STRUCTURAL AND FUNCTIONAL ORGANIZATION OF THE SYNAPSE

STRUCTURAL AND FUNCTIONAL ORGANIZATION OF THE SYNAPSE

Edited by

Johannes W. Hell
University of Iowa
Iowa City, IA, USA

and

Michael D. Ehlers
Duke University Medical Center
Durham, NC, USA

Springer

Editors
Johannes W. Hell
Department of Pharmacology 2-512 BSB
University of Iowa
Iowa City, IA 52242

Michael D. Ehlers
Department of Neurobiology
Duke University Medical Center
Box 3209
Durham, NC 27710

ISBN 978-0-387-77231-8 ISBN 978-0-387-77232-5 (eBook)
DOI 10.1007/978-0-387-77232-5

Library of Congress Control Number: 2007941249

Cover illustration: The cover illustration shows an immunofluorsecence micrograph of a hippocampal pyramidal neuron at two weeks in culture. The neuron was stained for the abundant calcium and calmodulin-dependent protein kinase CaMKII (green) and the presynaptic protein synapsin (red). The pictures was provided by Y. Chen and J. W. Hell, University of Iowa.

Printed on acid-free paper

9 8 7 6 5 4 3 2 1

springer.com

With special dedication to our late friend and colleague Alaa El-Husseini

Preface

The synapse is a fascinating structure for many reasons. Biologically, it is an exquisitely organized subcellular compartment that has a remarkable capacity for fidelity and endurance. Computationally, synapses play a central role in signal transmission and processing that represent evolution's solution to learning and memory. Nervous systems, including our own brains, possess an extraordinary capacity for adaptation and memory because the synapse, not the neuron, constitutes the basic unit for information storage.

Because the molecular complexities underlying signal processing and information storage must occur within the tiny space of the synapse, the precise molecular organization of proteins, lipids, and membranes at the synapse is paramount. Given the central role of the synapse in neuronal communication, it comes as no surprise that dysregulation of the synapse accounts for many, if not most, neurological and psychiatric disorders. Clinically, the synapse thus constitutes a prime target for treatments of these diseases.

It is for these reasons that we have chosen to focus our work on deciphering the structural and functional organization of the synapse. We have assembled leaders in the field of synapse biology to describe and distill the wonders and mysteries of the synapse. This book provides a fundamental description of the synapse developed over many decades by numerous investigators, paired with recent insight into new aspects of synapse structure and function that is still in flux and at the cutting edge of research.

This book grew out of a symposium and a research seminar at the University of Iowa that were sponsored, in large part, by the generous support of the Obermann Center for Advanced Studies. Obermann Seminars are specifically designed to gather international scholars and produce interdisciplinary research publications.

We are grateful for the exceptional efforts of our contributing authors, without whom this book would not have been possible. Their willingness to take time from their busy research schedules to share their insight and ideas with the breadth and depth that allow us to compile a collective work that is illuminating and useful, for both the general biologist and specialized neuroscientist, is very much appreciated. We express our gratitude to our assistants Ms. Susan Harward and Ms. Sue Birely for their professionalism and help with the book layout and proof reading. Lastly, we would like to thank our families (Laura and Henrik, Mary, Solon, Anselm, and Hans), who provided the support, encouragement, inspiration and comic relief, that in many ways helped to make this book possible.

Michael D. Ehlers and Johannes W. Hell
Editors

Contents

List of Contributors

Rana A. Al-Hallaq
Laboratory of Neurochemistry, National Institute on Deafness and Other Communication Disorders, National Institutes of Health, Bethesda, MD, USA, e-mail: wenthold@nidcd.nih.gov

I. Lorena Arancibia-Carcamo
Department of Pharmacology, University College London, Gower Street, London, WC1E 6BT, UK, e-mail: l.carcamo@ucl.ac.uk

Deniz Atasoy
Department of Neuroscience, U.T. Southwestern Medical Center, 5323 Harry Hines Boulevard, Dallas, TX 75390-9111, USA, e-mail: ege.kavalali@utsouthwestern.edu

George J. Augustine
Department of Neurobiology, Duke University Medical Center, Box 3209, Durham, NC 27710, USA, e-mail: georgea@neuro.duke.edu

Ben A. Barres
Department of Neurobiology, Stanford University School of Medicine, Stanford, CA 94305, USA, e-mail: barres@stanford.edu

Dwight E. Bergles
The Solomon H. Snyder Department of Neuroscience, Johns Hopkins School of Medicine, 725 N. Wolfe St., WBSB 1003, Baltimore, MD 21205, USA, e-mail: dbergles@jhmi.edu

Holly J. Carlisle
Division of Biology, California Institute of Technology, Pasadena, CA 91125, USA, e-mail: carlisle@caltech.edu

William A. Catterall
Department of Pharmacology, University of Washington, Seattle, WA 98195-7280, USA, e-mail: wcatt@u.washington.edu

Daniel Choquet
UMR 5091 CNRS, Université de Bordeaux 2, Physiologie Cellulaire de la Synapse, Institut François Magendie rue Camille Saint Saëns 33077 Bordeaux Cedex, France, e-mail: dchoquet@u-bordeaux2.fr

Robert H. Edwards
The Departments of Neurology and Physiology, UCSF School of Medicine, San Francisco, CA 94158-2517, USA, e-mail: robert.edwards@ucsf.edu

Michael D. Ehlers
Howard Hughes Medical Institute, Department of Neurobiology, Duke University Medical Center, Durham, NC 27710, USA, e-mail: ehlers@neuro.duke.edu

Alaa El-Husseini
University of British Columbia, Department of Psychiatry, Vancouver, British Columbia, Canada, e-mail: alaa@interchange.ubc.ca

Cagla Eroglu
Department of Neurobiology, Stanford University School of Medicine, Stanford, CA 94305, USA, e-mail: ceroglu@stanfordmedalumni.org

José A. Esteban
University of Michigan, Department of Pharmacology, Ann Arbor, MI 48109, USA, e-mail: estebanj@umich.edu

Craig C. Garner
Department of Psychiatry and Behavioral Science, Nancy Pritzker Laboratory, Stanford University, Palo Alto, CA 94304-5485, USA, e-mail: cgarner@stanford.edu

Matthew T. Geballe
Department of Chemistry, Emory University, Atlanta, GA 30322, USA, e-mail: strayne@emory.edu

Kasper B. Hansen
Department of Pharmacology, Emory University School of Medicine, Atlanta, GA 30322, USA, e-mail: strayne@emory.edu

Cyril Hanus
Howard Hughes Medical Institute, Department of Neurobiology, Duke University Medical Center, Durham, NC 27710, USA, e-mail: ehlers@neuro.duke.edu

Kristen M. Harris
Neurobiology Department, Center for Learning and Memory, University of Texas at Austin, Austin, Texas 78712, USA, e-mail: kharris@mail.clm.utexas.edu

Johannes W. Hell
Department of Pharmacology, University of Iowa, Iowa City, IA 52242-1109, USA, e-mail: johannes-hell@uiowa.edu

Thomas D. Helton
Department of Neurobiology, Duke University Medical Center, Howard Hughes Medical Institute, Durham, NC 27710, USA, e-mail: helton@neuro.duke.edu

Carrie L. Heusner
Department of Biological Chemistry and Department of Psychiatry and Biobehavioral Sciences, Brain Research Institute, Semel Institute for Neuroscience and Human Behavior, UCLA, Los Angeles, CA 90095, USA,
e-mail: kcmartin@mednet.ucla.edu

Yang Z. Huang
Duke University, Department of Neurobiology, Durham, NC 27710, USA,
e-mail: huang@neuro.duke.edu

Ege T. Kavalali
Department of Neuroscience, U.T. Southwestern Medical Center, 5323 Harry Hines Boulevard, Dallas, TX 75390-9111, USA, e-mail: ege.kavalali@utsouthwestern.edu

Mary B. Kennedy
Division of Biology, California Institute of Technology, Pasadena, CA 91125, USA,
e-mail: kennedym@its.caltech.edu

Josef T. Kittler
Department of Physiology, University College London, Gower Street, London, WC1E 6BT, UK, e-mail: j.kittler@ucl.ac.uk

Amy Lee
Department of Pharmacology, Emory University School of Medicine,
Atlanta, GA 30322, USA, e-mail: alee@pharm.emory.edu

John E. Lisman
Brandeis University, Biology Department and Volen Center for Complex Systems, Waltham, MA 02454, USA, e-mail: lisman@brandeis.edu

Edoardo Marcora
Division of Biology, California Institute of Technology, Pasadena, CA 91125, USA,
e-mail: marcora@caltech.edu

Kelsey C. Martin
Department of Biological Chemistry and Department of Psychiatry and Biobehavioral Sciences, Brain Research Institute, Semel Institute for Neuroscience and Human Behavior, UCLA, Los Angeles, CA 90095, USA,
e-mail: kcmartin@mednet.ucla.edu

Mark L. Mayer
Porter Neuroscience Research Center, Laboratory of Cellular & Molecular Neurophysiology, NICHD, NIH, Bethesda, MD 20892, USA,
e-mail: mayerm@mail.nih.gov

James O. McNamara
Duke University, Department of Neurobiology, Durham, NC 27710, USA,
e-mail: jmc@neuro.duke.edu

Peter S. McPherson
Department of Neurology and Neurosurgery, Montreal Neurological Institute, McGill University, 3801 Rue University, Montreal, Quebec, Canada, H3A 2B4,
e-mail: peter.mcpherson@mcgill.ca

Robert Meller
RS Dow Neurobiology Laboratory, Legacy Clinical Research and Technology Center, 1225 NE 2nd Ave, Portland, OR, 97232, USA,
e-mail: Rsimon@downeurobiology.org

Enhui Pan
Duke University, Department of Neurobiology, Durham, NC 27710, USA,
e-mail: enhui@neuro.duke.edu

Ronald S. Petralia
Laboratory of Neurochemistry, National Institute on Deafness and Other Communication Disorders, National Institutes of Health, Bethesda, MD, USA,
e-mail: wenthold@nidcd.nih.gov

Brigitte Ritter
Department of Neurology and Neurosurgery, Montreal Neurological Institute, McGill University, 3801 Rue University, Montreal, Quebec, Canada, H3A 2B4,
e-mail: peter.mcpherson@mcgill.ca

Bernardo L. Sabatini
Department of Neurobiology, Harvard Medical School, 220 Longwood Avenue, Boston, MA 02115, USA, e-mail: bsabatini@hms.haravard.edu

Michael W. Salter
Program in Neurosciences & Mental Health, The Hospital for Sick Children, and The University of Toronto Centre for the Study of Pain, Toronto, ON, Canada M5G 1X8, e-mail: mike.salter@utoronto.ca

Kang Shen
Department of Biological Sciences, Stanford University, Palo Alto, CA 94305-5020, USA, e-mail: kangshen@stanford.edu

Zu-Hang Sheng
Synaptic Functions Unit, National Institute of Neurological Disorders and Stroke, National Institutes of Health, Bethesda, MD 20892-3701, USA, e-mail: shengz@ninds.nin.gov

Roger P. Simon
RS Dow Neurobiology Laboratory, Legacy Clinical Research and Technology Center, 1225 NE 2nd Ave, Portland, OR, 97232, USA, e-mail: Rsimon@downeurobiology.org

Beth Stevens
Department of Neurobiology, Stanford University School of Medicine, Stanford, CA 94305, USA, e-mail: beths@stansfordmedalumni.org

Stefan Strack
Department of Pharmacology, University of Iowa, Iowa City, IA 52242-1109, USA, e-mail: stefan-strack@uiowa.edu

Karel Svoboda
Janelia Farm Research Campus, HHMI, 19700 Helix Drive, Ashburn, VA 20147, USA, e-mail: svobodak@janelia.hhmi.org

Catherine Croft Swanwick
Laboratory of Neurochemistry, National Institute on Deafness and Other Communication Disorders, National Institutes of Health, Bethesda, MD, USA, e-mail: wenthold@nidcd.nih.gov

Stephen F. Traynelis
Department of Pharmacology, Emory University School of Medicine, Atlanta, GA 30322, USA, e-mail: strayne@emory.edu

Antoine Triller
Inserm UR497, Ecole Normale Supérieure, Biologie Cellulaire de la Synapse N&P, 46, rue d'Ulm 75005 Paris, France, e-mail: triller@biologie.ens.fr

Gina Turrigiano
Brandeis University, Department of Biology and Center for Behavioral Genomics, Waltham, MA 02454, USA, e-mail: turrigiano@brandeis.edu

Michael J. Welsh
Departments of Internal Medicine and Molecular Physiology, and Howard Hughes Medical Institute, University of Iowa, Roy J. and Lucille A. Carver College of Medicine, Iowa City, IA 52242, USA, e-mail: michael-welsh@uiowa.edu

John A. Wemmie
Department of Psychiatry, Neuroscience Program, and Department of Veterans Affairs Medical Center, University of Iowa, Roy J. and Lucille A. Carver College of Medicine, Iowa City, IA,USA, e-mail: john-wemmie@uiowa.edu

Robert J. Wenthold
Laboratory of Neurochemistry, National Institute on Deafness and Other Communication Disorders, National Institutes of Health, Bethesda, MD, USA, e-mail: wenthold@nidcd.nih.gov

Hongjie Yuan
Department of Pharmacology, Emory University School of Medicine, Atlanta, GA 30322, USA, e-mail: strayne@emory.edu

Xiang-ming Zha
Department of Internal Medicine and Howard Hughes Medical Institute, University of Iowa, Roy J. and Lucille A. Carver College of Medicine, Iowa City, IA 52242, USA, e-mail: xiangming-zha@uiowa.edu

Diversity in Synapse Structure and Composition

Kristen M. Harris

Neurobiology Department, Center for Learning and Memory, University of Texas at Austin, Austin, Texas 78712, USA, kharris@mail.clm.utexas.edu

1 Introduction

This chapter describes diversity in the structure and composition of synapses at the resolution of serial section transmission electron microscopy (ssTEM). Section 1 introduces the synapse. Section 2 describes the structure and composition of presynaptic axons. Section 3 elucidates postsynaptic structure and composition. Section 4 discusses the impact of perisynaptic astroglial processes on synapses. Throughout all sections an effort is made to understand regularities in the relationships among these features that might contribute to the diversity in synapse size and number in systematic ways.

A synapse has a presynaptic component, usually an axon but sometimes a dendrite, and a postsynaptic component, usually part of a dendrite, cell soma, or axonal initial segment and occasionally an astroglial process. Perisynaptic astroglia represent a third component that occurs at some synapses. When perisynaptic astroglia are present, the structural complex has been referred to as a 'tripartite synapse' (45). Figure 1 illustrates an electron micrograph of an ultrathin (50 nm) section through dendrites, axons, astroglial processes an synapses in the rat hippocampus. This picture was chosen to open this chapter because it nicely illustrates the diversity in the composition of even a tiny segment of the neuropil.

J.W. Hell, M.D. Ehlers (eds.), *Structural and Functional Organization of the Synapse*,
DOI: 10.1007/978-0-387-77232-5_1,

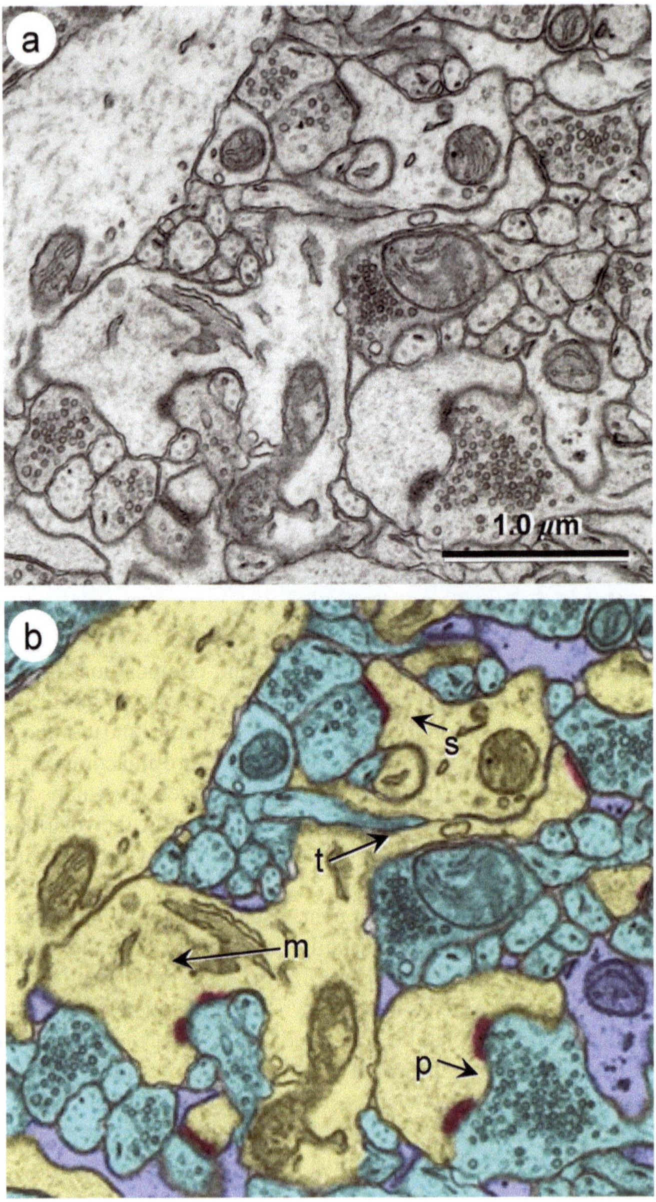

Fig. 1. A single thin section spanning approximately 10 μm^2 in the middle of the apical dendritic arbors of hippocampal area CA1 pyramidal cells. (**a**) *Gray* scale image obtained at the electron microscope. (**b**) Same section colorized to illustrate dendrites (*yellow*), axons and vesicle-filled axonal boutons (*green*), asymmetric postsynaptic densities of synapses located on diversely shaped dendritic spines (*red*), and astroglial processes (*lavender*). Stubby (s), thin (t) and mushroom (m) spines can be seen in longitudinal section emerging from dendrites. A large mushroom spine head has a perforated (p) postsynaptic density.

2 Presynaptic Structures and Composition

Three dimensional reconstructions of individual presynaptic axons that pass through the complex hippocampal neuropil show the diversity in their local trajectories (Fig. 2a). A three-dimensional reconstruction of a hippocampal dendritic segment that is approximately 10 microns long illustrates the more than 10 fold variation in the dimensions of neighboring dendritic spines and synapses (Fig. 2b).

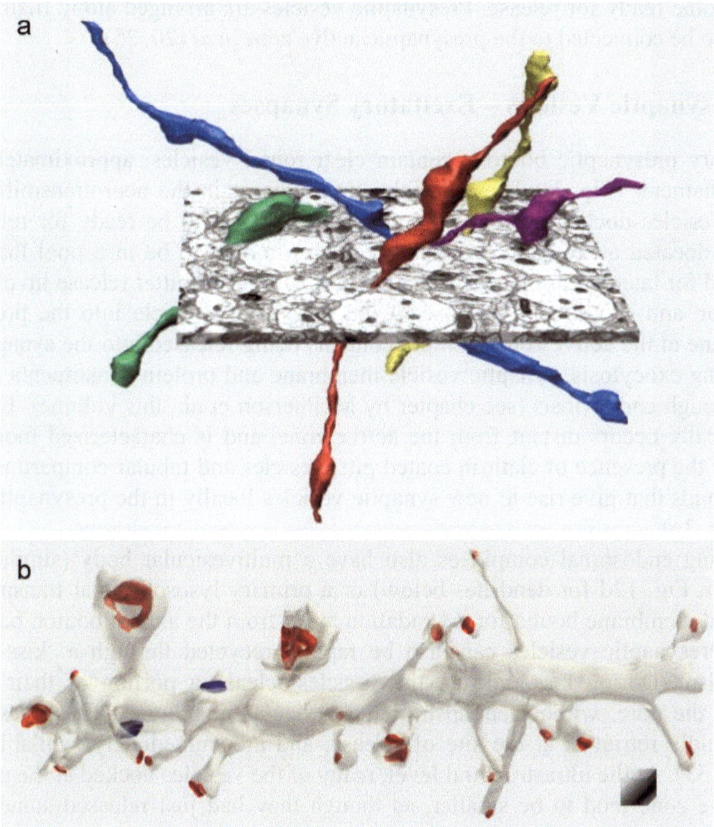

Fig. 2. Diversity in the trajectory of presynaptic axons and shapes of postsynaptic dendritic spines. (**a**) Three-dimensional reconstructions of a subset of axons passing through a single electron micrograph. (Adapted from (36)). (**b**) Three-dimensional reconstruction of a single dendritic segment, illustrating the diversity in dendritic spine shapes and their postsynaptic densities (*red*). Inhibitory shaft synapses are colorized in *blue*. (This is a recently surfaced image of dendrite 21 from (18) available at http://synapses.clm.utexas.edu/). Scale is approximately 0.5 μm3 for both reconstructions.

2.1 Presynaptic Active Zone

A presynaptic axonal bouton contains vesicles with a variety of shapes and sizes. These vesicles contain excitatory or inhibitory neurotransmitters, neuromodulatory peptides, proteins required to concentrate neurotransmitters, and a variety of proteins involved in the vesicle cycle or that are destined for the presynaptic active zone (11). The presynaptic active zone is a specialized region variously described as 'dense projections', the 'presynaptic grid' or mini-active zones (34) where vesicles dock and become ready for release. Presynaptic vesicles are arranged along filaments that appear to be connected to the presynaptic active zone area (20, 25).

2.2 Presynaptic Vesicles – Excitatory Synapses

Excitatory presynaptic boutons contain clear round vesicles, approximately 35–50 nm in diameter (Fig. 3). These vesicles usually contain the neurotransmitter glutamate. Vesicles docked at the active zones are thought to be ready for release and vesicles located away from the membrane are thought to be in a pool that can be recruited for later release. One mechanism for neurotransmitter release involves pore formation and subsequent collapse of the presynaptic vesicle into the presynaptic membrane at the active zone with the contents being released into the synaptic cleft. Following exocytosis, synaptic vesicle membrane and protein constituents are recycled through endocytosis (see chapter by McPherson et al., this volume). Endocytosis typically occurs distant from the active zone, and is characterized morphologically by the presence of clathrin coated pits, vesicles and tubular compartments with coated buds that give rise to new synaptic vesicles locally in the presynaptic bouton (e.g. Fig. 3g).

Sorting endosomal complexes also have a multivesicular body (similar to that shown in Fig. 12d for dendrites below) or a primary lysosome that transports proteins and membrane bound for degradation away from the axonal bouton back to the soma. Presynaptic vesicles can also be rapidly recycled through a 'kiss-and-run' mechanism. During kiss-and-run, the vesicles release a portion of their contents through the pore, without collapse of the vesicular membrane. These vesicles are then rapidly retrieved at the site of release, and are immediately available for re-release (33). At the ultrastructural level, many of the vesicles docked at the presynaptic active zone tend to be smaller, as though they had just released some of their contents at the time of fixation (19).

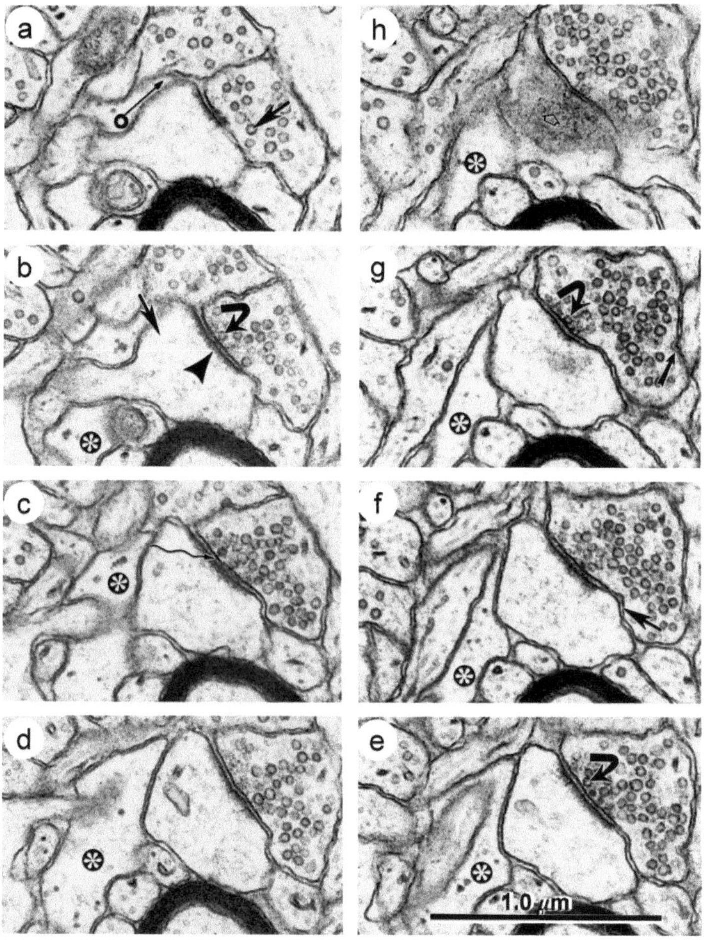

Fig. 3. Excitatory synapse revealed through ssTEM of a mushroom-shaped dendritic spine and its corresponding presynaptic axonal bouton. (**a**) A characteristic non-docked vesicle (*arrow*) in the presynaptic axonal bouton. (**b**) A docked vesicle (*curved arrow*) across from the postsynaptic density (PSD, chevron); on the spine head (*arrow*). (**c**) The synaptic cleft (*wiggly arrow*) located between the plasma membranes of the spine head and presynaptic bouton contains dense staining material, presumably composed of adhesion molecules and portions of receptors. (**d**) Serial section between (c) and (**e**) illustrates astroglial process (*) at the base of the spine head. (e) Curved arrow illustrates another docked vesicle. (**f**) Extracellular space (*arrow*) does not contain the dense-staining material found in the synaptic cleft. (**g**) Coated pit (*straight arrow*) at a site of endocytosis on side of the bouton away from the active zone; docked vesicle (*curved arrow*). (**h**) *Gray* surface of the plasma membrane (*open arrow*) viewed en face where it caps the head of the dendritic spine. In (a) the astroglial process labeled (°) is near the synaptic cleft where it might detect and control spillout of neurotransmitter; in all other sections the perisynaptic astroglial process is labeled with (*). (Adapted from (19)).

2.3 Presynaptic Vesicles – Inhibitory Synapses

Inhibitory presynaptic boutons contain smaller, pleiomorphic vesicles having both round and flattened shapes in aldehyde-fixed tissue (Fig. 4). The pleiomorphic vesicles usually contain the neurotransmitters GABA or glycine. Inhibitory synapses are most abundant at the neuronal soma (Fig. 4a) and proximal dendritic zones (see chapter by Arancibia-Carcamo, Triller and Kittler, this volume). In addition, inhibitory synapses can be interspersed among excitatory synapses (in the hippocampus about 1 inhibitory per 10–20 excitatory synapses) along a dendrite (Fig. 4b). Occasionally, inhibitory synapses are located at the axonal hillock, where activation of one or a small number of inhibitory synapses can regulate neuronal cell firing at their axons. In some brain regions an inhibitory synapse occurs on the necks of some dendritic spines, whether they veto excitatory activation likely depends on their frequency and the specific circuit involved (12, 24, 46). Neurosecretory peptides and some neurotransmitters are localized to the cytoplasm surrounding the pleiomorphic vesicles of inhibitory synapses, or in large dense core vesicles (~100 nm) (10, 43). If axons use these large secretory DCVs then more of them occur in each axonal bouton.

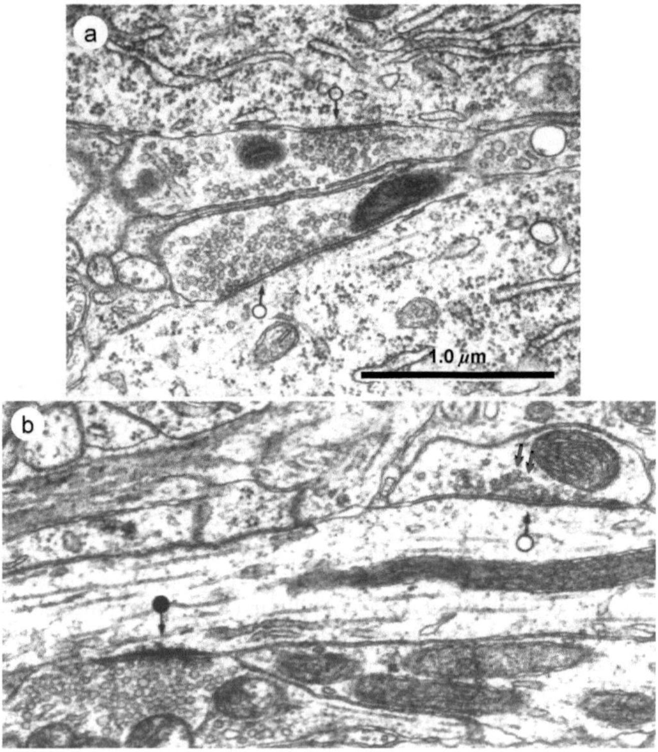

Fig. 4. Inhibitory symmetric synapses in mature hippocampus with thin pre- and postsynaptic densities and pleiomorphic vesicles. (**a**) Two inhibitory synapses on the pyramidal cell bodies (*open circles*). (**b**) Symmetric synapse (*open circle*) with flattened vesicles (*red arrows*) and asymmetric synapse (*closed circle*) with thicker PSD and larger rounder presynaptic vesicles. These synapses are located directly on the dendritic shaft of a nonspiny interneuron in mature hippocampal area CA1. (Modified from (16)).

2.4 Presynaptic Small Dense Core Vesicles as Active Zone Transporters

Small dense core vesicles (~80 nm) are distinct from large DCVs both in size and frequency (Fig. 5). Small DCVs are present in only about 20% of mature presynaptic axons, and when present, only 1–10 vesicles occur in a fully reconstructed axonal bouton. The outer membranes of small DCVs label with antibodies to proteins located at the presynaptic active zone, such as piccolo and bassoon, and they are prevalent along axons in the developing nervous system; hence the small DCVs are thought to be a local source of new presynaptic active zones (1, 50) (see chapter by Shen and Garner, this volume). Recent work has shown that there are fewer small DCVs during rapid synapse formation in the mature hippocampus in further support of their role in local delivery of presynaptic active-zones (39).

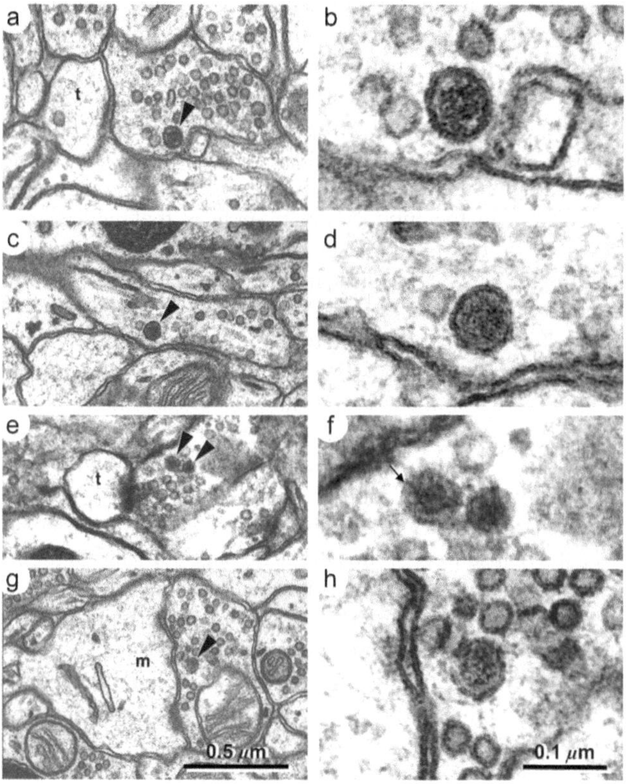

Fig. 5. Small dense core vesicles (*arrowheads*) at excitatory synapses in the apical dendritic field of mature hippocampus (CA1). (**a, b**) Low and higher power views of a small dense core vesicle (dcv) in typical location near plasma membrane but away from active zone. (**c, d**) Small DCV located near the beginning of an inter-varicosity region, suggesting it might be in transit. (**e, f**) Two small DCVs in a presynaptic axonal bouton. One of these vesicles clearly illustrates small 'spicules' emanating from its surface. (**g, h**) Small DCV located within one vesicle diameter of an active zone. DCVs are rarely located this close to an active zone, but show that they might also be involved in synapse enlargement, not just new synapse formation. Scale bar in (g) is for (a, c, and e). Scale bar in (h) is also for (b, d, and f). (Adapted from (38)).

2.5 Local Protein Synthesis in Presynaptic Boutons?

Although polyribosomes are not a prominent component of presynaptic axonal boutons in the central nervous system, mRNAs have been localized to them (32). In squid giant axons, local protein synthesis machinery appears to derive from the ensheathing glia (9, 14). Detailed three-dimensional reconstructions will be required to learn whether isolated polyribosomes are directed into vertebrate presynaptic axonal boutons to allow for local protein synthesis, similar to that observed in dendritic spines (see below).

2.6 Nonsynaptic, Single Synaptic and Multisynaptic Axonal Boutons

Axonal boutons containing clear synaptic vesicles, mitochondria, dense core vesicles, and multivesicular bodies occur both with and without postsynaptic partners (Fig. 6). In the mature hippocampus, about 96% of the vesicle-containing boutons have at least one postsynaptic partner. Single-synapse boutons predominate comprising about 75% of all vesicle-containing axonal boutons. About 21% of vesicle-containing boutons are multi-synaptic while about 4% are non-synaptic boutons in the mature hippocampus (36, 37). There are more multisynaptic boutons when rapid synaptogenesis occurs in the hippocampus, such as that which occurs during the estrus cycle (49) or following the preparation of mature hippocampal slices under ice-cold conditions (22, 23, 31). Multisynaptic dendritic spines, receiving input from more than one presynaptic axon occur relatively frequently during development and under conditions of synaptogenesis in the mature hippocampus. The axonal segment in Fig. 6c was from a hippocampal slice that had been prepared under ice-cold dissection conditions, which induces new synapses. These findings suggest that rapid synaptogenesis in the mature hippocampus does not require de novo formation of presynaptic axons. It is not known whether the nonsynaptic boutons also constitute a source of available presynaptic boutons to accommodate rapid synaptogenesis, if they represent vesicle clusters in transit between presynaptic varicosities, or if thery are sites of recent synapse loss. Similarly, presynaptic axonal boutons vary in structure along the axons from other brain regions such as cerebellar cortex (Fig. 7, (48)). Parallel fiber axons synapse with dendritic spines of Purkinje cell spiny branchlets (Fig. 7a–c; see also Figs. 8d, 11c and f below for further discussion of these postsynaptic spines). The climbing fiber axons that synapse along the proximal dendrite of the Purkinje cells have much larger, more irregularly shaped boutons (Fig. 7d) than parallel fiber axons.

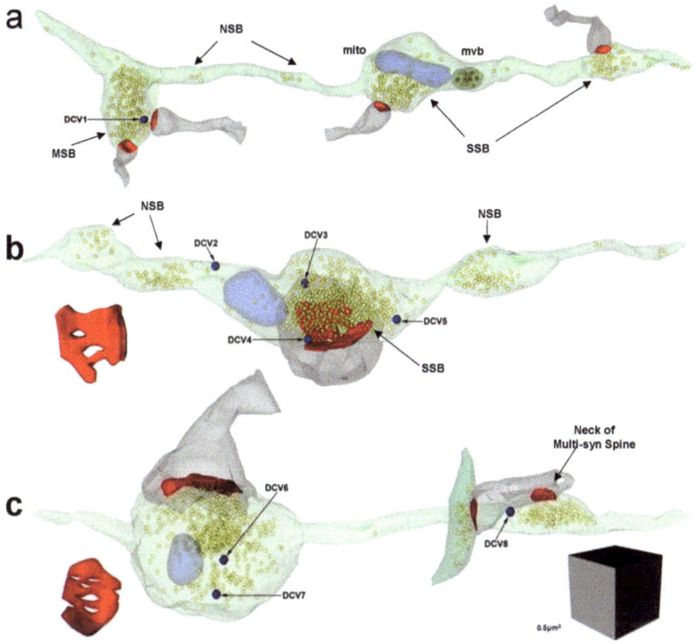

Fig. 6. Axonal segments from mature hippocampus (CA1). (**a**) This segment is 7.8 μm long with 2 single-synapse boutons (SSB), 2 nonsynaptic boutons (NSB), and 1 multiple synapse bouton (MSB) shared by 2 postsynaptic spines from different dendrites. (**b**) This axonal segment is 5.7 μm long with 1 SSB containing 3 small DCVs and a mitochondrion. It is surrounded by 3 NSBs. (**c**) This segment is 5.5 μm long with 1 SSB similar to that in (b) and another SSB in an unusual position along the neck of a multi-synaptic dendritic spine. (Axons – *green*, vesicles – *yellow*, mitochondria – *light blue*, DCVs – *dark blue*, PSDs – *red*; Adapted from (38)).

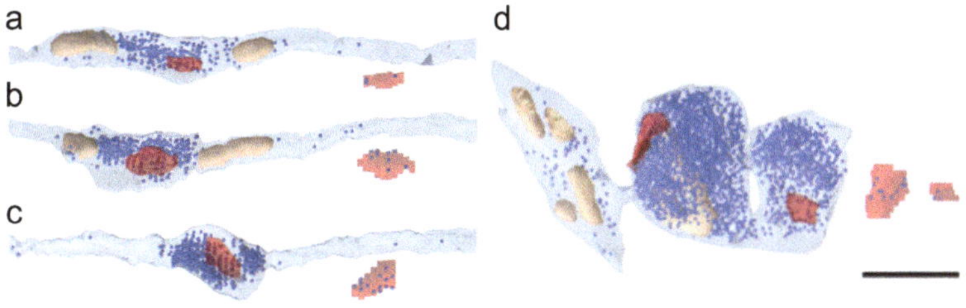

Fig. 7. Reconstructed axons from rat cerebellar cortex. (**a–c**) Parallel fiber axons. (**d**) Climbing fiber axon. Axons (*translucent light blue*); PSDs (*red*); vesicles (*dark blue*); mitochondria (*beige*). In all images, the locations of docked vesicles are superimposed on the 'enface' red PSD reconstructions to the right side of each axon. The scale bar is 1 micron for all 4 reconstructions. (Adapted from (48)).

3 Postsynaptic Structure and Composition

3.1 Diversity in Postsynaptic Dendritic Spine Structure

Postsynaptic structure is highly diverse among synapses on the same dendrite and across different cell types in the brain. For example, neighboring dendritic spines on a single hippocampal dendritic segment can vary more than 10 fold in their dimensions (e.g. Figs. 2b, 11d). Large highly branched dendritic spines are the postsynaptic partners of mossy fibers in the hippocampal area CA3 (Fig. 8a–c). The dendritic spines occurring along the spiny branchlets of cerebellar Purkinje cells vary widely in their size, yet they are more uniformly club-shaped (Fig. 8d). These cerebellar spines can be branched; however, the head of each branch on the spine is also club shaped (for more examples see (17)). Similarly, dendritic spines distributed along the same presynaptic axon vary greatly in their dimensions (Fig. 6). The size of the dendritic spine and its synapse correlates nearly perfectly with the number of vesicles in the presynaptic bouton for all brain regions tested so far (Fig. 9). Thus, it is of great interest to understand the rules that govern local changes in synaptic structure and how they are coordinated between the pre- and postsynaptic structures. One strategy has been to investigate the impact of altering the molecular composition of neurons to determine the impact on dendritic spine structure (reviewed in (4, 42)). Another strategy has been to compare the composition of subcellular organelles among dendritic spines and synapses of differing morphologies, during different stages of development, and during synaptic plasticity. Our focus in this chapter is on this second strategy.

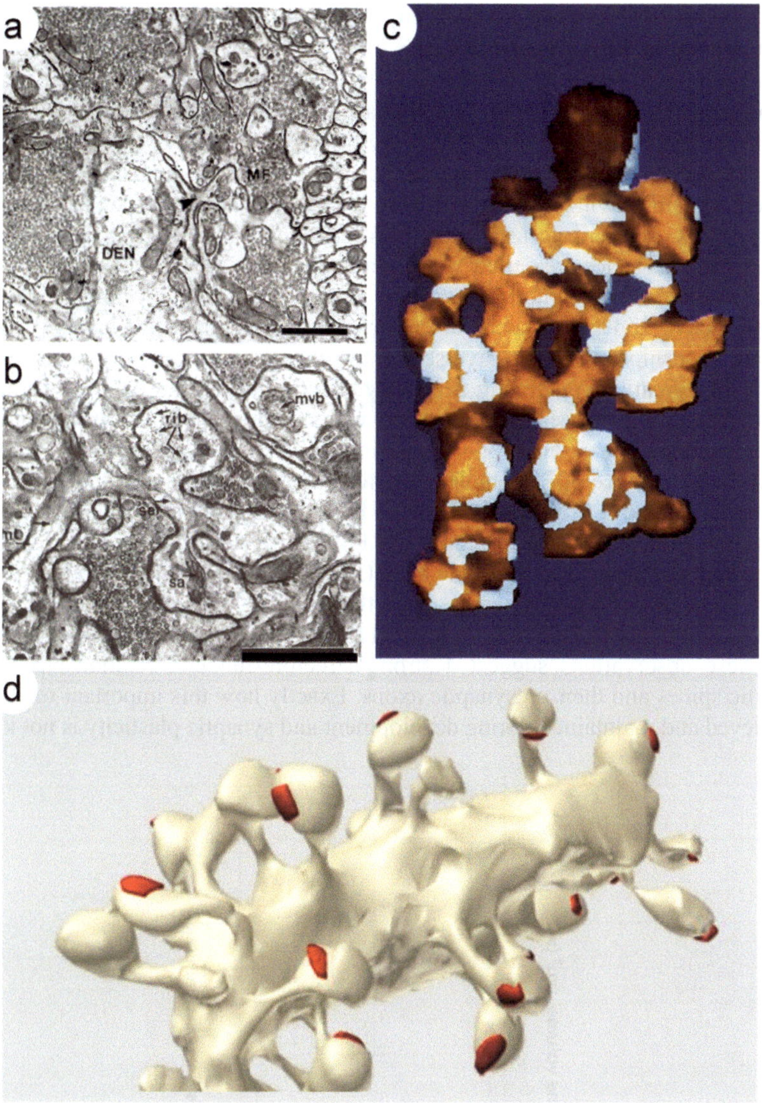

Fig. 8. 3D reconstructions of diverse dendritic spines. (**a**) Electron micrograph illustrating CA3 pyramidal cell dendritic shaft (den), the origin of a complex branched spine (*arrow head*) and mossy fiber (MF) bouton that synapses with spine heads from this and other dendrites. Scale = 1 μm. (**b**) CA3 spine which contains microtubules (mt), polyribosomes (rib), and a tubule of SER that becomes connected to a spine apparatus (sa). The fourth head connected to this spine on a different serial section, has a multivesicular body (mvb). (**c**) CA3 spine (*gold*) with multiple PSDs (*light blue*) distributed across its many heads. (**d**) Segment of spiney branchlet from a cerebellar Purkinje cell (PSDs *red*, dendrite and spines *beige*). Scale in (b) is 1 μm for (b, c, and d). (a–c) are Adapted from (5); and (d) is Adapted from http://synapses.clm. utexas.edu, courtesy J. Spacek.

3.2 Correlation Between PSD Area, Spine Head Volume, and Vesicles in Presynaptic Axon

The postsynaptic density (PSD) contains a host of important receptors, scaffolding proteins, and signaling complexes that vary from synapse to synapse (21). The diversity in molecular composition from synapse to synapse is great, with hundreds of different proteins having been identified in the PSD. In aldehyde-fixed tissue, excitatory synapses are characterized by a thick PSD relative to the thinner presynaptic active zone, and hence are called 'asymmetric synapses' (Figs. 1, 3, 5, 8, 10, 11, 12, 14, see also (15, 30)). Inhibitory synapses usually have equally thin PSDs and presynaptic active zones and hence are referred to as 'symmetric synapses' (Fig. 4). These relationships are highly diverse and the degree of pre- and postsynaptic thickening varies greatly even among excitatory or inhibitory synapses in the same brain region ((6), http://synapses.clm.utexas.edu/anatomy/chemical/colh.htm). Importantly, PSD thickness is also sensitive to experimental manipulations that are known to impact the abundance of PSD molecules, such as the calcium calmodulin dependent protein kinase type II (CaMKII), which are highly enriched in the PSD (28).

The reconstructed surface area of the PSD correlates nearly perfectly with the volume and surface area of the spine head (Fig. 9, (18, 26)). The PSD area and spine head volume also correlate nearly perfectly with the total number of presynaptic vesicles and the number of vesicles docked at the presynaptic active zone (18, 19, 26, 35). These observations suggest a strong structure-function relationship between dendritic spines and their presynaptic axons. Exactly how this important relationship is achieved and maintained during development and synaptic plasticity is not known.

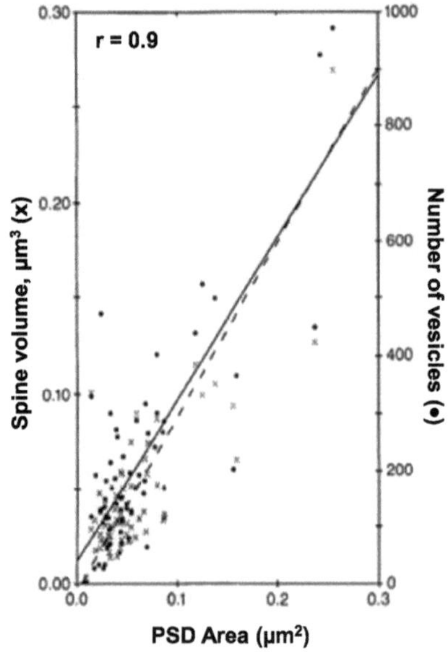

Fig. 9. Strong correlation between PSD area, spine volume, and t number of presynaptic vesicles. This dataset is from hippocampal area CA1, but these relationships hold true across all brain regions evaluated so far, including synapses in hippocampal area CA3, cerebellar cortex, striatum, and neocortex. (Adapted from (18, 26)).

3.3 Polyribosomes – Local Protein Synthesis in Dendrites and Spines

Polyribosomes occur in some, but certainly not all dendritic spines (41). They are sites where translation could have been occurring at the time of tissue fixation. Ribosomes are identified in dendrites and spines as 10–25 nm electron dense spheres surrounded by a gray halo; and polyribosomes are identified in ssTEM as having at least 3 ribosomes (Fig. 10ab). Free polyribosomes synthesize cytoplasmic proteins, like those in the PSD and have been found to increase in frequency during plasticity such as LTP (27). Bound polyribosomes are associated with endoplasmic reticulum and synthesize integral proteins, such as receptors. Occasionally polyribosomes have been detected in the vicinity of a spine apparatus (Fig. 10b, 11a). Quantitative ssTEM is required to ascertain the relative frequencies of free and bound polyribosomes in spines and at other synapse types in the developing and mature brain and during plasticity such as learning and memory which requires local protein synthesis.

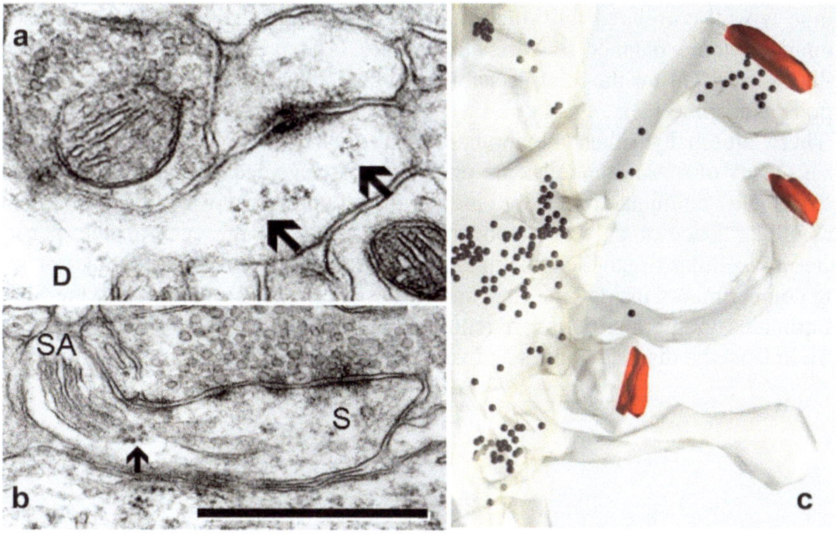

Fig. 10. Polyribosomes are sites of local protein synthesis present in some dendritic spines. (**a**) Head of a thin hippocampal dendritic spine with multiple polyribosomes (*arrows*; D, dendritic shaft). (**b**) Polyribosome (*arrow*) located amidst a hippocampal spine apparatus (SA). (**c**) Only one of the reconstructed spines on this dendrite had polyribosomes (ribosomes – *black spheres*, PSDs – *red*). (Adapted from http://synapses.clm.utexas.edu/anatomy/ribosome/ribo.stm courtesy J. Spacek). Scale in (b) is about 0.5 μm for all three images.

3.4 Diversity in Composition of Smooth Endoplasmic Reticulum and Endosomal Compartments in Dendrites and Spines

Some dendritic spines contain smooth endoplasmic reticulum (SER), which functions as an internal calcium store. In the hippocampus, only about 15% of all spines contain SER, and typically only the largest spines contain SER (40). In contrast, nearly all cerebellar dendritic spines contain SER (17). The spine apparatus is an enigmatic organelle variously thought to be involved in the regulation of calcium, synthesis of proteins, and post-translational modification of proteins like the Golgi apparatus (Fig. 11a–d). It contains SER-like membranes arranged in laminae separated by dense-staining bars that contain the actin-binding protein synaptopodin which is necessary for maintaining its laminar structure and for positioning it into spines (13). Only large hippocampal and cortical spines contain a spine apparatus; in contrast the SER of cerebellar spines forms a reticular membranous network without the dense-staining laminations found in the spine apparatus (Fig. 11ef). Total spine volume is well-correlated with the volume of SER, suggesting a functional role in maintaining ionic balance of cerebellar spines. The presence or absence of SER might also account for the relative stability of cerebellar spines, and plasticity among hippocampal spines (3).

There are many other membrane-bound organelles present in dendrites and spines. Many of these organelles are endosomes as revealed by tracking internalized gold particles conjugated with bovine serum albumin that were injected into the extracellular space of a PN21 hippocampal slice (7). The presence of gold particles in the intracellular organelles unambiguously identifies them as endosomal or recycling compartments inside dendrites and spines (Fig. 12a–d). In contrast, the SER is a continuous membrane-bound reticulum that is easily distinguished upon reconstruction from the discrete tubules and vesicles of endosomal compartments (Fig. 12e).

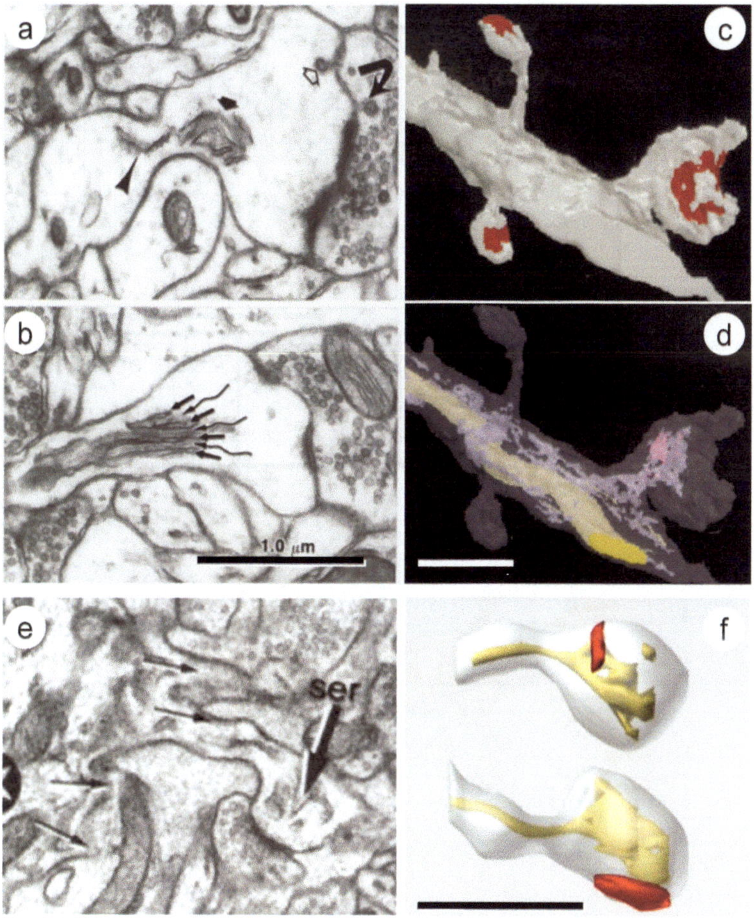

Fig. 11. Mushroom spines with spine apparatus. (**a**) A spine apparatus is characterized by the presence of SER laminated with dense staining material (*thick filled arrow*). At the base of this spine apparatus is rough endoplasmic reticulum (*thin arrowhead*). At the edge of the PSD there are reciprocal coated structures. The curved arrow indicates a double-walled coated vesicle, containing a small bit of the postsynaptic spine, or 'spinule', projecting from the spine into the presynaptic axon. The open *short arrow* indicates a coated pit, endocytosing into the postsynaptic dendritic spine. (**b**) Longitudinal section through a highly laminated spine apparatus. *Straight arrows* indicate SER; *wiggly arrows* indicate dense-staining material. (**c**) Reconstruction of a short hippocampal CA1 dendritic segment with 2 small thin spine and 1 large mushroom spine with surfaced PSDs (*red*). (**d**) Translucent surface of the dendrite and spine reveals the SER (*lavender*) and mitochondrion (*yellow*) in the dendritic shaft. Only the large mushroom spine has SER, which forms a spine apparatus (*pink*) in this example. (**e**) EM image of SER in dendrite and spines of a spiney branchlet from a cerebellar Purkinje cell. Other spine origins emerging from this dendrite are indicated with thin arrows on the left. (**f**) Reconstruction of cerebellar spines and their reticulum of SER inside. PSDs are *red* dendrite and spines are *gray* and ser is *yellow* in these spines. ((a–d) adapted from (40); (e) is adapted from (17); and (f) is from http://synapses.clm.utexas.edu/anatomy/compare/compare.stmcourtesy of J. Spacek).

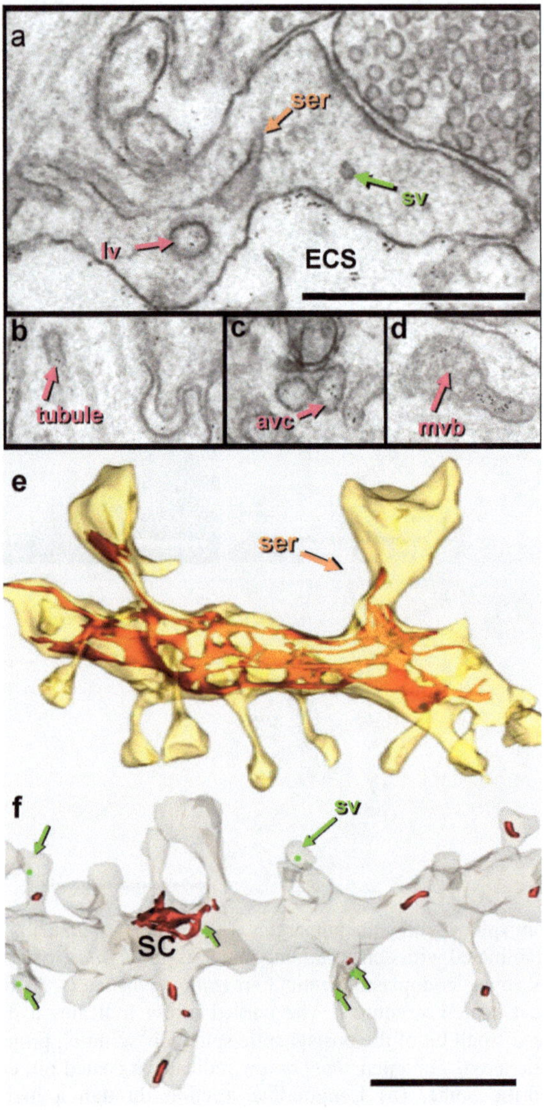

Fig. 12. Diversity among spines in composition of SER and endosomal compartments. (**a–d**) Large vesicles (lv), tubules with coated buds, amorphous vesicular clumps (avc) and multivesicular bodies with tubules all contained gold particles, endocytosed from the extracellular space. Smooth endoplasmic reticulum (ser) never contained gold particles, indicating that these two intracellular membranous compartments are not connected. (**e**) *Top*, reconstruction of SER in a dendrite and associated dendritic spines; about 14% of hippocampal dendritic spines contained SER. *Bottom*, reconstruction of a PN21 dendrite (*gray*) with endosomal structures (*red*) including a sorting complex (SC) which contains tubules, vesicles, a multivesicular body, and coated buds. *Green spheres* represent the locations of the smaller, smooth vesicles, presumably exocytic in function. (Adapted from (7)) Scale in (a) is 0.5 μm for (a–d), and in (**f**) it is 1 um for (e) and (f).

Quantitative work shows that about 50 percent of normal hippocampal dendritic spines from perfusion fixed brain contain no membrane bound organelles; and a distinction among spines that contain the different types of organelles (Fig. 13a). Interestingly, the amount of SER also varies along the shaft of the parent dendritic shaft (Fig. 13b). Recent work shows endosomal compartments to be dynamically regulated during synaptic plasticity (29). It will be interesting to learn whether the presence or absence of SER is also dynamically regulated in spines during synaptic plasticity.

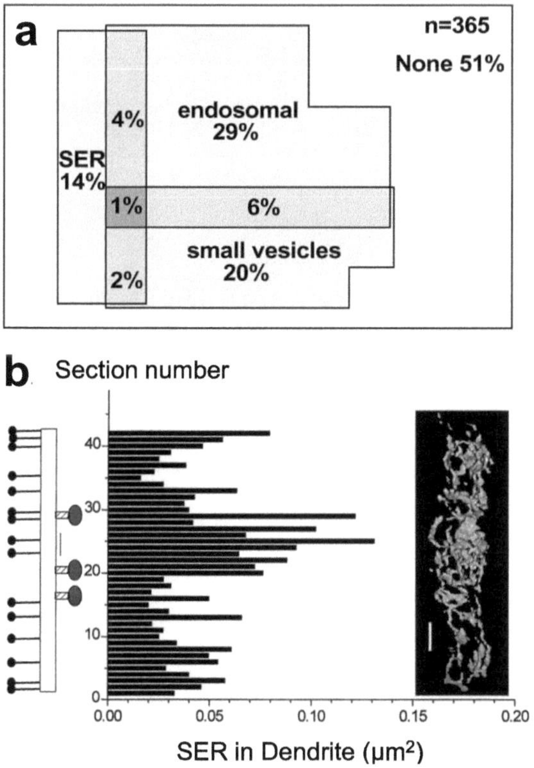

Fig. 13. Variation in the organization of SER and other membrane bound organelles in hippocampal dendrites. (**a**) Venn diagram showing the relative frequency of membrane-bound organelles inside hippocampal dendritic spines. About 49% of dendritic spines contain one or more membranous organelles. It represents an average of all of the dendritic spines that were reconstructed from PN15, PN21, and Adult area CA1 of the rat hippocampus. This relationship is about the same at all three ages in perfusion fixed brain, *in vivo*. (**b**) Relative volume of SER in the dendritic shaft varies by type and number of dendritic spines along the length. At the left is a schematic diagram of this dendrite with the location of thin spine origins along the dendritic segment illustrated to the left as small lollipops and mushroom spine locations illustrated by larger lollipops pointing to the right. The graph illustrates the cross-sectional area of SER in each section along the length of the dendrite. The three-dimensional reconstruction of the SER in *gray* also illustrates its non-uniform distribution in the dendritic shaft. (Diagram in (a) is Modified from (6); and in (b) from (40)).

4 Perisynaptic Astroglial Processes

In the cerebellum, nearly all synapses are enveloped by perisynaptic astroglial processes (Fig. 14ab). Astroglial processes are identified by their light cytoplasm and the presence of dark glycogen granules and astrocytic fibrils in larger processes. They end in thin or flattened processes that easily interdigitate among axons and dendrites throughout the brain. In contrast, less than 50% of neocortical and hippocampal dendritic spine synapses have astroglial processes at their perimeters (Fig. 14cd).

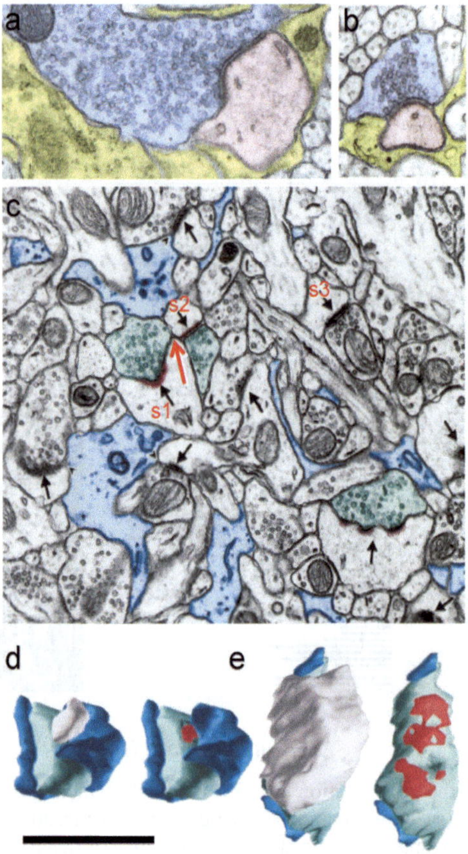

Fig. 14. Perisynaptic astroglial processes. In cerebellar cortex, astroglial processes surround the synaptic cleft of nearly all synapses made by (**a**) climbing fibers or (**b**) parallel fibers. In contrast, only about half of all hippocampal synapses have astroglial processes at the cleft, even between neighboring synapses (e.g. s1→ s2 *Red arrow*; s3 has no perisynaptic astroglia at the presynaptic bouton, the postsynaptic spine, or the perisynaptic cleft region. Scale bar = 1 um for all images). (a and b color schema: presynaptic axon (*blue*), postsynaptic spine (*pink*), perisynaptic astroglia (*yellow*); (**c**) and (**d**) color scheme: presynaptic axon (*green*); postsynaptic density (*red*); astroglia (*blue*)). (a, b adapted from (48)); (c, d adapted from (44)).

This differential arrangement of perisynaptic astroglia suggests that some synapses are highly regulated and supported by this important partner in the tripartite synapse, whereas other synapses are either more selective, unable to attract astroglial processes, or actively repel them (44). Exactly how this relationship is regulated by local synaptic plasticity remains to be determined, however, attraction to high concentrations of extracellular glutamate is a likely candidate mechanism (8). In the mature hippocampus, those synapses that have a perisynaptic astroglial process at its perimeter, are on average significantly larger than synapses without (47). Importantly, size is associated with the presence of an astroglial process juxtaposed to the postsynaptic spine and/or synaptic cleft; not the degree to which the astroglial process surrounds the synapse (Fig. 14de). In fact, very large hippocampal synapses might have only a small fraction of their perimeters surrounded by an astroglial process (compare Fig. 14de). These arrangements suggest that the perisynaptic astroglial processes do more than simply delineate boundaries between synapses to prevent spillover of neurotransmitter and crosstalk and are likely crucial partners in sustaining synapses (2). One intriguing possibility, small, new, relatively unstable synapses in hippocampus or cortex, share the extracellular products released during synaptic transmission from other synapses, until they grow big enough to attract their own stabilizing astroglial processes (see chapter by Eroglu, Barres and Stevens, this volume).

Summary

In this chapter I have described the highly diverse structure and composition of presynaptic axons, postsynaptic spines and dendrites, and perisynaptic astroglial processes. Postsynaptic size is proportional to presynaptic vesicle content and larger synaptic dendritic spines contain more of the subcellular organelles needed to remodel and sustain them. Perisynaptic astroglial processes also appear to be key players in determining synapse size. Although synapse structure may appear static in electron micrographs, it is highly dynamic, undergoing dramatic changes in gross morphology, as well as, the positioning of subcellular organelles during basic synaptic transmission and in response to conditions that elicit changes in synapse function or 'synaptic plasticity'. Similarly, the structure and composition of synapses can vary across brain regions, and change dynamically during development, maturation, normal aging, neurological disorders and trauma. Careful consideration of this structural diversity and plasticity is leading to new understanding about synapse formation, growth, maintenance and elimination.

References

1. Ahmari SE, Smith SJ. Knowing a nascent synapse when you see it. *Neuron* 2002; 34:333–6.
2. Allen NJ, Barres BA. Signaling between glia and neurons: focus on synaptic plasticity. *Curr. Opin. Neurobiol.* 2005; 15:542–8.
3. Bourne J, Harris KM. Do thin spines learn to be mushroom spines that remember? *Curr. Opin. Neurobiol.* 2007; 17(3):381–6.

4. Bourne JN, Harris KM. Balancing structure and function at hippocampal dendritic spines. *Annu. Rev. Neurosci.* 2008; 31:47–67.
5. Chicurel ME, Harris KM. Three-dimensional analysis of the structure and composition of CA3 branched dendritic spines and their synaptic relationships with mossy fiber boutons in the rat hippocampus. *J. Comp. Neurol.* 1992; 325:169–82.
6. Colonnier M. Synaptic patterns on different cell types in the different laminae of the cat visual cortex. An electron microscope study. *Brain Res.* 1968; 9:268–87.
7. Cooney JR, Hurlburt JL, Selig DK et al. Endosomal compartments serve multiple hippocampal dendritic spines from a widespread rather than a local store of recycling membrane. *J. Neurosci.* 2002; 22:2215–24.
8. Cornell-Bell AH, Thomas PG, Smith SJ. The excitatory neurotransmitter glutamate causes filopodia formation in cultured hippocampal astrocytes. *Glia* 1990; 3:322–34.
9. Crispino M, Kaplan BB, Martin R et al. Active polysomes are present in the large presynaptic endings of the synaptosomal fraction from squid brain. *J. Neurosci.* 1997; 17:7694–702.
10. Crivellato E, Nico B, Ribatti D. Ultrastructural evidence of piecemeal degranulation in large dense-core vesicles of brain neurons. *Anat. Embryol. (Berl)* 2005; 210:25–34.
11. De Camilli P, Slepnev VI, Shupliakov O et al. The structure of synapses. In: Cowen WM, Sudhof TC, Stevens CF, eds. *The structure of synapses*. Baltimore, MD: The Johns Hopkins University Press, 2000; 89–133.
12. Dehay C, Douglas RJ, Martin KA et al. Excitation by geniculocortical synapses is not 'vetoed' at the level of dendritic spines in cat visual cortex. *J. Physiol.* 1991; 440:723–34.
13. Deller T, Orth CB, Del TD et al. A role for synaptopodin and the spine apparatus in hippocampal synaptic plasticity. *Ann. Anat.* 2007; 189:5–16.
14. Eyman M, Cefaliello C, Ferrara E et al. Local synthesis of axonal and presynaptic RNA in squid model systems. *Eur. J. Neurosci.* 2007; 25:341–50.
15. Gray EG. Axo-somatic and axo-dendritic synapses of the cerebral cortex: An electron microscopic study. *J. Anat.* 1959; 93:420–33.
16. Harris KM, Landis DM. Membrane structure at synaptic junctions in area CA1 of the rat hippocampus. *Neuroscience* 1986; 19:857–72.
17. Harris KM, Stevens JK. Dendritic spines of rat cerebellar Purkinje cells: Serial electron microscopy with reference to their biophysical characteristics. *J. Neurosci.* 1988; 8:4455–69.
18. Harris KM, Stevens JK. Dendritic spines of CA1 pyramidal cells in the rat hippocampus: serial electron microscopy with reference to their biophysical characteristics. *J. Neurosci.* 1989; 9:2982–97.
19. Harris KM, Sultan P. Variation in number, location, and size of synaptic vesicles provides an anatomical basis for the non-uniform probability of release at hippocampal CA1 synapses. *J. Neuropharmacol.* 1995; 34:1387–95.
20. Heuser JE, Reese TS. Structure of the synapse. In: Kandel E, ed. *The handbook of physiology, the Nervous system I.* American physiological society, 1977; 261–94.
21. Kennedy MB. Signal-processing machines at the postsynaptic density. *Science* 2000; 290:750–4.
22. Kirov SA, Petrak LJ, Fiala JC et al. Dendritic spines disappear with chilling but proliferate excessively upon rewarming of mature hippocampus. *Neuroscience* 2004; 127: 69–80.
23. Kirov SA, Sorra KE, Harris KM. Slices have more synapses than perfusion-fixed hippocampus from both young and mature rats. *J. Neurosci.* 1999; 19:2876–86.
24. Knott GW, Quairiaux C, Genoud C et al. Formation of dendritic spines with GABAergic synapses induced by whisker stimulation in adult mice. *Neuron* 2002; 34:265–73.

25. Landis DM, Hall AK, Weinstein LA et al. The organization of cytoplasm at the presynaptic active zone of a central nervous system synapse. *Neuron* 1988; 1:201–9.
26. Lisman J, Harris KM. Quantal analysis and synaptic anatomy - integrating two views of hippocampal plasticity. *Trends Neurosci.* 1993; 16:141–7.
27. Ostroff LE, Fiala JC, Allwardt B et al. Polyribosomes redistribute from dendritic shafts into spines with enlarged synapses during LTP in developing rat hippocampal slices. *Neuron* 2002; 35:535–45.
28. Otmakhov N, Tao-Cheng JH, Carpenter S et al. Persistent accumulation of calcium/calmodulin-dependent protein kinase II in dendritic spines after induction of NMDA receptor-dependent chemical long-term potentiation. *J. Neurosci.* 2004; 24:9324–31.
29. Park M, Salgado JM, Ostroff L et al. Plasticity-induced growth of dendritic spines by exocytic trafficking from recycling endosomes. *Neuron* 2006; 52:817–30.
30. Peters A, Palay SL, Webster HD. *The fine structure of the nervous system: The neurons and supporting cells.* 3 ed. Philadelphia, London, Toronto: W.B. Saunders, Co., 1991; 1–406.
31. Petrak LJ, Harris KM, Kirov SA. Synaptogenesis on mature hippocampal dendrites occurs via filopodia and immature spines during blocked synaptic transmission. *J. Comp. Neurol.* 2005; 484:183–90.
32. Piper M, Holt C. RNA translation in axons. *Ann Rev. Cell Dev. Biol.* 2004; 20:505–523.
33. Rizzoli SO, Jahn R. Kiss-and-run, Collapse and 'Readily Retrievable' Vesicles. *Traffic* 2007; 8:1137–1144.
34. Rostaing P, Real E, Siksou L et al. Analysis of synaptic ultrastructure without fixative using high-pressure freezing and tomography. *Eur. J. Neurosci.* 2006; 24:3463–74.
35. Schikorski T, Stevens CF. Morphological correlates of functionally defined synaptic vesicle populations. *Nat. Neurosci.* 2001; 4:391–5.
36. Shepherd GM, Harris KM. Three-dimensional structure and composition of CA3-->CA1 axons in rat hippocampal slices: implications for presynaptic connectivity and compartmentalization. *J. Neurosci.* 1998; 18:8300–10.
37. Sorra KE, Harris KM. Occurrence and three-dimensional structure of multiple synapses between individual radiatum axons and their target pyramidal cells in hippocampal area CA1. *J. Neurosci.* 1993; 13:3736–48.
38. Sorra KE, Mishra A, Kirov SA et al. Dense core vesicles resemble active-zone transport vesicles and are diminished following synaptogenesis in mature hippocampal slices. *Neuroscience* 2006; 141:2097–106.
39. Sorra KE, Mishra A, Kirov SA et al. Dense core vesicles resemble active-zone transport vesicles and are diminished following synaptogenesis in mature hippocampal slices. *Neuroscience* 2006; 141:2097–106.
40. Spacek J, Harris KM. Three-dimensional organization of smooth endoplasmic reticulum in hippocampal CA1 dendrites and dendritic spines of the immature and mature rat. *J. Neurosci.* 1997; 17:190–203.
41. Steward O, Levy WB. Preferential localization of polyribosomes under the base of dendritic spines in granule cells of the dentate gyrus. *J. Neurosci.* 1982; 2:284–91.
42. Tada T, Sheng M. Molecular mechanisms of dendritic spine morphogenesis. *Curr. Opin. Neurobiol.* 2006; 16(1): 95–101.
43. Torrealba F, Carrasco MA. A review on electron microscopy and neurotransmitter systems. *Brain Res. Brain Res. Rev.* 2004; 47:5–17.
44. Ventura R, Harris KM. Three-dimensional relationships between hippocampal synapses and astrocytes. *J. Neurosci.* 1999; 19:6897–906.
45. Volterra A, Magistretti PJ, Haydon PG. *The Tripartite Synapse.* Oxford: Oxford University Press, 2002.

46. Wilson CJ, Groves PM, Kitai ST et al. Three dimensional structure of dendritic spines in rat striatum. *J. Neurosci.* 1983; 3:383–98.
47. Witcher MR, Kirov SA, Harris KM. Plasticity of perisynaptic astroglia during synaptogenesis in the mature rat hippocampus. *Glia* 2007; 55:13–23.
48. Xu-Friedman MA, Harris KM, Regehr WG. Three-dimensional comparison of ultrastructural characteristics at depressing and facilitating synapses onto cerebellar Purkinje cells. *J. Neurosci.* 2001; 21:6666–72.
49. Yankova M, Hart SA, Woolley CS. Estrogen increases synaptic connectivity between single presynaptic inputs and multiple postsynaptic CA1 pyramidal cells: a serial electron – microscopic study. *Proc. Natl. Acad. Sci. U.S.A* 2001; 98:3525–30.
50. Ziv NE, Garner CC. Cellular and molecular mechanisms of presynaptic assembly. *Nat. Rev. Neurosci.* 2004; 5:385–99.

The Role of Glutamate Transporters in Synaptic Transmission

Dwight E. Bergles[1] and Robert H. Edwards[2]

[1] The Solomon H. Snyder Department of Neuroscience, Johns Hopkins School of Medicine, 725 N. Wolfe St., WBSB 1003, Baltimore, MD 21205, USA, dbergles@jhmi.edu
[2] The Departments of Neurology and Physiology, UCSF School of Medicine, San Francisco, CA 94158-2517, USA, robert.edwards@ucsf.edu

1 Introduction

Exocytotic release of glutamate from preformed synaptic vesicles confers great speed on synaptic signaling, but also requires mechanisms to maintain the pool of available synaptic vesicles and remove glutamate from the extracellular space. Since the axon terminal is in general located at a great distance from the cell body, synapses rely on the local recycling of both synaptic vesicle membrane and their neurotransmitter contents. In this review, we address recycling of the excitatory transmitter glutamate, focusing on transport activities that operate at the plasma membrane and the synaptic vesicle, and the role that these transporters play in determining the amplitude and time course of synaptic responses.

2 Glutamate Transport into Synaptic Vesicles and the Presynaptic Regulation of Quantal Size

Although analysis of the neuromuscular junction originally suggested that the response to release of a single vesicle filled with neurotransmitter is fixed, and hence represents the elemental "quantum" of synaptic transmission (109), considerable work has now shown that quantal size can change as a function of activity, contributing to such forms of plasticity as long-term potentiation (124). Nonetheless, the locus for this regulation is postsynaptic, and involves changes in receptor number or sensitivity. More recently, it has become clear that changes in quantal size can also reflect presynaptic changes in vesicle filling.

It has long been appreciated that changes in the amount of neuromodulator released per vesicle can have profound consequences for the activation of receptors at a distance from the release site. Many G protein-coupled receptors have a relatively high affinity for their peptide and monoamine ligands, but the small

J.W. Hell, M.D. Ehlers (eds.), *Structural and Functional Organization of the Synapse*,
DOI: 10.1007/978-0-387-77232-5_2,

amounts of ligand that actually impinge on receptors are not likely to saturate binding. As a result, the release of more modulator activates more receptors, and considerable attention has focused on the regulation of quantal size for monoamines, taking advantage of electrochemical detection to measure dopamine release directly and in real time (181).

It has been less clear whether changes in vesicle filling with classical transmitters such as acetylcholine, GABA and glutamate make a difference in the postsynaptic response. These transmitters are generally released in close apposition to postsynaptic receptors, many of which are ionotropic and have a high affinity for ligand (such as NMDA receptors for glutamate). If receptors are normally saturated by the contents of a single vesicle, packaging more transmitter will have no effect on the postsynaptic response.

If receptors are saturated, this will tend to reduce the variation in postsynaptic response. However, quantal size exhibits considerable variation, particularly at central synapses. Although this might result from variation in the distance of different synapses from the recording electrode, due to differences in electrotonic filtering, as well as variation in release probability and the number of receptors at each synapse, a number of observations have demonstrated that variation in quantal size is intrinsic to a single synapse. Focal stimulation of one bouton, or localized dendritic recording, both show variation in quantal size similar to that observed from electrical stimulation of release from multiple boutons (17, 63, 120, 121). Increased cytosolic glutamate in the presynaptic terminal also increases quantal size at the calyx of Held in the auditory pathway (99), providing additional evidence against receptor saturation. Remarkably, a single vesicle filled with glutamate fails to saturate low-affinity AMPA receptors as well as high-affinity NMDA receptors (123, 132). Consistent with this, AMPA and NMDA responses are highly correlated at individual synapses, supporting a presynaptic locus for the variation. GABA receptors at many (but not all) inhibitory synapses also appear not to be saturated by a single vesicle (14, 67, 79).

How can synaptic release fail to saturate receptors? Although the concentration of transmitter achieved in the synaptic cleft is high, the receptors are closely apposed to the release site, and many are of high affinity, the peak concentration of transmitter is very brief, so that only a few receptors become activated. Regardless of the precise explanation, changes in the amount of transmitter per vesicle are thus predicted to have a major influence on the postsynaptic response.

The amount of neurotransmitter released from a synaptic vesicle may be controlled either before or after the fusion event. *After* fusion, premature closure of the pore may interrupt the full release of vesicle contents. Indeed, the exocytosis of large dense core vesicles frequently exhibits "kiss-and-run", but this mechanism remains controversial for small synaptic vesicles, and the topic has recently been reviewed elsewhere (60, 82). This review focuses on changes in quantal size *before* fusion with the plasma membrane, that involve direct changes in vesicle filling.

It is important to note that the mechanism of vesicular release poses several inherent problems. Large amounts of transmitter per vesicle will result in the activation of more receptors, but high rates of firing will also deplete transmitter from the terminal unless it is actively replaced by, for example, recycling or biosynthesis. At the same time, vesicular transport is generally slow, and may limit refilling if vesicles recycle quickly, even at concentrations of cytosolic transmitter that saturate the transport mechanism. Subsaturating cytosolic concentrations will further slow refilling and release. However, low cytosolic concentrations may be important to prevent the oxidation and toxicity of monoamines such as dopamine (136), and this is compensated for by the ability of the vesicular monoamine transporter to generate an extremely large concentration gradient up to 10^5 higher in the lumen than the cytoplasm. Other classical transmitters including glutamate produce toxicity through a specific interaction with cell surface receptors, and can therefore be tolerated at higher levels in the cytoplasm. This reduces the magnitude of the concentration gradient required to fill vesicles with glutamate, and presumably also speeds filling. We will therefore consider now the factors that influence vesicle filling with glutamate, from its cytosolic concentration to the H^+ electrochemical gradient that drives transport, glutamate transport itself, and finally, the physiological regulation of these mechanisms and their role in synaptic plasticity. The amount of transmitter achieved inside secretory vesicles indeed reflects the cytosolic concentration of transmitter, the driving force, transport into and non-specific leakage across the vesicle membrane.

2.1 Cytosolic Glutamate: Biosynthesis and Recycling

Like any enzymatic reaction, the concentration of lumenal transmitter (product) depends directly on the cytosolic concentration (substrate). In the case of monoamines, electrochemistry has recently made it possible to measure cytosolic concentrations directly, by inserting a small carbon fiber electrode into a patch pipette (136). However, glutamate does not oxidize with the properties required for electrochemical detection, and less direct methods have therefore been used to estimate cytosolic concentrations. Immunocytochemistry using specific antibodies to glutamate followed by electron microscopy originally suggested low millimolar levels in the cytoplasm (173, 179). More recently, dialysis of the calyx of Held with 50 mM glutamate increased the postsynaptic response of immature animals toward that observed by more mature, suggesting that cytosolic glutamate almost certainly exceeds 1 mM and might even approach 50 mM (99, 222).

2.1.1 Glutaminase

The cytosolic concentration of transmitter in turn depends on both biosynthesis and recycling. In the case of glutamate, classical studies have shown that the preponderance of glutamate released as transmitter derives from glutamine (81). The neuronal enzyme glutaminase (also referred to as the phosphate-activated or kidney glutaminase, PAG) converts glutamine to glutamate before transport into synaptic vesicles (42, 114), and a liver isoform may also contribute (45, 113). Interestingly, both isoforms can be regulated by inorganic phosphate (45), which converts the inactive monomer to active tetramer. However, the loss of kidney glutaminase has shown remarkably little defect in excitatory transmission (129). The animals do not survive past birth, but the enzyme also has roles in nitrogen metabolism and pH regulation that may account for the observed lethality. Indeed, excitatory neurotransmission appears grossly normal. The knock-out animals exhibit only a modest defect in the response to prolonged high frequency stimulation. It is possible that the liver enzyme compensates for loss of kidney PAG, but extracts from the knock-out mice show no residual glutaminase activity. On the other hand, the assay conditions may have precluded measurement of the liver isoform. Despite the early neurochemical and neuroanatomic studies, the physiological role of glutaminase in transmitter release thus remains unknown.

2.1.2 Glutamine-glutamate cycle

In contrast to the direct reuptake of monoamines, GABA, and glycine into the nerve terminals from which they were released, glutamate is cleared primarily by the Na^+-dependent excitatory amino acid transporters EAAT1 and 2 expressed by astrocytes (96, 164, 186, 191). After uptake by astrocytes, glutamate is converted into glutamine by glutamine synthetase (157, 158, 166, 215). The efflux of glutamine by amino acid system N transporters expressed in glia and the uptake of glutamine by closely related system A transporters expressed in neurons suggest a mechanism for glutamine transfer back to neurons, where glutaminase can produce glutamate (reviewed previously (35)).

Despite the circumstantial evidence, does the glutamine-glutamate cycle actually contribute to synaptic transmission? Partial blockade of system A with the high affinity, slowly transported substrate methyl-aminoisobutyric acid has been reported to decrease quantal size in autapses (6). In addition, disrupting glutamine synthesis and transport may reduce epileptic activity (11, 192). However, it has been very difficult to demonstrate a physiological role for glutamine uptake in glutamate release (105). On the other hand, the glutamine-glutamate cycle may have a more clear role at certain inhibitory synapses. Blocking glutamine production by astrocytes and the subsequent transfer to neurons decreases the size of both evoked and miniature inhibitory currents in the hippocampus, although the effect requires moderate synaptic activity (119). The glutamine-glutamate cycle may indeed be particularly important at inhibitory synapses that do not express a plasma membrane GABA transporter, and hence rely on other mechanisms for recycling.

2.2 H^+ Electrochemical Gradient

The release of neurotransmitter by exocytosis requires transport into secretory vesicles, a process that involves the exchange of lumenal H^+ for cytosolic transmitter. The packaging of all transmitters thus depends on a H^+ electrochemical gradient across the vesicle membrane which is generated by the vacuolar-type H^+-ATPase (Fig. 1). Indeed, the vacuolar H^+ pump acidifies endosomes, lysosomes and dense core vesicles as well as synaptic vesicles. (The yeast lysosome is a large membranous compartment called the vacuole). Why does the cell use H^+ rather than other ions such as Na^+, K^+ and Cl^- to drive transport into synaptic vesicles? Along with their small size, the rapid recycling of synaptic vesicles imposes a number of surprising constraints. Unlike the plasma membrane gradients of Na^+, K^+ and Cl^- that are large and stable, synaptic vesicles rely on a H^+ electrochemical gradient that is alternately dissipated and regenerated by exo- and endocytosis. The dependence on H^+ may serve to reduce the inappropriate activation of vesicle transporters on the cell surface. In addition, a vesicle pH of 6 (only 1 μM) translates into less than one free H^+ per vesicle, although buffering by vesicle proteins requires the entry of more H^+ to reach this lumenal pH. In contrast to the high millimolar concentrations of other ions, fewer H^+ are thus required to generate a gradient of the same magnitude as Na^+, K^+ and Cl^- at the plasma membrane (roughly two log units). The relatively low concentrations of H^+ under physiological conditions appear to confer flexibility, efficiency and the potential for rapid regulation.

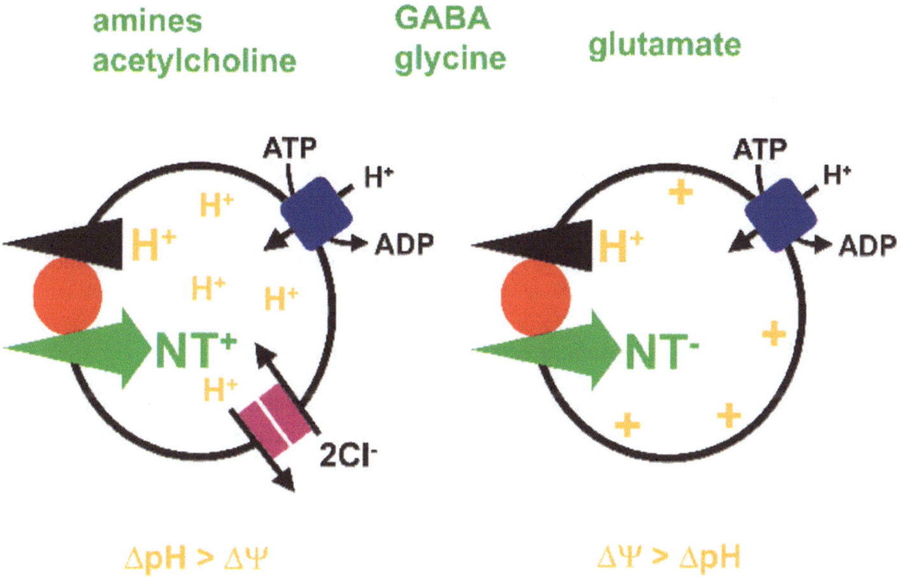

Fig. 1. Bioenergetics of vesicular neurotransmitter transport.
The vacuolar H^+-ATPase generates a H^+ electrochemical gradient that drives the uptake of neurotransmitter into secretory vesicles through a mechanism involving H^+ exchange. However, the transport of different transmitters depends to varying extents on the two components of the H^+ electrochemical gradient, the chemical gradient (ΔpH) and the membrane potential (ΔΨ). The transport of cationic transmitters such as monoamines and acetylcholinie depends primarily on ΔpH and the transport of anionic glutamate on ΔΨ, with zwitterionic GABA and glycine transport more equally dependent on both components. Nonetheless, all of the vesicular neurotransmitter transporters depend at least to some extent on both ΔpH and ΔΨ. The relative expression of ΔpH and ΔΨ is in turn influenced by the entry of chloride through Cl^-/H^+ exchangers, which dissipates ΔΨ and thus indirectly activates the H^+ pump to make a larger ΔpH.

2.2.1 H^+ Pump

The vacuolar H^+ pump that produces the principal ionic gradient across intracellular membranes resembles the F0/F1-ATPase from mitochondria and bacteria. The vacuolar pump and F0/F1-ATPase both move H^+ into a membranous compartment, but rather than using H^+ flux down its electrochemical gradient to produce ATP as in the case of the F0/F1-ATPase, the vacuolar pump generates a H^+ electrochemical gradient that depends on ATP hydrolysis. In addition, both proteins comprise two domains, a peripheral V1 domain that catalyzes either ATP production or hydrolysis, and a membrane-embedded V0 domain that moves H^+. Each of these domains in turn contains multiple subunits, whose specific functions remain poorly understood, although structural studies are beginning to provide suggestions (56, 62). Indeed, the v-ATPase uses a rotary mechanism very similar to the F0/F1-ATPase, with the V1 domain rotating relative to V0 during H^+ movement (138). In yeast, the vacuolar H^+ pump undergoes regulation at the level of trafficking and intrinsic transport activity (110, 125). Starvation produces disassembly of the V1 subunit, presumably to conserve ATP (106). However, direct regulation of the H^+ pump in the synaptic vesicle membrane has not yet been shown.

Interestingly, work in yeast and *Drosophila* has suggested a role for the vacuolar H^+ pump and specifically the V0 domain in membrane fusion, independent of its role in vesicle acidification (92, 152). A mutation in the V0 a1 subunit impairs synaptic vesicle exocytosis rather than acidification. Although the mechanism remains unknown, previous work has suggested that another subunit of V0 mediates the non-vesicular release of transmitter (57), and the same subunit contributes to homotypic vacuole fusion in yeast (152).

2.2.2 Chloride flux: pH gradient versus membrane potential

In contrast to the lack of information about direct regulation of the vacuolar H^+ pump, other ions have important roles in the indirect activation of pump activity. Indeed, purified synaptic vesicles (as well as endosomes and lysosomes) acidify very little with the addition of only ATP, and require the subsequent addition of a permeant anion such as chloride for massive acidification (Fig. 1). Presumably, the entry of chloride dissipates the inside positive membrane potential ($\Delta\Psi$) and so reduces one of the gradients against which the pump works, allowing it to generate a larger pH gradient (ΔpH). Consistent with this, chloride increases the ATPase activity of the H^+ pump (62). Chloride flux thus provides a mechanism to regulate independently the electrical ($\Delta\Psi$) and chemical (ΔpH) components of the H^+ electrochemical gradient.

A subset of the ClC channels mediate the flux of chloride across intracellular membranes required for vesicle acidification (102). The ClC-3 isoform is highly expressed in brain, and synaptic vesicles from the ClC-3 knock-out show impaired acidification (178). However, the effect on transmitter release and in particular quantal size was difficult to assess, because the knock-out also shows severe, early degeneration in the hippocampus and retina, with the remaining synapses presumably expressing a different ClC isoform and hence showing no impairment.

Originally considered channels, the intracellular ClCs appear to function as Cl^-/H^+ exchangers (1, 153, 170) (Fig. 1). In light of the H^+ exchange mechanism, how can Cl^- entry promote vesicle acidification? The stoichiometry of ionic coupling apparently provides the answer: the exchange of 2 Cl^- for one H^+ results in the movement of substantially more charge (3+) than H^+ (1). This ionic coupling in turn predicts more dissipation of $\Delta\Psi$ than ΔpH, producing secondary activation of the H^+ pump that exceeds the direct exchange of one lumenal H^+ and hence results in vesicle acidification. On the other hand, it is not clear why a Cl^-/H^+ exchanger would be selected for over a simple Cl^- channel.

2.2.3 Regulation of the H^+ electrochemical gradient

Several factors including chloride flux have the potential to regulate the vesicle H^+ electrochemical gradient, but does it undergo physiological regulation? In parafollicular cells of the thyroid, activation of a cell surface calcium receptor promotes vesicle acidification, apparently by increasing the chloride conductance (13, 188). Protracted stimulation of adrenal chromaffin cells also promotes vesicle acidification, increasing the proportion of larger "active" vesicles with a halo around the dense core, and quantal size (155). Indeed, chromaffin cells have been shown to increase quantal size with strong stimulation, and although this has been attributed to changes in the fusion pore, changes in the H^+ electrochemical gradient may in fact be responsible (58). These observations indicate the potential for physiological regulation of H^+ electrochemical gradient, but it is important to note that the mechanisms involved remain poorly understood.

3 Vesicular Glutamate Transport

The transport of all classical transmitters into synaptic vesicles requires a H^+ electrochemical gradient generated by the vacuolar H^+ pump, but transmitters differ in their reliance on the two components of the gradient, ΔpH and $\Delta\Psi$. The vesicular transport of monoamines and acetylcholine depend primarily on ΔpH, while the transport of glutamate depends primarily on $\Delta\Psi$ (33, 103, 131) (Fig. 1). Vesicular GABA transport lies somewhere in between, with more equal reliance on the chemical and electrical components of the H^+ electrochemical gradient (89, 111). Consistent with these differences in ionic coupling, the vesicular neurotransmitter transporters fall into three molecularly distinct families. However, all of the activities rely at least in part on $\Delta\Psi$ and are hence electrogenic. We will now focus on the vesicular glutamate transporters, describing their functional properties, identification, distribution and role in synaptic transmission.

3.1 Functional Characteristics

Vesicular glutamate transport exhibits several distinctive properties. Unlike the plasma membrane transporters, vesicular glutamate transport does not recognize aspartate (137). The primary dependence on $\Delta\Psi$ rather than ΔpH has also suggested that it functions as an ion channel rather than a H^+ exchanger (33, 131). To some extent, the apparent dependence of vesicular glutamate transport on $\Delta\Psi$ and ΔpH will reflect the relative magnitude of the two components under the conditions studied. However, the difference in mechanism between channel and transporter has profound consequences for the amount of glutamate achieved per vesicle. A channel would concentrate glutamate only ~10-fold for a $\Delta\Psi$ ~60 mV, but cytoplasmic glutamate ~10 mM would require a gradient of only ~10 to produce 100 mM glutamate inside the vesicle. On the other hand, since glutamate is an anion, the exchange of 1 lumenal H^+ would result in the net movement of 2+ charge. With this coupling, ΔpH ~2 and ΔΨ ~60 mV would produce a substantially larger glutamate gradient ~10^4. Although high cytosolic concentrations of glutamate indicate that a gradient of this magnitude is not needed to fill the vesicle, most of the evidence supports an H^+ exchange mechanism. Dissipation of ΔpH alone has minimal effect on vesicular glutamate transport, but the residual activity after dissipation of ΔΨ is abolished by the simultaneous dissipation of ΔpH (19, 182).

Glutamate transport into synaptic vesicles also exhibits unusual regulation by chloride. 2–10 mM Cl^- confers robust activity, with less but still detectable transport at both lower and higher concentrations (137). High chloride concentrations presumably dissipate the ΔΨ necessary for transport. Originally, it was thought that low chloride concentrations produce a small ΔpH that was also necessary for transport. However, the chloride dependence persists even with ΔpH clamped (218), and chloride apparently interacts with an allosteric site on the transporter that regulates both forward and reverse flux (84). Despite repeated demonstration of these findings, physiological analysis has recently shown that intracellular chloride concentration does not affect quantal size (160). At the calyx of Held, where receptors are not saturated by the transmitter in a single synaptic vesicle, the size of spontaneous events does not vary when the presynaptic terminal is dialyzed with a wide range of chloride concentrations from 5–100 mM. How can the physiological and biochemical results be reconciled? Although not proven, it seems most likely that the difference lies in the nature of the measurement. Radiotracer flux assesses only the earliest phase of uptake and is a kinetic measurement. In contrast, quantal size reflects a thermodynamic equilibrium. Thus, chloride concentration may affect only the rate of vesicle filling, not the final gradient achieved. Consistent with this, chloride has much less of an effect on transport assays using high concentrations of glutamate, which may reflect more of a steady-state (218).

3.2 Vesicular Glutamate Transporters

The vesicular glutamate transporters (VGLUTs) were originally identified as type 1 Na^+-dependent inorganic phosphate transporters (140), but it has since become clear that most members of this family perform a function other than phosphate transport. Sialin is a H^+ exchanger that exports sialic acid from lysosomes (135, 203, 219), and the founding member of the type I phosphate transporters NaPi-1 recognizes organic anions (28, 31). In addition, glutamate transport by the VGLUTs does not depend on Na^+, and inorganic phosphate does not inhibit glutamate uptake by vesicles expressing VGLUT1 (19). Rather, heterologous expression of the VGLUTs seems to up-regulate endogenous phosphate transport activities (which can be Na^+-independent or -dependent) (R.H.E., unpublished observations), and NaPi-1 behaves similarly (28). It is therefore remarkable that purified, reconstituted VGLUT2 has recently been shown to catalyze Na^+-dependent phosphate transport as well as glutamate/H^+ exchange (104). Using the reconstituted system, mutations that interfere with glutamate transport do not disrupt phosphate uptake, raising the possibility that the two activities use entirely different mechanisms. What role might phosphate uptake across the plasma membrane serve at the nerve terminal? Interestingly, phosphate activates both glutaminase isoforms responsible for glutamate production, and since the VGLUTs localize to synaptic vesicles, this might provide a mechanism to link glutamate production to release (18).

Consistent with a primary role in neurotransmitter packaging, heterologous expression of the VGLUTs reproduces all of the functional characteristics previously described using synaptic vesicles from the brain, including selective recognition of glutamate rather than aspartate, and biphasic regulation by chloride (19, 64, 66, 78, 90, 183–185, 201). In addition, the related NaPi-1 shows a large chloride conductance that is inhibited by organic anions (31), and VGLUT1 exhibits a chloride conductance which is inhibited by glutamate (but not aspartate) (19). The chloride conductance may indeed somehow contribute to the allosteric regulation of VGLUTs by chloride, but we still understand very little about the relationship between glutamate transport and chloride. Indeed, the available radiotracer flux assays, particularly after heterologous expression, currently limit our understanding of the VGLUTs. The vesicular monoamine transporters require only transfection into non-neural cells for robust activity, but the VGLUTs require more specialized conditions, and remain difficult to assay.

How do the properties of the VGLUTs contribute to glutamate release evoked by high frequency stimulation? In contrast to the vesicular monoamine transporters, which have an extremely high apparent affinity for their substrates (Km ~1 μM), essentially all of the other vesicular neurotransmitter transporters including the VGLUTs have a low apparent affinity (Km ~1–3 mM in the case of VGLUTs). This low apparent affinity presumably allows the VGLUTs to transport larger amounts of glutamate more rapidly, but we do not know the actual maximal turnover rate per transport protein. The turnover of VMATs for serotonin at 29° C is ~5/sec/protein, and presumably higher for the lower affinity substrate dopamine which has a 4-fold higher Vmax (151). The vesicular acetylcholine transporter has an even lower turnover of ~1/sec (200). However, the ability of VGLUTs to acidify synaptic vesicles as glutamate enters suggests that they transport much more rapidly. Indeed, glutamate behaves very much like chloride in promoting vesicle acidification, presumably by dissipating $\Delta\Psi$ and indirectly activating the H^+ pump.

3.2.1 Distribution and Synaptic Role

VGLUT1 and 2. The three characterized VGLUT isoforms show complementary distributions in the mature brain. Excitatory neurons in the cortex, hippocampus and cerebellar cortex (granule cells) express VGLUT1 whereas cells in the thalamus and brainstem express VGLUT2 (19, 64, 66, 78, 90, 184, 185, 201). Considering a possible alternative role in phosphate transport, how does the loss of VGLUT1 influence transmitter release? As predicted, the VGLUT1 knock-out shows no release at mature synapses that normally express VGLUT1 in adulthood (65). However, residual glutamatergic transmission persists at these synapses during early postnatal development, apparently due to the transient expression of VGLUT2. Further, the VGLUT1-independent transmission at these sites appears normal in many respects, suggesting that it originates from distinct synapses. On the other hand, VGLUT1-independent transmission in the hippocampus does exhibit more rapid synaptic depression than wild type, possibly because the vesicles at these synapses recycle more slowly than the more prevalent VGLUT1 synapses of wild type mice, or because VGLUT1 and 2 themselves differ in the rate of recycling. Consistent with differences in trafficking, recent work has shown that VGLUT1 interacts with endophilin, a protein involved in synaptic vesicle endocytosis, and this interaction influences the rate of recycling (205). Loss of VGLUT1 also produces abnormal endocytic structures (65), suggesting roles for the transporter in aspects of the synaptic vesicle cycle other than vesicle filling. Like vesicular monoamine and acetylcholine transporters (43, 149), however, inhibition of vesicular glutamate transport using the H^+ pump inhibitor bafilomycin does not impair vesicle recycling (227), and hippocampal neurons from the VGLUT1 knock-out can release empty vesicles (217). The VGLUT2 knock-out reported more recently cannot release glutamate at synapses in the brainstem involved in generation of the respiratory rhythm, and hence dies just after birth (134, 211).

VGLUT3. In contrast to the expression of VGLUT1 and 2 by known excitatory neurons, VGLUT3 is expressed by cells not traditionally considered glutamatergic, including serotonergic neurons in the raphe nuclei, cholinergic interneurons in striatum, and a subset of GABAergic interneurons in the cortex and hippocampus (64, 78, 169). VGLUT3 is also expressed outside the nervous system, and in contrast to VGLUT1 and 2, heterologous expression has not yet conferred detectable glutamate release (183), raising questions about the role of this isoform. Indeed, the glutamate packaged by VGLUT3 might simply influence the storage of the other classical transmitter released by VGLUT3$^+$ neurons (e.g., serotonin in the raphe). It is also possible that the glutamate stored by VGLUT3 has a role in metabolism. However, several observations indicate that VGLUT3 does contribute to glutamate release.

In contrast to VGLUT1 and 2, VGLUT3 can localize to the cell body and dendrites as well as the axon terminal. In projection neurons such as serotonergic cells in the raphe nuclei, VGLUT3 localizes almost exclusively to the axon terminal, but in interneurons such as cholinergic interneurons in the striatum and GABAergic interneurons in the cortex and hippocampus, VGLUT3 also localizes to the cell body and dendrites (64). In addition, VGLUT3 appears in the dendrites of pyramidal neurons in layer 2/3 of cortex, where the pharmacology indicates a role for the protein in dendritic release and retrograde synaptic signaling (83). Trafficking of VGLUT3 to different cellular compartments may therefore contribute to novel modes of signaling by glutamate.

VGLUT3 may have a particular role in plasticity. In the brainstem auditory pathway, synapses formed by neurons in the medial nucleus of the trapezoid body (MNTB) onto neurons of the lateral superior olive (LSO) switch from GABA to glycine during the tonotopic refinement and synapse strengthening that occurs in early postnatal development. These changes are presumably important for sound localization. Interestingly, MNTB neurons also transiently release glutamate that activates NMDA receptors, suggesting a role for the glutamate in plasticity, and the brief expression of VGLUT3 coincides temporally with this glutamate release (70). However, it remains unclear whether the early glutamate release actually contributes to synapse development and sound localization.

Other Vesicle Carriers. In addition to the H^+ pump, chloride carriers and neurotransmitter transporters, other factors may influence the contents of synaptic vesicles and hence quantal size. Extracellular ATP activates specific receptors but also promotes the co-storage of cationic transmitters such as serotonin (12) in the cytosol, although the vesicular ATP transporter remains unknown. Similarly, cationic compounds may promote the packaging of an anionic transmitter such as glutamate. A number of polytopic membrane proteins that localize specifically to synaptic vesicles indeed remain poorly understood. The SV2 family shows strong homology to a number of carbohydrate transporters, but its biochemical function also remains unknown. The loss of SV2A alone results in seizures and death shortly after birth, and physiological studies have suggested roles in both calcium regulation and the readily releasable pool of synaptic vesicles (44, 101). Remarkably, the SV2A isoform appears to be the direct target for the major clinical anticonvulsant levetiracetam (Keppra) (122).

4 The Presynaptic Regulation of Quantal Size

The amount of transmitter released per vesicle can be regulated in two distinct ways, by changes in the lumenal concentration, and by changes in vesicle size. We will now present several models for the presynaptic regulation of quantal size, each of which has the potential to influence one or both of these parameters.

4.1 Equilibrium

At thermodynamic equilibrium, the stoichiometry of ionic coupling, the H^+ electrochemical driving force, and the cytosolic concentration of transmitter should determine the concentration of transmitter inside a secretory vesicle. The number of transporters should influence vesicle filling only if there is not enough time to reach equilibrium, which could occur if the rate of vesicle recycling exceeds the rate of filling. The low turnover numbers for vesicular monoamine and ACh transport suggest that this might indeed be the case at aminergic and cholinergic synapses, particularly when stimulated at high frequency. On the other hand, the number of transporters expressed can influence quantal size even under resting conditions, implicating other mechanisms.

4.2 Transporter Expression and Leak

Initial studies of the vesicular monoamine transporter VMAT2 and the vesicular acetylcholine transporter (VAChT) have shown that changes in expression can influence quantal size independent of release rate. The over-expression of VMAT2 in chromaffin cells and neurons increases quantal size measured directly by amperometry (154), and these cells show little spontaneous activity in culture. Over-expression of VAChT at the developing neuromuscular junction also increases quantal size (175). Conversely, the loss of one functional VMAT2 allele reduces monoamine release (61, 196) and a reduction in VAChT by only ~45% both reduces quantal size and impairs cognitive function (159).

If the vesicles have enough time to fill, why does transporter expression make a difference? Multiple observations indicate a nonspecific (i.e., not mediated by the transporter) leak of transmitter across the vesicle membrane. Both native synaptic vesicles and the endosomes used for heterologous expression of the transporters exhibit substantial accumulation of monoamine driven by the H^+ electrochemical gradient, even without a functional transporter. Presumably, the unprotonated, neutral amine diffuses into the vesicle, then undergoes protonation that prevents its efflux. Nonspecific efflux must therefore also exist, and the efflux of monoamine from chromaffin granules triggered by dissipation of ΔpH is not blocked by VMAT inhibitors such as reserpine (128). More vesicle transporters may serve to offset this leak.

VGLUT expression may also influence quantal size. Using purified synaptic vesicles, inhibitors reduce both the steady-state amount and rate of glutamate accumulation (214). Over-expression of VGLUTs in transfected hippocampal neurons also appears to increase quantal size (214, 217). Conversely, the loss of either VGLUT1 or 2 can reduce quantal size (134, 217). The VGLUTs also undergo transcriptional regulation during development (25, 65, 133), and VGLUT1 and 2 were both originally identified in screens for differentially expressed genes (3, 140). Activity reduces VGLUT1 expression at both RNA and protein levels (49), suggesting that it is also under homeostatic control. However, VMAT2 is up-regulated in response to increased activity. Remarkably, the expression of VGLUT1 protein but not mRNA also appears to cycle with a circadian rhythm (223). These changes in transporter expression provide circumstantial evidence for the presynaptic regulation of quantal size.

Despite the evidence for regulation of neurotransmitter transport in vitro and transporter expression in vivo, other work suggests that VGLUT expression does not limit vesicle filling. Heterozygous VGLUT1 knock-out mice carrying a single functional VGLUT1 allele show no difference from wild type in hippocampal field potentials (65). Any change in quantal size should be reflected in this response, so the absence of a difference from wild type effectively rules out a change in quantal size with VGLUT reduction by 50%. Second, quantal size at the *Drosophila* neuromuscular junction does not change as VGLUT expression is reduced (47). Only the frequency of spontaneous events declines as the expression of VGLUT drops below 10–20% of the wild type level. The loss of miniature events presumably reflects the exocytosis of "empty" vesicles. Indeed, a single VGLUT apparently suffices to fill a vesicle, at least at this synapse. Although a modest reduction in VGLUT expression does not affect glutamate release, VGLUT over-expression in *Drosophila* appears to increase quantal size (48).

How can the results *in vivo* be reconciled with the biochemical and physiological studies in vitro? First, it is important to note that radiotracer flux assays measure the kinetics of transport, rather than the thermodynamic equilibrium. Second, the very low levels of residual transporter expression in autaptic cultures made from knock-out mice may well produce smaller quantal sizes than wild type, but the work from *Drosophila* suggests that more physiological reductions in expression might not have produced detectable changes. However, it is more difficult to understand how VGLUT over-expression increases quantal size when even a 50% reduction has no effect. VGLUT expression may influence vesicle size, as suggested by recent work in *Drosophila* (48). And if VGLUT expression does not influence vesicle filling, why would this differ from vesicular monoamine and acetylcholine transport (61, 154, 175)? One possibility is that cationic molecules may exhibit a greater nonspecific leak across the membrane, possibly due to abundant acidic phospholipid. In contrast, anions would have more difficulty permeating directly through this membrane. In addition, it is important to note that even if VGLUT expression does not affect the size of spontaneous events, it may become limiting with high rates of activity and vesicle recycling, when it is more difficult to measure quantal size directly. Indeed, reduced VGLUT2 expression in retinal ganglion cells of the zebrafish mutant *blumenkohl* has remarkably modest effects on neurotransmission, but dramatic effects on fine behavior guided by vision (174).

Differences in non-specific leak may also account for the distinct effects of vesicular transport inhibitors on vesicle storage. Without stimulation, an inhibitor should deplete stores only in the presence of a leak—in this case, the stores reflect a balance between leak and ongoing transport activity. In contrast, a transmitter that does not leak should require stimulation to deplete the stores. Permanently charged ACh should leak less than reversibly protonated monoamines, and thus require activity to deplete the stores. Indeed, the VAChT inhibitor vesamicol does not affect vesicular stores without stimulation (32), whereas the vesicular monoamine transport inhibitor reserpine rapidly depletes monoamines from resting chromaffin granules, without stimulation (112). Since there are very few if any potent, specific VGLUT inhibitors, it remains unclear how fast the vesicular stores would be depleted after transporter block. The H^+ pump inhibitor bafilomycin depletes stores relatively rapidly, but might simply promote reverse flux through the VGLUTs (227).

4.3 Set Point

Vesicular neurotransmitter transport has been suggested to undergo direct regulation at the level of the protein as well as transcription. A proteolytic fragment of the cytoskeletal protein fodrin was purified as an inhibitor of vesicular GABA and glutamate transport, but the mechanism for this regulation and its physiological role remain unknown (146, 189). In addition, heterotrimeric G proteins inhibit vesicular monoamine and glutamate transport. A variety of G proteins bind to secretory vesicles, and activation of $G\alpha o_2$ inhibits both monoamine and glutamate uptake by 30–50%, affecting Vmax rather than Km (2, 94, 147, 216). The regulation does not apparently involve a change in vesicle acidification, suggesting instead a direct interaction of α subunit with the transporter. In platelet granules that express VMAT2 and store serotonin, α_q has a similar role. Further, inhibition of transport seems to depend on lumenal transmitter: in mice lacking platelet tryptophan hydroxylase and hence serotonin, activated α_q does not inhibit transport, and the addition of monoamine restores inhibition (29). G proteins may thus convey a signal about vesicle filling back to the transporter, and so provide a set point. However, the lumenal sensor for any neurotransmitter remains unknown. Internalized plasma membrane G protein-coupled receptors might subserve this function, but it would require a major decrease in their affinity, which might occur at the low pH of secretory vesicles. In the case of vesicular glutamate transport, $G\alpha o_2$ also changes the chloride dependence, eliminating the activation by low chloride concentrations in αo_2 knock-out mice and shifting the peak activation to lower concentrations when stimulated by a non-hydrolyzable form of GTP (216). In both cases, the mechanism remains unclear.

4.4 The Relationship Between Neurotransmitter and Synaptic Vesicle Cycles

Is there a relationship between vesicle filling and the synaptic vesicle cycle of exo- and endocytosis? As mentioned above, the inhibition of filling does not appear to influence synaptic vesicle exocytosis or recycling in cholinergic, aminergic or glutamatergic neurons (43, 149, 217, 227), but the converse remains unexplored. Indeed, the allele-specific supression of a point mutation in the acetylcholine transporter by a point mutation in the v-SNARE synaptobrevin supports a connection between the synaptic vesicle and neurotransmitter cycles (167). However, similar mutations in other vesicle proteins partially rescue the transporter mutant, raising the alternative possibility that only the mutant but not wild type proteins interact.

4.5 Changes in Vesicle Volume

Secretory vesicles can vary in size as well as the concentration of lumenal transmitter. Changes in vesicle size were first recognized at the *Torpedo* electric organ, which has two biochemically distinct populations of synaptic vesicles (228). The larger, VP1 vesicles, contain 3-5-fold more acetylcholine than the smaller VP2, but VP2 expresses more acetylcholine transport activity (77). At rest, VP1 vesicles predominate, and the VP2 appear with activity (72). After stimulation, VP1 replaces VP2. VP1 may thus correspond to the reserve pool of vesicles, and VP2 to the recycling (162), although reserve and recycling pools at other synapses are functionally distinct. Considering that VP2 vesicles may simply swell to become VP1, it is interesting that the membrane of VP2 seems twice as thick as VP1 (198).

The analysis of monoamine storage and release has further indicated the potential of filling to influence vesicle size. In PC12 cells, administration of L-Dopa increases the size of dense core vesicles and the dopamine released per vesicle (detected by amperometry) (40, 74). Remarkably, however, the concentration of transmitter released (measured by voltammetry) does not change. Indeed, L-Dopa increases the size of the halo around the dense core, but not the dense core itself (40). Transport can thus regulate vesicle size without changing the concentration of lumenal transmitter. Consistent with the osmotic swelling of preformed vesicles, variation in quantal size in the leech correlates with variation in vesicle volume, and the average concentration of transmitter released per vesicle varies little despite large differences in vesicle size (30).

What accounts for the uniform lumenal concentration of transmitter? The cytosolic concentration of transmitter, the driving force for transport and the number of transporters per vesicle might generate vesicles of uniform concentration in a particular cell, but the development of an osmotic gradient provides an additional constraint that might account for the uniformity between cells. The H^+

electrochemical driving force for transport may thus produce both a concentration gradient of transmitter and an osmotic gradient which will eventually oppose any further increases in concentration. Indeed, assuming the approximate cytosolic concentrations of transmitter (1–10 mM) suggested for glutamate, the driving force and the coupling of VGLUTs to H^+ exchange (which could theoretically produce a concentration gradient of 10^4), the mechanism should produce extremely hyperosmolar lumenal contents. If these contents do not precipitate, they might be expected to distend the vesicle, presumably up to a limit determined by the physical properties of the membrane, and by the H^+ electrochemical driving force.

Changes in vesicle size may also reflect changes in biogenesis and hence intrinsic biochemical differences in the composition of vesicles. A number of *Drosophila* mutants show both an increase in quantal size and increased vesicle volume at the neuromuscular junction (107, 225). One of the mutations affects a *Drosophila* homologue of the clathrin adaptor AP180 (226) and another the lipid phosphatase synaptojanin that mediates clathrin uncoating (54), strongly suggesting that in these cases, the increased size reflects a defect in biogenesis. Chromaffin cells have also been reported to contain two distinct populations of dense core vesicles, and the adaptor AP-3 apparently controls their production (75, 76). 1b and 1s synapses at the wild type *Drosophila* neuromuscular junction also show differences in both quantal size and vesicle volume (107), supporting a role for the regulation of vesicle volume under physiological conditions. A recent study at the calyx of Held has tested the hypothesis that variation in vesicle volume underlies the variation in quantal size. The authors recorded both the capacitance change associated with fusion of a single synaptic vesicle, and the postsynaptic response (220). The results reveal no correlation between capacitance change and quantal size, indicating variation in the lumenal concentration of transmitter rather than vesicle size. However, capacitance detects only the surface area of the vesicle *after* fusion with the plasma membrane, and the role for an osmotic gradient can only operate *before* fusion. With exocytosis, any swollen vesicles presumably return to their original, undistended state.

4.6 Presynaptic Regulation of Quantal Size by Activity

Do changes in vesicle filling actually contribute to synaptic plasticity? At the neuromuscular junction, strong, protracted stimulation reduces quantal size by ~20%, and this change does not reflect alterations in postsynaptic sensitivity, implicating a presynaptic mechanism (55, 139). Inhibition of release delays this decline in quantal size, as if a pool of synaptic vesicles containing more transmitter releases first but is eventually depleted. In addition, a variety of intracellular signaling pathways acting presynaptically can influence quantal size by up to 4-fold (199). Further, blockade of neural activity for one week at the rat neuromuscular junction substantially increases quantal size, apparently through a presynaptic mechanism (212). The lack of correlation with activity in muscle further suggests that this mechanism does not involve retrograde signaling from the postsynaptic cell.

A number of other synapses also exhibit presynaptic regulation of quantal size. Stimulation increases the amount of stored glutamate in synaptosomal preparations from the rat brain (24) and quantal size detected electrochemically with adrenal chromaffin cells (155). Conversely, suppression of activity reduces the amplitude of spontaneous events through a presynaptic mechanism in hippocampal neurons, perhaps contributing to the homeostatic regulation of brain excitability (85). In *Drosophila*, quantal size fluctuates with the rate of foraging for food. Flies crawl faster just after the shift to a plate that lacks food, and the fastest subgroup exhibits a transient increase in both quantal size and evoked release that appears presynaptic (177). The increase in quantal size appears to reflect an increase in vesicle volume, but it remains unclear whether this reflects swelling of preformed vesicles or the recruitment of distinct vesicles with a different size.

4.7 Conclusions

Previous work has identified many of the proteins required for neurotransmitter uptake by the cell, and for the packaging of secretory vesicles, but we still understand relatively little about their regulation and their role in synaptic transmission. Future work in this area will focus on the physiological role of these mechanisms and their regulation in synaptic plasticity.

5 Glutamate Uptake at the Cell Surface

When a synaptic vesicle fuses with the presynaptic membrane, glutamate rapidly diffuses into the cleft where it has the first opportunity to encounter receptors. Although the peak concentration within the small volume of the cleft is estimated to reach several millimolar (39, 53), glutamate is rapidly diluted in the cleft and surrounding extrasynaptic spaces. The close proximity of low affinity AMPA and kainate receptors to release sites ensures that binding and activation occur reliably despite the rapid exit of glutamate. Although dilution and diffusion contribute substantially to the decline in the concentration of glutamate near receptors, ultimately glutamate must be removed to prevent accumulation and persistent signaling. Unlike the prototypical neurotransmitter acetylcholine, which is hydrolyzed in the extracellular space by acetylcholinesterase, extracellular enzymes capable of catalyzing the conversion or degradation of glutamate have not been identified. Instead, the removal of glutamate is catalyzed by a multigene family of integral membrane proteins termed EAATs (Excitatory Amino Acid Transporters) for the human variants, which transport glutamate from the extracellular space into the cytosol (46).

The EAAT family of transporters consists of five members (EAAT1-5), each a product of a unique gene (46), and for two members (EAAT1 and EAAT2), additional variants are generated through alternative splicing (97, 180). Highly homologous transporters are found in rodents, which have been termed GLAST (EAAT1), GLT-1 (EAAT2), and EAAC1 (EAAT3), EAAT4 and EAAT5. The diversity of glutamate transporters is unparalleled among neurotransmitter transporters (5). Although the precise role of these different transporter isoforms is still uncertain, it is likely that clearance of extracellular glutamate is parsed out among the different variants, with each transporter class responsible for removing glutamate over a distinct time frame or from a particular volume of extracellular space. By providing multiple routes for the exit of glutamate, EAATs may afford subtle control over glutamate dynamics, and thus receptor activation in and around synapses.

5.1 Clearance of Glutamate from Synapses

The impact of transporter-mediated clearance on glutamate transients and receptor activation at synapses depends on many factors, including the structure of the synapse, the number of release sites, the probability of release and patterns of activity experienced by the synapse, the proximity of neighboring synapses and their release properties, the extent of glial ensheathment, and the properties and locations of receptors. Given the vast structural and functional diversity of synapses in the CNS, it is not surprising that the involvement of transporters in synaptic signaling varies considerably among the population of excitatory synapses. It is also likely that the contribution of EAATs to glutamate clearance is not static, but changes with synaptic reorganization and shifting patterns of activity.

The involvement of EAATs in clearance of glutamate near synaptic receptors can be tested by applying transporter antagonists such as TBOA (threo-beta-benzyloxyaspartate), a substrate analog that is not transported (172), while monitoring spontaneous or evoked synaptic responses. If EAATs curtail synaptic glutamate transients, blocking transporters should increase the peak concentration of glutamate and slow the decay of each glutamate transient. Thus, transporter inhibition should increase the amplitude of synaptic currents and slow their decay; this assumes that under baseline conditions receptors are not saturated and the decay of the current is not already determined by the rate of receptor densensitization. At most excitatory synapses, AMPA receptor responses elicited following fusion of a single vesicle are not affected by transporter antagonists, indicating that quantal responses are determined primarily by the intrinsic properties of the receptors and a passive decrease in glutamate concentration through rapid diffusion and dilution in the extracellular space (91, 168). In contrast, EPSCs elicited in response to stimulation of multiple afferents typically decay more slowly in the presence of transporter antagonists, an effect that becomes more pronounced when trains of EPSCs are elicited at high frequency, or the activation of high affinity NMDA or metabotropic receptors (mGluRs) are monitored (7, 34, 36, 38, 145, 194). Such protocols raise the release probability at individual synapses and increase the likelihood that adjacent synapses will be active. These effects are partially due to the recruitment of receptors located outside the cleft (e.g. NMDA receptors and mGluRs) that are in proximity to transporters (see below). Together, these results suggest that repetitive release of glutamate from individual sites and "pooling" of glutamate from adjacent sites engage EAATs to a greater degree than isolated, univesicular release (16), allowing transporters shape synaptic responses.

In the continued presence of transporter blockers, extracellular glutamate levels rise (98, 100), which can decrease the amplitude of synaptic responses through occlusion and desensitization of receptors, or change frequency the frequency of spontaneous events through tonic activation of presynaptic mGluRs. Thus, EAATs are involved in regulating the ambient level of glutamate surrounding synaptic receptors, as well as the population of receptors activated following release of glutamate from nerve terminals. To understand how transporters are able to participate in synaptic events lasting only a few milliseconds, some of the characteristics of the proteins that catalyze glutamate translocation need to be considered.

5.1.1 How do transporters accomplish uptake?

The reliance on surface transporters to inactivate extracellular glutamate poses a number of challenges. Foremost among these is that the movement of glutamate into cells is opposed by considerable thermal and electrical barriers. As glutamate is negatively charged, movement through the hydrophobic interior of the membrane must be assisted. However, unlike channels, in which ion flow is dictated by electrochemical gradients, glutamate must be forced to enter the cell against a considerable opposing electrochemical gradient, as cells maintain a negative resting potential and the intracellular concentration of glutamate is likely to be much greater than that typically encountered by transporters (126, 222).

To force the influx of glutamate in the presence of these opposing gradients, EAATs harness the energy stored in Na^+ and K^+ gradients, coupling the movement of glutamate to the inward movement of Na^+ and the outward movement of K^+. Thus, while EAATs do not directly require ATP, they require gradients set up by ATP-dependent pumps. The stoichiometry for EAAT-dependent transport has been measured for GLT-1 (EAAT2) and EAAT3; for every cycle of transport, $3Na^+$, $1H^+$, and 1 glutamate$^-$ move inward, and one K^+ moves outward (118, 224). This stoichiometry predicts that EAATs are capable of establishing a concentration gradient of greater than a million, lowering extracellular glutamate to nanomolar levels at equilibrium when challenged with a single pulse of glutamate (224). Furthermore, due to this unbalanced stoichiometry, completion of a single transport cycle results in the net movement of two positive charges into the cytosol. Thus, EAATs are electrogenic molecules, which can induce macroscopic currents if a sufficient number of transporters are cycling at the same time (9, 15). It has been possible to resolve EAAT-mediated currents in a variety of cell types and even in cell-free, outside-out patches when their density is sufficiently high (10, 20, 22) (Fig. 2), which has given great insight into the capabilities of these transporters.

In addition to the movement of cations that are directly coupled to the movement of glutamate, EAATs also enable anions to move across the membrane. However, reversing the anion gradient, and thus the direction of anion movement, does not alter the next influx of glutamate (206). EAATs therefore contain an anion channel, but they do not require anions to transport their substrate, in contrast to the GABA, glycine and monoamine transporters (195). Under physiological conditions, the flux of Cl^- through all but EAAT5 is minute (8, 143, 206), and the physiological significance of this transporter-mediated anion flux has yet to be determined (see below). Nevertheless, this aspect of EAAT behavior has been exploited for physiological studies. In the presence of anions that are highly permeant through the associated anion conductance (e.g. thiocyanate, nitrate, perchlorate), glutamate-evoked transport currents become much larger, as many more anions flow per cycle than do cations, allowing transporter activity to be monitored when the density or total number of transporters is low (22, 144, 161) (Fig. 2).

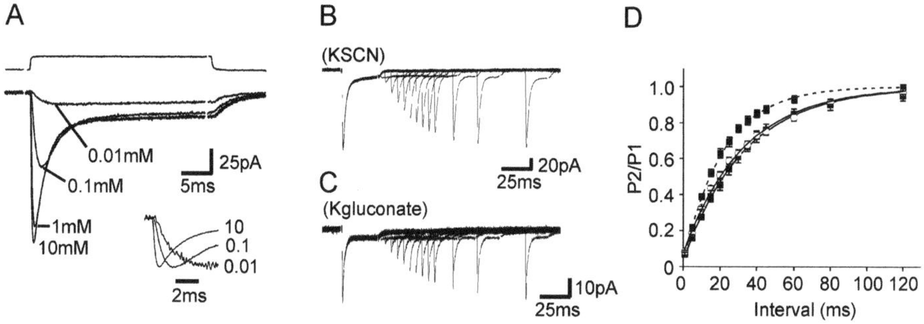

Fig. 2. GLT-1 transporter currents in outside-out patches.
(**a**) Response of GLT-1 transporters to a range of L-glutamate concentrations. The inset shows the peak responses to 0.01 and 0.1 mM L-glutamate, which are scaled to the peak of the response to 10 mM L-glutamate to illustrate the concentration dependence of the rise time of the transporter currents. The rise time to saturating glutamate was 248 μs, indicating that GLT-1 is capable of binding glutamate very rapidly. The trace above is the "open tip" response produced when the patch was removed, which shows the duration of glutamate application (KSCN-based internal solution). (**b**) Response of a patch to pairs of applications of 10 mM L-glutamate (KSCN-based internal solution). The interval between control (30 msec duration) and test (20 msec duration) applications was 1–120 msec. (**c**) Response of a patch to paired applications of 10 mM L-glutamate recorded without permeant anions in the internal solution (K-gluconate-based internal solution). (**d**) Plot of the ratio of the peak amplitude of the second (P2) to the first (P1) response to paired applications of 10 mM L-glutamate recorded with (filled squares) and without anions (open circles) in the internal solution (solid lines are single exponential fits to the data). Removal of TEA-Cl (10 mM) from the internal solution (open squares) speeded the decay by ~35% (dashed line), suggesting that TEA interacts with the internal K^+ binding site. The time constant of recovery, and thus the turnover rate of these transporters in the absence of TEA was 21.8 ± 0.8 ms (n = 10). All responses were recorded from outside-out patches removed from HEK cells expressing GLT-1. Reproduced from reference (183).

Although equilibrium measurements reveal the extraordinary concentrating power of EAATs, they provide little insight into how transporters will perform in the dynamic environment of the brain. In order to effectively restrict the spread of glutamate, transporters must operate in a concentration range similar to that required for binding to receptors. Measured affinities (K_m values) of the EAATs range from 1–20 μM (9), which are comparable to the affinities of NMDA, high affinity kainate, and mGluRs, as well as the glutamate levels that are able to induce desensitization of AMPA receptors (197). However, because coupled transport is a relatively slow process, glutamate clearance is subject to an additional constraint. To complete a single cycle, multiple binding and unbinding reactions have to occur in precise order to allow the conformational changes required for glutamate translocation; as a result, between 10–70 ms are required before each transporter is available to bind glutamate again (21, 23, 143, 207, 224). Because of this slow turnover rate, EAATs must be present at high densities to ensure that they are not saturated, as synapses can release at rates greater than 100 Hz and can be challenged by greater quantities of glutamate during episodes of multivesicular release (141, 208). Quantitiative immunogold measurements indicate that the density of EAATs near synapses can be as high as 10,000 per μm^2 (115), comparable to the density of nicotinic acetylcholine receptors at neuromuscular junctions (4), and physiological measurements have revealed that they are typically not saturated, even under high release conditions (52, 96).

The rates of glutamate diffusion within synaptic and extrasynaptic spaces have not been measured directly, but are likely to be rapid at conventional *en passant* synapses containing a single release site, given the speed with which glutamate transients decay (39) and the short delay between release and the onset of transporter currents in perisynaptic astrocytes (20, 22, 51). Although rapid diffusion will help dissipate glutamate transients though dilution, it could also allow glutamate to escape a zone enriched in transporters to reach receptors in extrasynaptic domains and adjacent synapses. However, EAATs seem well-designed to compete with receptors for glutamate. Glutamate-evoked transporter currents in outside-out patches reach a peak in a few hundred microseconds (10, 22, 144, 204, 210), indicating that they are also capable of binding glutamate rapidly (Fig. 2). Chemical-kinetic modeling suggests that binding to transporters (23, 143), like ionotropic receptors (87), is effectively limited only by the rate of glutamate diffusion. This rapid binding of glutamate may allow EAATs to quickly lower the concentration of glutamate below that required for receptor activation simply by binding. Thus, the rapid decline in the glutamate transient may be primarily due to the ability to EAATs to rapidly buffer glutamate as it exits the cleft (194).

The relative efficiency of EAAT-dependent transport can also influence the rate of glutamate clearance. Transport is accomplished when a molecule of glutamate binds to the transporter and is translocated and released into the cytosol; however, glutamate may unbind from the transporter before translocation, or following translocation it may fail to unbind or be replaced by another glutamate molecule and eventually delivered back to the extracellular space as the transporter completes a half cycle in reverse. This transport efficiency is dependent on the intracellular and extracellular concentrations of transported ions and substrate, as well as the transmembrane voltage. Under ideal conditions (0 Na^+, 0 glutamate inside), the efficiency of EAAT2 is estimated to be 65% (23), while EAAT4 has an estimated efficiency of 75% (143). However, under physiological conditions, the efficiency of EAAT4 is only about 50%, due to the high concentration of intracellular glutamate in neurons. By the same means, the cycling rate of these transporters slows down when Na^+ and K^+ gradients collapse. An increase in intracellular Na^+ and a rise in extracellular K^+ can even induce EAATs to cycle in reverse, a phenomenon that is likely to contribute to the prolonged elevation of extracellular glutamate in ischemia (163). Thus, the ability of EAATs to remove glutamate will be highly dependent on their local environment, and transport is likely to be compromised during high rates of activity, when extracellular K^+ increases, intracellular Na^+ and glutamate increase, and cells spend more time at depolarized potentials.

5.2 Localization of Glutamate Transporters

In order to understand the relative contribution of these transporters to the clearance of glutamate from synapses, the distribution of these transporters and their density around synapses must be determined. Anatomical studies have revealed that each EAAT variant exhibits a restricted distribution. EAAT1 is expressed only in glial cells, particularly astroglial cells in the brain and spinal cord (astrocytes, Bergmann glial cells, radial glial cells) and Müller glial cells in the retina (116, 165). Similarly, EAAT2 is, with few exceptions (37, 161), found only in the membranes of astrocytes (Fig. 3). EAAT3 is ubiquitously expressed by neurons throughout the nervous system, while EAAT4 and EAAT5 have a much more restricted distribution. EAAT4 is expressed only by Purkinje neurons in the cerebellum (59, 221), while EAAT5 is expressed by a variety of neurons in the retina (bipolar cells, photoreceptors), but nowhere else (156).

The ability of EAATs to affect the concentration of extracellular glutamate near synapses will depend on their density relative to sites of release. EAAT2 is the most abundant transporter in the forebrain, accounting for > 90% of all activity (164, 190), and the density of EAAT2 in hippocampal astrocytes has been estimated to reach 8,500 per μm^2 using quantitative immunogold analysis (115). EAAT1 reaches 2,300 per μm^2 in the same membranes, but the situation is reversed in the molecular layer of the cerebellum, where EAAT1 is more abundant than EAAT2 (115). Although early reports suggested that EAAC1 (EAAT3) may also be enriched in the postsynaptic membrane (88), recent studies suggest that indicate that it may be

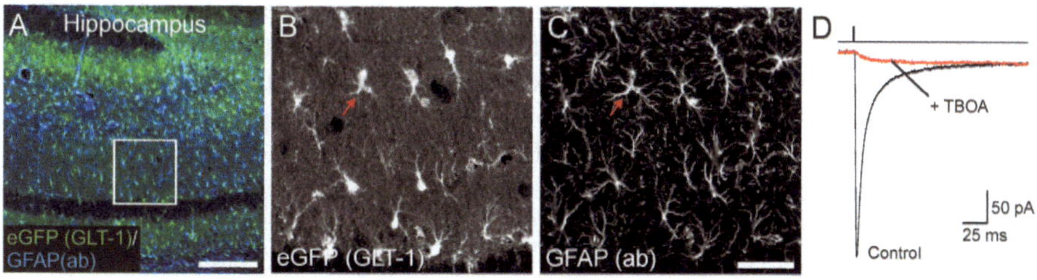

Fig. 3. Restricted expression of GLT-1 by astrocytes.
(**a**) Confocal fluorescence image of the CA1 region of the hippocampus from a GLT-1-eGFP transgenic mouse, in which eGTP is expressed under control of the GLT-1 promoter. This section was also immunolabeled with anti-GFAP antibodies (*blue*), showing that the GLT-1 promoter is only active in $GFAP^+$ astrocytes in this region. Scale = 300 μm. (**b, c**) Higher magnification images of the region delimited by the white box in *A*, showing expression of eGFP (**b**) by $GFAP^+$ astrocytes (**C**). Scale = 50 μm. (**d**) Whole-cell recording of glutamate transporter currents elicited in a stratum radiatum astrocyte (identified expression of eGFP) in response to photolysis of MNI-D-aspartate (125 μM). The red trace shows the response recorded in the presence of the glutamate transporter antagonist TBOA (300 μM). Reproduced from reference (180).

difficult to reliably determine the subcellular distribution of this transporter using immunogold techniques (93), due to the low abundance of this protein. These results suggest that the dominant route for clearance of synaptic glutamate in the forebrain is through uptake into astrocytes. This conclusion is supported by physiological studies which have shown that transporter currents can be recorded in astrocytes in response to photolysis of caged substrates (Fig. 3) and following stimulation of glutamatergic afferents (22), while glutamate transporter currents have not been observed in neurons (21), other than Purkinje cells and some neurons in the retina (see below).

To add to this complexity, the distribution of glutamate transporters is not uniform over the cell surface. High resolution immunogold labeling has revealed that the density of EAAT2 in astrocytes is higher in membranes found next to neuronal membranes than in membranes next to other astrocytes or blood vessels (116). Similarly, the highest density of EAAT4 is reached just outside the cleft in the perisynaptic region of the dendritic membrane (50). Together, these results indicate that there is a broad division of labor at synapses, with some of the glutamate transported into astroglial cells by EAAT2 and EAAT1, and at certain synapses in the cerebellum and retina by EAAT4 and EAAT5, respectively. These differences in transporter abundance suggest that EAATs are positioned near synapses to remove glutamate as it is released and shield receptors as it diffuses from the cleft.

5.3 Contributions of Astroglial Transport

Deciphering the particular roles of the different EAATs in synaptic function has proven difficult, in large part due to the lack of subtype specific antagonists. Nevertheless, comparisons between wild type and transporter-deficient mice have begun to reveal how different EAAT isoforms contribute to uptake at synapses.

The high density of EAAT2 near excitatory synapses in the forebrain indicates that it is likely to serve as the dominant route for glutamate uptake. This hypothesis is corroborated by the phenotype of GLT-1 (EAAT2) knockout mice, which have elevated glutamate in the cerebrospinal fluid, suffer from uncontrolled seizures and die within the first few weeks of life (190). These effects are mimicked by chronic administration of antisense oliginucleotides against GLT-1 to adult animals (164), indicating that this transporter isoform is essential for maintaining a low ambient level of glutamate. GLT-1 deficient mice also show a reduced threshold for the induction of long-term potentiation at Shaffer collateral-commissural synapses (108) and exhibit enhanced activation of mGluRs at recurrent synapses onto CA1 interneurons (96). Furthermore, acute application of dihydrokainate (DHK), a non-substrate antagonist that shows ~100-fold higher affinity for GLT-1, increases the amplitude of NMDA receptor-mediated responses in some CA1 pyramidal neurons (91), indicating that these transporters also shape glutamate transients near synaptic receptors in the hippocampus. However, GLAST (EAAT1) also contributes to uptake at these synapses, as DHK caused a greater enhancement of mGluR currents in EAAT1 (GLAST) knockout mice (96) and GLAST contributes to uptake currents in astrocytes induced following stimulation of Shaffer collateral axons (22). It is not yet clear why both transporters are expressed by astrocytes, as they have comparable affinities and turnover rates (9).

Although GLT-1 is not highly expressed in the molecular layer of the cerebellum, the important role of glial cells in glutamate clearance is conserved. Animals that lack GLAST, the primary transporter expressed by Bergmann glial cells, exhibit coordination deficits and an increased sensitivity to ischemic damage to the cerebellum (213). At the cellular level, slow climbing fiber mediated responses are visible in Purkinje neurons in GLAST knockout mice, which arise from spill over of glutamate from climbing fiber synapses formed onto adjacent Purkinje neurons (187, 213). These aberrant responses also appear in wild type mice in the presence of PMB-TBOA, an antagonist which is selective for GLAST and GLT-1 (187), indicating that GLAST is essential to maintain synapse specificity. Parallel fiber mediated EPSCs evoked by trains of stimuli are also greatly enhanced in GLAST-deficient mice (127), presumably due to enhanced pooling of glutamate from neighboring synapses.

In the mammalian cochlea, GLAST is also expressed by supporting cells (68) that surround synapses formed between inner hair cells and the dendrites of primary afferent neurons, the first synapse in the auditory pathway. Unlike central excitatory synapses, GLAST appears to be the only glutamate transporter present in abundance at these peripheral synapses, as transporter-mediated currents could be elicited in supporting cells, but not inner hair cells or afferent dendrites (73). GLAST knockout mice exhibit an increased sensitivity to acoustic overstimulation and a dramatic slowing of glutamate clearance from the perilymph following exposure to loud sound (80). Thus, even at these ribbon synapses where transmitter is released tonically, glutamate can be cleared effectively by transport into surrounding supporting cells.

The dependence on transporters in glial membranes suggests that the involvement of EAATs in removal of glutamate around synapses will vary depending on the extent to which synapses are ensheathed by the astroglial processes. In the CA1 region of the hippocampus, it has been estimated that only 57% of excitatory synapses are contacted by astrocytes (202), and most are only partially ensheathed, suggesting that transmitter uptake may vary even within a population of synapses formed by similar afferents. In contrast to the hippocampus, parallel and climbing fiber synapses on cerebellar Purkinje cells are tightly ensheathed by the processes of Bergmann glial cells (176), suggesting that these glial cells play an important role in the functional isolation of neighboring synapses in the cerebellar cortex. Recent anatomical studies indicate that in both the hippocampus and cerebellum, the extent of glial coverage is greater on the postsynaptic than presynaptic side (117). A simple prediction from this observation is that glutamate may have the opportunity to diffuse further in the presynaptic direction before being removed by transporters. Weather this anisotropy leads to differences in the spatial dynamics of glutamate diffusion and receptor activation around pre- and postsynaptic membranes remains to be determined.

Recent studies have shown that whisker stimulation enhanced the expression of GLT-1 and GLAST 2- to 4- fold in the corresponding cortical column, an effect that was accompanied by an increase in the extent of glial coverage (69). These provocative findings, and previous results showing that fiber lesions result in a rapid decrease in GLT-1 expression in the target field (71), suggest that the involvement of transporters in glutamate clearance may be highly dynamic. Studies of excitatory synapses in the hypothalamus have shown that physiological changes in the extent of astrocytic ensheathment can have profound effects on the profile of glutamate and the pattern of receptor activation at excitatory synapses. During lactation, astrocytes in the hypothalamus retract their processes from excitatory synapses (193), which results in a decrease in the density of EAAT2 near presynaptic mGluRs. Correspondingly, the absence of transporters in lactating animals allows enhanced tonic inhibition due to an elevation in ambient glutamate, and greater numbers of mGluR autoreceptors to be activated following activity-dependent glutamate release (142).

5.4 A Role for EAAT3 in Transmitter Clearance?

Simple conservation suggests that there would be advantages for neurons to recapture glutamate by transporting it back into the presynaptic terminal. The best candidate for neuronal uptake is EAAT3 (EAAC1), as it is expressed by neurons throughout the CNS (165). However, this transporter is found mainly in cell bodies and dendrites of principal neurons and occasionally in the terminals of GABAergic interneurons. Furthermore, in contrast to GLAST and GLT-1, which are primarily found at the cell surface, the majority of EAAC1 protein is found in the cytosol (41). In accordance with these findings, EAAT3-mediated transporter currents cannot be resolved in neurons (21), suggesting that it is expressed at a much lower level than GLAST or GLT-1, and CNS deficits have not been observed in EAAT3 knockout mice (150).

It is important to recognize that the threshold for detection of transporters using electrophysiological methods is quite high, even when transport-associated currents are amplified by recording in the presence of permeant anions. Thus, an inability to resolve transporter currents does not necessarily indicate that transporters are not present. Indeed, a density of transporters below the detection limit could likely still exert a significant effect on glutamate dynamics if they are localized near receptors. Although transporter currents have not been observed in CA1 pyramidal neurons, inhibition of postsynaptic uptake by substituting intracellular K^+ with N-methyl-D-glucamine, which does not support the final counter-transport step of the transport cycle, enhanced spillover of glutamate and prolonged NMDA receptor-mediated currents in these neurons. These findings raise the possibility that despite their low density, glutamate transporters in the postsynaptic membrane can influence extracellular glutamate dynamics.

EAAT3 may also be involved in a variety of metabolic processes. Adult animals administered antisense oligonucleotides against EAAT3 exhibited seizures (164). The change in neuronal excitability following acute loss of EAAT3 may result from a decrease in the GABA synthesis, as antisense treated mice exhibited a decrease in GABA levels and a reduction in the amplitude of miniature IPSCs (171), suggesting that less GABA was available for loading into synaptic vesicles. Furthermore, if glutamate transporters are inhibited during bouts of GABA release, the subsequent release of GABA is reduced (130), suggesting that glutamate uptake and GABA synthesis are tightly linked. These findings highlight the fact that glutamate is used in many other cellular processes, including protein synthesis, deamination of ammonia, and coping with reactive oxygen species through transport of cystine, a precursor of glutathione. It is likely that even high affinity glutamate transporters may play an essential role in providing glutamate for these metabolic processes.

5.5 EAAT4 Shapes Glutamate Transients at Cerebellar Synapses

A prominent role for neuronal uptake has been demonstrated at excitatory synapses in the cerebellum. EAAT4 is abundant in the perisynaptic membrane where mGluR1 densities are highest. If postsynaptic glutamate uptake is inhibited by perfusing Purkinje cells with a $Tris^+$ -based solution, which does not support transporter cycling, parallel fiber evoked mGluR currents are enhanced, and the threshold for the induction of long term depression (LTD) is lowered (27). Furthermore, Purkinje cells that express less EAAT4 show enhanced mGluR-dependent slow inward currents, greater LTD, and greater cannabinoid-mediated inhibition of IPSCs (209), suggesting that EAAT4 transporters normally shield mGluRs from synaptic glutamate. Parallel and climbing fiber mediated EPSCs also decay more slowly in the majority of Purkinje cells in EAAT4 knockout mice, supporting a role for EAAT4 in the rapid clearance of glutamate from these excitatory synapses (186). The dramatic slowing of climbing fiber EPSCs in EAAT4 knockout mice is surprising, as estimates of glutamate translocation based on the size of synaptically evoked transporter currents in Purkinje neurons, currents that are mediated exclusively by EAAT4 (95), indicate that EAAT4 removes less than 20% of the glutamate released from climbing fiber terminals (26, 95, 144).

5.6 EAAT5: A Glutamate-Gated Chloride Channel

The expression of EAAT5 in the terminal membranes of photoreceptors and bipolar cells in the retina (86, 148) suggest that this transporter may perform a unique role in synaptic function. Although all EAATs are permeable to anions (206), EAAT5 allows a much greater flux of Cl^- when glutamate is bound (8). This associated Cl^- conductance is so large that depolarization -induced release of glutamate from a bipolar cell terminal triggers an overt Cl^- current through autoactivation of EAAT5 transporters (148). Furthermore, in the presence of SCN^-, which is highly permeant through the EAAT5-associated anion channel, quantal transporter-mediated anion currents can be resolved in these cells (204). By lowering the membrane resistance (shunting) and by hyperpolarizing the terminal membrane, EAAT5 activation leads to a decrease in the release of glutamate onto amacrine cells. Thus, EAAT5 adds an additional feedback inhibitory influence on bipolar cells.

The presence of EAAT5 in terminal membranes suggests that it may also contribute to clearance. However, the number of glutamate molecules captured by these transporters, which has been estimated by taking into account the size of the EAAT5-mediated Cl^- current, the unitary conductance of the EAAT5 anion channel, and the number of vesicles released (measured through capacitance recordings), indicate that much less than 1% of the glutamate released is captured by EAAT5 (148). Therefore, at bipolar cell terminals, EAAT5 appears to function primarily as a glutamate-gated Cl^- channel rather than a sequestration mechanism. However, recent studies suggest that presynaptic EAAT5 may contribute significantly to uptake at ribbon synapses between rods and bipolar cells. Recordings from bipolar cells revealed that the prolongation of synaptic currents by TBOA could not be accounted for by clearance through GLT-1, GLAST or EAAC1 (86). In addition, EAAT5 was not found in the postsynaptic membrane of bipolar cells, pointing to a role for EAAT5 in rod terminal membrane in the clearance of glutamate. These results suggest that glutamate clearance mechanisms vary between ribbon synapses in sensory cells of the auditory and visual pathways.

6 Concluding Remarks

Studies of glutamate clearance at excitatory synapses indicate that the involvement of transporters varies considerably in different brain regions, paralleling the striking anatomical and functional diversity of synapses. Although there are risks to generalizing findings from a few model synapses, some general rules have emerged. At the majority of synapses, rapid diffusion and subsequent dilution of glutamate in the extracellular space is sufficient to drop the concentration of glutamate under conditions of isolated release. However, EAAT-dependent uptake plays an important role in restricting the diffusion of glutamate to extrasynaptic receptors and preventing glutamate transients from reaching neighboring synapses. Glutamate transport is accomplished outside the cleft; excluding transporters from the cleft presumably ensures that they do not compete with low affinity receptors for glutamate binding. The dominant route for glutamate transport is through EAAT2 at synapses in the forebrain, and EAAT1 and EAAT4 at synapses in the molecular layer of the cerebellum. Transport of synaptic glutamate into pre- and postsynaptic neurons appears to be restricted to a subset of synapses in the cerebellum and the retina. It is likely that EAAT-dependent uptake is dynamic, and changes in relation to the association of astroglial cells with synapses.

The overwhelming reliance on transporters in glial cells suggests that there is a clear division of labor at excitatory synapses, with neurons involved in packaging and releasing glutamate and astroglial cells involved in removing it from the extracellular space. It is clear that astrocytes provide an environment that is better suited for glutamate uptake. Astrocytes have a higher resting potential than neurons, which fluctuates little during activity. As the ability of EAATs to take up glutamate decreases with depolarization, this ensures that uptake will continue during high rates of activity. Repetitive depolarization, higher intracellular levels of glutamate, and repeated Na^+ influx during activity, would be expected to reduce the capacity of neurons for glutamate uptake. In addition, under extreme conditions, large numbers of glutamate transporters in neurons could be detrimental, due to the ability of EAATs to cycle in reverse. However, these interpretations assume that maximal influx of glutamate under all conditions is the ultimate goal. It is possible that activity dependent reductions in the efficiency or capacity of transport could be used to enhance the activation of perisynaptic receptors during periods of high release, in effect creating a frequency filter by placing transporters and receptors in close proximity.

Acknowledgements

The Bergles lab is supported by the NIH, the Deafness Research Foundation, and the Packard Center for ALS Research; and the Edwards lab is supported by NARSAD, National Parkinson Foundation, the Michael J. Fox Foundation, NIGMS, NINDS, NIMH and NIDA.

References

1. Accardi A and Miller C. Secondary active transport mediated by a prokaryotic homologue of ClC Cl- channels. Nature 427: 803–807, 2004.
2. Ahnert-Hilger G, Nurnberg B, Exner T, Schafer T, and Jahn R. The heterotrimeric G protein G02 regulates catecholamine uptake by secretory vesicles. EMBO J 17: 406–413, 1998.
3. Aihara Y, Mashima H, Onda H, Hisano S, Kasuya H, Hori T, Yamada S, Tomura H, Yamada Y, Inoue I, Kojima I, and Takeda J. Molecular cloning of a novel brain-type Na(+)-dependent inorganic phosphate cotransporter. J Neurochem 74: 2622–2625, 2000.
4. Albuquerque EX, Barnard EA, Porter CW, and Warnick JE. The density of acetylcholine receptors and their sensitivity in the postsynaptic membrane of muscle endplates. Proc Natl Acad Sci USA 71: 2818–2822, 1974.
5. Amara SG and Kuhar MJ. Neurotransmitter transporters: recent progress. Annu Rev Neurosci 16: 73–93, 1993.
6. Armano S, Coco S, Bacci A, Pravettoni E, Schenk U, Verderio C, Varoqui H, Erickson JD, and Matteoli M. Localization and functional relevance of system a neutral amino acid transporters in cultured hippocampal neurons. J Biol Chem 277: 10467–10473, 2002.
7. Arnth-Jensen N, Jabaudon D, and Scanziani M. Cooperation between independent hippocampal synapses is controlled by glutamate uptake. Nat Neurosci 5: 325–331, 2002.
8. Arriza JL, Eliasof S, Kavanaugh MP, and Amara SG. Excitatory amino acid transporter 5, a retinal glutamate transporter coupled to a chloride conductance. Proc Natl Acad Sci USA 94: 4155–4160, 1997.
9. Arriza JL, Fairman WA, Wadiche JI, Murdoch GH, Kavanaugh MP, and Amara SG. Functional comparisons of three glutamate transporter subtypes cloned from human motor cortex. J Neurosci 14: 5559–5569, 1994.
10. Auger C and Attwell D. Fast removal of synaptic glutamate by postsynaptic transporters. Neuron 28: 547–558, 2000.
11. Bacci A, Sancini G, Verderio C, Armano S, Pravettoni E, Fesce R, Franceschetti S, and Matteoli M. Block of glutamate-glutamine cycle between astrocytes and neurons inhibits epileptiform activity in hippocampus. J Neurophysiol 88: 2302–2310, 2002.
12. Bankston LA and Guidotti G. Characterization of ATP transport into chromaffin granule ghosts. Synergy of ATP and serotonin accumulation in chromaffin granule ghosts. J Biol Chem 271: 17132–17138, 1996.
13. Barasch J, Gershon MD, Nunez EA, Tamir H, and al-Awqati Q. Thyrotropin induces the acidification of the secretory granules of parafollicular cells by increasing the chloride conductance of the granular membrane. J Cell Biol 107: 2137–2147, 1988.
14. Barberis A, Petrini EM, and Cherubini E. Presynaptic source of quantal size variability at GABAergic synapses in rat hippocampal neurons in culture. Eur J Neurosci 20: 1803–1810, 2004.

15. Barbour B, Brew H, and Attwell D. Electrogenic uptake of glutamate and aspartate into glial cells isolated from the salamander (Ambystoma) retina. J Physiol (Lond) 436: 169–193, 1991.
16. Barbour B and Hausser M. Intersynaptic diffusion of neurotransmitter. Trends Neurosci 20: 377–384, 1997.
17. Bekkers JM, Richerson GB, and Stevens CF. Origin of variability in quantal size in cultured hippocampal neurons and hippocampal slices. Proc Natl Acad Sci USA 87: 5359–5362, 1990a.
18. Bellocchio EE, Hu H, Pohorille A, Chan J, Pickel VM, and Edwards RH. The localization of the brain-specific inorganic phosphate transporter suggests a specific presynaptic role in glutamatergic transmission. J Neurosci 18: 8648–8659, 1998.
19. Bellocchio EE, Reimer RJ, Fremeau RTJ, and Edwards RH. Uptake of glutamate into synaptic vesicles by an inorganic phosphate transporter. Science 289: 957–960, 2000.
20. Bergles DE, Dzubay JA, and Jahr CE. Glutamate transporter currents in bergmann glial cells follow the time course of extrasynaptic glutamate. Proc Natl Acad Sci USA 94: 14821–14825, 1997.
21. Bergles DE and Jahr CE. Glial contribution to glutamate uptake at Schaffer collateral-commissural synapses in the hippocampus. J Neurosci 18: 7709–7716, 1998.
22. Bergles DE and Jahr CE. Synaptic activation of glutamate transporters in hippocampal astrocytes. Neuron 19: 1297–1308, 1997.
23. Bergles DE, Tzingounis AV, and Jahr CE. Comparison of coupled and uncoupled currents during glutamate uptake by GLT-1 transporters. J Neurosci 22: 10153–10162, 2002.
24. Bole DG, Hirata K, and Ueda T. Prolonged depolarization of rat cerebral synaptosomes leads to an increase in vesicular glutamate content. Neurosci Lett 322: 17–20, 2002.
25. Boulland JL, Qureshi T, Seal RP, Rafiki A, Gundersen V, Bergersen LH, Fremeau RT, Jr., Edwards RH, Storm-Mathisen J, and Chaudhry FA. Expression of the vesicular glutamate transporters during development indicates the widespread corelease of multiple neurotransmitters. J Comp Neurol 480: 264–280, 2004.
26. Brasnjo G and Otis TS. Isolation of glutamate transport-coupled charge flux and estimation of glutamate uptake at the climbing fiber-Purkinje cell synapse. Proc Natl Acad Sci USA 101: 6273–6278, 2004.
27. Brasnjo G and Otis TS. Neuronal glutamate transporters control activation of postsynaptic metabotropic glutamate receptors and influence cerebellar long-term depression. Neuron 31: 607–616, 2001.
28. Bröer S, Schuster A, Wagner CA, Bröer A, Forster I, Biber J, Murer H, Werner A, Lang F, and Busch AE. Chloride conductance and Pi transport are separate functions induced by the expression of NaPi-1 in Xenopus oocytes. J Memb Biol 164: 71–77, 1998.
29. Brunk I, Blex C, Rachakonda S, Holtje M, Winter S, Pahner I, Walther DJ, and Ahnert-Hilger G. The first luminal domain of vesicular monoamine transporters mediates G-protein-dependent regulation of transmitter uptake. J Biol Chem 281: 33373–33385, 2006.
30. Bruns D, Riedel D, Klingauf J, and Jahn R. Quantal release of serotonin. Neuron 28: 205–220, 2000.
31. Busch AE, Schuster A, Waldegger S, Wagner CA, Zempel G, Broer S, Biber J, Murer H, and Lang F. Expression of a renal type I sodium/phosphate transporter (NaPi-1) induces a conductance in Xenopus oocytes permeable for organic and inorganic anions. Proc Natl Acad Sci USA 93: 5347–5351, 1996.
32. Cabeza R and Collier B. Acetylcholine mobilization in a sympathetic ganglion in the presence and absence of 2-(4-phenylpiperidino)cyclohexanol (AH5183). J Neurochem 50: 112–121, 1988.

33. Carlson MD, Kish PE, and Ueda T. Characterization of the solubilized and reconstituted ATP-dependent vesicular glutamate uptake system. J Biol Chem 264: 7369–7376, 1989a.
34. Carter AG and Regehr WG. Prolonged synaptic currents and glutamate spillover at the parallel fiber to stellate cell synapse. J Neurosci 20: 4423–4434, 2000.
35. Chaudhry FA, Reimer RJ, and Edwards RH. The glutamine commute: take the N line and transfer to the A. J Cell Biol 157: 349–355, 2002b.
36. Chen S and Diamond JS. Synaptically released glutamate activates extrasynaptic NMDA receptors on cells in the ganglion cell layer of rat retina. J Neurosci 22: 2165–2173, 2002.
37. Chen W, Mahadomrongkul V, Berger UV, Bassan M, DeSilva T, Tanaka K, Irwin N, Aoki C, and Rosenberg PA. The glutamate transporter GLT1a is expressed in excitatory axon terminals of mature hippocampal neurons. J Neurosci 24: 1136–1148, 2004.
38. Clark BA and Cull-Candy SG. Activity-dependent recruitment of extrasynaptic NMDA receptor activation at an AMPA receptor-only synapse. J Neurosci 22: 4428–4436, 2002.
39. Clements JD. Transmitter timecourse in the synaptic cleft: its role in central synaptic function. Trends Neurosci 19: 163–171, 1996.
40. Colliver TL, Pyott SJ, Achalabun M, and Ewing AG. VMAT-Mediated changes in quantal size and vesicular volume. J Neurosci 20: 5276–5282, 2000.
41. Conti F, DeBiasi S, Minelli A, Rothstein JD, and Melone M. EAAC1, a high-affinity glutamate tranporter, is localized to astrocytes and gabaergic neurons besides pyramidal cells in the rat cerebral cortex. Cereb Cortex 8: 108–116, 1998.
42. Conti F and Minelli A. Glutamate immuoreactivity in rat cerebral cortex is reversibly abolished by 6-diazo-5-oxo-L-norleucine. J Histochem Cytochem 42: 717–726, 1994.
43. Croft BG, Fortin GD, Corera AT, Edwards RH, Beaudet A, Trudeau LE, and Fon EA. Normal biogenesis and cycling of empty synaptic vesicles in dopamine neurons of vesicular monoamine transporter 2 knockout mice. Mol Biol Cell 16: 306–315, 2005.
44. Crowder KM, Gunther JM, Jones TA, Hale BD, Zhang HZ, Peterson MR, Scheller RH, Chavkin C, and Bajjalieh SM. Abnormal neurotransmission in mice lacking synaptic vesicle protein 2A (SV2A). Proc Natl Acad Sci USA 96: 15268–15273, 1999.
45. Curthoys NP and Watford M. Regulation of glutaminase activity and glutamine metabolism. Ann Rev Nutr 15: 133–159, 1995.
46. Danbolt NC. Glutamate uptake. Prog Neurobiol 65: 1–105, 2001.
47. Daniels RW, Collins CA, Chen K, Gelfand MV, Featherstone DE, and Diantonio A. A single vesicular glutamate transporter is sufficient to fill a synaptic vesicle. Neuron 49: 11–16, 2006.
48. Daniels RW, Collins CA, Gelfand MV, Dant J, Brooks ES, Krantz DE, and DiAntonio A. Increased expression of the Drosophila vesicular glutamate transporter leads to excess glutamate release and a compensatory decrease in quantal content. J Neurosci 24: 10466–10474, 2004.
49. De Gois S, Jeanclos E, Morris M, Grewal S, Varoqui H, and Erickson JD. Identification of endophilins 1 and 3 as selective binding partners for VGLUT1 and their co-localization in neocortical glutamatergic synapses: implications for vesicular glutamate transporter trafficking and excitatory vesicle formation. Cell Mol Neurobiol 26: 679–693, 2006.
50. Dehnes Y, Chaudhry FA, Ullensvang K, Lehre KP, Storm-Mathisen J, and Danbolt NC. The glutamate transporter EAAT4 in rat cerebellar Purkinje cells: a glutamate-gated chloride channel concentrated near the synapse in parts of the dendritic membrane facing astroglia. J Neurosci 18: 3606–3619, 1998.
51. Diamond JS. Deriving the glutamate clearance time course from transporter currents in CA1 hippocampal astrocytes: transmitter uptake gets faster during development. J Neurosci 25: 2906–2916, 2005.

52. Diamond JS and Jahr CE. Synaptically released glutamate does not overwhelm transporters on hippocampal astrocytes during high-frequency stimulation. J Neurophysiol 83: 2835–2843, 2000.
53. Diamond JS and Jahr CE. Transporters buffer synaptically released glutamate on a submillisecond time scale. J Neurosci 17: 4672–4687, 1997.
54. Dickman DK, Horne JA, Meinertzhagen IA, and Schwarz TL. A slowed classical pathway rather than kiss-and-run mediates endocytosis at synapses lacking synaptojanin and endophilin. Cell 123: 521–533, 2005.
55. Doherty P, Hawgood BJ, and Smith IC. Changes in miniature end-plate potentials after brief nervous stimulation at the frog neuromuscular junction. J Physiol 356: 349–358, 1984.
56. Drory O and Nelson N. The emerging structure of vacuolar ATPases. Physiology (Bethesda) 21: 317–325, 2006.
57. Dunant Y and Israel M. Neurotransmitter release at rapid synapses. Biochimie 82: 289–302, 2000.
58. Elhamdani A, Palfrey HC, and Artalejo CR. Quantal size is dependent on stimulation frequency and calcium entry in calf chromaffin cells. Neuron 31: 819–830, 2001.
59. Fairman WA, Vandenberg RJ, Arriza JL, Kavanaugh MP, and Amara SG. An excitatory amino-acid transporter with properties of a ligand-gated chloride channel. Nature 375: 599–603, 1995.
60. Fernandez-Peruchena C, Navas S, Montes MA, and Alvarez de Toledo G. Fusion pore regulation of transmitter release. Brain Res Brain Res Rev 49: 406–415, 2005.
61. Fon EA, Pothos EN, Sun B-C, Killeen N, Sulzer D, and Edwards RH. Vesicular transport regulates monoamine storage and release but is not essential for amphetamine action. Neuron 19: 1271–1283, 1997.
62. Forgac M. Structure, mechanism and regulation of the clathrin-coated vesicle and yeast vacuolar H(+)-ATPases. J Exp Biol 203 Pt 1: 71–80, 2000.
63. Forti L, Bossi M, Bergamaschi A, Villa A, and Malgaroli A. Loose-patch recordings of single quanta at individual hippocampal synapses. Nature 388: 874–878, 1997.
64. Fremeau RT, Jr., Burman J, Qureshi T, Tran CH, Proctor J, Johnson J, Zhang H, Sulzer D, Copenhagen DR, Storm-Mathisen J, Reimer RJ, Chaudhry FA, and Edwards RH. The identification of vesicular glutamate transporter 3 suggests novel modes of signaling by glutamate. Proc Natl Acad Sci USA 99: 14488–14493, 2002.
65. Fremeau RT, Jr., Kam K, Qureshi T, Johnson J, Copenhagen DR, Storm-Mathisen J, Chaudhry FA, Nicoll RA, and Edwards RH. Vesicular glutamate transporters 1 and 2 target to functionally distinct synaptic release sites. Science 304: 1815–1819, 2004.
66. Fremeau RT, Jr., Troyer MD, Pahner I, Nygaard GO, Tran CH, Reimer RJ, Bellocchio EE, Fortin D, Storm-Mathisen J, and Edwards RH. The expression of vesicular glutamate transporters defines two classes of excitatory synapse. Neuron 31: 247–260, 2001.
67. Frerking M and Wilson M. Effects of variance in mini amplitude on stimulus-evoked release: a comparison of two models. Biophys J 70: 2078–2091, 1996.
68. Furness DN and Lehre KP. Immunocytochemical localization of a high-affinity glutamate-aspartate transporter, GLAST, in the rat and guinea-pig cochlea. Eur J Neurosci 9: 1961–1969, 1997.
69. Genoud C, Quairiaux C, Steiner P, Hirling H, Welker E, and Knott GW. Plasticity of astrocytic coverage and glutamate transporter expression in adult mouse cortex. PLoS Biol 4: e343, 2006.
70. Gillespie DC, Kim G, and Kandler K. Inhibitory synapses in the developing auditory system are glutamatergic. Nat Neurosci 8: 332–338, 2005.

71. Ginsberg SD, Rothstein JD, Price DL, and Martin LJ. Fimbria-fornix transections selectively down-regulate subtypes of glutamate transporter and glutamate receptor proteins in septum and hippocampus. J Neurochem 67: 1208–1216, 1996.
72. Giompres PE, Zimmermann H, and Whittaker VP. Changes in the biochemical and biophysical parameters of cholinergic synaptic vesicles on transmitter release and during a subsequent period of rest. Neuroscience 6: 775–785, 1981.
73. Glowatzki E, Cheng N, Hiel H, Yi E, Tanaka K, Ellis-Davies GC, Rothstein JD, and Bergles DE. The glutamate-aspartate transporter GLAST mediates glutamate uptake at inner hair cell afferent synapses in the mammalian cochlea. J Neurosci 26: 7659–7664, 2006.
74. Gong LW, Hafez I, Alvarez de Toledo G, and Lindau M. Secretory vesicles membrane area is regulated in tandem with quantal size in chromaffin cells. J Neurosci 23: 7917–7921, 2003.
75. Grabner CP, Price SD, Lysakowski A, Cahill AL, and Fox AP. Regulation of large dense-core vesicle volume and neurotransmitter content mediated by adaptor protein 3. Proc Natl Acad Sci USA 103: 10035–10040, 2006.
76. Grabner CP, Price SD, Lysakowski A, and Fox AP. Mouse chromaffin cells have two populations of dense core vesicles. J Neurophysiol 94: 2093–2104, 2005.
77. Gracz LM, Wang W-C, and Parsons SM. Cholinergic synaptic vesicle heterogeneity: evidence for regulation of acetylcholine transport. Biochem 27: 5268–5274, 1988.
78. Gras C, Herzog E, Bellenchi GC, Bernard V, Ravassard P, Pohl M, Gasnier B, Giros B, and El Mestikawy S. A third vesicular glutamate transporter expressed by cholinergic and serotoninergic neurons. J Neurosci 22: 5442–5451, 2002.
79. Hajos N, Nusser Z, Rancz EA, Freund TF, and Mody I. Cell type- and synapse-specific variability in synaptic GABAA receptor occupancy. Eur J Neurosci 12: 810–818, 2000.
80. Hakuba N, Koga K, Gyo K, Usami S-i, and Tanaka K. Exacerbation of Noise-Induced Hearing Loss in Mice Lacking the Glutamate Transporter GLAST. J Neurosci 20: 8750–8753, 2000.
81. Hamberger A, Chiang GH, Sandoval E, and Cotman CW. Glutamate as a CNS transmitter. II. Regulation of synthesis in the releasable pool. Brain Res 168: 531–541, 1979b.
82. Harata NC, Aravanis AM, and Tsien RW. Kiss-and-run and full-collapse fusion as modes of exo-endocytosis in neurosecretion. J Neurochem 97: 1546–1570, 2006.
83. Harkany T, Holmgren C, Hartig W, Qureshi T, Chaudhry FA, Storm-Mathisen J, Dobszay MB, Berghuis P, Schulte G, Sousa KM, Fremeau RT, Jr., Edwards RH, Mackie K, Ernfors P, and Zilberter Y. Endocannabinoid-independent retrograde signaling at inhibitory synapses in layer 2/3 of neocortex: involvement of vesicular glutamate transporter 3. J Neurosci 24: 4978–4988, 2004.
84. Hartinger J and Jahn R. An anion binding site that regulates the glutamate transporter of synaptic vesicles. J Biol Chem 268: 23122–23127, 1993.
85. Hartman KN, Pal SK, Burrone J, and Murthy VN. Activity-dependent regulation of inhibitory synaptic transmission in hippocampal neurons. Nat Neurosci 9: 642–649, 2006.
86. Hasegawa J, Obara T, Tanaka K, and Tachibana M. High-density presynaptic transporters are required for glutamate removal from the first visual synapse. Neuron 50: 63–74, 2006.
87. Hausser M and Roth A. Dendritic and somatic glutamate receptor channels in rat cerebellar Purkinje cells. J Physiol (Lond) 501: 77–95, 1997.
88. He Y, Janssen WG, Rothstein JD, and Morrison JH. Differential synaptic localization of the glutamate transporter EAAC1 and glutamate receptor subunit GluR2 in the rat hippocampus. J Comp Neurol 418: 255–269, 2000.

89. Hell JW, Maycox PR, and Jahn R. Energy dependence and functional reconstitution of the gamma-aminobutyric acid carrier from synaptic vesicles. J Biol Chem 265: 2111–2117, 1990.
90. Herzog E, Bellenchi GC, Gras C, Bernard V, Ravassard P, Bedet C, Gasnier B, Giros B, and El Mestikaway S. The existence of a second vesicular glutamate transporter specifies subpopulations of glutamatergic neurons. J Neurosci 21: RC181, 2001.
91. Hestrin S, Sah P, and Nicoll RA. Mechanisms generating the time course of dual component excitatory synaptic currents recorded in hippocampal slices. Neuron 5: 247–253, 1990.
92. Hiesinger PR, Fayyazuddin A, Mehta SQ, Rosenmund T, Schulze KL, Zhai RG, Verstreken P, Cao Y, Zhou Y, Kunz J, and Bellen HJ. The v-ATPase V0 subunit a1 is required for a late step in synaptic vesicle exocytosis in Drosophila. Cell 121: 607–620, 2005.
93. Holmseth S, Dehnes Y, Bjornsen LP, Boulland JL, Furness DN, Bergles D, and Danbolt NC. Specificity of antibodies: unexpected cross-reactivity of antibodies directed against the excitatory amino acid transporter 3 (EAAT3). Neuroscience 136: 649–660, 2005.
94. Höltje M, von Jagow B, Pahner I, Lautenschlager M, Hörtnagl H, Nürnberg B, Jahn R, and Ahnert-Hilger G. The neuronal monoamine transporter VMAT2 is regulated by the trimeric GTPase Go(2). J Neurosci 20: 2131–2141, 2000.
95. Huang YH, Dykes-Hoberg M, Tanaka K, Rothstein JD, and Bergles DE. Climbing fiber activation of EAAT4 transporters and kainate receptors in cerebellar Purkinje cells. J Neurosci 24: 103–111, 2004.
96. Huang YH, Sinha SR, Tanaka K, Rothstein JD, and Bergles DE. Astrocyte glutamate transporters regulate metabotropic glutamate receptor-mediated excitation of hippocampal interneurons. J Neurosci 24: 4551–4559, 2004.
97. Huggett J, Vaughan-Thomas A, and Mason D. The open reading frame of the Na(+)-dependent glutamate transporter GLAST-1 is expressed in bone and a splice variant of this molecule is expressed in bone and brain. FEBS Lett 485: 13–18, 2000.
98. Isaacson JS and Nicoll RA. The uptake inhibitor L-trans-PDC enhances responses to glutamate but fails to alter the kinetics of excitatory synaptic currents in the hippocampus. J Neurophysiol 70: 2187–2191, 1993.
99. Ishikawa T, Sahara Y, and Takahashi T. A single packet of transmitter does not saturate postsynaptic glutamate receptors. Neuron 34: 613–621, 2002.
100. Jabaudon D, Shimamoto K, Yasuda-Kamatani Y, Scanziani M, Gahwiler BH, and Gerber U. Inhibition of uptake unmasks rapid extracellular turnover of glutamate of nonvesicular origin. Proc Natl Acad Sci USA 96: 8733–8738, 1999.
101. Janz R, Goda Y, Geppert M, Missler M, and Südhof TC. SV2A and SV2B function as redundant Ca2+ regulators in neurotransmitter release. Neuron 24: 1003–1016, 1999.
102. Jentsch TJ, Poet M, Fuhrmann JC, and Zdebik AA. Physiological functions of CLC Cl- channels gleaned from human genetic disease and mouse models. Annu Rev Physiol 67: 779–807, 2005.
103. Johnson RG. Accumulation of biological amines into chromaffin granules: a model for hormone and neurotransmitter transport. Physiol Rev 68: 232–307, 1988.
104. Juge N, Yoshida Y, Yatsushiro S, Omote H, and Moriyama Y. Vesicular glutamate transporter contains two independent transport machineries. J Biol Chem 281: 39499–39506, 2006.
105. Kam K and Nicoll R. Excitatory synaptic transmission persists independently of the glutamate-glutamine cycle. J Neurosci 27: 9192–9200, 2007.
106. Kane PM. Disassembly and reassembly of the yeast vacuolar H(+)-ATPase in vivo. J Biol Chem 270: 17025–17032, 1995.

107. Karunanithi S, Marin L, Wong K, and Atwood HL. Quantal size and variation determined by vesicle size in normal and mutant Drosophila glutamatergic synapses. J Neurosci 22: 10267–10276, 2002.
108. Katagiri H, Tanaka K, and Manabe T. Requirement of appropriate glutamate concentrations in the synaptic cleft for hippocampal LTP induction. Eur J Neurosci 14: 547–553, 2001.
109. Katz B. Quantal mechanism of neural transmitter release. Science 173: 123–126, 1971.
110. Kawasaki-Nishi S, Bowers K, Nishi T, Forgac M, and Stevens TH. The amino-terminal domain of the vacuolar proton-translocating ATPase a subunit controls targeting and in vivo dissociation, and the carboxyl-terminal domain affects coupling of proton transport and ATP hydrolysis. J Biol Chem 276: 47411–47420, 2001.
111. Kish PE, Fischer-Bovenkerk C, and Ueda T. Active transport of gamma-aminobutyric acid and glycine into synaptic vesicles. Proc Natl Acad Sci USA 86: 3877–3881, 1989.
112. Kozminski KD, Gutman DA, Davila V, Sulzer D, and Ewing AG. Voltammetric and pharmacological characterization of dopamine release from single exocytotic events at rat pheochromocytoma (PC12) cells. Anal Chem 70: 3123–3130, 1998.
113. Kvamme E, Torgner IA, and Roberg B. Kinetics and localization of brain phosphate activated glutaminase. J Neurosci Res 66: 951–958, 2001.
114. Laake JH, Slyngstad TA, Haug FM, and Ottersen OP. Glutamine from glial cells is essential for the maintenance of the nerve terminal pool of glutamate: immunogold evidence from hippocampal slice cultures. J Neurochem 65: 871–881, 1995.
115. Lehre KP and Danbolt NC. The number of glutamate transporter subtype molecules at glutamatergic synapses: chemical and stereological quantification in young adult rat brain. J Neurosci 18: 8751–8757, 1998.
116. Lehre KP, Levy LM, Ottersen OP, Storm-Mathisen J, and Danbolt NC. Differential expression of two glial glutamate transporters in the rat brain: quantitative and immunocytochemical observations. J Neurosci 15: 1835–1853, 1995.
117. Lehre KP and Rusakov DA. Asymmetry of glia near central synapses favors presynaptically directed glutamate escape. Biophys J 83: 125–134, 2002.
118. Levy LM, Warr O, and Attwell D. Stoichiometry of the glial glutamate transporter GLT-1 expressed inducibly in a Chinese hamster ovary cell line selected for low endogenous Na+-dependent glutamate uptake. J Neurosci 18: 9620–9628, 1998.
119. Liang SL, Carlson GC, and Coulter DA. Dynamic regulation of synaptic GABA release by the glutamate-glutamine cycle in hippocampal area CA1. J Neurosci 26: 8537–8548, 2006.
120. Liu G, Choi S, and Tsien RW. Variability of neurotransmitter concentration and nonsaturation of postsynaptic AMPA receptors at synapses in hippocampal cultures and slices see comments. Neuron 22: 395–409, 1999.
121. Liu G and Tsien RW. Properties of synaptic transmission at single hippocampal synaptic boutons. Nature 375: 404–408, 1995.
122. Lynch BA, Lambeng N, Nocka K, Kensel-Hammes P, Bajjalieh SM, Matagne A, and Fuks B. The synaptic vesicle protein SV2A is the binding site for the antiepileptic drug levetiracetam. Proc Natl Acad Sci USA 101: 9861–9866, 2004.
123. Mainen ZF, Malinow R, and Svoboda K. Synaptic calcium transients in single spines indicate that NMDA receptors are not saturated. Nature 399: 151–155, 1999.
124. Malinow R and Malenka RC. AMPA receptor trafficking and synaptic plasticity. Annu Rev Neurosci 25: 103–126, 2002.
125. Manolson MF, Wu B, Proteau D, Taillon BE, Roberts BT, Hoyt MA, and Jones EW. STV1 gene encodes functional homologue of 95-kDa yeast vacuolar H(+)-ATPase subunit Vph1p. J Biol Chem 269: 14064–14074, 1994.

126. Marc RE, Liu WL, Kalloniatis M, Raiguel SF, and van Haesendonck E. Patterns of glutamate immunoreactivity in the goldfish retina. J Neurosci 10: 4006–4034, 1990.
127. Marcaggi P, Billups D, and Attwell D. The role of glial glutamate transporters in maintaining the independent operation of juvenile mouse cerebellar parallel fibre synapses. J Physiol 552: 89–107, 2003.
128. Maron R, Stern Y, Kanner BI, and Schuldiner S. Functional asymmetry of the amine transporter from chromaffin granules. J Biol Chem 258: 11476–11481, 1983.
129. Masson J, Darmon M, Conjard A, Chuhma N, Ropert N, Thoby-Brisson M, Foutz AS, Parrot S, Miller GM, Jorisch R, Polan J, Hamon M, Hen R, and Rayport S. Mice lacking brain/kidney phosphate-activated glutaminase have impaired glutamatergic synaptic transmission, altered breathing, disorganized goal-directed behavior and die shortly after birth. J Neurosci 26: 4660–4671, 2006.
130. Mathews GC and Diamond JS. Neuronal glutamate uptake Contributes to GABA synthesis and inhibitory synaptic strength. J Neurosci 23: 2040–2048, 2003.
131. Maycox PR, Deckwerth T, Hell JW, and Jahn R. Glutamate uptake by brain synaptic vesicles. Energy dependence of transport and functional reconstitution in proteoliposomes. J Biol Chem 263: 15423–15428, 1988.
132. McAllister AK and Stevens CF. Nonsaturation of AMPA and NMDA receptors at hippocampal synapses. Proc Natl Acad Sci USA 97: 6173–6178, 2000.
133. Miyazaki T, Fukaya M, Shimizu H, and Watanabe M. Subtype switching of vesicular glutamate transporters at parallel fibre-Purkinje cell synapses in developing mouse cerebellum. Eur J Neurosci 17: 2563–2572, 2003.
134. Moechars D, Weston MC, Leo S, Callaerts-Vegh Z, Goris I, Daneels G, Buist A, Cik M, van der Spek P, Kass S, Meert T, D'Hooge R, Rosenmund C, and Hampson RM. Vesicular glutamate transporter VGLUT2 expression levels control quantal size and neuropathic pain. J Neurosci 26: 12055–12066, 2006.
135. Morin P, Sagne C, and Gasnier B. Functional characterization of wild-type and mutant human sialin. Embo J 23: 4560–4570, 2004.
136. Mosharov EV, Gong LW, Khanna B, Sulzer D, and Lindau M. Intracellular patch electrochemistry: regulation of cytosolic catecholamines in chromaffin cells. J Neurosci 23: 5835–5845, 2003.
137. Naito S and Ueda T. Characterization of glutamate uptake into synaptic vesicles. J Neurochem 44: 99–109, 1985.
138. Nakanishi-Matsui M and Futai M. Stochastic proton pumping ATPases: from single molecules to diverse physiological roles. IUBMB Life 58: 318–322, 2006.
139. Naves LA and Van der Kloot W. Repetitive nerve stimulation decreases the acetylcholine content of quanta at the frog neuromuscular junction. J Physiol 532: 637–647, 2001.
140. Ni B, Rosteck PR, Nadi NS, and Paul SM. Cloning and expression of a cDNA encoding a brain-specific Na+-dependent inorganic phosphate cotransporter. Proc Natl Acad Sci USA 91: 5607–5611, 1994.
141. Oertner TG, Sabatini BL, Nimchinsky EA, and Svoboda K. Facilitation at single synapses probed with optical quantal analysis. Nat Neurosci 5: 657–664, 2002.
142. Oliet SH, Piet R, and Poulain DA. Control of glutamate clearance and synaptic efficacy by glial coverage of neurons. Science 292: 923–926, 2001.
143. Otis TS and Jahr CE. Anion currents and predicted glutamate flux through a neuronal glutamate transporter. J Neurosci 18: 7099–7110, 1998.
144. Otis TS, Kavanaugh MP, and Jahr CE. Postsynaptic glutamate transport at the climbing fiber-Purkinje cell synapse. Science 277: 1515–1518, 1997.

145. Otis TS, Wu YC, and Trussell LO. Delayed clearance of transmitter and the role of glutamate transporters at synapses with multiple release sites. J Neurosci 16: 1634–1644, 1996.
146. Ozkan ED, Lee FS, and Ueda T. A protein factor that inhibits ATP-dependent glutamate and g-aminobutyric acid accumulation into synaptic vesicles: purification and initial characterization. Proc Natl Acad Sci USA 94: 4137–4142, 1997.
147. Pahner I, Holtje M, Winter S, Takamori S, Bellocchio EE, Spicher K, Laake P, Numberg B, Ottersen OP, and Ahnert-Hilger G. Functional G-protein heterotrimers are associated with vesicles of putative glutamatergic terminals: implications for regulation of transmitter uptake. Mol Cell Neurosci 23: 398–413, 2003.
148. Palmer MJ, Taschenberger H, Hull C, Tremere L, and von Gersdorff H. Synaptic activation of presynaptic glutamate transporter currents in nerve terminals. J Neurosci 23: 4831–4841, 2003.
149. Parsons RL, Calupca MA, Merriam LA, and Prior C. Empty synaptic vesicles recycle and undergo exocytosis at vesamicol-treated motor nerve terminals. J Neurophysiol 81: 2696–2700, 1999.
150. Peghini P, Janzen J, and Stoffel W. Glutamate transporter EAAC-1-deficient mice develop dicarboxylic aminoaciduria and behavioral abnormalities but no neurodegeneration. Embo J 16: 3822–3832, 1997.
151. Peter D, Jimenez J, Liu Y, Kim J, and Edwards RH. The chromaffin granule and synaptic vesicle amine transporters differ in substrate recognition and sensitivity to inhibitors. J Biol Chem 269: 7231–7237, 1994.
152. Peters C, Bayer MJ, Buhler S, Andersen JS, Mann M, and Mayer A. Trans-complex formation by proteolipid channels in the terminal phase of membrane fusion. Nature 409: 581–588, 2001.
153. Picollo A and Pusch M. Chloride/proton antiporter activity of mammalian CLC proteins ClC-4 and ClC-5. Nature 436: 420–423, 2005.
154. Pothos EN, Larsen KE, Krantz DE, Liu Y-J, Haycock JW, Setlik W, Gershon ME, Edwards RH, and Sulzer D. Synaptic vesicle transporter expression regulates vesicle phenotype and quantal size. J Neurosci 20: 7297–7306, 2000.
155. Pothos EN, Mosharov E, Liu KP, Setlik W, Haburcak M, Baldini G, Gershon MD, Tamir H, and Sulzer D. Stimulation-dependent regulation of the pH, volume and quantal size of bovine and rodent secretory vesicles. J Physiol 542: 453–476, 2002.
156. Pow DV and Barnett NL. Developmental expression of excitatory amino acid transporter 5: a photoreceptor and bipolar cell glutamate transporter in rat retina. Neurosci Lett 280: 21–24, 2000.
157. Pow DV and Crook DK. Direct immunocytochemical evidence for the transfer of glutamine from glial cells to neurons: use of specific antibodies directed against the d-stereoisomers of glutamate and glutamine. Neuroscience 70: 295–302, 1996.
158. Pow DV and Robinson SR. Glutamate in some retinal neurons is derived solely from glia. Neurosci 60: 355–366, 1994.
159. Prado VF, Martins-Silva C, de Castro BM, and Lima. Mice deficient for the vesicular acetylcholine transporter are myasthenic and have deficits in object and social recognition. Neuron 51, 2006.
160. Price GD and Trussell LO. Estimate of the chloride concentration in a central glutamatergic terminal: a gramicidin perforated-patch study on the calyx of Held. J Neurosci 26: 11432–11436, 2006.
161. Regan MR, Huang YH, Kim YS, Dykes-Hoberg MI, Jin L, Watkins AM, Bergles DE, and Rothstein JD. Variations in promoter activity reveal a differential expression and physiology of glutamate transporters by glia in the developing and mature CNS. J Neurosci 27: 6607–6619, 2007.

162. Rizzoli SO and Betz WJ. Synaptic vesicle pools. Nat Rev Neurosci 6: 57–69, 2005.
163. Rossi DJ, Oshima T, and Attwell D. Glutamate release in severe brain ischaemia is mainly by reversed uptake. Nature 403: 316–321, 2000.
164. Rothstein JD, Dykes-Hoberg M, Pardo CA, Bristol LA, Jin LA, Jin L, Kuncl RW, Kanai Y, Hediger MA, Wang Y, Schielke JP, and al e. Knockout of glutamate transporters reveals a major role for astroglial transport in excitotoxicity and clearance of glutamate. Neuron 16: 675–686, 1996.
165. Rothstein JD, Martin L, Levey AI, Dykes-Hoberg M, Jin L, Wu D, Nash N, and Kuncl RW. Localization of neuronal and glial glutamate transporters. Neuron 13: 713–725, 1994.
166. Rothstein JD and Tabakoff B. Alteration of striatal glutamate release after glutamine synthetase inhibition. J Neurochem 43: 1438–1446, 1984.
167. Sandoval GM, Duerr JS, Hodgkin J, Rand JB, and Ruvkun G. A genetic interaction between the vesicular acetylcholine transporter VAChT/UNC-17 and synaptobrevin/SNB-1 in C. elegans. Nat Neurosci 9: 599–601, 2006.
168. Sarantis M, Ballerini L, Miller B, Silver RA, Edwards M, and Attwell D. Glutamate uptake from the synaptic cleft does not shape the decay of the non-NMDA component of the synaptic current. Neuron 11: 541–549, 1993.
169. Schafer MK, Varoqui H, Defamie N, Weihe E, and Erickson JD. Molecular cloning and functional identification of mouse vesicular glutamate transporter 3 and its expression in subsets of novel excitatory neurons. J Biol Chem 277: 50734–50748, 2002.
170. Scheel O, Zdebik AA, Lourdel S, and Jentsch TJ. Voltage-dependent electrogenic chloride/proton exchange by endosomal CLC proteins. Nature 436: 424–427, 2005.
171. Sepkuty JP, Cohen AS, Eccles C, Rafiq A, Behar K, Ganel R, Coulter DA, and Rothstein JD. A neuronal glutamate transporter contributes to neurotransmitter GABA synthesis and epilepsy. J Neurosci 22: 6372–6379, 2002.
172. Shimamoto K, Lebrun B, Yasuda-Kamatani Y, Sakaitani M, Shigeri Y, Yumoto N, and Nakajima T. DL-threo-beta-benzyloxyaspartate, a potent blocker of excitatory amino acid transporters. Mol Pharmacol 53: 195–201, 1998.
173. Shupliakov O, Atwood HL, Ottersen OP, Storm-Mathisen J, and Brodin L. Presynaptic glutamate levels in tonic and phasic motor axons correlate with properties of synaptic release. J Neurosci 15: 7168–7180, 1995.
174. Smear MC, Tao HW, Staub W, Orger MB, Gosse NJ, Liu Y, Takahashi K, Poo MM, and Baier H. Vesicular glutamate transport at a central synapse limits the acuity of visual perception in zebrafish. Neuron 53: 65–77, 2007.
175. Song H-j, Ming G-l, Fon E, Bellocchio E, Edwards RH, and Poo M-m. Expression of a putative vesicular acetylcholine transporter facilitates quantal transmitter packaging. Neuron 18: 815–826, 1997.
176. Spacek J. Three-dimensional analysis of dendritic spines. III. Glial sheath. Anat Embryol (Berl) 171: 245–252, 1985.
177. Steinert JR, Kuromi H, Hellwig A, Knirr M, Wyatt AW, Kidokoro Y, and Schuster CM. Experience-dependent formation and recruitment of large vesicles from reserve pool. Neuron 50: 723–733, 2006.
178. Stobrawa SM, Breiderhoff T, Takamori S, Engel D, Schweizer M, Zdebik AA, Bösl MR, Ruether K, Jahn H, Draguhn A, Jahn R, and Jentsch TJ. Disruption of ClC-3, a chloride channel expressed on synaptic vesicles, leads to a loss of the hippocampus. Neuron 29: 185–196, 2001.
179. Storm-Mathisen J, Leknes AK, Bore AT, Vaaland JL, Edminson P, Haug FM, and Ottersen OP. First visualization of glutamate and GABA in neurones by immunocytochemistry. Nature 301: 517–520, 1983.

180. Sullivan R, Rauen T, Fischer F, Wiessner M, Grewer C, Bicho A, and Pow DV. Cloning, transport properties, and differential localization of two splice variants of GLT-1 in the rat CNS: implications for CNS glutamate homeostasis. Glia 45: 155–169, 2004.
181. Sulzer D and Pothos EN. Regulation of quantal size by presynaptic mechanisms. Rev Neurosci 11: 159–212, 2000.
182. Tabb JS, Kish PE, Van Dyke R, and Ueda T. Glutamate transport into synaptic vesicles. J Biol Chem 267: 15412–15418, 1992.
183. Takamori S, Malherbe P, Broger C, and Jahn R. Molecular cloning and functional characterization of human vesicular glutamate transporter 3. EMBO Rep 3: 798–803, 2002.
184. Takamori S, Rhee JS, Rosenmund C, and Jahn R. Identification of a vesicular glutamate transporter that defines a glutamatergic phenotype in neurons. Nature 407: 189–194, 2000.
185. Takamori S, Rhee JS, Rosenmund C, and Jahn R. Identification of differentiation-associated brain-specific phosphate transporter as a second vesicular glutamate transporter. J Neurosci 21: RC182, 2001.
186. Takayasu Y, Iino M, Kakegawa W, Maeno H, Watase K, Wada K, Yanagihara D, Miyazaki T, Komine O, Watanabe M, Tanaka K, and Ozawa S. Differential roles of glial and neuronal glutamate transporters in Purkinje cell synapses. J Neurosci 25: 8788–8793, 2005.
187. Takayasu Y, Iino M, Shimamoto K, Tanaka K, and Ozawa S. Glial glutamate transporters maintain one-to-one relationship at the climbing fiber-Purkinje cell synapse by preventing glutamate spillover. J Neurosci 26: 6563–6572, 2006.
188. Tamir H, Liu KP, Adlersberg M, Hsiung SC, and Gershon MD. Acidification of serotonin-containing secretory vesicles induced by a plasma membrane calcium receptor. J Biol Chem 271: 6441–6450, 1996.
189. Tamura Y, Ozkan ED, Bole DG, and Ueda T. IPF, a vesicular uptake inhibitory protein factor, can reduce the Ca(2+)-dependent, evoked release of glutamate, GABA and serotonin. J Neurochem 76: 1153–1164, 2001.
190. Tanaka K, Watase K, Manabe T, Yamada K, Watanabe M, Takahashi K, Iwama H, Nishikawa T, Ichihara N, Kikuchi T, Okuyama S, Kawashima N, Hori S, Takimoto M, and Wada K. Epilepsy and exacerbation of brain injury in mice lacking the glutamate transporter GLT-1. Science 276: 1699–1702, 1997.
191. Tanaka K, Watase K, Manabe T, Yamada K, Watanabe M, Takahashi K, Iwama H, Nishikawa T, Ichihara N, Kikuchi T, Okuyama S, Kawashima N, Hori S, Takimoto M, and Wada K. Epilepsy and exacerbation of brain injury in mice lacking the glutamate transporter GLT-1. Science 276: 1699–1702, 1997.
192. Tani H, Bandrowski AE, Parada I, Wynn M, Huguenard JR, Prince DA, and Reimer RJ. Modulation of epileptiform activity by glutamine and system A transport in a model of post-traumatic epilepsy. Neurobiol Dis 25: 230–238, 2007.
193. Theodosis DT and Poulain DA. Activity-dependent neuronal-glial and synaptic plasticity in the adult mammalian hypothalamus. Neuroscience 57: 501–535, 1993.
194. Tong G and Jahr CE. Block of glutamate transporters potentiates postsynaptic excitation. Neuron 13: 1195–1203, 1994.
195. Torres GE and Amara SG. Glutamate and monoamine transporters: new visions of form and function. Curr Opin Neurobiol 17: 304–312, 2007.
196. Travis ER, Wang YM, Michael DJ, Caron MG, and Wightman RM. Differential quantal release of histamine and 5-hydroxytryptamine from mast cells of vesicular monoamine transporter 2 knockout mice. Proc Natl Acad Sci USA 97: 162–167, 2000.
197. Trussell LO and Fischbach GD. Glutamate receptor desensitization and its role in synaptic transmission. Neuron 3: 209–218, 1989.

198. Van der Kloot W. Loading and recycling of synaptic vesicles in the Torpedo electric organ and the vertebrate neuromuscular junction. Prog Neurobiol 71: 269–303, 2003.
199. Van der Kloot W, Colasante C, Cameron R, and Molgo J. Recycling and refilling of transmitter quanta at the frog neuromuscular junction. J Physiol 523 Pt 1: 247–258, 2000.
200. Varoqui H and Erickson JD. Active transport of acetylcholine by the human vesicular acetylcholine transporter. J Biol Chem 271: 27229–27232, 1996.
201. Varoqui H, Schafer MK-H, Zhu H, Weihe E, and Erickson JD. Identification of the differentiation-associated Na+/Pi transporter as a novel vesicular glutamate transporter expressed in a distinct set of glutamatergic synapses. J Neurosci 22: 142–155, 2002.
202. Ventura R and Harris KM. Three-dimensional relationships between hippocampal synapses and astrocytes. J Neurosci 19: 6897–6906., 1999.
203. Verheijen FW, Verbeek E, Aula N, Beerens CE, Havelaar AC, Joosse M, Peltonen L, Aula P, Galjaard H, van der Spek PJ, and Mancini GM. A new gene, encoding an anion transporter, is mutated in sialic acid storage diseases. Nature Genetics 23: 462–465, 1999.
204. Veruki ML, Morkve SH, and Hartveit E. Activation of a presynaptic glutamate transporter regulates synaptic transmission through electrical signaling. Nat Neurosci 9: 1388–1396, 2006.
205. Voglmaier SM, Kam K, Yang H, Fortin DL, Hua Z, Nicoll RA, and Edwards RH. Distinct endocytic pathways control the rate and extent of synaptic vesicle protein recycling. Neuron 51: 71–84, 2006.
206. Wadiche JI, Amara SG, and Kavanaugh MP. Ion fluxes associated with excitatory amino acid transport. Neuron 15: 721–728, 1995.
207. Wadiche JI, Arriza JL, Amara SG, and Kavanaugh MP. Kinetics of a human glutamate transporter. Neuron 14: 1019–1027, 1995.
208. Wadiche JI and Jahr CE. Multivesicular release at climbing fiber-Purkinje cell synapses. Neuron 32: 301–313, 2001.
209. Wadiche JI and Jahr CE. Patterned expression of Purkinje cell glutamate transporters controls synaptic plasticity. Nat Neurosci 8: 1329–1334, 2005.
210. Wadiche JI and Kavanaugh MP. Macroscopic and microscopic properties of a cloned glutamate transporter/chloride channel. J Neurosci 18: 7650–7661, 1998.
211. Wallen-Mackenzie A, Gezelius H, Thoby-Brisson M, Nygard A, Enjin A, Fujiyama F, Fortin G, and Kullander K. Vesicular glutamate transporter 2 is required for central respiratory rhythm generation but not for locomotor central pattern generation. J Neurosci 26: 12294–12307, 2006.
212. Wang X, Li Y, Engisch KL, Nakanishi ST, Dodson SE, Miller GW, Cope TC, Pinter MJ, and Rich MM. Activity-dependent presynaptic regulation of quantal size at the mammalian neuromuscular junction in vivo. J Neurosci 25: 343–351, 2005.
213. Watase K, Hashimoto K, Kano M, Yamada K, Watanabe M, Inoue Y, Okuyama S, Sakagawa T, Ogawa S, Kawashima N, Hori S, Takimoto M, Wada K, and Tanaka K. Motor discoordination and increased susceptibility to cerebellar injury in GLAST mutant mice. Eur J Neurosci 10: 976–988, 1998.
214. Wilson NR, Kang J, Hueske EV, Leung T, Varoqui H, Murnick JG, Erickson JD, and Liu G. Presynaptic regulation of quantal size by the vesicular glutamate transporter VGLUT1. J Neurosci 25: 6221–6234, 2005.
215. Winkler BS, Kapousta-Bruneau N, Arnold MJ, and Green DG. Effects of inhibiting glutamine synthetase and blocking glutamate uptake on b-wave generation in the isolated rat retina. Vis Neurosci 16: 345–353, 1999.

216. Winter S, Brunk I, Walther DJ, Holtje M, Jiang M, Peter JU, Takamori S, Jahn R, Birnbaumer L, and Ahnert-Hilger G. Galphao2 regulates vesicular glutamate transporter activity by changing its chloride dependence. J Neurosci 25: 4672–4680, 2005.
217. Wojcik SM, Rhee JS, Herzog E, Sigler A, Jahn R, Takamori S, Brose N, and Rosenmund C. An essential role for vesicular glutamate transporter 1 (VGLUT1) in postnatal development and control of quantal size. Proc Natl Acad Sci USA 101: 7158–7163, 2004.
218. Wolosker H, de Souza DO, and de Meis L. Regulation of glutamate transport into synaptic vesicles by chloride and proton gradient. J Biol Chem 271: 11726–11731, 1996.
219. Wreden CC, Wlizla M, and Reimer RJ. Varied mechanisms underlie the free sialic acid storage disorders. J Biol Chem 280: 1408–1416, 2005.
220. Wu XS, Xue L, Mohan R, Paradiso K, Gillis KD, and Wu LG. The origin of quantal size variation: vesicular glutamate concentration plays a significant role. J Neurosci 27: 3046–3056, 2007.
221. Yamada K, Watanabe M, Shibata T, Tanaka K, Wada K, and Inoue Y. EAAT4 is a post-synaptic glutamate transporter at Purkinje cell synapses. Neuroreport 7: 2013–2017, 1996.
222. Yamashita T, Ishikawa T, and Takahashi T. Developmental increase in vesicular glutamate content does not cause saturation of AMPA receptors at the calyx of held synapse. J Neurosci 23: 3633–3638, 2003.
223. Yelamanchili SV, Pendyala G, Brunk I, Darna M, Albrecht U, and Ahnert-Hilger G. Differential sorting of the vesicular glutamate transporter 1 into a defined vesicular pool is regulated by light signaling involving the clock gene Period2. J Biol Chem 281: 15671–15679, 2006.
224. Zerangue N and Kavanaugh MP. Flux coupling in a neuronal glutamate transporter. Nature 383: 634–637, 1996.
225. Zhang B, Ganetzky B, Bellen HJ, and Murthy VN. Tailoring uniform coats for synaptic vesicles during endocytosis. Neuron 23: 419–422, 1999.
226. Zhang B, Koh YH, Beckstead RB, Budnik V, Ganetzky B, and Bellen HJ. Synaptic vesicle size and number are regulated by a clathrin adaptor protein required for endocytosis. Neuron 21: 1465–1475, 1998.
227. Zhou Q, Petersen CCH, and Nicoll RA. Effects of reduced vesicular filling on synaptic transmission in rat hippocampal neurones. J Physiol 525: 195–206, 2000.
228. Zimmermann H and Denston CR. Recycling of synaptic vesicles in the cholinergic synapses of the Torpedo electric organ during induced transmitter release. Neuroscience 2: 695–714, 1977.

Structure and Function of Vertebrate and Invertebrate Active Zones

Craig C. Garner[1] and Kang Shen[2]

[1] Department of Psychiatry and Behavioral Science, Nancy Pritzker Laboratory, Stanford University, Palo Alto, CA 94304-5485, USA, cgarner@stanford.edu
[2] Department of Biology, Stanford University, Palo Alto, CA 94305-5020, USA, kangshen@stanford.edu

1 Introduction

As a biologist, one is frequently struck by the synchrony between form and function and how it facilitates the ability of organisms to adapt to different environments. This concept extends from the gross anatomical features of plants, animals and aquatic creatures down to the shapes and attributes of individual cell types and their cellular organelles. Within the nervous system, form and function are essential driving elements in creating the myriad of neuronal and glial cell types, their captivating morphologies and mesmerizing networks of axons, dendrites and processes. However, the match between form and function can also be seen at the level of intercellular anatomy as oligodendrocytes and Schwann cells ensheath axons forming myelin or astrocytic end-feet envelope capillaries to create the blood brain barrier.

For those interested in inter-neuronal communication, nature's beauty in matching form and function is also present at the tiny sites of cell–cell contact formed between neuronal cells, called synapses. These micron size structures are the key, not only for encoding higher order cognitive functions, but also the simple reflex responsiveness of both large and small animals to adverse stimuli. Surveying the morphological and ultra-structural features of synapses across species and even within a single organism, one is immediately struck both by the similarities and differences in the morphology of synapses. At first, it is hard to understand why some synapses are large (hippocampal mossy fiber boutons) and others are small (excitatory pyramidal cell synapses), or why others exhibit elaborate convoluted and involuted membraneous substructures (neuromuscular junction).

However, as neuroscientists begin to define the functional properties of neurons and their synaptic connections, it becomes easier to see how once again synaptic form/morphology is tightly coupled to the functional properties of different types of synapses. In the present review, we consider both macro and microscopic aspects of

J.W. Hell, M.D. Ehlers (eds.), *Structural and Functional Organization of the Synapse*,
DOI: 10.1007/978-0-387-77232-5_3,

synaptic form and function from the point of view of presynaptic assembly and how complex ensembles of proteins can lead to presynaptic boutons with distinct features that are in the end ideally suited for the output characteristics of nerve cells and the networks within which they reside.

2 The Chemical Synapse

Chemical synapses are best characterized as asymmetric cellular junctions formed between neuronal cells as well as with target cells within glands, skin and muscles. They are composed of three compartments: the presynaptic bouton, the synaptic cleft and the postsynaptic reception apparatus. Presynaptic boutons generally found along axons are filled with anywhere from a few hundred to tens-of-thousands of 50 nm vesicles. These so called synaptic vesicles (SVs) are filled with neurotransmitter substances, e.g., small peptides or amino acids or their derivatives, that are often released into the synaptic cleft, a small space between the pre and postsynapse, in an activity dependent manner where they bind and activate neurotransmitter receptors embedded within the postsynaptic plasma membrane (Fig. 1). Importantly, the release and reception of neurotransmitters across the synaptic cleft is a highly regulated process involving the assembly of large numbers of proteins into macromolecular complexes (48, 64). These are not randomly distributed within the presynaptic bouton or postsynaptic membrane, but are organized and clustered into transsynaptic signaling elements called the synaptic junction. At the ultrastructural level, synaptic junctions are readily identifiable due to their avidity for heavy metals as electron densities (23, 28). When viewed in cross-section by electron microscopy (Fig. 1), these pre and postsynaptic electron densities are perfectly aligned and appear to be held in register by transsynaptic material, most likely extracellular matrix proteins, or cell adhesion molecules (see Chapter by Atasoy and Kavalali). Functionally, the presynaptic half of the synaptic junction, referred to as the active zone, is a specialized region of the presynaptic membrane where synaptic vesicles dock, fuse and release neurotransmitter in response to action potentials entering the bouton. Protein ensembles that comprise the postsynaptic density (PSD) translate the chemical signal back to electrical impulses via the activation of clusters of neurotransmitter receptors and downstream signaling complexes (see Chapter by Esteban). Although PSDs are fascinating in their own right, we will restrict our subsequent comments to features and properties of the different types of presynaptic boutons and refer the reader to a number of excellent previous reviews of this topic (48, 69).

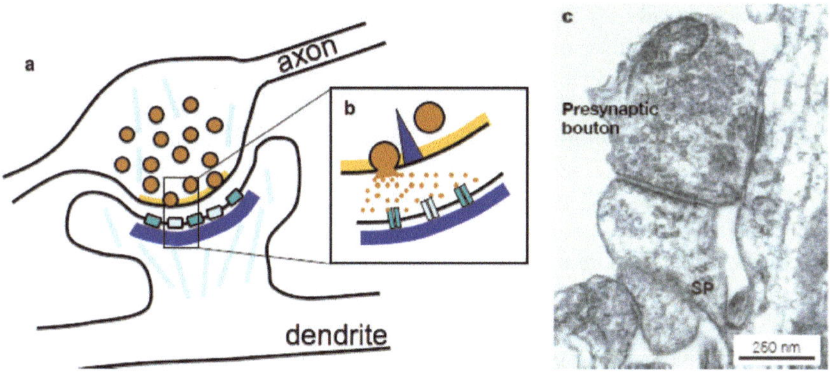

Fig. 1. *General structure of chemical synapses.* (**a**) Schematic diagram of an excitatory glutamatergic synapse formed on to a dendritic spine showing the typical structure of a chemical synapse, consisting a presynaptic bouton, a postsynaptic compartment and a synaptic cleft. Neurotransmitter is concentrated in small 50 nm synaptic vesicles (*orange circles*) that dock, fuse and released neurotransmitter at the active zone (*light orange band*) into the synaptic cleft upon the arrival of action potentials. Neurotransmitter receptors (*squares*) are localized on the postsynaptic membrane and tethered in place by an underlying cytoskeletal network called the PSD (*Dark blue band*). (**b**) Magnified view of a section of the synaptic junction showing a synaptic vesicle undergoing fusion and release of neurotransmitter into the synaptic cleft where they bind postsynaptic receptors. (**c**) A electron micrograph of an excitatory glutamatergic synapse formed on a dendritic spine of a hippocampal neuron grown in culture (Image provided by Joanne Buchanan, Stanford University). Visible are the numerous presynaptic vesicles, the flat synaptic junction and dendritic spine (SP).

3 Presynaptic Subcompartments

In attempting to translate presynaptic form into function, it is important to consider the most basic functions performed by presynaptic boutons before considering the nuances of synapses formed in different brain regions. Clearly the most fundamental function of the presynapse is the regulated release of neurotransmitter which is accomplished through the fusions of SVs with the presynaptic plasma membrane. SV fusion is but one step of the "SV Cycle" (29, 73). Essentially, this cycle is comprised of several discrete steps that begin with the biogenesis and maturation of SVs as they transit from Golgi membrane to newly forming presynaptic boutons. Upon their arrival at the presynaptic bouton, mature SVs are filled with neurotransmitters, such as glutamate, acetylcholine, glycine, dopamine, serotonin, adrenalin, noradrenalin or gamma amino butyric acid (GABA) depending on neuronal subtype. This filling requires specific vesicle transporters that utilize an electro-chemical gradient setup by an ATP dependent proton pump. Once filled these mature SVs then translocate to a region of the presynaptic plasma membrane, called the active zone (AZ) where they dock, become primed and undergo fusion in a calcium dependent manner. SV proteins deposited in the plasma membrane then undergo clathrin mediated endocytosis in a peri-active zonal region of the plasma membrane and are then processed through early endosomes present within the presynaptic bouton before budding as mature vesicles ready for the next round of the cycle. This regenerative process allows presynaptic boutons, millimeter to meters from the cell soma, to be self-reliant neuronal compartments that are only partially dependent on the soma for the biosynthesis of new synaptic vesicle proteins.

Functionally, SVs within presynaptic boutons can be divided into several pools. Those situated at the AZ are considered part of the readily releasable pool of SVs that are both docked and usually primed for fusion. Action potentials invading a bouton, opening voltage gated calcium channels, lead to the fusion of this population of SVs. SVs situated more distally from the AZ comprise the reserve pool of vesicle that translocate towards the AZ to refill these sites, restoring the readily releasable pool of SVs. For reasons that are poorly understood, SVs of the reserve pool can be split into two pools, one that participates in the SV cycle during a train of action potential and thus are considered part of the total recycling pool, and a second that fail to recycle upon stimulation. Whether this latter pool is part of a reserve–reserve pool of vesicles awaiting a special signal to enter into the total recycling pool or constitute a unique pool of vesicles with yet to be ascribed properties remains unclear. One intriguing possibility is that this pool represents a store-house of presynaptic proteins that contribute to the maintenance, stability and replenishment of AZs and other substructures of presynaptic boutons.

Additional membrane organelles within boutons include both early and late endosomal membrane and mitochondria. While the former are associated with the retrieval of synaptic vesicle proteins and the regeneration of mature SVs following vesicle exocytosis, the latter is essential for meeting the high energy demands of nerve terminal, e.g., the synthesis of ATP, as well as the sequestration of calcium following each train of action potentials. Although the characteristics of presynaptic endosomes are rather poorly understood, the study of clathrin mediated endocytosis has revealed an amazing complexity of proteins that identify, cluster and retrieve SV proteins deposited into the plasma membrane post calcium triggered fusion. For an excellent review on this topic see (15). Relevant to the current discussion is the observation that while SV exocytosis occurs within the confines of the AZ, clathrin mediate endocytosis appears to occur primarily in a peri-AZ location and requires a remarkable coordination between the formation of clathrin coated vesicles and actin filament assembly to direct SVs back towards the reserve pool (6, 15).

4 Distinct Pools of SVs Are Created by the Neuronal Cytoskeleton

A fundamental question in synaptic-biology is how presynaptic boutons create and maintain their reserve, readily releasable, endocytotic and recycling pools of SVs. Ultrastructural studies of presynaptic boutons using freeze fracture, deep-etch electron microscopy point to elements of the neuronal cytoskeleton as potential organizers of these pools (31, 32). The most important of these are the microtubules, microfilaments and cortical membrane associated cytoskeleton. Of these, microtubules are considered the main highways by which vesicular intermediates, mitochondria, endosomes and maturing SVs are transported from the cell soma to distant presynaptic sites. Microfilaments composed of filamentous actin and its associated proteins were initial thought to provide the guide wires within boutons mediating the translocation of SVs from the reserve pool to the readily releasable pool. This view is not well supported by functional studies demonstrating that neurotransmission occurs quite well in the absence of microfilaments (60). Instead, microfilaments have been functionally linked to endocytosis at a number of synaptic types (7) and in influencing SV release probably (49). The best candidate cytoskeletal protein implicated in maintaining the reserve pool SVs is Synapsin (24). This phospho-protein has several intriguing properties that make it well suited for this function. First, Synapsin can, not only bind SVs, but also microfilaments as well as themselves (24). The latter property endows them with the capacity to form short cross-bridges between SVs as seen by EM (32), and facilitating the sequestration of SVs within boutons preventing them for wandering off. Importantly, the association of Synapsin with SVs is regulated both by activity and phosphorylation (10) thus coupling the translocation of SV in the reserve and readily releasable pools, as the latter becomes depleted following the arrival of the first round of action potentials.

Perhaps the most remarkable aspect of presynaptic function is the regulated docking and fusion of SVs with the AZ plasma membrane within a few nanometers of voltage-gated calcium channels (VGCC). In contrast to constitutive membrane flow from the Golgi to the plasma membrane, which is not constrained temporally or spatially, SV fusion within presynaptic boutons is tightly controlled and appear to occur almost exclusively at the AZ. How is this accomplished? One notion, mentioned above, is that the spatial control of SV fusion is regulated by proteins that are incorporated into and thus define this patch of membrane (22, 64). Although many of the details are still lacking, several families of large multi-domain proteins have been identified that are selectively localized to AZs and thus meet the expected criteria of molecules potentially involved in AZ assembly (Fig. 2). Based on limited data, one can subdivide these families into several groups based on their hypothesized roles as either being structural or mediating SV docking and fusion.

The first group, which includes Piccolo, Bassoon, RIMs, ELKS (also known as ERCs or CAST), CASK, Velis, Mint and SYD-2/Liprin-α, has all the hallmarks of structural proteins, as they are composed primarily of domains characteristic for scaffolding proteins such as PDZ, Zinc fingers, coiled-coil, SAM, GUK and SH3 domains among others (64). Consistent with this concept, each has been found to be tightly associated with synaptic junctional preparations and to participate in intermolecular protein–protein interactions with other members of this group (see (63)). Many are conserved across evolution and are found at both vertebrate and invertebrate synapses. Two exceptions include Piccolo and Bassoon, which appear to have vertebrate specific functions. An important unresolved question is how they are woven together to create an AZ. At present, we have but low resolution models, based on protein–protein interaction studies, to suggest how they might come together to cluster calcium channels, tether SVs and position the SV release machinery (Fig. 2). Equally uncertain is whether specific members play essential roles in AZ assembly or whether such properties are masked by functional redundancy in the system. While some clues are coming from the study of invertebrate synapses (see below), most studies of vertebrate synapses, where loss of function mutation in many of these proteins has failed to show substantial impact on AZ or synapse assembly, indicate that the assembly of this structure involves numerous redundant features.

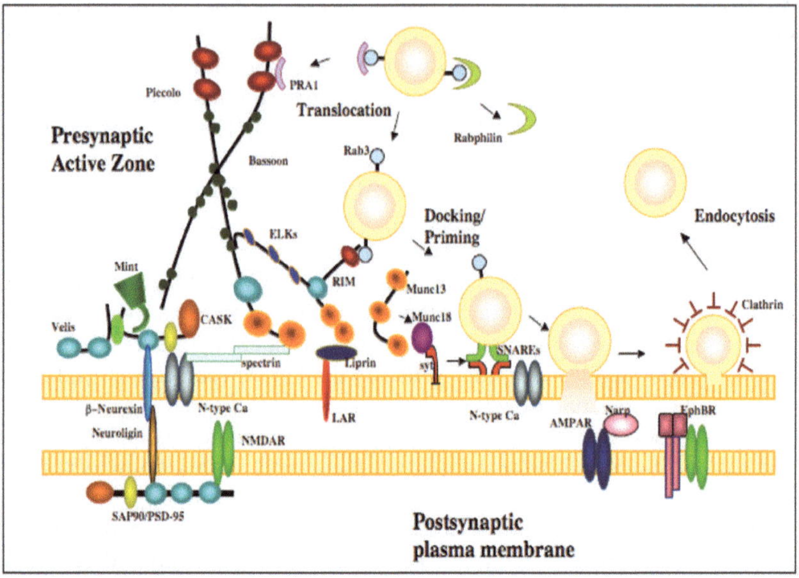

Fig. 2. *Molecular components of the presynaptic active zone.* A schematic diagram of a subset of proteins found at presynaptic active zones of vertebrate synapses. Note that many are structural molecules comprised of a modular domain structure that makes them suitable for their putative scaffolding functions. The relationships between these molecules draw in the diagram are derived primarily from in vitro binding data. In general, dark lines are meant to represent the backbone of the molecule and the colored elliptical objects to represent individual domains, such as, zinc fingers (*red*) C2 (*orange*) PDZ (*aqua*), PBH (*green*). Other proteins are named and represented as single object, e.g. neurexin, neuroligin, clathrin etc (for a more detailed description see (93)).

The second group of AZ proteins has been shown to physically interact with individual members of the first group, but have the auspicious responsibility of also participating in the docking, priming, or fusion of SVs (5). These include syntaxin, SNAP25, VAMP2, RIMs, Munc13, Munc18, voltage gated calcium channels (VGCC) and synaptotagmin to name a few. The first three are components of the SNARE complex and shown to be essential for the fusion of SVs with the plasma membrane (33). The next three appear to be critical for the priming of SVs and in facilitating the formation of the SNARE complex at the AZ plasma membrane (64). As such, they are emerging as key regulators of presynaptic plasticity as their activity can determine the probability that an action potential invading a bouton will lead to SV fusion and thus neurotransmitter release (20, 64). Finally, the last two, VGCC and synaptotagmin, have been shown to be critical for calcium triggered SV fusion in that they either allow the entry of calcium into a nerve terminal upon the arrival of an depolarizing action potential (eg. VGCC) or sense the presence of calcium within the bouton (synaptotagmin) and trigger SNARE mediate SV fusion (73).

The third group of AZ proteins are those involved in SV endocytosis. To call them AZ proteins is perhaps overstated at best, as they are only transiently associated with structural components of the AZ at times of rest. The remainder of the time this group, which includes clathrin, AP2, dynamin, synaptojanin and other components of the endocytic machinery, are working in a peri-active zonal region to collect, cluster and retrieve SV proteins deposited in the plasma membrane post SV fusion (7, 15). Although temporally slower than SV exocytosis (on the order of mins versus msec) (58, 65), SV endocytosis is tightly coupled to SV exocytosis using the influx of calcium via VGCC to activate kinases such as PKC and PI3Kinase to activate components of the endocytotic machinery (15).

5 Ultrastructural Organization of Active Zones

A survey of synapses from both vertebrate and invertebrate systems reveals a remarkable heterogeneity in the morphology and in particular in the ultrastructural features of presynaptic AZs (see (88)). At the seemingly simplest organized synapse, such as those found in C. elegans or Lamprey, the AZ appears as a thin patch of electron dense material with no apparent distinguishing features when viewed by transmission electron microscopy (TEM).

Similarly, most AZs of excitatory and inhibitory synapses of the vertebrate central nervous system appear in cross section to contain a small patch of electron dense material (23). Closer scrutiny of, in particular, excitatory spiny synapses has shown that these AZ are anything but simple. In fact, when viewed in cross-section the electron dense material between SVs appears as tuffs of material projecting some distance into the interior of the bouton, contacting both docked SVs as well as those sitting in next most proximal layer (Fig. 3). Furthermore when viewed on-face these tuffs appear to be organized in a regular array, or lattice, as if the AZ was assembled from geometric units that both creates slots and guides SVs to the active zonal plasma membrane (Fig. 3).

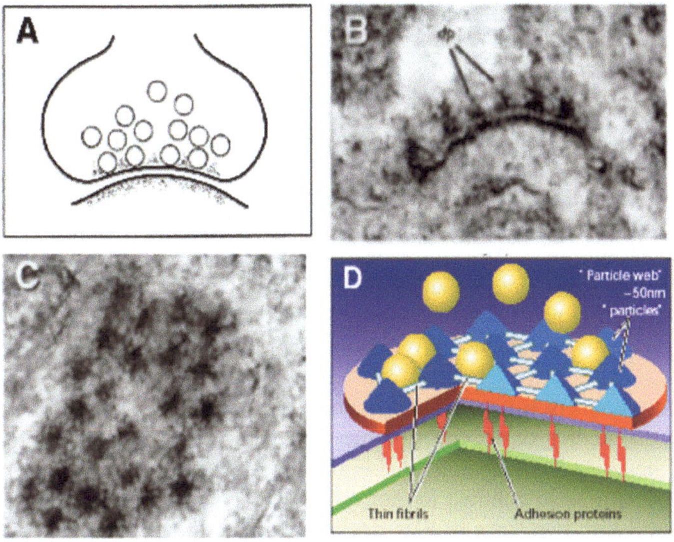

Fig. 3. *Ultrastructural organization of presynaptic active zones.* (**a**) A schematic diagram of a chemical synapse. (**b**) An electron micrograph of the cross section view of a presynaptic active zone with tuff like electron dense projections. (**c**) A on-face view of the presynaptic tuffs from an excitatory synapse. (**d**) A model of the organization of the active zone structures (images taken from (88)).

Although this view of an AZ as a highly organized geometric structure is quite attractive, recent studies using high pressure freezing and electron tomography indicate that part of this pattern could be a by-product of aldehyde fixation and the precipitation of soluble proteins on to the AZ surface between SVs (57, 71). In these newer studies that appear to leave the images less cluttered, tuffs of electron dense material can still be seen. However, the tuffs can now be resolved into a dense network of filamentous material projecting from the cytoplasmic face of the AZ membrane and extending several hundred nanometers into the interior of the bouton (Fig. 4). In these images, SVs appear to be linked to the plasma membrane by two types of the longer filaments. Some radiate singly from the plasma surface while others emerge as clusters from more organized patches of electron dense material at the AZ and appear to guide SV to the plasma membrane (71) (Fig. 4). As discussed below, these patches of material may represent clusters of the known AZ proteins predicted by genetic and functional studies to direct SV docking and fusion. If so, this structural element may represent a fundamental building block or functional unit of the AZ, that when duplicated many times or arranged in geometric array, could create AZs with multiple SV release sites.

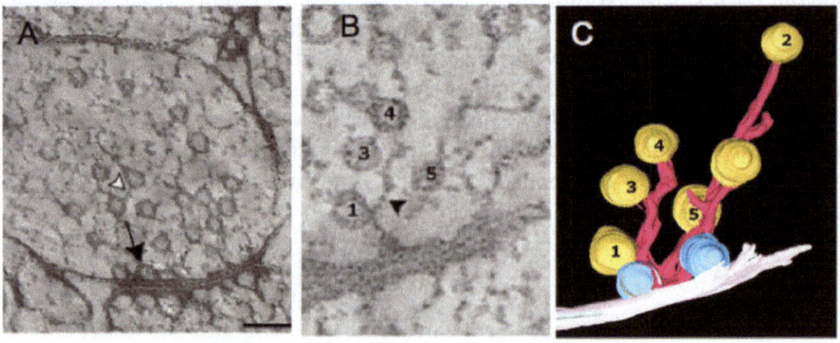

Fig. 4. *Ultrastructural features of the presynaptic active zone from a hippocampal CA1 synapse visualized by electron tomography* (See (71)). (**a**) EM micrograph of a single section through a chemical synapses using high pressure freezing protocol. (Arrows mark the presence of fine filaments connecting SV to the active zone). (**b**) A high magnification EM picture showing that filamentous material projecting from the cytoplasmic face of the AZ membrane and extending several hundred nm into the interior of the bouton making contact with SVs. (**c**) A model drawing of a presynaptic active zone (shown in b) illustrating how fibers link synaptic vesicles and to the presynaptic active zone.

This latter concept is supported by very elegant work by Jack McMahan at Stanford and their studies of the AZs of the frog neuromuscular junction (NMJ) (26). These data have lead to truly remarkable images, in part due to the choice of the frog NMJ as a specimen. In contrast to many AZs, those of the frog NMJ exhibit a striking symmetry with two rows of SVs docked at the plasma membrane along a single longitudinal axis, as if some underlying structure holds them in place (Fig. 5). When viewed in cross-section, these rows of SVs are invariably situated above the crest of the involuted postsynaptic muscle membrane, within which acetylcholine receptors are concentrated at high density (59). Clearly this geometric arrangement is an ideal match between form and function given that the fusion of these vesicles will lead to the deposition of neurotransmitter, acetylcholine, directly onto the underlying receptors. With regard to the AZ, the pairs of SVs regularly spaced along its length suggest some regularity or repetitiveness in the underlying molecular docking and fusion machinery. This impression is further reinforced by freeze fracture images of the extracellular face of the AZ plasma membrane into which two double rows of bumps are seen running the length of the AZ (26, 30). However, while this pattern and rhythmicity had been noted for more than 20 years, the structural underpinning has remained elusive.

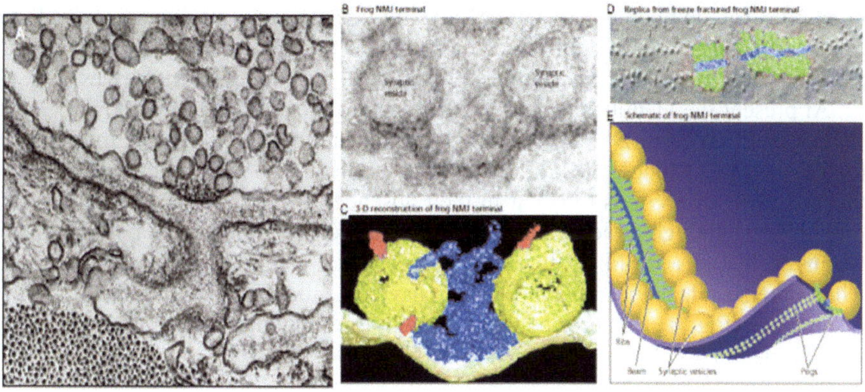

Fig. 5. *Ultrastructural organization of the frog neuromuscular junction.* (**a**) An electron micrograph of a frog neuromuscular junction taken by Dr. John Heuser. Electron dense proteineaous material associated with the cytoplasmic face of the active zone are associated with presynaptic SV clusters and are situated juxtaposed to postsynaptic membrane invaginations. (**b**) Tomographic image of a cross-section of an active zone showing SV docked at the plasma membrane and interacting with fine filamentous projection. (**c**) 3-D reconstruction of frog active zone shown in (**b**). (**d**) Replica from a freeze fractured frog NMJ showing both the presence of parallel rows of bumps (possibly Ca Channels) the relationship of the overlying ribs and beams that make up the presynaptic cytoskeletal matrix assembled at the active zone. (**e**) Schematic diagram of SVs organized into two parallel rows and tethered to the active zone via ribs, beams and pegs at the frog neuromuscular junction. (reproduced from (26)).

With the advent of tomography, this view has quickly changed. Of note, tomography is able to bring clarity to TEM images resolving out of focused material into distinct structural elements (Fig. 5). These included a set of beams running down the long axis of the AZ and a set of perpendicular ribs that reached up and make contact with the parallel rows of SVs. Furthermore, the ribs could be seen to tie into the plasma membrane through a structure called pegs that presumably represent the rows of bumps mentioned above to protrude through the AZ plasma membrane when viewed by freeze fracture EM (Fig. 5). Importantly, along the length of the AZs, the beams, ribs and pegs appeared in a regular repeated manner. In fact, they look as if the tuffs of electron dense material with their extending filaments seen at CNS synapses were polymerized into a linear array. These images provide strong support that cytoskeletal matrix assembled at AZ plays a fundamental role in guiding SVs to their penultimate docking and fusion sites.

Given these beautiful morphological images of AZs, which have also been collected from invertebrate synapses such as those in C. elegans (83), synaptic biologists are now challenged to translate the beam, pegs and filaments into molecules. Clearly one of the first questions is what exactly is the most fundamental building unit needed to create a single SV docking and fusion site? Part of the answer is coming from a combination of genetic and electrophysiological studies aimed at identifying critical functional components of the AZ necessary for SV docking and fusion and release. As discussed above, these are likely to include, among others, one or more VGCCs, components of the SV docking and priming machinery, as well as a few scaffold proteins to both tether these fusion components together and perhaps also provide a molecular link to trans-synaptic cell–cell adhesion molecules (Fig. 6).

At present the identities of these AZ molecules are only partially known. Genetic studies in *C. elegans*, *Drosophila* and mice indicate that both SNARE proteins and proteins such as Munc13, Munc18 and RIM involved in SV priming must be part of this complex, as SV fails to occur in their absence (55). Less clear are the key scaffold proteins, though proteins such as ELKS and liprins, which being conserved from worms to humans are likely to emerge as essential elements of this core complex, while others such as Piccolo and Bassoon have potential roles in guiding SVs to their fusion sites (Fig. 6). Features of these proteins and how they may contribute to synapse formation are discussed in greater detail below.

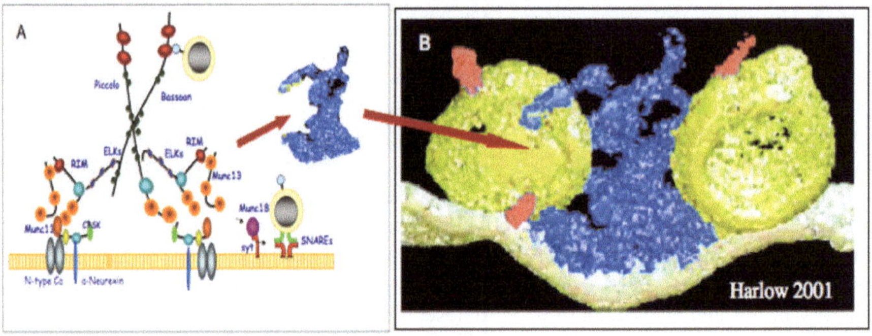

Fig. 6. *Unitary active zone element for SV exocytosis.* (**a**) Schematic drawing of how the known presynaptic active proteins might be assembled to create a pair of SV docking and fusion sites as seen at the frog NMJ. Presented are all of the key structural proteins and those involved in SV priming and fusion. Conceptually these proteins are assembled to create the core ribs, pegs and filaments (Blue amorphous material) that tether SV to the active zone illustrated in (**b**).

6 Cellular and Molecular Assembly of Presynaptic Structure

As discussed above, the primary function of the presynaptic terminal is to execute the SV cycle at specific subcellular location in a temporally controlled fashion. Therefore, the developmental program that builds presynaptic terminals must accumulate the repertoire of presynaptic proteins and organelles necessary for the SV cycle. It also has to do so at precise subcellular locations: directly opposing the postsynaptic specializations. In this section, we will consider the molecular and cellular mechanisms that cluster SVs, assemble presynaptic AZ proteins and restrict the assembly to appropriate locations. As a field, we are still at the very beginning of understanding how the presynaptic apparatus develop in general, not to mention how to explain the physiological and functional diversity of presynaptic structures described above. Nevertheless, forward genetics approaches in worms and flies, as well as elegant cell biology approaches using vertebrate neuronal cultures have begun to identify key players and provide insights into the cellular process organizing presynaptic assembly.

7 Trafficking and Maintenance of Synaptic Vesicles

SV precursors are formed in the cell body and transported to synaptic terminals via microtubule based molecular motors. Cell biological studies suggest that these transporting vesicles are pleiomorphic in size and shape. Therefore, it is possible that the endocytotic cycle at the presynaptic terminal generates the mature, 50 nm synaptic vesicles (1). In axons, microtubule filaments are oriented with their plus end pointing towards the distal axon. Because of this polarized orientation of the axonal microtubules (27), plus end-directed motors of the kinesin super-family are likely responsible for the trafficking of SV precursors from the cell body to synapses. With the large number of kinesin molecules in eukaryotic genome (75), it is an important question to understand which kinesin(s) are responsible for SV trafficking. The first clue to this question came from genetic analysis of a mutant in *C. elegans*. The *unc-104* mutant exhibits a variety of behavioral defects including drastically reduced locomotion and pharynx contraction. Electron microscopy (EM) analysis showed that SVs are almost completely absent at synapses in all neurons. Instead, SV precursors accumulate within neuronal cell bodies, suggestive of defects in intracellular anterograde transport (75). The mouse homologs of *unc-104*, KIF1a and KIF1b, were subsequently shown to be important for the transport of SV precursors. For example, in KIF1a and KIF1b null mice, the density of SVs at synapses is significantly reduced (86, 90). These experiments indicate that the evolutionarily conserved KIF1/UNC-104 family of motors are indispensable for the trafficking of SVs.

The microtubule "highway" enables KIF1/UNC-104 to transport SV precursors to synaptic sites. Once they arrived at synapses, there must be mechanisms to keep the SVs clustered at the presynaptic terminals. Synaptic vesicles associated with the readily releasable pool are most likely tethered directly to the presynaptic membrane via proteins involved in docking and priming. In contrast, the vast majority of the vesicles, present in the reserved pool, do not appear to directly contact the AZ. How are these SVs anchored at presynaptic terminals? As discussed above, synapsin is the most attractive candidate as it can bind both SVs and F-actin. This concept is supported by studies on the lamprey giant axon, where the injection of synapsin antibody causes the distal pool of SVs to dissipate without affecting a pool of SVs that are close to the presynaptic membrane (54). A similar result was observed at vertebrate synapses when the genes encoding synapsins were disrupted (24). A role for F-actin has also been suggested to play a role in restricting SV clusters, based primarily on its presynaptic localization, at the lamprey synapse, where it forms a ring that surrounds SVs and the AZ (49, 60). However disrupting presynaptic F-actin with Latrunculin A at vertebrate presynaptic terminal has little effect on the reserve pool of SVs, suggesting that F-actin might not be required for clustering of presynaptic SVs (60).

Further clues to the mechanisms regulating the clustering of SVs have come from Reichardt and colleagues. Here, the abrogation of β-catenin function, a protein interacting with the cytoplasmic domain of cadherin, led to a diffused appearance of SV markers in dissociated cultured neurons (4). Intriguingly, brain derived neurotrophic factor (BDNF) weakens the interaction between cadherin and β-catenin, allowing SV clusters to become more dynamic and frequently "splitting" into several new clusters (3). These data are consistent with a model where cadherin and β-catenin sets the fence to restrict the lateral movement of SV clusters at presynaptic terminals. However, it is unclear what creates the fence, as cadherin and β-catenin tightly associate with each other at the plasma membrane and, as discussed above, F-actin, a logical downstream player of cadherin mediated adhesion, is not essential for the retention of SVs within boutons.

A related question is what restricts the lateral mobility of SVs situated in close proximity to the AZ? As discussed above, tomographic studies suggest that this group of SVs may be held in place by their association with components of the AZ cytoskeletal matrix (25). From the SV perspective, one interesting candidate is Rab3a. This small ras like G-protein, which is associated with SVs, can directly interact with a number of presynaptic proteins in a GTP-dependent manner including rabphilin and the AZ protein RIM (80, 82). Of these, RIM is particularly attractive as a molecular tethering point between SVs, the exocytotic machinery and the AZ (82). Interestingly, mutations in the *rab-3* gene cause a mild redistribution of SVs at neuromuscular junctions in *C. elegans* (52). In these mutants, the number of presynaptic SVs is decreased, while more SVs can be detected in extrasynaptic regions of axons. These data suggest that Rab-3a functions to keep SVs at the terminals. However, a thorough analysis of the mouse homologs of rab-3 did not reveal similar phenotypes. Specifically quadruple knockout of Rab3A, B, C and D genes in mouse showed little effects on SV accumulation (62). Likewise, knocking out RIM1α in mice does not affect SV clustering (63).

8 Molecular Determinants of AZ Assembly

As described above, the cytoskeletal matrix assembled at the presynaptic plasma membrane is comprised of a complex matrix of proteins involved in defining the site where SVs dock and fuse. Given the complexity of this membrane specialization, it is interesting to consider how it is assembled. For example, is there a single master organizer that initiates the assembly of the entire presynaptic AZ? Alternatively, are their several organizers that facilitate the assembly of sub-complexes of proteins which then assemble into a stable presynaptic AZ? Different predictions can be made from these two models. The first model predicts that the entire presynaptic structure will dissipate if the master organizer is disrupted. The second model predicts that disruption of any single molecule would not give a dramatic phenotype in the assembly of AZ proteins, while only the simultaneous abrogation of multiple presynaptic proteins might lead to the collapse of presynaptic structures.

8.1 Is SYD-2/Liprin a Master Regulator of Presynaptic Development?

In order to distinguish the two models of presynaptic assembly, it is necessary to examine the localization of many presynaptic proteins in a variety of genetic mutant backgrounds where individual presynaptic proteins are mutated. When such experiments were performed in the *C. elegans* neuron HSNL, it was found that two presynaptic scaffold molecules, SYD-2/liprin and SYD-1 were essential for the presynaptic assembly of numerous AZ and peri-active zone components (53). In *syd-2* or *syd-1* mutant, numerous presynaptic markers fail to accumulate at synapses. Interestingly, single mutations in most of the other presynaptic molecules such as the calcium channels, *unc-13*/Munc13, *unc-18*/Munc18 and *unc-10*/RIM do not affect the synaptic accumulation of other presynaptic markers, although these mutants dramatically affect the presynaptic exocytosis function (55). At least in C. elegans, SYD-2/liprin is likely to be a key organizer as the overexpression of SYD-2/liprin rescues the assembly phenotype in *syd-1* mutants(13, 53). In contrast, the overexpression of SYD-1 fails to rescue the phenotypes of *syd-2*, suggesting that SYD-1 might be a positive regulator of SYD-2/liprin (53). These genetic experiments argue that SYD-2 has indispensable functions in the development of presynaptic apparatus and is responsible for clustering of numerous presynaptic proteins.

The biochemistry of liprin/SYD-2 supports this notion. Liprin has multiple coiled-coil domains at its N-terminus and three sterile alpha motif (SAM) domains at its C-terminus (92). In vitro binding experiments have shown that different AZ proteins directly associate with different domains in liprin. For example, liprin uses three distinct portions of its coiled-coil domains to bind to RIM, ELKS/ERC/CAST, and GIT (39). In addition, the SAM domains of liprin have been shown to bind to GRIP a scaffolding molecule for glutamate receptor trafficking and the postsynaptic density (84). Recent biochemistry experiments also suggest that SYD-2/ liprin not only can bind to itself but also forms large oligomer structures (M Patel and K Shen, unpublished results). Therefore, it is conceivable that SYD-2/ liprin is the key presynaptic scaffold, capable of forming a multimeric network and organizing the presynaptic terminal by direct binding to different presynaptic proteins.

Much of this model is supported by genetic data from analyzing the presynaptic HSNL neuron in *C. elegans*. Is SYD-2/ liprin the universal presynaptic organizer in all presynaptic terminals? The answer is likely to be much more complex. First, *syd-2* was first identified based on its presynaptic defects in another class of neurons, DD and VD in *C. elegans* (92). In the DD/VD neurons, the presynaptic defects of *syd-2* are more subtle than those of HSNL. Specifically, the number of SVs is only mildly decreased in the *syd-2* mutant DD/VD synapses. The most dramatic phenotype, which allowed the original identification of *syd-2,* is lengthened presynaptic AZs (92). Interestingly, the shape of AZs is also abnormal in Dliprin mutants when neuromuscular junctions were examined in Drosophila (35). These results argue that SYD-2/liprin is not absolutely required for the accumulation of SVs or de novo development of the AZ; instead, it regulates the size and shape of AZ in certain synapses. Analysis of more types of synapses will be necessary to gain a complete view of the importance of SYD-2/Liprin in presynaptic assembly.

Another intriguing observation on SYD-2/Liprin was that it affected the dynamics of KIF1a, the kinesin motor responsible for trafficking SV precursors from the cell body to synapses. In *syd-2/Liprin* mutants, the anterograde processivity of synaptic punctae decreases and the retrograde processivity increases, suggesting that SYD-2/Liprin regulates motor activity to ensure long distance transport away from the cell body (46). In support of this notion, direct binding between SYD-2/Liprin and KIF1a has been reported (46, 70). Thus, a more complete picture of SYD-2/Liprin's functions in presynaptic assembly is beginning to emerge. It is conceivable that one population of SYD-2/Liprin is localized at the developing presynaptic AZ, while another population of SYD-2/Liprin binds to the transporting vesicle precursor to stimulate the anterograde movement. One can imagine that once the transporting vesicles arrive at developing synaptic sites, SYD-2/Liprin might utilize its multimerization ability to incorporate itself into the growing presynaptic AZ network.

In summary, growing evidences suggest that SYD-2/Liprin is an important molecule for the presynaptic differentiation. In certain neurons such as HSNL in *C. elegans*, SYD-2/Liprin functions as a master regulator to trigger the assembly of many presynaptic molecules. In many other types of synapses, SYD-2/Liprin is also required, but probably function partially redundantly with other presynaptic molecules to orchestrate AZ assembly.

8.2 SYD-1 and SAD-1 Kinase, Presynaptic Molecules with Polarity Functions?

As discussed above, SYD-1 plays important roles in the assembly of presynaptic structures in HSNL. However, studies of DD/VD neurons and the chemosensory ASI neurons in *C. elegans* showed that SYD-1 also functions in establishing or maintaining axon-dendrite polarity. In *syd-1* mutants, SVs and presynaptic specializations are found in both axons and dendrites (25). The potential functional convergence of presynaptic assembly and axon-dendrite polarity is also evidenced from studies on SAD-1, which was originally isolated in two presynaptic morphology screens in *C. elegans* (12). In *sad-1* mutants, presynaptic markers in both ASI sensory neurons ASI and DD/VD motor neurons appear to be disorganized. SAD-1 encodes a serine/threonine kinase related to the mammalian PAR-1 kinase (12). In HSNL, SAD-1 is co-localized with SV markers as well as AZ markers and is partially required for the accumulation of normal level of SVs. *sad-1* mutants exhibit a dramatic loss of synaptic markers in HSNL, but the severity of the defect is less compared to those of *syd-1* and *syd-2* mutants. Localization and epistasis experiments suggest that SAD-1 functions downstream of SYD-1 and SYD-2 to cluster SVs and AZ proteins (53).

Similar to SYD-1, evidence suggest that SAD-1 is involved in axon-dendrite polarity, SAD-1 overexpression in the ASI neurons causes SVs to appear in sensory dendrites, which is normally devoid of presynaptic markers (12). When DD/VD neurons were examined, SV markers were also found in the dendritic process in addition to axons in *sad-1* loss-of-function mutants, suggesting that appropriate levels of *sad-1* are required for targeting of SVs to the correct subcellular compartments. The establishment of axon-dendrite polarity is more extensively studied in the vertebrate system. Indeed our understanding of SAD-1's function in synapses and polarity has benefited from genetic experiments in the vertebrate system, where both of the SAD-A and SAD-B genes were inactivated. Here, in SAD-A/B double knockout mice, many cortical neurons failed to establish clear axons and dendrites. The remaining neurites in the knockout animals tend to express both axonal and dendritic markers, suggestive of a polarity defect (36).

Although loss of SAD kinase in both *C. elegans* and in mouse neurons exhibited polarity defects, the nature of these defects might not be identical. In worms, the defect is primarily a SV targeting defect. Both ASI and DD/VD neurons in *sad-1* mutants are able to establish morphologically normal axons and dendrites, indicating that the initiation of the axon-dendrite program is intact without *sad-1*. However, in the SAD-A/B knockout mice, many neurons fail to even form morphologically distinct axons and dendrites, suggestive of a key role of SAD kinase in early polarity establishment. Therefore, the SAD kinase family might play multiple functions during the development of the nervous system. Alternatively, the proper targeting of SVs and establishment of the morphological features of axons and dendrites might be linked with each other at certain molecular level.

8.3 Dap160/Intersectin, an Endocytosis Periactive Zonal Organizer

As discussed above, presynaptic nerve terminals are organized into distinct zones, which carry out exocytosis and endocytosis. Early EM studies already indicated that vesicle exocytosis occurs in the middle of the AZ, while endocytosis takes place at the periactive zone (7, 29). What is the molecular basis for the periactive zonal endocytosis? Two studies on a particular scaffolding molecule, DAP160/intersectin, shed light on this question (40, 44).

Endocytosis at the presynaptic terminal requires multiple protein components including clathrin, dynamin, endophilin, synaptojanin and AP180 (15). Endocytosis must be very efficient to sustain the rapid exocytosis at the presynaptic terminal. An elegant way to enrich all the endocytosis components and cluster them at designated subsynaptic region is through using a specific scaffolding molecule. Dap160 contains protein interaction domains including coiled-coil domains and SH3 domains. Dap160 is present in the presynaptic boutons of Drosophila NMJ, and it is specifically localized to the periactive zone, a localization pattern that is identical to localization pattern of dynamin and endophilin (40, 44). In *dap160* mutant flies, the abundance of dynamin and endophilin at the presynaptic region is significantly reduced. As a result, the mutant NMJs fail to sustain high frequency SV release (40, 44). Importantly, the general synaptic structure and the localization of AZ markers were not affected by Dap160, suggesting that Dap160 is designated to be a scaffolding molecule for the endocytic machinery.

8.4 Bruchpilot/ELKs/CAST/ERCs, a Calcium Channel Anchor?

The cardinal feature of synaptic exocytosis is its temporal precision. Spikes that arrive at synapses ten milliseconds apart lead to dramatically different outcomes in synaptic plasticity (14). This precision of neurotransmitter release is achieved by the tight regulation of calcium concentration at the presynaptic terminals (34). Calcium channels are strategically placed in the proximity of the SNARE complex to ensure the rapid and precise onset of exocytosis upon the arrival of action potentials (9, 73). Physiological experiments suggest that there are a defined number of "calcium channel slots" at the AZ, suggesting that ordered interactions between the channels and AZ proteins must exist (8). How are calcium channels localized to the AZ? One potential link in vertebrate synapses is the presynaptic cell-adhesion molecule alpha-neurexin. Here, knockout mice of alpha neurexin have been found to exhibit a reduced calcium trigger neurotransmiter release (47).

A second candidate is Bruchpilot, which was biochemically identified in *Drosophila* as an active zone component of the body wall NMJ, and subsequently found to be important in assembling AZ and localizing calcium channels (37, 77). At the Drosophila NMJ, Bruchpilot was observed in donut-shaped structures centered at AZ of neuromuscular synapses. At *burp* mutant AZs, electron-dense projections (T-bars) were entirely lost, calcium channels were reduced in density, evoked vesicle release was depressed, and short-term plasticity was altered. BRP-like proteins might thus establish proximity between calcium channels and vesicles to allow efficient transmitter release and patterned synaptic plasticity. Therefore, Bruchpilot is required for structural and functional integrity of AZ in the fly NMJs.

The C-terminus of Bruchpilot is highly conserved within insects but not vertebrates, whereas the N-terminal half of Bruchpilot shows strong homology to a vertebrate family of proteins with different names including ELKS/CAST/ERC, as well as to the *C. elegans* AZ protein ELKS-1. These homologs have been shown to localize to the AZs of vertebrate CNS and worm synapses (16, 39, 81). In vertebrate synapses, ELKs have been shown to physically interact with other AZ proteins such RIM, Piccolo Bassoon and Liprin (63). Although the function of ELKS in vertebrate synapses is unclear, more is known about ELKS in *C. elegans*. Specifically, ELKS-1 appears to interact genetically with SYD-1 and SYD-2. ELKS-1 is required for the ability of a gain-of-function allele of *syd-2* to suppress a *syd-1* mutant (13). However, loss of function mutation in ELKs-1 has no obvious effect on the assembly or function of worm synapses (16) and thus its role in localizing calcium channels in other systems remains to be tested (13).

8.5 Bassoon Anchors Synaptic Ribbons to AZ Membrane

Ribbon synapses are highly specialized synapses found in the photoreceptor cells, retinal bipolar cells and hair cells. This type of synapse is characterized by an electron-dense "ribbon" structure that is associated with the AZ plasma membrane. This special type of AZ is thought to permit a high rate of continuous SV release (72, 76). Many AZ components have been localized to the ribbon structure (18). The distinctive size and shape of the ribbon provided a rare opportunity to study the sub-active zone localization of AZ proteins. Morphologically, the ribbon synapses can be divided into two parts. The first is the arciform density. This is a segment of the presynaptic plasma membrane where SVs dock and fuse and that sits juxtaposed to the PSD. The second is the synaptic ribbon itself. As the name implies, this is a ribbon of proteinaeous material with one edge attached to the arciform density. Tethered all along the two faces of the ribbon are large numbers of SVs. These are thought to either move progressively down along the ribbon until they reach the arciform density and fuse, or to engage in multi-vesicular fusion of a row of SVs (72, 76). Molecularly, the arciform density appears to be defined by molecules essential for SV fusion such as Munc13, RIM2, ELKS and calcium channels (74). In contrast, the ribbons are created by a variety of proteins including Ribeye as a core structural component and other AZ proteins such as Piccolo, CtBP1, RIM1 and Kif3a (74). Intriguingly, the large AZ protein Bassoon was found at the junction of the arciform density and the ribbon suggesting a potential role in linking ribbons to the AZ (74). To test this concept, the group of Helmut Brandstatter examined the impact of loss of function mutation in Bassoon on the formation of the mouse photoreceptor ribbon synapse (19). In these mice, the photoreceptor ribbons were found to form and contain rows of attached SVs. However, the ribbons failed to be tethered at the arciform density and instead were observed free floating in the cytoplasm (19). Interestingly while Piccolo, Ribeye, CtBP1, RIM1 and KIF3a were found on free floating ribbons, RIM2, Munc13-1 and CAST/ERC were associated with the arciform density at the plasma membrane (74). These results suggest that subfamily members of AZ proteins can be recruited into distinct substructures of an AZ to execute different functions. Furthermore, the data demonstrate that Bassoon plays a central role in anchoring ribbons to the arciform density. Molecularly this appears to be due to the ability of Bassoon to directly interact with both the ribbon protein Ribeye, as well as with the acriform density protein ELKS/CAST/ERC (74).

8.6 How are Presynaptic AZ Proteins Delivered to the Sites of Assembly?

Presynaptic differentiation transforms a patch of plasma membrane into highly organized presynaptic membrane characterized both by the presences of clusters of AZ proteins and SVs. A fundamental cell-biological question is how protein constituents of presynaptic boutons are delivered to both nascent and mature synapses. As discussed above, many vesicle-associated proteins such as those that make up mature SVs utilize membrane trafficking and active transport along microtubules to reach distant synapses. With regard to soluble/cytoskeletal proteins, two mechanisms appear to guide them to synapses: diffusion/slow axonal transport or in association with vesicular cargos. Either way, correctly sorting these vesicles or proteins into axons and getting them to nascent synapses in a timely manner and in the correct ratios is a major challenge for neuronal cells.

Studies during the last decade have begun to examine the timing of synapse formation and the cellular mechanisms underlying the delivery of different components. With regard to the AZ, the question has been how do mixtures of both soluble and integral membrane protein arrive at synapses. Two models have been considered: (a) that AZ proteins are co-transported as supermolecular complexes or (b) that they each have a specific transport/synaptic localization sequence and are sequentially recruited into nascent synapses.

Answers to these questions have largely come from studies of dissociated cultures of vertebrate neurons (93, 94). These studies have shown that functional AZ assembly can be rapid, occurrings within 20 min of initial axo-dendritic contact during which time key AZ proteins such as Bassoon and Piccolo are recruited (21, 66). Subsequent studies have shown that many AZ proteins, even those without transmembrane domains, appear to use vesicular intermediates for their delivery to nascent synapses (1, 87, 89). In fact, large clusters of vesicles of different size and shapes have been shown to accumulate within axonal varicosities at nascent synaptic sites supporting a general concept that vesicular transport is a fundamental aspect of AZ assembly (1). Moreover, emerging data indicate that small groups of AZ proteins, such as Piccolo and Bassoon may use specialized vesicles for their delivery to nascent synapses (66, 89) and that specific cell–cell adhesion molecules may play a role in their site specific fusion (78, 93, 94). This said little is known about the signaling pathways that lie between initial contact and the assembly of presynaptic AZs.

9 Regulation of Presynaptic Assembly by Ubiquitin Mediated Pathways

Many aspects of presynaptic assembly are regulated. For example, AZs usually adopt specific size and shape. The position and number of synapses formed between neurons are often stereotyped, suggesting that there must be mechanisms to control the molecular assembly program. Genetic analysis in model organisms has been instrumental in gaining insight into these regulatory mechanisms.

9.1 Rpm-1/highwire Controls Numbers of Synapses and AZ?

In search for mechanisms that regulates synaptic growth, Corey Goodman and colleagues tested the involvement of ubiquitin using a deubiquitinating protease, *fat facets* in the development of the *Drosophila* NMJ. Ubiquitination is a reversible modification stimulated by ubiquitin ligases and antagonized by deubiquitinating proteases. Overexpression of *fat facets* leads to massive overgrowth of the NMJ synapses, suggesting that ubiquitin mediated pathways are required during normal development to restrict the number of synapses forming at the developing *Drosophila* NMJ (17).

A genetic interaction screen of the *fat facets* overexpression phenotype identified a RING finger protein *highwire*, which enhance the *fat facets* gain-of-function phenotype. RING finger motifs are signatures for E3 ubiquitin ligases, suggesting that *highwire* represent the other side of the ubiquitin equation. Indeed, *highwire* loss of function mutants exhibit dramatic synaptic overgrowth of the NMJ (79). These experiments suggest that the number of synapses at the *Drosophila* NMJ is regulated by a delicate balance between the ubiquitin ligase activity and deubiquinating protease activity.

Independent studies in *C. elegans* synapses identified *rpm-1*, the worm ortholog of *highwire*, as a regulator of synaptic morphology. In *rpm-1* loss-of-function mutants, the morphology of DD/VD presynaptic morphology is disrupted. In mutants, individual presynaptic terminals often contain 2 or 3 AZs, instead of only one AZs found as typically found in wild type controls, indicating that *rpm-1* regulates the number of AZs in *C. elegans* (91). Studies in a different worm neuron PLM revealed an intriguing phenotype of *rpm-1*. In *rpm-1* mutants, SVs fail to accumulate at the presynaptic AZ (61). The reason for this distinct phenotype is not yet clear.

9.2 MAPK Pathway Is the Target of rpm-1?

The conserved RING finger domain in *rpm-1/highwire* suggests that it is an E3 ubiquitin ligase. Such ligases have been shown to catalyze the ubiquitination of protein substrates doomed for proteasome-mediated degradation (see Chapter by Helton and Ehlers, this volume). If *rpm-1/highwire* regulates AZ organization via this pathway, what are the targets for its E3 ligase? Forward genetic screens at the DD/VD synapses identified a kinase cascade that functions downstream of *rpm-1*. The p38 MAP kinase pathway is composed of the dual leucine zipper-bearing MAPKKK DLK-1, the MAPKK MKK-4, and the p38 MAP kinase PMK-3. Current data indicate that RPM-1 normally down-regulates the activity of this pathway. Furthermore, inactivation of this pathway suppresses the *rpm-1* phenotype, indicating that this pathway is the endogenous substrate for RPM-1 during synapse development (50). Another study showed that RPM-1 physically associates with an F-box protein FSN-1 and might function is a SCF (Skp/Cullin/F-box) complex (42). Similarly, the homolog of DLK-1, Wallenda was identified as the substrate of *highwire* in down-regulating synaptic growth in *Drosophila* (11). Hence, this evolutionarily conserved genetic pathway appears to regulate synaptic growth and the number of AZs.

9.3 A Specific SCF Complex Mediate Presynapse Elimination

Stereotyped synaptic connectivity can arise from both precise recognition between appropriate partners during synaptogenesis and selective synapse elimination after an initial phase of exuberant synapse formation. Synapse elimination is a hallmark of the developmental refinement of neural circuits in both the central and peripheral nervous systems of vertebrate animals (43). The molecular mechanisms that underlie selective synapse removal are largely unknown. Through time lapse experiments, it is apparent that stereotyped developmental elimination of *en passant* synapses in the *C. elegans* HSNL neuron gives rise to the synapse pattern of the mature animals. Exuberant synapses form at early stages of development, some of which are invariably lost as the animals mature. The stereotype synapse elimination is mediated by an SCF complex (E3 ubiquitin ligase) that is comprised of SKR-1 and the F-box protein SEL-10. SYG-1, a synaptic adhesion molecule, binds to SKR-1 directly and inhibits assembly of the SCF complex, thereby protecting nearby synapses. Since SYG-1 is localized to particular subcellular location, the regulated activity of SCF complex generates the stereotyped synaptic elimination (67, 68). Therefore, diverse ubiquitin E3 ligase complexes can negatively regulate presynaptic assembly programs in different modes: some leading to overall reduction of number of synapses, others giving rise to specific patterns of synaptic innervation.

10 What Initiates and Restricts the Presynaptic Assembly Program?

The molecular assembly program responsible for presynapse formation is likely to be regulated both spatially and temporally. Since synaptogenesis is initiated by contact between pre- and postsynaptic cells, it is probably safe to assume that interactions between transmembrane proteins on the synaptic partners trigger pre- and postsynaptic assembly program. Indeed several pairs of transmembrane proteins have been shown to trigger synapse assembly in cultured vertebrate neurons (reviewed in detailed in Chapter by Hanus and Ehlers of this book). Diffusible molecules have been demonstrated to promote synapse formation in the central nervous system, including ephrin/EphR, FGFs/FGFR, thrombospodin, Wnt/Frizzled and BDNF/TrkB (reviewed in (45)). It is not yet clear whether these extracellular molecules and their cognate receptors activation directly activate presynaptic assembly.

Many ion channels are present at the synaptic terminals. It has long been postulated that these ion channels might participate in synapse assembly. Ion channels can potentially directly interact with proteins in or across synaptic cleft using the extraceullar domains and interact with cytosolic scaffolding molecules with their intracellular domains. Therefore, the presynaptically localized ion channels are poised to bridge the extracellular signal and presynaptic assembly. The best experimental evidence to support this notion comes from studies on vertebrate NMJ.

During the development of vertebrate NMJs, it is well established that the basal lamina plays important roles for postsynaptic development. Agrin, a secreted protein found in high concentration at the basal lamina, is a key factor in maintaining postsynaptic acetylcholine receptor clusters (41). Much less is known about the importance of basal lamina on the development of presynaptic development. One recent study showed that another synaptic cleft basal lamina component, β2 laminin, can directly interact with the pore forming subunit of the presynaptic calcium channel. This interaction not only clusters voltage gated calcium channels, but also triggers the assembly of other presynaptic components. In β2 laminin -/- or $Ca_V2.1$ -/- mice, the development of AZs is partially compromised, suggesting this interaction is required in vivo for neuromuscular junction formation (51). Since β2 laminin is specifically localized to the basal lamina at the synaptic cleft, it is conceivable that it triggers the presynaptic development and ensures that the functional presynaptic specialization is directly juxtaposed to the postsynaptic apparatus. Alternatively, this pathway may not be inductive in nature, but may serve to place the AZ and postsynaptic apparatus in register with other signaling pathways regulating stability when this link is disrupted.

In *C. elegans*, a pair of transmembrane proteins SYG-1 and SYG-2, which are members of the immunoglobulin superfamily of adhesion molecules, specifies the synaptic location and target selection of the motor neuron HSNL. SYG-1 functions in the presynaptic neuron and is localized to the presynaptic region immediately before synaptogenesis in HSNL. Interestingly, SYG-2 is not made by the target cells of HSNL, instead is expressed by guidepost cells, which are essential for the specification of these synapses. SYG-2 clusters SYG-1 to the HSNL synaptic region through direct binding between their extracellular domains. In *syg-1* or *syg-2* mutants, HSNL fails to select its normal synaptic partners and fails to form synapses at the normal locations (67, 68). Therefore, it is hypothesized that SYG-1 initiates the assembly of presynaptic specializations in HSNL. If this is correct, one would predict that the localization of SVs or AZ markers would be abnormal in *syg-1* mutants. Indeed, every AZ and vesicle markers examined showed ectopic localization in *syg-1* mutants (53), suggesting that SYG-1 likely initiates many aspects of presynaptic assembly. However, no direct interaction between SYG-1 and SYD-1/SYD-2 has been observed (Patel and Shen, unpublished data). It is still not clear how SYG-1 activates critical assembly molecules like SYD-2.

Synaptic patterns can be generated by positive regulators of synapse formation, which promote local synaptic assembly. Alternatively, innervation patterns can also be created using negative regulators to restrict synapse formation to specific subcellular locations. When the synapses of the DA9 motor neuron was examined, it was found that presynapses are restricted to a specific segment of its axon. A Wnt *lin-44* localizes the Wnt receptor *lin-17*/Frizzled to a subdomain of the DA9 axon that is devoid of presynaptic specializations. When this signaling pathway, composed of the Wnts *lin-44* and *egl-20*, *lin-17*/Frizzled and *dsh-1*/Dishevelled, is compromised, synapses develop ectopically in this subdomain (38). These results suggest that morphogenetic signals can spatially regulate the patterning of synaptic connections by negative regulation of synaptogenesis and restricting the subcellular localization of synapses. Further experiments are required to understand how Wnt and Frizzled prevent location synaptic assembly.

11 All Roads Lead to Rome

The available data from different organisms and different synapses strongly suggest that presynaptic terminals are assembled via diverse mechanisms. This is probably not surprising considering the vastly diverse morphology of presynaptic terminals and AZs in the nervous system. The core components of the SV cycle are likely to be similar at all types of AZs. However, the size, shape and fine configuration of different AZs must be tailored to achieve differences in physiology observed at diverse presynaptic terminals.

11.1 Basal Laminae vs No Basal Laminae

As discussed above, β2 laminin triggers presynaptic assembly through direct binding to VGCC in the vertebrate NMJs (51). Many synapses do not have basal lamina structure, such as the vertebrate central synapses and worm synapses. It is therefore not likely that the same β2 laminin/VGCC interaction would be essential for presynaptic assembly in these synapses. Indeed, the loss of calcium channels do not appear to affect presynaptic markers in HSNL synapses in *C. elegans* (53). Interestingly, calcium channel mutants exhibit an undergrowth phenotype in *Drosophila* NMJ (56), where basal lamina is also absent, suggesting that calcium channel might interact with other cues.

11.2 Long Distance vs Short Distance

The length of axons varies significantly in the nervous system. In vertebrates, long axon projections can be as long as 1 m, while short axons in the retina are only tens of microns in length. It is possible that there are additional requirements for motor assisted intracellular transport in long axons to efficiently recruit presynaptic components. Some preliminary observations in *C. elegans* seem to agree with this notion. HSNL forms one group of synapses very close to its cell body (~30 μm away from the cell body). Genetic analysis on synaptic assembly showed that the presynaptic localization of SVs and vesicle associated markers strictly depends on UNC-104/KIF1A, a kinesin motor, while the assembly of AZ markers such as SYD-2/liprin and ELKS-1/CAST/ERC are not *unc-104* dependent. However, when similar AZ markers were examined in *C. elegans* neurons with longer axons such as DD/VD or DA9, both vesicle markers and AZ markers require UNC-104 to be localized to the presynaptic terminals (85) (K Shen unpublished data). It is possible that AZ markers can diffuse cross short distance and assembly at synapses near cell body, while the assembly of AZ far from cell body requires active vesicular transport mechanisms, such as the dense core vesicles as seen in the studies of nascent synapses in vertebrate cultured neurons (78).

11.3 Specialized AZs vs Non-specialized AZs

The specialized AZs are likely to have presynaptic assembly programs distinct from others. The ribbon synapses in the photoreceptor cells requires Bassoon to assume the normal synaptic "ribbon" AZ (19). In the Bassoon -/-, when hippocampal synapses were examined, no obvious morphological defects can be detected at the ultrastructural level, although the physiology of these synapses are compromised (2). These data suggest that Bassoon might be playing different roles in specialized AZs versus non-specialized AZs.

12 Conclusion

Significant progresses have been made towards an understanding of the structure, function and development of the presynaptic scaffold. The universal feature of the presynaptic scaffold is to organize the SV cycle. The core machinery that executes the cycle is found at all synapses. However, AZs of different size and shape have evolved presumably to achieve the diverse physiological task of different neurons. Much work is still ahead of us to understand how difference in molecular composition of the AZ give rise to distinctive morphological features and physiological signatures, and to understand how diverse developmental programs all lead to the functional assembly of the SV cycle.

References

1. Ahmari SE, Buchanan J, and Smith SJ. Assembly of presynaptic active zones from cytoplasmic transport packets. *Nature Neuroscience* 3: 445–451, 2000.
2. Altrock WD, tom Dieck S, Sokolov M, Meyer AC, Sigler A, Brakebusch C, Fassler R, Richter K, Boeckers TM, Potschka H, Brandt C, Loscher W, Grimberg D, Dresbach T, Hempelmann A, Hassan H, Balschun D, Frey JU, Brandstatter JH, Garner CC, Rosenmund C, and Gundelfinger ED. Functional inactivation of a fraction of excitatory synapses in mice deficient for the active zone protein bassoon. *Neuron* 37: 787–800, 2003.
3. Bamji SX, Rico B, Kimes N, and Reichardt LF. BDNF mobilizes synaptic vesicles and enhances synapse formation by disrupting cadherin-beta-catenin interactions. *The Journal of cell biology* 174: 289–299, 2006.
4. Bamji SX, Shimazu K, Kimes N, Huelsken J, Birchmeier W, Lu B, and Reichardt LF. Role of beta-catenin in synaptic vesicle localization and presynaptic assembly. *Neuron* 40: 719–731, 2003.
5. Becherer U and Rettig J. Vesicle pools, docking, priming, and release. *Cell and Tissue Research* 326: 393–407, 2006.
6. Brodin L, Low P, and Shupliakov O. Sequential steps in clathrin-mediated synaptic vesicle endocytosis. *Current Opinion in Neurobiology* 10: 312–320, 2000.
7. Brodin L and Shupliakov O. Giant reticulospinal synapse in lamprey: molecular links between active and periactive zones. *Cell and Tissue Research* 326: 301–310, 2006.
8. Cao YQ, Piedras-Renteria ES, Smith GB, Chen G, Harata NC, and Tsien RW. Presynaptic Ca2+ channels compete for channel type-preferring slots in altered neurotransmission arising from Ca2+ channelopathy. *Neuron* 43: 387–400, 2004.
9. Catterall WA, Hulme JT, Jiang X, and Few WP. Regulation of sodium and calcium channels by signaling complexes. *Journal of Receptor and Signal Transduction Research* 26: 577–598, 2006.
10. Chi P, Greengard P, and Ryan TA. Synaptic vesicle mobilization is regulated by distinct synapsin I phosphorylation pathways at different frequencies. *Neuron* 38: 69–78, 2003.
11. Collins CA, Wairkar YP, Johnson SL, and DiAntonio A. Highwire restrains synaptic growth by attenuating a MAP kinase signal. *Neuron* 51: 57–69, 2006.
12. Crump JG, Zhen M, Jin Y, and Bargmann CI. The SAD-1 kinase regulates presynaptic vesicle clustering and axon termination. *Neuron* 29: 115–129, 2001.
13. Dai Y, Taru H, Deken SL, Grill B, Ackley B, Nonet ML, and Jin Y. SYD-2 Liprin-alpha organizes presynaptic active zone formation through ELKS. *Nature Neuroscience* 9: 1479–1487, 2006.
14. Dan Y and Poo MM. Spike timing-dependent plasticity of neural circuits. *Neuron* 44: 23–30, 2004.
15. De Camilli P. Molecular mechanisms in membrane traffic at the neuronal synapse: role of protein-lipid interactions. *Harvey Lectures* 100: 1–28, 2004.
16. Deken SL, Vincent R, Hadwiger G, Liu Q, Wang ZW, and Nonet ML. Redundant localization mechanisms of RIM and ELKS in Caenorhabditis elegans. *The Journal of Neuroscience* 25: 5975–5983, 2005.
17. DiAntonio A, Haghighi AP, Portman SL, Lee JD, Amaranto AM, and Goodman CS. Ubiquitination-dependent mechanisms regulate synaptic growth and function. *Nature* 412: 449–452, 2001.
18. Dick O, Hack I, Altrock WD, Garner CC, Gundelfinger ED, and Brandstatter JH. Localization of the presynaptic cytomatrix protein Piccolo at ribbon and conventional synapses in the rat retina: comparison with Bassoon. *The Journal of Comparative Neurology* 439: 224–234, 2001.

19. Dick O, tom Dieck S, Altrock WD, Ammermuller J, Weiler R, Garner CC, Gundelfinger ED, and Brandstatter JH. The presynaptic active zone protein bassoon is essential for photoreceptor ribbon synapse formation in the retina. *Neuron* 37: 775–786, 2003.
20. Fejtova A and Gundelfinger ED. Molecular organization and assembly of the presynaptic active zone of neurotransmitter release. *Results and Probllems in Cell Differentiation* 43: 49–68, 2006.
21. Friedman HV, Bresler T, Garner CC, and Ziv NE. Assembly of new individual excitatory synapses: time course and temporal order of synaptic molecule recruitment. *Neuron* 27: 57–69, 2000.
22. Garner CC, Kindler S, and Gundelfinger ED. Molecular determinants of presynaptic active zones. *Current Opinion in Neurobiology* 10: 321–327, 2000.
23. Gray EG. Electron microscopy of presynaptic organelles of the spinal cord. *Journal of Anatomy* 97: 101–106, 1963.
24. Greengard P, Valtorta F, Czernik AJ, and Benfenati F. Synaptic vesicle phosphoproteins and regulation of synaptic function. *Science (New York, NY* 259: 780–785, 1993.
25. Hallam SJ, Goncharov A, McEwen J, Baran R, and Jin Y. SYD-1, a presynaptic protein with PDZ, C2 and rhoGAP-like domains, specifies axon identity in C. elegans. *Nature Neuroscience* 5: 1137–1146, 2002.
26. Harlow ML, Ress D, Stoschek A, Marshall RM, and McMahan UJ. The architecture of active zone material at the frog's neuromuscular junction. *Nature* 409: 479–484, 2001.
27. Heidemann SR, Landers JM, and Hamborg MA. Polarity orientation of axonal microtubules. *The Journal of Cell Biology* 91: 661–665, 1981.
28. Heuser J, Katz B, and Miledi R. Structural and functional changes of frog neuromuscular junctions in high calcium solutions. *Proceedings of the Royal Society of London Series B, Containing papers of a Biological character* 178: 407–415, 1971.
29. Heuser JE and Reese TS. Evidence for recycling of synaptic vesicle membrane during transmitter release at the frog neuromuscular junction. *The Journal of Cell Biology* 57: 315–344, 1973.
30. Heuser JE and Reese TS. Structural changes after transmitter release at the frog neuromuscular junction. *The Journal of Cell Biology* 88: 564–580, 1981.
31. Heuser JE, Reese TS, and Landis DM. Functional changes in frog neuromuscular junctions studied with freeze-fracture. *Journal of Neurocytology* 3: 109–131, 1974.
32. Hirokawa N, Sobue K, Kanda K, Harada A, and Yorifuji H. The cytoskeletal architecture of the presynaptic terminal and molecular structure of synapsin 1. *The Journal of Cell Biology* 108: 111–126, 1989.
33. Jahn R. Neuroscience. A neuronal receptor for botulinum toxin. *Science (New York, NY)* 312: 540–541, 2006.
34. Katz B and Miledi R. The Effect of Calcium on Acetylcholine Release from Motor Nerve Terminals. *Proceedings of the Royal Society of London Series B, Containing Papers of a Biological Character* 161: 496–503, 1965.
35. Kaufmann N, DeProto J, Ranjan R, Wan H, and Van Vactor D. Drosophila liprin-alpha and the receptor phosphatase Dlar control synapse morphogenesis. *Neuron* 34: 27–38, 2002.
36. Kishi M, Pan YA, Crump JG, and Sanes JR. Mammalian SAD kinases are required for neuronal polarization. *Science (New York, NY)* 307: 929–932, 2005.
37. Kittel RJ, Wichmann C, Rasse TM, Fouquet W, Schmidt M, Schmid A, Wagh DA, Pawlu C, Kellner RR, Willig KI, Hell SW, Buchner E, Heckmann M, and Sigrist SJ. Bruchpilot promotes active zone assembly, Ca2+ channel clustering, and vesicle release. *Science (New York, NY)* 312: 1051–1054, 2006.
38. Klassen MP and Shen K. Wnt Signaling Positions Neuromuscular Connectivity by Inhibiting Synapse Formation in C. elegans. *Cell* 130: 704–716, 2007.

39. Ko J, Na M, Kim S, Lee JR, and Kim E. Interaction of the ERC family of RIM-binding proteins with the liprin-alpha family of multidomain proteins. *The Journal of Biological Chemistry* 278: 42377–42385, 2003.
40. Koh TW, Verstreken P, and Bellen HJ. Dap160/intersectin acts as a stabilizing scaffold required for synaptic development and vesicle endocytosis. *Neuron* 43: 193–205, 2004.
41. Kummer TT, Misgeld T, and Sanes JR. Assembly of the postsynaptic membrane at the neuromuscular junction: paradigm lost. *Current Opinion in Neurobiology* 16: 74–82, 2006.
42. Liao EH, Hung W, Abrams B, and Zhen M. An SCF-like ubiquitin ligase complex that controls presynaptic differentiation. *Nature* 430: 345–350, 2004.
43. Lichtman JW and Colman H. Synapse elimination and indelible memory. *Neuron* 25: 269–278, 2000.
44. Marie B, Sweeney ST, Poskanzer KE, Roos J, Kelly RB, and Davis GW. Dap160/intersectin scaffolds the periactive zone to achieve high-fidelity endocytosis and normal synaptic growth. *Neuron* 43: 207–219, 2004.
45. McAllister AK. Dynamic Aspects of CNS Synapse Formation. *Annual Review of Neuroscience* 30: 425–450, 2007.
46. Miller KE, DeProto J, Kaufmann N, Patel BN, Duckworth A, and Van Vactor D. Direct observation demonstrates that Liprin-alpha is required for trafficking of synaptic vesicles. *Current Biology* 15: 684–689, 2005.
47. Missler M, Zhang W, Rohlmann A, Kattenstroth G, Hammer RE, Gottmann K, and Sudhof TC. Alpha-neurexins couple Ca2+ channels to synaptic vesicle exocytosis. *Nature* 424: 939–948, 2003.
48. Montgomery JM, Zamorano PL, and Garner CC. MAGUKs in synapse assembly and function: an emerging view. *Cellular and Molecular Life Science* 61: 911–929, 2004.
49. Morales M, Colicos MA, and Goda Y. Actin-dependent regulation of neurotransmitter release at central synapses. *Neuron* 27: 539–550, 2000.
50. Nakata K, Abrams B, Grill B, Goncharov A, Huang X, Chisholm AD, and Jin Y. Regulation of a DLK-1 and p38 MAP kinase pathway by the ubiquitin ligase RPM-1 is required for presynaptic development. *Cell* 120: 407–420, 2005.
51. Nishimune H, Sanes JR, and Carlson SS. A synaptic laminin-calcium channel interaction organizes active zones in motor nerve terminals. *Nature* 432: 580–587, 2004.
52. Nonet ML, Staunton JE, Kilgard MP, Fergestad T, Hartwieg E, Horvitz HR, Jorgensen EM, and Meyer BJ. Caenorhabditis elegans rab-3 mutant synapses exhibit impaired function and are partially depleted of vesicles. *Journal of Neuroscience* 17: 8061–8073, 1997.
53. Patel MR, Lehrman EK, Poon VY, Crump JG, Zhen M, Bargmann CI, and Shen K. Hierarchical assembly of presynaptic components in defined C. elegans synapses. *Nature Neuroscience* 9: 1488–1498, 2006.
54. Pieribone VA, Shupliakov O, Brodin L, Hilfiker-Rothenfluh S, Czernik AJ, and Greengard P. Distinct pools of synaptic vesicles in neurotransmitter release. *Nature* 375: 493–497, 1995.
55. Richmond JE and Broadie KS. The synaptic vesicle cycle: exocytosis and endocytosis in Drosophila and C. elegans. *Current Opinion in Neurobiology* 12: 499–507, 2002.
56. Rieckhof GE, Yoshihara M, Guan Z, and Littleton JT. Presynaptic N-type calcium channels regulate synaptic growth. *The Journal of Biological Chemistry* 278: 41099–41108, 2003.
57. Rostaing P, Real E, Siksou L, Lechaire JP, Boudier T, Boeckers TM, Gertler F, Gundelfinger ED, Triller A, and Marty S. Analysis of synaptic ultrastructure without fixative using high-pressure freezing and tomography. *European Journal of Neuroscience* 24: 3463–3474, 2006.

58. Ryan TA, Reuter H, Wendland B, Schweizer FE, Tsien RW, and Smith SJ. The kinetics of synaptic vesicle recycling measured at single presynaptic boutons. *Neuron* 11: 713–724, 1993.
59. Sanes JR and Lichtman JW. Induction, assembly, maturation and maintenance of a postsynaptic apparatus. *Nature Reviews. Neuroscience* 2: 791–805, 2001.
60. Sankaranarayanan S, Atluri PP, and Ryan TA. Actin has a molecular scaffolding, not propulsive, role in presynaptic function. *Nature Neuroscience* 6: 127–135, 2003.
61. Schaefer AM, Hadwiger GD, and Nonet ML. rpm-1, a conserved neuronal gene that regulates targeting and synaptogenesis in C. elegans. *Neuron* 26: 345–356, 2000.
62. Schluter OM, Schmitz F, Jahn R, Rosenmund C, and Sudhof TC. A complete genetic analysis of neuronal Rab3 function. *Journal of Neuroscience* 24: 6629–6637, 2004.
63. Schoch S, Castillo PE, Jo T, Mukherjee K, Geppert M, Wang Y, Schmitz F, Malenka RC, and Sudhof TC. RIM1alpha forms a protein scaffold for regulating neurotransmitter release at the active zone. *Nature* 415: 321–326, 2002.
64. Schoch S and Gundelfinger ED. Molecular organization of the presynaptic active zone. *Cell and Tissue Research* 326: 379–391, 2006.
65. Schweizer FE and Ryan TA. The synaptic vesicle: cycle of exocytosis and endocytosis. *Current Opinion in Neurobiology* 16: 298–304, 2006.
66. Shapira M, Zhai RG, Dresbach T, Bresler T, Torres VI, Gundelfinger ED, Ziv NE, and Garner CC. Unitary assembly of presynaptic active zones from Piccolo-Bassoon transport vesicles. *Neuron* 38: 237–252, 2003.
67. Shen K and Bargmann CI. The immunoglobulin superfamily protein SYG-1 determines the location of specific synapses in C. elegans. *Cell* 112: 619–630, 2003.
68. Shen K, Fetter RD, and Bargmann CI. Synaptic specificity is generated by the synaptic guidepost protein SYG-2 and its receptor, SYG-1. *Cell* 116: 869–881, 2004.
69. Sheng M and Hoogenraad CC. The Postsynaptic architecture of excitatory synapses: a more quantitative view. *Annual Review of Biochemistry* 76: 823–847, 2006.
70. Shin H, Wyszynski M, Huh KH, Valtschanoff JG, Lee JR, Ko J, Streuli M, Weinberg RJ, Sheng M, and Kim E. Association of the kinesin motor KIF1A with the multimodular protein liprin-alpha. *The Journal of Biological chemistry* 278: 11393–11401, 2003.
71. Siksou L, Rostaing P, Lechaire JP, Boudier T, Ohtsuka T, Fejtova A, Kao HT, Greengard P, Gundelfinger ED, Triller A, and Marty S. Three-dimensional architecture of presynaptic terminal cytomatrix. *Journal of Neuroscience* 27: 6868–6877, 2007.
72. Sterling P and Matthews G. Structure and function of ribbon synapses. *Trends in Neurosciences* 28: 20–29, 2005.
73. Sudhof TC. The synaptic vesicle cycle. *Annual Review of Neuroscience* 27: 509–547, 2004.
74. tom Dieck S, Altrock WD, Kessels MM, Qualmann B, Regus H, Brauner D, Fejtova A, Bracko O, Gundelfinger ED, and Brandstatter JH. Molecular dissection of the photoreceptor ribbon synapse: physical interaction of Bassoon and RIBEYE is essential for the assembly of the ribbon complex. *The Journal of Cell Biology* 168: 825–836, 2005.
75. Vale RD. The molecular motor toolbox for intracellular transport. *Cell* 112: 467–480, 2003.
76. von Gersdorff H and Matthews G. Electrophysiology of synaptic vesicle cycling. *Annual Review of Physiology* 61: 725–752, 1999.
77. Wagh DA, Rasse TM, Asan E, Hofbauer A, Schwenkert I, Durrbeck H, Buchner S, Dabauvalle MC, Schmidt M, Qin G, Wichmann C, Kittel R, Sigrist SJ, and Buchner E. Bruchpilot, a protein with homology to ELKS/CAST, is required for structural integrity and function of synaptic active zones in Drosophila. *Neuron* 49: 833–844, 2006.
78. Waites CL, Craig AM, and Garner CC. Mechanisms of vertebrate synaptogenesis. *Annual Review of Neuroscience* 28: 251–274, 2005.

79. Wan HI, DiAntonio A, Fetter RD, Bergstrom K, Strauss R, and Goodman CS. Highwire regulates synaptic growth in Drosophila. *Neuron* 26: 313–329, 2000.
80. Wang X, Hu B, Zimmermann B, and Kilimann MW. Rim1 and rabphilin-3 bind Rab3-GTP by composite determinants partially related through N-terminal alpha -helix motifs. *The Journal of Biological Chemistry* 276: 32480–32488, 2001.
81. Wang Y, Liu X, Biederer T, and Sudhof TC. A family of RIM-binding proteins regulated by alternative splicing: Implications for the genesis of synaptic active zones. *Proceedings of the National Academy of Sciences of the United States of America* 99: 14464–14469, 2002.
82. Wang Y, Okamoto M, Schmitz F, Hofmann K, and Sudhof TC. Rim is a putative Rab3 effector in regulating synaptic-vesicle fusion. *Nature* 388: 593–598, 1997.
83. Weimer RM, Gracheva EO, Meyrignac O, Miller KG, Richmond JE, and Bessereau JL. UNC-13 and UNC-10/rim localize synaptic vesicles to specific membrane domains. *Journal of Neuroscience* 26: 8040–8047, 2006.
84. Wyszynski M, Kim E, Dunah AW, Passafaro M, Valtschanoff JG, Serra-Pages C, Streuli M, Weinberg RJ, and Sheng M. Interaction between GRIP and liprin-alpha/SYD2 is required for AMPA receptor targeting. *Neuron* 34: 39–52, 2002.
85. Yeh E, Kawano T, Weimer RM, Bessereau JL, and Zhen M. Identification of genes involved in synaptogenesis using a fluorescent active zone marker in Caenorhabditis elegans. *Journal of Neuroscience* 25: 3833–3841, 2005.
86. Yonekawa Y, Harada A, Okada Y, Funakoshi T, Kanai Y, Takei Y, Terada S, Noda T, and Hirokawa N. Defect in synaptic vesicle precursor transport and neuronal cell death in KIF1A motor protein-deficient mice. *The Journal of Cell Biology* 141: 431–441, 1998.
87. Zhai R, Olias G, Chung WJ, Lester RA, tom Dieck S, Langnaese K, Kreutz MR, Kindler S, Gundelfinger ED, and Garner CC. Temporal appearance of the presynaptic cytomatrix protein bassoon during synaptogenesis. *Molecular and Cellular Neurosciences* 15: 417–428, 2000.
88. Zhai RG and Bellen HJ. The architecture of the active zone in the presynaptic nerve terminal. *Physiology (Bethesda)* 19: 262–270, 2004.
89. Zhai RG, Vardinon-Friedman H, Cases-Langhoff C, Becker B, Gundelfinger ED, Ziv NE, and Garner CC. Assembling the presynaptic active zone: a characterization of an active one precursor vesicle. *Neuron* 29: 131–143, 2001.
90. Zhao C, Takita J, Tanaka Y, Setou M, Nakagawa T, Takeda S, Yang HW, Terada S, Nakata T, Takei Y, Saito M, Tsuji S, Hayashi Y, and Hirokawa N. Charcot-Marie-Tooth disease type 2A caused by mutation in a microtubule motor KIF1Bbeta. *Cell* 105: 587–597, 2001.
91. Zhen M, Huang X, Bamber B, and Jin Y. Regulation of presynaptic terminal organization by C. elegans RPM-1, a putative guanine nucleotide exchanger with a RING-H2 finger domain. *Neuron* 26: 331–343, 2000.
92. Zhen M and Jin Y. The liprin protein SYD-2 regulates the differentiation of presynaptic termini in C. elegans. *Nature* 401: 371–375, 1999.
93. Ziv NE and Garner CC. Cellular and molecular mechanisms of presynaptic assembly. *Nature Reviews. Neuroscience* 5: 385–399, 2004.
94. Ziv NE and Garner CC. Principles of glutamatergic synapse formation: seeing the forest for the trees. *Current Opinion in Neurobiology* 11: 536–543, 2001.

79. Wan HI, DiAntonio A, Fetter RD, Bergstrom K, Strauss R, and Goodman CS. Highwire regulates synaptic growth in Drosophila. Neuron 26: 313–329, 2000.
80. Wang X, Hu B, Zimmermann B, and Kilimann MW. Rim1 and rabphilin-3 bind Rab3-GTP by composite determinants partially related through N-terminal alpha-helix motifs. Journal of Biological Chemistry 276: 32480–32488, 2001.
81. [illegible]
82. [illegible]
83. [illegible]
84. [illegible]
85. [illegible]
86. [illegible] Harada A, Okada Y, Funakoshi T, Kanai Y, Takei Y, Terada S, Noda T, and Hirokawa N. Defect in synaptic vesicle precursor transport and neuronal cell death [illegible]
87. [illegible]
88. [illegible]
89. [illegible]
90. [illegible] Nakata T, Takei Y, [illegible] Hirokawa N. Charcot-Marie-Tooth [illegible] Cell 105 [illegible]
[illegible]

Neurotransmitter Release Machinery: Components of the Neuronal SNARE Complex and Their Function

Deniz Atasoy and Ege T. Kavalali

Department of Neuroscience, U.T. Southwestern Medical Center, 5323 Harry Hines Boulevard, Dallas, TX 75390-9111, USA, ege.kavalali@utsouthwestern.edu

1 Introduction

Membrane fusion is a critical process for vesicle trafficking and intercellular communication in eukaryotes. Fusion of lipid bilayers is an endothermic reaction. It requires specialized proteins to bring two membranes in close proximity to overcome electrostatic forces derived from charged lipid head groups. Once the membranes are in close proximity, the boundary between hydrophilic and hydrophobic parts of the bilayer has to be destabilized. This destabilization may either lead to a complete merger of the two bilayers forming a fusion pore or may result in a hemifusion intermediate before complete opening of a fusion pore (61) (Fig. 1).

Synaptic transmission requires fusion of neurotransmitter filled synaptic vesicles with the plasma membrane. Synaptic vesicle fusion is typically driven by the Ca^{2+} influx triggered by action potential mediated depolarization of a presynaptic nerve terminal. Transmitter release is restricted to a specialized region on the nerve terminal known as the active zone, which can be easily identified in electron micrographs due to its high electron density. Synaptic vesicles dock at the active zone in the vicinity of voltage gated Ca^{2+} channels and upon action potential arrival, Ca^{2+} influx through these channels drive vesicle fusion with a time course of less than 100 microseconds (69). In addition, synaptic vesicles can also fuse spontaneously albeit at a very low probability (~1 vesicle per minute per active zone) (29, 52).

The last two decades have witnessed major leaps in our understanding of the mechanisms underlying neurotransmitter release (43, 66, 67, 83). These studies have identified several molecular components of the synaptic vesicle fusion machinery that are critical for neurotransmission. A core group of these proteins are called SNAREs (acronym for soluble N-ethylmaleimide-sensitive factor attachment protein

J.W. Hell, M.D. Ehlers (eds.), *Structural and Functional Organization of the Synapse*,
DOI: 10.1007/978-0-387-77232-5_4, © Springer Science+Business Media, LLC 2008

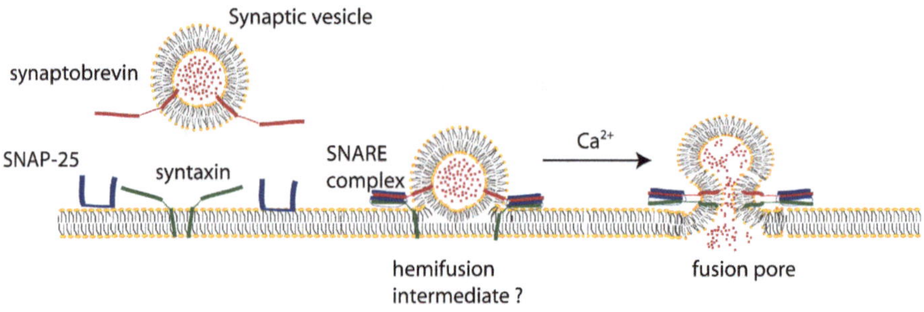

Fig. 1. Synaptic vesicle fusion is mediated by the formation of SNARE complexes from the SNARE proteins synaptobrevin/VAMP, syntaxin-1, and SNAP-25. Syntaxin and synaptobrevin are anchored on the plasma and the synaptic vesicle membranes, respectively, by a transmembrane region, whereas SNAP-25 is attached to the plasma membrane by palmitoylated cysteines. All SNAREs share a characteristic motif that consists of a stretch of approximately 60 amino acids called the SNARE motif. SNARE motifs associate into parallel four-helical bundles to form SNARE complexes. In the case of the synaptic SNARE complex, SNAP-25 contributes two SNARE motifs, and syntaxin and synaptobrevin each contributes one SNARE motif. Under physiological circumstances SNAREs closely cooperate with SM-proteins (Sec1/Munc18-like proteins) as well as synaptic active zone proteins munc13 and RIM to execute fusion. Studies so far suggest that these molecules act as regulators of SNARE-dependent fusion either by facilitating SNARE complex assembly or act in conjunction with SNAREs to promote fusion. In addition, rapidity and steep Ca^{2+}-dependence of synaptic vesicle fusion process is encoded by the interaction of SNAREs with synaptotagmin-1 and complexins. After fusion triggered by Ca^{2+}, SNARE complexes are disassembled by NSF, followed by recycling of used SNAREs and replenishment with vesicles from a reserve pool.

receptors) and belong to a family of proteins that mediate vesicle trafficking and fusion in the secretory pathway in all eukaryotes. In the brain, synaptobrevin-2/VAMP-2 is an abundant synaptic vesicle associated SNARE (v-SNARE) protein essential for normal synaptic transmission in central synapses. Its interaction with the target plasma membrane SNARE (t-SNARE) proteins, syntaxin-1 and SNAP-25, is critical for synaptic vesicle fusion and neurotransmitter release (Fig. 1). SNAREs alone appear to be sufficient to fuse lipid bilayer vesicles in vitro (95) as well as fibroblast plasma membranes when they are expressed on the cell surface in an inverted configuration (38), and therefore they may represent the minimal fusion machinery. However, SNARE-dependent fusion of lipid bilayer vesicles proceed with a time course of hours, an extremely slow rate compared to the less than millisecond time scale of synaptic vesicle fusion. Therefore, it is less clear whether under physiological circumstances SNAREs alone are sufficient to execute fusion. Increasing evidence suggest that SNAREs closely cooperate with SM-proteins (Sec1/Munc18-like proteins) to execute fusion. For instance, mouse knockout of munc18 shows complete loss of evoked as well as spontaneous neurotransmission (91). However, the exact mechanistic basis for the munc18-dependence of synaptic vesicle fusion and the molecular details of the cooperation between SM proteins and SNAREs during membrane fusion is poorly understood (66). In addition to munc18, genetic deletion of munc13 isoforms, a family of active zone proteins critical for priming synaptic vesicles for fusion, totally abolishes neurotransmission akin to loss of munc18 (90). Nevertheless, studies so far suggest that these molecules act as regulators of SNARE-dependent fusion either by facilitating SNARE complex assembly or act in conjunction with SNAREs to promote fusion (42). Therefore, our current understanding of the synaptic vesicle fusion process points to SNAREs as the core machinery for the fusion reaction, upon which other regulatory molecules act on. A key aspect of synaptic vesicle fusion is its rapidity and steep Ca^{2+}-dependence. This feature of the fusion process is encoded by the interaction of SNAREs with synaptotagmin-1 and complexins, which have specialized roles in Ca^{2+}-dependent regulation of fusion. This chapter will cover key properties of SNAREs and their interaction partners in regards to their role(s) in neurotransmitter release. For in depth information on the mechanisms of membrane fusion from a general perspective, the reader should refer to excellent recent review articles on this topic (41–43).

2 Overview of the Role of SNAREs in Synaptic Vesicle Fusion

Synaptic vesicle fusion is mediated by the formation of SNARE complexes (80) from the SNARE proteins synaptobrevin/VAMP, syntaxin-1, and SNAP-25 (42, 43). Syntaxin and synaptobrevin are anchored on the plasma and the synaptic vesicle membranes, respectively, by a transmembrane region, whereas SNAP-25 is attached to the plasma membrane by palmitoylated cysteines (37). All SNAREs share a characteristic motif that consists of a stretch of approximately 60 amino acids called the SNARE motif. SNARE motifs associate into parallel four-helical bundles to form SNARE complexes. In the case of the synaptic SNARE complex SNAP-25 contributes two SNARE motifs, and syntaxin and synaptobrevin each contributes one SNARE motif (85). SNARE complexes show exceptional stability, as they are typically resistant to treatment with the detergent SDS that would normally be sufficient to dissociate other macromolecular complexes. Under physiological conditions, this tight complex, referred to as the "core complex", can only be disassembled by the ATPase NSF (N-ethylmaleimide-sensitive factor) together with SNAPs (soluble NSF attachment proteins). Biochemical stability and molecular structure of the core complex suggest that formation of the complex can be sufficient to overcome the energy barrier to membrane fusion by bringing vesicular and target membranes into nanometer proximity (58). In this scheme, the association of the v-SNARE synaptobrevin with t-SNAREs SNAP-25 and syntaxin supplies the energy required for the fusion reaction in a process accordingly termed "zippering". During zippering SNARE complexes are thought to assemble by in an N- to C-terminal direction, thereby forcing their resident membranes close together (82). This proposal is supported by assays that monitor fluorescent dye mixing upon fusion of liposomes containing v- or t-SNAREs. As indicated above, in these reduced systems, complex formation itself is sufficient to drive fusion, albeit on a slow time scale (95). In this model, fusion reaction is followed by the dissociation and recycling of the involved SNARE proteins through action of NSF and its molecular partners. An alternative scenario for

SNARE function, which emerged from experiments performed in yeast, and is also supported by recent experiments in mouse knockouts, assigns a catalytic role for SNAREs in fusion. According to this view, core complex formation may bring the two membranes into close proximity short of actual fusion, and thus instead of driving fusion, this complex may serve as a substrate for other accessory proteins that drive fusion or render it Ca^{2+}-dependent (12, 77, 89).

In addition to the SNARE motif, syntaxin contains an N-terminal regulatory Habc domain. This region forms a three-helix bundle that folds independently and its crystal structure indicates that it folds back onto the SNARE motif forming a "closed conformation" which provides a high affinity binding site for munc18-1 (19). Whereas in the "open conformation" syntaxin-1 SNARE motif is disengaged from its interaction with the Habc regulatory domain to form core complex with other SNAREs. The open conformation is the only state when syntaxin can be engaged in a fusion reaction. There is biochemical and genetic evidence, which suggests that this switch between closed and open conformations of syntaxin-1 is mediated by munc13 and therefore is a key reaction that underlies synaptic vesicle priming for fusion (3, 65).

Although SNARE complex formation is primarily fueled by hydrophobic interactions, SNARE complexes contain a central "zero layer" composed of a hydrophilic electrostatic interaction mediated by three glutamine and one arginine residue (85). As a result, SNARE proteins are classified based on their SNARE motif sequences into types of Q-SNAREs and R-SNAREs (22). SNARE complex assembly requires the interaction of the 3 Q- and 1 R-SNARE motif, with the Q-SNAREs contribute glutamines and the R-SNARE arginine to the central ionic zero layer of the complex. The crystal structure of the synaptic core complex consists of a four helical bundle of SNARE motifs, with synaptobrevin-2 and syntaxin-1 each contributing one and SNAP-25 providing two motifs (1, 85). The ionic layer in the center of the core complex includes an arginine residue from synaptobrevin-2 and three glutamine residues originating from SNARE motifs of syntaxin and SNAP-25. The amino acids that contribute to the ionic layer are conserved among all SNAREs. In particular, recent evidence suggests that all core complexes contain a single SNARE motif with an arginine at this location (R-SNARE), and three SNARE motifs with a glutamine at the same locus (Q-SNARE). Single amino acid substitutions at these loci result in impairment of SNARE function (22, 44, 57, 93). In contrast, a previous classification of SNARE proteins with respect to their membrane localization as vesicle or target membrane SNAREs (v- and t-SNAREs) cannot account for all eukaryotic fusion reactions, such as homotypic fusion of functionally identical organelles (vacuoles) in yeast (89). Therefore, a new classification scheme into R- and Q-SNAREs has been proposed (22). In the case of synaptic vesicle fusion, the R-SNARE is the v-SNARE synaptobrevin-2 and the Q-SNAREs correspond to the t-SNAREs syntaxin and SNAP-25. It has also been shown that Q-SNAREs alone can form a core complex where synaptobrevin-2 is replaced by syntaxin complexing SNAP-25 with 2:1 stoichiometry. However, it is not known whether this complex exists in vivo.

3 Function of SNAREs in Synaptic Transmission Probed by Clostridial Neurotoxins

The role of SNARE proteins as essential components of normal synaptic vesicle fusion was initially supported by evidence from experiments that utilized proteolytic bacterial toxins. Tetanus and botulinum neurotoxins are zinc endopeptidases that cleave synaptobrevin, SNAP-25 and syntaxin at specific sites within the SNARE motif and cause potent inhibition of neurotransmitter release (76). Discovery of the specific proteolytic activities of tetanus and botulinum neurotoxins was instrumental in the elucidation of SNARE interactions as essential mediators of synaptic vesicle fusion (5–7, 74, 75). Several experimental strategies take advantage of the biochemical properties of SNARE core complexes to link distinct stages of the fusion reaction to stages of SNARE complex assembly. In this regard, the action of clostridial toxins are extremely informative as individual SNAREs within a SNARE complex, once formed, are resistant to proteolysis by these toxins (35). Microinjection of tetanus toxin light chain, which specifically cleaves synaptobrevin, into cultured synapses from leech Hirudo medicinalis resulted in inhibition of neurotransmitter release in strong correlation with cleavage of synaptobrevin (21). Injection of tetanus light chain into squid giant synapse, on the other hand, blocked neurotransmitter release without any effect on synaptic vesicle attachment to the active zone (40). In contrast to earlier suggestion of a role for SNAREs in physical attachment of synaptic vesicles at the active zone, these experiments revealed an increase in the number of docked vesicles. Similarly, targeted expression of tetanus toxin light chain in Drosophila motor nerve terminals abolished neurotransmitter release in response to action potential stimulation without affecting synaptic vesicle organization at the active zone (86). One puzzling finding in these studies was the persistence of spontaneous neurotransmitter release. This result was interpreted as the presence of alternative pathways presumably requiring a distinct set of SNARE proteins that mediate this unregulated form of release (39).

These earlier experiments supported the conclusion that SNAREs do not mediate morphological attachment (docking) of vesicles at the active zone, although exocytosis in response to action potential stimulation requires SNARE proteins. However, recent experiments in the nematode C. elegans took advantage of high-pressure rapid freezing electron microscopy to better preserve synapse morphology during fixation and challenged this notion. This study indicated that syntaxin and its active zone binding partner unc13 are required for establishing a physical contact between the synaptic vesicle membrane and the active zone. This docking defect in unc13 deficient terminals is only relieved by overexpression of a constitutively open form of syntaxin (33).

Analysis of SNARE function in mammalian central synapses also benefited from the availability of neurotoxins. In a comprehensive study of hippocampal synaptic transmission following treatment with clostridial and tetanus toxins, the rate of remaining spontaneous release after incubation of hippocampal slice cultures with tetanus toxin was unaffected by experimental maneuvers that increase Ca^{2+} influx or by application of hypertonic sucrose (+100 mOsm) (10). However, the same stimulation protocols could increase the frequency of neurotransmitter release after cleavage of syntaxin or SNAP-25 by botulinum neurotoxins-C and -A respectively. All toxins caused a significant decrease in spontaneous release, indicating a requirement for cognate synaptic SNAREs for this form of release. In this study, the absence of the well-characterized stimulatory effects of phorbol ester treatment on neurotransmitter release was interpreted to mean that there were few undamaged SNAREs under these circumstances. If so, cleavage of SNAREs with these toxins substantially compromises Ca^{2+} sensitivity of neurotransmitter release rather than abolishing fusion per se. Recent structure-function analyses in chromaffin cells (82) as well as mouse knockout of complexins, molecules that bind to the core complex (62), and synaptobrevin supported this conclusion. Taken together these results argued for a role of SNARE core complex itself and/or its molecular interaction partners in the regulation of Ca^{2+}-dependence of fusion.

Although mouse knockouts of major SNARE proteins provide the best setting for structure-function analysis (see below), clostridial toxins can also be quite useful for this type of analysis. For instance, a recent study used the botulinum neurotoxin E to impair function of endogenous SNAP-25. This maneuver preceded transfection of mutant SNAP-25s, which were expected to be defective in the "zippering" capability of SNARE core complex (26). In these experiments, partial rescue of action potential-dependent fusion, albeit with a very low release probability, argued against a direct role of core complex zippering in driving vesicle fusion.

4 Genetic Analysis of SNARE Function

Experiments that utilize clostridial toxins have been extremely useful in probing SNARE protein function. Yet this approach has limitations because these toxins cannot cleave SNARE proteins that are in tight complexes with other SNAREs (35). This fact may explain the observation that spontaneous neurotransmitter release persists after toxin treatment. Furthermore, some studies have suggested additional proteolytic targets for these toxins as well as additional enzymatic activities (2, 27). Therefore genetic inactivation of synaptobrevin-2 and other SNAREs are quite informative alternative experimental approaches to examine SNARE function.

Analyses of genetic deletions of individual SNARE proteins present a complex view of SNARE function in neurotransmitter release. Drosophila mutants of the R-SNARE synaptobrevin did not significantly affect spontaneous neurotransmitter release, although in the same mutants action potential-induced release was absent at low stimulation frequencies. Nevertheless, extensive Ca^{2+} influx during high K^+ induced depolarization or through Ca^{2+} ionophores could increase neurotransmitter release (16). Mutational inactivation of synaptobrevin in C. elegans resulted in a worm that retained some ability to move, indicating impaired but persistent synaptic function (55). In yeast, the deletion of the synaptobrevin homolog sncl/2p severely impairs health of cells. This phenotype, however, can be alleviated by alterations in lipid metabolism (13).

Physiological analysis of the synapses from knockout mice lacking synaptobrevin-2 provided further insight into SNARE function (77). Surprisingly, a form of regulated exocytosis and synaptic vesicle recycling, as detected by styryl dye methods, persists in hippocampal cultures extracted from these mice. Synaptobrevin-2 deficient synapses show a dramatic (more than 100-fold) reduction in neurotransmitter release in response to action potential stimulation, whereas there is only 10-fold reduction in fusion mediated by hypertonic sucrose. Moreover, both in flies and mice, synaptobrevin null mutants exhibit a facilitation of release during 10 Hz stimulation (14, 98). These results indicate that synaptobrevin-2 and synaptic core complex are partly dispensable for fusion although core complex formation may stabilize fusion intermediates that would facilitate triggering of synaptic vesicle exocytosis by Ca^{2+}. An alternative explanation for this result is the potential substitution of synaptobrevin-2 by other neuronal v-SNAREs, such as tetanus-toxin insensitive VAMP (Ti-VAMP) involved in vesicle trafficking in axonal growth cones (48). Other potential candidate v-SNAREs, synaptobrevin-1 and cellubrevin, do not appear to be present in synaptobrevin-2 deficient neurons (77).

In contrast to the relatively mild secretion phenotype of R-SNARE deletions, elimination of syntaxin, a Q-SNARE, revealed a more severe phenotype in Drosophila (79). In these mutants there was complete loss of evoked and spontaneous neurotransmission in addition to several developmental abnormalities. In C. elegans, syntaxin null mutants were almost completely paralyzed (70). These observations can be partly explained by the interaction of syntaxin with molecules other than its SNARE partners, such as munc18, which are known to be essential for fusion. Surprisingly, in mice, loss of syntaxin-1A results in normal basic neurotransmitter release, but there is a defect in hippocampal long-term potentiation as well as conditioned fear memory (28). This relatively weak phenotype might be due to compensation by syntaxin-1B. Similarly, in flies loss of SNAP-25 did not diminish neurotransmission substantially, partly due to potential compensation from SNAP-24, a protein closely related to SNAP-25 (54, 92). In contrast, in mice, SNAP-25 deletion leads to lethality at birth, and secretion, in particular stimulus evoked secretion, is severely impaired (9, 81, 87, 94). In SNAP-25 knockout neuronal cultures one can detect almost no calcium-evoked release. Even strong stimulation with elevated potassium could barely elicit responses. In contrast, spontaneous neurotransmission occurs reliably, albeit at a lower frequency than controls. In addition, SNAP-25 mutants always responded to hypertonic sucrose application, a calcium-independent form of stimulation. Furthermore, SNAP-25 deficient synapses are capable of synaptic vesicle recycling monitored by uptake and release of FM dyes in response to hypertonic sucrose stimulation. Although the mouse knock-out of SNAP-25 showed a phenotype reminiscent of R-SNARE deletions, taken together, these results suggest that SNAP-25 has a more significant role in calcium-secretion coupling than synaptobrevin-2. It is important to note that SNAP-25 contributes two SNARE motifs to the synaptic core complex, and, therefore, if the strong phenotype of Drosophila syntaxin null mutants were due to SNARE interactions alone, this would be expected to be mimicked by the loss of SNAP-25.

Altogether, these results support the notion that the SNARE core complex is a substrate upon which other regulatory molecules act on. SNAREs may perform an essential catalytic function that increases efficiency and robustness of regulated exocytosis. Therefore their impairment substantially diminishes neurotransmitter release without abolishing it completely. The alternative to this interpretation is the possibility that other cognate or non-cognate SNAREs could potentially substitute for the loss of a particular SNARE in the mutant neurons. This hypothesis agrees with the ability of non-cognate SNAREs (which do not function in synaptic vesicle fusion) to form tight core complexes. However, recruitment of non-cognate v-SNAREs into the synaptic vesicle-recycling pathway would violate functional specificity for SNAREs detected in other preparations (see below). Additionally, the fact that the remaining fusion in these mutants is not a simple scaled down version of the normal fusion with respect to its Ca^{2+} dependence suggests a functional inadequacy of the compensating SNAREs. This inadequacy would imply that SNAREs perform two separate functions in membrane fusion: an executive function in which different SNAREs are interchangeable, and a regulatory function for which a particular SNARE is specific.

5 Structural Requirements and Specificity of SNARE Function in Synaptic Vesicle Fusion

Availability of mouse knockouts for synaptobrevin-2 and SNAP-25 opened up the possibility for detailed structure-function analyses of these molecules to better understand their roles in synaptic transmission. Recent studies have addressed several aspects of SNARE function in rescue experiments performed on dissociated neuronal cultures prepared from synaptobrevin-2 or SNAP-25 deficient brains (15, 17). These studies examined three critical questions on SNARE function. First set of experiments tested whether SNAREs involved in distinct vesicle trafficking steps can substitute for each other and thus if SNAREs show functional specificity. Second, as stated above SNARE complexes contain a central "zero layer" composed of a hydrophilic electrostatic interaction mediated by three glutamine and one arginine residue which lead to the classification of SNARE proteins based on their SNARE motif sequences into types of Q-SNAREs and R-SNAREs (22). Although, SNARE complex assembly requires the interaction of the 3 Q- and 1 R-SNARE motifs, with the Q-SNAREs contributing the glutamines and the R-SNARE the arginine to the central ionic zero layer of the complex, the functional significance of this interaction in the synapse was unknown. Finally, these studies addressed whether there are differences in mechanisms that lead to spontaneous and evoked release. This is a major question raised by earlier SNARE loss of function studies using clostridial toxins or null mutants where impairment of SNARE function typically had only mild effects on spontaneous neurotransmission.

Aside from the premise that structural determinants of fusion may be embedded in the SNARE core complex, the ability of SNAREs to form stable complexes was initially thought to encode for the target specificity of vesicular transport (80). However, core complex formation among SNAREs originating from distinct subcellular compartments is promiscuous (23, 97). Therefore, the specificity of membrane trafficking cannot be explained by distinct propensities of cognate versus non-cognate SNAREs to form complexes or by the stability of the formed complexes alone. Nevertheless, recent experiments involving reconstitution of fusion by yeast SNAREs in liposomes argue for a compartmental specificity in the ability of SNAREs to drive fusion (50). Similarly, studies in whole PC-12 cells tested the ability of soluble

SNARE motifs to compete with interaction of endogenous SNAREs and subsequent fusion reaction (73). These results also indicated a high degree of functional specificity in SNARE interactions. In the case of "cracked" (mechanically dissected) PC-12 cell experiments, some of this specificity was attributed to factors other than SNAREs that could facilitate core complex formation in situ. In contrast to this premise, transgenic rescue experiments in Drosophila (4) argued for a functional interchangeability between non-synaptic and synaptic synaptobrevin isoforms. Recently, cellubrevin was shown to be responsible for the remaining release observed in synaptobrevin-deficient chromaffin cells (8). In the mean time, in hippocampal neurons cellubrevin expression could successfully rescue spontaneous and evoked synaptic vesicle fusion as well as efficient synaptic vesicle recycling (15). In contrast, double knockout of cellubrevin and synaptobrevin did not further reduce the fusion efficacy below what is seen in synaptobrevin knockout alone, consistent with the previous finding that cellubrevin is absent from hippocampal neurons (77). These data suggest a clear redundancy between the cellubrevin and synaptobrevin R-SNAREs role in fusion, in spite of their differences in N-terminal and C-terminal sequences. They also indicate that although synaptobrevin and cellubrevin are potentially functionally interchangeable, the remaining fusion in synaptobrevin deficient central synapses is not mediated by cellubrevin.

A recent study also examined the functional specificity of SNAP-25 isoforms in hippocampal neurons. Hippocampal neurons possess at least three isoforms of SNAP-25, these include two alternative splice variants of SNAP-25, SNAP-25a and SNAP-25b, as well as SNAP-23, which is also present in non-neuronal cells. In these experiments lentiviral expression of SNAP-25a, SNAP-25b, or SNAP-23 on SNAP-25 null background could fully rescue functional and anatomical phenotypes associated with loss of SNAP-25 (17). In contrast to nearly complete functional interchangeability of SNAP-25a and SNAP-25b, SNAP-23 could only rescue asynchronous release giving rise to a phenotype similar loss of synaptotagmin-1 (29). These findings suggest a level of functional specificity among closely related SNAP-25 isoforms, which does not exist between synaptobrevin and cellubrevin.

Second, to test the functional significance of the electrostatic interaction at the central ionic zero layer of the SNARE complex, hippocampal cultures from synaptobrevin-2 deficient mice were infected with synaptobrevin-2 mutants with a glutamine instead of an arginine in the central layer (R56Q), thus converting synaptobrevin-2 from an R-SNARE to a Q-SNARE. Expression of this mutant synaptobrevin-2 still resulted in full rescue of release, indicating that the 4Q combination is a functional core complex alternative. The lack of a major phenotype in synaptic exocytosis produced by the Q/R substitution agrees well with previous studies in yeast showing that similar substitutions did not abolish SNARE function (44, 57). Previous experiments in yeast, however, did not have the temporal resolution of synaptic exocytosis to detect changes in the kinetics or amplitude of exocytosis. This result thus extended these studies by showing that the 4Q SNARE complex is not only functional, but appears to act with a similar kinetics in exocytosis as the 3Q/1R SNARE complex. This is surprising given the tight interaction of the glutamines and arginine in the zero layer of the SNARE complex revealed in the crystal structure of the complex (85). In contrast, experiments with different R-Q mutations in the zero layer of the yeast SNARE proteins ykt6 and sec22 found a developmental arrest in the mutants (18, 31). These earlier findings suggest that a subtle change in the SNARE function can lead to a cumulative defect in a longer time frame. These defects may not detectable in relatively acute experimental settings. Furthermore, detection of neurotransmission in hippocampal cultures may not be an optimal setting to detect the consequences of R to Q substitution at the zero layer in exocytosis, as synaptic vesicle fusion is less sensitive to intracellular fusion steps within a short time frame.

Finally, recent work also addressed potential mechanisms underlying the distinct nature of spontaneous and evoked release. A new insight to this issue came from experiments that were designed to test whether the distance between the SNARE motif and transmembrane region is critical for fusion. This distance is expected to be an important determinant of a vesicle's fusion propensity if SNARE complexes force membranes into close proximity. To test this hypothesis, insertions of 12 or 24 residues were introduced between the SNARE motif and the transmembrane region of synaptobrevin-2 (15). These mutants revealed that the physical distance between the two regions of synaptobrevin-2 is indeed critical for the rescue of evoked fusion in agreement with the proposal that the assembly of SNARE complexes provides the energy for membrane fusion. Surprisingly, in contrast to the insertion of 24 amino acids, the 12 amino acids insertion mutant completely rescued spontaneous release suggesting that constraints on SNARE function during spontaneous fusion are more flexible than for evoked fusion. This finding argues against the traditional notion that spontaneous release events arise from the random low probability exocytosis of docked and fully primed vesicles in the readily releasable pool. According to this view, spontaneous fusion should possess the same structural requirements as evoked fusion. This finding could not be ascribed to a selective effect of synaptobrevin-2 in Ca^{2+}-dependence of evoked fusion (71, 99) because synaptobrevin carrying the 12-residue insertion was also largely unable to rescue hypertonic sucrose evoked fusion, which is Ca^{2+}-independent (68).

Differential requirement of membrane proximity in Ca^{2+}-evoked and spontaneous synaptic release raises the question whether the release machinery for spontaneous fusion is identical to the one required for evoked fusion. Clearly, both forms of fusion are impaired in the absence of the same SNARE molecules, albeit to different degrees (77, 94). Taken together with the earlier observations from the SNARE mutant studies, this finding suggest that spontaneous fusion may require an alternative fusion complex (with a different R-SNARE) or the same complex formed with less stringency (51, 96). Stronger dependency of evoked release on the close proximity of membranes is also more compatible with its strict Ca^{2+}-dependence. Paradoxically, the mechanism of evoked release appears to be closely related to the evolutionally conserved exocytosis because yeast vacuolar fusion was similarly strictly dependent on membrane proximity and the length of R-SNAREs (49).

6 Interaction Partners for Neuronal SNAREs that Underlie Ca^{2+}-Dependent Synaptic Vesicle Fusion

Despite the overwhelming evidence that SNAREs constitute the core machinery for fusion, it is also becoming increasingly clear that SNARE zippering alone cannot completely explain the speed of Ca^{2+} regulated neurotransmission in synapses. In addition to the question of speed, there are further reasons to investigate interactions of SNAREs with other molecules to fully account for the mechanism of synaptic vesicle fusion. First, in artificial fusion assays expression of SNAREs can cause fusion albeit at a very slow time scale that range between minutes to hours. Second, in these assays it is not clear whether vesicle contents are not exposed to outside, that is if the membrane integrity is protected during fusion. This is essential for synaptic fusion as most of the neurotransmitter would leak into cytosol if fusion pore does not remain intact during exocytosis. Finally, in the absence of SM proteins synaptic fusion does not occur at all, which suggest that SNAREs alone may not be sufficient to execute fusion.

A key aspect of synaptic vesicle fusion leading to neurotransmitter release is its steep Ca^{2+}-dependence. Synaptotagmin-1 is a major component of synaptic vesicle membrane and an essential target for Ca^{2+} to trigger rapid vesicle fusion. Synaptotagmins are a family of proteins characterized by two carboxyl Ca^{2+} binding C2-domains (C2A and C2B) and N-terminal transmembrane domain. Synaptotagmin-1 binds to 5 Ca^{2+} ions (three on C2A and two on C2B). The Ca^{2+} binding substantially increases synaptotagmin-1's affinity to phospholipids thus triggers its binding to the plasma membrane. In addition, synaptotagmin-1 can also bind to formed SNARE complexes. This feature of synaptotagmin-1 makes it an ideal candidate as the Ca^{2+} sensor for evoked fusion. In support of this hypothesis, synaptotagmin-1 knock out mice displayed severe impairments in rapid Ca^{2+}-dependent synchronous release (29). Lipid binding of C2 domains involves electrostatic interactions between the basic residues and phospholipids head groups, coordination of Ca^{2+} and insertion of hydrophobic residues into the membrane. Point mutation experiments supported this view, for instance R233Q mutation in the top loop of synaptotagmin-1 decreases its Ca^{2+} binding efficacy as well as the Ca^{2+} sensitivity of evoked fusion by 50% (24).

The other key players in regulation of Ca^{2+}-dependent fusion are complexins. Complexins are small soluble proteins that compete with synaptotagmins for binding to SNAREs. The crystal structure of complexins suggests that it forms an α-helix that interacts with SNAREs with 1:1 ratio in an anti-parallel manner (11). This interaction is thought to stabilize the C-terminal part of the coiled SNARE complex. In complexin knock-out mice Ca^{2+} triggered release is compromised but Ca^{2+}-independent hypertonic sucrose triggered release is normal (62). Enrichment of complexin at the SNARE complex sites by expression of a synaptobrevin-complexin fusion protein imitated synaptotagmin knockout phenotype (88). These results led to the hypothesis that complexins are essential for super-priming vesicles by stabilizing SNAREs and making them available for binding to synaptotagmin. According to this model, Ca^{2+} binding to synaptotagmin-1 leads to lipid as well as SNARE binding and displacement of complexin by synaptotagmin. In concert with the zippering of SNAREs, synaptotagmin's binding to the plasma membrane has been shown to induce high positive curvature on the membrane in a C2 domain dependent manner (47).

7 Interaction Partners for Neuronal SNAREs that Regulate Synaptic Vesicle Priming

Priming is a process that occurs after vesicle docking at the active zone and before the execution of fusion. Molecularly this process is thought to correspond to dissolution of SNARE interactions that typically hinder individual SNAREs from participating in SNARE complexes. Two specific SNARE interactions are the principal targets of priming reactions. First reaction is the disassembly of unproductive cis-SNARE complexes between plasma membrane SNAREs SNAP-25 and syntaxin, where two SNARE motifs of SNAP-25 forms a complex with two syntaxins. Such SNARE complexes can be rapidly dissociated by NSF and given the abundance of SNARE proteins in the synaptic terminals this activity of NSF might serve as a critical checkpoint. Second reaction involves the switch of syntaxin between its closed and open conformations. As indicated above, syntaxin contains an N-terminal regulatory Habc domain which forms a three-helix bundle that folds independently and folds back onto the SNARE motif forming a closed conformation (19). In the open conformation syntaxin-1 SNARE motif is disengaged from its interaction with the Habc regulatory domain to form complex with other SNAREs. Formation of SNARE complex would necessitate release of syntaxin from munc18 and to acquire an open conformation. The mechanism that switches syntaxin between the open and closed confirmation is not clear, yet it is thought to be regulated by active zone proteins munc13 and RIM. Munc13-1 and munc13-2 double knock out mice, as well as loss of function mutants of unc13 (munc13 homologue) in C. elegans show nearly complete loss of neurotransmitter release (64). Similarly in C. elegans RIM mutants neurotransmission is impaired whereas in mice there is a decrease in release probability (45, 78). Munc13 directly binds to the N-terminal of syntaxin, and in C.elegans this binding is sufficient to displace unc18 binding. In C. elegans, phenotypes of unc13 loss of function mutations can be rescued by constitutively open form of syntaxin supporting for role of munc13 in “opening” syntaxin (65). Similar observations were also made for C. elegans RIM mutants indicating a possible functional coordination between unc13 and RIM in regulation of SNARE complex assembly. Taken together with the function of SM proteins, which interact with syntaxin and also bind to SNARE complex (20), the switch between open and close conformations of syntaxin and its regulation by munc18, munc13 and RIM prevents nonspecific SNARE complex formation. The involvement of munc13 and RIM, two major components of the active zone, in this process implies that synaptic vesicle priming establishes spatial and temporal requirements on SNARE complex assembly thus restrict fusion to the active zone and potentially impede ectopic release.

8 After Fusion: SNAREs’ Role in Synaptic Vesicle Endocytosis and Recycling

The synaptic vesicle cycle is initiated by a rise in the intrasynaptic Ca^{2+} leading to vesicle fusion and neurotransmitter release at the active zone. According to the conventional picture of this process, upon fusion, synaptic vesicles completely collapse onto the plasma membrane and vesicle membrane components (lipids as well as proteins) intermix with their plasma membrane counterparts. Subsequently clathrin, through its adaptor proteins, is recruited to the membrane (typically within the periphery of the active zone) and forms coated vesicles by re-clustering vesicle membrane components which eventually bud off the plasma membrane with the help of the GTPase dynamin. Besides the plasma membrane, endocytosis of clathrin-coated vesicles may also occur from membrane infoldings or endosomal cisternae, which form upon accumulation of fused synaptic vesicles. A vacuolar type ATPase then lowers the pH in these vesicles and subsequently the vesicles are refilled with neurotransmitter. The re-acidification and neurotransmitter filling steps are thought to be rapid, and the entire process for exocytosed vesicles to be re-available for release occurs within 60 seconds (53).

Some evidence support the presence of an additional fast pathway of vesicle recycling referred to as "kiss-and-run" during which synaptic vesicles retain their identity and do not intermix with the plasma membrane or endosomal compartments. During this process, the fusion pore opens transiently without complete collapse of the vesicle and without intermixing of synaptic vesicle membrane components with the plasma membrane (25). In several models of cellular secretion, vesicle fusion process is easier to access technically through direct visualization or via membrane capacitance measurements. In these systems, there are several examples of transient vesicle fusion events that resemble what would be expected from kiss-and-run fusion. In central synapses, however, limited experimental access to the fusion process restricts our ability to assess the preponderance of kiss-and-run type fusion (36). In addition to limited morphological and functional characterization of the kiss-and-run process, the actual molecular mechanism of this pathway remains to be determined. There is evidence that this rapid component of endocytosis may operate independent of clathrin (59) and may be due to reversibility of fusion and therefore may not require independent fission machinery. However, it is currently a matter of debate whether kiss-and-run represents a reversible SNARE-mediated fusion event, or almost instantaneous fission of vesicle-plasma membrane interface upon fully executed SNARE-mediated fusion.

Several lines of evidence from previous studies suggest key functions for synaptobrevin-2 in the regulation of vesicle recycling at various steps following fusion, such as synaptic vesicle endocytosis and sorting. Therefore, synaptobrevin-2 can be involved both in kiss-and-run as well as in clathrin-mediated endocytosis. The involvement of synaptobrevin-2 in endocytotic trafficking can be of two types. First, it may be involved in fusion pore regulation. Second, synaptobrevin-2 could be involved in tagging synaptic vesicles for rapid retrieval machinery independent of SNARE function.

Biochemical evidence supports a role for synaptobrevin-2 in clathrin-mediated synaptic vesicle generation from endosomes that requires the AP3 clathrin adaptor complex (72). On the other hand, genetic evidence from C. elegans has shown a requirement for the clathrin adaptor, AP180, for proper sorting of synaptobrevin-2 to vesicles, suggesting an interaction between AP180 and synaptobrevin-2 (56). These observations are also supported by experiments in yeast where a deficiency of the synaptobrevin homologue, snc1, leads to a defect in endocytic trafficking (32). There is also evidence that endosomal recycling may require the function of synaptobrevin-2 as well as SNAP-25 in a complex with syntaxin-13 (84). Besides a potential role for synaptobrevin-2 in clathrin-mediated vesicle recycling, several lines of evidence also support a role for synaptobrevin-2 in kiss-and-run endocytosis presumably through the regulation of fusion pore kinetics. For instance, in a recent study, treatment of hippocampal synapses with tetanus toxin that cleaves synaptobrevin-2 decreased the glutamate concentration in the synaptic cleft per fusion event to a level that is insufficient to activate low affinity AMPA receptors but not high affinity NMDA receptors (63). This finding suggests a role for synaptobrevin-2 in the regulation of fusion pore dynamics and glutamate release. As indicated in previous sections, during fusion synaptobrevin-2, syntaxin-1 and SNAP-25 form an exceptionally stable complex. Biochemical stability and molecular structure of the core complex suggest that formation of the complex can be sufficient to overcome the energy barrier to membrane fusion by bringing vesicular and target membranes into nanometer proximity or even driving the fusion reaction itself (42, 43). In this scheme, the association of the synaptobrevin-2 with corresponding syntaxin and SNAP-25 supplies the energy required for the fusion reaction. Under physiological conditions, following vesicle fusion SNARE complex adopts a cis-configuration, which is recognized by a co-factor protein called SNAP. This interaction leads to the recruitment of NSF, a molecular chaperone that mediates the disassembly of this highly stable complex. As SNARE complex formation is an energetically highly favorable reaction and under normal conditions SNARE complex disassembly is an extremely rapidly process (46) it is difficult to envision how this reaction can be reversible. Nevertheless, kiss-and-run may still be due to very rapid reversibility of the fusion reaction mediated after SNARE complex disassembly presumably requiring fast action of the GTPase dynamin for fission of vesicles from the plasma membrane. In this way, vesicles can be endocytosed at the spot where they have fused without losing their curvature. This reversible fusion reaction may also involve regulation of core complex disassembly, providing a switch between complete fusion versus fusion pore flickering. Therefore, if synaptobrevin-2's involvement in kiss-and-run is an outcome of its function as a component of the SNARE complex then we would expect this function to be shared with other SNAREs. Indeed, in lamprey central synapses interaction of Gβγ subunits with SNAP-25 during serotonergic stimulation results in kiss-and-run type fusion events (30, 60) and restricts glutamate release suggesting a role for SNAP-25 in fusion pore regulation. In addition, in PC-12 cells the transmembrane domain of syntaxin is thought to line the fusion pore (34) and the regulation of plasma membrane SNARE interactions has a direct impact on fusion pore kinetics akin to regulation of ion channel openings (41). Along the same lines, recent studies also suggest a role for synaptobrevin-2 in rapid coupling of vesicle fusion and retrieval during activity presumably in kiss-and-run endocytosis. Loss of synaptobrevin-2 could alter the kinetics of FM dye release and also impair endocytosis that is tightly coupled to exocytosis (14). However, in contrast, similar experiments show that plasma membrane associated SNARE protein SNAP-25 does not appear to be essential for exo-endocytosis coupling in the same way that synaptobrevin-2 is (9). These findings strongly suggest a specific role for synaptobrevin-2 in dictating rapid endocytosis.

9 Concluding Remarks

In summary, our current understanding of the neurotransmitter release process points to SNAREs as the core machinery that mediates fusion of synaptic vesicles with the plasma membrane. However, a key aspect of synaptic vesicle fusion is its rapidity and steep Ca^{2+}-dependence, which is difficult to explain by the function of SNAREs alone. Therefore, accumulating evidence suggests that individual SNARE's availability to engage in SNARE complex formation as well as execution of the actual fusion reaction requires interaction of SNAREs with other key molecules. Among these synaptotagmin-1 and complexins are essential for assuring the phenomenal speed and synchronicity of synaptic vesicle fusion. Munc13 and munc18, on the other hand, appear to regulate SNARE complex assembly thus prime synaptic vesicles for fusion, although their exact role(s) in fusion are yet to be clarified. Furthermore, rapid coupling of vesicle fusion and retrieval during neurotransmission implies that SNARE proteins that drive rapid Ca^{2+}-dependent fusion may also be responsible for ensuring swift synaptic vesicle retrieval. However, despite extensive progress on the characterization of molecular interactions that involve SNAREs and their roles in fusion, the impact of SNARE interactions in synaptic vesicle trafficking events after fusion remains elusive. Taken together, regardless of all the remaining questions about the precise function of each player during the execution of synaptic vesicle fusion and recycling, it is becoming clear that the existing repertoire of molecules possess several key features that can account for all critical biophysical properties of neurotransmission.

References

1. Antonin W, Fasshauer D, Becker S, Jahn R and Schneider TR. Crystal structure of the endosomal SNARE complex reveals common structural principles of all SNAREs. *Nat Struct Biol* 9, 107–11, 2002.
2. Ashton AC, Li Y, Doussau F, Weller U, Dougan G, Poulain B and Dolly JO. Tetanus toxin inhibits neuroexocytosis even when its Zn(2+)-dependent protease activity is removed. *J Biol Chem* 270, 31386–90, 1995.
3. Betz A, Okamoto M, Benseler F and Brose N. Direct interaction of the rat unc-13 homologue Munc13-1 with the N terminus of syntaxin. *J Biol Chem* 272, 2520–6, 1997.
4. Bhattacharya S, Stewart BA, Niemeyer B, Burgess RW, McCabe B, Lin P, Boulianne G, O'Kane C and Schwarz TL. Members of the synaptobrevin/vesicle-associated membrane protein (VAMP) family in Drosophila are functionally interchangeable in vivo for neurotransmittter release and cell viability. *Proc Natl Acad Sci U S A* 99, 13867–72, 2002.
5. Binz T, Blasi J, Yamasaki S, Baumeister A, Link E, Sudhof TC, Jahn R and Niemann H. Proteolysis of SNAP-25 by types E and A botulinal neurotoxins. *J Biol Chem* 269, 1617–20, 1994.
6. Blasi J, Chapman ER, Link E, Binz T, Yamasaki S, De Camilli P, Sudhof TC, Niemann H and Jahn R. Botulinum neurotoxin A selectively cleaves the synaptic protein SNAP-25. *Nature* 365, 160–3, 1993.

7. Blasi J, Chapman ER, Yamasaki S, Binz T, Niemann H and Jahn R. Botulinum neurotoxin C1 blocks neurotransmitter release by means of cleaving HPC-1/syntaxin. *EMBO J* 12, 4821–8, 1993.
8. Borisovska M, Zhao Y, Tsytsyura Y, Glyvuk N, Takamori S, Matti U, Rettig J, Sudhof T and Bruns D. v-SNAREs control exocytosis of vesicles from priming to fusion. *EMBO J* 24, 2114–26, 2005.
9. Bronk P, Deak F, Wilson MC, Liu X, Sudhof TC and Kavalali ET. Differential Effects of SNAP-25 Deletion on Ca2+-Dependent and Ca2+-Independent Neurotransmission. *J Neurophysiol* 98, 794–806, 2007.
10. Capogna M, Mckinney RA, O'Connor V, Gahwiler BH and Thompson SM. Ca2+ and Sr2+ partially rescues synaptic transmission in hippocampal cultures treated with botulinum toxin A and C, but not tetanus toxin. *J Neurosci* 17, 7190–202, 1997.
11. Chen X, Tomchick DR, Kovrigin E, Arac D, Machius M, Sudhof TC and Rizo J. Three-dimensional structure of the complexin/SNARE complex. *Neuron* 33, 397–409, 2002.
12. Chen YA, Scales SJ, Jagath JR and Scheller RH. A discontinuous SNAP-25 C-terminal coil supports exocytosis. *J Biol Chem* 276, 28503–8, 2001.
13. David D, Sundarababu S and Gerst JE. Involvement of long chain fatty acid elongation in the trafficking of secretory vesicles in yeast. *J Cell Biol* 143, 1167–82, 1998.
14. Deak F, Schoch S, Liu X, Sudhof TC and Kavalali ET. Synaptobrevin is essential for fast synaptic-vesicle endocytosis. *Nat Cell Biol* 6, 1102–8, 2004.
15. Deak F, Shin OH, Kavalali ET and Sudhof TC. Structural determinants of synaptobrevin 2 function in synaptic vesicle fusion. *J Neurosci* 26, 6668–76, 2006.
16. Deitcher DL, Ueda A, Stewart BA, Burgess RW, Kidokoro Y and Schwarz TL. Distinct requirements for evoked and spontaneous release of neurotransmitter are revealed by mutations in the Drosophila gene neuronal-synaptobrevin. *J Neurosci* 18, 2028–39, 1998.
17. Delgado-Martinez I, Nehring RB and Sorensen JB. Differential abilities of SNAP-25 homologs to support neuronal function. *J Neurosci* 27, 9380–91, 2007.
18. Dilcher M, Kohler B and von Mollard GF. Genetic interactions with the yeast Q-SNARE VTI1 reveal novel functions for the R-SNARE YKT6. *J Biol Chem* 276, 34537–44, 2001.
19. Dulubova I, Sugita S, Hill S, Hosaka M, Fernandez I, Sudhof TC and Rizo J. A conformational switch in syntaxin during exocytosis: role of munc18. *EMBO J* 18, 4372–82, 1999.
20. Dulubova I, Khvotchev M, Liu S, Huryeva I, Sudhof TC and Rizo J. Munc18-1 binds directly to the neuronal SNARE complex. *Proc Natl Acad Sci U S A* 104, 2697–702, 2007.
21. Fasshauer D, Bruns D, Shen B, Jahn R and Brunger AT. A structural change occurs upon binding of syntaxin to SNAP-25. *J Biol Chem* 272, 4582–90, 1997.
22. Fasshauer D, Sutton RB, Brunger AT and Jahn R. Conserved structural features of the synaptic fusion complex: SNARE proteins reclassified as Q- and R-SNAREs. *Proc Natl Acad Sci U S A* 95, 15781–6, 1998.
23. Fasshauer D, Antonin W, Margittai M, Pabst S and Jahn R. Mixed and non-cognate SNARE complexes. Characterization of assembly and biophysical properties. *J Biol Chem* 274, 15440–6, 1999.
24. Fernandez-Chacon R, Konigstorfer A, Gerber SH, Garcia J, Matos MF, Stevens CF, Brose N, Rizo J, Rosenmund C and Sudhof TC. Synaptotagmin I functions as a calcium regulator of release probability. *Nature* 410, 41–9, 2001.
25. Fesce R, Grohovaz F, Valtorta F and Meldolesi J. Neurotransmitter release: fusion or 'kiss-and-run'? *Trends Cell Biol* 4, 1–4, 1994.

26. Finley MF, Patel SM, Madison DV and Scheller RH. The core membrane fusion complex governs the probability of synaptic vesicle fusion but not transmitter release kinetics. *J Neurosci* 22, 1266–72, 2002.
27. Foran P, Lawrence GW, Shone CC, Foster KA and Dolly JO. Botulinum neurotoxin C1 cleaves both syntaxin and SNAP-25 in intact and permeabilized chromaffin cells: correlation with its blockade of catecholamine release. *Biochemistry* 35, 2630–6, 1996.
28. Fujiwara T, Mishima T, Kofuji T, Chiba T, Tanaka K, Yamamoto A and Akagawa K. Analysis of knock-out mice to determine the role of HPC-1/syntaxin 1A in expressing synaptic plasticity. *J Neurosci* 26, 5767–76, 2006.
29. Geppert M, Goda Y, Hammer RE, Li C, Rosahl TW, Stevens CF and Sudhof TC. Synaptotagmin I: a major Ca2+ sensor for transmitter release at a central synapse. *Cell* 79, 717–27, 1994.
30. Gerachshenko T, Blackmer T, Yoon EJ, Bartleson C, Hamm HE and Alford S. Gbetagamma acts at the C terminus of SNAP-25 to mediate presynaptic inhibition. *Nat Neurosci* 8, 597–605, 2005.
31. Graf CT, Riedel D, Schmitt HD and Jahn R. Identification of functionally interacting SNAREs by using complementary substitutions in the conserved '0' layer. *Mol Biol Cell* 16, 2263–74, 2005.
32. Gurunathan S, Chapman-Shimshoni D, Trajkovic S and Gerst JE. Yeast exocytic v-SNAREs confer endocytosis. *Mol Biol Cell* 11, 3629–43, 2000.
33. Hammarlund M, Palfreyman MT, Watanabe S, Olsen S and Jorgensen EM. Open syntaxin docks synaptic vesicles. *PLoS Biol* 5, e198, 2007.
34. Han X, Wang CT, Bai J, Chapman ER and Jackson MB. Transmembrane segments of syntaxin line the fusion pore of Ca2+-triggered exocytosis. *Science* 304, 289–92, 2004.
35. Hayashi T, McMahon H, Yamasaki S, Binz T, Hata Y, Sudhof TC and Niemann H. Synaptic vesicle membrane fusion complex: action of clostridial neurotoxins on assembly. *EMBO J* 13, 5051–61, 1994.
36. He L, Wu XS, Mohan R and Wu LG. Two modes of fusion pore opening revealed by cell-attached recordings at a synapse. *Nature* 444, 102–5, 2006.
37. Hess DT, Slater TM, Wilson MC and Skene JH. The 25 kDa synaptosomal-associated protein SNAP-25 is the major methionine-rich polypeptide in rapid axonal transport and a major substrate for palmitoylation in adult CNS. *J Neurosci* 12, 4634–41, 1992.
38. Hu C, Ahmed M, Melia TJ, Sollner TH, Mayer T and Rothman JE. Fusion of cells by flipped SNAREs. *Science* 300, 1745–9, 2003.
39. Huang X, Wheeler MB, Kang YH, Sheu L, Lukacs GL, Trimble WS and Gaisano HY. Truncated SNAP-25 (1-197), like botulinum neurotoxin A, can inhibit insulin secretion from HIT-T15 insulinoma cells. *Mol Endocrinol* 12, 1060–70, 1998.
40. Hunt JM, Bommert K, Charlton MP, Kistner A, Habermann E, Augustine GJ and Betz H. A post-docking role for synaptobrevin in synaptic vesicle fusion. *Neuron* 12, 1269–79, 1994.
41. Jackson MB and Chapman ER. Fusion pores and fusion machines in Ca2+-triggered exocytosis. *Annu Rev Biophys Biomol Struct* 35, 135–60, 2006.
42. Jahn R, Lang T and Sudhof TC. Membrane fusion. *Cell* 112, 519–33, 2003.
43. Jahn R and Scheller RH. SNAREs–engines for membrane fusion. *Nat Rev Mol Cell Biol* 7, 631–43, 2006.
44. Katz L and Brennwald P. Testing the 3Q:1R "rule": mutational analysis of the ionic "zero" layer in the yeast exocytic SNARE complex reveals no requirement for arginine. *Mol Biol Cell* 11, 3849–58, 2000.
45. Koushika SP, Richmond JE, Hadwiger G, Weimer RM, Jorgensen EM and Nonet ML. A post-docking role for active zone protein Rim. *Nat Neurosci* 4, 997–1005, 2001.

46. Littleton JT, Chapman ER, Kreber R, Garment MB, Carlson SD and Ganetzky B. Temperature-sensitive paralytic mutations demonstrate that synaptic exocytosis requires SNARE complex assembly and disassembly. *Neuron* 21, 401–13, 1998.
47. Martens S, Kozlov MM and McMahon HT. How synaptotagmin promotes membrane fusion. *Science* 316, 1205–8, 2007.
48. Martinez-Arca S, Alberts P, Zahraoui A, Louvard D and Galli T. Role of tetanus neurotoxin insensitive vesicle-associated membrane protein (TI-VAMP) in vesicular transport mediating neurite outgrowth. *J Cell Biol* 149, 889–900, 2000.
49. McNew JA, Weber T, Engelman DM, Sollner TH and Rothman JE. The length of the flexible SNAREpin juxtamembrane region is a critical determinant of SNARE-dependent fusion. *Mol Cell* 4, 415–21, 1999.
50. McNew JA, Parlati F, Fukuda R, Johnston RJ, Paz K, Paumet F, Sollner TH and Rothman JE. Compartmental specificity of cellular membrane fusion encoded in SNARE proteins. *Nature* 407, 153–9, 2000.
51. Melia TJ, Weber T, McNew JA, Fisher LE, Johnston RJ, Parlati F, Mahal LK, Sollner TH and Rothman JE. Regulation of membrane fusion by the membrane-proximal coil of the t-SNARE during zippering of SNAREpins. *J Cell Biol* 158, 929–40, 2002.
52. Murthy VN and Stevens CF. Reversal of synaptic vesicle docking at central synapses. *Nat Neurosci* 2, 503–7, 1999.
53. Murthy VN and De Camilli P. Cell biology of the presynaptic terminal. *Annu Rev Neurosci* 26, 701–28, 2003.
54. Niemeyer BA and Schwarz TL. SNAP-24, a Drosophila SNAP-25 homologue on granule membranes, is a putative mediator of secretion and granule-granule fusion in salivary glands. *J Cell Sci* 113 (Pt 22), 4055–64, 2000.
55. Nonet ML, Saifee O, Zhao H, Rand JB and Wei L. Synaptic transmission deficits in Caenorhabditis elegans synaptobrevin mutants. *J Neurosci* 18, 70–80, 1998.
56. Nonet ML, Holgado AM, Brewer F, Serpe CJ, Norbeck BA, Holleran J, Wei L, Hartwieg E, Jorgensen EM and Alfonso A. UNC-11, a Caenorhabditis elegans AP180 homologue, regulates the size and protein composition of synaptic vesicles. *Mol Biol Cell* 10, 2343–60, 1999.
57. Ossig R, Schmitt HD, de Groot B, Riedel D, Keranen S, Ronne H, Grubmuller H and Jahn R. Exocytosis requires asymmetry in the central layer of the SNARE complex. *EMBO J* 19, 6000–10, 2000.
58. Otto H, Hanson PI and Jahn R. Assembly and disassembly of a ternary complex of synaptobrevin, syntaxin, and SNAP-25 in the membrane of synaptic vesicles. *Proc Natl Acad Sci U S A* 94, 6197–201, 1997.
59. Palfrey HC and Artalejo CR. Vesicle recycling revisited: rapid endocytosis may be the first step. *Neuroscience* 83, 969–89, 1998.
60. Photowala H, Blackmer T, Schwartz E, Hamm HE and Alford S. G protein betagamma-subunits activated by serotonin mediate presynaptic inhibition by regulating vesicle fusion properties. *Proc Natl Acad Sci U S A* 103, 4281–6, 2006.
61. Reese C, Heise F and Mayer A. Trans-SNARE pairing can precede a hemifusion intermediate in intracellular membrane fusion. *Nature* 436, 410–4, 2005.
62. Reim K, Mansour M, Varoqueaux F, McMahon HT, Sudhof TC, Brose N and Rosenmund C. Complexins regulate a late step in Ca2+-dependent neurotransmitter release. *Cell* 104, 71–81, 2001.
63. Renger JJ, Egles C and Liu G. A developmental switch in neurotransmitter flux enhances synaptic efficacy by affecting AMPA receptor activation. *Neuron* 29, 469–84, 2001.
64. Richmond JE, Davis WS and Jorgensen EM. UNC-13 is required for synaptic vesicle fusion in C. elegans. *Nat Neurosci* 2, 959–64, 1999.

65. Richmond JE, Weimer RM and Jorgensen EM. An open form of syntaxin bypasses the requirement for UNC-13 in vesicle priming. *Nature* 412, 338–41, 2001.
66. Rizo J and Sudhof TC. Snares and Munc18 in synaptic vesicle fusion. *Nat Rev Neurosci* 3, 641–53, 2002.
67. Rizo J, Chen X and Arac D. Unraveling the mechanisms of synaptotagmin and SNARE function in neurotransmitter release. *Trends Cell Biol* 16, 339–50, 2006.
68. Rosenmund C and Stevens CF. Definition of the readily releasable pool of vesicles at hippocampal synapses. *Neuron* 16, 1197–207, 1996.
69. Sabatini BL and Regehr WG. Timing of synaptic transmission. *Annu Rev Physiol* 61, 521–42, 1999.
70. Saifee O, Wei L and Nonet ML. The Caenorhabditis elegans unc-64 locus encodes a syntaxin that interacts genetically with synaptobrevin. *Mol Biol Cell* 9, 1235–52, 1998.
71. Sakaba T, Stein A, Jahn R and Neher E. Distinct Kinetic Changes in Neurotransmitter Release After SNARE Protein Cleavage. *Science* 309, 491–94, 2005.
72. Salem N, Faundez V, Horng JT and Kelly RB. A v-SNARE participates in synaptic vesicle formation mediated by the AP3 adaptor complex. *Nat Neurosci* 1, 551–6, 1998.
73. Scales SJ, Chen YA, Yoo BY, Patel SM, Doung YC and Scheller RH. SNAREs contribute to the specificity of membrane fusion. *Neuron* 26, 457–64, 2000.
74. Schiavo G, Benfenati F, Poulain B, Rossetto O, Polverino de Laureto P, DasGupta BR and Montecucco C. Tetanus and botulinum-B neurotoxins block neurotransmitter release by proteolytic cleavage of synaptobrevin. *Nature* 359, 832–5, 1992.
75. Schiavo G, Santucci A, Dasgupta BR, Mehta PP, Jontes J, Benfenati F, Wilson MC and Montecucco C. Botulinum neurotoxins serotypes A and E cleave SNAP-25 at distinct COOH-terminal peptide bonds. *FEBS Lett* 335, 99–103, 1993.
76. Schiavo G, Matteoli M and Montecucco C. Neurotoxins affecting neuroexocytosis. *Physiol Rev* 80, 717–66, 2000.
77. Schoch S, Deak F, Konigstorfer A, Mozhayeva M, Sara Y, Sudhof TC and Kavalali ET. SNARE function analyzed in Synaptobrevin/VAMP knockout mice. *Science* 294, 1117–22, 2001.
78. Schoch S, Castillo PE, Jo T, Mukherjee K, Geppert M, Wang Y, Schmitz F, Malenka RC and Sudhof TC. RIM1alpha forms a protein scaffold for regulating neurotransmitter release at the active zone. *Nature* 415, 321–6, 2002.
79. Schulze KL, Broadie K, Perin MS and Bellen HJ. Genetic and electrophysiological studies of Drosophila syntaxin-1A demonstrate its role in nonneuronal secretion and neurotransmission. *Cell* 80, 311–20, 1995.
80. Sollner T, Whiteheart SW, Brunner M, Erdjument-Bromage H, Geromanos S, Tempst P and Rothman JE. SNAP receptors implicated in vesicle targeting and fusion. *Nature* 362, 318–24, 1993.
81. Sorensen JB, Nagy G, Varoqueaux F, Nehring RB, Brose N, Wilson MC and Neher E. Differential control of the releasable vesicle pools by SNAP-25 splice variants and SNAP-23. *Cell* 114, 75–86, 2003.
82. Sorensen JB, Wiederhold K, Muller EM, Milosevic I, Nagy G, de Groot BL, Grubmuller H and Fasshauer D. Sequential N- to C-terminal SNARE complex assembly drives priming and fusion of secretory vesicles. *EMBO J* 25, 955–66, 2006.
83. Sudhof TC. The synaptic vesicle cycle. *Annu Rev Neurosci* 27, 509–47, 2004.
84. Sun W, Yan Q, Vida TA and Bean AJ. Hrs regulates early endosome fusion by inhibiting formation of an endosomal SNARE complex. *J Cell Biol* 162, 125–37, 2003.
85. Sutton RB, Fasshauer D, Jahn R and Brunger AT. Crystal structure of a SNARE complex involved in synaptic exocytosis at 2.4 A resolution. *Nature* 395, 347–53, 1998.

86. Sweeney ST, Broadie K, Keane J, Niemann H and O'Kane CJ. Targeted expression of tetanus toxin light chain in Drosophila specifically eliminates synaptic transmission and causes behavioral defects. *Neuron* 14, 341–51, 1995.
87. Tafoya LCR, Mameli M, Miyashita T, Guzowski JF, Valenzuela CF and Wilson MC. Expression and Function of SNAP-25 as a Universal SNARE component in GABAergic Neurons. *J Neurosci* 26, 7826–38, 2006.
88. Tang J, Maximov A, Shin OH, Dai H, Rizo J and Sudhof TC. A complexin/synaptotagmin 1 switch controls fast synaptic vesicle exocytosis. *Cell* 126, 1175–87, 2006.
89. Ungermann C and Wickner W. Vam7p, a vacuolar SNAP-25 homolog, is required for SNARE complex integrity and vacuole docking and fusion. *EMBO J* 17, 3269–76, 1998.
90. Varoqueaux F, Sigler A, Rhee JS, Brose N, Enk C, Reim K and Rosenmund C. Total arrest of spontaneous and evoked synaptic transmission but normal synaptogenesis in the absence of Munc13-mediated vesicle priming. *Proc Natl Acad Sci U S A* 99, 9037–42, 2002.
91. Verhage M, Maia AS, Plomp JJ, Brussaard AB, Heeroma JH, Vermeer H, Toonen RF, Hammer RE, van den Berg TK, Missler M, Geuze HJ and Sudhof TC. Synaptic assembly of the brain in the absence of neurotransmitter secretion. *Science* 287, 864–9, 2000.
92. Vilinsky I, Stewart BA, Drummond J, Robinson I and Deitcher DL. A Drosophila SNAP-25 null mutant reveals context-dependent redundancy with SNAP-24 in neurotransmission. *Genetics* 162, 259–71, 2002.
93. Wang Y, Dulubova I, Rizo J and Sudhof TC. Functional analysis of conserved structural elements in yeast syntaxin Vam3p. *J Biol Chem* 276, 28598–605, 2001.
94. Washbourne P, Thompson PM, Carta M, Costa ET, Mathews JR, Lopez-Bendito G, Molnar Z, Becher MW, Valenzuela CF, Partridge LD and Wilson MC. Genetic ablation of the t-SNARE SNAP-25 distinguishes mechanisms of neuroexocytosis. *Nat Neurosci* 5, 19–26, 2002.
95. Weber T, Zemelman BV, McNew JA, Westermann B, Gmachl M, Parlati F, Sollner TH and Rothman JE. SNAREpins: minimal machinery for membrane fusion. *Cell* 92, 759–72, 1998.
96. Xu T, Rammner B, Margittai M, Artalejo AR, Neher E and Jahn R. Inhibition of SNARE complex assembly differentially affects kinetic components of exocytosis. *Cell* 99, 713–22, 1999.
97. Yang B, Gonzalez L, Jr., Prekeris R, Steegmaier M, Advani RJ and Scheller RH. SNARE interactions are not selective. Implications for membrane fusion specificity. *J Biol Chem* 274, 5649–53, 1999.
98. Yoshihara M, Ueda A, Zhang D, Deitcher DL, Schwarz TL and Kidokoro Y. Selective effects of neuronal-synaptobrevin mutations on transmitter release evoked by sustained versus transient Ca2+ increases and by cAMP. *J Neurosci* 19, 2432–41, 1999.
99. Young SM, Jr. Proteolysis of SNARE proteins alters facilitation and depression in a specific way. *Proc Natl Acad Sci U S A* 102, 2614–9, 2005.

The Molecular Machinery for Synaptic Vesicle Endocytosis

Peter S. McPherson[1], Brigitte Ritter[1] and George J. Augustine[2]

[1] Department of Neurology and Neurosurgery, Montreal Neurological Institute, McGill University, 3801 Rue University, Montreal, Quebec, Canada, H3A 2B4, peter.mcpherson@mcgill.ca
[2] Department of Neurobiology, Duke University Medical Center, Box 3209, Durham, NC 27710, USA, georgea@neuro.duke.edu

1 Recycling of Synaptic Vesicles and Their Proteins

1.1 Introduction

Chemical neurotransmission is the major means of communication within the nervous system. Small molecule neurotransmitters are stored in synaptic vesicles (SVs) and their release following SV exocytosis is responsible for the fast, point-to-point signaling typical of the nervous system. SVs are defined by their small size (30–50 nm diameter), as well as their homogeneous morphology and their accumulation within the presynaptic nerve terminal. SVs have a defined and unique protein composition that allows them to package and store neurotransmitter and undergo rapid, Ca^{2+}-dependent fusion with the presynaptic plasma membrane. In fact, the protein components of SVs are present in defined stoichiometric ratios that must be maintained for normal function (153).

During exocytosis, SVs are thought to collapse into the plasma membrane of the presynaptic nerve terminal during neurotransmitter release (49, 59). There is also evidence for incomplete, transient fusion of the two membranes, which has been termed "kiss and run" fusion (52). Because presynaptic nerve terminals are located far away from the biosynthetic machinery in the cell body, and most nerve terminals have limited numbers of SVs, local mechanisms within the nerve terminal must reform SVs from the plasma membrane. Thus, SV exocytosis is followed by endocytosis.

J.W. Hell, M.D. Ehlers (eds.), *Structural and Functional Organization of the Synapse*,
DOI: 10.1007/978-0-387-77232-5_5, © Springer Science+Business Media, LLC 2008

It is widely thought that the major endocytic mechanism for retrieval of SV proteins and the recycling of functional SVs following exocytosis is based on the protein clathrin (4, 8, 49, 58). Such clathrin-mediated endocytosis (CME) at the synapse can be viewed as a specialized form of the "house-keeping" process that mediates internalization of receptors and other protein and lipid cargo in all cell types (115).

The main purpose of this chapter is to review the molecular mechanisms involved in CME of SVs. However, to understand the task faced by this synaptic endocytic machinery, it is first important to understand the fate of SV membranes and membrane proteins once they are transiently incorporated in the plasma membrane. At least three different scenarios are possible (Fig. 1): SV membranes and membrane proteins retain their unique identity within the plasma membrane; SV proteins diffuse but are captured by an endocytic scaffold that is organized around the active zone, the site of SV fusion; SV proteins diffuse broadly throughout the plasma membrane. We will discuss each model in turn.

2 Models of SV Fate

2.1 Maintenance of SV Integrity in the Plasma Membrane

Following exocytosis SVs may maintain their unique character within the presynaptic plasma membrane (Fig. 1, left). In support of this notion, Willig et al. (174) used STED (stimulated emission depletion) microscopy to follow the fate of synaptotagmin I, an integral protein of the SV membrane. STED microscopy reduces the focal spot of a light microscope to an order of magnitude below the diffraction limit, thereby allowing the analysis of individual SVs. Intriguingly, synaptotagmin I remained clustered in isolated patches on the plasma membrane whether SV fusion was triggered by mild or intense stimulation, suggesting that SVs maintain their integrity in the plasma membrane (174). A recent proteomic and lipidomic analysis of purified SVs revealed an extremely high protein packing density such that transmembrane helicies may occupy more than a quarter of the entire SV membrane volume (153). Because lateral diffusion of proteins within the plasma membrane is decreased at high protein density (43), it is possible that SVs could retain their identity as a lipid and protein microdomain that does not diffuse laterally in the plasma membrane. Moreover, clustering of SV proteins within the presynaptic membrane could be aided by cholesterol, which represents approximately 40% of the total lipid content of SVs (103). A proteomic analysis reveals that many components of SVs and the endocytic machinery are enriched in cholesterol-rich domains of synaptic membranes (71). The integrity of the SV microdomain need only be maintained for a few seconds before the endocytic machinery could form a clathrin coat to stabilize the cargo for CME. Maintaining the integrity of SVs within the plasma membrane

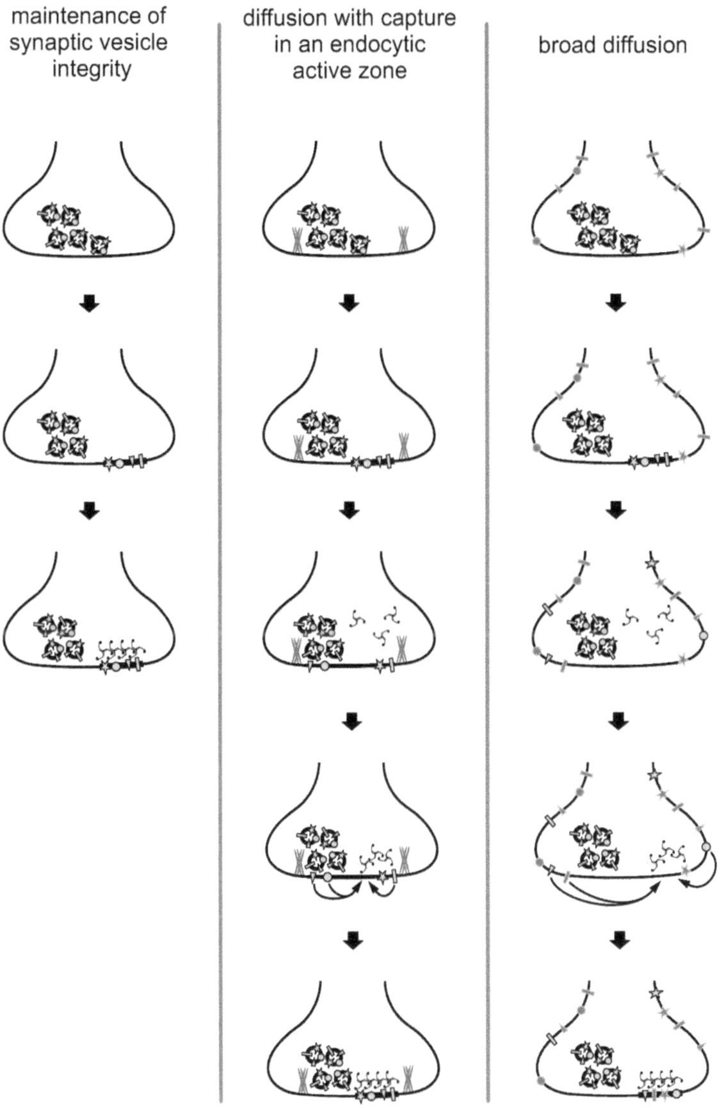

Fig. 1. Models of SV fate. In all three models, the five clustered SVs indicate the exocytic active zone. Each SV is modeled as having four distinct protein components and only one SV undergoes exocytosis. In the first model (*left panel*), the SV membrane (*black bar*) and SV proteins maintain their integrity in the plasma membrane following exocytosis. A clathrin lattice is then built on the patch of SV membrane. In the second model (*middle panel*), SV proteins diffuse locally upon fusion of the SV with the plasma membrane but are captured by the endocytic machinery organized in an endocytic active zone (indicated by *crossed lines*) surrounding the exocytic active zone. A clathrin coat assembles on the membrane and sorts the SV proteins into a CCP. In the third model (*right panel*), SV proteins diffuse broadly upon SV fusion and intermix with populations of SV proteins already present in the plasma membrane. A clathrin coat assembles on the membrane and sorts the SV proteins into a CCP.

would greatly simplify the task of reformation of SVs as the endocytic machinery, in theory, would need to target only one key protein in order to selectively retrieve all SV components.

2.2 Diffusion with Capture in an Endocytic Active Zone

A second model proposes that SV proteins diffuse within the plasma membrane following SV exocytosis, but are then captured in an endocytic active zone that surrounds the exocytic active zone (Fig. 1, center). One candidate scaffold protein for the organization of such a functional endocytic microdomain is dynamin-associated protein of 160 kDa (Dap160) (139). Dap160 is the *Drosophila* homologue of intersectin-short (intersectin-s), an endocytic protein of mammalian cells composed of two Eps15 homology (EH) domains, a coiled-coil region, and five tandem Src homology 3 (SH3) domains (176). Intersectin-s localizes to clathrin-coated pits (CCPs) and interacts with numerous components of the machinery for CME including the lipid phosphatase synaptojanin and the GTPase dynamin (63). In the neuromuscular synapse of *Drosophila*, Dap160/intersectin-s is concentrated in a peri-active zone region, which is also enriched in the adaptor protein 2 (AP-2) (47). As described below, AP-2 is a central factor in the nucleation of clathrin coats and the organization of the endocytic machinery. In *Drosophila*, loss-of-function mutants for Dap-160/intersectin-s have impaired SV endocytosis and are unable to sustain high-frequency neurotransmitter release (81, 94). Further, essential endocytic proteins are lost from the synapse in Dap-160 null *Drosophila*. Thus, the organization of the endocytic machinery around the active zone may provide a barrier to diffusion of SV proteins, such that the proteins would incorporate into nascent, perhaps preformed, CCPs followed by CME and SV reformation.

2.3 Broad Diffusion in the Plasma Membrane

An alternative model regarding the fate of SV proteins is that they diffuse broadly within the plasma membrane following exocytosis (Fig. 1, right). This would create the challenge of sorting and retrieving a large number of proteins at low concentrations. One possible mechanism to partially overcome this difficulty has been recently reported (39, 172). Using antibodies against the luminal domain of synaptotagmin I or tagged versions of synaptotagmin I and vesicle-associated membrane protein 2 (VAMP-2; also called synaptobrevin 2), it was demonstrated that significant pools of these integral SV proteins reside in the synaptic and axonal plasma membrane at rest and thus provide a reservoir of SV proteins that can be tapped during stimulation and CME (39, 152, 172) (Fig. 1). The surface reservoir would lessen the demands on the endocytic machinery by providing a high concentration of proteins for selection and inclusion in endocytic vesicles.

3 Clathrin and Endocytosis of Synaptic Vesicles

3.1 Clathrin-Coated Vesicles in Synaptic Vesicle Recycling

The endocytic machinery must select not only the appropriate proteins for inclusion in SVs, but it must select them at the correct molar ratios. It appears that clathrin-coated vesicles (CCVs) are the main organelles responsible for such protein recycling during endocytosis. A recent proteomic analysis of CCVs isolated from rat brain revealed SV proteins as the major cargo (8). In fact, 87% of the proteins from the CCV proteome assigned to the categories of SV and membrane trafficking were detected in a subsequent proteomic analysis of SVs; in sharp contrast, only 35% of the clathrin trafficking proteins from the CCVs were identified in SVs (8, 153). These data support an important role for CCVs in the recycling of SV components. Included in the CCV proteome were VAMP-1 and -2, which are v-SNARES and syntaxin 1A/B and synaptosomal associated protein of 25 kDa (SNAP-25), which are t-SNAREs. The pairing of these v- and t-SNAREs is crucial for SV exocytosis. The relative levels of the SNARE proteins were determined by quantitative peptide counting (8) revealing that their molar ratios were the same as previously reported for SVs (168). This suggests that CCVs directly recycle SVs without sorting out plasma membrane-derived t-SNAREs and other proteins via endosomes, consistent with physiological studies in neurons (49, 114). Thus, CCVs are critical players in SV endocytosis. In this chapter, we will provide a detailed description of the endocytic clathrin machinery of the presynaptic terminal and describe the role of this machinery in SV recycling.

3.2 A Clathrin-Based Protein Network

Over the last fifteen years there has been an explosion in the identification and functional characterization of the protein components of the clathrin-based endocytic machinery. In fact, until as late as the early 1990s, the activities of clathrin and AP-2 seemed sufficient to explain how cargo-laden CCVs were generated, at least in non-neuronal systems. We now recognize that clathrin and AP-2 are only two components of a large, complex and interconnected protein network that is necessary for CME. However, clathrin and AP-2 are key players as the globular ear (appendage) domains found at the C-terminal ends of the α- and β2-adaptin subunits of AP-2 (α-ear and β2-ear) and the globular N-terminal domain (TD) of the clathrin-heavy chain (CHC) are major hubs for the organization of the machinery. The components of the machinery work together to nucleate new CCPs, to recruit cargo, to stimulate and drive membrane curvature and eventually to mediate membrane fission and CCV uncoating. Much of the work that has led to the identification of the protein components of this machinery has stemmed from biochemical studies in the nervous system. This approach has been particularly fruitful because the proteins that regulate SV endocytosis are expressed in brain at much higher levels than are their counterparts that function in housekeeping forms of CME in non-neuronal cells. There are currently approximately 30 proteins implicated in CME of SVs (Fig. 2).

protein	function	binding motifs/ domains	binding partner	size (aa)
AAK1	ser/thr kinase, phosphorylates μ2-adaptin, regulates recruitment of YxxØ-based sorting motifs to AP-2	DPF WxxF-acidic	α-ear platform α-ear sandwich	961
amphiphysin I	recruits dynamin/synaptojanin to membranes, possibly coupled to membrane curvature	BAR domain SH3 domain DPF FxDxF PxxP clathrin box I and II	membranes (electrostatic), dimerization dynamin, synaptojanin α-ear platform α-ear platform, PHear domain endophilin A1 CHC TD	695
amphiphysin II	recruits dynamin/synaptojanin to membranes, possibly coupled to membrane curvature	BAR domain SH3 domain FxDxF PxxP clathrin box I and II	membranes (electrostatic), dimerization dynamin, synaptojanin α-ear platform, PHear domain endophilin A1 CHC TD	593
aftiphilin	unknown	FxxFxxF/L WxxF-acidic	α-ear platform α-ear sandwich	937
AP-2	recruits cargo, clathrin, and endocytic accessory proteins, clathrin assembly	α-ear platform α-ear sandwich β2-ear platform β2-ear sandwich C-term μ2 α-adaptin trunk	DPF/W, FxDxF, FxxFxxF/L WxxF-acidic FxxFxxxR CHC ankle PtdIns(4,5)P2 YxxØ PtdIns(4,5)P2	938 (α-adaptin) 937 (β2-adaptin) 433 (μ2-adaptin) 142 (σ2-adaptin)
AP180	clathrin membrane recruitment and assembly	ANTH domain DLL DPF FxDxF	PtdIns(4,5)P2 CHC TD α-ear platform α-ear platform, PHear domain	907
auxilin	recruits Hsc70 to CCVs, stimulates clathrin uncoating activity of Hsc70	DNAJ domain DPF FxDxF clathrin box	Hsc70 α-ear platform α-ear platform, PHear domain CHC TD	913
connecdenn	unknown	DENN domain FxDxF DPF WxxF-acidic PxxP	phospholipids α-ear platform, PHear domain α-ear platform α-ear sandwich endophilin, intersectin	1009
CHC	assembles into clathrin coats	TD domain	clathrin box type I and II, DLL	1675

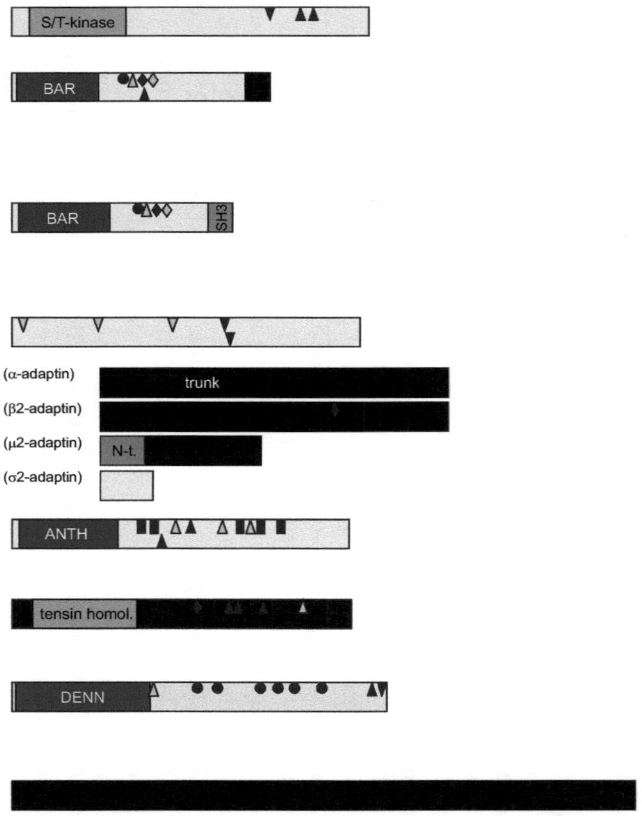

Fig. 2. (Continued)

dynamin 1	GTPase, drives fission reaction, release of CCVs from membrane, essential for SV formation following high frequency stimulation	PH domain PRD	inositol phospholipids multiple SH3 domain proteins	864
dynamin 2	GTPase, drives fission reaction, release of CCVs from membrane	PH domain PRD	inositol phospholipids multiple SH3 domain proteins	870
dynamin 3	GTPase, drives fission reaction, release of CCVs from membrane	PH domain PRD	inositol phospholipids SH3 domains?	863
endophilin A1	regulates membrane recruitment and phosphatase activity of synaptojanin, possibly coupled to membrane curvature	N-BAR domain SH3 domain	membranes (electrostatic), dimerization dynamin, synaptojanin, amphiphysins	352
epsin	clathrin recruitment and assembly, membrane curvature, adaptor for ubiquitinated cargo	ENTH domain DPW clathrin box NPF	PtdIns(4,5)P_2 α-ear platform CHC TD EH domains of eps15 and intersectin	550
eps15	regulation of clathrin assembly	DPF EH domain	α-ear platform NPF in epsin, AP180, and synaptojanin 170	762
HIP1	cargo adaptor for AMPA receptors	ANTH domain DPF FxDxF clathrin box helical domain THATCH domain	phospholipids α-ear platform α-ear platform, PHear domain CHC TD clathrin-light chain ?	1037
HIP1R/HIP12	productive linkage of CCPs to actin through clathrin light chain	ANTH domain clathrin box helical domain THATCH domain PRD	phospholipids CHC TD clathrin-light chain actin SH3 domain of cortactin	1068
Hsc70	uncoating ATPase	ATPase domain	auxilin	646
intersectin-s	scaffold at endocytic sites, couples endocytic and signaling pathways	EH domains helical domain SH3 domains	NPF motifs in epsin SNAP-25, eps15 dynamin, synaptojanin N-WASP, Sos, cdGAP, connecdenn	1220
intersectin-l	scaffold at endocytic sites, couples endocytic and signaling pathways, GEF for Cdc42, regulates actin assembly	EH domains helical domain SH3 domains DH/PH C2	NPF motifs in epsin SNAP-25, eps15 dynamin, synaptojanin, N-WASP, Sos, cdGAP, connecdenn Cdc42 ?	1721

Fig. 2. (Continued)

NECAP 1	recruits FxDxF motif proteins	PHear domain WxxF-acidic	FxDxF motifs α-ear sandwich	275
NECAP 2	recruits FxDxF motif proteins	PHear domain WxxF-acidic	FxDxF motifs α-ear sandwich	263
PIPkinase type 1γ	formation of PtdIns(4,5)P_2	kinase domain	PtdIns(4)P AP-2, Arf6, talin	668
PACSIN 1/syndapin I	links endocytosis and actin	F-BAR domain WxxF-acidic NPF SH3 domain	membranes (electrostatic), dimerization α-ear sandwich EHD proteins dynamin, N-WASP, Sos, synaptojanin, huntingtin	444
stonin 2	regulation of synaptic vesicle recycling	WxxF-acidic NPF stonin homology domain μ2 homology domain	α-ear sandwich EH domains of eps15 ? synaptotagmin I	905
synaptojanin 145	dephosphorylates inositol phospholipids, regulates coat stability	FxDxF WxxF-acidic PRD	α-ear platform, PHear domain α-ear sandwich amphiphysins, endophilin, intersectin, PACSIN	1311
synaptojanin 170	dephosphorylates inositol phospholipids, regulates coat stability	DPF FxDxF WxxF-acidic PRD	α-ear platform α-ear platform, PHear domain α-ear sandwich amphiphysins, endophilin, intersectin, PACSIN	1575
synaptotagmin I	regulates SV exo- and endocytosis	C2	phopholipids, AP-2 μ2, stonin 2	422

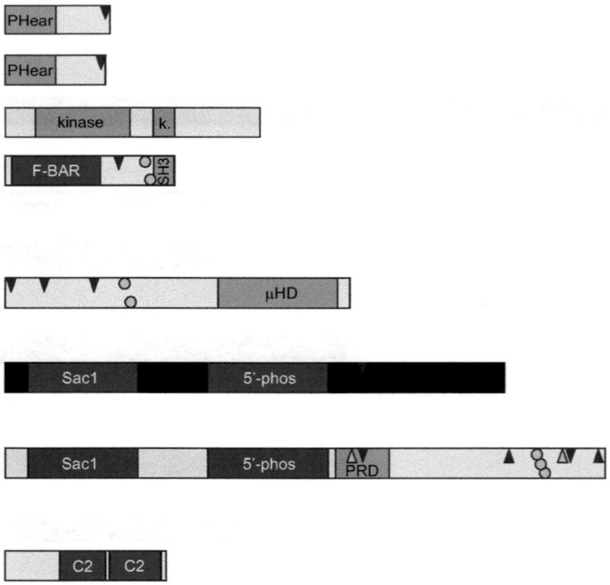

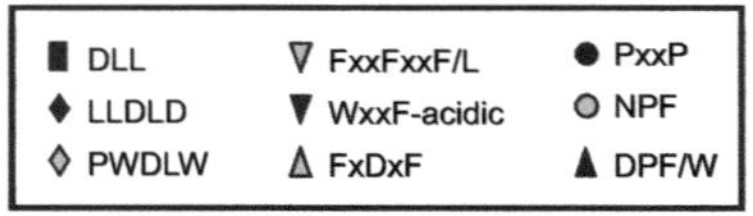

Fig. 2. (Continued) Protein machinery for SV endocytosis. Schematic representation of proteins associated with SV endocytosis and description of their function, binding motifs and domains, binding partners and their size. The abbreviations for most of the modular protein domains are defined in the text. Other abbreviations (5′-phos., 5′-phosphatase domain; C.-term, C-terminal; J, DnaJ domain; k., kinase domain; μHD, μ2-homology domain; N-t., N-terminal; S/T-kinase, serine/threonine-kinase domain; TD, terminal domain; trim., trimerization domain).

As identification and characterization of the protein components of the synaptic endocytic machinery have proceeded, several general lessons regarding CME have emerged. One is that the endocytic protein network is highly interconnected. In general, the proteins contain one or more of a variety of independently folded protein modules and additional, loosely structured or unstructured regions that may be referred to as peptide motif domains (Figs. 2 and 3). Examples of protein modules found in endocytic proteins include; PH (pleckstrin homology), PTB (protein tyrosine phosphate binding), SH3 (src homology 3), DNAJ, EH (Eps15 homology), E/ANTH (epsin/AP180 N-terminal homology), BAR (Bin1/amphiphysin/Rvs167), N-BAR and F-BAR, DENN (differentially expressed in normal versus neoplastic cells) and THATCH (talin-HIP1/R-Actin-Tethering C-terminal Homology) domains. Several of these domains mediate interactions with other proteins in the endocytic machinery through short, consensus peptide motifs found in the peptide motif domains (Fig. 3). These interactions are generally of low affinity. However, as the networks form, interconnections between the various components stabilize the overall network in a Velcro-like arrangement.

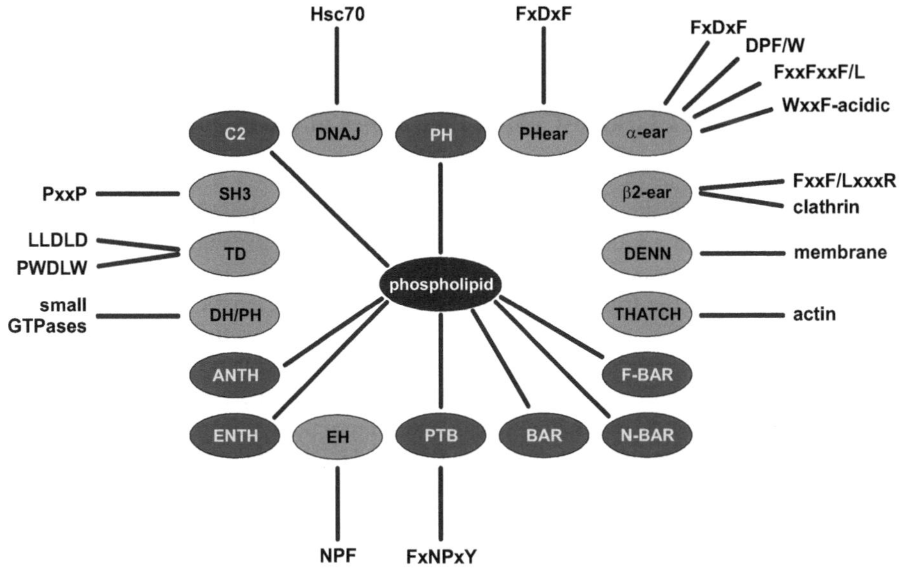

Fig. 3. Schematic presentation of protein modules involved in SV endocytosis and their binding partners. Phospholipids are depicted in *black*, lipid-binding modules are *dark grey*, and protein-binding modules are *light grey*. Protein binding partners and peptide motifs targeted by the domains are indicated for the specific modules.

Several of these protein modules interact with phospholipids. Most notable in this regard is phosphatidylinositol(4,5)P_2 [PtdIns(4,5)P_2], a key lipid in CME that is generated primarily on the inner leaflet of the plasma membrane. The interaction of the modules with phospholipids, which are of low affinity, help to anchor the protein network to membranes. This arrangement of multiple sites of low affinity contacts is ideally suited to the biology of CCVs in that the machinery needs to assemble rapidly to ensure vesicle formation, but it also needs to dissociate rapidly to ensure subsequent SV recycling. The specific sites of low affinity are readily subject to regulation by phosphorylation and dephosphorylation of proteins and lipids.

In addition to the formation of endocytic networks, the protein modules have other roles. For example, they bind to cargo and thus function in cargo recruitment to nascent CCPs. They interact with components of the cytoskeleton, notably actin linking CME to actin function. They also interact with signaling molecules allowing for the formation of signaling pathways on endocytic vesicles. Finally, the modules can directly modify membranes, critical to the formation of membrane curvature and vesicle formation.

4 Mechanisms of Clathrin-Mediated Endocytosis

The rest of this chapter describes the various components of the endocytic machinery with a focus on those involved in CME in presynaptic terminals. We will define what is currently known regarding their function and how these components integrate to form the overall endocytic machinery. Our discussion will be organized according to the postulated temporal sequence of the events involved in CME.

4.1 Nucleation of Clathrin-Coated Pits: Roles of Clathrin, AP-2 and AP180

The assembly unit of the clathrin coat is the triskelion, composed of three copies of CHC linked through a C-terminal trimerization domain (12) (Fig. 4). Clathrin light chains (CLCs) are associated with CHCs but their function is poorly understood. CHCs radiate from the trimerization domain with a characteristic curl such that the protein is divided into segments, the proximal leg, the knee, the distal leg and the ankle, ending in the globular TD (Fig. 4). As triskelia assemble into a clathrin coat or cage, the proximal leg segments interdigitate through numerous weak contacts to form a lattice of open hexagonal and pentagonal faces with a trimerization domain at each vertex (167). The inherent curvature of the CHC and thus, the triskelion means that when assembled, clathrin takes the form of a cage or basket. Electron cryomicroscopy of purified clathrin cages reveals that in the assembled coat, the C-terminal ends of the CHCs at the trimerization domain project inward (towards the membrane) and make contacts with the ankle regions of three additional triskelia, each centered two vertices away (42). These contacts appear to be invariant and likely provide stability to the lattice.

Clathrin triskelia do not bind directly to membranes. Thus, other proteins are needed to recruit clathrin and to mediate its membrane association and in fact nu merous proteins, collectively known as adaptors or clathrin-associated proteins have been identified that fulfill this role (161). These adaptors bind to both CHCs and phospholipids. In the case of SV endocytosis, a key clathrin adaptor is AP180 (Fig. 2), which was originally identified as a component of CCVs (1) and is enriched in presynaptic nerve terminals (179, 184). At its N-terminus, AP180 contains a globular

ANTH domain that binds to PtdIns(4,5)P_2 (40). Through its weakly structured C-terminal peptide motif domain (74), AP180 binds to the TD of the CHC (110) and purified AP180 is sufficient in vitro to recruit CHC to PtdIns(4,5)P_2-enriched lipid tubules (40). In addition, AP180 stimulates clathrin assembly and inhibition of the assembly activity in squid giant presynaptic nerve terminals disrupts formation of CCPs and SV endocytosis (110). Mutation of the AP180 gene of *Drosophila* (182) and *C. elegans* (119) also results in defects in synaptic vesicle trafficking. Thus, AP180 is a critical clathrin adaptor for SV endocytosis.

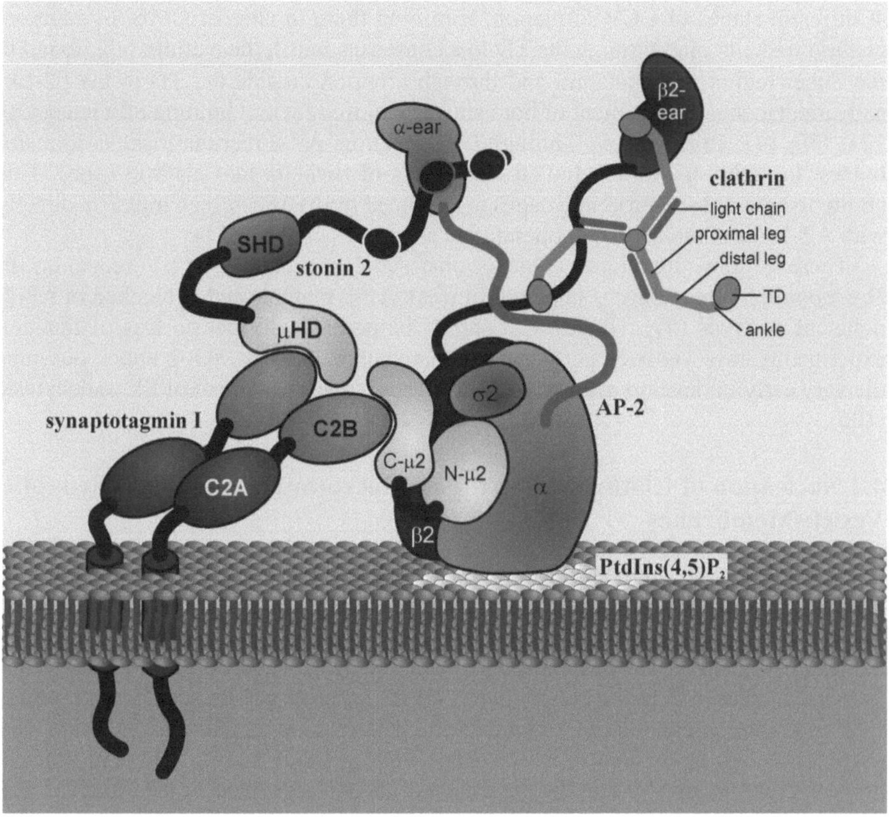

Fig. 4. Model of synaptotagmin I as cargo receptor for CME of SVs. Upon fusion of SVs with the plasma membrane synaptotagmin I could coordinate SV endocytosis by recruiting stonin 2 and AP-2 through its C2 domains. The endocytic protein network is stabilized by direct interaction of AP-2 with PtdIns(4,5)P2 in the plasma membrane and by concomitant interaction of stonin 2 and AP-2, which also allows clathrin recruitment and assembly. The individual subunits of the AP-2 complex are indicated as are the modules in stonin 2. The regions of the CHC are indicated as is the association of the clathrin-light chain. μHD, μ2 homology domain; SHD, stonin homology domain; TD, terminal domain.

Another key player in SV endocytosis is AP-2 (Figs. 2 and 4). AP-2 is enriched in nerve terminals (13, 178) and consists of a multi-subunit complex composed of two large subunits, α- and β2-adaptin and two smaller subunits, μ2- and σ2-adaptin (136) (Figs. 2 and 4). The large N-terminal regions of α- and β2-adaptin, along with the μ2 and σ2 subunits form the core of the AP-2 complex. The C-terminal regions of α and β2 contain the globular ear domains. The ear domains are in fact bi-lobed structures, composed of sandwich and platform subdomains that attach to the core via flexible linkers (122). Given the flexible nature of the linkers, it is thought that the α- and β2-ears project into the cytosol to bind to accessory proteins that function at different stages of CCV formation, recruiting them to sites of CME. In addition, clathrin triskelia bind through the TD to a consensus motif, the clathrin box, found in the linker region of β2-adaptin and through a region outside the TD to the β2-ear, and simultaneous engagement of both sites is required for full binding efficiency (30, 121, 130, 143, 159) (Fig. 4). Through CHC binding, AP-2 recruits triskelia to membranes. In addition, AP-2 stimulates assembly of triskelia into clathrin cages. This ability to assemble clathrin into cages is enhanced by AP180, which interacts directly with AP-2 in a complex that cooperatively assembles clathrin (51).

There is some functional evidence that AP-2 participates in SV recycling. In *Drosophila*, SV recycling is impaired in weak AP-2 mutants and is blocked in AP-2-deficient embryos (47). However, it should be pointed out that no loss of function experiments have yet been performed in mammalian neurons. AP-2 knock out mice die very early in development, which prevents reliable assessment of SV endocytosis (106).

4.2 Nucleation of Clathrin-Coated Pits: Recruitment of AP-2 to Synaptic Vesicle Membranes

Because clathrin recruitment requires the joint participation of AP-2 and AP180, the mechanisms that allow these adaptors to associate with synaptic membranes are fundamental to SV recycling. It appears that PtdIns(4,5)P_2 is crucial for the recruitment of AP-2 and AP180 to the plasma membrane (Fig. 4). In non-neuronal systems, transient increases in PtdIns(4,5)P_2 levels on the surface may be sufficient to initiate CCP formation at essentially random sites (31). However, in neurons, SV endocytosis must be tightly coupled to exocytosis – both in space and in time – in order to prevent depletion of SVs and the expansion of the synaptic plasma membrane. Thus, factors in addition to PtdIns(4,5)P_2 are needed to initiate CCP formation at specific sites on the presynaptic plasma membrane and at a time that follows exocytosis.

If SV proteins maintain their integrity within the plasma membrane, then a single SV protein could be sufficient to act as "the cargo" for AP-2 attachment, allowing this membrane-associated adaptor to concentrate and assemble clathrin on the SV membrane. Synaptotagmin I, a SV protein that serves as the Ca^{2+} sensor for exocytosis, is a strong candidate to play this role. A model for the mechanism by which PtdIns(4,5)P_2 and synaptotagmin may function co-operatively in nucleation of CCPs on SV membranes is presented in Fig. 4 and is described below.

Generation of PtdIns(4,5)P_2 mainly involves phosphorylation of another membrane lipid, PtdIns(4)P, by the type I PtdIns(4)P kinase (PIP kinase) (Fig. 2). PIP kinase type Iγ is the major PIP kinase in the synapse and knock out of this enzyme leads to decreased levels of PtdIns(4,5)P_2 and synaptic defects including delayed CME and slowing of SV recycling (28). GTP-bound Arf6 recruits PIP kinase type Iγ to the synaptic plasma membrane and stimulates its kinase activity leading to the formation of CCPs via increased PtdIns(4,5)P_2 (83). These results implicate GTP-Arf6 as an important upstream regulator of CCP nucleation. However, the mechanisms that localize Arf6 and switch it to the active, GTP-bound form during endocytosis remain unknown.

Given the high concentration of AP-2 in the synapse (8, 13), it is likely that this adaptor protein is constantly sampling the plasma membrane through kinetic action. Transient increases in PtdIns(4,5)P_2 levels – through activation of PIP kinase type Iγ – would then be detected by AP-2 through the low affinity (5–10 μM) PtdIns(4,5)P_2-binding site present in the α-adaptin subunit (62). The μ2-adaptin subunit of AP-2 also contains a PtdIns(4,5)P_2-binding site, but this site is conformationally hidden under steady-state conditions. Subsequent to membrane interaction, μ2 undergoes a conformational change that exposes the second site; having two PtdIns(4,5)P_2-binding sites in a single AP-2 complex greatly stabilizes its association with the membrane (62). Phosphorylation of a key μ2 residue, Thr156, by the AP-2-associated kinase (AAK1) that is associated with brain CCVs (19, 170) (Fig. 2) favors the open conformation of AP-2 (62). It is not yet known whether AAK1 activity has any functional role in SV endocytosis.

PtdIns(4,5)P_2 provides the mechanism for recruitment of AP-2/clathrin complexes to membranes. A recent study in yeast provides direct evidence that PtdIns(4,5)P_2 levels rise in conjunction with clathrin coat assembly and fall during disassembly (151). Moreover, in this system, PtdIns(4,5)P_2 is detected in patches on the plasma membrane that correlate precisely with sites of endocytosis (151). However, there is little evidence that PtdIns(4,5)P_2 within the presynaptic nerve terminal provides enough spatial information to guide clathrin cages specifically to SV membranes within the plasma membrane. Instead, this role may be played by synaptotagmin I. The first indication that synaptotagmin functions as a cargo for endocytic adaptors came with the observation that it binds to AP-2 (183), suggesting that synaptotagmin could mark SV membranes within the plasma membrane as the site to recruit clathrin coats. In fact functional studies in *C. elegans*, *Drosophila*, squid and hippocampal neurons have all demonstrated that synaptotagmin function is crucial for SV endocytosis (44, 72, 118, 126). The major binding site for synaptotagmin I within the AP-2 complex is the μ2 subunit (54), which is also involved in cargo recognition in non-neuronal cells. In fibroblasts, nascent CCPs will often collapse unless they incorporate cargo (31). This may occur because binding to cargo is needed to synergize with PtdIns(4,5)P_2 binding to provide sufficient stability to keep AP-2 attached to the membrane. Interestingly, cargo binding to AP-2 μ2 has also been reported to stimulate type I PIPKs, leading to increased production of PtdIns(4,5)P_2 (84). This could provide a positive feedback mechanism for CCP nucleation. Thus, synaptotagmin interactions with μ2 may stabilize AP-2 on membranes and in particular, at patches of SV membranes.

Although synaptotagmin I participates in SV endocytosis, synaptotagmin does not undergo endocytosis when transfected into fibroblasts (69). This suggests that neurons possess an endocytic adaptor that functions along with synaptotagmin. Stonin 1 and 2 may serve this role (Figs. 2 and 4). These proteins are mammalian homologues of the stoned proteins of *Drosophila*. Stoned proteins are located at the synapse and are essential for the recycling of SVs through direct interactions with synaptotagmin (37, 125). Stonin 2 binds to synaptotagmin I through a μ2-homology domain (29) and also binds to the α-ear of AP-2 through a WXXF-acidic motif (134, 135, 169), providing an additional bridge between synaptotagmin and clathrin coats. Thus, two individual synaptotagmin molecules (or a synaptotagmin dimer) could interact simultaneously with stonin 2 and with AP-2, which are bound to each other (Fig. 4). These interactions would provide the affinity needed for the formation of a stable clathrin coat on synaptotagmin-containing SV membranes. Consistent with this model, stonin 2/AP-2 endocytic complexes interact with synaptotagmin on the presynaptic plasma membrane face but not on intracellular SVs (77). However, it is not yet clear how the properties of synaptotagmin change when this protein resides on the plasma membrane. It also is not clear how these changes selectively allow recruitment of the endocytic coat. These and many related issues require further study.

4.3 Nucleation of Clathrin-Coated Pits: NECAPs at Work with AP-2

A proteomic analysis of highly enriched CCVs from brain revealed eight novel proteins (8, 170), seven of which have been subsequently linked to clathrin-mediated trafficking (19, 20, 132, 134, 170, our unpublished data). Two of these proteins were originally recognized as homologs but that apparently shared no homology or common domains with any previously characterized proteins. The proteins, which were named NECAP (adaptin ear binding coat associated protein) 1 and 2, are highly enriched on the coats of CCVs (134). NECAP 2 is ubiquitously distributed, while NECAP 1 is expressed predominantly in brain. Both NECAP proteins bind to the α-ear of AP-2 via their C-terminal regions (134) (Figs. 2 and 3). Within brain, NECAP 1 is present in the presynaptic compartment and overexpression of constructs that interfere with the interaction of NECAP 1 with AP-2 blocks CME of SVs (113). Thus, NECAP 1 may be a new component of the SV endocytic machinery.

The identification of the α-ear as a site for NECAP binding (134) was surprising: even though the α-ear is a key hub for organization of the endocytic machinery, NECAPs do not contain DPF/W or FXDXF motifs, which until recently were the only motifs known to bind to the α-ear (10, 120, 160). Through deletion analysis, the C-terminal sequence WVQF-COO^- of the NECAPs was identified as a novel AP-2 α-ear-binding motif (134). Interestingly, unlike DPF/W and FXDXF motifs, which bind to the platform subdomain of the α-ear, structural studies demonstrated that WVQF-COO^- binds to the sandwich subdomain (135). Subsequent database searches have identified a number of endocytic proteins that contain WVQF-like motifs, including a splice variant of synaptojanin and stonin 2, and the motifs within these proteins do indeed confer AP-2 binding (70, 169). Sequence alignment reveals a consensus motif of WXXF-acidic (where the acidic is provided by the free carboxy terminus or by one or more internal aspartic acid or glutamic acid residues) (135).

These results thus reveal a greater potential for the α-ear to function as a recruitment scaffold during CME.

The exact function of NECAP 1 in SV endocytosis is unknown. However, a recent structural study has revealed that the conserved N-terminal region of the protein forms a PH-like domain called the PHear domain (137) (Fig. 2). Remarkably, this domain binds to FXDXF motifs, which as described previously are binding motifs for the platform subdomain of the α-ear. In fact, the PHear domain functionally mimics the platform subdomain of the α-ear, hence the name (*PH* fold with *ear*-like function) (137). Thus, when NECAP is bound to the sandwich subdomain through the C-terminal WXXF-acidic motif, its N-terminal PHear domain would be positioned near the platform subdomain. Given that the α-ear is attached to the AP-2 core via a flexible linker region, the linker apparently acts as a "fishing line" with the α-ear serving as the "hook" needed to snare accessory proteins from the cytosol. With NECAP in place, this hook would have two barbs for interacting with FXDXF motif-bearing proteins. Because FXDXF interactions with the α-ear platform domain are of low affinity, in the 50 μM range, the combination of two domains should increase FXDXF protein binding. In summary, NECAPs appear to function cooperatively with AP-2 in the formation of CCPs.

4.4 Membrane Curvature: Eps15 and Epsin, EH and ENTH Domains

AP-2 and AP180 are responsible for recruiting clathrin triskelia to the synaptic plasma membrane and for stimulating their assembly into a lattice. The inherent curvature of triskelia leads to a curved lattice and eventually a spherical cage as the triskelia are assembled. In fact, assembly of purified triskelia in vitro leads to a clathrin cage with a diameter similar to that of a CCV in vivo. Thus, it was long thought that assembly of clathrin was sufficient to drive membrane budding and curvature in cells and some evidence in support of this idea has been reported by Hinrichsen et al. (60). However, it is now thought that, under normal physiological conditions, clathrin assembly is not the major mechanism to drive membrane budding. Instead, it appears that various components of the clathrin machinery act directly on lipid bilayers to stimulate and maintain curvature (98). A common theme is that the endocytic proteins stimulate curvature by insertion of amphipathic helices into membranes. Inserting such helices into the cytosolic leaflet of a membrane displaces lipid headgroups and thereby creates an asymmetry between the bilayer leaflets. Such leaflet asymmetry causes spontaneous membrane curvature toward the cytosolic side as predicted by the bilayer couple hypothesis (146).

The first example of such a mechanism for producing membrane curvature was provided by studies on the ENTH domain of the endocytic protein epsin (Fig. 2). Epsin (Eps15-interacting protein) was originally identified because of its ability to bind to the EH domains of Eps15 (16). Eps15 (EGFR pathway substrate clone no. 15) itself (Fig. 2) had been previously identified as a substrate for the EGFR kinase (36). Subsequent studies demonstrated that Eps15 binds to the α-ear of AP-2 (and the NECAP PHear domain via 15 DPF and 1 FXDXF motifs) and is a component of CCPs and CCVs, implicating Eps15 in CME (5, 6, 137, 158). At its N-terminus, Eps15 contains three copies of the EH domain (Fig. 2), which is found in other endocytic proteins including intersectin (109, 175). EH domains interact with a NPF tripeptide (Fig. 3) and epsin contains three copies of this motif (16). Both epsin and Eps15 are enriched in presynaptic terminals (16, 180). Disruption of interactions between EH domain proteins and their substrates in squid giant presynaptic nerve terminals inhibit the formation of CCPs and CCVs during SV endocytosis (112). Moreover, genetic studies have demonstrated that both the *C. elegans* and *Drosophila* homologues of Eps15 have crucial roles in SV recycling (92, 142). Together, these studies demonstrate the importance of EH domain/NPF-mediated networks for SV recycling (109).

In addition to the NPF motifs that interact with Eps15 and intersectin, the weakly structured C-terminal peptide motif domain of epsin (74) also contains eight copies of the DPW tripeptide that mediates binding to the α-ear of AP-2 (120, 160) and two distinct clathrin box motifs for binding the CHC TD (63, 140) (Figs. 2 and 3). At its N-terminus, epsin contains an ENTH domain, an approximately 150-residue region (76, 140) that has a globular structure composed of eight α-helices (40, 65, 67). A breakthrough in the understanding of epsin function and the role of endocytic regulatory proteins in membrane curvature came with the observation that the ENTH domain binds to PtdIns(4,5)P_2 through basic residues in helices 1, 3 and 4 and the loop between helix 1 and 2 (40, 67). More fundamentally, Ford et al. (41) demonstrated that binding to PtdIns(4,5)P_2 causes unstructured residues at the N-terminus of the ENTH domain to form a new α-helix with a series of hydrophobic residues on its outer surface. This new helix inserts into PtdIns(4,5)P_2-containing monolayers and this can be directly measured by assessing changes in membrane surface pressure (149). Interestingly, mutation of specific hydrophobic residues within the helix that prevented membrane insertion blocked the development of membrane curvature (41). Thus, epsin appears to serve a key role in the initial development of membrane curvature that is critical for the formation of CCPs and CCVs during SV recycling.

Interestingly, mutations in epsin that disrupt the ability of the ENTH domain to stimulate curvature block the progression of curvature even under conditions where clathrin has been recruited to membranes and has assembled into a lattice (41). This provided a key piece of evidence that clathrin assembly on its own is unlikely to be sufficient for the generation of curvature. Instead, clathrin may simply form a scaffold upon which the curvature machinery (ENTH domain and BAR domain proteins, see below) is organized. Additionally, the clathrin lattice may contribute to curvature by the formation of a mold into which the deformed membrane can be structured into a highly curved vesicle. Regardless of the precise mechanism, it has become apparent that generating a highly curved SV from an essentially flat synaptic plasma membrane is complex and likely involves a series of proteins working in concert.

4.5 Membrane Curvature: Endophilin and Amphiphysin, N-BAR and BAR Domains

Endophilin was originally identified in biochemical and yeast two-hybrid screens as a binding partner for the endocytic enzyme synaptojanin (25, 133). Endophilin is composed of an N-terminal BAR domain, a central domain and a C-terminal SH3 domain (Fig. 2). The SH3 domain mediates interactions with proline-rich sequences in the peptide motif domains of synaptojanin and dynamin (104, 133) (Fig. 3) and it is the ability to bind to these key endocytic regulatory enzymes that gave endophilin its name (104). Disruption of these interactions through injection of peptides and antibodies into the lamprey reticulospinal synapse impairs SV endocytosis, demonstrating a role for endophilin in SV recycling (45). Subsequent studies in *Drosophila* and *C. elegans* have revealed that endophilin mutants and synaptojanin mutants have nearly identical phenotypes: both have accumulations of CCVs in the synapse and an abnormal, clustered distribution of SVs away from the active zone (144, 164). Moreover, defects observed in synaptojanin and endophilin double mutants closely resemble those seen in either single mutant, demonstrating that the two proteins function on the same pathway (144, 164) as predicted from earlier biochemical studies (25, 133). Finally, synaptojanin was lost from the nerve terminals of endophilin mutants (144, 164), thus indicating that endophilin serves to recruit synaptojanin to nerve terminals during SV endocytosis. There also are indications that endophilin may play additional roles during endocytosis (26, 165)

Amphiphysin 1 and 2 are two proteins that have domain structures similar to that of endophilin. These proteins have an N-terminal BAR domain, a variable central domain and a C-terminal SH3 domain (Fig. 2). As is the case for endophilin, the SH3 domains of the amphiphysins bind to synaptojanin and dynamin (23, 99, 129). Amphiphysin 1 is expressed predominantly in brain, while amphiphysin 2, which is expressed as alternatively spliced isoforms, is more widely distributed (14, 23, 24, 129, 130). The predominant neuronal isoform of amphiphysin 2 forms heterodimers with amphiphysin 1 and this amphiphysin heterodimer is concentrated in presynaptic nerve terminals (129, 173). Both components of this heterodimer interact, via their central domain, with clathrin, NECAP and AP-2 (97, 129, 130, 137). Thus, amphiphysins are well positioned to function in the recruitment of synaptojanin and dynamin to sites of CCV formation during SV endocytosis. In fact, presynaptic microinjection of peptides that compete with the binding partners of the SH3 domains of the amphiphysins impairs SV endocytosis (148). Moreover, knock out of the amphiphysin 1 gene of mice leads to a concomitant loss of amphiphysin 2 and defects in SV recycling (27). Thus, the amphiphysins are key players in SV endocytosis.

Most recently, much of the focus on endophilin and amphiphysin has been on the BAR domain. The function of BAR domains was first suggested from the observation that the N-terminus of amphiphysin I, which mediates dimerization (131, 173) could bind to liposomes and cause them to form tubules (155). It was subsequently demonstrated that the N-terminus of endophilin, which is weakly homologous to the N-terminus of amphiphysin, also binds to membranes and converts lipid bilayers into narrow tubules (35). Structural studies of the BAR domain from arfaptin (157) and amphiphysin (124) revealed a dimer, with each monomer composed of three kinked α-helices that together form a six helix bundle. The BAR domain dimer has an elongated crescent structure, with positive charge along a concave surface that permits electrostatic interactions with negatively charged membranes (124). More recently, several groups have solved the structure of the BAR domain of endophilin (46, 96, 171). The structure is very similar to that of amphiphysin, except for the presence of an N-terminal amphipathic helix and two largely disordered domains that are inserted into helix 1, projecting from the concave face of the dimer. When the BAR domain engages the membrane, the N-terminal helix folds and is peripherally bound in the plane of the membrane, ideally situated to affect membrane curvature (46, 96). The combination of the N-terminal helix with the BAR domain has led to the designation of the N-BAR domain.

Thus, BAR domains are thought to have two distinct but overlapping functions. First, they can act as molecular switches. Different BAR domains can have different degrees of curvature on their concave faces and will thus best interact with membranes of a defined degree of curvature (98). They could thus serve to recruit proteins to newly formed vesicles only after a certain degree of curvature has been reached. For example, the amphiphysin BAR domain would best recognize a bilayer with curvature similar to the neck of a deeply invaginated CCP and could thus recruit dynamin specifically to this site. A second function for BAR domains, and in particular for N-BAR domains, is to contribute to the induction of curvature by forcing membranes to conform to the concave face of the module (98). In addition, insertion of the N-terminal helix of an N-BAR domain into the cytosolic leaflet stimulates curvature as descried for ENTH domains. In summary, amphiphysins and endophilin both recruit endocytic proteins, such as dynamin and synaptojanin, and contribute to curving of membranes during CCV formation.

4.6 Transition to Fission: F-BAR Domains Link Curvature to Fission Via Actin

The FCH (Fes/CIP4 homology) domain is a protein module found in a wide variety of proteins. Many of these proteins have been linked to regulation of the actin cytoskeleton while others have been implicated in CME, due to their SH3 domain-dependent binding to dynamin (68, 162). For example, the syndapin/PACSIN family of proteins (Fig. 2) functions in CME via interactions with dynamin and synaptojanin and also contributes to regulation of actin via binding to N-WASP (108, 127). Recent examination of sequences downstream of the FCH domain revealed a coiled-coil domain that, when considered as part of the FCH domain, is homologous to BAR domains. The extended module was thus renamed the extended FCH (EFC) or F-BAR domain (68, 162). Like BAR domains, the F-BAR domain tubulates lipid membranes both in liposomes and in cells. However, the tubules formed by F-BAR domains are larger in diameter than those produced by N-BAR domains (68, 162). Most recently, two groups have solved the crystal structure of the F-BAR domain (55, 147). The structures reveal a gently curving helical-bundle dimer with a distant relationship to BAR domains. The curved domain best fits a tubular membrane with a diameter of approximately 60 nm and generates tubules as large as 130 nm in diameter (55, 147). The module forms filaments through end-to-end interactions, leading to striations on tubular membranes after addition of the F-BAR domain (147). Thus, F-BAR domains may sense initial states of curvature of CCPs and then assemble into filaments around the invaginated membrane to drive further CCP invagination (147).

It thus appears that the membrane curvature needed for CCV formation is generated through the coordinated action of several lipid-binding protein modules. Initially, ENTH domains and N-BAR domains insert amphipathic helicies into membranes to create bilayer asymmetry and initiate curvature. This low degree of curvature of the membrane would then be sensed by F-BAR domain proteins, which would form filaments to further induce invagination. Once the membrane reaches a deeply invaginated state, BAR domain proteins would recognize the highly curved, small diameter membrane at the neck of the vesicle and recruit dynamin to drive membrane fission. In addition, many F-BAR domain proteins interact with regulatory proteins of the actin cytoskeleton, most notably N-WASP. Thus, as these proteins drive later stages of curvature they could also recruit N-WASP to endocytic sites. N-WASP stimulates actin assembly, which has been proposed to apply tension on the invaginated membrane for the function of dynamin (141). Through formation of such protein networks, which initiate, stabilize and resolve membrane curvature-dependent processes, early events and late events in the formation of CCPs and CCVs could be linked.

4.7 Transition to Fission: HIPs and Intersectin-l Provide Additional Links to Actin

Another link between the clathrin machinery and the actin cyoskeleton is provided by the huntingtin-interacting proteins (HIPs) (Fig. 2). HIP1 was initially identified in a yeast two-hybrid screen for proteins that interact with huntingtin, the protein product of the Huntington's disease gene (73). Another HIP1-related protein (HIP1R/HIP12) was identified by its sequence similarity to HIP1, although HIP1R does not bind directly to huntingtin (18). HIP1 and HIP1R have similar domain structures, with an N-terminal ANTH domain for phospholipid interaction (66), a helical domain that mediates dimerization and binding to CLCs, and a C-terminal THATCH domain (7, 11, 33, 56, 87) (Fig. 2). The interaction of HIPs with CLCs targets these proteins to clathrin-coated structures (17, 88), causing HIP1 and HIP1R to be major components of the coats of CCPs and CCVs (32, 33, 102, 105, 166).

The C-terminal THATCH domain allows HIP1R (but not HIP1) to bind to actin (11, 33, 87). Moreover, a proline-rich sequence of HIP1R binds to the SH3 domain of cortactin (85). While cortactin stimulates actin assembly via activation of N-WASP, interaction of cortactin with HIP1R inhibits actin assembly (85). Thus, knock down of HIP1R with small inhibitory RNA leads to over-assembly of actin and defects in CME (34, 85). HIP1R therefore appears to be necessary for the appropriate interaction of clathrin-coated structures with the actin cytoskeleton. Under normal physiological circumstances, HIP1R, perhaps in heterodimers with HIP1, could link actin to clathrin coats on deeply invaginated CCPs, providing the invaginating endocytic structure with the tension that is needed for dynamin to drive membrane fission (141).

Another actin regulatory protein that is also involved in CME is intersectin-long (intersectin-l). Intersectin-l, a splice variant of intersectin-s, has a C-terminal extension with a Dbl homology (DH), pleckstrin homology (PH) and C2 domain (63) (Fig. 2). The combination of DH and PH domains is characteristic of guanine-nucleotide exchange factors (GEFs) and allows intersectin-l to act as a Cdc42-specific GEF (64, 75, 181). The GEF activity of intersectin-l is stimulated by N-WASP binding to intersectin-l. This increases the amount of GTP-bound Cdc42, which in turn activates N-WASP and stimulates actin assembly (64). In addition, intersectin-l binds to dynamin (176). Thus, intersectin-l is ideally situated to stimulate actin assembly in proximity to CCPs and CCVs, particularly in the presence of dynamin, although whether or not intersectin-1 functions in CME of SVs has never been tested.

4.8 Fission of CCVs: Dynamin Drives Membrane Fission

Dynamin was one of the first endocytic accessory proteins to be identified and remains a central feature of all current models of endocytosis. Dynamin contains an N-terminal GTPase domain, a GED (GTPase effector domain), which acts as an intramolecular GTPase activating protein (GAP) (145), and a proline-rich domain that binds various SH3 domain-bearing proteins (48) (Fig. 2). A role for dynamin in CME was originally suggested by its identification as the product of the *Drosophila shibire* gene (163). Temperature-sensitive *Shibire* mutants display rapid paralysis at the restrictive temperature, due to impairment of SV recycling and subsequent block of synaptic transmission (80). This block of SV recycling at the non-permissive temperature is due to inhibition of membrane fission during endocytosis, evident as the appearance of endocytic vesicles that remain attached to the plasma membrane by narrow membrane stalks (82). Dynamin was subsequently demonstrated to oligomerize into rings with an inner diameter similar to that of the membrane stalks seen in *shibire* flies (61). Treatment of synaptosomes with GTPγS, which locks dynamin into a GTP-bound state and inhibits SV endocytosis (57, 177), leads to production of CCVs that remained attached to the plasma membrane via extended membrane necks (154). The necks are coated with dynamin spirals, leading to the hypothesis that dynamin functions in vesicle fission as a mechanochemical enzyme that utilizes GTP hydrolysis to provide a twisting force that severs the membrane tube (154).

Various models have been proposed to explain how dynamin functions in vesicle fission. In the GTP-bound form, dynamin appears as tightly packed rings on lipid nanotubes, with GTP hydrolysis increasing the space between the rings. This suggests that a spring-like expansion (conformational change) drives vesicle fission (150). The main alternative model is that dynamin does not function as a mechanochemical enzyme but – by analogy to other GTPases – functions in the recruitment of other factors required for fission (145). This model arose from the study of dynamin with mutations in the GED domain that inhibited the intramolecular GAP activity and thus held dynamin in the active, GTP-bound state. These mutants in fact stimulated CME, suggesting that GTP-bound dynamin recruits other (unidentified) proteins that produce fission (145). Thus, the exact mechanism of dynamin function has remained elusive.

A recent paper has provided support for the role of dynamin as a mechanochemical enzyme that drives fission (141). Specifically, the addition of GTP to dynamin-coated lipid tubules resulted in twisting of the tubules: this twisting motion was observed by using streptavidin beads conjugated to biotinylated dynamin, in which the beads were seen to swing around the tubules. Moreover, when lipid tubules were generated by the motor activity of kinesin on microtubules, and were thus subjected to longtitudinal tension, dynamin/GTP could drive fission, while no fission was observed in the absence of tension (141). Thus, the twisting force of dynamin appears to lead to constriction that causes fission on membranes under tension. In cells, it is likely that tension on the deeply invaginated CCPs is provided by the assembly of actin.

Work with the *shibire* mutant of *Drosophila* leaves no doubt that dynamin plays a crucial role in SV endocytosis. The function of dynamin in SV endocytosis in mammalian presynaptic terminals was recently clarified through the use of dynamin 1 knock out mice (38). Mammals express three dynamin genes: dynamin 1, which is brain specific and is the most abundant form, dynamin 2 that is expressed ubiquitously, and dynamin 3, which is expressed in brain and testis (15). In dynamin 1 null animals, SV endocytosis is highly compromised during strong exogenous stimulation. Somewhat surprisingly however, SV endocytosis is normal after strong stimulation stops (38). Expressing any of the three dynamin isoforms in knock-out neurons could compensate for the dynamin 1 phenotype, though dynamin 3 was more effective than dynamin 2. Thus, whereas dynamin 1 functions during high levels of neuronal activity, other mechanisms drive membrane fission during low levels of activity, with dynamin 3 being the most likely candidate for the fission that does not depend upon dynamin 1.

The proline-rich peptide motif domain of dynamin 1 has multiple sites for SH3 domain binding (Fig. 2), with endophilin, amphiphysin and syndapin/PACSIN being the three key interacting proteins. Each of these proteins contains an N-terminal domain that can sense membrane curvature: endophilin (N-BAR domain), amphiphysin (BAR domain) and syndapin (F-BAR domain). Interestingly, the degree of curvature that is best suited for each domain is unique (55, 124, 147). Thus, it is possible that these proteins contribute to the recruitment of dynamin 1 to the membrane at different stages of CCV formation. Although it is clear that dynamin functions late in CME, this protein also appears to function at earlier steps (91). For example, the F-BAR domain appears to best fit membrane curvatures consistent with a shallow CCP whereas the amphiphysin BAR domain recognizes highly curved membranes such as those at the neck of a deeply invaginated CCP (55, 124, 147).

Interestingly, the ability of dynamin 1 to interact with its SH3 domain partners is regulated by phosphorylation. In fact, dynamin 1 is one of a group of at least eight dephosphins, proteins that undergo a coordinated dephosphorylation reaction in nerve terminals during SV endocytosis (21). It is thought that the dephosphorylation promotes the assembly of protein complexes needed for SV endocytosis. The subsequent rephosphorylation of the proteins is necessary for maintaining the continuity of SV recycling (21). To date, only one dephosphin kinase has been identified, cyclin-dependent protein kinase 5 (Cdk5); disruption of its activity blocks CME of SVs (156). Careful analysis of the phosphorylation of dynamin 1 by Cdk5 has revealed that the major phosphorylation-regulated dynamin 1 partner is syndapin (3). Thus, although the functional role of the dynamin 1/syndapin interaction is unknown, it appears clear that these proteins form a highly regulated partnership that is critical for SV endocytosis and recycling.

4.9 Uncoating CCVs: Synaptojanin and Stability of Coats

Once CCVs have been liberated from the membrane by dynamin-dependent fission, they must rapidly uncoat before they can be reloaded with neurotransmitter and reenter the recycling pool of SVs. At least two types of proteins have been implicated in this process, synaptojanin and the uncoating ATPase Hsc70.

Synaptojanin was originally identified as a 145 kDa protein that bound to SH3 domains, was expressed predominantly in brain, and was enriched on endocytic intermediates of the synapse (99, 100). Subsequent cloning of the protein, which was later renamed synaptojanin 1-145, revealed a N-terminal Sac1 homology domain followed by a central inositol 5-phosphatase domain and a proline-rich C-terminus (101) (Fig. 2). The Sac1 domain dephosphorylates PtdIns(3)P and PtdIns(4)P and the 5-phosphatase domain dephosphorylates PtdIns(4,5)P_2 and PtdIns(3,4,5)P_3 at the 5′ position (50, 101). The identification of synaptojanin 1-145 was in fact instrumental in the early development of the idea that PtdIns(4,5)P_2 functions in SV endocytosis. The C-terminal peptide motif domain of synaptojanin 1-145 is primarily a protein interaction domain that binds to a variety of SH3 domain-containing proteins that are thought to be involved in targeting the protein to sites of CME. Through an alternative splice in the C-terminus, a second, longer synaptojanin 1 isoform, synaptojanin 1-170, is generated. This isoform is expressed in brain but has a much broader tissue distribution than synaptojanin 1-145 (128). Synaptojanin 1-170 contains additional peptide motifs for binding to clathrin, AP-2 and EH domains (Figs. 2 and 3). Interestingly, multicolor TIRF (total internal reflection fluorescence) microscopy has recently revealed that synaptojanin 1-145 is recruited to membranes as a burst at a late stage of CCP formation, whereas synaptojanin 1-170 is present at CCPs throughout their formation (123). Thus, dynamic phosphoinositide metabolism is likely to be important at all stages of CCP/CCV formation.

In neurons, synaptojanin 1 knock-out mice have increased overall levels of PtdIns(4,5)P_2 and an accumulation of uncoated CCVs in their presynaptic terminals (22). Moreover, mutations in unc-26, the *C. elegans* homologue of synaptojanin, also lead to defects in the budding of SVs from the plasma membrane and in the uncoating of CCVs (53). Thus, synaptojanin-dependent dephosphorylation of PtdIns(4,5)P_2 is likely important as a trigger to destabilize clathrin coats following SV endocytosis, allowing the coat to come off and the SV to re-enter the release pool. In yeast, patches of PtdIns(4,5)P_2 that form on the plasma membrane are rapidly lost coincident with the arrival of synaptojanin and disassembly of the clathrin coat (151). Indeed, synaptojanin knock-out mice show a delay in the reentry of recycling vesicles into the fusion-competent pool although SV recycling is not abolished (78). A second gene product, synaptojanin 2, has also been identified and at least one synaptojanin 2 splice variant is expressed in nerve terminals (116). The function of synaptojanin 2 in SV endocytosis has not been directly assessed but it is possible that it is functionally redundant with synaptojanin 1 and accounts for the residual SV endocytosis seen in synaptojanin 1 null animals.

4.10 Uncoating CCVs: Auxilin and Hsc70

While dephosphorylation of PtdIns(4,5)P_2 contributes to destabilization of the coat of CCVs, it is not sufficient to disassemble the clathrin lattice. In vitro, the assembly of clathrin triskelia into clathrin cages can proceed spontaneously, indicating that an energy-dependent step is necessary to dissociate the clathrin cage. The energy required for disassembly of clathrin cages appears to come from an ATPase, the heat shock cognate 70 (Hsc70) protein. Hsc70 is a critical factor in CCV uncoating in vitro, with the hydrolysis of three ATP molecules necessary for the release of one triskelia (9, 79) and dominant-negative forms of Hsc70 disrupt clathrin uncoating and multiple stages of clathrin-mediated trafficking in cells (117). Essential cofactors for Hsc70 are the auxilins. Auxilin 1, which is brain specific, contains an N-terminal region with sequence similarity to the phosphatase and C2 domains of PTEN (a PtdIns(3)-phosphatase), a central region containing motifs that bind to clathrin and AP-2, and a DNAJ domain at the C-terminus (Fig. 2). Auxilin 2, which is also known as cyclin G-associated kinase or GAK, has a ubiquitous tissue distribution and a similar domain structure as auxilin 1 but with an additional N-terminal ser/thr kinase domain (79). Auxilins bind directly to clathrin and AP-2 and through the DNAJ domain they bind Hsc70 in the ATP-bound form. Injection into presynaptic nerve terminals of auxilin 1 with a mutation in the conserved HPD motif of the DNAJ domain, which prevents auxilin 1 interaction with Hsc70, disrupts CCV uncoating and synaptic transmission (111). Thus, a key function for auxilin is to recruit Hsc70 to newly formed CCVs for uncoating and SV recycling.

Because auxilins should be able to bind to partially assembled lattices, it is important to understand what prevents Hsc70 from uncoating CCVs prematurely. One potential mechanism is that auxilin may be recruited to CCVs only late in their life cycle. Alternatively, Hsc70 could be recruited at any time but with the ATPase activity regulated to only activate uncoating at specific times. Two recent studies have examined the recruitment of auxilin over the life cycle of CCVs in non-neuronal systems (86, 95). Auxilin was found to be present at sites of CCV formation at early points, however, there is a major burst of auxilin recruitment following the peak of dynamin recruitment late in the CCV life cycle. Interestingly, the late recruitment of auxilin depends upon the PTEN-like domain. This domain binds specifically to inostiol phospholipids, most notably to PtdIns(3)P (86, 95). This result suggests that a switch in the levels of specific inositol phospholipid levels late in CCV formation may act as a trigger to recruit auxilin/Hsc70 for CCV uncoating. This is perhaps not surprising given the key role of inositol phospholipids in the nucleation of CCPs and the observation that knock out of synaptojanin leads to an accumulation of CCVs in nerve terminals (22).

Hsc70 has recently been identified as a binding partner of CHL1, a cell adhesion molecule of the immunoglobulin superfamily (89). Ablation of the CHL1 gene reduced the levels of Hsc70 in the presynaptic compartment and CHL1 deficiency or disruption of CHL1/Hsc70 interactions led to the accumulation of CCVs at synapses (89). Thus, CHL1 appears to function as a synapse-targeting factor for Hsc70. However, CHL1 was also found to undergo CME in an activity dependent manner and to interact with Hsc70 on CCVs in the synapse. Thus, CHL1 may represent an additional mechanism to recruit Hsc70 for CCV uncoating during SV recycling, which may be needed to meet the high demands placed on this specialized form of CME.

4.11 Post-Uncoating: Connecdenn and DENN Domains, a Late Function in Endocytosis?

Identification of the consensus WXXF-acidic α-ear-binding motif in the NECAP proteins led to a bioinformatics analysis that revealed a novel protein containing the C-terminal sequence WETFE-COO$^-$ (2). At its N-terminus, the protein contains a module referred to as the DENN (differentially expressed in normal versus neoplastic cells) domain, leading to the name connecdenn for this protein (2) (Fig. 2). Connecdenn is highly enriched on CCVs and, as expected, binds to the α-ear of AP-2. The protein is expressed predominantly in brain where it is enriched in presynaptic nerve terminals (2). In addition to the WXXF-acidic motif, connecdenn also contains single DPF and FXDXF motifs; all three of these motifs contribute to α-ear binding, although the WXXF-acidic motif is the strongest of the three. Disruption of the interaction of connecdenn with AP-2, through overexpression of C-terminal connecdenn fragments, blocks SV endocytosis in hippocampal neurons, as does connecdenn knock down using lentiviral delivery of connecdenn-specific shRNA (2). These experiments suggest that connecdenn is a component of the machinery for SV endocytosis.

DENN domains are composed of upstream (u)DENN and downstream (d)DENN modules flanking a DENN module, forming a unit that is readily detected using the conserved domain search of the National Center for Biotechnology Information (93). The domain is well conserved throughout evolution and is found in multiple proteins from diverse species including humans, *C. elegans* and *S. pombe*. Prominent DENN domain-containing proteins include myotubular myopathy related (MTMR) 5 and 13, which are homologues of the myotubularin lipid phosphatases that function at endosomes (138) and DENN/MADD, a guanine-nucleotide exchange factor for the SV protein Rab3 (107). However, the majority of DENN domain-containing proteins are uncharacterized and there is essentially no information regarding DENN domain structure or the biological context of DENN domain function (90). Connecdenn is thus important because it links DENN domain proteins to the process of SV endocytosis and introduces a new module into the endocytic machinery (Fig. 3).

Connecdenn has a unique association with CCVs in that it remains associated with the vesicles following CCV uncoating (2). In fact, the protein is tightly associated with the membranes and resists extraction even in the presence of high salt (2). Preliminary evidence demonstrates that the C-terminus of connecdenn guides the protein to CCVs but the DENN domain is responsible for the stable membrane association. It is likely that the stable association of connecdenn with CCV membranes ensures a post-uncoating function for the protein, likely in endosomal trafficking or morphology. Interestingly, knock-out of DENN/MADD leads to mice that have reduced numbers of SVs, suggesting a possible defect in SV reformation (107). Thus, DENN domain proteins may be involved generally in endocytic membrane trafficking and SV endocytosis/reformation.

5 Conclusions

Although it has been known for more than 30 years that endocytosis via CCVs is required for SV recycling, for most of these years our understanding of the molecular mechanisms involved in SV endocytosis has lagged far behind our appreciation of the molecular mechanisms of SV exocytosis. Fortunately, a surge in studies of CCVs and their molecular constituents has led to tremendous recent advances in our appreciation of the molecular underpinnings of SV endocytosis. One tangible result is that nearly 30 proteins now have been implicated in SV endocytosis (Fig. 2), though more endocytic proteins undoubtedly await discovery.

To date, the main conclusion is that the sequence of events that leads to removal of SV components from the plasma membrane and the recycling of SVs requires a complex web of interactions between proteins and lipids. This web is based on the interaction domains that are depicted in Fig. 3. This web of interactions must be regulated, both in time and in space, to yield a vectorial progression of CCVs through many molecular states.

The actual sequence of molecular events involved in endocytosis is not yet clear. Thus far, multiple proteins and protein-protein interactions have been implicated in each of the many partial reactions involved in endocytic retrieval of SVs. While the information presented in Section 4 can be assembled into a rudimentary model of SV endocytosis, further experimental work will be needed to turn these preliminary hypotheses into conclusive definitions of functional roles. An important step toward this goal will be to define the precise times at which each of these interactions occurs during SV endocytosis. Such work eventually will lead to a comprehensive and quantitative description of protein and lipid dynamics involved in SV endoytosis.

References

1. Ahle, S. and Ungewickell. E. Purification and properties of a new clathrin assembly protein. *EMBO J.* 5, 3143–3149, 1986.
2. Allaire, P.D., Ritter, B., Thomas, S., Burman, J.L., Denisov, A.Y., Legendre-Guillemin, V., Harper, S.Q., Davidson, B.L., Gehring, K. and McPherson, P.S. Connecdenn, a novel DENN domain-containing protein of neuronal clathrin-coated vesicles functioning in synaptic vesicle endocytosis. *J. Neurosci.* 26, 13202–13212, 2006.
3. Anggono, V., Smillie, K.J., Graham, M.E., Valova, V.A., Cousin, M.A. and Robinson, P.J. Syndapin I is the phosphorylation-regulated dynamin I partner in synaptic vesicle endocytosis. *Nat Neurosci.* 9, 752–760, 2006.
4. Augustine, G.J., Morgan, J.R., Villalba-Galea, C.A., Jin, S., Prasad, K. and Lafer, E.M. Clathrin and synaptic vesicle endocytosis: studies at the squid giant synapse. *Biochem. Soc. Trans.* 34, 68–72, 2006.
5. Benmerah, A., Gagnon, J., Begue, B., Megarbane, B., Dautry-Varsat, A. and Cerf-Bensussan, N. The tyrosine kinase substrate eps15 is constitutively associated with the plasma membrane adaptor AP-2. *J. Cell Biol.* 131, 1831–1838, 1995.
6. Benmerah, A., Begue, B., Dautry-Varsat, A. and Cerf-Bensussan, N. The ear of alpha-adaptin interacts with the COOH-terminal domain of the Eps 15 protein. *J. Biol. Chem.* 271, 12111–12116, 1996.

7. Bennett, E.M., Chen, C.Y., Engqvist-Goldstein, A.E., Drubin, D.G. and Brodsky, F.M. Clathrin hub expression dissociates the actin-binding protein Hip1R from coated pits and disrupts their alignment with the actin cytoskeleton. *Traffic* 2, 851–858, 2001.
8. Blondeau, F., Ritter, B., Allaire, P.D., Wasiak, S., Girard, M., Hussain, N.K., Angers, A., Legendre-Guillemin, V., Roy, L., Boismenu, D., Kearney, R.E., Bell, A.W., Bergeron, J.J. and McPherson PS. Tandem MS analysis of brain clathrin-coated vesicles reveals their critical involvement in synaptic vesicle recycling. *Proc. Natl. Acad. Sci. USA* 101, 3833–3838, 2004.
9. Braell, W.A., Schlossman, D.M., Schmid, S.L. and Rothman, J.E. Dissociation of clathrin coats coupled to the hydrolysis of ATP: role of an uncoating ATPase. *J. Cell Biol.* 99, 734–741, 1984.
10. Brett, T.J., Traub, L.M. and Fremont, D.H. Accessory protein recruitment motifs in clathrin-mediated endocytosis. *Structure* 10, 797–809, 2002.
11. Brett, T.J., Legendre-Guillemin, V., McPherson, P.S. and Fremont, D.H. Structural definition of the F-actin-binding THATCH domain from HIP1R. *Nat. Struct. Mol. Biol.* 13, 121–130, 2006.
12. Brodsky, F.M., Chen, C.Y., Knuehl, C., Towler, M.C. and Wakeham, D.E. Biological basket weaving: formation and function of clathrin-coated vesicles. *Annu. Rev. Cell Dev. Biol.* 17, 517–568, 2001.
13. Burman, J.L., Wasiak, S., Ritter, B., de Heuvel, E. and McPherson, P.S. Aftiphilin is a component of the clathrin machinery in neurons. *FEBS Lett.* 579, 2177–2184, 2005.
14. Butler, M.H., David, C., Ochoa, G.C., Freyberg, Z., Daniell, L., Grabs, D., Cremona, O. and De Camilli, P. Amphiphysin II (SH3P9; BIN1), a member of the amphiphysin/Rvs family, is concentrated in the cortical cytomatrix of axon initial segments and nodes of ranvier in brain and around T tubules in skeletal muscle. *J. Cell Biol.* 137, 1355–1367, 1997.
15. Cao, H., Garcia, F. and McNiven, M.A. Differential distribution of dynamin isoforms in mammalian cells. *Mol. Biol. Cell* 9, 2595–609, 1998.
16. Chen, H., Fre, S., Slepnev, V.I., Capua, M.R., Takei, K., Butler, M.H., Di Fiore, P.P. and De Camilli, P. Epsin is an EH-domain-binding protein implicated in clathrin-mediated endocytosis. *Nature* 394, 793–797, 1998.
17. Chen, C.Y. and Brodsky, F.M. Huntingtin-interacting protein 1 (Hip1) and Hip1-related protein (Hip1R) bind the conserved sequence of clathrin light chains and thereby influence clathrin assembly in vitro and actin distribution in vivo. *J. Biol. Chem.* 280, 6109–6117, 2005.
18. Chopra, V.S., Metzler, M., Rasper, D.M., Engqvist-Goldstein, A.E., Singaraja, R., Gan, L., Fichter, K.M., McCutcheon, K., Drubin, D., Nicholson, D.W. and Hayden, M.R. HIP12 is a non-proapoptotic member of a gene family including HIP1, an interacting protein with huntingtin. *Mamm. Genome* 11, 1006–1015, 2000.
19. Conner, S.D. and Schmid, S.L. Identification of an adaptor-associated kinase, AAK1, as a regulator of clathrin-mediated endocytosis. *J. Cell Biol.* 156, 921–929, 2002.
20. Conner, S.D. and Schmid, S.L. CVAK104 is a novel poly-L-lysine-stimulated kinase that targets the beta2-subunit of AP2. *J. Biol. Chem.* 280, 21539–21544, 2005.
21. Cousin, M.A. and Robinson, P.J. The dephosphins: dephosphorylation by calcineurin triggers synaptic vesicle endocytosis. *Trends Neurosci.* 24, 659–665, 2001.
22. Cremona, O., Di Paolo, G., Wenk, M.R., Luthi, A., Kim, W.T., Takei, K., Daniell, L., Nemoto, Y., Shears, S.B., Flavell, R.A., McCormick, D.A. and De Camilli, P. Essential role of phosphoinositide metabolism in synaptic vesicle recycling. *Cell* 99, 179–188, 1999.
23. David, C., McPherson, P.S., Mundigl, O. and De Camilli, P. A role of amphiphysin in synaptic vesicle endocytosis suggested by its binding to dynamin in nerve terminals. *Proc. Natl. Acad. Sci. USA* 93, 331–355, 1996.

24. De Camilli, P., Thomas, A., Cofiell, R., Folli, F., Lichte, B., Piccolo, G., Meinck, H.M., Austoni, M., Fassetta, G., Bottazzo, G., Bates, D., Cartlidge, N., Solimena, M. and Kilimann, M.W. The synaptic vesicle-associated protein amphiphysin is the 128-kD autoantigen of Stiff-Man syndrome with breast cancer. *J. Exp. Med.* 178, 2219–2223, 1993.
25. de Heuvel, E., Bell, A.W., Ramjaun, A.R., Wong, K., Sossin, W.S. and McPherson, P.S. Identification of the major synaptojanin-binding proteins in brain. *J. Biol. Chem.* 272, 8710–8716, 1997.
26. Dickman, D.K., Lu, Z., Meinertzhagen, I.A. and Schwarz, T.L. Altered synaptic development and active zone spacing in endocytosis mutants. *Curr. Biol.* 16, 591–598, 2006.
27. Di Paolo, G., Sankaranarayanan, S., Wenk, M.R., Daniell, L., Perucco, E., Caldarone, B.J., Flavell, R., Picciotto, M.R., Ryan, T.A., Cremona, O. and De Camilli, P. Decreased synaptic vesicle recycling efficiency and cognitive deficits in amphiphysin 1 knockout mice. *Neuron* 33, 789–804, 2002.
28. Di Paolo, G., Moskowitz, H.S., Gipson, K., Wenk, M.R., Voronov, S., Obayashi, M., Flavell, R., Fitzsimonds, R.M., Ryan, T.A. and De Camilli, P. Impaired PtdIns(4,5)P2 synthesis in nerve terminals produces defects in synaptic vesicle trafficking. *Nature* 431, 415–422, 2004.
29. Diril, M.K., Wienisch, M., Jung, N., Klingauf, J. and Haucke, V.. Stonin 2 is an AP-2-dependent endocytic sorting adaptor for synaptotagmin internalization and recycling. *Dev. Cell* 10, 233–244, 2006.
30. Edeling, M.A., Mishra, S.K., Keyel, P.A., Steinhauser, A.L., Collins, B.M., Roth, R., Heuser, J.E., Owen, D.J. and Traub, L.M. Molecular switches involving the AP-2 beta2 appendage regulate endocytic cargo selection and clathrin coat assembly. *Dev. Cell* 10, 329–342, 2006.
31. Ehrlich, M., Boll, W., Van Oijen, A., Hariharan, R., Chandran, K., Nibert, M.L. and Kirchhausen, T. Endocytosis by random initiation and stabilization of clathrin-coated pits. *Cell* 118, 591–605, 2004.
32. Engqvist-Goldstein, A.E., Kessels, M.M., Chopra, V.S., Hayden, M.R. and Drubin, D.G. An actin-binding protein of the Sla2/Huntingtin interacting protein 1 family is a novel component of clathrin-coated pits and vesicles. *J. Cell Biol.* 147, 1503–1518, 1999.
33. Engqvist-Goldstein, A.E., Warren, R.A., Kessels, M.M., Keen, J.H., Heuser, J. and Drubin, D.G. The actin-binding protein Hip1R associates with clathrin during early stages of endocytosis and promotes clathrin assembly in vitro. *J. Cell Biol.* 154, 1209–1223, 2001.
34. Engqvist-Goldstein, A.E., Zhang, C.X., Carreno, S., Barroso, C., Heuser, J.E. and Drubin, D.G. RNAi-mediated Hip1R silencing results in stable association between the endocytic machinery and the actin assembly machinery. *Mol. Biol. Cell* 15, 1666–1679, 2004.
35. Farsad, K., Ringstad, N., Takei, K., Floyd, S.R., Rose, K. and De Camilli, P. Generation of high curvature membranes mediated by direct endophilin bilayer interactions. *J. Cell Biol.* 155, 193–200, 2001.
36. Fazioli, F., Minichiello, L., Matoskova, B., Wong, W.T. and Di Fiore, P.P. eps15, a novel tyrosine kinase substrate, exhibits transforming activity. *Mol. Cell Biol.* 13, 5814–5828, 1993.
37. Fergestad, T., Davis, W.S. and Broadie, K. The stoned proteins regulate synaptic vesicle recycling in the presynaptic terminal. *J. Neurosci.* 19, 5847–5860, 1999.
38. Ferguson, S.M., Brasnjo, G., Hayashi, M., Wolfel, M., Collesi, C., Giovedi, S., Raimondi, A., Gong, L.W., Ariel, P., Paradise, S., O'toole, E., Flavell, R., Cremona, O., Miesenbock, G., Ryan, T.A. and De Camilli, P. A selective activity-dependent requirement for dynamin 1 in synaptic vesicle endocytosis. *Science* 316, 570–574, 2007.

39. Fernandez-Alfonso, T., Kwan, R. and Ryan, T.A. Synaptic vesicles interchange their membrane proteins with a large surface reservoir during recycling. *Neuron* 20, 179–186, 2006.
40. Ford, M.G., Pearse, B.M., Higgins, M.K., Vallis, Y., Owen, D.J., Gibson, A., Hopkins, C.R., Evans, P.R. and McMahon, H.T. Simultaneous binding of PtdIns(4,5)P2 and clathrin by AP180 in the nucleation of clathrin lattices on membranes. *Science* 291, 1051–1055, 2001.
41. Ford, M.G., Mills, I.G., Peter, B.J., Vallis, Y., Praefcke, G.J., Evans, P.R. and McMahon, H.T. Curvature of clathrin-coated pits driven by epsin. *Nature* 419, 361–366, 2002.
42. Fotin, A., Cheng, Y., Sliz, P., Grigorieff, N., Harrison, S.C., Kirchhausen, T. and Walz, T. Molecular model for a complete clathrin lattice from electron cryomicroscopy. *Nature* 432, 573–579, 2004.
43. Frick, M., Schmidt, K. and Nichols, B.J. Modulation of lateral diffusion in the plasma membrane by protein density. *Curr. Biol.* 6, 462–467, 2007.
44. Fukuda, M., Moreira, J.E., Lewis, F.M., Sugimori, M., Niinobe, M., Mikoshiba, K. and Llinas, R. Role of the C2B domain of synaptotagmin in vesicular release and recycling as determined by specific antibody injection into the squid giant synapse preterminal. *Proc. Natl. Acad. Sci. USA* 92, 10708–10712, 1995.
45. Gad, H., Ringstad, N., Low, P., Kjaerulff, O., Gustafsson, J., Wenk, M., Di Paolo, G., Nemoto, Y., Crun, J., Ellisman, M.H., De Camilli, P., Shupliakov, O. and Brodin, L. Fission and uncoating of synaptic clathrin-coated vesicles are perturbed by disruption of interactions with the SH3 domain of endophilin. *Neuron* 27, 301–312, 2000.
46. Gallop, J.L., Jao, C.C., Kent, H.M., Butler, P.J., Evans, P.R., Langen, R. and McMahon, H.T. Mechanism of endophilin N-BAR domain-mediated membrane curvature. *EMBO J.* 25, 2898–2910, 2006.
47. Gonzalez-Gaitan, M. and Jackle, H. Role of Drosophila {alpha}-adaptin in presynaptic vesicle recycling. *Cell* 88, 767–776, 1997.
48. Gout, I., Dhand, R., Hiles, I.D., Fry, M.J., Panayotou, G., Das, P., Truong, O., Totty, N.F., Hsuan, J., Booker, G.W., Campbell, I.D. and Waterfield, M.D. The GTPase dynamin binds to and is activated by a subset of SH3 domains. *Cell* 75, 25–36, 1993.
49. Granseth, B., Odermatt, B., Royle, S.J. and Lagnado, L. Clathrin-mediated endocytosis is the dominant mechanism of vesicle retrieval at hippocampal synapses. *Neuron* 51, 773–786, 2006.
50. Guo, S., Stolz, L.E., Lemrow, S.M. and York, J.D. SAC1-like domains of yeast SAC1, INP52, and INP53 and of human synaptojanin encode polyphosphoinositide phosphatases. *J. Biol. Chem.* 274, 12990–12995, 1999.
51. Hao, W., Luo, Z., Zheng, L., Prasad, K. and Lafer, E.M. AP180 and AP-2 interact directly in a complex that cooperatively assembles clathrin. *J. Biol. Chem.* 274, 22785–22794, 1999.
52. Harata, N.C., Aravanis, A.M. and Tsien, R.W. Kiss-and-run and full-collapse fusion as modes of exo-endocytosis in neurosecretion. *J. Neurochem.* 97, 1546–1570, 2006.
53. Harris, T.W., Hartwieg, E., Horvitz, H.R. and Jorgensen, E.M. Mutations in synaptojanin disrupt synaptic vesicle recycling. *J. Cell Biol.* 150, 589–600, 2000.
54. Haucke, V., Wenk, M.R., Chapman, E.R., Farsad, K. and De Camilli, P. Dual interaction of synaptotagmin with mu2- and alpha-adaptin facilitates clathrin-coated pit nucleation. *EMBO J.* 19, 6011–6019, 2000.
55. Henne, W.M., Kent, H.M., Ford, M.G., Hegde, B.G., Daumke, O., Butler, P.J., Mittal, R., Langen, R., Evans, P.R. and McMahon, H.T. Structure and Analysis of FCHo2 F-BAR Domain: A Dimerizing and Membrane Recruitment Module that Effects Membrane Curvature. *Structure* 15, 839–852.

56. Henry, K.R., D'Hondt, K., Chang, J., Newpher, T., Huang, K., Hudson, R.T., Riezman, H. and Lemmon, S.K. Scd5p and clathrin function are important for cortical actin organization, endocytosis, and localization of sla2p in yeast. *Mol. Biol. Cell* 13, 2607–2625, 2002.
57. Hess, S.D., Doroshenko, P.A. and Augustine, G.J. A functional role for GTP-binding proteins in synaptic vesicle cycling. *Science* 259, 1169–1172, 1993.
58. Heuser, J.E. and Reese, T.S. Evidence for recycling of synaptic vesicle membrane during transmitter release at the frog neuromuscular junction. *J. Cell Biol.* 57, 315–344, 1973.
59. Heuser, J.E., Reese, T.S., Dennis, M.J., Jan, Y., Jan, L. and Evans, L. Synaptic vesicle exocytosis captured by quick freezing and correlated with quantal transmitter release. *J. Cell Biol.* 81, 275–300, 1979.
60. Hinrichsen, L., Meyerholz, A., Groos, S. and Ungewickell, E.J. Bending a membrane: how clathrin affects budding. *Proc. Natl. Acad. Sci. USA* 103, 8715–8720, 2006.
61. Hinshaw, J.E. and Schmid, S.L. Dynamin self-assembles into rings suggesting a mechanism for coated vesicle budding. *Nature* 374, 190–192, 1995.
62. Honing, S., Ricotta, D., Krauss, M., Spate, K., Spolaore, B., Motley, A., Robinson, M., Robinson, C., Haucke, V. and Owen, D.J. Phosphatidylinositol-(4,5)-bisphosphate regulates sorting signal recognition by the clathrin-associated adaptor complex AP2. *Mol. Cell* 27, 519–531, 2005.
63. Hussain, N.K., Yamabhai, M., Ramjaun, A.R., Guy, A.M., Baranes, D., O'Bryan, J.P., Der, C.J., Kay, B.K. and McPherson, P.S. Splice variants of intersectin are components of the endocytic machinery in neurons and nonneuronal cells. *J. Biol. Chem.* 274, 15671–15677, 1999.
64. Hussain, N.K., Jenna, S., Glogauer, M., Quinn, C.C., Wasiak, S., Guipponi, M., Antonarakis, S.E., Kay, B.K., Stossel, T.P., Lamarche-Vane, N. and McPherson, P.S. Endocytic protein intersectin-l regulates actin assembly via Cdc42 and N-WASP. *Nat. Cell Biol.* 3, 927–932, 2001.
65. Hyman, J., Chen, H., Di Fiore, P.P., De Camilli, P. and Brunger, A.T. Epsin 1 undergoes nucleocytosolic shuttling and its eps15 interactor NH(2)-terminal homology (ENTH) domain, structurally similar to Armadillo and HEAT repeats, interacts with the transcription factor promyelocytic leukemia Zn(2)+ finger protein (PLZF*). J. Cell Biol.* 149, 537–546, 2000.
66. Hyun, T.S., Rao, D.S., Saint-Dic, D., Michael, L.E., Kumar, P.D., Bradley, S.V., Mizukami, I.F., Oravecz-Wilson, K.I. and Ross, T.S. HIP1 and HIP1r stabilize receptor tyrosine kinases and bind 3-phosphoinositides via epsin N-terminal homology domains. *J. Biol. Chem.* 279, 14294–14306, 2004.
67. Itoh, T., Koshiba, S., Kigawa, T., Kikuchi, A., Yokoyama, S. and Takenawa, T. Role of the ENTH domain in phosphatidylinositol-4,5-bisphosphate binding and endocytosis. *Science* 291, 1047–1051, 2001.
68. Itoh, T., Erdmann, K.S., Roux, A., Habermann, B., Werner, H. and De Camilli, P. Dynamin and the actin cytoskeleton cooperatively regulate plasma membrane invagination by BAR and F-BAR proteins. *Dev. Cell* 9, 791–804, 2005.
69. Jarousse, N. and Kelly, R.B. The AP2 binding site of synaptotagmin 1 is not an internalization signal but a regulator of endocytosis. *J. Cell Biol.* 154, 857–866, 2001.
70. Jha, A., Agostinelli, N.R., Mishra, S.K., Keyel, P.A., Hawryluk, M.J. and Traub, L.M. A novel AP-2 adaptor interaction motif initially identified in the long-splice isoform of synaptojanin 1, SJ170. *J. Biol. Chem.* 279, 2281–2290, 2004.
71. Jia, J.Y., Lamer, S., Schumann, M., Schmidt, M.R., Krause, E. and Haucke, V. Quantitative proteomics analysis of detergent-resistant membranes from chemical synapses: evidence for cholesterol as spatial organizer of synaptic vesicle cycling. *Mol. Cell Proteomics* 5, 2060–2071, 2006.

72. Jorgensen, E.M., Hartwieg, E., Schuske, K., Nonet, M.L., Jin, Y. and Horvitz, H.R. Defective recycling of synaptic vesicles in synaptotagmin mutants of *Caenorhabditis elegans*. *Nature* 378, 196–199, 1995.
73. Kalchman, M.A., Koide, H.B., McCutcheon, K., Graham, R.K., Nichol, K., Nishiyama, K., Kazemi-Esfarjani, P., Lynn, F.C., Wellington, C., Metzler, M., Goldberg, Y.P., Kanazawa, I., Gietz, R.D. and Hayden, M.R. HIP1, a human homologue of S. cerevisiae Sla2p, interacts with membrane-associated huntingtin in the brain. *Nat. Genet.* 16, 44–53, 1997.
74. Kalthoff, C., Alves, J., Urbanke, C., Knorr, R. and Ungewickell, E.J. Unusual structural organization of the endocytic proteins AP180 and epsin 1. *J. Biol. Chem.* 8, 8209–8216, 2002.
75. Karnoub, A.E., Worthylake, D.K., Rossman, K.L., Pruitt, W.M., Campbell, S.L., Sondek, J. and Der, C.J. Molecular basis for Rac1 recognition by guanine nucleotide exchange factors. *Nat. Struct. Biol.* 8, 1037–1041, 2001.
76. Kay, B.K., Yamabhai, M., Wendland, B. and Emr, S.D. Identification of a novel domain shared by putative components of the endocytic and cytoskeletal machinery. *Protein Sci.* 8, 435–438, 1999.
77. Khanna, R., Li, Q. and Stanley, E.F. Fractional recovery' analysis of a presynaptic synaptotagmin 1-anchored endocytic protein complex. *PLoS ONE* 1, e67, 2006.
78. Kim, W.T., Chang, S., Daniell, L., Cremona, O., Di Paolo, G. and De Camilli, P. Delayed reentry of recycling vesicles into the fusion-competent synaptic vesicle pool in synaptojanin 1 knockout mice. *Proc. Natl. Acad. Sci. USA* 99, 17143–17148, 2002.
79. Kirchhausen, T. Clathrin. *Annu. Rev. Biochem.* 69, 699–727, 2000.
80. Koenig, J.H., Saito, K. and Ikeda, K. Reversible control of synaptic transmission in a single gene mutant of *Drosophila melanogaster*. *J. Cell Biol.* 96, 1517–1522, 1983.
81. Koh, T.W., Verstreken, P. and Bellen, H.J. Dap160/intersectin acts as a stabilizing scaffold required for synaptic development and vesicle endocytosis. *Neuron* 43, 193–205, 2004.
82. Kosaka, T. and Ikeda, K. Possible temperature-dependent blockage of synaptic vesicle recycling induced by a single gene mutation in *Drosophila*. *J. Neurobiol.* 14, 207–225, 1983.
83. Krauss, M., Kinuta, M., Wenk, M.R., De Camilli, P., Takei, K. and Haucke, V. ARF6 stimulates clathrin/AP-2 recruitment to synaptic membranes by activating phosphatidylinositol phosphate kinase type Igamma. *J. Cell Biol.* 162, 113–124, 2003.
84. Krauss, M., Kukhtina, V., Pechstein, A. and Haucke, V. Stimulation of phosphatidylinositol kinase type I-mediated phosphatidylinositol (4,5)-bisphosphate synthesis by AP-2mu-cargo complexes. *Proc Natl Acad Sci USA* 103, 11934–11939, 2006.
85. Le Clainche, C., Pauly, B.S., Zhang, C.X., Engqvist-Goldstein, A.E., Cunningham, K. and Drubin, D.G. A Hip1R-cortactin complex negatively regulates actin assembly associated with endocytosis. *EMBO J.* 26, 1199–1210, 2007.
86. Lee, D.W., Wu, X., Eisenberg, E. and Greene, L.E. Recruitment dynamics of GAK and auxilin to clathrin-coated pits during endocytosis. *J. Cell Sci.* 119, 3502–3512, 2006.
87. Legendre-Guillemin, V., Metzler, M., Charbonneau, M., Gan, L., Chopra, V., Philie, J., Hayden, M.R. and McPherson, P.S. HIP1 and HIP12 display differential binding to F-actin, AP2, and clathrin. Identification of a novel interaction with clathrin light chain. *J. Biol. Chem.* 277, 19897–19904, 2002.
88. Legendre-Guillemin, V., Metzler, M., Lemaire, J.F., Philie, J., Gan, L., Hayden, M.R. and McPherson, P.S. Huntingtin interacting protein 1 (HIP1) regulates clathrin assembly through direct binding to the regulatory region of the clathrin light chain. *J. Biol. Chem.* 280, 6101–6108, 2005.

89. Leshchyns'ka, I., Sytnyk, V., Richter, M., Andreyeva, A., Puchkov, D. and Schachner, M. The adhesion molecule CHL1 regulates uncoating of clathrin-coated synaptic vesicles. *Neuron* 52, 1011–1025, 2006.
90. Levivier, E., Goud, B., Souchet, M., Calmels, T.P., Mornon, J.P. and Callebaut, I. uDENN, DENN, and dDENN: indissociable domains in Rab and MAP kinases signaling pathways. *Biochem. Biophys. Res. Commun.* 287, 688–695, 2001.
91. Macia, E., Ehrlich, M., Massol, R., Boucrot, E., Brunner, C. and Kirchhausen, T. Dynasore, a cell-permeable inhibitor of dynamin. *Dev. Cell* 10, 839–850, 2006.
92. Majumdar, A., Ramagiri, S. and Rikhy, R. *Drosophila* homologue of Eps15 is essential for synaptic vesicle recycling. *Exp. Cell Res.* 312, 2288–2298, 2006.
93. Marchler-Bauer, A. and Bryant, S.H. CD-Search: protein domain annotations on the fly. *Nucleic Acids Res.* 32, W327–W331, 2004.
94. Marie, B., Sweeney, S.T., Poskanzer, K.E., Roos, J., Kelly, R.B. and Davis, G.W. Dap160/intersectin scaffolds the periactive zone to achieve high-fidelity endocytosis and normal synaptic growth. *Neuron* 43, 207–219, 2004.
95. Massol, R.H., Boll, W., Griffin, A.M. and Kirchhausen, T. A burst of auxilin recruitment determines the onset of clathrin-coated vesicle uncoating. *Proc. Natl. Acad. Sci. USA* 103, 10265–10270, 2006.
96. Masuda, M., Takeda, S., Sone, M., Ohki, T., Mori, H., Kamioka, Y. and Mochizuki, N. Endophilin BAR domain drives membrane curvature by two newly identified structure-based mechanisms. *EMBO J.* 25, 2889–2897, 2006.
97. McMahon, H.T., Wigge, P. and Smith, C. Clathrin interacts specifically with amphiphysin and is displaced by dynamin. *FEBS Lett.* 413, 319–322, 1997.
98. McMahon, H.T. and Gallop, J.L. Membrane curvature and mechanisms of dynamic cell membrane remodelling. *Nature* 438, 590–596, 2005.
99. McPherson, P.S., Takei, K., Schmid, S.L. and De Camilli, P. p145, a major Grb2-binding protein in brain, is co-localized with dynamin in nerve terminals where it undergoes activity-dependent dephosphorylation. *J. Biol. Chem.* 269, 30132–30139, 1994.
100. McPherson, P.S., Czernik, A.J., Chilcote, T.J., Onofri, F., Benfenati, F., Greengard, P., Schlessinger, J. and De Camilli, P. Interaction of Grb2 via its Src homology 3 domains with synaptic proteins including synapsin I. *Proc. Natl. Acad. Sci. USA* 91, 6486–6490, 1994.
101. McPherson, P.S., Garcia, E.P., Slepnev, V.I., David, C., Zhang, X., Grabs, D., Sossin, W.S., Bauerfeind, R., Nemoto, Y. and De Camilli, P. A presynaptic inositol-5-phosphatase. *Nature* 379, 353–357, 1996.
102. Metzler, M., Legendre-Guillemin, V., Gan, L., Chopra, V., Kwok, A., McPherson, P.S. and Hayden, M.R. HIP1 functions in clathrin-mediated endocytosis through binding to clathrin and adaptor protein 2. *J. Biol. Chem.* 276, 39271–63927, 2001.
103. Michaelson, D.M., Barkai, G. and Barenholz, Y. Asymmetry of lipid organization in cholinergic synaptic vesicle membranes. *Biochem. J.* 211, 155–162, 1983.
104. Micheva, K.D., Kay, B.K. and McPherson, P.S. Synaptojanin forms two separate complexes in the nerve terminal. Interactions with endophilin and amphiphysin. *J. Biol. Chem.* 272, 27239–27245, 1997.
105. Mishra, S.K., Agostinelli, N.R., Brett, T.J., Mizukami, I., Ross, T.S. and Traub, L.M. Clathrin- and AP-2-binding sites in HIP1 uncover a general assembly role for endocytic accessory proteins. *J. Biol. Chem.* 276, 46230–26236, 2001.
106. Mitsunari, T., Nakatsu, F., Shioda, N., Love, P.E., Grinberg, A., Bonifacino, J.S. and Ohno, H.. Clathrin adaptor AP-2 is essential for early embryonal development. *Mol. Cell Biol.* 25: 9318–9323, 2005.
107. Miyoshi, J. and Takai, Y. Dual role of DENN/MADD (Rab3GEP) in neurotransmission and neuroprotection. *Trends Mol. Med.* 10, 476–480, 2004.

108. Modregger, J., Ritter, B., Witter, B., Paulsson, M. and Plomann, M. All three PACSIN isoforms bind to endocytic proteins and inhibit endocytosis. *J. Cell Sci.* 113, 4511–4521, 2000.
109. Montesinos, M.L., Castellano-Munoz, M., Garcia-Junco-Clemente, P. and Fernandez-Chacon, R. Recycling and EH domain proteins at the synapse. *Brain Res. Brain Res. Rev.* 49, 416–428, 2005.
110. Morgan, J.R., Prasad, K., Hao, W., Augustine, G.J. and Lafer, E.M. A conserved clathrin assembly motif essential for synaptic vesicle endocytosis. *J. Neurosci.* 20, 8667–8676, 2000.
111. Morgan, J.R., Prasad, K., Jin, S., Augustine, G.J. and Lafer, E.M. Uncoating of clathrin-coated vesicles in presynaptic terminals: roles for Hsc70 and auxilin. *Neuron* 32, 289–300, 2001.
112. Morgan, J.R., Prasad, K., Jin, S., Augustine, G.J. and Lafer, E.M. Eps15 homology domain-NPF motif interactions regulate clathrin coat assembly during synaptic vesicle recycling. *J. Biol. Chem.* 278, 33583–33592, 2003.
113. Murshid, A., Srivastava, A., Kumar, R. and Presley, J.F. Characterization of the localization and function of NECAP 1 in neurons. *J. Neurochem.* 98, 1746–1762, 2006.
114. Murthy, V.N. and Stevens, C.F. Synaptic vesicles retain their identity through the endocytic cycle. *Nature* 392, 497–501, 1998.
115. Murthy, V.N. and De Camilli, P. Cell biology of the presynaptic terminal. *Annu. Rev. Neurosci.* 26, 701–728, 2003.
116. Nemoto, Y., Wenk, M.R., Watanabe, M., Daniell, L., Murakami, T., Ringstad, N., Yamada, H., Takei, K. and De Camilli, P. Identification and characterization of a synaptojanin 2 splice isoform predominantly expressed in nerve terminals. *J. Biol. Chem.* 276, 41133–41142, 2001.
117. Newmyer, S.L. and Schmid, S.L. Dominant-interfering Hsc70 mutants disrupt multiple stages of the clathrin-coated vesicle cycle *in vivo*. *J. Cell Biol.* 152, 607–620, 2001.
118. Nicholson-Tomishima, K. and Ryan, T.A. Kinetic efficiency of endocytosis at mammalian CNS synapses requires synaptotagmin I. *Proc. Natl. Acad. Sci. USA* 101, 16648–16652, 2004.
119. Nonet, M.L., Holgado, A.M., Brewer, F., Serpe, C.J., Norbeck, B.A., Holleran, J., Wei, L., Hartwieg, E., Jorgensen, E.M., and Alfonso, A. UNC-11, a Caenorhabditis elegans AP180 homologue, regulates the size and protein composition of synaptic vesicles. *Mol. Biol. Cell.* 10, 2343–2360, 1999.
120. Owen, D.J., Vallis, Y., Noble, M.E., Hunter, J.B., Dafforn, T.R., Evans, P.R. and McMahon, H.T. A structural explanation for the binding of multiple ligands by the alpha-adaptin appendage domain. *Cell* 97, 805–815, 1999.
121. Owen, D.J., Vallis, Y., Pearse, B.M., McMahon, H.T. and Evans, P.R. The structure and function of the beta 2-adaptin appendage domain. *EMBO J.* 19, 4216–4227, 2000.
122. Owen, D.J., Collins, B.M. and Evans, P.R. Adaptors for clathrin coats: structure and function. *Annu. Rev. Cell Dev. Biol.* 20, 153–191, 2004.
123. Perera, R.M., Zoncu, R., Lucast, L., De Camilli, P. and Toomre, D. Two synaptojanin 1 isoforms are recruited to clathrin-coated pits at different stages. *Proc. Natl. Acad. Sci. USA* 103,19332–19337, 2006.
124. Peter, B.J., Kent, H.M., Mills, I.G., Vallis, Y., Butler, P.J., Evans, P.R. and McMahon, H.T. BAR domains as sensors of membrane curvature: the amphiphysin BAR structure. *Science* 303, 495–499, 2004.
125. Phillips, A.M., Smith, M., Ramaswami, M. and Kelly, L.E. The products of the Drosophila stoned locus interact with synaptic vesicles via synaptotagmin. *J. Neurosci.* 20, 8254–8261, 2000.

126. Poskanzer, K.E., Marek, K.W., Sweeney, S.T. and Davis, G.W. Synaptotagmin I is necessary for compensatory synaptic vesicle endocytosis *in vivo*. *Nature* 426, 559–563, 2003.
127. Qualmann, B., Roos, J., DiGregorio, P.J. and Kelly, R.B. Syndapin I, a synaptic dynamin-binding protein that associates with the neural Wiskott-Aldrich syndrome protein. *Mol. Biol. Cell* 10, 501–513, 1999.
128. Ramjaun, A.R. and McPherson, P.S. Tissue-specific alternative splicing generates two synaptojanin isoforms with differential membrane binding properties. *J. Biol. Chem.* 271, 24856–24861, 1996.
129. Ramjaun, A.R., Micheva, K.D., Bouchelet, I. and McPherson, P.S. Identification and characterization of a nerve terminal-enriched amphiphysin isoform. *J. Biol. Chem.* 272, 16700–16706, 1997.
130. Ramjaun, A.R. and McPherson, P.S. Multiple amphiphysin II splice variants display differential clathrin binding: identification of two distinct clathrin-binding sites. *J. Neurochem.* 70, 2369–2376, 1998.
131. Ramjaun, A.R., Philie, J., de Heuvel, E. and McPherson, P.S. The N terminus of amphiphysin II mediates dimerization and plasma membrane targeting. *J. Biol. Chem.* 274, 19785–19791, 1999.
132. Ridley, S.H., Ktistakis, N., Davidson, K., Anderson, K.E., Manifava, M., Ellson, C.D., Lipp, P., Bootman, M., Coadwell, J., Nazarian, A., Erdjument-Bromage, H., Tempst, P., Cooper, M.A., Thuring, J.W., Lim, Z.Y., Holmes, A.B., Stephens, L.R. and Hawkins, P.T. FENS-1 and DFCP1 are FYVE domain-containing proteins with distinct functions in the endosomal and Golgi compartments. *J. Cell Sci.* 114, 3991–4000, 2001.
133. Ringstad, N., Nemoto, Y. and De Camilli, P. The SH3p4/Sh3p8/SH3p13 protein family: binding partners for synaptojanin and dynamin via a Grb2-like Src homology 3 domain. *Proc. Natl. Acad. Sci. USA* 94, 8569–8574, 1997.
134. Ritter, B., Philie, J., Girard, M., Tung, E.C., Blondeau, F. and McPherson, P.S. Identification of a family of endocytic proteins that define a new alpha-adaptin ear-binding motif. *EMBO Rep.* 4, 1089–1095, 2003.
135. Ritter, B., Denisov, A.Y., Philie, J., Deprez, C., Tung, E.C., Gehring, K. and McPherson, P.S. Two WXXF-based motifs in NECAPs define the specificity of accessory protein binding to AP-1 and AP-2. *EMBO J.* 23, 3701–3710, 2004.
136. Ritter, B. and McPherson, P.S. *Molecular mechanisms in clathrin-mediated membrane budding. In Topics in Current Genetics*, Vol 10, pp 9–37. Sirkka Keränen and Jussi Jantti (Eds): Regulatory Mechanisms of Intracellular Membrane Transport. Springer-Verlag Berlin Heidelberg, 2004.
137. Ritter, B., Denisov, A. Yu., Philie, J., Allaire, P.D., Legendre-Guillemin, V., Zybergold, P., Gehring, K. and McPherson, P.S. The NECAP PHear domain increases clathrin accessory protein binding potential. *EMBO J.* 26, 4066–4077, 2007.
138. Robinson, F.L. and Dixon, J.E. Myotubularin phosphatases: policing 3-phosphoinositides. *Trends Cell Biol.* 16, 403–412, 2006.
139. Roos, J. and Kelly, R.B. Dap160, a neural-specific Eps15 homology and multiple SH3 domain-containing protein that interacts with Drosophila dynamin. *J. Biol. Chem.* 273, 19108–19119, 1998.
140. Rosenthal, J.A., Chen, H., Slepnev, V.I., Pellegrini, L., Salcini, A.E., Di Fiore, P.P. and De Camilli, P.. The epsins define a family of proteins that interact with components of the clathrin coat and contain a new protein module. *J. Biol. Chem.* 274, 33959–33965, 1999.
141. Roux, A., Uyhazi, K., Frost, A. and De Camilli, P. GTP-dependent twisting of dynamin implicates constriction and tension in membrane fission. *Nature* 441, 528–531, 2006.

142. Salcini, A.E., Hilliard, M.A., Croce, A., Arbucci, S., Luzzi, P., Tacchetti, C., Daniell, L., De Camilli, P., Pelicci, P.G., Di Fiore, P.P. and Bazzicalupo, P. The Eps15 C. elegans homologue EHS-1 is implicated in synaptic vesicle recycling. *Nat. Cell Biol.* 3, 755–760, 2001.
143. Schmid, E.M., Ford, M.G., Burtey, A., Praefcke, G.J., Peak-Chew, S.Y., Mills, I.G., Benmerah, A. and McMahon, H.T. Role of the AP2 beta-appendage hub in recruiting partners for clathrin-coated vesicle assembly. *PLoS Biol.* 4, e262, 2006.
144. Schuske, K.R., Richmond, J.E., Matthies, D.S., Davis, W.S., Runz, S., Rube, D.A., van der Bliek, A.M. and Jorgensen, E.M. Endophilin is required for synaptic vesicle endocytosis by localizing synaptojanin. *Neuron* 40, 749–762, 2003.
145. Sever, S., Muhlberg, A.B. and Schmid, S.L. Impairment of dynamin's GAP domain stimulates receptor-mediated endocytosis. *Nature* 398, 481–486, 1999.
146. Sheetz, M.P. and Singer, S.J. Biological membranes as bilayer couples. A molecular mechanism of drug-erythrocyte interactions. *Proc. Natl. Acad. Sci. USA* 71, 4457–4461, 1974.
147. Shimada, A., Niwa, H., Tsujita, K., Suetsugu, S., Nitta, K., Hanawa-Suetsugu, K., Akasaka, R., Nishino, Y., Toyama, M., Chen, L., Liu, Z.J., Wang, B.C., Yamamoto, M., Terada, T., Miyazawa, A., Tanaka, A., Sugano, S., Shirouzu, M., Nagayama, K., Takenawa, T. and Yokoyama, S. Curved EFC/F-BAR-domain dimers are joined end to end into a filament for membrane invagination in endocytosis. *Cell* 129, 761–772, 2007.
148. Shupliakov, O., Low, P., Grabs, D., Gad, H., Chen, H., David, C., Takei, K., De Camilli, P. and Brodin, L. Synaptic vesicle endocytosis impaired by disruption of dynamin-SH3 domain interactions. *Science* 276, 259–263, 1997.
149. Stahelin, R.V., Long, F., Peter, B.J., Murray, D., De Camilli, P., McMahon, H.T. and Cho, W. Contrasting membrane interaction mechanisms of AP180 N-terminal homology (ANTH) and epsin N-terminal homology (ENTH) domains. *J. Biol. Chem.* 278, 28993–28999, 2003.
150. Stowell, M.H., Marks, B., Wigge, P. and McMahon, H.T. Nucleotide-dependent conformational changes in dynamin: evidence for a mechanochemical molecular spring. *Nat. Cell Biol.* 1, 27–32, 1999.
151. Sun, Y., Carroll, S., Kaksonen, M., Toshima, J.Y. and Drubin, D.G. PtdIns(4,5)P2 turnover is required for multiple stages during clathrin- and actin-dependent endocytic internalization. *J. Cell Biol.* 177, 355–367, 2007.
152. Tabares, L., Ruiz, R., Linares-Clemente, P., Gaffield, M.A., Alvarez de Toledo, G., Fernandez-Chacon, R. and Betz, W.J. Monitoring synaptic function at the neuromuscular junction of a mouse expressing synaptopHluorin. *J. Neurosci.* 27, 5422–5430, 2007.
153. Takamori, S., Holt, M., Stenius, K., Lemke, E.A., Gronborg, M., Riedel, D., Urlaub, H., Schenck, S., Brugger, B., Ringler, P., Muller, S.A., Rammner, B., Grater, F., Hub, J.S., De Groot, B.L., Mieskes, G., Moriyama, Y., Klingauf, J., Grubmuller, H., Heuser, J., Wieland, F. and Jahn, R. Molecular anatomy of a trafficking organelle. *Cell* 17, 831–846, 2006.
154. Takei, K., McPherson, P.S., Schmid, S.L. and De Camilli, P. Tubular membrane invaginations coated by dynamin rings are induced by GTP-gamma S in nerve terminals. *Nature* 374, 186–190, 1995.
155. Takei, K., Slepnev, V.I., Haucke, V. and De Camilli, P. Functional partnership between amphiphysin and dynamin in clathrin-mediated endocytosis. *Nat. Cell Biol.* 1, 33–39, 1999.
156. Tan, T.C., Valova, V.A., Malladi, C.S., Graham, M.E., Berven, L.A., Jupp, O.J., Hansra, G., McClure, S.J., Sarcevic, B., Boadle, R.A., Larsen, M.R., Cousin, M.A. and Robinson, P.J. Cdk5 is essential for synaptic vesicle endocytosis. *Nat. Cell Biol.* 5, 701–710, 2003.

157. Tarricone, C., Xiao, B., Justin, N., Walker, P.A., Rittinger, K., Gamblin, S.J. and Smerdon, S.J. The structural basis of Arfaptin-mediated cross-talk between Rac and Arf signalling pathways. *Nature* 411, 215–219, 2001.
158. Tebar, F., Sorkina, T., Sorkin, A., Ericsson, M. and Kirchhausen, T. Eps15 is a component of clathrin-coated pits and vesicles and is located at the rim of coated pits. *J. Biol. Chem.* 271, 28727–28730, 1996.
159. ter Haar, E., Harrison, S.C. and Kirchhausen, T. Peptide-in-groove interactions link target proteins to the beta-propeller of clathrin. *Proc. Natl. Acad. Sci. USA* 97, 1096–1100, 2000.
160. Traub, L.M., Downs, M.A., Westrich, J.L. and Fremont, D.H. Crystal structure of the alpha appendage of AP-2 reveals a recruitment platform for clathrin-coat assembly. *Proc. Natl. Acad. Sci. USA* 96, 8907–8912, 1999.
161. Traub, L.M. Common principles in clathrin-mediated sorting at the Golgi and the plasma membrane. *Biochim. Biophys. Acta.* 1744, 415–437, 2005.
162. Tsujita, K., Suetsugu, S., Sasaki, N., Furutani, M., Oikawa, T. and Takenawa, T. Coordination between the actin cytoskeleton and membrane deformation by a novel membrane tubulation domain of PCH proteins is involved in endocytosis. *J. Cell Biol.* 172, 269–279, 2006.
163. van der Bliek, A.M. and Meyerowitz, E.M. Dynamin-like protein encoded by the Drosophila shibire gene associated with vesicular traffic. *Nature* 351, 411–414, 1991.
164. Verstreken, P., Koh, T.W., Schulze, K.L., Zhai, R.G., Hiesinger, P.R., Zhou, Y., Mehta, S.Q., Cao, Y., Roos, J. and Bellen, H.J. Synaptojanin is recruited by endophilin to promote synaptic vesicle uncoating. *Neuron* 40, 733–748, 2003.
165. Voglmaier, S.M., Kam, K., Yang, H., Fortin, D.L., Hua, Z., Nicoll, R.A. and Edwards, R.H. Distinct endocytic pathways control the rate and extent of synaptic vesicle protein recycling. *Neuron* 51, 71–84, 2006.
166. Waelter, S., Scherzinger, E., Hasenbank, R., Nordhoff, E., Lurz, R., Goehler, H., Gauss, C., Sathasivam, K., Bates, G.P., Lehrach, H. and Wanker, E.E. The huntingtin interacting protein HIP1 is a clathrin and alpha-adaptin-binding protein involved in receptor-mediated endocytosis. *Hum. Mol. Genet.* 10, 1807–1817, 2001.
167. Wakeham, D.E., Chen, C.Y., Greene, B., Hwang, P.K. and Brodsky, F.M. Clathrin self-assembly involves coordinated weak interactions favorable for cellular regulation. *EMBO J.* 22, 4980–4990, 2003.
168. Walch-Solimena, C., Blasi, J., Edelmann, L., Chapman, E.R., von Mollard, G.F. and Jahn, R. The t-SNAREs syntaxin 1 and SNAP-25 are present on organelles that participate in synaptic vesicle recycling. *J. Cell Biol.* 128, 637–645, 1995.
169. Walther, K., Diril, M.K., Jung, N. and Haucke, V. Functional dissection of the interactions of stonin 2 with the adaptor complex AP-2 and synaptotagmin. *Proc. Natl. Acad. Sci. USA* 101, 964–969, 2004.
170. Wasiak, S., Legendre-Guillemin, V., Puertollano, R., Blondeau, F., Girard, M., de Heuvel, E., Boismenu, D., Bell, A.W., Bonifacino, J.S. and McPherson, P.S. Enthoprotin: a novel clathrin-associated protein identified through subcellular proteomics. *J. Cell Biol.* 158, 855–862, 2002.
171. Weissenhorn, W. Crystal structure of the endophilin-A1 BAR domain. *J. Mol. Biol.* 351, 653–661, 2005.
172. Wienisch, M. and Klingauf, J. Vesicular proteins exocytosed and subsequently retrieved by compensatory endocytosis are nonidentical. *Nat. Neurosci.* 9, 1019–1027, 2006.
173. Wigge, P., Kohler, K., Vallis, Y., Doyle, C.A., Owen, D., Hunt, S.P. and McMahon, H.T. Amphiphysin heterodimers: potential role in clathrin-mediated endocytosis. *Mol. Biol. Cell* 8, 2003–2015, 1997.

174. Willig, K.I., Rizzoli, S.O., Westphal, V., Jahn, R. and Hell, S.W. STED microscopy reveals that synaptotagmin remains clustered after synaptic vesicle exocytosis. *Nature* 13, 935–939, 2006.
175. Wong, W.T., Schumacher, C., Salcini, A.E., Romano, A., Castagnino, P., Pelicci, P.G. and Di Fiore, P.P. A protein-binding domain, EH, identified in the receptor tyrosine kinase substrate Eps15 and conserved in evolution. *Proc. Natl. Acad. Sci. USA* 92, 9530–9534, 1995.
176. Yamabhai, M., Hoffman, N.G., Hardison, N.L., McPherson, P.S., Castagnoli, L., Cesareni, G. and Kay, B.K. Intersectin, a novel adaptor protein with two Eps15 homology and five Src homology 3 domains. *J. Biol. Chem.* 273, 31401–31407, 1998.
177. Yamashita, T., Hige, T. and Takahashi, T. Vesicle endocytosis requires dynamin-dependent GTP hydrolysis at a fast CNS synapse. *Science* 307, 124–127, 2005.
178. Yao, P.J., Zhang, P., Mattson, M.P. and Furukawa, K. Heterogeneity of endocytic proteins: distribution of clathrin adaptor proteins in neurons and glia. *Neuroscience* 121, 25–37, 2003.
179. Yao, P.J., Petralia, R.S., Bushlin, I., Wang, Y. and Furukawa, K. Synaptic distribution of the endocytic accessory proteins AP180 and CALM. *J. Comp. Neurol.* 481, 58–69, 2005.
180. Yao, P.J., Bushlin, I. and Petralia, R.S. Partially overlapping distribution of epsin1 and HIP1 at the synapse: analysis by immunoelectron microscopy. *J. Comp. Neurol.* 494, 368–379, 2006.
181. Zamanian, J.L. and Kelly, R.B. Intersectin 1L guanine nucleotide exchange activity is regulated by adjacent src homology 3 domains that are also involved in endocytosis. *Mol. Biol. Cell* 14, 1624–1637, 2003.
182. Zhang, B., Koh, Y.H., Beckstead, R.B., Budnik, V., Ganetzky, B. and Bellen, H.J. Synaptic vesicle size and number are regulated by a clathrin adaptor protein required for endocytosis. *Neuron* 21, 1465–1475, 1998.
183. Zhang, J.Z., Davletov, B.A., Sudhof, T.C. and Anderson, R.G. Synaptotagmin I is a high affinity receptor for clathrin AP-2: implications for membrane recycling. *Cell* 78, 751–760, 1994.
184. Zhou, S., Sousa, R., Tannery, N.H. and Lafer, E.M. Characterization of a novel synapse-specific protein. II. cDNA cloning and sequence analysis of the F1-20 protein. *J. Neurosci.* 12, 2144–2155, 1992.

Initiation and Regulation of Synaptic Transmission by Presynaptic Calcium Channel Signaling Complexes

Zu-Hang Sheng[1], Amy Lee[2], and William A. Catterall[3]

[1] Synaptic Function Section, National Institute of Neurological Disorders and Stroke, National Institutes of Health, Bethesda, MD 20892-3701, USA, shengz@ninds.nin.gov

[2] Department of Pharmacology, Emory University School of Medicine, Atlanta, GA 30322, USA, alee@pharm.emory.edu

[3] Department of Pharmacology, University of Washington, Seattle, WA 98195-7280, USA, wcatt@u.washington.edu

1 Introduction

Ca^{2+} entry through presynaptic voltage-gated Ca^{2+} (Ca_v) channels initiates release of neurotransmitters. Multiple mechanisms directly or indirectly modulate the function of these presynaptic Ca^{2+} channels and thereby regulate synaptic transmission (66, 17, 7). Ca^{2+} channels that have been characterized biochemically are complexes of a pore-forming α_1 subunit of approximately 250 kDa with auxiliary β, $\alpha2\delta$, and γ subunits (77, 7, 18). Ten α_1 subunits are known in vertebrates, which fall into three structurally and functionally distinct subfamilies (18). The $Ca_V2.1$ and $Ca_V2.2$ channels are the primary Ca^{2+} entry pathways at conventional fast synapses, whereas ribbon synapses in specialized sensory cells have $Ca_V1.3$ or $Ca_V1.4$ channels (17, 7, 51). Neuromodulators affect the ability of Ca_v channels to open, close, or inactivate in response to membrane depolarization, and alter the response of these channels to repetitive stimuli (7). This chapter reviews the structure and function of presynaptic Ca^{2+} channels and the progress toward characterization of the cellular and molecular mechanisms that modulate their activity and regulate synaptic transmission.

2 Calcium Currents in Excitable Cells

Ionic currents conducted by Ca^{2+} in different cell types have diverse physiological roles and pharmacological properties, and an alphabetical nomenclature has evolved for the distinct classes of Ca^{2+} currents (80). L-type Ca^{2+} currents typically require a strong depolarization for activation, have slow voltage-dependent inactivation, and are blocked by the organic L-type Ca^{2+} channel antagonists, including dihydropyridines, phenylalkylamines, and benzothiazepines (79). In muscle and endocrine cells, L-type currents initate contraction and secretion, respectively.

J.W. Hell, M.D. Ehlers (eds.), *Structural and Functional Organization of the Synapse*,
DOI: 10.1007/978-0-387-77232-5_6, © Springer Science+Business Media, LLC 2008

L-type currents are also prominent in cell bodies and dendrites of neurons, where they are important for integration of synaptic inputs and regulation of many intracellular events. In retinal photorecepetors and cochlear hair cells, presynaptic L-type Ca^{2+} currents trigger synaptic transmission at specialized ribbon synapses.

N-type, P/Q-type, and R-type Ca^{2+} currents also require strong depolarization for activation (79). They are relatively unaffected by L-type Ca^{2+} channel antagonists but are blocked by specific polypeptide toxins from snail and spider venoms (47). They are expressed primarily in neurons, where they initiate neurotransmission at most fast synapses and also mediate Ca^{2+} entry into cell bodies and dendrites (17, 7).

T-type Ca^{2+} currents are activated by weak depolarization and are transient due to rapid voltage-dependent inactivation (6, 52). They are resistant to both organic Ca^{2+} channel antagonists and to most snake and spider toxins used to define the L-, N-, P/Q-, and R-type Ca^{2+} currents. T-type Ca^{2+} currents regulate action potential properties and control patterns of repetitive firing, but they have not been shown to have a major role in presynaptic nerve terminals.

3 Calcium Channel Subunit Structure

The Ca^{2+} channels that have been characterized biochemically are composed of four or five distinct subunits (Fig. 1 (77, 7)). The α_1 subunit of 190–250 kDa is the largest subunit, and it incorporates the conduction pore, the voltage sensors and gating apparatus, and most of the known sites of channel regulation by second messengers, drugs, and toxins. The α_1 subunits are composed of about 2000 amino acid residues organized in four homologous domains (I–IV). Each domain of the α1 subunits consists of six transmembrane alpha helices (S1 through S6) and a membrane-associated P loop between S5 and S6. Intensive studies of the structure and function of the related pore-forming subunits of Na^+, Ca^{2+}, and K^+ channels have led to identification of their principal functional domains (8, 9). The S1 through S4 segments serve as the voltage sensor module containing the gating charges in the S4 segments (Fig. 1a, *yellow*), whereas transmembrane segments S5 and S6 in each domain and the P loop between them form the pore module (Fig. 1a, *green*). The three-dimensional structure of a vertebrate voltage-gated K^+ channel, $K_V1.2$, has been determined by x-ray crystallography (45), and a model based on that structure is illustrated in Fig. 1b (91). The narrow external pore is lined by the P loop, which contains a glutamate residue in each domain of Ca_V channels that is required for Ca^{2+} selectivity (27). The inner pore is lined by the S6 segments. The S4 segment in each voltage-sensing module serves as the voltage sensor. Four to eight positively charged amino acids at three-residue intervals move across the membrane in response to depolarization according to a 'sliding helix' model, conducting the gating charge, and opening the S5-S6 pore module by pulling on the connecting S4-S5 segment (91). The voltage-sensing module of one domain interacts primarily with the pore-forming module of the adjacent domain in the clockwise direction (45). It is likely that Ca_V channels have the same functional architecture as the $K_V1.2$ channel.

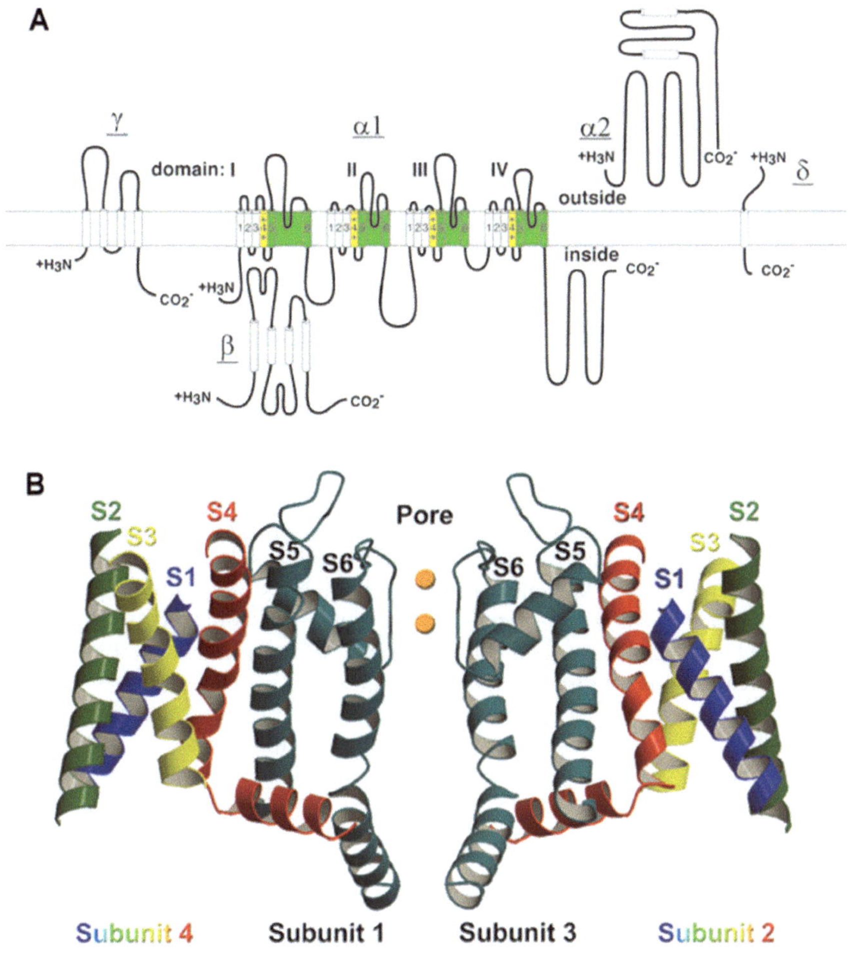

Fig. 1. Subunit structure of Ca_V1 channels. (**a**) The subunit composition and structure of high-voltage-activated Ca^{2+} channels are illustrated. Predicted helices are depicted as cylinders. The lengths of lines correspond approximately to the lengths of the polypeptide segments represented. The voltage-sensing module is illustrated in *yellow* and the pore-forming module in *green.* (**b**) Structural model (91) based on the x-ray crystallographic structure of the $K_V1.2$ channel (45). The voltage-sensing module is illustrated in multiple colors to highlight the different transmembrane segments, and the pore-forming module in *green.* forming module of the adjacent domain in the clockwise direction (Fig. 1b (45)). It is likely that Ca_V channels have the same functional architecture as the $K_V1.2$ channel.

The α_1 subunits are associated with four distinct auxiliary protein subunits (Fig. 1a). The intracellular β subunit is a hydrophilic protein of 50–65 kDa. The transmembrane, disulfide-linked $\alpha_2\delta$ subunit complex is encoded by a single gene, but the resulting pre-polypeptide is posttranslationally cleaved and disulfide-bonded to yield the mature covalently linked α_2 and δ subunits. A γ subunit having four transmembrane segments is a component of skeletal muscle Ca^{2+} channels, and related subunits are expressed in heart and brain. The auxiliary subunits of Ca^{2+} channels have an important influence on their function (28, 16). $Ca_V\beta$ subunits greatly enhance cell surface expression of $\alpha 1$ subunits and shift their kinetics and voltage dependence of activation and inactivation. Most $Ca_V\beta$ subunits accelerate activation and voltage-dependent inactivation and shift the voltage dependence of gating to more negative membrane potentials, but the $Ca_V\beta 2a$ subunit has the opposite effect in most cases. The $\alpha 2\delta$ subunits also enhance cell surface expression of $\alpha 1$ subunits, but have smaller and less consistent effects on the kinetics and voltage dependence of gating (13). The γ subunits do not increase cell surface expression of Ca_V channels, and in some cases reduce it substantially. The functional role of these subunits of Ca^{2+} channels is the least well-defined. Although these four auxiliary subunits modulate the functional properties of the Ca^{2+} channel complex, the pharmacological and physiological diversity of Ca^{2+} channels arises primarily from the existence of multiple α_1 subunits.

Ca^{2+} channel α_1 subunits are encoded by ten distinct genes in mammals (66, 7, 18). The Ca_V1 subfamily of channels ($Ca_V1.1$ to $Ca_V1.4$) conduct L-type Ca^{2+} currents. The Ca_V2 subfamily members ($Ca_V2.1$ to $Ca_V2.3$) conduct P/Q-type, N-type, and R-type Ca^{2+} currents, respectively. The Ca_V3 subfamily members ($Ca_V3.1$ to $Ca_V3.3$) conduct T-type Ca^{2+} currents. The amino acid sequences of these α_1 subunits are more than 70% identical within a subfamily but less than 40% identical among the three subfamilies. Division of Ca^{2+} channels into these three subfamilies is phylogenetically ancient, as single representatives of each are found in the *C. elegans* genome. Evidently, the molecular properties of these three different channel families enable them to participate in distinct physiological pathways.

4 Physiological Roles of Calcium Channels in Neurons

Ca^{2+} channels convert depolarization of the cell surface membrane during the action potential into a transient, local increase in intracellular Ca^{2+}, which serves as a second messenger to initiate many physiological events. In neurons, three of the key physiological events initiated by Ca^{2+} channels are synaptic integration, synaptic transmission, and gene regulation. All three subfamilies of Ca^{2+} channels participate in Ca^{2+} signaling in dendrites and cell bodies and in integration of synaptic signals. In contrast, Ca_V1 and Ca_V2 channels are specialized for synaptic transmission at ribbon synapses and fast conventional synapses, respectively (66, 17, 55, 7, 51). Ca^{2+} entering neurons through $Ca_V2.1$ and $Ca_V2.2$ channels is primarily responsible for initiating synaptic transmission at fast conventional synapses in the peripheral and central nervous systems (17, 47). $Ca_V2.2$ channels, which conduct N-type Ca^{2+} current, are most important at synapses formed by neurons of the peripheral nervous system. In contrast, $Ca_V2.1$ channels, which conduct P/Q-type Ca^{2+} currents, play the major role at most synapses formed by neurons of the mammalian central nervous system. Both $Ca_V2.2$ channels and $Ca_V2.3$ channels also have a significant role in initiating transmitter release at central synapses. In some neurons, including a subset of inhibitory interneurons of the hippocampus (55), $Ca_V2.2$ channels are predominant in neurotransmitter release at central synapses. On the other hand, $Ca_V2.3$ channels seem always to play a secondary role at central synapses compared to $Ca_V2.1$ and/or $Ca_V2.2$ channels. The molecular basis for the prominent role of Ca_V2 channels in neurotransmission is not completely understood, but one factor is their direct interactions with the neurotransmitter release machinery in Ca^{2+} channel signaling complexes.

5 Calcium Channel Signaling Complexes

Ca^{2+} entering neurons through Ca^{2+} channels forms a transient microdomain of high Ca^{2+} concentration in the presynaptic nerve terminal (44, 65, 71). Intracellular proteins that must bind Ca^{2+} to initiate or regulate essential physiological processes must be located near Ca^{2+} channels in order to receive the Ca^{2+} signal. In many cases, this close localization is achieved by direct interaction with the intracellular domains of Ca^{2+} channels, which serve as signal transduction platforms for cytosolic Ca^{2+} signaling (7). Similarly, proteins that regulate Ca^{2+} channels in response to second messenger signaling are also often directly associated with Ca^{2+} channels (7). The signaling complexes of presynaptic Ca^{2+} channels include interactions with the SNARE proteins involved in exocytosis and many Ca^{2+}-binding proteins involved in regulation of channel activity and initiation of Ca^{2+}-dependent responses.

6 Presynaptic Calcium Channels and SNARE Proteins

6.1 Tight Spatial Co-Localization of Presynaptic Ca^{2+} Channels and the Synaptic Vesicle Release Machinery

Synaptic transmission relies on local Ca^{2+} transients generated near the mouth of presynaptic Ca^{2+} channels, and even Ca^{2+} entry through a single Ca^{2+} channel can trigger vesicular release (70). Ca^{2+}-triggered synaptic vesicle exocytosis depends on the assembly of the SNARE complex, the basic fusion machinery, in which the vesicle-associated v-SNARE protein synaptobrevin (VAMP) interacts with two plasma membrane-associated t-SNARE proteins, SNAP-25 and syntaxin-1 (67, 1, 75). Maturation into a release-ready synaptic vesicle requires synaptotagmin, an integral Ca^{2+}-binding protein of the synaptic vesicle membrane that provides the Ca^{2+}-dependent regulation of the fusion machinery. Ca^{2+} influx into the presynaptic terminal triggers the Ca^{2+} sensor of the fusion machinery, upon which the SNARE complex binding progresses from a trans- to a cis-state, resulting in the fusion of apposing membranes and the release of neurotransmitter.

Neurotransmitter release is initiated within 200 μs after the arrival of the action potential and critically depends on Ca^{2+} influx through high-voltage-activated Ca^{2+} channels. Exocytosis of synaptic vesicles requires high Ca^{2+} concentration, with a threshold of 10–50 μM and near maximal activation at 200 μM (59). Given that the intracellular Ca^{2+} concentration falls off steeply as a function of the distance away from the Ca^{2+} channels, the brief rise in the Ca^{2+} concentration to the level necessary for exocytosis is thought to occur only in close proximity to the presynaptic Ca^{2+} channels (44) (Fig. 2). Thus, the tightly localized rise in intracellular Ca^{2+} and the subsequent activation of synaptic vesicle fusion suggests a direct physical link between the Ca^{2+}-channels and the synaptic vesicle release machinery (70, 62) (Fig. 2).

Biochemical and immunocytochemical studies of brain neurons indicate association of syntaxin and synaptotagmin with Ca_v2 channels. Both $Ca_V2.1$ and $Ca_V2.2$ channels localize densely with syntaxin-1 at the presynaptic nerve terminals of many neurons and are crucial components in neuronal excitation-secretion coupling (11, 85, 86). Furthermore, these channels can be isolated as a complex with SNARE proteins from brain homogenates, suggesting that both types of Ca_v2 channels are part of the synaptic vesicle fusion complexes (2, 94, 41). These findings sparked investigation into what molecular determinants allow these channels to bind SNARE proteins and how this interaction regulates synaptic transmission and channel function.

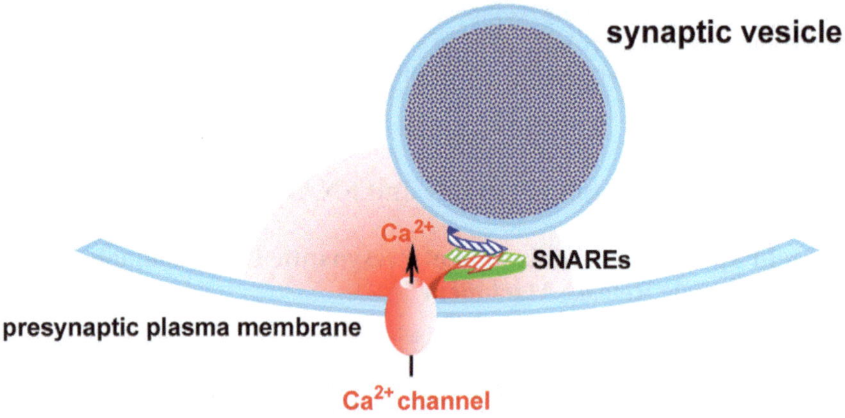

Fig. 2. Physical and functional coupling of presynaptic Ca_V2 channels and synaptic vesicle fusion machinery. A synaptic vesicle docked near an open Ca_V2 channel within the microdomain of high Ca^{2+}. Because the intracellular Ca^{2+} concentration drops steeply as a function of distance away from the Ca^{2+} entry sites, fast exocytosis occurs only in proximity to the Ca^{2+} channels. Binding of the SNARE proteins to the *synprint* site of Ca_V2 channels contributes to the efficiency of neurotransmitter release by localizing docked synaptic vesicles near sites of Ca^{2+} entry and maximizing Ca^{2+} influx through channels with nearby docked vesicles. Following vesicle exocytosis, the interaction between SNARE proteins and Ca_V2 channels is disrupted to allow recovery of vesicle components and re-priming of the synaptic vesicle for release.

6.2 SNARE Protein Interaction Site on Presynaptic Ca_v2 Channels

The large intracellular N- and C-terminal domains and the three loops that connect the four homologous domains of the α1 subunits of Ca_V2 channels are the most attractive candidates for synaptic protein interactions. In vitro biochemical analysis demonstrated that the SNARE proteins syntaxin-1A and SNAP-25, but not VAMP2, specifically interact with the $Ca_v2.2$ channel by binding to an amino acid sequence (residues 718–963) within the intracellular loop between domains II and III ($L_{II\text{-}III}$) of $\alpha_12.2$ subunit (Fig. 3a, (61)), which comprises the ***syn**aptic **pr**otein **int**eraction site, or* *synprint* site. The *synprint* peptide blocks co-immunoprecipitation of native $Ca_V2.2$ channels with syntaxin-1, indicating that this binding site is required for the stable interaction of these two proteins. Interestingly, this interaction is Ca^{2+}-dependent, with maximal binding at 20 μM Ca^{2+} and reduced binding at low, resting Ca^{2+} concentrations and at concentrations above 50 μM (60). Although the physiological

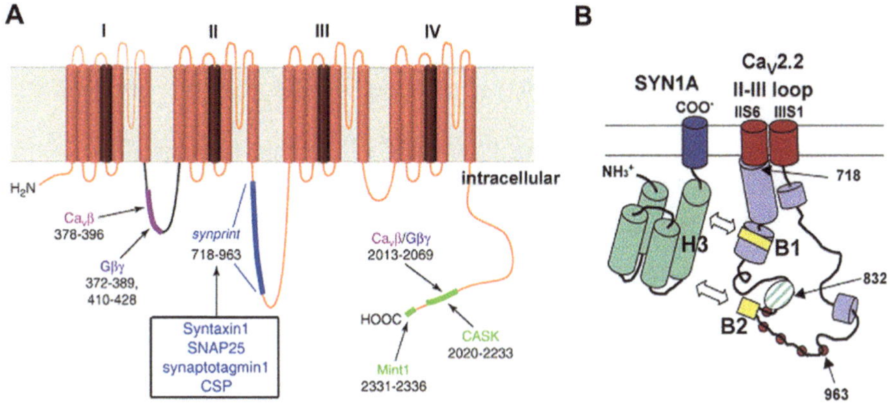

Fig. 3. Structural model of the $Ca_V2.2$ *synprint* site. (**a**) Protein interaction sites of the $Ca_V2.2$ channels. The α1 subunit is composed of four homologous transmembrane domains, which are formed by six membrane-spanning helices and connected by large cytoplasmic loops. $Ca_V\beta$ subunits and G protein βγ subunits bind to adjacent sites in the intracellular loop connecting domains I and II. The domain II-III loop contains the *synprint* site, which binds syntaxin-1, SNAP-25, cysteine string protein (CPS), and synaptotagmin I. $Ca_V\beta$ subunits and G protein βγ subunits bind to secondary sites in the C-terminal domain, and the scaffolding proteins CASK and Mint also bind to specific sequences there. Adapted from (69). (**b**) A structural model of the interaction between the $Ca_V2.2$ *synprint* site and syntaxin-1A. Based on a Chou-Fasman secondary structural prediction algorithm, an ~80 amino acid length bundle with high probability of forming a helical domain is depicted extending from IIS6 into the loop LII-III. This helical segment is interrupted by a short random coil that precedes a second alpha helix containing the first binding site (B1; *yellow box*). Following this helical segment, a coil domain with collagen-like sequence homology (aa 823–848; hatched oval) is adjacent to the second binding site (B2; *yellow box*). Red balls mark the arginine- and histidine-rich clusters, which may serve as membrane surface interaction sites. The synprint B1 and B2 segments provide potential binding contact sites for the carboxyl-terminal coiled-coil (H3) of syntaxin-1 in a closed confirmation (*white arrows*). Adapted from (92).

significance of this biphasic Ca^{2+}–dependent interaction is not yet known, the reduction in binding affinity at high Ca^{2+} levels suggests sequential steps of association and dissociation of Ca_V2 channels with the SNARE complex as a function of Ca^{2+} concentration in order for fusion to proceed (62).

Further structure-function studies defined the molecular determinants of the *synprint* site responsible for binding to syntaxin-1A and SNAP-25 more completely (92). Two different peptide segments, $Ca_v2.2(773–786)$ and $Ca_v2.2(872–901)$, are independently required to bind syntaxin-1A and SNAP-25 to two halves of the *synprint* site, which are separated by a linker sequence that contains a collagen-like strand with repeated proline and glycine residues (Fig. 3b). This structural motif is predicted to form an extended, flexible linker connecting the two SNARE-binding segments. Furthermore, the *synprint* sequence exhibits a high charge density with a charged residue found every 2.5 residues on average and repeated arginine-histidine-rich sequences. This hydrophilic structure of the *synprint* site may hold it near the intracellular surface membrane to facilitate interactions with the hydrophilic surface of the amphipathic alpha helices of SNARE proteins and provide points of contact with phospholipids in the plasma membrane or in the docked synaptic vesicles, thus contributing to structural tethering of synaptic vesicles near presynaptic $Ca_v2.2$ channels.

$Ca_v2.1$ channels, which conduct P/Q-type Ca^{2+} currents, also contain a *synprint* site in their $L_{II\text{-}III}$. Two isoforms of the $\alpha_12.1$ subunit of $Ca_v2.1$ channels, initially named BI and rbA, have different patterns of interactions with the SNARE proteins and exhibit different Ca^{2+}-dependence (58, 34). The BI isoform interacts with both syntaxin-1A and SNAP-25, while the rbA isoform interacts with SNAP-25 but not with syntaxin-1. Given that these Ca^{2+} channels are differentially distributed in synapses in the central nervous system, the observed differences in the interactions between the channels and the SNARE proteins may produce subtle variations in the efficacy of the initiation of transmitter release in response to Ca^{2+} influx through these different channels and may confer specialized regulatory properties that contribute to synaptic modulation.

The *synprint* site may be unique to $Ca_v2.1$ and $Ca_v2.2$ channels, because Ca_v1 channels lack a similar *synprint* motif in the corresponding intracellular $L_{II\text{-}III}$ region (50). However, functional interactions of $Ca_V1.2$ channels with SNARE proteins have been reported, suggesting additional interaction mechanisms that do not involve the *synprint* site (89).

Both $Ca_v2.1$ and $Ca_v2.2$ channels interact with the C-terminal coiled-coil domain or H3 helix of syntaxin-1A (181–288), suggesting that Ca^{2+} channels bind to syntaxin-1 at a C-terminal site near the intracellular surface of the plasma membrane (Fig. 3b (61)). The N-terminal sequence (1–69) of syntaxin-1A also binds the *synprint* site in vitro (31), though it is not clear whether this binding contributes to functional coupling between the SNARE proteins and Ca^{2+} channels at nerve terminals.

6.3 Synaptotagmin Binds to Presynaptic Ca^{2+} Channels

Synaptotagmin has been widely accepted as the Ca^{2+}-sensor for the SNARE-based release machinery (21, 74), with synaptotagmins-1, -2, and -9 playing the major role at fast synapses (90). Synaptotagmin-1 contains two homologous C2 domains, which bind Ca^{2+} to initiate synchronous transmitter release (74). Biochemical studies have revealed a complex of synaptotagmin-1 and both $Ca_V2.1$ and $Ca_V2.2$ channels in brain homogenates (63). The *synprint* site directly binds to the same region in the C2B domain of synaptotagmin-I required for synaptotagmin oligomerization and for AP-2 binding. Two adjacent lysine residues (K326 and K327) within this region are critical for synaptotagmin oligomerization, AP-2, and *synprint* binding (10). Moreover, syntaxin interacts exclusively with either *synprint* or synaptotagmin in a Ca^{2+}-dependent manner, such that at low Ca^{2+} concentrations syntaxin-1 binds *synprint* whereas at higher concentrations (>30 μM) its association with synaptotagmin increases, providing a potential biochemical correlate for the sequential events occurring at the *synprint* site of presynaptic Ca^{2+} channels during synaptic vesicle exocytosis (60). Collectively, these studies suggest a versatile physical link of the neuronal $Ca_V2.1$ and $Ca_V2.2$ channels with the synaptic vesicle fusion machinery, enabling tight structural and functional coupling of the sites for Ca^{2+} entry and neurotransmitter release through sequential protein-protein interactions (62).

6.4 Ca^{2+} Channel-SNARE Interaction Through the Synprint Site is Important for Efficient Release of Neurotransmitters

The Ca^{2+} concentration at the nerve terminal drops steeply as a function of the distance from Ca^{2+} entry sites, requiring close association between presynaptic Ca^{2+} channels and the Ca^{2+} sensor of the release machinery for efficient initiation of exocytosis (72). Maintenance of such a critical intermolecular distance must involve anchoring of the release machinery to the source of the Ca^{2+} influx. Thus, the *synprint*-mediated interaction of Ca^{2+} channels and SNARE proteins provides an attractive model for interaction with the docked synaptic vesicles spatially localized within Ca^{2+} microdomains.

Peptides derived from the *synprint* site competitively inhibit interactions between SNARE proteins and Ca_V2 channels in vitro, and therefore provide a potential test of the physiological significance of these interactions. Injection of *synprint* peptides from $Ca_v2.2$ channels into presynaptic superior cervical ganglion neurons (SCGNs) in culture significantly reduced the excitatory postsynaptic response by competitive uncoupling of the endogenous Ca^{2+} channel-SNARE interaction at nerve terminals (48). Rapid, synchronous synaptic transmission was selectively inhibited following the injection, while late asynchronous release and paired-pulse facilitation were increased. This is consistent with the assumption that synaptic vesicles are shifted from an optimally positioned pool for synchronous release to a more distant pool located too far away for rapid Ca^{2+} sensor response. Evidently, $Ca_V2.2$ channels have two distinct functional roles in neurotransmitter release: conducting Ca^{2+} entry and serving as a signaling platform for SNARE proteins and docked synaptic vesicles to bind and receive the local Ca^{2+} signal and initiate exocytosis (48).

The hypothesis that the inhibition of neurotransmitter release by the *synprint* peptides is due to the detachment of $Ca_V2.2$ channels from release sites was further tested in *Xenopus* nerve-muscle co-cultures (57). Injection of the *synprint* peptides into embryonic *Xenopus* spinal neurons reduced transmitter release by approximately 50% when cells were stimulated in an extracellular solution containing physiological Ca^{2+} concentration. Increasing the external Ca^{2+} concentrations effectively rescued this inhibition, implying that the Ca^{2+} channels are competitively displaced away from docked synaptic vesicles by the injected *synprint* peptides and this effect can be overcome by flooding the presynaptic terminal with Ca^{2+} from the higher concentration in the extracellular pool.

Deletion of the *synprint* site in the $Ca_v2.1$ and $Ca_v2.2$ channels also affects synaptic vesicle release at nerve terminals and exocytosis in secretory cells. Deletion of the *synprint* site on the $Ca_v2.1$ P/Q-type channel results in a substantial reduction in neurotransmitter release in transfected SCG neurons, correlated with reduced localization of the mutant channels to presynaptic terminals (49, 50). Transfer of the *synprint* sequence from $Ca_v2.1$ to $Ca_v1.2$ channels, which conduct L-type Ca^{2+} currents and are ineffective in supporting synaptic transmission, is sufficient to allow these channels to initiate synaptic transmission. A related study addressed this issue using mouse pheochromocytoma (MPC) 9/3L cells, which lack voltage-gated Ca^{2+} channels (25). Compared to cells transiently expressing wild-type $Ca_v2.2$ channels, cells expressing the *synprint* deletion mutant showed a substantial reduction in rate of secretion determined by capacitance measurements, even though Ca^{2+} currents were similar regardless of whether *synprint* is deleted. In addition, in transfected neurons in cell culture, localization of Ca_V2 channels to nerve terminals is substantially reduced when the synprint site is partially or completely deleted (76). These studies further support the concept that the *synprint* site is necessary for efficient presynaptic localization and functional coupling of $Ca_v2.1$ and $Ca_v2.2$ channels to vesicle exocytosis (Figs. 2 and 3).

6.5 Additional Interactions that may Localize the Exocytotic Machinery Near Ca^{2+} Channels

While the studies above indicate that interaction of the *synprint* site of presynaptic Ca^{2+} channels with SNARE proteins enhances the efficiency of neurotransmitter release in vertebrate neurons, several lines of evidence suggest that this interaction is not absolutely required for synaptic vesicle exocytosis. First, inhibition of *synprint* interaction or deletion of the *synprint* site on both $Ca_v2.1$ and $Ca_v2.2$ channels reduces the efficiency of synaptic transmission but does not completely abolish synaptic transmission and vesicle exocytosis (48, 57, 50, 25). Second, although invertebrate Ca_v2 channels effectively initiate synaptic transmission, they lack a *synprint* site (68). In invertebrates, other protein-protein interactions may substitute for interaction with the *synprint* site of Ca^{2+} channels in spatial organization of the exocytosis process (68), or neurotransmitter release may be less dependent on local organization than in vertebrates. Splice variants of the human $Ca_V2.2$ channels lacking the *synprint* region have been identified, but their localization in nerve terminals and their physiological relevance in supporting synaptic transmission are not known (32). These findings raise the question of whether these splice variants couple to the synaptic vesicle fusion machinery or have other physiological roles (69).

The SNARE proteins are not selectively localized at nerve terminals. Therefore, while interaction with SNARE proteins may be required for efficient presynaptic localization, it seems unlikely that interactions with SNARE proteins can serve as a primary targeting mechanism for Ca_V2 channels. Consistent with this idea, the active zone scaffolding proteins Mint and CASK also play critical roles in the presynaptic targeting of $Ca_V2.2$ channels via direct interactions with the C-terminal tail of the channels (46). Moreover, in invertebrate Ca^{2+} channels that lack a synprint site, Mint and CASK are required for targeting of presynaptic Ca^{2+} channels to the synapse (68, 69). Interactions with Mint and CASK, and perhaps with other unidentified targeting proteins, may represent the primary targeting signals required for the localization of the presynaptic Ca^{2+} channels in the nerve terminals.

7 Regulation of Presynaptic Ca^{2+} Channels by SNARE Proteins

7.1 Biphasic Regulation of Calcium Channel Inactivation by SNARE Proteins

In addition to linking presynaptic Ca^{2+} channels to the vesicle release machinery, syntaxin-1A and SNAP-25 also regulate channel function. This retrograde regulation serves as a negative feedback mechanism for the modulation of synaptic transmission. Co-expression of syntaxin-1A and/or SNAP-25 with $Ca_v2.1$ or $Ca_v2.2$ channels reduces the availability of the channels to open and shifts the voltage dependence of steady-state inactivation towards more negative membrane potentials (3, 87, 95). These inhibitory effects of syntaxin on $Ca_V2.2$ channels can be reversed by co-expressing SNAP-25 (87, 31), and the inhibitory effects of SNAP-25 on $Ca_V2.1$ channels can be relieved by co-expressing synaptotagmin-1 (88, 95). These physiological data correlate with biochemical observations that syntaxin interacts exclusively with either *synprint* or synaptotagmin in a Ca^{2+}-dependent manner (63). This relief of inhibition of Ca^{2+} channels by formation of a complete synaptotagmin/SNARE complex favors Ca^{2+} influx through Ca_V2 channels near docked synaptic vesicles that are ready for release. Thus, this regulatory mechanism can provide selective Ca^{2+} entry through Ca_V2 channels with associated docked vesicles, whereas channels associated with only one t-SNARE protein would remain relatively inactive. This regulatory mechanism also allows the modulatory interaction of the channels to be switched on and off at appropriate stages of formation of nearby complexes of the SNARE proteins and synaptic vesicles. The inhibitory role of syntaxin-1 on channel opening is also abolished in the presence of nSec-1, a protein that blocks the assembly of the SNARE complex by locking syntaxin-1 in a closed conformation (31). Altogether, these findings support a biphasic model for the SNARE-mediated modulation of presynaptic Ca^{2+} channels based on the availability of docked synaptic vesicles nearby that are ready for rapid release.

7.2 Separate Anchoring and Regulatory Sites on Syntaxin-1

By combining site-directed mutagenesis with functional expression in *Xenopus* oocytes, Bezprozvanny et al. (4) further addressed whether the anchoring and modulatory interactions between syntaxin-1 and Ca^{2+} channels are mediated by the same molecular determinants. Their studies showed that the transmembrane region and a short segment within the H3 helical cytoplasmic domain of syntaxin-1A are critical for channel modulation but not for binding to the *synprint* site of the N-type Ca^{2+} channels. Given that this short segment is buried within the core of the four-helix structure of the SNARE complex, the modulatory action of syntaxin-1A can be enabled only after disassembly of the SNARE complex following synaptic vesicle exocytosis. The *synprint* site binds to the whole H3 region of syntaxin-1A (4), in agreement with the previous biochemical findings (61, 60). Deletion of the *synprint* site weakened the modulation of the channels by syntaxin-1A, but did not abolish it, arguing that the *synprint* sequence acts as an important anchor in facilitating channel modulation but is not absolutely essential for the modulatory action. Identification of partially separate anchoring and modulation sites on Ca_V2 channels provides a means to test the relative roles of these *synprint* functions in synaptic transmission.

7.3 Functional Roles of Vesicle Tethering and Ca_V2 Channel Regulation Via the Synprint Site in Synaptic Transmission in vivo

At first glance, it seems that interactions of Ca_V2 channels with SNARE proteins have two opposing effects: tethering synaptic vesicles near the point of Ca^{2+} entry would increase synaptic transmission whereas enhancing Ca_V2 channel inactivation would reduce synaptic transmission. These effects were dissected by use of competing *synprint* peptides and mutant syntaxin in *Xenopus* neuromuscular junctions in vivo (33). Injection of competing *synprint* peptides into developing neuromuscular junctions reduced the basal efficiency of synaptic transmission, as reflected in increased paired-pulse facilitation and reduced quantal content of synaptic transmission. Evidently, the effect of the *synprint* peptide to reduce linkage of docked synaptic vesicles to Ca_V2 channels is predominant, because its effect to relieve inhibition of Ca^{2+} channels by individual plasma membrane SNARE proteins would primarily affect channels with no docked vesicles that do not effectively initiate transmitter release. In contrast, over-expression of a syntaxin mutant that is unable to regulate $Ca_V2.2$ channels but still binds to them increased the efficiency of synaptic transmission, as reflected in reduced paired-pulse facilitation and increased quantal content (33). In this case, the syntaxin mutant relieves enhanced inactivation of $Ca_V2.2$ channels caused by endogenous syntaxin, thereby increasing Ca^{2+} entry and synaptic transmission, but does not alter linkage of docked synaptic vesicles to $Ca_V2.2$ channels. These results demonstrate a bidirectional regulation of synaptic transmission in vivo by interactions of SNARE proteins with $Ca_V2.2$ channels.

7.4 Crosstalk Between Syntaxin and G-Protein Regulation of $Ca_v2.2$ Channels

Further regulation of the $Ca_v2.2$ channels involves the interplay between the *synprint* site interactions and the second messenger modulation of calcium channels by G-proteins. Stanley and Mirotznik (73) revealed that syntaxin-1A is required for G-protein inhibition of presynaptic calcium channels in intact nerve terminals. Botulinum toxin, which cleaves syntaxin-1A, prevents G-protein modulation of presynaptic Ca^{2+} channels in chick calyx synapses, suggesting a crosstalk mechanism between syntaxin-1A and the G-protein effects on the N-type Ca^{2+} channels. Further studies by Jarvis et al. (30) showed that the *synprint* site and Gβ subunit bind to distinct portions of syntaxin-1A and that expression of syntaxin-1A appears to be a prerequisite for tonic G-protein inhibition of the N-type Ca^{2+} channels. These results support the hypothesis that syntaxin-1A acts as a chaperone by tethering Gβ to $Ca_V2.2$ channels in synapses, thus facilitating G-protein regulation.

7.5 Additional SNARE-Binding Proteins add Complexity to Regulation of Ca_V2 Channels

The SNARE complex together with synaptotagmin is thought to be the minimal essential machinery required for Ca^{2+}-dependent synaptic vesicle fusion. However, a multitude of presynaptic proteins have been identified that directly participate in exocytosis or regulate the synaptic vesicle cycle via interactions with the SNARE proteins. These interactions have a potential impact on the modulation of the SNARE-Ca_V2 channel complex. RIM (Rab-Interacting Molecule), an active zone protein that is required for synaptic transmission and is implicated in short- and long-term synaptic plasticity, has also been reported to interact with the *synprint* region in vitro (12). In addition, RIM interacts with $Ca_V\beta$ subunits and shifts the voltage dependence of inactivation to more positive membrane potentials, increasing Ca^{2+} channel activity (35). In the neuroendocrine cell line PC12, interaction of RIM with $Ca_V2.2$ channels increases docking of neurotransmitter-containing vesicles (35). Cysteine string protein (CSP) was identified as a synaptic vesicle-associated protein with the ability to modify Ca^{2+} influx in *Drosophila* and *Torpedo* (82). Subsequent studies demonstrated that CSP interacts with the *synprint* site of Ca_V2 channels and that this interaction results in a robust tonic inhibition of channel activity by G-protein βγ-subunits (42). Regulation of presynaptic Ca^{2+} channel function and vesicle docking by RIM and CSP provides additional pathways for control of Ca^{2+} channel activity in response to the state of nearby docked synaptic vesicles.

8 Modulation of the SNARE-*Synprint* Interaction by Protein Phosphorylation

Second messenger-activated regulation of neurotransmitter release via modulation of protein interactions within the exocytotic apparatus has a potentially important role in synaptic plasticity. Several protein kinases, including PKA, PKC, PKG, and

CaMKII, are expressed in presynaptic terminals and phosphorylate Ca^{2+} channels and SNARE proteins. Phosphorylation of the *synprint* peptide by PKC and CaMKII in vitro strongly inhibits its binding to recombinant syntaxin-1A and SNAP-25 and prevents its association with the native SNARE complex isolated from rat brain homogenates (93). Moreover, PKC phosphorylation blocks the negative shift of steady-state inactivation of $Ca_v2.2$ channels, possibly by reducing binding of syntaxin-1 to the *synprint* site (31). A more detailed analysis of these phosphorylation sites reveals a bipartite mode of regulation of SNARE protein binding, in which the two separate domains of the *synprint* site that each bind syntaxin-1 and SNAP-25 in vitro are regulated by PKC phosphorylation at serines 774 and 898 and by CaMKII phosphorylation at serines 784 and 896 (92) (Fig. 3b). Each phosphorylation site can control syntaxin-1 and SNAP-25 binding to half of the *synprint* site. However, phosphorylation of the intact $Ca_v2.2$ channels before exposure to syntaxin-1A reduces the binding affinity to syntaxin-1A, but does not completely prevent binding (92). These observations are consistent with functional studies, in which PKC activation in transfected tsA-201 cells does not completely dissociate syntaxin-1A from the channels but does completely reverse the negative shift of the voltage dependence of inactivation caused by syntaxin-1A (92). These studies suggest that phosphorylation of the *synprint* site by PKC or CaMKII may serve as a biochemical switch for the SNARE-*synprint* interaction and regulation of channel function.

9 Ca^{2+}-Dependent Regulation of Presynaptic Ca^{2+} Channels

Ca^{2+} entering presynaptic nerve terminals through Ca^{2+} channels forms a local microdomain of high concentration near the intracellular mouth of the channel (44, 65, 71). In order to respond to this local Ca^{2+} signal, Ca^{2+}-dependent signaling proteins must be located near the Ca^{2+} channel in order to sense this Ca^{2+} transient. Therefore, many proteins that respond to local Ca^{2+} transients in presynaptic terminals in order to initiate or regulate neurotransmission bind directly to the intracellular domains of Ca_V2 channels.

9.1 Ca^{2+}-dependent Regulation of Ca_V1 and Ca_V2 Channels

Although all Ca_v channels readily conduct both Ca^{2+} and Ba^{2+} ions, L-type and P/Q-type Ca^{2+} currents inactivate strikingly faster than Ba^{2+} currents (7). This process, Ca^{2+}-dependent inactivation, prevents Ca^{2+} overload and refines activity-dependent Ca^{2+} signals in excitable cells. Ca^{2+}-dependent inactivation of L-type Ca^{2+} currents conducted by $Ca_V1.2$ channels in cardiac myocytes and other cell types is observed in the presence of high concentrations of intracellular EGTA in the recording pipette, indicating that the inactivation process binds entering Ca^{2+} rapidly and with high affinity, before Ca^{2+} ions are chelated by EGTA (29).

Ca^{2+}-dependent inactivation was long considered specific for L-type Ca^{2+} currents, in part because most whole-cell patch clamp recordings employ high concentrations of EGTA or BAPTA (5–10 mM) in the intracellular recording solution, which are permissive for Ca^{2+}-dependent inactivation of L-type currents. However, with EGTA reduced to 0.5 mM, a level of buffering more similar to intracellular conditions, Ca^{2+}-dependent inactivation for $Ca_v2.1$ channels transfected in noneuronal cells is robust (39, 37) (Fig. 4a, b), similar to presynaptic P/Q-type currents at the calyx of Held, a giant synapse in the brainstem (20). During trains of depolarizations, P/Q-type Ca^{2+} currents increase in size during the first pulses due to facilitation and then inactivate faster during repeated pulses than Ba^{2+} currents (37) (Fig. 4c). Inactivation, but not facilitation, is prevented by a high intracellular concentration of EGTA (10 mM) (37). These results indicate that the facilitation process has higher affinity and/or more rapid binding of Ca^{2+} than the inactivation process. This dual feedback regulation may permit activity-dependent sharpening of presynaptic Ca^{2+} signals underlying synaptic plasticity, providing a boost of synaptic efficiency during short trains of impulses and then reduction of synaptic efficacy during long trains. With low concentrations of EGTA (0.5 mM), Ca^{2+}-dependent inactivation is also observed for $Ca_v2.2$ and $Ca_v2.3$ channels (43). Therefore, all three Ca_v2 family channels show Ca^{2+}-dependent feedback regulation that is sensitive to the Ca^{2+} buffering conditions of the cell, unlike Ca_V1 channels.

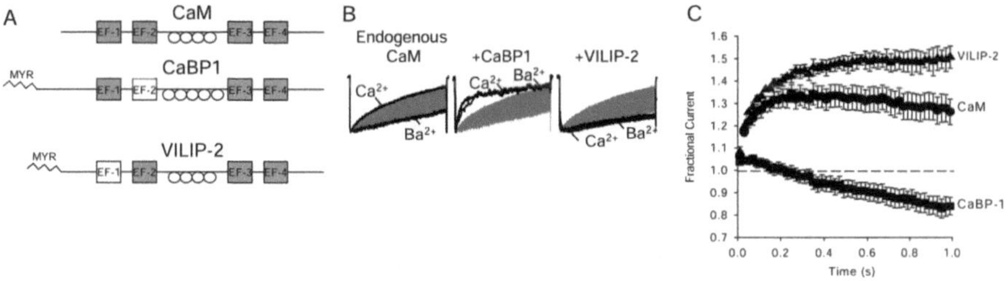

Fig. 4. Ca^{2+}-dependent regulation of $Ca_V2.1$ channels. (**a**) Structures of calmodulin (CaM), CaBP1, and VILIP-2. EF-hands: active, gray; inactive, white. MYR, myristoylation. Circles, alpha helix. (**b**) Ca^{2+}-dependent inactivation of transfected $Ca_V2.1$ channels. Ca^{2+} and Ba^{2+} currents were measured in whole-cell voltage clamp for $Ca_V2.1$ channels containing $\alpha2\delta$ and β_{2a} subunits with an intracellular Ca^{2+} buffer of 0.5 mM EGTA. Recordings are compared for cells expressing only endogenous calmodulin, transfected CaBP-1, or VILIP-2. (**c**) Facilitation and inactivation of transfected $Ca_V2.1$ shaded area represents Ca^{2+}-dependent inactivation channels during trains of stimuli. Ca^{2+} and Ba^{2+} currents were measured in whole-cell voltage clamp for $Ca_V2.1$ channels containing $\alpha2\delta$ and β_{2a} subunits with an intracellular Ca^{2+} buffer of 0.5 mM EGTA. Cells were stimulated with 5 ms depolarizations to +10 mV at 100 Hz. The normalized mean peak Ca^{2+} or Ba^{2+} current during each stimulation is plotted. Recordings are compared for cells expressing only endogenous calmodulin, transfected CaBP1, or transfected VILIP-2.

9.2 Regulation of Ca_v1 Channels by Calmodulin

Ca^{2+}-feedback for Ca_v1 channels involves the ubiquitous Ca^{2+}-binding protein calmodulin (100, 53, 56, 98). For $Ca_v1.2$ channels, calmodulin binds to a consensus site (IQ-domain) in the C-terminal domain of $\alpha_1 1.2$ ((99). Ca^{2+} ions permeating the channel bind EF-hand motifs in the N- and C-terminal lobes of calmodulin, causing conformational changes that favor inactivation. Ba^{2+} ions bind poorly to calmodulin, and so cannot support rapid inactivation of Ba^{2+} currents. Calmodulin interactions with the $\alpha_1 1.2$ IQ-domain are functionally complex: mutation of the initial isoleucine to glutamate abolishes Ca^{2+}-dependent inactivation, but substitution of alanine for this residue reveals Ca^{2+}-dependent facilitation, which is normally masked by intense Ca^{2+}-dependent inactivation for $Ca_v1.2$ channels expressed in recombinant systems (99, 54). Structural and functional studies demonstrate positive and negative Ca^{2+}-feedback dependent on the N- and C-terminal lobes of calmodulin, respectively, which make contacts with different residues in the IQ-domain. Among different Ca_v1 channels, the IQ-domain and other sequences (Pre-IQ and EF-hand) required for Ca^{2+}-dependent inactivation are nearly identical, illustrating the essential nature of these determinants for Ca_v1 channel function.

9.3 Ca^{2+}-Dependent Facilitation and Inactivation of $Ca_V2.1$ Channels

Both facilitation and inactivation of $Ca_V2.1$ channels are dependent on calmodulin (39, 37, 14). In the C-terminal domain of the $\alpha_1 2.1$ subunit, calmodulin interacts with a modified IQ-like domain, which begins with the sequence isoleucine-methionine (IM) rather than isoleucine-glutamine (IQ), and with a nearby downstream second site, the calmodulin binding domain (CBD), both of which are involved in Ca^{2+}-dependent feedback regulation ((14, 40); Fig. 5). Ca^{2+}-dependent facilitation is impaired by mutations in the IQ-like domain of $\alpha_1 2.1$ or by calmodulin mutants unable to bind Ca^{2+} at the C-terminal EF-hands. In contrast, Ca^{2+}-dependent inactivation is inhibited by deletion of the CBD of $\alpha_1 2.1$ or by mutations of the Ca^{2+}-binding sites in the N-terminal lobe of calmodulin (14, 40). Since the C-terminal lobe of calmodulin binds Ca^{2+} with higher affinity than the N-terminal lobe, the dependence of Ca^{2+}-dependent facilitation of $Ca_V2.1$ channels on the C-terminal of calmodulin may be responsible for its ability to respond to local increases in Ca^{2+} in the presence of a high concentration of EGTA. Similarly, the increased Ca^{2+}-buffer sensitivity of Ca^{2+}-dependent inactivation for $Ca_v2.1$ channels compared to $Ca_v1.2$ channels may reflect Ca^{2+} binding to the lower affinity N-terminal lobe for inactivation of $Ca_V2.1$ channels versus the higher affinity C-terminal lobe for $Ca_V2.1$ channels.

The two lobes of calmodulin interact differentially with the two calmodulin-binding subsites in the C-terminal domain of $Ca_V2.1$ channels (40). Mutations of the IQ-like domain primarily impair facilitation, indicating that they interact primarily with the C-terminal lobe of calmodulin (14, 40). In contrast, mutations of the CBD primarily impair Ca^{2+}-dependent inactivation (40), suggesting that they interact primarily with the lower affinity N-terminal lobe of calmodulin. These results lead to a model in which rapid, high-affinity binding of Ca^{2+} to the C-terminal lobe of calmodulin and interaction with the IQ-like motif of $Ca_V2.1$ channels causes facilitation, whereas subsequent slower and/or lower affinity binding of Ca^{2+} to the N-terminal lobe of calmodulin and interaction with the CBD of $Ca_V2.1$ channels causes inactivation (Fig. 5).

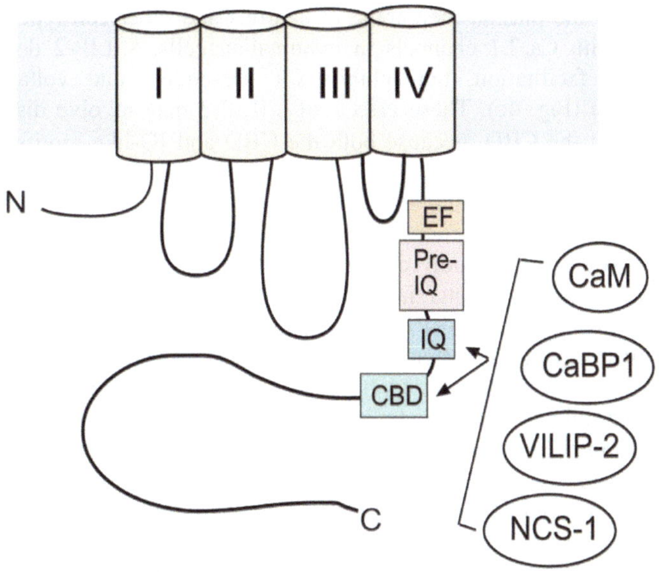

Fig. 5. Binding sites for Ca^{2+}-binding proteins mediating Ca^{2+}-dependent modulation of $Ca_v2.1$ channels. Each homologous domain of the $Ca_v2.1$ channels is illustrated as a cylinder. Binding sites in the C-terminus of the $\alpha_12.1$ subunit for calmodulin (CaM), CaBP1, and VILIP-2, and potentially NCS-1 are indicated.

10 Regulation of Presynaptic Ca_V Channels by Calmodulin-Like Ca^{2+}-binding Proteins

Calmodulin is the primordial member of a large family of EF-hand Ca^{2+} binding proteins, some of which are expressed specifically in neurons (nCaBPs; (5, 23)). Like calmodulin, these CaBPs possess four EF-hand Ca^{2+} binding sites organized in two lobes connected by an α-helix. At least one of the N-terminal EF-hands of the nCaBPs is nonfunctional in Ca^{2+} binding due to changes in amino acid sequence, and some of the nCaBPs are myristolyated at their N-terminus (5). nCaBPs are similar enough to displace calmodulin from shared binding sites in the α_1 subunits of Ca_V channels, but different enough to confer distinct forms of regulation.

10.1 Regulation of $Ca_v2.1$ Channels by CaBP

CaBP1 is a member of a subfamily of nCaBPs highly expressed in the brain and retina (24) and is colocalized with presynaptic $Ca_v2.1$ channels in some synapses (38). Like calmodulin, CaBP1 binds to the CBD of $\alpha_12.1$, but its binding is Ca^{2+}-independent (38). It causes rapid inactivation that is independent of Ca^{2+}, and it does not support Ca^{2+}-dependent facilitation (38) (Fig. 4c). A second nCaBP that modulates $Ca_v2.1$ channels is visinin-like protein-2 (VILIP-2), which is highly expressed in the neocortex and hippocampus and undergoes Ca^{2+}-dependent association with the plasma membrane in neurons and other cell-types (36). When cotransfected with $Ca_v2.1$ channels in mammalian cells, VILIP-2 does not affect Ca^{2+}-dependent facilitation, but inhibits Ca^{2+}-dependent and voltage-dependent inactivation (36) (Fig. 4c). These effects of VILIP-2 may involve displacement of calmodulin from the CBD, because both the CBD and IQ-like motifs of $\alpha_12.1$ are required for binding of VILIP-2. Thus, CaBP-1 and VILIP-2 bind to the same site as calmodulin but have differential effects on $Ca_V2.1$ channel activity. In a presynaptic terminal, these differential effects on facilitation and inactivation of the P/Q-type Ca^{2+} current would substantially change the encoding properties of the synapse in response to trains of action potentials (97).

How can VILIP-2 and CaBP1 have such opposing effects on $Ca_v2.1$ function? The mechanism for these effects is not yet clear, but both proteins must be myristolyated on the N-termini to have their distinctive regulatory effects (19). Because calmodulin is not myristoylated at its N-terminal and does not have an inactive N-terminal EF-hand, it is tempting to think that differential interactions of the inactive EF-hands in the N-terminal domains of these Ca^{2+}-binding proteins with the IQ-like domain and CBD of $Ca_V2.1$ channels are responsible for their differences in action. The divergent actions of nCaBPs on $Ca_v2.1$ channels may fine-tune the function and regulatory properties of presynaptic P/Q-type Ca^{2+} currents, allowing a greater range of input-output relationships and short-term plasticity at different synapses.

$Ca_v2.1$ channels also interact with other CaBPs, including neuronal Ca^{2+} sensor-1 (NCS-1), which has inhibitory effects on the amplitude of P/Q-type Ca^{2+} currents in adrenal chromaffin cells that are mediated by a tyrosine kinase signaling pathway (83, 5, 84). However, injection of purified NCS-1 protein into presynaptic nerve terminals at the calyx of Held synapse promotes facilitation of P/Q-type currents, and activity-dependent facilitation of P/Q-type currents at this synapse can be prevented by injection of NCS-1-inhibitor peptides (81). Similarly, NCS-1 can enhance facilitation of synaptic transmission in cultured hippocampal neurons (64). Therefore, the effects of NCS-1 on $Ca_v2.1$ channels may depend on the cellular context in which they are co-expressed, but in synapses it appears to act as an enhancer of Ca^{2+} channel activity and synaptic transmission.

10.2 Regulation of Ca_v1 Channels by CaBP

The Ca_v1 family of Ca^{2+} channels are also differentially regulated by CaBPs (96). Unlike its inhibitory effects on $Ca_v2.1$ channels, CaBP1 prolongs $Ca_v1.2$ Ca^{2+} currents and completely abolishes Ca^{2+}/calmodulin-dependent inactivation. The opposing actions of calmodulin and CaBP1 on $Ca_v1.2$ depend on differences in how the two interact with the channel (96). CaBP1 binds to the IQ-domain as well as to a site in the N-terminal domain of the $\alpha_11.2$ subunit. Deletion of the N-terminal site prevents the effects of CaBP1 but not calmodulin on $Ca_v1.2$ inactivation. In contrast, IQ-EE substitutions in the IQ-domain prevent calmodulin-dependent inactivation, but partially spare the effects of CaBP1. Interestingly, a CaBP1 variant, caldendrin, causes more modest suppression of $Ca_v1.2$ Ca^{2+}-dependent inactivation than CaBP1 through interactions solely with the IQ-domain and not the N-terminal site in $\alpha_11.2$

(78). CaBP1 and caldendrin associate and colocalize with $Ca_v1.2$ in somatodendritic domains of neurons, and so may help fine-tune and boost postsynaptic Ca^{2+} signals conducted by $Ca_v2.1$, which regulate gene transcription and neuronal excitability.

$Ca_V1.4$ channels conduct L-type Ca^{2+} currents that initiate transmitter release at specialized ribbon synapses in sensory cells in the retinal photoreceptors and bipolar cells (26). $Ca_V1.4$ channels interact with the calmodulin-like Ca^{2+} binding protein CaBP4, which is essential for normal formation and function of the photoreceptor synapse. Interaction with CaBP-4 increases $Ca_V1.4$ channel activity by negatively shifting its voltage dependence of activation. This regulatory effect of CaBP4 may regulate synaptic transmission at the ribbon synapses of photoreceptor cells (22).

11 Regulation of Ca_V2 Channels and Synaptic Plasticity

Short-term synaptic plasticity of neurotransmitter release from presynaptic terminals shapes the response of postsynaptic neurons to bursts of impulses and is crucial for fine-grained encoding of information in the nervous system (97). This form of synaptic plasticity is thought to be mediated by residual Ca^{2+} in the cytosol of the presynaptic terminal remaining from previous stimuli, but the molecular mechanisms responsible for this form of synaptic plasticity are unknown. During trains of stimuli at many synapses, facilitation of neurotransmitter release increases the postsynaptic response, and subsequent depression of neurotransmitter release reduces it. Activity-dependent regulation of presynaptic calcium channels is predicted to have profound effects on short-term synaptic plasticity because the amount of neurotransmitter released at a synapse is steeply dependent on calcium entry, approximately proportional to $n^{3.5}$ of extracellular Ca^{2+} (15). In order to critically test the role of activity-dependent regulation of presynaptic calcium channels via calmodulin and nCaBPs in short-term synaptic plasticity, it is necessary to compare synaptic transmission initiated by wild-type $Ca_V2.1$ channels and by mutant channels with impaired activity-dependent regulation. This is a challenging experiment because endogenous Ca^{2+} channels are present in high concentration in presynaptic active zones, and replacement of them with exogenously expressed mutant Ca^{2+} channels is difficult. Cultured superior cervical ganglion neurons can be transfected successfully with cDNA encoding $Ca_V2.1$ channels by microinjection, and these transfected neurons then express functional $Ca_V2.1$ channels in their cell bodies and synapses (49). Whole-cell voltage clamp recordings of transfected $Ca_V2.1$ channels at the cell body show that they undergo Ca^{2+}-dependent facilitation like $Ca_V2.1$ channels expressed in non-neuronal cells (50a). The contribution of these transfected $Ca_V2.1$ channels to initiation of synaptic transmission can then be determined by blocking the endogenous N-type calcium current specifically with ω-conotoxin GVIA. Excitatory postsynaptic potentials (EPSPs) recorded in the postsynaptic neurons in response to action potentials elicited in the presynaptic neuron are approximately 30–40% of the size of those initiated by endogenous N-type Ca^{2+} currents (49). In contrast, transfected $Ca_V1.2$ channels and $Ca_V2.3$ channels are much less effective (49). Recent studies using this model synaptic system show that activity-dependent facilitation and inactivation of transfected $Ca_V2.1$ channels mediated by nCaBPs induces short-term synaptic plasticity in the cultured superior cervical ganglion neurons, including paired-pulse facilitation and depression as well as synaptic facilitation, augmentation, post-tetanic potentiation, and depression during trains of stimuli (50a). Presynaptic expression of the pore-forming $\alpha_1$2.1 subunit containing a mutation in the IQ-like motif in the C-terminal domain that blocks Ca^{2+}/calmodulin-dependent facilitation of P/Q-type calcium currents prevented paired-pulse facilitation and markedly reduced synaptic facilitation and augmentation during trains of stimuli (50a). Deletion of the nearby calmodulin-binding domain (CBD), which reduces Ca^{2+}-dependent inactivation in heterologous expression systems, prevented paired-pulse depression and also substantially reduced depression of EPSPs during trains of action potentials (50a). These results establish a new molecular mechanism for the action for residual Ca^{2+} in presynaptic terminals, long known to be important in short-term synaptic plasticity. Residual Ca^{2+} induces synaptic facilitation and augmentation through Ca^{2+}/calmodulin-dependent facilitation of the activity of presynaptic Ca_V2.1 channels. Furthermore, Ca^{2+}/calmodulin-dependent inactivation of presynaptic calcium channels, not vesicle depletion, is the primary cause of the rapid phase of synaptic depression at this model synapse. These results point to activity-dependent dual regulation of presynaptic Ca^{2+} channels by calmodulin and nCaBP proteins as a primary determinant of short-term synaptic plasticity and information-processing in the nervous system.

12 Conclusion

Signaling complexes formed by presynaptic Ca_V channels have three crucial functions. First, the presynaptic Ca^{2+} channel provides the rapid, spatially focused Ca^{2+} entry that initiates synaptic transmission. Second, through specific protein-protein interactions with SNARE proteins and scaffolding proteins, the presynaptic Ca^{2+} channels bring docked synaptic vesicles close to the source of Ca^{2+} entry, allow them to respond efficiently to the micordomain of high Ca^{2+} concentration, and mediate regulatory effects that serve to focus Ca^{2+} entry on sites with docked synaptic vesicles. Third, through specific protein-protein interactions with Ca^{2+}-binding proteins, the presynaptic Ca^{2+} channels respond to residual Ca^{2+} in nerve terminals and mediate important aspects of short-term synaptic plasticity, including paired-pulse facilitation and depression, and facilitation and depression during trains of stimuli.

References

1. Bajjalieh, S.M. and Scheller, R.H. (1995) The biochemistry of neurotransmitter secretion. J Biol Chem 270, 1971–1974.
2. Bennett, M.K., Calakos, N. and Scheller, R.H. (1992) Syntaxin: A synaptic protein implicated in docking of synaptic vesicles at presynaptic active zones. Science 257, 255–259.

3. Bezprozvanny, I., Scheller, R.H. and Tsien, R.W. (1995) Functional impact of syntaxin on gating of N-type and Q-type calcium channels. Nature 378, 623–626.
4. Bezprozvanny, I., Zhong, P., Scheller, R.H. and Tsien, R.W. (2000) Molecular determinants of the functional interaction between syntaxin and N-type calcium channel gating. Proc Natl Acad Sci USA 97, 13943–13948.
5. Burgoyne, R.D. and Weiss, J.L. (2001) The neuronal calcium sensor family of calcium-binding proteins. Biochem J 353, 1–12.
6. Carbone, E. and Swandulla, D. (1989) Neuronal calcium channels: Kinetics, blockade and modulation. Prog Biophys Mol Biol 54, 31–58.
7. Catterall, W.A. (2000) Structure and regulation of voltage-gated calcium channels. Annu Rev Cell Dev Bio 16, 521–555.
8. Catterall, W.A., Goldin, A.L. and Waxman, S.G. (2005) International Union of Pharmacology. XLVII. Nomenclature and structure-function relationships of voltage-gated sodium channels. Pharmacol Rev 57, 397–409.
9. Catterall, W.A., Perez-Reyes, E., Snutch, T.P. and Striessnig, J. (2005) International Union of Pharmacology. XLVIII. Nomenclature and structure-function relationships of voltage-gated calcium channels. Pharmacol Rev 57, 411–425.
10. Chapman, E.R., Desai, R.C., Davis, A.F. and Tornehl, C.K. (1998) Delineation of the oligomerization, AP-2 binding, and synprint binding region of the C2B domain of synaptotagmin. J Biol Chem 273, 32966–32972.
11. Cohen, M.W., Jones, O.T. and Angelides, K.J. (1991) Distribution of Ca^{2+} channels on frog motor nerve terminals revealed by fluorescent omega-conotoxin. J Neurosci 11, 1032–1039.
12. Coppola, T., Magnin-Luthi, S., Perret-Menoud, V., Gattesco, S., Schiavo, G. and Regazzi, R. (2001) Direct interaction of the Rab3 effector RIM with Ca^{2+} channels, SNAP-25, and synaptotagmin. J Biol Chem 276, 32756–32762.
13. Davies, A., Hendrich, J., Van Minh, A.T., Wratten, J., Douglas, L. and Dolphin, A.C. (2007) Functional biology of the alpha(2)delta subunits of voltage-gated calcium channels. Trends Pharmacol Sci 28, 220–228.
14. DeMaria, C.D., Soong, T.W., Alseikhan, B.A., Alvania, R.S. and Yue, D.T. (2001) Calmodulin bifurcates the local calcium signal that modulates P/Q-type calcium channels. Nature 411, 484–489.
15. Dodge, F.A. and Rahamimoff, R. (1967) Co-operative action of calcium ions in transmitter release at the neuromuscular junction. J Physiol (Lond) 193, 419–432.
16. Dolphin, A.C. (2003) Beta subunits of voltage-gated calcium channels. J Bioenerg Biomembr 35, 599–620.
17. Dunlap, K., Luebke, J.I. and Turner, T.J. (1995) Exocytotic Ca^{2+} channels in mammalian central neurons. TINS 18, 89–98.
18. Ertel, E.A., Campbell, K.P., Harpold, M.M., Hofmann, F., Mori, Y., Perez-Reyes, E., Schwartz, A., Snutch, T.P., Tanabe, T., Birnbaumer, L., Tsien, R.W. and Catterall, W.A. (2000) Nomenclature of voltage-gated calcium channels. 25, 533–535.
19. Few, A.P., Lautermilch, N.J., Westenbroek, R.E., Scheuer, T. and Catterall, W.A. (2005) Differential regulation of $Ca_v2.1$ channels by calcium-binding protein 1 and visinin-like protein-2 requires N-terminal myristoylation. J Neurosci 25, 7071–7080.
20. Forsythe, I.D., Tsujimoto, T., Barnes-Davies, M., Cuttle, M.F. and Takahashi, T. (1998) Inactivation of presynaptic calcium current contributes to synaptic depression at a fast central synapse. Neuron 20, 797–807.
21. Geppert, M., Goda, Y., Hammer, R.E., Li, C., Rosahl, T.W., Stevens, C.F. and Sudhof, T.C. (1994) Synaptotagmin I: a major Ca^{2+} sensor for transmitter release at a central synapse. Cell 79, 717–727.

22. Haeseleer, F., Imanishi, Y., Maeda, T., Possin, D.E., Maeda, A., Lee, A., Rieke, F. and Palczewski, K. (2004) Essential role of Ca^{2+}-binding protein 4, a $Ca_v1.4$ channel regulator, in photoreceptor synaptic function. Nat Neurosci 7, 1079–1087.
23. Haeseleer, F., Imanishi, Y., Sokal, I., Filipek, S. and Palczewski, K. (2002) Calcium-binding proteins: intracellular sensors from the calmodulin superfamily. Biochem Biophys Res Commun 290, 615–623.
24. Haeseleer, F., Sokal, I., Verlinde, C.L., Erdjument-Bromage, H., Tempst, P., Pronin, A.N., Benovic, J.L., Fariss, R.N. and Palczewski, K. (2000) Five members of a novel Ca^{2+}-binding protein (CABP) subfamily with similarity to calmodulin. J Biol Chem 275, 1247–1260.
25. Harkins, A.B., Cahill, A.L., Powers, J.F., Tischler, A.S. and Fox, A.P. (2004) Deletion of the synaptic protein interaction site of the N-type ($Ca_v2.2$) calcium channel inhibits secretion in mouse pheochromocytoma cells. Proc Natl Acad Sci USA 101, 15219–15224.
26. Heidelberger, R., Thoreson, W.B. and Witkovsky, P. (2005) Synaptic transmission at retinal ribbon synapses. Prog Retin Eye Res 24, 682–720.
27. Heinemann, S.H., Terlau, H., Stühmer, W., Imoto, K. and Numa, S. (1992) Calcium channel characteristics conferred on the sodium channel by single mutations. Nature 356, 441–443.
28. Hofmann, F., Lacinová, L. and Klugbauer, N. (1999) Voltage-dependent calcium channels: from structure to function. Rev Physiol Biochem Pharmacol 139, 33–87.
29. Imredy, J.P. and Yue, D.T. (1992) Submicroscopic Ca^{2+} diffusion mediates inhibitory coupling between individual Ca^{2+} channels. Neuron 9, 197–207.
30. Jarvis, S.E., Magga, J.M., Beedle, A.M., Braun, J.E. and Zamponi, G.W. (2000) G protein modulation of N-type calcium channels is facilitated by physical interactions between syntaxin 1A and Gβγ. J Biol Chem 275, 6388–6394.
31. Jarvis, S.E. and Zamponi, G.W. (2001) Distinct molecular determinants govern syntaxin 1A-mediated inactivation and G-protein inhibition of N-type calcium channels. J Neurosci 21, 2939–2948.
32. Kaneko, S., Cooper, C.B., Nishioka, N., Yamasaki, H., Suzuki, A., Jarvis, S.E., Akaike, A., Satoh, M. and Zamponi, G.W. (2002) Identification and characterization of novel human $Ca_v2.2$ (alpha 1B) calcium channel variants lacking the synaptic protein interaction site. J Neurosci 22, 82–92.
33. Keith, R.K., Poage, R.E., Yokoyama, C.T., Catterall, W.A. and Meriney, S.D. (2007) Bidirectional modulation of transmitter release by calcium channel/syntaxin interactions in vivo. J Neurosci 27, 265–269.
34. Kim, D.K. and Catterall, W.A. (1997) Ca^{2+}-dependent and -independent interactions of the isoforms of the α1A subunit of brain Ca^{2+} channels with presynaptic SNARE proteins. Proc Natl Acad Sci USA 94, 14782–14786.
35. Kiyonaka, S., Wakamori, M., Miki, T., Uriu, Y., Nonaka, M., Bito, H., Beedle, A.M., Mori, E., Hara, Y., De Waard, M., Kanagawa, M., Itakura, M., Takahashi, M., Campbell, K.P. and Mori, Y. (2007) RIM1 confers sustained activity and neurotransmitter vesicle anchoring to presynaptic Ca^{2+} channels. Nat Neurosci 10, 691–701.
36. Lautermilch, N.J., Few, A.P., Scheuer, T. and Catterall, W.A. (2005) Modulation of $Ca_v2.1$ channels by the neuronal calcium-binding protein visinin-like protein-2. J Neurosci 25, 7062–7070.
37. Lee, A., Scheuer, T. and Catterall, W.A. (2000) Ca^{2+}-Calmodulin dependent inactivation and facilitation of P/Q-type Ca^{2+} channels. J. Neurosci 20, 6830–6838.
38. Lee, A., Westenbroek, R.E., Haeseleer, F., Palczewski, K., Scheuer, T. and Catterall, W.A. (2002) Differential modulation of $Ca_v2.1$ channels by calmodulin and calcium-binding protein 1. Nat Neurosci 5, 210–217.

39. Lee, A., Wong, S.T., Gallagher, D., Li, B., Storm, D.R., Scheuer, T. and Catterall, W.A. (1999) Ca^{2+}/calmodulin binds to and modulates P/Q-type calcium channels. Nature 399, 155–159.
40. Lee, A., Zhou, H., Scheuer, T. and Catterall, W.A. (2003) Molecular determinants of Ca^{2+}/calmodulin-dependent regulation of $Ca_v2.1$ channels. Proc Natl Acad Sci USA 100, 16059–16064.
41. Leveque, C., El Far, O., Martin-Moutot, N., Sato, K., Kato, R., Takahashi, M. and Seagar, M.J. (1994) Purification of the N-type calcium channel associated with syntaxin and synaptotagmin: a complex implicated in synaptic vesicle exocytosis. J Biol Chem 269, 6306–6312.
42. Leveque, C., Pupier, S., Marqueze, B., Geslin, L., Kataoka, M., Takahashi, M., De Waard, M. and Seagar, M. (1998) Interaction of cysteine string proteins with the α1A subunit of the P/Q-type calcium channel. J Biol Chem 273, 13488–13492.
43. Liang, H., DeMaria, C.D., Erickson, M.G., Mori, M.X., Alseikhan, B.A. and Yue, D.T. (2003) Unified mechanisms of Ca^{2+} regulation across the Ca^{2+} channel family. Neuron 39, 951–960.
44. Llinás, R., Sugimori, M. and Silver, R.B. (1992) Microdomains of high calcium concentration in a presynaptic terminal. Science 256, 677–679.
45. Long, S.B., Campbell, E.B. and Mackinnon, R. (2005) Crystal structure of a mammalian voltage-dependent Shaker family K^+ channel. Science 309, 897–903.
46. Maximov, A. and Bezprozvanny, I. (2002) Synaptic targeting of N-type calcium channels in hippocampal neurons. J Neurosci 22, 6939–6352.
47. Miljanich, G.P. and Ramachandran, J. (1995) Antagonists of neuronal calcium channels: structure, function, and therapeutic implications. Annu Rev Pharmacol Toxicol 35, 707–734.
48. Mochida, S., Sheng, Z.H., Baker, C., Kobayashi, H. and Catterall, W.A. (1996) Inhibition of neurotransmission by peptides containing the synaptic protein interaction site of N-type Ca^{2+} channels. Neuron 17, 781–788.
49. Mochida, S., Westenbroek, R.E., Yokoyama, C.T., Itoh, K. and Catterall, W.A. (2003) Subtype-selective reconstitution of synaptic transmission in sympathetic ganglion neurons by expression of exogenous calcium channels. Proc Natl Acad Sci USA 100, 2813–2818.
50. Mochida, S., Westenbroek, R.E., Yokoyama, C.T., Zhong, H., Myers, S.J., Scheuer, T., Itoh, K. and Catterall, W.A. (2003) Requirement for the synaptic protein interaction site for reconstitution of synaptic transmission by P/Q-type calcium channels. Proc Natl Acad Sci USA 100, 2819–2824.
50a. Mochida, S., Few, A.P., Scheuer, T., and Catterall, W.A. (2008) Regulation of presynaptic $Ca_v2.1$ channels by Ca^{2+} sensor proteins mediates short-term synaptic plasticity. Neuron 24, 210–216.
51. Morgans, C.W., Bayley, P.R., Oesch, N.W., Ren, G., Akileswaran, L. and Taylor, W.R. (2005) Photoreceptor calcium channels: insight from night blindness. Vis Neurosci 22, 561–568.
52. Perez-Reyes, E. (2003) Molecular physiology of low-voltage-activated t-type calcium channels. Physiol Rev 83, 117–161.
53. Peterson, B.Z., DeMaria, C.D. and Yue, D.T. (1999) Calmodulin is the Ca^{2+} sensor for Ca^{2+}-dependent inactivation of L-type calcium channels. Neuron 22, 549–558.
54. Pitt, G.S., Zuhlke, R.D., Hudmon, A., Schulman, H., Reuter, H. and Tsien, R.W. (2001) Molecular basis of calmodulin tethering and calcium-dependent inactivation of L-type calcium channels. J Biol Chem 276, 30794–30802.
55. Poncer, J.C., McKinney, R.A., Gähwiler, B.H. and Thompson, S.M. (1997) Either N- or P-type calcium channels mediate GABA release at distinct hippocampal inhibitory synapses. Neuron 18, 463–472.

56. Qin, N., Olcese, R., Bransby, M., Lin, T. and Birnbaumer, L. (1999) Ca^{2+} induced inhibition of the cardiac Ca^{2+} channel depends on calmodulin. Proc Natl Acad Sci USA 96, 2435–2438.
57. Rettig, J., Heinemann, C., Ashery, U., Sheng, Z.H., Yokoyama, C.T., Catterall, W.A. and Neher, E. (1997) Alteration of Ca^{2+} dependence of neurotransmitter release by disruption of Ca^{2+} channel/syntaxin interaction. J Neurosci 17, 6647–6656.
58. Rettig, J., Sheng, Z.H., Kim, D.K., Hodson, C.D., Snutch, T.P. and Catterall, W.A. (1996) Isoform-specific interaction of the α1A subunits of brain Ca^{2+} channels with the presynaptic proteins syntaxin and SNAP-25. Proc Natl Acad Sci USA 93, 7363–7368.
59. Schneggenburger, R. and Neher, E. (2005) Presynaptic calcium and control of vesicle fusion. Curr Opin Neurobiol 15, 266–274.
60. Sheng, Z.H., Rettig, J., Cook, T. and Catterall, W.A. (1996) Calcium-dependent interaction of N-type calcium channels with the synaptic core-complex. Nature 379, 451–454.
61. Sheng, Z.H., Rettig, J., Takahashi, M. and Catterall, W.A. (1994) Identification of a syntaxin-binding site on N-type calcium channels. Neuron 13, 1303–1313.
62. Sheng, Z.H., Westenbroek, R.E. and Catterall, W.A. (1998) Physical link and functional coupling of presynaptic calcium channels and the synaptic vesicle docking/fusion machinery. J Bioeng Biomem 30, 335–345.
63. Sheng, Z.H., Yokoyama, C. and Catterall, W.A. (1997) Interaction of the synprint site of N-type Ca^{2+} channels with the C2B domain of synaptotagmin I. Proc Natl Acad Sci USA 94, 5405–5410.
64. Sippy, T., Cruz-Martin, A., Jeromin, A. and Schweizer, F.E. (2003) Acute changes in short-term plasticity at synapses with elevated levels of neuronal calcium sensor-1. Nat Neurosci 6, 1031–1038.
65. Smith, S.J., Buchanan, J., Osses, L.R., Charlton, M.P. and Augustine, G.J. (1993) The spatial distribution of calcium signals in squid presynaptic terminals. J Physiol 472, 573–593.
66. Snutch, T.P. and Reiner, P.B. (1992) Ca^{2+} channels: diversity of form and function. Curr Opin Neurobiol 2, 247–253.
67. Sollner, T., Whiteheart, S.W., Brunner, M., Erdjument-Bromage, H., Geromanos, S., Tempst, P. and Rothman, J.E. (1993) SNAP receptors implicated in vesicle targeting and fusion. Nature 362, 297–298.
68. Spafford, J.D., Munno, D.W., Van Nierop, P., Feng, Z.P., Jarvis, S.E., Gallin, W.J., Smit, A.B., Zamponi, G.W. and Syed, N.I. (2003) Calcium channel structural determinants of synaptic transmission between identified invertebrate neurons. J Biol Chem 278, 4258–4267.
69. Spafford, J.D. and Zamponi, G.W. (2003) Functional interactions between presynaptic calcium channels and the neurotransmitter release machinery. Curr Opin Neurobiol 13, 308–314.
70. Stanley, E.F. (1993) Single calcium channels and acetylcholine release at a presynaptic nerve terminal. Neuron 11, 1007–1011.
71. Stanley, E.F. (1997) The calcium channel and the organization of the presynaptic transmitter release face. Trends Neurosci 20, 404–409.
72. Stanley, E.F. (1993) Presynaptic calcium channels and the transmitter release mechanism. Ann NY Acad Sci 681, 368–372.
73. Stanley, E.F. and Mirotznik, R.R. (1997) Cleavage of syntaxin prevents G-protein regulation of presynaptic calcium channels. Nature 385, 340–343.
74. Sudhof, T.C. (2004) The synaptic vesicle cycle. Annu Rev Neurosci 27, 509–547.
75. Sudhof, T.C. (1995) The synaptic vesicle cycle: a cascade of protein-protein interactions. Nature 375, 645–653.

76. Szabo, Z., Obermair, G.J., Cooper, C.B., Zamponi, G.W. and Flucher, B.E. (2006) Role of the synprint site in presynaptic targeting of the calcium channel $Ca_v2.2$ in hippocampal neurons. Eur J Neurosci 24, 709–718.
77. Takahashi, M., Seagar, M.J., Jones, J.F., Reber, B.F. and Catterall, W.A. (1987) Subunit structure of dihydropyridine-sensitive calcium channels from skeletal muscle. Proc Natl Acad Sci USA 84, 5478–5482.
78. Tippens, A.L. and Lee, A. (2007) Caldendrin, a neuron-specific modulator of $Ca_v/1.2$ (L-type) Ca^{2+} channels. J Biol Chem 282, 8464–8473.
79. Tsien, R.W., Elinor, P.T. and Horne, W.A. (1991) Molecular diversity of voltage-dependent calcium channels. Trends Neurosci 12, 349–354.
80. Tsien, R.W., Lipscombe, D., Madison, D.V., Bley, K.R. and Fox, A.P. (1988) Multiple types of neuronal calcium channels and their selective modulation. Trends Neurosci 11, 431–438.
81. Tsujimoto, T., Jeromin, A., Saitoh, N., Roder, J.C. and Takahashi, T. (2002) Neuronal calcium sensor 1 and activity-dependent facilitation of P/Q-type calcium currents at presynaptic nerve terminals. Science 295, 2276–2279.
82. Umbach, J.A. and Gundersen, C.B. (1997) Evidence that cysteine string proteins regulate an early step in the Ca^{2+}-dependent secretion of neurotransmitter at Drosophila neuromuscular junctions. J Neurosci 17, 7203–7209.
83. Weiss, J.L., Archer, D.A. and Burgoyne, R.D. (2000) Neuronal Ca^{2+} sensor-1/frequenin functions in an autocrine pathway regulating Ca^{2+} channels in bovine adrenal chromaffin cells. J Biol Chem 275, 40082–40087.
84. Weiss, J.L. and Burgoyne, R.D. (2001) Voltage-independent inhibition of P/Q-type Ca^{2+} channels in adrenal chromaffin cells via a neuronal Ca^{2+} sensor-1-dependent pathway involves Src family tyrosine kinase. J Biol Chem 276, 44804–44811.
85. Westenbroek, R.E., Hell, J.W., Warner, C., Dubel, S.J., Snutch, T.P. and Catterall, W.A. (1992) Biochemical properties and subcellular distribution of an N-type calcium channel $\alpha1$ subunit. Neuron 9, 1099–1115.
86. Westenbroek, R.E., Sakurai, T., Elliott, E.M., Hell, J.W., Starr, T.V.B., Snutch, T.P. and Catterall, W.A. (1995) Immunochemical identification and subcellular distribution of the $\alpha1A$ subunits of brain calcium channels. J Neurosci 15, 6403–6418.
87. Wiser, O., Bennett, M.K. and Atlas, D. (1996) Functional interaction of syntaxin and SNAP-25 with voltage-sensitive L- and N-type Ca^{2+} channels. EMBO J 15, 4100–4110.
88. Wiser, O., Tobi, D., Trus, M. and Atlas, D. (1997) Synaptotagmin restores kinetic properties of a syntaxin-associated N-type voltage sensitive calcium channel. FEBS Lett 404, 203–207.
89. Wiser, O., Trus, M., Hernandez, A., Renstrom, E., Barg, S., Rorsman, P. and Atlas, D. (1999) The voltage-sensitive Lc-type Ca^{2+} channel is functionally coupled to the exocytotic machinery. Proc Natl Acad Sci USA 96, 248–253.
90. Xu, J., Mashimo, T. and Sudhof, T.C. (2007) Synaptotagmin-1, -2, and -9: $Ca(^{2+})$ sensors for fast release that specify distinct presynaptic properties in subsets of neurons. Neuron 54, 567–581.
91. Yarov-Yarovoy, V., Baker, D. and Catterall, W.A. (2006) Voltage sensor conformations in the open and closed states in ROSETTA structural models of K^+ channels. Proc Natl Acad Sci USA 103, 7292–7297.
92. Yokoyama, C.T., Myers, S.J., Fu, J., Mockus, S.M., Scheuer, T. and Catterall, W.A. (2005) Mechanism of SNARE protein binding and regulation of Ca_v2 channels by phosphorylation of the synaptic protein interaction site. Mol Cell Neurosci 28, 1–17.
93. Yokoyama, C.T., Sheng, Z.H. and Catterall, W.A. (1997) Phosphorylation of the synaptic protein interaction site on N-type calcium channels inhibits interactions with SNARE proteins. J Neurosci 17, 6929–6938.

94. Yoshida, A., Oho, C., Omori, A., Kawahara, R., Ito, T. and Takahashi, M. (1992) HPC-1 is associated with synaptotagmin and ω-conotoxin receptor. J Biol Chem 267, 24925–24928.
95. Zhong, H., Yokoyama, C., Scheuer, T. and Catterall, W.A. (1999) Reciprocal regulation of P/Q-type Ca^{2+} channels by SNAP-25, syntaxin and synaptotagmin. Nat Neurosci 2, 939–941.
96. Zhou, H., Kim, S.A., Kirk, E.A., Tippens, A.L., Sun, H., Haeseleer, F. and Lee, A. (2004) Ca^{2+}-binding protein-1 facilitates and forms a postsynaptic complex with $Ca_v1.2$ (L-type) Ca^{2+} channels. J Neurosci 24, 4698–4708.
97. Zucker, R.S. and Regehr, W.G. (2002) Short-term synaptic plasticity. Annu Rev Physiol 64, 355–405.
98. Zühlke, R.D., Pitt, G.S., Deisseroth, K., Tsien, R.W. and Reuter, H. (1999) Calmodulin supports both inactivation and facilitation of L-type calcium channels. Nature 399, 159–162.
99. Zühlke, R.D., Pitt, G.S., Tsien, R.W. and Reuter, H. (2000) Ca^{2+}-sensitive inactivation and facilitation of L-type Ca^{2+} channels both depend on specific amino acid residues in a consensus calmodulin-binding motif in the alpha1C subunit. J Biol Chem 275, 21121–21129.
100. Zühlke, R.D. and Reuter, H. (1998) Ca^{2+}-sensitive inactivation of L-type Ca^{2+} channels depends on multiple cytoplasmic amino acid sequences of the α1C subunit. Proc Natl Acad Sci USA 95, 3287–3294.

Adhesion Molecules at the Synapse

Alaa El-Husseini

University of British Columbia, Department of Psychiatry, Vancouver, British Columbia, Canada alaa@interchange.ubc.ca

Abstract. Communication between neurons is mediated mainly by chemical synapses, at which neurotransmitters are released by the presynaptic neuron, diffuse across the narrow cleft, and activate receptors located on postsynaptic neurons. Synapses provide the structural and functional basis for the formation and maintenance of the complex neural networks that exist in the brain. This process is governed by the number, type, and location of synapses formed. Therefore, specialized molecular machinery must exist to ensure assembly of appropriate synaptic partners. There is considerable evidence that synaptic contact formation and remodeling is regulated by adhesion molecules that modulate the stabilization and maintenance of initial contacts between neuronal cells, as well as recruitment of other molecules that control synapse maturation, specificity and function. Thus, disruption of the adhesive properties of these molecules not only compromises synaptic transmission, but also many aspects of brain function that control memory formation and behavior. Here, I will describe the various stages of synaptogenesis, with focus on adhesion molecules that regulate contact initiation, synapse maturation/stabilization or elimination, and synaptic plasticity. Emphasis will be on the potential mechanisms that regulate adhesion molecule recruitment and function at the synapse. I will also discuss how dysfunction in specific adhesive systems may contribute to synaptic imbalance and the development of brain disorders.

1 Introduction

1.1 Mechanisms That Govern Synapse Development

Synapses form when axons reach their target cells and establish stable contacts with them. This process requires spatially and temporally controlled changes in contact structure as well as recruitment of specific proteins to contact sites (18, 26, 39, 45, 74, 75, 154, 191, 197). The axon terminal, the so-called presynaptic side, is engorged with synaptic vesicles containing neurotransmitters, the chemical basis of neuron to neuron communication. These vesicles are docked on the presynaptic membrane, attached by a complex cytomatrix of proteins, primed for the proper signal for release. At postsynaptic sites, neurotransmitter receptors and a wide

J.W. Hell, M.D. Ehlers (eds.), *Structural and Functional Organization of the Synapse*,
DOI: 10.1007/978-0-387-77232-5_7, © Springer Science+Business Media, LLC 2008

array of transmembrane, cytoskeletal and signaling proteins are clustered and thereby positioned to rapidly respond to the neurotransmitter released from the presynaptic terminal (112). This asymmetrical assembly of proteins at the synapse makes synapses distinct from other known cell–cell junctions.

The formation of a new synapse is thought to involve several characteristic steps: contact initiation, recruitment of presynaptic and postsynaptic proteins, and contact stabilization. As discussed earlier, neuronal contact formation relies on the ability of axons to extend to and reach their target fields and contact the correct postsynaptic cell. To achieve this, axons often travel long and arduous paths before finding their final target. Before reaching the target neuron, axons encounter a multitude of postsynaptic cells along the way, yet despite this, do not establish synapses with inappropriate cells. Thus, precise axon guidance by cues alone are not sufficient for target-specific synapse formation (187). Developing dendrites lined with filopodia containing proper adhesion molecules interact with extending axons and are thus thought to promote initial contact stabilization (64, 201). Given the large number of neuronal contacts formed during early stages of synaptogenesis, polymorphic adhesion systems must exist to offer sufficient combinatorial possibilities to ensure the formation and stabilization of contact sites between specific neurons. This 'lock and key' mechanism was first hypothesized by Sperry (1963), implying the existence of specific adhesion molecules that pair axons with their targets for the specification of synaptic connections (169). The second step in synaptogenesis is thought to involve the recruitment of presynaptic and postsynaptic components to sites of initial contact. At this stage, the content and morphology of pre- and postsynaptic sites develop in a coordinated fashion. The number of synaptic vesicles in the presynaptic compartment increases, as does the size and content of the postsynaptic density (141, 186). The correlation between changes in the size of the pre- and postsynaptic compartments suggests the involvement of trans-synaptic adhesion complexes in this process. These trans-synaptic signals would require specificity, being able to recruit the appropriate neurotransmitter on the presynaptic side as well as its cognate receptors on the postsynaptic side, to avoid mismatching. After the recruitment of presynaptic and postsynaptic proteins, newly formed synapses are further stabilized or lost, in an effort to fine-tune neuronal circuitry, a process which also requires modulation of the adhesive properties of synapses (78). Adhesion systems important for these processes are likely to be modulated by synaptic activity, leading to changes in protein levels or adhesive properties (83, 121). Overall, each of the steps of synaptogenesis outlined above requires crosstalk between two neurons, a fundamental property of all cell adhesion molecules. Developmental regulation of expression, as well as mechanisms that govern localization of adhesion molecules to particular contact sites, are two processes that potentially control contact initiation, stabilization and plasticity (Fig. 1).

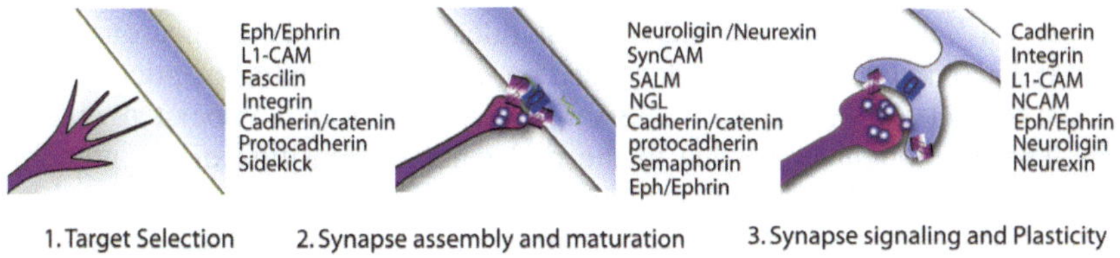

Fig. 1. Putative adhesion molecules involved in synaptogenesis. Synapse formation involves several characteristic steps that include: Target selection (**1**), and Synapse assembly and maturation (**2**). These steps involve the recruitment of pre- and postsynaptic proteins to newly formed contacts, followed by contact stabilization or elimination. At mature synapses, adhesion molecules modulate retrograde signaling to control synaptic strength and plasticity (**3**). Examples of some of the tentative roles of specific adhesion molecules at various stages of synaptogenesis are indicated. Functional redundancy between these adhesive systems has been observed, suggesting that synapse formation and maturation rely on the collective action of several adhesive systems to ensure proper connectivity and recruitment of appropriate neurotransmitter receptors to contact sites.

1.2 Heterogeneity of CNS Synapses

Despite the identification of several molecules that modulate synapse maturation, mechanisms of synaptogenesis in the central nervous system (CNS) are poorly understood (27, 38, 98, 100, 144, 162, 163). This process is complicated by the enormous heterogeneity of neuronal cells, the timing of their development and the type of neurotransmitter released. In addition, the effect of a particular neurotransmitter on the postsynaptic neuron is context-specific, depending on several factors, including the developmental stage, the type and number of receptors present on the postsynaptic membrane, and neuronal activity (194).

Chemical synapses are divided mainly into excitatory and inhibitory subtypes, depending on the type of neurotransmitter involved, and their ability to depolarize or hyperpolarize the postsynaptic membrane. Excitatory synaptic transmission in the mammalian brain is primarily mediated by the neurotransmitter glutamate. As such, most of our knowledge of synaptogenesis in the CNS is based on research done on glutamatergic synapses. Excitatory synapse maturation involves the transformation of an immature glutamatergic postsynaptic compartment into a mature spine, an actin-based dendritic protrusion with a bulbous head (64, 82, 103–105, 137, 139, 168). This postsynaptic compartment is characterized by an electron-dense organelle called the postsynaptic density (PSD), which contains the two main types of glutamate-receptor channels, N-methyl-D-aspartate (NMDA) and α-amino-3-hydroxy-5-methyl-4-isoxazole propionic acid (AMPA) receptors. The majority of inhibitory neurotransmission in the vertebrate CNS is mediated by γ-aminobutyric acid (GABA), a neurotransmitter responsible for modulation of neuronal excitability and function. The action of GABA is mediated by ionotropic ($GABA_A$ or $GABA_C$) and metabotropic ($GABA_B$) receptors, which are ubiquitously expressed in the CNS (8, 14). GABAergic function is fine-tuned at multiple levels, including transmitter synthesis, catalyzed by two isoforms of glutamic acid decarboxylase (GAD); vesicular storage; Ca^{2+} dependent and independent release; re-uptake in neurons and glial cells; and activation of receptors localized pre-, post-, and extrasynaptically (106). In contrast to excitatory synapses, GABAergic synapses are not associated with a clear PSD. This is thought to be due to the molecular composition of the postsynaptic compartment being less elaborate. Excitatory and inhibitory synapses are therefore clearly distinguishable on the basis of morphology and molecular composition of the postsynaptic compartment (Fig. 2).

A critical aspect of synaptogenesis is the recruitment of factors that determine whether a nascent contact develops into an excitatory or an inhibitory synapse. This is a difficult task to achieve considering that a single neuron may receive tens of thousands of heterogeneous synaptic inputs, yet virtually no mismatches between pre- and postsynaptic elements occur in the case of mature synapses (2, 144). How do neurons regulate this process? An ensemble of hundreds of proteins containing several adhesion molecules has been identified at the PSD of excitatory synapses (138, 188).

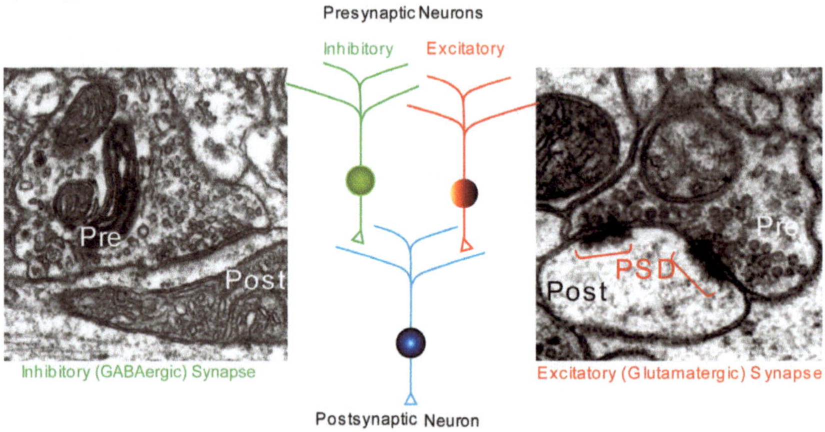

Fig. 2. Heterogeneity of synaptic contacts. Electron microscopy images highlighting the differences between the two major types of synapses; excitatory glutamatergic synapses and inhibitory GABAergic synapses. The postsynaptic (Post) sites are in perfect opposition to neurotransmitter containing synaptic vesicles located at presynaptic (Pre) terminals. The postsynaptic sites of excitatory glutamatergic synapses form a bulbous protrusion known as a spine, which is enriched in glutamate receptors and associated proteins. These protrusions are characterized by an electron-dense organelle called the postsynaptic density (PSD). In contrast to excitatory synapses, GABAergic synapses are not associated with a clear PSD.

Most likely, the coordinated recruitment and modes of action of these molecules on contact stabilization and maturation help establish the identity of newly formed glutamatergic neuronal contacts. In contrast, few adhesion molecules have been identified at inhibitory synapses. The limited information available on the constituents of inhibitory contacts has hindered our understanding of the mechanisms that govern inhibitory synapse formation and maturation. The recent discovery of some of the adhesion molecules that influence maturation of either excitatory or inhibitory contacts will help unravel the early events involved in dictating synaptic specificity and function. In this chapter, the effects of several newly discovered adhesion systems will be further discussed to highlight some of the identified mechanisms involved in synaptogenesis. A large body of work has been conducted recently to characterize the role of neuroligins and neurexins in excitatory and inhibitory synapse development. Hence, the effects of these molecules on synapse induction, maturation and function will be discussed in detail to elaborate on some of the newly emerging mechanisms involved in initial contact formation to maturation and plasticity.

2 Adhesion Molecules That Modulate Synapse Induction

2.1 Effects of Neuroligins and Neurexins on Excitatory Synapse Maturation and Function

Neuroligins, a family of synaptic cell adhesion molecules, are coming to the forefront as critical modulators of synapse maturation and plasticity (18, 46, 77, 90, 91, 118, 122, 157, 158). Functions of neuroligins at the synapse have been linked to their interaction with neurexins, a family of highly polymorphic, brain-specific proteins, identified while cloning a receptor for the black widow spider toxin, α-latrotoxin (Fig. 3) (182).

While neuroligins are enriched in dendrites specifically at postsynaptic membranes, neurexins are thought to be present at both pre- and postsynaptic sites (23, 127, 166). The first insights into a role for neuroligins in synapse development came form a pioneering study from Scheiffele and colleagues, that developed a coculture assay to study the molecular machinery sufficient for synapse induction at contact sites between non-neuronal and neuronal cells (157). In these experiments, presentation of neuroligin-1 or -2 on the surface of heterologous cells was sufficient to induce morphological and functional presynaptic differentiation in contacting axons of co-cultured neurons (157). In addition, neuroligins were found to induce the accumulation of the synaptic vesicle protein, synapsin, in pontine axons or granules cells. Moreover, these vesicles could undergo exocytosis in a depolarization-dependent manner (157). In this study, the ability to induce presynaptic differentiation was unique to neuroligins, as other adhesion molecules tested, including agrin, N-Cadherin, ephrinB1, TAG-1, and L1, failed to do so (157). Electrophysiological recordings of the artificial contacts formed between neurons and co-cultured heterologous cells expressing neuroligin-1 and NMDAR or AMPAR subunits provided further evidence that neuroligins can drive functional synapse formation (70, 155). Further studies showed that the synaptogenic effects of neuroligin depend on interaction with neurexins (47, 157). Consistent with these results, overexpression of neuroligin-1 in neurons induces presynaptic differentiation, triggering clustering of endogenous neurexin and increasing the size of synapsin clusters (47).

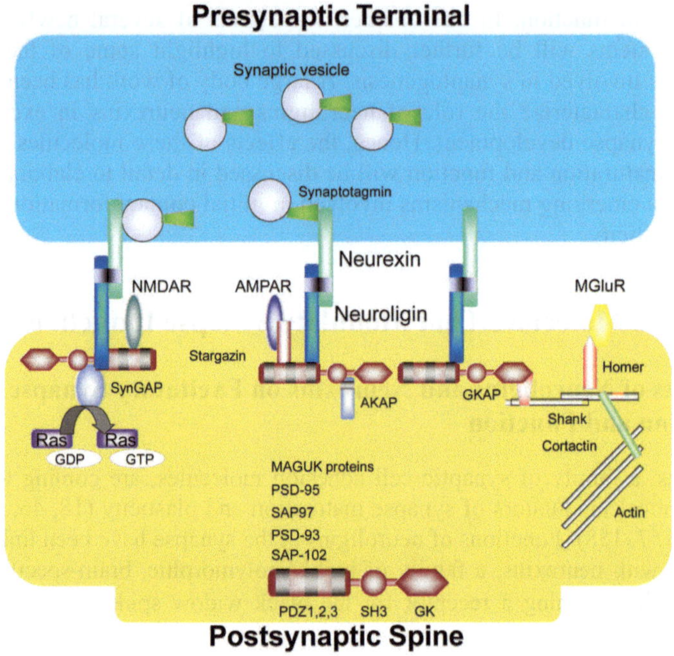

Fig. 3. Trans-synaptic signaling at glutamatergic synapses mediated by neuroligins and neurexins. Clustering of neurexins by neuroligins is thought to induce recruitment of presynaptic proteins associated with vesicle release machinery, including synapsin, synaptotagmin, synaptobrevin/VAMP, and VGLUT. Neuroligin clustering by β-neurexin triggers accumulation of postsynaptic scaffolding proteins including members of the MAGUK family, GKAP, and SynGAP, as well as the excitatory neurotransmitter receptor subunit NR1. Some of the molecules thought to regulate neuroligin actions in excitatory synapse maturation as well as molecules involved in modulating synaptic strength, retrograde signaling and plasticity are indicated. For a more comprehensive overview of the molecules that regulate neuroligins and neurexins action at the synapse see reference (122).

2.2 Molecular and Functional Determinants of Neuroligin/Neurexin Interaction

Neuroligins belong to a large family of adhesion molecules, known as cholinesterase-like adhesion molecules (CLAMs), which include the invertebrate proteins, glutactin, neurotactin and gliotactin (77). The hallmark of CLAMs distinguishing them from cholinesterases is the lack of various residues in the catalytic triad of the extracellular cholinesterase-like domain, rendering these proteins enzymatically inactive. Four genes encoding neuroligin family members have been identified in rodents, with at least five genes coding for neuroligins in humans (12). Neuroligins are type I transmembrane proteins containing three major domains that regulate their function: a cholinesterase-like domain (CLD), an O-glycosylation cassette, and a short C-terminal tail containing a type I PDZ binding motif (90). The extracellular region of neuroligins is thought to mediate receptor/ligand-like interaction implicated in cell–cell adhesion. Comparison of rat sequences revealed that the intracellular regions are the least conserved, with only 31% identity. As such, this divergence may underlie the differential localizations and/or functions of neuroligin family members. Neuroligin-1, the most characterized member of the family, was first purified from brain lysates by affinity chromatography on immobilized neurexin 1β lacking an insert in the alternatively spliced G domain, also referred as splice site 4 (90). Additional neuroligin family members have subsequently been identified as splice-site specific ligands for β-neurexins (91).

Neurexins are the products of three genes, from which two classes of mRNAs are generated: long mRNAs encoding the α-neurexins, and short mRNAs which encode the β-neurexins. Thus, at least 6 principal neurexin isoforms are expressed in the brain; three α-neurexins (Iα, IIα, IIIα) and three β-neurexins (Iβ, IIβ, IIIβ) (180, 181). Both α- and β-neurexins are single transmembrane proteins, with distinct N-terminal sequences that contain the LNS domain, named after repeated sequences of Laminin A, Neurexin, and Sex hormone-binding protein. LNS domains are also referred to as G domains, in reference to domain repeats present in laminin A, agrin, and slit, proteins implicated in cell recognition processes during development of the nervous system (182, 148). In α-neurexin, the LNS domains are part of a three-repeat cassette, each repeat containing 2 LNS domains (A and B) separated by an epidermal growth factor-like (EGF) sequence. In comparison, β-neurexins contain only the last LNS unit, which is preceded by a β-neurexin-specific sequence resulting from the N-terminal truncation of a long, atypical signal peptide (181, 182). In all neurexins, the last LNS unit is followed by a carbohydrate attachment sequence, a single transmembrane region and a short cytoplasmic tail containing a type II PDZ recognition motif. Neurexins also have multiple alternatively spliced regions making them highly polymorphic. Neuroligins can bind to both α- and β-neurexins, depending on alternative splicing of neuroligins at splice site B (13).

2.3 Balancing Excitation and Inhibition by Neuroligins and Neurexins

An important finding emerging from recent studies on the effects of neuroligins on synapse induction, is their ability to influence not only excitatory synapse development, but also inhibitory synapse differentiation and function (31, 33, 119, 143, 184). The first evidence for a role for neuroligins in driving maturation of inhibitory synapses came from work performed by Prange et al. (143), showing that neuroligin-1 not only induces excitatory presynaptic contact formation, but also the maturation of GABAergic synapses when overexpressed in hippocampal neurons (143). Later, it was discovered that these effects can be reproduced to varying extents by all members of the neuroligin family (30, 80, 117). Endogenously however, neuroligin-1 and -3 are enriched at excitatory synapses, whereas neuroligin-2 is found mainly at inhibitory synapses (31, 117, 166, 185).

The effects on excitatory and inhibitory contacts seem to require coupling to neurexin, since application of soluble β-neurexin extracellular domain reduced frequency of excitatory and inhibitory currents (118). Other studies showed that β-neurexin has the ability to induce the clustering of neuroligins and other postsynaptic proteins normally present at both types of synapses. Moreover, while neuroligin-1, -3, and –4 co-aggregated mainly with PSD-95, neuroligin-2 co-aggregated significantly more with gephyrin, a scaffolding protein found mainly at inhibitory postsynaptic sites (80). Studies of a similar design showed that coexpression of neuroligin-2 and $GABA_A$ receptors in non-neuronal cells is sufficient for reconstitution of functional GABAergic synapses (53). Interestingly, while presentation of β-neurexin induces clustering of both PSD-95 and NMDA receptors, no clustering of AMPA type glutamate receptors was observed (80, 131). Additional experiments demonstrated that AMPA receptors are recruited upon application of glutamate or overexpression of CaMKII in neurons (131). Thus, synapses induced by neuroligins seem to resemble silent synapses which become functional at later stages of synapse development, via recruitment of AMPA receptors. However, despite the ability of neuroligins to induce recruitment of excitatory and inhibitory synaptic proteins to contact sites, their physiological role seems to be more to exert specific effects on synaptic activity. For example, neuroligin-1 expressed in both non-neuronal and neuronal cells enhances excitatory synaptic activity, while expression of neuroligin-2 enhances inhibitory synaptic currents (33, 53, 70, 131, 165). Further work must be carried out to clarify mechanisms that control targeting and function of specific neuroligins at different synapse types.

In addition to modulation of glutamatergic and GABAergic synapse maturation, recent evidence indicates that neuroligins also enhance nicotinic synapse maturation (36). Namely, overexpression of neuroligins in neurons induces presynaptic specializations in adjacent cholinergic neurons. In addition, manipulation of endogenous neuroligin function with dominant negative constructs and siRNA compromises nicotinic synaptic function. How neuroligins are able to drive the maturation of heterogeneous types of synapses using the same presynaptic partner remains unclear. Secondary presynaptic binding partners or crosstalk with other adhesion systems may account for this phenomenon, but this remains to be determined.

2.4 Regulation of Neuroligin/Neurexin Trafficking and Assembly

Recent studies revealed that neuroligins and neurexins have the inherent ability to affect both excitatory and inhibitory synapse maturation. The question then arises, if neuroligins and neurexins have the ability to modulate both types of contacts, what accounts for their proper sorting to a particular synapse in vivo? One potential mechanism that influences the recruitment of neuroligins to particular synapse types, is alternative splicing (90, 91). Although the N-terminal regions of neuroligins share a relatively high degree of sequence similarity, the splice sites found within these regions provide differing sequences which may be used to direct targeting of these molecules to excitatory versus inhibitory synapses (13, 79, 80). Indeed, a recent study has shown that neuroligins lacking insert at splice site B target more efficiently to inhibitory synapses, while those containing an insert at this splice site localize mainly to excitatory synapses. Thus, this splice code guides the specific binding of particular neuroligin and neurexin variants (39, 122). A 30-residue insert at splice site 4 of β-neurexins reduces the affinity of interaction specifically with neuroligin-1 that contains an insert at splice site B, and promotes GABAergic synapse development. In contrast, the presence of an insert at site B of neuroligin-1, which is the predominantly expressed neuroligin-1 protein, promotes targeting to glutamatergic synapses (32). The basis for the binding affinity of alternatively spliced neuroligin for neurexins appears to be protein glycosylation. N-glycosylation at Asn 303 located in splice site B containing insert hinders binding of neuroligin-1 to neurexin 1β. In contrast, neuroligin-1 lacking the insert at splice site B binds both β- and α-neurexins (13). These results suggest that differential combinations of neuroligin and neurexin isoforms at contact sites activate distinct signaling codes that specify particular synaptic properties. Whether the spatial and temporal expression of these splice variants control synapse specificity remains an open question.

Intracellular binding partners of neuroligins and neurexins also influence the differential targeting of specific neuroligin family members, which would direct the effects of a particular neuroligin to one synapse type over another. In addition to possessing PDZ-binding motifs, the C-termini of neuroligins show the highest degree of sequence divergence, and may therefore provide a basis for different family members trafficking to distinct synapse types. For instance, coexpression of neuroligin-1 or -2 with the excitatory scaffold protein, PSD-95, restricts the presynaptic-inducing effects of neuroligins to excitatory contacts (117, 143). By assembling a core complex containing stargazin, AMPA receptors, and cell adhesion molecules, PSD-95 and other scaffolding proteins found in the PSD may therefore determine excitatory synapse identity, whereas other unidentified adaptor proteins may control sorting and stabilization of cell adhesion molecules at inhibitory postsynaptic sites (16, 59).

Other molecules that potentially contribute to differential retention of neuroligins at excitatory and inhibitory synapses include shank, GRIP, gephyrin and dystroglycan (39, 75, 115, 116, 122, 124). Shank is a PDZ protein enriched at glutamatergic synapses, and has been shown to influence spine morphology and presynaptic function (76, 145, 152, 153, 163). The GRIP proteins (GRIP1 and GRIP2, also known as ABP; AMPA receptor binding protein) represents another important family of scaffolding molecules, whose members contain up to seven PDZ domains and have been implicated in clustering of glutamate receptors (22, 28, 44, 49, 52, 87, 99, 120, 135, 136). GRIP family members were initially identified by their association with the GluR2/GluR3 subunits of the AMPA receptor, which bind to the fourth and fifth PDZ domains of GRIP. Disruption of the GluR2-GRIP interaction inhibits accumulation of AMPA receptors at the synapse. In addition, other studies have shown that GRIP is also present at inhibitory synapses. Thus, members of the GRIP family comprise another class of proteins potentially involved in sequestration of neuroligins to excitatory and inhibitory contacts. Gephyrin is a molecule concentrated within the postsynaptic domain of GABAergic synapses, that has been implicated in clustering of $GABA_A$ receptor subunits in hippocampal neurons (19, 42, 65, 107, 108, 114, 116). A direct interaction between gephyrin and neuroligins has not been assessed, and thus it remains a possibility that gephyrin directly influences neuroligin trafficking, or does so indirectly, through other factors that influence the function of neuroligins at inhibitory synapses. Another molecule enriched at inhibitory synapses is dystroglycan, a protein composed of an extracellular α subunit and a transmembrane β subunit. β-Dystroglycan binds dystrophin and utrophin intracellularly, while α-dystroglycan binds several matrix molecules including agrin, laminin, and perlecan extracellularly (115). The binding of dystroglycan to neurexin makes it a good candidate for modulating the effects of neuroligin/neurexin on synapse development (172). Thus, interplay between these molecules influences assembly of neuroligins at both excitatory and inhibitory synapses.

Mechanisms that control early sorting and delivery of individual neuroligin molecules to nascent neuronal contacts are less understood. To address this issue, we have monitored the mobility of neuroligin molecules at synaptic sites (Fig. 4).

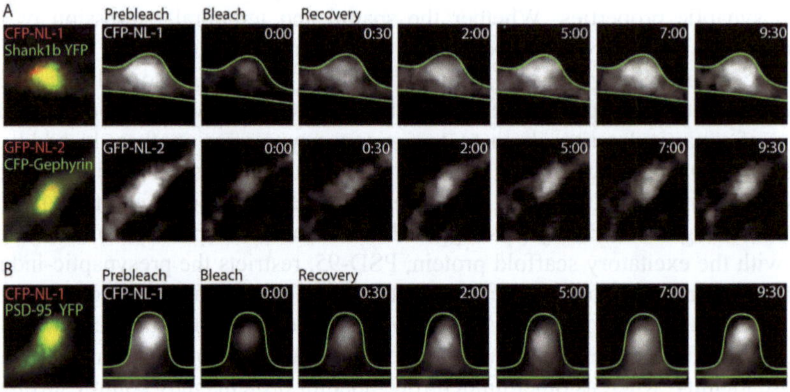

Fig. 4. Differential mobility of neuroligins at synaptic sites. (**a**) Fluorescence recovery after photobleaching anslysis shows rapid recovery of fluorescene intensity of neuroligin-1 clusters present at excitaory synpases (Shank1b-YFP-positive), suggesting the existence of a dynamic pool of neuroligin-1 at contact sites. Neuroligin-1 clusters exhibit faster recovery than those of neuroligin-2 at inhbitory (CFP-Gephyrin-positive) sites, revealing differential mobility and/or dynamics of recruitment of these proteins at synapses. (**b**) Renewal of neuroligin-1 clusters is influenced by overexpression of PSD-95, which prolongs the fluorescence recovery time of neuroligin-1, indicating that PSD-95 stabilizes the diffusible pool of neuroligin-1 at the synapse. Time is indicated in seconds.

To detect cell surface receptors, a biotinylation-based approach that allows the visualization of surface proteins in live neurons was utilized (88). This analysis revealed that clusters of both neuroligin-1 and -2 show rapid changes in fluorescence intensity, suggesting continous remodeling of neuroligin clusters at the cell surface. Fluorescence recovery after photobleaching anslysis shows similar recovery dynamics for GFP- (representative of the total pool) and AP-tagged (representative of the surface pool) neuroligins. This suggests that the accumulation of neuroligins at existing clusters involves the retention of a diffusible surface pool, rather than direct insertion from an intracellular pool. However, neuroligin-1 clusters exhibit faster recovery than those of neuroligin-2, and this process is further depressed in the case of neuroligin-2 in more mature neurons.

The renewal of neuroligin-1 clusters is also influenced by expression of PSD-95, which reduces the recovery time of neuroligin-1, indicating that PSD-95 interaction plays a role in stabilizing the diffusible pool of neuroligin-1 at the synapse. Consistent with these findings, PSD-95 enhances recruitment of endogenous neuroligin-1 to excitatory contacts, and this process correlates with an increase in the size of presynaptic terminals (143). Recent data also show that the assembly of a complex of scaffold proteins containing neuroligin-1 may precede the recruitment of presynaptic vesicle proteins and the machinery required for active recycling of neurotransmitter. In this case, the presence of postsynaptic proteins such as neuroligins may facilitate recruitment of presynaptic molecules important for vesicular release (76). Previous studies, however, demonstrated that recruitment of PSD-95 can also occur after the establishment of an active presynaptic terminal (1, 17, 69). Although the role of adhesion complexes was not examined in these studies, the presumed role of neuroligin would be to recruit PSD-95 to these sites. Another study looking at nascent contacts made by axonal filopodia highlighted the rapid recruitment of NMDA-type glutamate receptors independently of PSD-95, before the establishment of an active presynaptic terminal (190). Thus, it will be important to distinguish between trafficking of newly synthesized neuroligin molecules and retention of pre-existing protein at the cell surface and synaptic sites. Perhaps the role of PSD-95 relates more to the former at earlier stages of synaptic development during the assembly of postsynaptic proteins. At mature synapses, PSD-95 may restrict the mobility of neuroligins at the synapse and thus modulate their retention at particular contact sites.

Less is known about binding partners that influence neurexin targeting and function at presynaptic sites. The interaction between α-neurexins and neurexophilins, a family of secreted glycoproteins, has been suggested to mediate cell signaling (126, 140). Another extracellular ligand of both α- and β-neurexins is dystroglycan (172). Although the functional relevance of this interaction is uncertain, it has been suggested that it may regulate cell adhesion at inhibitory contacts. Intracellularly, neurexins are coupled to CASK, which forms a stoichiometric complex with Mint-1 (also called X11 or Lin-10) and Velis (also called MALS or Lin-7) potentially involved in organizing presynaptic vesicle release machinery (5, 23). In CASK-deficient neurons, glutamatergic spontaneous synaptic release events are increased, and GABAergic synaptic release events are decreased, indicating that coupling of neurexins to these scaffolding molecules may modulate presynaptic release.

2.5 Other Mechanisms that Regulate Neuroligin/Neurexin Assembly and Function

Other mechanisms that control neuroligin/neurexin interaction and function include calcium binding, glycosylation, and oligomerization (13, 32, 34, 35, 90, 181). Neuroligin/β-neurexin binding requires only low Ca^{2+} concentrations, and thus under physiological conditions, changes in Ca^{2+} levels at the synaptic cleft following synaptic activation may not influence association of these molecules at the synapse. Neuronal depolarization, however, may modulate splicing of specific isoforms of α-neurexins, particularly at splice sites 1 and 3 (146). Another mechanism that potentially modulates the function of neuroligins is protein oligomerization. Neuroligin-1 mutants which fail to cluster neurexins and induce presynaptic differentiation also show disrupted formation of neuroligin-1 homo-oligomers (47). These results highlight the functional importance of neuroligin oligomerization with respect to its ability to signal to the presynaptic side. It is possible that intermolecular binding occurs between different neuroligin family members, resulting in the formation of neuroligin hetero-oligomers. This may have implications in modulating neuroligin-directed recruitment of specific molecules at the synapse.

Recent evidence shows that postsynaptically localized β-neurexin inhibits trans-binding of neuroligin-1 to presynaptic neurexins, blocking synaptogenic activity (177). These results suggest that postsynaptic interactions between neuroligins and neurexins provide a mechanism for silencing the activity of this adhesion complex. This finding presents yet further insight into mechanisms potentially involved in modulating excitation and inhibition via neuroligin-neurexin interaction. Other studies showed that retrograde signaling through the neuroligin/neurexin protein complex modulates the release probability of synaptic transmitter vesicles due to alterations in presynaptic short-term plasticity (72).

2.6 In Vivo Assessment of the Function of Neurexins and Neuroligins

Analysis of α-neurexin triple knockout mice revealed an essential role of α-neurexins in Ca^{2+}-regulated neurotransmitter release, potentially through coupling N- and P/Q-type calcium channels to the presynaptic exocytotic machinery (54, 55, 125, 128, 167, 200). Mice lacking all three α-neurexins show reduced neurotransmitter release at excitatory and inhibitory synapses and die at early postnatal stages. However, no gross anatomical defects or changes in the distribution of synaptic proteins or axonal projections were observed. Structural alterations observed include a 20% reduction of neuropil area in many brain regions, due to shortened distal dendritic branches and fewer spines. A moderate decrease (30%) in inhibitory synapse number in the neocortex was also observed, indicating that neurexin function is likely more critical for inhibitory synapse maturation. Overall, these studies illustrate that α-neurexins are not essential for the formation of the majority of synapses in vivo. However, it is possible that the expression of β-neurexins or other adhesion systems may have compensated for the loss of α-neurexins in these mice. In contrast, studies on flies mutant for α-neurexin show that this adhesion molecule is required for synapse formation and associative learning (199). Neurexin-1 mutant flies are viable and fertile, but exhibit a decrease in synapse number and have shortened lifespan. Similar investigations on mice lacking neuroligins show that neuroligins are essential for survival but not for CNS synapse formation (184). Despite a significant reduction in the expression of many presynaptic proteins, there were no obvious changes in synapse number or morphology in many brain areas examined. Interestingly, mice lacking neuroligins 1–3 die shortly after birth due to respiratory failure (184). In these mutant animals, excitatory and inhibitory synaptic transmission is impaired in the brainstem respiratory center. A shift in the balance between excitation and inhibition most likely accounts for the defects in the respiratory control system, indicating that neuroligins and neurexins are not necessary for synapse induction, but rather for survival and for the regulation of specific aspects of synapse maturation that control synaptic strength, signaling and function. Together, the current data indicate that neuroligins and neurexins may take part in multiple molecular interactions to regulate a host of neuronal processes, ranging from synapse maturation to neurotransmitter exocytosis and cellular signaling (Fig. 5). Future studies are required to further elucidate the mechanisms that control their action at the synapse.

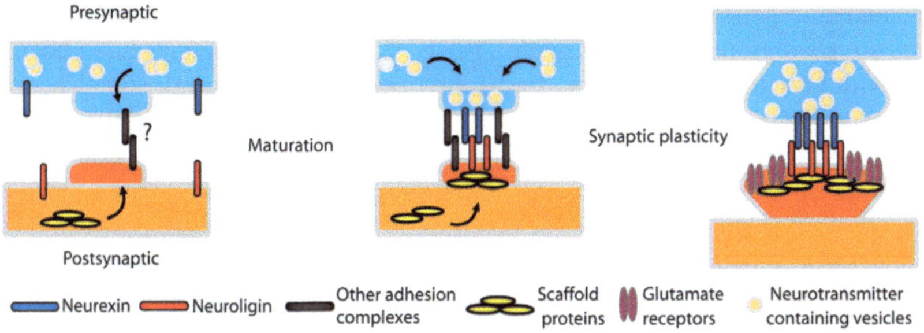

Fig. 5. Potential functions mediated by the neuroligin/neurexin adhesion complex. Putative functions proposed for the neuroligin/neurexin complex. A combination of specific isoforms of neuroligins and neurexins may allow the recognition of a particular type of neuron or even a specific location on a target neuron, thus determining the type of synaptic contact formed (e.g. excitatory versus inhibitory; *left panel*). After contact formation between pre- and postsynaptic neurons, recruitment of neuroligin and neurexin may trigger coordinated recruitment of presynaptic protein packets and postsynaptic scaffold complexes to control synapse stabilization and maturation (*middle panel*). The recruitment and/or stabilization of the neuroligin/neurexin complex at mature synapses may also influence synaptic strength. Enhanced association of neuroligin/neurexin by increased protein delivery to/retention at the synapse will strengthen trans-synaptic signaling (*right panel*). In contrast, uncoupling of the neuroligin/neurexin complex may result in reduced synaptic strength.

2.7 Potential Link to Neurodevelopmental Disorders

Recent studies indicate that altered synapse formation and maturation is the leading cause of neurodevelopmental disorders such as autism and schizophrenia (202). Defects in spine maturation have been implicated in many forms of mental retardation and memory impairment. Autism and Asperger syndrome are classified as pervasive developmental disorders, also referred to as autism-spectrum disorders (ASD). These syndromes are characterized by varying degrees of abnormality in communication ability and social interactions, as well as repetitive and stereotyped patterns of behavior. A number of twin and other family studies have supported a strong genetic component in autism. Although intensive genetic research in recent years has identified a number of genes that may be involved in the pathology of autism, the biological mechanisms underlying neurological deficits remain poorly understood. It has been hypothesized that the behavioral and cognitive deficits found in autistic patients may result from alterations in neuronal circuitry implicated in sensory, mnemonic, social and emotional information processing. Neuroimaging and neuropathological studies of these circuits in autistic patients have provided important insights into the etiology of autism, leading to a model that could explain the observed dysfunctions. This model postulates that an increase in the ratio of excitation/inhibition in neural circuits involved in information processing may underlie the dysfunctions characteristic of autism (147). Considering the role of neuroligins in controlling the function of excitatory and inhibitory synapses, these proteins are good candidate genes for being affected in these disorders. Indeed, rearrangement of chromosomal regions harboring neuroligin-1, -2 and PSD-95 genes have been associated with autism (6, 111, 202). Two mutations in the X-chromosomal neuroligin genes, neuroligin-3 and -4, have also been linked to some forms of autism, including Asperger syndrome (94). The associated mutation in neuroligin-3 consisted of a substitution of Arg451 for Cys (R451C), resulting in a missense mutation. In the case of neuroligin-4, a novel insertion at position 396 (D396X) resulted in a frame shift and premature termination of the transcript. Support for the involvement of neuroligin-4 in autism was provided by another genetic study of a large French family with members affected by non-specific X-linked mental retardation with or without autism or pervasive developmental disorders. In the latter study, another frame-shift mutation in neuroligin-4 was identified (113). In this case, a 2-base-pair deletion was found to generate a stop codon in the middle of the protein, leading to the formation of a truncated protein lacking sequences crucial for neuroligin dimerization required for binding to β-neurexins, as well as the transmembrane and intracellular domains. The fact that this mutation was found in both autistic and non-autistic mentally retarded patients indicates that neuroligin-4 may be implicated in both autism and mental retardation. Since neuroligin-associated disorders seem to include both autism and mental retardation, it suggests these two syndromes may share a common synaptically modulated mechanism (63).

Recent linkage and copy number variation analyses also implicate chromosome 11p12–p13 and neurexins, respectively, among other candidate associated with autism (175). In addition to neuroligins and neurexins, mutations in the adhesion molecule NrCam (151) and the scaffolding protein Shank3 (56) have been linked to autism. Dysfunction in adhesion molecule expression appears to be a common defect in language and/or social communication disorders.

3 Synapse Induction by Members of the IgG Superfamily

3.1 SynCAMs

Similar to neuroligins, SynCAM1 (Synaptic Cell Adhesion Molecule) has been shown to promote presynaptic differentiation at contacts between axons and heterologous cells, and to drive the recruitment of synaptic proteins at sites of contact (10, 47, 157). SynCAM1 is localized to both sides of the synaptic cleft and shares similarities with the invertebrate molecules fascicilin II and apCAM. SynCAM1 is a member of the Ig superfamily that shows homophilic, calcium-independent binding. SynCAM1 contain 3 Ig-domains, and a C-terminal PDZ-binding motif which mediates interactions with the synaptic scaffolding proteins CASK and syntenin. Loss-of-function studies indicate that PDZ domain containing adaptor proteins are involved in this process. There are three other genes encoding SynCAM, and heterophilic adhesion between the various SynCAM isoforms has been demonstrated (164). However, the function of other SynCAM isoforms and the consequences of heterophilic binding between them have not been studied in neurons. At the presynaptic terminal, the known protein interaction motifs in the cytosolic tails of SynCAM1 and β-neurexins are highly conserved. They therefore presumably bind the same adaptor proteins. Although SynCAMs and neuroligins are strong inducers of synapses, it appears that synapses induced by these adhesion molecules are functionally distinct (156). The artificial presynaptic terminals induced by these molecules on heterologous cells are identical in all parameters measured, and to true synapses. Overexpression in neurons however, produced different effects. SynCAM1 specifically promotes excitatory synaptic transmission. Moreover, only expression of SynCAM1 increased synaptic function by electrophysiological measures, whereas morphological analysis revealed the opposite, only neuroligin increased synapse number (156). In contrast, other studies have used electrophysiology to show increased synaptic function in neuroligin overexpressing neurons (117, 143), and that knockdown of neuroligin diminishes synaptic function and activity (30). The differences between these observations may be explained by variations in levels of protein overexpression and neuronal culture preparation. Despite the fact that SynCAM and neuroligin are potent synapse inducers *in vitro*, neither has been shown to be absolutely required for synapse induction in vivo. Mice lacking SynCAM1 are viable and do not exhibit major synapse abnormalities. It remains unclear whether synaptogenesis in mice lacking SynCAM1 is not altered due to functional compensation by other members of the SynCAM family (71).

3.2 NGLs and SALMS

Neuronal proteins containing leucine-rich repeats (LRRs), such as the Nogo-66 receptor, Slit, AMIGO, LINGO, NGL, and NLRR, have been implicated in neurite outgrowth and cell migration (102, 109). However, recent studies revealed that particular members of LRR containing proteins regulate synaptogenesis. Examples include the netrin G ligands (NGLs) and synaptic adhesion-like molecules (SALMs), molecules that belong to the immunoglobulin superfamily of proteins (109, 110, 189). NGLs and SALMs have similar domain structure and share the ability to bind PSD-95. The NGL family has three known members, NGL1-3, which are expressed at higher levels in postnatal brains than in embryonic tissue, supporting an important role for NGLs in synapse development. Recent studies have shown that NGL2, expressed in nonneuronal cells, induces the formation of functional presynaptic terminals in contacting axons of co-cultured neurons. Moreover, overexpression of NGL2 in neurons increases the number of excitatory synapses and clustering of postsynaptic proteins such as PSD-95 and NMDA receptors. These results suggest that NGL2 modulates excitatory synapse development through its association with PSD-95. The SALM family includes five members, each contain six LRRs, an Ig domain, and a fibronectin III domain, followed by a transmembrane domain and a C-terminal PDZ-binding motif. When overexpressed in neurons, SALMs promote neurite outgrowth, excitatory synapse maturation and clustering of NMDA receptors. However, unlike NGLs, SALMs expression in non-neuronal cells is not sufficient for synapse induction.

4 Other Adhesion Molecules Involved in Synapse Maturation

4.1 Cadherins, Catenins, and Protocadherins

Members of cadherins and protocadherins have been implicated in controlling specific stages of synaptogenesis that involve target recognition. Due to their wide expression in the brain and the existence of multiple isoforms, matching cadherins in axons and dendrites is believed to promote selective adhesion between appropriate partners. All three major subtypes of cadherins are expressed in the brain: classical cadherins, cadherin-related proteins, and protocadherins. The expression and subcellular distribution of classical cadherins, and their intracellular binding partners, catenins, varies with neuronal type and developmental stage. This distinct distribution of each of these adhesive molecules potentially provides the molecular basis for specificity of synapse connectivity (161).

At individual synapses, classical cadherins are detected at pre- and postsynaptic sites (9, 95, 178). In development, N-cadherin and β-catenin are localized diffusely in motile dendritic filopodia. Upon contact with an axon, the cadherin complex accumulates at the tip of filopodia at points of contact (95, 178). Cadherins signal via catenins to regulate dendritic spine morphology and motility. In addition, in vivo studies revealed that the loss of particular catenin molecules results in abnormal synapse formation and/or maturation. Studies using a dominant negative approach (N-cadherin lacking ectodomain), demonstrates that blockade of N-cadherin function results in a conversion of mature spines to filopodia-like spines and loss of β-catenin (178). These effects are more dramatic in younger neurons, further emphasizing the importance of cadherins at early stages of synaptogenesis. Interfering with cadherin function suggests that these molecules are more likely involved in the regulation of target specificity, rather than induction. For example, initial synaptic assembly is delayed, but not blocked, in neurons that are transfected with dominant-negative cadherin (15). Studies in the retinotectal system show that the addition of antibodies that block N-cadherin function causes retinal ganglion cells to overshoot their target and form exuberant synapses (92). Thus, evidence indicates that cadherins are important for target recognition, but they are not essential for synapse differentiation. Blocking N-cadherin adhesion causes the formation of widely separated synaptic clefts in the chick retinotectal tract (196). Mislocalization of β-catenin, through disruption of its interaction with cadherin, alters synapsin and PSD-95 localization and affects presynaptic vesicle release (7, 130). Thus, the recruitment of the cadherin-catenin complex appears to modulate maturation of pre- and postsynaptic sites.

The cadherin complex mediates profound changes in the actin cytoskeleton through several effectors including cortactin, Arp1/3, formin-1, and the Rho family of GTPases (27, 85). Several adhesion complexes have been shown to affect spine morphology or the actin cytoskeleton, including neuroligins, cadherins, integrins, syndecans, and numerous members of the immunoglobulin superfamily (61). Also, interaction between adhesion molecules with the extracellular matrix is believed to regulate spine maturation. Two integrin ligands have been demonstrated to affect dendritic spines: laminin increases spine density, whereas reelin promotes spine stability (61). In addition, the transmembrane proteoglycan syndecan has been shown to induce early transformation of filopodia protrusions into mature mushroom-shaped spines (60).

The large sequence diversity of protocadherins has also led to the suggestion of their role in target selection and specifying neuronal circuits. About 70 protocadherins are known in mouse and human, and many of these are expressed in the nervous system in largely overlapping patterns, with some regional and cell-type differences in strength of expression (67, 68). Further analysis revealed that protocadherin-γ lacking neurons form significantly fewer synapses than wild type neurons (192). Correlating expression pattern of certain protocadherin isoforms, to neuronal subsets for which the pattern of connectivity is known, may help us determine how synapse specificity is controlled.

4.2 Sidekick

The expression pattern and the consequences of its mis-expression on connectivity have also been demonstrated for the adhesion molecule sidekick. Sidekick is a member of the immunoglobulin superfamily of proteins and contains 6 Ig domains and 13 fibronectin III repeats. Sidekick was identified based on its ability to regulate photoreceptor differentiation in *Drosophila* (132). In mammals, sidekick-1 and -2 concentrate at pre- and postsynaptic sites respectively, in non-overlapping subsets of retinal neurons (198). Cells expressing sidekick-1 and -2 form separate aggregates. Deletion of the domains required for homophilic association abolishes adhesion. Moreover, the generation of chimeras reverses adhesion specificity (84). Since sidekick is enriched at synapses in narrow sub-laminae of the inner plexiform layer, this suggests that sidekick-mediated adhesion may promote laminae specific connectivity.

4.3 Ephrins and Ephrin Receptors

Ephrin receptors (Eph's) are a large family of 14 receptors and 8 ligands known as ephrins. Eph receptors and ephrin ligands are expressed through out the developing and mature nervous system and were initially characterized based on their ability to guide axons to their proper targets, and later determined to be present at synapses (20, 66, 179). Eph receptors are divided into two classes: EphA and EphB, which bind ephrin A and ephrin B respectively. Ephrins must be clustered to bind and activate their cognate receptor (43). EphB receptor tyrosine kinases have been shown to regulate excitatory synaptogenesis, including the clustering of NMDARs and AMPARs, dendritic spine formation, and presynaptic differentiation (60, 97). EphB-mediated signaling seems to be particularly significant for a subset of synaptic inputs in vivo (61).

4.4 NCAM and L1-CAM

The neural cell adhesion molecule (NCAM) has been shown to regulate synapse maturation through homo- and heterophilic interactions with recognition molecules (21, 40, 48, 50, 51, 133, 149, 160, 171, 174, 183). In particular, NCAM is thought to regulate the early events that control contact formation, and fasciculation of axons to form bundles, such as the retinotectal and mossy fiber projections (40, 149). NCAM is encoded by a single gene located on chromosome 11, and as a result of alternative splicing, is synthesized in three main membrane-bound isoforms (NCAM-120, NCAM-140, and NCAM-180) which differ in their C-terminals. NCAM-180 and -140 are transmembrane proteins, with different length intracellular domains, whereas NCAM-120 is attached to the membrane by a glycosylphosphalitdylinositol (GPI) anchor. All three isoforms have a conserved extracellular domain consisting of 5 Ig domains followed by 2 fibronectin type III repeats. All three isoforms of NCAM can be released from primary hippocampal neurons in culture (89). The significance of proteolytic processing of membrane-bound NCAM is not well known, however, soluble NCAM forms have been implicated in synaptic plasticity, including long-term potentiation (LTP) (62, 101). Live imaging studies have shown that NCAM translocates to contact sites within minutes after initial contact formation (174). Thus, NCAM is potentially involved in early synaptogenesis and in synaptic maturation. At later stages, the relative levels of post-, but not presynaptic, NCAM expression regulates the strength of excitatory synapses in a NMDA receptor-dependent manner (50). This mechanism involves the interaction between the polysialylated form of NCAM and heparan sulfate proteoglycans, and requires fibroblast growth factor receptor (FGFR)-mediated signaling. The importance of mammalian NCAM for synapse dynamics is further supported by activity-dependent expression of NCAM (159) and impaired LTP in mice deficient in NCAM (21, 57, 129).

L1-CAMs are expressed by almost all post-mitotic neurons in the CNS at the onset of differentiation, but not by glia. During development L1-CAM has been shown to regulate neuronal migration, neurite outgrowth, fasciculation and guidance, survival, and myelination, and some of their roles in the CNS, particularly at synapses, have recently been elucidated. Synaptic plasticity in hippocampal slices from adult constitutively L1-deficient mice do not have significant changes in LTP induced by theta-burst or high frequency stimulation in the CA1 region (11). These findings were recapitulated in juvenile mice, but interestingly, perisomatic inhibition was noticeably impaired, with a decrease in the frequency and amplitude of inhibitory postsynaptic currents, as well as a decrease in the number of inhibitory synapses (150).

The repertoire of molecules that influence synapse induction and maturation is rapidly increasing. An RNAi-based approach identified other candidate molecules involved in glutamatergic and GABAergic synapse development. A survey of 22 cadherin family members demonstrated distinct roles for cadherin-11 and cadherin-13 in synapse development. This screen also revealed roles for the class 4 Semaphorins, Sema4B and Sema4D, in the development of glutamatergic and GABAergic synapses. Sema4D influences GABAergic, but not glutamatergic synapse development. These experiments identify new components that influence the genetic program that regulate synapse formation and/or maintenance.

5 Adhesion Molecules Involved in Inhibitory Synapse Development

A role for neuroligins and protocadherins in inhibitory synapse maturation and function has been established (75). A few other adhesion molecules have been implicated in inhibitory synaptogenesis including dystroglycan, L1, and N-CAM. Dystroglycan was the first identified adhesive macromolecule at mature GABA synapses, clustering at synaptic loci after synaptic vesicles, $GABA_A$ receptors, and gephyrin (108, 115). Dystroglycan is composed of an extracellular α-subunit and a transmembrane β-subunit derived from proteolytic cleavage and glycosylation of a single precursor protein, and binds several extracellular matrix molecules such as agrin, laminin and perlecan. In addition, dystroglycan binds to the intracellular proteins, dystrophin and utrophin (193). Importantly, dystroglycan also binds to two cell surface proteins involved in synaptogenesis, namely agrin and β-neurexin (172, 173). Despite the potential of dystroglycan as an inhibitory synaptic adhesion protein, studies on neurons obtained from mice mutant for dystroglycan indicate that this protein is not essential, since clustering of many proteins that localize to inhibitory synapses is not affected in these mice (115). These findings suggest that dystroglycans may play a more prominent role in the maturation of inhibitory contacts.

A clue for one of the functions of L1-CAM in inhibitory synapse development comes from a study on a member of the family known as neurofascin186 (3). The precise localization of neurofascin186 by the membrane adaptor protein ankyrin G is responsible for establishment of a gradient of this adhesion molecule at the initial axon segment of cerebellar Purkinje cells. Disruption of this gradient results in disruption of the subcellular localization of GABAergic synapses on the axonal initial segment (3). These data suggest that L1 may be important in directing formation of inhibitory synapses to specific sites on the target neurons. Protocadherins also appear to be important for inhibitory synapse development in the spinal cord. Spinal cord neurons prepared from protocadherin mutant mice show a decrease in the number of GABAergic terminals. Moreover, electrophysiological recordings revealed a decrease in both spontaneous and evoked inhibitory currents (192). Whether protocadherins are important for inhibitory synapse development in other areas of the brain remains to be explored.

6 Role for Adhesion Molecules in Synaptic Plasticity

Expression of many adhesion molecules persists into adulthood, and as such, these molecules are thought to play a role in the activity-dependent remodeling of synaptic contacts, a process associated with learning and memory paradigms. The expression level of several adhesion molecules that participate in the modification of existing synapses is also regulated by neuronal activity. For instance, N-cadherin, NCAM and L1 are up-regulated by distinct action potential firing patterns (93). Moreover, the promoter for NCAM is sensitive to AMPA receptor activity (86). Functional blocking approaches and gene knockouts also support a role for several families of cell adhesion molecules, including integrins, cadherins, ephrins and NCAM, in the modulation of synaptic plasticity and animal behavior.

Integrins potentially play multiple roles in adhesion, signal transduction and cytoskeletal organization (58, 134, 142). Blocking integrin function in the adult rat hippocampus with an RGD sequence peptide, which interferes with integrin binding to its endogenous ligands in the extracellular matrix, causes a decay of LTP, an enduring enhancement of excitatory synaptic transmission (170, 195). Furthermore, chronic blockade by RGD peptides or by functional antibodies against the β3 integrin subunit prevents activity-dependent reduction in the probability of glutamate release (29). Members of the Ig superfamily, in particular NCAM and L1, also control synaptic plasticity. Antibodies against L1 or NCAM infused into the hippocampus of rats decreases efficiency of spatial learning tasks such as the Morris water maze (4). L1 knockout mice show similar learning impairments (96). *In vitro*, L1 antibodies and soluble oligomannosides that disrupt interaction with NCAM, significantly reduce LTP induced by theta-burst stimulation (123).

Several adhesion complexes regulate retrograde signaling between pre- and postsynaptic compartments. The EphB-ephrinB receptor-ligand pair is of particular interest due to its capacity for bidirectional signaling (66), which is thought to coordinate changes in both sides of the synapse during LTP. On the postsynaptic membrane, addition of ephrinB induces interaction between EphB and NMDA-type glutamate receptors, an increase in NMDA-mediated calcium influx and gene expression (41, 176). Disruption of Eph/ephrin signaling by interfering peptides inhibits retrograde signaling through ephrinB, blocking LTP at mossy fibers in the cerebellum (37). Thus, multiple adhesion systems exist to modulate communication between pre- and postsynaptic cells to induce long lasting changes in synaptic efficacy. Both LTP and LTD have been reported at inhibitory synapses in a variety of brain regions, including the hippocampus, cortex, cerebellum and brain stem (24, 25, 73, 81). However, it remains to be determined whether changes in plasticity of inhibitory synapses are controlled by similar mechanisms that involve retrograde signaling through adhesive complexes.

7 Conclusions and Future Directions

Numerous studies indicate that several of the recently identified adhesion molecules can modulate synaptic contact number, morphology and function. However, strong evidence that any of these molecules are indispensable for synapse formation in vivo is lacking, suggesting functional redundancy. Thus, one can envisage that synapse formation and maturation rely on the collective action of these adhesive systems. Most likely, crosstalk between the different adhesive systems ensures proper connectivity and recruitment of appropriate neurotransmitter receptors and associated protein complexes. Whether these adhesion molecules act in parallel or in a hierarchical manner remains unknown. Future studies are needed to determine how these various adhesion systems orchestrate the establishment and function of synapses. Time-lapse imaging studies monitoring the behavior of several molecules implicated in synapse development will help address some of these issues. Moreover, future in vivo experiments are needed to define molecules essential for synapse induction, contact initiation, differentiation, and plasticity.

Acknowledgements

Many thanks to Rochelle Hines for providing the electron micrographs for illustration, and to Joshua Levinson for insightful comments. A.E.H. research is supported by funds from CIHR, EJLB, MSFHR, and Neuroscience Canada.

References

1. Ahmari SE, Buchanan J, and Smith SJ. Assembly of presynaptic active zones from cytoplasmic transport packets. *Nat Neurosci* 3: 445–451, 2000.
2. Anderson TR, Shah PA, and Benson DL. Maturation of glutamatergic and GABAergic synapse composition in hippocampal neurons. *Neuropharmacology* 47: 694–705, 2004.
3. Ango F, di Cristo G, Higashiyama H, Bennett V, Wu P, and Huang ZJ. Ankyrin-based subcellular gradient of neurofascin, an immunoglobulin family protein, directs GABAergic innervation at purkinje axon initial segment. *Cell* 119: 257–272, 2004.
4. Arami S, Jucker M, Schachner M, and Welzl H. The effect of continuous intraventricular infusion of L1 and NCAM antibodies on spatial learning in rats. *Behav Brain Res* 81: 81–87, 1996.
5. Atasoy D, Schoch S, Ho A, Nadasy KA, Liu X, Zhang W, Mukherjee K, Nosyreva ED, Fernandez-Chacon R, Missler M, Kavalali ET, and Sudhof TC. Deletion of CASK in mice is lethal and impairs synaptic function. *Proc Natl Acad Sci U S A* 104: 2525–2530, 2007.
6. Auranen M, Vanhala R, Varilo T, Ayers K, Kempas E, Ylisaukko-Oja T, Sinsheimer JS, Peltonen L, and Jarvela I. A genomewide screen for autism-spectrum disorders: evidence for a major susceptibility locus on chromosome 3q25-27. *Am J Hum Genet* 71: 777–790, 2002.
7. Bamji SX, Shimazu K, Kimes N, Huelsken J, Birchmeier W, Lu B, and Reichardt LF. Role of beta-catenin in synaptic vesicle localization and presynaptic assembly. *Neuron* 40: 719–731, 2003.
8. Barnard EA, Skolnick P, Olsen RW, Mohler H, Sieghart W, Biggio G, Braestrup C, Bateson AN, and Langer SZ. International Union of Pharmacology. XV. Subtypes of gamma-aminobutyric acidA receptors: classification on the basis of subunit structure and receptor function. *Pharmacol Rev* 50: 291–313, 1998.
9. Benson DL and Tanaka H. N-cadherin redistribution during synaptogenesis in hippocampal neurons. *J Neurosci* 18: 6892–6904, 1998.
10. Biederer T, Sara Y, Mozhayeva M, Atasoy D, Liu X, Kavalali ET, and Sudhof TC. SynCAM, a synaptic adhesion molecule that drives synapse assembly. *Science* 297: 1525–1531, 2002.
11. Bliss T, Errington M, Fransen E, Godfraind JM, Kauer JA, Kooy RF, Maness PF, and Furley AJ. Long-term potentiation in mice lacking the neural cell adhesion molecule L1. *Curr Biol* 10: 1607–1610, 2000.
12. Bolliger MF, Frei K, Winterhalter KH, and Gloor SM. Identification of a novel neuroligin in humans which binds to PSD-95 and has a widespread expression. *Biochem J* 356: 581–588, 2001.

13. Boucard AA, Chubykin AA, Comoletti D, Taylor P, and Sudhof TC. A Splice Code for trans-Synaptic Cell Adhesion Mediated by Binding of Neuroligin 1 to alpha- and beta-Neurexins. *Neuron* 48: 229–236, 2005.
14. Bowery NG, Bettler B, Froestl W, Gallagher JP, Marshall F, Raiteri M, Bonner TI, and Enna SJ. International Union of Pharmacology. XXXIII. Mammalian gamma-aminobutyric acid(B) receptors: structure and function. *Pharmacol Rev* 54: 247–264, 2002.
15. Bozdagi O, Valcin M, Poskanzer K, Tanaka H, and Benson DL. Temporally distinct demands for classic cadherins in synapse formation and maturation. *Mol Cell Neurosci* 27: 509–521, 2004.
16. Bredt DS and Nicoll RA. AMPA receptor trafficking at excitatory synapses. *Neuron* 40: 361–379, 2003.
17. Bresler T, Shapira M, Boeckers T, Dresbach T, Futter M, Garner CC, Rosenblum K, Gundelfinger ED, and Ziv NE. Postsynaptic density assembly is fundamentally different from presynaptic active zone assembly. *J Neurosci* 24: 1507–1520, 2004.
18. Brose N. Synaptic cell adhesion proteins and synaptogenesis in the mammalian central nervous system. *Naturwissenschaften* 86: 516–524, 1999.
19. Brunig I, Suter A, Knuesel I, Luscher B, and Fritschy JM. GABAergic terminals are required for postsynaptic clustering of dystrophin but not of GABA(A) receptors and gephyrin. *J Neurosci* 22: 4805–4813, 2002.
20. Buchert M, Schneider S, Meskenaite V, Adams MT, Canaani E, Baechi T, Moelling K, and Hovens CM. The Junction-associated protein AF-6 interacts and clusters with specific Eph receptor tyrosine kinases at specialized sites of cell–cell contact in the brain. *J Cell Biol* 144: 361–371, 1999.
21. Bukalo O, Fentrop N, Lee AY, Salmen B, Law JW, Wotjak CT, Schweizer M, Dityatev A, and Schachner M. Conditional ablation of the neural cell adhesion molecule reduces precision of spatial learning, long-term potentiation, and depression in the CA1 subfield of mouse hippocampus. *J Neurosci* 24: 1565–1577, 2004.
22. Burette A, Khatri L, Wyszynski M, Sheng M, Ziff EB, and Weinberg RJ. Differential cellular and subcellular localization of ampa receptor-binding protein and glutamate receptor-interacting protein. *J Neurosci* 21: 495–503, 2001.
23. Butz S, Okamoto M, and Sudhof TC. A tripartite protein complex with the potential to couple synaptic vesicle exocytosis to cell adhesion in brain. *Cell* 94: 773–782, 1998.
24. Caillard O, Ben-Ari Y, and Gaiarsa J-L. Long-term potentiation of GABAergic synaptic transmission in neonatal rat hippocampus. *J Physiol (Lond)* 518: 109–119, 1999.
25. Caillard O, Ben-Ari Y, and Gaiarsa J-L. Mechanisms of Induction and Expression of Long-Term Depression at GABAergic Synapses in the Neonatal Rat Hippocampus. *J Neurosci* 19: 7568–7577, 1999.
26. Cantallops I and Cline HT. Synapse formation: if it looks like a duck and quacks like a duck. *Curr Biol* 10: R620–R623, 2000.
27. Carlisle HJ and Kennedy MB. Spine architecture and synaptic plasticity. *Trends Neurosci* 28: 182–187, 2005.
28. Charych EI, Yu W, Li R, Serwanski DR, Miralles CP, Li X, Yang BY, Pinal N, Walikonis R, and De Blas AL. A four PDZ domain-containing splice variant form of GRIP1 is localized in GABAergic and glutamatergic synapses in the brain. *J Biol Chem* 279: 38978–38990, 2004.
29. Chavis P and Westbrook G. Integrins mediate functional pre- and postsynaptic maturation at a hippocampal synapse. 411: 317–321, 2001.
30. Chih B, Engelman H, and Scheiffele P. Control of excitatory and inhibitory synapse formation by neuroligins. *Science* 307: 1324–1328, 2005.

31. Chih B, Engelman H, and Scheiffele P. Control of excitatory and inhibitory synapse formation by neuroligins. *Science* 307: 1324–1328, 2005.
32. Chih B, Gollan L, and Scheiffele P. Alternative splicing controls selective trans-synaptic interactions of the neuroligin-neurexin complex. *Neuron* 51: 171–178, 2006.
33. Chubykin AA, Atasoy D, Etherton MR, Brose N, Kavalali ET, Gibson JR, and Sudhof TC. Activity-dependent validation of excitatory versus inhibitory synapses by neuroligin-1 versus neuroligin-2. *Neuron* 54: 919–931, 2007.
34. Comoletti D, Flynn R, Jennings LL, Chubykin A, Matsumura T, Hasegawa H, Sudhof TC, and Taylor P. Characterization of the interaction of a recombinant soluble neuroligin-1 with neurexin-1beta. *J Biol Chem* 278: 50497–50505, 2003.
35. Comoletti D, Flynn RE, Boucard AA, Demeler B, Schirf V, Shi J, Jennings LL, Newlin HR, Sudhof TC, and Taylor P. Gene selection, alternative splicing, and post-translational processing regulate neuroligin selectivity for beta-neurexins. *Biochemistry* 45: 12816–12827, 2006.
36. Conroy WG, Nai Q, Ross B, Naughton G, and Berg DK. Postsynaptic neuroligin enhances presynaptic inputs at neuronal nicotinic synapses. *Dev Biol* 307: 79–91, 2007.
37. Contractor A, Rogers C, Maron C, Henkemeyer M, Swanson GT, and Heinemann SF. Trans-synaptic Eph receptor-ephrin signaling in hippocampal mossy fiber LTP. *Science* 296: 1864–1869, 2002.
38. Craig AM and Boudin H. Molecular heterogeneity of central synapses: afferent and target regulation. *Nat Neurosci* 4: 569–578, 2001.
39. Craig AM and Kang Y. Neurexin-neuroligin signaling in synapse development. *Curr Opin Neurobiol* 17: 43–52, 2007.
40. Cremer H, Chazal G, Goridis C, and Represa A. NCAM is essential for axonal growth and fasciculation in the hippocampus. *Mol Cell Neurosci* 8: 323–335, 1997.
41. Dalva MB, Takasu MA, Lin MZ, Shamah SM, Hu L, Gale NW, and Greenberg ME. EphB receptors interact with NMDA receptors and regulate excitatory synapse formation. *Cell* 103: 945–956, 2000.
42. Danglot L, Triller A, and Bessis A. Association of gephyrin with synaptic and extrasynaptic GABA receptors varies during development in cultured hippocampal neurons. *Mol Cell Neurosci* 23: 264–278, 2003.
43. Davis S, Gale NW, Aldrich TH, Maisonpierre PC, Lhotak V, Pawson T, Goldfarb M, and Yancopoulos GD. Ligands for EPH-related receptor tyrosine kinases that require membrane attachment or clustering for activity. *Science* 266: 816–819, 1994.
44. Daw MI, Chittajallu R, Bortolotto ZA, Dev KK, Duprat F, Henley JM, Collingridge GL, and Isaac JT. PDZ proteins interacting with C-terminal GluR2/3 are involved in a PKC-dependent regulation of AMPA receptors at hippocampal synapses. *Neuron* 28: 873–886, 2000.
45. Dean C and Dresbach T. Neuroligins and neurexins: linking cell adhesion, synapse formation and cognitive function. *Trends Neurosci* 29: 21–29, 2006.
46. Dean C and Dresbach T. Neuroligins and neurexins: linking cell adhesion, synapse formation and cognitive function. *Trends Neurosci* 29: 21-29, 2006.
47. Dean C, Scholl FG, Choih J, DeMaria S, Berger J, Isacoff E, and Scheiffele P. Neurexin mediates the assembly of presynaptic terminals. *Nat Neurosci* 6: 708–716, 2003.
48. Delling M, Wischmeyer E, Dityatev A, Sytnyk V, Veh RW, Karschin A, and Schachner M. The neural cell adhesion molecule regulates cell-surface delivery of G-protein-activated inwardly rectifying potassium channels via lipid rafts. *J Neurosci* 22: 7154–7164, 2002.
49. DeSouza S, Fu J, States BA, and Ziff EB. Differential palmitoylation directs the AMPA receptor-binding protein ABP to spines or to intracellular clusters. *J Neurosci* 22: 3493–3503, 2002.

50. Dityatev A, Dityateva G, and Schachner M. Synaptic strength as a function of post- versus presynaptic expression of the neural cell adhesion molecule NCAM. *Neuron* 26: 207–217, 2000.
51. Dityatev A, Dityateva G, Sytnyk V, Delling M, Toni N, Nikonenko I, Muller D, and Schachner M. Polysialylated neural cell adhesion molecule promotes remodeling and formation of hippocampal synapses. *J Neurosci* 24: 9372–9382, 2004.
52. Dong H, O'Brien RJ, Fung ET, Lanahan AA, Worley PF, and Huganir RL. GRIP: a synaptic PDZ domain-containing protein that interacts with AMPA receptors. *Nature* 386: 279–284, 1997.
53. Dong N, Qi J, and Chen G. Molecular reconstitution of functional GABAergic synapses with expression of neuroligin-2 and GABA(A) receptors. *Mol Cell Neurosci* 35: 14–23, 2007.
54. Dudanova I, Sedej S, Ahmad M, Masius H, Sargsyan V, Zhang W, Riedel D, Angenstein F, Schild D, Rupnik M, and Missler M. Important contribution of alpha-neurexins to Ca2+-triggered exocytosis of secretory granules. *J Neurosci* 26: 10599–10613, 2006.
55. Dudanova I, Tabuchi K, Rohlmann A, Sudhof TC, and Missler M. Deletion of alpha-neurexins does not cause a major impairment of axonal pathfinding or synapse formation. *J Comp Neurol* 502: 261–274, 2007.
56. Durand CM, Betancur C, Boeckers TM, Bockmann J, Chaste P, Fauchereau F, Nygren G, Rastam M, Gillberg IC, Anckarsater H, Sponheim E, Goubran-Botros H, Delorme R, Chabane N, Mouren-Simeoni MC, de Mas P, Bieth E, Roge B, Heron D, Burglen L, Gillberg C, Leboyer M, and Bourgeron T. Mutations in the gene encoding the synaptic scaffolding protein SHANK3 are associated with autism spectrum disorders. *Nat Genet* 39: 25–27, 2007.
57. Eckhardt M, Bukalo O, Chazal G, Wang L, Goridis C, Schachner M, Gerardy-Schahn R, Cremer H, and Dityatev A. Mice deficient in the polysialyltransferase ST8SiaIV/PST-1 allow discrimination of the roles of neural cell adhesion molecule protein and polysialic acid in neural development and synaptic plasticity. *J Neurosci* 20: 5234–5244, 2000.
58. Einheber S, Schnapp LM, Salzer JL, Cappiello ZB, and Milner TA. Regional and ultrastructural distribution of the alpha 8 integrin subunit in developing and adult rat brain suggests a role in synaptic function. *J Comp Neurol* 370: 105–134, 1996.
59. El-Husseini AE, Schnell E, Chetkovich DM, Nicoll RA, and Bredt DS. PSD-95 involvement in maturation of excitatory synapses. *Science* 290: 1364–1368, 2000.
60. Ethell IM, Irie F, Kalo MS, Couchman JR, Pasquale EB, and Yamaguchi Y. EphB/syndecan-2 signaling in dendritic spine morphogenesis. *Neuron* 31: 1001–1013, 2001.
61. Ethell IM and Pasquale EB. Molecular mechanisms of dendritic spine development and remodeling. *Prog Neurobiol* 75: 161–205, 2005.
62. Fazeli MS, Breen K, Errington ML, and Bliss TV. Increase in extracellular NCAM and amyloid precursor protein following induction of long-term potentiation in the dentate gyrus of anaesthetized rats. *Neurosci Lett* 169: 77–80, 1994.
63. Feng J, Schroer R, Yan J, Song W, Yang C, Bockholt A, Cook EH, Jr., Skinner C, Schwartz CE, and Sommer SS. High frequency of neurexin 1beta signal peptide structural variants in patients with autism. *Neurosci Lett* 409: 10–13, 2006.
64. Fiala JC, Feinberg M, Popov V, and Harris KM. Synaptogenesis via dendritic filopodia in developing hippocampal area CA1. *J Neurosci* 18: 8900–8911, 1998.
65. Fischer F, Kneussel M, Tintrup H, Haverkamp S, Rauen T, Betz H, and Wassle H. Reduced synaptic clustering of GABA and glycine receptors in the retina of the gephyrin null mutant mouse. *J Comp Neurol* 427: 634–648, 2000.
66. Flanagan JG and Vanderhaeghen P. The Ephrins and Eph Receptors in Neural Deveopment. *Annu Rev Neurosci* 21: 309–345, 1998.

67. Frank M, Ebert M, Shan W, Phillips GR, Arndt K, Colman DR, and Kemler R. Differential expression of individual gamma-protocadherins during mouse brain development. *Mol Cell Neurosci Aug;* 29: 603–616, 2005.
68. Frank M and Kemler R. Protocadherins. *Curr Opin Cell Biol* 14: 557–562, 2002.
69. Friedman HV, Bresler T, Garner CC, and Ziv NE. Assembly of new individual excitatory synapses: time course and temporal order of synaptic molecule recruitment. *Neuron* 27: 57–69, 2000.
70. Fu Z, Washbourne P, Ortinski P, and Vicini S. Functional excitatory synapses in HEK293 cells expressing neuroligin and glutamate receptors. *J Neurophysiol* 90: 3950–3957, 2003.
71. Fujita E, Kouroku Y, Ozeki S, Tanabe Y, Toyama Y, Maekawa M, Kojima N, Senoo H, Toshimori K, and Momoi T. Oligo-astheno-teratozoospermia in mice lacking RA175/TSLC1/SynCAM/IGSF4A, a cell adhesion molecule in the immunoglobulin superfamily. *Mol Cell Biol* 26: 718–726, 2006.
72. Futai K, Kim MJ, Hashikawa T, Scheiffele P, Sheng M, and Hayashi Y. Retrograde modulation of presynaptic release probability through signaling mediated by PSD-95-neuroligin. *Nat Neurosci* 10: 186–195, 2007.
73. Gaiarsa J-L, Caillard O, and Ben-Ari Y. Long-term plasticity at GABAergic and glycinergic synapses: mechanisms and functional significance. *Trends Neurosci* 25: 564–570, 2002.
74. Garner CC, Zhai RG, Gundelfinger ED, and Ziv NE. Molecular mechanisms of CNS synaptogenesis. *Trends Neurosci* 25: 243–251, 2002.
75. Gerrow K and El-Husseini A. Cell adhesion molecules at the synapse. *Front Biosci* 11: 2400–2419, 2006.
76. Gerrow K, Romorini S, Nabi SM, Colicos MA, Sala C, and El-Husseini A. A preformed complex of postsynaptic proteins is involved in excitatory synapse development. *Neuron* 49: 547–562, 2006.
77. Gilbert MM and Auld VJ. Evolution of clams (cholinesterase-like adhesion molecules): structure and function during development. *Front Biosci* 10: 2177–2192, 2005.
78. Goda Y and Davis GW. Mechanisms of synapse assembly and disassembly. *Neuron* 40: 243–264, 2003.
79. Graf ER, Kang Y, Hauner AM, and Craig AM. Structure function and splice site analysis of the synaptogenic activity of the neurexin-1 beta LNS domain. *J Neurosci* 26: 4256–4265, 2006.
80. Graf ER, Zhang X, Jin SX, Linhoff MW, and Craig AM. Neurexins Induce Differentiation of GABA and Glutamate Postsynaptic Specializations via Neuroligins. *Cell* 119: 1013–1026, 2004.
81. Gubellini P, Ben-Ari Y, and Gaiarsa J-L. Activity- and age-dependent GABAergic synaptic plasticity in the developing rat hippocampus. *Eur J Neurosci* 14: 1937–1946, 2001.
82. Harris KM, Jensen FE, and Tsao B. Three-dimensional structure of dendritic spines and synapses in rat hippocampus (CA1) at postnatal day 15 and adult ages: implications for the maturation of synaptic physiology and long-term potentiation. *J Neurosci* 12: 2685–2705, 1992.
83. Hashimoto K and Kano M. Functional differentiation of multiple climbing fiber inputs during synapse elimination in the developing cerebellum. *Neuron* 38: 785–796, 2003.
84. Hayashi K, Kaufman L, Ross MD, and Klotman PE. Definition of the critical domains required for homophilic targeting of mouse sidekick molecules. *FASEB J* 19: 614–616, 2005.
85. Hering H and Sheng M. Dendritic spines: structure, dynamics and regulation. *Nat Rev Neurosci* 2: 880–888, 2001.

86. Holst BD, Vanderklish PW, Krushel LA, Zhou W, Langdon RB, McWhirter JR, Edelman GM, and Crossin KL. Allosteric modulation of AMPA-type glutamate receptors increases activity of the promoter for the neural cell adhesion molecule, N-CAM. *Proc Natl Acad Sci U S A* 95: 2597–2602, 1998.
87. Hoogenraad CC, Milstein AD, Ethell IM, Henkemeyer M, and Sheng M. GRIP1 controls dendrite morphogenesis by regulating EphB receptor trafficking. *Nat Neurosci* 8: 906–915, 2005.
88. Howarth M.J.,- Chinnapen F, Dorrestein P.C., Gerrow K., Grandy M.R., Kelleher NL, El-Husseini A, and Ting. AY. A monovalent streptavidin with a single femtomolar biotin binding site. *Nature Methods* 3:267–273, 2006.
89. Hübschmann MV, Skladchikova G, Bock E, and Berezin V. Neural cell adhesion molecule function is regulated by metalloproteinase-mediated ectodomain release. *J Neurosci Res* 80: 826–837, 2005.
90. Ichtchenko K, Hata Y, Nguyen T, Ullrich B, Missler M, Moomaw C, and Sudhof TC. Neuroligin 1: a splice site-specific ligand for beta-neurexins. *Cell* 81: 435–443, 1995.
91. Ichtchenko K, Nguyen T, and Sudhof TC. Structures, alternative splicing, and neurexin binding of multiple neuroligins. *J Biol Chem* 271: 2676–2682, 1996.
92. Inoue A and Sanes JR. Lamina-specific connectivity in the brain: regulation by N-cadherin, neurotrophins, and glycoconjugates. *Science* 276: 1428–1431, 1997.
93. Itoh K, Ozaki M, Stevens B, and Fields RD. Activity-dependent regulation of N-cadherin in DRG neurons: differential regulation of N-cadherin, NCAM, and L1 by distinct patterns of action potentials. *J Neurobiol* 33: 735–748, 1997.
94. Jamain S, Quach H, Betancur C, Rastam M, Colineaux C, Gillberg IC, Soderstrom H, Giros B, Leboyer M, Gillberg C, and Bourgeron T. Mutations of the X-linked genes encoding neuroligins NLGN3 and NLGN4 are associated with autism. *Nat Genet* 34: 27–29, 2003.
95. Jontes JD, Emond MR, and Smith SJ. In vivo trafficking and targeting of N-cadherin to nascent presynaptic terminals. *J Neurosci* 24: 9027–9034, 2004.
96. Kamiguchi H, Hlavin ML, and Lemmon V. Role of L1 in Neural development: what the knockouts tell us. *Mol Cell Neurosci* 12: 48–55, 1998.
97. Kayser MS, McClelland AC, Hughes EG, and Dalva MB. Intracellular and trans-synaptic regulation of glutamatergic synaptogenesis by EphB receptors. *J Neurosci* 26: 12152–12164, 2006.
98. Kennedy MB. The postsynaptic density at glutamatergic synapses. *Trends Neurosci* 20: 264–268, 1997.
99. Kim CH, Chung HJ, Lee HK, and Huganir RL. Interaction of the AMPA receptor subunit GluR2/3 with PDZ domains regulates hippocampal long-term depression. *Proc Natl Acad Sci U S A* 98: 11725–11730, 2001.
100. Kim E and Sheng M. PDZ domain proteins of synapses. *Nat Rev Neurosci* 5: 771–781, 2004.
101. Kim JS, Kim HY, Kim JH, Shin HK, Lee SH, Lee YS, and Son H. Enhancement of rat hippocampal long-term potentiation by 17[beta]-estradiol involves mitogen-activated protein kinase-dependent and -independent components. *Neurosci Lett* 332: 65–69, 2002.
102. Kim S, Burette A, Chung HS, Kwon SK, Woo J, Lee HW, Kim K, Kim H, Weinberg RJ, and Kim E. NGL family PSD-95-interacting adhesion molecules regulate excitatory synapse formation. *Nat Neurosci* 9: 1294–1301, 2006.
103. Kirov SA, Goddard CA, and Harris KM. Age-dependence in the homeostatic upregulation of hippocampal dendritic spine number during blocked synaptic transmission. *Neuropharmacology* 47: 640–648, 2004.
104. Kirov SA and Harris KM. Dendrites are more spiny on mature hippocampal neurons when synapses are inactivated. *Nat Neurosci* 2: 878–883, 1999.

105. Kirov SA, Petrak LJ, Fiala JC, and Harris KM. Dendritic spines disappear with chilling but proliferate excessively upon rewarming of mature hippocampus. *Neuroscience* 127: 69–80, 2004.
106. Kneussel M and Betz H. Clustering of inhibitory neurotransmitter receptors at developing postsynaptic sites: the membrane activation model. *Trends Neurosci* 23: 429–435, 2000.
107. Kneussel M, Brandstatter JH, Gasnier B, Feng G, Sanes JR, and Betz H. Gephyrin-independent clustering of postsynaptic GABA(A) receptor subtypes. *Mol Cell Neurosci* 17: 973–982, 2001.
108. Kneussel M, Brandstatter JH, Laube B, Stahl S, Muller U, and Betz H. Loss of postsynaptic GABA(A) receptor clustering in gephyrin-deficient mice. *J Neurosci* 19: 9289–9297, 1999.
109. Ko J and Kim E. Leucine-rich repeat proteins of synapses. *J Neurosci Res* 85: 2829–2832, 2007.
110. Ko J, Kim S, Chung HS, Kim K, Han K, Kim H, Jun H, Kaang BK, and Kim E. SALM synaptic cell adhesion-like molecules regulate the differentiation of excitatory synapses. *Neuron* 50: 233–245, 2006.
111. Konstantareas MM and Homatidis S. Chromosomal abnormalities in a series of children with autistic disorder. *J Autism Dev Disord* 29: 275–285, 1999.
112. Kornau HC, Seeburg PH, and Kennedy MB. Interaction of ion channels and receptors with PDZ domain proteins. *Curr Opin Neurobiol* 7: 368–373, 1997.
113. Laumonnier F, Bonnet-Brilhault F, Gomot M, Blanc R, David A, Moizard MP, Raynaud M, Ronce N, Lemonnier E, Calvas P, Laudier B, Chelly J, Fryns JP, Ropers HH, Hamel BC, Andres C, Barthelemy C, Moraine C, and Briault S. X-linked mental retardation and autism are associated with a mutation in the NLGN4 gene, a member of the neuroligin family. *Am J Hum Genet* 74: 552–557, 2004.
114. Lee SH and Sheng M. Development of neuron-neuron synapses. *Curr Opin Neurobiol* 10: 125–131, 2000.
115. Levi S, Grady RM, Henry MD, Campbell KP, Sanes JR, and Craig AM. Dystroglycan is selectively associated with inhibitory GABAergic synapses but is dispensable for their differentiation. *J Neurosci* 22: 4274–4285, 2002.
116. Levi S, Logan SM, Tovar KR, and Craig AM. Gephyrin is critical for glycine receptor clustering but not for the formation of functional GABAergic synapses in hippocampal neurons. *J Neurosci* 24: 207–217, 2004.
117. Levinson JN, Chery N, Huang K, Wong TP, Gerrow K, Kang R, Prange O, Wang YT, and El-Husseini A. Neuroligins Mediate Excitatory and Inhibitory Synapse Formation: INVOLVEMENT OF PSD-95 AND NEUREXIN-1{beta} IN NEUROLIGIN-INDUCED SYNAPTIC SPECIFICITY. *J Biol Chem* 280: 17312–17319, 2005.
118. Levinson JN, Chery N, Huang K, Wong TP, Gerrow K, Kang R, Prange O, Wang YT, and El-Husseini A. Neuroligins mediate excitatory and inhibitory synapse formation: involvement of PSD-95 and neurexin-1beta in neuroligin-induced synaptic specificity. *J Biol Chem* 280: 17312–17319, 2005.
119. Levinson JN and El-Husseini A. Building excitatory and inhibitory synapses: balancing neuroligin partnerships. *Neuron* 48: 171–174, 2005.
120. Li RW, Serwanski DR, Miralles CP, Li X, Charych E, Riquelme R, Huganir RL, and de Blas AL. GRIP1 in GABAergic synapses. *J Comp Neurol* 488: 11–27, 2005.
121. Lichtman JW and Colman H. Synapse elimination and indelible memory. *Neuron* 25: 269–278, 2000.
122. Lise MF and El-Husseini A. The neuroligin and neurexin families: from structure to function at the synapse. *Cell Mol Life Sci* 63: 1833–1849, 2006.

123. Luthi A, Mohajeri H, Schachner M, and Laurent JP. Reduction of hippocampal long-term potentiation in transgenic mice ectopically expressing the neural cell adhesion molecule L1 in astrocytes. *J Neurosci Res* 46: 1–6, 1996.
124. Meyer G, Varoqueaux F, Neeb A, Oschlies M, and Brose N. The complexity of PDZ domain-mediated interactions at glutamatergic synapses: a case study on neuroligin. *Neuropharmacology* 47: 724–733, 2004.
125. Missler M, Fernandez-Chacon R, and Sudhof TC. The making of neurexins. *J Neurochem* 71: 1339–1347, 1998.
126. Missler M, Hammer RE, and Sudhof TC. Neurexophilin binding to alpha-neurexins. A single LNS domain functions as an independently folding ligand-binding unit. *J Biol Chem* 273: 34716–34723, 1998.
127. Missler M and Sudhof TC. Neurexins: three genes and 1001 products. *Trends Genet* 14: 20–26, 1998.
128. Missler M, Zhang W, Rohlmann A, Kattenstroth G, Hammer RE, Gottmann K, and Sudhof TC. Alpha-neurexins couple Ca2+ channels to synaptic vesicle exocytosis. *Nature* 423: 939–948, 2003.
129. Muller D, Wang C, Skibo G, Toni N, Cremer H, Calaora V, Rougon G, and Kiss JZ. PSA-NCAM is required for activity-induced synaptic plasticity. *Neuron* 17: 413–422, 1996.
130. Murase S, Mosser E, and Schuman EM. Depolarization drives beta-Catenin into neuronal spines promoting changes in synaptic structure and function. *Neuron* 35: 91–105, 2002.
131. Nam CI and Chen L. Postsynaptic assembly induced by neurexin-neuroligin interaction and neurotransmitter. *Proc Natl Acad Sci U S A* 102: 6137–6142, 2005.
132. Nguyen DN, Liu Y, Litsky ML, and Reinke R. The sidekick gene, a member of the immunoglobulin superfamily, is required for pattern formation in the Drosophila eye. *Development* 124: 3303–3312, 1997.
133. Niethammer P, Delling M, Sytnyk V, Dityatev A, Fukami K, and Schachner M. Cosignaling of NCAM via lipid rafts and the FGF receptor is required for neuritogenesis. *J Cell Biol* 157: 521–532, 2002.
134. Nishimura SL, Boylen KP, Einheber S, Milner TA, Ramos DM, and Pytela R. Synaptic and glial localization of the integrin alphavbeta8 in mouse and rat brain. *Brain Res* 791: 271–282, 1998.
135. O'Brien RJ, Lau LF, and Huganir RL. Molecular mechanisms of glutamate receptor clustering at excitatory synapses. *Curr Opin Neurobiol* 8: 364–369, 1998.
136. Osten P, Khatri L, Perez JL, Kohr G, Giese G, Daly C, Schulz TW, Wensky A, Lee LM, and Ziff EB. Mutagenesis reveals a role for ABP/GRIP binding to GluR2 in synaptic surface accumulation of the AMPA receptor. *Neuron* 27: 313–325, 2000.
137. Park M, Salgado JM, Ostroff L, Helton TD, Robinson CG, Harris KM, and Ehlers MD. Plasticity-induced growth of dendritic spines by exocytic trafficking from recycling endosomes. *Neuron* 52: 817–830, 2006.
138. Peng J, Kim MJ, Cheng D, Duong DM, Gygi SP, and Sheng M. Semiquantitative proteomic analysis of rat forebrain postsynaptic density fractions by mass spectrometry. *J Biol Chem* 279: 21003–21011, 2004.
139. Petrak LJ, Harris KM, and Kirov SA. Synaptogenesis on mature hippocampal dendrites occurs via filopodia and immature spines during blocked synaptic transmission. *J Comp Neurol* 484: 183–190, 2005.
140. Petrenko AG, Ullrich B, Missler M, Krasnoperov V, Rosahl TW, and Sudhof TC. Structure and evolution of neurexophilin. *J Neurosci* 16: 4360–4369, 1996.
141. Pierce JP and Mendell LM. Quantitative ultrastructure of Ia boutons in the ventral horn: scaling and positional relationships. *J Neurosci* 13: 4748–4763, 1993.

142. Pinkstaff JK, Detterich J, Lynch G, and Gall C. Integrin subunit gene expression is regionally differentiated in adult brain. *J Neurosci* 19: 1541–1556, 1999.
143. Prange O, Wong TP, Gerrow K, Wang YT, and El-Husseini A. A balance between excitatory and inhibitory synapses is controlled by PSD-95 and neuroligin. *Proc Natl Acad Sci U S A* 101: 13915–13920, 2004.
144. Rao A, Kim E, Sheng M, and Craig AM. Heterogeneity in the molecular composition of excitatory postsynaptic sites during development of hippocampal neurons in culture. *J Neurosci* 18: 1217–1229, 1998.
145. Romorini S, Piccoli G, Jiang M, Grossano P, Tonna N, Passafaro M, Zhang M, and Sala C. A functional role of postsynaptic density-95-guanylate kinase-associated protein complex in regulating Shank assembly and stability to synapses. *J Neurosci* 24: 9391–9404, 2004.
146. Rozic-Kotliroff G and Zisapel N. Ca2+ -dependent splicing of neurexin IIalpha. *Biochem Biophys Res Commun* 352: 226–230, 2007.
147. Rubenstein JL and Merzenich MM. Model of autism: increased ratio of excitation/inhibition in key neural systems. *Genes Brain Behav* 2: 255–267, 2003.
148. Rudenko G, Nguyen T, Chelliah Y, Sudhof TC, and Deisenhofer J. The structure of the ligand-binding domain of neurexin Ibeta: regulation of LNS domain function by alternative splicing. *Cell* 99: 93–101, 1999.
149. Rutishauser U. Influences of the neural cell adhesion molecule on axon growth and guidance. *J Neurosci Res* 13: 123–131, 1985.
150. Saghatelyan AK, Nikonenko AG, Sun M, Rolf B, Putthoff P, Kutsche M, Bartsch U, Dityatev A, and Schachner M. Reduced GABAergic transmission and number of hippocampal perisomatic inhibitory synapses in juvenile mice deficient in the neural cell adhesion molecule L1. *Mol Cell Neurosci* 26: 191–203, 2004.
151. Sakurai T, Ramoz N, Reichert JG, Corwin TE, Kryzak L, Smith CJ, Silverman JM, Hollander E, and Buxbaum JD. Association analysis of the NrCAM gene in autism and in subsets of families with severe obsessive-compulsive or self-stimulatory behaviors. *Psychiatr Genet* 16: 251–257, 2006.
152. Sala C, Piech V, Wilson NR, Passafaro M, Liu G, and Sheng M. Regulation of dendritic spine morphology and synaptic function by Shank and Homer. *Neuron* 31: 115–130, 2001.
153. Sala C, Roussignol G, Meldolesi J, and Fagni L. Key role of the postsynaptic density scaffold proteins Shank and Homer in the functional architecture of Ca2+ homeostasis at dendritic spines in hippocampal neurons. *J Neurosci* 25: 4587–4592, 2005.
154. Sanes JR and Lichtman JW. Development of the vertebrate neuromuscular junction. *Annu Rev Neurosci* 22: 389–442, 1999.
155. Sara Y, Biederer T, Atasoy D, Chubykin A, Mozhayeva MG, Sudhof TC, and Kavalali ET. Selective capability of SynCAM and neuroligin for functional synapse assembly. *J Neurosci* 25: 260–270, 2005.
156. Sara Y, Biederer T, Atasoy D, Chubykin A, Mozhayeva MG, Sudhof TC, and Kavalali ET. Selective capability of SynCAM and neuroligin for functional synapse assembly. *J Neurosci* 25: 260–270, 2005.
157. Scheiffele P, Fan J, Choih J, Fetter R, and Serafini T. Neuroligin expressed in nonneuronal cells triggers presynaptic development in contacting axons. *Cell* 101: 657–669, 2000.
158. Scholl FG and Scheiffele P. Making connections: cholinesterase-domain proteins in the CNS. *Trends Neurosci* 26: 618–624, 2003.
159. Schuster T, Krug M, Hassan H, and Schachner M. Increase in proportion of hippocampal spine synapses expressing neural cell adhesion molecule NCAM180 following long-term potentiation. *J Neurobiol* 37: 359–372, 1998.

160. Senkov O, Sun M, Weinhold B, Gerardy-Schahn R, Schachner M, and Dityatev A. Polysialylated neural cell adhesion molecule is involved in induction of long-term potentiation and memory acquisition and consolidation in a fear-conditioning paradigm. *J Neurosci* 26: 10888–109898, 2006.
161. Serafini T. An old friend in a new home: cadherins at the synapse. *Trends Neurosci* 20: 322–323, 1997.
162. Sheng M and Kim MJ. Postsynaptic signaling and plasticity mechanisms. *Science* 298: 776–780, 2002.
163. Sheng M and Sala C. PDZ domains and the organization of supramolecular complexes. *Annu Rev Neurosci* 24: 1–29, 2001.
164. Shingai T, Ikeda W, Kakunaga S, Morimoto K, Takekuni K, Itoh S, Satoh K, Takeuchi M, Imai T, Monden M, and Takai Y. Implications of nectin-like molecule-2/IGSF4/RA175/SgIGSF/TSLC1/SynCAM1 in cell–cell adhesion and transmembrane protein localization in epithelial cells. *J Biol Chem* 278: 35421–35427, 2003.
165. Sia GM, Beique JC, Rumbaugh G, Cho R, Worley PF, and Huganir RL. Interaction of the N-Terminal domain of the AMPA receptor GluR4 subunit with the neuronal pentraxin NP1 mediates GluR4 synaptic recruitment. *Neuron* 55: 87–102, 2007.
166. Song JY, Ichtchenko K, Sudhof TC, and Brose N. Neuroligin 1 is a postsynaptic cell-adhesion molecule of excitatory synapses. *Proc Natl Acad Sci U S A* 96: 1100–1105, 1999.
167. Sons MS, Busche N, Strenzke N, Moser T, Ernsberger U, Mooren FC, Zhang W, Ahmad M, Steffens H, Schomburg ED, Plomp JJ, and Missler M. alpha-Neurexins are required for efficient transmitter release and synaptic homeostasis at the mouse neuromuscular junction. *Neuroscience* 138: 433–446, 2006.
168. Spacek J and Harris KM. Trans-endocytosis via spinules in adult rat hippocampus. *J Neurosci* 24: 4233–4241, 2004.
169. Sperry RW. Chemoaffinity in the orderly growth of nerve fibre patterns and connections. *Proc Natl Acad Sci U S A* 50: 703–710, 1963.
170. Staubli U, Chun D, and Lynch G. Time-dependent reversal of long-term potentiation by an integrin antagonist. *J Neurosci* 18: 3460–3469, 1998.
171. Stoenica L, Senkov O, Gerardy-Schahn R, Weinhold B, Schachner M, and Dityatev A. In vivo synaptic plasticity in the dentate gyrus of mice deficient in the neural cell adhesion molecule NCAM or its polysialic acid. *Eur J Neurosci* 23: 2255–2264, 2006.
172. Sugita S, Saito F, Tang J, Satz J, Campbell K, and Sudhof TC. A stoichiometric complex of neurexins and dystroglycan in brain. *J Cell Biol* 154: 435–445, 2001.
173. Sugiyama J, Bowen DC, and Hall ZW. Dystroglycan binds nerve and muscle agrin. *Neuron* 13: 103–115, 1994.
174. Sytnyk V, Leshchyns'ka I, Dityatev A, and Schachner M. Trans-Golgi network delivery of synaptic proteins in synaptogenesis. *J Cell Sci* 117: 381–388, 2004.
175. Szatmari P, Paterson AD, Zwaigenbaum L, Roberts W, Brian J, Liu XQ, Vincent JB, Skaug JL, Thompson AP, Senman L, Feuk L, Qian C, Bryson SE, Jones MB, Marshall CR, Scherer SW, Vieland VJ, Bartlett C, Mangin LV, Goedken R, Segre A, Pericak-Vance MA, Cuccaro ML, Gilbert JR, Wright HH, Abramson RK, Betancur C, Bourgeron T, Gillberg C, Leboyer M, Buxbaum JD, Davis KL, Hollander E, Silverman JM, Hallmayer J, Lotspeich L, Sutcliffe JS, Haines JL, Folstein SE, Piven J, Wassink TH, Sheffield V, Geschwind DH, Bucan M, Brown WT, Cantor RM, Constantino JN, Gilliam TC, Herbert M, Lajonchere C, Ledbetter DH, Lese-Martin C, Miller J, Nelson S, Samango-Sprouse CA, Spence S, State M, Tanzi RE, Coon H, Dawson G, Devlin B, Estes A, Flodman P, Klei L, McMahon WM, Minshew N, Munson J, Korvatska E, Rodier PM, Schellenberg GD, Smith M, Spence MA, Stodgell C, Tepper PG, Wijsman EM, Yu CE, Roge B, Mantoulan C, Wittemeyer K, Poustka A, Felder B, Klauck SM, Schuster C,

Poustka F, Bolte S, Feineis-Matthews S, Herbrecht E, Schmotzer G, Tsiantis J, Papanikolaou K, Maestrini E, Bacchelli E, Blasi F, Carone S, Toma C, Van Engeland H, de Jonge M, Kemner C, Koop F, Langemeijer M, et al. Mapping autism risk loci using genetic linkage and chromosomal rearrangements. *Nat Genet* 39: 319–328, 2007.

176. Takasu MA, Dalva MB, Zigmond RE, and Greenberg ME. Modulation of NMDA receptor-dependent calcium influx and gene expression through EphB receptors. *Science* 295: 491–495, 2002.
177. Taniguchi H, Gollan L, Scholl FG, Mahadomrongkul V, Dobler E, Limthong N, Peck M, Aoki C, and Scheiffele P. Silencing of neuroligin function by postsynaptic neurexins. *J Neurosci* 27: 2815–2824, 2007.
178. Togashi H, Abe K, Mizoguchi A, Takaoka K, Chisaka O, and Takeichi M. Cadherin regulates dendritic spine morphogenesis. *Neuron* 35: 77–89, 2002.
179. Torres R, Firestein BL, Dong H, Staudinger J, Olson EN, Huganir RL, Bredt DS, Gale NW, and Yancopoulos GD. PDZ proteins bind, cluster, and synaptically colocalize with Eph receptors and their ephrin ligands. *Neuron* 21: 1453–1463, 1998.
180. Ullrich B, Ushkaryov YA, and Sudhof TC. Cartography of neurexins: more than 1000 isoforms generated by alternative splicing and expressed in distinct subsets of neurons. *Neuron* 14: 497–507, 1995.
181. Ushkaryov YA, Hata Y, Ichtchenko K, Moomaw C, Afendis S, Slaughter CA, and Sudhof TC. Conserved domain structure of beta-neurexins. Unusual cleaved signal sequences in receptor-like neuronal cell-surface proteins. *J Biol Chem* 269: 11987–11992, 1994.
182. Ushkaryov YA, Petrenko AG, Geppert M, and Sudhof TC. Neurexins: synaptic cell surface proteins related to the alpha-latrotoxin receptor and laminin. *Science* 257: 50–56, 1992.
183. Vaithianathan T, Matthias K, Bahr B, Schachner M, Suppiramaniam V, Dityatev A, and Steinhauser C. Neural cell adhesion molecule-associated polysialic acid potentiates alpha-amino-3-hydroxy-5-methylisoxazole-4-propionic acid receptor currents. *J Biol Chem* 279: 47975–47984, 2004.
184. Varoqueaux F, Aramuni G, Rawson RL, Mohrmann R, Missler M, Gottmann K, Zhang W, Sudhof TC, and Brose N. Neuroligins determine synapse maturation and function. *Neuron* 51: 741–754, 2006.
185. Varoqueaux F, Jamain S, and Brose N. Neuroligin 2 is exclusively localized to inhibitory synapses. *Eur J Cell Biol* 83: 449–456, 2004.
186. Vaughn JE. Fine structure of synaptogenesis in the vertebrate central nervous system. *Synapse* 3: 255–285, 1989.
187. Vogt AK, Brewer GJ, Decker T, Bocker-Meffert S, Jacobsen V, Kreiter M, Knoll W, and Offenhausser A. Independence of synaptic specificity from neuritic guidance. *Neuroscience* 134: 783–790, 2005.
188. Walikonis RS, Jensen ON, Mann M, Provance DW, Jr., Mercer JA, and Kennedy MB. Identification of proteins in the postsynaptic density fraction by mass spectrometry. *J Neurosci* 20: 4069–4080, 2000.
189. Wang CY, Chang K, Petralia RS, Wang YX, Seabold GK, and Wenthold RJ. A novel family of adhesion-like molecules that interacts with the NMDA receptor. *J Neurosci* 26: 2174–2183, 2006.
190. Washbourne P, Bennett JE, and McAllister AK. Rapid recruitment of NMDA receptor transport packets to nascent synapses. *Nat Neurosci* 5: 751–759, 2002.
191. Washbourne P, Dityatev A, Scheiffele P, Biederer T, Weiner JA, Christopherson KS, and El-Husseini A. Cell Adhesion Molecules in Synapse Formation. *J Neurosci* 24: 9244–9249, 2004.

192. Weiner JA, Wang X, Tapia JC, and Sanes JR. Gamma protocadherins are required for synaptic development in the spinal cord. *Proc Natl Acad Sci U S A* 102: 8–14, 2005.
193. Williamson RA, Henry MD, Daniels KJ, Hrstka RF, Lee JC, Sunada Y, Ibraghimov-Beskrovnaya O, and Campbell KP. Dystroglycan is essential for early embryonic development: disruption of Reichert's membrane in Dag1-null mice. *Hum Mol Genet* 6: 831–841, 1997.
194. Wilson DA, Best AR, and Sullivan RM. Plasticity in the olfactory system: lessons for the neurobiology of memory. *Neuroscientist* 10: 513–524, 2004.
195. Xiao P, Bahr BA, Staubli U, Vanderklish PW, and Lynch G. Evidence that matrix recognition contributes to stabilization but not induction of LTP. *Neuroreport* 2: 461–464, 1991.
196. Yamagata M and Sanes JR. Target-independent diversification and target-specific projection of chemically defined retinal ganglion cell subsets. *Development* 121: 3763–3776, 1995.
197. Yamagata M, Sanes JR, and Weiner JA. Synaptic adhesion molecules. *Curr Opin Cell Biol* 15: 621–632, 2003.
198. Yamagata M, Weiner JA, and Sanes JR. Sidekicks: synaptic adhesion molecules that promote lamina-specific connectivity in the retina. *Cell* 110: 649–660, 2002.
199. Zeng X, Sun M, Liu L, Chen F, Wei L, and Xie W. Neurexin-1 is required for synapse formation and larvae associative learning in Drosophila. *FEBS Lett* 581: 2509–2516, 2007.
200. Zhang W, Rohlmann A, Sargsyan V, Aramuni G, Hammer RE, Sudhof TC, and Missler M. Extracellular domains of alpha-neurexins participate in regulating synaptic transmission by selectively affecting N- and P/Q-type Ca2+ channels. *J Neurosci* 25: 4330–4342, 2005.
201. Ziv NE and Smith SJ. Evidence for a role of dendritic filopodia in synaptogenesis and spine formation. *Neuron* 17: 91–102, 1996.
202. Zoghbi HY. Postnatal neurodevelopmental disorders: meeting at the synapse? *Science* 302: 826–830, 2003.

Dendritic Organelles for Postsynaptic Trafficking

Cyril Hanus and Michael D. Ehlers

Howard Hughes Medical Institute, Department of Neurobiology, Duke University Medical Center, Durham, NC 27710, USA, ehlers@neuro.duke.edu

Abstract. The development, maintenance, and plasticity of synapses require the trafficking of molecular components to postsynaptic specializations distributed across elaborate dendritic arbors. The large size and geometric complexity of dendrites necessitate specialized mechanisms that can both deliver postsynaptic cargo over extended distances and regulate dendritic composition on a submicron scale. Despite the fundamental importance of membrane trafficking in neuronal development and synaptic plasticity, the cellular rules governing postsynaptic trafficking are only now emerging. We review here current knowledge on the organization of dendritic organelles, and describe cell biological mechanisms that allow long-range transport of postsynaptic components that are then locally processed at individual synapses.

1 Introduction

With long and highly branched processes extending up to several hundred microns from the cell body, neurons have surface areas up to 10,000 times greater than typical mammalian cells and display a remarkable degree of locally specified physiological properties (106). These spatially restricted properties (e.g. action potential generation and neurotransmitter release in the axon, postsynaptic response to neurotransmitters in dendrites, synapse-specific modification for long-lasting plasticity) are due to a high degree of molecular compartmentalization, the basis for which is only now emerging.

While inhibitory synapses are typically localized on the dendritic shaft (190), most excitatory synapses in the brain are localized on micron-sized membranous protrusions present along the entire length of dendrites (85). These dendritic spines contact one or more presynaptic terminals, and the sites of contact contain a dense array of receptors, scaffold proteins, and signaling molecules at the postsynaptic density (PSD) (96). The molecular content of dendritic branches, individual spines, and the PSD is subject to rapid and long-lasting modification based on developmental stage and patterns of neural activity (45, 122, 290). This multi-layered compartmental organization requires long-range transport of postsynaptic proteins to reach the most

J.W. Hell, M.D. Ehlers (eds.), *Structural and Functional Organization of the Synapse*,
DOI: 10.1007/978-0-387-77232-5_8, © Springer Science+Business Media, LLC 2008

distal dendrites through geometrically complex branches. Once delivered to spines, synaptic components including postsynaptic receptors are subject to local trafficking rules, which determine whether they are incorporated into synapses or removed from the plasma membrane, and whether they are recycled or degraded.

Many aspects of postsynaptic receptor trafficking are regulated by neuronal activity, allowing rapid changes of receptor amount at the postsynapse and thereby the strength of postsynaptic responses (24). Dendritic membrane protein trafficking has emerged as a key postsynaptic mechanism underlying synapse development and various forms of synaptic plasticity, such as long term potentiation (LTP) and depression (LTD) (45). Because individual PSDs contain only tens to a few hundred neurotransmitter receptors, the addition or subtraction of just a few molecules can alter neurotransmission (39, 182, 189, 277). Moreover, adjacent dendritic spines can have strikingly different steady-state levels of synaptic proteins (13, 64, 183), implying that postsynaptic trafficking is subject to stringent spatial regulation.

In the present chapter, we describe the fundamental organization of neuronal organelles, emphasizing that, while many basic principles established in model systems such as yeast and fibroblasts also apply to neurons, neurons have many of their own rules to orchestrate protein trafficking. Our discussion will be limited largely to protein trafficking in dendrites. After presenting formal principles of protein trafficking in dendrites, we review recent data explaining how specific mechanisms known to control the dynamics of postsynaptic components can be integrated in a larger perspective taking into account specific properties of neuronal organelles. Due to limitations of space, local mRNA translation in dendrites, which is an important aspect of synaptic function and plasticity (272, 274), will not be reviewed here.

2 Dendritic Organelles

2.1 ER, Golgi and the Neuronal Secretory Pathway

2.1.1 Fundamental Organization of the Secretory Pathway

Maintenance of the dendritic membrane requires continuous delivery of newly synthesized membrane proteins through the secretory pathway. The secretory pathway consists of an organized array of membrane bound organelles and tubulo-vesicular intermediates that are required for the synthesis, maturation and transport of proteins destined for the cell surface (210). The main organelles of the secretory pathway are the endoplasmic reticulum (ER), the ER-Golgi intermediate compartment (ERGIC), the Golgi apparatus (GA) and the trans-Golgi-network (TGN) (Fig. 1) (163). Newly synthesized proteins typically pass through these organelles in a sequential manner, with transport between compartments mediated by vesicular carriers moving along microtubules (163). The machinery for eukaryotic secretory trafficking has been largely conserved during evolution, and many of the basic principles first established in yeast apply to higher eukaryotic cells including neurons (107, 157).

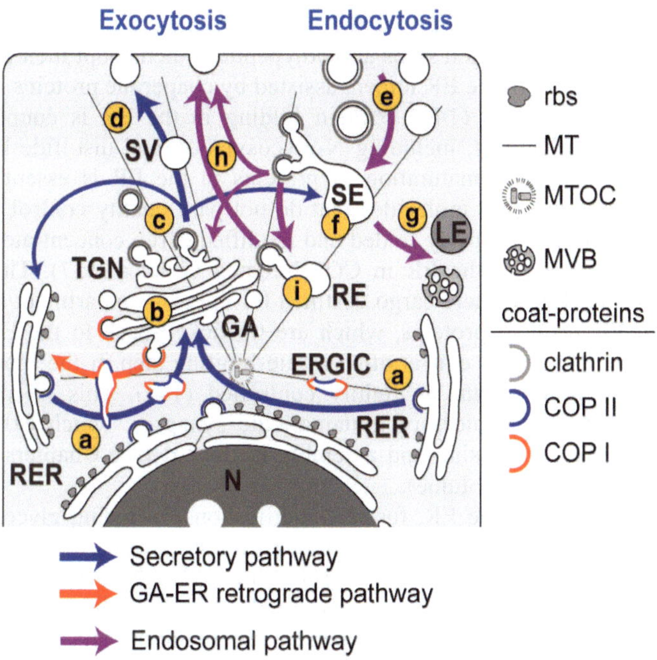

Fig. 1. Membrane trafficking pathways in eukaryotic cells. *Secretory pathway*. Newly synthesized membrane proteins exit the endoplasmic reticulum (ER) in COP-II-coated vesicles (**a**), which form at specialized locations termed ER-exit sites (ERES). Cargo is then trafficked to the Golgi apparatus (GA) for post-translational modifications (**b**) via ER-Golgi intermediate compartments (ERGIC). ER-resident proteins are retrieved from the ERGIC or GA and trafficked back to the ER in COP-I coated-vesicles. Cargo progresses through the GA is sorted at the TGN (**c**), and trafficked either directly or via endosomes to the plasma membrane for exocytosis (**d**). *Endosomal pathway*. Membrane proteins are internalized by means of clathrin-dependent or clathrin-independent endocytosis (**e**). Newly internalized vesicles fuse to form early/sorting endosomes (SE) where cargo is sorted (**f**) and trafficked to late endosomes/multivesicular bodies (MVBs) for degradation (**g**), to the TGN, or back to the plasma membrane (**h**), either directly or via recycling endosomes (**i**). *Abbreviations*. N, nucleus; RER, rough endoplasmic reticulum; ERGIC, endoplasmic reticulum-Golgi intermediate compartment; GA, Golgi apparatus; TGN, trans-Golgi network; SV, secretory vesicle; SE, sorting endosome; LE, late endosome; RE, recycling endosome; rbs, ribosome; MT, microtubule; MTOC, microtubule organizing center; MVB, multivesicular body.

While nuclear and cytoplasmic proteins are synthesized in the cytoplasm, the secretory pathway is the obligatory biosynthetic pathway for most surface and extracellular proteins, as well as for transmembrane and intraluminal proteins destined for all membrane bound organelles other than mitochondria, chloroplasts and peroxisomes (157, 163). The entry of proteins into the secretory pathway is coupled to their synthesis and occurs at the rough ER (RER) where actively translating ribosome-mRNA complexes associate with the ER membrane (246). Ribosomes are targeted to the ER via the signal peptide recognition particle (SRP) which associates with the translocon, the protein channel complex through which nascent polypeptides are translocated across the ER membrane (132, 246). Ribosomes are released as protein elongation stops and polypeptides then adopt their proper three-dimensional structure in the ER lumen, assisted by chaperone proteins such as BiP, calnexin, and calreticulin (144). Protein folding in the ER is coupled to post-translational modifications, including N-glycosylation and disulfide bond formation. Proper folding and maturation of proteins in the ER is essential for their secretory progression, and molecules that do not pass quality control are targeted for degradation (144). Properly folded and modified cargo concentrates at ER-exit sites (ERES) and leaves the ER in COPII-coated vesicles (157). These vesicles merge with the ERGIC where cargo destined for the cis-compartment of the GA is sorted from ER resident proteins, which are trafficked back to the ER in COPI-coated vesicles (157). ER exit is often a rate-limiting step in the biosynthesis of transmembrane proteins and is tightly controlled (144). This is true for both AMPA and NMDA receptors of glutamatergic synapses, which exhibit several forms of regulated processing and assembly in the ER (see Chapters by Esteban and Wenthold et al., this volume).

Following exit from the ER, further modifications, including glycosylation and proteolysis, occur as cargo progresses through the cis, medial and trans compartments of the GA (160). Finally, cargo reaches the TGN where it is sorted for either transport to the plasma membrane or endosomal membranes. In most cell types, the ER, ERES and ERGIC elements are dispersed throughout the cell, whereas the Golgi network is located in the pericentriolar area close to the nucleus (160). The anterograde progression of cargo through the secretory pathway therefore involves centrifugal and centripetal displacements of intermediate carriers along microtubule tracks as cargo moves from the ER to the Golgi, and from the TGN to the plasma membrane, respectively (Fig. 1).

2.1.2 Spatial Organization and Unique Features of the Neuronal Secretory Pathway

Although membrane trafficking organelles were observed in neurons over a century ago (81), the fundamental organization of the neuronal secretory pathway is only beginning to emerge. The unique shape and size of neurons dictates a specialized secretory system capable of transporting cargo up to several hundred microns from the cell body to a network of extremely complex dendritic arbors. The neuronal secretory pathway has evolved a unique spatial distribution to allow trafficking of lipids and proteins to the most distal regions of dendrites.

As documented by electron microscopy (EM) studies, the ER extends widely throughout dendrites in addition to the soma (Fig. 2a) (27, 75, 268). Although smooth ER (SER), devoid of ribosomes, dominates in distal dendrites, RER is also found distally (221). Immunogold labeling has shown that the translocon protein Sec61a is present in both the dendritic shaft and in some spines (220, 221). Active ER exit sites also distribute throughout the somato-dendritic compartment as first indicated by direct functional visualization of emerging cargo (105), as well as by the concentration of COPII machinery at discrete puncta in the soma and dendrites (9, 105).

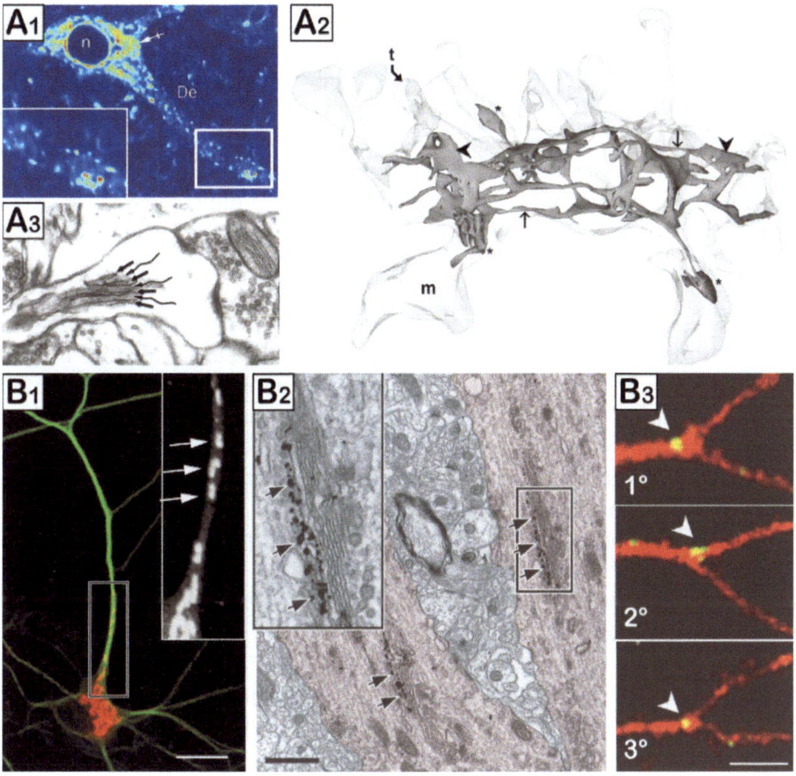

Fig. 2. Dendritic organelles I. Endoplasmic reticulum and Golgi apparatus. (**a**) Endoplasmic Reticulum (ER). (**A1**) Confocal section of a spinal cord neuron immunostained for the ER chaperone protein BiP (*pseudocolored*) illustrating the presence of ER elements throughout dendrites. Adapted from (75); reproduced with permission from the Society for Neuroscience, copyright 2001. (**A2**) Three-dimensional reconstruction of serial electron micrographs showing the distribution of smooth ER (*SER, dark grey*) in dendrites and spines of a CA1 hippocampal neuron. Large flat compartments (*arrowheads*) are linked by thin extensions (*thin arrows*). Note the extension of SER elements in several spines (*asterisks*). Adapted from (47); reprinted with permission from the Society for Neuroscience, copyright 2002. (**A3**) Electron micrograph of the spine apparatus showing the lamination of cisternae (*straight arrows*) between regions of high electron density (*wavy arrows*). Adapted from (268); reproduced with permission from the Society for Neuroscience, copyright 1997. (**b**) Golgi apparatus (GA). (**B1**) Cultured hippocampal neuron stained for MAP2 (*green*), and the cis-Golgi marker GM130 (*red*). Inset: higher magnification of GM130 labeling documenting Golgi outposts dispersed in the apical dendrite (*arrows*). (**B2**) Immunogold labeling for GM130 in adult rat hippocampus demonstrating the presence of mini Golgi-stacks in the apical dendrite of a CA1 pyramidal neuron in vivo. Scale bar, 1 μm. (**B3**) Golgi outposts at dendritic branch points. VSVGts045 (*green*) accumulates at primary, secondary and tertiary dendritic branch points following release from the ER at 20°C. Scale bar, 5 μm. B2–3 adapted from (108); reprinted with permission from Elsevier, copyright 2005.

In addition to its function in lipid biogenesis (21), the SER plays a crucial role in neuronal calcium homeostasis. The SER sequesters calcium by the action of the ER-associated Ca^{2+}-ATPase (SERCA) and releases it in the cytoplasm upon activation of inositol trisphosphate receptors (IP3R) or ryanodine receptors (236, 240). The dendritic SER often extends into the neck of mature spines, providing a conduit to the synapse (Fig. 2a).

In addition to a role as a Ca^{2+} reservoir important for postsynaptic signaling (236), spine SER may also be involved in lipid and protein secretion to the spine surface. A specialized SER-like derivative composed of stacked cisternae, termed the spine apparatus, is found in some dendritic spines (Fig. 2a) (84–86, 268). The presence of vesicles near the tip of the spine apparatus raises the possibility that it could supply membrane for spine growth, and possibly traffic important synaptic proteins, such as AMPA receptors and NMDA receptors, which can be localized to this organelle (200, 227, 268). Though defined largely on morphological criteria, the precise compartmental identity of the spine apparatus has yet to be determined. Notably, the spine apparatus is absent in mice lacking synaptopodin (55). These mice display deficits in LTP and spatial learning, indicating a potential link between the spine apparatus and mechanisms of synaptic plasticity (55).

Spine SER membranes occasionally extend to the periphery of the PSD (288). This tight association of intracellular calcium stores and the PSD may depend on the direct binding of inositol trisphosphate receptors (IP3Rs) in the ER to the postsynaptic scaffold molecule Homer (242). SER membranes are absent from Purkinje neuron spines in *dilute* mice, which lack functional myosin Va (276). Parallel fiber synapses onto Purkine neurons in *dilute* mice have impaired long-term depression, supporting a requirement of spine SER calcium in cerebellar LTD (188).

The organization of the GA in neurons is quite different from what is observed in most non-neuronal cells. Immunofluorescence labeling shows that the neuronal GA consists not only of the perinuclear membrane array found in most cell types, but also of discrete structures dispersed in dendrites, termed Golgi outposts (Fig. 2b) (52, 75, 105, 108, 168, 285). This organization has been confirmed by EM, where miniature Golgi-stacks and intracellular membranes immuno-reactive for Golgi membranes markers such as α-mannosidase II, giantin, GM130, and Rab6, have been observed in dendrites (75, 108, 220). Dendritic Golgi membranes can be found close to synapses but seem to be only localized to the dendritic shaft. Moreover, not all dendrites contain morphological or molecular markers of the GA (108), suggesting dendrite-specific compartmentalization of secretory trafficking.

The dendritic localization of RER, ERES and Golgi membranes suggests that "satellite" secretory systems exist in some dendrites that could locally synthesize and process secretory proteins at sites remote from the cell body. Numerous mRNA species are found in dendrites, including mRNAs encoding postsynaptic receptors such as NMDA receptors, AMPA receptors and glycine receptors (272). Although the local translation of membrane proteins in dendrites is still not as clearly delineated as that of cytoplasmic proteins, protein synthesis and enzymatic activities associated with the secretory pathway, such as glycosylation, persist in hippocampal dendrites isolated from the soma (284, 285). The capacity of isolated dendrites to synthesize exogenously expressed integral membrane proteins (e.g., glutamate receptors) and deliver them to the plasma membrane has been reported (127, 128).

The involvement of dendritic Golgi ouposts in post-ER membrane trafficking was recently established by imaging hippocampal neurons expressing the ts045 thermosensitive mutant of the vesicular stomatitis viral glycoprotein (VSVG-ts) (105). VSVG-ts is retained in the ER at 39.5°C, but is rapidly released by reducing the temperature to 32°C, allowing control and monitoring of the synchronous progression of cargo through the secretory pathway (16). Upon release from the ER, a fraction of pre-Golgi carriers containing VSVG-ts merge with dendritic compartments immunoreactive for the Golgi markers galactosyltransferase and GM130, demonstrating that dendritic Golgi outposts are functional trafficking platforms (105). Golgi outposts also engage in the local dendritic trafficking of BDNF, confirming their role in processing endogenous proteins (105). The neural cell adhesion molecule (NCAM) promotes the accumulation of TGN-derived vesicles at neuron contact sites (275), indicating that post-Golgi compartments can be directed by intercellular signaling.

Interestingly, at 20°C, a temperature which blocks vesicle budding from the TGN (181), newly released VSVG-ts frequently accumulates at Golgi outposts located at dendritic branch points (Fig. 2b) (105). These outposts engage in ongoing post-Golgi trafficking after release from TGN-exit blockade (105), and appear ideally positioned to regulate the identity or quantity of cargo that is trafficked to each branch, a possibility that awaits further investigation.

Although dendritic Golgi outposts are present in some dendrites of hippocampal and cortical neurons, not all neurons possess detectable dendritic Golgi membranes (105, 108). Moreover, even in those dendrites containing Golgi outposts, the majority of ER-to-Golgi carriers originating in the dendrite are trafficked all the way back to the somatic Golgi in the cell body (105). Thus, dual modes of early secretory trafficking exist in dendrites. The major mode of ER-to-Golgi trafficking is directed long distances to the somatic Golgi apparatus and likely represents the exclusive mode of early secretory trafficking in those dendrites lacking Golgi outposts. A second mode of ER-to-Golgi trafficking occurs locally in dendrites containing Golgi outposts, and may be specialized for the processing of specific cargo or the control of dendritic secretion.

2.2 Dendritic Endosomes

Spines or dendritic segments need to regulate the surface expression of neurotransmitter receptors, adhesion molecules, and ion channels. Often separated by only a few microns, adjacent spines on a dendrite can have strikingly different steady-state levels of synaptic proteins (13, 64, 95, 183), requiring different rates of insertion, removal, and subsequent recycling. The endosomal system provides a primary mechanism for highly localized regulation of spine membrane composition.

The endosomal system is an interconnected and dynamic network of membrane bound structures that receive endocytic vesicles from the plasma membrane (Fig. 1). Newly internalized vesicles formed by means of clathrin-dependent or clathrin-independent endocytosis fuse to form carly/sorting endosomes (185). In much the same way that the TGN functions for secretory proteins, sorting endosomes act as a station from which internalized cargo can be targeted for degradation, transport to the TGN, or recycling back to the cell surface, either directly or via recycling endosomes (Fig. 1) (19). Sorting endosomes display a characteristic tubulo-vesicular morphology directly related to their function. Molecules destined for reinsertion into the plasma membrane or to the TGN exit the sorting endosome on vesicles pinched off from small-diameter tubules (19). Cargo targeted for degradation remains in sorting endosomes which become progressively more acidic as they mature into late endosomes, with degradative cargo budding intraluminally to form multivesicular bodies (MVBs) prior to fusion with lysosomes where cargo degradation is completed within minutes (185). Endocytic organelles exhibit a complex morphological organization in the form of membrane vacuoles, cisternae, tubules, and multilamellar or multivesicular bodies, and differ in their biochemical composition and localization within the cell. In most non-neuronal cells, early endosomes are localized close to the cell periphery while recycling endosomes (REs) are generally found in the peri-centriolar area (Fig. 1) (185). This spatial segregation is likely crucial for the sequential processing of molecules as they progress from one compartment to the other (187). In neurons however, both early and REs are localized in dendrites remote from cell body, potentially facilitating local processing of internalized cargo (63).

The observation of an intracellular accumulation of various plasma proteins in Purkinje neuron dendrites was some of the first evidence for dendritic endocytosis (72). As shown by immunofluorescent labeling of a broad cast of endosome associated proteins, virtually all elements of the endosomal network are present in hippocampal neurons dendrites, including early endosomes, recycling endosomes, late endosomes, and lysosomes (17, 63), although spatial variations in endosome abundance in different dendrites or different neuronal classes has not been thoroughly investigated.

By combining three-dimensional electron microscopy reconstruction of dendrites in vivo with the tracking of internalized gold-conjugated endocytic cargo in acute hippocampal slices, Harris and colleagues have provided important insight into the fundamental organization of the dendritic endosomal network (47). Internal endosomal structures are present in dendrites, including coated and uncoated vesicles, tubular structures and multi-vesicular bodies (Fig. 3a) (47). Interestingly, nearly two thirds of dendritic endosomal structures are situated within or at the base of spines (213), suggesting a direct involvement in subsynaptic membrane trafficking. Endosome-like structures generally extend over stretches of dendrite encompassing one to three spines, suggesting that multiple spines share the same endocytic organelles (47). Clathrin-like coats are present at the tip of some tubular protrusions emerging from these structures, indicating that they may generate smaller trafficking intermediates within the dendritic endosomal pathway (47).

Other studies showed that syntaxin 13, a SNARE protein enriched in early and recycling endosomes, is found in dendritic tubulo-vesicular structures containing transferrin receptors, which are known to be recycled through the endosomal pathway (225). Consistent with local endosomal trafficking, the transferrin receptor continuously cycles back and forth between the dendritic surface and internal structures

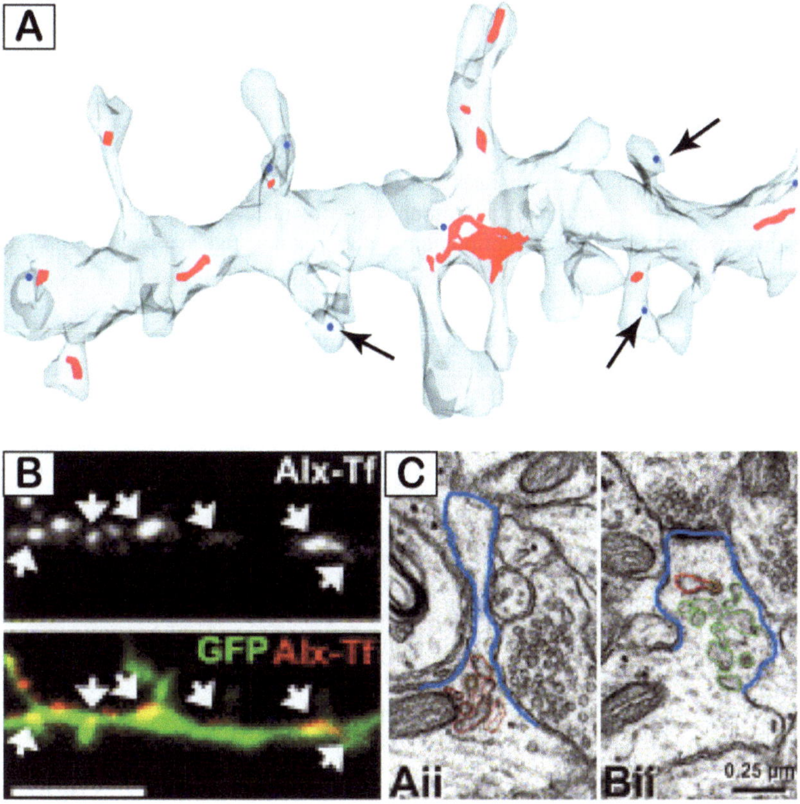

Fig. 3. Dendritic organelles II. Endosomal network. (**a**) Three-dimensional reconstruction of serial electron micrographs showing the distribution of dendritic endosomes in a postnatal day 21 rat hippocampal neuron. Endosomes (*red*) and small vesicles (*blue, arrows*) are distributed throughout dendrites. Adapted from (47); reprinted with permission from the Society for Neuroscience, copyright 2002. (**b**) A cultured hippocampal neuron expressing green fluorescent protein (*GFP, green*) after loading with fluorescent transferrin (*Alx-Tf, red*), showing the presence of recycling endosomes (REs) at the base of dendritic spines (*arrows*). (**c**) Electron micrographs of hippocampal dendritic spines *in vivo* after tracing of the spine surface (*blue lines*), recycling endosomes (*red lines*) and amorphous vesicular clumps (*green lines*). b and c were adapted from (213); reprinted with permission from Elsevier, copyright 2006.

(32, 191, 300). Interestingly, both measurements of transferrin uptake and immunolabeling of hippocampal neuron dendrites indicate that REs are often localized at the base of the spine neck and occasionally in the spine head (Fig. 3b,c) (212, 213). Notably, REs translocate into spines upon LTP-inducing stimuli where they contribute exocytic membrane for spine growth during synapse potentiation (213). These observations demonstrate the presence of a dendritic endosomal network in close proximity to glutamatergic synapses and support a model for local synaptic protein recycling and degradation via endosomes. This issue is discussed in more detail in Sections 3.2. and 3.3.

2.3 Dendritic Mitochondria

Synaptic remodeling requires membrane fission, fusion, protein degradation and local protein synthesis, all of which are ATP-dependent processes. Indeed mitochondria, strongly impregnated by common contrast agents used in electron microscopy, were one of the first dendritic organelles to be identified (217). Dendrite mitochondria are mainly located in the dendritic shaft, but can occasionally be found associated with spines (1, 31). Although dendritic mitochondria usually appear as individual units, three-dimensional EM reconstructions have documented the existence of continuous mitochondrial networks extending over 10–30 μm in dendrites of the CA1 region of rodent hippocampus (222).

Dendritic mitochondria are quite dynamic. In addition to fusion and fission, mitochondria can move over long distances in dendrites in an activity-dependent manner (161). Synaptic stimulation decreases the mobility of mitochondria and induces their translocation into spines (161). Decreased mitochondrial mobility near active synapses and increased mobility in the absence of neural activity would be predicted to distribute mitochondria near highly active, high ATP-utilizing, dendritic regions. Consistent with this notion, reducing the number of dendritic mitochondria by dominant-negative overexpression of the GTPase Drp1, a dynamin-like protein involved in mitochondrial fission, decreased the number of synapses. Conversely, increasing the number of dendritic mitochondria by overexpressing wild type Drp1 or increasing mitochondrial function by treating cells with creatine, nearly doubled synapse number, demonstrating that synapse formation or maintenance is normally limited by mitochondrial activity (161).

3 Polarized Trafficking and Long-Range Targeting to Synapses

The maintenance of synaptic structures with distinct compositions and properties requires continuous and accurate targeting of newly synthesized proteins to their final destination. Like any polarized cell, but amplified due to their morphological complexity, neurons require sorting and targeting mechanisms to ensure the delivery of molecular components to the appropriate compartment. Cell compartmentalization is often most obvious at the plasma membrane where diffusion barriers prevent integral proteins from being freely exchanged from one compartment to another. In neurons, such barriers are present at the axon initial segment, preventing mixing of components of the axonal and somatodendritic membranes (145, 197, 301, 302). On a finer (and possibly more impressive) scale, the thousands of synaptic inputs that a given neuron receives further segment the somatodendritic membrane into an array of postsynaptic microdomains, each with specific compositions. This raises fundamental questions on how such an extreme degree of compartmentalization can be maintained.

3.1 Asymmetric Protein Trafficking in Neurons

Much of what is known about protein trafficking in polarized cells comes from studies in epithelial cells, most prominently the Madin-Darby canine kidney (MDCK) cell-line, which differentiates in vitro into mono-layered epithelia with a well-defined apical/basolateral asymmetry (235). Various modes of polarized trafficking have been described in this system, including selective sorting of cargo into specialized post-Golgi carriers directly targeted to specific cellular domains, and non-specific delivery to the plasma membrane followed by endocytosis and transport to the appropriate destination (235, 289).

Studies in MDCK cells have demonstrated that cargo sorting into distinct post-Golgi carriers relies on intrinsic sequence determinants, with basolateral sorting sequences generally located in the cytoplasmic tail, and apical sorting sequences found frequently in the transmembrane domain (134, 235). Sorting into apical carriers can also result from post-translational modifications, such as N- or O- glycosylation, or addition of a glycosylphosphatidylinositol (GPI) anchor (235). When expressed in neurons, several basolateral and apical proteins are sorted preferentially to the somatodendritic compartment and axon respectively, including VSVG and LDL receptor (basolateral/somatodendritic), and influenza HA protein (apical/axonal) (60, 61, 123). However, more recent experiments suggest that neurons do not recognize some dihydrophobic motifs directing cargo to the basolateral domain in epithelial cells (260). Additionally, the sequence determinant that directs transferrin receptor to the basolateral domain in epithelial cells overlaps only partially with the somatodendritic targeting motif (300). Therefore, while many of the general principles of protein sorting established in epithelial cells apply to neurons, neurons have many of their own rules for establishing protein asymmetry.

In addition to direct delivery of proteins to their final membrane destination, a more circuitous mode of polarized trafficking is observed in certain epithelial and endothelial cells, where cargo destined for the apical plasma membrane is first exocytosed to the basolateral membrane and then endocytosed and transported to the apical membrane (289). Such transcytosis is also observed in neurons and accounts for the trafficking of certain proteins (e.g. VAMP2 and NgCAM/L1), which are initially delivered to the somatodendritic surface but ultimately localize to the presynaptic terminal or axonal membrane (243, 303). Mutations disrupting VAMP2 internalization leave VAMP2 stranded at the somatodendritic surface, indicating that its proper localization requires a redistribution occurring after endocytosis. While NgCAM is enriched in the axonal plasma membrane, intracellular NgCAM containing vesicles are found in both dendrites and axons (30), indicating that the steady state distribution of the molecule does not result from a vectorial targeting of post-Golgi carriers. Monitoring NgCAM insertion in the cell-surface following a synchronous release from the ER revealed that this protein first appears at the somatodendritic plasma membrane, and is subsequently internalized to be trafficked to the axon (243, 303). Interestingly, a single tyrosine point mutation (Y33A) results in direct axonal targeting of NgCAM (303), demonstrating that NgCAM can follow different routes to reach the same final destination. The direct targeting to the axon thus likely represents a default route for some axonal integral proteins.

Another example of polarized trafficking is neuronal potassium channels. Neurons express a wide array of potassium channels, which regulate crucial aspects of cellular ion homeostasis and excitability. Although displaying very similar primary structures within particular subclasses, given channel isoforms can have distinct distribution in the neuronal plasma membrane as well as specific trafficking itineraries (101, 171). For example, while voltage-gated potassium (Kv) channels Kv1.2, Kv1.3 and Kv1.4 are primarily found in the axonal plasma membrane, Kv4.2 is selectively expressed at the somato-dendritic surface (256). The conserved structure and specific surface expression patterns of Kv channels allowed the identification of molecular determinants of asymmetric protein trafficking in neurons. Analysis of chimeric molecules led to the identification of a C-terminal 16 amino acid di-leucine motif required for the proper targeting of Kv4.2 and sufficient to target the axonally localized channels Kv1.3 and Kv1.4 to dendrites (233). Conversely, interaction of the cytoplasmic N-terminal T1 tetramerization domain of Kv1 channels with Kvβ subunits is required for Kv1 expression at the axonal membrane and is sufficient to target dendritic molecules such as the transferrin receptor to the axonal surface (91).

Similar approaches led to the characterization of the molecular address targeting the AMPA receptor GluR1 subunit to the somatodendritic compartment. This determinant resides in the proximal segment of the C-terminal cytoplasmic tail of GluR1 and is sufficient to redirect otherwise axonally expressed proteins such as influenza virus hemagglutinin to the somatodendritic membrane (239).

Finally, of particular interest is the tendency of certain receptors and ion channels to have unusual reverse gradient concentration along dendrites, with distal dendrites displaying a higher concentration of receptors/channels than proximal ones. This is notably the case of AMPA receptors (7, 262) and HCN1 channels (167), whose distance-dependent enrichment along apical dendrites is important for the scaling of dendritic currents and their integration in the soma of hippocampal pyramidal neurons (73). This suggests differential regulation of protein trafficking or stability as a function of dendritic distance, a possibility that still awaits further investigation.

3.2 Polarized Secretory Trafficking for Asymmetric Dendrite Outgrowth

In addition to membrane protein processing, the ER and the Golgi are the primary site of lipid biosynthesis (163), a metabolic function subject to considerable challenge during the massive expansion of the plasma membrane that occurs during neurite outgrowth. Consistent with an immediate requirement for membrane trafficking through the secretory pathway in dendrite growth, disrupting Golgi function with brefeldin A in developing hippocampal or cortical pyramidal neurons dramatically reduces dendritic growth and branching (108). Perhaps more surprisingly, in mature cortical pyramidal neurons the same treatment causes a dramatic simplification of dendrite morphology, demonstrating the importance of an ongoing membrane flux through the secretory pathway for the maintenance of dendritic arbors (108).

The site of membrane insertion in growing dendrites is uncertain, but the spatial organization of the neuronal secretory pathway provides some clues. Somatic Golgi is nearly always oriented toward the apical dendrite in hippocampal pyramidal cells in vivo, and toward the longest and most branched dendrite of hippocampal neurons in culture (Fig. 2b) (108). As a consequence, most post-Golgi flux originating from the somatic Golgi is directed toward the apical dendrite. Disrupting the polarization of the somatic Golgi by overexpressing GRASP65, a Golgi matrix protein required for cisternal stacking (6), blocks the specification of the apical dendrite without diminishing the overall rate and extent of total dendrite growth (108), demonstrating that polarized post-Golgi trafficking sustains asymmetric dendritic growth.

In another study, the centrosome and associated organelles were found to cluster at sites opposite of the plane of final mitotic division of neural progenitors at a location where the first neurite emerges and becomes the axon (51). However, at slightly later developmental stages after initial neurite outgrowth but before either polarized growth or the appearance of spatially restricted axonal markers (hippocampal neuron polarity stage 1 in the scheme of Banker) (62), the somatic Golgi shows no polarization towards the axon (108). One possibility is that post-Golgi trafficking is redirected from axons to dendrites at different developmental stages, during migration, or in different in vivo contexts. Interestingly, disrupting the secretory pathway by overexpressing a kinase-dead form of protein kinase D1 that prevents cargo budding from the TGN (14), results in cessation of dendritic growth while axonal growth persists over days (108), indicating that distinct mechanisms govern membrane addition to the growing axon. Thus, akin to what occurs during polarization of non-neuronal cell architecture (69, 148), reorganization of the secretory pathway is essential for the asymmetric growth of dendrites. In this context, it is interesting to note that several molecules implicated in neurite polarization and outgrowth, such as PI3K, LIMK1, and Cdk5, are localized to the Golgi apparatus (6, 208, 238), suggesting close coordination between dendritic compartmentalization, polarized cell growth, and directed membrane trafficking through the neuronal Golgi.

3.3 Microtubule Transport

The cytoskeleton provides anchoring points for the stabilization and transport of cellular structures, and is a major determinant of cell architecture. The microtubule network provides tracks along which cargo including organelles, vesicles, and mRNA granules is transported over long distances. In some instances, stabilization of microtubule plus ends allows direct vectorial delivery of cargo to subcompartments of the plasma membrane, as observed for the targeting of gap junction components to epithelial cell adherens junctions (255). In a comparable manner, the establishment of a subsynaptic secretory pathway at the neuromuscular junction (NMJ) or in *Torpedo* electrocytes also involves a local reorganization of microtubules (33, 124–126, 229).

The organization of neuronal microtubules presents interesting specializations. In model non-neuronal cell types, most microtubules share the same polarity and are arranged in a radial manner with their minus end anchored at the microtubule organizing center (MTOC) and their plus end radiating towards the cell periphery (280). In contrast, neuronal microtubules in proximal dendrites adopt both plus end-out and plus end-in orientations while axons and distal dendrites contain microtubules with their plus ends pointing away from the cell body (12). Although the bidirectional organization of microtubules predicts significant complexity for dendritic membrane trafficking, the precise physiological consequences remain largely unknown.

Active movement of cargo along microtubules requires two classes of motor proteins – the large family of kinesins and cytoplasmic dynein. These microtubule motors move, in most cases, towards the plus and minus ends of microtubules, respectively (293). Indeed, the first kinesin to be described was characterized as the molecular motor responsible for fast axonal transport (23, 294). Subsequently, at least 45 members of the kinesin family have been discovered in mammals (102). Some of these kinesins play distinct roles in dendrites. For example, KIFC2 and KIF5 (the conventional kinesin, also known as kinesin1), localize to the somatodendritic compartment, where they are required for trafficking multivesicular bodies and mRNA granules (137, 241).

In the last ten years, biochemical characterization established that postsynaptic receptors are found with their interacting partners in large protein complexes containing an unexpectedly broad cast of molecular motors, especially kinesins, indicating that receptor microtubule transport is a fundamental aspect of postsynaptic receptor biology (138). This notion has been directly demonstrated for KIF17 which is required for the delivery of NR2B-containing NMDA receptors to synapses (92, 251). Interestingly, interaction with different cargo can "steer" kinesins to specific cellular domains. For example, the glutamate receptor interacting protein 1 (GRIP1), which binds the AMPA receptor subunit GluR2, directs KIF5 primarily to the somatodendritic domain, while a different kinesin binding protein, JSAP1, routes KIF5 to the axon (252). In addition to participating in AMPA receptor trafficking, the GRIP1/KIF5 interaction has important functions in dendrite development and maintenance. Decreasing GRIP1 expression by RNA interference (RNAi) causes a loss of dendrites which can be rescued by overexpression of EphB2 (104), a receptor tyrosine kinase known to affect spine development and synaptic plasticity (117). Conversely, disrupting the GRIP1/KIF5 interaction impairs EphB2 trafficking to dendrites and inhibits dendritic growth, supporting a model where GRIP1 acts as a kinesin adaptor important for dendritic trafficking (104, 252).

In a recent study, Banker and colleagues showed that the behavior of specific kinesins changes during early stages of neuronal morphogenesis. Whereas truncated kinesin 3 accumulates at the tip of all neurites throughout development, truncated kinesin 1 localizes to only a subset of neurites and rapidly redistributes in the emerging axon (119). This event represents one of the earliest molecular readouts of axon specification, and demonstrates that neurite differentiation involves a tight control of the microtubule network and associated kinesin-based transport.

3.4 Synaptic Targeting: Postsynaptic Receptors as Model Cargo

Postsynaptic receptors are highly enriched at synapses, where they form microdomains apposed to presynaptic active zones releasing cognate neurotransmitters. This accumulation depends on postsynaptic scaffolds, which stabilize receptors and connect them to downstream signaling effectors (see Chapter by Kennedy et al., this volume). Time-lapse imaging and photobleaching of GFP-tagged receptors (10, 11, 147, 254, 258) as well as real-time imaging of single fluorescent probes bound to surface receptors has revealed that receptors continuously diffuse in and out synapses, occasionally exploring multiple synapses (20, 50, 64, 278) (see Chapter by Choquet and Triller, this volume). These latter experimental approaches highlight the highly dynamic and probabilistic nature of receptor stabilization at synapses, emphasizing the importance of lateral diffusion (287). Indeed, upon expressing recombinant glycine receptor subunits under conditions where receptor exocytosis was detected only in the soma, it was found that lateral diffusion is sufficient for receptor accumulation at synapses on distal dendrites (237).

The relative contribution of vesicular trafficking and lateral diffusion to the trafficking of postsynaptic receptors remains a topic of active investigation. In some instances, both modes of trafficking have been found to contribute for a given receptor or membrane cargo. For example, while immunogold labeling and biochemical data suggest that microtubule dependent transport is a major mode of NMDA receptor targeting to synapses, live cell imaging indicates that synapses either receive quantal "packets" of NMDA receptors (299), or can gradually acquire NMDA receptors (25). While the latter study supports a "diffusional" accumulation of NMDA receptors, the former supports a direct delivery by discrete intracellular carriers. Monitoring the recovery of NMDA receptor-mediated synaptic currents following irreversible inactivation with the open-channel blocker MK801 demonstrated that synaptic NMDA receptors exchange with their extrasynaptic counterparts (286). Comparable approaches applied to $GABA_A$ receptors (281) and AMPA receptors (2) support the same general conclusion.

Although lateral diffusion in the plasma membrane is a major aspect of receptor movement and exchange, endocytosis and exocytosis are also crucial for rapid variations of receptor amount at synapses (see Chapters by Esteban, Wenthold et al., Lorena Arancibia-Carcamo et al., and Lisman and Hell, this volume). However, if exo- and endocytosis can occur locally in close proximity to excitatory synapses, membrane fusion and budding have not been demonstrated directly at the PSD (see Section 4 below), implying that lateral diffusion is an obligatory step for receptors to enter or exit synapses (287).

In the case of AMPA receptors, local activity at single synapses reduces the lateral mobility of GluR1, leading to accumulation of receptors by a diffusional trap (64), while local exocytosis from recycling endosomes supplies receptors for LTP (147, 212). Exocytosis and lateral diffusion are thus likely subject to distinct forms of regulation for different classes of receptors and membrane cargo (38, 203). Reconciliation of these two modes of synaptic trafficking (lateral diffusion versus vectorial vesicular targeting) awaits further investigation. Yet, the architecture of the dendritic membrane provides some interesting clues.

Most central excitatory synapses are localized on dendritic spines, and are thus isolated from the dendrite proper by the narrow spine neck, which limits exchange of cytoplasmic and integral membrane proteins between the spine head and the dendritic shaft (10, 11, 18, 232). Local diffusible pools of membrane cargo inside spines would thus provide a reserve pool for synaptic exchange (103). Utilization of vesicular traffic locally within spines, in turn, requires the positioning or mobilization of vesicle budding and trafficking machinery. The combination of a limited local pool of reserve cargo together with spine-localized membrane trafficking potentially affords exquisite control or "tuning" of molecular abundance at synapses via the balance of exocytosis, endocytosis and diffusional exchange. On the other hand, synapses onto dendritic shafts, including most GABAergic synapses, have direct access to a much larger surface pool of extrasynaptic receptors and are likely less affected by short term changes in the balance between exo- and endocytosis. Such constraints of geometry might help explain the distinct mechanisms for targeting and maintaining postsynaptic membrane components at excitatory spine synapses versus inhibitory shaft synapses (see Chapter by Lorena Arancibia-Carcamo et al., this volume).

3.5 Pre-assembly of Postsynaptic Membranes in the Secretory Pathway

The function of the ER and the Golgi apparatus in membrane trafficking extends beyond transmembrane protein processing. These organelles also direct the trafficking of cytoplasmic proteins modified by acylation. Indeed, acyltransferases, the core enzymes mediating protein palmitoylation localize to the Golgi apparatus (231, 263). The addition of a lipid moiety to proteins increases their hydrophobicity and facilitates their incorporation in lipid rafts (263), which participate in the nano-organization of cellular membranes by stabilizing multi-protein complexes (57, 115). Interestingly, numerous synaptic proteins, including cytoplasmic molecules, are palmitoylated. As shown directly for GRIP, PSD-95, AMPA receptors (GluR1-GluR4), ACh receptors, and $GABA_A$ receptors, palmitoylation regulates synaptic targeting and stabilization (56, 66, 67, 71, 97). PSD-95, the major postsynaptic scaffold protein at glutamateric synapses, is palmitoylated by HIP14, while specific AMPA receptor and $GABA_A$ receptor subunits are palmitoylated by GODZ, two palmitoylating enzymes primarily localized to the Golgi (71, 74, 113, 133, 230, 261).

As described at central synapses (56, 66, 67, 94, 172, 245), and in *Torpedo* electrocytes (175, 176), several postsynaptic scaffold proteins are localized to intracellular membranes containing receptors that likely correspond to compartments of the secretory pathway. Live cell imaging experiments have revealed the appearance and transport of PSD scaffold proteins such as PSD-95 as discrete clusters in dendrites (78, 177, 204). In the case of PSD-95, dendritic transport occurs through the secretory pathway (66) and can be stimulated by BDNF (307). The origin of these "pre-packaged" secretory carriers and the adaptor molecules or coat proteins involved in cargo sorting from earlier stages in the secretory pathway are not yet clear. However, many of these transported scaffold proteins are lipid modified, suggesting that acylation at the Golgi apparatus is an important step in scaffold protein targeting, and may direct the pre-assembly of postsynaptic membranes in the secretory patway. PSD95 is particularly interesting in this regard since its palmitoylation is regulated by synaptic activity (67), and is essential for its co-trafficking with AMPA receptors (66), its targeting to synapses (49, 66), and its ability to cluster transmembrane proteins (41).

4 Local Trafficking of Postsynaptic Components

4.1 Dendritic Exocytosis

4.1.1 Regulated Exocytosis Near the Postsynaptic Membrane

Membrane fusion requires the concerted action of numerous factors to first tether a vesicle to the membrane, prime it for release, and finally to fuse the separate lipid bilayers into a contiguous membrane (120). The eukaryotic membrane fusion machinery has been conserved during evolution, and consists of a core complex of the soluble N-ethylmaleimide sensitive factor attachment protein receptor (SNARE) family of proteins. As elucidated in detail for exocytosis in presynaptic nerve terminals (see Chapter by Atasoy and Kavalali, this volume), vesicle-bound SNARES (v-SNARES) form a complex with target membrane associated SNARES (t-SNARES) through coiled-coil motifs in a process facilitated by members of the Rab family of small GTPases (120). Accessory proteins superimposed on this core machinery allow for regulation of exocytosis by coupling membrane fusion to diverse signaling pathways. Many of the principles governing regulated exocytosis have been established by studies on the Ca^{2+}-dependent release of synaptic vesicles (see Chapter by Atasoy and Kavalali, this volume).

Although less well understood than synaptic vesicle release, dendritic exocytosis has emerged as a key mechanism for regulating synaptic physiology. Among the first indications of regulated exocytosis in dendrites came from experiments demonstrating that the styryl dye FM1-43 accumulates in dendritic TGN-derived compartments and can be rapidly unloaded in response to Ca^2 influx (174). The finding that this activity-dependent dendritic exocytosis required CaMKII (173) pointed towards a direct link between postsynaptic exocytosis and synaptic plasticity signaling pathways. Additional evidence for such a link came from observations that brefeldin A, which disrupts the secretory pathway, impairs synaptic potentiation in hippocampal slices, as does postsynaptic loading of N-ethylmaleimide or botulinum toxin B, two inhibitors of membrane fusion (28, 166). In conjunction with the discovery of silent synapses lacking AMPA receptors (see Chapters by Esteban and Lisman and Hell, this volume), these observations suggested that the rapid insertion of AMPA receptors in the postsynaptic membrane accounted for an early component of the expression of LTP. Subsequent imaging studies demonstrated that AMPA receptors rapidly appear and accumulate near the postsynaptic plasma membrane and are incorporated into synapses in response to stimuli triggering synapse potentiation (45, 147, 212). However, the core machinery for postsynaptic exocytosis and the mechanisms coupling NMDA receptor-induced Ca^{2+} influx to dendritic membrane fusion remain poorly understood. Moreover, the exact location of receptor insertion in the dendritic membrane is unclear, although NMDA receptor-induced exocytic events have recently been visualized in spines (213). For AMPA receptors, either the site of exocytosis or the location of rapid receptor accumulation after exocytosis, depends on subunit composition, since newly inserted GluR2 receptors accumulate more rapidly at synapses than do GluR1 receptors (214, 257). This suggests that a subclass of exocytic vesicles containing distinct populations of receptors may be inserted in close

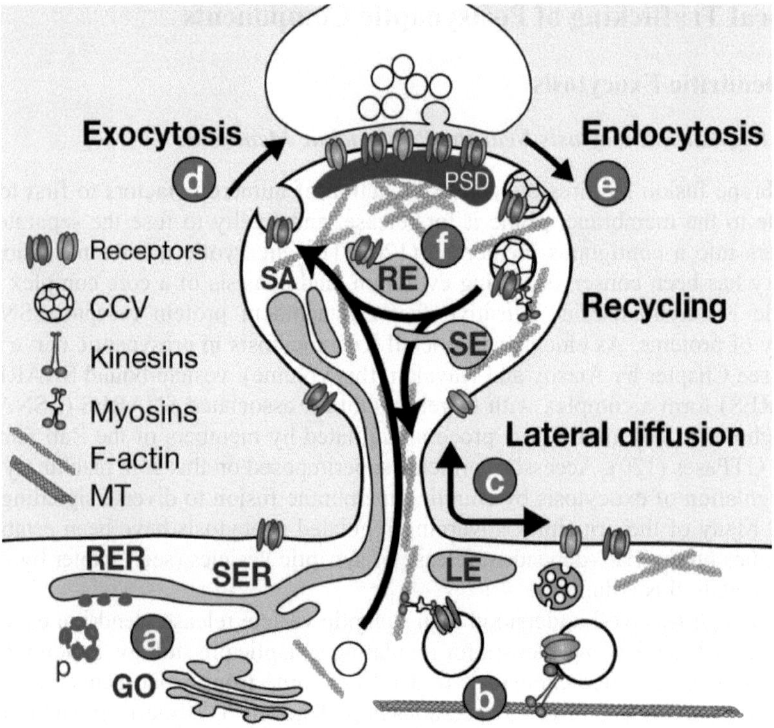

Fig. 4. Model for local trafficking in dendritic spines. Postsynaptic membrane cargo is provided by local synthesis in the secretory pathway (**a**), vesicular transport along microtubules (*dendritic shaft*) and F-actin (*spines*) (**b**), and lateral diffusion in the plasma membrane (**c**). Vesicles are translocated through the spine neck and can release cargo directly in the spine head by exocytosis (**d**). Cargo molecules are stabilized at the postsynaptic membrane but can diffuse in and out the postsynaptic density (PSD). Released receptors and other integral membrane proteins are captured and internalized by the spine endocytic zone located laterally to the PSD (**e**). Internalized receptors are sorted in early/sorting endosomes (SE) and sent either to late endosomes (LE) for degradation, or to recycling endosomes (RE) for return to the spine surface (**f**). *Abbreviations.* SER, smooth endoplasmic reticulum; RER, rough endoplasmic reticulum; GO, Golgi outpost; SA, spine apparatus; PSD, postsynaptic density; CCV, clathrin coated vesicle; SE, sorting endosome; RE, recycling endosome; LE, late endosome; MT, microtubule; p, polysome.

proximity to synapses (Fig. 4). Such exocytic vesicles may derive from recycling endosomes, which traffic into spines and fuse with the spine membrane during LTP-induced potentiation and spine growth (213).

Although initially described at vertebrate NMJs or in *Torpedo* electrocytes (33, 124, 125, 229), sub-synaptic trafficking compartments have been characterized in detail in larval *Drosophila* neuromuscular synapses (179, 180, 266). Genetic approaches in this system have provided important clues to the molecular mechanisms accounting for postsynaptic exocytosis. For example, the SNARE protein Gtaxin interacts with the postsynaptic scaffolding protein Dlg and is required for the targeted delivery of membranes to the postsynapse (82). In addition, synaptotagmin 4 serves as a Ca^{2+} sensor for postsynaptic exocytosis at the fly neuromuscular synapse (306), suggesting that mechanisms responsible for regulated postsynaptic exocytosis may be very similar to those operating at presynaptic terminals (see Chapter by Atasoy and Kavalali, this volume). Despite these initial clues, many questions remain, such as the identity and requirement for specific molecules that tether postsynaptic vesicles to the plasma membrane, and the nature of machinery linking postsynaptic exocytosis to neural activity.

4.1.2 Rab Proteins and Dendritic Exocytosis

In addition to SNARE proteins, Rab family GTPases are major regulators of membrane trafficking and vesicular fusion. Rab proteins localize to distinct cellular compartments and control trafficking specificity by facilitating interactions between donor and acceptor membranes (90). In general, Rab proteins bind to donor membranes in their GTP-bound state and recruit accessory factors that interact with proteins on the target membrane. When the appropriate combination of vesicle and target membrane proteins is assembled, Rab proteins hydrolyze their bound GTP, locking the vesicle in place for subsequent fusion (90).

Two Rab proteins with well-defined function in non-neuronal cells have been shown to regulate postsynaptic receptor exocytosis. Rab11a controls trafficking through recycling endosomes (291) and is required for NMDA receptor-dependent synapse potentiation including LTP (212). Rab8 is involved in trafficking from the TGN to the cell surface (114) and is also required for the postsynaptic delivery of AMPA receptor subunits (76). Rab11 resides in recycling endosomes, a compartment through which trafficking is mediated by vesicular fusion involving syntaxin 13 (226). Acute disruption of syntaxin 13-dependent trafficking abolishes stimulus-dependent insertion of AMPA receptors and accompanying LTP within minutes (212), suggesting that this membrane trafficking pathway is immediately proximal to NMDA receptor-induced dendritic exocytosis. Still to be determined are the molecular mechanisms underlying synaptic activity-dependent regulation of Rab8 and Rab11. Rab-specific guanine nucleotide exchange factors (GEFs) and GTPase activating proteins (GAPs) orchestrate Rab protein function by controlling their activation and inactivation (79, 250), and are themselves controlled by numerous signaling pathways (250), making them particularly attractive candidates for coupling neural activity to postsynaptic exocytosis.

4.1.3 Myosin Motors and Micron-Range Trafficking of Spine Cargo

In non-neuronal cells, post-TGN transport of secretory cargo to the plasma membrane occurs in two general stages. Vesicles are first transported along microtubules towards the cell-periphery by kinesins, and are then transferred to myosin motors, most prominently class V myosins, which translocate vesicles through the F-actin network of the cell cortex to bring them close to the plasma membrane (152). Myosins are multisubunit motor molecules that move cargo along F-actin in a unidirectional manner. Myosin motors have diverse functions in vesicle trafficking. At the plasma membrane, myosins control movements of vesicles during exocytosis and endocytosis (152). Many types of myosins are expressed in neurons, and although their individual roles are only beginning to emerge, they seem to be especially important for synaptic receptor recycling (26) (Fig. 4). This is particularly true for class V and class VI myosins, which are enriched in dendritic spines (165, 196, 206). While myosin VI regulates AMPA receptor endocytosis by virtue of its ability to move vesicles inward toward the pointed or minus-ends of actin filaments (206), members of the myosin V family facilitate delivery of AMPA receptors to the dendritic membrane (165).

Interestingly, a light chain common to myosin V and dynein indirectly interacts with AMPA receptors and NMDA receptors via GKAP/PSD95 (196), and class V myosins also associate with specific kinesins (152). Glutamate receptors are thus part of multi-protein complexes containing both microtubule and actin motors, a feature that is likely important for the transfer of receptor containing vesicles from the microtubule-rich dendritic shaft to spines, whose cytoskeleton primarily consists of actin. Recent in vitro experiments demonstrated that myosin Va can traverse filamentous junctions and displays diffusive "scanning" along microtubules (4), emphasizing the ability of this motor to navigate complex cytoskeletal structures.

Class V myosins are calmodulin-binding molecules and are regulated by calcium-dependent conformational changes (279). Increasing calcium triggers a conformational change of myosin Va from a closed, inactive, conformation to an open conformation that exposes the cargo binding domain. Although not yet fully characterized in neurons, this calcium-dependent regulation of class V myosins could play a role in activity-dependent trafficking of cargo to dendritic spines. Indeed myosin Va is highly enriched in the PSD (196, 297). However, *dilute* mutant mice lacking myosin Va have no defects in basal synaptic transmission or LTP at CA1 hippocampal synapses (248) and exhibit no quantitative difference in AMPA receptor abundance at synapses when assessed by immungold electron microscopy (218).

4.1.4 The Exocyst

Genetic studies of targeted membrane trafficking in *S. cerevisiae* led to the identification of conserved proteins important for the late processing of post-Golgi vesicles. Among these proteins were components of the exocyst, an octameric protein complex whose subunits (Sec3, Sec5, Sec6, Sec8, Sec10, Sec15, Exo 70 and Exo84) are required for tethering vesicles to distinct cellular domains (164).

In *Drosophila*, Sec5 and Sec6 mutants display defective neurite outgrowth resulting from impaired exocytosis (192, 193). However, synaptic vesicle exocytosis at the neuromuscular junction is normal in Sec5 mutants, indicating that the exocyst is not involved in all forms of exocytosis (192). Other studies in flies revealed a specific role for Sec15 in axon outgrowth and delivery of a subset of surface proteins to the plasma membrane (e.g. fasciclin II and chaoptin) (186), suggesting that particular components of the exocyst complex play distinct roles in neuronal trafficking.

Components of the exocyst complex are expressed in the mammalian nervous system (112), and notably in the developing brain where Sec6/8 is present in layers displaying ongoing synaptogenesis (98). At later stages, the exocyst is broadly expressed in the adult brain and is enriched in synaptic membranes (112, 244). Consistent with a role in directing exocytosis to discrete domains, Sec6/8 is located at sites of membrane addition, including neurites, filopodia, and growth cones in cultured hippocampal neurons (98, 296). More recently, the exocyst has been shown to participate in the trafficking of glutamate receptors to the synapse. Specifically, Sec8 contains a PDZ-binding motif that interacts with the synaptic scaffolding proteins PSD-95 and SAP102 (244). This interaction thus connects the exocyst complex to NMDA receptors and potentially AMPA receptor/TARP complexes, and is required for NMDA receptor trafficking to the cell surface (244).

The exocyst also plays an important role during late stages of receptor delivery to the plasma membrane, including in the spine head itself. Overexpression of truncated forms of Sec8 and Exo70 uncovered two successive exocyst-dependent steps of AMPA receptor exocytosis required for the turnover of synaptic receptors. Whereas impairment of Sec8 leads to the accumulation of AMPA receptor-containing vesicles in the dendritic shaft, overexpression of Exo70 mutant results in an accumulation of these vesicles in the spine head, leaving them stranded very close to the PSD (77). These data support the notion that AMPA receptor exocytosis can occur directly in spines, and that exocyst subunits play distinct and synergistic functions in the local targeting or tethering of receptor-laden vesicles.

4.2 Dendritic Endocytosis and Post-Endocytic Sorting

Eukaryotic cells continuously invaginate and internalize portions of their plasma membrane to take up extracellular nutrients, trophic factors, and pathogens, as well as to regulate their surface area and control surface levels of membrane proteins. Clathrin-dependent endocytosis is the best characterized mechanism of membrane protein internalization (139), and has been widely studied in presynaptic terminals where it mediates the retrieval of synaptic vesicles (see Chapter by McPherson et al., this volume). Clathrin-mediated endocytosis is initiated by adaptor proteins, such as AP-2, AP-180, and epsin, which associate with phosphoinositides and membrane protein cargo and nucleate the oligomerization of clathrin into a lattice, or coat, apposed to the cytoplasmic face of the plasma membrane (215). Upon acquisition of membrane curvature and invagination, clathrin-coated pits bud from the membrane to form clathrin-coated vesicles (118, 215). The fission reaction is mediated by members of the dynamin family of large GTPases, which assemble into oligomeric rings or spirals around the necks of coated pits (223). All three dynamin family members (dynamins 1–3) are expressed in neurons. While dynamin 2 is ubiquitious and mediates membrane fission in all cells of the body, dynamin 1 is specialized for membrane recycling of synaptic vesicles at the presynaptic terminal following exocytosis (273). Like dynamin 1, dynamin 3 is enriched in brain, but also localizes postsynaptically to dendritic spine heads and may play a role in postsynaptic membrane trafficking (87, 292).

Exocytosis, endocytosis and post-endocytic sorting represent important checkpoints for controlling cell-surface signaling. Although relying on the same basic machinery, turnover rates of surface proteins vary greatly, with internalization time constants ranging from seconds for membrane retrival at presynaptic terminals (273), to minutes in the case of G-protein coupled receptors (GPCRs) (267) and postsynaptic receptors (63, 141, 162). In addition to simply controlling the number of cell-surface receptors, endocytosis is intimately coupled to receptor signaling. A classic example of this is provided by GPCRs and receptor tyrosine kinases (RTKs), whose endocytosis is stimulated by agonist binding, and desensitizes the cell toward the receptor ligand (267). In some cases, endocytosis itself is required for signaling since some receptors initiate signal transduction from internal endosomes (54, 99, 110, 305).

Endocytosis and endosomal sorting of postsynaptic receptors have emerged as key mechanisms for tuning synaptic responses rapidly within minutes, or more slowly during homeostatic scaling and synapse development (45, 290). It is now clear that postsynaptic compartments possess an exquisite degree of specialization for receptor internalization and post-endocytic trafficking (Fig. 4). This specialization is reflected by the diverse stimuli that trigger postsynaptic endocytosis, the complex biochemical interactions that link synaptic receptors to endocytic effectors, and the organization of the spine membrane itself.

4.2.1 Activity-Dependent Regulation of Endocytosis and Post-endocytic Sorting

One indication of an intimate relationship between postsynaptic proteins and endocytic machinery comes from biochemical data showing that postsynaptic receptors and scaffolds directly interact with the endocytic machinery. For example, Homer and mGluR5 interact with dynamin 3 (87), and both the AMPA receptor subunit GluR2 and the NMDA receptor subunit NR2B interact with the AP-2 clathrin adaptor complex (130, 155, 158, 234). The $GABA_A$ receptor subunits β1-3 and γ2 also bind AP-2, indicating that similar mechanisms operate at inhibitory synapses (141). Consistently, interfering with receptor internalization leads to modifications of both excitatory and inhibitory synaptic responses, sometime within minutes (45). Indeed, the regulation of postsynaptic receptor endocytosis is a key set-point for various form of synaptic plasticity.

As initially established for AMPA receptors, internalization rates and post-endocytic sorting are directly modulated by synaptic activity (63, 159, 162). Various mechanisms account for the regulation of AMPA receptor endocytosis, and endocytosis can be triggered either directly by agonist binding, or indirectly by Ca^{2+}- and phosphatase-dependent signaling cascades. Control of postsynaptic receptor internalization involves a broad cast of proteins, including hippocalcin (211), Rab5 (29), CPG2 (48), Arc/Arg3.1. (40), and a significant array of receptor binding partners (45) (see Chapters by Esteban, Wenthold et al., and Lorena Arancibia-Carcamo et al., this volume).

The actin cytoskeleton is a critical determinant of the local dynamics of membrane proteins (150), and actin dynamics are tightly controlled at all steps of endocytosis (68). Accordingly, actin polymerization and depolymerization are required for stimulus-dependent endocytosis of AMPA receptors (309), and actin plays a critical role in the structural maintenance and plasticity of dendritic spines. It is thus not surprising that spines contain numerous actin regulatory proteins (58). Some of these proteins, notably cortactin (88, 195, 228), Arc/Arg3.1 (40, 170) and myosin VI (206) participate in AMPA receptor endocytosis. In the case of NMDA receptors, the endocytic adaptor PACSIN-1/syndapin-1, which interacts with dynamin and N-WASP and thereby couples actin dynamics to endocytosis (136), directly binds the noncanonical subunit NR3A and drives receptor internalization (216). That actin dynamics is itself regulated by synaptic activity (43, 205, 269) suggests that cytoskeletal remodeling links activity-dependent signaling to the endocytic machinery.

Following endocytosis, the fate of internalized postsynaptic cargo is governed by extrinsic factors such as synaptic activity or metabotropic receptors. For example, while AMPA receptors internalized in response to AMPA stimulation are trafficked to dendritic lysosomes and degraded, receptors internalized in response to NMDA stimulation are sorted into recycling endosomes in a PKA-dependent manner (63). This endosomal sorting is regulated by the duration of the stimulus (265) and can in turn be altered by long-term changes in network activity (159, 201). In a comparable manner, synaptic activity also determines the fate of internalized kainate receptors (178). The subunit composition within a given receptor subtype also modifies the behavior of internalized receptors. In the case of NMDA receptors, NR2A and NR2B subunits contain distinct post-endocytic sorting motifs, which direct them along degradative and recycling pathways, respectively (154). In addition, both NR1 and NR2 subunits contain conserved membrane proximal motifs, which direct internalized receptors for degradation by default if not counteracted by the C-terminal recycling motifs present in NR2B (249).

4.2.2 The Spine Endocytic Zone

By directly visualizing clathrin coats, Blanpied et al. (2002) showed that clathrin puncta localize to most hippocampal dendritic spines in close proximity to the PSD (Fig. 5ab) (17). Live cell imaging together with fluorescence recovery after photobleaching (FRAP) indicated that clathrin repeatedly assembles and disassembles at specific endocytic sites in dendrites and spines (17). The concentration and internalization of labeled transferrin at these spine clathrin puncta established that they are functional, thus demonstrating the existence of stable endocytic zones in close association with the PSD at glutamatergic synapses (17). Ultrastructural analysis confirmed the presence of coated structures in spines in vivo recapitulating all the early stages of the endocytic process (Fig. 5c) (219, 228). Immunogold labeling of its core components revealed an exquisite level of spatial organization of the clathrin endocytic machinery in spines, where AP-2, clathrin and dynamin are localized on the lateral side of the spine surface, with AP-2 closest and dynamin furthest from the PSD (228) (Fig. 5d). In some cases, clathrin and AP-2 are detected at the spine plasma membrane in the absence of a morphologically identified coated pit (228) (Fig. 5d), suggesting a pre-organization of endocytic molecules at the lateral spine membrane.

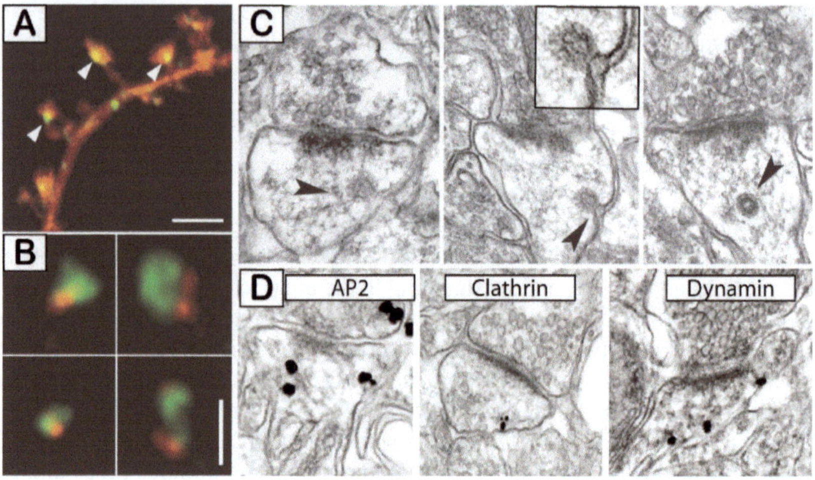

Fig. 5. The spine endocytic zone. (**a**) A cultured hippocampal neuron expressing a cell-fill (*red*) and clathrin-YFP (*green*), illustrating the localization of clathrin puncta to dendritic spine heads (*arrows*). Scale bar, 2 μm. (**b**) Examples of the close association of the spine endocytic zone (*red*) and the PSD (*green*) in hippocampal neurons expressing Clathrin-DsRed and PSD95-GFP. Scale bar, 1 μm. a and b adapted from (17); reprinted with permission from Elsevier, copyright 2002. (**c–d**) Electron micrographs showing the different stages of endocytosis (c) and the distribution of the endocytic machinery (d) in dendritic spine heads of CA1 pyramidal neurons in vivo. c: a clathrin coated invagination prior to scission (*left panel*), a coated invagination (*middle panel*), and a coated vesicle post-scission (*right panel*). d: note the presence of AP-2 (*left panel*), clathrin heavy chain (*middle panel*) and dynamin (*right panel*) immunoreactivity on the lateral side of spines. Adapted from (228); reprinted with permission from Nature Publishing Group, copyright 2004.

These data demonstrate that endocytosis occurs in spines, raising important questions regarding the subsynaptic trafficking via endosomes. Functional studies of the clathrin- and early endosome-associated small GTPase Rab5 showed that it is localized to the perimeter of the PSD and facilitates AMPA receptor internalization in response to LTD-inducing stimuli (29). Rab5 overexpression does not alter the spine/shaft ratio of total (internal and surface) AMPA receptors, but rather selectively decreases the number of receptors at the spine surface (29). This indicates that some internalized receptors remain in the spine, supporting a model where receptors internalized at endocytic zones on the spine surface are retained in a local endosomal pool (Fig. 4).

While the presence of endocytic zones at inhibitory synapses has not been extensively studied, immunolabeling of newly endocytosed $GABA_A$ receptors documented the presence of a stable pool of internalized receptors in close proximity to inhibitory synapses (295). Although the identity of these compartments awaits further investigation, these findings suggest that specialized endocytic machinery may also exist at inhibitory synapses. In this context, it is interesting to note that GRIP1/ABP, a molecule involved in the control of AMPA receptor intracellular trafficking, is also localized to inhibitory synapses and is found in internal membrane compartments containing inhibitory receptors (56, 59, 140), suggesting that the control of inhibitory and excitatory receptor endocytosis might share common effectors in addition to AP-2 and clathrin.

4.2.3 Recycling Endosomes as a Mobilizable Source of Postsynaptic Cargo

Quantitative analysis of AMPA receptor internalization and re-insertion in the plasma membrane in cultured cortical neurons has provided important insight into the involvement of post-endocytic recycling in synaptic remodeling (63, 162). Internalization of cell-surface AMPA receptors occurs with a time constant of approximately 10 min under basal conditions, and reaches a steady state that plateaus after 15–20 min, a time point at which approximately 15–20% of initially labeled receptors are internalized (63, 162). This apparent rate represents a balance between endocytosis itself and cycling of molecules (initially labeled when at the cell-surface) between endosomes and the plasma membrane.

Focusing selectively on the exocytosis of internalized molecules, it was shown that almost 90% of internalized receptors are reinserted in the cell-surface within 60 min under basal conditions, with a similar time constant of 10 min (63). As with internalization, the rate or reinsertion is directly modulated by neuronal activity. This demonstrates that both endocytosis and reinsertion are controlled by activity, and also indicates that the kinetics of reinsertion to the cell-surface can determine the speed and timing of receptor turnover. Adding to this rather complex regulation of trafficking kinetics, selective activation of synaptic AMPA receptors and NMDA receptors can change the compartmental itinerary of internalized cargo by differentially routing internalized AMPA receptors for lysosomal degradation, which in turn leads to changes in the number of receptors in the recycling pool (63, 159, 162). Altogether, these data strongly suggest that recycling endosomes function as a kinetic trap whereby activity-dependent tuning of receptor recycling kinetics and endosomal sorting controls the abundance of postsynaptic receptors in an online fashion (Fig. 4).

As previously mentioned, morphological studies have demonstrated the presence of an endosomal network throughout dendrites, and have specifically shown that recycling endosomes (REs) reside at the base of, and in some instance within, dendritic spines (Fig. 3) (47, 63, 213). This spatial distribution of REs is particularly surprising given that these compartments are usually found in the peri-nuclear area in non-neuronal cells (Fig. 1) (187), and suggests that dendritic REs may play a general role in a local trafficking of postsynaptic cargo. Indeed, whereas some dendritic AMPA receptors are transported to the cell body after endocytosis, an important fraction of internalized receptors remain in dendrites (63, 159, 162). Mechanisms by which different classes of endosomes are positioned in dendrites and spines, and the machinery that determines whether internalized postsynaptic cargo molecules are degraded or locally recycled back to the dendritic surface are only beginning to emerge.

The neuron-enriched endosomal protein of 21 kDa (Neep21) localizes to early endosomes and facilitates recycling of surface receptors (53, 271). In hippocampal neurons Neep21 loss of function decreases recycling of internalized GluR1 and GluR2 (271). This effect is mediated, at least for GluR2, by an activity-dependent interaction between Neep21 and GRIP1, since disrupting this interaction causes GluR2, but not GluR1 or transferrin receptor, to accumulate in early endosomes and lysosomes (270). Consistent with a function of Neep21 in activity-dependent recycling of AMPA receptor, disrupting Neep21 expression in hippocampal slices impairs LTP (3). The activity dependence of the Neep21/GRIP1 interaction suggests that synaptic activity defines the fraction of internalized receptor that is sorted for recycling at a given synapse, and conversely that this sorting step might tune the amount of receptor available for activity-dependent synaptic potentiation.

Overexpressing dominant-inhibitory mutants of Rab11a, Rme1/EHD1, or syntaxin-13, three molecules required form endocytic recycling, Park et al. (2004) (212) directly tested the role of REs in excitatory receptor trafficking (Fig. 6a). An important finding of this study was that blocking post-endocytic recycling not only decreases the basal level of surface AMPA receptors, but also prevents the rapid NMDA receptor-dependent insertion of GluR1 at hippocampal synapses, as well as LTP expression at Schaffer collateral/CA1 synapses in hippocampal slices. Finally, by selectively tagging or masking those receptors originally at the plasma membrane or those receptors cycling between the cell-surface and endocytic structures, this study established that AMPA receptors rapidly recruited at synapses are mobilized from a pool of internalized molecules, demonstrating the fundamental importance of REs in synapse plasticity.

4.2.4 Endosomal Recycling as a Cellular Basis for Spine Structural Plasticity

While mature neurons maintain their overall size and architecture over years, dendritic spines can be highly dynamic, growing and shrinking over minutes (184, 198). At the scale of the entire cell, endocytosis and exocytosis balance each other to maintain cellular surface area. However, the continuous appearance and disappearance of some dendritic spines necessarily requires increases or decreases in local membrane area. One attractive mechanism for refining spine morphology is local addition or removal of membrane by exocytosis and endocytosis. Indeed, the observation that spine volume correlates with the size of the PSD, the number of postsynaptic AMPA receptors and the abundance of spine endomembranes (100, 169, 182, 283) suggests a quantitative coupling between membrane trafficking, synaptic strength, and spine size. Consistent with this notion, LTP- and LTD-inducing stimuli trigger an increase or decrease in spine size, respectively (151, 183, 194, 207, 308).

The source of membrane for local spine growth has been a matter of debate since the dendritic shaft, and in some cases the spine head itself, house a wide array of endomembranes (135), and remodeling of the actin cytoskeleton alone can expand cellular membranes (70). The recently recognized contribution of endosomal recycling in glutamatergic synapse maintenance and plasticity prompted experiments to test whether REs contribute to spine growth. Combining three-dimensional serial reconstruction electron microscopy and high resolution fluorescence timelapse imaging, Park et al. (2006) (213) showed that spine growth is correlated with both the translocation of endosomal compartments in growing spines and the local exocytosis of RE cargo, occurring in some instances directly in the spine head (Fig. 6b). Furthermore, Rab11a and syntaxin 13 mutants that block endosomal recycling impair stimulus-induced spine growth, thus demonstrating that recycling endosomes contribute membrane material for spine growth (213). By supplying both AMPA receptors (212) and membrane material (213) during LTP, activity-induced cargo transport from recycling endosomes provides an attractive mechanism for coupling functional and structural plasticity at glutamatergic synapses.

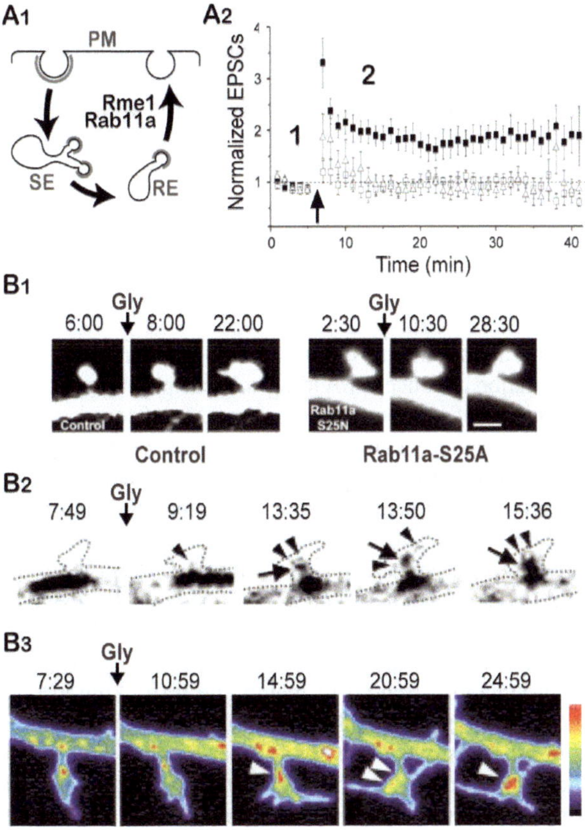

Fig. 6. Recycling endosomes as reservoirs for functional and structural spine plasticity. (**a**) Endosomal recycling is required for the expression of LTP. (**A1**) Schematic diagram showing trafficking pathways between the plasma membrane (PM), sorting endosomes (SE) and recycling endosomes (RE), and the transport step controlled by the regulatory proteins Rab11a and Rme1. (**A2**) Normalized EPSCs at Schaffer collateral/CA1 synapses in hippocampal slices before (1) or after (2) high frequency stimulation (*Arrow*) demonstrating the absence of LTP in cells expressing either Rme1 or Rab11a mutants (open symbols) compared to control cells (black squares). Adapted from (212); reprinted with permission from AAAS, copyright 2004. (**b**) Transport from recycling endosomes is required for activity-induced spine growth. (**B1**) Hippocampal neurons expressing a fluorescent cell fill (tdTomato) before and after a chemical LTP stimulation (Gly). Blocking recycling endosome transport by expression of a Rab11a dominant inhibitory mutant (Rab11a-S25N) prevented stimulus-induced spine growth (*right panel*) as compared to cells expressing GFP (*left panel*). Times indicated in min:sec. (**B2**) Time-lapse sequence of a dendritic spine of a hippocampal neuron expressing a GFP-tagged tranferrin receptor (inverted contrast) before and after a chemical LTP stimulation (Gly) documenting the stimulus-dependent recruitment of recycling endosomes into the spine head. (**B3**) Similar experiment in a neuron expressing a pHluorin-tagged transferrin receptor (TfR) to detect exocytic events in spine (*arrowheads*). B1–3 adapted from (213); reprinted with permission from Elsevier, copyright 2006.

Intriguingly, the importance of Rab11-dependent membrane recycling in the control of cell geometry seems to extend beyond plastic changes of spine morphology, as indicated by the recent characterization of protrudin, a mammalian Rab11-GDP binding protein which promotes and controls neurite formation (259). Spatially directed membrane trafficking from recycling endosomes may thus be a general mechanism for neuronal growth and structural plasticity during the elaboration and remodeling of diverse neural circuits.

5 Concluding Remarks

The discovery that membrane trafficking underlies diverse forms of neuronal plasticity represents a major advance, providing a focal point for experiments addressing the basic mechanisms of learning and memory. Many core issues of neuronal development and circuit plasticity can now be framed as cell biological problems, raising fundamental questions as to how conserved cellular machinery (Fig. 1) has evolved and adapted to serve specialized functions in the mammalian brain. Though far from complete, a picture is emerging of a previously unanticipated and remarkable degree of molecular dynamics at synapses orchestrated by conserved trafficking machinery and diverse dendritic organelles which are regulated by synaptic activity. These molecular dynamics function to maintain reliable, but still exquisitely plastic, communication between highly interconnected cells (Fig. 4). While general rules coupling changes of synaptic strength to membrane trafficking are becoming clarified, our understanding of how cell biological events coordinate neuronal physiology is still constrained by unresolved questions. What controls the organization and distribution of dendritic organelles themselves? To what extent are changes in synaptic strength confined to specific synapses, and under which circumstances? What is the molecular machinery that couples neural activity to dendritic membrane trafficking?

While much recent work has focused on the trafficking of postsynaptic receptors to and from the synapse, it is clear that the trafficking of other synaptic proteins, such as adhesion molecules (83, 224, 247), CaMKII (80), PSD-95 (44, 65, 78, 254), and gephyrin (94, 172) is equally important in tuning synaptic properties (116). A detailed description of the fundamental organization of neuronal organelles will be essential to understand how the regulation of multiprotein complexes produces the requisite integration of diverse signals for coordinating functional and structural plasticity. The relative contribution of long-range versus local trafficking in the development, maintenance, and plasticity of single synapses is not established. Within a given cell, mechanisms accounting for the turnover and plastic changes of synapses vary with synapse localization. For example, recent data suggest that endosomal trafficking and lateral diffusion may differentially contribute to postsynaptic receptors at synapses localized near the soma and those localized to distal dendrites (2). In pyramidal neurons, Golgi outposts are present primarily in apical dendrites (108), suggesting distinct mechanisms in apical versus basolateral dendrites. Indeed synapses onto these dendritic compartments contain different subclasses of NMDA receptors (131), and perhaps other molecules. One possibility is that a heterogeneous spatial restriction of neuronal organelles such as the Golgi apparatus, post-Golgi compartments, or various classes of endosomes contributes to spatial variation in synapse composition.

At the level of single spines, the actin cytoskeleton is a central effector that couples morphological and functional synaptic plasticity. Although actin can directly affect the organization of the PSD itself (5, 37, 43, 149), its contribution to local spine membrane trafficking is only beginning to emerge. An intriguing observation is that while synaptic activity triggers the recruitment of membrane bound structures in the spine head (213, 282, 283), the diffusional coupling between the spine head and the dendritic shaft is simultaneously decreased (18). A detailed description of spine neck properties will be crucial to understand how calcium signaling, actin dynamics, and membrane trafficking are coordinated to control spine size and molecular composition.

Finally, from a clinical perspective, recent findings illustrate how core issues of dendritic membrane trafficking open new perspectives for therapeutic strategies. For example, the Huntington's disease protein huntingtin binds to proteins such as HIP (huntingtin interacting protein) 1 and HIP14 which participate in the control of inhibitory (142, 143, 156, 209) and excitatory receptor (113) turnover, respectively. The role of postsynaptic trafficking in Alzheimer's disease (AD) is particularly interesting. It has been known for some time that the processing of the amyloid precursor protein during β-amyloid (Aβ) production occurs during trafficking through secretory and endocytic compartments (93), and swollen neuronal endosomes correlate with disease in late-onset AD (34–36, 89, 199). Recent data indicate that Aβ release is activity-dependent, and liberated Aβ in turn attenuates synaptic strength (129) through mechanisms that share similarities with LTD, including increased endocytosis of postsynaptic glutamate receptors (111, 129, 202, 253, 264). Conversely, Aβ oligomers block LTP (298) and presumably AMPA receptor insertion. Thus, it is tempting to speculate that at its early stages, AD is a disorder of postsynaptic trafficking and synapse dysfunction. Beyond neurodegenerative diseases, alterations in glutamate receptor trafficking and defects in synaptic properties have been reported in animal models of neuropsychiatric disorders including addiction (22, 304) and schizophrenia (153), as well as in models of congenital mental retardation (8, 15, 109, 146). Autism has been associated with mutations in specific neuroligin isoforms (121). These mutations impair the expression of these adhesion molecules at the plasma membrane (42, 46), and are likely to have important consequences on synaptic properties given the importance of neuroligins in the clustering of postsynaptic proteins (83, 247). Regulation of dendritic organelles and trafficking machinery is thus requisite for modifying neural circuits during development and plasticity, and provides a novel venue for restoring neuronal function in injury and disease.

Acknowledgements

We thank Morgan Sheng for insightful comments on the manuscript. We also thank Matthew J. Kennedy, and Ian G. Davison for their critical readings. We apologize to those whose work we could not cite due to space limitations. MDE is an Investigator of the Howard Hughes Medical Institute. Work in the lab of MDE is supported by the NIH.

References

1. Adams I and Jones DG. Quantitative ultrastructural changes in rat cortical synapses during early-, mid- and late-adulthood. *Brain Res* 239: 349–363, 1982.
2. Adesnik H, Nicoll RA, and England PM. Photoinactivation of native AMPA receptors reveals their real-time trafficking. *Neuron* 48: 977–985, 2005.
3. Alberi S, Boda B, Steiner P, Nikonenko I, Hirling H, and Muller D. The endosomal protein NEEP21 regulates AMPA receptor-mediated synaptic transmission and plasticity in the hippocampus. *Mol Cell Neurosci* 29: 313–319, 2005.
4. Ali MY, Krementsova EB, Kennedy GG, Mahaffy R, Pollard TD, Trybus KM, and Warshaw DM. Myosin. Va maneuvers through actin intersections and diffuses along microtubules. *Proc Natl Acad Sci USA* 104: 4332–4336, 2007.
5. Allison DW, Chervin AS, Gelfand VI, and Craig AM. Postsynaptic scaffolds of excitatory and inhibitory synapses in hippocampal neurons: maintenance of core components independent of actin filaments and microtubules. *J Neurosci* 20: 4545–4554, 2000.
6. Altan-Bonnet N, Sougrat R, and Lippincott-Schwartz J. Molecular basis for Golgi maintenance and biogenesis. *Curr Opin Cell Biol* 16: 364–372, 2004.
7. Andrasfalvy BK and Magee JC. Distance-dependent increase in AMPA receptor number in the dendrites of adult hippocampal CA1 pyramidal neurons. *J Neurosci* 21: 9151–9159, 2001.
8. Antar LN, Afroz R, Dictenberg JB, Carroll RC, and Bassell GJ. Metabotropic glutamate receptor activation regulates fragile × mental retardation protein and FMR1 mRNA localization differentially in dendrites and at synapses. *J Neurosci* 24: 2648–2655, 2004.
9. Aridor M, Guzik AK, Bielli A, and Fish KN. Endoplasmic reticulum export site formation and function in dendrites. *J Neurosci* 24: 3770–3776, 2004.
10. Ashby MC, De La Rue SA, Ralph GS, Uney J, Collingridge GL, and Henley JM. Removal of AMPA receptors (AMPARs) from synapses is preceded by transient endocytosis of extrasynaptic AMPARs. *J Neurosci* 24: 5172–5176, 2004.
11. Ashby MC, Maier SR, Nishimune A, and Henley JM. Lateral diffusion drives constitutive exchange of AMPA receptors at dendritic spines and is regulated by spine morphology. *J Neurosci* 26: 7046–7055, 2006.
12. Baas PW, Deitch JS, Black MM, and Banker GA. Polarity orientation of microtubules in hippocampal neurons: uniformity in the axon and nonuniformity in the dendrite. *Proc Natl Acad Sci USA* 85: 8335–8339, 1988.
13. Bagal AA, Kao JP, Tang CM, and Thompson SM. Long-term potentiation of exogenous glutamate responses at single dendritic spines. *Proc Natl Acad Sci USA* 102: 14434–14439, 2005.
14. Bard F and Malhotra V. The formation of TGN-to-plasma-membrane transport carriers. *Annu Rev Cell Dev Biol* 22: 439–455, 2006.
15. Bear MF, Huber KM, and Warren ST. The mGluR theory of fragile × mental retardation. *Trends Neurosci* 27: 370–377, 2004.
16. Bergmann JE. Using temperature-sensitive mutants of VSV to study membrane protein biogenesis. *Methods Cell Biol* 32: 85–110, 1989.
17. Blanpied TA, Scott DB, and Ehlers MD. Dynamics and regulation of clathrin coats at specialized endocytic zones of dendrites and spines. *Neuron* 36: 435–449, 2002.
18. Bloodgood BL and Sabatini BL. Neuronal activity regulates diffusion across the neck of dendritic spines. *Science* 310: 866–869, 2005.
19. Bonifacino JS and Rojas R. Retrograde transport from endosomes to the trans-Golgi network. *Nat Rev Mol Cell Biol* 7: 568–579, 2006.
20. Borgdorff AJ and Choquet D. Regulation of AMPA receptor lateral movements. *Nature* 417: 649–653, 2002.

21. Borgese N, Francolini M, and Snapp E. Endoplasmic reticulum architecture: structures in flux. *Curr Opin Cell Biol* 18: 358–364, 2006.
22. Boudreau AC and Wolf ME. Behavioral sensitization to cocaine is associated with increased AMPA receptor surface expression in the nucleus accumbens. *J Neurosci* 25: 9144–9151, 2005.
23. Brady ST. A novel brain ATPase with properties expected for the fast axonal transport motor. *Nature* 317: 73–75, 1985.
24. Bredt DS and Nicoll RA. AMPA receptor trafficking at excitatory synapses. *Neuron* 40: 361–379, 2003.
25. Bresler T, Shapira M, Boeckers T, Dresbach T, Futter M, Garner CC, Rosenblum K, Gundelfinger ED, and Ziv NE. Postsynaptic density assembly is fundamentally different from presynaptic active zone assembly. *J Neurosci* 24: 1507–1520, 2004.
26. Bridgman PC. Myosin-dependent transport in neurons. *J Neurobiol* 58: 164–174, 2004.
27. Broadwell RD and Cataldo AM. The neuronal endoplasmic reticulum: its cytochemistry and contribution to the endomembrane system. I. Cell bodies and dendrites. *J Histochem Cytochem* 31: 1077–1088, 1983.
28. Broutman G and Baudry M. Involvement of the secretory pathway for AMPA receptors in NMDA-induced potentiation in hippocampus. *J Neurosci* 21: 27–34, 2001.
29. Brown TC, Tran IC, Backos DS, and Esteban JA. NMDA receptor-dependent activation of the small GTPase Rab5 drives the removal of synaptic AMPA receptors during hippocampal LTD. *Neuron* 45: 81–94, 2005.
30. Burack MA, Silverman MA, and Banker G. The role of selective transport in neuronal protein sorting. *Neuron* 26: 465–472, 2000.
31. Cameron HA, Kaliszewski CK, and Greer CA. Organization of mitochondria in olfactory bulb granule cell dendritic spines. *Synapse* 8: 107–118, 1991.
32. Cameron PL, Sudhof TC, Jahn R, and De Camilli P. Colocalization of synaptophysin with transferrin receptors: implications for synaptic vesicle biogenesis. *J Cell Biol* 115: 151–164, 1991.
33. Camus G, Jasmin BJ, and Cartaud J. Polarized sorting of nicotinic acetylcholine receptors to the postsynaptic membrane in Torpedo electrocyte. *Eur J Neurosci* 10: 839–852, 1998.
34. Cataldo AM, Barnett JL, Pieroni C, and Nixon RA. Increased neuronal endocytosis and protease delivery to early endosomes in sporadic Alzheimer's disease: neuropathologic evidence for a mechanism of increased beta-amyloidogenesis. *J Neurosci* 17: 6142–6151, 1997.
35. Cataldo AM, Hamilton DJ, Barnett JL, Paskevich PA, and Nixon RA. Properties of the endosomal-lysosomal system in the human central nervous system: disturbances mark most neurons in populations at risk to degenerate in Alzheimer's disease. *J Neurosci* 16: 186–199, 1996.
36. Cataldo AM, Petanceska S, Peterhoff CM, Terio NB, Epstein CJ, Villar A, Carlson EJ, Staufenbiel M, and Nixon RA. App gene dosage modulates endosomal abnormalities of Alzheimer's disease in a segmental trisomy 16 mouse model of down syndrome. *J Neurosci* 23: 6788–6792, 2003.
37. Charrier C, Ehrensperger MV, Dahan M, Levi S, and Triller A. Cytoskeleton regulation of glycine receptor number at synapses and diffusion in the plasma membrane. *J Neurosci* 26: 8502–8511, 2006.
38. Chen L, Chetkovich DM, Petralia RS, Sweeney NT, Kawasaki Y, Wenthold RJ, Bredt DS, and Nicoll RA. Stargazin regulates synaptic targeting of AMPA receptors by two distinct mechanisms. *Nature* 408: 936–943, 2000.
39. Cheng D, Hoogenraad CC, Rush J, Ramm E, Schlager MA, Duong DM, Xu P, Wijayawardana SR, Hanfelt J, Nakagawa T, Sheng M, and Peng J. Relative and absolute quan-

tification of postsynaptic density proteome isolated from rat forebrain and cerebellum. *Mol Cell Proteomics* 5: 1158–1170, 2006.

40. Chowdhury S, Shepherd JD, Okuno H, Lyford G, Petralia RS, Plath N, Kuhl D, Huganir RL, and Worley PF. Arc/Arg3.1 interacts with the endocytic machinery to regulate AMPA receptor trafficking. *Neuron* 52: 445–459, 2006.
41. Christopherson KS, Sweeney NT, Craven SE, Kang R, El-Husseini Ael D, and Bredt DS. Lipid- and protein-mediated multimerization of PSD-95: implications for receptor clustering and assembly of synaptic protein networks. *J Cell Sci* 116: 3213–3219, 2003.
42. Chubykin AA, Liu X, Comoletti D, Tsigelny I, Taylor P, and Sudhof TC. Dissection of synapse induction by neuroligins: effect of a neuroligin mutation associated with autism. *J Biol Chem* 280: 22365–22374, 2005.
43. Colicos MA, Collins BE, Sailor MJ, and Goda Y. Remodeling of synaptic actin induced by photoconductive stimulation. *Cell* 107: 605–616, 2001.
44. Colledge M, Snyder EM, Crozier RA, Soderling JA, Jin Y, Langeberg LK, Lu H, Bear MF, and Scott JD. Ubiquitination regulates PSD-95 degradation and AMPA receptor surface expression. *Neuron* 40: 595–607, 2003.
45. Collingridge GL, Isaac JT, and Wang YT. Receptor trafficking and synaptic plasticity. *Nat Rev Neurosci* 5: 952–962, 2004.
46. Comoletti D, De Jaco A, Jennings LL, Flynn RE, Gaietta G, Tsigelny I, Ellisman MH, and Taylor P. The Arg451Cys-neuroligin-3 mutation associated with autism reveals a defect in protein processing. *J Neurosci* 24: 4889–4893, 2004.
47. Cooney JR, Hurlburt JL, Selig DK, Harris KM, and Fiala JC. Endosomal compartments serve multiple hippocampal dendritic spines from a widespread rather than a local store of recycling membrane. *J Neurosci* 22: 2215–2224, 2002.
48. Cottrell JR, Borok E, Horvath TL, and Nedivi E. CPG2: a brain- and synapse-specific protein that regulates the endocytosis of glutamate receptors. *Neuron* 44: 677–690, 2004.
49. Craven SE, El-Husseini AE, and Bredt DS. Synaptic targeting of the postsynaptic density protein PSD-95 mediated by lipid and protein motifs. *Neuron* 22: 497–509, 1999.
50. Dahan M, Levi S, Luccardini C, Rostaing P, Riveau B, and Triller A. Diffusion dynamics of glycine receptors revealed by single-quantum dot tracking. *Science* 302: 442–445, 2003.
51. de Anda FC, Pollarolo G, Da Silva JS, Camoletto PG, Feiguin F, and Dotti CG. Centrosome localization determines neuronal polarity. *Nature* 436: 704–708, 2005.
52. De Camilli P, Moretti M, Donini SD, Walter U, and Lohmann SM. Heterogeneous distribution of the cAMP receptor protein RII in the nervous system: evidence for its intracellular accumulation on microtubules, microtubule-organizing centers, and in the area of the Golgi complex. *J Cell Biol* 103: 189–203, 1986.
53. Debaigt C, Hirling H, Steiner P, Vincent JP, and Mazella J. Crucial role of neuron-enriched endosomal protein of 21 kDa in sorting between degradation and recycling of internalized G-protein-coupled receptors. *J Biol Chem* 279: 35687–35691, 2004.
54. Delcroix JD, Valletta JS, Wu C, Hunt SJ, Kowal AS, and Mobley WC. NGF signaling in sensory neurons: evidence that early endosomes carry NGF retrograde signals. *Neuron* 39: 69–84, 2003.
55. Deller T, Korte M, Chabanis S, Drakew A, Schwegler H, Stefani GG, Zuniga A, Schwarz K, Bonhoeffer T, Zeller R, Frotscher M, and Mundel P. Synaptopodin-deficient mice lack a spine apparatus and show deficits in synaptic plasticity. *Proc Natl Acad Sci USA* 100: 10494–10499, 2003.
56. DeSouza S, Fu J, States BA, and Ziff EB. Differential palmitoylation directs the AMPA receptor-binding protein ABP to spines or to intracellular clusters. *J Neurosci* 22: 3493–3503, 2002.

57. Di Paolo G and De Camilli P. Phosphoinositides in cell regulation and membrane dynamics. *Nature* 443: 651–657, 2006.
58. Dillon C and Goda Y. The actin cytoskeleton: integrating form and function at the synapse. *Annu Rev Neurosci* 28: 25–55, 2005.
59. Dong H, Zhang P, Liao D, and Huganir RL. Characterization, expression, and distribution of GRIP protein. *Ann NY Acad Sci* 868: 535–540, 1999.
60. Dotti CG, Parton RG, and Simons K. Polarized sorting of glypiated proteins in hippocampal neurons. *Nature* 349: 158–161, 1991.
61. Dotti CG and Simons K. Polarized sorting of viral glycoproteins to the axon and dendrites of hippocampal neurons in culture. *Cell* 62: 63–72, 1990.
62. Dotti CG, Sullivan CA, and Banker GA. The establishment of polarity by hippocampal neurons in culture. *J Neurosci* 8: 1454–1468, 1988.
63. Ehlers MD. Reinsertion or degradation of AMPA receptors determined by activity-dependent endocytic sorting. *Neuron* 28: 511–525, 2000.
64. Ehlers MD, Heine M, Groc L, Lee MC, and Choquet D. Diffusional trapping of GluR1 AMPA receptors by input-specific synaptic activity. *Neuron* 54: 447–460, 2007.
65. Ehrlich I and Malinow R. Postsynaptic density 95 controls AMPA receptor incorporation during long-term potentiation and experience-driven synaptic plasticity. *J Neurosci* 24: 916–927, 2004.
66. El-Husseini AE, Craven SE, Chetkovich DM, Firestein BL, Schnell E, Aoki C, and Bredt DS. Dual palmitoylation of PSD-95 mediates its vesiculotubular sorting, postsynaptic targeting, and ion channel clustering. *J Cell Biol* 148: 159–172, 2000.
67. El-Husseini Ael D, Schnell E, Dakoji S, Sweeney N, Zhou Q, Prange O, Gauthier-Campbell C, Aguilera-Moreno A, Nicoll RA, and Bredt DS. Synaptic strength regulated by palmitate cycling on PSD-95. *Cell* 108: 849–863, 2002.
68. Engqvist-Goldstein AE and Drubin DG. Actin assembly and endocytosis: from yeast to mammals. *Annu Rev Cell Dev Biol* 19: 287–332, 2003.
69. Etienne-Manneville S and Hall A. Integrin-mediated activation of Cdc42 controls cell polarity in migrating astrocytes through PKCzeta. *Cell* 106: 489–498, 2001.
70. Etienne-Manneville S and Hall A. Rho GTPases in cell biology. *Nature* 420: 629–635, 2002.
71. Fang C, Deng L, Keller CA, Fukata M, Fukata Y, Chen G, and Luscher B. GODZ-mediated palmitoylation of GABA(A) receptors is required for normal assembly and function of GABAergic inhibitory synapses. *J Neurosci* 26: 12758–12768, 2006.
72. Fishman PS, Farrand DA, and Kristt DA. Internalization of plasma proteins by cerebellar Purkinje cells. *J Neurol Sci* 100: 43–49, 1990.
73. Frick A, Magee J, Koester HJ, Migliore M, and Johnston D. Normalization of Ca2+ signals by small oblique dendrites of CA1 pyramidal neurons. *J Neurosci* 23: 3243–3250, 2003.
74. Fukata M, Fukata Y, Adesnik H, Nicoll RA, and Bredt DS. Identification of PSD-95 palmitoylating enzymes. *Neuron* 44: 987–996, 2004.
75. Gardiol A, Racca C, and Triller A. Dendritic and postsynaptic protein synthetic machinery. *J Neurosci* 19: 168–179, 1999.
76. Gerges NZ, Backos DS, and Esteban JA. Local control of AMPA receptor trafficking at the postsynaptic terminal by a small GTPase of the Rab family. *J Biol Chem* 279: 43870–43878, 2004.
77. Gerges NZ, Backos DS, Rupasinghe CN, Spaller MR, and Esteban JA. Dual role of the exocyst in AMPA receptor targeting and insertion into the postsynaptic membrane. *EMBO J* 25: 1623–1634, 2006.

78. Gerrow K, Romorini S, Nabi SM, Colicos MA, Sala C, and El-Husseini A. A preformed complex of postsynaptic proteins is involved in excitatory synapse development. *Neuron* 49: 547–562, 2006.
79. Geyer M and Wittinghofer A. GEFs, GAPs, GDIs and effectors: taking a closer (3D) look at the regulation of Ras-related GTP-binding proteins. *Curr Opin Struct Biol* 7: 786–792, 1997.
80. Gleason MR, Higashijima S, Dallman J, Liu K, Mandel G, and Fetcho JR. Translocation of CaM kinase II to synaptic sites in vivo. *Nat Neurosci* 6: 217–218, 2003.
81. Golgi C. On the structure of nerve cells. 1898. *J Microsc* 155: 3–7, 1989.
82. Gorczyca D, Ashley J, Speese S, Gherbesi N, Thomas U, Gundelfinger E, Gramates LS, and Budnik V. Postsynaptic membrane addition depends on the Discs-Large-interacting t-SNARE Gtaxin. *J Neurosci* 27: 1033–1044, 2007.
83. Graf ER, Zhang X, Jin SX, Linhoff MW, and Craig AM. Neurexins induce differentiation of GABA and glutamate postsynaptic specializations via neuroligins. *Cell* 119: 1013–1026, 2004.
84. Gray EG. Axo-somatic and axo-dendritic synapses of the cerebral cortex: an electron microscope study. *J Anat* 93: 420–433, 1959.
85. Gray EG. Electron microscopy of synaptic contacts on dendrite spines of the cerebral cortex. *Nature* 183: 1592–1593, 1959.
86. Gray EG and Guillery RW. A Note on the Dendritic Spine Apparatus. *J Anat* 97: 389–392, 1963.
87. Gray NW, Fourgeaud L, Huang B, Chen J, Cao H, Oswald BJ, Hemar A, and McNiven MA. Dynamin 3 is a component of the postsynapse, where it interacts with mGluR5 and Homer. *Curr Biol* 13: 510–515, 2003.
88. Gray NW, Kruchten AE, Chen J, and McNiven MA. A dynamin-3 spliced variant modulates the actin/cortactin-dependent morphogenesis of dendritic spines. *J Cell Sci* 118: 1279–1290, 2005.
89. Grbovic OM, Mathews PM, Jiang Y, Schmidt SD, Dinakar R, Summers-Terio NB, Ceresa BP, Nixon RA, and Cataldo AM. Rab5-stimulated up-regulation of the endocytic pathway increases intracellular beta-cleaved amyloid precursor protein carboxyl-terminal fragment levels and Abeta production. *J Biol Chem* 278: 31261–31268, 2003.
90. Grosshans BL, Ortiz D, and Novick P. Rabs and their effectors: achieving specificity in membrane traffic. *Proc Natl Acad Sci USA* 103: 11821–11827, 2006.
91. Gu C, Jan YN, and Jan LY. A conserved domain in axonal targeting of Kv1 (Shaker) voltage-gated potassium channels. *Science* 301: 646–649, 2003.
92. Guillaud L, Setou M, and Hirokawa N. KIF17 dynamics and regulation of NR2B trafficking in hippocampal neurons. *J Neurosci* 23: 131–140, 2003.
93. Haass C and Selkoe DJ. Soluble protein oligomers in neurodegeneration: lessons from the Alzheimer's amyloid beta-peptide. *Nat Rev Mol Cell Biol* 8: 101–112, 2007.
94. Hanus C, Vannier C, and Triller A. Intracellular association of glycine receptor with gephyrin increases its plasma membrane accumulation rate. *J Neurosci* 24: 1119–1128, 2004.
95. Harms KJ, Tovar KR, and Craig AM. Synapse-specific regulation of AMPA receptor subunit composition by activity. *J Neurosci* 25: 6379–6388, 2005.
96. Harris KM. Structure, development, and plasticity of dendritic spines. *Curr Opin Neurobiol* 9: 343–348, 1999.
97. Hayashi T, Rumbaugh G, and Huganir RL. Differential regulation of AMPA receptor subunit trafficking by palmitoylation of two distinct sites. *Neuron* 47: 709–723, 2005.
98. Hazuka CD, Foletti DL, Hsu SC, Kee Y, Hopf FW, and Scheller RH. The sec6/8 complex is located at neurite outgrowth and axonal synapse-assembly domains. *J Neurosci* 19: 1324–1334, 1999.

99. Heerssen HM and Segal RA. Location, location, location: a spatial view of neurotrophin signal transduction. *Trends Neurosci* 25: 160–165, 2002.
100. Hering H and Sheng M. Dendritic spines: structure, dynamics and regulation. *Nat Rev Neurosci* 2: 880–888, 2001.
101. Heusser K and Schwappach B. Trafficking of potassium channels. *Curr Opin Neurobiol* 15: 364–369, 2005.
102. Hirokawa N and Takemura R. Molecular motors and mechanisms of directional transport in neurons. *Nat Rev Neurosci* 6: 201–214, 2005.
103. Holcman D and Triller A. Modeling synaptic dynamics driven by receptor lateral diffusion. *Biophys J* 91: 2405–2415, 2006.
104. Hoogenraad CC, Milstein AD, Ethell IM, Henkemeyer M, and Sheng M. GRIP1 controls dendrite morphogenesis by regulating EphB receptor trafficking. *Nat Neurosci* 8: 906–915, 2005.
105. Horton AC and Ehlers MD. Dual modes of endoplasmic reticulum-to-Golgi transport in dendrites revealed by live-cell imaging. *J Neurosci* 23: 6188–6199, 2003.
106. Horton AC and Ehlers MD. Neuronal polarity and trafficking. *Neuron* 40: 277–295, 2003.
107. Horton AC and Ehlers MD. Secretory trafficking in neuronal dendrites. *Nat Cell Biol* 6: 585–591, 2004.
108. Horton AC, Racz B, Monson EE, Lin AL, Weinberg RJ, and Ehlers MD. Polarized secretory trafficking directs cargo for asymmetric dendrite growth and morphogenesis. *Neuron* 48: 757–771, 2005.
109. Hou L, Antion MD, Hu D, Spencer CM, Paylor R, and Klann E. Dynamic translational and proteasomal regulation of fragile × mental retardation protein controls mGluR-dependent long-term depression. *Neuron* 51: 441–454, 2006.
110. Howe CL, Valletta JS, Rusnak AS, and Mobley WC. NGF signaling from clathrin-coated vesicles: evidence that signaling endosomes serve as a platform for the Ras-MAPK pathway. *Neuron* 32: 801–814, 2001.
111. Hsieh H, Boehm J, Sato C, Iwatsubo T, Tomita T, Sisodia S, and Malinow R. AMPAR removal underlies Abeta-induced synaptic depression and dendritic spine loss. *Neuron* 52: 831–843, 2006.
112. Hsu SC, Ting AE, Hazuka CD, Davanger S, Kenny JW, Kee Y, and Scheller RH. The mammalian brain rsec6/8 complex. *Neuron* 17: 1209–1219, 1996.
113. Huang K, Yanai A, Kang R, Arstikaitis P, Singaraja RR, Metzler M, Mullard A, Haigh B, Gauthier-Campbell C, Gutekunst CA, Hayden MR, and El-Husseini A. Huntingtin-interacting protein HIP14 is a palmitoyl transferase involved in palmitoylation and trafficking of multiple neuronal proteins. *Neuron* 44: 977–986, 2004.
114. Huber LA, Pimplikar S, Parton RG, Virta H, Zerial M, and Simons K. Rab8, a small GTPase involved in vesicular traffic between the TGN and the basolateral plasma membrane. *J Cell Biol* 123: 35–45, 1993.
115. Ikonen E. Roles of lipid rafts in membrane transport. *Curr Opin Cell Biol* 13: 470–477, 2001.
116. Inoue A and Okabe S. The dynamic organization of postsynaptic proteins: translocating molecules regulate synaptic function. *Curr Opin Neurobiol* 13: 332–340, 2003.
117. Irie F and Yamaguchi Y. EphB receptors regulate dendritic spine development via intersectin, Cdc42 and N-WASP. *Nat Neurosci* 5: 1117–1118, 2002.
118. Itoh T and De Camilli P. BAR, F-BAR (EFC) and ENTH/ANTH domains in the regulation of membrane-cytosol interfaces and membrane curvature. *Biochim Biophys Acta* 1761: 897–912, 2006.

119. Jacobson C, Schnapp B, and Banker GA. A change in the selective translocation of the Kinesin-1 motor domain marks the initial specification of the axon. *Neuron* 49: 797–804, 2006.
120. Jahn R, Lang T, and Sudhof TC. Membrane fusion. *Cell* 112: 519–533, 2003.
121. Jamain S, Quach H, Betancur C, Rastam M, Colineaux C, Gillberg IC, Soderstrom H, Giros B, Leboyer M, Gillberg C, and Bourgeron T. Mutations of the X-linked genes encoding neuroligins NLGN3 and NLGN4 are associated with autism. *Nat Genet* 34: 27–29, 2003.
122. Jan YN and Jan LY. The control of dendrite development. *Neuron* 40: 229–242, 2003.
123. Jareb M and Banker G. The polarized sorting of membrane proteins expressed in cultured hippocampal neurons using viral vectors. *Neuron* 20: 855–867, 1998.
124. Jasmin BJ, Antony C, Changeux JP, and Cartaud J. Nerve-dependent plasticity of the Golgi complex in skeletal muscle fibres: compartmentalization within the subneural sarcoplasm. *Eur J Neurosci* 7: 470–479, 1995.
125. Jasmin BJ, Cartaud J, Bornens M, and Changeux JP. Golgi apparatus in chick skeletal muscle: changes in its distribution during end plate development and after denervation. *Proc Natl Acad Sci USA* 86: 7218–7222, 1989.
126. Jasmin BJ, Changeux JP, and Cartaud J. Compartmentalization of cold-stable and acetylated microtubules in the subsynaptic domain of chick skeletal muscle fibre. *Nature* 344: 673–675, 1990.
127. Ju W, Morishita W, Tsui J, Gaietta G, Deerinck TJ, Adams SR, Garner CC, Tsien RY, Ellisman MH, and Malenka RC. Activity-dependent regulation of dendritic synthesis and trafficking of AMPA receptors. *Nat Neurosci* 7: 244–253, 2004.
128. Kacharmina JE, Job C, Crino P, and Eberwine J. Stimulation of glutamate receptor protein synthesis and membrane insertion within isolated neuronal dendrites. *Proc Natl Acad Sci USA* 97: 11545–11550, 2000.
129. Kamenetz F, Tomita T, Hsieh H, Seabrook G, Borchelt D, Iwatsubo T, Sisodia S, and Malinow R. APP processing and synaptic function. *Neuron* 37: 925–937, 2003.
130. Kastning K, Kukhtina V, Kittler JT, Chen G, Pechstein A, Enders S, Lee SH, Sheng M, Yan Z, and Haucke V. Molecular determinants for the interaction between AMPA receptors and the clathrin adaptor complex AP-2. *Proc Natl Acad Sci USA* 104: 2991–2996, 2007.
131. Kawakami R, Shinohara Y, Kato Y, Sugiyama H, Shigemoto R, and Ito I. Asymmetrical allocation of NMDA receptor epsilon2 subunits in hippocampal circuitry. *Science* 300: 990–994, 2003.
132. Keenan RJ, Freymann DM, Stroud RM, and Walter P. The signal recognition particle. *Annu Rev Biochem* 70: 755–775, 2001.
133. Keller CA, Yuan X, Panzanelli P, Martin ML, Alldred M, Sassoe-Pognetto M, and Luscher B. The gamma2 subunit of GABA (A) receptors is a substrate for palmitoylation by GODZ. *J Neurosci* 24: 5881–5891, 2004.
134. Keller P and Simons K. Post-Golgi biosynthetic trafficking. *J Cell Sci* 110 (Pt 24): 3001–3009, 1997.
135. Kennedy MJ and Ehlers MD. Organelles and trafficking machinery for postsynaptic plasticity. *Annu Rev Neurosci* 29: 325–362, 2006.
136. Kessels MM and Qualmann B. The syndapin protein family: linking membrane trafficking with the cytoskeleton. *J Cell Sci* 117: 3077–3086, 2004.
137. Kiebler MA and Bassell GJ. Neuronal RNA granules: movers and makers. *Neuron* 51: 685–690, 2006.
138. Kim E and Sheng M. PDZ domain proteins of synapses. *Nat Rev Neurosci* 5: 771–781, 2004.
139. Kirchhausen T. Clathrin. *Annu Rev Biochem* 69: 699–727, 2000.

140. Kittler JT, Arancibia-Carcamo IL, and Moss SJ. Association of GRIP1 with a GABA(A) receptor associated protein suggests a role for GRIP1 at inhibitory synapses. *Biochem Pharmacol* 68: 1649–1654, 2004.
141. Kittler JT, Delmas P, Jovanovic JN, Brown DA, Smart TG, and Moss SJ. Constitutive endocytosis of GABAA receptors by an association with the adaptin AP2 complex modulates inhibitory synaptic currents in hippocampal neurons. *J Neurosci* 20: 7972–7977, 2000.
142. Kittler JT and Moss SJ. Modulation of GABAA receptor activity by phosphorylation and receptor trafficking: implications for the efficacy of synaptic inhibition. *Curr Opin Neurobiol* 13: 341–347, 2003.
143. Kittler JT, Thomas P, Tretter V, Bogdanov YD, Haucke V, Smart TG, and Moss SJ. Huntingtin-associated protein 1 regulates inhibitory synaptic transmission by modulating gamma-aminobutyric acid type A receptor membrane trafficking. *Proc Natl Acad Sci USA* 101: 12736–12741, 2004.
144. Kleizen B and Braakman I. Protein folding and quality control in the endoplasmic reticulum. *Curr Opin Cell Biol* 16: 343–349, 2004.
145. Kobayashi T, Storrie B, Simons K, and Dotti CG. A functional barrier to movement of lipids in polarized neurons. *Nature* 359: 647–650, 1992.
146. Koekkoek SK, Yamaguchi K, Milojkovic BA, Dortland BR, Ruigrok TJ, Maex R, De Graaf W, Smit AE, VanderWerf F, Bakker CE, Willemsen R, Ikeda T, Kakizawa S, Onodera K, Nelson DL, Mientjes E, Joosten M, De Schutter E, Oostra BA, Ito M, and De Zeeuw CI. Deletion of FMR1 in Purkinje cells enhances parallel fiber LTD, enlarges spines, and attenuates cerebellar eyelid conditioning in Fragile × syndrome. *Neuron* 47: 339–352, 2005.
147. Kopec CD, Li B, Wei W, Boehm J, and Malinow R. Glutamate receptor exocytosis and spine enlargement during chemically induced long-term potentiation. *J Neurosci* 26: 2000–2009, 2006.
148. Kupfer A, Dennert G, and Singer SJ. Polarization of the Golgi apparatus and the microtubule-organizing center within cloned natural killer cells bound to their targets. *Proc Natl Acad Sci USA* 80: 7224–7228, 1983.
149. Kuriu T, Inoue A, Bito H, Sobue K, and Okabe S. Differential control of postsynaptic density scaffolds via actin-dependent and -independent mechanisms. *J Neurosci* 26: 7693–7706, 2006.
150. Kusumi A, Nakada C, Ritchie K, Murase K, Suzuki K, Murakoshi H, Kasai RS, Kondo J, and Fujiwara T. Paradigm shift of the plasma membrane concept from the two-dimensional continuum fluid to the partitioned fluid: high-speed single-molecule tracking of membrane molecules. *Annu Rev Biophys Biomol Struct* 34: 351–378, 2005.
151. Lang C, Barco A, Zablow L, Kandel ER, Siegelbaum SA, and Zakharenko SS. Transient expansion of synaptically connected dendritic spines upon induction of hippocampal long-term potentiation. *Proc Natl Acad Sci USA* 101: 16665–16670, 2004.
152. Langford GM. Myosin-V, a versatile motor for short-range vesicle transport. *Traffic* 3: 859–865, 2002.
153. Lau CG and Zukin RS. NMDA receptor trafficking in synaptic plasticity and neuropsychiatric disorders. *Nat Rev Neurosci* 8: 413–426, 2007.
154. Lavezzari G, McCallum J, Dewey CM, and Roche KW. Subunit-specific regulation of NMDA receptor endocytosis. *J Neurosci* 24: 6383–6391, 2004.
155. Lavezzari G, McCallum J, Lee R, and Roche KW. Differential binding of the AP-2 adaptor complex and PSD-95 to the C-terminus of the NMDA receptor subunit NR2B regulates surface expression. *Neuropharmacology* 45: 729–737, 2003.

156. Le Clainche C, Pauly BS, Zhang CX, Engqvist-Goldstein AE, Cunningham K, and Drubin DG. A Hip1R-cortactin complex negatively regulates actin assembly associated with endocytosis. *EMBO J*, 26: 1199–1210, 2007.
157. Lee MC, Miller EA, Goldberg J, Orci L, and Schekman R. Bi-directional protein transport between the ER and Golgi. *Annu Rev Cell Dev Biol* 20: 87–123, 2004.
158. Lee SH, Liu L, Wang YT, and Sheng M. Clathrin adaptor AP2 and NSF interact with overlapping sites of GluR2 and play distinct roles in AMPA receptor trafficking and hippocampal LTD. *Neuron* 36: 661–674, 2002.
159. Lee SH, Simonetta A, and Sheng M. Subunit rules governing the sorting of internalized AMPA receptors in hippocampal neurons. *Neuron* 43: 221–236, 2004.
160. Levine CG, Mitra D, Sharma A, Smith CL, and Hegde RS. The efficiency of protein compartmentalization into the secretory pathway. *Mol Biol Cell* 16: 279–291, 2005.
161. Li Z, Okamoto K, Hayashi Y, and Sheng M. The importance of dendritic mitochondria in the morphogenesis and plasticity of spines and synapses. *Cell* 119: 873–887, 2004.
162. Lin JW, Ju W, Foster K, Lee SH, Ahmadian G, Wyszynski M, Wang YT, and Sheng M. Distinct molecular mechanisms and divergent endocytotic pathways of AMPA receptor internalization. *Nat Neurosci* 3: 1282–1290, 2000.
163. Lippincott-Schwartz J, Roberts TH, and Hirschberg K. Secretory protein trafficking and organelle dynamics in living cells. *Annu Rev Cell Dev Biol* 16: 557–589, 2000.
164. Lipschutz JH and Mostov KE. Exocytosis: the many masters of the exocyst. *Curr Biol* 12: R212–214, 2002.
165. Lise MF, Wong TP, Trinh A, Hines RM, Liu L, Kang R, Hines DJ, Lu J, Goldenring JR, Wang YT, and El-Husseini A. Involvement of myosin Vb in glutamate receptor trafficking. *J Biol Chem* 281: 3669–3678, 2006.
166. Lledo PM, Zhang X, Sudhof TC, Malenka RC, and Nicoll RA. Postsynaptic membrane fusion and long-term potentiation. *Science* 279: 399–403, 1998.
167. Lorincz A, Notomi T, Tamas G, Shigemoto R, and Nusser Z. Polarized and compartment-dependent distribution of HCN1 in pyramidal cell dendrites. *Nat Neurosci* 5: 1185–1193, 2002.
168. Lowenstein PR, Morrison EE, Bain D, Shering AF, Banting G, Douglas P, and Castro MG. Polarized distribution of the trans-Golgi network marker TGN38 during the in vitro development of neocortical neurons: effects of nocodazole and brefeldin A. *Eur J Neurosci* 6: 1453–1465, 1994.
169. Luscher C, Nicoll RA, Malenka RC, and Muller D. Synaptic plasticity and dynamic modulation of the postsynaptic membrane. *Nat Neurosci* 3: 545–550, 2000.
170. Lyford GL, Yamagata K, Kaufmann WE, Barnes CA, Sanders LK, Copeland NG, Gilbert DJ, Jenkins NA, Lanahan AA, and Worley PF. Arc, a growth factor and activity-regulated gene, encodes a novel cytoskeleton-associated protein that is enriched in neuronal dendrites. *Neuron* 14: 433–445, 1995.
171. Ma D and Jan LY. ER transport signals and trafficking of potassium channels and receptors. *Curr Opin Neurobiol* 12: 287–292, 2002.
172. Maas C, Tagnaouti N, Loebrich S, Behrend B, Lappe-Siefke C, and Kneussel M. Neuronal cotransport of glycine receptor and the scaffold protein gephyrin. *J Cell Biol* 172: 441–451, 2006.
173. Maletic-Savatic M, Koothan T, and Malinow R. Calcium-evoked dendritic exocytosis in cultured hippocampal neurons. Part II: mediation by calcium/calmodulin-dependent protein kinase II. *J Neurosci* 18: 6814–6821, 1998.
174. Maletic-Savatic M and Malinow R. Calcium-evoked dendritic exocytosis in cultured hippocampal neurons. Part I: trans-Golgi network-derived organelles undergo regulated exocytosis. *J Neurosci* 18: 6803–6813, 1998.

175. Marchand S, Bignami F, Stetzkowski-Marden F, and Cartaud J. The myristoylated protein rapsyn is cotargeted with the nicotinic acetylcholine receptor to the postsynaptic membrane via the exocytic pathway. *J Neurosci* 20: 521–528, 2000.
176. Marchand S, Stetzkowski-Marden F, and Cartaud J. Differential targeting of components of the dystrophin complex to the postsynaptic membrane. *Eur J Neurosci* 13: 221–229, 2001.
177. Marrs GS, Green SH, and Dailey ME. Rapid formation and remodeling of postsynaptic densities in developing dendrites. *Nat Neurosci* 4: 1006–1013, 2001.
178. Martin S and Henley JM. Activity-dependent endocytic sorting of kainate receptors to recycling or degradation pathways. *EMBO J* 23: 4749–4759, 2004.
179. Mathew D, Ataman B, Chen J, Zhang Y, Cumberledge S, and Budnik V. Wingless signaling at synapses is through cleavage and nuclear import of receptor DFrizzled2. *Science* 310: 1344–1347, 2005.
180. Mathew D, Popescu A, and Budnik V. Drosophila amphiphysin functions during synaptic Fasciclin II membrane cycling. *J Neurosci* 23: 10710–10716, 2003.
181. Matlin KS and Simons K. Reduced temperature prevents transfer of a membrane glycoprotein to the cell surface but does not prevent terminal glycosylation. *Cell* 34: 233–243, 1983.
182. Matsuzaki M, Ellis-Davies GC, Nemoto T, Miyashita Y, Iino M, and Kasai H. Dendritic spine geometry is critical for AMPA receptor expression in hippocampal CA1 pyramidal neurons. *Nat Neurosci* 4: 1086–1092, 2001.
183. Matsuzaki M, Honkura N, Ellis-Davies GC, and Kasai H. Structural basis of long-term potentiation in single dendritic spines. *Nature* 429: 761–766, 2004.
184. Matus A. Growth of dendritic spines: a continuing story. *Curr Opin Neurobiol* 15: 67–72, 2005.
185. Maxfield FR and McGraw TE. Endocytic recycling. *Nat Rev Mol Cell Biol* 5: 121–132, 2004.
186. Mehta SQ, Hiesinger PR, Beronja S, Zhai RG, Schulze KL, Verstreken P, Cao Y, Zhou Y, Tepass U, Crair MC, and Bellen HJ. Mutations in Drosophila sec15 reveal a function in neuronal targeting for a subset of exocyst components. *Neuron* 46: 219–232, 2005.
187. Miaczynska M, Pelkmans L, and Zerial M. Not just a sink: endosomes in control of signal transduction. *Curr Opin Cell Biol* 16: 400–406, 2004.
188. Miyata M, Finch EA, Khiroug L, Hashimoto K, Hayasaka S, Oda SI, Inouye M, Takagishi Y, Augustine GJ, and Kano M. Local calcium release in dendritic spines required for long-term synaptic depression. *Neuron* 28: 233–244, 2000.
189. Momiyama A, Silver RA, Hausser M, Notomi T, Wu Y, Shigemoto R, and Cull-Candy SG. The density of AMPA receptors activated by a transmitter quantum at the climbing fibre-Purkinje cell synapse in immature rats. *J Physiol* 549: 75–92, 2003.
190. Moss SJ and Smart TG. Constructing inhibitory synapses. *Nat Rev Neurosci* 2: 240–250, 2001.
191. Mundigl O, Matteoli M, Daniell L, Thomas-Reetz A, Metcalf A, Jahn R, and De Camilli P. Synaptic vesicle proteins and early endosomes in cultured hippocampal neurons: differential effects of Brefeldin A in axon and dendrites. *J Cell Biol* 122: 1207–1221, 1993.
192. Murthy M, Garza D, Scheller RH, and Schwarz TL. Mutations in the exocyst component Sec5 disrupt neuronal membrane traffic, but neurotransmitter release persists. *Neuron* 37: 433–447, 2003.
193. Murthy M, Ranjan R, Denef N, Higashi ME, Schupbach T, and Schwarz TL. Sec6 mutations and the Drosophila exocyst complex. *J Cell Sci* 118: 1139–1150, 2005.
194. Nagerl UV, Eberhorn N, Cambridge SB, and Bonhoeffer T. Bidirectional activity-dependent morphological plasticity in hippocampal neurons. *Neuron* 44: 759–767, 2004.

195. Naisbitt S, Kim E, Tu JC, Xiao B, Sala C, Valtschanoff J, Weinberg RJ, Worley PF, and Sheng M. Shank, a novel family of postsynaptic density proteins that binds to the NMDA receptor/PSD-95/GKAP complex and cortactin. *Neuron* 23: 569–582, 1999.
196. Naisbitt S, Valtschanoff J, Allison DW, Sala C, Kim E, Craig AM, Weinberg RJ, and Sheng M. Interaction of the postsynaptic density-95/guanylate kinase domain- associated protein complex with a light chain of myosin-V and dynein. *J Neurosci* 20: 4524–4534, 2000.
197. Nakada C, Ritchie K, Oba Y, Nakamura M, Hotta Y, Iino R, Kasai RS, Yamaguchi K, Fujiwara T, and Kusumi A. Accumulation of anchored proteins forms membrane diffusion barriers during neuronal polarization. *Nat Cell Biol* 5: 626–632, 2003.
198. Nimchinsky EA, Sabatini BL, and Svoboda K. Structure and function of dendritic spines. *Annu Rev Physiol* 64: 313–353, 2002.
199. Nixon RA. Endosome function and dysfunction in Alzheimer's disease and other neurodegenerative diseases. *Neurobiol Aging* 26: 373–382, 2005.
200. Nusser Z, Lujan R, Laube G, Roberts JD, Molnar E, and Somogyi P. Cell type and pathway dependence of synaptic AMPA receptor number and variability in the hippocampus. *Neuron* 21: 545–559, 1998.
201. O'Brien RJ, Kamboj S, Ehlers MD, Rosen KR, Fischbach GD, and Huganir RL. Activity-dependent modulation of synaptic AMPA receptor accumulation. *Neuron* 21: 1067–1078, 1998.
202. Oddo S, Caccamo A, Shepherd JD, Murphy MP, Golde TE, Kayed R, Metherate R, Mattson MP, Akbari Y, and LaFerla FM. Triple-transgenic model of Alzheimer's disease with plaques and tangles: intracellular Abeta and synaptic dysfunction. *Neuron* 39: 409–421, 2003.
203. Oh MC, Derkach VA, Guire ES, and Soderling TR. Extrasynaptic membrane trafficking regulated by GluR1 serine 845 phosphorylation primes AMPA receptors for long-term potentiation. *J Biol Chem* 281: 752–758, 2006.
204. Okabe S, Kim HD, Miwa A, Kuriu T, and Okado H. Continual remodeling of postsynaptic density and its regulation by synaptic activity. *Nat Neurosci* 2: 804–811, 1999.
205. Okamoto K, Nagai T, Miyawaki A, and Hayashi Y. Rapid and persistent modulation of actin dynamics regulates postsynaptic reorganization underlying bidirectional plasticity. *Nat Neurosci* 7: 1104–1112, 2004.
206. Osterweil E, Wells DG, and Mooseker MS. A role for myosin VI in postsynaptic structure and glutamate receptor endocytosis. *J Cell Biol* 168: 329–338, 2005.
207. Ostroff LE, Fiala JC, Allwardt B, and Harris KM. Polyribosomes redistribute from dendritic shafts into spines with enlarged synapses during LTP in developing rat hippocampal slices. *Neuron* 35: 535–545, 2002.
208. Paglini G, Peris L, Diez-Guerra J, Quiroga S, and Caceres A. The Cdk5-p35 kinase associates with the Golgi apparatus and regulates membrane traffic. *EMBO Rep* 2: 1139–1144, 2001.
209. Pal A, Severin F, Lommer B, Shevchenko A, and Zerial M. Huntingtin-HAP40 complex is a novel Rab5 effector that regulates early endosome motility and is up-regulated in Huntington's disease. *J Cell Biol* 172: 605–618, 2006.
210. Palade G. Intracellular aspects of the process of protein synthesis. *Science* 189: 347–358, 1975.
211. Palmer CL, Lim W, Hastie PG, Toward M, Korolchuk VI, Burbidge SA, Banting G, Collingridge GL, Isaac JT, and Henley JM. Hippocalcin functions as a calcium sensor in hippocampal LTD. *Neuron* 47: 487–494, 2005.
212. Park M, Penick EC, Edwards JG, Kauer JA, and Ehlers MD. Recycling endosomes supply AMPA receptors for LTP. *Science* 305: 1972–1975, 2004.

213. Park M, Salgado JM, Ostroff L, Helton TD, Robinson CG, Harris KM, and Ehlers MD. Plasticity-induced growth of dendritic spines by exocytic trafficking from recycling endosomes. *Neuron* 52: 817–830, 2006.
214. Passafaro M, Piech V, and Sheng M. Subunit-specific temporal and spatial patterns of AMPA receptor exocytosis in hippocampal neurons. *Nat Neurosci* 4: 917–926, 2001.
215. Pearse BM, Smith CJ, and Owen DJ. Clathrin coat construction in endocytosis. *Curr Opin Struct Biol* 10: 220–228, 2000.
216. Perez-Otano I, Lujan R, Tavalin SJ, Plomann M, Modregger J, Liu XB, Jones EG, Heinemann SF, Lo DC, and Ehlers MD. Endocytosis and synaptic removal of NR3A-containing NMDA receptors by PACSIN1/syndapin1. *Nat Neurosci* 9: 611–621, 2006.
217. Peters A, Palay SL, Webster HF. The fine structure of the nervous system. Oxford: Oxford University Press, 1991.
218. Petralia RS, Wang YX, Sans N, Worley PF, Hammer JA, 3rd, and Wenthold RJ. Glutamate receptor targeting in the postsynaptic spine involves mechanisms that are independent of myosin Va. *Eur J Neurosci* 13: 1722–1732, 2001.
219. Petralia RS, Wang YX, and Wenthold RJ. Internalization at glutamatergic synapses during development. *Eur J Neurosci* 18: 3207–3217, 2003.
220. Pierce JP, Mayer T, and McCarthy JB. Evidence for a satellite secretory pathway in neuronal dendritic spines. *Curr Biol* 11: 351–355, 2001.
221. Pierce JP, van Leyen K, and McCarthy JB. Translocation machinery for synthesis of integral membrane and secretory proteins in dendritic spines. *Nat Neurosci* 3: 311–313, 2000.
222. Popov V, Medvedev NI, Davies HA, and Stewart MG. Mitochondria form a filamentous reticular network in hippocampal dendrites but are present as discrete bodies in axons: a three-dimensional ultrastructural study. *J Comp Neurol* 492: 50–65, 2005.
223. Praefcke GJ and McMahon HT. The dynamin superfamily: universal membrane tubulation and fission molecules? *Nat Rev Mol Cell Biol* 5: 133–147, 2004.
224. Prange O, Wong TP, Gerrow K, Wang YT, and El-Husseini A. A balance between excitatory and inhibitory synapses is controlled by PSD-95 and neuroligin. *Proc Natl Acad Sci USA* 101: 13915–13920, 2004.
225. Prekeris R, Foletti DL, and Scheller RH. Dynamics of tubulovesicular recycling endosomes in hippocampal neurons. *J Neurosci* 19: 10324–10337, 1999.
226. Prekeris R, Klumperman J, Chen YA, and Scheller RH. Syntaxin 13 mediates cycling of plasma membrane proteins via tubulovesicular recycling endosomes. *J Cell Biol* 143: 957–971, 1998.
227. Racca C, Stephenson FA, Streit P, Roberts JD, and Somogyi P. NMDA receptor content of synapses in stratum radiatum of the hippocampal CA1 area. *J Neurosci* 20: 2512–2522, 2000.
228. Racz B, Blanpied TA, Ehlers MD, and Weinberg RJ. Lateral organization of endocytic machinery in dendritic spines. *Nat Neurosci* 7: 917–918, 2004.
229. Ralston E, Lu Z, and Ploug T. The organization of the Golgi complex and microtubules in skeletal muscle is fiber type-dependent. *J Neurosci* 19: 10694–10705, 1999.
230. Rathenberg J, Kittler JT, and Moss SJ. Palmitoylation regulates the clustering and cell surface stability of GABAA receptors. *Mol Cell Neurosci* 26: 251–257, 2004.
231. Resh MD. Fatty acylation of proteins: new insights into membrane targeting of myristoylated and palmitoylated proteins. *Biochim Biophys Acta* 1451: 1–16, 1999.
232. Richards DA, De Paola V, Caroni P, Gahwiler BH, and McKinney RA. AMPA-receptor activation regulates the diffusion of a membrane marker in parallel with dendritic spine motility in the mouse hippocampus. *J Physiol* 558: 503–512, 2004.

233. Rivera JF, Ahmad S, Quick MW, Liman ER, and Arnold DB. An evolutionarily conserved dileucine motif in Shal K+ channels mediates dendritic targeting. *Nat Neurosci* 6: 243–250, 2003.
234. Roche KW, Standley S, McCallum J, Dune Ly C, Ehlers MD, and Wenthold RJ. Molecular determinants of NMDA receptor internalization. *Nat Neurosci* 4: 794–802, 2001.
235. Rodriguez-Boulan E, Kreitzer G, and Musch A. Organization of vesicular trafficking in epithelia. *Nat Rev Mol Cell Biol* 6: 233–247, 2005.
236. Rose CR and Konnerth A. Stores not just for storage. intracellular calcium release and synaptic plasticity. *Neuron* 31: 519–522, 2001.
237. Rosenberg M, Meier J, Triller A, and Vannier C. Dynamics of glycine receptor insertion in the neuronal plasma membrane. *J Neurosci* 21: 5036–5044, 2001.
238. Rosso S, Bollati F, Bisbal M, Peretti D, Sumi T, Nakamura T, Quiroga S, Ferreira A, and Caceres A. LIMK1 regulates Golgi dynamics, traffic of Golgi-derived vesicles, and process extension in primary cultured neurons. *Mol Biol Cell* 15: 3433–3449, 2004.
239. Ruberti F and Dotti CG. Involvement of the proximal C terminus of the AMPA receptor subunit GluR1 in dendritic sorting. *J Neurosci* 20: RC78, 2000.
240. Sabatini BL, Maravall M, and Svoboda K. Ca(2+) signaling in dendritic spines. *Curr Opin Neurobiol* 11: 349–356, 2001.
241. Saito N, Okada Y, Noda Y, Kinoshita Y, Kondo S, and Hirokawa N. KIFC2 is a novel neuron-specific C-terminal type kinesin superfamily motor for dendritic transport of multivesicular body-like organelles. *Neuron* 18: 425–438, 1997.
242. Sala C, Piech V, Wilson NR, Passafaro M, Liu G, and Sheng M. Regulation of dendritic spine morphology and synaptic function by Shank and Homer. *Neuron* 31: 115–130, 2001.
243. Sampo B, Kaech S, Kunz S, and Banker G. Two distinct mechanisms target membrane proteins to the axonal surface. *Neuron* 37: 611–624, 2003.
244. Sans N, Prybylowski K, Petralia RS, Chang K, Wang YX, Racca C, Vicini S, and Wenthold RJ. NMDA receptor trafficking through an interaction between PDZ proteins and the exocyst complex. *Nat Cell Biol* 5: 520–530, 2003.
245. Sans N, Racca C, Petralia RS, Wang YX, McCallum J, and Wenthold RJ. Synapse-associated protein 97 selectively associates with a subset of AMPA receptors early in their biosynthetic pathway. *J Neurosci* 21: 7506–7516, 2001.
246. Schatz G and Dobberstein B. Common principles of protein translocation across membranes. *Science* 271: 1519–1526, 1996.
247. Scheiffele P, Fan J, Choih J, Fetter R, and Serafini T. Neuroligin expressed in nonneuronal cells triggers presynaptic development in contacting axons. *Cell* 101: 657–669, 2000.
248. Schnell E and Nicoll RA. Hippocampal synaptic transmission and plasticity are preserved in myosin Va mutant mice. *J Neurophysiol* 85: 1498–1501, 2001.
249. Scott DB, Michailidis I, Mu Y, Logothetis D, and Ehlers MD. Endocytosis and degradative sorting of NMDA receptors by conserved membrane-proximal signals. *J Neurosci* 24: 7096–7109, 2004.
250. Seabra MC and Wasmeier C. Controlling the location and activation of Rab GTPases. *Curr Opin Cell Biol* 16: 451–457, 2004.
251. Setou M, Nakagawa T, Seog DH, and Hirokawa N. Kinesin superfamily motor protein KIF17 and mLin-10 in NMDA receptor-containing vesicle transport. *Science* 288: 1796–1802, 2000.
252. Setou M, Seog DH, Tanaka Y, Kanai Y, Takei Y, Kawagishi M, and Hirokawa N. Glutamate-receptor-interacting protein GRIP1 directly steers kinesin to dendrites. *Nature* 417: 83–87, 2002.

253. Shankar GM, Bloodgood BL, Townsend M, Walsh DM, Selkoe DJ, and Sabatini BL. Natural oligomers of the Alzheimer amyloid-beta protein induce reversible synapse loss by modulating an NMDA-type glutamate receptor-dependent signaling pathway. *J Neurosci* 27: 2866–2875, 2007.
254. Sharma K, Fong DK, and Craig AM. Postsynaptic protein mobility in dendritic spines: long-term regulation by synaptic NMDA receptor activation. *Mol Cell Neurosci* 31: 702–712, 2006.
255. Shaw RM, Fay AJ, Puthenveedu MA, von Zastrow M, Jan YN, and Jan LY. Microtubule plus-end-tracking proteins target gap junctions directly from the cell interior to adherens junctions. *Cell* 128: 547–560, 2007.
256. Sheng M, Tsaur ML, Jan YN, and Jan LY. Subcellular segregation of two A-type K+ channel proteins in rat central neurons. *Neuron* 9: 271–284, 1992.
257. Shi S, Hayashi Y, Esteban JA, and Malinow R. Subunit-specific rules governing AMPA receptor trafficking to synapses in hippocampal pyramidal neurons. *Cell* 105: 331–343, 2001.
258. Shi SH, Hayashi Y, Petralia RS, Zaman SH, Wenthold RJ, Svoboda K, and Malinow R. Rapid spine delivery and redistribution of AMPA receptors after synaptic NMDA receptor activation. *Science* 284: 1811–1816, 1999.
259. Shirane M and Nakayama KI. Protrudin induces neurite formation by directional membrane trafficking. *Science* 314: 818–821, 2006.
260. Silverman MA, Peck R, Glover G, He C, Carlin C, and Banker G. Motifs that mediate dendritic targeting in hippocampal neurons: a comparison with basolateral targeting signals. *Mol Cell Neurosci* 29: 173–180, 2005.
261. Singaraja RR, Hadano S, Metzler M, Givan S, Wellington CL, Warby S, Yanai A, Gutekunst CA, Leavitt BR, Yi H, Fichter K, Gan L, McCutcheon K, Chopra V, Michel J, Hersch SM, Ikeda JE, and Hayden MR. HIP14, a novel ankyrin domain-containing protein, links huntingtin to intracellular trafficking and endocytosis. *Hum Mol Genet* 11: 2815–2828, 2002.
262. Smith MA, Ellis-Davies GC, and Magee JC. Mechanism of the distance-dependent scaling of Schaffer collateral synapses in rat CA1 pyramidal neurons. *J Physiol* 548: 245–258, 2003.
263. Smotrys JE and Linder ME. Palmitoylation of intracellular signaling proteins: regulation and function. *Annu Rev Biochem* 73: 559–587, 2004.
264. Snyder EM, Nong Y, Almeida CG, Paul S, Moran T, Choi EY, Nairn AC, Salter MW, Lombroso PJ, Gouras GK, and Greengard P. Regulation of NMDA receptor trafficking by amyloid-beta. *Nat Neurosci* 8: 1051–1058, 2005.
265. Snyder EM, Philpot BD, Huber KM, Dong X, Fallon JR, and Bear MF. Internalization of ionotropic glutamate receptors in response to mGluR activation. *Nat Neurosci* 4: 1079–1085, 2001.
266. Sone M, Suzuki E, Hoshino M, Hou D, Kuromi H, Fukata M, Kuroda S, Kaibuchi K, Nabeshima Y, and Hama C. Synaptic development is controlled in the periactive zones of Drosophila synapses. *Development* 127: 4157–4168, 2000.
267. Sorkin A and Von Zastrow M. Signal transduction and endocytosis: close encounters of many kinds. *Nat Rev Mol Cell Biol* 3: 600–614, 2002.
268. Spacek J and Harris KM. Three-dimensional organization of smooth endoplasmic reticulum in hippocampal CA1 dendrites and dendritic spines of the immature and mature rat. *J Neurosci* 17: 190–203, 1997.
269. Star EN, Kwiatkowski DJ, and Murthy VN. Rapid turnover of actin in dendritic spines and its regulation by activity. *Nat Neurosci* 5: 239–246, 2002.
270. Steiner P, Alberi S, Kulangara K, Yersin A, Sarria JC, Regulier E, Kasas S, Dietler G, Muller D, Catsicas S, and Hirling H. Interactions between NEEP21, GRIP1 and GluR2

regulate sorting and recycling of the glutamate receptor subunit GluR2. *EMBO J* 24: 2873–2884, 2005.
271. Steiner P, Sarria JC, Glauser L, Magnin S, Catsicas S, and Hirling H. Modulation of receptor cycling by neuron-enriched endosomal protein of 21 kD. *J Cell Biol* 157: 1197–1209, 2002.
272. Steward O and Schuman EM. Compartmentalized synthesis and degradation of proteins in neurons. *Neuron* 40: 347–359, 2003.
273. Sudhof TC. The synaptic vesicle cycle. *Annu Rev Neurosci* 27: 509–547, 2004.
274. Sutton MA and Schuman EM. Dendritic protein synthesis, synaptic plasticity, and memory. *Cell* 127: 49–58, 2006.
275. Sytnyk V, Leshchyns'ka I, Delling M, Dityateva G, Dityatev A, and Schachner M. Neural cell adhesion molecule promotes accumulation of TGN organelles at sites of neuron-to-neuron contacts. *J Cell Biol* 159: 649–661, 2002.
276. Takagishi Y, Oda S, Hayasaka S, Dekker-Ohno K, Shikata T, Inouye M, and Yamamura H. The dilute-lethal (dl) gene attacks a Ca2+ store in the dendritic spine of Purkinje cells in mice. *Neurosci Lett* 215: 169–172, 1996.
277. Tanaka J, Matsuzaki M, Tarusawa E, Momiyama A, Molnar E, Kasai H, and Shigemoto R. Number and density of AMPA receptors in single synapses in immature cerebellum. *J Neurosci* 25: 799–807, 2005.
278. Tardin C, Cognet L, Bats C, Lounis B, and Choquet D. Direct imaging of lateral movements of AMPA receptors inside synapses. *EMBO J* 22: 4656–4665, 2003.
279. Taylor KA. Regulation and recycling of myosin V. *Curr Opin Cell Biol* 19: 67–74, 2007.
280. Thery M and Bornens M. Cell shape and cell division. *Curr Opin Cell Biol* 18: 648–657, 2006.
281. Thomas P, Mortensen M, Hosie AM, and Smart TG. Dynamic mobility of functional GABAA receptors at inhibitory synapses. *Nat Neurosci* 8: 889–897, 2005.
282. Toni N, Buchs PA, Nikonenko I, Bron CR, and Muller D. LTP promotes formation of multiple spine synapses between a single axon terminal and a dendrite. *Nature* 402: 421–425, 1999.
283. Toni N, Buchs PA, Nikonenko I, Povilaitite P, Parisi L, and Muller D. Remodeling of synaptic membranes after induction of long-term potentiation. *J Neurosci* 21: 6245–6251, 2001.
284. Torre ER and Steward O. Demonstration of local protein synthesis within dendrites using a new cell culture system that permits the isolation of living axons and dendrites from their cell bodies. *J Neurosci* 12: 762–772, 1992.
285. Torre ER and Steward O. Protein synthesis within dendrites: glycosylation of newly synthesized proteins in dendrites of hippocampal neurons in culture. *J Neurosci* 16: 5967–5978, 1996.
286. Tovar KR and Westbrook GL. Mobile NMDA receptors at hippocampal synapses. *Neuron* 34: 255–264, 2002.
287. Triller A and Choquet D. Surface trafficking of receptors between synaptic and extrasynaptic membranes: and yet they do move! *Trends Neurosci* 28: 133–139, 2005.
288. Tu JC, Xiao B, Naisbitt S, Yuan JP, Petralia RS, Brakeman P, Doan A, Aakalu VK, Lanahan AA, Sheng M, and Worley PF. Coupling of mGluR/Homer and PSD-95 complexes by the Shank family of postsynaptic density proteins. *Neuron* 23: 583–592, 1999.
289. Tuma PL and Hubbard AL. Transcytosis: crossing cellular barriers. *Physiol Rev* 83: 871–932, 2003.
290. Turrigiano GG and Nelson SB. Homeostatic plasticity in the developing nervous system. *Nat Rev Neurosci* 5: 97–107, 2004.
291. Ullrich O, Reinsch S, Urbe S, Zerial M, and Parton RG. Rab11 regulates recycling through the pericentriolar recycling endosome. *J Cell Biol* 135: 913–924, 1996.

292. Urrutia R, Henley JR, Cook T, and McNiven MA. The dynamins: redundant or distinct functions for an expanding family of related GTPases? *Proc Natl Acad Sci USA* 94: 377–384, 1997.
293. Vale RD. The molecular motor toolbox for intracellular transport. *Cell* 112: 467–480, 2003.
294. Vale RD, Reese TS, and Sheetz MP. Identification of a novel force-generating protein, kinesin, involved in microtubule-based motility. *Cell* 42: 39–50, 1985.
295. van Rijnsoever C, Sidler C, and Fritschy JM. Internalized GABA-receptor subunits are transferred to an intracellular pool associated with the postsynaptic density. *Eur J Neurosci* 21: 327–338, 2005.
296. Vega IE and Hsu SC. The exocyst complex associates with microtubules to mediate vesicle targeting and neurite outgrowth. *J Neurosci* 21: 3839–3848, 2001.
297. Walikonis RS, Jensen ON, Mann M, Provance DW, Jr., Mercer JA, and Kennedy MB. Identification of proteins in the postsynaptic density fraction by mass spectrometry. *J Neurosci* 20: 4069–4080, 2000.
298. Walsh DM, Klyubin I, Fadeeva JV, Cullen WK, Anwyl R, Wolfe MS, Rowan MJ, and Selkoe DJ. Naturally secreted oligomers of amyloid beta protein potently inhibit hippocampal long-term potentiation in vivo. *Nature* 416: 535–539, 2002.
299. Washbourne P, Liu XB, Jones EG, and McAllister AK. Cycling of NMDA receptors during trafficking in neurons before synapse formation. *J Neurosci* 24: 8253–8264, 2004.
300. West AE, Neve RL, and Buckley KM. Identification of a somatodendritic targeting signal in the cytoplasmic domain of the transferrin receptor. *J Neurosci* 17: 6038–6047, 1997.
301. Winckler B, Forscher P, and Mellman I. A diffusion barrier maintains distribution of membrane proteins in polarized neurons. *Nature* 397: 698–701, 1999.
302. Winckler B and Mellman I. Neuronal polarity: controlling the sorting and diffusion of membrane components. *Neuron* 23: 637–640, 1999.
303. Wisco D, Anderson ED, Chang MC, Norden C, Boiko T, Folsch H, and Winckler B. Uncovering multiple axonal targeting pathways in hippocampal neurons. *J Cell Biol* 162: 1317–1328, 2003.
304. Wolf ME, Sun X, Mangiavacchi S, and Chao SZ. Psychomotor stimulants and neuronal plasticity. *Neuropharmacology* 47(Suppl 1): 61–79, 2004.
305. Ye H, Kuruvilla R, Zweifel LS, and Ginty DD. Evidence in support of signaling endosome-based retrograde survival of sympathetic neurons. *Neuron* 39: 57–68, 2003.
306. Yoshihara M, Adolfsen B, Galle KT, and Littleton JT. Retrograde signaling by Syt 4 induces presynaptic release and synapse-specific growth. *Science* 310: 858–863, 2005.
307. Yoshii A and Constantine-Paton M. BDNF induces transport of PSD-95 to dendrites through PI3K-AKT signaling after NMDA receptor activation. *Nat Neurosci* 10: 702–711, 2007.
308. Zhou Q, Homma KJ, and Poo MM. Shrinkage of dendritic spines associated with long-term depression of hippocampal synapses. *Neuron* 44: 749–757, 2004.
309. Zhou Q, Xiao M, and Nicoll RA. Contribution of cytoskeleton to the internalization of AMPA receptors. *Proc Natl Acad Sci USA* 98: 1261–1266, 2001.

292. [illegible] and McNiven MA. The dynamins: redundant or distinct [illegible] of related GTPases? Proc Natl Acad Sci USA [illegible]

[illegible]

299. Washbourne P, [illegible] Trafficking of NMDA receptors [illegible] synapse formation. J Neurosci 24: 8253–8264, 2004.

300. West AE, Neve RL, and Buckley KM. Identification of a somatodendritic targeting signal in the cytoplasmic domain of the transferrin receptor. J Neurosci 17: 6038–6047, 1997.

[illegible]

Structure and Mechanism of Action of AMPA and Kainate Receptors

Mark L. Mayer

Porter Neuroscience Research Center, Laboratory of Cellular & Molecular Neurophysiology, NICHD, NIH, Bethesda, MD 20892, USA, mayerm@mail.NIH.gov

1 Introduction

Structural studies on glutamate receptor ion channels (iGluRs) have reached a level of sophistication that was impossible to imagine when the individual receptor genes were cloned in the 1980s. Because these membrane proteins mediate excitatory responses at approximately 60% of synapses in the brain, they have attracted enormous interest due to their role in synaptic plasticity, learning and memory, development of the CNS, and disease. There has also been immense interest in understanding how these molecules work. In this chapter I will highlight some of the key advances in our understanding of the function of glutamate receptor ion channels which have arisen from structural work, and address some of the most important issues which remain unanswered. The results discussed focus almost entirely on AMPA and kainate receptors, while NMDA receptors are discussed in the chapter by Yuan et al.

2 Molecular Architecture

Our initial ideas about the structure of iGluRs were strongly influenced by prior work on nicotinic acetylcholine receptors and related members of the cys loop family of ligand gated ion channels. These assemble as pentamers, with a barrel stave architecture in which the ion channel lies at the center of a 5-fold axis of symmetry (37). Five years after the first iGluR was cloned (14) a series of experiments using principally biochemical and bioinformantic approaches revealed that iGluRs have a strikingly different architecture. Biochemical experiments using glycosidase sensitivity, and chimeric constructs to identify the extracellular ligand binding domain, unequivocally established the correct membrane topology of iGluRs (13,

J.W. Hell, M.D. Ehlers (eds.), *Structural and Functional Organization of the Synapse*, DOI: 10.1007/978-0-387-77232-5_9,

40) and introduced the now widely used S1S2 nomenclature for labeling the pre and post ion channel segments of the ligand binding core. Bioinformantic studies played a key role in this work by revealing that iGluRs were modular proteins in which the individual domains showed sequence, and thus most likely structural homology with a series of bacterial proteins (20, 26, 40). Such approaches revealed that the ion channel of iGluRs has a pore loop motif like that in cyclic nucleotide and voltage-gated ion channels (21), and that its structure must closely resemble that of KcsA, the first ion channel protein to be crystallized (10). This was reinforced by discovery of GluR0, a bacterial glutamate gated potassium selective ion channel with high sequence homology to KcsA (7). Likewise, the ligand binding domain was shown to be closely related in sequence and thus structure to a family of bacterial periplasmic proteins exemplified by glutamine binding protein (40). Finally, the 1st 400 or so amino acids of an iGluR subunit, which form the amino terminal domain (ATD), show sequence homology to a second family of periplasmic binding proteins of which leucine, isoleucine, valine binding protein (LIVBP) is the canonical member (28). A scheme illustrating how these domains are assembled in an individual iGluR subunit is shown in Fig. 1a.

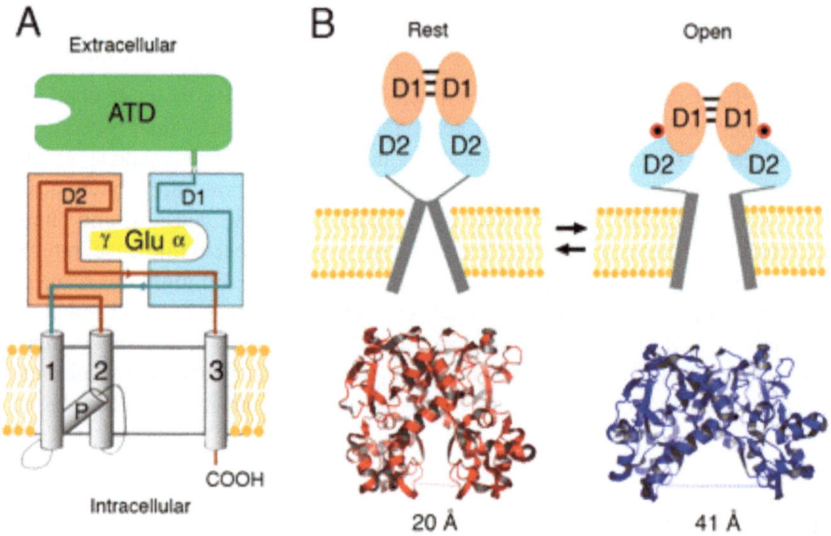

Fig. 1. Domain organization and activation mechanism for iGluR gating. (**a**) shows a cartoon representation of a single iGluR submit, four of which coassemble to form an ion channel. The extracellular N-terminal is at the top and consists of the amino terminal domain (ATD), the ligand binding core composed of domains 1 and 2 (D1 and D2), with the ion channel linkers to the membrane embedded pore loop emerging from domain 2; the C-terminus is in the cytoplasm. (**b**) shows a cartoon representation of an iGluR dimmer in the resting and active conformations (*top*) and corresponding iGluR crystal structures (*bottom*) with the separation of the residues leading to the ion channel linkers in the intact protein given in A (PDB 2F34 and 2F36)

This assembly scheme reveals the surprising fact that the ligand binding domain is interrupted by insertion of the ion channel. This initially made no sense when the membrane topology of iGluRs was thought to be similar to that of the cys-loop family of ligand gated ion channels, since one half of the putative ligand binding domain was on the extracellular side of the post synaptic membrane, while the second half was on the intracellular side. Assignment of the correct membrane topology resolved this problem, but raised the question as to how the ligand binding domain could function when split into two halves by the membrane embedded ion channel segment. The answer to this question lies at the heart of how iGluRs use the conformational change induced by the binding of glutamate to gate the ion channel, and reveal a beautiful example of molecular evolution. The solution to the problem came from the combination of crystallography, patch clamp recording, and biophysical chemistry, and as a result we now have substantial insight into the mechanisms underlying iGluR activation, desensitization, allosteric modulation, and partial agonist activity. As important, we can now explain in atomic detail the selective binding properties of all of the major iGluR families, and several groups are using this information for the development subtype selective ligands, the availability of which will allow unprecedented functional studies to address the role of iGluRs in synaptic function.

3 Mechanism of Activation by Glutamate

The ligand binding domains of iGluRs are evolutionarily related to bacterial periplasmic binding proteins, the structures of numerous examples of which had been solved prior to the beginning of similar work on iGluRs. Thus, the idea that the individual ligand binding cores of iGluRs use a 'Venus fly trap' mechanism to bind glutamate had precedent from prior structural studies on bacterial periplasmic proteins, which crystallize in an 'open' conformation in the absence of substrate, and a 'closed' conformation when substrate is bound (34). The extent of domain closure in iGluRs, while not as large as observed in the fly trap of insectivorous plants, is nonetheless substantial, between 20–30°, and differs for individual ligands and iGluR subtypes. The 'fly trap' has two domains (D1 and D2) which are connected by a flexible hinge formed by antiparallel beta strands. The first crystal structure of an iGluR ligand binding domain revealed that insertion of the ion channel segment does not destroy its function because the linkers leading to the ion channel proper all emerge from the bottom surface of D2 (5).

From a mechanistic perspective the most striking observation to emerge from structural studies was the discovery that the isolated ligand binding cores assemble as dimers (2), since this immediately suggested experimentally testable models for activation (Fig. 1b). Also important was the discovery that agonists, partial agonists, and competitive antagonists stabilize the ligand binding domain in different conformations. The dimers are formed by a back to back assembly, in which the intermolecular contacts stabilizing the dimer assembly are made exclusively by the rear surface of domain 1, so that the entrance to the ligand binding site faces outwards, and is accessible to the extracellular solution in the postsynaptic membrane, ready to bind glutamate. Ion channel activation occurs because, following the binding of glutamate, D2 in the pair of subunits in a dimer assembly moves in a scissors-like motion (Fig. 1b). This exerts strain on the ion channel, 'pulling' it into the open conformation because the outward movement of D2 and its associated linkers applies force to the TM helices, which favors the open channel conformation. This model is bolstered by crystallographic studies on potassium channels, which reveal that in the closed conformation the channel has a teepee like structure (10), and that in the open conformation the smoke hole opens (17).

Although this model for iGluR gating was developed using crystal structures of the isolated ligand binding domain, which constitutes roughly 25% of the mass of an intact iGluR, there is essentially no doubt that its is correct, although important details will only be resolved when we have crystal structures of full length iGluRs. Several experimental results give us confidence in this model. First, deletion of the ATD, which removes approximately 40% of the mass of an iGluR, leaves activation and desensitization intact, and in both AMPA and kainate receptors gating occurs with essentially wild type kinetics (15, 31, 33). Second, structural studies on potassium channels, some of which share a related gating mechanism, provide clear precedent for a mechanical linkage between the ion channel pore and the ligand binding domain. Third, iGluR dimers, chemically cross-linked via introduction of intermolecular disulfide bridges to trap them in the permissive conformation for activation by glutamate, continue to show normal ion channel activation (46). Key details missing from this picture include the lack of any real understanding of how the pair of dimers in a tetrameric iGluR assembly are packed with respect to each other, and knowledge of the extent of movement that occurs in the ion channel during the transition from the resting to open conformation. Although crystallographic studies with agonists and antagonists clearly reveal that the ligand binding domains can move large distances, it is by no means certain that the ion channel moves as much. The extent of movement will be determined by the structure of the linkers which connect the ion channel proper with D2; rigid linkers would impose large movements on the ion channel TM domains; flexible, perhaps disordered linkers, would absorb large movements, and impose much less force on the ion channel.

An understanding of the structure of the dimer of dimers assembly that must occur in the extracellular segments of a tetrameric iGluR is critical to several issues. First, we have no real understanding of the extent of cross talk, if any, between the dimers; this will of course have a large impact on the cooperativity of ion channel activation. Second, lack of information about the connectivity between the ATD dimers and the ligand binding domain dimers prevents us from understanding how allosteric modulators act on the ATD to modulate ion channel gating. This is particularly important in the case of NMDA receptors (see chapter by Yuan et al.) Third, the transition from a dimer of dimers assembly to the ion channel segment, which must contain elements of 4-fold symmetry, presents the conundrum that, at least in homomeric iGluR assemblies, chemically identical peptide chains will need to assume different conformations to accommodate the symmetry mismatch. However, it is possible that in iGluRs the ion channel does not show perfect 4-fold symmetry, and recent work shows that the symmetry transition, if present, must occur below the extracellular entrance to the channel (39).

4 Mechanism of Desensitization

The dimer assembly of the iGluR ligand binding domains also suggests a mechanism for desensitization (Fig. 2a). Clues about the mechanism arose from a series of prior 'structure'-function studies which gave tantalizing hints about important sites in the protein but did not elucidate the underlying molecular events. The first corner stone was the observation that the desensitized state binds glutamate with high affinity, even though the ion channel is in a non-conducting conformation; thus the ligand binding domains must be in their closed cleft active conformation, even though the ion channel is also closed. The second clue came from studies which revealed that mutation of amino acids, which were subsequently mapped to the dimer interface observed in iGluR crystal structures, had profound effects on both the extent and

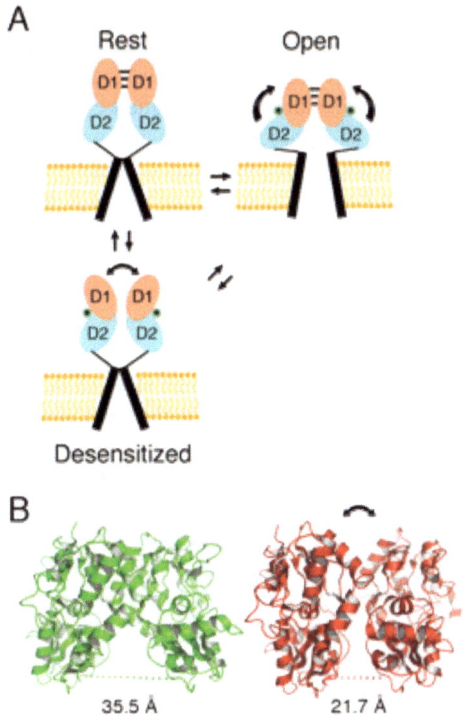

Fig. 2. The mechanism of iGluR desensitization. (a) shows a cartoon representation of an iGluR dinner assembly in the resting, active and agonist bound desensitized states. (b) shows crystals structures for the GluR2 ligand binding cores in the active (PDB 1FTJ) and desensitized (PDB 2I3V) conformations with the separation of the residues leading to the ion channel linkers in the intact protein given in Å.

kinetics of onset of desensitization (41), as well as on allosteric modulation by drugs like cyclothiazide and aniracetam (29, 30). The dimer assembly observed in AMPA receptor crystal structures provided an explanation for these results, and suggested that a pivoting motion of the pair of subunits in a dimer assembly, around an axis running close to the boundary between domains 1 and 2, would allow the ion channel linkers to relax back towards their resting position, shutting the ion channel, even though the individual ligand binding cores remained in their active closed cleft conformation (42). Direct support for this model came from combining the results of measurement of the affinity for dimer formation by a series of mutants of the isolated ligand binding cores, using analytical centrifugation, with functional measurements of the extent of desensitization measured in intact receptors, using rapid perfusion techniques and outside-out patch recording (42). This revealed that the rate of onset of desensitization was inversely correlated with the Kd for dimer formation by the isolated ligand binding cores: mutations that disrupted dimer formation accelerated desensitization, and mutations that stabilized the dimer slowed desensitization. Consistent with this, allosteric modulators that attenuate desensitization were found to stabilize dimer assemblies. Moreover, crystal structures of AMPA receptor complexes with cyclothiazide and aniracetam revealed that the modulators bind at the base of the dimer assembly, forming cross-links between each subunit, and sterically obstructing the rotation required to allow each of the subunits in a dimer pair to pivot into the desensitized state (19, 42). One key aspect of this model is that it implies that desensitization is a process that occurs within a dimer, and is not the result of a conformational change in individual iGluR subunits. This was further reinforced by disulfide cross linking experiments in which pairs of cysteine residues engineered into the dimer interface, to covalently trap the dimer in a conformation permissive for activation by glutamate, were shown to abolish desensitization (46). This result unambiguously establishes the dimer model for desensitization, and raises the question of how much cross talk, if any, occurs between dimer partners in AMPA and kainate receptors.

Hints about the structure of the desensitized state came from a clever series of experiments in which the pair of subunits in a glutamate bound GluR2 dimer were individually docked onto the domain 2 coordinates of the GluR2 apo crystal structure to generate a 1st approximation of the conformation of the desensitized state. The resulting model structure suggested the location of positions at which cysteine residues could be introduced to trap the receptor in the desensitized state (3). These mutants were then tested in functional assays using intact receptors, searching for positions at which reducing agents produced a large potentiation of glutamate responses following the introduction of cysteine residues. Once these sites were identified, the corresponding cross-linked dimer was crystallized to reveal the conformation of the desensitized state (Fig. 2b). As expected, both ligand binding cores remain in their closed cleft glutamate bound conformation, but due to pivoting around an axis parallel to the dimer interface, the separation of the domain 2 linkers that lead to the ion channel, decreases by nearly 14Å, which is sufficient to allow the ion channel to enter a non-conducting state, which however is not necessarily the same as the resting closed state (3). Further support for a partial collapse of the dimer interface during entry into the desensitized state came from measuring the state dependence of accessibility of additional positions for which introduction of mutant cysteine residues did not form spontaneous disulfide bonds in the resting, active or desensitized states, but which showed large increases in rate of reactivity when the receptor was desensitized. These positions mapped to structures that are buried in the resting and active conformations, but which become solvent exposed when the dimer collapses into its desensitized conformation.

The conformational changes underlying activation and desensitization described in above sections are likely to apply in broad detail to all subtype of iGluRs, including NMDA receptors, although the extent of the allosteric coupling between the ligand binding domain and the ion channel clearly differs between receptor families. The basis for this is not well understood at present, and will likely only be revealed when structures are solved for intact receptors from members of several iGluR subtypes. Thus, despite the spectacular advances which have revealed the general mechanisms by which iGluRs work, our understanding of how individual receptor subtypes gain their unique functional properties is still quite vague.

5 The Agonist Binding Site

The ligand binding cores have been the subject of substantial structural work, and crystal structures have been solved for GluR2, GluR5, GluR6, NR1, NR2A, NR3A and NR3B, delta 2, as well as for GluR0, a prokaryotic glutamate receptor ion channel that plausibly represents an ancestral precursor to the large family of diverse iGluRs that have evolved in eukaryotes. At the time of writing the only missing members are the KA subtype kainate receptors. Strikingly different from the cys-loop family of ligand gated ion channels, in which agonists bind at the interface between subunits, in iGluRs the agonist binding site is a self contained structure, four copies of which are present in each subunit. Thus, for AMPA and kainate receptors there are four glutamate binding sites per receptor, while for NMDA receptors there are two glycine and two glutamate binding sites (see chapter by Yuan et al.)

The neurotransmitter L-glutamate is a bifunctional molecule in which the alpha carboxylate and ammonium groups interact principally with domain 1, while the gamma carboxylate group interacts exclusively with domain 2 (Fig. 3). A conserved arginine side chain forms a salt bridge with the agonist alpha carboxylate group, while main chain amide and carbonyl groups form hydrogen bonds with the alpha carboxylate and ammonium groups respectively. Coordination of the alpha ammonium group is in addition mediated via a salt bridge with a conserved glutamate or aspartate side chain (Glu707 in Fig. 3), but this interaction is conformation specific, occurs only in the closed cleft active conformation, and is not possible in the apo state because the acidic side chain is blocked by formation of a salt bridge with a lysine side chain that projects up from one of the interdomain beta strands that form the base of the ligand binding domain (2). Thus, during the initial encounter of glutamate with the receptor, the ligand alpha carbon carboxylate and amino groups most likely dock with the domain 1 binding site, displacing bound water molecules, and only after this complex is formed does cleft closure and glutamate interaction with domain 2 occur (1, 4). This picture has been reinforced by

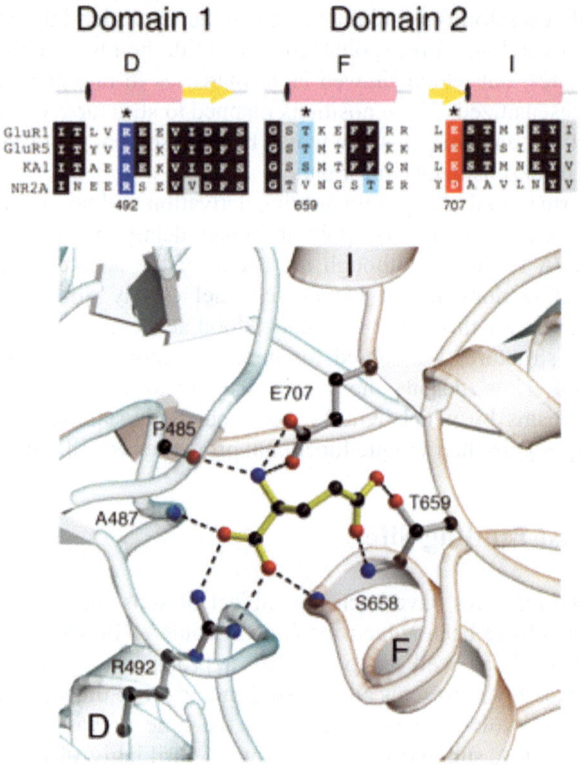

Fig. 3. Conserved binding motif for the glutamate binding site of AMPA, kainite and NMDA receptons. The secondary structure assignment for GluR6 and the sequence alignment for GluR1, GluR5, KA1 and NR2A, illustrate conserved residues in the agonist binding site, as revealed in the crystal structure for GluR6 (PDB 1S7Y). Labels indicated the positions of helices D, F and I; conserved amnio acids in the binding site which interact with the glutamate (*yellow*) are drawn in *ball and stick* represtation.

studies using time-resolved vibrational spectroscopy which reveals that formation of the initial iGluR agonist complex starts with the docking of glutamate at the alpha-carboxylate moiety and ends with the formation of interactions between the gamma carboxylate of glutamate and the protein (9).Within domain 2, a conserved threonine or serine side chain near the tip of helix F acts as the hydrogen bond donor for the agonist gamma carboxylate group, while main chain amides in helix F act as hydrogen bond donors for both the alpha and gamma agonist carboxyl groups. The large conformational difference between the apo and glutamate bound structures has the consequence that agonist interactions with helix F occur only in the active, closed-cleft conformation.

The formation of protein-protein bonds between domains 1 and 2 which are conformation specific, and formed only in the closed cleft conformation, also help to stabilize the active state of the agonist receptor complex (35). Amino acid sequence differences between AMPA and kainate receptors involved in such contacts is one of the mechanisms that contributes to the different kinetics of AMPA and kainate receptor gating. The stability of the closed cleft glutamate bound conformation is determined in part by residues at the lip of the clam shell, which differ in GluR2 and GluR6. Exchange of these tunes the kinetics of deactivation and recovery from desensitization (45). Such experiments produced both AMPA receptor gain of function mutants, in which the time constant of recovery from desensitization for GluR2 was slowed 5-fold, by introducing pairs of aspartate residues from kainate receptors in place of the alanine and serine residues in wild type GluR2, as well as kainate receptor loss of function mutants in which the time constant of recovery from desensitization was increased up to six-fold by breaking interdomain contacts in GluR6 (45).

At the time of writing only two iGluR apo structures have been solved, those for GluR2 and delta 2 and these give us our sole view of the resting state of an iGluR ligand binding domain dimer assembly. For other receptor subtypes we have a series of competitive antagonist structures, which might be considered equivalent to the apo state, and indeed clearly define the mechanism of action of competitive antagonists as due to steric occlusion via a foot in the door effect, which prevents the domain closure necessary to trigger ion channel activation. These structures reveal a surprisingly large range of differences in domain closure relative to the GluR2 apo structure, and indicate that in order to be able to bind antagonists of different structure the iGluR ligand binding cores must be extremely flexible proteins which can sample a wide range of conformations in solution, and also in the intact protein. Without additional apo structures it is not possible to determine if some of these antagonist structures are hyper-extended compared to the apo conformation, and even when this information becomes available, we will need to consider that the conformation of the apo state observed in crystal structures is likely to some unknown degree to be influenced by contacts with the crystal lattice and/or NCS related molecules. One solution to this problem would be to obtain high resolution NMR structures of the isolated ligand binding domain, but even this leaves unaddressed a second concern that excision from the ion channel frees the ligand binding domain from steric constraints which might limit its range of motion in an intact receptor. Even with all of these caveats in mind, it is remarkable how much insight has been gained into iGluR function from study of the isolated ligand binding domains.

6 Chemistry of Ligand Binding and Subtype Pharmacology

Because the structure of the domain 1 binding site is highly conserved in different iGluRs (Fig. 3), virtually all of the selectivity for drugs which bind to the discrete subtypes of iGluRs arises from amino acid sequence variations in the domain 2 surface of the ligand binding site. Thus, with crystal structures in hand, the structural basis for the rich pharmacology of iGluRs can now be understood and exploited for drug design. There is a surprisingly large variation in volume of the ligand binding cavity in individual iGluRs, which ranges from about 50Å^3 in the NR1, NR3A and NR3B subunits, to 218 Å^3 for GluR2, 255 Å^3 for GluR6 and 305Å^3 for GluR5. In all of these structures agonists are buried in the interior of the protein and are completely desolvated with respect to bulk solvent. In the case of AMPA and kainate receptors, the ligand binding cavity is much larger than necessary to bind glutamate, and contains well ordered water molecules, which essentially act in combination with glutamate as surrogate ligand atoms. Thus, although glutamate interacts with highly conserved structures in the ligand binding domains of the 9 AMPA and kainate receptor genes, in the surrounding protein landscape there exist cavities and protrusions, with unique chemical profiles. These permit the high affinity binding of AMPA to GluR1-4, but absolutely exclude AMPA binding to GluR6 by steric occlusion (23). Similar steric exclusion mechanisms underlie the selective binding of 5-substituted willardiine agonists to GluR5 but not GluR6 subtype kainate receptors, and the high selectivity of the competitive antagonist UBP310 for GluR5 versus AMPA receptors, and other subtypes of kainate receptor (24). As discussed later, steric effects also play a key role in the mechanism of action of partial agonists. In the case of the NR1, NR3A and NR3B subunits, which bind glycine, the small size of the ligand binding cavity cannot accommodate glutamate due to steric clash with the gamma carboxylate group, and this simple fact explains the exquisite discrimination by these subunits between glutamate and glycine.

The chemistry of ligand interaction with iGluRs is enormously rich, and much more complex than we can explain. That is, in most cases we are unable to predict the affinity of ligand binding with any degree of accuracy. The reasons for this are complex, but include the complex electrostatic environment of the ligand binding site; the need to model solvent structure in the ligand binding site and how this is altered upon ligand binding; the need to describe the change in entropy of ligands with highly mobile groups on transfer from solution to the conformation adopted when bound to the protein; the change in solvation energy of ligand upon transfer from solution into the ligand binding site; and of course the energy difference between the apo and ligand bound conformational states of the protein. As a 'simple' example it is interesting to consider the case of 2S,4R-4-methylglutamate, a potent kainate receptor agonist that has been widely used in radioligand binding assays because of its high affinity compared to glutamate. The addition of a single methyl group to the $C\gamma$ atom of glutamate, to yield 2S,4R-4-methylglutamate, increases affinity for GluR6 by 40-fold, from a Kd of 1.4 μM to 36 nM, but decreases affinity for GluR2 by 33-fold, from a Kd of 0.8 μM to 27 μM. How does addition of a single methyl group to glutamate have such a dramatic and selective effect? Crystal structures of the GluR6 ligand complexes reveal that the mechanism of binding of 2S,4R-4-methylglutamate to GluR6 is identical to that for glutamate, with the exception that the 4-methyl group makes additional van der Waals contacts with Tyrosine 457 and Valine 654. Least squares superposition of $C\alpha$ positions for the GluR6 glutamate and 2S,4R-4-methylglutamate structures gave rmsds of 0.12 Å indicating essentially identical structures. The extent of domain closure in the 2S,4R-4-methylglutamate and glutamate structures, 26.4° and 26.6° is also identical within the limits of experimental error (23).

The following three differences between glutamate and its 4-methyl derivative are sufficient to explain with quite good accuracy the difference in binding energy of the two compounds. First, addition of the methyl group to give the 4R enantiomer introduces a steric barrier that favors the conformation glutamate adopts when bound to GluR6; this gives about a 1.34 kcal per mole increase in binding energy. Second, the methyl group is shielded from bulk solvent in the GluR6 complex, and this transfer from an aqueous environment to a hydrophobic cavity yields about 1.2 kcal per mole; third, van der Waals contacts of the ligand methyl group with the greasy methylene atoms of the Tyrosine 457 and Valine 654 side chains gives a boost in affinity of somewhat less than a kcal per mole. In contrast, for AMPA receptors, the substitution of a larger leucine side chain at the position equivalent to valine 654, introduces a steric clash with the 4-methyl group, which requires the receptor binding site to expand to relieve the steric constraint, explaining why glutamate binds with higher affinity than its 4-methyl derivative. From this example one begins to see that computational docking experiments that ignore the complex chemistry of ligand binding are unlikely to accurately predict ligand affinity, although of course they are still very valuable tools for drug design, since they can eliminate many poor candidates, and aid in the design of new drugs that can be tested experimentally with increased anticipation of a successful outcome as a result of rational drug design.

7 Mechanism of Partial Agonist Activity

A large number of AMPA and kainate receptor agonist complexes have been solved for GluR2, GluR5, GluR6 and these revealed the striking observation that for a given receptor species and crystal system, the extent of domain closure differed by less than 1° for most agonists. In contrast, there is a much greater extent of variation in the range of domain closure observed for competitive antagonists, even though the number of antagonist structures solved is substantially less. This difference in extent of domain closure for antagonists clearly results from steric effects, and suggests that the ligand binding cores can expand and contract to bind antagonists of different size and shape without activation of ion channel gating. Thus, the ligand binding domains do not switch between single open and closed conformations, as required by the simplest Monod Wyman Changeux models of allostery, and instead exhibit induced fit to accommodate a range of ligands sizes. This naturally raises the question of whether there is a threshold for domain closure, the crossing of which triggers ion channel gating. This is indeed true and in fact is one of the major mechanisms which underlies the action of AMPA and kainate receptor partial agonists. Most work has been done for AMPA receptors, and the results suggest that even the 12° domain closure produced by the partial agonist kainate produces at most a few percent of the activation produced by the full agonists glutamate, AMPA, and quisqualate (4).

Two sets of experiments provide unequivocal support for a mechanism by which the extent of domain closure directly controls the extent of receptor activation by partial agonists. The first involved measurement of both the extent of domain closure for a series of 5-substituted willardiines, for which crystal structures were solved for the hydro-, fluoro-, bromo-, and iodo-derivatives, as well as measurement of the extent of activation of GluR2 by these agonists in the presence and absence of the allosteric modulator cyclothiazide (18). When desensitization was blocked the results revealed a strong correlation between domain closure and efficacy. Due to steric constraints, the extent of domain closure was determined by size of the 5-substituent, and followed the sequence H > F > Br > I, with the same sequence for peak current amplitudes. Conversely, when desensitization was intact, there was an inverse correlation between domain closure and the amplitude of equilibrium responses; this occurs because the least efficacious agonists also produce the smallest extent of desensitization. The logical continuation of this sequence extends to competitive antagonists, which of course produce neither activation nor desensitization. Studies with additional compounds confirm and extend this model for partial agonist activity and reveal that it is common to both AMPA and kainate receptors (12, 23, 27). Additional support for this mechanism came from protein engineering studies in which steric interactions with the ligand isopropenyl group were reduced by replacing the wild type Leu650 side chain with the smaller threonine residue (4). As expected this increased the efficacy of kainate, from 2% to 24% relative to the response to glutamate without altering the maximum response to quisqualate. Crystallization of the L650T mutant complex with kainate revealed a 3° increase in domain closure relative to the wild type complex, but this small change in domain closure has large effects on the efficiency of ion channel activation.

A surprising result of these studies is the extreme sensitivity of receptor activation to very small changes in domain closure, which reinforces two ideas discussed above. First, there is a threshold extent of domain closure required to trigger ion channel gating, and this requires quite large movements, about a 10° domain closure for AMPA receptors. Second, it is likely that the ion channel linkers are not rigid structures, and instead can absorb large movements of the ligand binding domain without disturbing the closed conformation of the ion channel. A surprising advantage come from this seemingly inefficient design: domain closure due to thermal vibration of the ligand binding domains, which could cause spontaneous ion channel activation, and degrade the fidelity of synaptic transmission, would be substantially reduced. This view presupposes that the membrane embedded ion channel domain is quite rigid, but this is consistent with the view that has emerged from crystallographic studies on potassium channels (16). Structural studies on intact iGluRs are required to test these ideas.

The above picture leaves unaddressed some very important issues, which complicate any analysis of iGluR gating mechanisms. The first is that measurements of macroscopic current amplitudes, which average out the activity of single ion channels, hide the fact that for AMPA receptors the ion channel opens to a number of substates of lower conductance, and that the relative substate occupancy and life time of these varies with concentration of glutamate (36, 38). This behavior has been successfully incorporated into a kinetic model for partial agonist activity in which a coupling constant, which varies with domain closure, was used to describe substate occupancy as a function of ligand concentration (18). This model builds on prior work suggesting that each ligand binding core can act somewhat independently to activate ion channel gating, but obtaining a realistic kinetic model of the single channel activity of AMPA receptors remains a formidable goal. Surprisingly, it is less clear that the same general model can be used to describe kainate receptor gating, since the available evidence suggests that concentration dependent ion channel substates may not occur (38). If true this would further complicate our limited understanding of partial agonist activity, since crystal structures for GluR6 reveal that as for AMPA receptors, kainic and domoic acid produce intermediate degrees of domain closure consistent with their action as kainate receptor partial agonists.

A second major issue largely ignored by current models for iGluR gating arises from the fact that following the binding of glutamate, the ion channel oscillates between open and closed states, producing bursts of openings while the agonist remains bound. The molecular identity of the glutamate bound state when the ion channel is closed is not known, but it seems unlikely that it reflects transitions between the open and closed cleft conformations observed in crystal structures for the apo and glutamate bound structures. It is possible that the clam shell partially opens, to an extent sufficient to permit ion channel closure, perhaps to a conformation like that observed in the AMPA bound complex of the GluR2 L650T mutant (4). The extremely short open times, and relatively low conductance of AMPA receptor substates, complicates the experiments needed to explore these issues, and as discussed in the chapter by Yuan et al., much better data is starting to emerge for NMDA receptors, which open for longer times, and to fewer subconductance states, than AMPA and kainate receptors.

Although to a first approximation the domain closure model accounts for partial agonist action quite well, additional details are required to accurately describe iGluR gating. It is clear from the results of FTIR spectroscopy that the energy of interaction of the alpha carbon amine group with the receptor differs for individual ligands, and although steric mechanisms dictate the extent of domain closure possible for many AMPA and kainate receptor agonists, binding energy is also an important determinant of agonist efficacy (22). Recent work showing that mutations of amino acids in the ligand binding core of AMPA receptors that contribute to stabilization of the closed cleft conformation alter agonist efficacy illustrates another example of how receptor activity can be controlled (35).

Any discussion of AMPA receptor gating is now incomplete without description of the modulatory effect of a recently discovered family of proteins named TARPs, short for transmembrane AMPA receptor regulatory proteins (44). There is a growing body of evidence, including studies with single particle electron microscopy (25), which indicate that TARPs are integral components of native AMPA receptors in at least some cell types. Whether this is true for all AMPA receptors is not yet established. The issue is complicated by the fact that TARPs have two distinct functions. First, in some cell types, and especially cerebellar granule cells where their biological role was first discovered (8), association of AMPA receptors with TARPs is essential for both cell surface expression and targeting to synapses. However, in addition to this, coexpression with TARPs modulates AMPA receptor gating, increasing the efficacy of activation by agonists, and slowing deactivation and desensitization (43). At the time of writing no TARP-like partners have been found for other iGluR subtypes, but the extreme mismatch in kinetics for synaptically activated kainate receptor responses with those for kainate receptors expressed in heterologous systems such as HEK cells is striking, and might be resolved if such proteins were discovered.

8 Modulation by Extracellular Ions

AMPA and kainate receptors are cation permeable ion channels which under physiological conditions use the transmembrane sodium ion concentration gradient to generate depolarizing responses at synapses. The extracellular and intracellular concentrations of the major biological cations and anions, Na^+, K^+ and Cl^-, are quite tightly regulated, albeit subject to transient changes during intense bursts of synaptic activity or during spreading depression. It is thus surprising that kainate receptors in particular, and AMPA receptors to a lesser extent, show profound modulation by changes in monovalent ion concentration and species (6, 32). Recent structural work has identified the site and mechanism of action of anions on kainate receptors, and reveals, unexpectedly that chloride ions are integral components of a kainate receptor dimer in its active conformation (33). The anion binding site shows exquisite selectivity, and strongly prefers chloride and bromide over fluoride and iodide; larger organic anions, like nitrate, methanesulfonate and sulfate also do not substitute for Cl^-. All kainate receptor subunits, i.e. GluR5, GluR6 and GluR7, expressed either alone or in combination with the KA1 or KA2 subunits, require chloride for normal gating. Crystallographic studies reveal that the anion binding site is formed at the D1 interface between a pair of subunits in a dimer assembly, and that chloride acts as a counter ion to a positively charged cavity, and thus provides electrostatic glue which contributes binding energy to dimer stabilization. Consistent with this, in functional experiments removal of chloride has two effects. First it accelerates the rate of onset of desensitization, reflecting reduced stability of the dimer assembly. Second, it reduces the peak current amplitude, by lowering open channel probability and the number of receptors competent for activation by glutamate. This effect was interpreted as a consequence of binding of glutamate to anion free receptors resulting in the triggering of either of two events: dimer rearrangement before the ion channel opens, or normal activation, but with a reduced stability of the active conformation.

Crystal structures of the chloride and bromide bound complexes reveal in detail the mechanism of anion binding (Fig. 4). The Cl^- ion shows six-fold coordination by the ammonium groups of Lys516 (3.67 Å), the guanidinium groups of Arg760 (3.4 Å), and a pair of equatorial water molecules (2.81 Å) which form hydrogen bonds with the side chain hydroxyl groups of Thr764. The bound anion is located on the molecular two-fold axis of symmetry at the dimer interface, 10 Å below the top surface of domain 1, defined by the ridge of methyl groups formed by Pro758. A single Cl^- or Br^- ion together with two water molecules is trapped in a cavity of volume 70 $Å^3$ located between the pair of subunits in a dimer assembly. The lid of the cavity is formed by a pair of intermolecular salt bridges between the side chains of Arg760 and Asp761. The carboxyl group of Asp761 is only 4.4 Å from the Cl^- ion, and is presumably shielded by formation of bidentate intermolecular salt bridges with Arg760 of the dimer partner (2.84 and 2.91 Å) and hydrogen bond formation with the trapped water molecules (2.66 Å) and solvent outside the anion binding cavity. The base of the cavity is formed by another pair of salt bridges formed between the side chains of Glu509 and Lys516. The structure is remarkable for the close apposition of multiple charged side chains which, to a first approximation, form counterbalanced rings of charge around the top of the dimer interface.

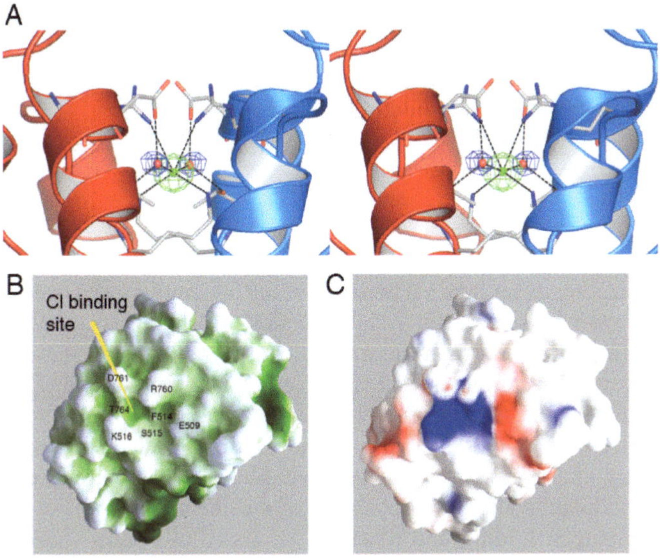

Fig. 4. The anion binding site in kainate receptors. (**a**) shows a stereo figure illustrating the six-fold coordination of chloride (*green*) by Arg and Lys side chains, and by water molecules (*blue*) trapped in the dimer interface. (**b**)and (**c**) show the molecular surface for the D1 dimer interface for a single GluR5 subunit, and illustrate the location of the anion binding site (b), and the surface electrostatic potential calculated for an intact dimer with the bound ion and solvent omitted from the calculation (c).

Although these structural results are compelling, kainate receptor crystals grown in different conditions did not consistently exhibit the mode of anion binding described above. For crystals grown in the presence of high concentrations of sulfate, the structural data make sense and the $\chi 4$ dihedral angle of the pair of Arg760 residues in the dimer complex is rotated by 94° compared to the Cl^- complex, so that the guanidinium groups point upwards towards the solvent exposed SO_4^{2-} ion, thus breaking the intermolecular salt bridges with Asp761. This is consistent with our functional studies, for which substitution of 150 mM NaCl by either 75 or 150 mM Na_2SO_4 increased the rate of onset of desensitization for GluR6 and reduced the peak amplitude of responses to glutamate. However, in other cases structures of anion bound complexes appear to have no biological significance.

To validate identification of the anion binding site revealed by structural studies, mutants were designed to perturb the anion binding site and examined for changes in gating kinetics and sensitivity to anions. The predicted results from these experiments were loss of function phenotypes, since perturbing the binding site would be expected to destabilize anion binding, and thus accelerate the rate of onset of desensitization. The results obtained with even conservative mutations to anion binding site residues were consistent with this prediction and the R744K and T748N mutations in GluR6 accelerated the rate of onset of desensitization 2 and 6-fold respectively. Of more interest, these mutations changed the rank order of sensitivity to anion modulation, which is consistent with the functional effects of anions being mediated by the binding site identified in structural experiments. Consistent with this mechanism, anion modulation was not altered in the GluR6 ATD(-) deletion construct, but was blocked when the D1 interfaces were locked in their active conformation by introduction of intermolecular cross links.

9 Conclusion

Clearly, with just 25% of the mass of a typical iGluR crystallized, understanding the domain organization of an intact tetrameric iGluR assembly, even at low resolution, would be a tremendous advance. Achieving this will be a heroic undertaking, but ultimately without this information we will never really understand how iGluRs work. Adding to the complexity of this picture, structures with TARPs will be necessary to understand AMPA receptor function, and perhaps similar proteins will be identified for other iGluR subtypes in the future. Structural studies are also the only way to try and begin to understand how phosphorylation, and other forms of cytoplasmic modulation alter the probability of ion channel gating. An understanding of the mechanism of folding and assembly of an iGluR, and its coassembly with chaperones and transport partners is beginning to emerge (11), but the molecular mechanisms are a long way from being understood in structural terms. In the near future it is to be anticipated that the rational design of subtype selective ligands will be achieved with increasing frequency, which will finally provide tools to begin to cleanly answer questions about the diverse biology of iGluRs in synaptic pathways throughout the CNS and spinal cord. Even with the ligand binding cores many important issues remain to be sorted out, and likely this will result in the identification of new sites of action of allosteric modulators.

Acknowledgements

Work in the authors laboratory is supported by the Intramural Research Program of NICHD, NIH.

References

1. Abele R, Keinanen K, and Madden DR. Agonist-induced isomerization in a glutamate receptor ligand-binding domain. A kinetic and mutagenetic analysis. J Biol Chem 275: 21355–21363, 2000.
2. Armstrong N and Gouaux E. Mechanisms for activation and antagonism of an AMPA-sensitive glutamate receptor: Crystal structures of the GluR2 ligand binding core. Neuron 28: 165–181, 2000.
3. Armstrong N, Jasti J, Beich-Frandsen M, and Gouaux E. Measurement of conformational changes accompanying desensitization in an ionotropic glutamate receptor. Cell 127: 85–97, 2006.
4. Armstrong N, Mayer M, and Gouaux E. Tuning activation of the AMPA-sensitive GluR2 ion channel by genetic adjustment of agonist-induced conformational changes. Proc Natl Acad Sci USA 100: 5736–5741, 2003.

5. Armstrong N, Sun Y, Chen GQ, and Gouaux E. Structure of a glutamate-receptor ligand-binding core in complex with kainate. Nature 395: 913–917, 1998.
6. Bowie D. External anions and cations distinguish between AMPA and kainate receptor gating mechanisms. J Physiol 539: 725–733, 2002.
7. Chen GQ, Cui C, Mayer ML, and Gouaux E. Functional characterization of a potassium-selective prokaryotic glutamate receptor. Nature 402: 817–821, 1999.
8. Chen L, Chetkovich DM, Petralia RS, Sweeney NT, Kawasaki Y, Wenthold RJ, Bredt DS, and Nicoll RA. Stargazin regulates synaptic targeting of AMPA receptors by two distinct mechanisms. Nature 408: 936–943, 2000.
9. Cheng Q, Du M, Ramanoudjame G, and Jayaraman V. Evolution of glutamate interactions during binding to a glutamate receptor. Nat Chem Biol 1: 329–332, 2005.
10. Doyle DA, Cabral JM, Pfuetzner RA, Kuo A, Gulbis JM, Cohen SL, Chait BT, and MacKinnon R. The structure of the potassium channel: molecular basis of K+ conduction and selectivity. Science 280: 69–77, 1998.
11. Greger IH and Esteban JA. AMPA receptor biogenesis and trafficking. Curr Opin Neurobiol, 2007.
12. Hogner A, Kastrup J, Jin R, Liljefors T, Mayer M, Egebjerg J, Larsen I, and Gouaux E. Structural Basis for AMPA Receptor Activation and Ligand Selectivity: Crystal Structures of Five Agonist Complexes with the GluR2 Ligand-binding Core. J Mol Biol 322: 93, 2002.
13. Hollmann M, Maron C, and Heinemann S. N-Glycosylation site tagging suggests a three transmembrane domain topology for the glutamate receptor GluR1. Neuron 13: 1331–1343, 1994.
14. Hollmann M, O'Shea-Greenfield A, Rogers SW, and Heinemann S. Cloning by functional expression of a member of the glutamate receptor family. Nature 342: 643–648, 1989.
15. Horning MS and Mayer ML. Regulation of AMPA receptor gating by ligand binding core dimers. Neuron 41: 379–388, 2004.
16. Jiang Y, Lee A, Chen J, Cadene M, Chait BT, and MacKinnon R. Crystal structure and mechanism of a calcium-gated potassium channel. Nature 417: 515–522, 2002.
17. Jiang Y, Lee A, Chen J, Cadene M, Chait BT, and MacKinnon R. The open pore conformation of potassium channels. Nature 417: 523–526, 2002.
18. Jin R, Banke TG, Mayer ML, Traynelis SF, and Gouaux E. Structural basis for partial agonist action at ionotropic glutamate receptors. Nature Neurosci 6: 803–810, 2003.
19. Jin R, Clark S, Weeks AM, Dudman JT, Gouaux E, and Partin KM. Mechanism of positive allosteric modulators acting on AMPA receptors. J Neurosci 25: 9027–9036, 2005.
20. Kuner T, Seeburg PH, and Guy HR. A common architecture for K+ channels and ionotropic glutamate receptors? Trends Neurosci 26: 27–32, 2003.
21. MacKinnon R. Pore loops: an emerging theme in ion channel structure. Neuron 14: 889–892, 1995.
22. Mankiewicz KA, Rambhadran A, Du M, Ramanoudjame G, and Jayaraman V. Role of the chemical interactions of the agonist in controlling alpha-amino-3-hydroxy-5-methyl-4-isoxazolepropionic acid receptor activation. Biochemistry 46: 1343–1349, 2007.
23. Mayer ML. Crystal structures of the GluR5 and GluR6 ligand binding cores: molecular mechanisms underlying kainate receptor selectivity. Neuron 45: 539–552, 2005.
24. Mayer ML, Ghosal A, Dolman NP, and Jane DE. Crystal structures of the kainate receptor GluR5 ligand binding core dimer with novel GluR5-selective antagonists. J Neurosci 26: 2852–2861, 2006.
25. Nakagawa T, Cheng Y, Ramm E, Sheng M, and Walz T. Structure and different conformational states of native AMPA receptor complexes. Nature 433: 545–549, 2005.

26. Nakanishi N, Shneider NA, and Axel R. A family of glutamate receptor genes: evidence for the formation of heteromultimeric receptors with distinct channel properties. Neuron 5: 569–581, 1990.
27. Nanao MH, Green T, Stern-Bach Y, Heinemann SF, and Choe S. Structure of the kainate receptor subunit GluR6 agonist-binding domain complexed with domoic acid. Proc Natl Acad Sci USA 103: 1708–1713, 2005.
28. Paoletti P, Perin-Dureau F, Fayyazuddin A, Le Goff A, Callebaut I, and Neyton J. Molecular organization of a zinc binding n-terminal modulatory domain in a NMDA receptor subunit. Neuron 28: 911–925, 2000.
29. Partin KM, Bowie D, and Mayer ML. Structural determinants of allosteric regulation in alternatively spliced AMPA receptors. Neuron 14: 833–843, 1995.
30. Partin KM, Fleck MW, and Mayer ML. AMPA receptor flip/flop mutants affecting deactivation, desensitization, and modulation by cyclothiazide, aniracetam, and thiocyanate. J Neurosci 16: 6634–6647, 1996.
31. Pasternack A, Coleman SK, Jouppila A, Mottershead DG, Lindfors M, Pasternack M, and Keinanen K. Alpha-amino-3-hydroxy-5-methyl-4-isoxazolepropionic acid (AMPA) receptor channels lacking the N-terminal domain. J Biol Chem 277: 49662–49667, 2002.
32. Paternain AV, Cohen A, Stern-Bach Y, and Lerma J. A role for extracellular Na+ in the channel gating of native and recombinant kainate receptors. J Neurosci 23: 8641–8648, 2003.
33. Plested AJ and Mayer ML. Structure and mechanism of kainate receptor modulation by anions. Neuron 53: 829–841, 2007.
34. Quiocho FA and Ledvina PS. Atomic structure and specificity of bacterial periplasmic receptors for active transport and chemotaxis: variation of common themes. Mol Microbiol 20: 17–25, 1996.
35. Robert A, Armstrong N, Gouaux JE, and Howe JR. AMPA receptor binding cleft mutations that alter affinity, efficacy, and recovery from desensitization. J Neurosci 25: 3752–3762, 2005.
36. Rosenmund C, Stern-Bach Y, and Stevens CF. The tetrameric structure of a glutamate receptor channel. Science 280: 1596–1599, 1998.
37. Sine SM and Engel AG. Recent advances in Cys-loop receptor structure and function. Nature 440: 448–455, 2006.
38. Smith TC and Howe JR. Concentration-dependent substate behavior of native AMPA receptors. Nat Neurosci 3: 992–997, 2000.
39. Sobolevsky AI, Yelshansky MV, and Wollmuth LP. The outer pore of the glutamate receptor channel has 2-fold rotational symmetry. Neuron 41: 367–378, 2004.
40. Stern-Bach Y, Bettler B, Hartley M, Sheppard PO, O'Hara PJ, and Heinemann SF. Agonist-selectivity of glutamate receptors is specified by two domains structurally related to bacterial amino acid binding proteins. Neuron 13: 1345–1357, 1994.
41. Stern-Bach Y, Russo S, Neuman M, and Rosenmund C. A point mutation in the glutamate binding site blocks desensitization of AMPA receptors. Neuron 21: 907–918, 1998.
42. Sun Y, Olson R, Horning M, Armstrong N, Mayer M, and Gouaux E. Mechanism of glutamate receptor desensitization. Nature 417: 245–253, 2002.
43. Tomita S, Adesnik H, Sekiguchi M, Zhang W, Wada K, Howe JR, Nicoll RA, and Bredt DS. Stargazin modulates AMPA receptor gating and trafficking by distinct domains. Nature 435: 1052–1058, 2005.
44. Tomita S, Chen L, Kawasaki Y, Petralia RS, Wenthold RJ, Nicoll RA, and Bredt DS. Functional studies and distribution define a family of transmembrane AMPA receptor regulatory proteins. J Cell Biol 161: 805–816, 2003.

45. Weston MC, Gertler C, Mayer ML, and Rosenmund C. Interdomain interactions in AMPA and kainate receptors regulate affinity for glutamate. J Neurosci 26: 7650–7658, 2006.
46. Weston MC, Schuck P, Ghosal A, Rosenmund C, and Mayer ML. Conformational restriction blocks glutamate receptor desensitization. Nat Struct Mol Biol 13: 1120–1127, 2006.

Cellular Biology of AMPA Receptor Trafficking and Synaptic Plasticity

José A. Esteban

University of Michigan, Department of Pharmacology, Ann Arbor, MI 48109, USA, estebanj@umich.edu

1 Introduction

Intracellular membrane trafficking is an essential process in all eukaryotic cells, but it is particularly critical at synaptic terminals, where a large number of specific ion channels, scaffolding molecules and multiple signal transduction regulators have to be precisely targeted to ensure proper synaptic function (80, 126). At the level of the postsynaptic terminal, local membrane trafficking is now appreciated as major factor controlling synaptic function (58). In particular, the regulation of neurotransmitter receptor transport and targeting is crucial for the maintenance of synaptic strength, and for the activity-dependent changes associated to synaptic plasticity (19).

Most excitatory transmission in the central nervous system is mediated by two types of glutamate receptors: γ-amino-3-hydroxy-5-methylisoxazole-4-proprionic acid (AMPA) and N-methyl-D-aspartate (NMDA) receptors. These two types of receptors have very different roles in synaptic function (22, 51). AMPA receptors (AMPARs) mediate most excitatory (depolarizing) currents in conditions of basal neuronal activity. Hence, they have a major influence in the strength of the synaptic response. NMDA receptors (NMDARs), on the other hand, remain silent at resting membrane potential (85), but they are crucial for the induction of specific forms of synaptic plasticity, such as long-term potentiation (LTP) and long-term depression (LTD) (6). Although AMPARs and NMDARs reside at the same synapses in most brain regions, they traffic to the synaptic membrane through quite different programs. In the brain, soon after birth, most excitatory synapses in the hippocampus and other brain regions contain only NMDARs, whereas the prevalence of AMPARs increases gradually over development (24, 53, 55, 73, 92). In fact, the delivery of AMPARs into synapses is a regulated process that depends on NMDAR activation and underlies some forms of synaptic plasticity (78).

J.W. Hell, M.D. Ehlers (eds.), *Structural and Functional Organization of the Synapse*, DOI: 10.1007/978-0-387-77232-5_10, © Springer Science+Business Media, LLC 2008

Synaptic plasticity is thought to underlie higher cognitive functions, such as learning and memory (9, 90, 119), and is also critical for neural development (20). Thus, it is not surprising that alterations in synaptic plasticity have been implicated in the pathology of several neurological disorders, including Alzheimer's disease (98), schizophrenia (110), Down's syndrome (32) and other forms of mental retardation (82). Consequently, there is considerable interest in understanding the underlying mechanisms of synaptic plasticity, among which the regulation of AMPAR trafficking plays a prominent role.

This review will summarize our current knowledge of the membrane trafficking pathways that steer AMPARs from their biosynthesis at the endoplasmic reticulum (ER) to their destination at excitatory synapses, with special emphasis in the regulatory steps that contribute to synaptic plasticity. Most of the experimental observations that are the basis for this chapter have been obtained from hippocampal principal neurons, although it is expected that most of the principles described here will be applicable to the regulation of AMPAR trafficking in multiple brain regions.

2 AMPA Receptor Assembly and Exit from the Endoplasmic Reticulum

AMPARs are hetero-tetramers (96), composed of different combinations of GluR1, GluR2, GluR3, and GluR4 subunits (51). In the mature hippocampus, most AMPARs are composed of GluR1–GluR2 or GluR2–GluR3 combinations (118), whereas GluR4-containing AMPARs are expressed mainly in early postnatal development (125). These oligomeric combinations are formed in the endoplasmic reticulum (ER) (40), possibly assembling as dimers of dimers (112) via interactions between the luminal, N-terminal domains of the subunits (43, 66, 69). After assembly, exit from the ER is tightly regulated by quality control mechanisms that monitor the competency of newly synthesized receptors for ligand binding and gating (30).

Interestingly, AMPAR trafficking through the ER is subunit-specific. Thus, GluR1–GluR2 hetero-oligomers exit the ER rapidly, and traffic to the Golgi compartment where they become fully glycosylated (42). In contrast, GluR2–GluR3 heteromers take much longer to exit (i.e., are retained longer in) the ER. In fact, a fraction of the GluR2 subunits remains unassembled within the ER, in a manner that depends on the presence of an edited arginine residue (R607) at the channel pore region (41, 42). These immature AMPAR subunits associate with molecular chaperones residing at the ER (31, 42). Interactions with cytosolic proteins also control trafficking through the ER. For example, the GluR2 C-terminus has a PDZ (PSD-95/Discs-Large/ZO-1) consensus motif (-SVKI) that interacts with the PDZ domain-containing protein PICK1 (Protein Interacting with C-Kinase 1) (21, 122). This interaction is required for GluR2's exit from the ER (42).

Additionally, export of AMPARs from the ER and surface expression is also facilitated by direct interaction with a family of "Transmembrane AMPAR regulatory proteins" (TARPs) (117). In fact, TARPs may well be considered auxiliary subunits of AMPARs (31), which assist in their proper folding and affect channel kinetics (7, 113, 116).

3 AMPA Receptor Trafficking along the Microtubular Cytoskeleton in Dendrites

Messenger RNAs for AMPAR subunits are present in both cell body and dendritic compartments (44), and dendritic synthesis of AMPARs has been reported (56). Nevertheless, most newly synthesized receptors are likely to travel long distances from their point of biosynthesis to their final synaptic targets. The long-range dendritic transport of AMPARs is likely to depend on the microtubular cytoskeleton that runs along dendritic shafts. The transport of membrane organelles on microtubule tracks is an active process powered mainly by motor proteins of the kinesin and dynein superfamilies (37). Therefore, membrane compartments bearing AMPARs are likely to be recognized and transported by some of these motor proteins. The molecular mechanisms underlying these processes are still being elucidated.

The PDZ domain-containing protein Glutamate Receptor Interacting Protein 1 (GRIP1) interacts directly with the heavy chain of conventional kinesin, as revealed by yeast two-hybrid screening (104). GRIP binds to the C-terminal PDZ motif of GluR2 and GluR3 (23, 108), and hence, may serve as the link between AMPARs and microtubular motor proteins. In fact, the ternary complex formed by GluR2, GRIP1, and kinesin can be immunoprecipitated from brain lysates, and dominant-negative versions of kinesin reduce the presence of AMPAR at synapses (104).

AMPARs have also been shown to associate with a different neuron-specific kinesin motor: KIF1 (107). In this case, the adaptor molecule is liprin-α, which interacts with GluR2–GRIP (121) and with KIF1 (107). Another member of the liprin-α – AMPAR–GRIP complex is GIT1, which is also involved in AMPAR trafficking (61). Based on these combined studies, the GRIP1–AMPAR complex is likely to be transported along dendrites by multiple types of kinesin motors.

In addition to this microtubular-dependent transport, it has recently been reported that the export of AMPARs from the cell body into the dendritic surface is powered by a specific actin-based motor protein: myosin Vb (71). Interestingly, this transport system was specific for the GluR1 subunit, and required the small GTPase Rab11, possibly acting as a linker between the motor protein and its membrane cargo. From these combined studies, it seems likely that multiple links between AMPARs and cytoskeletal motor proteins will be discovered in the future, possibly mediated by specific scaffolding molecules.

4 Actin-Dependent Trafficking in Spines

Most excitatory synapses in the adult brain occur on small dendritic protuberances called spines (50). Dendritic spines lack microtubular cytoskeleton, but they are rich in highly motile actin filaments (29). Therefore, at some point, AMPAR-containing organelles, trafficking along microtubular tracks, must be transferred to the actin-based cytoskeleton for their final delivery into synapses. The importance of the actin cytoskeleton for local AMPAR trafficking is underscored by the observation that pharmacological depolymerization of actin filaments leads to the removal of AMPARs from dendritic spines (3) and from synapses (60).

The molecular mechanisms that may mediate the actin-based movement of AMPARs are largely unknown. Nevertheless, AMPARs can be linked to the actin cytoskeleton through several scaffolding proteins, such as 4.1N (105) and RIL (reversion-induced LIM protein) (102). Members of the protein 4.1 family are known to couple the spectrin–actin cytoskeleton to different membrane-associated proteins (52). In particular, the neuronal isoform 4.1N interacts directly with GluR1 (105) and GluR4 (18) through the juxtamembrane region of their cytoplasmic C-terminal tails. The other potential actin linker for AMPARs, RIL, is a multi-functional protein that interacts with an internal region of the GluR1 C-terminus through its LIM (Lin11/rat Isl-1/Mec3) domain, and with α-actinin through its PDZ domain (102). Interestingly, only AMPAR subunits with long C-tails (GluR1 and GluR4) have been shown so far to couple with the actin cytoskeleton. Since these long-tail subunits are the ones involved in regulated (activity-dependent) delivery at the synapse (77), it is tempting to speculate that actin-dependent transport may be particularly critical for AMPAR synaptic insertion during plasticity.

The transport of AMPARs along the spine actin cytoskeleton is likely to be bidirectional, since AMPARs are known to move in and out of synapses in a very dynamic manner. This expectation has been recently confirmed by the identification of an actin-based motor protein, myosin VI, as a mediator of the endocytic removal of AMPARs from synapses (86). Myosin VI interacts with the GluR1-binding protein SAP97 (Synapse-Associated Protein 97) (120), providing a mechanistic link between AMPARs (again through a long-tail subunit) and the motor protein that drives their internalization. Undoubtedly, further studies will be required to unravel what is likely to be a network of interactions mediating the transport of AMPARs along the actin cytoskeleton in synapses.

5 Subunit-Specificity for Constitutive and Regulated Synaptic Delivery of AMPA Receptors

It is now well established that the final steps in the synaptic trafficking of AMPARs depend on their subunit composition, and specifically, on cis-signals contained within their cytosolic carboxy termini (89, 106). In hippocampus, hetero-tetramers formed by GluR1-GluR2 and GluR2-GluR3 subunits, together with a smaller contribution from GluR1 homomers, represent the most common combinations in excitatory synapses (118). Based on experiments expressing recombinant AMPAR subunits in hippocampal neurons, it has been shown that GluR2/GluR3 hetero-tetramers continuously cycle in and out of synapses in a manner largely independent from synaptic activity (89, 106). This process (constitutive pathway) preserves the total number of receptor at synapses, and therefore, it has been proposed to help maintaining synaptic strength in the face of protein turnover (77). This constitutive cycling is very fast (half-time of minutes) and it requires a direct interaction between GluR2 and NSF (N-ethyl-maleimide Sensitive Factor) (83). The precise role of NSF in this trafficking pathway is not fully understood, but it may involve NSF-assisted dissociation of GluR2 from the PDZ domain protein PICK1, allowing AMPARs to cycle back into synapses (47). The continuous synaptic cycling of AMPARs also requires the molecular chaperon Hsp90 (35), although the mechanistic link between AMPARs and Hsp90 has not been elucidated yet.

The constitutive synaptic cycling of AMPARs was originally described to involve endo- and exocytic trafficking, based on experiments using whole-cell infusion of peptides in hippocampal slices (76). However, this interpretation has been recently challenged by experiments involving acute inactivation of surface AMPARs in dissociated neurons, where fast exchange between synaptic and extrasynaptic receptors occurred exclusively on the neuronal surface (2). The ongoing controversy between surface and intracellular delivery is further discussed in the following section on TARPs and AMPAR surface trafficking.

In contrast with the constitutive trafficking, AMPARs containing GluR1 (49), GluR2-long (63) (a splice variant of GluR2 (62)) or GluR4 (125) are added into synapses in an activity-dependent manner during synaptic plasticity. This regulated pathway is triggered transiently upon induction of long-term potentiation (LTP), and results in a net increase in the number of AMPARs present at synapses (77). The synaptic delivery of GluR1 is also regulated by physiological stimulation in living animals, as it has been reported for neocortical neurons upon sensory stimulation (111), and in the lateral amygdala after fear conditioning (99). The subunit composition of the endogenous AMPARs that participate in regulated synaptic delivery has been more difficult to establish. Thus, both GluR2-lacking receptors (presumably GluR1 homomers) (94) and GluR2-containing receptors (presumably GluR1/GluR2 heteromers) (1, 4) have been proposed to be rapidly inserted into synapses upon NMDA receptor activation in hippocampal slices. Although the details remain to be clarified, the importance of subunit composition for the regulation of synaptic delivery is well established. This has been recently corroborated by in vivo studies, which demonstrated that sensory stimulation (17) or deprivation (36), as well as cocaine administration (8), can alter the prevalence of AMPARs with different subunit assemblies at synapses.

The activity-dependent synaptic delivery of AMPARs is regulated by several protein kinases, such as Ca2+/calmodulin-dependent protein kinase II (CaMKII; reviewed in (72)), PKA (25, 28, 39), PKC (10, 38, 70) and phosphoinositide 3-kinase (PI3K) (79). Interestingly, the signaling cascades controlling the delivery of AMPARs to synapses, as well as the AMPAR subunits involved, change during development. Thus, early in postnatal development of the hippocampus, the regulated delivery of AMPARs involves GluR4-containing receptors (125), and PKA-mediated phosphorylation of this subunit triggers receptor delivery (28). Around the second postnatal week, LTP is mostly mediated by the synaptic delivery of GluR2-long (63). Then, later in development, the regulated addition of AMPARs requires both GluR1 phosphorylation by PKA and CaMKII activation (28). These developmental changes in the regulation of AMPAR synaptic delivery fit very well with the switch in signaling cascades that are required for LTP induction at different postnatal ages (123).

6 TARPs and AMPA Receptor Surface Trafficking

TARPs are the only known transmembrane proteins found to be associated with AMPARs. The first TARP to be identified was stargazin, which was found as a spontaneous mutation in the stargazer mouse (68). By sequence and structural homology, stargazin belongs to a large group of proteins that includes γ-subunits of Ca2+ channels and the claudin family of cell-adhesion molecules. Nevertheless, only five of these proteins have been described to bind AMPARs and affect their trafficking: stargazin (γ-2), γ-3, γ-4, γ-8 (114), and more recently, γ-7 (57). Therefore, these are the proteins collectively known as TARPs.

Different TARPs display specific expression patterns in brain, which are to some extent complementary (114). The most striking property of TARPs is their critical role for the expression of AMPARs at the extrasynaptic neuronal surface. Genetic ablation of stargazin (γ-2), the TARP member most abundantly expressed in cerebellum, results in a virtual depletion of AMPARs from the extrasynaptic surface in granule cells (14). Similarly, removal of γ-8, a TARP member that is almost exclusively expressed in hippocampus, precludes AMPAR surface expression in hippocampal pyramidal neurons (97). Interestingly, TARPs are a limiting factor for AMPAR cell surface delivery in these neurons, since overexpression of the corresponding TARP leads to a marked increase in the number of AMPARs expressed on the plasma membrane (14, 97). The role of these extrasynaptic surface receptors is still debated, although morphological evidence indicates that they are highly mobile and can reach the postsynaptic membrane through lateral diffusion (15, 115).

TARPs also participate in the trafficking of AMPARs into the synaptic membrane. TARPs contain a PDZ consensus sequence at the C-terminus, which can bind the PDZ domain of MAGUK (membrane-associated guanylate kinase) proteins, such as PSD95 and PSD93 (Post-Synaptic Density Proteins 95 and 93) (14). MAGUKs are synaptic scaffolding molecules, which have been shown to be critical regulators of AMPAR delivery and/or stabilization at synapses (26, 27). Therefore, TARPs are thought to be the molecular linkers between AMPARs and MAGUKs. In particular, the association between TARPs and MAGUKs has been recently shown to be critical to retain AMPARs at synapses. Thus, impairment of the PDZ interaction between stargazin (TARP) and PSD95 (MAGUK) leads to increased receptor diffusion out of the synaptic membrane (5). Based on these results, it is concluded that a major function of the TARP-MAGUK interaction is the stabilization/anchoring of AMPARs at synapses.

The dual role of TARPs in extrasynaptic surface expression and in receptor stabilization at synapses has led to the hypothesis that AMPAR synaptic delivery occurs in two steps: insertion in the extrasynaptic surface followed by lateral diffusion and synaptic trapping. Indeed, there are morphological (89) and electrophysiological (2) observations supporting this scenario. However, there are also indications that extrasynaptic surface receptors are not a necessary source for synaptic delivery. For example, genetic ablation of the hippocampal TARP (γ-8) produced a virtual depletion of extrasynaptic AMPARs, with only a modest effect on the accumulation of AMPARs at synapses (97). Conversely, TARP overexpression produces a massive increase in extrasynaptic AMPARs without any detectable effect on AMPAR-mediated synaptic transmission (97, 101).

The controversy about the subcellular pathway for AMPAR synaptic delivery is far from being resolved. Nevertheless, it is worth mentioning that the use of different experimental systems may underlie some of the reported discrepancies. Thus, extrasynaptic surface delivery is typically demonstrated in dissociated neuronal cultures, whereas most evidence for intracellular trafficking has been obtained in organotypic hippocampal slices. On the other hand, endogenous receptors are usually monitored on dissociated neurons, while overexpressed proteins are often the basis for the experiments usings hippocampal slices. Clearly, more work will be required to decipher the anatomical details of AMPAR synaptic trafficking. It is also worth keeping in mind that the precise mechanism of AMPAR delivery may vary among different synapse types and developmental stages.

7 Local Intracellular Trafficking of AMPARs. Role of Rab Proteins and the Exocyst

Although it is difficult to evaluate the relative contribution of surface versus intracellular trafficking, there is concurrent evidence that rapid exocytic events can mediate the activity-dependent delivery of AMPARs into synapses (64, 74, 87, 93). In this sense, it may come as a surprise that very little is known about the subcellular organization of the membrane trafficking machinery that mediates AMPAR synaptic delivery. This picture has started to change recently, with the identification of local endosomal compartments in close proximity to synapses, or even within dendritic spines, that mediate the delivery of AMPARs into the synaptic membrane (33, 87, 88). These new reports are starting to offer a glimpse of the complexity of the membrane trafficking machinery operating at postsynaptic terminals, and how it may relate to the subunit-specific synaptic delivery of AMPARs.

Most intracellular membrane sorting in eukaryotic cells is governed by small GTPases of the Rab family (124). Therefore, the identification of specific Rab proteins involved in AMPAR trafficking may give us some clues as to how the intracellular sorting and synaptic targeting of AMPARs is organized in neurons. It was recently proposed that recycling endosomes driven by the small GTPase Rab11 mediate the activity-dependent delivery of GluR1-containing AMPARs into synapses (87). In addition, Rab8, which controls trans-Golgi network trafficking (54) and a separate endosomal population (48), is also required for GluR1 synaptic insertion and LTP (33). Therefore, it seems that the activity-dependent delivery of AMPARs involves a relay of at least two distinct membrane compartments, whose sorting is controlled by Rab11 and Rab8 possibly acting in separate trafficking steps. Rab11-containing endosomes have recently been localized at the base of dendritic spines (88), whereas ultrastructural studies have detected Rab8 in close proximity to the postsynaptic membrane (33). According to these morphological observations, we propose a model in which AMPARs enter spines through Rab11-dependent endosomes. Subsequently, an additional endosomal population, controlled by Rab8, would drive their insertion into the synaptic membrane (see model in Fig. 1).

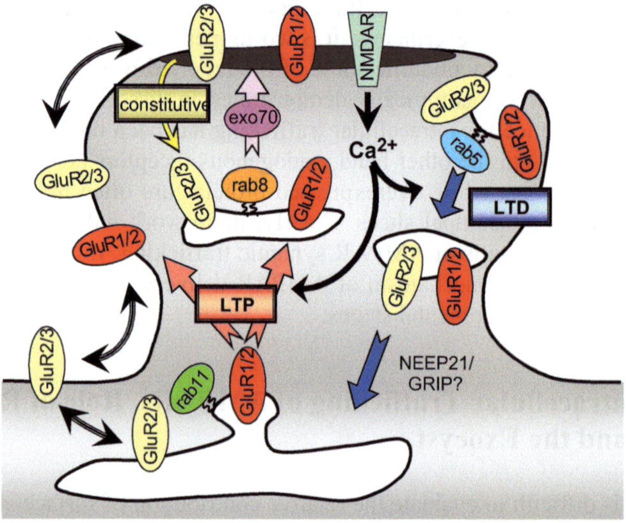

Fig. 1. Schematic model for the synaptic trafficking of AMPA receptors and the endosomal machinery operating at postsynaptic termianls. The activity-dependent entry of GluR1-containing AMPARs into spines is controlled by Rab11 upon LTP induction (*red upward arrows*). Once within the spine, both GluR1-GluR2 (*red*) and GluR2-GluR3 (*yellow*) AMPARs are driven into synapses in an exocytic process controlled by Rab8 and the exocyst subunit Exo70. In addition, GluR2-GluR3 receptors are engaged in constitutive cycling in and out of the postsynaptic membrane (*yellow arrow*). In addition to this intracellular pathway, AMPARs can also reach the synaptic membrane through lateral diffusion from the extrasynaptic surface (*double-headed arrows on the left*). The relative contribution from these two pathways is currently being investigated. The activity-dependent internalization of AMPARs is mediated by Rab5, acting on the lateral (extrasynaptic) membrane within the spine (*blue downward arrows*). Re-entry of internalized receptors into the Rab11-Rab8 delivery circuit may require the participation of NEEP21 and GRIP.

As mentioned above, the constitutive trafficking of AMPARs in and out of synapses may involve surface exchange and/or endo-exocytic events (Fig. 1). Indeed, very little is known about the intracellular machinery that controls this process. It has been shown that Rab proteins typically associated with recycling endosomes, such as Rab4 and Rab11, do not participate in constitutive AMPAR synaptic cycling (33). However, Rab8 appears to be critically required (33). Since Rab8 is also involved in activity-dependent trafficking (see above), these results indicate that there is a partial overlap between the endosomal machinery mediating constitutive and regulated delivery of AMPARs at synapses (see model in Fig. 1). The potential endocytic arm of this continuous cycling of receptors is even less characterized. The prototypic Rab protein for endocytosis, Rab5 (12), does not participate in constitutive AMPAR internalization (11). Dynamin was proposed to be required for this process (76), and the role of clathrin has not been directly tested yet. Obviously, more work will be required to elucidate the cellular basis of this very dynamic aspect of the intracellular trafficking of AMPARs.

The final step in the intracellular trafficking of AMPARs involves their functional insertion and stabilization at the postsynaptic membrane. As mentioned before, several members of the MAGUK family of scaffolding proteins are critical factors for the synaptic targeting of AMPARs (27). Interestingly, these synaptic scaffolding molecules associate with the exocyst (95, 100), a known effector of Rab-dependent exocytic trafficking (45, 84). Therefore, the exocyst may act as a link between incoming AMPAR-containing vesicles and the synaptic scaffold. In agreement with this scenario, it has recently been shown that the exocyst acts within the dendritic spine to mediate the insertion of AMPARs into the postsynaptic membrane (34). In particular, interference with the Exo70 subunit of the exocyst leads to the accumulation of AMPARs within the postsynaptic density, before fusion with the synaptic membrane (34). This observation suggests that AMPAR membrane insertion occurs directly at the level of the postsynaptic density (see model in Fig. 1).

8 Activity-Dependent Internalization and Sorting of AMPA Receptors

Synaptic AMPARs are internalized in an activity-dependent manner, leading to long-term synaptic depression (LTD). This process does require clathrin-mediated endocytosis (see review in (13)). Interestingly, and in contrast with constitutive endocytosis, the small GTPase Rab5 drives the regulated internalization of AMPARs during LTD (11) (see model in Fig. 1). In fact, Rab5 is rapidly and transiently activated upon NMDAR activation during LTD induction (11). Therefore, these results suggest that constitutive and regulated AMPAR internalization may engage different components of the endocytic machinery.

In contrast to the subunit-specific rules for AMPAR delivery, the contribution of different receptor populations to activity-dependent removal still remains controversial. Hippocampal neurons lacking both GluR2 and GluR3 subunits display normal LTD, suggesting that GluR1 removal contributes to synaptic depression (81). On the other hand, GluR2 subunits are removed during LTD in hippocampal neurons (103), and cerebellar LTD requires PKC phosphorylation of GluR2 (16). Therefore, both GluR1- and GluR2-containing receptors seem to participate in the synaptic trafficking associated with LTD. In fact, most experimental evidence is compatible with an initial indiscriminate internalization of all AMPAR populations upon LTD induction. However, it is increasingly appreciated that AMPARs undergo complicated intracellular sorting and recycling events after synaptic removal, which may involve significant subunit-specificity (67).

The molecular mechanisms that organize post-endocytic sorting of AMPARs and potential reinsertion into synaptic and/or extra-synaptic membranes are still far from clear. Nevertheless, the balance between GRIP/ABP (AMPA receptor Binding

Protein) and PICK1 interactions with GluR2 has been shown to be a critical factor (59, 91). In hippocampal and parallel fiber-Purkinje cell synapses, PICK1 is required for the synaptic removal of phosphorylated GluR2 receptors (16, 59). This role is facilitated by the calcium-dependent interactions between GluR2 and PICK1 (46). Subsequently, a fraction of these internalized GluR2 subunits recycles back into synaptic sites, in a process likely mediated by direct GRIP/ABP-PICK1 interactions (75) and NSF-mediated dissociation of the GluR2-PICK1 complex (47). The connection between these AMPAR binding proteins and the intracellular membrane trafficking machinery are still being elucidated, but it has been recently proposed that the return of AMPARs to synaptic sites may be mediated by phosphorylation-regulated interactions between GRIP/ABP and the endosomal protein NEEP21 (Neuron-Enriched Endosomal Protein of 21 KDa) (65, 109) (see model in Fig. 1).

9 Conclusions

The field of AMPAR trafficking is advancing at a fast pace. New proteins interacting with AMPARs or with the AMPAR trafficking machinery are constantly being identified. These new investigations are uncovering an intricate choreography, in which AMPARs are assembled, sorted and targeted throughout the neuronal secretory pathway. We are starting to identify the core cellular machinery that transports AMPARs, as well as the regulatory molecules that orchestrate their dynamic behavior close to the synapse, where bidirectional AMPAR movement results in long-lasting changes in synaptic strength.

Much remains to be known about the regulation of the membrane trafficking machinery and how it is linked to the signal transduction pathways triggered during synaptic plasticity. For example, it has been reported that induction of LTD leads to the activation of the endocytic GTPase Rab5 (11). Conversely, exocytic trafficking associated to recycling endosomes is enhanced upon LTP induction (87, 88). However, in neither case is obvious how activation of NMDARs and calcium entry is mechanistically connected to the trafficking machinery. Nevertheless, it is reasonable to expect that the more pieces of the puzzle we are able to properly place, the better chances we will have to find the direct targets and effectors that execute AMPAR transport in response to neuronal activity.

These are exciting times, when the fields of AMPAR trafficking and synaptic plasticity are started to be integrated within the realm of cellular biology.

References

1. Adesnik H and Nicoll RA. Conservation of glutamate receptor 2-containing AMPA receptors during long-term potentiation. J Neurosci 27: 4598–4602, 2007.
2. Adesnik H, Nicoll RA, and England PM. Photoinactivation of native AMPA receptors reveals their real-time trafficking. Neuron 48: 977–985, 2005.
3. Allison DW, Gelfand VI, Spector I, and Craig AM. Role of actin in anchoring postsynaptic receptors in cultured hippocampal neurons: differential attachment of NMDA versus AMPA receptors. J Neurosci 18: 2423–2436, 1998.

4. Bagal AA, Kao JP, Tang CM, and Thompson SM. Long-term potentiation of exogenous glutamate responses at single dendritic spines. Proc Natl Acad Sci USA 102: 14434–14439, 2005.
5. Bats C, Groc L, and Choquet D. The interaction between Stargazin and PSD-95 regulates AMPA receptor surface trafficking. Neuron 53: 719–734, 2007.
6. Bear MF and Malenka RC. Synaptic plasticity: LTP and LTD. Curr Opin Neurobiol 4: 389–399, 1994.
7. Bedoukian MA, Weeks AM, and Partin KM. Different domains of the AMPA receptor direct stargazin-mediated trafficking and stargazin-mediated modulation of kinetics. J Biol Chem 281: 23908–23921, 2006.
8. Bellone C and Luscher C. Cocaine triggered AMPA receptor redistribution is reversed in vivo by mGluR-dependent long-term depression. Nat Neurosci 9: 636–641, 2006.
9. Bliss TV and Collingridge GL. A synaptic model of memory: long-term potentiation in the hippocampus. Nature 361: 31–39, 1993.
10. Boehm J, Kang MG, Johnson RC, Esteban J, Huganir RL, and Malinow R. Synaptic incorporation of AMPA receptors during LTP is controlled by a PKC phosphorylation site on GluR1. Neuron 51: 213–225, 2006.
11. Brown TC, Tran IC, Backos DS, and Esteban JA. NMDA receptor-dependent activation of the small GTPase Rab5 drives the removal of synaptic AMPA receptors during hippocampal LTD. Neuron 45: 81–94, 2005.
12. Bucci C, Parton RG, Mather IH, Stunnenberg H, Simons K, Hoflack B, and Zerial M. The small GTPase rab5 functions as a regulatory factor in the early endocytic pathway. Cell 70: 715–728., 1992.
13. Carroll RC, Beattie EC, von Zastrow M, and Malenka RC. Role of AMPA receptor endocytosis in synaptic plasticity. Nat Rev Neurosci 2: 315–324, 2001.
14. Chen L, Chetkovich DM, Petralia RS, Sweeney NT, Kawasaki Y, Wenthold RJ, Bredt DS, and Nicoll RA. Stargazin regulates synaptic targeting of AMPA receptors by two distinct mechanisms. Nature 408: 936–943, 2000.
15. Choquet D and Triller A. The role of receptor diffusion in the organization of the postsynaptic membrane. Nat Rev Neurosci 4: 251–265, 2003.
16. Chung HJ, Steinberg JP, Huganir RL, and Linden DJ. Requirement of AMPA receptor GluR2 phosphorylation for cerebellar long-term depression. Science 300: 1751–1755, 2003.
17. Clem RL and Barth A. Pathway-specific trafficking of native AMPARs by in vivo experience. Neuron 49: 663–670, 2006.
18. Coleman SK, Cai C, Mottershead DG, Haapalahti JP, and Keinanen K. Surface expression of GluR-D AMPA receptor is dependent on an interaction between its C-terminal domain and a 4.1 protein. J Neurosci 23: 798–806, 2003.
19. Collingridge GL, Isaac JT, and Wang YT. Receptor trafficking and synaptic plasticity. Nat Rev Neurosci 5: 952–962, 2004.
20. Constantine-Paton M. NMDA receptor as a mediator of activity-dependent synaptogenesis in the developing brain. Cold Spring Harb Symp Quant Biol 55: 431–443, 1990.
21. Dev KK, Nishimune A, Henley JM, and Nakanishi S. The protein kinase C alpha binding protein PICK1 interacts with short but not long form alternative splice variants of AMPA receptor subunits. Neuropharmacology 38: 635–644, 1999.
22. Dingledine R, Borges K, Bowie D, and Traynelis SF. The glutamate receptor ion channels. Pharmacol Rev 51: 7–61, 1999.
23. Dong H, O'Brien RJ, Fung ET, Lanahan AA, Worley PF, and Huganir RL. GRIP: a synaptic PDZ domain-containing protein that interacts with AMPA receptors. Nature 386: 279–284, 1997.

24. Durand GM, Kovalchuk Y, and Konnerth A. Long-term potentiation and functional synapse induction in developing hippocampus. Nature 381: 71–75, 1996.
25. Ehlers MD. Reinsertion or degradation of AMPA receptors determined by activity-dependent endocytic sorting. Neuron 28: 511–525, 2000.
26. El-Husseini AE, Schnell E, Chetkovich DM, Nicoll RA, and Bredt DS. PSD-95 involvement in maturation of excitatory synapses. Science 290: 1364–1368, 2000.
27. Elias GM, Funke L, Stein V, Grant SG, Bredt DS, and Nicoll RA. Synapse-specific and developmentally regulated targeting of AMPA receptors by a family of MAGUK scaffolding proteins. Neuron 52: 307–320, 2006.
28. Esteban JA, Shi SH, Wilson C, Nuriya M, Huganir RL, and Malinow R. PKA phosphorylation of AMPA receptor subunits controls synaptic trafficking underlying plasticity. Nat Neurosci 6: 136–143, 2003.
29. Fischer M, Kaech S, Knutti D, and Matus A. Rapid actin-based plasticity in dendritic spines. Neuron 20: 847–854, 1998.
30. Fleck MW. Glutamate receptors and endoplasmic reticulum quality control: looking beneath the surface. Neuroscientist 12: 232–244, 2006.
31. Fukata Y, Tzingounis AV, Trinidad JC, Fukata M, Burlingame AL, Nicoll RA, and Bredt DS. Molecular constituents of neuronal AMPA receptors. J Cell Biol 169: 399–404, 2005.
32. Galdzicki Z and Siarey RJ. Understanding mental retardation in Down's syndrome using trisomy 16 mouse models. Genes Brain Behav 2: 167–178, 2003.
33. Gerges NZ, Backos DS, and Esteban JA. Local control of AMPA receptor trafficking at the postsynaptic terminal by a small GTPase of the Rab family. J Biol Chem 279: 43870–43878, 2004.
34. Gerges NZ, Backos DS, Rupasinghe CN, Spaller MR, and Esteban JA. Dual role of the exocyst in AMPA receptor targeting and insertion into the postsynaptic membrane. Embo J 25: 1623–1634, 2006.
35. Gerges NZ, Tran IC, Backos DS, Harrell JM, Chinkers M, Pratt WB, and Esteban JA. Independent functions of hsp90 in neurotransmitter release and in the continuous synaptic cycling of AMPA receptors. J Neurosci 24: 4758–4766, 2004.
36. Goel A, Jiang B, Xu LW, Song L, Kirkwood A, and Lee HK. Cross-modal regulation of synaptic AMPA receptors in primary sensory cortices by visual experience. Nat Neurosci 9: 1001–1003, 2006.
37. Goldstein LS and Yang Z. Microtubule-based transport systems in neurons: the roles of kinesins and dyneins. Annu Rev Neurosci 23: 39–71, 2000.
38. Gomes AR, Correia SS, Esteban JA, Duarte CB, and Carvalho AL. PKC anchoring to GluR4 AMPA receptor subunit modulates PKC-driven receptor phosphorylation and surface expression. Traffic 8: 259–269, 2007.
39. Gomes AR, Cunha P, Nuriya M, Faro CJ, Huganir RL, Pires EV, Carvalho AL, and Duarte CB. Metabotropic glutamate and dopamine receptors co-regulate AMPA receptor activity through PKA in cultured chick retinal neurones: effect on GluR4 phosphorylation and surface expression. J Neurochem 90: 673–682, 2004.
40. Greger IH and Esteban JA. AMPA receptor biogenesis and trafficking. Curr Opin Neurobiol 17: 289–297, 2007.
41. Greger IH, Khatri L, Kong X, and Ziff EB. AMPA receptor tetramerization is mediated by Q/R editing. Neuron 40: 763–774, 2003.
42. Greger IH, Khatri L, and Ziff EB. RNA editing at arg607 controls AMPA receptor exit from the endoplasmic reticulum. Neuron 34: 759–772, 2002.
43. Greger IH, Ziff EB, and Penn AC. Molecular determinants of AMPA receptor subunit assembly. Trends Neurosci 30: 407–416, 2007.

44. Grooms SY, Noh KM, Regis R, Bassell GJ, Bryan MK, Carroll RC, and Zukin RS. Activity bidirectionally regulates AMPA receptor mRNA abundance in dendrites of hippocampal neurons. J Neurosci 26: 8339–8351, 2006.
45. Guo W, Roth D, Walch-Solimena C, and Novick P. The exocyst is an effector for Sec4p, targeting secretory vesicles to sites of exocytosis. Embo J 18: 1071–1080, 1999.
46. Hanley JG and Henley JM. PICK1 is a calcium-sensor for NMDA-induced AMPA receptor trafficking. Embo J 24: 3266–3278, 2005.
47. Hanley JG, Khatri L, Hanson PI, and Ziff EB. NSF ATPase and alpha-/beta-SNAPs disassemble the AMPA receptor-PICK1 complex. Neuron 34: 53–67, 2002.
48. Hattula K, Furuhjelm J, Tikkanen J, Tanhuanpaa K, Laakkonen P, and Peranen J. Characterization of the Rab8-specific membrane traffic route linked to protrusion formation. J Cell Sci 119: 4866–4877, 2006.
49. Hayashi Y, Shi SH, Esteban JA, Piccini A, Poncer JC, and Malinow R. Driving AMPA receptors into synapses by LTP and CaMKII: requirement for GluR1 and PDZ domain interaction. Science 287: 2262–2267, 2000.
50. Hering H and Sheng M. Dendritic spines: structure, dynamics and regulation. Nat Rev Neurosci 2: 880–888, 2001.
51. Hollmann M and Heinemann S. Cloned glutamate receptors. Annu Rev Neurosci 17: 31–108, 1994.
52. Hoover KB and Bryant PJ. The genetics of the protein 4.1 family: organizers of the membrane and cytoskeleton. Curr Opin Cell Biol 12: 229–234, 2000.
53. Hsia AY, Malenka RC, and Nicoll RA. Development of excitatory circuitry in the hippocampus. J Neurophysiol 79: 2013–2024, 1998.
54. Huber LA, Pimplikar S, Parton RG, Virta H, Zerial M, and Simons K. Rab8, a small GTPase involved in vesicular traffic between the TGN and the basolateral plasma membrane. J Cell Biol 123: 35–45, 1993.
55. Isaac JT, Crair MC, Nicoll RA, and Malenka RC. Silent synapses during development of thalamocortical inputs. Neuron 18: 269–280, 1997.
56. Ju W, Morishita W, Tsui J, Gaietta G, Deerinck TJ, Adams SR, Garner CC, Tsien RY, Ellisman MH, and Malenka RC. Activity-dependent regulation of dendritic synthesis and trafficking of AMPA receptors. Nat Neurosci 7: 244–253, 2004.
57. Kato AS, Zhou W, Milstein AD, Knierman MD, Siuda ER, Dotzlaf JE, Yu H, Hale JE, Nisenbaum ES, Nicoll RA, and Bredt DS. New transmembrane AMPA receptor regulatory protein isoform, gamma-7, differentially regulates AMPA receptors. J Neurosci 27: 4969–4977, 2007.
58. Kennedy MJ and Ehlers MD. Organelles and trafficking machinery for postsynaptic plasticity. Annu Rev Neurosci 29: 325–362, 2006.
59. Kim CH, Chung HJ, Lee HK, and Huganir RL. Interaction of the AMPA receptor subunit GluR2/3 with PDZ domains regulates hippocampal long-term depression. Proc Natl Acad Sci USA 98: 11725–11730, 2001.
60. Kim CH and Lisman JE. A role of actin filament in synaptic transmission and long-term potentiation. J Neurosci 19: 4314–4324, 1999.
61. Ko J, Kim S, Valtschanoff JG, Shin H, Lee JR, Sheng M, Premont RT, Weinberg RJ, and Kim E. Interaction between liprin-alpha and GIT1 is required for AMPA receptor targeting. J Neurosci 23: 1667–1677, 2003.
62. Kohler M, Kornau HC, and Seeburg PH. The organization of the gene for the functionally dominant alpha-amino-3-hydroxy-5-methylisoxazole-4-propionic acid receptor subunit GluR-B. J Biol Chem 269: 17367–17370, 1994.
63. Kolleker A, Zhu JJ, Schupp BJ, Qin Y, Mack V, Borchardt T, Kohr G, Malinow R, Seeburg PH, and Osten P. Glutamatergic plasticity by synaptic delivery of GluR-B(long)-containing AMPA receptors. Neuron 40: 1199–1212, 2003.

64. Kopec CD, Li B, Wei W, Boehm J, and Malinow R. Glutamate receptor exocytosis and spine enlargement during chemically induced long-term potentiation. J Neurosci 26: 2000–2009, 2006.
65. Kulangara K, Kropf M, Glauser L, Magnin S, Alberi S, Yersin A, and Hirling H. Phosphorylation of glutamate receptor interacting protein 1 regulates surface expression of glutamate receptors. J Biol Chem 282: 2395–2404, 2007.
66. Kuusinen A, Abele R, Madden DR, and Keinanen K. Oligomerization and ligand-binding properties of the ectodomain of the alpha-amino-3-hydroxy-5-methyl-4-isoxazole propionic acid receptor subunit GluRD. J Biol Chem 274: 28937–28943, 1999.
67. Lee SH, Simonetta A, and Sheng M. Subunit rules governing the sorting of internalized AMPA receptors in hippocampal neurons. Neuron 43: 221–236, 2004.
68. Letts VA, Felix R, Biddlecome GH, Arikkath J, Mahaffey CL, Valenzuela A, Bartlett FS, 2nd, Mori Y, Campbell KP, and Frankel WN. The mouse stargazer gene encodes a neuronal Ca2+-channel gamma subunit. Nat Genet 19: 340–347, 1998.
69. Leuschner WD and Hoch W. Subtype-specific assembly of alpha-amino-3-hydroxy-5-methyl-4-isoxazole propionic acid receptor subunits is mediated by their n-terminal domains. J Biol Chem 274: 16907–16916, 1999.
70. Ling DS, Benardo LS, and Sacktor TC. Protein kinase Mzeta enhances excitatory synaptic transmission by increasing the number of active postsynaptic AMPA receptors. Hippocampus 16: 443–452, 2006.
71. Lise MF, Wong TP, Trinh A, Hines RM, Liu L, Kang R, Hines DJ, Lu J, Goldenring JR, Wang YT, and El-Husseini A. Involvement of myosin Vb in glutamate receptor trafficking. J Biol Chem 281: 3669–3678, 2006.
72. Lisman JE and Zhabotinsky AM. A model of synaptic memory: a CaMKII/PP1 switch that potentiates transmission by organizing an AMPA receptor anchoring assembly. Neuron 31: 191–201, 2001.
73. Losi G, Prybylowski K, Fu Z, Luo JH, and Vicini S. Silent synapses in developing cerebellar granule neurons. J Neurophysiol 87: 1263–1270, 2002.
74. Lu W, Man H, Ju W, Trimble WS, MacDonald JF, and Wang YT. Activation of synaptic NMDA receptors induces membrane insertion of new AMPA receptors and LTP in cultured hippocampal neurons. Neuron 29: 243–254, 2001.
75. Lu W and Ziff EB. PICK1 interacts with ABP/GRIP to regulate AMPA receptor trafficking. Neuron 47: 407–421, 2005.
76. Luscher C, Xia H, Beattie EC, Carroll RC, von Zastrow M, Malenka RC, and Nicoll RA. Role of AMPA receptor cycling in synaptic transmission and plasticity. Neuron 24: 649–658, 1999.
77. Malinow R, Mainen ZF, and Hayashi Y. LTP mechanisms: from silence to four-lane traffic. Curr Opin Neurobiol 10: 352–357, 2000.
78. Malinow R and Malenka RC. AMPA receptor trafficking and synaptic plasticity. Annu Rev Neurosci 25: 103–126, 2002.
79. Man HY, Wang Q, Lu WY, Ju W, Ahmadian G, Liu L, D'Souza S, Wong TP, Taghibiglou C, Lu J, Becker LE, Pei L, Liu F, Wymann MP, MacDonald JF, and Wang YT. Activation of PI3-kinase is required for AMPA receptor insertion during LTP of mEPSCs in cultured hippocampal neurons. Neuron 38: 611–624, 2003.
80. McGee AW and Bredt DS. Assembly and plasticity of the glutamatergic postsynaptic specialization. Curr Opin Neurobiol 13: 111–118, 2003.
81. Meng Y, Zhang Y, and Jia Z. Synaptic transmission and plasticity in the absence of AMPA glutamate receptor GluR2 and GluR3. Neuron 39: 163–176, 2003.
82. Newey SE, Velamoor V, Govek EE, and Van Aelst L. Rho GTPases, dendritic structure, and mental retardation. J Neurobiol 64: 58–74, 2005.

83. Nishimune A, Isaac JT, Molnar E, Noel J, Nash SR, Tagaya M, Collingridge GL, Nakanishi S, and Henley JM. NSF binding to GluR2 regulates synaptic transmission. Neuron 21: 87–97, 1998.
84. Novick P, Medkova M, Dong G, Hutagalung A, Reinisch K, and Grosshans B. Interactions between Rabs, tethers, SNAREs and their regulators in exocytosis. Biochem Soc Trans 34: 683–686, 2006.
85. Nowak L, Bregestovski P, Ascher P, Herbet A, and Prochiantz A. Magnesium gates glutamate-activated channels in mouse central neurones. Nature 307: 462–465, 1984.
86. Osterweil E, Wells DG, and Mooseker MS. A role for myosin VI in postsynaptic structure and glutamate receptor endocytosis. J Cell Biol 168: 329–338, 2005.
87. Park M, Penick EC, Edwards JG, Kauer JA, and Ehlers MD. Recycling endosomes supply AMPA receptors for LTP. Science 305: 1972–1975, 2004.
88. Park M, Salgado JM, Ostroff L, Helton TD, Robinson CG, Harris KM, and Ehlers MD. Plasticity-induced growth of dendritic spines by exocytic trafficking from recycling endosomes. Neuron 52: 817–830, 2006.
89. Passafaro M, Piech V, and Sheng M. Subunit-specific temporal and spatial patterns of AMPA receptor exocytosis in hippocampal neurons. Nat Neurosci 4: 917–926, 2001.
90. Pastalkova E, Serrano P, Pinkhasova D, Wallace E, Fenton AA, and Sacktor TC. Storage of spatial information by the maintenance mechanism of LTP. Science 313: 1141–1144, 2006.
91. Perez JL, Khatri L, Chang C, Srivastava S, Osten P, and Ziff EB. PICK1 targets activated protein kinase Calpha to AMPA receptor clusters in spines of hippocampal neurons and reduces surface levels of the AMPA-type glutamate receptor subunit 2. J Neurosci 21: 5417–5428, 2001.
92. Petralia RS, Esteban JA, Wang YX, Partridge JG, Zhao HM, Wenthold RJ, and Malinow R. Selective acquisition of AMPA receptors over postnatal development suggests a molecular basis for silent synapses. Nat Neurosci 2: 31–36, 1999.
93. Pickard L, Noel J, Duckworth JK, Fitzjohn SM, Henley JM, Collingridge GL, and Molnar E. Transient synaptic activation of NMDA receptors leads to the insertion of native AMPA receptors at hippocampal neuronal plasma membranes. Neuropharmacology 41: 700–713, 2001.
94. Plant K, Pelkey KA, Bortolotto ZA, Morita D, Terashima A, McBain CJ, Collingridge GL, and Isaac JT. Transient incorporation of native GluR2-lacking AMPA receptors during hippocampal long-term potentiation. Nat Neurosci 9: 602–604, 2006.
95. Riefler GM, Balasingam G, Lucas KG, Wang S, Hsu SC, and Firestein BL. Exocyst complex subunit sec8 binds to postsynaptic density protein-95 (PSD-95): a novel interaction regulated by cypin (cytosolic PSD-95 interactor). Biochem J 373: 49–55, 2003.
96. Rosenmund C, Stern-Bach Y, and Stevens CF. The tetrameric structure of a glutamate receptor channel. Science 280: 1596–1599, 1998.
97. Rouach N, Byrd K, Petralia RS, Elias GM, Adesnik H, Tomita S, Karimzadegan S, Kealey C, Bredt DS, and Nicoll RA. TARP gamma-8 controls hippocampal AMPA receptor number, distribution and synaptic plasticity. Nat Neurosci 8: 1525–1533, 2005.
98. Rowan MJ, Klyubin I, Cullen WK, and Anwyl R. Synaptic plasticity in animal models of early Alzheimer's disease. Philos Trans R Soc Lond B Biol Sci 358: 821–828, 2003.
99. Rumpel S, LeDoux J, Zador A, and Malinow R. Postsynaptic receptor trafficking underlying a form of associative learning. Science 308: 83–88, 2005.
100. Sans N, Prybylowski K, Petralia RS, Chang K, Wang YX, Racca C, Vicini S, and Wenthold RJ. NMDA receptor trafficking through an interaction between PDZ proteins and the exocyst complex. Nat Cell Biol 5: 520–530, 2003.

101. Schnell E, Sizemore M, Karimzadegan S, Chen L, Bredt DS, and Nicoll RA. Direct interactions between PSD-95 and stargazin control synaptic AMPA receptor number. Proc Natl Acad Sci USA 99: 13902–13907, 2002.
102. Schulz TW, Nakagawa T, Licznerski P, Pawlak V, Kolleker A, Rozov A, Kim J, Dittgen T, Kohr G, Sheng M, Seeburg PH, and Osten P. Actin/alpha-actinin-dependent transport of AMPA receptors in dendritic spines: role of the PDZ-LIM protein RIL. J Neurosci 24: 8584–8594, 2004.
103. Seidenman KJ, Steinberg JP, Huganir R, and Malinow R. Glutamate receptor subunit 2 Serine 880 phosphorylation modulates synaptic transmission and mediates plasticity in CA1 pyramidal cells. J Neurosci 23: 9220–9228, 2003.
104. Setou M, Seog DH, Tanaka Y, Kanai Y, Takei Y, Kawagishi M, and Hirokawa N. Glutamate-receptor-interacting protein GRIP1 directly steers kinesin to dendrites. Nature 417: 83–87, 2002.
105. Shen L, Liang F, Walensky LD, and Huganir RL. Regulation of AMPA receptor GluR1 subunit surface expression by a 4. 1N-linked actin cytoskeletal association. J Neurosci 20: 7932–7940, 2000.
106. Shi S, Hayashi Y, Esteban JA, and Malinow R. Subunit-specific rules governing AMPA receptor trafficking to synapses in hippocampal pyramidal neurons. Cell 105: 331–343, 2001.
107. Shin H, Wyszynski M, Huh KH, Valtschanoff JG, Lee JR, Ko J, Streuli M, Weinberg RJ, Sheng M, and Kim E. Association of the Kinesin Motor KIF1A with the Multimodular Protein Liprin-alpha. J Biol Chem 278: 11393–11401, 2003.
108. Srivastava S, Osten P, Vilim FS, Khatri L, Inman G, States B, Daly C, DeSouza S, Abagyan R, Valtschanoff JG, Weinberg RJ, and Ziff EB. Novel anchorage of GluR2/3 to the postsynaptic density by the AMPA receptor-binding protein ABP. Neuron 21: 581–591, 1998.
109. Steiner P, Alberi S, Kulangara K, Yersin A, Sarria JC, Regulier E, Kasas S, Dietler G, Muller D, Catsicas S, and Hirling H. Interactions between NEEP21, GRIP1 and GluR2 regulate sorting and recycling of the glutamate receptor subunit GluR2. Embo J 24: 2873–2884, 2005.
110. Stephan KE, Baldeweg T, and Friston KJ. Synaptic plasticity and dysconnection in schizophrenia. Biol Psychiatry 59: 929–939, 2006.
111. Takahashi T, Svoboda K, and Malinow R. Experience strengthening transmission by driving AMPA receptors into synapses. Science 299: 1585–1588, 2003.
112. Tichelaar W, Safferling M, Keinanen K, Stark H, and Madden DR. The Three-dimensional Structure of an Ionotropic Glutamate Receptor Reveals a Dimer-of-dimers Assembly. J Mol Biol 344: 435–442, 2004.
113. Tomita S, Adesnik H, Sekiguchi M, Zhang W, Wada K, Howe JR, Nicoll RA, and Bredt DS. Stargazin modulates AMPA receptor gating and trafficking by distinct domains. Nature 435: 1052–1058, 2005.
114. Tomita S, Chen L, Kawasaki Y, Petralia RS, Wenthold RJ, Nicoll RA, and Bredt DS. Functional studies and distribution define a family of transmembrane AMPA receptor regulatory proteins. J Cell Biol 161: 805–816, 2003.
115. Triller A and Choquet D. Surface trafficking of receptors between synaptic and extrasynaptic membranes: and yet they do move! Trends Neurosci 28: 133–139, 2005.
116. Turetsky D, Garringer E, and Patneau DK. Stargazin modulates native AMPA receptor functional properties by two distinct mechanisms. J Neurosci 25: 7438–7448, 2005.
117. Vandenberghe W, Nicoll RA, and Bredt DS. Interaction with the unfolded protein response reveals a role for stargazin in biosynthetic AMPA receptor transport. J Neurosci 25: 1095–1102, 2005.

118. Wenthold RJ, Petralia RS, Blahos J, II, and Niedzielski AS. Evidence for multiple AMPA receptor complexes in hippocampal CA1/CA2 neurons. J Neurosci 16: 1982–1989, 1996.
119. Whitlock JR, Heynen AJ, Shuler MG, and Bear MF. Learning induces long-term potentiation in the hippocampus. Science 313: 1093–1097, 2006.
120. Wu H, Nash JE, Zamorano P, and Garner CC. Interaction of SAP97 with minus-end-directed actin motor myosin VI. Implications for AMPA receptor trafficking. J Biol Chem 277: 30928–30934, 2002.
121. Wyszynski M, Kim E, Dunah AW, Passafaro M, Valtschanoff JG, Serra-Pages C, Streuli M, Weinberg RJ, and Sheng M. Interaction between GRIP and liprin-alpha/SYD2 is required for AMPA receptor targeting. Neuron 34: 39–52, 2002.
122. Xia J, Zhang X, Staudinger J, and Huganir RL. Clustering of AMPA receptors by the synaptic PDZ domain-containing protein PICK1. Neuron 22: 179–187, 1999.
123. Yasuda H, Barth AL, Stellwagen D, and Malenka RC. A developmental switch in the signaling cascades for LTP induction. Nat Neurosci 6: 15–16, 2003.
124. Zerial M and McBride H. Rab proteins as membrane organizers. Nat Rev Mol Cell Biol 2: 107–117., 2001.
125. Zhu JJ, Esteban JA, Hayashi Y, and Malinow R. Postnatal synaptic potentiation: delivery of GluR4-containing AMPA receptors by spontaneous activity. Nat Neurosci 3: 1098–1106, 2000.
126. Ziv NE and Garner CC. Cellular and molecular mechanisms of presynaptic assembly. Nat Rev Neurosci 5: 385–399, 2004.

Structure and Function of the NMDA Receptor

Hongjie Yuan[1], Matthew T. Geballe[2], Kasper B. Hansen[1], and Stephen F. Traynelis[1]

[1] Department of Pharmacology, Emory University School of Medicine, Atlanta, GA 30322, USA, strayne@emory.edu
[2] Department of Chemistry, Emory University, Atlanta, GA 30322, USA, strayne@emory.edu

1 Introduction

Ionotropic glutamate receptors are ligand-gated ion channels that mediate excitatory synaptic transmission in the central nervous system. Based on their structural and pharmacological properties, ionotropic glutamate receptors can be divided into three groups, which include α-amino-3-hydroxy-5-methyl-4-isoxasolepropionic acid (AMPA) receptors, kainate receptors, and N-methyl-D-aspartate (NMDA) receptors. Historically, these groups were named on the basis of the activating agonist. AMPA and kainate receptors are discussed in the chapter by Mayer. NMDA receptors are distinguished from other ionotropic glutamate receptors by their unique features including requirement for simultaneous binding of the co-agonists glycine and glutamate, voltage-dependent Mg^{2+} block, and high permeability to Ca^{2+}. NMDA receptors contribute to the slow component of the excitatory postsynaptic current (EPSC) and play key roles in neuronal development, synaptic plasticity, learning, and memory, as well as in a number of pathophysiological conditions including epilepsy, stroke, neurodegenerative diseases (e.g. Parkinson's disease, Huntington's disease and Alzheimer's disease), and psychiatric disorders (e.g. schizophrenia). Therefore, understanding the relationship between structure and function of the NMDA receptor will provide valuable insights into the mechanisms of synaptic transmission, as well as pathophysiology of a number of disorders in the central nervous system. Understanding this relationship will also facilitate the development of therapeutically useful compounds.

1.1 Subunit Diversity of NMDA Receptors

There are seven NMDA receptor subunits that have been identified including the NR1 subunit, four different NR2 subunits (NR2A-D), and two NR3 subunits (NR3A-B). Functional NMDA receptors are tetrameric assemblies composed of

J.W. Hell, M.D. Ehlers (eds.), *Structural and Functional Organization of the Synapse*,
DOI: 10.1007/978-0-387-77232-5_11, © Springer Science+Business Media, LLC 2008

two glycine-binding NR1 subunits combined with two glutamate-binding NR2 subunits (31, 39). NR3 subunits can assemble with NR1-NR2 to decrease NMDA receptor current amplitudes, or with NR1 subunits alone to form glycine-activated receptors (16, 30, 102).

NR1 subunits, which are distributed ubiquitously in the brain, have eight different functional splice variants that arise from alternative splicing of three exons: one in the N-terminus (exon 5, named N1 cassette) and two in the C-terminus (exon 21, named C1 cassette; exon 22, named C2 and C2' cassettes). Different NR1 splice variants display distinct regional and developmental expression patterns and NR1 splicing controls several different aspects of NMDA receptor function, such as agonist affinity, receptor trafficking, and allosteric regulation. Receptors lacking exon 5 (i.e. lacking the N1-cassette) show increased agonist potency (i.e. decreased EC_{50}; 35) with enhanced sensitivity to hydrogen ions, zinc, and polyamines (35, 36, 52, 137, 140). Consistent with the effects of exon 5 on agonist potency, the presence of N1 cassette accelerates deactivation of L-glutamate responses (119).

The four different NR2 subunits (NR2A-D) share 55% overall amino acid sequence identity and show both developmental and regional differences in expression. The four NR2 subunits play different roles during development of the central nervous system and endow NMDA receptors with different biophysical and pharmacological properties, as well as different sensitivity to Mg^{2+} and extracellular modulators. NMDA receptors that contain different NR2 subunits show different agonist potency (39, 42). For example, the EC_{50} for glutamate activating heterodimeric NR1/NR2D receptors (0.5 μM) is lower (i.e. glutamate is more potent) than that for NR1/NR2A receptors (3–5 μM), with NR1/NR2B and NR1/NR2C subunits showing intermediate EC50 values. NR1/NR2A and NR1/NR2B have higher single channel conductance levels (50, 40 pS) than NR1/NR2C and NR1/NR2D receptors (36, 22 pS; and 36, 16 pS) (Fig. 1a,b) (9, 44, 131). In addition, heterodimeric NR1/NR2 NMDA receptors have different peak open probability, which in excised membrane patches is ~0.5 for recombinant NR1/NR2A, ~0.1 for NR1/NR2B (40), and <0.05 for NR1/NR2C and NR1/NR2D (Fig. 1) (34, 155). The time constants of deactivation (tau decay) also depends on the identity of NR2 subunits; tau decay for NR1/NR2A is ~50 ms, for NR1/NR2B ~ 400 ms, for NR1/NR2C ~290 ms, and for NR1/NR2D > 1 s (Fig. 1c) (97, 141, 155).

In addition to heterodimeric complexes, NMDA receptors are also thought to assemble into tri-heteromeric receptors containing more than one type of NR2 subunit, or both NR2 and NR3 subunits. The existence of tri-heteromeric assemblies in native neurons has been reported (13, 63, 111) although it is difficult to distinguish between the heterotrimeric and mixed populations of heterodimeric NMDA receptor assemblies in the brain. Despite their potential importance, little is known about how combinations of subunits with dissimilar properties impact receptor function. Recent single-channel recording studies suggest that possible heterotrimeric NMDA receptors that consist of two different types of NR2 subunits show distinct and intermediate conductance (NR1/NR2A/NR2D; (18)). Heterotrimeric receptors have been suggested to retain responses to subunit specific allosteric modulators, but with somewhat modified properties (51).

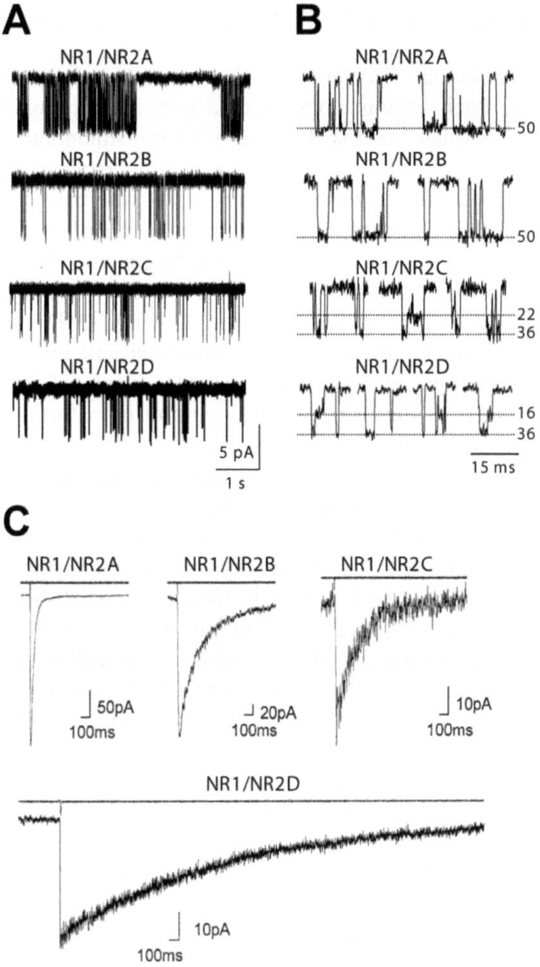

Fig. 1. Functional properties of recombinant NMDA receptor subtypes. **(a)** Single-channel recordings of currents at NMDA receptor subtypes expressed in HEK293 cells (outside-out membrane patches). Open probability is ~0.5 for NR1/NR2A, ~0.1 for NR1/NR2B, and <0.05 for NR1/NR2C and NR1/NR2D. **(b)** Highlights of the individual openings of the NMDA receptor subtypes. NR1/NR2A and NR1/NR2B have higher channel conductance (~50 pS) compared to NR1/NR2C (~22 and ~36 pS) and NR1/NR2D (~16 and ~36 pS). Experiments were performed using 0.5 mM Ca^{2+} in the extracellular buffer, which eliminates the low conductance open state of NR1/NR2A and NR1/NR2B. Dotted lines indicate open states. Numbers are conductance in picosiemens (pS). **(c)** Whole-cell patch-clamp recordings of macroscopic responses from brief application of glutamate (1 ms of 1 mM glutamate) to HEK293 cells transfected with different NMDA receptor subtypes. Upper trace depicts the open tip current used to measure the duration of the drug application. Notice the pronounced difference in deactivation rate between NR1/NR2A and NR1/NR2D. Adapted from (141); reprinted with permission from American Physiological Society, copyright 1998.

1.2 Molecular Architecture of NMDA Receptor Subtypes

Similar to AMPA and kainate receptors, NMDA receptor subunits share a common membrane topology. Each subunit consists of a large extracellular amino terminal domain, the bi-lobed agonist binding domain, three transmembrane domains (membrane associated regions M1, M3, and M4), one cytoplasmic re-entrant pore loop (referred to as the P-region for pore forming region, or alternatively M2 for the second membrane-associated domain), and an intracellular carboxyl terminal domain (Fig. 2). The ion channel pore region of NMDA receptors, which bears strong resemblance to an inverted potassium channel architecture, is composed of the re-entrant pore forming loop from all four subunits. Among NMDA receptor subunits, the sequences of the pore region are highly conserved and this region determines basic permeation properties. One of the key determinants of ion permeation for NMDA receptors resides at the apex of the re-entrant pore forming loop, which is sometimes referred to as the Q/R/N site on the basis of amino acid residues found at this position in AMPA, kainate, and NMDA receptors. This site partially defines the Ca^{2+} permeability as well as the sensitivity to Mg^{2+}.

Two modular semi-autonomous extracellular domains are thought to be encoded by each subunit. The amino terminal domain is composed of approximately the first 350 amino acids of the protein and appears to form a clamshell-like structure, which shows sequence homology with bacterial leucine/isoleucine/valine binding domain as well as the glutamate binding site of the G-protein coupled receptor mGluR1. The amino terminal domain interacts with various extracellular allosteric modulators such as zinc and polyamines, plays an important role in subunit assembly, and influences deactivation (see below). This domain can also bind allosteric regulators such as ifenprodil. Although no crystallographic structural information exists describing the amino terminal domain, important insights have been derived from structural models (Fig. 2) (88, 109).

The glutamate and glycine binding domains are formed by two lobes referred to as D1 and D2, which are composed jointly of S1 (the polypeptide chain between the amino terminal domain and the first transmembrane domain) and S2 (the extracellular polypeptide chain between the second and third transmembrane domains) (see Fig. 2). The NR1 subunit binds glycine, whereas the NR2 subunit binds glutamate. These regions fold into a hinged bi-lobed, clamshell-like structure that contains the agonist binding site. This domain has been crystallized ((46, 47, 56) see below), and shows sequence homology with bacterial periplasmic amino acid binding proteins such as the glutamine binding protein (QBP) and lysine/arginine/ornithnine-binding protein (LAOBP). The agonist binding site likely exists as a pair of NR1/NR2 heterodimers that form an extracellular portion of the tetramer (47).

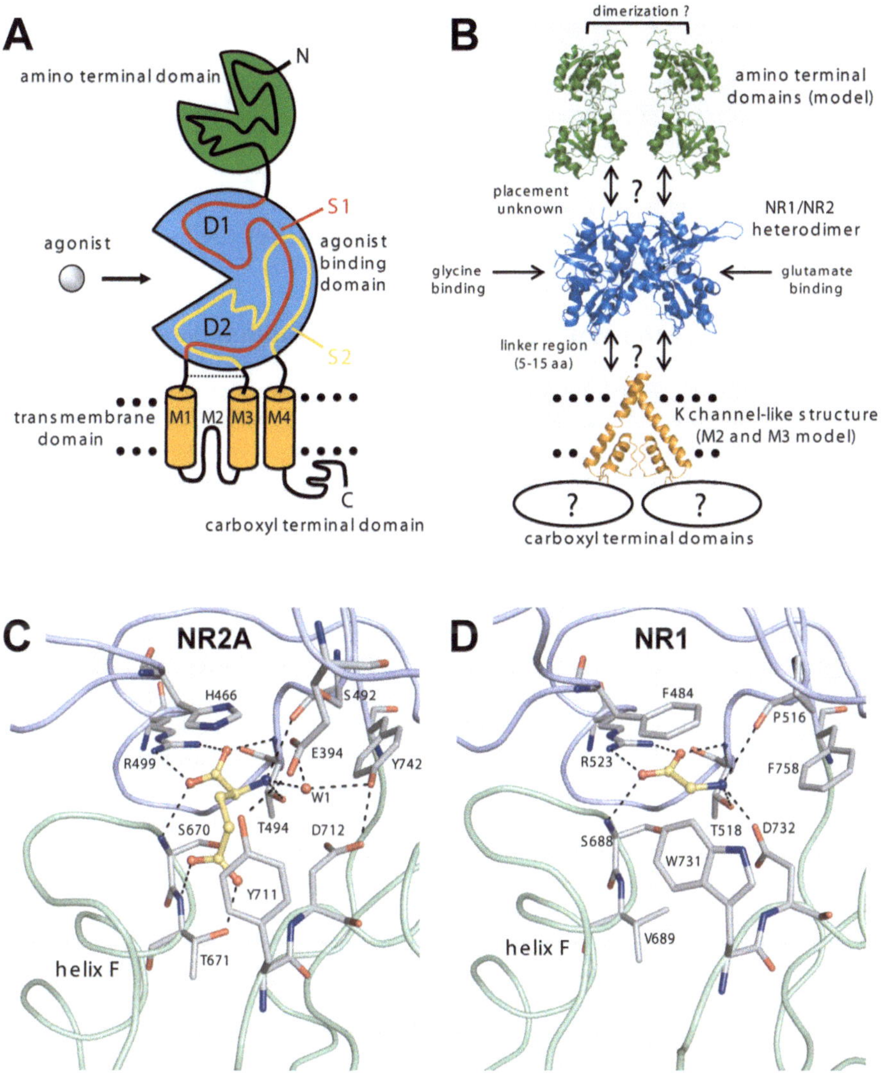

Fig. 2. Structural arrangement of NMDA receptor subunits. (**a**) The iGluRs are tetramers assembled from subunits with a highly modular structure comprised of four discrete domains: two large extracellular domains referred to as the amino terminal domain and the agonist binding domain, a transmembrane domain that forms part of the ion channel pore, and an intracellular carboxyl terminal domain. The agonist binding domain is defined by two segments of amino acids termed S1 and S2. The transmembrane domain contains four membrane-associated regions: three membrane-spanning helices (M1, M3, and M4, sometimes referred to as TM1, TM2, TM3) and a membrane re-entrant loop (M2). Soluble recombinant proteins containing the isolated S1 and S2 segments have been constructed by deleting the amino terminal domain along with the transmembrane domain and joining S1 and S2 with at hydrophilic linker (*dashed line*). In these structures, the agonist binding domain exists in a bi-lobed clamshell-like arrangement with the agonist binding pocket located deep within the cleft

2 Structure and Function of Agonist Binding Domain

While much has been learned about the structure and function of AMPA receptors in the past 10 years, crystal structures of the glutamate binding domain of NMDA receptors have only recently been solved. The release of these structures has provided a great deal of information about the agonist binding domains of NMDA receptors, and provided a means by which to interpret a large amount of mutagenesis work. A number of excellent reviews have summarized the structure-function relationship derived from this crystallographic data (22, 50, 67, 91, 94).

2.1 Structural Features of the NR2 Glutamate Binding Site

When the structure of the NR2A agonist binding domain in complex with the NR1 agonist binding domain was described (47), it provided the first experimental structure of the glutamate binding site of the NMDA receptor as well as the first structural information about the heterodimer interface within the NMDA receptor multimeric complex. The existence of this interface between NR1 and NR2 provides a possible structural explanation for allosteric interactions observed between glutamate and glycine. This allosteric interaction is shown by NR2 control of glycine potency at NR1 (98) as well as glycine-dependent desensitization (see below). The agonist glutamate binds in the cleft between the two clearly defined lobes (D1 and D2) of the agonist binding domain, near the hinge region (Fig. 2). The lower half of the binding pocket is formed largely by the ends of helix F and helix I. Residues from loops within the upper lobe (D1) form most of the upper half of the binding pocket, although an arginine (Arg499 in NR2A) is located on a more distant helix with the length of its side chain allowing it to reach the ligand. The pocket is largely lined with polar amino acid side chains, although some aromatic character is provided by a histidine (His466 in NR2A) and two tyrosines (Tyr711 and Tyr742 in NR2A) (Fig. 2). Virtually all of the contact residues in NR2A are analogous to contact residues in

Fig. 2. (Continued) between the two lobes. The S1 and S2 segments contribute to both the two lobes, which are referred to as D1 and D2. Adapted from (50); reprinted with permission from Elsevier Limited, copyright 2007. **(b)** The crystal structure of the agonist binding domains as a NR1/NR2 heterodimer (blue; PDB code 2A5T) with models of the amino terminal domains (green) and the transmembrane domain (M2 and M3; orange). The positioning of amino terminal domains and the transmembrane domains relative to the agonist binding domains is unknown. The amino terminal domain is a homology model build from the crystal structure of the glutamate binding site of mGluR1 (PDB code 1EWK). M2 and M3 of the transmembrane domain is a homology model based on KcsA contributed by Johnson and Beth Siegler-Retchless. **(c)** Binding of glutamate in the ligand binding pocket of the NR2A agonist binding domain crystal structure (PDB code 2A5S). The α-amino group of glutamate forms water-mediated hydrogen bonds to E394 (Glu394) and Y742 (Tyr742). Numbering of NR2A residues is according to the mature protein excluding the signal peptide. (d) Binding of glycine in the ligand binding pocket of the NR1 agonist binding domain crystal structure (PDB code 1PB7). Specificity of NR1 for glycine is likely mediated by the hydrophobic environment created by V689 (Val689) and the steric barrier formed by W731 (Trp731). Numbering of NR1 residues is according to the total protein including the signal peptide.

AMPA and kainate receptor binding pockets. Despite being tucked away between the lobes, the binding site is fairly accessible to water, with 6 water oxygens found within 4 Å of the NR2A-bound glutamate. Two of these waters form hydrogen bonds with bound glutamate, one with the amino group and one with the γ-carboxyl group (47).

In addition to this contact with water, the γ-carboxyl group of the glutamate agonist also makes hydrogen bond contact to the side chain and backbone amide of a threonine (Thr671 in NR2A) (Fig. 2). The amino group of glutamate forms hydrogen bonds with the backbone carbonyl of a serine (Ser492 in NR2A), the side chain of a threonine (Thr494 in NR2A), and the previously mentioned water molecule. In NR2A, the α-carboxyl group of glutamate forms a bifurcated salt bridge with Arg499, as well as hydrogen bonds with the amide backbones of both Ser670 and Thr494.

2.2 Determinants of Glutamate Binding and Function within the NR2 Subunit

Prior to the crystallographic data, many residues were suggested to influence glutamate binding on the basis of mutagenesis studies (summarized in (22, 32)). For example, Anson et al. (3) probed the glutamate binding site of NR2A using site-directed mutagenesis. Three of ten mutations (H466A, G669A, and T671A) were found to increase the EC_{50} of glutamate by over 100-fold, with the effect of the NR2A(T671A) mutation increasing EC_{50} over 1000-fold. His466 and Tyr671 contact glutamate in the crystallographic structure and Gly669 is proximal to glutamate and mutation of Gly669 likely disturbs the structure of the binding pocket. More recently, effects of mutations at agonist contact residues were examined for several different agonists and partial agonists of NR2A (21).

Although only structural information for the NR2A agonist binding domain exists, investigations on other NR2 subtypes have resulted in important data about the function of these. Recent studies have examined the effects of mutants in the agonist binding site of NR2B, and suggested structural roles for several of these mutants on the basis of changes in potency and efficacy of glutamate and several other ligands (31, 50, 78). A molecular dynamics study of NR1/NR2A and a homology model for NR1/NR2D ligand binding domains showed similar contact residues in the NR2A and NR2D glutamate binding pockets, with the exception of Tyr711 (Tyr732 in NR2D), which forms an inter-domain bridge in NR2A and a direct ligand contact in NR2D (42). These similar contacts occurred although simulations showed differences between NR2A and NR2D in the placement of the D2 region, specifically in helix F. Mutagenesis studies support the proposed difference in the NR2D pocket, and also predict that a number of additional inter-domain hydrogen bonds in NR2D enhance the affinity of that subunit for glutamate. Interestingly, 4-methyl glutamate can distinguish between the NR2A and NR2D binding pockets, with the methyl group clashing with Tyr711 in NR2A because of its position as an inter-domain contact. However, the corresponding residue in NR2D (Tyr732) bends out of the way as it makes contact with the agonist, creating space for the 4-methyl group. This leads to a 50-fold increase in potency of (*2S,4R*)-4-methyl glutamate (SYM2081; (42)) for NR2D over NR2A.

2.3 Structural Features of the NR1 Glycine Binding Site

Glycine was first found to potentiate NMDA receptors, and later argued to be a required co-agonist for NMDA receptor activation (60, 70, 81). Mutagenesis studies provided the initial evidence that the NR1 subunit harbors the glycine binding site (77, 145, 150), which was subsequently confirmed when structures of the NR1 agonist binding domain were solved by Furukawa and Gouaux (46). The binding pocket for the co-agonist glycine in NR1 is situated similarly within the agonist binding domain as the glutamate pocket was in NR2, primarily formed by residues at the end of two D2 helices and loops from D1. The pocket is smaller and more hydrophobic than that found in NR2A, with Phe484, Trp731, and Phe758 replacing the three more hydrophilic aromatic residues in NR2A (His466, Tyr711, and Tyr742) (Fig. 2). Furthermore, Thr671 in NR2A is replaced by a valine in NR1 (Val689). The hydrophobic Val689 and the larger hydrophobic residues Phe484, Trp731, and Phe758 likely occlude binding of glutamate to NR1. Glycine binds deeper between the two lobes, and is less exposed to water than glutamate in NR2A. The amino group of glycine forms hydrogen bonds with the carbonyl backbone of Pro516 and the side chains of Thr518 and Asp732. The alpha-carboxyl of glycine forms a similar hydrogen bonding interaction as seen with glutamate, a bifurcated salt bridge with Arg523, and interactions with the backbone amides of Thr518 and Ser688. Structures are also available for the endogenous agonist D-serine, which shows additional interactions between the hydroxyl of D-serine and residues Thr518 and Ser688.

A pair of cysteine residues within the NR1 subunit are conserved across the glutamate receptor family (NR1 Cys744 and Cys798), and have been proposed to form a site that is sensitive to redox potential (1, 133). These two residues are oxidized in the crystal structure, and the resulting disulfide bond imparts some conformational constraints on the receptor. Mutagenesis studies suggest that relief of this constraint in NR1 (but not NR2) by reduction of the Cys744-Cys798 disulfide bond enhances receptor function (133), apparently by increasing channel opening frequency with little effect on single channel conductance or channel open time (135). Interestingly, NR1(C744A) and NR1(C798A) mutations enhance the potency of NMDA acting at the NR2 subunit by an unknown mechanism (133). Several other disulfide bonds exist within the crystallized NR1 and NR2A subunits, although functional effects of their reduction or oxidation have not yet been described (46, 47).

2.4 Determinants of Glycine Binding and Function within the NR1 Subunit

Multiple ligands have each been crystallized in the NR1 agonist binding domain (46, 47, 56), providing a structural basis for understanding the effects of partial agonists, agonists, and antagonists. Similar to previous structures of related glutamate receptors as described in the chapter by Mayer, the antagonist 5,7-dichlorokynurenate stabilizes an open-cleft conformation of the NR1 agonist binding domain, resulting in a loss of 24 degrees of domain closure when compared to the glycine-bound structure. The stabilization of an open-cleft conformation of the agonist binding domain is similar to what has been found for the soluble agonist binding core of AMPA and kainate receptors. However, despite this similarity, the structure of NR1 bound by partial agonists was unlike that observed for AMPA and kainate receptors. While previous structures of AMPA receptor glutamate binding domains showed partial domain closure that correlated with the efficacy of the partial agonist (e.g. (58)), no such relationship was seen in the NR1 agonist binding domain. Multiple structures showed that the partial agonists D-cycloserine, ACPC, and ACBC bound with virtually identical degrees of domain closure of NR1 compared to the glycine complex. By contrast, a more subtle effect is revealed in the incremental shift in the position of helix F and the hinge region between the two lobes of the clamshell structure (56, 71). These and other data (see (71) suggest that the position of helix F may influence efficacy, or perhaps may be involved in translating the effects of agonist binding to gating.

Interestingly, the heterodimer interface between NR1 and NR2A may also impact receptor function. Three separate areas of contact between NR1 and NR2A were identified from the dimer crystal structure (47). Mutagenesis in the two outer regions (sites I and III) was shown to influence dimerization of the soluble agonist binding domains of NR1 and NR2A. Modification of site II, which is analogous to a binding site for small ligands that modulate deactivation in AMPA receptors (59, 96, 107), impacts deactivation with different mutants of Tyr535 in NR1 capable of increasing or decreasing the rate of deactivation. These results suggest domain interface stability can control NMDA receptor function, similar to what has been found for AMPA and kainate receptors as discussed in the chapter by Mayer.

2.5 Agonist Binding to the NR3 Subunit

Several groups simultaneously isolated cDNAs that encoded proteins with similarity to the NMDA receptor subunits NR1 and NR2 (27, 132). The two subunits comprising this new subfamily (NR3A and NR3B) appear to co-assemble with the NR1 subunit alone to produce glycine-activated excitatory currents (16). In addition, the NR3 subunit can also co-assemble with the NR1 and NR2 subunits as heterotrimeric receptors. NMDA receptors containing the NR3 subunit show reduced Ca^{2+} permeability, reduced sensitivity to voltage dependent block by extracellular Mg^{2+}, and reduced single channel conductance (16, 27, 90, 102, 108, 120, 132). A soluble NR3A agonist binding domain has been purified, which shows high affinity for glycine (40 nM) and pharmacology distinct from the NR1 subunit (157). The NR3A subunit also binds D-serine with lower affinity than glycine. Interestingly, glycine binding to the NR1 subunit appears to play an auto-inhibitory role in NR1/NR3A heterodimers, since NR1-selective antagonists can enhance glycine-evoked current (87). This suggests a strikingly different role of the NR1 subunit in NR1/NR3 receptors compared to NR1/NR2 receptors, which require glycine binding to NR1 for activation.

3 Structural Basis of Channel Gating, Ion Permeation, and Channel Block

3.1 Receptor Gating

NMDA receptor activation is a complex series of events that start with agonist binding, followed by multiple protein conformational changes, opening of the channel pore, block and unblock by Mg^{2+}, and cation permeation. The structural basis of the mechanisms that control the opening and closing of the channel are not fully understood. The activation of NMDA receptors requires the binding of glutamate and glycine to the receptor. Although the structure of the *apo* form of the NR1 or NR2A agonist binding domain has not been described, we will assume NMDA receptors generally bind glutamate as AMPA or kainate receptors do. In this scenario, the initial interactions of glutamate with its binding site are associated with a rapid rearrangement of the agonist binding domain that involves closure of the angle between the two lobes that comprise the clamshell binding site. This leads to association of the γ-carboxyl of glutamate with the lower portion of the bi-lobed binding site, and the ultimate formation of intra-lobe hydrogen bonds. The resulting energy associated with closure of the clamshell structure is conveyed through long range intra-protein interactions to cause the structure of the channel to undergo a conformational change that opens a permeation pore. There is as yet insufficient structural or functional data to delineate the temporal sequence of protein conformational changes or long range intra-protein interactions that lead to gating, although a number of indirect experiments have suggested the identity of some key residues. A major advance in our understanding of gating will occur as new structural and functional information identifies the sequence of structural changes that connect binding to gating.

Site-directed mutagenesis studies show that all three transmembrane domains (M1, M3, and M4) as well as the reentrant pore forming loop are involved in the gating. Among these regions, the second transmembrane domain is thought to be an important component of channel gating, since activation of the channel causes this domain to move. Within this domain, there is a highly conserved nine amino acids motif (SYTANLAAF). Site-directed mutagenesis of residues on this motif significantly alters the channel function (8, 62, 85). The NMDA receptor second transmembrane domain is presumably a rigid helix that physically contacts the reentrant pore forming loop. One hypothetical scheme suggests that the second transmembrane domain of both subunits moves during the gating, which directly results in the movement of the pore forming loop, and in turn, the channel pore. The narrowest constriction within the channel pore, which consists of the reentrant loop tip, might thus control opening or closing of the pore.

The first widely accepted kinetic model for NMDA receptor activation was proposed by Lester and Jahr (82) (Fig. 3a). This model consists of two independent but identical glutamate binding sites, one open state, one closed state, and one desensitized state. This simple model describes the macroscopic time course for NMDA responses remarkably well, although it does not capture the complexity seen in single channel records. This model predicts the response with two decay phases of the EPSC and thus accounts for the synaptic response (82). There are several limitations

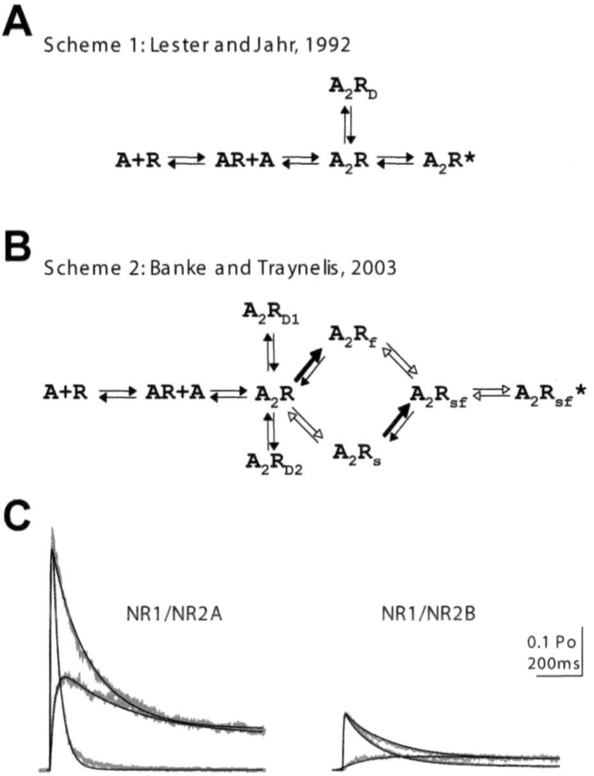

Fig. 3. Kinetic models for NMDA receptor activation. (**a**) Scheme 1: The first widely accepted kinetic model for NMDA receptor activation proposed by Lester and Jahr (82). This model consists of two independent but identical glutamate binding sites, one open state (A_2R^*), one closed state (A_2R), and one desensitized state (A_2R_D). Glutamate binding is described by the agonist (A) binding, while all glycine binding sites are assumed to be fully occupied. This simple model describes the macroscopic time course for NMDA responses remarkably well, although it does not capture the complexity seen in single channel records. (**b**) Scheme 2: The model proposed by Banke and Traynelis (7) and similar models include two new states (A_2R_f and A_2R_s) that reflect the fast gating transitions controlled by the NR1 subunit and slower gating transitions of the NR2 subunit. These conformational changes are hypothesized to occur in either order to arrive at the A_2R_{sf} state that is then followed by channel opening. This multi-state gating model reconciled microscopic and macroscopic properties of channel function, and provides an example of how different subunits within the same receptor complex can uniquely contribute to complex channel properties. (**c**) The kinetic model proposed by Banke and Traynelis (7) (Scheme 2) for NMDA receptor activation has been used to describe activation, desensitization, and deactivation of NMDA receptors. Here, the model is fitted to macroscopic currents recorded in outside-out patches excised from HEK293 cells expressing NR1/NR2A (*left*) and NR1/NR2B (*right*) in response to 1 mM glutamate brief pulse, 1 mM glutamate long pulse, and 5 μM glutamate long pulse. Fits are in black and currents are in grey. Traces are normalized to the open probability (Po). Adapted from (40); reprinted with permission from Wiley-Blackwell Publishing, copyright 2005.

to this model, such as the disregard of the glycine binding steps, which are important for channel activation. To overcome these limitations, newer models, which incorporate both glutamate and glycine binding steps, have been proposed (11, 83, 100).

A key feature of several recently published conceptual models of function that provide a better description of single channel data is the incorporation of multiple steps between binding and gating (7, 41, 112, 114, 123). A recent study of single channel and macroscopic responses to full and partial agonists on recombinant NR1/NR2B receptors (7) suggested that agonist binding to different subunits in some fashion controls different steps in the receptor activation scheme, suggesting a working hypothesis that the binding of glutamate and glycine on separate subunits controls different, perhaps subunit specific, conformational changes that are required for channel opening (Fig. 3b). These conformational changes are hypothesized to occur in either order to arrive at an intermediate state that is then followed by channel opening. Similar models have been proposed by Schorge et al. (123) and Auerbach and Zhou (4). A number of additional models (39, 113) can also account for microscopic and macroscopic properties by incorporating multiple gating steps in a linear reaction scheme in which a specific sequence of transitions is implicit in the reaction mechanism. An important point made by all of these studies is that faithful representation of single channel data as well as macroscopic responses requires multiple pre-gating steps as well as multiple open states (Fig. 3c).

3.2 Structural Determinants of Ion Permeation

The channel pore of NMDA receptors consists of an outer extracellular vestibule, an inner intracellular vestibule and a narrow constriction, which likely lies just past the midpoint traversing the membrane from extra- to intracellular. The extracellular vestibule is formed by the residues on the N-terminal side of the first transmembrane domain segment, the C-terminal portion of the second transmembrane domain, and the N-terminal portion of the third transmembrane domain of both NR1 and NR2 subunits. In NR1/NR2 receptors, both NR1 and NR2 subunits contribute to the extracellular vestibule; the contribution of NR1 and NR2 may be staggered by one helical turn in the vertical axis perpendicular to the plane of the membrane (130). The second transmembrane domain of NR1 and NR2 appears to be the only transmembrane helix that contributes to the deep voltage-dependent portion of the vestibule (129). The re-entrant pore-forming loop lines an intracellular vestibule with the tip of the loop forming the narrowest constriction of the channel. Both extracellular and intracellular vestibules are thought to be water-filled (8, 76). The narrow constriction, which may act as a molecular sieve, is referred to as the selectivity filter, and is typically a major determinant of ion selectivity in channels. By using the relative permeability of differently sized organic cations as an indicator of the dimension of the narrow constriction, the cross-sectional diameter of the narrow constriction of wild type recombinant NR1/NR2A receptor was suggested to be approximately 0.55 nm (142).

An accumulating body of evidence suggests that ionotropic glutamate receptors and potassium channels share a common architecture in their transmembrane pores. Sequence comparison, topology profiling, and analogy to a bacterial glutamate-activated receptor with a potassium-selective channel pore (92) provides compelling evidence for structural similarity between glutamate receptors and potassium channels. The primary differences in the pore-forming region between potassium channels and glutamate receptors is the orientation of the pore loop relative to the membrane as well as the lack of a selectivity filter for NMDA receptors; NMDA receptors are permeable to Na^+, K^+, Ca^{2+}, whereas potassium channels are selectively permeable to K^+. Both ends of the re-entrant pore forming loop in glutamate receptors face the cytoplasmic side of the membrane, whereas both ends in potassium channels such as KcsA face the extracellular side. The structures of the KcsA and MthK channels have created a template for understanding glutamate receptor transmembane structure (33, 57).

One feature that distinguishes NMDA receptors from other ligand gated channels is their unique subconductance levels and permeability to Ca^{2+}. Multi-sublevel conductance of NMDA receptors has been observed not only on recombinant NMDA receptors, but also in native neurons. The unitary conductance roughly correlates with Ca^{2+} permeability among NMDA receptors containing different NR2 subunits. In contrast to studies of AMPA receptors (117, 128), the frequency of subconductance level transitions in NMDA receptors are concentration-independent. The channel of NMDA receptors is more selective for Ca^{2+} than for monovalent ions such as Na^+ and K^+. The NMDA receptor-mediated Ca^{2+} influx not only alters intracellular Ca^{2+} concentration, but also triggers many biological functions, such as changes in synaptic efficacy, gene expression, development of synaptic connections, as well as pathological processes. A number of lines of investigation suggest that the apex of the reentrant pore forming loop reaches towards the extracellular vestibule and forms the narrow region, which contributes to the ion selectivity (153). Several residues that determine the ion permeability have been identified (15, 55, 66, 126). The same asparagine residue in the pore forming reentrant loop of both NR1 and NR2 subunits that plays a key role in Mg^{2+} block also is involved in the Ca^{2+} permeation (15). This residue is analogous to the Q/R-site in AMPA receptors, which also controls Ca^{2+} permeation.

3.3 Structural Determinants of Channel Block

One unique feature of NMDA receptors is their voltage-dependent channel block by extracellular Mg^{2+}. Both native and recombinant NMDA receptors show a strong voltage-dependent block by physiological concentrations of Mg^{2+}, which acts to prevent influx of Ca^{2+} and monovalent cations. This Mg^{2+} blockade is relieved by the depolarization of the membrane potential, providing the receptor with the ability to detect coincident release of glutamate and postsynaptic membrane depolarization. If the membrane potential is sufficiently depolarized to relieve voltage-dependent channel block by extracellular Mg^{2+} at the time NMDA receptors bind glutamate and glycine, the channel will open and allow ion permeation that can lead to a significant influx of Ca^{2+}. This in turn will trigger a series of intracellular events under both physiological and pathophysiological conditions that can alter the strength of synaptic signaling. The degree of the Mg^{2+} blockade is different on the different NR2-containing NMDA receptors. NR2A- and NR2B-subunit containing NMDA receptors are more sensitive to Mg^{2+} block than the NR2C- and NR2D-subunit containing NMDA receptors. In addition, NR2A- and NR2B-containing receptors exhibit a slower component of Mg^{2+} unblock than NR2C- and NR2D-containing receptors (29).

Although the structural determinants of Mg^{2+} block are becoming clearer, more information is still needed before a specific model of Mg^{2+} block at atomic resolution can be proposed. An N-site asparagine has been located as a key determinant of Mg^{2+} block. Several mutagenesis studies suggest that channels containing amino acid substitutions at the N-site or adjacent asparagines in the NR2A subunit show an increased Mg^{2+} permeability, suggesting that these adjacent asparagine residues form a barrier for inward Mg^{2+} influx (152). Other than the N-site asparagines, several additional regions for blockade by Mg^{2+} have been identified. Studies of chimeric receptors suggest that multiple structural elements control the NR2 subunit specificity of Mg^{2+} block, including the first transmembrane domain, the small linker connecting pore forming reentrant loop and the second transmembrane domain, and the third transmembrane domain (74).

Besides external Mg^{2+}, a number of other rigid organic cations of diverse chemical structures are capable of blocking the NMDA pore when the channel opens. Most of these compounds are positively charged at physiological pH and function as uncompetitive inhibitors. A defining property of these channel blockers is that they bind to and block agonist-gated open channels more rapidly than the closed channels, a feature termed use-dependence. Based on the difference in the interaction with the channel gate, the open channel blockers are further classified into three categories: (1) trapping blockers, such as MK-801, phencyclidine (PCP) and ketamine, which are trapped inside the pore as the channel returns to the closed state and the agonist unbinds; (2) "sequential" or "foot-in-the-door" blockers, such as aminoacridine derivatives and tetraetylammonium (TEA), which bind to the channel only when it is open and prevent channel closure; (3) partial trapping blockers, such as memantine and amantadine, which exhibit intermediate properties between above two categories, impede the channel closure without completely preventing it.

Numerous studies show that NMDA channel blockers are neuroprotective in animal models of neurological disorders that involve excessive stimulation of the NMDA receptors, such as traumatic brain injury, epilepsy, stroke, neuropathic pain, as well as neurodegenerative diseases, including Alzheimer disease, Parkinson's disease and Huntington's disease. However, clinical trials have largely been disappointing, as dose-limiting side effects and a narrow temporal window for intervention have likely confounded interpretation and derailed positive results. Interestingly, low affinity channel blockers are well tolerated, and memantine, which shows fast blocking/unblocking kinetics, has been approved for clinical use in the United States for treating Alzheimer's disease (84). With more attention to subunit selectivity, rapid block and unblock kinetics, additional channel blockers may show promise as clinically useful agents.

4 Regulation of NMDA Receptor Function

NMDA receptors are regulated by phosphorylation, extracellular ions, and intracellular binding proteins. We will focus primarily on regulation of NMDA receptor function by extracellular ions and molecules, with particular attention given to mechanisms for which structural information exists.

4.1 Desensitization of NMDA Receptors

Desensitization is operational defined as a decrement in receptor response in the continued presence of a stimulus. Native and recombinant NMDA receptors desensitize by several different mechanisms, including glycine-dependent desensitization, Zn^{2+}-dependent desensitization, Ca^{2+} dependent desensitization, and glycine/Ca^{2+}/Zn^{2+}-independent desensitization. We will consider each in turn.

Mayer and colleagues (93) were the first to describe an apparent desensitization of neuronal NMDA receptors in the presence of sub-saturating glycine concentration, which abated with increases in the concentration of extracellular glycine. They postulated that a negative allosteric interaction occurred between the glutamate- and glycine binding domains such that the binding of glutamate decreased the affinity for glycine (10, 83). The idea embodied in this working hypothesis was simple—when glutamate binds, the affinity for the glycine binding site becomes lower, and in the absence of high concentrations of glycine, the current diminishes with the relaxation to a new equilibrium as glycine unbinds from the receptor. Since glycine is a required co-agonist of the NMDA receptor, sub-saturating concentrations of glycine results in a desensitization of the current. The time course for the desensitization thus reflects the unbinding of glycine from the receptor, and is within the range of synaptic NMDA receptor time course, suggesting it could impact synaptic signaling. While this mechanism was proposed purely from functional data, subsequent structural studies (47) showed that the glycine binding domain of the NR1 subunit crystallizes as a heterodimer with the glutamate binding domain of NR2, providing a plausible structural model for negative allosteric coupling between the glutamate and glycine binding sites.

Extracellular Zn^{2+} has long been known to inhibit neuronal NMDA receptors (110, 146). A number of investigators have reported high affinity regulation of NR2A-containing recombinant NMDA receptors by nanomolar concentrations of extracellular Zn^{2+} (20, 104, 137, 150). However, Raymond and colleagues observed the appearance of a rapid component of desensitization in the presence of extracellular Zn^{2+}, and postulated that Zn^{2+} accelerated receptor desensitization (20). Subsequent work showed that Zn^{2+}-induced desensitization occurred by a similar mechanism as glycine-dependent desensitization. Zheng et al. (159) proposed that a positive allosteric intra-subunit interaction exists between the glutamate binding domain and the Zn^{2+} binding domain, which likely resides in the NR2A amino terminal domain (26, 45, 86, 106). The working hypothesis, supported by a range of experimentation, suggests that binding of glutamate enhances Zn^{2+} binding, such that in the presence of subsaturating concentrations of Zn^{2+}, the system relaxes to a new equilibrium with Zn^{2+} binding to the receptor in a concentration-dependent fashion.

Thus, the desensitization time course reflects the time course for Zn^{2+} binding to the receptors following a shift of the Zn^{2+} binding site into a high affinity state (42).

NMDA receptors additionally undergo a Ca^{2+}-dependent inhibition that is proportional to the magnitude of the current, and is often referred to as Ca^{2+} dependent desensitization or inactivation (28, 80, 116, 143). Ca^{2+}-dependent desensitization develops slowly over seconds, and can be blocked by introduction of Ca^{2+} chelators into the cell. The magnitude of Ca^{2+}-dependent desensitization varies among NR2 subunits, being more prominent for NR2A than for NR2D, and absent for NR2B and NR2C receptors (72, 95). The working hypothesis underlying this form of regulation is that an increase in the concentration of intracellular Ca^{2+} occurs in the immediate vicinity of the intracellular portion of the channel, and this increased Ca^{2+} inhibits the receptor by stimulating the uncoupling of the receptor from filamentous actin in a manner sensitive to second messenger systems (118). Additional data suggest a role for calmodulin binding to the NR1 subunit of the receptor in this form of desensitization. NR1 splice variants can remove one of a pair of calmodulin binding sites from the receptor and abolish Ca^{2+}-dependent desensitization (37, 38). Furthermore, mutations within one of these NR1 calmodulin binding sites disrupt Ca^{2+}-dependent desensitization (73, 158). Interestingly, alpha-actinin2 binds to the intracellular C-terminal of the NR1 subunit in a manner that is antagonized by calmodulin (2, 156), suggesting alpha-actinin2 may be involved in Ca^{2+}-dependent desensitization.

A wide range of ligand-gated channels undergo a form of desensitization that appears to reflect conformational changes within the receptor. Independent of the forms of desensitization discussed above, the NMDA receptor can additional desensitize in the continued presence of agonist by a mechanism that is independent of Ca^{2+}, Zn^{2+} and glycine; such desensitization is more prominent in excised outside-out membrane patches, suggesting a regulation by a diffusible factor (121, 122). This form of desensitization can be reduced by mutations both in the conserved SYTANLAAF motif in the second transmembrane domain and deep within the channel pore, suggesting that it may be coupled to the gating mechanism (19, 53). Interestingly, glycine and Ca^{2+}-independent desensitization of certain subunits are also controlled by the presence of CamKII, which accelerates the rate of desensitization of NR1/NR2B, but slows the rate of desensitization of NR1/NR2A receptors (125).

4.2 Allosteric Regulation of NMDA Receptor Function by Protons

A number of studies have described inhibition of NMDA receptors by extracellular protons (49, 134, 139, 144). Although virtually all receptors show some form of pH sensitivity (reviewed by (136)), NMDA receptors appeared somewhat unique in that protons potently (IC_{50} 50 nM) and completely inhibited receptor function with a concentration-effect relationship that approximately followed a single binding site isotherm. These reports suggested that neuronal NMDA receptors are tonically inhibited by physiological concentrations of protons with an IC_{50} value that corresponds to about pH 7.4, or approximately 50 nM protons. Proton inhibition depends on alternative RNA splicing of the NR1 subunit and the identity of the NR2 subunit (140). Under physiological conditions, the NMDA receptor resides under tonic inhibition, and can respond to small changes in extracellular pH, which is dynamic and changes with neuronal activity (24). Changes in pH driven by pathological situations, such as seizure or ischemia, are also more than sufficient to inhibit NMDA receptor function (5, 24, 25, 101, 127).

A number of experiments all show that extracellular (but not intracellular) protons act in a voltage-independent manner that is also independent of actions at the agonist binding site. At the single channel level, protons appear to reduce open probability for NR1/NR2B recombinant receptors, with only modest effects on the open time and no effects on single channel conductance (6, 138). Mutagenesis data, while not implicating any single residue as causative for proton inhibition, shows a clustering of residues at which mutagenesis alters pH sensitivity that is near the linkers coupling the agonist binding domain to the transmembrane helices (85). Interestingly, inward rectifier potassium channels, which share a number of common features with the glutamate receptor family (75), are also pH sensitive (e.g. (124)), raising the possibility that the two channels share some structural determinants or mechanisms of proton inhibition.

Several recent studies suggest that the downstream effects of a number of modulators may reflect a subtle change in the pKa of the proton sensor that either enhances or reduces tonic proton inhibition at physiological pH; enhancement of proton sensitivity will appear as inhibition, whereas a reduction in sensitivity to protons appears as a potentiation. For example, regulation of NMDA receptor function by extracellular polyamines such as spermine (61, 148) occurs in a subunit-dependent fashion, with only the NR2B subunit showing potentiation by 100's of micromolar spermine (35, 36). The potentiation by polyamines only occurs at pH values that produce tonic inhibition, suggesting that the binding of polyamines such as spermine shifts the pKa of the proton sensor to reduce tonic inhibition. Consistent with this idea, spermine can shift the experimentally-determined proton inhibition curve, and there is a strong correlation between the degree of proton inhibition and the degree of spermine potentiation, consistent with a relief of proton inhibition by polyamines (140). There is also a strong correlation between the ability of various mutations to shift the proton sensitivity and their effects on polyamine potentiation (64, 65, 89, 140, 151). These data all suggest that the full potentiating effect of polyamine binding (in the presence of saturating glutamate and glycine) can be accounted for by relief of tonic proton inhibition. This finding created two important precedents. First, this conclusion (spermine relieves proton inhibition by shifting the pKa of a proton sensor; (140)) established the idea that exogenous molecules could regulate NMDA receptor function through actions on the proton sensor. Second, implicit in this straightforward mechanism is the idea that agents that enhance or diminish proton sensitivity can cause inhibition and potentiation of NMDA receptor function, respectively.

4.3 Allosteric Regulation of NMDA Receptor Function by Extracellular Zn^{2+}

The isolation of cDNA's encoding the NMDA receptor subunits in the 1990's provided an opportunity to pursue a molecular understanding of Zn^{2+} inhibition. One of the first observations made was that the NR2A subunit was exquisitely sensitive to extracellular Zn^{2+}, with variable IC_{50} values reported in the nM range (e.g. (20, 147)). Paoletti et al. (104) used buffered Zn^{2+} to unambiguously determine the IC_{50} value for the high affinity Zn^{2+} site, and further showed that hundreds of nanomolar Zn^{2+} often persists as a contaminant of physiological saline solutions made from commercially available salts. One interesting feature to emerge from these studies is that Zn^{2+} inhibition curves for NR1/NR2A receptors are biphasic. The interpretation of this was that saturating concentrations of Zn^{2+} at the nM high affinity site do not impart full inhibition, which could only be observed at higher concentrations of Zn^{2+} (10's of micromolar), which cause a voltage-dependent channel block. Studies of chimeric NR2A-NR2C receptors suggested that the amino terminal domain of NR2A controlled the high affinity Zn^{2+} site (45). Site-directed mutagenesis further isolated a series of histidine residues that were candidates for chelating Zn^{2+} within its binding site (26, 45, 86). The amino terminal domain harboring the Zn^{2+} site shows a number of similarities to both bacterial periplasmic amino acid binding proteins as well as the glutamate binding site of mGluR1, allowing structural models of amino terminal domain to be developed (105). These models and supporting experimentation show convincingly that the amino terminal domain folds into a bi-lobed clamshell structure, with the Zn^{2+} binding site residing within the cleft.

Following work on ifenprodil ((99), see below) and spermine control of NMDA receptor proton sensitivity of receptors (140), Choi and Lipton (26) proposed that the binding of extracellular Zn^{2+} to the amino terminal domain enhances the sensitivity of the NR1/NR2A receptor to protons. The leftward shift of the proton inhibition curve for Zn^{2+}-bound receptors means that at pH 7.4, the extent of inhibition by protons increases from 50% to perhaps 80%. The result of this is an apparent Zn^{2+}-induced inhibition that reflects an increase in the proportion of protonated and non-functional receptors. A number of lines of evidence support this hypothesis. At alkaline pH values where the proton inhibition is absent, Zn^{2+} causes no inhibition. Furthermore, the close functional interactions of Zn^{2+} binding and proton inhibition implies a strong correlation between the ability of mutations to shift the proton sensitivity and their effects on the Zn^{2+} inhibition (86). Lastly, kinetic modeling of Zn^{2+}-induced enhancement of proton inhibition can quantitatively account for the actions of Zn^{2+}on proton binding rates, providing additional support to this hypothesis (43).

Zn^{2+} inhibits the NR2B-containing NMDA receptors with an IC_{50} value near 1 micromolar. Rachline et al. (115) have recently extended the structural hypothesis developed for Zn^{2+} inhibition at NR2A to the NR2B subunit. Their ideas are consistent with functional model of inhibition originally proposed by Choi & Lipton (26) for Zn^{2+} inhibition of both NR2A and NR2B containing receptors. This idea is also consistent with mutagenesis studies showing similar sensitivity to mutations of Zn^{2+} and proton IC_{50} values (137). Interestingly, the high affinity Zn^{2+}site is partially maintained in heterotrimeric NR1/NR2A/NR2B receptors, but shows shallower inhibition (51).

4.4 NR2B Selective Allosteric Modulators

The phenylethanolamine ifenprodil was the first subunit-selective antagonist identified at any NMDA receptor subunit (149). Characterization of this compound and a host of analogues (67) have provided an invaluable set of tools for dissecting the functional role of the NR2B subunit in a range of neurophysiological, developmental, and pathological situations. Most data support the idea that ifenprodil binds to the amino terminal domain of an NR2B subunit in a fashion analogous to Zn^{2+} binding to NR2A (14, 48, 109, 154), although residues in the NR1 amino terminal domain also have important effects on ifenprodil binding (89). Molecular modeling studies support an intra-cleft binding site within the NR2B bi-lobed amino terminal domain (54, 88, 109), analogous to that proposed for Zn^{2+} binding to the NR2A amino terminal domain. In addition, ifenprodil and its analogues exert a range of actions including a positive allosteric enhancement of submaximal glutamate responses (68, 69). At higher concentrations, ifenprodil induces a non-competitive, voltage-independent inhibition of NR1/NR2B receptors that is incomplete. Extensive studies of the mechanism suggest ifenprodil promotes entry of the receptor into a long-lived shut state that is analogous to desensitization (68, 69). A large number of structural classes of NR2B-selective ligands have been described, which show a surprising degree of diversity in structure (12, 17, 23, 79).

Mott et al. (99) have proposed that ifenprodil binding causes a unique action on the proton sensor, which enhances the proton sensitivity of the receptor. Evaluation of the proton inhibition curve in the absence and presence of ifenprodil shows a leftward shift, with ifenprodil enhancing proton inhibition of receptor function at all pH values except alkaline levels. This enhancement of proton sensitivity increases the level of tonic inhibition at physiological pH, which appears as inhibition. The potency of ifenprodil is pH sensitive, and there is a strong correlation with the potency shift induced in ifenprodil and proton sensitivity by a series of mutants (99, 103). Perhaps the strongest evidence supporting the idea of enhancement of proton sensitivity by ifenprodil comes from experiments at alkaline pH values where there is little proton inhibition; under this condition there is no effect of ifenprodil even though it remains bound to the receptor. These data provide a link between mechanisms of regulation by ligands of the amino terminal domain, and also suggest that the tight regulation of gating by protons could be a downstream substrate for several regulatory pathways.

5 Conclusions

The remarkable advances in our understanding of all features of the NMDA receptors suggest that their modular receptor design impacts function and regulation. That is, autonomous domains have clear and distinct effects on receptor function. It is also clear that key features of receptor structure are conserved across the glutamate receptor family, although different subunit families appear to have perhaps somewhat different mechanisms of activation and desensitization. What seems clear is that the first steps of channel activation are similar between AMPA, kainate and NMDA receptors, which all presumably undergo agonist-induced domain closure. After this point, it remains unclear how the similar molecular themes of conformational changes in the ligand binding domain, stability of the dimer interface, and intra-subunit conformational changes shared in all families work to provide unique activation mechanisms. Reconciling the structural features with gating mechanisms will be an important step forward in advancing understanding of gating. In addition, differences between subunits impact their pharmacology and function, providing a wide range of properties that can be utilized by cells for different functions at different times. Understanding the molecular basis of receptor pharmacology will provide further insight into receptor function. Moreover, the various autonomous domains provide a host of therapeutic targets, and the best tolerated NMDA receptor antagonists so far are subunit-selective allosteric modulators. It seems highly likely that as additional efforts are focused on modulation, more subunit selective modulators will emerge for clinical evaluation.

Acknowledgments

We thank M. Mayer for critical review of the manuscript. We apologize to those whose work we could not cite due to space limitations. Work in SFT's lab is supported by the NIH, the Michael J. Fox Foundation for Parkinson's Research, and NARSAD, the Mental Health Research Association.

References

1. Aizenman E, Lipton SA and Loring RH. Selective modulation of NMDA responses by reduction and oxidation. *Neuron* 2: 1257–1263, 1989.
2. Allison DW, Gelfand VI, Spector I and Craig AM. Role of actin in anchoring postsynaptic receptors in cultured hippocampal neurons: differential attachment of NMDA versus AMPA receptors. *J Neurosci* 18: 2423–2436, 1998.
3. Anson LC, Chen PE, Wyllie DJ, Colquhoun D and Schoepfer R. Identification of amino acid residues of the NR2A subunit that control glutamate potency in recombinant NR1/NR2A NMDA receptors. *J Neurosci* 18: 581–589, 1998.
4. Auerbach A and Zhou Y. Gating reaction mechanisms for NMDA receptor channels. *J Neurosci* 25: 7914–7923, 2005.
5. Balestrino M and Somjen GG. Concentration of carbon dioxide, interstitial pH and synaptic transmission in hippocampal formation of the rat. *J Physiol* 396: 247–266, 1988.
6. Banke TG, Dravid SM and Traynelis SF. Protons trap NR1/NR2B NMDA receptors in a nonconducting state. *J Neurosci* 25: 42–51, 2005.
7. Banke TG and Traynelis SF. Activation of NR1/NR2B NMDA receptors. *Nat Neurosci* 6: 144–152, 2003.
8. Beck C, Wollmuth LP, Seeburg PH, Sakmann B and Kuner T. NMDAR channel segments forming the extracellular vestibule inferred from the accessibility of substituted cysteines. *Neuron* 22: 559–570, 1999.
9. Béhé P, Colquhoun D and Wyllie DJ. Activation of single AMPA- and NMDA-type glutamate-receptor channels. In: *Ionotropic glutamate receptors in the CNS*, edited by Jonas P and Monyer H. Springer, New York, 1999, pp. 175–218.
10. Benveniste M, Clements J, Vyklicky L, Jr. and Mayer ML. A kinetic analysis of the modulation of N-methyl-D-aspartic acid receptors by glycine in mouse cultured hippocampal neurones. *J Physiol* 428: 333–357, 1990.

11. Benveniste M and Mayer ML. Kinetic analysis of antagonist action at N-methyl-D-aspartic acid receptors. Two binding sites each for glutamate and glycine. *Biophys J* 59: 560–573, 1991.
12. Borza I and Domany G. NR2B selective NMDA antagonists: the evolution of the ifenprodil-type pharmacophore. *Curr Top Med Chem* 6: 687–695, 2006.
13. Brickley SG, Misra C, Mok MH, Mishina M and Cull-Candy SG. NR2B and NR2D subunits coassemble in cerebellar Golgi cells to form a distinct NMDA receptor subtype restricted to extrasynaptic sites. *J Neurosci* 23: 4958–4966, 2003.
14. Brimecombe JC, Gallagher MJ, Lynch DR and Aizenman E. An NR2B point mutation affecting haloperidol and CP101,606 sensitivity of single recombinant N-methyl-D-aspartate receptors. *J Pharmacol Exp Ther* 286: 627–634, 1998.
15. Burnashev N, Schoepfer R, Monyer H, Ruppersberg JP, Gunther W, Seeburg PH and Sakmann B. Control by asparagine residues of calcium permeability and magnesium blockade in the NMDA receptor. *Science* 257: 1415–1419, 1992.
16. Chatterton JE, Awobuluyi M, Premkumar LS, Takahashi H, Talantova M, Shin Y, Cui J, Tu S, Sevarino KA, Nakanishi N, Tong G, Lipton SA and Zhang D. Excitatory glycine receptors containing the NR3 family of NMDA receptor subunits. *Nature* 415: 793–798, 2002.
17. Chazot PL. The NMDA receptor NR2B subunit: a valid therapeutic target for multiple CNS pathologies. *Curr Med Chem* 11: 389–396, 2004.
18. Cheffings CM and Colquhoun D. Single channel analysis of a novel NMDA channel from Xenopus oocytes expressing recombinant NR1a, NR2A and NR2D subunits. *J Physiol* 526 Pt 3: 481–491, 2000.
19. Chen N, Li B, Murphy TH and Raymond LA. Site within N-Methyl-D-aspartate receptor pore modulates channel gating. *Mol Pharmacol* 65: 157–164, 2004.
20. Chen N, Moshaver A and Raymond LA. Differential sensitivity of recombinant N-methyl-D-aspartate receptor subtypes to zinc inhibition. *Mol Pharmacol* 51: 1015–1023, 1997.
21. Chen PE, Geballe MT, Stansfeld PJ, Johnston AR, Yuan H, Jacob AL, Snyder JP, Traynelis SF and Wyllie DJ. Structural features of the glutamate binding site in recombinant NR1/NR2A N-methyl-D-aspartate receptors determined by site-directed mutagenesis and molecular modeling. *Mol Pharmacol* 67: 1470–1484, 2005.
22. Chen PE and Wyllie DJ. Pharmacological insights obtained from structure-function studies of ionotropic glutamate receptors. *Br J Pharmacol* 147: 839–853, 2006.
23. Chenard BL and Menniti FS. Antagonists selective for NMDA receptors containing the NR2B subunit. *Curr Pharm Des* 5: 381–404, 1999.
24. Chesler M. Regulation and modulation of pH in the brain. Physiol Rev 83: 1183–1221, 2003.
25. Chesler M and Kaila K. Modulation of pH by neuronal activity. *Trends Neurosci* 15: 396–402, 1992.
26. Choi YB and Lipton SA. Identification and mechanism of action of two histidine residues underlying high-affinity Zn2+ inhibition of the NMDA receptor. *Neuron* 23: 171–180, 1999.
27. Ciabarra AM, Sullivan JM, Gahn LG, Pecht G, Heinemann S and Sevarino KA. Cloning and characterization of chi-1: a developmentally regulated member of a novel class of the ionotropic glutamate receptor family. *J Neurosci* 15: 6498–6508, 1995.
28. Clark GD, Clifford DB and Zorumski CF. The effect of agonist concentration, membrane voltage and calcium on N-methyl-D-aspartate receptor desensitization. *Neuroscience* 39: 787–797, 1990.
29. Clarke RJ and Johnson JW. NMDA receptor NR2 subunit dependence of the slow component of magnesium unblock. *J Neurosci* 26: 5825 5834, 2006.

30. Das S, Sasaki YF, Rothe T, Premkumar LS, Takasu M, Crandall JE, Dikkes P, Conner DA, Rayudu PV, Cheung W, Chen HS, Lipton SA and Nakanishi N. Increased NMDA current and spine density in mice lacking the NMDA receptor subunit NR3A. *Nature* 393: 377–381, 1998.
31. Dingledine R, Borges K, Bowie D and Traynelis SF. The glutamate receptor ion channels. *Pharmacol Rev* 51: 7–61, 1999.
32. Dingledine R and McBain CJ. Glutamate and Aspartate. In: *Basic Neurochemistry*: Molecular, Cellular and Medical Aspects, edited by Siegel GJ, Agranoff BW, Albers RW, Fisher SK and Uhler MD. 1999, p. 315–334.
33. Doyle DA, Morais CJ, Pfuetzner RA, Kuo A, Gulbis JM, Cohen SL, Chait BT and MacKinnon R. The structure of the potassium channel: molecular basis of K+ conduction and selectivity. *Science* 280: 69–77, 1998.
34. Dravid SM and Traynelis SF. Activation of NR1/NR2C NMDA receptors. *The 35th Annual Meeting Society for Neuroscience*, 2005.
35. Durand GM, Bennett MV and Zukin RS. Splice variants of the N-methyl-D-aspartate receptor NR1 identify domains involved in regulation by polyamines and protein kinase C. *Proc Natl Acad Sci U S A* 90: 6731–6735, 1993.
36. Durand GM, Gregor P, Zheng X, Bennett MV, Uhl GR and Zukin RS. Cloning of an apparent splice variant of the rat N-methyl-D-aspartate receptor NMDAR1 with altered sensitivity to polyamines and activators of protein kinase C. *Proc Natl Acad Sci U S A* 89: 9359–9363, 1992.
37. Ehlers MD, Fung ET, O'Brien RJ and Huganir RL. Splice variant-specific interaction of the NMDA receptor subunit NR1 with neuronal intermediate filaments. *J Neurosci* 18: 720–730, 1998.
38. Ehlers MD, Zhang S, Bernhadt JP and Huganir RL. Inactivation of NMDA receptors by direct interaction of calmodulin with the NR1 subunit. *Cell* 84: 745–755, 1996.
39. Erreger K, Chen PE, Wyllie DJ and Traynelis SF. Glutamate receptor gating. *Crit Rev Neurobiol* 16: 187–224, 2004.
40. Erreger K, Dravid SM, Banke TG, Wyllie DJ and Traynelis SF. Subunit-specific gating controls rat NR1/NR2A and NR1/NR2B NMDA channel kinetics and synaptic signalling profiles. *J Physiol* 563: 345–358, 2005.
41. Erreger K, Geballe MT, Dravid SM, Snyder JP, Wyllie DJ and Traynelis SF. Mechanism of partial agonism at NMDA receptors for a conformationally restricted glutamate analog. *J Neurosci* 25: 7858–7866, 2005.
42. Erreger K, Geballe MT, Kristensen AS, Chen PE, Hansen KB, Lee CJ, Yuan H, Le P, Lyuboslavsky P, Micale N, Jørgensen L, Clausen RP, Wyllie DJ, Snyder JP and Traynelis SF. Subunit-specific agonist activity for glutamate analogs at NR2A, NR2B, NR2C, and NR2D N-methyl-D-aspartate glutamate receptors. *Mol Pharmacol* 72: 907–920, 2007.
43. Erreger K and Traynelis SF. Allosteric interaction between zinc and glutamate binding domains on NR2A causes desensitization of NMDA receptors. *J Physiol* 569: 381–393, 2005.
44. Farrant M, Feldmeyer D, Takahashi T and Cull-Candy SG. NMDA-receptor channel diversity in the developing cerebellum. *Nature* 368: 335–339, 1994.
45. Fayyazuddin A, Villarroel A, Le GA, Lerma J and Neyton J. Four residues of the extracellular N-terminal domain of the NR2A subunit control high-affinity Zn2+ binding to NMDA receptors. *Neuron* 25: 683–694, 2000.
46. Furukawa H and Gouaux E. Mechanisms of activation, inhibition and specificity: crystal structures of the NMDA receptor NR1 ligand-binding core. *EMBO J* 22: 2873–2885, 2003.

47. Furukawa H, Singh SK, Mancusso R and Gouaux E. Subunit arrangement and function in NMDA receptors. *Nature* 438: 185–192, 2005.
48. Gallagher MJ, Huang H, Pritchett DB and Lynch DR. Interactions between ifenprodil and the NR2B subunit of the N-methyl-D-aspartate receptor. *J Biol Chem* 271: 9603–9611, 1996.
49. Giffard RG, Monyer H, Christine CW and Choi DW. Acidosis reduces NMDA receptor activation, glutamate neurotoxicity, and oxygen-glucose deprivation neuronal injury in cortical cultures. *Brain Res* 506: 339–342, 1990.
50. Hansen KB, Yuan H and Traynelis SF. Structural aspects of AMPA receptor activation, desensitization, and deactivation. *Curr Opin Neurobiol* 17: 281–288, 2007.
51. Hatton CJ and Paoletti P. Modulation of triheteromeric NMDA receptors by N-terminal domain ligands. *Neuron* 46: 261–274, 2005.
52. Hollmann M, Boulter J, Maron C, Beasley L, Sullivan J, Pecht G and Heinemann S. Zinc potentiates agonist-induced currents at certain splice variants of the NMDA receptor. *Neuron* 10: 943–954, 1993.
53. Hu B and Zheng F. Molecular determinants of glycine-independent desensitization of NR1/NR2A receptors. *J Pharmacol Exp Ther* 313: 563–569, 2005.
54. Huggins DJ and Grant GH. The function of the amino terminal domain in NMDA receptor modulation. *J Mol Graph Model* 23: 381–388, 2005.
55. Hume RI, Dingledine R and Heinemann SF. Identification of a site in glutamate receptor subunits that controls calcium permeability. *Science* 253: 1028–1031, 1991.
56. Inanobe A, Furukawa H and Gouaux E. Mechanism of Partial Agonist Action at the NR1 Subunit of NMDA Receptors. *Neuron* 47: 71–84, 2005.
57. Jiang Y, Lee A, Chen J, Cadene M, Chait BT and MacKinnon R. Crystal structure and mechanism of a calcium-gated potassium channel. *Nature* 417: 515–522, 2002.
58. Jin R, Banke TG, Mayer ML, Traynelis SF and Gouaux E. Structural basis for partial agonist action at ionotropic glutamate receptors. *Nat Neurosci* 6: 803–810, 2003.
59. Jin R, Clark S, Weeks AM, Dudman JT, Gouaux E and Partin KM. Mechanism of positive allosteric modulators acting on AMPA receptors. *J Neurosci* 25: 9027–9036, 2005.
60. Johnson JW and Ascher P. Glycine potentiates the NMDA response in cultured mouse brain neurons. *Nature* 325: 529–531, 1987.
61. Johnson TD. Modulation of channel function by polyamines. *Trends Pharmacol Sci* 17: 22–27, 1996.
62. Jones KS, VanDongen HM and VanDongen AM. The NMDA receptor M3 segment is a conserved transduction element coupling ligand binding to channel opening. *J Neurosci* 22: 2044–2053, 2002.
63. Jones S and Gibb AJ. Functional NR2B- and NR2D-containing NMDA receptor channels in rat substantia nigra dopaminergic neurones. *J Physiol* 569: 209–221, 2005.
64. Kashiwagi K, Fukuchi J, Chao J, Igarashi K and Williams K. An aspartate residue in the extracellular loop of the N-methyl-D-aspartate receptor controls sensitivity to spermine and protons. *Mol Pharmacol* 49: 1131–1141, 1996.
65. Kashiwagi K, Pahk AJ, Masuko T, Igarashi K and Williams K. Block and modulation of N-methyl-D-aspartate receptors by polyamines and protons: role of amino acid residues in the transmembrane and pore-forming regions of NR1 and NR2 subunits. *Mol Pharmacol* 52: 701–713, 1997.
66. Kawajiri S and Dingledine R. Multiple structural determinants of voltage-dependent magnesium block in recombinant NMDA receptors. *Neuropharmacology* 32: 1203–1211, 1993.
67. Kew JN and Kemp JA. Ionotropic and metabotropic glutamate receptor structure and pharmacology. *Psychopharmacology* (Berl) 179: 4–29, 2005.

68. Kew JN, Trube G and Kemp JA. A novel mechanism of activity-dependent NMDA receptor antagonism describes the effect of ifenprodil in rat cultured cortical neurones. *J Physiol* 497 (Pt 3): 761–772, 1996.
69. Kew JN, Trube G and Kemp JA. State-dependent NMDA receptor antagonism by Ro 8-4304, a novel NR2B selective, non-competitive, voltage-independent antagonist. *Br J Pharmacol* 123: 463–472, 1998.
70. Kleckner NW and Dingledine R. Requirement for glycine in activation of NMDA-receptors expressed in Xenopus oocytes. *Science* 241: 835–837, 1988.
71. Kristensen AS, Geballe MT, Snyder JP and Traynelis SF. Glutamate receptors: variation in structure-function coupling. *Trends Pharmacol Sci* 27: 65–69, 2006.
72. Krupp JJ, Vissel B, Heinemann SF and Westbrook GL. Calcium-dependent inactivation of recombinant N-methyl-D-aspartate receptors is NR2 subunit specific. *Mol Pharmacol* 50: 1680–1688, 1996.
73. Krupp JJ, Vissel B, Thomas CG, Heinemann SF and Westbrook GL. Interactions of calmodulin and alpha-actinin with the NR1 subunit modulate Ca2+-dependent inactivation of NMDA receptors. *J Neurosci* 19: 1165–1178, 1999.
74. Kuner T and Schoepfer R. Multiple structural elements determine subunit specificity of Mg2+ block in NMDA receptor channels. *J Neurosci* 16: 3549–3558, 1996.
75. Kuner T, Seeburg PH and Guy HR. A common architecture for K+ channels and ionotropic glutamate receptors? *Trends Neurosci* 26: 27–32, 2003.
76. Kuner T, Wollmuth LP, Karlin A, Seeburg PH and Sakmann B. Structure of the NMDA receptor channel M2 segment inferred from the accessibility of substituted cysteines. *Neuron* 17: 343–352, 1996.
77. Kuryatov A, Laube B, Betz H and Kuhse J. Mutational analysis of the glycine-binding site of the NMDA receptor: structural similarity with bacterial amino acid-binding proteins. *Neuron* 12: 1291–1300, 1994.
78. Laube B, Schemm R and Betz H. Molecular determinants of ligand discrimination in the glutamate-binding pocket of the NMDA receptor. *Neuropharmacology* 47: 994–1007, 2004.
79. Layton ME, Kelly MJ, III and Rodzinak KJ. Recent advances in the development of NR2B subtype-selective NMDA receptor antagonists. *Curr Top Med Chem* 6: 697–709, 2006.
80. Legendre P, Rosenmund C and Westbrook GL. Inactivation of NMDA channels in cultured hippocampal neurons by intracellular calcium. *J Neurosci* 13: 674–684, 1993.
81. Lerma J, Zukin RS and Bennett MV. Glycine decreases desensitization of N-methyl-D-aspartate (NMDA) receptors expressed in Xenopus oocytes and is required for NMDA responses. *Proc Natl Acad Sci U S A* 87: 2354–2358, 1990.
82. Lester RA and Jahr CE. NMDA channel behavior depends on agonist affinity. *J Neurosci* 12: 635–643, 1992.
83. Lester RA, Tong G and Jahr CE. Interactions between the glycine and glutamate binding sites of the NMDA receptor. *J Neurosci* 13: 1088–1096, 1993.
84. Lipton SA. The molecular basis of memantine action in Alzheimer's disease and other neurologic disorders: low-affinity, uncompetitive antagonism. *Curr Alzheimer Res* 2: 155–165, 2005.
85. Low CM, Lyuboslavsky P, French A, Le P, Wyatte K, Thiel WH, Marchan EM, Igarashi K, Kashiwagi K, Gernert K, Williams K, Traynelis SF and Zheng F. Molecular determinants of proton-sensitive N-methyl-D-aspartate receptor gating. *Mol Pharmacol* 63: 1212–1222, 2003.
86. Low CM, Zheng F, Lyuboslavsky P and Traynelis SF. Molecular determinants of coordinated proton and zinc inhibition of N-methyl-D-aspartate NR1/NR2A receptors. *Proc Natl Acad Sci U S A* 97: 11062–11067, 2000.

87. Madry C, Mesic I, Bartholomaus I, Nicke A, Betz H and Laube B. Principal role of NR3 subunits in NR1/NR3 excitatory glycine receptor function. *Biochem Biophys Res Commun* 354: 102–108, 2007.
88. Malherbe P, Mutel V, Broger C, Perin-Dureau F, Kemp JA, Neyton J, Paoletti P and Kew JN. Identification of critical residues in the amino terminal domain of the human NR2B subunit involved in the RO 25-6981 binding pocket. *J Pharmacol Exp Ther* 307: 897–905, 2003.
89. Masuko T, Kuno T, Kashiwagi K, Kusama T, Williams K and Igarashi K. Stimulatory and inhibitory properties of aminoglycoside antibiotics at N-methyl-D-aspartate receptors. *J Pharmacol Exp Ther* 290: 1026–1033, 1999.
90. Matsuda K, Kamiya Y, Matsuda S and Yuzaki M. Cloning and characterization of a novel NMDA receptor subunit NR3B: a dominant subunit that reduces calcium permeability. *Brain Res Mol Brain Res* 100: 43–52, 2002.
91. Mayer ML and Armstrong N. Structure and function of glutamate receptor ion channels. *Annu Rev Physiol* 66: 161–181, 2004.
92. Mayer ML, Olson R and Gouaux E. Mechanisms for ligand binding to GluR0 ion channels: crystal structures of the glutamate and serine complexes and a closed apo state. *J Mol Biol* 311: 815–836, 2001.
93. Mayer ML, Vyklicky L, Jr. and Clements J. Regulation of NMDA receptor desensitization in mouse hippocampal neurons by glycine. *Nature* 338: 425–427, 1989.
94. McFeeters RL and Oswald RE. Emerging structural explanations of ionotropic glutamate receptor function. *FASEB J* 18: 428–438, 2004.
95. Medina I, Filippova N, Charton G, Rougeole S, Ben-Ari Y, Khrestchatisky M and Bregestovski P. Calcium-dependent inactivation of heteromeric NMDA receptor-channels expressed in human embryonic kidney cells. *J Physiol* 482 (Pt 3): 567–573, 1995.
96. Mitchell NA and Fleck MW. Targeting of AMPA receptor gating processes by allosteric modulators and mutations. *Biophys J* 92: 2392–2402, 2007.
97. Monyer H, Burnashev N, Laurie DJ, Sakmann B and Seeburg PH. Developmental and regional expression in the rat brain and functional properties of four NMDA receptors. *Neuron* 12: 529–540, 1994.
98. Monyer H, Sprengel R, Schoepfer R, Herb A, Higuchi M, Lomeli H, Burnashev N, Sakmann B and Seeburg PH. Heteromeric NMDA receptors: molecular and functional distinction of subtypes. *Science* 256: 1217–1221, 1992.
99. Mott DD, Doherty JJ, Zhang S, Washburn MS, Fendley MJ, Lyuboslavsky P, Traynelis SF and Dingledine R. Phenylethanolamines inhibit NMDA receptors by enhancing proton inhibition. *Nat Neurosci* 1: 659–667, 1998.
100. Nahum-Levy R, Lipinski D, Shavit S and Benveniste M. Desensitization of NMDA receptor channels is modulated by glutamate agonists. *Biophys J* 80: 2152–2166, 2001.
101. Nedergaard M, Kraig RP, Tanabe J and Pulsinelli WA. Dynamics of interstitial and intracellular pH in evolving brain infarct. *Am J Physiol* 260: R581–R588, 1991.
102. Nishi M, Hinds H, Lu HP, Kawata M and Hayashi Y. Motoneuron-specific expression of NR3B, a novel NMDA-type glutamate receptor subunit that works in a dominant-negative manner. *J Neurosci* 21: RC185, 2001.
103. Pahk AJ and Williams K. Influence of extracellular pH on inhibition by ifenprodil at N-methyl-D-aspartate receptors in Xenopus oocytes. *Neurosci Lett* 225: 29–32, 1997.
104. Paoletti P, Ascher P and Neyton J. High-affinity zinc inhibition of NMDA NR1-NR2A receptors. *J Neurosci* 17: 5711–5725, 1997.
105. Paoletti P and Neyton J. NMDA receptor subunits: function and pharmacology. *Curr Opin Pharmacol* 7: 39–47, 2007.

106. Paoletti P, Perin-Dureau F, Fayyazuddin A, Le GA, Callebaut I and Neyton J. Molecular organization of a zinc binding n-terminal modulatory domain in a NMDA receptor subunit. *Neuron* 28: 911–925, 2000.
107. Partin KM, Fleck MW and Mayer ML. AMPA receptor flip/flop mutants affecting deactivation, desensitization, and modulation by cyclothiazide, aniracetam, and thiocyanate. *J Neurosci* 16: 6634–6647, 1996.
108. Perez-Otano I, Schulteis CT, Contractor A, Lipton SA, Trimmer JS, Sucher NJ and Heinemann SF. Assembly with the NR1 subunit is required for surface expression of NR3A-containing NMDA receptors. *J Neurosci* 21: 1228–1237, 2001.
109. Perin-Dureau F, Rachline J, Neyton J and Paoletti P. Mapping the binding site of the neuroprotectant ifenprodil on NMDA receptors. *J Neurosci* 22: 5955–5965, 2002.
110. Peters S, Koh J and Choi DW. Zinc selectively blocks the action of N-methyl-D-aspartate on cortical neurons. *Science* 236: 589–593, 1987.
111. Pina-Crespo JC and Gibb AJ. Subtypes of NMDA receptors in new-born rat hippocampal granule cells. *J Physiol* 541: 41–64, 2002.
112. Popescu G and Auerbach A. Modal gating of NMDA receptors and the shape of their synaptic response. *Nat Neurosci* 6: 476–483, 2003.
113. Popescu G and Auerbach A. The NMDA receptor gating machine: lessons from single channels. *Neuroscientist* 10: 192–198, 2004.
114. Popescu G, Robert A, Howe JR and Auerbach A. Reaction mechanism determines NMDA receptor response to repetitive stimulation. *Nature* 430: 790–793, 2004.
115. Rachline J, Perin-Dureau F, Le GA, Neyton J and Paoletti P. The micromolar zinc-binding domain on the NMDA receptor subunit NR2B. *J Neurosci* 25: 308–317, 2005.
116. Rosenmund C, Feltz A and Westbrook GL. Calcium-dependent inactivation of synaptic NMDA receptors in hippocampal neurons. *J Neurophysiol* 73: 427–430, 1995.
117. Rosenmund C, Stern-Bach Y and Stevens CF. The tetrameric structure of a glutamate receptor channel. *Science* 280: 1596–1599, 1998.
118. Rosenmund C and Westbrook GL. Calcium-induced actin depolymerization reduces NMDA channel activity. *Neuron* 10: 805–814, 1993.
119. Rumbaugh G, Prybylowski K, Wang JF and Vicini S. Exon 5 and spermine regulate deactivation of NMDA receptor subtypes. *J Neurophysiol* 83: 1300–1306, 2000.
120. Sasaki YF, Rothe T, Premkumar LS, Das S, Cui J, Talantova MV, Wong HK, Gong X, Chan SF, Zhang D, Nakanishi N, Sucher NJ and Lipton SA. Characterization and comparison of the NR3A subunit of the NMDA receptor in recombinant systems and primary cortical neurons. *J Neurophysiol* 87: 2052–2063, 2002.
121. Sather W, Dieudonne S, MacDonald JF and Ascher P. Activation and desensitization of N-methyl-D-aspartate receptors in nucleated outside-out patches from mouse neurones. *J Physiol* 450: 643–672, 1992.
122. Sather W, Johnson JW, Henderson G and Ascher P. Glycine-insensitive desensitization of NMDA responses in cultured mouse embryonic neurons. *Neuron* 4: 725–731, 1990.
123. Schorge S, Elenes S and Colquhoun D. Maximum likelihood fitting of single channel NMDA activity with a mechanism composed of independent dimers of subunits. *J Physiol* 569: 395–418, 2005.
124. Schulte U, Hahn H, Konrad M, Jeck N, Derst C, Wild K, Weidemann S, Ruppersberg JP, Fakler B and Ludwig J. pH gating of ROMK (K(ir)1.1) channels: control by an Arg-Lys-Arg triad disrupted in antenatal Bartter syndrome. *Proc Natl Acad Sci U S A* 96: 15298–15303, 1999.
125. Sessoms-Sikes S, Honse Y, Lovinger DM and Colbran RJ. CaMKIIalpha enhances the desensitization of NR2B-containing NMDA receptors by an autophosphorylation-dependent mechanism. *Mol Cell Neurosci* 29: 139–147, 2005.

126. Sharma G and Stevens CF. A mutation that alters magnesium block of N-methyl-D-aspartate receptor channels. *Proc Natl Acad Sci U S A* 93: 9259–9263, 1996.
127. Siesjo BK. Acid-base homeostasis in the brain: physiology, chemistry, and neurochemical pathology. *Prog Brain Res* 63: 121–154, 1985.
128. Smith TC and Howe JR. Concentration-dependent substate behavior of native AMPA receptors. *Nat Neurosci* 3: 992–997, 2000.
129. Sobolevsky AI, Prodromou ML, Yelshansky MV and Wollmuth LP. Subunit-specific contribution of pore-forming domains to NMDA receptor channel structure and gating. *J Gen Physiol* 129: 509–525, 2007.
130. Sobolevsky AI, Rooney L and Wollmuth LP. Staggering of subunits in NMDAR channels. *Biophys J* 83: 3304–3314, 2002.
131. Stern P, Behe P, Schoepfer R and Colquhoun D. Single-channel conductances of NMDA receptors expressed from cloned cDNAs: comparison with native receptors. *Proc Biol Sci* 250: 271–277, 1992.
132. Sucher NJ, Akbarian S, Chi CL, Leclerc CL, Awobuluyi M, Deitcher DL, Wu MK, Yuan JP, Jones EG and Lipton SA. Developmental and regional expression pattern of a novel NMDA receptor-like subunit (NMDAR-L) in the rodent brain. *J Neurosci* 15: 6509–6520, 1995.
133. Sullivan JM, Traynelis SF, Chen HS, Escobar W, Heinemann SF and Lipton SA. Identification of two cysteine residues that are required for redox modulation of the NMDA subtype of glutamate receptor. *Neuron* 13: 929–936, 1994.
134. Tang CM, Dichter M and Morad M. Modulation of the N-methyl-D-aspartate channel by extracellular H+. *Proc Natl Acad Sci U S A* 87: 6445–6449, 1990.
135. Tang LH and Aizenman E. The modulation of N-methyl-D-aspartate receptors by redox and alkylating reagents in rat cortical neurones in vitro. *J Physiol* 465: 303–323, 1993.
136. Traynelis SF. pH modulation of ligand-gated channels. In: *pH and brain function*, edited by Kaila K and Ransom.BR. NY: Wiley Liss, 1998.
137. Traynelis SF, Burgess MF, Zheng F, Lyuboslavsky P and Powers JL. Control of voltage-independent zinc inhibition of NMDA receptors by the NR1 subunit. *J Neurosci* 18: 6163–6175, 1998.
138. Traynelis SF and Cull-Candy SG. Pharmacological properties and H+ sensitivity of excitatory amino acid receptor channels in rat cerebellar granule neurones. *J Physiol* 433: 727–763, 1991.
139. Traynelis SF and Cull-Candy SG. Proton inhibition of N-methyl-D-aspartate receptors in cerebellar neurons. *Nature* 345: 347–350, 1990.
140. Traynelis SF, Hartley M and Heinemann SF. Control of proton sensitivity of the NMDA receptor by RNA splicing and polyamines. *Science* 268: 873–876, 1995.
141. Vicini S, Wang JF, Li JH, Zhu WJ, Wang YH, Luo JH, Wolfe BB and Grayson DR. Functional and pharmacological differences between recombinant N-methyl-D-aspartate receptors. *J Neurophysiol* 79: 555–566, 1998.
142. Villarroel A, Burnashev N and Sakmann B. Dimensions of the narrow portion of a recombinant NMDA receptor channel. *Biophys J* 68: 866–875, 1995.
143. Vyklicky L, Jr. Calcium-mediated modulation of N-methyl-D-aspartate (NMDA) responses in cultured rat hippocampal neurones. *J Physiol* 470: 575–600, 1993.
144. Vyklicky L, Jr., Vlachova V and Krusek J. The effect of external pH changes on responses to excitatory amino acids in mouse hippocampal neurones. *J Physiol* 430: 497–517, 1990.
145. Wafford KA, Kathoria M, Bain CJ, Marshall G, Le BB, Kemp JA and Whiting PJ. Identification of amino acids in the N-methyl-D-aspartate receptor NR1 subunit that contribute to the glycine binding site. *Mol Pharmacol* 47: 374–380, 1995.

146. Westbrook GL and Mayer ML. Micromolar concentrations of Zn2+ antagonize NMDA and GABA responses of hippocampal neurons. *Nature* 328: 640–643, 1987.
147. Williams K. Separating dual effects of zinc at recombinant N-methyl-D-aspartate receptors. *Neurosci Lett* 215: 9–12, 1996.
148. Williams K. Interactions of polyamines with ion channels. *Biochem J* 325 (Pt 2): 289–297, 1997.
149. Williams K. Ifenprodil discriminates subtypes of the N-methyl-D-aspartate receptor: selectivity and mechanisms at recombinant heteromeric receptors. *Mol Pharmacol* 44: 851–859, 1993.
150. Williams K, Chao J, Kashiwagi K, Masuko T and Igarashi K. Activation of N-methyl-D-aspartate receptors by glycine: role of an aspartate residue in the M3-M4 loop of the NR1 subunit. *Mol Pharmacol* 50: 701–708, 1996.
151. Williams K, Kashiwagi K, Fukuchi J and Igarashi K. An acidic amino acid in the N-methyl-D-aspartate receptor that is important for spermine stimulation. *Mol Pharmacol* 48: 1087–1098, 1995.
152. Wollmuth LP, Kuner T and Sakmann B. Adjacent asparagines in the NR2-subunit of the NMDA receptor channel control the voltage-dependent block by extracellular Mg2+. *J Physiol* 506 (Pt 1): 13–32, 1998.
153. Wollmuth LP, Kuner T, Seeburg PH and Sakmann B. Differential contribution of the NR1- and NR2A-subunits to the selectivity filter of recombinant NMDA receptor channels. *J Physiol* 491 (Pt 3): 779–797, 1996.
154. Wong E, Ng FM, Yu CY, Lim P, Lim LH, Traynelis SF and Low CM. Expression and characterization of soluble amino-terminal domain of NR2B subunit of N-methyl-D-aspartate receptor. *Protein Sci* 14: 2275–2283, 2005.
155. Wyllie DJ, Behe P and Colquhoun D. Single-channel activations and concentration jumps: comparison of recombinant NR1a/NR2A and NR1a/NR2D NMDA receptors. *J Physiol* 510 (Pt 1): 1–18, 1998.
156. Wyszynski M, Lin J, Rao A, Nigh E, Beggs AH, Craig AM and Sheng M. Competitive binding of alpha-actinin and calmodulin to the NMDA receptor. *Nature* 385: 439–442, 1997.
157. Yao Y and Mayer ML. Characterization of a soluble ligand binding domain of the NMDA receptor regulatory subunit NR3A. *J Neurosci* 26: 4559–4566, 2006.
158. Zhang S, Ehlers MD, Bernhardt JP, Su CT and Huganir RL. Calmodulin mediates calcium-dependent inactivation of N-methyl-D-aspartate receptors. *Neuron* 21: 443–453, 1998.
159. Zheng F, Erreger K, Low CM, Banke T, Lee CJ, Conn PJ and Traynelis SF. Allosteric interaction between the amino terminal domain and the ligand binding domain of NR2A. *Nat Neurosci* 4: 894–901, 2001.

Molecular Properties and Cell Biology of the NMDA Receptor

Robert J. Wenthold, Rana A. Al-Hallaq, Catherine Croft Swanwick, and Ronald S. Petralia

Laboratory of Neurochemistry, National Institute on Deafness and Other Communication Disorders, National Institutes of Health, Bethesda, MD, USA, wenthold@nidcd.nih.gov

Abstract. NMDA receptors (NMDARs) play a distinct role at excitatory glutamatergic synapses, where they are usually localized with other ionotropic glutamate receptors, including AMPA receptors (AMPARs) and kainate receptors, as well as metabotropic glutamate receptors. Two features are essential to their specialized roles in synaptic plasticity and the excitotoxic cell death that results from their deregulation (66). First, NMDARs exhibit a voltage-dependent magnesium block, the removal of which requires depolarization of the membrane potential. Second, upon activation, the NMDAR channel passes sodium and, importantly, calcium into the neuron. Calcium is the universal second messenger in numerous intracellular signaling cascades and is critical in synaptic plasticity and mechanisms of neurotoxicity (288).

1 NMDAR Structure and Subunits

The functional NMDAR is a tetrameric protein complex with subunits comprised of three gene families: NR1, NR2, and NR3. All NMDAR subunits have similar membrane topology: an extracellular amino-terminal domain (ATD), 3 transmembrane domains (TM1, 3, 4), a re-entrant P loop at TM2 that lines the pore of the channel, and an intracellular C-terminal tail. The ATD is implicated in subunit assembly (210, 213) and in receptor sensitivity to modulators such as protons, polyamines, Zn^{2+}, and ifenprodil (76, 198, 256) (for review, see (122)). The extracellular S1 (N-terminal) and S2 (between TM3 and TM4) regions comprise the glycine-binding domains of NR1 (124, 162, 350) and NR3 (10, 40), and the glutamate-binding domains of the NR2 subunits (172, 318). The intracellular C-terminal tail, the sequence and length of which varies considerably among the different subunits, interacts with numerous binding partners (252, 364) and contains multiple phosphorylation sites (145, 283, 364).

NR1 subunits contain three alternatively spliced exons- one near the N-terminus (N1; exon 5) and two in the C-terminal tail (C1 and C2/C2'; exons 21 and 22). Thus, there are eight different splice forms, seven of which have been identified in vivo (174), that exhibit distinct regional distributions over development (175, 269, 392).

J.W. Hell, M.D. Ehlers (eds.), *Structural and Functional Organization of the Synapse*, DOI: 10.1007/978-0-387-77232-5_12, © Springer Science+Business Media, LLC 2008

The most common isoforms either contain the C1 and C2 cassettes (i.e., NR1-1a/b) or lack C1 but contain C2 (i.e. NR1-2a/b) with "a" and "b" denoting the absence or presence of the N1 cassette, respectively (174, 175). Importantly, NR1 isoforms differ in properties such as deactivation kinetics (280) and sensitivity to H^+, polyamines, and other modulators (77, 337). Knockout of NR1 leads to death shortly after birth (88), indicating that the NMDAR plays an essential role in normal nervous system function after birth.

Four genes constitute the NR2 subunit family: NR2A-D. Expression of these subunits is regulated developmentally and regionally (135, 173, 226). Briefly, NR2A and NR2B are the most widely expressed NR2 subunits in the forebrain. NR2B is expressed early in development, and decreases with age as NR2A expression increases. NR2C is abundant in adult cerebellar granule cells (324), and NR2D expression peaks at postnatal day (P)7 in the diencephalon and brainstem (75, 226). Physiological properties, including probability of opening, single channel conductance, and sensitivity to block by magnesium, differ significantly among the subunits (66). Additionally, subunit composition specifies binding partners and sensitivity to regulation by phosphorylation (145, 283, 364).

Transgenic mice have revealed important roles of the NR2 subunits. Mice lacking NR2A exhibit impaired synaptic maturation (91, 336) and synaptic plasticity (264). The NR2B knockout is fatal, indicating that NR2B is essential for normal neuronal function after birth. Mice overexpressing NR2B show enhanced learning and memory (326) and altered behavioral responses to inflammatory pain (361). Mice lacking both NR2A and NR2C exhibit a dramatic loss of spontaneous and evoked EPSCs in cerebellar granule cells and deficits in motor coordination (139). Mice lacking NR2D are less sensitive to stress and have altered monoaminergic neuronal function (222); overexpression of NR2D impairs NMDAR-dependent long-term potentiation (LTP) (243).

NR3A and NR3B are the most recently identified NMDAR subunits. NR3A expression is highest in early postnatal development (4, 50, 321). In contrast, NR3B remains high in adult brains and is expressed predominantly in motor neurons of the brainstem and spinal cord (19, 206). Numerous studies have reported incorporation of NR3 subunits into NR1- and NR2-containing NMDAR complexes (4, 254), resulting in decreased channel conductance and calcium permeability. Additionally, NR3 subunits associate with NR1 to form an excitatory glycinergic receptor (40). NR3A knockout mice exhibit enhanced NMDA responses and an increased density of dendritic spines in the cerebral cortex (58), consistent with a role for NR3A in depressing NMDAR activity during postnatal development.

Additional diversity of the NMDAR response arises from the complexity of subunit composition and localization. For example, different NMDARs may be expressed at different synapses within the same neuron (for review, see (154)). Furthermore, evidence exists for the incorporation of more than one type of NR2 subunit in each complex (i.e., di-heteromeric NR1/NR2X or tri-heteromeric NR1/NR2X/NR2Y), such as NR1/NR2A/NR2B receptors in hippocampal neuron synapses, NR1/NR2A/NR2C in cerebellar granule cell synapses, and NR1/NR2B/NR2D in substantia nigra dopaminergic neurons (for review, see (55)). Additionally, subcellular localization differs as some receptors are targeted to synaptic sites, both pre- and postsynaptic, while others are localized extrasynaptically (183, 319, 333) (reviewed in (154)). Thus, NMDAR subunits assemble in complexes that exhibit a range of physiological properties, binding partners, and subcellular localization.

2 Assembly of Functional NMDARs

2.1 Processing in the Endoplasmic Reticulum

The NMDAR is a typical integral membrane protein that is thought to follow a standard pathway in its biosynthesis. Most integral membrane proteins are synthesized by ribosomes that bind to the membrane of the endoplasmic reticulum (ER) shortly after the initiation of translation. This initial interaction with the ER occurs at sites known as translocons where a complex of proteins facilitates the translocation of the polypeptide across the ER membrane. As it is being translated, the growing polypeptide undergoes a complex process that will lead to its correct folding and orientation in the membrane. For multi-subunit complexes, assembly may occur co-translationally (86) or after translation is complete (45).

The NR1 subunit is made in large stoichiometric excess relative to NR2 (42, 127). Such a mechanism would ensure that sufficient amounts of NR1 partners are available for all newly synthesized NR2 subunits, and is consistent with the fact that the NR2 subunit is the main determinant of the function of the receptor complex (364). Unassembled NR1 is rapidly degraded with a half-life of one to two hours (127); this is a common fate for unneeded proteins in the ER. For the NMDAR, the first step is expected to be assembly of the NR1/NR2 dimer, followed by the formation of the tetrameric complex. Evidence for the initial formation of an NR1/NR2 dimer is based on the crystal structure of the mature protein (93), although the formation of such intermediates has not been detected biochemically. The mechanism underlying NR1/NR2 assembly remains largely unexplored. Interestingly, biochemical studies have shown that unassembled NR1 is present primarily as a homodimer (213) raising the question of whether this is simply a default route after failure to assemble with NR2 or if NR1 dimer formation is functionally important, perhaps as an intermediate in the interaction with NR2. NR1 can assemble with NR3 to form a glycine receptor (254), and the homodimerization of NR1 may play a role in the formation of NR1/NR3 complexes.

2.2 Retention Mechanisms for NMDARs

To prevent use of unassembled, misfolded and otherwise malformed complexes, the ER uses a rigorous quality control mechanism for proteins exported from the ER (83). All NR2 subunits and some NR1 splice variants are retained in the ER unless assembled (209). The NR1 subunit has at least one identified ER retention motif in its C-terminus. Of the NR1 splice variants, only NR1-1, one of the most abundant splice variants, does not reach the cell surface when expressed alone in heterologous cells (244). Using chimeric proteins of the C-terminus and single transmembrane proteins (tac = interleukin-2 receptor α subunit), a consensus ER retention motif, RXR, was identified in the C1 cassette (295, 314, 373). Retention by this site can be modulated by phosphorylation of nearby serine residues by PKA and PKC (293). The NR1-3 splice variant, which contains the C1 and C2' cassettes, is also expressed on the cell surface due to an export signal in the distal part of the C2' cassette (295, 314, 373). This site (-STVV) contains a PDZ binding domain as well as a binding site for COPII, which is involved in the export of proteins from the ER. In addition to regulating retention and export of the unassembled subunits, the same sites may affect the processing rate of the assembled complex, as discussed below. Assembly with NR2 negates the retention sites, and the assembled complex leaves the ER. Most of these studies have been done on chimeric proteins, so the role of these retention and export sites in full-length NR1 or the assembled NR1/NR2 complex, is not clear. However, the general principle of ER retention has been demonstrated in the intact animal. Mice in which the NR1 subunit is deleted in the hippocampus show an accumulation of NR2 subunits in the ER (92). While less is known about NR3, it may also use RXR motifs for ER retention and require association with NR1 for export from the ER (206).

The ER retention of the NR2 subunits is less studied and likely more complex. None of the four NR2 subunits has been shown to exit the ER unless assembled with NR1. The isolated C-terminus of NR2B appended to tac is retained in the ER, suggesting an ER retention site (119). Efforts to find a specific motif responsible for the retention have been unsuccessful thus far, although truncation of the C-terminus appended to tac leads to increased surface expression in constructs containing the C-terminus up to residue 1070 of NR2B (119). The NR2 subunit, truncated a few amino acids after TM4, is retained in the ER suggesting that there are one or more ER retention signals in the remaining part of the molecule. While NR1 and NR2 subunits can assemble and form functional receptors without most of the C-terminus, a short segment of the C-terminus that includes the motif (HLFY), which begins with the second amino acid after TM4 in NR2A and NR2B, is required to form functional receptors (119). Constructs lacking this motif or containing mutations in this motif are not functional. This seems to be due to the fact that without this motif, the assembled complex does not leave the ER, since the complex appears to be properly folded and assembled based on its ability to bind MK801 (119). Whether the HLFY motif functions as an export signal or is simply an area of the molecule critical to its proper folding remains to be determined. A later study (376) showed that the HLFY motif is not necessary, but can be replaced by alanines if the remainder of the C-terminus is absent. This may indicate the HLFY provides more of a structural role to ensure the proper orientation of the C-terminus, rather than an export motif that is involved in overriding ER retention. Interestingly, the NR2 subunit can form functional receptors when expressed in two distinct pieces (291). Two segments split shortly before TM4 can form functional receptors when expressed in heterologous cells with NR1. Presumably the C-terminus segment can assemble with the remainder of the molecule either in the ER or after exit from the ER. In either case, the retention mechanisms for both segments are overridden.

2.3 Preferential Assembly of Complexes

While the ER is an essential station through which nearly all integral membrane proteins must pass and is required for their assembly and folding, very little is known about how these events occur. In neurons expressing the two most common NR2 subunits, NR2A and NR2B, three types of receptor are formed, NR1/NR2A, NR1/NR2B and NR1/NR2A/NR2B. Since the functional properties of these three receptors are dramatically different, it is interesting to speculate that their formation may not be due simply to the random association of the subunits, but may be influenced by other factors. In the young adult CA1/CA2 neurons, where both NR2 subunits are relatively abundantly expressed, about 60% of NR2A and 70% of NR2B are present as diheteromeric complexes (NR1/NR2B and NR1/NR2A) based on co-immunoprecipitation results using detergent solubilized extracts (3). Interestingly, at P7, when NR2A is just beginning to increase in abundance, the diheteromeric complexes are present at the same relative amounts, suggesting that although NR2B is present in much greater abundance, only a fraction of NR2A assembles with NR2B. This would argue that formation of complexes is not dependent solely on the relative abundance of the NR2 subunits. It is interesting to speculate that assembly is influenced by factors such as the developmental stage or neuronal activity and that the proportion of receptors containing 2A, 2B or both 2A and 2B can be modulated. Similar effects could be expected for the NR1 splice variants. Although they have a less important effect on receptor function than NR2, there is evidence that they can influence receptor trafficking. Variants containing the C2' domain show an accelerated trafficking from the ER (228, 244) probably through the interaction with COPII, as noted above. Neuronal activity preferentially leads to increased expression of C2'-containing variants (228), and the combined effect would be an increased production of mature NMDARs. It has also been reported that there may be a preference for pairing NR2 subunits with different NR1 splice variants (300). The high level of regulation involving the ER retention of both NR1 and NR2 as well as the exit of the assembled receptor complex reflects the critical role of this receptor in neurons.

3 Post Golgi Sorting and Dendritic Trafficking

3.1 Packaging into Organelles

Following release from the ER, integral membrane proteins enter the Golgi apparatus and the *trans*-Golgi network (TGN) for additional processing and packaging into vesicular and tubulovesicular carriers for transport to the plasma membrane. In neurons many proteins are segregated into either axonal or dendritic compartments, which may involve the packaging of specific proteins into transport vesicles destined for either axons or dendrites either at the level of the TGN or at endosomal compartments soon after release from the TGN. The sorting may also occur later, and the mechanisms by which proteins are sorted into axonal and dendritic compartments can vary. For example, vesicle-associated membrane protein 2 (VAMP2) is delivered to both axons and dendrites but is preferentially removed from dendrites through endocytosis, leading to an accumulation in axons (284). These findings would imply that trafficking in neurons cannot be accomplished with only two transport vesicles, one for axons and one for dendrites, but that multiple routes of delivery must exist for each domain. It is equally unlikely that each protein has its own transport vesicle, but how many there are and how proteins are sorted for packaging into specific vesicle populations remains largely unexplored. Another complication involves explaining sorting of proteins that are present in both axons and dendrites in subsets of neurons, as is the case for NMDARs (discussed below). This would require that different mechanisms be used in different neurons. While evidence suggests that most axonal and dendritic membrane proteins are delivered while associated with some sort of intracellular organelle, proteins can move by lateral diffusion within the plasma membrane. NMDARs and AMPARs can both move rapidly within the plasma membrane ((338); described more in Section 5), and one study showed that the first appearance of new AMPARs following irreversible inactivation of existing receptors is at the cell body (2).

NMDARs can be shown in adult neurons with immunocytochemistry to be associated with intracellular tubulovesicular structures near the TGN and in dendrites (Fig. 1). In young (3 days in vitro (DIV)) cultures of cortical neurons, mobile vesicular structures containing NMDARs can be identified through live imaging (358). In these studies, the NR1 subunit was tagged with EGFP or DsRed and transfected into cultured neurons. Live imaging showed that organelles containing the tagged NR1 move along microtubules at rates averaging about 4 μm/min, somewhat slower than that seen with fast axonal transport (about 30 μm/min). Using co-transfection with tagged AMPARs, about 61% of the NR1 clusters also contained AMPARs. However, when only mobile clusters were analyzed, 28% of the NR1 was associated with AMPARs; this latter figure gives a more accurate estimate of actual transport vesicles that contain both receptors since the total contains stationary clusters that may be nascent synapses, which contain both receptors. Nevertheless, this modest overlap (28%) indicates neither a specific exclusion of the two populations nor an active mechanism for co-localization. Rather, it suggests that the packaging is poorly controlled or entirely random. An important caveat, however, is that these studies were done using over-expressed proteins, and aberrant packaging due to artificially high levels of proteins may have occurred. In a subsequent study (359), the same authors found that transport of NMDARs involved a continuous recycling of receptors from the cell surface to intracellular organelles. Interestingly, these receptors were associated with SAP102, which appeared to remain associated with the receptor complex. In their earlier study (358), these authors had found that PSD-95 was not associated with NMDARs during their transport.

3.2 Association with Proteins During Transport

Some studies have addressed the molecular nature of transport vesicles and motors that may be involved in the movement of NMDARs in dendrites. These studies show that the NMDAR is part of a large molecular complex, although the different studies identify different associated proteins; such results would fit with the idea that the

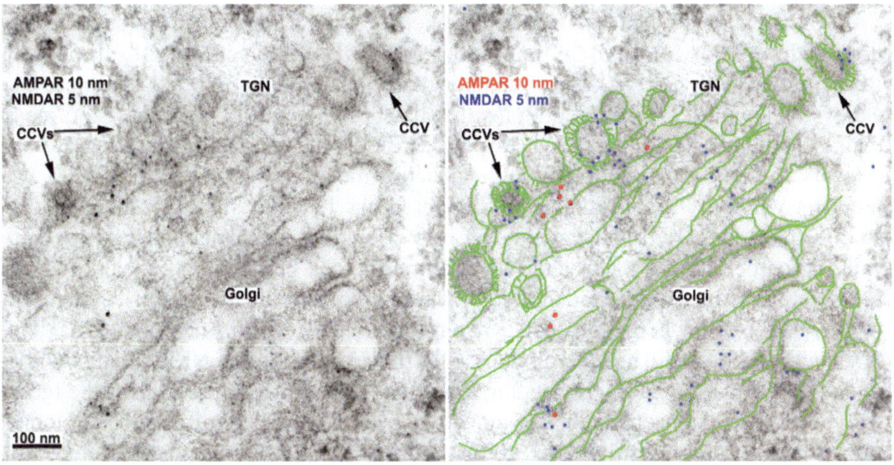

Fig. 1. NMDARs can exit the Golgi/TGN via clathrin-coated vesicles. Double-labeling using rabbit polyclonal antibodies to AMPARs (combination of 3 antibodies: GluR1, GluR2, GluR2/3) and 10 nm immunogold, and mouse monoclonal antibodies to NR1 (3 antibodies) and 5 nm immunogold, in a section of a neuron soma in the CA1 stratum pyramidale of the hippocampus from a P10 rat. In these studies, identification of 5 nm gold particles, where unclear, was confirmed by detailed examination and photography at higher magnifications. On the right, the micrograph is repeated and major structures are highlighted in color (*green* for membranous structures and clathrin coats; *red* for AMPAR-10 nm gold; *blue* for NR1-5 nm gold). Note that AMPA and NMDA receptor labeling is localized mainly separately in both the Golgi apparatus and trans-Golgi network (TGN), and NR1 is found alone in several clathrin-coated vesicles (CCVs) budding off of the TGN (R.S. Petralia, Y.-X. Wang, and R.J. Wenthold, unpublished data; (259, 260)).

receptor is associated with more than one type of protein complex and transport vesicle during its transport to and from synapses and extrasynaptic locations. KIF17, a dendrite-specific microtubule-dependent molecular motor binds directly to the first PDZ domain of mLin-10 and transports a large protein complex containing the NMDAR (297). The NMDAR is linked to the complex through a PDZ interaction of NR2B and mLin-7, which then binds to mLin-2, which forms a link with mLin-10. KIF17 complexes move in dendrites at an average speed of 46 μm/min, which is similar to fast axonal transport but significantly faster than the movement of NR1 (4 μm/min) as described above. The expression of KIF17 and NR2B are closely linked since decreasing KIF17 with antisense oligonucleotides leads to a corresponding decrease in NR2B (110). Interestingly, this decrease in NR2B is offset by a corresponding increase in NR2A, which suggests that the KIF17 model applies only to NR2B-containing receptors. Mice overexpressing KIF17 also up-regulate NR2B expression and showed enhanced learning and memory (368). Further evidence for a microtubule-based movement of NMDARs is that microtubule-destabilizing drugs decrease the number of NMDARs on the surface of dendrites (383).

Transport of NMDARs may also involve association with membrane-associated guanylate kinases (MAGUKs), particularly SAP102. With its multiple domains,

SAP102 is capable of interacting simultaneously with a number of proteins in addition to the NMDARs. SAP102 (and other MAGUKs) interacts directly through its PDZ domain with Sec8 (285), a member of the mammalian exocyst complex. The exocyst is a complex of eight proteins that has been studied in both yeast and mammalian cells and has been shown to play a role in the intracellular trafficking and delivery to the plasma membrane of a subset of membrane proteins (126, 378). It is not clear why only some proteins are delivered via the exocyst route, although some studies have suggested that the exocyst may provide a mechanism to more specifically target delivery to a particular location on the cell membrane (215). Expression of a Sec8 construct with a mutated PDZ binding domain blocks the surface delivery of the NMDAR in both transfected heterologous cells and in neurons (285). This effect can be negated in heterologous cells by expressing an NR2 subunit with a mutated PDZ binding domain, leading to normal cell surface expression of the receptor, indicating that a separate pathway can be used for delivery of non-PDZ interacting proteins. These results suggest that NMDARs are routinely associated with SAP102 (or another MAGUK) and the exocyst complex at some point in their biosynthetic pathway. Expression of Sec8 with a mutated PDZ binding domain leads to a buildup of NMDAR at several intracellular sites, including the TGN, but the exact point where the exocyst interaction begins and ends is not known.

Another protein that interacts with SAP102 (and other MAGUKs) and influences NMDAR trafficking is mPins (the mammalian homologue of the *Drosophila melanogaster* partner of inscuteable), also referred to as LGN (286). mPins, a multiple domain protein (26, 69), interacts with SAP102 through its domain known as its linker region and the SH3/GK domain of SAP102. Unlike the exocyst complex, which interacts with all NMDARs, mPins plays a more modulatory role in regulating NMDAR trafficking; while NMDAR surface expression can be nearly eliminated by dominant-negative Sec8, similar constructs of mPins have less dramatic effects. Over-expression of mPins or expression of individual domains of mPins, which may act as dominant-negative constructs, influences the number of NMDARs at synapses, as well as spine size and number. One binding partner of mPins is the G-protein subunit, $G_{\alpha i}$ (367), raising the interesting possibility that NMDAR trafficking could be influenced through mPins by G-protein signaling. A complex consisting of $G_{\alpha i}$, SAP102 and the NMDAR can be identified in brain by co-immunoprecipitation, and over-expression of $G_{\alpha i}$ enhances NMDARs on dendrites of cultured hippocampal neurons (286).

A particular motor that could move a SAP102 based complex has not been identified, although KIF1Bα interacts directly with several MAGUKs and could fill this purpose (224). While microtubule-based motors are expected to be involved throughout transport in the dendrite, myosin motors are likely involved in the delivery to and from the plasma membrane. NMDARs are regulated by myosin light chain kinase and interact directly with myosin regulatory light chain (6, 179).

The studies discussed above point to two different mechanisms for the delivery of NMDARs, one SAP102-based and the other mLin-7/mLin-2/mLin-10-based. They have very different complexes of proteins that associate directly or indirectly with the NMDAR, but an important common feature is that both rely on PDZ interactions for the formation of the complex, pointing out the importance of PDZ proteins not only to NMDAR organization at the synapse, but also in NMDAR trafficking. With the identification of at least two distinct mechanisms for dendritic transport of NMDARs, a challenge is to determine at which points in the trafficking of NMDARs the two mechanisms are functioning. Given the complex nature of the trafficking of membrane proteins through multiple organelles on their way to and from the cell surface, it is likely that there are additional mechanisms with other associated proteins involved. Like most other membrane proteins, NMDARs are endocytosed and recycled to the plasma membrane. As discussed above, this may be a normal process for their delivery in young neurons. Are new and recycled receptors simply mixed together or are their deliveries separated? Perhaps this is one role for the exocyst, which may associate only with newly-synthesized proteins.

4 NMDARs at the Synapse

NMDARs are present at most glutamatergic synapses in the brain (Fig. 2). Although the NMDAR is relatively stable at the synapse, several studies have shown that a number of conditions can cause rapid changes in synaptic NMDARs. For example, activation of type 1 metabotropic receptors causes a rapid internalization of synaptic NMDARs (307), and some forms of LTP and long-term depression (LTD) involve changes in synaptic NMDARs (14, 22, 106, 117, 225, 238). Synaptic NMDARs recycle and exchange with cytoplasmic and extrasynaptic receptors (267, 279, 334). NMDARs may also be altered during homeostatic plasticity (255) as neurons maintain a normal firing rate by scaling their synaptic strength (340).

As described in the section on structure, most NMDARs form tetramers containing two obligatory NR1 subunits and two NR2 subunits. Subunit composition of synaptic NMDARs shapes the kinetics of NMDAR-mediated currents. For instance, the presence of the NR2A subunit confers a faster decay time for NMDAR-mediated currents than the presence of the NR2B subunit (347). Therefore, mature synapses exhibit shorter periods of NMDAR activity than immature synapses, which may lead to greater precision of coincidence detection from converging synaptic inputs. Both the number and composition of synaptic NMDARs are closely regulated by factors including (1) their addition to, and removal from, the synapse, (2) their posttranslational modifications such as phosphorylation, and (3) their interaction with associated proteins; all of which can be modified in response to neuronal activity. Each of these factors will be discussed in further detail.

4.1 Regulation of Synaptic NMDARs

4.1.1 Addition and Removal of NMDARs

The insertion of NR2A- and NR2B-containing NMDARs into synapses occurs under different conditions. In hippocampal slices cultured from rats at P7, NR2B-containing receptors are added constitutively, whereas NR2A-containing receptors are delivered after activation of pre-existing synaptic NMDARs (12). Similar results were found in the visual cortex, where visual experience leads to NMDAR activation that is responsible for the rapid insertion of NR2A-containing receptors (272) and is reversed after visual deprivation (263, 271). Likewise, an initial phase of learning in

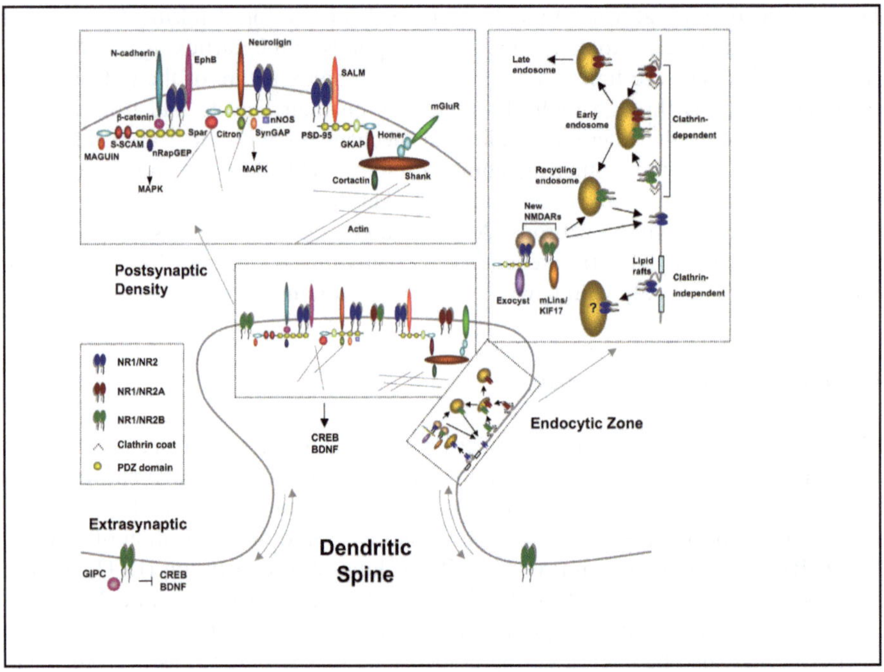

Fig. 2. Summary of NMDAR trafficking at the synapse in mature forebrain. NMDARs are clustered at the postsynaptic membrane. Their presence in this region is regulated by interactions with PDZ proteins and adhesion molecules that can serve as scaffolds to bring them into contact with signaling and regulatory proteins (see sections on interacting proteins). The number of NMDARs at the synapse can also be modulated by their addition or removal from the synaptic membrane (see Section 4.1.1). To be internalized, NMDARs move to an endocytic zone located in the perisynaptic membrane to undergo clathrin-dependent endocytosis. NMDARs may also be internalized by a clathrin-independent pathway, although this mechanism is not well defined. The insertion of newly synthesized or recycled NMDARs may occur in the endocytic zone or the postsynaptic density. In addition, mobile pools of NMDARs may diffuse laterally between the extrasynaptic and synaptic membranes (see Section 5). Downstream signaling of extrasynaptic and synaptic NMDARs may differ, resulting in deactivation or activation, respectively, of CREB and BDNF.

the olfactory system requires NMDAR activation and enhances the expression of NR2A-containing receptors relative to NR2B-containing receptors in the piriform cortex (270). However, there also may be regional differences in the mechanisms of synaptic insertion for NR2A- and NR2B-containing receptors. In hippocampal slices prepared from rats at P6-7 and cultured for approximately 2.5 days, the insertion of NR2B-containing receptors does not change the kinetics of synaptic NMDARs (12), whereas the insertion of NR2A-containing receptors induces a faster decay time (12, 272). This suggests that both are replacing pre-existing NR2B-containing receptors. In contrast, in cerebellar granule cells harvested at P6 and cultured for 6-8 DIV, overexpression of NR2B slows the decay of NMDAR-mediated currents and overexpression of NR2A quickens their kinetics (268), indicating that both types of NMDARs can replace the other form.

The mechanism that determines the number of synaptic NMDARs is unknown, although it appears to be distinct from the mechanism regulating extrasynaptic receptors. In cultured cerebellar granule cells, overexpression of NR2A or NR2B subunits causes an increase in the total number of functional NMDARs, whereas overexpression of the NR1 subunit does not (268). However, overexpression of NR2 subunits does not change the number of synaptic receptors, indicating that synaptic receptors are not regulated solely by availability. On the other hand, extrasynaptic receptors appear to be determined, at least to some extent, by the number of receptors synthesized. It is important, however, to point out that this latter property may vary among different neuron types.

The removal of synaptic receptors likely requires at least three stages. First, the receptor is released from its anchor that holds it at the synapse. Second, for clathrin-dependent internalization, the receptor moves in the plasma membrane to the edge of the synapse either by diffusion or some form of active transport (see Section 5). Finally, the receptor is removed from the plasma membrane. Most current evidence shows that NMDARs are endocytosed through a clathrin-dependent mechanism (176, 237, 258, 279, 348). For clathrin-dependent internalization from the plasma membrane, proteins often contain an endocytic motif that is recognized by the AP-2 adaptor complex. Interestingly, NR2A and NR2B subunits use different motifs for removal. The NR2B subunit has a tyrosine-based YXXΦ consensus motif, YEKL, near its C-terminus that can bind to AP-2 and promote clathrin-dependent internalization (279). In contrast, a homologous tyrosine-based YKKM motif on the NR2A subunit is not involved with NMDAR endocytosis, but rather, a dileucine motif (LL), which also binds to the AP-2 complex, is involved in endocytosis of NR2A subunits (176). Another YXXΦ consensus motif of the NR2A subunit near its last transmembrane domain binds to AP-2 and leads to use-dependent rundown of NMDAR current (348). In fact, tyrosine-containing motifs present near the last transmembrane domain of all NR2 subunits are involved in NMDAR internalization and may target NMDARs to late endosomes for degradation (294). Therefore, multiple sites on the C-termini of both NR1 and NR2 subunits are involved in endocytosis of NMDARs, and additional sites are likely to be identified.

Recent evidence suggests that NMDARs may also be removed by nonclathrin-mediated pathways. For instance, prolonged activation of NMDARs leads to their degradation by calpain in a process that is not dependent on clathrin-mediated endocytosis (371). In clathrin-independent pathways, endocytosis occurs via uncoated invaginations in lipid rafts (149, 234), microdomains of the membrane proposed to act as platforms for neurotransmitter signaling (5). Although little is known about clathrin-independent internalization of NMDARs, it may occur in concert with clathrin-mediated pathways. Similar to the paradigm for internalization of epidermal growth factor receptors (302), higher levels of glutamate in the synapse may trigger clathrin-independent removal of NMDARs. The site of clathrin-independent endocytosis is undefined.

Once NMDARs are removed from the synaptic membrane, the intracellular pathways of NR2A- and NR2B-containing receptors diverge. Both encounter early endosomes, but then most NR2B-containing receptors are sent to recycling endosomes, whereas most NR2A-containing receptors travel to late endosomes (176). This implies that although NR2B-containing receptors are endocytosed more readily than NR2A-containing receptors (176), they are more likely to recycle and return to the synapse.

4.1.2 Modification of NMDARs by Phosphorylation

Members of the protein tyrosine kinase (PTK) family include Src, Fyn, Lyn, Lck, and Yes. Phosphorylation of tyrosine residues on NR2A and NR2B subunits potentiates NMDAR-mediated currents (357). Tyrosine residues exist on NR2A and NR2B subunits (283). The YEKL internalization motif on the NR2B subunit discussed above contains the tyrosine residue Tyr1472 that is a major substrate for Fyn kinase (231). In cerebellar granule cells, phosphorylation of this residue by Fyn increases the synaptic localization of NMDARs by preventing their internalization (267). Also, in striatal tissue of Fyn knockout mice, basal tyrosine phosphorylation of NR2A and NR2B is reduced and NMDARs fail to redistribute to the synapse upon stimulation of dopamine receptors (72).

Calcium influx through NMDARs generates the production of cyclic AMP, which activates cyclic AMP-dependent protein kinase (PKA). PKA can phosphorylate NR1, NR2A, and NR2B subunits. This phosphorylation can increase NMDAR activity by promoting synaptic targeting (54) or enhancing calcium permeability of NMDARs (305). PKA also can associate indirectly with NMDARs via Yotiao, allowing it to overcome the constitutive activity of protein phosphatase 1 (366). In addition, PKA can have indirect effects on NMDARs, such as in its phosphorylation of the immediate early gene cAMP-responsive element-binding protein (CREB), which can then activate other signaling cascades such as the MAP kinase pathway or induce new protein synthesis (352). Due to its calcium-dependent role, PKA is important for the induction of LTP and spatial long-term memory (1).

Protein kinase C (PKC) also phosphorylates NR1, NR2A, and NR2B subunits. PKC activation increases the amplitude of NMDAR-mediated currents and channel open probability (43, 101, 374), in part by increasing the number of NMDARs inserted into the synaptic membrane (167). However, these enhancements are probably due to indirect signaling through second messenger cascades (390). Direct phosphorylation of NMDARs by PKC induces rapid dispersal of NMDARs from the synaptic to extrasynaptic membrane (87, 329) and increases their sensitivity to inactivation by intracellular Ca^{2+} (202).

Activation of casein kinase II (CKII) potentiates NMDAR activity by prolonging channel open time (187) and may contribute to the induction of LTP (39). CKII may also facilitate the replacement of NR2B-containing receptors with NR2A-containing receptors because it phosphorylates serine residue Ser1480 in the PDZ binding domain of the NR2B subunit, thereby blocking the interaction of NR2B-containing receptors with PDZ proteins and decreasing their surface expression in an activity-dependent manner (49).

Cyclin-dependent kinase 5 (Cdk5) phosphorylates NR2A subunits at serine residue Ser1232 (184). This modification is important for mediating NMDA-evoked synaptic currents during the induction of LTP (184), as well as ischemic degeneration of hippocampal CA1 pyramidal neurons (355).

Activity-dependent calcium influx induces autophosphorylation of Ca^{2+}/calmodulin-dependent protein kinase II (CaMKII), leading to its increased localization at the postsynaptic density (67, 299). CaMKII can bind NR1, NR2A, or NR2B subunits (100, 180), although its interaction with NR2B-containing receptors is much stronger than with NR2A-containing receptors (320). The interaction of CaMKII and NR2B-containing receptors locks CaMKII in an active conformation (16), enabling it to phosphorylate AMPARs and mediate LTP (7, 13, 63).

4.1.3 Other Modifications

Ubiquitin conjugation involves the sequential actions of ubiquitin-activating enzymes (E1), ubiquitin-conjugating enzymes (E2), and ubiquitin ligases (E3) (362). The E3 ubiquitin ligase confers protein specificity, in part by forming a complex with F-box proteins. Many postsynaptic proteins undergo ubiquitination and subsequent degradation by proteasomes, and this process may be dependent on neuronal activity (78). It is unclear whether NMDARs undergo ubiquitination. In synaptosomal membranes from cortical neurons, ubiquitination of the NR1, NR2A, and NR2B subunits was not detectable (78). However, in a separate study, expression cloning identified F-box protein 2 (Fbx2) as an interactor of the NR1 subunit (143). Fbx2 was shown to induce the ubiquitination of the NR1 subunit in HEK293T cells through binding to high-mannose glycans on the extracellular domain of the NR1 subunit. Interestingly, effects of Fbx2 inhibition on NMDAR ubiquitination were only observed under conditions of enhanced neuronal activity (143).

NMDARs are also modified by S-nitrosylation of cysteines of the NR2A subunit by nitric oxide, which can down regulate ion channel activity (47).

4.1.4 Interactions with PDZ Proteins

NMDARs can bind to the PDZ domains of all MAGUKs (156, 171, 235). MAGUKs are abundant at all excitatory synapses, although their expression varies during development. In the hippocampus, the expression of SAP102 peaks during early development whereas the expressions of PSD-95 and PSD-93 are highest during adulthood (82, 287). This parallels the developmental shift from NR2B-containing receptors to NR2A-containing receptors at synapses, suggesting that SAP102 may preferentially bind to NR2B-containing receptors and PSD-95 and PSD-93 may selectively bind to NR2A-containing receptors (287, 336).

MAGUKs may function as anchors that retain and cluster proteins such as NMDARs at a particular cellular location. Studies on animal models have shown that the NR2 C-terminus is required for synaptic localization of NMDARs (312, 317). However, these results, in which nearly the entire C-terminus was deleted, could be due to the loss of other components of the C-terminus rather than only the PDZ binding domain. Studies using transfection of wild type and mutant NMDAR subunits have shown that NR2B-containing receptors depend on the PDZ binding domain for entry/retention at the synapse (267), while NR2A-containing receptors do not (12, 267, 327). Overexpression studies are conflicting, for in hippocampal neurons PSD-95 overexpression does not affect the synaptic clustering or function of NMDARs (81, 290), although in cerebellar granule cells it does promote the synaptic insertion of NR2A-containing receptors (196). Knockout mice lacking PSD-95 (218) or PSD-93 (208) have no significant loss of synaptic NMDARs. However, double knockout mice lacking both PSD-95 and PSD-93 show a reduction in NMDAR-mediated currents (82), indicating that compensatory mechanisms may stabilize NMDARs at synapses. NMDARs are less affected than AMPARs by changes in MAGUK expression. Acute knockdown studies using shRNA for PSD-95 show either no change in NMDARs while AMPARs decrease (82) or a moderate reduction in NMDA EPSCs (79).

MAGUKs also function as scaffolds that bring NMDARs into contact with regulatory and signaling molecules. For example, interactions with MAGUKs may inhibit internalization of NR2B-containing NMDARs by promoting the phosphorylation of Tyr1472 by Fyn kinase (72, 267), thus preventing the interaction of the NR2B subunit with AP-2 (see Section 4.1.1). Some proteins can be linked to NMDARs by binding to adjacent PDZ domains of MAGUKs. Examples include neuronal nitric oxide synthase (nNOS), which can mediate NMDAR-induced excitotoxicity (36), and synaptic Ras-GTPase activating protein (SynGAP), which couples NMDARs to the MAP kinase pathway (148). Other proteins may be linked to NMDARs by guanylate kinase-associated protein (GKAP), a central molecule that binds to the GK domain of MAGUKs (146). For instance, GKAP binds to Shank, which binds to Homer dimers that then bind to various other proteins including metabotropic glutamate receptors or IP3 receptors (230, 339).

NMDARs are linked to the actin cytoskeleton through interactions (147) such as PSD-95-SPAR-actin, GKAP-Shank-cortactin-actin, or PSD-95-citron, which is a target of Rho (94). NMDARs can also associate with actin filaments via actinin (372) or α1-chimerin, an inhibitor of Rac1 (344). This suggests a mechanism for NMDAR activity to control overall spine structure (see Section 6.2).

Other PDZ proteins that can also act as scaffolding molecules for NMDARs include synaptic scaffolding molecule (S-SCAM) (125) and channel interacting PDZ domain protein (CIPP) (161). CIPP is highly expressed in thalamus, colliculi, cerebellum, and brainstem, and can also bind to adhesion molecules (161). S-SCAM can link the MAP kinase pathway to NMDARs by binding to nRap GEP (neural GDP/GTP exchange protein for Rap1 small G-protein). S-SCAM (or PSD-95) also binds to membrane-associated guanylate kinase interacting protein (MAGUIN) (377), which may affect cell polarity (242).

4.1.5 Interactions with Adhesion Molecules

NMDARs interact directly and indirectly with a variety of adhesion molecules at the synapse, including some that act transynaptically to link NMDARs to the presynaptic terminal. Adhesion molecules play a role in the formation, maturation, function, and plasticity of synapses (56). Similar to PDZ proteins, one of their functions may be to cluster NMDARs at synapses. Recent evidence suggests that PDZ proteins and adhesion molecules work together (95, 278), perhaps in a compensatory manner.

The binding of presynaptic ephrin-B to postsynaptic EphB receptors induces a direct interaction of EphB receptors with the extracellular domain of NR1 subunits (57) that enhances NMDAR-mediated synaptic function (108, 120). Activated EphB receptors also have tyrosine kinase activity that can indirectly potentiate NMDAR-mediated Ca^{2+} influx (325) and lead to synapse formation (57). EphB is critical for the formation of the postsynaptic density and dendritic spines (121).

A family of synaptic adhesion-like molecules (SALMs) was recently discovered on the basis of its interactions with MAGUKs (153, 353). Of the five members, SALMS1-3 have PDZ binding domains which could allow SALMs to be linked indirectly with NMDARs at the synapse. However, SALM1 can also bind directly to the extracellular domain of the NR1 subunit when expressed in heterologous cells (353). Although a presynaptic ligand has not yet been found, SALM1 and SALM2 can induce synapse formation and neurite outgrowth (153, 353).

Other families of transynaptic adhesion molecules may act indirectly to affect synaptic NMDARs ((56); see also the Section 6.1 below). Presynaptic neurexin binds to postsynaptic neuroligin, which interacts with the MAGUKs. Neuroligin localization at synapses may help recruit NMDARs to synapses and determine if developing synapses become excitatory or inhibitory (182). Similarly, homophilic interactions between presynaptic and postsynaptic cadherins induce interaction with postsynaptic β-catenin, which can bind directly to the PDZ scaffold S-SCAM and may control its synaptic targeting (236). NrCAM binds directly to SAP102 (59) and has been implicated in neuronal positioning and dendrite orientation (62). NCAM180 may interact directly with the NR2A subunit (96) to help mediate synaptic plasticity (31). Furthermore, integrins mediate the developmental switch from NR2B-containing receptors to NR2A-containing receptors at synapses (41) and potentiate NMDAR-mediated currents by activating src tyrosine kinases (21, 190).

4.2 Presynaptic NMDARs

NMDARs are also found at the presynaptic terminal. The effect of presynaptic NMDARs on synaptic transmission varies among regions of the central nervous system. In some areas they negatively regulate neuronal excitability. They can inhibit glutamate release in spinal cord (11), enhance GABA release in cerebellum (71), or mediate LTD in the cerebellum (38), cortex (304), and retinotectal projections (188). However, in other brain regions they positively impact synaptic transmission. Activation of presynaptic NMDARs enhances glutamate release in hippocampal CA1 pyramidal neurons during early postnatal development (203) and is involved in associative LTP of the lateral amygdala (128). Interestingly, the function of presynaptic NMDARs in the entorhinal cortex decreases during development but can be elevated during epileptic states (375). Presynaptic NMDARs may preferentially contain NR2B (138, 375) or NR2D subunits (11, 203, 328) along with NR1 subunits with exon 5 inserts in the N-terminal domain (356).

5 Extrasynaptic NMDARs

5.1 Distribution

NMDARs in extrasynaptic membrane locations, i.e., anywhere except within the synaptic junction, consist largely of NR1/NR2B and NR1/NR2D receptor complexes (30, 183, 333); in fact, NR1/NR2D may never enter the synapse (30). In much of the adult brain, typical mature synapses contain mainly NMDARs with NR2A, while NR2B-containing NMDARs are common in the extrasynaptic membrane (255, 333, 336). One explanation for this pattern is that NR2B-containing NMDARs are removed more readily from synapses than NR2A-containing NMDARs (176, 255, 336); in contrast, NR2A-containing NMDARs form more stable associations with PSD-95 at synapses (12, 255, 336). Nevertheless, this separation of NR2A in synapses and NR2B in the extrasynaptic membrane is not absolute, since some NR2B occurs in adult synapses ((257, 333); either as NR1/NR2B or NR1/NR2A/NR2B). Conversely, extrasynaptic NMDARs can contain NR2A (183, 281, 327). Unlike hippocampal neurons, cerebellar granule cells lose all of their NR2B, both synaptic and extrasynaptic, in later maturation and have both NR2A and NR2C in the adult (91, 281).

5.2 Function

The function of the predominant separation of NMDARs into synaptic NR2A-containing and extrasynaptic NR2B-containing receptors in adults is not clear, but may relate to a separation of short distance (synaptic) and long distance (extrasynaptic) signals. Thus, a precise release of glutamate at the synapse could activate synaptic receptors, while extensive release of glutamate followed by spillover into the extrasynaptic spaces could be required to activate extrasynaptic receptors. In this way, extrasynaptic NMDARs could elicit plastic changes to compensate for synapse overactivity. NR2A has been reported to be associated mainly with LTP and NR2B mainly with LTD (193, 369) although several studies show that both NR2A and NR2B can induce LTP (20, 363). However, problems in specificity of the antagonist used in the former studies make their results for NR2A problematic (90, 233, 363). Control of plasticity may depend more on location (synaptic versus extrasynaptic) rather than NR2 type. Thus, extrasynaptic NMDARs show a greater tendency for rundown (decline in peak current with repeated, sustained agonist application) than synaptic NMDARs, apparently for both NR2A- and NR2B-containing NMDARs (183). Rundown may protect neurons from damage following hyperactivity leading to glutamate spillover. In addition, activation of extrasynaptic NMDARs generates inhibitory currents via BK-type calcium-activated potassium channels in rat olfactory bulb granule cells (134). Also, synaptic and extrasynaptic NMDARs can mediate opposite long-term changes in neuron gene expression, probably due to differences in their local, associated signaling complexes (reviewed by (214, 346, 388)). The signaling pathways of synaptic NMDARs selectively activate a regulatory cascade involving Ras, the kinase ERK, and CREB, while that of extrasynaptic NMDARs inactivate them. Thus, synaptic NMDARs promote CREB activation and induction of BDNF gene expression, while extrasynaptic ones shut off CREB and inhibit BDNF gene expression (346). Activation of synaptic NMDARs may up-regulate pro-survival genes and down-regulate pro-death genes, while activation of extrasynaptic NMDARs does the exact opposite (214, 388). Lastly, a more general function of extrasynaptic NMDARs is to form a mobile reserve pool of receptors for interchange with those in the synapse (338) (see below).

5.3 Localization, Mobility, and Relation to Synaptic Receptors

Relatively little is known about the arrangement and distribution of extrasynaptic NMDARs on the neuron cell membrane; in fact, relatively little is known about general protein distribution on the plasma membrane (163, 189). Some of the functional studies noted above indicate the presence of some extrasynaptic NMDARs that may be stabilized in extrasynaptic domains, but other extrasynaptic NMDARs may be moving to sites of endocytosis or moving out of sites of exocytosis. In some cases, extrasynaptic NMDARs may associate with specialized non-synaptic junctions such as the cadherin-based attachment plaques of the cerebellar glomeruli. These contain abundant NMDARs and associated PSD-95, and thus glutamate spillover from adjacent synapses in the glomerulus may control the overall stability of the glomerulus (262). Other examples of specialized extrasynaptic locations of NMDARs are those on or near exocytic or endocytic zones or in the perisynaptic region directly surrounding the synapse. In addition, there are extrasynaptic NMDARs of presynaptic membranes.

Clathrin-mediated endocytosis of NMDARs (and glutamate receptors in general) may be localized mainly to the extrasynaptic membrane on the side of a spine (24, 258, 273). While this location is typical for endocytosis in mature synaptic spines, in immature synapses and early postnatal ages, the site of endocytosis often is located at the edge of the postsynaptic density (PSD) in the perisynaptic region (24, 258). Only a few examples of labeling for NMDARs in clathrin-coated pits have been found ((258) and Petralia et al., unpublished data). Also, RNAi knockdown of CPG2 (candidate plasticity gene 2), which is localized specifically to the postsynaptic endocytotic zone of excitatory synapses, increases the number of postsynaptic clathrin-coated vesicles, including some that contain NMDARs, and increases the number of surface NR1 and AMPAR GluR2 molecules (53).

Targeting of the NMDARs to sites of endocytosis may be influenced by stabilization of NMDARs on the surface, especially in extrasynaptic locations. GIPC is associated with endocytosis and contains a PDZ domain that binds to the C-terminal ESDV domain of NR2B. Preliminary studies indicate that GIPC helps regulate the surface stabilization, endocytosis, and recycling of NMDARs, but the localization of extrasynaptic GIPC/NMDAR associations is not clear (379, 380). The adaptor PACSIN1/syndapin1 binds to the C-terminal of the NR3A, as well as to dynamin and the actin-organizing protein N-WASP, and mediates the selective endocytosis of NR3A-containing NMDARs during postnatal development. This mechanism is activity-dependent and may help regulate synaptic maturation (253). In this case, NR3A is localized largely in the outer zone of the synapse and in the adjacent perisynaptic area; it is suggested that PACSIN1 might facilitate movement of NR3A out of synapses and into the endocytic zone on the side of the spine.

It is not clear exactly where NMDARs and other glutamate receptors enter the plasma membrane via exocytosis (105, 152). Labeling for AMPARs is often seen in pit-like structures, without evident clathrin coats, on the sides of spines in adults (258, 259, 332). These structures probably correspond to the noncoated pits seen in structural studies in these locations (311) and could be exocytic sites or noncoated endocytic sites. Preliminary data indicate that the soluble NSF attachment protein receptor (SNARE), SNAP-23, is concentrated in this area, suggesting that these are sites of exocytosis or sites of lipid raft/caveoli, which might be involved in receptor regulation (64, 123, 141). In addition to these possible sites of exocytosis of AMPARs on the sides of spines (248), we also find immunogold-labeled vesicles and pits, labeled with either AMPARs or NMDARs, in the perisynaptic region and possibly within the PSD itself (102, 259, 260). Consistent with this, some studies have suggested that NMDARs could be incorporated more directly into synapses, presumably via actin/myosin-mediated transport (110), although it also has been reported that newly-exocytosed NMDARs form extrasynaptic clusters first (276).

In any case, wherever they reach the surface, NMDARs can be very mobile in the surface membrane, and thus may move readily to and from sites of function (334). In hippocampal neuron cultures, at least 65% of synaptic NMDARs are mobile and can be exchanged rapidly (334). Extrasynaptic NMDARs are not as mobile as extrasynaptic AMPARs, while mobility for these two kinds of receptors is similar within the synapse (104, 338). Also, while changes in neuronal activity modify AMPAR mobility but not NMDAR mobility, activation of PKC modifies mobility for both (104, 338). These latter studies employed live surface labeling of native NMDARs on cultured hippocampal neurons using either single molecule-fluorescence microscopy tracking of receptors labeled with Cy3-coupled antibodies or single particle-tracking of semiconductor quantum dot-coupled antibodies. Similar methods reveal that NR2A-containing NMDARs are less mobile than NR2B-containing NMDARs (105). In addition, overexpression of NR2A subunits reduces surface diffusion of NR2B-containing NMDARs, possibly due to formation of NR1/NR2A/NR2B triheteromers that retain the stabilizing effect of NR2A. One caveat for these studies is that they are limited by the resolution of light microscopy; their definitions of synaptic versus extrasynaptic areas cannot be as accurate as those obtained with electron microscopy (EM). EM using markers such as DAB and immunogold may be useful for showing surface localization of glutamate receptors including AMPARs (258) and NMDARs (Petralia et al., unpublished data). Nevertheless, such EM studies may be affected by artifacts such as DAB "bleeding" and restrictions of movements of the relatively large gold particles, which may have difficulty entering into the synaptic cleft (for surface labeling protocols;(258) and Petralia et al., unpublished data); similar difficulties are indicated in the above light microscope studies comparing Cy3 tracking versus the larger quantum dots.

6 NMDARs in Developing Neurons

6.1 Subunit Changes During Development

NMDARs are already present throughout the brain in embryos and include mainly NR1, NR2B, NR2D and NR3A (discussed below in relation to neuronal migration), and maintain this pattern in early postnatal development (see also Section 1). Overall, NR2B, NR2D and NR3A tend to decrease during later postnatal development and maturation; conversely, NMDARs that contain NR2A and/or NR2C tend to become more abundant in the maturing brain (50, 150, 173, 360, 365); relative levels of different NR1 splice variants also change with age (175). This has been studied best in the forebrain and cerebellum (see below), and exceptions occur in other areas. Thus, motoneurons of the hypoglossal, facial and trigeminal motor nuclei in the mouse express abundant mRNA for NR2A early in development (245). In the hypoglossal nucleus, NR2A mRNA is high at least by embryonic day 13 (E13) and peaks at P1-P7; the protein is evident at least by P1. It is likely that this early appearance of NR2A in these motoneurons reflects their function in the suckling reflex at birth. However, NR2A knockout mice manage to develop normally after birth, perhaps due to substitution of NR2B for NR2A in these circuits.

As noted above, the major change in NMDAR subunits in development is the switch from primarily NR2B-containing receptors to those having mainly NR2A subunits. This switchover within synapses has been confirmed specifically with ultrastructural studies in the thalamus and cerebral cortex (84, 194) and hippocampus (257). The latter study examined postnatal development in the CA1 region of the hippocampus using EM immunogold analysis, and results correlate well with previous immunoblot analyses (287). Synaptic labeling for NR2B is highest at P2 and decreases gradually to approximately half that level by P35. In contrast, immunogold labeling for NR2A at synapses is present but very low at P2, and there is about twelve times as much present by P35; this is consistent with mRNA studies in the mouse, where the first distinct signal for NR2A is in the CA1 region of the hippocampus at P1, and this is followed by a substantial increase throughout the brain over the next 2 postnatal weeks (360). Interestingly, AMPARs are not common at these synapses at P2, and like NR2A, they increase during maturation (261). Such synapses are probably "silent synapses" (133) because, in the absence of AMPARs, the NMDAR channels remain blocked by magnesium during neurotransmitter release; activation of these channels normally requires activation of AMPARs and the subsequent change in membrane potential necessary to release the magnesium block. Without AMPARs, activation of these early NMDAR-containing synapses might require initial activation of excitatory GABA receptors, which are prevalent in the early postnatal brain ((342); but see also (250)). Alternatively or in addition, activation of NR2B-containing NMDARs could be preceded by activation of NMDARs with low sensitivity to magnesium, including those with NR2C, NR2D, or NR3A (252) (see also Sections 4.2, 5, and 6.3). Several studies show that the switch from NR2B- to NR2A-containing NMDARs is linked to learning experiences (255), including visual experience (271, 272) and rule learning in odor discrimination (270).

This change is also accompanied by an increase in PSD-95 (255, 287, 381) as discussed in other sections of this chapter.

Another major switch in NR2 subunits of NMDARs occurs in cerebellar granule cells, where NR2C replaces NR2B. Once the granule cells complete migration and are innervated by the mossy fibers, they down-regulate NR2B and begin to express NR2C (246), developing in a caudal to rostral progression through the cerebellum (142); note also that NR2A is present during this period (91, 281). Expression of NR2C by granule cells is regulated by neuregulin secreted by mossy fiber terminals. Neuregulin can interact with ErbB2 and ErbB4 receptors on granule cells, inducing NR2C expression. In this case, expression of NR2C may involve MAGUKs that can form an indirect structural link between neuregulin and NMDARs (98, 111).

6.2 Role in Synapse Formation

NMDARs probably are not present in the earliest nascent excitatory synaptic contacts, which likely are formed largely by various adhesion factors ((186, 257, 391); see also above in Section 4.1.5). One example is NCAM, which associates with both sides of nascent synaptic contacts and may play a major role in synaptogenesis and synapse maturation (266, 323). Other adhesion factors implicated in the earliest contacts are neuroligin, EphB receptors, and cadherins/catenins (discussed above in the Section 4.1.5). In the CA1 stratum radiatum of the hippocampus, immunogold labeling for catenin shows a preference for the more immature-appearing synaptic contacts, while in comparison, labeling for NR2B, SAP102, and Homer 1b/c is more prevalent on the mature-appearing synapses, again suggesting a progression with adhesion proteins preceding entry of other proteins at early synaptic contacts (257).

After the initial contact mediated by adhesion factors, presynaptic components can be delivered in packets, followed by delivery of postsynaptic MAGUKs and NMDARs (89, 186). This process is surprisingly rapid; a distinctive PSD containing multiple proteins linked directly or indirectly to NMDARs is formed in less than two hours (28, 89). Consistent with this, ultrastructural studies of the developing CA1 region of the hippocampus at P2 reveal that many PSDs appear to be relatively mature and they label with immunogold for many of the proteins found in adult synapses (257). Also, the spatial arrangements of these proteins resemble those found in the adult. Thus, at all ages studied (P2, P10 and adult), positions of gold labeling for PSD proteins reflect their molecular associations. For example, NMDARs at the postsynaptic membrane link to MAGUKs, which link to GKAP and subsequently to Shank and Homer; these latter two are positioned below the density where they can form further links to proteins in internal storage organelles and to other proteins in the perisynaptic membrane surrounding the synapse.

In these developing synapses of the hippocampus, while this basic chain of molecules remains the same throughout postnatal development, the forms of the first three components, i.e., NMDARs, MAGUKs, and GKAP, change from types prevalent in early development to those more prevalent in the adult. Thus, as noted above: (1) the predominant NMDAR changes from NR2B- to mainly NR2A-containing, (2) the predominant MAGUK changes from SAP102 to mainly PSD-95, and also (3) the predominant GKAP changes from mainly a higher molecular weight form (130 kD) to a more even mixture of two forms (130 and 95 kD) in the adult (257, 287, 345, 381). Note, however, that in adults, NR2B and SAP102 are still fairly common at synapses in the hippocampus, but the overall pattern of NMDAR/MAGUK/GKAP has changed dramatically during maturation. The importance of these changes is reflected in studies of NR2A knockout mice (336). In visual neurons in these mice, PSD-95 may not form an effective scaffold for NR2B-containing NMDARs, since there is a selective loss of spontaneous NMDAR currents in these mice during development. Again, normally there would be a general replacement of NR2B-containing NMDARs that are associated with SAP102, with NR2A-containing NMDARs that are associated with PSD-95. However, it is not clear how selective this NMDAR/MAGUK association may be; at least in adults, NR2A and NR2B appear to associate equally with SAP102, PSD-95 and PSD-93 (3). Another change during synapse maturation is the appearance of CaMKIIα, which is usually absent from early postnatal synapses (257); although other isoforms such as CaMKIIβ may be present early on. Switching from NR2B-containing NMDARs, which bind CaMKII with high affinity, to NR2A-containing NMDARs, which bind CaMKII with low affinity, results in a substantial reduction in LTP and may be associated with the reduced plasticity of glutamatergic synapses of the mature forebrain (13).

The final stage in the maturation of excitatory synapses, especially those on principal output neurons, is the development of the postsynaptic spine. A number of proteins that bind to NMDARs or to NMDAR-associated proteins help regulate actin formation, which is a major component of spine formation. The NR2A-binding protein α1-chimerin contains a GTPase-activating (GAP) domain that can inhibit the Rho GTPase family member Rac1; Rac1 promotes actin polymerization, increases dendrite arbor complexity, and stimulates spine formation (344). EphB2 receptors can bind to NMDARs and play a role in Rac regulation and spine formation ((221, 330, 331); see also Sections 4.1.4 and 4.1.5). SPAR is a Rap-specific GAP that binds to the GK domain of PSD-95 and regulates spine morphology, both via a direct interaction with actin and perhaps also through Rap signaling (247). Finally, citron is a PSD-95-associated protein that may mediate forms of NMDAR-dependent synaptic plasticity in some specialized neurons; it is a target of Rho, which regulates organization of the actin skeleton (94, 389). In spite of these studies indicating that NMDARs play roles in spine formation, mice with CA1-targeted NR1-knockout (NR1 is lost gradually, mainly during late postnatal development) nevertheless can maintain spines and even may produce new spines in the adult after environmental enrichment (92, 275).

6.3 Neuronal Migration and Neurite Outgrowth

NMDARs are expressed on the surface of young neurons prior to synapse formation and can modulate neuronal migration and neurite outgrowth and differentiation. In mice, mRNAs for NR1 and NR2B are widespread throughout the brain (along with a more restricted distribution of NR2D) by E13 (360). Also, NR3A mRNA is present in the rat brain at least by E15 (50), and its earliest distribution is somewhat similar to that of NR2D in mice. In rats, functional NMDARs that respond to endogenous neurotransmitters appear on neurons of the cortical plate at least by E16 (within one day of the first appearance of this structure; (197)). Neuronal migration appears to involve glutamate and GABA receptors to varying degrees, depending on neuron type, region, and stage in development. In the mouse embryonic cortex, glutamate acts at NMDARs to stimulate neuronal migration (18). In the hippocampus, GABA and NMDA receptors modulate radial migration of pyramidal neurons, while, in contrast, tangential migration of interneurons involves AMPARs but not GABA or NMDARs (204, 205). In the mouse olfactory bulb, tangentially migrating neurons express GABA receptors first, then calcium-impermeable (presumably containing GluR2) AMPARs, and express NMDARs only after changing to radial migration (37). Similarly, migration of gonadotropin-releasing hormone neurons from their origin in the nose and then into the brain is enhanced by AMPAR (including all four AMPAR subunits) activation (303). Once in the brain, these neurons begin expressing NMDARs, which may slow migration. In the mouse cerebellum, the rate of granule cell migration is regulated by changes in calcium influx through N-type calcium channels and NMDARs; completion of migration involves the loss of these transient calcium elevations (155, 159). Note also that neuronal migration can involve various locomotor mechanisms that differ in sensitivity to NMDAR and other receptor activation. Thus, in mouse cortical interneurons, nucleokinesis (movement of the nucleus through the neuron) involves a calcium elevation in the proximal region of the leading process, and this does not involve responses to agonists including NMDA (227). Finally, overall neuroanatomy at birth in full NR1 knockout mice appears normal, suggesting that at least embryonic neuronal migration can proceed in the absence of functional NMDARs, presumably due to some compensatory mechanisms (88).

Once a neuron migrates to its final destination, it undergoes substantial differentiation involving neurite outgrowth, including both axon and dendrites, and dendrite arborization, involving formation of specific branching patterns. Early studies implicated NMDARs in neurite outgrowth. In rat cerebellar granule cell cultures, NMDA application increases the rate of neurite outgrowth and the proportion of neuronal processes with growth cones, while the NMDAR antagonist APV causes a retraction of existing processes (251, 277). Similarly, in rat hippocampal dentate gyrus granule cell cultures, NMDA increases total neurite length and the number of branch points per neuron, although it does not affect the number of major neurite branches off of the soma (29). Interestingly, the effect of NMDARs is different for growing axons and dendrites. In developing neurons of *Xenopus* optic tectum, during early development of the dendritic arbor, APV treatment decreases the number of branch extensions and new branch additions (274). These effects are independent of the generation of sodium-dependent action potentials since they are not affected by TTX. Thus, they may involve NMDARs that are either extrasynaptic or are synaptic but activated by inputs that are below the threshold needed for firing action potentials. Later in development, when the dendritic arbor is more complex, AMPARs and NMDARs appear to work together to control the rate of branch length extension, and antagonists to both AMPARs and NMDARs decrease growth rates (274). However, antagonists cannot alter rates of branch additions or retractions in these more mature arbors. Others have found evidence that NMDARs can suppress neurite sprouting and elongation in cultured *Xenopus* optic tectal neurons (191, 220). This is dependent on calcium influx through NMDARs at the earliest neurite contacts between young neurons; these NMDARs show low magnesium sensitivity and their activation may be mediated by co-activation of type II metabotropic glutamate receptors. Essentially, this mechanism allows the neurite to stop growing when it has found a suitable contact, and appears to apply at least to axonal neurites (effects on dendrites are less clear). These controls of axon growth in *Xenopus* optic tectal neurons were clarified by Ruthazer et al. (282) in which they show that NMDARs help mediate axon branch elimination from inappropriate locations and selective stabilization of branches to appropriate locations. Similarly, in rat hippocampal slice cultures, NMDARs can suppress axonal outgrowth (211). Also, Colonnese et al. (52) examined axon sprouting in the superficial visual layers of the superior colliculus of the rat and found that during early postnatal ages, activation of NMDARs seems to reduce early axon sprouting and synapse formation. In conclusion, it appears that NMDARs in immature neurons can stimulate the growth of the dendrite arbor and also regulate the stabilization of early axon-dendrite contacts, by modulating axonal growth and synapse formation.

7 NMDA Receptor Trafficking and Disease

Excessive calcium influx through the NMDAR may trigger excitotoxic cell death (55, 201). Interestingly, increased intracellular calcium resulting from other channel or receptor activities does not lead to excitotoxic cell death, implying that NMDAR-mediated calcium influx mediates specific downstream signaling cascades that may ultimately lead to cell death (341). While excitotoxic cell death has been regarded to be the major mechanism by which NMDARs contribute to these disease states, mounting evidence suggests that disruptions in NMDAR trafficking and targeting may play a distinct role. Loss of synaptic function, involving disruptions in protein complexes and/or NMDAR dysfunction, may mediate a collapse of the synaptic functional unit in the absence of calcium-mediated cell death. Furthermore, while loss of synaptic function was believed to result from excitotoxic cell death, evidence suggests that these changes may actually occur before cell death and, in some diseases, pre-symptomatically (85, 107, 296, 306). Thus, the complex nature of NMDAR function in disease states has expanded beyond the traditional view of excitotoxic cell death to include NMDAR deregulation. Extensive reviews have been published that thoroughly detail the role of NMDARs in multiple diseases; therefore, our goal is to review recent developments addressing NMDAR mistargeting in neuropathological conditions (for a recent review, see (170)).

7.1 NMDAR Subunit-Specific Determinants of Excitotoxic Cell Death

Mechanisms of NMDAR deregulation may be common to various diseases. Firstly, NMDAR subunit composition is an important determinant in NMDAR-mediated excitotoxicity: cells transfected with NR1/NR2B or NR1/NR2A subunits are more susceptible to cell death than those transfected with NR1/NR2C or NR1/NR2D (201). These subunit-specific effects may be determined by higher calcium permeability of NR1/NR2A and NR1/NR2B NMDARs. Furthermore, increases in toxicity occur with development as NR2B decreases and NR2A increases (46), consistent with subunit-specific vulnerability to excitotoxic cell death.

Secondly, subcellular localization may determine subunit-specific effects. NR2A-containing receptors are enriched at the synapse while NR2B-containing receptors are found extrasynaptically and are coupled to distinct downstream signaling cascades (see Section 5). Extrasynaptic NMDARs are associated with the induction of cell death mechanisms, while synaptic receptors mediate induction of synaptic plasticity mechanisms (116, 310). Thus, multiple factors such as subunit composition and localization are involved in determining whether NMDAR activation will lead to synaptic plasticity or to neurotoxicity (115).

7.2 Huntington's Disease

Huntington's disease (HD) is an inherited neurodegenerative disorder that results in cognitive deficits, motor decline, and mood dysfunction (118). Cell death occurs most extensively in GABAergic projection medium-sized spiny neurons (MSNs) of the neostriatum and, to a lesser extent, in the cortex (349). Expansion of the polyglutamine repeat in the gene encoding the protein huntingtin (htt) has been identified as the cause of the disease (168).

Numerous lines of evidence implicate the NMDAR in HD neuropathology (for review, see (85)). A disproportionate loss of NMDAR binding sites occurs in the striatum of HD patients compared to GABAergic or cholinergic binding sites, suggesting that NMDAR expression increases vulnerability of the neurons to cell death (382). Administration of NMDAR agonists such as quinolinic acid into the striatum in rodents and primates induces HD-like behavioral and neuropathological symptoms (114, 292). In addition, in cultured striatal neurons, NMDAR agonists induce neurotoxicity more effectively than agonists for other glutamate receptor subclasses (reviewed in (65)). In HD, cognitive deficits often precede motor symptoms (137, 177, 249). Similarly, presymptomatic alterations in glutamatergic transmission occur in dissociated cortical neurons from a mouse model of HD (8), suggesting a role for the NMDARs in the disease independent of excitotoxicity. In a transgenic Huntington's mouse model, NMDAR physiological properties were altered in a population of MSNs, exhibiting an increased response to glutamate and decreased sensitivity to magnesium (315). Mouse models of HD display behavioral and electrophysiological alterations in the absence of cell loss (181), suggesting that the observed neuropathology is not merely the result of cell death.

While numerous studies report inconsistent results regarding changes in NMDAR subunit expression (for review, see (85)), a specific role for NR2B has been identified (44). The striatum is enriched in NR2B (160, 169) with a significantly higher ratio of NR2B to NR2A than in cortex or hippocampus (185). Furthermore, the majority of NMDARs in MSNs are sensitive to the NR2B-specific antagonist ifenprodil, which is as effective at blocking NMDAR-mediated cell death as the NMDAR antagonist MK-801 (387), indicating a high population of NR2B-containing receptors. The current model suggests that striatal neurons expressing NR2B are more susceptible to excitotoxic cell death (387) (see above in the Section 7.1). Co-expression of NR2B with mutant htt in HEK293 cells leads to larger

NMDAR currents than with wild-type htt, while no differences are seen when NR2A is co-expressed with mutant versus wild-type htt (44). Apoptotic cell death is also significantly greater in cells co-transfected with mutant htt and NR2B, but not NR2A (386). Furthermore, the NR2B subunit is enriched extrasynaptically, and a role for extrasynaptic NR2B-containing receptors in mediating cell death has been suggested (116, 310).

NMDAR function may be affected by other factors, such as receptor trafficking. NR1 localization on the cell surface is increased in cultured MSNs from transgenic mice expressing htt with moderately expanded polyglutamine lengths, and NMDAR may be linked to htt by cytoskeletal proteins (85), supporting a trafficking component. Additionally, the SH3 domain of PSD-95, which plays a role in NMDAR synaptic localization (see above in Section 4.1.4), binds htt in transfected HEK293 cells and in human cortical tissue (322). The PSD-95/htt association decreases in the presence of mutant htt, presumably increasing the availability of unbound PSD-95 to bind to NMDARs (322) and stabilize them on the cell surface (279). Thus, wild-type htt serves a neuroprotective role by blocking excessive NMDAR activity (178). Furthermore, PSD-95 links NMDARs to downstream signaling molecules, including neuronal nitric oxide synthase (nNOS) (48); in a transgenic model of HD, a decrease in membrane-associated nNOS was observed in addition to a decrease in PSD-95 (136, 200). Therefore, the presence of mutant htt may disrupt NMDAR function not only by increasing calcium influx, but also by disrupting normal synaptic function.

In addition to protein associations, NMDAR function may be altered in HD by changes in phosphorylation of the NMDAR and its downstream signaling partners. Mutant htt, but not wild-type htt, increases phosphorylated Src and targeting of activated phospho-Src and PSD-95 to membrane fractions (309). As mentioned above, Src regulates NMDAR channel activity. Additionally, changes in PKC and PKA phosphorylation may be affected by mutant htt (85). Studies on transgenic mouse models of HD report decrease of NR1 phosphorylation at Ser897 (136), which results in decreased dopamine D1-mediated CREB activation (70). Inhibition of CRE-mediated gene transcription has been linked to HD (168), and decreased CREB activity has been linked to increased NMDAR-mediated cell death (116). Further support for a role for NMDAR-mediated downstream signaling events in HD neuropathology comes from a study reporting increased CRE-mediated transcription in transgenic mice over-expressing mutant htt compared with mice overexpressing wild-type htt (239).

7.3 Alzheimer's Disease

Alzheimer's disease (AD) is a progressive neurodegenerative disease characterized by impairments in memory and cognition, likely caused by the pathogenic accumulation of a hydrophobic 42-amino-acid form of the β-amyloid peptide. While much of the earlier work focused on the role of the cholinergic system, mounting evidence suggests AD also exhibits a glutamatergic synaptic dysfunction in early disease stages (296, 306). Memantine, a non-competitive antagonist that more effectively blocks highly activated NMDARs, such as occurs in pathological conditions, but not during normal physiological activation (192), has been approved for treatment of AD (289).

One early observation linking synaptic changes to AD is the reported loss of synaptic spines in biopsied and postmortem AD brains, accompanied by a decrease in the number of synapses per cortical neuron (60). These findings have since been confirmed (165, 301), and more recent studies report that loss of synapses occurs by an NMDAR-dependent mechanism (298). Furthermore, β-amyloid oligomers were found to interrupt synaptic plasticity in vivo and in slices (166) and to transiently impair learned behavior in rats (51, 335, 351). Interestingly, non-toxic amounts of secreted β-amyloid reduce synaptic plasticity and glutamatergic transmission (140), suggesting an effect on NMDAR transmission that is independent of cell death.

Alterations of NMDAR subunit expression have been reported in analyses of postmortem AD brain. NR1 isoforms containing the N1 alternatively spliced cassette are decreased in AD affected patients in brain regions most susceptible to pathological damage (129, 130). Furthermore, while levels of NR2C and NR2D mRNA are not altered, NR2A and NR2B mRNA and protein show a decrease in susceptible brain regions (131). In a study investigating subunit expression in hippocampal subregions, NR1 and NR2B mRNA and protein levels were reduced, while NR2A mRNA and protein levels were unchanged (219), suggesting an altered neuronal vulnerability. Additional studies investigating subunit expression have reported inconsistent results, suggesting that additional mechanisms may be involved in the neuropathological changes (reviewed in (132)).

Protein trafficking may also play a role in the neuropathology of AD. Interestingly, β-amyloid decreases NMDAR localization at synaptic sites (308). Additionally, the 42-amino-acid form of β-amyloid increases endocytosis of NMDAR in cultured cortical neurons and in transgenic mice expressing mutant amyloid precursor protein (APP), with no affect on the $GABA_A$ β 2/3 subunit (308). Furthermore, an increase in the tyrosine phosphatase function of Striatal-Enriched Phosphatase (STEP) is responsible for the decrease in phosphorylation of NR2B at Tyr1472 (308), which results in increased endocytosis of the receptor (279). One might speculate that prolonged depression of NMDAR-mediated transmission may initiate the pathological changes. Alternatively, a depression of NMDAR function may be neuroprotective based on the finding that preconditioning with β-amyloid reduces glutamate-induced neurotoxicity by promoting endocytosis of NMDARs (103).

7.4 Parkinson's Disease

Parkinson's disease (PD) is a progressive disorder characterized by degeneration of the dopaminergic neurons projecting from the substantia nigra to the striatum. Animal models of PD, including nigrostriatal denervation induced by 6-hydroxydopamine (6-OHDA) injections, mimic this dopamine deficit (112) (for review, see (99, 113)). Treatment with L-DOPA (levodopa) alleviates the characteristic symptoms of PD, which include resting tremors, rigidity, and other motor symptoms. However, chronic treatment results in dyskinesia, i.e., abnormal, involuntary spasmodic movements (113). NMDAR antagonists, including the NR2B antagonists ifenprodil and CP-101,606, are effective in reducing the neuropathology in experimental models (195, 232) as well as in humans (343), suggesting a role for the glutamatergic system in PD neuropathology.

Alterations in the subcellular organization and function of NMDARs have been reported in animal models of PD, as well as following chronic L-DOPA treatment (112) (for review, see (99, 113)). In the unilateral 6-OHDA model of PD, alterations in synaptic function include reorganization of spines (34), decrease in synapse number (9), and reduction of LTP and LTD (32, 33). Also, changes in NMDAR binding has been reported in experimental models of PD and human postmortem PD brain (reviewed in (113)). In human postmortem brain of PD patients, NR1 expression is not altered (35, 217); however, in a nonhuman primate model of dyskinesia, NR2A is upregulated (112).

Changes in subunit composition, subcellular localization, and phosphorylation states of NMDARs have been reported in 6-OHDA mouse models of PD (74, 216, 265). Similarly, in L-DOPA-induced dyskinesia, NR2B redistributes from a synaptic to an extrasynaptic localization (99). While no changes were observed in total NMDAR subunit expression ((74, 216); but see (240)), the abundance of NR1 and NR2B in synaptosomal membrane fractions was found to decrease in the lesioned striatum relative to unlesioned striatum, with no change in NR2A (74), indicating decreased synaptic localization. Significantly, chronic L-DOPA treatment of 6-OHDA-lesioned rats restores the abundance of NR1, NR2A, and NR2B subunits in homogenate and synaptosomal membrane fractions (74, 240).

While the exact mechanisms by which NR2B-containing receptors redistribute following 6-OHDA lesions are not known, these may be affected by changes in phosphorylation. NR2A and NR2B subunits in the synaptosomal membrane, but not the light membrane of synaptic vesicle-enriched fractions, are tyrosine phosphorylated (73). While no changes were measured in the phosphorylation of NR2A or NR2B in total homogenates (74, 216), changes in NR2B tyrosine phosphorylation were reported in the PD 6-OHDA lesion model (216, 240). This decrease in NR2B phosphorylation is specific for the subpopulation of NMDARs in the membrane fraction (74), suggesting that changes in phosphorylation alter subcellular localization in PD. In support of this model, the dopamine D1 agonist increases NR1, NR2A, and NR2B localization to the synaptosomal membrane fraction and tyrosine phosphorylation of the subunits located in that fraction (73).

Furthermore, dopamine depletion decreases phosphorylation of NR1 at Ser890 and Ser896 in a unilateral 6-OHDA rat model of PD (74). While Ser897 was not affected in this study, increased phosphorylation was reported in another study utilizing the 6-OHDA model, as well as in an MPTP primate model of PD (23). Phosphorylation of Ser890 results in the dispersion of surface NMDARs (329), and increased phosphorylation at Ser897 results in increased dopamine D1-mediated CREB activation (see Section 7.2 above) (70). Also, an increase in serine phosphorylation of NR2A, but not NR2B, was reported (241). Interestingly, chronic L-DOPA treatment enhances serine phosphorylation of NR1 at Ser890, Ser896, and Ser897 and of NR2A, as well as tyrosine phosphorylation of both NR2A and NR2B (74, 240, 241). These results indicate that surface delivery and/or stabilization can be regulated by dopamine via alterations in NMDAR subunit phosphorylation.

7.5 Schizophrenia and Mood Disorders

Investigations into therapies for schizophrenia and mood disorders have more recently focused on the glutamate dysfunction hypothesis (313). In the case of schizophrenia, reduction of NMDAR function by non-competitive antagonists such as ketamine or phencyclidine result in dopaminergic hyperactivity and behavioral changes characteristic of schizophrenia (199). Altered NR1, NR2A, NR2B, and, interestingly, NR3A expression has been reported in postmortem tissue of schizophrenic patients (68, 97, 212, 229). Additionally, mice expressing only 5% of normal levels of NR1 display behavioral abnormalities that correlate with schizophrenia, which can be reversed by treatment with the antipsychotics haloperidol or clozapine (223). Decreased striatal expression of SAP102 was reported (158), suggesting possible changes in trafficking or localization of NMDARs. Interestingly, schizophrenia is associated with hypofunction of the glutamatergic system (313); notably, this is strong evidence that NMDARs play a role in neuropathologies such as schizophrenia by mechanisms other than excitotoxic cell death.

Neuregulins, which are widely expressed growth and differentiation factors, have been genetically linked to schizophrenia (316). Neuregulin function has been associated with changes in glutamatergic synapse function (109, 111), perhaps providing a link between neuregulins, glutamatergic hypofunction, and schizophrenia. The neuregulin-β isoform has been shown to induce NR2C mRNA expression in cultured cerebellar granule cells in an NMDAR activity-dependent manner (246). Furthermore, in the prefrontal cortex, neuregulin 1 promotes actin-dependent NMDAR internalization and decreases NMDAR-mediated EPSCs (109). While similar studies in hippocampal neurons report neuregulin-1-mediated decreases in AMPAR-mediated EPSCs, not NMDAR EPSCs, and a corresponding increase in AMPAR endocytosis (164), these data support a role for glutamatergic regulation by neuregulins. Furthermore, analysis in postmortem brain tissue of schizophrenic patients identified a more tightly coupled complex of the neuregulin receptor ErbB4, PSD-95, and NMDARs than in normal controls (111). Neuregulin 1 activity suppresses NMDAR activation, an effect that is more pronounced in schizophrenic subjects than in normal controls (111). These data, therefore, suggest that the effects of neuregulins on NMDAR function may be related to the glutamatergic hypofunction model.

Major depression and bipolar disorder have been linked to the glutamatergic system, largely as a result of effective treatment of these diseases with antagonists of NMDARs (384). Notably, subjects with major depressive disorder exhibited significant improvements following a single dose of the NMDAR-specific antagonist ketamine (385). Furthermore, mice lacking the NR2A subunit exhibit anxiolytic and antidepressant-like effects (27). Exercise in NR2A knockout mice does not enhance neurogenesis as in wild-type mice (151), in agreement with the model that depression may result from impaired neurogenesis (17). Treatment of rats with lithium, an effective treatment for bipolar disorder, results in block of NMDAR-mediated signaling via phopholipase A_2 activation and arachadonic acid release (15). As in schizophrenia, alterations in NR1, NR2A, NR2B, and, interestingly, NR3A subunit expression have been reported for bipolar disorder and depression (158, 207, 229, 370).

Decreased striatal expression of PSD-95 and SAP102, or SAP102 only, has been reported in postmortem tissue of patients with bipolar disorder or depression, respectively (158).

7.6 Other Disorders

NMDARs have also been implicated in other neuropathological conditions, including lupus (61, 157), neuropathic pain (25), ischemia (354), epilepsy (99), ALS (80), and HIV dimentia (144).

References

1. Abel, T., P. V. Nguyen, M. Barad, T. A. Deuel, E. R. Kandel, and R. Bourtchouladze. Genetic demonstration of a role for PKA in the late phase of LTP and in hippocampus-based long-term memory. *Cell* 88: 615–626, 1997.
2. Adesnik, H., R. A. Nicoll, and P. M. England. Photoinactivation of native AMPA receptors reveals their real-time trafficking. *Neuron* 48: 977–985, 2005.
3. Al-Hallaq, R. A., T. P. Conrads, T. D. Veenstra, and R. J. Wenthold. NMDA diheteromeric receptor populations and associated proteins in rat hippocampus. *J Neurosci* 27: 8334–8343, 2007.
4. Al-Hallaq, R. A., B. R. Jarabek, Z. Fu, S. Vicini, B. B. Wolfe, and R. P. Yasuda. Association of NR3A with the N-methyl-D-aspartate receptor NR1 and NR2 subunits. *Mol Pharmacol* 62: 1119–1127, 2002.
5. Allen, J. A., R. A. Halverson-Tamboli, and M. M. Rasenick. Lipid raft microdomains and neurotransmitter signalling. *Nat Rev Neurosci* 8: 128–140, 2007.
6. Amparan, D., D. Avram, C. G. Thomas, M. G. Lindahl, J. Yang, G. Bajaj, and J. E. Ishmael. Direct interaction of myosin regulatory light chain with the NMDA receptor. *J Neurochem* 92: 349–361, 2005.
7. Andrasfalvy, B. K., and J. C. Magee. Changes in AMPA receptor currents following LTP induction on rat CA1 pyramidal neurones. *J Physiol* 559: 543–554, 2004.
8. Andre, V. M., C. Cepeda, A. Venegas, Y. Gomez, and M. S. Levine. Altered cortical glutamate receptor function in the R6/2 model of Huntington's disease. *J Neurophysiol* 95: 2108–2119, 2006.
9. Arbuthnott, G. W., C. A. Ingham, and J. R. Wickens. Dopamine and synaptic plasticity in the neostriatum. *J Anat* 196: 587–596, 2000.
10. Awobuluyi, M., J. Yang, Y. Ye, J. E. Chatterton, A. Godzik, S. A. Lipton, and D. Zhang. Subunit-specific roles of glycine-binding domains in activation of NR1/NR3 N-methyl-D-aspartate receptors. *Mol Pharmacol* 71: 112–122, 2007.
11. Bardoni, R., C. Torsney, C. K. Tong, M. Prandini, and A. B. MacDermott. Presynaptic NMDA receptors modulate glutamate release from primary sensory neurons in rat spinal cord dorsal horn. *J Neurosci* 24: 2774–2781, 2004.
12. Barria, A., and R. Malinow. Subunit-specific NMDA receptor trafficking to synapses. *Neuron* 35: 345–353, 2002.
13. Barria, A., and R. Malinow. NMDA receptor subunit composition controls synaptic plasticity by regulating binding to CaMKII. *Neuron* 48: 289–301, 2005.
14. Bashir, Z. I., S. Alford, S. N. Davies, A. D. Randall, and G. L. Collingridge. Long-term potentiation of NMDA receptor-mediated synaptic transmission in the hippocampus. *Nature* 349: 156–158, 1991.

15. Basselin, M., L. Chang, J. M. Bell, and S. I. Rapoport. Chronic lithium chloride administration attenuates brain NMDA receptor-initiated signaling via arachidonic acid in unanesthetized rats. *Neuropsychopharmacology* 31: 1659–1674, 2006.
16. Bayer, K. U., K. De, P., A. S. Leonard, J. W. Hell, and H. Schulman. Interaction with the NMDA receptor locks CaMKII in an active conformation. *Nature* 411: 801–805, 2001.
17. Becker, S., and J. M. Wojtowicz. A model of hippocampal neurogenesis in memory and mood disorders. *Trends Cogn Sci* 11: 70–76, 2007.
18. Behar, T. N., C. A. Scott, C. L. Greene, X. Wen, S. V. Smith, D. Maric, Q. Y. Liu, C. A. Colton, and J. L. Barker. Glutamate acting at NMDA receptors stimulates embryonic cortical neuronal migration. *J Neurosci* 19: 4449–4461, 1999.
19. Bendel, O., B. Meijer, Y. Hurd, and G. von Euler. Cloning and expression of the human NMDA receptor subunit NR3B in the adult human hippocampus. *Neurosci Lett* 377: 31–36, 2005.
20. Berberich, S., P. Punnakkal, V. Jensen, V. Pawlak, P. H. Seeburg, O. Hvalby, and G. Kohr. Lack of NMDA receptor subtype selectivity for hippocampal long-term potentiation. *J Neurosci* 25: 6907–6910, 2005.
21. Bernard-Trifilo, J. A., E. A. Kramar, R. Torp, C. Y. Lin, E. A. Pineda, G. Lynch, and C. M. Gall. Integrin signaling cascades are operational in adult hippocampal synapses and modulate NMDA receptor physiology. *J Neurochem* 93: 834–849, 2005.
22. Berretta, N., F. Berton, R. Bianchi, M. Brunelli, M. Capogna, and W. Francesconi. Long-term Potentiation of NMDA Receptor-mediated EPSP in Guinea-pig Hippocampal Slices. *Eur J Neurosci* 3: 850–854, 1991.
23. Betarbet, R., O. Poisik, T. B. Sherer, and J. T. Greenamyre. Differential expression and ser897 phosphorylation of striatal N-methyl-D-aspartate receptor subunit NR1 in animal models of Parkinson's disease. *Exp Neurol* 187: 76–85, 2004.
24. Blanpied, T. A., D. B. Scott, and M. D. Ehlers. Dynamics and regulation of clathrin coats at specialized endocytic zones of dendrites and spines. *Neuron* 36: 435–449, 2002.
25. Bleakman, D., A. Alt, and E. S. Nisenbaum. Glutamate receptors and pain. *Semin Cell Dev Biol* 17: 592–604, 2006.
26. Blumer, J. B., L. J. Chandler, and S. M. Lanier. Expression analysis and subcellular distribution of the two G-protein regulators AGS3 and LGN indicate distinct functionality. Localization of LGN to the midbody during cytokinesis. *J Biol Chem* 277: 15897–15903, 2002.
27. Boyce-Rustay, J. M., and A. Holmes. Genetic inactivation of the NMDA receptor NR2A subunit has anxiolytic- and antidepressant-like effects in mice. *Neuropsychopharmacology* 31: 2405–2414, 2006.
28. Bresler, T., M. Shapira, T. Boeckers, T. Dresbach, M. Futter, C. C. Garner, K. Rosenblum, E. D. Gundelfinger, and N. E. Ziv. Postsynaptic density assembly is fundamentally different from presynaptic active zone assembly. *J Neurosci* 24: 1507–1520, 2004.
29. Brewer, G. J., and C. W. Cotman. NMDA receptor regulation of neuronal morphology in cultured hippocampal neurons. *Neurosci Lett* 99: 268–273, 1989.
30. Brickley, S. G., C. Misra, M. H. Mok, M. Mishina, and S. G. Cull-Candy. NR2B and NR2D subunits coassemble in cerebellar Golgi cells to form a distinct NMDA receptor subtype restricted to extrasynaptic sites. *J Neurosci* 23: 4958–4966, 2003.
31. Bukalo, O., N. Fentrop, A. Y. Lee, B. Salmen, J. W. Law, C. T. Wotjak, M. Schweizer, A. Dityatev, and M. Schachner. Conditional ablation of the neural cell adhesion molecule reduces precision of spatial learning, long-term potentiation, and depression in the CA1 subfield of mouse hippocampus. *J Neurosci* 24: 1565–1577, 2004.
32. Calabresi, P., D. Centonze, and G. Bernardi. Electrophysiology of dopamine in normal and denervated striatal neurons. *Trends Neurosci* 23: S57–S63, 2000.

33. Calabresi, P., P. Giacomini, D. Centonze, and G. Bernardi. Levodopa-induced dyskinesia: a pathological form of striatal synaptic plasticity? *Ann Neurol* 47: S60–S68; discussion S68–S69, 2000.
34. Calabresi, P., N. B. Mercuri, G. Sancesario, and G. Bernardi. Electrophysiology of dopamine-denervated striatal neurons. Implications for Parkinson's disease. *Brain* 116: 433–452, 1993.
35. Calon, F., A. H. Rajput, O. Hornykiewicz, P. J. Bedard, and T. Di Paolo. Levodopa-induced motor complications are associated with alterations of glutamate receptors in Parkinson's disease. *Neurobiol Dis* 14: 404–416, 2003.
36. Cao, J., J. I. Viholainen, C. Dart, H. K. Warwick, M. L. Leyland, and M. J. Courtney. The PSD95-nNOS interface: a target for inhibition of excitotoxic p38 stress-activated protein kinase activation and cell death. *J Cell Biol* 168: 117–126, 2005.
37. Carleton, A., L. T. Petreanu, R. Lansford, A. Alvarez-Buylla, and P. M. Lledo. Becoming a new neuron in the adult olfactory bulb. *Nat Neurosci* 6: 507–518, 2003.
38. Casado, M., P. Isope, and P. Ascher. Involvement of presynaptic N-methyl-D-aspartate receptors in cerebellar long-term depression. *Neuron* 33: 123–130, 2002.
39. Charriaut-Marlangue, C., S. Otani, C. Creuzet, Y. Ben-Ari, and J. Loeb. Rapid activation of hippocampal casein kinase II during long-term potentiation. *Proc Natl Acad Sci USA* 88: 10232–10236, 1991.
40. Chatterton, J. E., M. Awobuluyi, L. S. Premkumar, H. Takahashi, M. Talantova, Y. Shin, J. Cui, S. Tu, K. A. Sevarino, N. Nakanishi, G. Tong, S. A. Lipton, and D. Zhang. Excitatory glycine receptors containing the NR3 family of NMDA receptor subunits. *Nature* 415: 793–798, 2002.
41. Chavis, P., and G. Westbrook. Integrins mediate functional pre- and postsynaptic maturation at a hippocampal synapse. *Nature* 411: 317–321, 2001.
42. Chazot, P. L., and F. A. Stephenson. Biochemical evidence for the existence of a pool of unassembled C2 exon-containing NR1 subunits of the mammalian forebrain NMDA receptor. *J Neurochem* 68: 507–516, 1997.
43. Chen, L., and L. Y. Huang. Protein kinase C reduces Mg2+ block of NMDA-receptor channels as a mechanism of modulation. *Nature* 356: 521–523, 1992.
44. Chen, N., T. Luo, C. Wellington, M. Metzler, K. McCutcheon, M. R. Hayden, and L. A. Raymond. Subtype-specific enhancement of NMDA receptor currents by mutant huntingtin. *J Neurochem* 72: 1890–1898, 1999.
45. Chen, W., and A. Helenius. Role of ribosome and translocon complex during folding of influenza hemagglutinin in the endoplasmic reticulum of living cells. *Mol Biol Cell* 11: 765–772, 2000.
46. Cheng, C., D. M. Fass, and I. J. Reynolds. Emergence of excitotoxicity in cultured forebrain neurons coincides with larger glutamate-stimulated [Ca(2+)](i) increases and NMDA receptor mRNA levels. *Brain Res* 849: 97–108, 1999.
47. Choi, Y. B., L. Tenneti, D. A. Le, J. Ortiz, G. Bai, H. S. Chen, and S. A. Lipton. Molecular basis of NMDA receptor-coupled ion channel modulation by S-nitrosylation. *Nat Neurosci* 3: 15–21, 2000.
48. Christopherson, K. S., B. J. Hillier, W. A. Lim, and D. S. Bredt. PSD-95 assembles a ternary complex with the N-methyl-D-aspartic acid receptor and a bivalent neuronal NO synthase PDZ domain. *J Biol Chem* 274: 27467–27473, 1999.
49. Chung, H. J., Y. H. Huang, L. F. Lau, and R. L. Huganir. Regulation of the NMDA receptor complex and trafficking by activity-dependent phosphorylation of the NR2B subunit PDZ ligand. *J Neurosci* 24: 10248–10259, 2004.
50. Ciabarra, A. M., J. M. Sullivan, L. G. Gahn, G. Pecht, S. Heinemann, and K. A. Sevarino. Cloning and characterization of chi-1: a developmentally regulated member of a novel class of the ionotropic glutamate receptor family. *J Neurosci* 15: 6498–6508, 1995.

51. Cleary, J., J. M. Hittner, M. Semotuk, P. Mantyh, and E. O'Hare. Beta-amyloid(1–40) effects on behavior and memory. *Brain Res* 682: 69–74, 1995.
52. Colonnese, M. T., J. P. Zhao, and M. Constantine-Paton. NMDA receptor currents suppress synapse formation on sprouting axons in vivo. *J Neurosci* 25: 1291–1303, 2005.
53. Cottrell, J. R., E. Borok, T. L. Horvath, and E. Nedivi. CPG2: a brain- and synapse-specific protein that regulates the endocytosis of glutamate receptors. *Neuron* 44: 677–690, 2004.
54. Crump, F. T., K. S. Dillman, and A. M. Craig. cAMP-dependent protein kinase mediates activity-regulated synaptic targeting of NMDA receptors. *J Neurosci* 21: 5079–5088, 2001.
55. Cull-Candy, S., S. Brickley, and M. Farrant. NMDA receptor subunits: diversity, development and disease. *Curr Opin Neurobiol* 11: 327–335, 2001.
56. Dalva, M. B., A. C. McClelland, and M. S. Kayser. Cell adhesion molecules: signalling functions at the synapse. *Nat Rev Neurosci* 8: 206–220, 2007.
57. Dalva, M. B., M. A. Takasu, M. Z. Lin, S. M. Shamah, L. Hu, N. W. Gale, and M. E. Greenberg. EphB receptors interact with NMDA receptors and regulate excitatory synapse formation. *Cell* 103: 945–956, 2000.
58. Das, S., Y. F. Sasaki, T. Rothe, L. S. Premkumar, M. Takasu, J. E. Crandall, P. Dikkes, D. A. Conner, P. V. Rayudu, W. Cheung, H. S. Chen, S. A. Lipton, and N. Nakanishi. Increased NMDA current and spine density in mice lacking the NMDA receptor subunit NR3A. *Nature* 393: 377–381, 1998.
59. Davey, F., M. Hill, J. Falk, N. Sans, and F. J. Gunn-Moore. Synapse associated protein 102 is a novel binding partner to the cytoplasmic terminus of neurone-glial related cell adhesion molecule. *J Neurochem* 94: 1243–1253, 2005.
60. Davies, C. A., D. M. Mann, P. Q. Sumpter, and P. O. Yates. A quantitative morphometric analysis of the neuronal and synaptic content of the frontal and temporal cortex in patients with Alzheimer's disease. *J Neurol Sci* 78: 151–164, 1987.
61. DeGiorgio, L. A., K. N. Konstantinov, S. C. Lee, J. A. Hardin, B. T. Volpe, and B. Diamond. A subset of lupus anti-DNA antibodies cross-reacts with the NR2 glutamate receptor in systemic lupus erythematosus. *Nat Med* 7: 1189–1193, 2001.
62. Demyanenko, G. P., M. Schachner, E. Anton, R. Schmid, G. Feng, J. Sanes, and P. F. Maness. Close homolog of L1 modulates area-specific neuronal positioning and dendrite orientation in the cerebral cortex. *Neuron* 44: 423–437, 2004.
63. Derkach, V., A. Barria, and T. R. Soderling. Ca2+/calmodulin-kinase II enhances channel conductance of alpha-amino-3-hydroxy-5-methyl-4-isoxazolepropionate type glutamate receptors. *Proc Natl Acad Sci USA* 96: 3269–3274, 1999.
64. deSouza, S., and E. B. Ziff. AMPA receptors do the electric slide. *Sci STKE* 2002: PE45, 2002.
65. DiFiglia, M. Excitotoxic injury of the neostriatum: a model for Huntington's disease. *Trends Neurosci* 13: 286–289, 1990.
66. Dingledine, R., K. Borges, D. Bowie, and S. F. Traynelis. The glutamate receptor ion channels. *Pharmacol Rev* 51: 7–61, 1999.
67. Dosemeci, A., J. H. Tao-Cheng, L. Vinade, C. A. Winters, L. Pozzo-Miller, and T. S. Reese. Glutamate-induced transient modification of the postsynaptic density. *Proc Natl Acad Sci USA* 98: 10428–10432, 2001.
68. Dracheva, S., S. A. Marras, S. L. Elhakem, F. R. Kramer, K. L. Davis, and V. Haroutunian. N-methyl-D-aspartic acid receptor expression in the dorsolateral prefrontal cortex of elderly patients with schizophrenia. *Am J Psychiatry* 158: 1400–1410, 2001.
69. Du, Q., P. T. Stukenberg, and I. G. Macara. A mammalian partner of inscuteable binds NuMA and regulates mitotic spindle organization. *Nat Cell Biol* 3: 1069–1075, 2001.

70. Dudman, J. T., M. E. Eaton, A. Rajadhyaksha, W. Macias, M. Taher, A. Barczak, K. Kameyama, R. Huganir, and C. Konradi. Dopamine D1 receptors mediate CREB phosphorylation via phosphorylation of the NMDA receptor at Ser897-NR1. *J Neurochem* 87: 922–934, 2003.
71. Duguid, I. C., and T. G. Smart. Retrograde activation of presynaptic NMDA receptors enhances GABA release at cerebellar interneuron-Purkinje cell synapses. *Nat Neurosci* 7: 525–533, 2004.
72. Dunah, A. W., A. C. Sirianni, A. A. Fienberg, E. Bastia, M. A. Schwarzschild, and D. G. Standaert. Dopamine D1-dependent trafficking of striatal N-methyl-D-aspartate glutamate receptors requires Fyn protein tyrosine kinase but not DARPP-32. *Mol Pharmacol* 65: 121–129, 2004.
73. Dunah, A. W., and D. G. Standaert. Dopamine D1 receptor-dependent trafficking of striatal NMDA glutamate receptors to the postsynaptic membrane. *J Neurosci* 21: 5546–5558, 2001.
74. Dunah, A. W., Y. Wang, R. P. Yasuda, K. Kameyama, R. L. Huganir, B. B. Wolfe, and D. G. Standaert. Alterations in subunit expression, composition, and phosphorylation of striatal N-methyl-D-aspartate glutamate receptors in a rat 6-hydroxydopamine model of Parkinson's disease. *Mol Pharmacol* 57: 342–352, 2000.
75. Dunah, A. W., R. P. Yasuda, Y. H. Wang, J. Luo, M. Davila-Garcia, M. Gbadegesin, S. Vicini, and B. B. Wolfe. Regional and ontogenic expression of the NMDA receptor subunit NR2D protein in rat brain using a subunit-specific antibody. *J Neurochem* 67: 2335–2345, 1996.
76. Durand, G. M., M. V. Bennett, and R. S. Zukin. Splice variants of the N-methyl-D-aspartate receptor NR1 identify domains involved in regulation by polyamines and protein kinase C. *Proc Natl Acad Sci USA* 90: 6731–6735, 1993.
77. Durand, G. M., P. Gregor, X. Zheng, M. V. Bennett, G. R. Uhl, and R. S. Zukin. Cloning of an apparent splice variant of the rat N-methyl-D-aspartate receptor NMDAR1 with altered sensitivity to polyamines and activators of protein kinase C. *Proc Natl Acad Sci USA* 89: 9359–9363, 1992.
78. Ehlers, M. D. Activity level controls postsynaptic composition and signaling via the ubiquitin-proteasome system. *Nat Neurosci* 6: 231–242, 2003.
79. Ehrlich, I., M. Klein, S. Rumpel, and R. Malinow. PSD-95 is required for activity-driven synapse stabilization. *Proc Natl Acad Sci USA* 104: 4176–4181, 2007.
80. Eisen, A., and M. Weber. Treatment of amyotrophic lateral sclerosis. *Drugs Aging* 14: 173–196, 1999.
81. El-Husseini, A. E., E. Schnell, D. M. Chetkovich, R. A. Nicoll, and D. S. Bredt. PSD-95 involvement in maturation of excitatory synapses. *Science* 290: 1364–1368, 2000.
82. Elias, G. M., L. Funke, V. Stein, S. G. Grant, D. S. Bredt, and R. A. Nicoll. Synapse-specific and developmentally regulated targeting of AMPA receptors by a family of MAGUK scaffolding proteins. *Neuron* 52: 307–320, 2006.
83. Ellgaard, L., and A. Helenius. ER quality control: towards an understanding at the molecular level. *Curr Opin Cell Biol* 13: 431–437, 2001.
84. Erisir, A., and J. L. Harris. Decline of the critical period of visual plasticity is concurrent with the reduction of NR2B subunit of the synaptic NMDA receptor in layer 4. *J Neurosci* 23: 5208–5218, 2003.
85. Fan, M. M., and L. A. Raymond. N-Methyl-D-aspartate (NMDA) receptor function and excitotoxicity in Huntington's disease. *Prog Neurobiol* 81: 272–293, 2007.
86. Fiebiger, E., D. Tortorella, M. H. Jouvin, J. P. Kinet, and H. L. Ploegh. Cotranslational endoplasmic reticulum assembly of FcepsilonRI controls the formation of functional IgE-binding receptors. *J Exp Med* 201: 267–277, 2005.

87. Fong, D. K., A. Rao, F. T. Crump, and A. M. Craig. Rapid synaptic remodeling by protein kinase C: reciprocal translocation of NMDA receptors and calcium/calmodulin-dependent kinase II. *J Neurosci* 22: 2153–2164, 2002.
88. Forrest, D., M. Yuzaki, H. D. Soares, L. Ng, D. C. Luk, M. Sheng, C. L. Stewart, J. I. Morgan, J. A. Connor, and T. Curran. Targeted disruption of NMDA receptor 1 gene abolishes NMDA response and results in neonatal death. *Neuron* 13: 325–338, 1994.
89. Friedman, H. V., T. Bresler, C. C. Garner, and N. E. Ziv. Assembly of new individual excitatory synapses: time course and temporal order of synaptic molecule recruitment. *Neuron* 27: 57–69, 2000.
90. Frizelle, P. A., P. E. Chen, and D. J. Wyllie. Equilibrium constants for (R)-[(S)-1-(4-bromophenyl)-ethylamino]-(2,3-dioxo-1,2,3,4-tetrahydroquino xalin-5-yl)-methyl]-phosphonic acid (NVP-AAM077) acting at recombinant NR1/NR2A and NR1/NR2B N-methyl-D-aspartate receptors: implications for studies of synaptic transmission. *Mol Pharmacol* 70: 1022–1032, 2006.
91. Fu, Z., S. M. Logan, and S. Vicini. Deletion of the NR2A subunit prevents developmental changes of NMDA-mEPSCs in cultured mouse cerebellar granule neurones. *J Physiol* 563: 867–881, 2005.
92. Fukaya, M., A. Kato, C. Lovett, S. Tonegawa, and M. Watanabe. Retention of NMDA receptor NR2 subunits in the lumen of endoplasmic reticulum in targeted NR1 knockout mice. *Proc Natl Acad Sci USA* 100: 4855–4860, 2003.
93. Furukawa, H., S. K. Singh, R. Mancusso, and E. Gouaux. Subunit arrangement and function in NMDA receptors. *Nature* 438: 185–192, 2005.
94. Furuyashiki, T., K. Fujisawa, A. Fujita, P. Madaule, S. Uchino, M. Mishina, H. Bito, and S. Narumiya. Citron, a Rho-target, interacts with PSD-95/SAP-90 at glutamatergic synapses in the thalamus. *J Neurosci* 19: 109–118, 1999.
95. Futai, K., M. J. Kim, T. Hashikawa, P. Scheiffele, M. Sheng, and Y. Hayashi. Retrograde modulation of presynaptic release probability through signaling mediated by PSD-95-neuroligin. *Nat Neurosci* 10: 186–195, 2007.
96. Fux, C. M., M. Krug, A. Dityatev, T. Schuster, and M. Schachner. NCAM180 and glutamate receptor subtypes in potentiated spine synapses: an immunogold electron microscopic study. *Mol Cell Neurosci* 24: 939–950, 2003.
97. Gao, X. M., K. Sakai, R. C. Roberts, R. R. Conley, B. Dean, and C. A. Tamminga. Ionotropic glutamate receptors and expression of N-methyl-D-aspartate receptor subunits in subregions of human hippocampus: effects of schizophrenia. *Am J Psychiatry* 157: 1141–1149, 2000.
98. Garcia, R. A., K. Vasudevan, and A. Buonanno. The neuregulin receptor ErbB-4 interacts with PDZ-containing proteins at neuronal synapses. *Proc Natl Acad Sci USA* 97: 3596–3601, 2000.
99. Gardoni, F., and M. Di Luca. New targets for pharmacological intervention in the glutamatergic synapse. *Eur J Pharmacol* 545: 2–10, 2006.
100. Gardoni, F., L. H. Schrama, D. van, J. J., W. H. Gispen, F. Cattabeni, and L. Di, M. AlphaCaMKII binding to the C-terminal tail of NMDA receptor subunit NR2A and its modulation by autophosphorylation. *FEBS Lett* 456: 394–398, 1999.
101. Gerber, G., I. Kangrga, P. D. Ryu, J. S. Larew, and M. Randic. Multiple effects of phorbol esters in the rat spinal dorsal horn. *J Neurosci* 9: 3606–3617, 1989.
102. Gerges, N. Z., D. S. Backos, C. N. Rupasinghe, M. R. Spaller, and J. A. Esteban. Dual role of the exocyst in AMPA receptor targeting and insertion into the postsynaptic membrane. *EMBO J* 25: 1623–1634, 2006.
103. Goto, Y., T. Niidome, A. Akaike, T. Kihara, and H. Sugimoto. Amyloid beta-peptide preconditioning reduces glutamate-induced neurotoxicity by promoting endocytosis of NMDA receptor. *Biochem Biophys Res Commun* 351: 259–265, 2006.

104. Groc, L., M. Heine, L. Cognet, K. Brickley, F. A. Stephenson, B. Lounis, and D. Choquet. Differential activity-dependent regulation of the lateral mobilities of AMPA and NMDA receptors. *Nat Neurosci* 7: 695–696, 2004.
105. Groc, L., M. Heine, S. L. Cousins, F. A. Stephenson, B. Lounis, L. Cognet, and D. Choquet. NMDA receptor surface mobility depends on NR2A-2B subunits. *Proc Natl Acad Sci USA* 103: 18769–18774, 2006.
106. Grosshans, D. R., D. A. Clayton, S. J. Coultrap, and M. D. Browning. LTP leads to rapid surface expression of NMDA but not AMPA receptors in adult rat CA1. *Nat Neurosci* 5: 27–33, 2002.
107. Grossman, S. D., L. J. Rosenberg, and J. R. Wrathall. Temporal-spatial pattern of acute neuronal and glial loss after spinal cord contusion. *Exp Neurol* 168: 273–282, 2001.
108. Grunwald, I. C., M. Korte, D. Wolfer, G. A. Wilkinson, K. Unsicker, H. P. Lipp, T. Bonhoeffer, and R. Klein. Kinase-independent requirement of EphB2 receptors in hippocampal synaptic plasticity. *Neuron* 32: 1027–1040, 2001.
109. Gu, Z., Q. Jiang, A. K. Fu, N. Y. Ip, and Z. Yan. Regulation of NMDA receptors by neuregulin signaling in prefrontal cortex. *J Neurosci* 25: 4974–4984, 2005.
110. Guillaud, L., M. Setou, and N. Hirokawa. KIF17 dynamics and regulation of NR2B trafficking in hippocampal neurons. *J Neurosci* 23: 131–140, 2003.
111. Hahn, C. G., H. Y. Wang, D. S. Cho, K. Talbot, R. E. Gur, W. H. Berrettini, K. Bakshi, J. Kamins, K. E. Borgmann-Winter, S. J. Siegel, R. J. Gallop, and S. E. Arnold. Altered neuregulin 1-erbB4 signaling contributes to NMDA receptor hypofunction in schizophrenia. *Nat Med* 12: 824–828, 2006.
112. Hallett, P. J., A. W. Dunah, P. Ravenscroft, S. Zhou, E. Bezard, A. R. Crossman, J. M. Brotchie, and D. G. Standaert. Alterations of striatal NMDA receptor subunits associated with the development of dyskinesia in the MPTP-lesioned primate model of Parkinson's disease. *Neuropharmacology* 48: 503–516, 2005.
113. Hallett, P. J., and D. G. Standaert. Rationale for and use of NMDA receptor antagonists in Parkinson's disease. *Pharmacol Ther* 102: 155–174, 2004.
114. Hantraye, P., D. Riche, M. Maziere, and O. Isacson. A primate model of Huntington's disease: behavioral and anatomical studies of unilateral excitotoxic lesions of the caudate-putamen in the baboon. *Exp Neurol* 108: 91–104, 1990.
115. Hardingham, G. E., and H. Bading. The Yin and Yang of NMDA receptor signalling. *Trends Neurosci* 26: 81–89, 2003.
116. Hardingham, G. E., Y. Fukunaga, and H. Bading. Extrasynaptic NMDARs oppose synaptic NMDARs by triggering CREB shut-off and cell death pathways. *Nat Neurosci* 5: 405–414, 2002.
117. Harney, S. C., M. Rowan, and R. Anwyl. Long-term depression of NMDA receptor-mediated synaptic transmission is dependent on activation of metabotropic glutamate receptors and is altered to long-term potentiation by low intracellular calcium buffering. *J Neurosci* 26: 1128–1132, 2006.
118. Harper, P. S. The epidemiology of Huntington's disease. *Hum Genet* 89: 365–376, 1992.
119. Hawkins, L. M., K. Prybylowski, K. Chang, C. Moussan, F. A. Stephenson, and R. J. Wenthold. Export from the endoplasmic reticulum of assembled N-methyl-D-aspartic acid receptors is controlled by a motif in the c terminus of the NR2 subunit. *J Biol Chem* 279: 28903–28910, 2004.
120. Henderson, J. T., J. Georgiou, Z. Jia, J. Robertson, S. Elowe, J. C. Roder, and T. Pawson. The receptor tyrosine kinase EphB2 regulates NMDA-dependent synaptic function. *Neuron* 32: 1041–1056, 2001.
121. Henkemeyer, M., O. S. Itkis, M. Ngo, P. W. Hickmott, and I. M. Ethell. Multiple EphB receptor tyrosine kinases shape dendritic spines in the hippocampus. *J Cell Biol* 163: 1313–1326, 2003.

122. Herin, G. A., and E. Aizenman. Amino terminal domain regulation of NMDA receptor function. *Eur J Pharmacol* 500: 101–111, 2004.
123. Hering, H., C. C. Lin, and M. Sheng. Lipid rafts in the maintenance of synapses, dendritic spines, and surface AMPA receptor stability. *J Neurosci* 23: 3262–3271, 2003.
124. Hirai, H., J. Kirsch, B. Laube, H. Betz, and J. Kuhse. The glycine binding site of the N-methyl-D-aspartate receptor subunit NR1: identification of novel determinants of co-agonist potentiation in the extracellular M3-M4 loop region. *Proc Natl Acad Sci USA* 93: 6031–6036, 1996.
125. Hirao, K., Y. Hata, N. Ide, M. Takeuchi, M. Irie, I. Yao, M. Deguchi, A. Toyoda, T. C. Sudhof, and Y. Takai. A novel multiple PDZ domain-containing molecule interacting with N-methyl-D-aspartate receptors and neuronal cell adhesion proteins. *J Biol Chem* 273: 21105–21110, 1998.
126. Hsu, S. C., A. E. Ting, C. D. Hazuka, S. Davanger, J. W. Kenny, Y. Kee, and R. H. Scheller. The mammalian brain rsec6/8 complex. *Neuron* 17: 1209–1219, 1996.
127. Huh, K. H., and R. J. Wenthold. Turnover analysis of glutamate receptors identifies a rapidly degraded pool of the N-methyl-D-aspartate receptor subunit, NR1, in cultured cerebellar granule cells. *J Biol Chem* 274: 151–157, 1999.
128. Humeau, Y., H. Shaban, S. Bissiere, and A. Luthi. Presynaptic induction of heterosynaptic associative plasticity in the mammalian brain. *Nature* 426: 841–845, 2003.
129. Hynd, M. R., H. L. Scott, and P. R. Dodd. Glutamate(NMDA) receptor NR1 subunit mRNA expression in Alzheimer's disease. *J Neurochem* 78: 175–182, 2001.
130. Hynd, M. R., H. L. Scott, and P. R. Dodd. Selective loss of NMDA receptor NR1 subunit isoforms in Alzheimer's disease. *J Neurochem* 89: 240–247, 2004.
131. Hynd, M. R., H. L. Scott, and P. R. Dodd. Differential expression of N-methyl-D-aspartate receptor NR2 isoforms in Alzheimer's disease. *J Neurochem* 90: 913–919, 2004.
132. Hynd, M. R., H. L. Scott, and P. R. Dodd. Glutamate-mediated excitotoxicity and neurodegeneration in Alzheimer's disease. *Neurochem Int* 45: 583–595, 2004.
133. Isaac, J. T., R. A. Nicoll, and R. C. Malenka. Evidence for silent synapses: implications for the expression of LTP. *Neuron* 15: 427–434, 1995.
134. Isaacson, J. S., and G. J. Murphy. Glutamate-mediated extrasynaptic inhibition: direct coupling of NMDA receptors to Ca(2+)-activated K+ channels. *Neuron* 31: 1027–1034, 2001.
135. Ishii, T., K. Moriyoshi, H. Sugihara, K. Sakurada, H. Kadotani, M. Yokoi, C. Akazawa, R. Shigemoto, N. Mizuno, M. Masu, et al. Molecular characterization of the family of the N-methyl-D-aspartate receptor subunits. *J Biol Chem* 268: 2836–2843, 1993.
136. Jarabek, B. R., R. P. Yasuda, and B. B. Wolfe. Regulation of proteins affecting NMDA receptor-induced excitotoxicity in a Huntington's mouse model. *Brain* 127: 505–516, 2004.
137. Jason, G. W., O. Suchowersky, E. M. Pajurkova, L. Graham, M. L. Klimek, A. T. Garber, and D. Poirier-Heine. Cognitive manifestations of Huntington disease in relation to genetic structure and clinical onset. *Arch Neurol* 54: 1081–1088, 1997.
138. Jourdain, P., L. H. Bergersen, K. Bhaukaurally, P. Bezzi, M. Santello, M. Domercq, C. Matute, F. Tonello, V. Gundersen, and A. Volterra. Glutamate exocytosis from astrocytes controls synaptic strength. *Nat Neurosci* 10: 331–339, 2007.
139. Kadotani, H., T. Hirano, M. Masugi, K. Nakamura, K. Nakao, M. Katsuki, and S. Nakanishi. Motor discoordination results from combined gene disruption of the NMDA receptor NR2A and NR2C subunits, but not from single disruption of the NR2A or NR2C subunit. *J Neurosci* 16: 7859–7867, 1996.
140. Kamenetz, F., T. Tomita, H. Hsieh, G. Seabrook, D. Borchelt, T. Iwatsubo, S. Sisodia, and R. Malinow. APP processing and synaptic function. *Neuron* 37: 925–937, 2003.

141. Kanzaki, M., and J. E. Pessin. Insulin signaling: GLUT4 vesicles exit via the exocyst. *Curr Biol* 13: R574–R576, 2003.
142. Karavanova, I., K. Vasudevan, J. Cheng, and A. Buonanno. Novel regional and developmental NMDA receptor expression patterns uncovered in NR2C subunit-beta-galactosidase knock-in mice. *Mol Cell Neurosci* 34: 468–480, 2007.
143. Kato, A., N. Rouach, R. A. Nicoll, and D. S. Bredt. Activity-dependent NMDA receptor degradation mediated by retrotranslocation and ubiquitination. *Proc Natl Acad Sci USA* 102: 5600–5605, 2005.
144. Kaul, M., and S. A. Lipton. Signaling pathways to neuronal damage and apoptosis in human immunodeficiency virus type 1-associated dementia: chemokine receptors, excitotoxicity, and beyond. *J Neurovirol* 10(Suppl 1): 97–101, 2004.
145. Kennedy, M. B., and P. Manzerra. Telling tails. *Proc Natl Acad Sci USA* 98: 12323–12324, 2001.
146. Kim, E., S. Naisbitt, Y. P. Hsueh, A. Rao, A. Rothschild, A. M. Craig, and M. Sheng. GKAP, a novel synaptic protein that interacts with the guanylate kinase-like domain of the PSD-95/SAP90 family of channel clustering molecules. *J Cell Biol* 136: 669–678, 1997.
147. Kim, E., and M. Sheng. PDZ domain proteins of synapses. *Nat Rev Neurosci* 5: 771–781, 2004.
148. Kim, J. H., D. Liao, L. F. Lau, and R. L. Huganir. SynGAP: a synaptic RasGAP that associates with the PSD-95/SAP90 protein family. *Neuron* 20: 683–691, 1998.
149. Kirkham, M., and R. G. Parton. Clathrin-independent endocytosis: new insights into caveolae and non-caveolar lipid raft carriers. *Biochim Biophys Acta* 1746: 349–363, 2005.
150. Kirson, E. D., C. Schirra, A. Konnerth, and Y. Yaari. Early postnatal switch in magnesium sensitivity of NMDA receptors in rat CA1 pyramidal cells. *J Physiol* 521(Pt 1): 99–111, 1999.
151. Kitamura, T., M. Mishina, and H. Sugiyama. Enhancement of neurogenesis by running wheel exercises is suppressed in mice lacking NMDA receptor epsilon 1 subunit. *Neurosci Res* 47: 55–63, 2003.
152. Kneussel, M. Postsynaptic scaffold proteins at non-synaptic sites. The role of postsynaptic scaffold proteins in motor-protein-receptor complexes. *EMBO Rep* 6: 22–27, 2005.
153. Ko, J., S. Kim, H. S. Chung, K. Kim, K. Han, H. Kim, H. Jun, B. K. Kaang, and E. Kim. SALM synaptic cell adhesion-like molecules regulate the differentiation of excitatory synapses. *Neuron* 50: 233–245, 2006.
154. Kohr, G. NMDA receptor function: subunit composition versus spatial distribution. *Cell Tissue Res* 326: 439–446, 2006.
155. Komuro, H., and P. Rakic. Modulation of neuronal migration by NMDA receptors. *Science* 260: 95–97, 1993.
156. Kornau, H. C., L. T. Schenker, M. B. Kennedy, and P. H. Seeburg. Domain interaction between NMDA receptor subunits and the postsynaptic density protein PSD-95. *Science* 269: 1737–1740, 1995.
157. Kowal, C., L. A. Degiorgio, J. Y. Lee, M. A. Edgar, P. T. Huerta, B. T. Volpe, and B. Diamond. Human lupus autoantibodies against NMDA receptors mediate cognitive impairment. *Proc Natl Acad Sci USA* 103: 19854–19859, 2006.
158. Kristiansen, L. V., and J. H. Meador-Woodruff. Abnormal striatal expression of transcripts encoding NMDA interacting PSD proteins in schizophrenia, bipolar disorder and major depression. *Schizophr Res* 78: 87–93, 2005.
159. Kumada, T., and H. Komuro. Completion of neuronal migration regulated by loss of Ca(2+) transients. *Proc Natl Acad Sci USA* 101: 8479–8484, 2004.

160. Kuppenbender, K. D., D. G. Standaert, T. J. Feuerstein, J. B. J. Penney, A. B. Young, and G. B. Landwehrmeyer. Expression of NMDA receptor subunit mRNAs in neurochemically identified projection and interneurons in the human striatum. *J Comp Neurol* 419: 407–421, 2000.
161. Kurschner, C., P. G. Mermelstein, W. T. Holden, and D. J. Surmeier. CIPP, a novel multivalent PDZ domain protein, selectively interacts with Kir4.0 family members, NMDA receptor subunits, neurexins, and neuroligins. *Mol Cell Neurosci* 11: 161–172, 1998.
162. Kuryatov, A., B. Laube, H. Betz, and J. Kuhse. Mutational analysis of the glycine-binding site of the NMDA receptor: structural similarity with bacterial amino acid-binding proteins. *Neuron* 12: 1291–1300, 1994.
163. Kusumi, A., and K. Suzuki. Toward understanding the dynamics of membrane-raft-based molecular interactions. *Biochim Biophys Acta* 1746: 234–251, 2005.
164. Kwon, O. B., M. Longart, D. Vullhorst, D. A. Hoffman, and A. Buonanno. Neuregulin-1 reverses long-term potentiation at CA1 hippocampal synapses. *J Neurosci* 25: 9378–9383, 2005.
165. Lacor, P. N., M. C. Buniel, P. W. Furlow, A. S. Clemente, P. T. Velasco, M. Wood, K. L. Viola, and W. L. Klein. Abeta oligomer-induced aberrations in synapse composition, shape, and density provide a molecular basis for loss of connectivity in Alzheimer's disease. *J Neurosci* 27: 796–807, 2007.
166. Lambert, M. P., A. K. Barlow, B. A. Chromy, C. Edwards, R. Freed, M. Liosatos, T. E. Morgan, I. Rozovsky, B. Trommer, K. L. Viola, P. Wals, C. Zhang, C. E. Finch, G. A. Krafft, and W. L. Klein. Diffusible, nonfibrillar ligands derived from Abeta1–42 are potent central nervous system neurotoxins. *Proc Natl Acad Sci USA* 95: 6448–6453, 1998.
167. Lan, J. Y., V. A. Skeberdis, T. Jover, S. Y. Grooms, Y. Lin, R. C. Araneda, X. Zheng, M. V. Bennett, and R. S. Zukin. Protein kinase C modulates NMDA receptor trafficking and gating. *Nat Neurosci* 4: 382–390, 2001.
168. Landles, C., and G. P. Bates. Huntingtin and the molecular pathogenesis of Huntington's disease. Fourth in molecular medicine review series. *EMBO Rep* 5: 958–963, 2004.
169. Landwehrmeyer, G. B., D. G. Standaert, C. M. Testa, J. B. J. Penney, and A. B. Young. NMDA receptor subunit mRNA expression by projection neurons and interneurons in rat striatum. *J Neurosci* 15: 5297–5307, 1995.
170. Lau, C. G., and R. S. Zukin. NMDA receptor trafficking in synaptic plasticity and neuropsychiatric disorders. *Nat Rev Neurosci* 8: 413–426, 2007.
171. Lau, L. F., A. Mammen, M. D. Ehlers, S. Kindler, W. J. Chung, C. C. Garner, and R. L. Huganir. Interaction of the N-methyl-D-aspartate receptor complex with a novel synapse-associated protein, SAP102. *J Biol Chem* 271: 21622–21628, 1996.
172. Laube, B., H. Hirai, M. Sturgess, H. Betz, and J. Kuhse. Molecular determinants of agonist discrimination by NMDA receptor subunits: analysis of the glutamate binding site on the NR2B subunit. *Neuron* 18: 493–503, 1997.
173. Laurie, D. J., I. Bartke, R. Schoepfer, K. Naujoks, and P. H. Seeburg. Regional, developmental and interspecies expression of the four NMDAR2 subunits, examined using monoclonal antibodies. *Brain Res Mol Brain Res* 51: 23–32, 1997.
174. Laurie, D. J., J. Putzke, W. Zieglgansberger, P. H. Seeburg, and T. R. Tolle. The distribution of splice variants of the NMDAR1 subunit mRNA in adult rat brain. *Brain Res Mol Brain Res* 32: 94–108, 1995.
175. Laurie, D. J., and P. H. Seeburg. Regional and developmental heterogeneity in splicing of the rat brain NMDAR1 mRNA. *J Neurosci* 14: 3180–3194, 1994.
176. Lavezzari, G., J. McCallum, C. M. Dewey, and K. W. Roche. Subunit-specific regulation of NMDA receptor endocytosis. *J Neurosci* 24: 6383–6391, 2004.

177. Lawrence, A. D., B. J. Sahakian, J. R. Hodges, A. E. Rosser, K. W. Lange, and T. W. Robbins. Executive and mnemonic functions in early Huntington's disease. *Brain* 119: 1633–1645, 1996.
178. Leavitt, B. R., J. M. van Raamsdonk, J. Shehadeh, H. Fernandes, Z. Murphy, R. K. Graham, C. L. Wellington, L. A. Raymond, and M. R. Hayden. Wild-type huntingtin protects neurons from excitotoxicity. *J Neurochem* 96: 1121–1129, 2006.
179. Lei, S., E. Czerwinska, W. Czerwinski, M. P. Walsh, and J. F. MacDonald. Regulation of NMDA receptor activity by F-actin and myosin light chain kinase. *J Neurosci* 21: 8464–8472, 2001.
180. Leonard, A. S., I. A. Lim, D. E. Hemsworth, M. C. Horne, and J. W. Hell. Calcium/calmodulin-dependent protein kinase II is associated with the N-methyl-D-aspartate receptor. *Proc Natl Acad Sci USA* 96: 3239–3244, 1999.
181. Levine, M. S., C. Cepeda, M. A. Hickey, S. M. Fleming, and M. F. Chesselet. Genetic mouse models of Huntington's and Parkinson's diseases: illuminating but imperfect. *Trends Neurosci* 27: 691–697, 2004.
182. Levinson, J. N., N. Chery, K. Huang, T. P. Wong, K. Gerrow, R. Kang, O. Prange, Y. T. Wang, and A. El-Husseini. Neuroligins mediate excitatory and inhibitory synapse formation: involvement of PSD-95 and neurexin-1beta in neuroligin-induced synaptic specificity. *J Biol Chem* 280: 17312–17319, 2005.
183. Li, B., N. Chen, T. Luo, Y. Otsu, T. H. Murphy, and L. A. Raymond. Differential regulation of synaptic and extra-synaptic NMDA receptors. *Nat Neurosci* 5: 833–834, 2002.
184. Li, B. S., M. K. Sun, L. Zhang, S. Takahashi, W. Ma, L. Vinade, A. B. Kulkarni, R. O. Brady, and H. C. Pant. Regulation of NMDA receptors by cyclin-dependent kinase-5. *Proc Natl Acad Sci USA* 98: 12742–12747, 2001.
185. Li, L., M. Fan, C. D. Icton, N. Chen, B. R. Leavitt, M. R. Hayden, T. H. Murphy, and L. A. Raymond. Role of NR2B-type NMDA receptors in selective neurodegeneration in Huntington disease. *Neurobiol Aging* 24: 1113–1121, 2003.
186. Li, Z., and M. Sheng. Some assembly required: the development of neuronal synapses. *Nat Rev Mol Cell Biol* 4: 833–841, 2003.
187. Lieberman, D. N., and I. Mody. Casein kinase-II regulates NMDA channel function in hippocampal neurons. *Nat Neurosci* 2: 125–132, 1999.
188. Lien, C. C., Y. Mu, M. Vargas-Caballero, and M. M. Poo. Visual stimuli-induced LTD of GABAergic synapses mediated by presynaptic NMDA receptors. *Nat Neurosci* 9: 372–380, 2006.
189. Lillemeier, B. F., J. R. Pfeiffer, Z. Surviladze, B. S. Wilson, and M. M. Davis. Plasma membrane-associated proteins are clustered into islands attached to the cytoskeleton. *Proc Natl Acad Sci USA* 103: 18992–18997, 2006.
190. Lin, B., A. C. Arai, G. Lynch, and C. M. Gall. Integrins regulate NMDA receptor-mediated synaptic currents. *J Neurophysiol* 89: 2874–2878, 2003.
191. Lin, S. Y., and M. Constantine-Paton. Suppression of sprouting: an early function of NMDA receptors in the absence of AMPA/kainate receptor activity. *J Neurosci* 18: 3725–3737, 1998.
192. Lipton, S. A. The molecular basis of memantine action in Alzheimer's disease and other neurologic disorders: low-affinity, uncompetitive antagonism. *Curr Alzheimer Res* 2: 155–165, 2005.
193. Liu, L., T. P. Wong, M. F. Pozza, K. Lingenhoehl, Y. Wang, M. Sheng, Y. P. Auberson, and Y. T. Wang. Role of NMDA receptor subtypes in governing the direction of hippocampal synaptic plasticity. *Science* 304: 1021–1024, 2004.
194. Liu, X. B., K. D. Murray, and E. G. Jones. Switching of NMDA receptor 2A and 2B subunits at thalamic and cortical synapses during early postnatal development. *J Neurosci* 24: 8885–8895, 2004.

195. Loschmann, P. A., C. De Groote, L. Smith, U. Wullner, G. Fischer, J. A. Kemp, P. Jenner, and T. Klockgether. Antiparkinsonian activity of Ro 25–6981, a NR2B subunit specific NMDA receptor antagonist, in animal models of Parkinson's disease. *Exp Neurol* 187: 86–93, 2004.
196. Losi, G., K. Prybylowski, Z. Fu, J. Luo, R. J. Wenthold, and S. Vicini. PSD-95 regulates NMDA receptors in developing cerebellar granule neurons of the rat. *J Physiol* 548: 21–29, 2003.
197. LoTurco, J. J., M. G. Blanton, and A. R. Kriegstein. Initial expression and endogenous activation of NMDA channels in early neocortical development. *J Neurosci* 11: 792–799, 1991.
198. Low, C. M., F. Zheng, P. Lyuboslavsky, and S. F. Traynelis. Molecular determinants of coordinated proton and zinc inhibition of N-methyl-D-aspartate NR1/NR2A receptors. *Proc Natl Acad Sci USA* 97: 11062–11067, 2000.
199. Luby, E. D., B. D. Cohen, G. Rosenbaum, J. S. Gottlieb, and R. Kelley. Study of a new schizophrenomimetic drug; sernyl. *AMA Arch Neurol Psychiatry* 81: 363–369, 1959.
200. Luthi-Carter, R., B. L. Apostol, A. W. Dunah, M. M. DeJohn, L. A. Farrell, G. P. Bates, A. B. Young, D. G. Standaert, L. M. Thompson, and J. H. Cha. Complex alteration of NMDA receptors in transgenic Huntington's disease mouse brain: analysis of mRNA and protein expression, plasma membrane association, interacting proteins, and phosphorylation. *Neurobiol Dis* 14: 624–636, 2003.
201. Lynch, D. R., and R. P. Guttmann. Excitotoxicity: perspectives based on N-methyl-D-aspartate receptor subtypes. *J Pharmacol Exp Ther* 300: 717–723, 2002.
202. MacDonald, J. F., S. A. Kotecha, W. Y. Lu, and M. F. Jackson. Convergence of PKC-dependent kinase signal cascades on NMDA receptors. *Curr Drug Targets* 2: 299–312, 2001.
203. Mameli, M., M. Carta, L. D. Partridge, and C. F. Valenzuela. Neurosteroid-induced plasticity of immature synapses via retrograde modulation of presynaptic NMDA receptors. *J Neurosci* 25: 2285–2294, 2005.
204. Manent, J. B., M. Demarque, I. Jorquera, C. Pellegrino, Y. Ben-Ari, L. Aniksztejn, and A. Represa. A noncanonical release of GABA and glutamate modulates neuronal migration. *J Neurosci* 25: 4755–4765, 2005.
205. Manent, J. B., I. Jorquera, Y. Ben-Ari, L. Aniksztejn, and A. Represa. Glutamate acting on AMPA but not NMDA receptors modulates the migration of hippocampal interneurons. *J Neurosci* 26: 5901–5909, 2006.
206. Matsuda, K., M. Fletcher, Y. Kamiya, and M. Yuzaki. Specific assembly with the NMDA receptor 3B subunit controls surface expression and calcium permeability of NMDA receptors. *J Neurosci* 23: 10064–10073, 2003.
207. McCullumsmith, R. E., L. V. Kristiansen, M. Beneyto, E. Scarr, B. Dean, and J. H. Meador-Woodruff. Decreased NR1, NR2A, and SAP102 transcript expression in the hippocampus in bipolar disorder. *Brain Res* 1127: 108–118, 2007.
208. McGee, A. W., J. R. Topinka, K. Hashimoto, R. S. Petralia, S. Kakizawa, F. W. Kauer, A. Aguilera-Moreno, R. J. Wenthold, M. Kano, and D. S. Bredt. PSD-93 knock-out mice reveal that neuronal MAGUKs are not required for development or function of parallel fiber synapses in cerebellum. *J Neurosci* 21: 3085–3091, 2001.
209. McIlhinney, R. A., B. Le Bourdelles, E. Molnar, N. Tricaud, P. Streit, and P. J. Whiting. Assembly intracellular targeting and cell surface expression of the human N-methyl-D-aspartate receptor subunits NR1a and NR2A in transfected cells. *Neuropharmacology* 37: 1355–1367, 1998.
210. McIlhinney, R. A., E. Philipps, B. Le Bourdelles, S. Grimwood, K. Wafford, S. Sandhu, and P. Whiting. Assembly of N-methyl-D-aspartate (NMDA) receptors. *Biochem Soc Trans* 31: 865–868, 2003.

211. McKinney, R. A., A. Luthi, C. E. Bandtlow, B. H. Gahwiler, and S. M. Thompson. Selective glutamate receptor antagonists can induce or prevent axonal sprouting in rat hippocampal slice cultures. *Proc Natl Acad Sci USA* 96: 11631–11636, 1999.
212. Meador-Woodruff, J. H., and D. J. Healy. Glutamate receptor expression in schizophrenic brain. *Brain Res Brain Res Rev* 31: 288–294, 2000.
213. Meddows, E., B. Le Bourdelles, S. Grimwood, K. Wafford, S. Sandhu, P. Whiting, and R. A. McIlhinney. Identification of molecular determinants that are important in the assembly of N-methyl-D-aspartate receptors. *J Biol Chem* 276: 18795–18803, 2001.
214. Medina, I. Extrasynaptic NMDA receptors reshape gene ranks. *Sci STKE* 2007: pe23, 2007.
215. Mehta, S. Q., P. R. Hiesinger, S. Beronja, R. G. Zhai, K. L. Schulze, P. Verstreken, Y. Cao, Y. Zhou, U. Tepass, M. C. Crair, and H. J. Bellen. Mutations in Drosophila sec15 reveal a function in neuronal targeting for a subset of exocyst components. *Neuron* 46: 219–232, 2005.
216. Menegoz, M., L. F. Lau, D. Herve, R. L. Huganir, and J. A. Girault. Tyrosine phosphorylation of NMDA receptor in rat striatum: effects of 6-OH-dopamine lesions. *Neuroreport* 7: 125–128, 1995.
217. Meoni, P., B. H. Bunnemann, A. E. Kingsbury, D. G. Trist, and N. G. Bowery. NMDA NR1 subunit mRNA and glutamate NMDA-sensitive binding are differentially affected in the striatum and pre-frontal cortex of Parkinson's disease patients. *Neuropharmacology* 38: 625–633, 1999.
218. Migaud, M., P. Charlesworth, M. Dempster, L. C. Webster, A. M. Watabe, M. Makhinson, Y. He, M. F. Ramsay, R. G. Morris, J. H. Morrison, T. J. O'Dell, and S. G. Grant. Enhanced long-term potentiation and impaired learning in mice with mutant postsynaptic density-95 protein. *Nature* 396: 433–439, 1998.
219. Mishizen-Eberz, A. J., R. A. Rissman, T. L. Carter, M. D. Ikonomovic, B. B. Wolfe, and D. M. Armstrong. Biochemical and molecular studies of NMDA receptor subunits NR1/2A/2B in hippocampal subregions throughout progression of Alzheimer's disease pathology. *Neurobiol Dis* 15: 80–92, 2004.
220. Miskevich, F., W. Lu, S. Y. Lin, and M. Constantine-Paton. Interaction between metabotropic and NMDA subtypes of glutamate receptors in sprout suppression at young synapses. *J Neurosci* 22: 226–238, 2002.
221. Misra, C., and E. B. Ziff. EphB2 gets a GRIP on the dendritic arbor. *Nat Neurosci* 8: 848–850, 2005.
222. Miyamoto, Y., K. Yamada, Y. Noda, H. Mori, M. Mishina, and T. Nabeshima. Lower sensitivity to stress and altered monoaminergic neuronal function in mice lacking the NMDA receptor epsilon 4 subunit. *J Neurosci* 22: 2335–2342, 2002.
223. Mohn, A. R., R. R. Gainetdinov, M. G. Caron, and B. H. Koller. Mice with reduced NMDA receptor expression display behaviors related to schizophrenia. *Cell* 98: 427–436, 1999.
224. Mok, H., H. Shin, S. Kim, J. R. Lee, J. Yoon, and E. Kim. Association of the kinesin superfamily motor protein KIF1Balpha with postsynaptic density-95 (PSD-95), synapse-associated protein-97, and synaptic scaffolding molecule PSD-95/discs large/zona occludens-1 proteins. *J Neurosci* 22: 5253–5258, 2002.
225. Montgomery, J. M., and D. V. Madison. State-dependent heterogeneity in synaptic depression between pyramidal cell pairs. *Neuron* 33: 765–777, 2002.
226. Monyer, H., N. Burnashev, D. J. Laurie, B. Sakmann, and P. H. Seeburg. Developmental and regional expression in the rat brain and functional properties of four NMDA receptors. *Neuron* 12: 529–540, 1994.
227. Moya, F., and M. Valdeolmillos. Polarized increase of calcium and nucleokinesis in tangentially migrating neurons. *Cereb Cortex* 14: 610–618, 2004.

228. Mu, Y., T. Otsuka, A. C. Horton, D. B. Scott, and M. D. Ehlers. Activity-dependent mRNA splicing controls ER export and synaptic delivery of NMDA receptors. *Neuron* 40: 581–594, 2003.
229. Mueller, H. T., and J. H. Meador-Woodruff. NR3A NMDA receptor subunit mRNA expression in schizophrenia, depression and bipolar disorder. *Schizophr Res* 71: 361–370, 2004.
230. Naisbitt, S., E. Kim, J. C. Tu, B. Xiao, C. Sala, J. Valtschanoff, R. J. Weinberg, P. F. Worley, and M. Sheng. Shank, a novel family of postsynaptic density proteins that binds to the NMDA receptor/PSD-95/GKAP complex and cortactin. *Neuron* 23: 569–582, 1999.
231. Nakazawa, T., S. Komai, T. Tezuka, C. Hisatsune, H. Umemori, K. Semba, M. Mishina, T. Manabe, and T. Yamamoto. Characterization of Fyn-mediated tyrosine phosphorylation sites on GluR epsilon 2 (NR2B) subunit of the N-methyl-D-aspartate receptor. *J Biol Chem* 276: 693–699, 2001.
232. Nash, J. E., S. H. Fox, B. Henry, M. P. Hill, D. Peggs, S. McGuire, Y. Maneuf, C. Hille, J. M. Brotchie, and A. R. Crossman. Antiparkinsonian actions of ifenprodil in the MPTP-lesioned marmoset model of Parkinson's disease. *Exp Neurol* 165: 136–142, 2000.
233. Neyton, J., and P. Paoletti. Relating NMDA receptor function to receptor subunit composition: limitations of the pharmacological approach. *J Neurosci* 26: 1331–1333, 2006.
234. Nichols, B. J., and J. Lippincott-Schwartz. Endocytosis without clathrin coats. *Trends Cell Biol* 11: 406–412, 2001.
235. Niethammer, M., E. Kim, and M. Sheng. Interaction between the C terminus of NMDA receptor subunits and multiple members of the PSD-95 family of membrane-associated guanylate kinases. *J Neurosci* 16: 2157–2163, 1996.
236. Nishimura, W., I. Yao, J. Iida, N. Tanaka, and Y. Hata. Interaction of synaptic scaffolding molecule and Beta -catenin. *J Neurosci* 22: 757–765, 2002.
237. Nong, Y., Y. Q. Huang, and M. W. Salter. NMDA receptors are movin' in. *Curr Opin Neurobiol* 14: 353–361, 2004.
238. O'Connor, J. J., M. J. Rowan, and R. Anwyl. Long-lasting enhancement of NMDA receptor-mediated synaptic transmission by metabotropic glutamate receptor activation. *Nature* 367: 557–559, 1994.
239. Obrietan, K., and K. R. Hoyt. CRE-mediated transcription is increased in Huntington's disease transgenic mice. *J Neurosci* 24: 791–796, 2004.
240. Oh, J. D., D. S. Russell, C. L. Vaughan, and T. N. Chase. Enhanced tyrosine phosphorylation of striatal NMDA receptor subunits: effect of dopaminergic denervation and L-DOPA administration. *Brain Res* 813: 150–159, 1998.
241. Oh, J. D., C. L. Vaughan, and T. N. Chase. Effect of dopamine denervation and dopamine agonist administration on serine phosphorylation of striatal NMDA receptor subunits. *Brain Res* 821: 433–442, 1999.
242. Ohtakara, K., M. Nishizawa, I. Izawa, Y. Hata, S. Matsushima, W. Taki, H. Inada, Y. Takai, and M. Inagaki. Densin-180, a synaptic protein, links to PSD-95 through its direct interaction with MAGUIN-1. *Genes Cells* 7: 1149–1160, 2002.
243. Okabe, S., C. Collin, J. M. Auerbach, N. Meiri, J. Bengzon, M. B. Kennedy, M. Segal, and R. D. McKay. Hippocampal synaptic plasticity in mice overexpressing an embryonic subunit of the NMDA receptor. *J Neurosci* 18: 4177–4188, 1998.
244. Okabe, S., A. Miwa, and H. Okado. Alternative splicing of the C-terminal domain regulates cell surface expression of the NMDA receptor NR1 subunit. *J Neurosci* 19: 7781–7792, 1999.
245. Oshima, S., M. Fukaya, N. Masabumi, T. Shirakawa, H. Oguchi, and M. Watanabe. Early onset of NMDA receptor GluR epsilon 1 (NR2A) expression and its abundant

postsynaptic localization in developing motoneurons of the mouse hypoglossal nucleus. *Neurosci Res* 43: 239–250, 2002.
246. Ozaki, M., M. Sasner, R. Yano, H. S. Lu, and A. Buonanno. Neuregulin-beta induces expression of an NMDA-receptor subunit. *Nature* 390: 691–694, 1997.
247. Pak, D. T., S. Yang, S. Rudolph-Correia, E. Kim, and M. Sheng. Regulation of dendritic spine morphology by SPAR, a PSD-95-associated RapGAP. *Neuron* 31: 289–303, 2001.
248. Passafaro, M., V. Piech, and M. Sheng. Subunit-specific temporal and spatial patterns of AMPA receptor exocytosis in hippocampal neurons. *Nat Neurosci* 4: 917–926, 2001.
249. Paulsen, J. S., R. E. Ready, J. M. Hamilton, M. S. Mega, and J. L. Cummings. Neuropsychiatric aspects of Huntington's disease. *J Neurol Neurosurg Psychiatry* 71: 310–314, 2001.
250. Pavlov, I., R. Riekki, and T. Taira. Synergistic action of GABA-A and NMDA receptors in the induction of long-term depression in glutamatergic synapses in the newborn rat hippocampus. *Eur J Neurosci* 20: 3019–3026, 2004.
251. Pearce, I. A., M. A. Cambray-Deakin, and R. D. Burgoyne. Glutamate acting on NMDA receptors stimulates neurite outgrowth from cerebellar granule cells. *FEBS Lett* 223: 143–147, 1987.
252. Perez-Otano, I., and M. D. Ehlers. Learning from NMDA receptor trafficking: clues to the development and maturation of glutamatergic synapses. *Neurosignals* 13: 175–189, 2004.
253. Perez-Otano, I., R. Lujan, S. J. Tavalin, M. Plomann, J. Modregger, X. B. Liu, E. G. Jones, S. F. Heinemann, D. C. Lo, and M. D. Ehlers. Endocytosis and synaptic removal of NR3A-containing NMDA receptors by PACSIN1/syndapin1. *Nat Neurosci* 9: 611–621, 2006.
254. Perez-Otano, I., C. T. Schulteis, A. Contractor, S. A. Lipton, J. S. Trimmer, N. J. Sucher, and S. F. Heinemann. Assembly with the NR1 subunit is required for surface expression of NR3A-containing NMDA receptors. *J Neurosci* 21: 1228–1237, 2001.
255. Perez-Otano, I., and M. D. Ehlers. Homeostatic plasticity and NMDA receptor trafficking. *Trends Neurosci* 28: 229–238, 2005.
256. Perin-Dureau, F., J. Rachline, J. Neyton, and P. Paoletti. Mapping the binding site of the neuroprotectant ifenprodil on NMDA receptors. *J Neurosci* 22: 5955–5965, 2002.
257. Petralia, R. S., N. Sans, Y. X. Wang, and R. J. Wenthold. Ontogeny of postsynaptic density proteins at glutamatergic synapses. *Mol Cell Neurosci* 29: 436–452, 2005.
258. Petralia, R. S., Y. X. Wang, and R. J. Wenthold. Internalization at glutamatergic synapses during development. *Eur J Neurosci* 18: 3207–3217, 2003.
259. Petralia, R. S., L. A. Dunbar, Y.-X. Wang, and R. J. Wenthold. Morphological correlates of glutamate receptor trafficking to synapses. *Soc Neurosci Abstr* 843.2, 2004.
260. Petralia, R. S., L. A. Dunbar, Y.-X. Wang, and R. J. Wenthold. Specific endosomal associations during the trafficking of synaptic glutamate receptors. *Soc Neurosci Abs* 949.1, 2005.
261. Petralia, R. S., J. A. Esteban, Y. X. Wang, J. G. Partridge, H. M. Zhao, R. J. Wenthold, and R. Malinow. Selective acquisition of AMPA receptors over postnatal development suggests a molecular basis for silent synapses. *Nat Neurosci* 2: 31–36, 1999.
262. Petralia, R. S., Y. X. Wang, and R. J. Wenthold. NMDA receptors and PSD-95 are found in attachment plaques in cerebellar granular layer glomeruli. *Eur J Neurosci* 15: 583–587, 2002.
263. Philpot, B. D., A. K. Sekhar, H. Z. Shouval, and M. F. Bear. Visual experience and deprivation bidirectionally modify the composition and function of NMDA receptors in visual cortex. *Neuron* 29: 157–169, 2001.
264. Philpot, B. D., K. K. Cho, and M. F. Bear. Obligatory Role of NR2A for Metaplasticity in Visual Cortex. *Neuron* 53: 495–502, 2007.

265. Picconi, B., F. Gardoni, D. Centonze, D. Mauceri, M. A. Cenci, G. Bernardi, P. Calabresi, and M. Di Luca. Abnormal Ca2+-calmodulin-dependent protein kinase II function mediates synaptic and motor deficits in experimental parkinsonism. *J Neurosci* 24: 5283–5291, 2004.
266. Polo-Parada, L., C. M. Bose, F. Plattner, and L. T. Landmesser. Distinct roles of different neural cell adhesion molecule (NCAM) isoforms in synaptic maturation revealed by analysis of NCAM 180 kDa isoform-deficient mice. *J Neurosci* 24: 1852–1864, 2004.
267. Prybylowski, K., K. Chang, N. Sans, L. Kan, S. Vicini, and R. J. Wenthold. The synaptic localization of NR2B-containing NMDA receptors is controlled by interactions with PDZ proteins and AP-2. *Neuron* 47: 845–857, 2005.
268. Prybylowski, K., Z. Fu, G. Losi, L. M. Hawkins, J. Luo, K. Chang, R. J. Wenthold, and S. Vicini. Relationship between availability of NMDA receptor subunits and their expression at the synapse. *J Neurosci* 22: 8902–8910, 2002.
269. Prybylowski, K. L., and B. B. Wolfe. Developmental differences in alternative splicing of the NR1 protein in rat cortex and cerebellum. *Brain Res Dev Brain Res* 123: 143–150, 2000.
270. Quinlan, E. M., D. Lebel, I. Brosh, and E. Barkai. A molecular mechanism for stabilization of learning-induced synaptic modifications. *Neuron* 41: 185–192, 2004.
271. Quinlan, E. M., D. H. Olstein, and M. F. Bear. Bidirectional, experience-dependent regulation of N-methyl-D-aspartate receptor subunit composition in the rat visual cortex during postnatal development. *Proc Natl Acad Sci USA* 96: 12876–12880, 1999.
272. Quinlan, E. M., B. D. Philpot, R. L. Huganir, and M. F. Bear. Rapid, experience-dependent expression of synaptic NMDA receptors in visual cortex in vivo. *Nat Neurosci* 2: 352–357, 1999.
273. Racz, B., T. A. Blanpied, M. D. Ehlers, and R. J. Weinberg. Lateral organization of endocytic machinery in dendritic spines. *Nat Neurosci* 7: 917–918, 2004.
274. Rajan, I., and H. T. Cline. Glutamate receptor activity is required for normal development of tectal cell dendrites in vivo. *J Neurosci* 18: 7836–7846, 1998.
275. Rampon, C., Y. P. Tang, J. Goodhouse, E. Shimizu, M. Kyin, and J. Z. Tsien. Enrichment induces structural changes and recovery from nonspatial memory deficits in CA1 NMDAR1-knockout mice. *Nat Neurosci* 3: 238–244, 2000.
276. Rao, A., E. Kim, M. Sheng, and A. M. Craig. Heterogeneity in the molecular composition of excitatory postsynaptic sites during development of hippocampal neurons in culture. *J Neurosci* 18: 1217–1229, 1998.
277. Rashid, N. A., and M. A. Cambray-Deakin. N-methyl-D-aspartate effects on the growth, morphology and cytoskeleton of individual neurons in vitro. *Brain Res Dev Brain Res* 67: 301–308, 1992.
278. Regalado, M. P., R. T. Terry-Lorenzo, C. L. Waites, C. C. Garner, and R. C. Malenka. Transsynaptic signaling by postsynaptic synapse-associated protein 97. *J Neurosci* 26: 2343–2357, 2006.
279. Roche, K. W., S. Standley, J. McCallum, L. Dune, C., M. D. Ehlers, and R. J. Wenthold. Molecular determinants of NMDA receptor internalization. *Nat Neurosci* 4: 794–802, 2001.
280. Rumbaugh, G., K. Prybylowski, J. F. Wang, and S. Vicini. Exon 5 and spermine regulate deactivation of NMDA receptor subtypes. *J Neurophysiol* 83: 1300–1306, 2000.
281. Rumbaugh, G., and S. Vicini. Distinct synaptic and extrasynaptic NMDA receptors in developing cerebellar granule neurons. *J Neurosci* 19: 10603–10610, 1999.
282. Ruthazer, E. S., C. J. Akerman, and H. T. Cline. Control of axon branch dynamics by correlated activity in vivo. *Science* 301: 66–70, 2003.
283. Salter, M. W., and L. V. Kalia. Src kinases: a hub for NMDA receptor regulation. *Nat Rev Neurosci* 5: 317–328, 2004.

284. Sampo, B., S. Kaech, S. Kunz, and G. Banker. Two distinct mechanisms target membrane proteins to the axonal surface. *Neuron* 37: 611–624, 2003.
285. Sans, N., K. Prybylowski, R. S. Petralia, K. Chang, Y. X. Wang, C. Racca, S. Vicini, and R. J. Wenthold. NMDA receptor trafficking through an interaction between PDZ proteins and the exocyst complex. *Nat Cell Biol* 5: 520–530, 2003.
286. Sans, N., P. Y. Wang, Q. Du, R. S. Petralia, Y. X. Wang, S. Nakka, J. B. Blumer, I. G. Macara, and R. J. Wenthold. mPins modulates PSD-95 and SAP102 trafficking and influences NMDA receptor surface expression. *Nat Cell Biol* 7: 1179–1190, 2005.
287. Sans, N., R. S. Petralia, Y. X. Wang, J. Blahos, 2nd, J. W. Hell, and R. J. Wenthold. A developmental change in NMDA receptor-associated proteins at hippocampal synapses. *J Neurosci* 20: 1260–1271, 2000.
288. Sattler, R., and M. Tymianski. Molecular mechanisms of calcium-dependent excitotoxicity. *J Mol Med* 78: 3–13, 2000.
289. Scarpini, E., P. Scheltens, and H. Feldman. Treatment of Alzheimer's disease: current status and new perspectives. *Lancet Neurol* 2: 539–547, 2003.
290. Schnell, E., M. Sizemore, S. Karimzadegan, L. Chen, D. S. Bredt, and R. A. Nicoll. Direct interactions between PSD-95 and stargazin control synaptic AMPA receptor number. *Proc Natl Acad Sci USA* 99: 13902–13907, 2002.
291. Schorge, S., and D. Colquhoun. Studies of NMDA receptor function and stoichiometry with truncated and tandem subunits. *J Neurosci* 23: 1151–1158, 2003.
292. Schwarcz, R., and C. Kohler. Differential vulnerability of central neurons of the rat to quinolinic acid. *Neurosci Lett* 38: 85–90, 1983.
293. Scott, D. B., T. A. Blanpied, and M. D. Ehlers. Coordinated PKA and PKC phoshorylation suppresses RXR-mediated ER retention and regulates the surface delivery of NMDA receptors. *Neuropharmacology* 45: 755–767, 2003.
294. Scott, D. B., I. Michailidis, Y. Mu, D. Logothetis, and M. D. Ehlers. Endocytosis and degradative sorting of NMDA receptors by conserved membrane-proximal signals. *J Neurosci* 24: 7096–7109, 2004.
295. Scott, D. B., T. A. Blanpied, G. T. Swanson, C. Zhang, and M. D. Ehlers. An NMDA receptor ER retention signal regulated by phosphorylation and alternative splicing. *J Neurosci* 21: 3063–3072, 2001.
296. Selkoe, D. J. Alzheimer's disease is a synaptic failure. *Science* 298: 789–791, 2002.
297. Setou, M., T. Nakagawa, D. H. Seog, and N. Hirokawa. Kinesin superfamily motor protein KIF17 and mLin-10 in NMDA receptor-containing vesicle transport. *Science* 288: 1796–1802, 2000.
298. Shankar, G. M., B. L. Bloodgood, M. Townsend, D. M. Walsh, D. J. Selkoe, and B. L. Sabatini. Natural oligomers of the Alzheimer amyloid-beta protein induce reversible synapse loss by modulating an NMDA-type glutamate receptor-dependent signaling pathway. *J Neurosci* 27: 2866–2875, 2007.
299. Shen, K., and T. Meyer. Dynamic control of CaMKII translocation and localization in hippocampal neurons by NMDA receptor stimulation. *Science* 284: 162–166, 1999.
300. Sheng, M., J. Cummings, L. A. Roldan, Y. N. Jan, and L. Y. Jan. Changing subunit composition of heteromeric NMDA receptors during development of rat cortex. *Nature* 368: 144–147, 1994.
301. Shrestha, B. R., O. V. Vitolo, P. Joshi, T. Lordkipanidze, M. Shelanski, and A. Dunaevsky. Amyloid beta peptide adversely affects spine number and motility in hippocampal neurons. *Mol Cell Neurosci* 33: 274–282, 2006.
302. Sigismund, S., T. Woelk, C. Puri, E. Maspero, C. Tacchetti, P. Transidico, F. Di, P. P., and S. Polo. Clathrin-independent endocytosis of ubiquitinated cargos. *Proc Natl Acad Sci USA* 102: 2760–2765, 2005.

303. Simonian, S. X., and A. E. Herbison. Differing, spatially restricted roles of ionotropic glutamate receptors in regulating the migration of gnrh neurons during embryogenesis. *J Neurosci* 21: 934–943, 2001.
304. Sjostrom, P. J., G. G. Turrigiano, and S. B. Nelson. Neocortical LTD via coincident activation of presynaptic NMDA and cannabinoid receptors. *Neuron* 39: 641–654, 2003.
305. Skeberdis, V. A., V. Chevaleyre, C. G. Lau, J. H. Goldberg, D. L. Pettit, S. O. Suadicani, Y. Lin, M. V. Bennett, R. Yuste, P. E. Castillo, and R. S. Zukin. Protein kinase A regulates calcium permeability of NMDA receptors. *Nat Neurosci* 9: 501–510, 2006.
306. Small, D. H., S. S. Mok, and J. C. Bornstein. Alzheimer's disease and Abeta toxicity: from top to bottom. *Nat Rev Neurosci* 2: 595–598, 2001.
307. Snyder, E. M., B. D. Philpot, K. M. Huber, X. Dong, J. R. Fallon, and M. F. Bear. Internalization of ionotropic glutamate receptors in response to mGluR activation. *Nat Neurosci* 4: 1079–1085, 2001.
308. Snyder, E. M., Y. Nong, C. G. Almeida, S. Paul, T. Moran, E. Y. Choi, A. C. Nairn, M. W. Salter, P. J. Lombroso, G. K. Gouras, and P. Greengard. Regulation of NMDA receptor trafficking by amyloid-beta. *Nat Neurosci* 8: 1051–1058, 2005.
309. Song, C., Y. Zhang, C. G. Parsons, and Y. F. Liu. Expression of polyglutamine-expanded huntingtin induces tyrosine phosphorylation of N-methyl-D-aspartate receptors. *J Biol Chem* 278: 33364–33369, 2003.
310. Soriano, F. X., S. Papadia, F. Hofmann, N. R. Hardingham, H. Bading, and G. E. Hardingham. Preconditioning doses of NMDA promote neuroprotection by enhancing neuronal excitability. *J Neurosci* 26: 4509–4518, 2006.
311. Spacek, J., and K. M. Harris. Three-dimensional organization of smooth endoplasmic reticulum in hippocampal CA1 dendrites and dendritic spines of the immature and mature rat. *J Neurosci* 17: 190–203, 1997.
312. Sprengel, R., B. Suchanek, C. Amico, R. Brusa, N. Burnashev, A. Rozov, O. Hvalby, V. Jensen, O. Paulsen, P. Andersen, J. J. Kim, R. F. Thompson, W. Sun, L. C. Webster, S. G. Grant, J. Eilers, A. Konnerth, J. Li, J. O. McNamara, and P. H. Seeburg. Importance of the intracellular domain of NR2 subunits for NMDA receptor function in vivo. *Cell* 92: 279–289, 1998.
313. Stahl, S. M. Beyond the dopamine hypothesis to the NMDA glutamate receptor hypofunction hypothesis of schizophrenia. *CNS Spectr* 12: 265–268, 2007.
314. Standley, S., K. W. Roche, J. McCallum, N. Sans, and R. J. Wenthold. PDZ domain suppression of an ER retention signal in NMDA receptor NR1 splice variants. *Neuron* 28: 887–898, 2000.
315. Starling, A. J., V. M. Andre, C. Cepeda, M. de Lima, S. H. Chandler, and M. S. Levine. Alterations in N-methyl-D-aspartate receptor sensitivity and magnesium blockade occur early in development in the R6/2 mouse model of Huntington's disease. *J Neurosci Res* 82: 377–386, 2005.
316. Stefansson, H., E. Sigurdsson, V. Steinthorsdottir, S. Bjornsdottir, T. Sigmundsson, S. Ghosh, J. Brynjolfsson, S. Gunnarsdottir, O. Ivarsson, T. T. Chou, O. Hjaltason, B. Birgisdottir, H. Jonsson, V. G. Gudnadottir, E. Gudmundsdottir, A. Bjornsson, B. Ingvarsson, A. Ingason, S. Sigfusson, H. Hardardottir, R. P. Harvey, D. Lai, M. Zhou, D. Brunner, V. Mutel, A. Gonzalo, G. Lemke, J. Sainz, G. Johannesson, T. Andresson, D. Gudbjartsson, A. Manolescu, M. L. Frigge, M. E. Gurney, A. Kong, J. R. Gulcher, H. Petursson, and K. Stefansson. Neuregulin 1 and susceptibility to schizophrenia. *Am J Hum Genet* 71: 877–892, 2002.
317. Steigerwald, F., T. W. Schulz, L. T. Schenker, M. B. Kennedy, P. H. Seeburg, and G. Kohr. C-Terminal truncation of NR2A subunits impairs synaptic but not extrasynaptic localization of NMDA receptors. *J Neurosci* 20: 4573–4581, 2000.

318. Stern-Bach, Y., B. Bettler, M. Hartley, P. O. Sheppard, P. J. O'Hara, and S. F. Heinemann. Agonist selectivity of glutamate receptors is specified by two domains structurally related to bacterial amino acid-binding proteins. *Neuron* 13: 1345–1357, 1994.
319. Stocca, G., and S. Vicini. Increased contribution of NR2A subunit to synaptic NMDA receptors in developing rat cortical neurons. *J Physiol* 507: 13–24, 1998.
320. Strack, S., and R. J. Colbran. Autophosphorylation-dependent targeting of calcium/calmodulin-dependent protein kinase II by the NR2B subunit of the N-methyl-D-aspartate receptor. *J Biol Chem* 273: 20689–20692, 1998.
321. Sucher, N. J., S. Akbarian, C. L. Chi, C. L. Leclerc, M. Awobuluyi, D. L. Deitcher, M. K. Wu, J. P. Yuan, E. G. Jones, and S. A. Lipton. Developmental and regional expression pattern of a novel NMDA receptor-like subunit (NMDAR-L) in the rodent brain. *J Neurosci* 15: 6509–6520, 1995.
322. Sun, Y., A. Savanenin, P. H. Reddy, and Y. F. Liu. Polyglutamine-expanded huntingtin promotes sensitization of N-methyl-D-aspartate receptors via post-synaptic density 95. *J Biol Chem* 276: 24713–24718, 2001.
323. Sytnyk, V., I. Leshchyns'ka, M. Delling, G. Dityateva, A. Dityatev, and M. Schachner. Neural cell adhesion molecule promotes accumulation of TGN organelles at sites of neuron-to-neuron contacts. *J Cell Biol* 159: 649–661, 2002.
324. Takahashi, T., D. Feldmeyer, N. Suzuki, K. Onodera, S. G. Cull-Candy, K. Sakimura, and M. Mishina. Functional correlation of NMDA receptor epsilon subunits expression with the properties of single-channel and synaptic currents in the developing cerebellum. *J Neurosci* 16: 4376–4382, 1996.
325. Takasu, M. A., M. B. Dalva, R. E. Zigmond, and M. E. Greenberg. Modulation of NMDA receptor-dependent calcium influx and gene expression through EphB receptors. *Science* 295: 491–495, 2002.
326. Tang, Y. P., E. Shimizu, G. R. Dube, C. Rampon, G. A. Kerchner, M. Zhuo, G. Liu, and J. Z. Tsien. Genetic enhancement of learning and memory in mice. *Nature* 401: 63–69, 1999.
327. Thomas, C. G., A. J. Miller, and G. L. Westbrook. Synaptic and extrasynaptic NMDA receptor NR2 subunits in cultured hippocampal neurons. *J Neurophysiol* 95: 1727–1734, 2006.
328. Thompson, C. L., D. L. Drewery, H. D. Atkins, F. A. Stephenson, and P. L. Chazot. Immunohistochemical localization of N-methyl-D-aspartate receptor subunits in the adult murine hippocampal formation: evidence for a unique role of the NR2D subunit. *Brain Res Mol Brain Res* 102: 55–61, 2002.
329. Tingley, W. G., M. D. Ehlers, K. Kameyama, C. Doherty, J. B. Ptak, C. T. Riley, and R. L. Huganir. Characterization of protein kinase A and protein kinase C phosphorylation of the N-methyl-D-aspartate receptor NR1 subunit using phosphorylation site-specific antibodies. *J Biol Chem* 272: 5157–5166, 1997.
330. Tolias, K. F., J. B. Bikoff, A. Burette, S. Paradis, D. Harrar, S. Tavazoie, R. J. Weinberg, and M. E. Greenberg. The Rac1-GEF Tiam1 couples the NMDA receptor to the activity-dependent development of dendritic arbors and spines. *Neuron* 45: 525–538, 2005.
331. Tolias, K. F., J. B. Bikoff, C. G. Kane, C. S. Tolias, L. Hu, and M. E. Greenberg. The Rac1 guanine nucleotide exchange factor Tiam1 mediates EphB receptor-dependent dendritic spine development. *Proc Natl Acad Sci USA* 104: 7265–7270, 2007.
332. Tomita, S., L. Chen, Y. Kawasaki, R. S. Petralia, R. J. Wenthold, R. A. Nicoll, and D. S. Bredt. Functional studies and distribution define a family of transmembrane AMPA receptor regulatory proteins. *J Cell Biol* 161: 805–816, 2003.
333. Tovar, K. R., and G. L. Westbrook. The incorporation of NMDA receptors with a distinct subunit composition at nascent hippocampal synapses in vitro. *J Neurosci* 19: 4180–4188, 1999.

334. Tovar, K. R., and G. L. Westbrook. Mobile NMDA receptors at hippocampal synapses. *Neuron* 34: 255–264, 2002.
335. Townsend, M., J. P. Cleary, T. Mehta, J. Hofmeister, S. Lesne, E. O'Hare, D. M. Walsh, and D. J. Selkoe. Orally available compound prevents deficits in memory caused by the Alzheimer amyloid-beta oligomers. *Ann Neurol* 60: 668–676, 2006.
336. Townsend, M., A. Yoshii, M. Mishina, and M. Constantine-Paton. Developmental loss of miniature N-methyl-D-aspartate receptor currents in NR2A knockout mice. *Proc Natl Acad Sci USA* 100: 1340–1345, 2003.
337. Traynelis, S. F., M. Hartley, and S. F. Heinemann. Control of proton sensitivity of the NMDA receptor by RNA splicing and polyamines. *Science* 268: 873–876, 1995.
338. Triller, A., and D. Choquet. Surface trafficking of receptors between synaptic and extrasynaptic membranes: and yet they do move!. *Trends Neurosci* 28: 133–139, 2005.
339. Tu, J. C., B. Xiao, S. Naisbitt, J. P. Yuan, R. S. Petralia, P. Brakeman, A. Doan, V. K. Aakalu, A. A. Lanahan, M. Sheng, and P. F. Worley. Coupling of mGluR/Homer and PSD-95 complexes by the Shank family of postsynaptic density proteins. *Neuron* 23: 583–592, 1999.
340. Turrigiano, G. G., K. R. Leslie, N. S. Desai, L. C. Rutherford, and S. B. Nelson. Activity-dependent scaling of quantal amplitude in neocortical neurons. *Nature* 391: 892–896, 1998.
341. Tymianski, M., M. P. Charlton, P. L. Carlen, and C. H. Tator. Source specificity of early calcium neurotoxicity in cultured embryonic spinal neurons. *J Neurosci* 13: 2085–2104, 1993.
342. Tyzio, R., A. Represa, I. Jorquera, Y. Ben-Ari, H. Gozlan, and L. Aniksztejn. The establishment of GABAergic and glutamatergic synapses on CA1 pyramidal neurons is sequential and correlates with the development of the apical dendrite. *J Neurosci* 19: 10372–10382, 1999.
343. Uitti, R. J., A. H. Rajput, J. E. Ahlskog, K. P. Offord, D. R. Schroeder, M. M. Ho, M. Prasad, A. Rajput, and P. Basran. Amantadine treatment is an independent predictor of improved survival in Parkinson's disease. *Neurology* 46: 1551–1556, 1996.
344. Van de Ven, T. J., H. M. VanDongen, and A. M. VanDongen. The nonkinase phorbol ester receptor alpha 1-chimerin binds the NMDA receptor NR2A subunit and regulates dendritic spine density. *J Neurosci* 25: 9488–9496, 2005.
345. van Zundert, B., A. Yoshii, and M. Constantine-Paton. Receptor compartmentalization and trafficking at glutamate synapses: a developmental proposal. *Trends Neurosci* 27: 428–437, 2004.
346. Vanhoutte, P., and H. Bading. Opposing roles of synaptic and extrasynaptic NMDA receptors in neuronal calcium signalling and BDNF gene regulation. *Curr Opin Neurobiol* 13: 366–371, 2003.
347. Vicini, S., J. F. Wang, J. H. Li, W. J. Zhu, Y. H. Wang, J. H. Luo, B. B. Wolfe, and D. R. Grayson. Functional and pharmacological differences between recombinant N-methyl-D-aspartate receptors. *J Neurophysiol* 79: 555–566, 1998.
348. Vissel, B., J. J. Krupp, S. F. Heinemann, and G. L. Westbrook. A use-dependent tyrosine dephosphorylation of NMDA receptors is independent of ion flux. *Nat Neurosci* 4: 587–596, 2001.
349. Vonsattel, J. P., and M. DiFiglia. Huntington disease. *J Neuropathol Exp Neurol* 57: 369–384, 1998.
350. Wafford, K. A., M. Kathoria, C. J. Bain, G. Marshall, B. Le Bourdelles, J. A. Kemp, and P. J. Whiting. Identification of amino acids in the N-methyl-D-aspartate receptor NR1 subunit that contribute to the glycine binding site. *Mol Pharmacol* 47: 374–380, 1995.
351. Walsh, D. M., and D. J. Selkoe. Deciphering the molecular basis of memory failure in Alzheimer's disease. *Neuron* 44: 181–193, 2004.

352. Waltereit, R., and M. Weller. Signaling from cAMP/PKA to MAPK and synaptic plasticity. *Mol Neurobiol* 27: 99–106, 2003.
353. Wang, C. Y., K. Chang, R. S. Petralia, Y. X. Wang, G. K. Seabold, and R. J. Wenthold. A novel family of adhesion-like molecules that interacts with the NMDA receptor. *J Neurosci* 26: 2174–2183, 2006.
354. Wang, C. X., and A. Shuaib. NMDA/NR2B selective antagonists in the treatment of ischemic brain injury. *Curr Drug Targets CNS Neurol Disord* 4: 143–151, 2005.
355. Wang, J., S. Liu, Y. Fu, J. H. Wang, and Y. Lu. Cdk5 activation induces hippocampal CA1 cell death by directly phosphorylating NMDA receptors. *Nat Neurosci* 6: 1039–1047, 2003.
356. Wang, J. K., and V. Thukral. Presynaptic NMDA receptors display physiological characteristics of homomeric complexes of NR1 subunits that contain the exon 5 insert in the N-terminal domain. *J Neurochem* 66: 865–868, 1996.
357. Wang, Y. T., and M. W. Salter. Regulation of NMDA receptors by tyrosine kinases and phosphatases. *Nature* 369: 233–235, 1994.
358. Washbourne, P., J. E. Bennett, and A. K. McAllister. Rapid recruitment of NMDA receptor transport packets to nascent synapses. *Nat Neurosci* 5: 751–759, 2002.
359. Washbourne, P., X. B. Liu, E. G. Jones, and A. K. McAllister. Cycling of NMDA receptors during trafficking in neurons before synapse formation. *J Neurosci* 24: 8253–8264, 2004.
360. Watanabe, M., Y. Inoue, K. Sakimura, and M. Mishina. Developmental changes in distribution of NMDA receptor channel subunit mRNAs. *Neuroreport* 3: 1138–1140, 1992.
361. Wei, F., G. D. Wang, G. A. Kerchner, S. J. Kim, H. M. Xu, Z. F. Chen, and M. Zhuo. Genetic enhancement of inflammatory pain by forebrain NR2B overexpression. *Nat Neurosci* 4: 164–169, 2001.
362. Weissman, A. M. Themes and variations on ubiquitylation. *Nat Rev Mol Cell Biol* 2: 169–178, 2001.
363. Weitlauf, C., Y. Honse, Y. P. Auberson, M. Mishina, D. M. Lovinger, and D. G. Winder. Activation of NR2A-containing NMDA receptors is not obligatory for NMDA receptor-dependent long-term potentiation. *J Neurosci* 25: 8386–8390, 2005.
364. Wenthold, R. J., K. Prybylowski, S. Standley, N. Sans, and R. S. Petralia. Trafficking of NMDA receptors. *Annu Rev Pharmacol Toxicol* 43: 335–358, 2003.
365. Wenzel, A., D. Benke, H. Mohler, and J. M. Fritschy. N-methyl-D-aspartate receptors containing the NR2D subunit in the retina are selectively expressed in rod bipolar cells. *Neuroscience* 78: 1105–1112, 1997.
366. Westphal, R. S., S. J. Tavalin, J. W. Lin, N. M. Alto, I. D. Fraser, L. K. Langeberg, M. Sheng, and J. D. Scott. Regulation of NMDA receptors by an associated phosphatase-kinase signaling complex. *Science* 285: 93–96, 1999.
367. Willard, F. S., R. J. Kimple, and D. P. Siderovski. Return of the GDI: the GoLoco motif in cell division. *Annu Rev Biochem* 73: 925–951, 2004.
368. Wong, R. W., M. Setou, J. Teng, Y. Takei, and N. Hirokawa. Overexpression of motor protein KIF17 enhances spatial and working memory in transgenic mice. *Proc Natl Acad Sci USA* 99: 14500–14505, 2002.
369. Woo, N. H., H. K. Teng, C. J. Siao, C. Chiaruttini, P. T. Pang, T. A. Milner, B. L. Hempstead, and B. Lu. Activation of p75NTR by proBDNF facilitates hippocampal long-term depression. *Nat Neurosci* 8: 1069–1077, 2005.
370. Woo, T. U., J. P. Walsh, and F. M. Benes. Density of glutamic acid decarboxylase 67 messenger RNA-containing neurons that express the N-methyl-D-aspartate receptor subunit NR2A in the anterior cingulate cortex in schizophrenia and bipolar disorder. *Arch Gen Psychiatry* 61: 649–657, 2004.

371. Wu, H. Y., E. Y. Yuen, Y. F. Lu, M. Matsushita, H. Matsui, Z. Yan, and K. Tomizawa. Regulation of N-methyl-D-aspartate receptors by calpain in cortical neurons. *J Biol Chem* 280: 21588–21593, 2005.
372. Wyszynski, M., J. Lin, A. Rao, E. Nigh, A. H. Beggs, A. M. Craig, and M. Sheng. Competitive binding of alpha-actinin and calmodulin to the NMDA receptor. *Nature* 385: 439–442, 1997.
373. Xia, H., Z. D. Hornby, and R. C. Malenka. An ER retention signal explains differences in surface expression of NMDA and AMPA receptor subunits. *Neuropharmacology* 41: 714–723, 2001.
374. Xiong, Z. G., R. Raouf, W. Y. Lu, L. Y. Wang, B. A. Orser, E. M. Dudek, M. D. Browning, and J. F. MacDonald. Regulation of N-methyl-D-aspartate receptor function by constitutively active protein kinase C. *Mol Pharmacol* 54: 1055–1063, 1998.
375. Yang, J., G. L. Woodhall, and R. S. Jones. Tonic facilitation of glutamate release by presynaptic NR2B-containing NMDA receptors is increased in the entorhinal cortex of chronically epileptic rats. *J Neurosci* 26: 406–410, 2006.
376. Yang, W., C. Zheng, Q. Song, X. Yang, S. Qiu, C. Liu, Z. Chen, S. Duan, and J. Luo. A three amino acid tail following the TM4 region of the N-methyl-D-aspartate receptor (NR) 2 subunits is sufficient to overcome endoplasmic reticulum retention of NR1-1a subunit. *J Biol Chem* 282: 9269–9278, 2007.
377. Yao, I., Y. Hata, N. Ide, K. Hirao, M. Deguchi, H. Nishioka, A. Mizoguchi, and Y. Takai. MAGUIN, a novel neuronal membrane-associated guanylate kinase-interacting protein. *J Biol Chem* 274: 11889–11896, 1999.
378. Yeaman, C., K. K. Grindstaff, J. R. Wright, and W. J. Nelson. Sec6/8 complexes on trans-Golgi network and plasma membrane regulate late stages of exocytosis in mammalian cells. *J Cell Biol* 155: 593–604, 2001.
379. Yi, Z., R. S. Petralia, K. Prybylowski, N. Sans, Y.-X. Wang, and R. J. Wenthold. GIPC, a single PDZ domain-containing protein, interacts with the NMDA receptor and regulates its surface expression. *Soc Neurosci Abs* 31.14, 2006.
380. Yi, Z., R. S. Petralia, N. Sans, and W. RJ. NMDA receptors interact with GIPC. *Soc Neurosci Abs* 487.10, 2005.
381. Yoshii, A., M. H. Sheng, and M. Constantine-Paton. Eye opening induces a rapid dendritic localization of PSD-95 in central visual neurons. *Proc Natl Acad Sci USA* 100: 1334–1339, 2003.
382. Young, A. B., J. T. Greenamyre, Z. Hollingsworth, R. Albin, C. D'Amato, I. Shoulson, and J. B. Penney. NMDA receptor losses in putamen from patients with Huntington's disease. *Science* 241: 981–983, 1988.
383. Yuen, E. Y., Q. Jiang, J. Feng, and Z. Yan. Microtubule regulation of N-methyl-D-aspartate receptor channels in neurons. *J Biol Chem* 280: 29420–29427, 2005.
384. Zarate, C. A., J. Quiroz, J. Payne, and H. K. Manji. Modulators of the glutamatergic system: implications for the development of improved therapeutics in mood disorders. *Psychopharmacol Bull* 36: 35–83, 2002.
385. Zarate, C. A. J., J. B. Singh, P. J. Carlson, N. E. Brutsche, R. Ameli, D. A. Luckenbaugh, D. S. Charney, and H. K. Manji. A randomized trial of an N-methyl-D-aspartate antagonist in treatment-resistant major depression. *Arch Gen Psychiatry* 63: 856–864, 2006.
386. Zeron, M. M., N. Chen, A. Moshaver, A. T. Lee, C. L. Wellington, M. R. Hayden, and L. A. Raymond. Mutant huntingtin enhances excitotoxic cell death. *Mol Cell Neurosci* 17: 41–53, 2001.
387. Zeron, M. M., O. Hansson, N. Chen, C. L. Wellington, B. R. Leavitt, P. Brundin, M. R. Hayden, and L. A. Raymond. Increased sensitivity to N-methyl-D-aspartate receptor-mediated excitotoxicity in a mouse model of Huntington's disease. *Neuron* 33: 849–860, 2002.

388. Zhang, S. J., M. N. Steijaert, D. Lau, G. Schutz, C. Delucinge-Vivier, P. Descombes, and H. Bading. Decoding NMDA Receptor Signaling: identification of Genomic Programs Specifying Neuronal Survival and Death. *Neuron* 53: 549–562, 2007.
389. Zhang, W., L. Vazquez, M. Apperson, and M. B. Kennedy. Citron binds to PSD-95 at glutamatergic synapses on inhibitory neurons in the hippocampus. *J Neurosci* 19: 96–108, 1999.
390. Zheng, X., L. Zhang, A. P. Wang, M. V. Bennett, and R. S. Zukin. Protein kinase C potentiation of N-methyl-D-aspartate receptor activity is not mediated by phosphorylation of N-methyl-D-aspartate receptor subunits. *Proc Natl Acad Sci USA* 96: 15262–15267, 1999.
391. Ziv, N. E., and C. C. Garner. Cellular and molecular mechanisms of presynaptic assembly. *Nat Rev Neurosci* 5: 385–399, 2004.
392. Zukin, R. S., and M. V. Bennett. Alternatively spliced isoforms of the NMDARI receptor subunit. *Trends Neurosci* 18: 306–313, 1995.

Surface Trafficking of Membrane Proteins at Excitatory and Inhibitory Synapses

Daniel Choquet[1] and Antoine Triller[2]

[1] UMR 5091 CNRS, Université de Bordeaux 2, Physiologie Cellulaire de la Synapse, Institut François Magendie rue Camille Saint Saëns 33077 Bordeaux Cedex, France, dchoquet@u-bordeaux2.fr

[2] Inserm UR497, Ecole Normale Supérieure, Biologie Cellulaire de la Synapse N&P, 46, rue d'Ulm 75005 Paris, France, triller@biologie.ens.fr

1 Introduction

Until recently, synapse organization was seen as rather stable, being modified only during plastic processes over time courses ranging from minutes to hours. Bulk approaches, i.e. probing large numbers (fifty to few hundreds) of molecules simultaneously has permitted the observation of dynamic reorganization of synaptic components, and this has challenged the notion that synapses are stable entities. However, these approaches did not allow the behavior of individual molecules to be followed. The first single molecule technique to be used in synaptic biology was patch-clamp recordings measuring the electrophysiological properties of individual ion channels. This approach, though powerful, does not allow for a determination of the spatial distribution of a given molecule over time. As trafficking of molecular components has emerged as a major pathway in the regulation of synapse function, the field was searching for means to investigate the microscopic behavior of molecules with high spatio-temporal resolution. Once the concept was established that synaptic molecules are in movement, it was implicit that they are in reversible interactions in a short time scale (i.e. receptor scaffold interactions).

The recent developments in single molecule imaging technologies are now leading to wide applications in cellular biology allowing for the unraveling of new mechanisms related to molecular movements. These advances were made possible by a combination of improvements in optical detectors combined with new molecular tagging strategies and the advent of biocompatible nanotechnological tools.

Single molecule imaging has been critical for synaptic biology since molecules traffic between microdomains with specific molecular composition. In recent years, fluorescence recovery after photobleaching (FRAP, e.g. (3, 143)) and single particle tracking (SPT, see (142)) have revealed that synaptic molecules move and change diffusive behavior over short time scales. A major advantage of tracking a single molecule over time is that it allows a direct access to the characteristic times of its

J.W. Hell, M.D. Ehlers (eds.), *Structural and Functional Organization of the Synapse*, DOI: 10.1007/978-0-387-77232-5_13,

stochastic molecular interactions. These measurements are beginning to uncover nanoscopic structural plasticity. However, bulk and single molecule methods are complementary to access the molecular physiology of synapses because single molecule approaches bear their own limitations related to sampling and observation times.

In this chapter, we will review recent data on the surface trafficking of receptors at inhibitory and excitatory synapses. We will put this data in the context of the general notion that the synaptic molecular composition results from a dynamic equilibrium between different subcellular compartments and that plasticity processes largely result from changes in the equilibrium set point.

2 Evidence and Necessity for Excitatory and Inhibitory Receptors to Diffuse on the Neuron Surface

One key step in synapse formation is the recruitment and the stabilization of neurotransmitter receptors in the postsynaptic membrane. Once assembled in the endoplasmic reticulum (ER), neurotransmitter receptors traffic from the ER to the postsynaptic membrane (Fig. 1) (see Chapter by Hanus and Ehlers, this volume). However, the routes used by neurotransmitter receptors to reach the synapse remain unclear. Over the last decade, two paths have emerged that are likely complementary. On the one hand, neurotransmitter receptors cycle to and from the plasma membrane through exocytosis and endocytosis processes, respectively. On the other hand, neurotransmitter receptors diffuse at the surface of neurons (lateral diffusion), and may thus enter and leave the synapse by diffusing from and to the extrasynaptic membrane (Fig. 1).

Since it appears that most cycling events take place in the extrasynaptic membranes, both cycling and surface trafficking processes are likely involved in the regulation of the number of synaptic neurotransmitter receptors. The first report of neurotransmitter receptor lateral diffusion came from studies in the early seventies in which acetylcholine (ACh) receptor trafficking was investigated at the surface of cultured muscle fibers (4, 5). But it was only 30 years later that the lateral trafficking of neurotransmitter receptors at the surface of neurons was reported (20, 88, 121, 141). Since then, continuous improvement in the tracking technique using live imaging has been achieved, giving to neurobiologists the tools to localize fluorescent molecules with sub-wavelength precision and to identify receptor subpopulations. The use of nanometer-sized ligand-fluorophore complexes has made it possible to track targets within confined cellular compartments, such as the synaptic cleft. Since excitatory as well as inhibitory receptors exchange between extrasynaptic and synaptic membranes, the regulation of receptor numbers at postsynaptic sites is therefore likely to also depend on lateral diffusion.

3 Receptor Exocytosis and Endocytosis

Most studies of ionotropic neurotransmitter receptor endo and exocytosis have concentrated on AMPARs. For discussion of GABAA and glycine receptor endo and

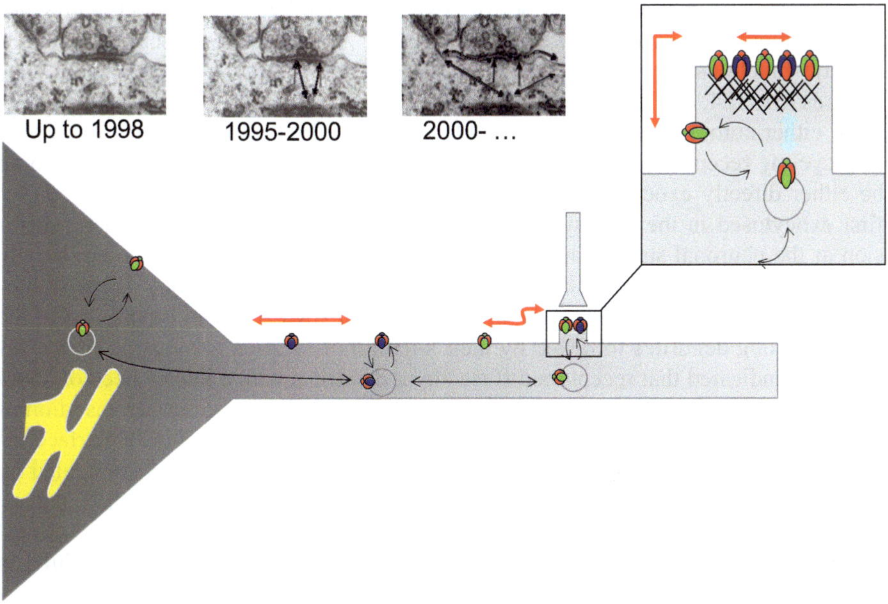

Fig. 1. Receptor trafficking from synthesis to insertion in the synapse. Receptors are synthesized mainly in the cell body but possibly also in dendrites (67) and then trafficked intracellularly along dendrites by microtubule based vesicular transport. There is debate (1, 100–102) on the site of insertion of receptors that could occur at the cell body (1), along dendrites (102), within the spine (100, 101), or even directly at the PSD (52). Receptor endocytosis occurs peri-synaptically. On the other hand, most receptors constantly move in and out synapses by lateral diffusion within the plane of the membrane. Receptors trafficking into and out of synapses thus rely on sequential and complementary routes involving diffusion, endo- and exocytosis.

exocytosis, see Chapter by Lorena Arancibia-Carcamo et al., this volume. The wealth of data on metabotropic receptor endocytosis is not reviewed here as beyond the scope of this chapter.

Multiple studies have shown that AMPAR localization is a very dynamic process, with AMPARs cycling in and out of synapses under a variety of situations related to synaptic plasticity and development (e.g. (12, 46, 78, 84, 86, 129, 145)). Similarly, NMDARs also cycle in and out synapses, albeit at slower rates (reviewed in (74)).

3.1 Exocytosis

One of the last steps in the long journey of receptors to the synapse is their delivery into the specialized dendritic membrane that constitutes the postsynaptic terminal. Exocytosis of receptors is not only required for insertion of new receptors in basal conditions but appears to bc also required for LTP (81, 100). Dendritic recycling endosomes contain the reserve pool of AMPARs mobilized during LTP and available for modifying synaptic strength (100, 101). As previously described, subunit-specific functions govern the synaptic delivery of AMPARs in hippocampal slice cultures (61, 124, 148) and in dissociated cultures (102). The sites of receptor exocytosis, either coming from a pool of newly synthesized receptors (67) or from a pool of recycling receptors, still remains elusive. Excitatory or inhibitory receptors could be either directly exocytozed at the synapse, as suggested by recent work (52), or first exocytosed in the extrasynaptic membrane and followed by their lateral diffusion at the neuronal surface and trapping at synaptic sites. This was shown initially by using cleavable extracellular epitope tags on transfected receptors. These studies (102, 113) have suggested that glycine receptor and AMPARs can be inserted at the soma or along dendrites followed by their lateral translocation to synaptic sites. This approach indicated that receptors diffuse along dendrites with a linear speed of about 1 μm/min (113). For AMPA receptors, it was established that the surface insertion of AMPARs occurs along dendrites in a subunit-dependent manner (102). Surface insertion of the GluR1 subunit occurs slowly in basal conditions and is stimulated by NMDAR activation (see also (105)) and insulin, whereas GluR2 exocytosis is constitutively rapid. Although both subunits ultimately concentrate in synapses, GluR2 accumulation at synapses is faster than that of GluR1 suggesting that either GluR1 is inserted initially at extrasynaptic sites while GluR2 is inserted more directly at synapses or that both receptors are inserted at extrasynaptic sites but that GluR2 diffuses faster in the membrane and can thus accumulate faster in synapses. The rapid accumulation of newly exocytosed GluR2 subunits at synaptic sites was found to depend on the NSF binding sites in the C terminus of GluR2 (14). Similar dendritic insertion of AMPAR subunits was found in other studies using different types of epitope tags (bungarotoxin binding site or bi-arsenical dyes) (67, 120).

Studies on native AMPARs found that LTP was accompanied by a rapid insertion of native AMPARs and by increased clustering of AMPARs at the surface of dendritic membranes. AMPAR insertion was blocked by intracellular tetanus toxin, providing evidence that AMPARs are inserted into excitatory synapses via a SNARE-dependent exocytosis during LTP (82). Real-time measurements of receptor exocytosis onto the surface of neurons have recently been made possible through the use of green fluorescent protein (GFP) variants that display strong pH-dependent fluorescence (ecliptic pHluorins) (91). Using this assay, it has been shown that chemically induced LTP rapidly enhances spine size with subsequent surface incorporation of GluR1 AMPARs by exocytosis from intracellular sites. GluR1 receptors increase on spine surfaces with no significant net change on the nearby dendrite. GluR2 receptors increase moderately and transiently on the spine surface (69). Although the exact site of exocytosis could not be determined, the absence of change in net dendritic signal may suggest that there is no significant net shift of receptors from dendritic surface to spine surface. However, a small amount of lateral movement from dendrite to spines could have been missed due to the large amount of dendritic surface receptors. A recent study on the role of components of the exocyst complex in AMPAR delivery have suggested that Exo70 mediates AMPAR insertion directly within the postsynaptic density, rather than at extrasynaptic membranes (52). Most interestingly, recent results indicate that LTP-inducing stimuli promote the mobilization of recycling endosomes and vesicles into spines, thus suggesting that AMPARs therein contained could be secondarily exocytosed in spines (101).

However such model has been recently challenged. Adesnik et al. used a photoreactive AMPAR antagonist (6-azido-7-nitro-1,4-dihydroquinoxaline-2,3-dione, ANQX) (1, 29). Its global or focal photoactivation (by UV light) allows for measurement of the recovery of currents after specific inactivation of all or a subpopulation of surface AMPARs. The turn-over rate of synaptic receptors with intracellular pools was found to be surprisingly slow, in the time scale of several hours, differing from previous results. In contrast, AMPARs were shown to be rapidly exocytosed to the plasma membrane in the somatic compartment. Importantly it was also found that somatic surface AMPARs diffuse laterally at high rate. Thus, it is possible that new functional AMPARs in the dendrites come from receptors exocytosed at the soma and traveling down dendrites by lateral diffusion. This model is however hard to reconcile with the wealth of data that has shown AMPARs recycling in dendrites and it is improbable that receptors are delivered to distal dendrites purely by diffusion at the neuronal surface as this would take extended periods of time (a few to several hours per 100 μm is an optimistic estimate). Moreover, it is not clear how this model fits with electrophysiological measurements indicating an increasing gradient of AMPA receptors as one moves proximally to distally in CA1 pyramidal neurons.

In contrast to the flurry of information on AMPARs, much less is known regarding the exocytosis of NMDARs. Interactions between SAP102 and Sec8 are involved in the delivery of NMDARs to the cell surface in heterologous cells and neurons (115).

3.2 Endocytosis

AMPARs undergo rapid constitutive internalization that is regulated by synaptic activity (46, 78). For NMDARs, constitutive internalization in mature neurons is slow relative to AMPARs and not regulated by activity but is rapid in immature neurons (46, 78, 112). Although NMDA receptors are thought to be relatively stable during synaptic plasticity, certain stimuli can induce their acute internalization (96, 129). Mounting evidence suggests that the internalization of glutamate receptors is a primary mechanism of long-term depression (LTD) (12, 78, 86, 129, 145), wherein a specific stimulus causes a decrease in synaptic strength (11). Upon internalization, AMPARs are sorted in early endosomes either to a specialized recycling endosome compartment for reinsertion to the plasma membrane or to late endosomes and lysosomes for degradation (46, 75).

Where are the endocytosis sites? Glutamate receptor internalization is thought to occur through clathrin-mediated endocytosis (12, 28, 46, 78, 86, 145). An endocytic zone at the postsynaptic side of excitatory synapses, segregated from the postsynaptic density (PSD), is thought to be the site of internalization for synaptic proteins, including glutamate receptors (17). Electron microscopy has shown the presence of clathrin-coated pits and vesicles in dendritic spines (36, 104, 131, 139). In young neurons, clathrin assembly and disassembly occur rapidly, locally, and repeatedly at "hot spots" throughout dendrites and at the tips of dendritic filopodia (17). In contrast, clathrin coats in mature dendrites reside in stable, long-lasting zones at sites of endocytosis, where clathrin undergoes continuous exchange with local cytosolic pools (17). In dendritic spines, endocytic zones lie laterally to the PSD where they develop and persist independently of synaptic activity, akin to the PSD itself (17).

Proteins mediating endocytosis are systematically arrayed within dendritic spines, tangential to the synapse (108). While the disruption of components of the general endocytic machinery has been shown to disrupt glutamate receptor internalization (27, 83, 89, 147), CPG2 is a protein that was recently shown to be specifically involved in the internalization of postsynaptic proteins. CPG2 is a component of the postsynaptic endocytic pathway that mediates both constitutive and activity-regulated glutamate receptor internalization (37). CPG2 localizes to an endocytic zone of excitatory synapses. CPG2 knockdown disrupts constitutive AMPAR and NMDAR internalization and activity-induced AMPAR internalization. Since CPG2 is primarily localized to the postsynaptic endocytic zone of excitatory synapses, it is not a constitutive part of the ubiquitous endocytic machinery. In CPG2 knockdown neurons, there is an increase in synaptic glutamate receptors, which is likely due to the disruption of glutamate receptor endocytosis. Although some reports have indicated that endocytosis inhibition does not affect the number of surface AMPARs (81, 86), this data is consistent with results showing that acute blockade of endocytosis causes an increase in synaptic transmission (83). These ultrastructural data are corroborated by direct imaging of removal of AMPARs from synapses showing that it is preceded by transient endocytosis of extrasynaptic AMPARs. Ecliptic pHluorin-tagged GluR2 has been used to visualize changes in AMPAR surface expression in real time. Synaptic and extrasynaptic AMPARs respond very differently to NMDAR activation; there is a rapid internalization of extrasynaptic AMPARs that precedes the delayed removal of synaptic AMPARs (2).

4 Approaches to Track Receptor Mobility

Experimental studies of diffusion of particles have originally been carried out by using methods that measure the average behaviour of an entire population, such as dynamic light scattering (15) or neutron scattering (62) in physics, or FRAP (110) or fluorescence correlation spectroscopy (FCS) (6) commonly used in biology. Bulk measurements using FRAP demonstrated long ago that biological membranes are fluid mosaics (5). Early on, this was recognized to be also true for neuronal membranes (e.g. (68)). However, single molecule techniques have only recently been applied in neuroscience (39, 133).

For twenty years, optical methods have been developed for observing the motion of molecules through the use of single particle imaging (51, 117). These methods, referred to as SPT methods, rely on following the trajectory of a marker attached to the diffusing molecule. Transport properties of the particle are then derived through a statistical analysis of the trajectory, which includes measurement of the mean square displacement.

Visualisation of the diffusive behaviour of single membrane proteins in living cells has revealed that these molecules undergo a variety of motions, such as Brownian, confined, or directed (70, 117). In this context, SPT is a powerful tool for investigating the membrane structure and the mechanisms responsible for these motions, which ultimately influence reactions and interactions between biomolecules. Current analyses of trajectories allow the determination of the translational diffusion coefficient of a symmetric molecule which undergoes isotropic diffusion.

4.1 Fluorescence Recovery After Photobleaching

The first experimental approach used to measure surface protein trafficking was the fluorescence recovery after photobleaching technique (110). In this approach, molecules of interest are tagged with a fluorophore (e.g. fluorescent irreversible antagonist or antibody). A small area of the cell is quickly photobleached using an intense excitation light and the rate of fluorescence recovery in the excited spot is monitored over time. The fluorescence recovery depends on protein mobility from the area surrounding the bleached spot. This technique only gives averaged estimates of protein mobility, as the movements of many molecules are measured simultaneously and derivation of diffusion coefficients are based on assumption on bleaching beam profile and membrane isotropy. Further, it has limited spatial resolution (~250 nm). The relationship between FRAP and single molecule tracking is discussed below.

4.2 Single Particle and Molecule Tracking

The emergence of single particle tracking experiments allowed real-time monitoring of the movement of individual or small groups of proteins (Fig. 2). In SPT approaches, submicrometer-sized particles are bound to the protein of interest, such as neurotransmitter receptors, through ligands that recognize the extracellular domain (7, 21, 33, 70, 117). Particles vary in composition, size and in the optical method used for their detection.

The advent of fluorescence imaging in the 1990s together with the improvement of charge-coupled device (CCD) camera sensitivity allowed the detection of single fluorophores in living cells. Organic dyes were the first fluorescent probes to be used and present the advantages of being small (<1 nm), easily coupled to ligands by simple chemistry and amenable to multicolor detection. However their use is limited by their rapid (few seconds) photobleaching.

To gather sufficient photons to image one dye molecule, the preparation must be excited at saturation through a defocused laser. However, this substantially accelerates the photobleaching, and thus single dye molecules can usually be imaged for only few seconds (at 30 Hz rate acquisition). Thus, the single molecule detection approach provides the advantage of small probe size that enters confined spaces but has the disadvantage of a limited tracking time. An alternative approach, FCS, has been developed. Using FCS, the mobility of individual fluorescent molecules is measured as their residence time in a small illumination volume. The faster molecules diffuse, the shorter their residence time is in the excitation volume. As with FRAP, this approach gives measurements of the population mobility. Furthermore, it is limited to the measurement of rapidly moving molecules, as the fluorophore must remain active during the whole period of residence in the measurement volume and not photobleach.

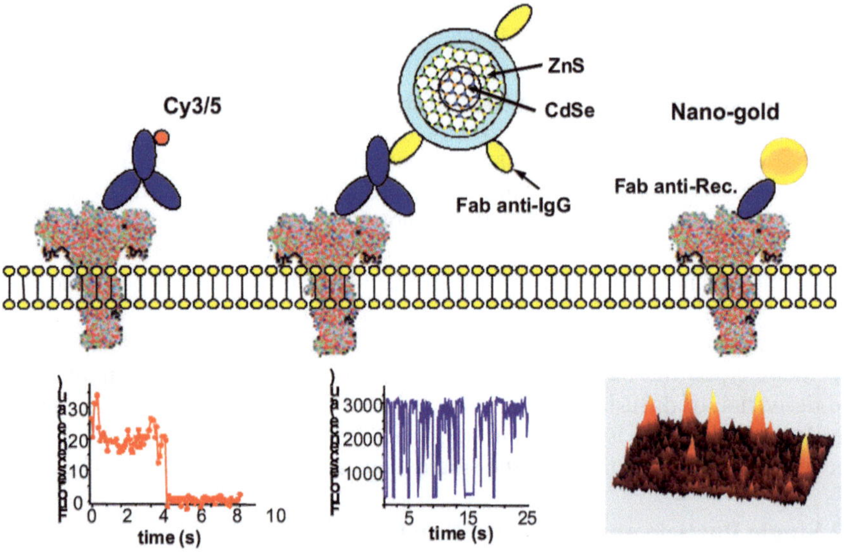

Fig. 2. Labeling methods for tracking single receptors. Labeling receptors relies on attaching a probe specifically to its extracellular domain. For probes, single molecule studies rely largely on fluorescent probes such as single organic dyes (*top left*) that photobleach rapidly (*bottom left*), or quantum dot nanocrystals (7) that are bulkier (*top middle*) but extremely photoresistant (*top bottom*). The new photothermal imaging approach (72) can image and track nano-gold particles as small as 1.4 nm and may provide an alternative nano-probe allowing tracking in very confined environments such as a living brain.

New developments may reboost the use of organic fluorescent dyes by improving the time during which they can be imaged. Either by using antibleach reagents, by the development of new photoresistant dyes (45), or by increasing fluorescence signals through relaxation of transient molecular dark states, such as the triplet states, between two molecular absorption events (44).

Small (nanometer range) non-organic fluorescent probes with virtually no photobleaching, have been developed to circumvent the limitation of organic dyes. Quantum dots (QD) which are nanometer-sized semiconductor fluorescent particles provide long-lasting fluorescence emission (10, 24, 90). The core fluorescent moiety of QD is a small CdSe crystal surrounded by a ZnS shell. It is hydrophobic and has to be encapsulated to ensure its biocompatibility, it then reaches a hydrodynamic size of 15–25 nm. Their fluorescent properties include: (i) a high absorption cross section that allows the use of standard illumination sources and high signal to noise ratio allowing recordings at high rates (up to the kHz); (ii) a narrow emission wave length depending on the QD size; (iii) high photoresistance; (iv) non-ergodic blinking between active and dark states that allows easy identification of individual QDs in temporal image series.

However, despite these attempts to minimize the particle size, even the smallest ones (e.g. QDs) used to track receptors within confined spaces have shown limitations, likely because of the complex size. Actual improvements in the QD structure, such as reduction of the core (few nanometers) or the organic shell sizes, will likely overcome this issue (106).

An alternative imaging method has been developed that allows detection of nanometer sized metallic particles (Fig. 2). It relies on the use of a photothermal detection method (13, 23, 34, 72). Its main advantages over QDs are the absence of photobleaching or blinking. Gold nanoparticles are heat modulated at high frequency (MHz) at the plasmon resonance (530 nm) and the resulting change in the diffraction index is detected using a probe laser beam and a lock in amplifier. This allows detection of particles down to 5 nm at 100 Hz with heating of less than 1°C. This approach has been used to track the movement of AMPARs in living neurons for over 20 min (72).

4.3 Surface Versus Intracellular Tracking

Up to now, the vast majority of single nano-object tracking studies have targeted the movement of cell surface molecules. A few exceptions lie in studies that have investigated the intracellular trafficking of viruses (119), molecular motors (25), endocytosed receptors (26, 77) or whole single vesicles (38, 76). The absence of studies on the tracking of intracellular cytosolic proteins at the single molecule level mostly arises from the intrinsic difficulty in exogenously labelling them. On the one hand, most exogenous labels such as QDots are not membrane permeable (see (41)), and on the other hand intracellularly injected probes cannot be easily washed out. Expanding the range of application of single particle/molecule studies to intracellular molecules will require the design of very high affinity (subnanomolar) cell permeable probes.

4.4 Ligands

Organic or inorganic probes, dyes or particles, have to be bound to the protein to be tracked through a specific linker ligand. Specificity can be achieved using natural ligands, peptides, chemical linkage, biotin, or antibodies. These links can be used to recognize endogenous unmodified receptors, or receptor subunits modified genetically or chemically to add a target tag.

Adequate linkers between the probes and the studied molecules must fulfill some prerequisites: linkers have to bind extracellular domains of receptors in living cells without modifying their function; linker affinity has to be in the nanomolar range to ensure stability of the receptor-linker-probe complex. One of the most difficult requirements is control over the stoichiometry of the receptor-to-probe interaction. This usually requires controlling not only the number of linkers on the probe, but also the number of acceptor sites on the studied molecule.

These requisites are met differently for the various ligands. In the central nervous system, there are only few natural ligands which can be used, mainly because their affinities are too low or because chemical modifications needed for their coupling to the probes destroys their binding properties. In most cases only antibodies are available for recognition. Specific tagging of receptors is best achieved using artificial tags added chemically or genetically allowing the use of excellent ligands. Various strategies have been used with success: epitope tags (88, 121), biotin tags (64), toxin binding sites (120), polyhistidine-NTA (73), cysteine modification (130). The linking to the probe (a dye or a nano particle) is performed either by passive adsorption or by covalent linkage. For single molecule tracking it is usually straightforward to chemically link an organic fluophore. Fluorescent proteins (XFPs) represent a specific case in which the fluorophores are directly genetically encoded with the studied protein. However, the poor signal to noise ratio yielded by these natural fluorophores make them of difficult use for SMT (32). The recent progress in fluorescence excitation procedures (44) may open the track to a new era for SMT of XFPs.

In the case of particles, the situation is more complex as complexation of the ligand to the probe can alter the physical properties of one or the other. Interesting new directions lie is the direct binding of peptides to gold or QD nanoparticles, allowing for marked decrease in the hydrodynamic radii of the probe-linker complex. In any case, the main hurdles appear to be the nonspecific binding of the probe to biological sample and the control of the stoichiometry of the probe-linker-target complex. Indeed, while it is relatively easy to ascertain a one-to-one labelling ratio of receptor to organic dyes, it is much more difficult to do it with functionalized particles.

4.5 Effect of Label Size on Diffusion and Trafficking of Surface Receptors

Particle or molecule movement, and therefore diffusion of the underlying receptors, is imaged by video-enhanced differential interference contrast microscopy or fluorescence imaging. The relative position of the particle is then measured with good accuracy (in the 10 nm range). In open spaces, the size of the tag only marginally influences the membrane protein movements, which are dominated by Brownian and viscous forces, because membrane viscosity is approximately 500-fold greater than that of extracellular fluids. The membrane-anchored receptors thus slow down the particles and not the opposite. This is of course only true when the probe diffusion is not hindered by extracellular molecular obstacles or narrow spaces between cells. Another issue lies in the difficulty to determine whether a given tracked particle remains at the cell surface and/or interferes with the endocytosis of the bound receptor. On the one hand, the available particles do not report a surface versus intracellular localization and the optical resolution does not permit an unambiguous determination of this localization. Incorporation of pH sensitive dyes in particles may allow detecting internalization events through the normal acidification of the luminal domain within the endocytotic vesicles (128). On the other hand, there are few published data that have studied the impact of particles on the endocytic process (26).

4.6 Labelling of Surface Receptors

Tracking individual receptors in primary neuronal cultures and other cultured cell types has now been achieved in many laboratories, and all the studies have confirmed the dynamic nature of membrane protein organization. However access to more intact systems (tissue slices and in vivo) has still resisted investigations at the single molecule level. A few recent reports have analyzed the dynamics of synaptic molecules in slices or in vivo using the FRAP approach (53). Interestingly, the dynamics which have been measured are only mildly quantitatively different from those found in neuronal culture. Two elements are detrimental to measurements of single molecules in tissue: the size of the probe, and light diffraction which imposes the need for a large number of photons. Measurement of diffusion volumes in brain slices indicate that probes with hydrodynamic diameters of 5–10 nm may be used to track receptors (136). The need for a high photon yield being already satisfied by QDs or metallic nanoparticles, development for small functionalizing linkers remain the main obstacle. In addition, rapid 3-D tracking is still to be developed. Altogether, tracking individual receptors in tissue slices is the next challenge that now seems to be within reach.

4.7 Pointing Accuracy

An important advantage of SPT and SMT approaches lies in their high pointing accuracy that allows localisation of the tracked object below the optical diffraction limits reaching in the best case few nanometers. This property results from the fact that pixel sizes are images of areas smaller than the point spread function of the optical pathway (PSF) and are intensity coded. The center of gravity of the image of the PSF on the CCD is a good approximation of the localization of the source and can be computed with a precision where signal variance is dominated by photon noise and varies according to the inverse square root of photon numbers. This implies that the more photon is gathered the better is the pointing accuracy. The latter being given by:

$$\frac{\lambda_{em}}{2\pi n_a \sqrt{\gamma A t}}$$

where λ_{em}, n_a, γ, A, and t denote the emission wavelength of the single molecule, the numerical aperture of the objective, the efficiency of the optical system, the emission rate of the single molecule and the acquisition time, respectively (98)

Trajectories are reconstructed from sequences of images of single particles or molecules. First, single objects must be detected in the images. That a single object is detected is ascertained according to the adequate characteristics of each probe. In the case of micron-sized latex beads, the center of gravity of the image is used to position the object. In the case of single fluorophores such as organic dyes or QDs, the object is smaller than the diffraction limit and thus its image corresponds to the point spread function of the optical system. Single organic dyes are recognized by one step photo-bleaching while single QD are identified by their fluorescence intermittency. Fluorescent spots are detected by cross-correlating the image with a Gaussian model of the point spread function. A least-squares Gaussian fit is applied (around the local maximum above a threshold) to determine the center of each spot with a spatial accuracy of 5–10 nm (depending on the signal-to-noise ratio). This is applied successively to each frame of the sequence.

The pointing accuracy is different from the resolution, which is the ability to separate two objects. Single nano object imaging display the standard optical resolution given by half the light wavelength. Other techniques (see below) aim at breaking this diffraction limit. This is at the origin of the difficulty to assert colocalization of a single molecule with a bulk labelling (e.g. a synaptic stain) because strictly speaking the resolution is not improved. In case of multiple single nano-object markers, the use of pointing accuracy allows for the indirect improvement in resolution, breaking the diffraction limit, as recently applied in the PALM technique (16).

4.8 Trajectory Reconstruction and Analysis

After nano-object detection, trajectories are assembled automatically by linking, from frame to frame, the centers of fluorescent spots likely arising from the same probe (7, 21, 32). The association criterion is based on the assumption of free Brownian diffusion, and takes into account short blinking events in the case of QDs. For analysis, the following parameters can be calculated from the reconstructed trajectories: (1) The mean square displacement (117), which allows one to characterize the movement (directed, free Brownian or confined), (2) the diffusion coefficient (71) which characterizes the speed of molecular displacement; (3) the confinement area (71, 30) which measure the space in which a molecule diffuses with a given motion type; and (4) the rotational diffusion (58) that depend on the shape and interactions of the diffusing object. These physical parameters give access to the properties of the followed molecule that diffuses in a given molecular environment. It allows important elements to be characterized such as molecular interactions (e.g. receptors with scaffold molecules), or diffusional restriction due to local molecular crowding.

4.9 Transition Between Compartments and Dwell Time Within a Compartment

Special compartments such as synapses can be labelled with fluorescent dyes. Overlaying the image of a compartment marker with the QD-trajectory allows the transitions of a QD-labelled molecule in and out synapses to be directly observed (7, 30, 39). Transitions between compartments, dwell time, and the average diffusion coefficient within a compartment can thus be estimated from the trajectory (30).

It is immediately observable from individual trajectories that molecule diffusion displays spatial and temporal heterogeneities (Fig. 3). Refining the analysis indicate that molecules alternate between states with different diffusional properties following Markovian rules. This results either from intrinsic properties of the receptor or from reversible interactions with its immediate environment. This was first established for glycine receptors using latex beads and optical tweezers (88). Glycine receptors diffused in and out of gephyrin clusters with mean residency times of about

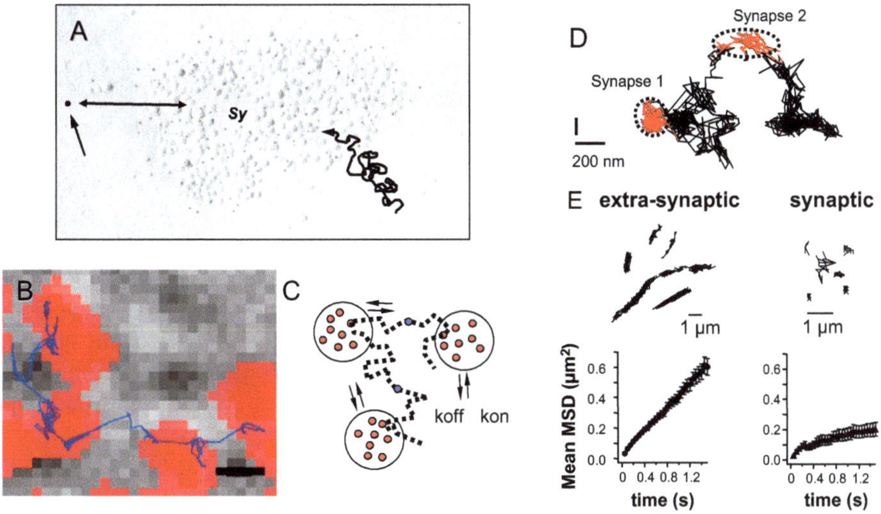

Fig. 3. Surface trafficking of receptors into and out synapses. (**a**) AMPA receptor distribution over a dendrite revealed by SDS-digested freeze-fracture replica labeling (132). Note that although receptors are accumulated at the postsynaptic membrane (Sy) some at low density are extrasynaptic (*arrows*). Diffusion trajectory is more likely to be tortuous than direct. (**b**) Diffusion trajectory (*blue*) of a glycine receptor diffusing from one synaptic contact to another. Active synaptic boutons are labeled with the amphiphilic dye FM4-64. (**c**) Schematic representation of a receptor trajectory from one postsynaptic density to another. The entry-exit kinetic allow to define a global on (kon) and off (koff) rate. (**d**) Trajectory of a single QD bound AMPAR diffusing sequentially in and out two synapses (*dotted circles*). (**e**) Sample trajectories of AMPARs moving inside and outside synapses. The mean square displacement is drastically different in both compartments, receptors displaying free Brownian movement outside synapses and confined movement inside.

20 s. Interestingly, the distribution of dwell times over clusters followed complex kinetics that could not result from a simple bi-molecular interaction. SPT analysis with quantum dots provided direct evidence for multiple association states between glycine receptors and gephyrin (48). Similar results were found for mGluR (121) or AMPAR-stargazin complexes (8) diffusion on homer and PSD-95 clusters respectively.

The cell membrane (i) contains various membrane proteins that can either act as traps by transiently immobilizing the diffusing particle or as obstacles by hindering its diffusion (70, 93) (ii) contains lipid microdomains, the size of which can range from tens to hundreds of nanometers (43), and (iii) is further compartmentalized with domains as large as a few hundreds nanometers in diameter (49). This latter compartmentalization results from the existence of corrals formed by membrane proteins anchored to a submembranous cytoskeleton network. This heterogeneity, whether structured or not, dynamic or not, affects the diffusion of proteins in the membrane. This heterogeneity can be quantified by a characteristic length below which the membrane is homogenous and does not contain any obstacle/trap for the diffusing protein.

Properties of diffusion within a compartment are characterized by several parameters such as the instantaneous diffusion coefficient, the degree of confinement and the dwell time. In neurons, the instantaneous diffusion of receptors spans several orders of magnitude but is within the same range inside and outside synapses, indicating that receptors can reach comparable slow and high instantaneous speeds in both compartments. Yet, the averaged movements are strikingly different in synaptic and extrasynaptic domains. Movements in synapses are confined within submicron domains while they are free and Brownian outside synapses. As a result, extrasynaptic receptors can travel much longer distances.

4.10 Bulk Versus Single Molecule Measurements

Although detailed analysis of the trajectories of individual molecules can yield considerably precise information, the representativeness of the measures can always be biased by sampling and labeling procedures. Therefore, single molecule data cannot be interpreted without knowledge of the population behavior. The most common bulk approach to receptor movements is FRAP. Both FRAP and SPT are different views of the same phenomenon which is related to the movement and immobilization of receptors entering and exiting a defined subdomain. However the interpretation of FRAP data and extraction of pertinent parameters such as diffusion coefficients, dwell times and confinement requires multiple and hidden assumptions while single molecule tracking directly provides these parameters (63).

FRAP experiments consist of bleaching tagged receptors in a specific region, such as the PSD, and measuring the time it takes for unbleached receptors to fill the bleached area. The recovery process depends on different molecular events such as the lateral diffusion of unbleached receptors into the PSD, which depends on both their availability and mobility, and the exit of bleached receptors from the PSD, which depends on both their mobility and unbinding from scaffolding molecule.

The recovery curves depend on different parameters such as the geometrical characteristics of the bleached spot, the flux of incoming receptors in the bleached area, and the total number of binding sites. From the recovery curves, two main parameters can be recovered, the time constant of recovery and the amount of recovery once steady state is reached, otherwise called mobile fraction.

The time constant of recovery depends on the incoming flux which itself depends on gradients creating a chemical potential and on the mobility of the molecules. The mobile fraction is harder to interpret as it depends on the reservoir of receptors outside the bleached zone (e.g. microdomains limit the amount of available receptors) and the number of available binding sites in the bleached zone, which itself depend on the dissociation kinetics of the bleached receptors from the binding sites. One can understand from these considerations that reconciling single molecule and FRAP data is not straightforward. Variations of FRAP approaches such as FRAP at varying radius (114) or FLIP (fluorescence loss in photobleaching (79)) can however facilitate this correlation.

4.11 The Next Challenge: Live Imaging of Receptor Trafficking in Slices

Tracking individual receptors in primary neuronal cultures has now been achieved in many labs, and all the studies have confirmed the highly dynamic nature of membrane protein organization. However the access to more intact systems (slices and in vivo) has resisted single molecule investigations. A few recent reports have analyzed the dynamic of synaptic molecules in slices or in vivo using FRAP approach (53). Interestingly the dynamics which have been measured are nearly identical to those found in culture. Two elements are detrimental to measurements of single molecules in tissues: the size of the probe, and the light diffraction which impose large number of photons. Further improvements in probe size and quantum yield are likely to allow single particle tracking in tissue slices in a foreseeable future.

5 Diffusion Properties of Neuronal Membranes: Differences and Similarities Between Synaptic and Extrasynaptic Membranes

Neuronal membranes are not homogeneous environments for diffusion and barriers to diffusion have been described at several levels. The distribution of proteins, such as the neurotransmitter receptors and voltage-gated channels, at the neuronal surface is highly compartmentalized. The most striking example of such compartmentalization in neurons is between the somatodentritic and axonal compartments. A diffusion barrier is present in the cell membrane of the axonal initial segment (AIS), which separates the somatodendritic domain and the axon domains. It is formed by the accumulation of proteins anchored to the dense membrane skeleton which function as rows of pickets, which even stop the overall diffusion of phospholipids (94, 146). Neurons control polarized delivery of membrane proteins to each domain, and in addition, this diffusion barrier could block diffusion mixing of proteins between these domains. At a smaller scale, the synapse is a sub-compartment where neurotransmitter receptors are concentrated and confined within the postsynaptic membrane.

Because of the high heterogeneity of the neuronal membrane, that varies in space and time, diffusion coefficients of receptors span several orders of magnitude, ranging from the immobile within detection limits (below 10^{-3} $\mu m^2/s$) to very mobile (1 $\mu m^2/s$). It is noteworthy that this value is still lower than that expected form a transmembrane protein diffusing in a pure lipid bilayer.

Early on, it was recognized that diffusion rates of excitatory and inhibitory receptors varies extensively over time (88, 121), switching often (within seconds) and abruptly between states of rapid sow movements as the receptors explored the dendritic space. These observations provided additional evidence for the spatial heterogeneity of the neuronal membrane and indicated that molecules can exchange between compartments. Further studies indicated that states of low mobilities were preferentially, although not exclusively, associated with the presence of receptors over clusters of scaffolding molecules or synaptic sites. An important point to raise is that analysis of the exchange of receptors between optically resolved compartments is limited by the diffraction limit of visible light used in optical microscopy. As a consequence, receptor trajectories can only be ascertained to correspond to an optically identified compartment with a precision of around 250 nm. Importantly, the dynamic signature of movements that is obtained with a sub-diffraction resolution in each compartment verifies that receptors diffuse in specific environments. The diffusional properties that characterize a compartment are (1) the distribution of the speeds (instantaneous diffusion rate) at which molecules move within the compartment (2) the size of the compartment, (3) the dwell time of the molecule within the compartment. The global diffusion rate, averaged over all time intervals and obtained by fitting the whole MSD curve instead of the initial slope, provides a mean value of receptor diffusion and can be related to the rate of recovery obtained in FRAP experiments. All these parameters are calculated from trajectories and the derived MSD function.

To date, most studies have concentrated on the specific diffusional differences that exist between synaptic and extrasynaptic domains (3, 30, 39, 47, 56, 133). Generally speaking, and at first glance, the differences between these two domains appear surprisingly minor. For AMPARs, the percentage of immobile receptors is only about two fold larger in synaptic versus extrasynaptic domains. Similarly, the median instantaneous diffusion coefficients are about two fold smaller in synapses. In striking contrast, the level of confinement is very different in both compartments. While receptors display free Brownian diffusion in extrasynaptic domains, traveling several microns per minute, they display confined diffusion in synapses, their movement being rapid, but restricted to a sub-domain of a few hundreds of nanometer perfectly co-localized with synaptic markers (Fig. 3). As a consequence, receptors dwell proportionally for longer times in a synaptic than an extra-synaptic area. These findings obtained by single molecule measurements are translated in FRAP experiments by slower rates of recovery and a higher immobile fraction in synapses.

Spine shape might be another regulator of receptor diffusion. Spine necks likely act as passive filters that help maintaining recycling receptors in the spine head (100). In mammalian excitatory neurons, dendritic spines are separated from dendrites by thin necks. Diffusion across the neck limits the chemical and electrical isolation of each spine. The neck also represents a barrier to membrane (111) and volume (18) diffusion and these barriers seem to be bi-directionally regulated by neuronal activity. Glutamate receptors activation regulates the mobility of molecules in the inner leaflet of the plasma membrane through an action upon the actin cytoskeleton, suggesting a mechanism for the regulation of postsynaptic receptor density and composition. The regulation of spine/dendrite diffusional equilibration may have several functional consequences as the susceptibility of individual synapses to plasticity induction may be influenced by the ability of signalling molecules and receptors to move into and out of the spine head (3). This reinforces the control of the autonomy of the synapses on spines by isolating the spine head as an individual diffusion domain (63).

What might be the biological basis for corralling in synapses? Obviously, intracellular scaffold and cytoskeletal elements, transmembrane adhesion proteins, and extracellular matrix are candidates. The differentiating feature between these obstacle molecules lies in whether they act as specific binding sites for the receptors (e.g. PSD-95 for stargazin or gephyrin for GlyRs) or participate in overall crowding of the synapse acting as posts to slow receptor movements. As for elements participating in specific binding, the affinity and avidity of the receptor-target interaction will determine the residency time of the interaction. When the target is itself immobile, for example if linked to cytoskeletal elements, this interaction time will set the percentage of time receptors remain immobile within the PSD. Interestingly, residency times of receptors in clusters or synapses, whether AMPARs, mGluRs, GlyRs, or GABARs, follow distributions that cannot be fit by monoexponential functions and thus very likely do not follow a simple reaction rate kinetic with one bound and one unbound state (88, 121). This further suggests that receptor-binder interactions generate multiple states of distinct stability and could also indicate that receptor subpopulations move in and out synapses with different rate constants.

The PSD itself is most likely to have a spatially heterogenous structure. Immunoelectron microscopy has suggested that some molecules display anisotropic colocalization in the PSD. For example the adhesion protein cadherin and mGluR1/5 have been reported to lie laterally in the PSD. Classical EM however has a limited labeling efficacy, thus questioning the significance of these observations. Recent advanced cryo-immunoelectron microscopy associated with partial solubilization of freeze-fracture replicates (87) has demonstrated a near 100% labeling efficacy (132) and has demonstrated a concentration of AMPARs in a few tens of nanometer sized within the PSD at most cerebellar synapses, while the remaining ones display diffuse staining for AMPARs. Interestingly, high resolution single QD tracking has revealed that AMPARs are confined in 50 nm areas at active synapses while they explore the whole PSD at silenced synapses (47). Altogether, these data indicate that the PSD may bear subdomains in which receptors are more stable than others. In principle, these could be linked to presynaptic release sites which may also occur at more specific sites than previously thought (126).

6 Receptor Surface Trafficking at Rest: General Rules and Subunit Specific Properties

All neurotransmitter receptors studied so far display some level of lateral diffusion and alternate between states of free diffusion, confinement and immobilization (Fig. 4). The percentage of time and the properties of each individual state vary considerably from one receptor to the other. Generally speaking, one can assume that the state of free diffusion corresponds to receptors unbound from stabilizing elements and moving in relatively non-crowded membrane areas, such as the extrasynaptic membrane. The state of free diffusion is characterized by a linear MSD indicating that the area explored by the receptors increases proportionally with time. The diffusion coefficient of this "freely" diffusive state can however vary sizeably, likely proportionally to the size of the diffusing molecular complex that can be affected by

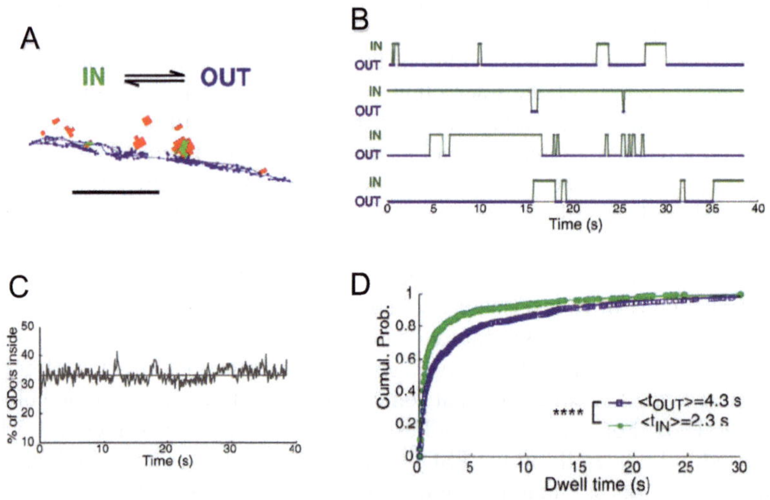

Fig. 4. Properties of the equilibrium between GlyRs inside and outside clusters. (**a**) Simple scheme of the dynamic equilibrium (*top panel*). (*Bottom panel*) Example of a trajectory (in neurons) alternating between regions inside (*gray*) and outside (*black*) gephyrin clusters (*dark gray*). Scale bar, 5 μm. (**b**) Examples of localization (IN and OUT) as a function of time. The upper plot corresponds to the trajectory in (a). (**c**) Proportion of QDots within gephyrin clusters as a function of time, computed for "swapping" GlyRs. (**d**) Comparison between the cumulative probability of in (*green*) and out (*blue*) dwell times. Mean dwell times inside (t_{in}) and outside (t_{out}) scaffold clusters. Modified from (48).

large intracellular tails or bound scaffold molecules (137). Indeed, freely diffusing large complexes are nevertheless slowed down by membrane viscosity. Measured values of this state range from very slow to nearly as rapid as that of transmembrane proteins in a pure membrane bilayer (diffusion coefficient from less than 0.005 $\mu m^2/s$ to 1 $\mu m^2/s$). The confined state corresponds to receptors diffusing in a crowded or corralled environment. This state is characterized by the area of the confinement domain, the diffusion coefficient, and the residency time within the domain. These values are highly variable from one receptor to the other and have been reported to range from 20 to 200 nm for domains and 0.005 to 0.1 $\mu m^2/s$ for diffusion. Finally, the immobile state corresponds to receptors bound to stabilizing elements and is mainly characterized by the residency time in the immobile state. We will here summarize the main differences that have been reported between receptor subtypes.

6.1 Glutamate Receptors

In basal conditions, AMPARs diffuse ten to hundred times faster than NMDARs, as observed by tracking the movement of GluR1- or GluR2-containing native AMPARs versus NR1-containing NMDARs. While basal AMPAR diffusion does not seem to depend on subunit composition, NMDARs diffusion strikingly depend on the presence of NR2A or NR2B subunits (57). While NR2A containing NMDARs are largely immobilized at synaptic sites, with only 20% being in a mobile state at any given time, NR2B containing NMDARs are rather mobile, albeit at slower rates than AMPARs. Triheteromeric NR2A/NR2B/NR1 containing NMDARs likely behave as diheteromeric NR2A/NR1 NMDARs (57). The synaptic composition of NMDARs changes over maturation with an increase in the NR2A/NR2B subunit ratio (92, 122, 134, 140). Interestingly, the developmental switch in the synaptic NR2A- and NR2B-containing NMDAR surface distribution correlates with developmental changes in the time spent by subunits within synapses (54). This suggests that the switch in synaptic NMDAR subtypes depends on the regulation of receptor surface trafficking. Regarding AMPARs, a global decrease in surface diffusion is observed during synaptic development (20, 55), but receptors can still diffuse in and out of mature spines (3, 8).

6.2 Glycine, GABA and Other Receptors

Using latex beads and SPT, glycine receptors (88) where the first to be shown to be mobile in plane of the neuronal plasma membrane and to exchange between clusters of gephyrin (its scaffold protein) through lateral diffusion. Quantum dots were used to track individual glycine receptors (GlyRs) and to analyze their lateral dynamics in the neuronal membrane of living cells for periods ranging from milliseconds to minutes (39). This allowed the characterization of multiple diffusion domains in relation to the synaptic, perisynaptic, or extrasynaptic GlyR localization. The median diffusion rate in the plane of the plasma membrane was in the range of 10^{-1}–10^{-2} μm^2/s in the extracellular space, and 10^{-3} μm^2/s or below at synapses. The entry of GlyRs into the synapse by diffusion could also be observed and further confirmed by electron microscopy imaging of QD-tagged receptors. A careful analysis of the diffusive properties further revealed that the receptor scaffold interaction is rather complex and results from multiple association states (48).

The dynamics of $GABA_A$ receptor mobility between synaptic and extrasynaptic locations in hippocampal pyramidal neurons has also been established by FRAP using fluorescent $GABA_A$ receptors that are capable of binding α-bungarotoxin (Bgt) (19) and by introducing a binding site for an irreversible inhibitor into the $GABA_A$ receptor alpha1 subunit channel lining region that can be accessed only when the receptor is activated (135). These studies established that the synaptic receptor population turns over rapidly by the import of new functional receptors within minutes through receptor diffusion within the plane of the membrane rather than trafficking from intracellular pools. Imaging with Bgt in hippocampal neurons revealed that $GABA_A$ receptor endocytosis occurred exclusively at extrasynaptic sites, consistent with the preferential colocalization of extrasynaptic receptors with the AP2 clathrin adaptor complex.

Receptor insertion into the plasma membrane was also predominantly extrasynaptic, and pulse-chase analysis revealed that these newly inserted receptors are then able to access directly synaptic sites. These results demonstrate that synaptic $GABA_A$ receptors are directly recruited from their extrasynaptic counterparts.

Diffusion of molecules and transient stabilization may also be instrumental to create local asymmetry. During the development of the nervous system, the tip of a growing axon, the growth cone (GC), must respond accurately to stimuli that direct its growth. This axonal navigation depends on extracellular concentration gradients of numerous guidance cues, including GABA. GCs can detect even weak directional signals, yet the mechanisms underlying this sensitivity remain unclear. Past studies in other eukaryotic chemotactic systems have pointed to the role of the spatial reorganization of the transduction pathway in their sensitive response. Using single molecule assay it was possible to observe individual $GABA_A$ receptors ($GABA_ARs$) in the plasma membrane of nerve GCs subjected to directional stimuli. In the presence of an external GABA gradient $GABA_ARs$ redistributed asymmetrically across the GC toward the gradient source. Single-particle tracking of $GABA_ARs$ showed that the redistribution resulted from transient interactions between receptors and microtubules. Moreover, the relocalization was accompanied by an enhancement in the asymmetry of intracellular calcium concentration. Altogether, these results revealed a microtubule-dependent polarized reorganization of chemoreceptors at the cell surface and suggested that this polarization serves as an amplification step in GABA gradient sensing by nerve GCs (22).

The surface diffusion properties of other neuronal receptors have been analyzed by ensemble or single molecule dynamic imaging. Using organotypic culture from rat striatum transfected with D1R fused to a fluorescent protein, the majority of dendritic D1R has been shown to diffuse laterally (118). NMDA receptors can act as a scaffold to recruit laterally diffusing dopamine D1 receptors (D1R) to neuronal spines. Human mu opiod receptors have been shown to diffuse in permeable submicrometer sized domains (40, 116). The cell surface dynamics of the 5HT1A receptor is modulated in a G protein-dependent manner (107). The movement of the human odorant receptor OR17-40 has been monitored in living cells by ensemble and single-molecule imaging. Although mobile receptors initially diffused either freely or within domains of various sizes, binding of an agonist or an antagonist increased partitioning of receptors into small domains of approximately 190 nm, which likely are precursors of clathrin-coated pits (66).

7 Receptor Scaffold Interactions and Cytoskeleton as Key Determinants of Surface Trafficking

Synapses contain over 300 types of proteins (65, 103), including neurotransmitter receptors, adhesion proteins and intracellular scaffolding, cytoskeleton or signaling elements. These various proteins can bind one with the other following a complex set of interactions that is far from worked out. These interactions can regulate the stabilization, function or trafficking of receptors and a number of them have been shown to regulate receptor surface diffusion and exchange between synaptic and extrasynaptic sites (Figs. 5,and 6).

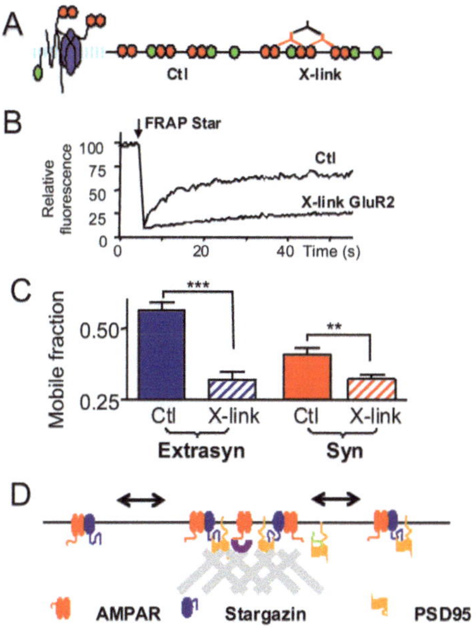

Fig. 5. AMPARs and stargazin traffic together in and out synapses. FRAP experiments demonstrate that stargazin and AMPARs are associated both in the synaptic and extrasynaptic membrane. Hippocampal neurons were transfected with mRFP-tagged AMPAR and GFP-tagged stargazin. FRAP was performed on stargazin-GFP in and out synapses either in control conditions or after cross-linking induced immobilization of AMPARs (**a**). (**b**) Recordings of stargazin-GFP fluorescence on a spot before and after photobleaching (*arrow*). Note that AMPAR crosslinking (X-link) induces immobilization of stargazin. (**c**) Quantification of the mobile fraction in and out synapses. (**d**) Schematic drawing of AMPAR/stargazin complex diffusion in and out synapses. Modified from (8, 48).

The fast reversibility of receptor-scaffold interactions were first demonstrated in live neurons using single latex particle tracking both for the glycine receptor/gephyrin interaction (88) and the group I metabotropic glutamate receptor 5 (mGluR5)/Homer interaction (121). Single-particle tracking was used to follow in real time, movements of the glycine receptor (GlyR) on the cell surface with or without the GlyR stabilizing protein gephyrin. GlyR alternate within seconds between diffusive and confined states. In the absence of gephyrin, GlyRs are mostly freely diffusing. Gephyrin induced long confinement periods spatially associated with submembranous clusters of gephyrin. However, even when most receptors are stabilized, they still frequently make transitions to the diffusive state. Analysis of the diffusion properties of two GlyR subunit types, either able or unable to bind gephyrin, gave access to the association states of GlyRs with its scaffolding protein (48). This indicated that an important portion of GlyR could be linked to a few molecules of gephyrin outside gephyrin clusters. The presence of extrasynaptic GlyR-gephyrin

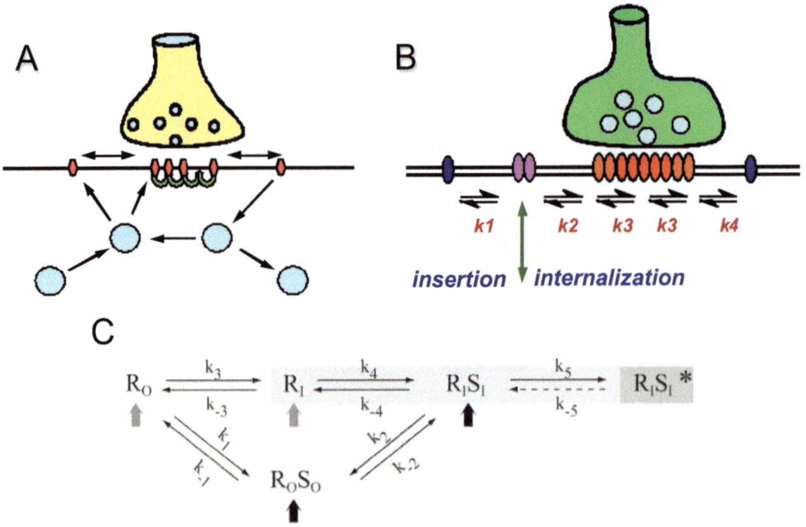

Fig. 6. Trafficking pathways leading to stabilization of receptors at synapses. (**a**) The classical route to postsynaptic differentiation now includes diffusion in the plane of the plasmamembrane. (**b**) Various kinetics can be derived from receptor diffusion properties depending on receptor localization with respect to the synaptic domain. (**c**) Schematic view of the different paths leading to stabilization of GlyR by gephyrin clusters. Receptor (R) and its scaffolding protein gephyrin (S) may be pre-assembled before being inserted into the cell membrane (*black arrow*) or they may reach the membrane separately (*gray arrow*). Receptor-scaffold complexes may be formed outside (equilibrium 1) or inside (equilibrium 4) gephyrin clusters. Both exchange of receptors (equilibrium 3) and of receptor-scaffold assemblies (equilibrium 2) may occur inside (suffix I) or outside (suffix O) gephyrin clusters. Once within clusters, receptor-scaffold complexes may reach a higher level of stabilization (equilibrium 5, *). Modified from (48).

complexes is not a surprise since the receptor associates with gephyrin intracellularly, and this insertion increases its plasma membrane accumulation rate (60). This emphasizes the role of scaffolding proteins in the extrasynaptic membrane and supports the implication of gephyrin-gephyrin interactions in the stabilization of GlyR at synapses.

The kinetic parameters controlling the equilibrium between GlyR inside and outside clusters were also characterized, and even within clusters, two subpopulations of GlyR with distinct degrees of stabilization between receptors and scaffolding proteins could be identified. At present, these subpopulations cannot be ascribed to specific molecular entities. At excitatory synapses, the receptor scaffold interaction is also a complex matter. mGluR5 is concentrated in a domain around the postsynaptic density but also found in large amounts in the extrasynaptic membrane (9, 97, 144). Receptor activation by glutamate increases receptor diffusion, whereas the scaffolding protein Homer favours confinement of receptor movements within clusters of Homer-mGluR5. However, this stabilization is reversible, because even in the presence of Homer, receptors still enter and exit from clusters at fast rates. Furthermore, clusters themselves are highly dynamic both in their movements and in their composition, which can vary within tens of seconds. These results first demonstrated that the exchange of receptors between dispersed and clustered states is fast and regulated during physiological processes and that receptor number in a cluster results from a dynamic equilibrium between the pools of stabilized and freely mobile receptors. This suggested that modification of this equilibrium could be involved in regulation of the number of receptors at synapses and may explain certain fast changes in receptor composition.

More recently, AMPAR stabilization at excitatory synapses has been shown to be mediated by interactions with the AMPAR-associated protein stargazin (95, 149) via the PDZ-domain scaffold protein PSD-95 (8). Using single quantum dot imaging and FRAP in live hippocampal neurons, AMPARs were shown to diffuse in the neuronal membrane together with Stargazin and to be trapped reversibly at synapses through the specific binding of stargazin to PSD-95 (Fig. 5). Disruption of the interaction between stargazin and PSD-95 strongly increased AMPAR surface diffusion, removing AMPARs from postsynaptic sites. Interestingly, and in contrast, the C-terminal domain of GluR2 does not seem to be involved in controlling AMPAR lateral diffusion. The direct interaction of the GluR2 C-terminus with the PDZ-containing proteins ABP/GRIP and PICK1 has been shown to play an important role in the regulation of AMPARs expression at synaptic sites (reviewed in (35)). These proteins may be involved primarily in modulating the surface expression of AMPARs rather than anchoring them at synaptic sites. The GluR2 PDZ-binding motif could be either required for AMPAR retention in plasma membrane (99) or for receptor recycling at the surface (102).

Cytoskeletal elements have also been shown to control receptor diffusion (30). Pharmacological disruption of F-actin and microtubules decreases the amount of GlyR and gephyrin, the backbone of the inhibitory postsynaptic scaffold, at synapses. F-actin and microtubule disruption increases GlyR exchange between the synaptic and extrasynaptic membranes and decreases receptor dwell time at synapses (Fig. 7). GlyR lateral diffusion is predominantly controlled by microtubules in the extrasynaptic membrane and by actin at synapses. Both diffusion coefficients and confinement at synapses are affected after F-actin disruption. These results indicate that receptor exchange between synaptic and extrasynaptic compartments depends on the properties of both the postsynaptic specialization and the extrasynaptic membrane. Consequently, GlyR number at synapses may be rapidly modulated by the cytoskeleton through the regulation of lateral diffusion in the plasma membrane and of receptor stabilization at synapses. In fact, the receptor is not diffusing above a stable matrix of gephyrins, and the postsynaptic scaffolds display a submicrometric rapid lateral motion continuously moving on the dendritic shaft (59). This dynamic

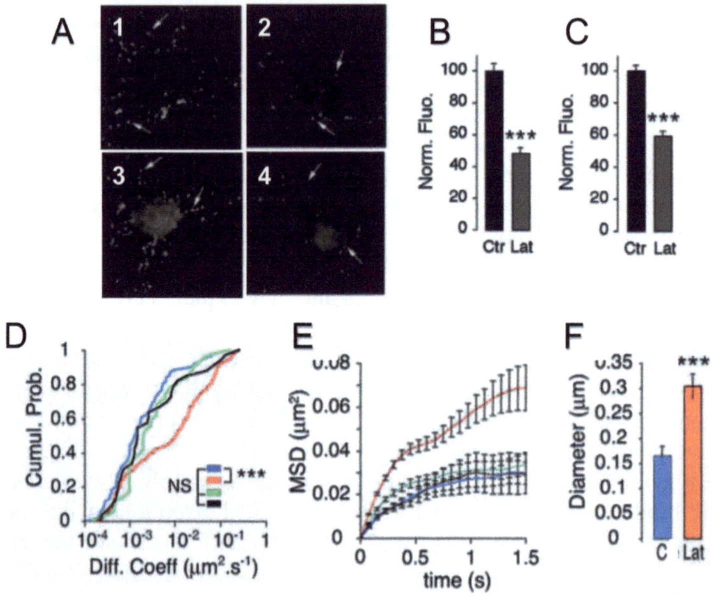

Fig. 7. Role of the cytoskeleton in controlling receptor surface trafficking. Example of regulation by actin of glycine receptor number and diffusion at synapses. (**a**) Effects of latrunculin-induced F-actin disruption (A2,4) on GlyR (A1,2) and gephyrin (A3,4) clusters. (**b,c**) Decrease of GlyR (B) and gephyrin (C) fluorescence at synapses after F-actin disruption. (**d**) Cumulative probabilities of GlyR-QD diffusion coefficients when tracked in the synaptic membrane. Latrunculin (*red*), but not nocodazole (*green*), led to an acceleration of GlyR lateral diffusion (control, *black*; gephyrin cluster, *blue*). (**e**) Average MSD as a function of time for GlyR-QDs at synapses in control (Ctr) (*blue*), latrunculin (*red*), or nocodazole (*green*) conditions. The black curve corresponds to GlyR-QDs tracked on neurons at identified inhibitory synapses. Values are means ± SEM. (**f**) Size of the domain of confinement at synapses (means ± SEM). Latrunculin decreased the confinement. Modified from (30).

behaviour is calcium dependent and also controlled by the cytoskeleton. Minute rearrangements within the gephyrin scaffold as well as the scaffold lateral displacements are F-actin dependent. The lateral movements are counteracted by microtubules. Moreover, the action of the potassium channel blocker 4-aminopyridine and receptor antagonists indicate that the dynamics of postsynaptic gephyrin scaffolds are controlled by synaptic activity. This set of data emphasizes once more that receptor-scaffold complexes at synapses should be considered as a multimolecular assemblies were mutual interactions together with individual molecule behaviour determine the set point of an equilibrium.

8 Receptor Surface Trafficking and Synaptic Plasticity

Rapid or long-term modifications in the efficacy of synaptic transmission are referred to as processes of synaptic plasticity. These are observed during development (synaptogenesis and synaptic maturation) and in the adult. In the last decade, activity-dependent synaptic plasticity has been shown to rely in large part on changes in the number and subtype of receptors concentrated in synapses (reviewed in (42, 54, 80, 85)). Surface diffusion of receptors in and out synaptic sites has emerged as a major mechanism to control these changes and this has fostered the analysis of activity-dependent regulation of receptor lateral diffusion (31, 142).

Although this field is only in its infancy, it is becoming apparent that both excitatory and inhibitory receptor diffusion is strongly and tightly regulated. It was initially shown that local rises in intracellular calcium or glutamate uncaging induce a rapid and reversible immobilization of AMPARs associated with a local accumulation of receptors (20). Although the molecular mechanisms underlying this phenomenon are still unclear, this is likely an important process to locally stabilize AMPARs during activity dependent processes. In contrast, global neuronal depolarization or bath application of glutamate induce an increase in AMPAR mobility (56, 133). This could underlie the loss of synaptic receptors observed during synaptic depression, although this remains to be established.

Basal synaptic activity was also recently shown to regulate AMPAR accumulation at synapses (47). Using single-molecule tracking together with the selective silencing of individual presynaptic boutons, local synaptic activity was shown to reduce diffusional exchange of GluR1 between synaptic and extrasynaptic domains, resulting in the postsynaptic accumulation of GluR1 (Fig. 8). At neighboring inactive synapses, GluR1 is highly mobile with individual receptors frequently escaping the synapse. Within the synapse, spontaneous activity confines the small-scale diffusional movement of GluR1 to restricted subregions of the postsynaptic membrane. Thus, local activity restricts GluR1 diffusional mobility on a submicron scale, defining a novel input-specific mechanism for regulating AMPA receptor composition and abundance.

The molecular mechanisms underlying short- or long-term activity-dependent regulation of AMPARs surface trafficking are still unknown but most likely involve regulation of interactions between the AMPAR/TARP complex and intracellular scaffold proteins. The C-termini of GluR1 and GluR2 have both been extensively shown to be involved in activity-dependent regulation of AMPAR numbers at synapses (reviewed in (35, 149)). Alternatively, phosphorylation of the C-terminus of TARPs has been shown to bi-directionally regulate synaptic plasticity (138). Deciphering the respective roles of AMPARs and TARPs in synaptic plasticity will undoubtedly require analyzing their differential involvement in the various trafficking pathways (lateral diffusion versus recycling) to and from the synapse.

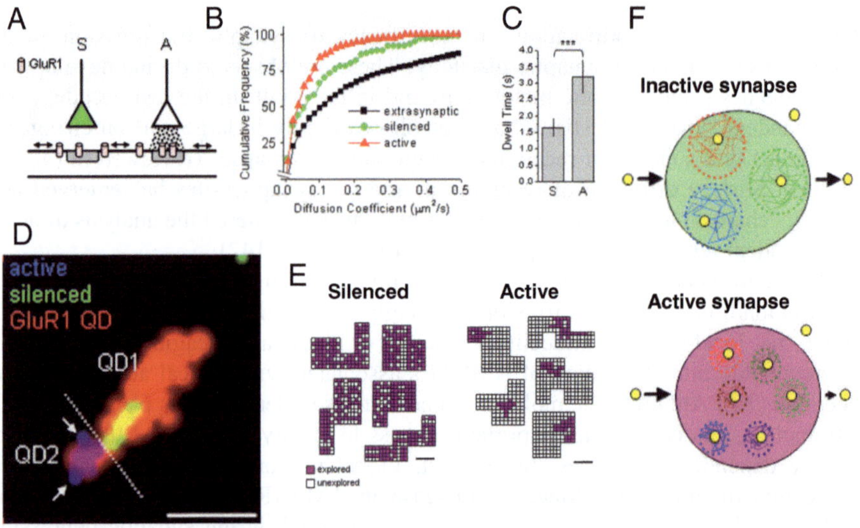

Fig. 8. Neuronal activity regulates AMPAR diffusion in synapses. (**a**) Individual synaptic terminals are simultaneously silenced and visualized by co-expression of tetanus toxin light chain and synaptophysin-GFP, respectively. Thus on the same postsynaptic dendrite are present active (A) and silenced (S) synapses. (**b**) The diffusion rate of GluR1-containing AMPARs is higher in inactive than active synapses, but still lower than in the extrasynaptic membrane. (**c**) The residency time of GluR1-containing AMPARs is smaller in silenced (S) than active (A) synapses. (**d**) Maximum projections of two QD-labeled GluR1 AMPARs (GluR1 QD, *red*) at adjacent silenced (*green*) and active (*blue*) synapses. Scale bar, 1 µm. (**e**) Single GluR1-QDs explore large areas within inactive silenced synapses and only small subregions within active synapses. Shown are five silenced and five active synaptic regions. Each pixel was divided into 0.0016 µm^2 subdomains and coded based on the presence (*pink*) or absence (*white*) of the GluR1-QD at any time during during the imaging period. Scale bar, 0.2 µm. (**f**) A schematic model for GluR1 lateral diffusion at active and inactive synapses viewed en face. Input-specific spontaneous synaptic activity reduces receptor mobility, limits exchange with the extrasynaptic membrane, and confines GluR1 within small subdomains of the postsynaptic membrane. This diffusional trap leads to GluR1 accumulation at active synapses. Modified from (47).

9 Concepts Which Have Emerged from Single Molecule Imaging in Neuroscience

9.1 New Imaging Technologies Together with Single Channel Recordings Have Favoured the Emergence of Molecular Physiology

Technological developments have allowed the investigation of dynamic processes which are hidden by the averaging inherent to bulk experiments. Numerous physical parameters are only accessible through nonstationary dynamics such as forward and reverse rate constants with respect to affinity. At steady state, molecules (receptor and scaffold proteins) enter and exit synapses at rates much higher than previously thought. This leads to the concept that synapses are multimolecular assemblies whose components are in a dynamic equilibrium with the reserve pools of molecules in the extrasynaptic membrane and in intracellular compartments (Figs. 6 and 9). The new concept is that synaptic plasticity results from changes in the equilibrium set points rather than all or none addition or removal of receptors. Equilibrium set points are likely to be part of a continuum rather than representing discrete states. The molecular basis for the changes in set points likely lies at several levels such as of receptor/scaffold affinities, flux rate of molecules, as well as their absolute levels of expression. In any case, the transition between equilibrium states relies on diffusion of molecules (including receptors and scaffold proteins) in and out synapses and the characteristic residence time of individual molecules or molecular assemblies in specific subcellular compartments. Diffusion provides a rapid sorting mechanism as already postulated for other cellular systems. Given the size of PSDs (diameter around 300 nm), the density of synapses (about 1 per μm^2), and the diffusion coefficients measured in the synaptic and extrasynaptic compartment, reasonable assumptions allow one to estimate the time needed for a receptor to exit (100 msec) or encounter a synapse (in the order of the minute) (63). This scale difference results from the large ratio of extrasynaptic to synaptic surface. Interestingly at excitatory synapses, local delivery of receptors to the spine head (101) which itself constitute a confined space, may allow an increase of the probability of diffusing receptors to be trapped by the PSD. Changes in the shape of spines may regulate this trapping efficiency (3).

That molecules only reside transiently in synapses implies that the molecular interactions underlying their stabilization are reversible over a short time scale. Time resolved measurements indicate that receptor exchange between the outside and inside of the PSD results from multiple kinetic pathways. This suggests multiple association states between receptors and scaffolding molecules as described above for the GlyR-gephyrin interaction (48). In a minimal model, these states can be defined as bound synaptic, bound extrasynaptic, superbound, etc. The latter state corresponds to receptors which stay for a long period (>30 s) and a higher confinement. Allocating the different states to identified structural features and determining the on and off rates between the states will allow access to the molecular biochemistry of receptor binding to scaffolding proteins and scaffolding proteins binding to other scaffolding proteins in living cells.

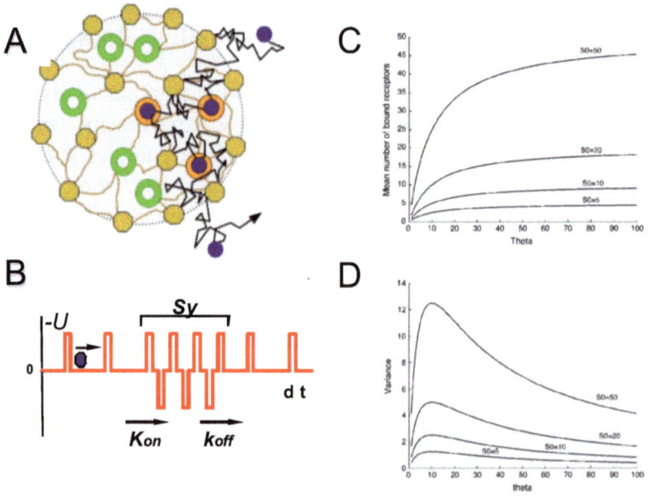

Fig. 9. Energy potential fluctuations seen by an itinerant receptor diffusing in the plane of the plasma membrane. (**a**) Schematic representation of a PSD domain containing free (*green donuts*) and bound (*orange donuts*) scaffolding molecules and many other structures such as transmembrane molecules (*brown dots*), submembranous cytoskeleton constituting obstacles, and fences (*brown lines*). (**b**) An itinerant receptor experiences potential fluctuations (-U) which are barriers (obstacles) such as adhesion molecules or other receptors already bound to scaffolds or potential wells corresponding to binding sites (scaffolding molecules) below the plasma membrane. Combination of barriers and wells at synapses tune the entry (K_{on}) and exit (K_{off}) receptor fluxes. (**c d**) Modeling of parameters to determine the mean (C) and variance (D) of the number of bound receptors at a single synapse allows the modeling of FRAP experiments derived from single molecule diffusion properties. The mean and the variance are plotted as a function of an equilibrium parameter (theta) defined for various numbers of scaffolding molecules with a given number of obstacles. Number of scaffolding molecules: S0 = 5, 10, 20, 50. See (63) for formula.

Thus, from a general view point, probing molecular interactions in living neurons will be of utmost importance. It is of particular interest to go beyond in vitro (i.e. in a test tube) studies which cannot take into account biochemical and biophysical properties due to the two dimensional nature of the membrane, inhomogeneities of protein concentrations or interfering interactions with other molecules such as lipids. The advent of single molecule and FRET techniques in live cells now provides a way to study in situ biochemical reactions. Such powerful tools will allow for a more complete deciphering of the kinetic properties of biological assemblies in relation to pharmacological properties.

9.2 Adaptation and Formation of Synapses

Another major outcome of the proposed model is that it accounts for changes during synaptic plasticity or even during synapse formation, which may result from modifications of receptor concentration in the plasma membrane and/or from changes in the concentration of scaffold proteins in the cytosol. It explains how changes in concentrations, i.e. chemical potentials, of receptors and scaffolds, lead to new steady state of the postsynaptic molecular assembly: the cooperativity at the origin of the discontinuous change of density distributions allows the system to switch from one set point of equilibrium to another one by small changes of key parameters. These chemical kinetics will need to be reconciled with specific biological mechanisms. This can now be achieved because the behaviour of individual molecules can be monitored, therefore allowing access to mechanisms normally hidden in the convoluted statistics of the behaviour of large numbers of molecules.

9.3 Reconciling Component Turnover and Synapse Structure Stability at Steady State

Stability of the synaptic structure in light of mobile receptors is a complex matter, because the local turnover (at synapses) of constitutive elements of the synapse is shorter than the lifetime of the structure (the synapse) (123). Recently, a new model has been proposed (125) in which the stability of receptor density is compatible with receptor individual turnover (Fig. 9). It accounts for the key idea that molecular interactions and cooperativity can be at the origin of a metastability of clusters and that their lifetime could be several orders of magnitude larger than that of their constitutive molecules. This quasi-equilibrium maintaining a stable density of receptor is also shared by our model.

The question of how much time an individual receptor spends in the synaptic zone is an important issue (Fig. 9). At steady state, the fraction of time spent by a particular receptor on a particular synaptic contact should be proportional to the density of the receptors at the contact. This is true if every receptor is well mixed so that there are no separation between the permanently stagnant receptors and mobile receptors. However, if some receptors are immobilized irreversibly (superbound) within the synaptic zone, the model can be modified to incorporate the contribution of these receptors. In order to test the hypothesis of mixing using single particle tracking, care must be taken so that the ensemble of trajectories tracked are not biased: if only those receptors which are initially diffusing in the extrasynaptic zone are sampled, then the statistics will be biased because of undetected receptors which stay within a synaptic site throughout the tracking experiment. Single-particle-tracking measurement has established at central excitatory synapses that about half of the receptors are mobile (39, 133). In contrast, FRAP (fluorescence recovery after photo-breaching) experiments of glutamate receptors at Drosophila neuromuscular junctions suggests that they are immobilized once they enter into the postsynaptic domain (109). However, one should stress that the stabilization of receptor density in the synaptic zone with indefinite lifetime is compatible with a finite residence time of an individual receptor on a synaptic site. Recently, Fusi et al. proposed a cascade mechanism to generate different timescales of synaptically stored memories (50), which sheds light on the quasi-equilibrium approach that we propose. Actually, different timescales can coexist to account for the dynamic turnover of constituting molecules in the postsynaptic density.

Another concept, which arises from the present model, is that stabilization is a reciprocal and cooperative mechanism. In other words, scaffold proteins stabilize receptors, and receptors stabilize scaffold proteins. This means that the local turnover of a given protein is not likely to determine by itself the turnover of the structure (see above). Indeed, the characteristic time of the individual molecule is independent from that of the structure, but one should also consider the molecular interactions. In biological terms, scaffolding molecules are not stabilizing the structure because they would have a longer turnover in the structure. Actually, reciprocity can either attenuate or amplify the amplitude of stochastic fluctuations of the receptor numbers at each synaptic site, depending on phase matching and consequently at the origin of a phase locking by reciprocity. This introduces robustness against the fluctuations of total receptor number associated with exo-/endocytosis at extrasynaptic sites.

10 Conclusion

Do receptor movements also exist in vivo in the freely behaving animal? This is a technical challenge. However in slices, FRAP experiments have revealed that cell surface receptors such as AMPAR and NMDAR also display diffusive behaviors. The basic biophysical properties of membranes have no reason to be fundamentally different from those characterized in cultured cells. Until now, nobody has challenged the fluid mosaic membrane model (127) in vivo. Nevertheless, extracellular and intercellular environments are likely different in vivo than in vitro and thus may impose supplementary constraints on receptor diffusion. These measurements represent the next frontier in this field.

Acknowledgements

We wish to thank Cecile Bats, Cédric Bouzigues, Cécile Charrier, Laurent Cognet, Maxime Dahan, Marie-Virginie Erhensperger, Laurent Groc, Cyril Hanus, Martin Heine, David Holcman, Sabine Lévi, Brahim Lounis and Ken Sekimoto whose work is presented here and without whom this chapter would not be possible.

References

1. Adesnik H, Nicoll RA, and England PM. Photoinactivation of native AMPA receptors reveals their real-time trafficking. *Neuron* 48: 977–985, 2005.
2. Ashby MC, De La Rue SA, Ralph GS, Uney J, Collingridge GL, and Henley JM. Removal of AMPA receptors (AMPARs) from synapses is preceded by transient endocytosis of extrasynaptic AMPARs. *J Neurosci* 24: 5172–5176, 2004.

3. Ashby MC, Maier SR, Nishimune A, and Henley JM. Lateral diffusion drives constitutive exchange of AMPA receptors at dendritic spines and is regulated by spine morphology. *J Neurosci* 26: 7046–7055, 2006.
4. Axelrod D, Koppel DE, Schlessinger J, Elson E, and Webb WW. Mobility measurement by analysis of fluorescence photobleaching recovery kinetics. *Biophys J* 16: 1055–1069, 1976.
5. Axelrod D, Ravdin P, Koppel DE, Schlessinger J, Webb WW, Elson EL, and Podleski TR. Lateral motion of fluorescently labeled acetylcholine receptors in membranes of developing muscle fibers. *Proc Natl Acad Sci USA* 73: 4594–4598, 1976.
6. Bacia K and Schwille P. A dynamic view of cellular processes by in vivo fluorescence auto- and cross-correlation spectroscopy. *Methods* 29: 74–85, 2003.
7. Bannai H, Levi S, Schweizer C, Dahan M, and Triller A. Imaging the lateral diffusion of membrane molecules with quantum dots. *Nat Protoc* 1: 2628–2634, 2006.
8. Bats C, Groc L, and Choquet D. The interaction between Stargazin and PSD-95 regulates AMPA receptor surface trafficking. *Neuron* 53: 719–734, 2007.
9. Baude A, Nusser Z, Roberts JDB, Mulvihill E, McIhinney RAJ, and Somogyi P. The metabotropic glutamate receptors (mGluR1α) is concentrated at perisynaptic membrane of neuronal subpopulations as detected by immunoglod reaction. *Neuron* 11: 771–787, 1993.
10. Bawendi MG, Wilson WL, Rothberg L, Carroll PJ, Jedju TM, Steigerwald ML, and Brus LE. Electronic structure and photoexcited-carrier dynamics in nanometer-size CdSe clusters. *Phys Rev Lett* 65: 1623–1626, 1990.
11. Bear MF and Malenka RC. Synaptic plasticity: LTP and LTD. *Curr Opin Neurobiol* 4: 389–399, 1994.
12. Beattie EC, Carroll RC, Yu X, Morishita W, Yasuda H, von Zastrow M, and Malenka RC. Regulation of AMPA receptor endocytosis by a signaling mechanism shared with LTD. *Nat Neurosci* 3: 1291–1300, 2000.
13. Berciaud S, Cognet L, Blab GA, and Lounis B. Photothermal heterodyne imaging of individual nonfluorescent nanoclusters and nanocrystals. *Phys Rev Lett* 93: 257402, 2004.
14. Beretta F, Sala C, Saglietti L, Hirling H, Sheng M, and Passafaro M. NSF interaction is important for direct insertion of GluR2 at synaptic sites. *Mol Cell Neurosci* 28: 650–660, 2005.
15. Berne BJ and Pecora R. Dynamic Light Scattering: With Applications to Chemistry, Biology, and Physics. New York: Dover, 2000.
16. Betzig E, Patterson GH, Sougrat R, Lindwasser OW, Olenych S, Bonifacino JS, Davidson MW, Lippincott-Schwartz J, and Hess HF. Imaging intracellular fluorescent proteins at nanometer resolution. *Science* 313: 1642–1645, 2006.
17. Blanpied TA, Scott DB, and Ehlers MD. Dynamics and regulation of clathrin coats at specialized endocytic zones of dendrites and spines. *Neuron* 36: 435–449, 2002.
18. Bloodgood BL and Sabatini BL. Neuronal activity regulates diffusion across the neck of dendritic spines. *Science* 310: 866–869, 2005.
19. Bogdanov Y, Michels G, Armstrong-Gold C, Haydon PG, Lindstrom J, Pangalos M, and Moss SJ. Synaptic GABAA receptors are directly recruited from their extrasynaptic counterparts. *EMBO J* 25: 4381–4389, 2006.
20. Borgdorff AJ and Choquet D. Regulation of AMPA receptor lateral movements. *Nature* 417: 649–653, 2002.
21. Bouzigues C, Levi S, Triller A, and Dahan M. Single quantum dot tracking of membrane receptors. *Methods Mol Biol* 374: 81–92, 2007.

22. Bouzigues C, Morel M, Triller A, and Dahan M. Asymmetric redistribution of GABA receptors during GABA gradient sensing by nerve growth cones analyzed by single quantum dot imaging. *Proc Natl Acad Sci USA*, 104: 11251–11256, 2007.
23. Boyer D, Tamarat P, Maali A, Lounis B, and Orrit M. Photothermal imaging of nanometer-sized metal particles among scatterers. *Science* 297: 1160–1163, 2002.
24. Bruchez M, Jr., Moronne M, Gin P, Weiss S, and Alivisatos AP. Semiconductor nanocrystals as fluorescent biological labels. *Science* 281: 2013–2016, 1998.
25. Cai D, Verhey KJ, and Meyhofer E. Tracking single Kinesin molecules in the cytoplasm of mammalian cells. *Biophys J* 92: 4137–4144, 2007.
26. Cambi A, Lidke DS, Arndt-Jovin DJ, Figdor CG, and Jovin TM. Ligand-conjugated quantum dots monitor antigen uptake and processing by dendritic cells. *Nano Lett* 7: 970–977, 2007.
27. Carroll RC, Beattie EC, Xia H, Luscher C, Altschuler Y, Nicoll RA, Malenka RC, and von Zastrow M. Dynamin-dependent endocytosis of ionotropic glutamate receptors. *Proc Natl Acad Sci USA* 96: 14112–14117, 1999.
28. Carroll RC, Lissin DV, von Zastrow M, Nicoll RA, and Malenka RC. Rapid redistribution of glutamate receptors contributes to long-term depression in hippocampal cultures. *Nat Neurosci* 2: 454–460, 1999.
29. Chambers JJ, Gouda H, Young DM, Kuntz ID, and England PM. Photochemically knocking out glutamate receptors in vivo. *J Am Chem Soc* 126: 13886–13887, 2004.
30. Charrier C, Ehrensperger MV, Dahan M, Levi S, and Triller A. Cytoskeleton regulation of glycine receptor number at synapses and diffusion in the plasma membrane. *J Neurosci* 26: 8502–8511, 2006.
31. Choquet D and Triller A. The role of receptor diffusion in the organization of the postsynaptic membrane. *Nat Rev Neurosci* 4: 251–265, 2003.
32. Cognet L, Coussen F, Choquet D, and Lounis B. Fluorescence microscopy of single autofluorescent proteins for cellular biology. *C R Phys* 3: 645–656, 2002.
33. Cognet L, Lounis B, and Choquet D. Tracking receptors by optical imaging of single molecules. In: *Imaging in Neuroscience and Development: A Laboratory Manual* (Yuste R and Konnerth A ed.). Cold Spring Harbor, NY: Cold Spring Harbor Lab Press, 2005, p. 521.
34. Cognet L, Tardin C, Boyer D, Choquet D, Tamarat P, and Lounis B. Single metallic nanoparticle imaging for protein detection in cells. *Proc Natl Acad Sci USA* 100: 11350–11355, 2003.
35. Collingridge GL, Isaac JT, and Wang YT. Receptor trafficking and synaptic plasticity. *Nat Rev Neurosci* 5: 952–962, 2004.
36. Cooney JR, Hurlburt JL, Selig DK, Harris KM, and Fiala JC. Endosomal compartments serve multiple hippocampal dendritic spines from a widespread rather than a local store of recycling membrane. *J Neurosci* 22: 2215–2224, 2002.
37. Cottrell JR, Borok E, Horvath TL, and Nedivi E. CPG2: a brain- and synapse-specific protein that regulates the endocytosis of glutamate receptors. *Neuron* 44: 677–690, 2004.
38. Courty S, Luccardini C, Bellaiche Y, Cappello G, and Dahan M. Tracking individual kinesin motors in living cells using single quantum-dot imaging. *Nano Lett* 6: 1491–1495, 2006.
39. Dahan M, Levi S, Luccardini C, Rostaing P, Riveau B, and Triller A. Diffusion dynamics of glycine receptors revealed by single-quantum dot tracking. *Science* 302: 442–445, 2003.
40. Daumas F, Destainville N, Millot C, Lopez A, Dean D, and Salomé L. Confined diffusion without fences of a G protein coupled receptor as revealed by single particle tracking. *Biophys J* 84, 2003.

41. Delehanty JB, Medintz IL, Pons T, Brunel FM, Dawson PE, and Mattoussi H. Self-assembled quantum dot-peptide bioconjugates for selective intracellular delivery. *Bioconjug Chem* 17: 920–927, 2006.
42. Derkach VA, Oh MC, Guire ES, and Soderling TR. Regulatory mechanisms of AMPA receptors in synaptic plasticity. *Nat Rev Neurosci* 8: 101–113, 2007.
43. Dietrich C, Yang B, Fujiwara T, Kusumi A, and Jacobson K. Relationship of lipid rafts to transient confinement zones detected by single particle tracking. *Biophys J* 82: 274–284, 2002.
44. Donnert G, Eggeling C, and Hell SW. Major signal increase in fluorescence microscopy through dark-state relaxation. *Nat Methods* 4: 81–86, 2007.
45. Eggeling C, Widengren J, Brand L, Schaffer J, Felekyan S, and Seidel CA. Analysis of photobleaching in single-molecule multicolor excitation and Forster resonance energy transfer measurements. *J Phys Chem A Mol Spectrosc Kinet Environ Gen Theory* 110: 2979–2995, 2006.
46. Ehlers MD. Reinsertion or degradation of AMPA receptors determined by activity-dependent endocytic sorting. *Neuron* 28: 511–525, 2000.
47. Ehlers MD, Heine M, Groc L, Lee MC, and Choquet D. Diffusional trapping of GluR1 AMPA receptors by input-specific synaptic activity. *Neuron* 54: 447–460, 2007.
48. Ehrensperger MV, Hanus C, Vannier C, Triller A, and Dahan M. Multiple association states between glycine receptors and gephyrin identified by SPT analysis. *Biophys J* 92: 3706–3718, 2007.
49. Fujiwara T, Ritchie K, Murakoshi H, Jacobson K, and Kusumi A. Phospholipids undergo hop diffusion in compartmentalized cell membrane. *J Cell Biol* 157: 1071–1081, 2002.
50. Fusi S, Drew PJ, and Abbott LF. Cascade models of synaptically stored memories. *Neuron* 45: 599–611, 2005.
51. Geerts H, De Brabander M, Nuydens R, Geuens S, Moeremans M, De Mey J, and Hollenbeck P. Nanovid tracking: a new automatic method for the study of mobility in living cells based on colloidal gold and video microscopy. *Biophys J* 52: 775–782, 1987.
52. Gerges NZ, Backos DS, Rupasinghe CN, Spaller MR, and Esteban JA. Dual role of the exocyst in AMPA receptor targeting and insertion into the postsynaptic membrane. *EMBO J* 25: 1623–1634, 2006.
53. Gray NW, Weimer RM, Bureau I, and Svoboda K. Rapid redistribution of synaptic PSD-95 in the neocortex in vivo. *PLoS Biol* 4: e370, 2006.
54. Groc L and Choquet D. AMPA and NMDA glutamate receptor trafficking: multiple roads for reaching and leaving the synapse. *Cell Tissue Res* 326: 423–438, 2006.
55. Groc L, Gustafsson B, and Hanse E. AMPA signalling in nascent glutamatergic synapses: there and not there! *Trends Neurosci* 29: 132–139, 2006.
56. Groc L, Heine M, Cognet L, Brickley K, Stephenson FA, Lounis B, and Choquet D. Differential activity-dependent regulation of the lateral mobilities of AMPA and NMDA receptors. *Nat Neurosci* 7: 695–696, 2004.
57. Groc L, Heine M, Cousins SL, Stephenson FA, Lounis B, Cognet L, and Choquet D. NMDA receptor surface mobility depends on NR2A-2B subunits. *Proc Natl Acad Sci USA* 103: 18769–18774, 2006.
58. Han Y, Alsayed AM, Nobili M, Zhang J, Lubensky TC, and Yodh AG. Brownian motion of an ellipsoid. *Science* 314: 626–630, 2006.
59. Hanus C, Ehrensperger MV, and Triller A. Activity-dependent movements of postsynaptic scaffolds at inhibitory synapses. *J Neurosci* 26: 4586–4595, 2006.
60. Hanus C, Vannier C, and Triller A. Intracellular association of glycine receptor with gephyrin increases its plasma membrane accumulation rate. *J Neurosci* 24: 1119–1128, 2004.

61. Hayashi Y, Shi SH, Esteban JA, Piccini A, Poncer JC, and Malinow R. Driving AMPA receptors into synapses by LTP and CaMKII: requirement for GluR1 and PDZ domain interaction. *Science* 287: 2262–2267, 2000.
62. Higgins JS and Benoit HC. *Polymers and Neutron Scattering*. Oxford: Oxford University Press, 1997.
63. Holcman D and Triller A. Modeling synaptic dynamics driven by receptor lateral diffusion. *Biophys J* 91: 2405–2415, 2006.
64. Howarth M, Takao K, Hayashi Y, and Ting AY. Targeting quantum dots to surface proteins in living cells with biotin ligase. *Proc Natl Acad Sci USA* 102: 7583–7588, 2005.
65. Husi H and Grant SG. Proteomics of the nervous system. *Trends Neurosci* 24: 259–266, 2001.
66. Jacquier V, Prummer M, Segura JM, Pick H, and Vogel H. Visualizing odorant receptor trafficking in living cells down to the single-molecule level. *Proc Natl Acad Sci USA* 103: 14325–14330, 2006.
67. Ju W, Morishita W, Tsui J, Gaietta G, Deerinck TJ, Adams SR, Garner CC, Tsien RY, Ellisman MH, and Malenka RC. Activity-dependent regulation of dendritic synthesis and trafficking of AMPA receptors. *Nat Neurosci* 7: 244–253, 2004.
68. Koda LY and Partlow LM. Membrane marker movement on sympathetic axons in tissue culture. *J Neurobiol* 7: 157–172, 1976.
69. Kopec CD, Li B, Wei W, Boehm J, and Malinow R. Glutamate receptor exocytosis and spine enlargement during chemically induced long-term potentiation. *J Neurosci* 26: 2000–2009, 2006.
70. Kusumi A, Nakada C, Ritchie K, Murase K, Suzuki K, Murakoshi H, Kasai RS, Kondo J, and Fujiwara T. Paradigm shift of the plasma membrane concept from the two-dimensional continuum fluid to the partitioned fluid: high-speed single-molecule tracking of membrane molecules. *Annu Rev Biophys Biomol Struct* 34: 351–378, 2005.
71. Kusumi A, Sako Y, and Yamamoto M. Confined lateral diffusion of membrane receptors as studied by single particle tracking (nanovid microscopy). Effects of calcium-induced differentiation in cultured epithelial cells. *Biophys J* 65: 2021–2040, 1993.
72. Lasne D, Blab GA, Berciaud S, Heine M, Groc L, Choquet D, Cognet L, and Lounis B. Single nanoparticle photothermal tracking (SNaPT) of 5-nm gold beads in live cells. *Biophys J* 91: 4598–4604, 2006.
73. Lata S, Gavutis M, Tampe R, and Piehler J. Specific and stable fluorescence labeling of histidine-tagged proteins for dissecting multi-protein complex formation. *J Am Chem Soc* 128: 2365–2372, 2006.
74. Lau CG and Zukin RS. NMDA receptor trafficking in synaptic plasticity and neuropsychiatric disorders. *Nat Rev Neurosci* 8: 413–426, 2007.
75. Lee SH, Simonetta A, and Sheng M. Subunit rules governing the sorting of internalized AMPA receptors in hippocampal neurons. *Neuron* 43: 221–236, 2004.
76. Levitan ES, Lanni F, and Shakiryanova D. In vivo imaging of vesicle motion and release at the Drosophila neuromuscular junction. *Nat Protoc* 2: 1117–1125, 2007.
77. Lidke DS, Lidke KA, Rieger B, Jovin TM, and Arndt-Jovin DJ. Reaching out for signals: filopodia sense EGF and respond by directed retrograde transport of activated receptors. *J Cell Biol* 170: 619–626, 2005.
78. Lin JW, Ju W, Foster K, Lee SH, Ahmadian G, Wyszynski M, Wang YT, and Sheng M. Distinct molecular mechanisms and divergent endocytotic pathways of AMPA receptor internalization. *Nat Neurosci* 3: 1282–1290, 2000.
79. Lippincott-Schwartz J, Snapp E, and Kenworthy A. Studying protein dynamics in living cells. *Nat Rev Mol Cell Biol* 2: 444–456, 2001.

80. Lisman J and Raghavachari S. A unified model of the presynaptic and postsynaptic changes during LTP at CA1 synapses. *Sci STKE* 2006: re11, 2006.
81. Lledo PM, Zhang X, Sudhof TC, Malenka RC, and Nicoll RA. Postsynaptic membrane fusion and long-term potentiation. *Science* 279: 399–403, 1998.
82. Lu W, Man H, Ju W, Trimble WS, MacDonald JF, and Wang YT. Activation of synaptic NMDA receptors induces membrane insertion of new AMPA receptors and LTP in cultured hippocampal neurons. *Neuron* 29: 243–254, 2001.
83. Luscher C, Xia H, Beattie EC, Carroll RC, von Zastrow M, Malenka RC, and Nicoll RA. Role of AMPA receptor cycling in synaptic transmission and plasticity. *Neuron* 24: 649–658, 1999.
84. Luthi A, Chittajallu R, Duprat F, Palmer MJ, Benke TA, Kidd FL, Henley JM, Isaac JT, and Collingridge GL. Hippocampal LTD expression involves a pool of AMPARs regulated by the NSF-GluR2 interaction. *Neuron* 24: 389–399, 1999.
85. Malinow R and Malenka RC. AMPA receptor trafficking and synaptic plasticity. *Annu Rev Neurosci* 25: 103–126, 2002.
86. Man HY, Lin JW, Ju WH, Ahmadian G, Liu L, Becker LE, Sheng M, and Wang YT. Regulation of AMPA receptor-mediated synaptic transmission by clathrin-dependent receptor internalization. *Neuron* 25: 649–662, 2000.
87. Masugi-Tokita M, Tarusawa E, Watanabe M, Molnar E, Fujimoto K, and Shigemoto R. Number and density of AMPA receptors in individual synapses in the rat cerebellum as revealed by SDS-digested freeze-fracture replica labeling. *J Neurosci* 27: 2135–2144, 2007.
88. Meier J, Vannier C, Sergé A, Triller A, and Choquet D. Fast and reversible trapping of surface glycine receptors by gephyrin. *Nat Neurosci* 4: 253–260, 2001.
89. Metzler M, Li B, Gan L, Georgiou J, Gutekunst CA, Wang Y, Torre E, Devon RS, Oh R, Legendre-Guillemin V, Rich M, Alvarez C, Gertsenstein M, McPherson PS, Nagy A, Wang YT, Roder JC, Raymond LA, and Hayden MR. Disruption of the endocytic protein HIP1 results in neurological deficits and decreased AMPA receptor trafficking. *EMBO J* 22: 3254–3266, 2003.
90. Michalet X, Pinaud FF, Bentolila LA, Tsay JM, Doose S, Li JJ, Sundaresan G, Wu AM, Gambhir SS, and Weiss S. Quantum dots for live cells, in vivo imaging, and diagnostics. *Science* 307: 538–544, 2005.
91. Miesenbock G, De Angelis DA, and Rothman JE. Visualizing secretion and synaptic transmission with pH-sensitive green fluorescent proteins. *Nature* 394: 192–195, 1998.
92. Monyer H, Burnashev N, Laurie DJ, Sakmann B, and Seeburg PH. Developmental and regional expression in the rat brain and functional properties of four NMDA receptors. *Neuron* 12: 529–540, 1994.
93. Murase K, Fujiwara T, Umemura Y, Suzuki K, Iino R, Yamashita H, Saito M, Murakoshi H, Ritchie K, and Kusumi A. Ultrafine membrane compartments for molecular diffusion as revealed by single molecule techniques. *Biophys J* 86: 4075–4093, 2004.
94. Nakada C, Ritchie K, Oba Y, Nakamura M, Hotta Y, Iino R, Kasai RS, Yamaguchi K, Fujiwara T, and Kusumi A. Accumulation of anchored proteins forms membrane diffusion barriers during neuronal polarization. *Nat Cell Biol* 5: 626–632, 2003.
95. Nicoll RA, Tomita S, and Bredt DS. Auxiliary subunits assist AMPA-type glutamate receptors. *Science* 311: 1253–1256, 2006.
96. Nong Y, Huang YQ, Ju W, Kalia LV, Ahmadian G, Wang YT, and Salter MW. Glycine binding primes NMDA receptor internalization. *Nature* 422: 302–307, 2003.
97. Nusser Z, Mulvihill E, Streit P, and Somogyi P. Subsynaptic segragation of metabotropic and ionotropic glutamate receptors as revealed by immunogold localization. *Neuroscience* 61: 421–427, 1994.

98. Ober RJ, Ram S, and Ward ES. Localization accuracy in single-molecule microscopy. *Biophys J* 86: 1185–1200, 2004.
99. Osten P, Khatri L, Perez JL, Kohr G, Giese G, Daly C, Schulz TW, Wensky A, Lee LM, and Ziff EB. Mutagenesis reveals a role for ABP/GRIP binding to GluR2 in synaptic surface accumulation of the AMPA receptor. *Neuron* 27: 313–325, 2000.
100. Park M, Penick EC, Edwards JG, Kauer JA, and Ehlers MD. Recycling endosomes supply AMPA receptors for LTP. *Science* 305: 1972–1975, 2004.
101. Park M, Salgado JM, Ostroff L, Helton TD, Robinson CG, Harris KM, and Ehlers MD. Plasticity-induced growth of dendritic spines by exocytic trafficking from recycling endosomes. *Neuron* 52: 817–830, 2006.
102. Passafaro M, Piech V, and Sheng M. Subunit-specific temporal and spatial patterns of AMPA receptor exocytosis in hippocampal neurons. *Nat Neurosci* 4: 917–926, 2001.
103. Peng J, Kim MJ, Cheng D, Duong DM, Gygi SP, and Sheng M. Semiquantitative proteomic analysis of rat forebrain postsynaptic density fractions by mass spectrometry. *J Biol Chem* 279: 21003–21011, 2004.
104. Petralia RS, Wang YX, and Wenthold RJ. Internalization at glutamatergic synapses during development. *Eur J Neurosci* 18: 3207–3217, 2003.
105. Pickard L, Noel J, Duckworth JK, Fitzjohn SM, Henley JM, Collingridge GL, and Molnar E. Transient synaptic activation of NMDA receptors leads to the insertion of native AMPA receptors at hippocampal neuronal plasma membranes. *Neuropharmacology* 41: 700–713, 2001.
106. Pinaud F, King D, Moore HP, and Weiss S. Bioactivation and cell targeting of semiconductor CdSe/ZnS nanocrystals with phytochelatin-related peptides. *J Am Chem Soc* 126: 6115–6123, 2004.
107. Pucadyil TJ and Chattopadhyay A. The human serotonin(1A) receptor exhibits G-protein-dependent cell surface dynamics. *Glycoconj J* 24: 25–31, 2007.
108. Racz B, Blanpied TA, Ehlers MD, and Weinberg RJ. Lateral organization of endocytic machinery in dendritic spines. *Nat Neurosci* 7: 917–918, 2004.
109. Rasse TM, Fouquet W, Schmid A, Kittel RJ, Mertel S, Sigrist CB, Schmidt M, Guzman A, Merino C, Qin G, Quentin C, Madeo FF, Heckmann M, and Sigrist SJ. Glutamate receptor dynamics organizing synapse formation in vivo. *Nat Neurosci* 8: 898–905, 2005.
110. Reits EA and Neefjes JJ. From fixed to FRAP: measuring protein mobility and activity in living cells. *Nat Cell Biol* 3: E145–E147, 2001.
111. Richards DA, De Paola V, Caroni P, Gahwiler BH, and McKinney RA. AMPA-receptor activation regulates the diffusion of a membrane marker in parallel with dendritic spine motility in the mouse hippocampus. *J Physiol* 558: 503–512, 2004.
112. Roche KW, Standley S, McCallum J, Dune Ly C, Ehlers MD, and Wenthold RJ. Molecular determinants of NMDA receptor internalization. *Nat Neurosci* 4: 794–802, 2001.
113. Rosenberg M, Meier J, Triller A, and Vannier C. Dynamics of glycine receptor insertion in the neuronal plasma membrane. *J Neurosci* 21: 5036–5044, 2001.
114. Salome L, Cazeils JL, Lopez A, and Tocanne JF. Characterization of membrane domains by FRAP experiments at variable observation areas. *Eur Biophys J* 27: 391–402, 1998.
115. Sans N, Prybylowski K, Petralia RS, Chang K, Wang YX, Racca C, Vicini S, and Wenthold RJ. NMDA receptor trafficking through an interaction between PDZ proteins and the exocyst complex. *Nat Cell Biol* 5: 520–530, 2003.
116. Sauliere A, Gaibelet G, Millot C, Mazeres S, Lopez A, and Salome L. Diffusion of the mu opioid receptor at the surface of human neuroblastoma SH-SY5Y cells is restricted to permeable domains. *FEBS Lett* 580: 5227–5231, 2006.
117. Saxton MJ and Jacobson K. Single-particle tracking: applications to membrane dynamics. *Annu Rev Biophys Biomol Struct* 26: 373–399, 1997.

118. Scott L, Zelenin S, Malmersjo S, Kowalewski JM, Markus EZ, Nairn AC, Greengard P, Brismar H, and Aperia A. Allosteric changes of the NMDA receptor trap diffusible dopamine 1 receptors in spines. *Proc Natl Acad Sci USA* 103: 762–767, 2006.
119. Seisenberger G, Ried MU, Endre{beta} T, Buning H, Hallek M, and Brauchle C. Real-time single-molecule imaging of the infection pathway of an adeno-associated virus. *Science* 294: 1929–1932, 2001.
120. Sekine-Aizawa Y and Huganir RL. Imaging of receptor trafficking by using alpha-bungarotoxin-binding-site-tagged receptors. *Proc Natl Acad Sci USA* 101: 17114–17119, 2004.
121. Serge A, Fourgeaud L, Hemar A, and Choquet D. Receptor activation and homer differentially control the lateral mobility of metabotropic glutamate receptor 5 in the neuronal membrane. *J Neurosci* 22: 3910–3920, 2002.
122. Sheng M, Cummings J, Roldan LA, Jan YN, and Jan LY. Changing subunit composition of heteromeric NMDA receptors during development of rat cortex. *Nature* 368: 144–147, 1994.
123. Sheng M and Nakagawa T. Neurobiology: glutamate receptors on the move. *Nature* 417: 601–602, 2002.
124. Shi S, Hayashi Y, Esteban JA, and Malinow R. Subunit-specific rules governing AMPA receptor trafficking to synapses in hippocampal pyramidal neurons. *Cell* 105: 331–343, 2001.
125. Shouval HZ. Clusters of interacting receptors can stabilize synaptic efficacies. *Proc Natl Acad Sci USA* 102: 14440–14445, 2005.
126. Siksou L, Rostaing P, Lechaire J, Boudier T, Ohtsuka T, Fejtová A, Kao H, Greengard P, Gundelfinger E, Triller A, and Marty S. Three-dimensional architecture of presynaptic terminal cytomatrix. *J Neurosci* 27: 6868–6877, 2007.
127. Singer SJ and Nicolson GL. The fluid mosaic model of the structure of cell membranes. *Science* 175: 720–731, 1972.
128. Snee PT, Somers RC, Nair G, Zimmer JP, Bawendi MG, and Nocera DG. A ratiometric CdSe/ZnS nanocrystal pH sensor. *J Am Chem Soc* 128: 13320–13321, 2006.
129. Snyder EM, Philpot BD, Huber KM, Dong X, Fallon JR, and Bear MF. Internalization of ionotropic glutamate receptors in response to mGluR activation. *Nat Neurosci* 4: 1079–1085, 2001.
130. Sonnleitner A, Mannuzzu LM, Terakawa S, and Isacoff EY. Structural rearrangements in single ion channels detected optically in living cells. *Proc Natl Acad Sci USA* 99: 12759–12764, 2002.
131. Spacek J and Harris KM. Three-dimensional organization of smooth endoplasmic reticulum in hippocampal CA1 dendrites and dendritic spines of the immature and mature rat. *J Neurosci* 17: 190–203, 1997.
132. Tanaka J, Matsuzaki M, Tarusawa E, Momiyama A, Molnar E, Kasai H, and Shigemoto R. Number and density of AMPA receptors in single synapses in immature cerebellum. *J Neurosci* 25: 799–807, 2005.
133. Tardin C, Cognet L, Bats C, Lounis B, and Choquet D. Direct imaging of lateral movements of AMPA receptors inside synapses. *EMBO J* 22: 4656–4665, 2003.
134. Thomas CG, Miller AJ, and Westbrook GL. Synaptic and extrasynaptic NMDA receptor NR2 subunits in cultured hippocampal neurons. *J Neurophysiol* 95: 1727–1734, 2006.
135. Thomas P, Mortensen M, Hosie AM, and Smart TG. Dynamic mobility of functional GABAA receptors at inhibitory synapses. *Nat Neurosci* 8: 889–897, 2005.
136. Thorne RG and Nicholson C. In vivo diffusion analysis with quantum dots and dextrans predicts the width of brain extracellular space. *Proc Natl Acad Sci USA* 103: 5567–5572, 2006.

137. Thoumine O, Saint-Michel E, Dequidt C, Falk J, Rudge R, Galli T, Faivre-Sarrailh C, and Choquet D. Weak effect of membrane diffusion on the rate of receptor accumulation at adhesive contacts. *Biophys J* 89: L40–L42, 2005.
138. Tomita S, Stein V, Stocker TJ, Nicoll RA, and Bredt DS. Bidirectional synaptic plasticity regulated by phosphorylation of stargazin-like TARPs. *Neuron* 45: 269–277, 2005.
139. Toni N, Buchs PA, Nikonenko I, Povilaitite P, Parisi L, and Muller D. Remodeling of synaptic membranes after induction of long-term potentiation. *J Neurosci* 21: 6245–6251, 2001.
140. Tovar KR and Westbrook GL. The incorporation of NMDA receptors with a distinct subunit composition at nascent hippocampal synapses in vitro. *J Neurosci* 19: 4180–4188, 1999.
141. Tovar KR and Westbrook GL. Mobile NMDA receptors at hippocampal synapses. *Neuron* 34: 255–264, 2002.
142. Triller A and Choquet D. Surface trafficking of receptors between synaptic and extrasynaptic membranes: and yet they do move! *Trends Neurosci* 28: 133–139, 2005.
143. Tsuriel S, Geva R, Zamorano P, Dresbach T, Boeckers T, Gundelfinger ED, Garner CC, and Ziv NE. Local sharing as a predominant determinant of synaptic matrix molecular dynamics. *PLoS Biol* 4: e271, 2006.
144. Vidnyánszky Z, Hámori J, Négyessy L, Rüegg D, Knöpfel T, Kuhn R, and Görcs T. Cellular and subecellular localization of the mGluR5 metabotropicglutamate receptor in rat spinal cord. *NeuroReport* 6: 209–213, 1994.
145. Wang YT and Linden DJ. Expression of cerebellar long-term depression requires postsynaptic clathrin-mediated endocytosis. *Neuron* 25: 635–647, 2000.
146. Winckler B, Forscher P, and Mellman I. A diffusion barrier maintains distribution of membrane proteins in polarized neurons. *Nature* 397: 698–701, 1999.
147. Zhou Q, Xiao M, and Nicoll RA. Contribution of cytoskeleton to the internalization of AMPA receptors. *Proc Natl Acad Sci USA* 98: 1261–1266, 2001.
148. Zhu JJ, Esteban JA, Hayashi Y, and Malinow R. Postnatal synaptic potentiation: delivery of GluR4-containing AMPA receptors by spontaneous activity. *Nat Neurosci* 3: 1098–1106, 2000.
149. Ziff EB. TARPs and the AMPA receptor trafficking paradox. *Neuron* 53: 627–633, 2007.

Scaffold Proteins in the Postsynaptic Density

Mary B. Kennedy [1], Edoardo Marcora [2] and Holly J. Carlisle [3]

[1] Division of Biology, California Institute of Technology, Pasadena, CA 91125, USA, kennedym@its.caltech.edu

[2] Division of Biology, California Institute of Technology, Pasadena, CA 91125, USA, marcora@caltech.edu

[3] Division of Biology, California Institute of Technology, Pasadena, CA 91125, USA, carlisle@caltech.edu

1 Introduction

Many intractable neurological and mental diseases, including epilepsy, depression, and schizophrenia, are believed to result, in part, from derangements of regulation of synaptic transmission in the brain. For this reason, much effort has been made to discover how the delicate mechanisms of signal transduction at synapses lead to modification of synaptic strength. One fruitful area of research over the last twenty years has been the postsynaptic signaling apparatus in glutamatergic spines (8, 97, 99, 106). Spines contain clusters of receptors and signaling proteins located in a dense submembranous structure that can be seen in the electron microscope and is called the postsynaptic density or PSD (For review of early work see 98).

In the 1970's, Philip Siekevitz (40) and Carl Cotman (42) worked out subcellular fractionation methods to purify tiny disks from brain that contain many of the proteins that are most tightly bound to the PSD. Their purification methods were later used to obtain material from which several ubiquitous PSD proteins were identified and sequenced, including the prototypical scaffold protein, PSD-95 (37, 109, 113), synGAP (35), the shank/proSAP family (20), and densin (6). At the same time, and independently, some of these proteins, and additional new ones from the PSD fraction, were identified by their interactions in yeast two hybrid screens, including GKAP (104), synGAP (107), shank/proSAP (158), and homer (24). Continuing proteomic analyses of the PSD fraction or related "NMDA-receptor complexes" (89, 127, 177, 235) have produced a catalog of putative PSD-associated proteins that comprises a few hundred individual proteins. More recent studies have begun to address the crucial question of the relative abundance of each protein in the PSD fraction and their stoichiometric ratios (126, 127, 177).

However, we are still in the early stages of understanding which core protein complexes are present in most excitatory synapses, and which proteins are present only in a subset of synapses, perhaps conferring specialized properties.

J.W. Hell, M.D. Ehlers (eds.), *Structural and Functional Organization of the Synapse*,
DOI: 10.1007/978-0-387-77232-5_14,

The molecular picture of the PSD machinery that has emerged suggests it is constructed to direct biochemical signaling reactions in time and space, and to organize complex mechanisms to modulate biochemical regulation of spine functions at several intermediate steps. The molecular organization of the PSD is maintained by an array of scaffold proteins that bind signaling proteins into spatially restricted complexes, and that also interact with each other to form the "superstructure" of the PSD. Signaling enzymes are positioned to respond appropriately to Ca^{2+} influx through activated NMDA-type glutamate receptors (NMDARs) or to messages generated by activation of metabotropic glutamate receptors (mGluRs). The enzymes can catalyze cascades of reactions that lead to modulation of receptors and channels within the spine, and to insertion or removal of AMPA receptors from the postsynaptic membrane. In this way, the signaling machinery controls changes in synaptic strength such as long-term potentiation or long-term depression (e.g., see Chapter by Lisman and Hell, this volume). The scaffold of the PSD is attached to the larger actin-based cytoskeleton of the spine (146). The signaling enzymes also help to coordinate rearrangements of actin filaments that underlie changes in spine shape accompanying alterations in synaptic strength (34).

2 Major Classes of Postsynaptic Scaffold Proteins

Scaffold proteins of the PSD fall into seven classes (Fig. 1); the MAGUK family (Membrane-Associated Guanylate Kinase-like proteins [e.g. PSD-95]), the GRIP/ABP and GRIP-like family (proteins containing a string of PDZ domains [e.g. PICK]), the GKAP/SAPAP family (Guanylate Kinase-Associated Proteins), the shank/proSAP family, the homer family (EPV domain-containing proteins), TARPs (Transmembrane AMPA receptor Regulatory Proteins), and the cytosolic tails of transmembrane, PSD-specific proteins (the NR2 subunits of the NMDAR and densin).

One useful way to envision the molecular architecture of the PSD is to imagine it as a top-down hierarchy starting from the three major classes of glutamate receptors; NMDARs, AMPA-type glutamate receptors (AMPARs), and mGluRs, and continuing with scaffold proteins that bind directly to these receptors. These "proximal" scaffold proteins include PSD-95 and other members of the MAGUK family, SAP97(134, 157), PSD-93/Chapsyn-110 (28, 103), and SAP102 (119, 156); as well as TARPs. Proximal scaffold proteins, in turn, link to more distal scaffold proteins such as GKAP/SAPAP, shank/proSAP, and the Homer family, which form links among the protein complexes associated with each class of glutamate receptor. At every level (receptors, proximal scaffolds, and distal scaffolds), other effector proteins, regulatory enzymes, and adaptors are recruited into specific signal transduction modules that enable specialized adaptive functions. Although this general organization of the PSD is understood, we still have much to learn about the precise organization of each part of the scaffold, and how the assembly of these parts is controlled. Here, we will consider the specific roles of the scaffold proteins in the context of their interaction with each receptor subtype.

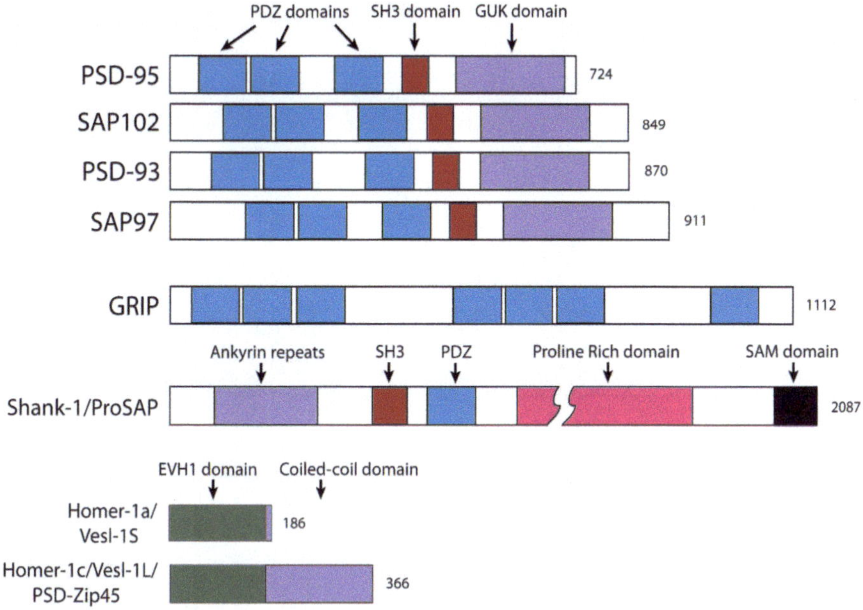

Fig. 1. Domain structures of major families of scaffold proteins in the PSD. Each protein is depicted by a bar scaled to its size. The size in amino acid residues is given at the end of the bar. PSD-95, PSD-93, SAP102, and SAP97 are the four principal MAGUK proteins (*M*embrane-*a*ssociated *gu*anylate *k*inase). Each has three PDZ domains, followed by an SH3 domain and a GK-domain. The figure includes one example of the GRIP/APB family, GRIP, which has six PDZ domains; one example of the shank/proSAP family, shank-1; and two examples of the homer family, a "short" isomer, homer-1a, and a long isomer, homer-1c. Domain structures for the other families are less well understood. TARPs (320 to 420 residues in length) contain four transmembrane domains and are homologous to γ-subunits of Ca^{2+} channels. The domain structures of GKAPs (666 residues in length) and of the tails of the NR2 subunits of the NMDA receptor (approx. 600 residues in length) are not yet well understood.

3 Scaffold Proteins Associated with the NMDA Receptor

The NMDA-type glutamate receptor (see Chapter by Traynelis et al.) is a ligand-gated Ca^{2+}-permeable ion channel that plays a critical role in several forms of synaptic plasticity, including long-term potentiation and long-term depression. It is a tetramer comprised of a pair of NR1 subunits and a pair of NR2 subunits. The NR2 subunits have unusually long cytosolic tails of 500 to 600 amino acids. These tails extend into the cytosol and interact with a variety of scaffold proteins and signaling enzymes. Of the four known NR2 subunits, NR2A and NR2B are the most abundant in forebrain neurons where the protein interactions of the tails have been most thoroughly studied. Individual receptors in forebrain neurons can contain two copies of NR2A, two of NR2B, or one copy of each (50, 86, 147, 148).

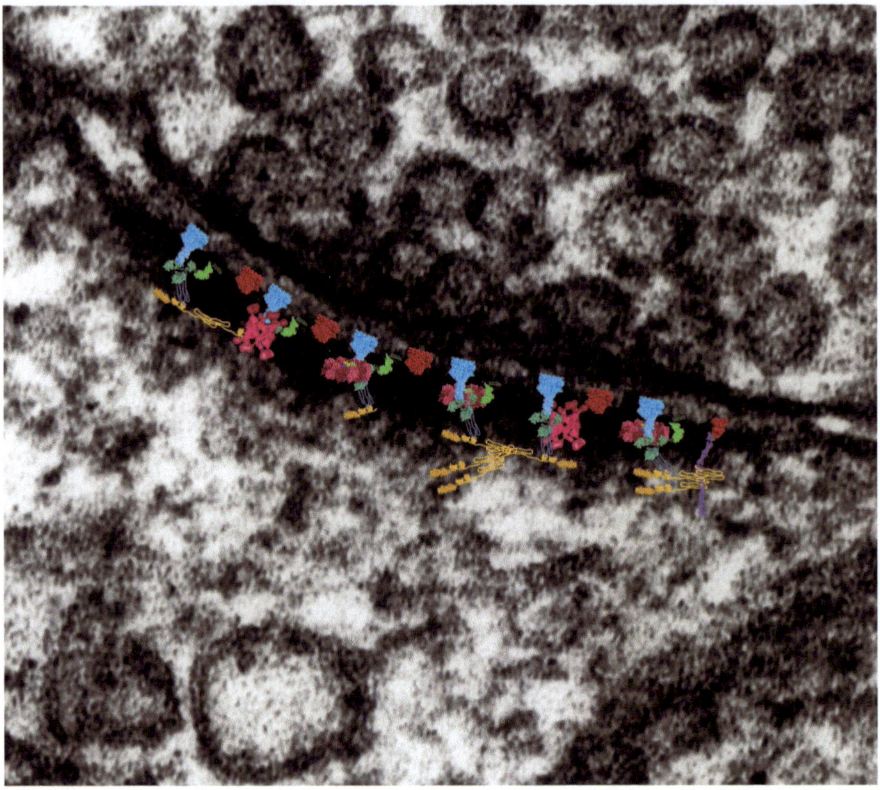

Fig. 2. Scaled cartoon of selected protein machinery in the postsynaptic density. Actual or hypothetical structures of protein complexes are overlaid to scale on an electron micrograph of a glutamatergic postsynaptic density (~450 nm in length) from the hippocampus. Light blue and red orange proteins represent NMDA and AMPA receptors, respectively. The hypothetical structures of the receptors were assembled as described in (150) from X-ray structures of the ATD domain of mGluR1 (1EWV), the ligand binding core complex of NR1/NR2A (2A5T), and the KcsA channel (POA334). The tails of the two NR2 subunits of the NMDA receptor are completely fictional as no information is available about their structures; however, the cartoon structure is drawn approximately to scale to depict the length of the tails with respect to the rest of the receptor. The aqua structure represents PSD-95 assembled from X-ray structures of its various domains (1IU2, 1QLC, 1BE9, and 1TP5). The domains have been slightly separated for clarity; the actual structure is likely more compact then shown here (see 114). The green protein represents synGAP; the GAP domain is that of p120rasGAP (1WER); other domains of synGAP with uncertain structures are depicted as beaded strings. The magenta proteins represent two different versions of the structure of the CaMKII holoenzyme. The flat, disk-shaped structure is based on the X-ray structure of Rosenberg et al. (192) and represents CaMKII in its inactive form. The open, crown-like structure is a lower resolution structure obtained by cryo-electron microscopy (see 112). This structure may represent CaMKII in its fully active form. The gold elongated proteins represent multimers of members of the shank/proSAP family. The amino terminal ankyrin repeats, SH3 domain, and PDZ domain are constructed from homologous X-ray structures of similar domains in other proteins (2PNN,

3.1 The PSD-95 Family

PSD-95 (also known as SAP90) was first identified as an abundant component of the PSD (37, 109). The first hint that it might be a scaffold protein came from the presence in its sequence of three repeated 90 residue domains and an SH3 domain. A fifth domain, now called a GK-domain (guanylate kinase-like) was at first thought to have an enzymatic function; but it has been shown to be enzymatically silent and instead to act as a protein-binding domain (104, 108). The three repeated domains are now called PDZ domains. Indeed, PDZ domains were first defined in PSD-95 (37, 113). A PDZ domain is a consensus protein-binding motif that is found in a wide variety of membrane-associated proteins. They bind specific sequences (e.g. a "tSXV motif") that are usually located at the carboxyl terminus of a membrane protein (114, 164) or they can associate through PDZ-PDZ interactions (164). In NMDARs, the C-terminal consensus sequence S-I/L-E-S-D/E-V in NR2 subunits binds to the first and second PDZ domains of PSD-95 (113). This tSXV motif can also bind to the corresponding PDZ domains of the other MAGUKs, SAP97, PSD-93, and SAP102. The PDZ domains of the PSD-95 family are relatively promiscuous and have been shown to bind several other synaptic proteins. Thus, they serve as "multifunctional" scaffold proteins. It is still not fully understood how the precise associations of PSD-95 and the other MAGUKs are controlled inside cells in situations when many binding partners are available.

Overexpression of PSD-95 family proteins, NMDARs, or other target proteins in heterologous cells induces their clustering (103). Studies of NMDARs expressed with PSD-95 in Xenopus oocytes suggest that PSD-95 can also modulate surface expression and biophysical properties of NMDARs (131, 252). These findings were interpreted to mean that PSD-95 might be important for insertion into the plasma membrane and postsynaptic clustering of NMDARs. However, mutant mice in which the first and second PDZ domains of PSD-95 are deleted have normal levels of surface synaptic NMDARs (154). They do, however display abnormal synaptic plasticity, suggesting that the crucial role of PSD-95 is organization of signaling machinery near the NMDAR.

Interestingly, PSD-95 and its homologues play a role in targeting of AMPA receptors to synapses in hippocampal neurons (36, 122, 199) (see below). When both PSD-95 and PSD-93 are deleted, AMPA-receptor trafficking, but not NMDAR trafficking, is altered in many synapses (12, 59). Taken together these data suggest that

Fig. 2. (Continued) 1SRL, and 1BE9, respectively). The long proline rich domains and terminal SAM domains are depicted as beaded strings. The blue beaded strings represent the approximate size and position of GKAP linking shank with PSD-95. The bright red protein on the right represents one copy of mGluR, assembled from 1EWV, 1EWK, and POA334. The violet proteins represent the long forms of homer shown as dimers that interact with each other via their EVH1 domains. The terminal EVH1 domains are taken from 1DDW and the coiled coil domains are represented by a similar region in actinin (1QUU). The cartoon is meant to provide a picture of the approximate size, relative to the electron micrograph, of each protein and of the complexes they can form in the PSD. The structures are based on data of various kinds, but are still hypothetical.

PSD-95 may be important for control of AMPA receptor trafficking; but its role in relation to NMDA receptors is more complex. The interaction between the NMDAR and PSD-95 appears to be important for proper alignment and interaction of signaling proteins within the spine but it is not crucial for NMDAR clustering at the synapse.

3.1.1 Complexes Formed by PSD-95

The PDZ domains of PSD-95 have been shown to bind to several proteins that are located in spines. However, it is still not known which of these associations actually occurs in vivo. Nor is it clear whether a particular set of associations predominates in the spine, or whether the range of associations shifts depending on regulatory cues. The first and second PDZ domains of PSD-95 can bind to the cytosolic carboxyl termini of the receptor tyrosine kinase ErbB4 (71, 88), semaphorin adhesion proteins (31, 201), the tails of shaker-type voltage-gated K^+ channels (105), and several inwardly rectifying potassium channels (161). The third PDZ domain binds the postsynaptic adhesion protein neuroligin (91, 212) and thus influences synaptic development (185) and retrograde signaling (69). It also can bind to synGAP and kalirin7, both of which influence signaling through small GTPases. Like EphB2, PSD-95 can bind to Src-family kinases and thus may localize and/or regulate them in the PSD (93, 94, 224).

3.1.2 Ca^{2+}-Regulated Enzymes

One important role for PSD-95 is as a scaffold for Ca^{2+}-activated signaling enzymes, placing them in proximity to Ca^{2+} flowing through the NMDAR and, thus, forming a signaling module whose activity is spatially and temporally restricted (64, 97). PSD-95 can bind to neuronal nitric oxide synthase (nNOS) through an atypical interaction with the first and second PDZ domains (27, 38) and to synGAP, an abundant ras-GTPase activating protein in the PSD fraction, through the third PDZ domain (35, 107). Both nNOS and synGAP are regulated by Ca^{2+} entry through the NMDAR (38, 167).

3.1.3 Small GTPases

PSD-95 has been reported to bind to several regulators of small GTPase proteins, including synGAP (35, 107), kalirin7 (179), SPAR (174), and citron (259), suggesting that it may play a significant role in the orchestration of GTPase signaling in the postsynaptic spine.

SynGAP is a ras and rap GTPase that is concentrated in the PSD presumably by virtue of its interaction with the third PDZ domain of PSD-95 (35, 116). Its GAP activity would be expected to increase the rate of inactivation of ras that has been activated by rasGRF (see below), or by postsynaptic TrkB receptors stimulated by BDNF. This activity is enhanced by phosphorylation by CaMKII that is activated by stimulation of NMDARs (167). Deletion of the synGAP gene is lethal in mice a few days after birth, indicating that it is a particularly crucial regulatory protein (110, 111, 234). Mice heterozygous for the synGAP deletion have behavioral learning deficits and defects in generation of LTP (111). Neurons cultured from mouse embryos homozygous for the synGAP deletion show derangements in cytoskeletal regulation (234) and in AMPA receptor trafficking (116, 192). The association of synGAP with PSD-95 is important for regulation of the cytoskeleton, as evidenced by the fact that a mutant synGAP protein with a deletion of its tSXV domain cannot rescue the cytoskeletal defects (234). Taken together, these studies suggest that PSD-95 is important for proper coordination of signaling through small GTPases in the PSD.

Kalirin7, a guanine nucleotide exchange factor (GEF) for the Rho family of small GTPases, associates with the third PDZ domain of PSD-95. By virtue of this association, PSD-95 may help to coordinate regulation by EphB2 of spine morphology and NMDA receptor current. Kalirin7 has been shown to regulate formation and maintenance of dendrites and spines during development (138, 179, 180, 187). When it is activated by binding of ephrin to the EphB2 receptor, kalirin7 activates Rac which then activates PAK1, an enzyme that controls the actin cytoskeleton. During development, activation of PAK1 in dendrites leads to formation of mature dendritic spines (178). Activation of EphB2 also triggers an independent structural rearrangement by promoting association between EphB2 itself and the NR1 subunit of the NMDAR (45), as well as, with postsynaptic PDZ domain-containing scaffold proteins such as GRIP/ABP (51, 215) and PICK1 (227). In this way, activated EphB2 is recruited to the postsynaptic membrane. There it, in turn, recruits and activates members of the Src family of non-receptor tyrosine kinases which phosphorylate NR2 subunits of the NMDA receptor, resulting in an increase in conductance of Ca^{2+} upon glutamate stimulation (221, 237, 256).

The GK-domain of PSD-95 can bind directly to a spine-associated Rap-GAP termed SPAR and recruit it into the NMDAR complex. SPAR also binds to actin and contributes to regulation of the actin cytoskeleton of the spine by down-regulating the activity of Rap (174). It also acts as a scaffold through its association with the master scaffold protein shank/proSAP (discussed below) via an intermediate protein (proSAPiP1) (241).

Finally, the third PDZ domain of PSD-95 can associate with citron-N, a splice variant of citron-kinase that lacks the kinase domain, and is situated in neuronal cell bodies, as well as concentrated at glutamatergic synapses in inhibitory neurons in the hippocampus and cortex and excitatory neurons in the thalamus and basal ganglia (68, 258, 259). Citron-N is a target of activated Rho and participates in organization of the Golgi apparatus in neurons (32). It binds to the Golgi apparatus, and controls actin filaments locally by assembling a complex of the Rho effector ROCK-II and the neuron-specific actin-binding protein profilin-IIa. The precise role of citron-N at synapses is still unknown. However, mutations in the gene encoding citron were recently implicated in genetic predisposition toward affective disorders (137).

3.1.4 AMPA Receptors

As we will discuss in a later section, the PDZ domains of PSD-95 and its homologues, including SAP97 can associate with complexes surrounding AMPA receptors.

3.1.5 The Guanylate Kinase-like (GK) Domain and Linkage to the Shank/proSAP Scaffold

The C-terminal GK-domain of PSD-95 and its homologues also acts as a specific protein interaction site. One important family of PSD proteins called GK-associated proteins (GKAPs) (104, 198, 222) bind to the GK-domain and link the PSD-95 scaffold to members of the shank/proSAP family (21, 158). Shank proteins are large scaffold proteins that contain a series of protein interaction motifs and lack identified catalytic domains. The single PDZ domain of shank binds to the C-terminus of GKAP. Shank proteins interact with a series of other scaffold proteins linking together protein complexes associated with all three major classes of glutamate receptors.

The protein kinase scaffold AKAP-79/150 binds to the SH3 and GK-domains of PSD-95 or SAP97 (41). AKAP-79/150 contains binding sites for cAMP-dependent protein kinase, C-kinase, and the Ca^{2+}-regulated phosphatase, calcineurin. Its association with MAGUKs may help to properly position these enzymes near the postsynaptic membrane.

How and when each of these possible complexes is formed at different synapses, and how they are regulated, are subjects for future study.

3.1.6 Dynamics of Association of PSD-95 with the PSD

Because PSD-95 and the other MAGUK proteins are capable of a large number of high affinity associations with synaptic proteins, it is easy to imagine that individual PSD-95 proteins remain stably bound within a particular PSD once they arrive there. However, one study of tagged PSD-95 proteins in vivo reveals quite a different situation. Although the size of individual PSDs is generally stable, individual PSD-95 proteins exist in a dynamic equilibrium moving rapidly among neighboring spines with an average retention time in a single spine of 1–2 hr (78). The behavior suggests that individual spines compete for limiting amounts of PSD-95 and that larger PSDs contain more PSD-95 binding sites. This rapid dynamic equilibrium may permit correspondingly rapid remodeling of individual PSDs in response to biochemical changes in PSD-95 binding proteins triggered by synaptic activity.

Structural studies of PSD-95 itself suggest possible regulatory mechanisms that could control the availability of its binding sites. The SH3 domain of PSD-95 interacts intimately with the GK domain such that a portion of the GK domain is necessary to complete the SH3 fold (149, 150, 223). Point mutations that disrupt the interaction between the two domains interfere with the ability of PSD-95 to cluster potassium channels in heterologous cells, suggesting the conformation and/or availability of PDZ domains has been altered (209). Furthermore, patch analysis of the tertiary structure of PSD-95 suggests that it can exist in two distinct stable conformations; one that obscures the binding cleft of the third PDZ domain and another that exposes it (112). Future studies of regulation of the conformation of PSD-95 may reveal new mechanisms that contribute to control and remodeling of signaling complexes in the spine.

3.2 Direct Association of Signaling Proteins with the Tails of NMDA Receptor Subunits

The structure of the NMDA receptor is unusual in several ways. One of its most striking structural features is the extended carboxyl termini of the NR2 subunits. These tails comprise approximately 600 residues and extend into the cytosol where they interact with PSD-95 (113); but also bind directly to the signaling enzymes CaMKII (11) and rasGRFI (115). The tails can also associate with the actin-associated protein, α-actinin (248), and the scaffold protein yotiao which binds protein phosphatase 1 (PP1) and protein kinase A (PKA) (242). Thus, these tails appear to serve a scaffolding function of their own that helps to arrange the NMDA receptor signaling complex.

3.2.1 CaMKII

The most thoroughly studied enzyme that binds to the tail of NR2 subunits is Ca^{2+}/calmodulin-dependent protein kinase II (CaMKII) (14, 190). It is one of several critical determinants of synaptic plasticity that are activated by influx of Ca^{2+} through the NMDAR (65, 99, 132, 193, 261). Its positioning near the NMDAR would place it in an ideal location to influence synaptic plasticity.

CaMKII has an elaborate structure. Individual catalytic subunits are activated by the Ca^{2+}-bound form of calmodulin (Ca^{2+}/CaM), which binds to a single CaM-binding domain within each subunit, and in so doing opens up the catalytic site so that it is able to interact with substrate proteins (14, 15, 176). The catalytic subunits assemble into a large holoenzyme made up of 12 catalytic units held together by interactions among association domains in the carboxyl third of their sequence (14, 189, 190). The holoenzymes are randomly assembled from two closely related catalytic subunits termed α and β (14, 29, 100). In the forebrain, α-subunits predominate over β-subunits by a ratio of 3 to 1. When all of the subunits are in an inactive state, the holoenzyme takes the form of six dimers because individual pairs of catalytic subunits bind to each other by a coiled coil interaction between their calmodulin binding domains (190). Activation of subunits by binding of Ca^{2+}/CaM is accompanied by autophosphorylation of a key residue in each subunit, threonine-286. Phosphorylation of this site prevents the catalytic site from closing when CaM later dissociates; thus, it holds the subunit in an active conformation (155). In order for this autophosphorylation to occur, Ca^{2+}/CaM must be bound to two neighboring subunits, one that acts as the "kinase" and a second that acts as the "substrate" (79). Based on their crystal structure, Rosenberg et al. (190) proposed that binding of Ca^{2+}/CaM dissociates the members of an inactive dimer pair, exposing their active sites and permitting autophosphorylation to occur between them.

CaMKII is an abundant constituent of the PSD fraction (101) which implies that the PSD contains many sites at which it can bind. In turn, each holoenzyme likely contains at least 12 possible sites that could bind to docking proteins in the PSD. CaMKII is an abundant protein in the forebrain, constituting ~ 1–2% of the total protein content (60). Therefore, it is likely to be present in excess of its individual binding sites in the PSD. At least three proteins in the PSD have been identified as potential docking sites for CaMKII; the tails of the NR2 subunits of the NMDA receptor, the NR1 subunit of the NMDA receptor, and densin, a sialylated transmembrane protein that has a short cytosolic domain and is highly enriched in the PSD (6). Three other proteins located near the PSD also bind CaMKII. α-actinin binds to CaMKII directly and forms a stable ternary complex with CaMKII and densin (236), and also with CaMKII and NR2B (121). F-actin, which makes up the spine cytoskeleton, binds the less abundant β-subunit of CaMKII, but does not bind the more abundant α-subunit (168, 205). Finally, the PDZ-domain-containing scaffold protein MUPP-1 has been reported to bind directly to CaMKII (116). The best studied of all of these potential interactions is the one with the tails of the NR2A and NR2B subunits.

The two most prominent NR2 subunits in the forebrain, NR2A and NR2B, both bind CaMKII at sites located in their long cytosolic carboxyl termini (73, 123, 216). Binding to either one requires activation of CaMKII by Ca^{2+}/CaM; however, the binding sites in NR2A and NR2B are not identical. The tail of NR2B binds CaMKII with high affinity at two distinct sites (11, 217); a site distal to the membrane (residues 1259–1310) that requires binding of Ca^{2+}/CaM to CaMKII, and a more proximal site (residues 829–1120) that only binds CaMKII after autophosphorylation at Thr286 (11, 200). Binding to the distal site stabilizes the activated form of CaMKII (11). However, phosphorylation by CaMKII of Ser1303 on NR2B reduces the affinity of the distal site and leads to slow dissociation of the kinase from NR2B (217). The tail of NR2A binds CaMKII with a lower affinity than NR2B. Binding to a site in NR2A between residues 1244 and 1464 is enhanced by activation by Ca^{2+}/CaM and by autophosphorylation (74). Phosphorylation by protein kinase C of Ser1416 in NR2A inhibits binding of CaMKII and promotes dissociation of the CaMKII/NR2A complex, providing a mechanism by which activation of mGluRs could decrease binding of CaMKII to the NMDA receptor (72). These interactions are likely important for appropriate positioning of CaMKII in the spine, although their precise functional roles, in particular, the significance of the difference in binding bwtween NR2A and NR2B, have not yet been dissected.

The NR1 subunit, which contains a 30 residue membrane proximal region termed C0 in its short cytosolic tail (121), can bind to CaMKII. Like the binding to the membrane proximal domain of the tail of NR2B, binding to NR1 requires autophosphorylation of CaMKII at Thr286. Both α-actinin and Ca^{2+}/CaM also bind to the C0 domain, and binding of CaMKII is competitive with their binding (121). Binding of α-actinin has been shown to increase the open probability of the NMDAR. For this reason, it has been suggested that the competitive binding of autophosphorylated CaMKII may contribute to Ca^{2+}-dependent inactivation of NMDARs (117, 257).

Translocation of CaMKII into spines following stimulation of glutamate receptors has been observed with the use of GFP-labeled CaMKII transfected into cultured neurons. In young, cultured hippocampal neurons, pharmacological stimulation of NMDARs caused a rapid, transient redistribution of CaMKII from the actin cytoskeleton to synaptic sites (203, 204). In another paradigm, chemical induction of LTP in hippocampal slices by application of forskolin in medium containing no Mg^{2+} caused translocation of CaMKII into spines (172). Hypoxia can also cause accumulation of CaMKII in the PSD (53, 218), so there was some doubt about whether physiological stimulation in vivo produces translocation of CaMKII to the PSD. However, translocation under normal physiological circumstances in vivo was reported in living zebrafish (76) where repeated sensory stimulation produced reproducible and reversible translocation of GFP-CaMKII to the PSD in an identified interneuron in the sensorimotor circuit.

The complex regulation of localization of CaMKII in the spine observed in cells and in the PSD, studied biochemically, illustrates how scaffold proteins may help to dynamically configure the location of regulatory proteins in the spine. For example, binding of Ca^{2+}/CaM to CaMKII causes its release from F-actin (168) and from the scaffold protein MUPP1 (116). On the other hand, binding of Ca^{2+}/CaM and autophosphorylation dramatically increase binding of CaMKII to the NMDAR. Thus, Ca^{2+} influx into the spine would be predicted to promote release of CaMKII from the spine actin cytoskeleton and subsequent binding to the NMDAR (2, 152). As another example, binding of CaMKII to the NMDAR or to densin is facilitated by concomitant binding of these proteins to α-actinin (121, 236). However, densin has a much greater affinity for the α-subunit of CaMKII than for the β-subunit (236); whereas, the NMDAR does not show this selectivity. The α-subunit, and not the β-subunit, can be synthesized from mRNA in dendrites in response to synaptic stimuli that induce potentiation (30, 173). Thus, densin in the PSD may preferentially concentrate holoenzymes of CaMKII that are newly synthesized in dendrites.

3.2.2 RasGRF1

Activation of ras is critical for regulation of insertion and removal of AMPA receptors in the spine (260) and for activation of the protein kinase cascade leading to activation of MAP kinases Erk 1 and 2 (219). Ras can be activated at synapses via several different receptors, including the Trk receptor protein tyrosine kinases. However, rasGRF1, a ras-specific GDP/GTP exchange factor (GEF) that is regulated by Ca^{2+}/CaM, provides a means by which activation of the NMDAR can directly activate ras and its downstream pathways (62, 80, 115). RasGRF1 has been shown to bind to the cytosolic tail of the NR2B subunit of the NMDA receptor (115). This binding would locate rasGRF1 near the mouth of the NMDAR and thus enable it to efficiently sense a local rise in Ca^{2+} concentration around the activated receptor. Learning and memory are impaired in rasGRF1 knockout mice (25, 75), confirming the importance of this protein in synaptic plasticity. Furthermore, disruption of the interaction between rasGRF1 and the NMDA receptor by introduction of a specific peptide blocker into neurons disupts activation of the ERK1/2 pathway following application of NMDA (115).

The finding that rasGRF1 associates specifically with the NR2B subunit of the NMDAR, and not with NR2A, suggests that the presence of rasGRF1 might be an important difference between signaling complexes formed near receptors containing NR2A subunits and those containing only NR2B. There may be competition between rasGRF1 and CaMKII for binding to NR2B. The binding sites for the two proteins on NR2B overlap extensively; however, it hasn't yet been demonstrated that they compete for binding. SynGAP, a ras inactivating GTPase that binds to PSD-95 is stimulated by phosphorylation by CaMKII that is activated by Ca^{2+} flux through the

NMDA receptor. The relative timing of activation of rasGRF and synGAP would determine the time course of ras activation near NMDARs. It will be interesting to learn how interactions among the scaffold proteins associated with the NMDAR in the PSD regulate this timing. Since ras can drive insertion of AMPA receptors into the synaptic plasma membrane, and also leads to changes in the actin cytoskeleton through activation of rac, the answer will have implications for the orchestration of activity-dependent synaptic plasticity in the spine.

4 Scaffold Proteins Associated with AMPA-Type Glutamate Receptors

AMPARs mediate rapid excitatory glutamatergic transmission in the brain. They are composed of four homologous subunits (GluR1-4, also called GluRA-D) which assemble in combinations to form the distinct AMPAR subtypes (50, 86, 147, 148). Their trafficking to, insertion into, and removal from the postsynaptic membrane is tightly regulated at all stages of development. In mature neurons, this regulation underlies activity-dependent changes in synaptic strength such as LTP and LTD (7, 9, 16, 26, 43, 82, 92, 128, 140, 141, 145, 175, 207, 208, 210). In pyramidal neurons of the hippocampus, where their regulation is most often studied, AMPARs are primarily hetero-oligomers of GluR1 and 2 subunits (GluR1/2) or of GluR 2 and 3 subunits (GluR2/3). The insertion of GluR1/2 AMPARs into the postsynaptic membrane is regulated by synaptic activity (207); whereas, GluR2/3 AMPARs cycle constitutively in and out of the postsynaptic membrane. The detailed mechanisms of AMPAR trafficking are discussed in the Chapter by Esteban. Here we focus on interactions of GluR subunits with scaffold proteins that contribute to their localization at the postsynaptic membrane.

Five scaffold proteins interact directly with GluR subunits and participate in regulation of AMPAR trafficking: Stargazin (a TARP protein), GRIP (GluR-interacting protein)/ABP (AMPAR-binding protein), PICK1 (protein interacting with C kinase 1), and NSF (N-ethylmaleimide-sensitive factor). PSD-95 participates indirectly in trafficking of AMPARs through its interaction with Stargazin. In addition, PSD-95 binds directly to the GluR6 and KA2 subunits of kainate receptors (70, 151).

4.1 GRIP/ABP, PICK1, and NSF

Certain scaffold proteins bind selectively to the GluR2/3 AMPARs and are believed to regulate their insertion into, and removal from, synaptic sites (133). Two of these, the homologous proteins GRIP (GRIP1) (51) and ABP (GRIP2) (52, 213, 249) contain a total of six to seven PDZ domains and no catalytic domains. The cytosolic tail of GluR2 can bind to the fourth or fifth PDZ domain of either protein. Both GRIP and ABP are palmitoylated. The palmitoylated form is targeted to the postsynaptic membrane; whereas, the nonpalmitoylated form is targeted to intracellular membranes (48, 253). Thus, regulated palmitoylation of GRIP and ABP is believed to anchor AMPARs to one or the other of these membrane compartments to modulate AMPAR trafficking (23, 47, 48, 66, 85, 102, 169, 202). However, disruption of the interaction between the AMPAR and GRIP/ABP interferes more severely with synaptic localization of the receptors than with internal trafficking (51). Thus, we still have an incomplete understanding of the role of GRIP and ABP in receptor localization.

A third scaffold protein that binds to GluR2 is PICK1, which contains a PDZ domain that binds selectively to protein kinase C-α (PKCα) (214) and a BAR domain that binds preferentially to curved membranes (182). PICK1 and GRIP/ABP appear to work together to regulate trafficking and surface expression of GluR2/3 receptors. In addition to PKCα, the single PDZ domain of PICK1 also associates with the cytosolic tail of the short forms of GluR2 and GluR3 subunits (49). Through formation of homomultimers, PICK1 is believed to position PKCα near the tail of GluR2 (39, 181) where it can phosphorylate GluR2 on serine-880 (144, 181). This phosphorylation releases GluR2 from GRIP/APB, but leaves it bound to PICK1.

Dissociation of GluR2 from GRIP/ABP appears to be required for both insertion and removal of AMPARs at synaptic sites (23, 47, 66, 85, 102, 169, 202). The proposed mechanism begins with PICK1 promoting dissociation of GluR2 from GRIP/ABP when it binds to the GRIP/ABP/GluR2 complex and induces phosphorylation of GluR2 on serine-880 by PKC. As GRIP/ABP unbinds from the phosphorylated form of GluR2, PDZ domains on PICK1 multimers bind to the freed PDZ-binding motif on GluR2. The BAR domain on PICK1 then promotes association of GluR2 with budding endo- and exocytotic vesicles that mediate trafficking of AMPARs to and from the postsynaptic membrane. This mechanism is consistent with the findings that PICK1 promotes both internalization and recycling of GluR2 (133), and that disruption of interaction of GluR2/3 with PICK1 or GRIP/ABP impairs expression of LTD (102, 202).

Although they are not strictly scaffold proteins, NSF (N-ethylmaleimide sensitive fusion protein) and the clathrin adaptor AP2 interact with the cytosolic tail of the GluR2 subunit at a membrane proximal region different from the C-terminal PDZ-binding domain that interacts with GRIP/ABP and PICK1. They are required for regulation of AMPAR trafficking and surface expression and thus they also have been shown to be important for normal expression of synaptic plasticity, in particular LTD (120, 136, 163, 170, 211).

Scaffold proteins generally interact with several proteins, sometimes at the same sites. In this regard, PSD scaffold proteins are no exception. In addition to the AMPAR, GRIP/ABP binds to the cytosolic tail of the receptor tyrosine kinase EphB2 (87), the scaffold protein Liprin-α (247), and the neuronal ras-GEF GRASP-1 (255). Interactions with these proteins likely help to regulate other, less well understood, aspects of postsynaptic function.

4.2 SAP97 and SAP102

The PSD-95 homologues, SAP97 and SAP102, have functions distinct from, but related to, those of PSD-95. SAP97 binds specifically to the tail of the GluR1 subunit of AMPARs (122, 197). It can form a trimeric complex with GluR1 and the motor protein myosin VI (244). The interaction of GluR1 and SAP97 occurs early in the biosynthetic pathway of the receptor, either in the endoplasmic reticulum or cis-Golgi apparatus; on the other hand, few synaptic receptors are associated with

SAP97 (197). These results suggest that SAP97 may mediate trafficking of GluR1 to the synapse via a myosin VI motor pathway. SAP102 associates with the tails of NMDA receptor subunits, but is most prominently expressed early in development (184). In mice, the levels of SAP102 decrease during the second week after birth, while levels of PSD-95 increase. Thus, SAP102 may play an important role in transport or targeting of NMDARs in the early stages of synaptic development.

4.3 Stargazin and PSD-95

Stargazin is a member of a family of proteins, collectively referred to as TARPs. The family comprises Stargazin (also called γ-2) and three homologous proteins (γ-3, γ-4, and γ-8), all of which are transmembrane auxiliary subunits of AMPARs (162, 171, 225). Stargazin/γ-2 is structurally related to the γ-1 subunit of the voltage-dependent calcium channel (VDCC) and was initially believed to be a calcium channel subunit (124). It was soon recognized that the TARPs are not part of the VDCC; rather they bind to subunits of the AMPA receptor (36, 226) and influence their trafficking, targeting, and biophysical properties (67, 159, 233).

TARPs contain a C-terminal consensus binding motif that binds selectively to the first two PDZ domains of PSD-95 (67). This interaction is required for activity-dependent translocation of AMPA receptors to postsynaptic sites (199). Disruption of the synaptic localization of PSD-95 dramatically reduces the number of synaptic AMPARs (58). In contrast, when PSD-95 is over-expressed the number of synaptic AMPARs increases (199), producing an enhancement of AMPA receptor-mediated synaptic transmission that occludes LTP and increases the amplitude of LTD (215). These findings suggest that interaction of AMPAR/TARP complexes with PSD-95 recruits extrasynaptic AMPAR/TARP complexes to postsynaptic sites in order to increase synaptic strength (57). The mechanisms of AMPA receptor trafficking and its role in synaptic plasticity are discussed in more detail in Chapters by Esteban, Choquet and Triller, and Lisman and Hell.

5 Protein Complexes Associated with the Metabotropic Glutamate Receptor Through the Scaffold Protein Homer

Glutamate activates two groups of G-protein coupled metabotropic receptors that modulate neuronal excitability and synaptic strength. Group I mGluRs, mGluR1 and mGluR5, regulate phospholipase C which produces the second messengers IP3 and diacylglycerol. Group I mGluRs are concentrated at the periphery of the PSD (10, 135, 165) and can influence synaptic plasticity (22). Their linkage to the PSD is mediated by the homer family of scaffold proteins (24, 250). The family comprises homer1, 2, and 3, each of which is alternatively spliced into long and short isoforms (termed "a" isoforms) (95, 250). All of the homers contain an N-terminal EVH1 domain which binds to a proline-rich motif (PPXXF) in mGluRs and in several other proteins (13). The long homer isoforms contain a C-terminal coiled-coil (CC) domain which promotes their self-association into tetramers (81, 220). In contrast, the short isoforms do not contain the CC domain and thus are monomeric (95). The short homers, for example homer1a, were originally discovered as "immediate early genes" whose expression is rapidly induced by synaptic activity (24). In contrast, long homers, for example homer1b/c, are constitutively expressed (61, 251).

5.1 Linkage to the IP3R and Shank/proSAP

In addition to group 1 mGluRs, the EVH1 domain of homer binds to the type 1 inositol trisphosphate receptor (IP3R) (229) and shank, a PSD scaffold protein (228). Because the long homer isoforms can form tetramers, they act as hubs that cross-link the different EVH1 domain-interacting proteins (228, 229). In contrast, the short monomeric homers antagonize this function and thus act as endogeneous "dominant negatives", breaking up linkages between mGluRs and other EVH1-interacting proteins when they are rapidly and transiently expressed upon synaptic stimulation (24, 96).

An interesting example of this dynamic is the coupling between group 1 mGluRs and IP3Rs (229). The IP3R is an IP3-sensitive Ca^{2+} channel localized in the smooth endoplasmic reticulum (SER). Activation of group 1 mGluRs results in production of the second messenger IP3. IP3 binds to the nearby IP3R causing it to open and release Ca^{2+} from the SER, which functions as an intracellular calcium store. The mGluR/homer/IP3R complex thus couples synaptic activation of group 1 mGluRs to release of Ca^{2+} from SER located in dendritic spines (196). In the electron microscope, the SER of dendritic spines can be seen juxtaposed to the postsynaptic membrane at the periphery of the PSD where group I mGluRs and homer are located (250). Activity-induced expression of short isoforms of homer apparently interferes with the coupling between mGluRs and the IP3R, because over-expression of a short homer impairs release of intracellular calcium evoked by quisqualate, an mGluR agonist (229). Binding of the short form homer1a to receptor isoforms mGluR1a and mGluR5a and b also induces constitutive receptor activation (5). When the long homer3 is in excess, it masks the constitutive activity conferred by homer1a (5). Thus, expression of homer1a would be expected to reduce release of Ca^{2+} from intracellular stores following activation of mGluRs; but, for some receptor subtypes, increase production of IP3 and diacylglycerol.

6 The Shank/proSAP Scaffold of Scaffolds

The distribution of scaffold proteins in the PSD, measured by immunoelectron microscopy, suggests a laminar organization (183, 232). The peak concentration of PSD-95 occurs at an average distance of 12 nm from the synaptic plasma membrane; whereas, the peak of GKAP-shank was detected at an average distance of 24–26 nm from the membrane. Cript and dynein, which may help to link PSD proteins to cytoskeletal elements, lie even further away, 29–31 nm from the membrane. The shank/ProSAP family of scaffold proteins (20, 21, 129, 158, 228, 254) acts as a scaffold of scaffolds, linking together PSD-95-associated proteins, homer-associated proteins, and the actin cytoskeleton (8, 19, 206).

6.1 Shank/proSAP Multimerization

Shank forms homomultimers through interactions of its SAM domains with each other and with sharpin (130, 158). Its single PDZ domain can also dimerize in an anti-parallel configuration, contributing to multimerization (90). The SAM domains form sheets of helical rods in vitro suggesting that shank multimers may form a platform that stabilizes the organization of the PSD (8). By virtue of shank's rich collection of protein-interaction domains (6-7 ankyrin repeats, an SH3 domain, a PDZ domain, a proline-rich region) this two-dimensional protein lattice could likely accommodate several ligands including homer and GKAP/SAPAP, which links to PSD-95 (8). The shank platform could thus cross-link the NMDA and metabotropic glutamate receptor complexes. It may also link to AMPAR complexes either via direct binding to the GluR1 subunit (230) or through interactions between the TARP proteins and PSD-95 (162).

6.2 Interaction with Homer

The physical interaction between group 1 mGluRs and the shank scaffold is mediated by the long isoforms of homer (56). This structural coupling may underlie functional coupling between group 1 mGluRs and the NMDAR (166), in which mGluR1 can upregulate current through the NMDAR by stimulating phosphorylation of the receptor by the proline-rich tyrosine kinase 2 (Pyk2) and Src-family kinases that are part of the NMDAR complex (83). The long isoforms of homer and shank also appear to cooperate in morphogenesis of dendritic spines. Accumulation of shank in synapses depends on its ability to bind to Homer (194) and also to GKAP/SAPAP (158). When over-expressed alone or in combination, the long isoforms of homer and shank synergistically increase the number and size of spines (195). As expected, overexpression of the short isoforms of homer have the opposite effect on spine morphogenesis, reducing synaptic targeting of shank and other PSD proteins, and inhibiting synaptic transmission (194).

6.3 Interaction with the Actin Cytoskeleton

Through its interactions with cortactin, Abp1 and spectrin/fodrin, the shank scaffold mediates binding of the PSD to the underlying spine actin cytoskeleton (18, 158, 186) and thus facilitates a functional connection between activation of postsynaptic receptors and changes in the spine cytoskeleton, as discussed below.

7 Attachment of the Cytoskeleton to the PSD by Actin-Associated Proteins

Dendritic spines contain a higher density of actin filaments than any other neuronal structure, and the arrangement of these filaments gives spines their characteristic shapes (63, 143). The narrow neck contains longitudinally bundled filaments, whereas the head contains a lattice-like mesh that supports its bulbous shape (118).

In mature synapses, the actin cytoskeleton immobilizes some, but not all, components of the PSD. For example, treatment of cultured neurons with latrunculin A, a toxin that causes depolymerization of actin by sequestering G-actin, results in a decrease in spines of AMPA and NMDA receptors, α-actinin-2, drebrin (an actin binding protein that can cluster actin), and CaMKII. Thus, the attachment of F-actin to the PSD contributes to immobilization of receptors and other proteins at the synapse and helps to regulate their function. In contrast, PSD-95 remains in spines even after the depolymerization of actin (3, 4), suggesting that PSD-95 may form a stable core scaffold that does not require the cytoskeleton for its aggregation.

Three actin binding proteins physically connect the cytoskeleton to the PSD: α-actinin, cortactin, and spectrin.

7.1 α-Actinin

α-Actinin is an actin binding protein that can cross-link actin filaments (153). The α-actinin family comprises 4 members (α-actinin1-4), of which isoforms 1, 2, and 4 are expressed in the brain (235, 236, 246). Proteins of the α-actinin family contain an actin-binding domain at the N-terminus, four spectrin repeats in their central region (153), and a Ca^{2+}-binding EF hand motif (243) and PDZ domain-binding sequence at the C-terminus (236). α-Actinin self-associates into antiparallel dimers that can bind actin at both ends, which gives it the ability to crosslink F-actin. Overexpression of α-actinin-2 has been shown to increase the length and density of dendritic protrusions in cultured hippocampal neurons (160).

α-actinin also plays an important role in the localization and function of NMDA receptors. α-actinin-2 colocalizes with NMDA receptors in spines and binds directly to both the NR1 and NR2 subunits (55, 248). This interaction is antagonized by both calcium-bound CaM and autophosphorylated CaMKII (121, 248). Interestingly, dissociation of the NMDA receptor from F-actin causes rapid inactivation of the channel (248, 257), while association with the cytoskeleton enhances NMDA receptor activity (191). Thus, NMDA receptors are probably tethered to F-actin under resting conditions but dissociate and become inactivated in response to calcium influx through the channel.

Another possible point of attachment in the PSD for F-actin via α-actinin is densin. The conserved C-terminal region of α-actinin-4 has been shown to bind to the PDZ domain of densin. α-actinin-4 can also bind directly to CaMKII; whereas, densin binds both α-actinin-4 and CaMKII. Thus, the three proteins form a highly stable ternary complex. Autophosphorylation of CaMKII increases its affinity for densin, indicating that formation of the complex probably depends on the level of intracellular calcium (236).

7.2 Cortactin

Cortactin drives de novo nucleation of actin filaments and creates branch points on preexisting filaments by activating the ARP2/3 complex. In addition to promoting extension of actin filaments, cortactin stabilizes existing F-actin. Cortactin contains an acidic domain at its N-terminus, as do other actin nucleation promoting proteins, including WASP, N-WASP, and SCAR/WAVE; all of which interact with the ARP2/3 complex (231, 238, 240, 245). The central region of cortactin comprises 6.5 tandem copies of a 37 amino acid repeat that are the sites of interaction with F-actin (245). Its C-terminus contains a helical domain, a proline-rich domain, and an SH3 domain (46). The SH3 domain of cortactin interacts with a PSD protein originally named CortBP1 (54), which is a splice variant of shank2/proSAP-1. This interaction has been shown to physically link the PSD scaffold to the cytoskeleton (54, 158).

Cortactin is highly enriched in spines where it colocalizes with F-actin (84, 188). Electron microscopic studies of immunogold labeled cortactin show that approximately 10% of cortactin in the spine is located in the PSD, while the bulk of the staining occurs within the core of the spine, with the peak average concentration 100–150 nm away from the PSD (188). This finding indicates that there are at least two pools of cortactin in the spine, a small one interacting with the PSD, and a larger one interacting with the actin cytoskeleton of the spine. Interestingly, NMDA receptor activation induces translocation of cortactin from spines into the dendrite (84), suggesting that movement of cortactin might be important for activity-dependent remodeling of the spine cytoskeleton.

As predicted, the level of cortactin is a strong determinant of spine morphology in cultured hippocampal neurons. Knockdown of endogenous cortactin in mature cultured hippocampal neurons by siRNA drastically reduces the total number of dendritic protrusions (84). This result means that cortactin is required for the maintenance of mature spines perhaps because it stabilizes filaments or promotes overall growth of the actin network. Overexpression of cortactin in cultured neurons produces abnormally long spines, similar to those observed when α-actinin is overexpressed. This elongation requires both the N-terminal region of cortactin that interacts with F-actin and the Arp2/3 complex (84).

7.3 Spectrin/fodrin

Spectrin is the main component of the cytoskeleton that underlies cell membranes and gives them their shape. It was first identified in erythrocytes where it forms a filamentous network required for red blood cells to maintain their shape and elasticity (142). The spectrin family of proteins includes α- and β-spectrin and α- and β-fodrin. Fodrin is the predominant form of spectrin in the brain and is also referred to as spectrin II or brain spectrin. The α and β subunits of fodrin differ in their domain composition: the α subunit contains 21 triple helical repeats with a segment containing an SH3 domain in the 10th repeat, while the β subunit contains only 17 triple helical repeats and does not contain an SH3 domain. The β subunit also contains an actin-binding region at its N-terminus and some splice variants of β subunits contain a PH domain. Within the membrane cytoskeleton, the α and β subunits typically form heterotetramers by first forming antiparallel heterodimers which then associate head to head to form a tetramer. Fodrin connects the actin cytoskeleton to the cell membranes by interacting with transmembrane proteins such as ankyrin (44, 77).

In spines, fodrin is also present in the PSD where it binds directly to NR1, NR2A and NR2B subunits of NMDA receptors in an activity-dependent manner (33, 125,

239). The interaction between fodrin and the NMDAR is regulated by calcium and by various protein kinases. Such regulation provides a potential mechanism by which the motility of the NMDA receptor in the membrane may be regulated by synaptic activity. CaM inhibits binding between fodrin and NR1 and Ca^{2+} antagonizes binding between fodrin and NR2B (239). Binding of fodrin is also inhibited by phosphorylation of the NR1 C-terminal domain by PKA and PKC and the NR2A C-terminal domain by the tyrosine kinase fyn (239). In the neuromuscular junction, removal of actin and spectrin from myotube membranes results in dispersal of AChRs (17). By analogy, fodrin may be important for immobilizing receptors at the synapse (1).

8 Conclusion

The postsynaptic density provides a quintessential example of the organization of signaling proteins to provide specificity of regulatory control. Three classes of proteins; the PSD-95 family of MAGUK proteins, the homer family, and the shank/proSAP family, have evolved to provide a highly organized structural framework to which signaling molecules can bind to carry out their functions. Future studies will allow us to understand how this scaffold structure provides for accurate regulation of synaptic plasticity, and how the structure is maintained and sculpted in a dynamic equilibrium.

Acknowledgements

We thank the members of the Kennedy laboratory for many useful discussions. This work was supported by P.H.S. grants NS44306, NS17660, and NS028710 (MBK), and NS047894 (HJC); the Hereditary Disease Foundation (EM) and the Huntington's Disease Society of America (EM, MBK).

References

1. Adam G and Matus A. Role of actin in the organisation of brain postsynaptic densities. *Brain Res Mol Brain Res* 43: 246–250, 1996.
2. Ahmed R, Zha XM, Green SH, and Dailey ME. Synaptic activity and F-actin coordinately regulate CaMKIIalpha localization to dendritic postsynaptic sites in developing hippocampal slices. *Mol Cell Neurosci* 31: 37–51, 2006.
3. Allison DW, Chervin AS, Gelfand VI, and Craig AM. Postsynaptic scaffolds of excitatory and inhibitory synapses in hippocampal neurons: maintenance of core components independent of actin filaments and microtubules. *J Neurosci* 20: 4545–4554, 2000.
4. Allison DW, Gelfand VI, Spector I, and Craig AM. Role of actin in anchoring postsynaptic receptors in cultured hippocampal neurons: differential attachment of NMDA versus AMPA receptors. *J Neurosci* 18: 2423–2436, 1998.
5. Ango F, Prezeau L, Muller T, Tu JC, Xiao B, Worley PF, Pin JP, Bockaert J, and Fagni L. Agonist-independent activation of metabotropic glutamate receptors by the intracellular protein Homer. *Nature* 411: 962–965, 2001.

6. Apperson ML, Moon I-S, and Kennedy MB. Characterization of densin-180, a new brain-specific synaptic protein of the O-sialoglycoprotein family. *J Neurosci* 16: 6839–6852, 1996.
7. Bagal AA, Kao JP, Tang CM, and Thompson SM. Long-term potentiation of exogenous glutamate responses at single dendritic spines. *Proc Natl Acad Sci USA* 102: 14434–14439, 2005.
8. Baron MK, Boeckers TM, Vaida B, Faham S, Gingery M, Sawaya MR, Salyer D, Gundelfinger ED, and Bowie JU. An architectural framework that may lie at the core of the postsynaptic density. *Science* 311: 531–535, 2006.
9. Barry MF and Ziff EB. Receptor trafficking and the plasticity of excitatory synapses. *Curr Opin Neurobiol* 12: 279–286, 2002.
10. Baude A, Nusser Z, Roberts JD, Mulvihill E, McIlhinney RA, and Somogyi P. The metabotropic glutamate receptor (mGluR1 alpha) is concentrated at perisynaptic membrane of neuronal subpopulations as detected by immunogold reaction. *Neuron* 11: 771–787, 1993.
11. Bayer KU, De Koninck P, Leonard AS, Hell JW, and Schulman H. Interaction with the NMDA receptor locks CaMKII in an active conformation. *Nature* 411: 801–805, 2001.
12. Beique JC, Lin DT, Kang MG, Aizawa H, Takamiya K, and Huganir RL. Synapse-specific regulation of AMPA receptor function by PSD-95. *Proc Natl Acad Sci US* 103: 19535–19540, 2006.
13. Beneken J, Tu JC, Xiao B, Nuriya M, Yuan JP, Worley PF, and Leahy DJ. Structure of the Homer EVH1 domain-peptide complex reveals a new twist in polyproline recognition. *Neuron* 26: 143–154, 2000.
14. Bennett MK, Erondu NE, and Kennedy MB. Purification and characterization of a calmodulin-dependent protein kinase that is highly concentrated in brain. *J Biol Chem* 258: 12735–12744, 1983.
15. Bennett MK and Kennedy MB. Deduced primary structure of the β subunit of brain type II Ca^{2+}/calmodulin-dependent protein kinase determined by molecular cloning. *Proc Natl Acad Sci USA* 84: 1794–1798, 1987.
16. Bliss TVP and Collingridge GL. A synaptic model of memory: long-term potentiation in the hippocampus. *Nature* 361: 31–39, 1993.
17. Bloch RJ. Actin at receptor-rich domains of isolated acetylcholine receptor clusters. *J Cell Biol* 102: 1447–1458, 1986.
18. Bockers TM, Mameza MG, Kreutz MR, Bockmann J, Weise C, Buck F, Richter D, Gundelfinger ED, and Kreienkamp HJ. Synaptic scaffolding proteins in rat brain. Ankyrin repeats of the multidomain Shank protein family interact with the cytoskeletal protein alpha-fodrin. *J Biol Chem* 276: 40104–40112, 2001.
19. Boeckers TM, Bockmann J, Kreutz MR, and Gundelfinger ED. ProSAP/Shank proteins – a family of higher order organizing molecules of the postsynaptic density with an emerging role in human neurological disease. *J Neurochem* 81: 903–910, 2002.
20. Boeckers TM, Kreutz MR, Winter C, Zuschratter W, Smalla KH, Sanmarti-Vila L, Wex H, Langnaese K, Bockmann J, Garner CC, and Gundelfinger ED. Proline-rich synapse-associated protein-1/cortactin binding protein 1 (ProSAP1/CortBP1) is a PDZ-domain protein highly enriched in the postsynaptic density. *J Neurosci* 19: 6506–6518, 1999.
21. Boeckers TM, Winter C, Smalla KH, Kreutz MR, Bockmann J, Seidenbecher C, Garner CC, and Gundelfinger ED. Proline-rich synapse-associated proteins ProSAP1 and ProSAP2 interact with synaptic proteins of the SAPAP/GKAP family. *Biochem Biophys Res Commun* 264: 247–252, 1999.
22. Bortolotto ZA, Fitzjohn SM, and Collingridge GL. Roles of metabotropic glutamate receptors in LTP and LTD in the hippocampus. *Curr Opin Neurobiol* 9: 299–304, 1999.

23. Braithwaite SP, Xia H, and Malenka RC. Differential roles for NSF and GRIP/ABP in AMPA receptor cycling. *Proc Natl Acad Sci USA* 99: 7096–7101, 2002.
24. Brakeman PR, Lanahan AA, O'Brien R, Roche K, Barnes CA, Huganir RL, and Worley PF. Homer: a protein that selectively binds metabotropic glutamate receptors. *Nature* 386: 284–288, 1997.
25. Brambilla R, Gnesutta N, Minichiello L, White G, Roylance AJ, Herron CE, Ramsey M, Wolfer DP, Cestari V, Rossi -A, C., Grant SGN, Chapman PF, Lipp H-P, Sturani E, and Klein R. A role for the Ras signalling pathway in synaptic transmission and long-term memory. *Nature* 390: 281–286, 1997.
26. Bredt DS and Nicoll RA. AMPA receptor trafficking at excitatory synapses. *Neuron* 40: 361–379, 2003.
27. Brenman JE, Chao DS, Gee SH, McGee AW, Craven SE, Santillano DR, Wu Z, Huang F, Xia H, Peters MF, Froehner SC, and Bredt DS. Interaction of nitric-oxide synthase with the postsynaptic density protein PSD-95 and a-1-syntrophin mediated by PDZ domains. *Cell* 84: 757–767, 1996.
28. Brenman JE, Christopherson KS, Craven SE, McGee AW, and Bredt DS. Cloning and characterization of postsynaptic density-93, a nitric oxide synthase interacting protein. *J Neurosci* 16: 7407–7415, 1996.
29. Bulleit RF, Bennett MK, Molloy SS, Hurley JB, and Kennedy MB. Conserved and variable regions in the subunits of brain type II Ca^{2+}/calmodulin-dependent protein kinase. *Neuron* 1: 63–72, 1988.
30. Burgin KE, Waxham MN, Rickling S, Westgate SA, Mobley WC, and Kelly PT. In situ hybridization histochemistry of Ca^{2+} calmodulin-dependent protein kinase in developing rat brain. *J Neurosci* 10: 1788–1798, 1990.
31. Burkhardt C, Muller M, Badde A, Garner CC, Gundelfinger ED, and Puschel AW. Semaphorin 4B interacts with the post-synaptic density protein PSD-95/SAP90 and is recruited to synapses through a C-terminal PDZ-binding motif. *FEBS Lett* 579: 3821–3828, 2005.
32. Camera P, da Silva JS, Griffiths G, Giuffrida MG, Ferrara L, Schubert V, Imarisio S, Silengo L, Dotti CG, and Di Cunto F. Citron-N is a neuronal Rho-associated protein involved in Golgi organization through actin cytoskeleton regulation. *Nat Cell Biol* 5: 1071–1078, 2003.
33. Carlin RK, Bartelt D, and Siekevitz P. Identification of fodrin as a major calmodulin-binding protein in postsynaptic density preparations. *J Cell Biol* 96: 443–448, 1983.
34. Carlisle HJ and Kennedy MB. Spine Architecture and Synaptic Plasticity. *Trends in Neurosci* 28: 182–187, 2005.
35. Chen H-J, Rojas-Soto M, Oguni A, and Kennedy MB. A synaptic Ras-GTPase activating protein (p135 SynGAP) inhibited by CaM Kinase II. *Neuron* 20: 895–904, 1998.
36. Chen L, Chetkovich DM, Petralia RS, Sweeney NT, Kawasaki Y, Wenthold RJ, Bredt DS, and Nicoll RA. Stargazin regulates synaptic targeting of AMPA receptors by two distinct mechanisms. *Nature* 408: 936–943., 2000.
37. Cho K-O, Hunt CA, and Kennedy MB. The rat brain postsynaptic density fraction contains a homolog of the Drosophila discs-large tumor suppressor protein. *Neuron* 9: 929–942, 1992.
38. Christopherson KS, Hillier BJ, Lim WA, and Bredt DS. PSD-95 assembles a ternary complex with the N-methyl-D-aspartic acid receptor and a bivalent neuronal NO synthase PDZ domain. *J Biol Chem* 274: 27467–27473, 1999.
39. Chung HJ, Xia J, Scannevin RH, Zhang X, and Huganir RL. Phosphorylation of the AMPA receptor subunit GluR2 differentially regulates its interaction with PDZ domain-containing proteins. *J Neurosci* 20: 7258–7267, 2000.

40. Cohen RS, Blomberg F, Berzins K, and Siekevitz P. The structure of postsynaptic densities isolated from dog cerebral cortex I. overall morphology and protein composition. *J Cell Biol* 74: 181–203, 1977.
41. Colledge M, Dean RA, Scott GK, Langeberg LK, Huganir RL, and Scott JD. Targeting of PKA to glutamate receptors through a MAGUK-AKAP complex. *Neuron* 27: 107–119, 2000.
42. Cotman CW, Banker B, Churchill L, and Taylor D. Isolation of postsynaptic densities from rat brain. *J Cell Biol* 63: 441–455, 1974.
43. Cummings JA, Mulkey RM, Nicoll RA, and Malenka RC. Ca2+ signaling requirements for long-term depression in the hippocampus. *Neuron* 16: 825–833, 1996.
44. Czogalla A and Sikorski AF. Spectrin and calpain: a 'target' and a 'sniper' in the pathology of neuronal cells. *Cell Mol Life Sci* 62: 1913–1924, 2005.
45. Dalva MB, Takasu MA, Lin MZ, Shamah SM, Hu L, Gale NW, and Greenberg ME. EphB receptors interact with NMDA receptors and regulate excitatory synapse formation. *Cell* 103: 945–956, 2000.
46. Daly RJ. Cortactin signalling and dynamic actin networks. *Biochem J* 382: 13–25, 2004.
47. Daw MI, Chittajallu R, Bortolotto ZA, Dev KK, Duprat F, Henley JM, Collingridge GL, and Isaac JT. PDZ proteins interacting with C-terminal GluR2/3 are involved in a PKC-dependent regulation of AMPA receptors at hippocampal synapses. *Neuron* 28: 873–886, 2000.
48. DeSouza S, Fu J, States BA, and Ziff EB. Differential palmitoylation directs the AMPA receptor-binding protein ABP to spines or to intracellular clusters. *J Neurosci* 22: 3493–3503, 2002.
49. Dev KK, Nishimune A, Henley JM, and Nakanishi S. The protein kinase C alpha binding protein PICK1 interacts with short but not long form alternative splice variants of AMPA receptor subunits. *Neuropharmacology* 38: 635–644, 1999.
50. Dingledine R, Borges K, Bowie D, and Traynelis SF. The glutamate receptor ion channels. *Pharmacol Rev* 51: 7–61, 1999.
51. Dong H, O'Brien RJ, Fung ET, Lanahan AA, Worley PF, and Huganir RL. GRIP: a synaptic PDZ domain-containing protein that interacts with AMPA receptors [see comments]. *Nature* 386: 279–284, 1997.
52. Dong H, Zhang P, Song I, Petralia RS, Liao D, and Huganir RL. Characterization of the glutamate receptor-interacting proteins GRIP1 and GRIP2. *J Neurosci* 19: 6930–6941, 1999.
53. Dosemeci A, Tao-Cheng JH, Vinade L, Winters CA, Pozzo-Miller L, and Reese TS. Glutamate-induced transient modification of the postsynaptic density. *Proc Natl Acad Sci USA* 98: 10428–10432, 2001.
54. Du Y, Weed SA, Xiong WC, Marshall TD, and Parsons JT. Identification of a novel cortactin SH3 domain-binding protein and its localization to growth cones of cultured neurons. *Mol Cell Biol* 18: 5838–5851, 1998.
55. Dunah AW, Wyszynski M, Martin DM, Sheng M, and Standaert DG. alpha-actinin-2 in rat striatum: localization and interaction with NMDA glutamate receptor subunits. *Brain Res Mol Brain Res* 79: 77–87, 2000.
56. Ehlers MD. Synapse structure: glutamate receptors connected by the shanks. *Curr Biol* 9: R848–850, 1999.
57. Ehrlich I and Malinow R. Postsynaptic density 95 controls AMPA receptor incorporation during long-term potentiation and experience-driven synaptic plasticity. *J Neurosci* 24: 916–927, 2004.
58. El-Husseini Ael D, Schnell E, Dakoji S, Sweeney N, Zhou Q, Prange O, Gauthier-Campbell C, Aguilera-Moreno A, Nicoll RA, and Bredt DS. Synaptic strength regulated by palmitate cycling on PSD-95. *Cell* 108: 849–863, 2002.

59. Elias GM, Funke L, Stein V, Grant SG, Bredt DS, and Nicoll RA. Synapse-specific and developmentally regulated targeting of AMPA receptors by a family of MAGUK scaffolding proteins. *Neuron* 52: 307–320, 2006.
60. Erondu NE and Kennedy MB. Regional distribution of type II Ca^{2+}/calmodulin-dependent protein kinase in rat brain. *J Neurosci* 5: 3270–3277, 1985.
61. Fagni L, Worley PF, and Ango F. Homer as both a scaffold and transduction molecule. *Sci STKE* 2002: RE8, 2002.
62. Farnsworth CL, Freshney NW, Rosen LB, Ghosh A, Greenberg ME, and Feig LA. Calcium activation of Ras mediated by neuronal exchange factor Ras-GRF. *Nature* 376: 524–527, 1995.
63. Fifkova E and Delay RJ. Cytoplasmic actin in neuronal processes as a possible mediator of synaptic plasticity. *J Cell Biol* 95: 345–350, 1982.
64. Franks KM, Keller DX, Bartol TMJ, and Sejnowski TJ. Subcellular spatial compartmentalization of calcium sources within a dendritic spine. *submitted for publication*, 2006.
65. Franks KM and Sejnowski TJ. Complexity of calcium signaling in synaptic spines. *Bioessays* 24: 1130–1144, 2002.
66. Fu J, deSouza S, and Ziff EB. Intracellular membrane targeting and suppression of Ser880 phosphorylation of glutamate receptor 2 by the linker I-set II domain of AMPA receptor-binding protein. *J Neurosci* 23: 7592–7601, 2003.
67. Fukata Y, Tzingounis AV, Trinidad JC, Fukata M, Burlingame AL, Nicoll RA, and Bredt DS. Molecular constituents of neuronal AMPA receptors. *J Cell Biol* 169: 399–404, 2005.
68. Furuyashiki T, Fujisawa K, Fujita A, Madaule P, Uchino S, Mishina M, Bito H, and Narumiya S. Citron, a rho-target, interacts with PSD-95/SAP-90 at glutamatergic synapses in the thalamus. *J Neurosci* 19: 109–118, 1999.
69. Futai K, Kim MJ, Hashikawa T, Scheiffele P, Sheng M, and Hayashi Y. Retrograde modulation of presynaptic release probability through signaling mediated by PSD-95-neuroligin. *Nat Neurosci* 10: 186–195, 2007.
70. Garcia EP, Mehta S, Blair LA, Wells DG, Shang J, Fukushima T, Fallon JR, Garner CC, and Marshall J. SAP90 binds and clusters kainate receptors causing incomplete desensitization. *Neuron* 21: 727–739, 1998.
71. Garcia RA, Vasudevan K, and Buonanno A. The neuregulin receptor ErbB-4 interacts with PDZ-containing proteins at neuronal synapses. *Proc Natl Acad Sci USA* 97: 3596–3601, 2000.
72. Gardoni F, Bellone C, Cattabeni F, and Di Luca M. Protein kinase C activation modulates alpha-calmodulin kinase II binding to NR2A subunit of N-methyl-D-aspartate receptor complex. *J Biol Chem* 276: 7609–7613, 2001.
73. Gardoni F, Caputi A, Cimino M, Pastorino L, Cattabeni F, and Di Luca M. Calcium/calmodulin-dependent protein kinase II is associated with NR2A/B subunits of NMDA receptor in postsynaptic densities. *J Neurochem* 71: 1733–1741, 1998.
74. Gardoni F, Schrama LH, van Dalen JJ, Gispen WH, Cattabeni F, and Di Luca M. AlphaCaMKII binding to the C-terminal tail of NMDA receptor subunit NR2A and its modulation by autophosphorylation. *FEBS Lett* 456: 394–398, 1999.
75. Giese KP, Friedman E, Telliez JB, Fedorov NB, Wines M, Feig LA, and Silva AJ. Hippocampus-dependent learning and memory is impaired in mice lacking the Ras-guanine-nucleotide releasing factor 1 (Ras-GRF1). *Neuropharmacology* 41: 791–800, 2001.
76. Gleason MR, Higashijima S, Dallman J, Liu K, Mandel G, and Fetcho JR. Translocation of CaM kinase II to synaptic sites in vivo. *Nat Neurosci* 6: 217–218, 2003.
77. Goodman SR, Zimmer WE, Clark MB, Zagon IS, Barker JE, and Bloom ML. Brain spectrin: of mice and men. *Brain Res Bull* 36: 593–606, 1995.

78. Gray NW, Weimer RM, Bureau I, and Svoboda K. Rapid redistribution of synaptic PSD-95 in the neocortex in vivo. *PLoS Biol* 4: e370, 2006.
79. Hanson PI, Meyer T, Stryer L, and Schulman H. Dual role of calmodulin in autophosphorylation of multifunctional CaM kinase may underlie decoding of calcium signals. *Neuron* 12: 943–956, 1994.
80. Hardingham GE, Arnold FJ, and Bading H. A calcium microdomain near NMDA receptors: on switch for ERK-dependent synapse-to-nucleus communication. *Nat Neurosci* 4: 565–566, 2001.
81. Hayashi MK, Ames HM, and Hayashi Y. Tetrameric hub structure of postsynaptic scaffolding protein homer. *J Neurosci* 26: 8492–8501, 2006.
82. Hayashi Y, Shi SH, Esteban JA, Piccini A, Poncer JC, and Malinow R. Driving AMPA receptors into synapses by LTP and CaMKII: requirement for GluR1 and PDZ domain interaction. *Science* 287: 2262–2267, 2000.
83. Heidinger V, Manzerra P, Wang XQ, Strasser U, Yu S-P, Choi DW, and Behrens MM. Metabotropic glutamate receptor 1-induced upregulation of NMDA receptor current: mediation through the Pyk2/Src-family kinase pathway in cortical neurons. *J Neurosci* 22: 5452–5461, 2002.
84. Hering H and Sheng M. Activity-dependent redistribution and essential role of cortactin in dendritic spine morphogenesis. *J Neurosci* 23: 11759–11769, 2003.
85. Hirbec H, Francis JC, Lauri SE, Braithwaite SP, Coussen F, Mulle C, Dev KK, Coutinho V, Meyer G, Isaac JT, Collingridge GL, and Henley JM. Rapid and differential regulation of AMPA and kainate receptors at hippocampal mossy fibre synapses by PICK1 and GRIP. *Neuron* 37: 625–638, 2003.
86. Hollmann M and Heinemann S. Cloned glutamate receptors. *Annu Rev Neurosci* 17: 31–108, 1994.
87. Hoogenraad CC, Milstein AD, Ethell IM, Henkemeyer M, and Sheng M. GRIP1 controls dendrite morphogenesis by regulating EphB receptor trafficking. *Nat Neurosci* 8: 906–915, 2005.
88. Huang YZ, Won S, Ali DW, Wang Q, Tanowitz M, Du QS, Pelkey KA, Yang DJ, Xiong WC, Salter MW, and Mei L. Regulation of neuregulin signaling by PSD-95 interacting with ErbB4 at CNS synapses. *Neuron* 26: 443–455, 2000.
89. Husi H, Ward MA, Choudhary JS, Blackstock WP, and Grant SGN. Proteomic analysis of NMDA receptor-adhesion protein signaling complexes. *Nat Neurosci* 3: 661–669, 2000.
90. Im YJ, Lee JH, Park SH, Park SJ, Rho SH, Kang GB, Kim E, and Eom SH. Crystal structure of the Shank PDZ-ligand complex reveals a class I PDZ interaction and a novel PDZ-PDZ dimerization. *J Biol Chem* 278: 48099–48104, 2003.
91. Irie M, Hata Y, Takeuchi M, Ichtchenko A, Toyoda A, Hirao K, Takai Y, Rosahl TW, and Sudhof TC. Binding of neuroligins to PSD-95. *Science* 277: 1511–1515, 1997.
92. Isaac JTR, Nicoll RA, and Malenka RC. Evidence for silent synapses – implications for the expression of LTP. *Neuron* 15: 427–434, 1995.
93. Kalia LV, Pitcher GM, Pelkey KA, and Salter MW. PSD-95 is a negative regulator of the tyrosine kinase Src in the NMDA receptor complex. *EMBO J* 25: 4971–4982, 2006.
94. Kalia LV and Salter MW. Interactions between Src family protein tyrosine kinases and PSD-95. *Neuropharmacology* 45: 720–728, 2003.
95. Kato A, Ozawa F, Saitoh Y, Fukazawa Y, Sugiyama H, and Inokuchi K. Novel members of the Vesl/Homer family of PDZ proteins that bind metabotropic glutamate receptors. *J Biol Chem* 273: 23969–23975, 1998.
96. Kato A, Ozawa F, Saitoh Y, Hirai K, and Inokuchi K. vesl, a gene encoding VASP/Ena family related protein, is upregulated during seizure, long-term potentiation and synaptogenesis. *FEBS Lett* 412: 183–189, 1997.

97. Kennedy MB. Signal-processing machines at the postsynaptic density. *Science* 290: 750–754, 2000.
98. Kennedy MB. The postsynaptic density at glutamatergic synapses. *Trends Neurosci* 20: 264–268, 1997.
99. Kennedy MB, Beale HC, Carlisle HJ, and Washburn LR. Integration of biochemical signalling in spines. *Nat Rev Neurosci* 6: 423–434, 2005.
100. Kennedy MB, Bennett MK, Bulleit RF, Erondu NE, Jennings VR, Miller SM, Molloy SS, Patton BL, and Schenker LJ. Structure and regulation of type II calcium/calmodulin-dependent protein kinase in central nervous system neurons. *Cold Spring Harb Symp Quant Biol* 55: 101–110, 1990.
101. Kennedy MB, Bennett MK, and Erondu NE. Biochemical and immunochemical evidence that the "major postsynaptic density protein" is a subunit of a calmodulin-dependent protein kinase. *Proc Natl Acad Sci USA* 80: 7357–7361, 1983.
102. Kim CH, Chung HJ, Lee HK, and Huganir RL. Interaction of the AMPA receptor subunit GluR2/3 with PDZ domains regulates hippocampal long-term depression. *Proc Natl Acad Sci USA* 98: 11725–11730, 2001.
103. Kim E, Cho KO, Rothschild A, and Sheng M. Heteromultimerization and NMDA receptor-clustering activity of chapsyn-110, a member of the PSD-95 family of proteins. *Neuron* 17: 103–113, 1996.
104. Kim E, Naisbitt S, Hsueh YP, Rao. A, Rothschild A, Craig AM, and Sheng M. GKAP, a novel synaptic protein that interacts with the guanylate kinase-like domain of the PSD-95/SAP90 family of channel clustering molecules. *J Cell Biol* 136: 669–678, 1997.
105. Kim E, Niethammer M, Rothschild A, Jan YN, and Sheng M. Clustering of shaker-type K+ channels by interaction with a family of membrane-associated guanylate kinases. *Nature* 378: 85–88, 1995.
106. Kim E and Sheng M. PDZ domain proteins of synapses. *Nat Rev Neurosci* 5: 771–781, 2004.
107. Kim JH, Liao D, Lau L-F, and Huganir RL. SynGAP: a synaptic RasGAP that associates with the PSD-95/SAP90 protein family. *Neuron* 20: 683–691, 1998.
108. Kistner U, Garner CC, and Linial M. Nucleotide-binding by the synapse associated protein sap90. *FEBS Lett* 359: 159–163, 1995.
109. Kistner U, Wenzel BM, Veh RW, Cases-Langhoff C, Garner AM, Appeltauer U, Voss B, Gundelfinger ED, and Garner CC. SAP90, a rat presynaptic protein related to the product of the Drosophila tumor suppressor gene dlg-A. *J Biol Chem* 268: 4580–4583, 1993.
110. Knuesel I, Elliott A, Chen HJ, Mansuy IM, and Kennedy MB. A role for synGAP in regulating neuronal apoptosis. *Eur J Neurosci* 21: 611–621, 2005.
111. Komiyama NH, Watabe AM, Carlisle HJ, Porter K, Charlesworth P, Monti J, Strathdee DJ, O'Carroll CM, Martin SJ, Morris RG, O'Dell TJ, and Grant SG. SynGAP regulates ERK/MAPK signaling, synaptic plasticity, and learning in the complex with postsynaptic density 95 and NMDA receptor. *J Neurosci* 22: 9721–9732, 2002.
112. Korkin D, Davis FP, Alber F, Luong T, Shen MY, Lucic V, Kennedy MB, and Sali A. Structural modeling of protein interactions by analogy: application to PSD-95. *PLoS Comput Biol* 2: e153, 2006.
113. Kornau H-C, Schenker LT, Kennedy MB, and Seeburg PH. Domain interaction between NMDA receptor subunits and the postsynaptic density protein PSD-95. *Science* 269: 1737–1740, 1995.
114. Kornau H-C, Seeburg PH, and Kennedy MB. Interaction of ion channels and receptors with PDZ domain proteins. *Curr Opin Neurobiol* 7: 368–373, 1997.

115. Krapivinsky G, Krapivinsky L, Manasian Y, Ivanov A, Tyzio R, Pellegrino C, Ben-Ari Y, Clapham DE, and Medina I. The NMDA receptor is coupled to the ERK pathway by a direct interaction between NR2B and RasGRF1. *Neuron* 40: 775–784, 2003.
116. Krapivinsky G, Medina I, Krapivinsky L, Gapon S, and Clapham DE. SynGAP-MUPP1-CaMKII synaptic complexes regulate p38 MAP kinase activity and NMDA receptor-dependent synaptic AMPA receptor potentiation. *Neuron* 43: 563–574, 2004.
117. Krupp JJ, Vissel B, Thomas CG, Heinemann SF, and Westbrook GL. Interactions of calmodulin and alpha-actinin with the NR1 subunit modulate Ca2+-dependent inactivation of NMDA receptors. *J Neurosci* 19: 1165–1178, 1999.
118. Landis DMD and Reese TS. Cytoplasmic organization in cerebellar dendritic spines. *J Cell Biol* 97: 1169–1178, 1983.
119. Lau LF, Mammen A, Ehlers MD, Kindler S, Chung WJ, Garner CC, and Huganir RL. Interaction of the N-methyl-D-aspartate receptor complex with a novel synapse-associated protein, sap102. *J Biol Chem* 271: 21622–21628, 1996.
120. Lee SH, Liu L, Wang YT, and Sheng M. Clathrin adaptor AP2 and NSF interact with overlapping sites of GluR2 and play distinct roles in AMPA receptor trafficking and hippocampal LTD. *Neuron* 36: 661–674, 2002.
121. Leonard AS, Bayer KU, Merrill MA, Lim IA, Shea MA, Schulman H, and Hell JW. Regulation of calcium/calmodulin-dependent protein kinase II docking to N-methyl-D-aspartate receptors by calcium/calmodulin and alpha-actinin. *J Biol Chem* 277: 48441–48448, 2002.
122. Leonard AS, Davare MA, Horne MC, Garner CC, and Hell JW. SAP97 is associated with the alpha-amino-3-hydroxy-5-methylisoxazole-4- propionic acid receptor GluR1 subunit. *J Biol Chem* 273: 19518–19524, 1998.
123. Leonard AS, Lim IA, Hemsworth DE, Horne MC, and Hell JW. Calcium/calmodulin-dependent protein kinase II is associated with the N- methyl-D-aspartate receptor. *Proc Natl Acad Sci USA* 96: 3239–3244, 1999.
124. Letts VA, Felix R, Biddlecome GH, Arikkath J, Mahaffey CL, Valenzuela A, Bartlett FS, II, Mori Y, Campbell KP, and Frankel WN. The mouse stargazer gene encodes a neuronal Ca2+-channel gamma subunit. *Nat Genet* 19: 340–347, 1998.
125. LeVine H and Sahyoun NE. Involvement of fodrin-binding proteins in the structure of the neuronal postsynaptic density and regulation by phosphorylation. *Biochem Biophys Res Commun* 138: 59–65, 1986.
126. Li K, Hornshaw MP, van Minnen J, Smalla KH, Gundelfinger ED, and Smit AB. Organelle proteomics of rat synaptic proteins: correlation-profiling by isotope-coded affinity tagging in conjunction with liquid chromatography-tandem mass spectrometry to reveal post-synaptic density specific proteins. *J Proteome Res* 4: 725–733, 2005.
127. Li KW, Hornshaw MP, Van Der Schors RC, Watson R, Tate S, Casetta B, Jimenez CR, Gouwenberg Y, Gundelfinger ED, Smalla KH, and Smit AB. Proteomics analysis of rat brain postsynaptic density. Implications of the diverse protein functional groups for the integration of synaptic physiology. *J Biol Chem* 279: 987–1002, 2004.
128. Liao DZ, Hessler NA, and Malinow R. Activation of postsynaptically silent synapses during pairing-induced LTP in CA1 region of hippocampal slice. *Nature* 375: 400–404, 1995.
129. Lim S, Naisbitt S, Yoon J, Hwang JI, Suh PG, Sheng M, and Kim E. Characterization of the Shank family of synaptic proteins. Multiple genes, alternative splicing, and differential expression in brain and development. *J Biol Chem* 274: 29510–29518, 1999.
130. Lim S, Sala C, Yoon J, Park S, Kuroda S, Sheng M, and Kim E. Sharpin, a novel postsynaptic density protein that directly interacts with the shank family of proteins. *Mol Cell Neurosci* 17: 385–397, 2001.

131. Lin Y, Jover-Mengual T, Wong J, Bennett MV, and Zukin RS. PSD-95 and PKC converge in regulating NMDA receptor trafficking and gating. *Proc Natl Acad Sci USA* 103: 19902–19907, 2006.
132. Lisman J, Schulman H, and Cline H. The molecular basis of CaMKII function in synaptic and behavioural memory. *Nat Rev Neurosci* 3: 175–190, 2002.
133. Lu W and Ziff EB. PICK1 interacts with ABP/GRIP to regulate AMPA receptor trafficking. *Neuron* 47: 407–421, 2005.
134. Lue RA, Marfatia SM, Branton D, and Chishti AH. Cloning and characterization of hdlg – the human homolog of the drosophila disks large tumor-suppressor binds to protein-4.1. *Proc Natl Acad Sci USA* 91: 9818–9822, 1994.
135. Lujan R, Roberts JD, Shigemoto R, Ohishi H, and Somogyi P. Differential plasma membrane distribution of metabotropic glutamate receptors mGluR1 alpha, mGluR2 and mGluR5, relative to neurotransmitter release sites. *J Chem Neuroanat* 13: 219–241, 1997.
136. Luscher C, Xia H, Beattie EC, Carroll RC, von Zastrow M, Malenka RC, and Nicoll RA. Role of AMPA receptor cycling in synaptic transmission and plasticity. *Neuron* 24: 649–658, 1999.
137. Lyons-Warren A, Chang JJ, Balkissoon R, Kamiya A, Garant M, Nurnberger J, Scheftner W, Reich T, McMahon F, Kelsoe J, Gershon E, Coryell W, Byerley W, Berrettini W, Depaulo R, McInnis M, and Sawa A. Evidence of association between bipolar disorder and Citron on chromosome 12q24. *Mol Psychiatry* 10: 807–809, 2005.
138. Ma XM, Huang J, Wang Y, Eipper BA, and Mains RE. Kalirin, a multifunctional Rho guanine nucleotide exchange factor, is necessary for maintenance of hippocampal pyramidal neuron dendrites and dendritic spines. *J Neurosci* 23: 10593–10603, 2003.
139. Malenka RC and Bear MF. LTP and LTD: an embarrassment of riches. *Neuron* 44: 5–21, 2004.
140. Malinow R. AMPA receptor trafficking and long-term potentiation. *Philos Trans R Soc Lond B Biol Sci* 358: 707–714, 2003.
141. Malinow R and Malenka RC. AMPA receptor trafficking and synaptic plasticity. *Annu Rev Neurosci* 25: 103–126, 2002.
142. Marchesi VT and Steers E, Jr. Selective solubilization of a protein component of the red cell membrane. *Science* 159: 203–204, 1968.
143. Markham JA and Fifkova E. Actin filament organization within dendrites and dendritic spines during development. *Brain Res* 392: 263–269, 1986.
144. Matsuda S, Mikawa S, and Hirai H. Phosphorylation of serine-880 in GluR2 by protein kinase C prevents its C terminus from binding with glutamate receptor-interacting protein. *J Neurochem* 73: 1765–1768, 1999.
145. Matsuzaki M, Honkura N, Ellis-Davies GC, and Kasai H. Structural basis of long-term potentiation in single dendritic spines. *Nature* 429: 761–766, 2004.
146. Matus A. Actin-based plasticity in dendritic spines. *Science* 290: 754–758., 2000.
147. Mayer ML. Glutamate receptor ion channels. *Curr Opin Neurobiol* 15: 282–288, 2005.
148. Mayer ML and Armstrong N. Structure and function of glutamate receptor ion channels. *Annu Rev Physiol* 66: 161–181, 2004.
149. McGee AW and Bredt DS. Identification of an intramolecular interaction between the SH3 and guanylate kinase domains of PSD-95. *J Biol Chem* 274: 17431–17436, 1999.
150. McGee AW, Dakoji SR, Olsen O, Bredt DS, Lim WA, and Prehoda KE. Structure of the SH3-guanylate kinase module from PSD-95 suggests a mechanism for regulated assembly of MAGUK scaffolding proteins. *Mol Cell* 8: 1291–1301, 2001.
151. Mehta S, Wu H, Garner CC, and Marshall J. Molecular mechanisms regulating the differential association of kainate receptor subunits with SAP90/PSD-95 and SAP97. *J Biol Chem* 276: 16092–16099, 2001.

152. Merrill MA, Chen Y, Strack S, and Hell JW. Activity-driven postsynaptic translocation of CaMKII. *Trends Pharmacol Sci* 26: 645–653, 2005.
153. Meyer RK and Aebi U. Bundling of actin filaments by alpha-actinin depends on its molecular length. *J Cell Biol* 110: 2013–2024, 1990.
154. Migaud M, Charlesworth P, Dempster M, Webster LC, Watabe AM, Makhinson M, He Y, Ramsay MF, Morris RG, Morrison JH, O'Dell TJ, and Grant SG. Enhanced long-term potentiation and impaired learning in mice with mutant postsynaptic density-95 protein [see comments]. *Nature* 396: 433–439, 1998.
155. Miller SG and Kennedy MB. Regulation of brain type II Ca^{2+}/calmodulin-dependent protein kinase by autophosphorylation: a Ca^{2+}-triggered molecular switch. *Cell* 44: 861–870, 1986.
156. Muller BM, Kistner U, Kindler S, Chung WJ, Kuhlendahl S, Fenster SD, Lau LF, Veh RW, Huganir RL, Gundelfinger ED, and Garner CC. Sap102, a novel postsynaptic protein that interacts with NMDA receptor complexes in-vivo. *Neuron* 17: 255–265, 1996.
157. Muller BM, Kistner U, Veh RW, Caseslanghoff C, Becker B, Gundelfinger ED, and Garner CC. Molecular characterization and spatial-distribution of sap97, a novel presynaptic protein homologous to sap90 and the drosophila disks-large tumor-suppressor protein. *J Neurosci* 15: 2354–2366, 1995.
158. Naisbitt S, Kim E, Tu JC, Xiao B, Sala C, Valtschanoff J, Weinberg RJ, Worley PF, and Sheng M. Shank, a novel family of postsynaptic density proteins that binds to the NMDA receptor/PSD-95/GKAP complex and cortactin. *Neuron* 23: 569–582, 1999.
159. Nakagawa T, Cheng Y, Ramm E, Sheng M, and Walz T. Structure and different conformational states of native AMPA receptor complexes. *Nature* 433: 545–549, 2005.
160. Nakagawa T, Engler JA, and Sheng M. The dynamic turnover and functional roles of alpha-actinin in dendritic spines. *Neuropharmacology* 47: 734–745, 2004.
161. Nehring RB, Wischmeyer E, Doring F, Veh RW, Sheng M, and Karschin A. Neuronal inwardly rectifying K(+) channels differentially couple to PDZ proteins of the PSD-95/SAP90 family. *J Neurosci* 20: 156–162, 2000.
162. Nicoll RA, Tomita S, and Bredt DS. Auxiliary subunits assist AMPA-type glutamate receptors. *Science* 311: 1253–1256, 2006.
163. Nishimune A, Isaac JT, Molnar E, Noel J, Nash SR, Tagaya M, Collingridge GL, Nakanishi S, and Henley JM. NSF binding to GluR2 regulates synaptic transmission. *Neuron* 21: 87–97, 1998.
164. Nourry C, Grant SG, and Borg JP. PDZ domain proteins: plug and play! *Sci STKE* 2003: RE7, 2003.
165. Nusser Z, Mulvihill E, Streit P, and Somogyi P. Subsynaptic segregation of metabotropic and ionotropic glutamate receptors as revealed by immunogold localization. *Neuroscience* 61: 421–427, 1994.
166. O'Connor JJ, Rowan MJ, and Anwyl R. Long-lasting enhancement of NMDA receptor-mediated synaptic transmission by metabotropic glutamate receptor activation. *Nature* 367: 557–559, 1994.
167. Oh JS, Manzerra P, and Kennedy MB. Regulation of the neuron-specific Ras GTPase activating protein, synGAP, by Ca2+/calmodulin-dependent protein kinase II. *J Biol Chem* 279: 17980–17988, 2004.
168. Ohta Y, Nishida E, and Sakai H. Type II Ca2+/calmodulin-dependent protein kinase binds to actin filaments in a calmodulin-sensitive manner. *FEBS Lett* 208: 423–426, 1986.
169. Osten P, Khatri L, Perez JL, Kohr G, Giese G, Daly C, Schulz TW, Wensky A, Lee LM, and Ziff EB. Mutagenesis reveals a role for ABP/GRIP binding to GluR2 in synaptic surface accumulation of the AMPA receptor. *Neuron* 27: 313–325, 2000.

170. Osten P, Srivastava S, Inman GJ, Vilim FS, Khatri L, Lee LM, States BA, Einheber S, Milner TA, Hanson PI, and Ziff EB. The AMPA receptor GluR2 C terminus can mediate a reversible, ATP-dependent interaction with NSF and alpha- and beta-SNAPs. *Neuron* 21: 99–110, 1998.
171. Osten P and Stern-Bach Y. Learning from stargazin: the mouse, the phenotype and the unexpected. *Curr Opin Neurobiol* 16: 275–280, 2006.
172. Otmakhov N, Tao-Cheng JH, Carpenter S, Asrican B, Dosemeci A, Reese TS, and Lisman J. Persistent accumulation of calcium/calmodulin-dependent protein kinase II in dendritic spines after induction of NMDA receptor-dependent chemical long-term potentiation. *J Neurosci* 24: 9324–9331, 2004.
173. Ouyang Y, Rosenstein A, Kreiman G, Schuman EM, and Kennedy MB. Tetanic stimulation leads to increased accumulation of Ca(2+)/calmodulin-dependent protein kinase II via dendritic protein synthesis in hippocampal neurons. *J Neurosci* 19: 7823–7833, 1999.
174. Pak DT, Yang S, Rudolph-Correia S, Kim E, and Sheng M. Regulation of dendritic spine morphology by SPAR, a PSD-95-associated RapGAP. *Neuron* 31: 289–303, 2001.
175. Park M, Penick EC, Edwards JG, Kauer JA, and Ehlers MD. Recycling endosomes supply AMPA receptors for LTP. *Science* 305: 1972–1975, 2004.
176. Payne ME, Fong YL, Ono T, Colbran RJ, Kemp BE, Soderling TR, and Means AR. Calcium/calmodulin-dependent protein kinase II. Characterization of distinct calmodulin binding and inhibitory domains. *J Biol Chem* 263: 7190–7195, 1988.
177. Peng J, Kim MJ, Cheng D, Duong DM, Gygi SP, and Sheng M. Semiquantitative proteomic analysis of rat forebrain postsynaptic density fractions by mass spectrometry. *J Biol Chem* 279: 21003–21011, 2004.
178. Penzes P, Beeser A, Chernoff J, Schiller MR, Eipper BA, Mains RE, and Huganir RL. Rapid induction of dendritic spine morphogenesis by trans-synaptic ephrinB-EphB receptor activation of the Rho-GEF kalirin. *Neuron* 37: 263–274, 2003.
179. Penzes P, Johnson RC, Alam MR, Kambampati V, Mains RE, and Eipper BA. An isoform of kalirin, a brain-specific GDP/GTP exchange factor, is enriched in the postsynaptic density fraction. *J Biol Chem* 275: 6395–6403, 2000.
180. Penzes P, Johnson RC, Sattler R, Zhang X, Huganir RL, Kambampati V, Mains RE, and Eipper BA. The neuronal Rho-GEF Kalirin-7 interacts with PDZ domain-containing proteins and regulates dendritic morphogenesis. *Neuron* 29: 229–242, 2001.
181. Perez JL, Khatri L, Chang C, Srivastava S, Osten P, and Ziff EB. PICK1 targets activated protein kinase Calpha to AMPA receptor clusters in spines of hippocampal neurons and reduces surface levels of the AMPA-type glutamate receptor subunit 2. *J Neurosci* 21: 5417–5428, 2001.
182. Peter BJ, Kent HM, Mills IG, Vallis Y, Butler PJ, Evans PR, and McMahon HT. BAR domains as sensors of membrane curvature: the amphiphysin BAR structure. *Science* 303: 495–499, 2004.
183. Petersen JD, Chen X, Vinade L, Dosemeci A, Lisman JE, and Reese TS. Distribution of postsynaptic density (PSD)-95 and Ca2+/calmodulin-dependent protein kinase II at the PSD. *J Neurosci* 23: 11270–11278, 2003.
184. Petralia RS, Sans N, Wang YX, and Wenthold RJ. Ontogeny of postsynaptic density proteins at glutamatergic synapses. *Mol Cell Neurosci* 29: 436–452, 2005.
185. Prange O, Wong TP, Gerrow K, Wang YT, and El-Husseini A. A balance between excitatory and inhibitory synapses is controlled by PSD-95 and neuroligin. *Proc Natl Acad Sci USA* 101: 13915–13920, 2004.
186. Qualmann B, Boeckers TM, Jeromin M, Gundelfinger ED, and Kessels MM. Linkage of the actin cytoskeleton to the postsynaptic density via direct interactions of Abp1 with the ProSAP/Shank family. *J Neurosci* 24: 2481–2495, 2004.

187. Rabiner CA, Mains RE, and Eipper BA. Kalirin: a dual Rho guanine nucleotide exchange factor that is so much more than the sum of its many parts. *Neuroscientist* 11: 148–160, 2005.
188. Racz B and Weinberg RJ. The subcellular organization of cortactin in hippocampus. *J Neurosci* 24: 10310–10317, 2004.
189. Rosenberg OS, Deindl S, Comolli LR, Hoelz A, Downing KH, Nairn AC, and Kuriyan J. Oligomerization states of the association domain and the holoenyzme of Ca/CaM kinase II. *Febs J* 273: 682–694, 2006.
190. Rosenberg OS, Deindl S, Sung RJ, Nairn AC, and Kuriyan J. Structure of the autoinhibited kinase domain of CaMKII and SAXS analysis of the holoenzyme. *Cell* 123: 849–860, 2005.
191. Rosenmund C and Westbrook GL. Calcium-induced actin depolymerization reduces NMDA channel activity. *Neuron* 10: 805–814, 1993.
192. Rumbaugh G, Adams JP, Kim JH, and Huganir RL. SynGAP regulates synaptic strength and mitogen-activated protein kinases in cultured neurons. *Proc Natl Acad Sci USA* 103: 4344–4351, 2006.
193. Sabatini BL, Maravall M, and Svoboda K. Ca^{2+} signaling in dendritic spines. *Curr Opin Neurobiol* 11: 349–356, 2001.
194. Sala C, Futai K, Yamamoto K, Worley PF, Hayashi Y, and Sheng M. Inhibition of dendritic spine morphogenesis and synaptic transmission by activity-inducible protein Homer1a. *J Neurosci* 23: 6327–6337, 2003.
195. Sala C, Piech V, Wilson NR, Passafaro M, Liu G, and Sheng M. Regulation of dendritic spine morphology and synaptic function by Shank and Homer. *Neuron* 31: 115–130, 2001.
196. Sala C, Roussignol G, Meldolesi J, and Fagni L. Key role of the postsynaptic density scaffold proteins Shank and Homer in the functional architecture of Ca2+ homeostasis at dendritic spines in hippocampal neurons. *J Neurosci* 25: 4587–4592, 2005.
197. Sans N, Racca C, Petralia RS, Wang YX, McCallum J, and Wenthold RJ. Synapse-associated protein 97 selectively associates with a subset of AMPA receptors early in their biosynthetic pathway. *J Neurosci* 21: 7506–7516, 2001.
198. Satoh K, Yanai H, Senda T, Kohu K, Nakamura T, Okumura N, Matsumine A, Kobayashi S, Toyoshima K, and Akiyama T. DAP-1, a novel protein that interacts with the guanylate kinase-like domains of hDLG and PSD-95. *Genes Cells* 2: 415–424, 1997.
199. Schnell E, Sizemore M, Karimzadegan S, Chen L, Bredt DS, and Nicoll RA. Direct interactions between PSD-95 and stargazin control synaptic AMPA receptor number. *Proc Natl Acad Sci USA* 99: 13902–13907, 2002.
200. Schulman H. Activity-dependent regulation of calcium/calmodulin-dependent protein kinase II localization. *J Neurosci* 24: 8399–8403, 2004.
201. Schultze W, Eulenburg V, Lessmann V, Herrmann L, Dittmar T, Gundelfinger ED, Heumann R, and Erdmann KS. Semaphorin4F interacts with the synapse-associated protein SAP90/PSD-95. *J Neurochem* 78: 482–489, 2001.
202. Seidenman KJ, Steinberg JP, Huganir R, and Malinow R. Glutamate receptor subunit 2 Serine 880 phosphorylation modulates synaptic transmission and mediates plasticity in CA1 pyramidal cells. *J Neurosci* 23: 9220–9228, 2003.
203. Shen K and Meyer T. Dynamic control of CaMKII translocation and localization in hippocampal neurons by NMDA receptor stimulation. *Science* 284: 162–166, 1999.
204. Shen K, Teruel MN, Connor JH, Shenolikar S, and Meyer T. Molecular memory by reversible translocation of calcium/calmodulin-dependent protein kinase II. *Nat Neurosci* 3: 881–886, 2000.

205. Shen K, Teruel MN, Subramanian K, and Meyer T. CaMKIIbeta functions as an F-actin targeting module that localizes CaMKIIalpha/beta heterooligomers to dendritic spines. *Neuron* 21: 593–606, 1998.
206. Sheng M and Kim E. The Shank family of scaffold proteins. *J Cell Sci* 113: 1851–1856, 2000.
207. Shi S, Hayashi Y, Esteban JA, and Malinow R. Subunit-specific rules governing AMPA receptor trafficking to synapses in hippocampal pyramidal neurons. *Cell* 105: 331–343, 2001.
208. Shi SH, Hayashi Y, Petralia RS, Zaman SH, Wenthold RJ, Svoboda K, and Malinow R. Rapid spine delivery and redistribution of AMPA receptors after synaptic NMDA receptor activation [see comments]. *Science* 284: 1811–1816, 1999.
209. Shin H, Hsueh YP, Yang F-C, Kim E, and Sheng M. An intramolecular interaction between src homology 3 domain and guanylate kinase-like domain required for channel clustering by postsynaptic density-95/SAP90. *J Neurosci* 20: 3580–3587, 2000.
210. Song I and Huganir RL. Regulation of AMPA receptors during synaptic plasticity. *Trends Neurosci* 25: 578–588, 2002.
211. Song I, Kamboj S, Xia J, Dong H, Liao D, and Huganir RL. Interaction of the N-ethylmaleimide-sensitive factor with AMPA receptors. *Neuron* 21: 393–400, 1998.
212. Song JY, Ichtchenko K, Sudhof TC, and Brose N. Neuroligin 1 is a postsynaptic cell-adhesion molecule of excitatory synapses. *Proc Natl Acad Sci USA* 96: 1100–1105, 1999.
213. Srivastava S, Osten P, Vilim FS, Khatri L, Inman G, States B, Daly C, DeSouza S, Abagyan R, Valtschanoff JG, Weinberg RJ, and Ziff EB. Novel anchorage of GluR2/3 to the postsynaptic density by the AMPA receptor-binding protein ABP. *Neuron* 21: 581–591, 1998.
214. Staudinger J, Lu J, and Olson EN. Specific interaction of the PDZ domain protein PICK1 with the COOH terminus of protein kinase C-alpha. *J Biol Chem* 272: 32019–32024, 1997.
215. Stein V, House DR, Bredt DS, and Nicoll RA. Postsynaptic density-95 mimics and occludes hippocampal long-term potentiation and enhances long-term depression. *J Neurosci* 23: 5503–5506, 2003.
216. Strack S and Colbran RJ. Autophosphorylation-dependent targeting of calcium/calmodulin- dependent protein kinase II by the NR2B subunit of the N-methyl- D- aspartate receptor. *J Biol Chem* 273: 20689–20692, 1998.
217. Strack S, McNeill RB, and Colbran RJ. Mechanism and regulation of calcium/calmodulin-dependent protein kinase II targeting to the NR2B subunit of the N-methyl-D-aspartate receptor. *J Biol Chem* 275: 23798–23806, 2000.
218. Suzuki T, Okumuranoji K, Tanaka R, and Tada T. Rapid translocation of cytosolic Ca^{2+}/calmodulin-dependent protein kinase II into postsynaptic density after decapitation. *J Neurochem* 63: 1529–1537, 1994.
219. Sweatt JD. Mitogen-activated protein kinases in synaptic plasticity and memory. *Curr Opin Neurobiol* 14: 311–317, 2004.
220. Tadokoro S, Tachibana T, Imanaka T, Nishida W, and Sobue K. Involvement of unique leucine-zipper motif of PSD-Zip45 (Homer 1c/vesl-1L) in group 1 metabotropic glutamate receptor clustering. *Proc Natl Acad Sci USA* 96: 13801–13806, 1999.
221. Takasu MA, Dalva MB, Zigmond RE, and Greenberg ME. Modulation of NMDA receptor-dependent calcium influx and gene expression through EphB receptors. *Science* 295: 491–495, 2002.
222. Takeuchi M, Hata Y, Hirao K, Toyoda A, Irie M, and Takai Y. SAPAPs. A family of PSD-95/SAP90-associated proteins localized at postsynaptic density. *J Biol Chem* 272: 11943–11951, 1997.

223. Tavares GA, Panepucci EH, and Brunger AT. Structural characterization of the intramolecular interaction between the SH3 and guanylate kinase domains of PSD-95. *Mol Cell* 8: 1313–1325, 2001.
224. Tezuka T, Umemori H, Akiyama T, Nakanishi S, and Yamamoto T. PSD-95 promotes Fyn-mediated tyrosine phosphorylation of the N-methyl-D- aspartate receptor subunit NR2A. *Proc Natl Acad Sci USA* 96: 435–440, 1999.
225. Tomita S, Chen L, Kawasaki Y, Petralia RS, Wenthold RJ, Nicoll RA, and Bredt DS. Functional studies and distribution define a family of transmembrane AMPA receptor regulatory proteins. *J Cell Biol* 161: 805–816, 2003.
226. Tomita S, Fukata M, Nicoll RA, and Bredt DS. Dynamic interaction of stargazin-like TARPs with cycling AMPA receptors at synapses. *Science* 303: 1508–1511, 2004.
227. Torres R, Firestein BL, Dong H, Staudinger J, Olson EN, Huganir RL, Bredt DS, Gale NW, and Yancopoulos GD. PDZ proteins bind, cluster, and synaptically colocalize with Eph receptors and their ephrin ligands. *Neuron* 21: 1453–1463, 1998.
228. Tu JC, Xiao B, Naisbitt S, Yuan JP, Petralia RS, Brakeman P, Doan A, Aakalu VK, Lanahan AA, Sheng M, and Worley PF. Coupling of mGluR/Homer and PSD-95 complexes by the Shank family of postsynaptic density proteins. *Neuron* 23: 583–592, 1999.
229. Tu JC, Xiao B, Yuan JP, Lanahan AA, Leoffert K, Li M, Linden DJ, and Worley PF. Homer binds a novel proline-rich motif and links group 1 metabotropic glutamate receptors with IP3 receptors. *Neuron* 21: 717–726, 1998.
230. Uchino S, Wada H, Honda S, Nakamura Y, Ondo Y, Uchiyama T, Tsutsumi M, Suzuki E, Hirasawa T, and Kohsaka S. Direct interaction of post-synaptic density-95/Dlg/ZO-1 domain-containing synaptic molecule Shank3 with GluR1 alpha-amino-3-hydroxy-5-methyl-4-isoxazole propionic acid receptor. *J Neurochem* 97: 1203–1214, 2006.
231. Uruno T, Liu J, Zhang P, Fan Y, Egile C, Li R, Mueller SC, and Zhan X. Activation of Arp2/3 complex-mediated actin polymerization by cortactin. *Nat Cell Biol* 3: 259–266, 2001.
232. Valtschanoff JG and Weinberg RJ. Laminar organization of the NMDA receptor complex within the postsynaptic density. *J Neurosci* 21: 1211–1217, 2001.
233. Vandenberghe W, Nicoll RA, and Bredt DS. Stargazin is an AMPA receptor auxiliary subunit. *Proc Natl Acad Sci USA* 102: 485–490, 2005.
234. Vazquez LE, Chen HJ, Sokolova I, Knuesel I, and Kennedy MB. SynGAP regulates spine formation. *J Neurosci* 24: 8796–8805, 2004.
235. Walikonis RS, Jensen ON, Mann M, Provance DWJ, Mercer JA, and Kennedy MB. Identification of proteins in the postsynaptic density fraction by mass spectrometry. *J Neurosci* 20: 4069–4080, 2000.
236. Walikonis RS, Oguni A, Khorosheva EM, Jeng C-J, Asuncion FJ, and Kennedy MB. Densin-180 forms a ternary complex with the α-subunit of CaMKII and α-actinin. *J Neurosci* 21: 423–433, 2001.
237. Wang YT and Salter MW. Regulation of NMDA receptors by tyrosine kinases and phosphatases. *Nature* 369: 233–235, 1994.
238. Weaver AM, Karginov AV, Kinley AW, Weed SA, Li Y, Parsons JT, and Cooper JA. Cortactin promotes and stabilizes Arp2/3-induced actin filament network formation. *Curr Biol* 11: 370–374, 2001.
239. Wechsler A and Teichberg VI. Brain spectrin binding to the NMDA receptor is regulated by phosphorylation, calcium and calmodulin. *EMBO J* 17: 3931–3939, 1998.
240. Weed SA, Karginov AV, Schafer DA, Weaver AM, Kinley AW, Cooper JA, and Parsons JT. Cortactin localization to sites of actin assembly in lamellipodia requires interactions with F-actin and the Arp2/3 complex. *J Cell Biol* 151: 29–40, 2000.
241. Wendholt D, Spilker C, Schmitt A, Dolnik A, Smalla KH, Proepper C, Bockmann J, Sobue K, Gundelfinger ED, Kreutz MR, and Boeckers TM. ProSAP-interacting protein 1

(ProSAPiP1), a novel protein of the postsynaptic density that links the spine-associated Rap-Gap (SPAR) to the scaffolding protein ProSAP2/Shank3. *J Biol Chem* 281: 13805–13816, 2006.
242. Westphal RS, Tavalin SJ, Lin JW, Alto NM, Fraser ID, Langeberg LK, Sheng M, and Scott JD. Regulation of NMDA receptors by an associated phosphatase-kinase signaling complex. *Science* 285: 93–96, 1999.
243. Witke W, Hofmann A, Koppel B, Schleicher M, and Noegel AA. The Ca(2+)-binding domains in non-muscle type alpha-actinin: biochemical and genetic analysis. *J Cell Biol* 121: 599–606, 1993.
244. Wu H, Nash JE, Zamorano P, and Garner CC. Interaction of SAP97 with minus-end-directed actin motor myosin VI. Implications for AMPA receptor trafficking. *J Biol Chem* 277: 30928–30934, 2002.
245. Wu H and Parsons JT. Cortactin, an 80/85-kilodalton pp60src substrate, is a filamentous actin-binding protein enriched in the cell cortex. *J Cell Biol* 120: 1417–1426, 1993.
246. Wyszynski M, Kharazia V, Shanghvi R, Rao A, Beggs AH, Craig AM, Weinberg R, and Sheng M. Differential regional expression and ultrastructural localization of alpha-actinin-2, a putative NMDA receptor-anchoring protein, in rat brain. *J Neurosci* 18: 1383–1392, 1998.
247. Wyszynski M, Kim E, Dunah AW, Passafaro M, Valtschanoff JG, Serra-Pages C, Streuli M, Weinberg RJ, and Sheng M. Interaction between GRIP and liprin-alpha/SYD2 is required for AMPA receptor targeting. *Neuron* 34: 39–52, 2002.
248. Wyszynski M, Lin J, Rao A, Nigh E, Beggs AH, Craig AM, and Sheng M. Competitive binding of α-actinin and calmodulin to the NMDA receptor. *Nature* 385: 439–442, 1997.
249. Wyszynski M, Valtschanoff JG, Naisbitt S, Dunah AW, Kim E, Standaert DG, Weinberg R, and Sheng M. Association of AMPA receptors with a subset of glutamate receptor-interacting protein in vivo. *J Neurosci* 19: 6528–6537, 1999.
250. Xiao B, Tu JC, Petralia RS, Yuan JP, Doan A, Breder CD, Ruggiero A, Lanahan AA, Wenthold RJ, and Worley PF. Homer regulates the association of group 1 metabotropic glutamate receptors with multivalent complexes of homer-related, synaptic proteins. *Neuron* 21: 707–716, 1998.
251. Xiao B, Tu JC, and Worley PF. Homer: a link between neural activity and glutamate receptor function. *Curr Opin Neurobiol* 10: 370–374, 2000.
252. Yamada Y, Chochi Y, Takamiya K, Sobue K, and Inui M. Modulation of the channel activity of the epsilon2/zeta1-subtype N-methyl D-aspartate receptor by PSD-95. *J Biol Chem* 274: 6647–6652, 1999.
253. Yamazaki M, Fukaya M, Abe M, Ikeno K, Kakizaki T, Watanabe M, and Sakimura K. Differential palmitoylation of two mouse glutamate receptor interacting protein 1 forms with different N-terminal sequences. *Neurosci Lett* 304: 81–84, 2001.
254. Yao I, Hata Y, Hirao K, Deguchi M, Ide N, Takeuchi M, and Takai Y. Synamon, a novel neuronal protein interacting with synapse-associated protein 90/postsynaptic density-95-associated protein. *J Biol Chem* 274: 27463–27466, 1999.
255. Ye B, Liao D, Zhang X, Zhang P, Dong H, and Huganir RL. GRASP-1: a neuronal RasGEF associated with the AMPA receptor/GRIP complex. *Neuron* 26: 603–617, 2000.
256. Yu XM, Askalan R, Keil GJ, 2nd, and Salter MW. NMDA channel regulation by channel-associated protein tyrosine kinase Src. *Science* 275: 674–678, 1997.
257. Zhang S, Ehlers MD, Bernhardt JP, Su CT, and Huganir RL. Calmodulin mediates calcium-dependent inactivation of N-methyl-D- aspartate receptors. *Neuron* 21: 443–453, 1998.
258. Zhang W and Benson DL. Targeting and clustering citron to synapses. *Mol Cell Neurosci* 31: 26–36, 2006.

259. Zhang W, Vazquez L, Apperson M, and Kennedy MB. Citron binds to PSD-95 at glutamatergic synapses on inhibitory neurons in the hippocampus. *J Neurosci* 19: 96–108, 1999.
260. Zhu JJ, Qin Y, Zhao M, Van Aelst L, and Malinow R. Ras and Rap control AMPA receptor trafficking during synaptic plasticity. *Cell* 110: 443–455, 2002.
261. Zucker RS. Calcium- and activity-dependent synaptic plasticity. *Curr Opin Neurobiol* 9: 305–313, 1999.

Ca^{2+} Signaling in Dendritic Spines

Bernardo L. Sabatini[1] and Karel Svoboda[2]

[1] Department of Neurobiology, Harvard Medical School, 220 Longwood Avenue, Boston, MA 02115, USA, bsabatini@hms.harvard.edu

[2] Janelia Farm Research Campus, HHMI, 19700 Helix Drive, Ashburn, VA 20147, USA, svobodak@janelia.hhmi.org

Abstract. Dendritic spines are neuronal microcompartments that are diffusionally isolated from their parent dendrites and neighboring spines. Optical studies of spine Ca^{2+} dynamics have revealed the Ca^{2+} sources and Ca^{2+} clearance mechanisms under a variety of experimental conditions and in diverse experimental systems. Here we review our understanding of Ca^{2+} signaling in spines, point out gaps in our knowledge, and discuss the limits of the existing experimental techniques.

1 Introduction

Postsynaptic calcium (Ca^{2+}) signals couple synaptic transmission and neuronal excitation to complex intracellular biochemical signals and transduction pathways. Of particular interest are Ca^{2+} accumulations in dendrites and dendritic spines. Spine Ca^{2+} accumulations trigger long-term-potentiation (LTP) and depression (LTD) of synaptic transmission (75). They also trigger structural plasticity of dendritic spines (123), activate transcriptional programs (30), and induce cytotoxicity (114). How does one second messenger participate in these disparate functions?

The diverse functions of Ca^{2+} can likely be attributed in part to the diverse molecular pathways underlying Ca^{2+} influx. Depending on the stimulus conditions, Ca^{2+} can enter the spine cytoplasm through voltage-sensitive Ca^{2+} channels (VSCCs), synaptic glutamate receptors, Ca-induced Ca^{2+} release and other pathways. In addition, Ca^{2+} signals with different amplitudes and durations likely have different meanings for postsynaptic signal transduction. The dynamics of spine Ca^{2+} signals depend on the properties of the Ca^{2+} sources, on the concentration and affinity of endogenous Ca^{2+} buffers, the efficiency of Ca^{2+} extrusion, and the local morphology of the spine and surrounding dendrite.

The advent of two-photon excitation laser scanning microscopy (TPLSM) (16) has allowed high-resolution imaging in intact brain slices and even in the intact brain. TPLSM together with Ca^{2+} imaging and two-photon photorelease (TPPR) of

J.W. Hell, M.D. Ehlers (eds.), *Structural and Functional Organization of the Synapse*,
DOI: 10.1007/978-0-387-77232-5_15, © Springer Science+Business Media, LLC 2008

glutamate have allowed the analysis of the factors shaping Ca^{2+} signals in dendritic spines (Fig. 1). We begin by reviewing general and cell-type specific aspects of Ca^{2+} signals in mammalian central nervous system neurons. We then discuss the methods used for measuring Ca^{2+} signals in neuronal microcompartments, with a particular emphasis on their limitations.

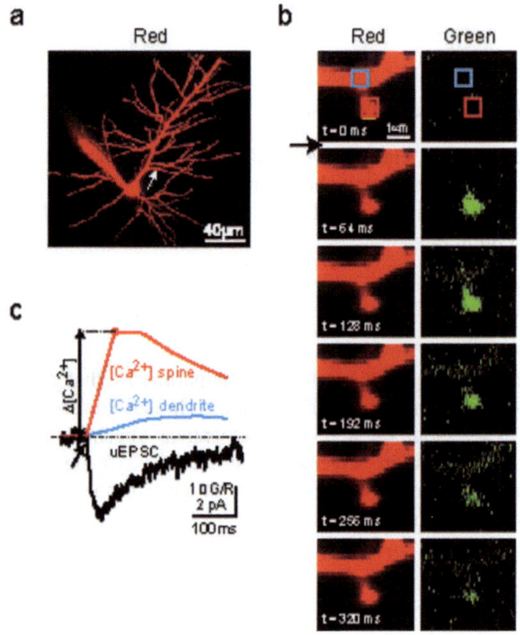

Fig. 1. Simultaneous TPLSM $[Ca^{2+}]$ Imaging and TPPR of Glutamate at Single Spines. Ca^{2+} influx was through NMDARs. (**a**) A neuron loaded with a high-affinity $[Ca^{2+}]$ indicator (Fluo-5F; *green*) and a Ca^{2+}-insensitive fluorophore (Alexa 594; *red*). (**b**) Time-lapse image of a region of interest (white arrow in (A)). TPPR of glutamate is indicated by the white bar (*top, left*) and occurred immediately after the first image. Two regions of interest are also shown. (**c**) NMDA-receptor-mediated current measured using a somatic whole cell pipette (*black*) and changes in calcium concentration (boxes in **b**). Under these conditions $\Delta G/R = 1$ corresponds to $[Ca^{2+}] = 220$ nM. Note that the dendritic Ca^{2+} accumulations are due to the presence of the high-affinity indicator, which slows the extrusion of Ca^{2+} and facilitates the spread of Ca^{2+} (Equation (1)). Adapted from (100).

2 General Classes of Postsynaptic Ca2+ Signals

2.1 Back-Propagating Action Potentials

In most vertebrate neurons, action potentials are generated near the soma and actively propagate down the axon. In many classes of neurons, the action potential also invades and propagates through at least the proximal portion of the dendritic arbor. These "back-propagating action potentials (bAPs)" open voltage-sensitive

Ca^{2+} channels (VSCCs), triggering Ca^{2+} entry into the dendrite and spines. The magnitude and kinetics of bAP-mediated Ca^{2+} transients as well as the distance that the bAP propagates into the dendrite varies among different classes of neurons. Furthermore, the classes of VSCCs that are expressed differ between the dendrite shaft and spine, between basal and apical dendrites, and between neurons classes.

Nevertheless, bAP-evoked Ca^{2+} transients share common features across neurons. Because the duration of the bAP is typically brief and because VSCC activation and deactivation kinetics are fast, bAPs evoke short-lived Ca^{2+} currents (~1 ms). Indeed, the rising phases of bAP-evoked Ca^{2+} transients in dendrite shafts and in spines is so brief that they must result from the opening of voltage-sensitive Ca^{2+} channels (VSCCs) directly in both compartments (e.g. (124)). Pharmacological analysis has identified multiple classes of VSCCs in spines with the exact complement of channels varying both by cell type and species (8, 12, 21, 40, 89, 95, 122). The functional relevance of bAP triggered activation of VSCCs is unclear, but, as discussed below, opening of VSCCs on the spine head can activate Ca-dependent signaling cascades that regulate other ion channels expressed in the spine (8, 122).

2.2 Synaptically-Evoked Ca2+ Transients

Glutamatergic synapses typically contain at least one class of ionotropic glutamate receptor that is Ca^{2+} permeable. This can be NMDA-type glutamate receptors (NMDARs), which efficiently conduct Ca, or certain isoforms of AMPA (AMPARs) and Kainate-type glutamate receptors (KRs) that, depending on their subunit composition, can be Ca^{2+} permeable. In addition, metabotropic-glutamate receptors (mGluRs) may also be expressed at synapses. These G-protein coupled receptors (GPCRs) typically signal through G_q-type G-proteins and phospholipase C (PLC) to trigger the release of Ca^{2+} from intracellular stores.

Synaptically-evoked Ca^{2+} transients are spatially and temporally complex compared to those evoked by bAPs (17, 124). Since Ca^{2+} entry is spatially inhomogeneous – i.e. it occurs at active synapses but neither in the remainder of the dendrite nor in inactive spines – [Ca] gradients are present along the dendrite. In spiny neurons, the thin spine neck (see Chapter by Harris) posses a significant barrier to diffusion that restricts synaptically-evoked Ca^{2+} increases to the head of the active spine (59, 87, 104). In aspiny neurons, synaptically-evoked Ca^{2+} increases are restricted to small stretches of dendrites in the immediate vicinity of active synapses due to the action of low-mobility Ca-binding proteins, efficient Ca^{2+} clearance, and other specializations of the dendrite (31, 102).

Because of its relevance to the induction of synaptic plasticity, NMDAR-dependent Ca^{2+} signaling has been extensively studied (Fig. 1). NMDARs typically bind glutamate with high-affinity and therefore stay open for ~100 ms (50), generating prolonged Ca^{2+} increases in the spine head (58, 125). NMDAR opening depends not only on glutamate binding but is also regulated by extracellular Mg ions that transiently enter and exit the conduction pore and provide a flickering, voltage-dependent block of the channel (64, 80). For this reason, NMDAR conductance at typical resting potentials is reduced ~20 fold compared that in the absence of extracellular Mg. However, because the block by Mg is incomplete and the driving force for Ca^{2+} entry is very large, NMDAR mediated Ca^{2+} transients in active spines reach ~ 1 μM even in neurons at resting potentials and in the presence of normal extracellular Mg (90). Furthermore, the voltage-dependence of Mg block makes Ca^{2+} influx through NMDARs dependent on the local, subthreshold changes in membrane potential and sensitive to the activity of AMPARs, voltage-sensitive Na and Ca^{2+} channels, and Ca-activated K channels near or in the postsynaptic terminal (8, 21, 48, 76, 78). Lastly, if synaptic stimulation is followed shortly by a bAP, Mg block of the NMDAR is further relieved and Ca^{2+} influx through the receptor is accentuated (12, 46, 57, 76, 77, 124). However, because the bAP is brief and the kinetics of Mg block are rapid, this effect is transient and contributes a rapid, pulse-like increase in Ca^{2+} influx through the NMDAR. Conversely, a bAP that arrives shortly before a synaptic stimulus can suppress synaptically-evoked Ca^{2+} signals in spines (46, 76), although the effect is small.

These nonlinear effects on Ca^{2+} signals of closely timed synaptic stimuli and bAPs may be necessary for "spike-timing dependent plasticity" (STDP) generated by pairing protocols (reviewed (14)). This form of plasticity has been demonstrated in a variety of neurons including hippocampal (7, 15), layer 5 and 2/3 (25, 29, 61) pyramidal neurons as well as in striatal medium spiny (28), cortical layer 4 (20), and dorsal cochlear nucleus neurons (113). In general, LTP is triggered if the synaptic activity consistently precedes bAPs by <10 ms, whereas LTD occurs if the bAP precedes the synaptic activity by less than ~100 ms (25, 55, 57, 61, 118, 119, 126) (however see (20, 113) for exceptions).

Ca-permeable AMPARs may also contribute to synaptically-evoked Ca^{2+} signals in a variety of non-pyramidal cell types (12, 102, 116). AMPARs lacking the GluR2 subunit have high Ca^{2+} permeability compared to receptors that contain this subunit (37, 39, 42, 96). Ca-permeable AMPARs are not found in all neurons and are even differentially expressed in synapses of individual neurons (81, 110). Furthermore, even in neurons that express GluR2, highly Ca-permeable GluR2-lacking AMPARs may still form and endow spines with AMPAR-mediated synaptic Ca^{2+} signals. For example, in striatal medium spiny neurons, which express GluR2, a subpopulation of Ca-permeable AMPARs contribute a small fraction of the total synaptic current but mediate a large fraction of synaptically-evoked Ca^{2+} signals (12).

2.3 General Mechanisms of Ca2+ Clearance and Sequestration

The resting level of free Ca^{2+} in neurons is on the order of 100 nM and is set by the balance of Ca^{2+} influx and efflux. When Ca^{2+} is increased beyond the resting level net extrusion of Ca^{2+} from the cytoplasm occurs through one of several mechanisms. ATP-utilizing transporters pump Ca^{2+} across the plasmamembrane and out of the cell (plasmamembrane Ca^{2+} ATPase, PMCA) or across the ER membrane and into its lumen (smooth endoplasmic reticulum Ca^{2+} ATPase, SERCA). In addition, the Ca/Na exchanger uses net influx of Na to power the extrusion of Ca^{2+} and may contribute not only to the clearance of Ca^{2+} but to the accumulation of Na following strong activity (84). Once Ca^{2+} enters the cytoplasm, most of it is rapidly bound by Ca^{2+} binding proteins. The concentrations and affinities of Ca^{2+} binding proteins vary from neuron to neuron and even within different regions of the cell. For example, in apical dendritic spines of hippocampal CA1 pyramidal neurons ~25 ions of Ca^{2+} are bound by protein for every one that remains free whereas in the cell body and apical dendrite the ratio is closer to 100:1 (60, 90). As explained in the *Perturbations due to Ca^{2+} indicators* section below, the ratio of bound to free Ca^{2+} is referred to as the "Ca^{2+} binding ratio" and affects the amplitude and time course of intracellular Ca^{2+} transients as well as the diffusion of Ca^{2+} within the cytoplasm.

3 Cell-Type Specific Features of Ca2+ Signaling

3.1 Pyramidal Neurons

In pyramidal neurons of the hippocampus and cortex, activation of the synapse associated with an individual dendritic spine leads to Ca^{2+} accumulation that is limited to the spine head (48, 58, 70, 79, 87, 100, 124, 125). These synaptically-evoked Ca^{2+} signals are principally mediated by NMDARs and are largely eliminated by pharmacological antagonists of these receptors (48, 58, 87, 125). Synaptically-evoked Ca^{2+} signals have been most extensively examined in spines of CA1 hippocampal pyramidal neurons and results from studies of these cells will be used for the discussion in this section. In these cells, the amplitude and kinetics of synaptically-evoked spine head Ca^{2+} signals are determined by the properties and numbers of NMDARs present in the postsynaptic terminal as well as by the membrane potential, the activity of kinases and phosphatases, and the opening of other ion channels in the active spine.

NMDARs are heteromers of NR1 and NR2 subunits, and the subunit composition of the receptor strongly influences synaptic Ca^{2+} signals. In rat CA1 pyramidal neurons, NMDARs containing NR1 and NR2A or NR2B subunits are typically found in spines and at synapses. The contributions of NR2A and NR2B containing receptors to NMDAR-dependent currents and spine head Ca^{2+} transients vary widely from spine to spine (100). Since NR2B-containing NMDARs deactivate more slowly (67), contribute relatively more Ca^{2+} to synaptically-evoked Ca^{2+} transients (100), and are coupled to different downstream signaling systems than NR2A-containing receptors (3, 101), this heterogeneity may allow NMDAR opening to have different functional implications for plasticity induction (4, 52, 62) (but see (5, 69)). Once incorporated into the synapse, the Ca^{2+} permeability of NMDARs is also regulated on a fast timescale by phosphorylation. For example, PKA activity enhances the Ca^{2+} permeability of both NR2A and NR2B containing receptors (99) and repetitive activation of NR2B-containing NMDARs activates a serine/threonine phosphatase that acts within an inhibitory regulatory loop to decreases Ca^{2+} permeability of NMDARs (101).

Synaptically-evoked Ca^{2+} signals in hippocampal CA1 pyramidal neurons cells are also shaped by the activity and properties of ion channels other than NMDARs. Several studies have shown that blocking AMPARs reduces synaptically-evoked Ca^{2+} signals in active spines (23, 48) (although see (58)). Since Ca-permeable AMPARs are typically not found at synapses made onto pyramidal neurons (although see (83, 109)), these effects suggest that the spine is sufficiently depolarized during the synaptic potential to reduce the efficacy of Mg-block of the NMDAR, thereby enhancing Ca^{2+} entry.

Depolarization in the spine head also activates VSCCs that contribute directly to spine head Ca^{2+} transients and indirectly shape synaptic signals. In mouse hippocampal CA1 pyramidal neurons blockade of L-type VSCCs reduces spine head Ca^{2+} transients evoked by 2-photon glutamate uncaging, suggesting that these channels contribute to synaptically-evoked Ca^{2+} transients (8). In these same cells, R-type VSCCs are found selectively on the spine and not in the dendrite and their blockade leads to a counter-intuitive increase in the amplitude of spine head Ca^{2+} transients and somatic synaptic potentials. This phenomenon is explained by the presence of an inhibitory regulatory mechanism in which Ca^{2+} influx through R-type VSCCs activates SK-type Ca-activated K channels which repolarize the spine and truncate Ca^{2+} influx through NMDARs (8, 78). Interestingly, the properties of R and L-type VSCCs in the spine head can be quickly modulated and may offer points of regulation of synaptic signals. In rat CA1 pyramidal neurons L-type VSCCs opening is revealed by activation of noradrenergic receptors (38) whereas activation of L-type VSCCs during prolonged bAP trains reduces the opening probability of R-type VSCCs (122).

However, conclusions drawn from the analysis of one class of pyramidal neuron in one species should not be considered to be generally applicable. Thus, although spines of mouse hippocampal CA1 pyramidal neurons express L-type VSCCs whose opening contributes to both bAP- and synaptically-evoked Ca^{2+} transients (8), Ca^{2+} influx through these channels is not detectable in spines of rat CA1 pyramidal neurons in the basal state (38, 89). Similarly, contributions of N and P/Q-type VSCCs to spine Ca^{2+} signals are seen in rat cortical layer 5 and mouse CA1 hippocampal pyramidal neurons but not in rat CA1 hippocampal pyramidal neurons (8, 89, 95, 122). Furthermore, the contribution of Ca^{2+} release from intracellular stores to synaptic signals in pyramidal neurons in controversial. Endoplasmic reticulum (ER) is a prominent internal Ca^{2+} store found in many spines in hippocampal pyramidal neurons (2, 103), with ryanodine receptors (RyRs) found in both dendrites and spines, and IP3 receptors (IP3Rs) restricted to dendrites (6, 97). A subset of studies have found that NMDAR activation triggers Ca^{2+} release from internal Ca^{2+} stores (1, 23), possibly via Ca-induced Ca^{2+} release (CICR). However, this contribution may depend on the recording conditions or experimental preparation, having been described in organotypic slices (23) but not in acute slices (48, 58, 87). Other studies have found that combinations of strong synaptic stimulation, bAPs trains, and GPCR activation can trigger waves of Ca^{2+} release from intracellular stores in the apical dendrite (72, 73).

3.2 Purkinje Cells

Cerebellar Purkinje Cells (PCs) have large dendrites with a very high density of spines and receive >100,000 glutamatergic synapses from parallel fibers (PFs), the axons of granule cells (98). In addition, mature PCs receive a powerful input from a single climbing fiber, the terminal axon of a neuron in the inferior olive.

Synaptically-evoked Ca^{2+} signals in PCs dendrites have been extensively studied and found to be fundamentally different than those in pyramidal neurons. Mature PCs lack functional NMDARs (24, 82), and PF-evoked Ca^{2+} signals are instead mediated by activation of synaptic AMPARs and mGluRs. Furthermore, PF-evoked Ca^{2+} signals in PC dendrites vary dramatically in amplitude and spatial extent depending on the strength of stimulation (35, 65). Weak synaptic stimulation triggers fast Ca^{2+} signals that are localized to individual spines and are mediated either by direct Ca^{2+} influx through AMPARs or by activation of VSCCs in the spine triggered by the synaptic potential (17). Stronger stimulation, such as brief trains of subthreshold PF input, produces biphasic Ca^{2+} transients in individual spines or small stretches of dendrites and the additional slow component is blocked by mGluR or IP3R antagonists (27, 107). Similar, prolonged dendritic Ca^{2+} transients can be evoked by direct activation of mGluRs or intracellular generation of IP3 and are mediated by Ca^{2+} release from internal stores (44, 53). Purkinje cells have ER in dendrites and spines and express IP3Rs and RyRs in dendrites as well as IP3Rs in spines (6, 34, 93, 115). Brief trains of PF activity also induce, in a Ca-dependent manner, production of endocannabinoids in the PC dendrite which signal retrogradely to reduce probability of glutamate release from the presynaptic bouton (10, 11). Lastly, strong parallel fiber stimulation triggers dendritic spikes and large Ca^{2+} signals that are mediated by VSCCs and extend through larger portions of the dendrite (22).

In contrast to the graded nature of PF responses, activation of the CF input elicits a complex spike in the PC and a widespread Ca^{2+} signal in the proximal dendrites (85). CF-evoked Ca^{2+} signals are blocked by AMPAR antagonists, and are mediated by VSCCs (45, 47, 65). CF stimulation, when paired with stimulation of PF stimulation, nonlinearly boosts synaptically-evoked Ca^{2+} signals in the regions of dendrite containing active PF inputs (10, 117). Although CF stimulation alone is insufficient to trigger endocannibinoid release, this pairing protocol elevates intracellular Ca^{2+} to levels that trigger production and release of endocannabinoids from the PC (9, 10). The mechanism by which pairing CF inputs with weak PF stimulation generates supralinear Ca^{2+} signals is unclear and may include saturation of Ca^{2+} binding proteins in the dendrite (56). A contribution of release of Ca^{2+} release from internal stores has been reported in a study that focused on signaling within spines (27, 117) whereas other studies have reported no contribution of Ca^{2+} release (10).

Repetitive, paired activation of CF and PF induces long-term depression (LTD) of active PF to PC synapses (41, 51). LTD induction requires Ca^{2+} signaling, is abolished by blocking mGluRs, depleting internal Ca^{2+} stores, or interfering with endocannabinoid mediated signaling (27, 43, 91, 92, 117).

3.3 Other Spiny Neurons

Synaptically-evoked Ca^{2+} signals in several other cell types have also been examined, including layer 4 spiny stellate cells (76), striatal medium spiny neurons (12), olfactory bulb granule cells (21) and lateral amygdala projection neurons (40). In the somatosensory cortex, the spines of layer 4 spiny stellate cells display synaptically-evoked Ca^{2+} signals mediated by NMDARs (76). While blocking

AMPARs in these cells also reduces the synaptically-evoked Ca^{2+} signals, this is thought to be due to the elimination of relief of Mg block of the NMDARs. In contrast, medium spiny neurons in the striatum have synaptically-evoked Ca^{2+} signals in spines that are mediated by both Ca-permeable AMPARs and NMDARs (12). In these cells, both AMPAR and NMDAR Ca^{2+} signals are strongly influenced subthreshold shifts in the resting membrane potential. These depolarizations decrease the AMPAR Ca^{2+} signal due a decreased driving force for Ca^{2+} entry, and increase the NMDAR Ca^{2+} signal via relief of Mg block. Finally, granule cells in the olfactory bulb have synaptically-evoked Ca^{2+} signals at dendrodendritic synapses mediated by a variety of Ca^{2+} sources, including NMDARs, T-type channels and release from internal Ca^{2+} stores (21)

4 Measuring Spine Ca2+ Signals

Ca^{2+} imaging has revolutionized our understanding of Ca^{2+} signaling in neurons. Ca^{2+} imaging in spines has been made possible by the development of fluorescent Ca^{2+} indicators (reviewed in (112)) and microscopy methods that allow high-resolution and high-sensitivity imaging (reviewed in (19)). Cooled CCD cameras and confocal microscopy for the first time allowed the imaging of Ca^{2+} changes in single dendritic spines in cultured neurons (71). However, synapses ultimately need to be studied in intact tissues, such as brain slices or whole animals. Scattering by neural tissue severely degrades contrast and resolution in wide-field and confocal microscopy. The problems posed by scattering can be overcome by two-photon excitation laser scanning microscopy (TPLSM) (16) (reviewed in (105)). TPLSM can resolve individual spines in brain slices (19) and even in the intact brain (18, 49). TPLSM can be combined with fluorescent Ca^{2+} indicators to image Ca^{2+} dynamics in neuronal microcompartments, including spines (124) and presynaptic terminals (13) in intact tissues.

Ca^{2+} imaging has been used to measure the time-course of Ca^{2+} signals in response to physiological stimuli. However, since Ca^{2+} indicators are Ca^{2+} buffers, their presence – distorts the Ca^{2+} signals (36, 60, 74, 88). Furthermore, Ca^{2+} indicators are non-linear Ca^{2+} sensors with limited dynamic ranges. A critical assessment of results derived with Ca^{2+} imaging in dendritic spines must take the limitations of Ca^{2+} imaging into account.

4.1 Resolution Limits

TPLSM can resolve objects ~0.5 micrometers apart, sufficient to isolate fluorescence signals arising from most spines without crosstalk with other spines and the parent dendrite (105). For example, Ca^{2+} signals in "mushroom" and "thin" spines can be isolated (definitions of spine types are from (33)). However, the length of "stubby spines", which lack spine necks, is on the order of the resolution limit of TPLSM, making them impossible to resolve from their parent dendrite. In addition, the lack of a spine neck makes the delineation of spine head and dendrite arbitrary. For stubby spines, fluorescence signals from spine and dendrite will thus be mixed.

TPLSM averages signals over the two-photon excitation volume (~0.5×0.5×1.5 μm^3) (105). The excitation volume is on the order of the size of typical spine heads (33). Using standard synthetic Ca^{2+} indicators, TPLSM is therefore insensitive to the strong Ca^{2+} gradients that must exist in dendritic spines. Over the last few years it has become clear that Ca^{2+} gradients, over length scales of nanometers, may be a critical determinant of Ca-dependent signaling. In particular, molecular Ca^{2+} sensors, such as calmodulin, are often fixed in Ca^{2+} microdomains close to a particular Ca^{2+} sources, such as voltage-sensitive calcium channels (VSCCs). For example, these molecular level interactions determine the VSCC response to intracellular Ca^{2+} (68). The length scales of Ca^{2+} microdomains (nanometers) are therefore 10–100 times smaller than the resolution limit of optical microscopy.

4.2 Ca2+ Indicator Calibration

Standard Ca^{2+} indicators consist of a fluorophore coupled to a Ca^{2+} buffer, such as ethylenedioxybis(o-phenylenenitrilo) tetraacetic acid (BAPTA). Ca^{2+} binding changes the fluorescence properties of the indicator (typically the fluorescence yield increases). A large number of Ca^{2+} indicators, with differing K_d and dynamic range, have become available. The choice of the particular indicator must be made on the basis of the amplitude and time-course of the Ca^{2+} signal to be studied.

For quantitative experiments it is important to operate in the linear range of the indicator. Nonlinearities cause difficulties in quantifying [Ca], buffering capacities, and biologically interesting fluctuations in [Ca] (89). The fractional nonlinearity (NL) is given by ~ [Ca]/Kd. Therefore, when [Ca] $\ll K_D$, the fluorescence intensity is proportional to [Ca]. Similarly, to limit *NL* to less than 20%, [Ca] should not exceed ~ $K_D/5$. It is best practice to express Ca^{2+} transient amplitudes in fractional units, where "1" corresponds to saturating Ca^{2+} (8, 94, 101).

Action potential-evoked evoked Δ[Ca] can be measured using a variety of indicators, but only low affinity indicators (e.g. Fluo-4FF, X-Rhod-FF) are appropriate for quantitative measurements of the larger NMDA-mediated [Ca] accumulations. Higher Ca^{2+} indicator concentrations lead to smaller changes in [Ca], and therefore also tend to improve linearity, but at the expense of decreasing signal-to-noise ratio (SNR). Supposedly quantitative Ca^{2+} imaging studies that do not treat indicator non-linearities explicitly should be viewed with skepticism.

4.3 Perturbations Due to Ca2+ Indicators

Ca^{2+} signals in cells are shaped by the Ca^{2+} sources, Ca^{2+} buffers, and Ca^{2+} clearance mechanisms (36, 60, 74, 88, 108). Since Ca^{2+} indicators themselves are Ca^{2+} buffers their presence perturbs Ca^{2+} signals and downstream signals. Measuring Ca^{2+} therefore invariably perturbs Ca^{2+} signals and Ca^{2+} signaling.

A variety of Ca^{2+} indicators are available to measure Ca^{2+} over many orders of magnitude in concentration. The perturbing effects of Ca^{2+} buffers can be summarized by the buffer binding ratio (κ), the fraction of the total Ca^{2+} entering the cell that is bound to buffers. For buffer X, κ_x is approximately equal to the concentration of the buffer divided by its dissociation constant for Ca^{2+} (κ_x = [X]/Kd,x).

The binding ratios for endogenous buffers (κ_B) are on the order of 100, but can range from 20 in small pyramidal dendrites and spines (90) up to more than 1,000 in Purkinje neurons (26). The buffer added by the indicator (κ_i) is typically on the order of $\kappa_i \sim 500$ (100 µM of Oregon Green Bapta-1 or Fluo-4). The relative magnitudes of κ_i and κ_B determine the extent to which the presence of the Ca^{2+} indicator perturbs the measured Ca^{2+} signal. In most experiments the indicator dominates intracellular buffers, markedly reducing the amplitudes and prolonging the durations of [Ca] transients (86, 90, 121). A variety of schemes have been used to correct for the presence of Ca^{2+} indicators and thereby calculate Ca^{2+} dynamics in the absence of Ca^{2+} indicators (36, 60, 90). However, these schemes generally are applicable only to conditions in which spatial Ca^{2+} gradients are negligible. This might apply for situations in which Ca^{2+} signals are relatively global and in which Ca^{2+} dynamics can be described by single-compartment models (e.g. action potential-evoked Ca^{2+} dynamics). The vast majority of Ca^{2+} imaging studies are performed in the regime $\kappa_I >> \kappa_B$. This means that nearly all Ca^{2+} ions entering the cell bind to the indicator, and changes in fluorescence are proportional to the total amount of Ca^{2+} entering the cell (i.e. the charge). Very few studies have seriously attempted to correct for the indicator and extrapolate to $\kappa_I = 0$ (e.g. 87).

In the presence of Ca^{2+} gradients, Ca^{2+} indicators can have highly perturbative and non-intuitive effects. In particular, Ca^{2+} indicators can dramatically accelerate the transport of Ca. Transport of Ca^{2+} is characterized by an effective diffusion coefficient, D_{eff}, which depends on the diffusion coefficients of free Ca^{2+} (D_{Ca}), and Ca^{2+} bound to mobile buffers (D_{m}) and indicator (D_{i}):

$$D_{eff} = \frac{D_{Ca} + D_m \kappa_m + D_i \kappa_i}{1 + \kappa_B + \kappa_m + \kappa_i} \tag{1}$$

Here κ_B represents the strength of immobile buffers ($D_B = 0$). The mobilities of Ca^{2+} dyes can be orders of magnitude larger than those of cellular buffers, since endogenous buffers seem to be largely immobile ($\kappa_m \approx 0$). Equation (1) therefore implies that even in situations of "low added buffer" ($\kappa_i \leq \kappa_B$), where [Ca] amplitudes and extrusion rates are relatively unperturbed, the spread of Ca^{2+} can still be highly distorted. The presence of indicator rapidly breaks down micrometer-level dendritic [Ca] gradients and thus accelerates equilibration of [Ca] between spine and dendrite (Fig. 1c). Conclusions about the spread of Ca^{2+} based on Ca^{2+} imaging therefore need to be viewed with great skepticism. For example, the slight spread of Ca^{2+} observed in Fig. 1b,c can be attributed to the presence of the indicator.

4.4 Genetically Encoded Ca2+ Indicators

Genetically encoded Ca^{2+} indicators (GECIs) have the potential to overcome some of the limitations of synthetic indicators (66). GECIs could be used to reduce the effect of Ca^{2+} indicators on Ca^{2+} spread. For example, GECIs could be fused to actin, which is concentrated in spines and stable over the time scales relevant to Ca^{2+} accumulations. Alternatively, GECIs could be used to selectively read out Ca^{2+} entering neurons through particular channels. For example, low-affinity GECIs could be fused to particular channels or associated scaffolds (32). Under the appropriate conditions (distance between GECI and the mouth of the channel, affinity between GECI and Ca^{2+}) GECIs would only respond to Ca^{2+} entering the cell through the tagged channel. However, current GECIs are inferior Ca^{2+} sensors compared to synthetic indicators, producing smaller and slower responses, limiting their use in small compartments such as dendritic spines. Furthermore, their responses to Ca^{2+} are complex and non-linear, which may make quantitative Ca^{2+} measurements challenging (121).

4.5 Two-Photon Glutamate Uncaging

Analysis of spine Ca^{2+} signaling has benefited from combining TPLSM and two-photon photorelease (TPPR) of caged glutamate (63). One laser is used to trigger photolysis of caged glutamate while a second is used for Ca^{2+} imaging (12, 54; 100) (Fig. 1). Using this approach, the postsynaptic terminal on a spine can be stimulated while the evoked signals are monitored. Furthermore, combined TPLSM and TPPR allow analysis of postsynaptic signaling using perturbations that might affect the presynaptic terminal. TPPR on single spines can produce postsynaptic currents with amplitudes and time-courses that mimic unitary synaptic currents (12, 63).

However, several limitations of this approach must be considered. Glutamate is uncaged throughout the two-photon excitation volume and may spread further by diffusion before binding to glutamate receptors. The effective volume of TPPR (~1 fL) is therefore far larger than the synaptic cleft. A different set of receptors, including extrasynaptic receptors, may therefore be activated with TPPR compared to synaptic transmission. This problem is exacerbated if the duration of the uncaging laser pulse is more than a few hundred microseconds.

A related issue is that the efficiency of glutamate receptor activation falls off relatively slowly with distance from the spine, with a length constant of ~1 μm (100). Dendritic spines need to be chosen for stimulation that are well-separated from other spines, biasing the sample. When stimulating short spines, such as stubby spines, glutamate receptors on the shaft may also be activated.

Lastly, control of the laser pulse duration and power allows the production of a broad range of spatio-temporal glutamate concentration profiles. It is important to avoid strong stimuli that might uncover phenomena that have no physiological relevance. It is critical to calibrate and standardize the amount of glutamate released onto spines at different depths and in varying optical environments (e.g. (8)).

5 Conclusions and Outlook

The development of Ca^{2+} indicators and high-resolution imaging methods have revolutionized our understanding of neuronal Ca^{2+} signaling. In a number of systems, and under a variety of physiologically relevant conditions, we understand how Ca^{2+} enters spines. We also understand what fraction of Ca^{2+} binds to buffers and how Ca^{2+} is eventually cleared from the neuronal cytoplasm. However, we know relatively little about the dynamics and roles of the fast nanometer sized Ca^{2+}

gradients that are known to exit close to the mouth of Ca^{2+} channels. There is mounting indirect evidence that such microdomains are critical to trigger specific Ca-dependent signaling in dendritic spines (3, 8, 122). Targeting of Ca^{2+} sensors to particular molecular scaffolds might allow the study of these Ca^{2+} microdomains (32, 111) in dendritic spines. In addition, we know little about how Ca^{2+} activates different Ca-dependent pathways. To answer these questions, it will be important to develop imaging probes that provide sensitivities similar to Ca^{2+} indicators to track Ca-dependent signaling in neuronal microcompartments (106, 120).

References

1. Alford S, Frenguelli BG, Schofield JG, and Collingridge GL. Characterization of Ca2+ signals induced in hippocampal CA1 neurones by the synaptic activation of NMDA receptors. *J Physiol* 469: 693–716, 1993.
2. Bannai H, Inoue T, Nakayama T, Hattori M, and Mikoshiba K. Kinesin dependent, rapid, bi-directional transport of ER sub-compartment in dendrites of hippocampal neurons. *J Cell Sci* 117: 163–175, 2004.
3. Barria A and Malinow R. NMDA receptor subunit composition controls synaptic plasticity by regulating binding to CaMKII. *Neuron* 48: 289–301, 2005.
4. Bartlett TE, Bannister NJ, Collett VJ, Dargan SL, Massey PV, Bortolotto ZA, Fitzjohn SM, Bashir ZI, Collingridge GL, and Lodge D. Differential roles of NR2A and NR2B-containing NMDA receptors in LTP and LTD in the CA1 region of two-week old rat hippocampus. *Neuropharmacology* 52: 60–70, 2007.
5. Berberich S, Punnakkal P, Jensen V, Pawlak V, Seeburg PH, Hvalby O, and Kohr G. Lack of NMDA receptor subtype selectivity for hippocampal long-term potentiation. *J Neurosci* 25: 6907–6910, 2005.
6. Berridge MJ. Neuronal calcium signaling. *Neuron* 21: 13–26, 1998.
7. Bi GQ and Poo MM. Synaptic modifications in cultured hippocampal neurons: dependence on spike timing, synaptic strength, and postsynaptic cell type. *J Neurosci* 18: 10464–10472, 1998.
8. Bloodgood BL and Sabatini BL. Nonlinear regulation of unitary synaptic signals by CaV(2.3) voltage-sensitive calcium channels located in dendritic spines. *Neuron* 53: 249–260, 2007.
9. Brenowitz SD, Best AR, and Regehr WG. Sustained elevation of dendritic calcium evokes widespread endocannabinoid release and suppression of synapses onto cerebellar Purkinje cells. *J Neurosci* 26: 6841–6850, 2006.
10. Brenowitz SD and Regehr WG. Associative short-term synaptic plasticity mediated by endocannabinoids. *Neuron* 45: 419–431, 2005.
11. Brown SP, Brenowitz SD, and Regehr WG. Brief presynaptic bursts evoke synapse-specific retrograde inhibition mediated by endogenous cannabinoids. *Nat Neurosci* 6: 1048–1057, 2003.
12. Carter AG and Sabatini BL. State-dependent calcium signaling in dendritic spines of striatal medium spiny neurons. *Neuron* 44: 483–493, 2004.
13. Cox CL, Denk W, Tank DW, and Svoboda K. Action potentials reliably invade axonal arbors of rat neocortical neurons. *Proc Natl Acad Sci U S A* 97: 9724–9728, 2000.
14. Dan Y and Poo M-M. Spike timing-dependent plasticity: from synapse to perception. *Physiol Rev* 86: 1033–1048, 2006.

15. Debanne D, Gahwiler BH, and Thompson SM. Long-term synaptic plasticity between pairs of individual CA3 pyramidal cells in rat hippocampal slice cultures. *J Physiol* 507 (Pt 1): 237–247, 1998.
16. Denk W, Strickler JH, and Webb WW. Two-photon laser scanning fluorescence microscopy. *Science* 248: 73–76, 1990.
17. Denk W, Sugimori M, and Llinas R. Two types of calcium response limited to single spines in cerebellar Purkinje cells. *Proc Natl Acad Sci U S A* 92: 8279–8282, 1995.
18. Denk W and Svoboda K. Photon upmanship: why multiphoton imaging is more than a gimmick. *Neuron* 18: 351–357, 1997.
19. Denk W, Yuste R, Svoboda K, and Tank DW. Imaging calcium dynamics in dendritic spines. *Curr Opin Neurobiol* 6: 372–378, 1996.
20. Egger V, Feldmeyer D, and Sakmann B. Coincidence detection and changes of synaptic efficacy in spiny stellate neurons in rat barrel cortex. *Nat Neurosci* 2: 1098–1105, 1999.
21. Egger V, Svoboda K, and Mainen ZF. Dendrodendritic synaptic signals in olfactory bulb granule cells: local spine boost and global low-threshold spike. *J Neurosci* 25: 3521–3530, 2005.
22. Eilers J, Augustine GJ, and Konnerth A. Subthreshold synaptic Ca2+ signalling in fine dendrites and spines of cerebellar Purkinje neurons. *Nature* 373: 155–158, 1995.
23. Emptage N, Bliss TV, and Fine A. Single synaptic events evoke NMDA receptor-mediated release of calcium from internal stores in hippocampal dendritic spines. *Neuron* 22: 115–124, 1999.
24. Farrant M and Cull-Candy SG. Excitatory amino acid receptor-channels in Purkinje cells in thin cerebellar slices. *Proc Biol Sci* 244: 179–184, 1991.
25. Feldman DE. Timing-based LTP and LTD at vertical inputs to layer II/III pyramidal cells in rat barrel cortex. *Neuron* 27: 45–56, 2000.
26. Fierro L and Llano I. High endogenous calcium buffering in Purkinje cells from rat cerebellar slices. *J Physiol* 496 (Pt 3): 617–625, 1996.
27. Finch EA and Augustine GJ. Local calcium signalling by inositol-1,4,5-trisphosphate in Purkinje cell dendrites. *Nature* 396: 753–756, 1998.
28. Fino E, Glowinski J, and Venance L. Bidirectional activity-dependent plasticity at corticostriatal synapses. *J Neurosci* 25: 11279–11287, 2005.
29. Froemke RC and Dan Y. Spike-timing-dependent synaptic modification induced by natural spike trains. *Nature* 416: 433–438, 2002.
30. Ghosh A, Ginty DD, Bading H, and Greenberg ME. Calcium regulation of gene expression in neuronal cells. *J Neurobiol* 25: 294–303, 1994.
31. Goldberg JH, Tamas G, Aronov D, and Yuste R. Calcium microdomains in aspiny dendrites. *Neuron* 40: 807–821, 2003.
32. Guerrero G, Reiff DF, Agarwal G, Ball RW, Borst A, Goodman CS, and Isacoff EY. Heterogeneity in synaptic transmission along a Drosophila larval motor axon. *Nat Neurosci* 8: 1188–1196, 2005.
33. Harris KM, Jensen FE, and Tsao B. Three-dimensional structure of dendritic spines and synapses in rat hippocampus (CA1) at postnatal day 15 and adult ages: implications for the maturation of synaptic physiology and long-term potentiation. *J Neurosci* 12: 2685–2705, 1992.
34. Harris KM and Stevens JK. Dendritic spines of rat cerebellar Purkinje cells: serial electron microscopy with reference to their biophysical characteristics. *J Neurosci* 8: 4455–4469, 1988.
35. Hartell NA. Strong activation of parallel fibers produces localized calcium transients and a form of LTD that spreads to distant synapses. *Neuron* 16: 601–610, 1996.
36. Helmchen F, Imoto K, and Sakmann B. Ca2+ buffering and action potential-evoked Ca2+ signaling in dendrites of pyramidal neurons. *Biophys J* 70: 1069–1081, 1996.

37. Hollmann M, Hartley M, and Heinemann S. Ca2+ permeability of KA-AMPA–gated glutamate receptor channels depends on subunit composition. *Science* 252: 851–853, 1991.
38. Hoogland TM and Saggau P. Facilitation of L-type Ca2+ channels in dendritic spines by activation of beta2 adrenergic receptors. *J Neurosci* 24: 8416–8427, 2004.
39. Hume RI, Dingledine R, and Heinemann SF. Identification of a site in glutamate receptor subunits that controls calcium permeability. *Science* 253: 1028–1031, 1991.
40. Humeau Y, Herry C, Kemp N, Shaban H, Fourcaudot E, Bissiere S, and Luthi A. Dendritic spine heterogeneity determines afferent-specific Hebbian plasticity in the amygdala. *Neuron* 45: 119–131, 2005.
41. Ito M and Kano M. Long-lasting depression of parallel fiber-Purkinje cell transmission induced by conjunctive stimulation of parallel fibers and climbing fibers in the cerebellar cortex. *Neurosci Lett* 33: 253–258, 1982.
42. Jonas P and Burnashev N. Molecular mechanisms controlling calcium entry through AMPA-type glutamate receptor channels. *Neuron* 15: 987–990, 1995.
43. Khodakhah K and Armstrong CM. Induction of long-term depression and rebound potentiation by inositol trisphosphate in cerebellar Purkinje neurons. *Proc Natl Acad Sci U S A* 94: 14009–14014, 1997.
44. Khodakhah K and Ogden D. Functional heterogeneity of calcium release by inositol trisphosphate in single Purkinje neurones, cultured cerebellar astrocytes, and peripheral tissues. *Proc Natl Acad Sci U S A* 90: 4976–4980, 1993.
45. Knopfel T, Vranesic I, Staub C, and Gahwiler BH. Climbing fibre responses in olivo-cerebellar slice cultures. II. Dynamics of cytosolic calcium in purkinje cells. *Eur J Neurosci* 3: 343–348, 1991.
46. Koester HJ and Sakmann B. Calcium dynamics in single spines during coincident pre- and postsynaptic activity depend on relative timing of back-propagating action potentials and subthreshold excitatory postsynaptic potentials. *Proc Natl Acad Sci U S A* 95: 9596–9601, 1998.
47. Konnerth A, Dreessen J, and Augustine GJ. Brief dendritic calcium signals initiate long-lasting synaptic depression in cerebellar Purkinje cells. *Proc Natl Acad Sci U S A* 89: 7051–7055, 1992.
48. Kovalchuk Y, Eilers J, Lisman J, and Konnerth A. NMDA receptor-mediated subthreshold Ca(2+) signals in spines of hippocampal neurons. *J Neurosci* 20: 1791–1799, 2000.
49. Lendvai B, Stern E, Chen B, and Svoboda K. Experience-dependent plasticity of dendritic spines in the developing rat barrel cortex in vivo. *Nature* 404: 876–881, 2000.
50. Lester RA, Clements JD, Westbrook GL, and Jahr CE. Channel kinetics determine the time course of NMDA receptor-mediated synaptic currents. *Nature* 346: 565–567, 1990.
51. Linden DJ and Connor JA. Long-term synaptic depression. *Annu Rev Neurosci* 18: 319–357, 1995.
52. Liu L, Wong TP, Pozza MF, Lingenhoehl K, Wang Y, Sheng M, Auberson YP, and Wang YT. Role of NMDA receptor subtypes in governing the direction of hippocampal synaptic plasticity. *Science* 304: 1021–1024, 2004.
53. Llano I, Dreessen J, Kano M, and Konnerth A. Intradendritic release of calcium induced by glutamate in cerebellar Purkinje cells. *Neuron* 7: 577–583, 1991.
54. Losonczy A and Magee JC. Integrative properties of radial oblique dendrites in hippocampal CA1 pyramidal neurons. *Neuron* 50: 291–307, 2006.
55. Lynch G, Larson J, Kelso S, Barrionuevo G, and Schottler F. Intracellular injections of EGTA block induction of hippocampal long-term potentiation. *Nature* 305: 719–721, 1983.

56. Maeda H, Ellis-Davies GC, Ito K, Miyashita Y, and Kasai H. Supralinear Ca2+ signaling by cooperative and mobile Ca2+ buffering in Purkinje neurons. *Neuron* 24: 989–1002, 1999.
57. Magee JC and Johnston D. A synaptically controlled, associative signal for Hebbian plasticity in hippocampal neurons. *Science* 275: 209–213, 1997.
58. Mainen ZF, Malinow R, and Svoboda K. Synaptic calcium transients in single spines indicate that NMDA receptors are not saturated. *Nature* 399: 151–155, 1999.
59. Majewska A, Brown E, Ross J, and Yuste R. Mechanisms of calcium decay kinetics in hippocampal spines: role of spine calcium pumps and calcium diffusion through the spine neck in biochemical compartmentalization. *J Neurosci* 20: 1722–1734, 2000.
60. Maravall M, Mainen ZF, Sabatini BL, and Svoboda K. Estimating intracellular calcium concentrations and buffering without wavelength ratioing. *Biophys J* 78: 2655–2667, 2000.
61. Markram H, Lubke J, Frotscher M, and Sakmann B. Regulation of synaptic efficacy by coincidence of postsynaptic APs and EPSPs. *Science* 275: 213–215, 1997.
62. Massey PV, Johnson BE, Moult PR, Auberson YP, Brown MW, Molnar E, Collingridge GL, and Bashir ZI. Differential roles of NR2A and NR2B-containing NMDA receptors in cortical long-term potentiation and long-term depression. *J Neurosci* 24: 7821–7828, 2004.
63. Matsuzaki M, Ellis-Davies GC, Nemoto T, Miyashita Y, Iino M, and Kasai H. Dendritic spine geometry is critical for AMPA receptor expression in hippocampal CA1 pyramidal neurons. *Nat Neurosci* 4: 1086–1092, 2001.
64. Mayer ML, Westbrook GL, and Guthrie PB. Voltage-dependent block by Mg2+ of NMDA responses in spinal cord neurones. *Nature* 309: 261–263, 1984.
65. Miyakawa H, Lev-Ram V, Lasser-Ross N, and Ross WN. Calcium transients evoked by climbing fiber and parallel fiber synaptic inputs in guinea pig cerebellar Purkinje neurons. *J Neurophysiol* 68: 1178–1189, 1992.
66. Miyawaki A, Nagai T, and Mizuno H. Engineering fluorescent proteins. *Adv Biochem Eng Biotechnol* 95: 1–15, 2005.
67. Monyer H, Burnashev N, Laurie DJ, Sakmann B, and Seeburg PH. Developmental and regional expression in the rat brain and functional properties of four NMDA receptors. *Neuron* 12: 529–540, 1994.
68. Mori MX, Erickson MG, and Yue DT. Functional stoichiometry and local enrichment of calmodulin interacting with Ca2+ channels. *Science* 304: 432–435, 2004.
69. Morishita W, Lu W, Smith GB, Nicoll RA, Bear MF, and Malenka RC. Activation of NR2B-containing NMDA receptors is not required for NMDA receptor-dependent long-term depression. *Neuropharmacology* 52: 71–76, 2007.
70. Muller W and Connor JA. Dendritic spines as individual neuronal compartments for synaptic Ca2+ responses. *Nature* 354: 73–76, 1991.
71. Murphy TH, Blatter LA, Wier WG, and Baraban JM. Spontaneous synchronous synaptic calcium transients in cultured cortical neurons. *J Neurosci* 12: 4834–4845, 1992.
72. Nakamura T, Lasser-Ross N, Nakamura K, and Ross WN. Spatial segregation and interaction of calcium signalling mechanisms in rat hippocampal CA1 pyramidal neurons. *J Physiol* 543: 465–480, 2002.
73. Nakamura T, Nakamura K, Lasser-Ross N, Barbara JG, Sandler VM, and Ross WN. Inositol 1,4,5-trisphosphate (IP3)-mediated Ca2+ release evoked by metabotropic agonists and backpropagating action potentials in hippocampal CA1 pyramidal neurons. *J Neurosci* 20: 8365–8376, 2000.
74. Neher E and Augustine GJ. Calcium gradients and buffers in bovine chromaffin cells. *J Physiol* 450: 273–301, 1992.

75. Neveu D and Zucker RS. Postsynaptic levels of [Ca2+]i needed to trigger LTD and LTP. *Neuron* 16: 619–629, 1996.
76. Nevian T and Sakmann B. Single spine Ca2+ signals evoked by coincident EPSPs and backpropagating action potentials in spiny stellate cells of layer 4 in the juvenile rat somatosensory barrel cortex. *J Neurosci* 24: 1689–1699, 2004.
77. Nevian T and Sakmann B. Spine Ca2+ signaling in spike-timing-dependent plasticity. *J Neurosci* 26: 11001–11013, 2006.
78. Ngo-Anh TJ, Bloodgood BL, Lin M, Sabatini BL, Maylie J, and Adelman JP. SK channels and NMDA receptors form a Ca2+-mediated feedback loop in dendritic spines. *Nat Neurosci* 8: 642–649, 2005.
79. Nimchinsky EA, Yasuda R, Oertner TG, and Svoboda K. The number of glutamate receptors opened by synaptic stimulation in single hippocampal spines. *J Neurosci* 24: 2054–2064, 2004.
80. Nowak L, Bregestovski P, Ascher P, Herbet A, and Prochiantz A. Magnesium gates glutamate-activated channels in mouse central neurones. *Nature* 307: 462–465, 1984.
81. Ogoshi F and Weiss JH. Heterogeneity of Ca2+-permeable AMPA/kainate channel expression in hippocampal pyramidal neurons: fluorescence imaging and immunocytochemical assessment. *J Neurosci* 23: 10521–10530, 2003.
82. Perkel DJ, Petrozzino JJ, Nicoll RA, and Connor JA. The role of Ca2+ entry via synaptically activated NMDA receptors in the induction of long-term potentiation. *Neuron* 11: 817–823, 1993.
83. Plant K, Pelkey KA, Bortolotto ZA, Morita D, Terashima A, McBain CJ, Collingridge GL, and Isaac JT. Transient incorporation of native GluR2-lacking AMPA receptors during hippocampal long-term potentiation. *Nat Neurosci* 9: 602–604, 2006.
84. Regehr WG. Interplay between sodium and calcium dynamics in granule cell presynaptic terminals. *Biophys J* 73: 2476–2488, 1997.
85. Ross WN and Werman R. Mapping calcium transients in the dendrites of Purkinje cells from the guinea-pig cerebellum in vitro. *J Physiol* 389: 319–336, 1987.
86. Sabatini BL, Maravall M, and Svoboda K. Ca^{2+} signaling in dendritic spines. *Curr Opin Neurobiol* 11: 349–356, 2001.
87. Sabatini BL, Oertner TG, and Svoboda K. The life cycle of Ca(2+) ions in dendritic spines. *Neuron* 33: 439–452, 2002.
88. Sabatini BL and Regehr WG. Optical measurement of presynaptic calcium currents. *Biophys J* 74: 1549–1563, 1998.
89. Sabatini BL and Svoboda K. Analysis of calcium channels in single spines using optical fluctuation analysis. *Nature* 408: 589–593, 2000.
90. Sabatini BS, Oertner TG, and Svoboda K. The life-cycle of Ca^{2+} ions in spines. *Neuron* 33: 439–452, 2002.
91. Safo PK and Regehr WG. Endocannabinoids control the induction of cerebellar LTD. *Neuron* 48: 647–659, 2005.
92. Sakurai M. Calcium is an intracellular mediator of the climbing fiber in induction of cerebellar long-term depression. *Proc Natl Acad Sci U S A* 87: 3383–3385, 1990.
93. Satoh T, Ross CA, Villa A, Supattapone S, Pozzan T, Snyder SH, and Meldolesi J. The inositol 1,4,5,-trisphosphate receptor in cerebellar Purkinje cells: quantitative immunogold labeling reveals concentration in an ER subcompartment. *J Cell Biol* 111: 615–624, 1990.
94. Scheuss V, Yasuda R, Sobczyk A, and Svoboda K. Nonlinear [Ca2+] signaling in dendrites and spines caused by activity-dependent depression of Ca2+ extrusion. *J Neurosci* 26: 8183–8194, 2006.

95. Schiller J, Schiller Y, and Clapham DE. NMDA receptors amplify calcium influx into dendritic spines during associative pre- and postsynaptic activation. *Nat Neurosci* 1: 114–118, 1998.
96. Schneggenburger R, Zhou Z, Konnerth A, and Neher E. Fractional contribution of calcium to the cation current through glutamate receptor channels. *Neuron* 11: 133–143, 1993.
97. Sharp AH, McPherson PS, Dawson TM, Aoki C, Campbell KP, and Snyder SH. Differential immunohistochemical localization of inositol 1,4,5-trisphosphate- and ryanodine-sensitive Ca2+ release channels in rat brain. *J Neurosci* 13: 3051–3063, 1993.
98. Shepherd GM. The Synaptic Organization of the Brain (4th ed.). New York: Oxford University Press, 1998, p. 638.
99. Skeberdis VA, Chevaleyre V, Lau CG, Goldberg JH, Pettit DL, Suadicani SO, Lin Y, Bennett MV, Yuste R, Castillo PE, and Zukin RS. Protein kinase A regulates calcium permeability of NMDA receptors. *Nat Neurosci* 9: 501–510, 2006.
100. Sobczyk A, Scheuss V, and Svoboda K. NMDA receptor subunit-dependent [Ca2+] signaling in individual hippocampal dendritic spines. *J Neurosci* 25: 6037–6046, 2005.
101. Sobczyk A and Svoboda K. Activity-dependent plasticity of the NMDA-receptor fractional Ca2+ current. *Neuron* 53: 17–24, 2007.
102. Soler-Llavina GJ and Sabatini BL. Synapse-specific plasticity and compartmentalized signaling in cerebellar stellate cells. *Nat Neurosci* 9: 798–806, 2006.
103. Spacek J and Harris KM. Three-dimensional organization of smooth endoplasmic reticulum in hippocampal CA1 dendrites and dendritic spines of the immature and mature rat. *J Neurosci* 17: 190–203, 1997.
104. Svoboda K, Tank DW, and Denk W. Direct measurement of coupling between dendritic spines and shafts. *Science* 272: 716–719, 1996.
105. Svoboda K and Yasuda R. Principles of two-photon excitation microscopy and its applications to neuroscience. *Neuron* 50: 823–839, 2006.
106. Takao K, Okamoto K, Nakagawa T, Neve RL, Nagai T, Miyawaki A, Hashikawa T, Kobayashi S, and Hayashi Y. Visualization of synaptic Ca2+ /calmodulin-dependent protein kinase II activity in living neurons. *J Neurosci* 25: 3107–3112, 2005.
107. Takechi H, Eilers J, and Konnerth A. A new class of synaptic response involving calcium release in dendritic spines. *Nature* 396: 757–760, 1998.
108. Tank DW, Regehr WG, and Delaney KR. A quantitative analysis of presynaptic calcium dynamics that contribute to short-term enhancement. *J Neurosci* 15: 7940–7952, 1995.
109. Thiagarajan TC, Lindskog M, and Tsien RW. Adaptation to synaptic inactivity in hippocampal neurons. *Neuron* 47: 725–737, 2005.
110. Toth K, Suares G, Lawrence JJ, Philips-Tansey E, and McBain CJ. Differential mechanisms of transmission at three types of mossy fiber synapse. *J Neurosci* 20: 8279–8289, 2000.
111. Tour O, Adams SR, Kerr RA, Meijer RM, Sejnowski TJ, Tsien RW, and Tsien RY. Calcium green FlAsH as a genetically targeted small-molecule calcium indicator. *Nat Chem Biol* 3: 423–431, 2007.
112. Tsien RY. Fluorescent probes of cell signaling. *Annu Rev Neurosci* 12: 227–253, 1989.
113. Tzounopoulos T, Kim Y, Oertel D, and Trussell LO. Cell-specific, spike timing-dependent plasticities in the dorsal cochlear nucleus. *Nat Neurosci* 7: 719–725, 2004.
114. Vanhoutte P and Bading H. Opposing roles of synaptic and extrasynaptic NMDA receptors in neuronal calcium signalling and BDNF gene regulation. *Curr Opin Neurobiol* 13: 366–371, 2003.
115. Walton PD, Airey JA, Sutko JL, Beck CF, Mignery GA, Sudhof TC, Deerinck TJ, and Ellisman MH. Ryanodine and inositol trisphosphate receptors coexist in avian cerebellar Purkinje neurons. *J Cell Biol* 113: 1145–1157, 1991.

116. Wang S, Jia Z, Roder J, and Murphy TH. AMPA receptor-mediated miniature synaptic calcium transients in GluR2 null mice. *J Neurophysiol* 88: 29–40, 2002.
117. Wang SS, Denk W, and Hausser M. Coincidence detection in single dendritic spines mediated by calcium release. *Nat Neurosci* 3: 1266–1273, 2000.
118. Wigstrom H, Gustafsson B, Huang YY, and Abraham WC. Hippocampal long-term potentiation is induced by pairing single afferent volleys with intracellularly injected depolarizing current pulses. *Acta Physiol Scand* 126: 317–319, 1986.
119. Wigstrom H, Swann JW, and Andersen P. Calcium dependency of synaptic long-lasting potentiation in the hippocampal slice. *Acta Physiol Scand* 105: 126–128, 1979.
120. Yasuda R, Harvey CD, Zhong H, Sobczyk A, van Aelst L, and Svoboda K. Supersensitive Ras activation in dendrites and spines revealed by two-photon fluorescence lifetime imaging. *Nat Neurosci* 9: 283–291, 2006.
121. Yasuda R, Nimchinsky EA, Scheuss V, Pologruto TA, Oertner TG, Sabatini BL, and Svoboda K. Imaging calcium concentration dynamics in small neuronal compartments. *Sci STKE* 2004: pl5, 2004.
122. Yasuda R, Sabatini BL, and Svoboda K. Plasticity of calcium channels in dendritic spines. *Nat Neurosci* 6: 948–955, 2003.
123. Yuste R and Bonhoeffer T. Morphological changes in dendritic spines associated with long-term synaptic plasticity. *Annu Rev Neurosci* 24: 1071–1089, 2001.
124. Yuste R and Denk W. Dendritic spines as basic functional units of neuronal integration. *Nature* 375: 682–684, 1995.
125. Yuste R, Majewska A, Cash SS, and Denk W. Mechanisms of calcium influx into hippocampal spines: heterogeneity among spines, coincidence detection by NMDA receptors, and optical quantal analysis. *J Neurosci* 19: 1976–1987, 1999.
126. Zhang LI, Tao HW, Holt CE, Harris WA, and Poo M. A critical window for cooperation and competition among developing retinotectal synapses. *Nature* 395: 37–44, 1998.

Postsynaptic Targeting of Protein Kinases and Phosphatases

Stefan Strack and Johannes W. Hell

Department of Pharmacology, University of Iowa, Iowa City, IA 52242-1109, USA, stefan-strack@uiowa.edu, johannes-hell@uiowa.edu

1 Introduction

The discovery of an unsuspected number of protein-protein interactions during the past 15 years was an eye opener for the unanticipated detailed spatial and functional organization of proteins inside cells. Colocalization of protein kinases and phosphatases with their regulators and substrates is of special interest because it is critical for swift, effective, and specific signaling (189). At the plasma membrane a number of extracellular signals are relayed with high selectivity to specific intracellular signaling pathways. Synapses are rather small subcellular compartments that fulfill multiple functions: signal transmission between neurons, processing of neuronal signals, and storage of information. Targeting mechanisms for kinases and phosphatases are thus particularly important at synapses to ensure spatial and temporal fidelity of signal transduction. Such targeting is not limited to kinases but often includes upstream regulators and downstream effectors of kinases. For instance, PKA forms a signaling complex with the postsynaptic L-type Ca^{2+} channel $Ca_v1.2$ that contains all elements of the classic β-adrenergic – PKA signaling cascade including the β_2 adrenergic receptor, the trimeric G protein G_s, and adenylyl cyclase (55, 93). Upon activation of the β_2 adrenergic receptor and the ensuing cAMP production, PKA phosphorylates $Ca_v1.2$ to increase its Ca^{2+} channel activity. Despite involving the diffusible second messenger cAMP, signaling from the β_2 adrenergic receptor via PKA to $Ca_v1.2$ is spatially highly restricted fostered by the formation of a supramolecular signaling complex (55).

This chapter describes the molecular basis and functional relevance of postsynaptic kinase and phosphatase anchoring at glutamatergic synapses. Glutamate is the main excitatory neurotransmitter in the brain. Most synaptic events are mediated by AMPA-type glutamate receptors. A temporary increase in transmission frequency leads to activation of NMDA-type glutamate receptors. The resulting Ca^{2+} influx through NMDA receptors causes long-term-depression (LTD) at more modest frequencies (1–3 Hz) and long-term potentiation (LTP) at higher frequencies.

J.W. Hell, M.D. Ehlers (eds.), *Structural and Functional Organization of the Synapse*, DOI: 10.1007/978-0-387-77232-5_16, © Springer Science+Business Media, LLC 2008

AMPA and NMDA receptors are described in Chapters by Mayer, Esteban, Yuan et al., Wenthold et al., and Choquet and Triller, this volume and LTP in Chapter by Lisman and Hell, this volume. Detailed information is available for postsynaptic localization of the protein kinases CaMKII, PKA, PKC, Pyk2, Src, and Fyn, and of the phosphatases PP1, PP2A, and PP2B. We will also describe upstream regulators and downstream effectors and substrates of these kinases and phosphatases in regards to their spatial co-distribution.

2 Molecular Basis and Function of Postsynaptic Kinase Targeting

2.1 CaMKII

CaMKII is the most prevalent kinase and the most abundant protein at postsynaptic sites (113). It plays a central role in synaptic plasticity and learning and postsynaptic CaMKII targeting will be described first. It phosphorylates the AMPA receptor subunit GluR1 and stargazin, which is important for postsynaptic targeting of AMPA receptors (see Fig. 1b) to increase postsynaptic responses during LTP (Chapter by Lisman and Hell, this volume).

2.1.1 Role of CaMKII in LTP, Learning, and Memory

LTP is a lasting increase in synaptic transmission following intense synaptic stimulation (34, 50, 135, 146, 148). Because of its synapse specificity and permanency and because a wide variety of manipulations that affect LTP also affect memory, LTP is thought to underlie learning and memory (153). Ca^{2+} influx through the NMDA receptor and the ensuing stimulation of CaMKII plays a central role in LTP (137, 147, 149, 218) and in learning and memory (73, 84, 136, 217).

2.1.2 Biochemical Properties and Activation of CaMKII

CaMKII is a large complex that consists of 12 catalytic subunits accounting for 1–2% of total brain protein (46, 107, 160, 202). CaMKIIα and CaMKIIβ are encoded by two homologous genes and are the two main isoforms in brain forming mixed α/β holoenzymes. Ca^{2+}-occupied calmodulin (CaM) binds downstream of and to some extent overlapping with the autoinhibitory pseudosubstrate region, which otherwise contacts and occludes the substrate binding site (S site) in the catalytic domain (Fig. 1a). Immediately N-terminal to the pseudosubstrate region of CaMKIIα is threonine 286 (T286; T287 in CaMKIIβ and *Drosophila melanogaster* CaMKII). Under resting conditions T286 contacts a hydrophobic pocket (T site) that includes valine 208 and tryptophan 237 near the S site. This interaction fosters binding of the pseudosubstrate region to the catalytic site. When Ca^{2+}/CaM displaces T286 and the pseudosubstrate region from the T- and S-site, respectively, in two neighboring subunits, T286 of one subunit can contact the S-site in the catalytic domain of the other subunit to become phosphorylated. This intersubunit autophosphorylation blocks re-association of T286 with the T site, thereby preserving kinase activity beyond removal of Ca^{2+}/CaM (46, 107, 160). Autophosphorylation of its 12 subunits allows CaMKII to translate Ca^{2+} spike frequency into graded kinase activity (57).

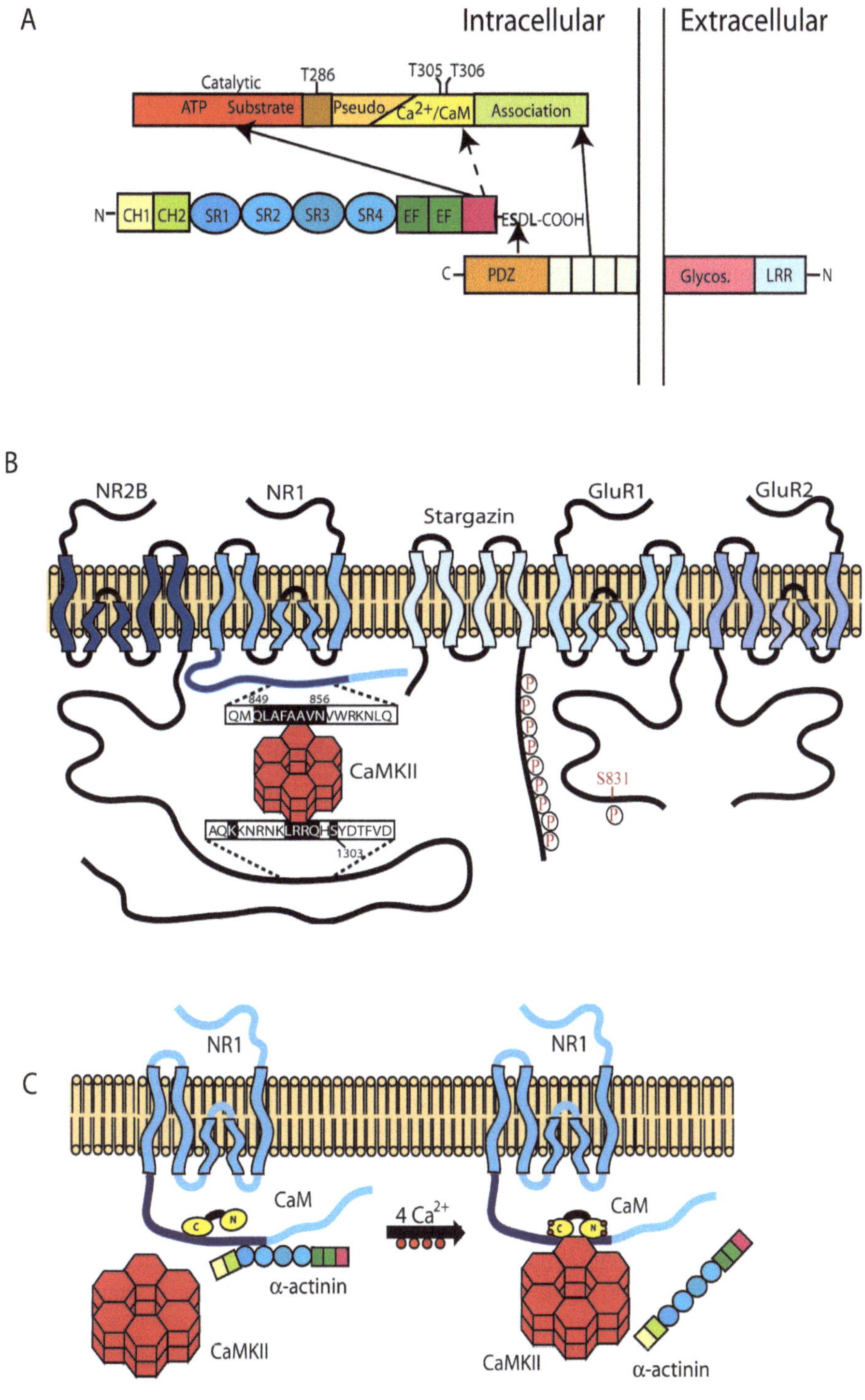
A
Intracellular
Extracellular
Catalytic
T286
T305 T306
ATP Substrate
Pseudo
Ca2+/CaM
Association
N
CH1
CH2
SR1
SR2
SR3
SR4
EF
EF
ESDL-COOH
C
PDZ
Glycos.
LRR
N
B
NR2B
NR1
Stargazin
GluR1
GluR2
849
856
QMQLAFAAVNVWRKNLQ
CaMKII
S831
AQKKNRNKLRRQHSYDTFVD
1303
C
NR1
NR1
CaM
4 Ca2+
CaM
α-actinin
CaMKII
CaMKII
α-actinin

2.1.3 Molecular Mechanisms of Postsynaptic CaMKII Accumulation

CaMKII binding proteins that are specifically concentrated at postsynaptic sites are the NMDA receptor, densin-180, F-actin, and α-actinin (Table 1). The NMDA receptor consists of two NR1 and two NR2 subunits, the latter encoded by four genes (NR2A – NR2D) (Chapters by Yuan et al. and Wenthold et al., this volume). The prevailing NR2 subunits in mammalian forebrain are NR2A and NR2B. CaMKII interacts with NR1 (19, 122, 124, 161), NR2A (81, 83), and NR2B (19, 122, 124, 228, 229). The CaMKII binding site on NR1 is formed by residues 847–861 constituting the C-terminal half of the so-called C0 domain with residues 849–857 being most critical for the interaction (122) (Leonard and Hell, unpublished) (Fig. 1b). CaMKII independently binds to the membrane-proximal residues 839–1120 (NR2B/P) and residues 1290–1310 in the 644 amino acid-long C-terminus of NR2B (Fig. 1b) (19, 124, 228, 229). The NR2A region that is nearly identical to NR2B/C does not bind CaMKII because of the insertion of two vicinal residues (Ile-Asp) right in the middle of this motif (15, 83, 229). However, a region further downstream has been reported to interact with CaMKII, with residues 1412–1419 forming the core of this interaction site (80, 82). Phosphorylation of S1303 in NR2B/C by CaMKII and of S1415 in NR2A by PKC inhibits CaMKII association with these sites (80, 229).

Fig. 1. (Continued) **(a)** Linear model of CaMKII and binding to α-actinin and densin-180. The catalytic domain of CaMKII (*red*) contains the ATP binding site, the substrate binding site (S site), and the hydrophobic pocket (T site), which interacts with T286 under resting conditions. The T286 segment (*brown*; T287 in CaMKIIβ) and pseudosubstrate region (*orange*) form the autoinhibitory domain, which binds to the T and S site, respectively, in the basal inactive state. 12 subunits interact with their C-terminal association domains (*green*) to form one CaMKII complex. The pseudosubstrate region (*orange*) overlaps with the Ca^{2+}/CaM binding site (*yellow*). Autophosphorylation of T305 or T306 blocks Ca^{2+}/CaM binding. α-actinin consists of two calponin-homology domains (CH1, CH2), four spectrin repeats (SR1-4), two EF hands and a C-terminal region. It binds with its C-terminal region to the catalytic domain and possibly the Ca^{2+}/CaM binding site of CaMKII. Densin-180 carries a PDZ domain at its N-terminus that interacts with the very C-terminal ESDL sequence of α-actinin. The membrane-proximal region of densin-180 is differentially spliced and binds to the association domain of CaMKII. It is followed by a transmembrane segment, a glycosylated region, and a leucine-rich region (LRR). **(b)** CaMKII binding sites on NR1 and NR2B/C and Phosphorylation Targets stargazin and GluR1. CaMKII binds to the region surrounding Ser1303 in the C-terminus of NR2B and to the C0 region in the C-terminus of NR1 (*dark blue*). Critical residues defined by binding studies with peptide libraries of overlapping sequences of NR1 C0 and by interaction studies with NR2B/C constructs carrying point mutations are given in white on black background. CaMKII can phosporylate S1303 in NR2B, S831 in GluR1 (*blue*) and 9 diferent serines in stargazin (*light blue*). **(c)** Interactions of CaM, α-actinin, and CaMKII with NR1 C0. Under resting conditions, NR1 C0 simultaneously binds to α-actinin and the C-domain of apo-CaM, with α-actinin preventing CaMKII binding. Upon addition of Ca^{2+} as occurring during Ca^{2+} influx Ca^{2+}/CaM displaces α-actinin, thereby promoting CaMKII binding.

Table 1. CaMKII binding proteins

CaMKAP	Target Domain	CaMKII Domain	Requirements	References
NR1	849–856 nec. 847–858 supp.	unknown	T286 phos. nec.	(122, 124)
NR2A	1412–1419 nec. 1420–1464	catalytic domain	T286 phos. supp.	(80, 83)
NR2B/P	839–1120	unknown	T286 phos. nec.	(19, 122, 124)
NR2B/C	1290–1309	cat. domain (Ile205)	Ca^{2+}/CaM or T286 phos. nec.[1]	(19, 228, 229)
Densin-180	1355–1382 nec. 1331–1404 supp.	assn. domain (α homomers only[2])	T286 phos. augments >10x	(198, 230, 255)
α-actinin-2	819–894	cat. domain (1–380; autoinh. segment?)	T305/6 phos. or Ca^{2+}/CaM inhibits	(197, 198)
α-actinin-4	806–871	cat. domain (1–316)		(255)
F-actin	unknown	variable domain (CaMKIIβ only)	Ca^{2+}/CaM and T287 phos. inh.	(212, 214)
Camguk/ CASK	unknown	autoinh. domain (275–310)	T306/7 phos.[3]	(138)
Eag K^+ channel	772–794	cat. domain	Ca^{2+}/CaM or T286 phos. nec.[1]	(232)

[1] Ca^{2+}/CaM induced binding might require adenosine nucleotides; binding can maintain kinase activity under certain conditions

[2] CaMKIIα homomers are likely formed by translation of CaMKIIα mRNA in dendrites, which do not contain CaMKIIβ mRNA

[3] if Ca^{2+}/CaM is absent Camguk promotes T306/7 phosphosrylation and its dissociation

CaMKII binding to NR1 and NR2B/P requires T286 autophosphorylation. CaMKII binding to NR2B/C can be triggered either by Ca^{2+}/CaM or T286 autophosphorylation in vitro (19, 122). These biochemical characteristics immediately suggest that CaMKII binding to the NMDA receptor depends on Ca^{2+} influx to allow Ca^{2+}/CaM binding to CaMKII and T286 autophosphorylation of CaMKII (Section 2.1.4). The K_d values for CaMKII binding to NR1 and NR2B are in the high nM to low μM range (124, 225, 228), well below the average concentration of CaMKII in neurons. Binding to NR2A is much weaker than to NR1 and NR2B and often hard to detect (124, 228, 229).

The C-terminal half of the NR1 C0 region interacts not only with CaMKII but also with CaM and α-actinin (122). The CaMKII binding site partially overlaps with the CaM/α-actinin binding site, the latter being C-terminally shifted by 3–4 residues (122). Ca^{2+} free (apo) CaM and α-actinin concurrently bind to NR1 C0 under basal conditions (2, 161). Addition of Ca^{2+} leads to displacement of α-actinin by Ca^{2+}/CaM from NR1 C0 (161, 266), which causes inactivation of the NMDA receptor (118, 277), likely to limit excessive Ca^{2+} influx. It is currently unknown what role CaMKII binding to NR1 C0 plays. In vitro, NR1 C0 can simultaneously bind Ca^{2+}/CaM and CaMKII but not α-actinin and CaMKII. We propose that under basal conditions with low intracellular Ca^{2+}, apo-CaM and α-actinin interact with NR1 C0. Ca^{2+} influx induces dislocation of α-actinin by Ca^{2+}/CaM, which then permits binding of activated CaMKII to this site (Fig. 1c).

Densin-180 selectively binds to CaMKIIα but not CaMKIIβ. Forebrain CaMKII holoenzymes, which contain on average 9α and 3β subunits, show little if any binding to densin-180 (198, 255). Homomeric CaMKIIα holoenzymes could arise from local translation of CaMKIIα mRNA, which is present throughout dendrites in contrast to CaMKIIβ mRNA (36). Association of CaMKIIα with densin-180 is stimulated by T286 autophosphorylation but does not strictly dependent on it (K_d is 75–250 nM for T286-autophosphorylated CaMKII and likely more than 10 fold higher for unphosphorylated CaMKII) (197, 230, 255). The association domain of CaMKIIα binds to residues 1331–1404 immediately upstream of the PDZ domain that is located at the C-terminal end of densin-180 (198, 230, 255) (Fig. 1a).

α-actinin is an F-actin binding protein concentrated at various cell adhesion sites including dendritic spines (182, 190, 254, 255, 266). The α-actinin family comprises 4 genes encoding the highly homologous α-actinin-1 through -4. In forebrain neurons mRNA and protein are detectable for α-actinin-1, -2, and -4 but not -3 (Schnizler, Hall, and Hell, unpublished results) (190, 254, 255, 266). α-actinins consist of two calponin homology domains, followed by four spectrin repeats, two EF hand motifs, and a C-terminal tail (Fig. 1a). A fragment of CaMKII that includes the catalytic and autoinhibitory domain binds to the C-terminal 74–76 residues of α-actinin-2 and -4 immediately downstream of the two EF hand motifs (198, 255) (Fig. 1a). This CaMKII binding site on α-actinin is likely composed of non-contiguous segments (198). Whether the extreme C-terminus of α-actinin is required for CaMKII binding is a matter of debate (198, 255). This segment and especially the last four residues of α-actinin constitute a classical PDZ interaction motif (E**S**D**L-COO**$^-$; moieties critical for PDZ binding are in bold). This motif binds to the PDZ domain at the C-terminal end of densin-180 (198, 255). Densin-180 can thus provide an additional link that acts in a synergistic manner to promote CaMKII – α-actinin binding (198).

T286 autophosphorylation does not modulate CaMKII binding to α-actinin. However, autophosphorylation of T305 and T306 in the Ca^{2+}/CaM binding site of the autoinhibitory domain of CaMKII abrogates CaMKII binding for α-actinin (197) as it does for Ca^{2+}/CaM (47, 187). T305/306 phosphorylation can occur upon Ca^{2+}/CaM dissociation from T286 autophosphorylation CaMKII (161) or association with Camguk/CASK (138) (see below). Furthermore, Ca^{2+}/CaM binding to this autoinhibitory CaMKII segment and Ca^{2+}/CaM-dependent CaMKII activity is competitively blocked by α-actinin indicating that α-actinin and Ca^{2+}/CaM compete for the same binding site on CaMKII (EC_{50} for α-actinin is ~1 μM) (197). Because the affinity of Ca^{2+}/CaM for T286-phosphorylated CaMKII is more than 1000 fold higher than for unphosphorylated CaMKII (162), Ca^{2+}/CaM competes most effectively with α-actinin for T286-phosphorylated CaMKII and much less so for unphosphorylated CaMKII (197).

NR2B/C and α-actinin do not compete with each other or with densin-180 for CaMKII binding. Furthermore, monomeric CaMKII lacking the association domain, which mediates oligomerization of CaMKII and association with densin-180, is capable of simultaneously binding NR2B/C and α-actinin-2. By forming dodecameric holoenzyme CaMKII can simultaneously bind multiple NR2B/C, densin-180 or α-actinin-2 polypeptides; it can concomitantly interact with NR2B/C plus densin-180 or α-actinin and with densin-180 plus α-actinin (198, 255). CaMKII could thus play a structural role at the postsynaptic site by forming multimeric complexes with NR2B/C, α-actinin, and densin-180 (198, 255). A non-catalytic role for CaMKIIα was recently demonstrated in presynaptic events underlying short-term potentiation (99) and for CaMKIIβ in bundling F-actin in spines (177).

CaMKIIβ binds with its variable region located between the autoinhibitory and association domain to F-actin, which is concentrated in dendritic spines (78, 214). This segment is not conserved in CaMKIIα, which thus does not directly interact with F-actin but can be recruited to cortical F-actin through CaMKIIβ if co-assembled into holomers. The F-actin binding segment of CaMKIIβ is located immediately downstream of the Ca^{2+}/CaM binding site in the CaMKII autoinhibitory domain. Ca^{2+}/CaM binding and T287 autophosphorylation leads to displacement of CaMKIIβ from F-actin (212, 214). As this interaction is very dynamic (212, 214) and inhibited by CaMKII activation, it cannot mediate activity-induced postsynaptic CaMKII accumulation. Nevertheless, it recruits CaMKIIβ to F-actin in spine shafts (176, 212), which could function to provide a readily available CaMKII pool for swift translocation to the PSD. Furthermore, recent work indicates that CaMKIIβ is important for maintaining dendritic spine structure by bundling F-actin with its F-actin binding domain in conjunction with the CaMKII association domain but independent of the kinase domain (177). It seems F-actin and CaMKIIβ mutually support each others accumulation in spines. However, the strong rise in postsynaptic CaMKII signal following Ca^{2+} influx can only be explained by translocation of extrasynaptic CaMKII to spines (183, 212, 213) (see below and Fig. 2).

Another postsynaptic binding partner for CaMKII is Camguk (138). It is the Drosophila homolog of CASK, which consists of an N-terminal CaMKII-like domain, followed by SH3, GK, and PDZ domains. The interaction of CaMKII with Camguk is enhanced by ATP if Ca^{2+}/CaM is present and by non-hydrolizable ATP analogues (138). Perhaps binding of ATP to CaMKII promotes initial Camguk binding by fostering exposure of a Camguk binding site on CaMKII. With Ca^{2+}/CaM present, autophosphorylation of T287 in Drosophila CaMKII occurs in this complex. In the absence of Ca^{2+}/CaM, Camguk-associated CaMKII autophosphorylates on T306/307 without T287 autophosphorylation, which is normally required for T306/307 phosphorylation (T287/T306/T307 in Drosophila CaMKII are equivalent to T286/T305/T306 in mammalian CaMKIIα). T306/307 phosphorylation displaces CaMKII from Camguk. T306/307 phosphorylated CaMKII cannot bind Ca^{2+}/CaM. At quiescent synapses with little Ca^{2+}/CaM present, Camguk can thus induce silencing of CaMKII (138). At active synapses Ca^{2+}/CaM can bind to T306/307, thereby inhibiting T306/307 phosphorylation and stabilizing the CaMKII-Camguk complex. CASK might perform a similar function in mammals but no evidence is available.

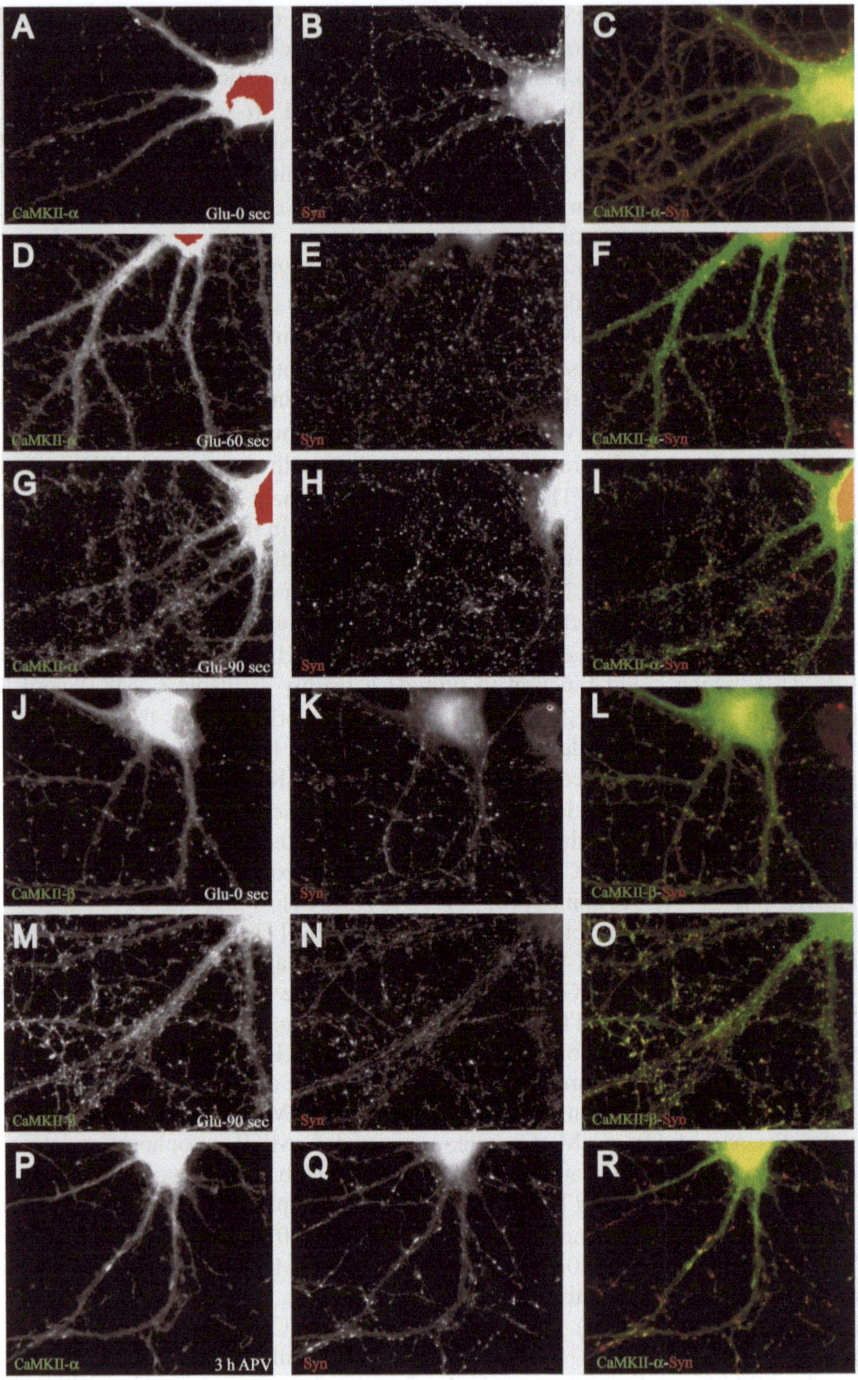
A
CaMKII-α
Glu-0 sec
B
Syn
C
CaMKII-α-Syn
D
CaMKII-α
Glu-60 sec
E
Syn
F
CaMKII-α-Syn
G
CaMKII-α
Glu-90 sec
H
Syn
I
CaMKII-α-Syn
J
CaMKII-β
Glu-0 sec
K
Syn
L
CaMKII-β-Syn
M
CaMKII-β
Glu-90 sec
N
Syn
O
CaMKII-β-Syn
P
CaMKII-α
3 h APV
Q
Syn
R
CaMKII-α-Syn

2.1.4 Activity-Dependent Postsynaptic Accumulation of CaMKII

Inhibition of K^+ channels with tetraethylammonium (TEA) causes depolarization, glutamate release, postsynaptic Ca^{2+} influx through NMDA receptors and Ca^{2+} channels, and synaptic potentiation in hippocampal slices. This chemically induced form of LTP increases the amount of CaMKII in a subcellular fraction that is enriched in postsynaptic densities (PSD) (227). Specific activation of Ca^{2+} influx through the NMDA receptor by NMDA in the presence of the Na^+ channel blocker tetrodotoxin (97) augments CaMKII binding to the NMDA receptor NR1 and NR2B subunit (124). NMDA receptor-mediated Ca^{2+} influx stimulates postsynaptic accumulation of ectopically expressed GFP-CaMKIIα as observed by live imaging in hippocampal cultures (183, 212, 213) and in zebrafish in vivo (87). GFP-tagged CaMKIIα and β have a pronounced tendency to cluster in neuronal somata and dendritic shafts under basal conditions (Chen and Hell, unpublished data). We tested whether endogenous CaMKII undergoes activity-driven postsynaptic targeting. CaMKII is detectable in synapses under resting conditions likely due to basal NMDA receptor activity as postsynaptic CaMKII clustering is reduced by the NMDA receptor antagonist AP5 (Fig. 2). Short (60–90 s) application of glutamate augments synaptic accumulation of endogenous CaMKIIα and CaMKIIβ in dendritic spines as obvious from the enhanced CaMKII staining intensity at synapses compared to the intensity in the dendritic shaft (Fig. 2). Glutamate-induced CaMKII accumulations colocalize with the presynaptic marker synapsin (Fig. 2) and with AMPA receptors, NMDA receptors, and PSD-95 (Chen and Hell, unpublished). CaMKII clustering is blocked by removal of extracellular Ca^{2+} and by the NMDA receptor antagonist MK-801 (Chen and Hell, unpublished). An increase in endogenous CaMKIIα immunoreactivity at the PSD is also detectable by immunogold electron microscopy upon glutamate treatment of hippocampal cultures for 90 s (69).

Optimal CaMKII binding to the NMDA receptor and densin-180 requires Ca^{2+} for Ca^{2+}/CaM binding to CaMKII and ensuing T286 autophosphorylation and for displacement of α-actinin from NR1C0 (Section 2.1.3). Therefore, the NMDA receptor and densin-180 are the prime candidates for mediating activity-driven postsynaptic CaMKII accumulation. These interactions strategically target CaMKII to postsynaptic sites that experience increased activation of NMDA receptors as occurring during LTP and other forms of synaptic plasticity. CaMKII thus accumulates at postsynaptic sites where Ca^{2+} influx through the NMDA receptor is

Fig. 2. (Continued) Postsynaptic accumulation of endogenous CaMKIIα and β upon stimulation with glutamate. Primary hippocampal cultures (18 DIV) were incubated with vehicle (**a**–**c**, **j**–**l**), 0.1 mM glutamate for 60 s (**d**–**f**) or 90 s (**g**–**i**, **m**–**o**) to induce NMDA receptor-mediated Ca^{2+} influx, or 0.1 mM APV for 3 h to block basal Ca^{2+} influx through NMDA receptor (**p**–**r**). Cultures were fixed (4% paraformaldehyde, 15 min), permeabilized (0.3% Triton X-100), and double stained for CaMKIIα or β plus the presynaptic marker synapsin as indicated (131). Glutamate treatment increased synaptic CaMKII staining within 90 s. Prolonged APV treatment reduced synaptic CaMKII staining suggesting that postsynaptic targeting of CaMKII under basal conditions is at least in part due to spontaneous neuronal activity and the ensuing NMDA receptor activation.

highest for further activation of CaMKII and for phosphorylation of nearby AMPA receptors and of other proteins critical for LTP. LTP specifically occurs at those synapses that receive increased stimulation. Ca^{2+}-induced binding supports CaMKII accumulation specifically at those synapses that undergo LTP. Such activity-driven accumulation is not only economical but likely also constitutes positive feedback that ensures reliable and specific potentiation of stimulated synapses.

Postsynaptic CaMKII accumulation might foster phosphorylation of the AMPA receptor GluR1 subunit at S831 (Fig. 1b), which is elevated during LTP (16, 121). This phosphorylation increases ion currents through homomeric GluR1 but not heteromeric GluR1/R2 receptors (61, 174). However, this phosphorylation likely contributes only to a limited degree to LTP and only in mice that are 2 and perhaps more than 7 weeks but not 3–4 weeks old. The reason for this limitation is that only at the former ages are GluR2-lacking, presumably GluR1 homomeric, AMPA receptors functionally present and only transiently so during the first 20 min of LTP (1, 142, 193) (Chapter by Lisman and Hell, this volume). This constraint was met in previous work showing an increase in AMPA receptor single channel conductivity following LTP induction (25). It reduces the time window during which this regulation could contribute to potentiation and thereby establishment of LTP during its early phase.

CaMKII also contributes to postsynaptic accumulation of GluR1-containing AMPA receptors (95, 201). CaMKII phosphorylation of stargazin (see Fig. 1b) (and potentially of its relatives γ_3, γ_4, γ_7, γ_8) recently emerged as an important molecular mechanism in LTP by fostering postsynaptic accumulation of AMPA receptors (244). Stargazin also positively modulates gating of AMPA receptors (194, 243, 250) but it is unclear whether CaMKII regulates gating by phosphorylating stargazin.

2.1.5 Regulation of Postsynaptic Accumulation by CaMKII Autophosphorylation

Upon Ca^{2+} influx T286A mutant CaMKIIα translocates like wild-type CaMKIIα to postsynaptic sites. However, in contrast to wild-type CaMKII, the T286A mutant disperses within a few seconds rather than minutes following cessation of Ca^{2+} influx (212, 213). Furthermore, Ca^{2+} influx also induces postsynaptic accumulation of the phosphomimetic T286D mutant, which requires more than 1 h for relocation after termination of Ca^{2+} influx (213). Accordingly, T286 phosphorylation is not sufficient for activity-induced translocation but does stabilize CaMKII at postsynaptic sites. The resulting prolonged postsynaptic CaMKII accumulation can be reversed by phosphorylation of T305/306 and dephosphorylation of T286: mutating T305 and T306 to alanine or application of the phosphatase inhibitor calyculin A, like the T286D mutation, slow postsynaptic CaMKII dispersal in cultured neurons (212) whereas mutating T305 to aspartate reduces the CaMKII content of PSDs in knock-in mice under basal conditions (73). If PKA is stimulated in parallel with Ca^{2+} influx postsynaptically clustered CaMKII is retained for extended time periods (183, 213). PKA phosphorylates inhibitor-1, which then inhibits the phosphatase PP1 and thereby dephosphorylation of CaMKII on T286 to promote LTP (28). PKA could also be involved in prolonged postsynaptic CaMKII accumulation upon extended stimulation with glutamate (20), as the ensuing Ca^{2+} influx can activate Ca^{2+}-stimulated adenylyl cyclase 1 and 8.

As T286 autophosphorylation is not sufficient for postsynaptic CaMKII accumulation, initial binding might be triggered by Ca^{2+}/CaM, which can induce CaMKII binding to NR2B/C in the absence of ATP in vitro (19, 122). In support of an important role for binding of Ca^{2+}/CaM to CaMKII in postsynaptic targeting, blocking Ca^{2+}/CaM binding by mutating CaMKII A302 to arginine prevents glutamate-induced translocation of CaMKIIα (212). A second potential mechanism involves displacement of α-actinin from NR1 C0 (Fig. 1c): binding to this site requires both T286 phosphorylation and dislocation of α-actinin by Ca^{2+}/CaM (161). Accordingly, CaMKII T286D will not be able to bind to NR1 C0 under basal conditions as long as α-actinin is associated with NR1 C0. Such a mechanism would also explain the finding that translocation of CaMKII to postsynaptic sites requires specifically Ca^{2+} influx through NMDA receptors (241).

2.1.6 Binding-Induced CaMKII Activity

Once formed, Ca^{2+}/CaM-triggered CaMKII binding to NR2B/C is stable even after Ca^{2+} chelation to remove CaM (19). Notably, a complex between non-phosphorylated CaMKII and NR2B possesses autonomous activity similar to T286-autophosphorylated CaMKII (19). Binding of CaMKII to the *Drosophila* eag K^+ channel leads to an analogous constitutive activation of CaMKII (232). The amino acid sequence of the NR2B/C and the eag binding site are similar to the T286 region. We hypothesize that displacement of T286 from the T site (Section 2.1.2) allows stable binding of NR2B/C or eag with this site to prevent re-association of not only T286 with the T site but also the pseudosubstrate region with the S site. These interactions would in this way confer autophosphorylation-independent autonomous activity, which in turn could promote T286 phosphorylation of a neighboring Ca^{2+}/CaM-bound subunit in the CaMKII holoenzyme. Binding to NR2B/C or eag could, thereby, locally spark autonomous activity when a limited amount of Ca^{2+}/CaM is available. However, S1303 in NR2B/C is a CaMKII phosphorylation site (180) and NR2B/C can also interact with catalytic sites in a not fully understood complex manner to inhibit CaMKII activity under certain conditions (197).

2.2 Postsynaptic Targeting of PKA by AKAPs

PKA is a tetramer that consists of two dimerized regulatory (R) and two catalytic (C) subunits. Four genes encode RIα, RIβ, RIIα, and RIIβ and three genes Cα, Cβ, and Cγ (33, 65). Cγ is only detectable in testis. Binding to R dimers inhibits catalytic activity of C subunits until their release by cAMP binding to R. RII and to a lesser extent RI associate with a number of substrates via A kinase anchor proteins (AKAPs) (37, 204, 264). AKAPs are diverse proteins that range from 15 to >300 kDa in size but share 14–18 residue long amphipathic α helices that bind the R dimer.

Full activation of C subunits requires their release from R subunits by cAMP, which should displace C from AKAP – RII complexes and thereby reduce phosphorylation of the targeted proteins. In fact, prolonged stimulation via cAMP of $Ca_V1.1$ reduces its potentiation by PKA consistent with dislocation of the C subunit from the channel – AKAP (here AKAP7 previously know as AKAP15/18) – R subunit complex (110). However, C subunits can be activated by cAMP without fully dissociating from RII subunits (252, 274). Incomplete dissociation might keep C subunits anchored near their substrates. RI isoforms, which are anchored to a lesser degree than RII, appear to more fully release their C subunit during substrate phosphorylation in vitro (252).

2.2.1 Postsynaptic Localization of AKAP5 and its Function

Rodent AKAP150, human AKAP79, and bovine AKAP75 are orthologs and were recently re-named AKAP5. AKAP150 differs from AKAP75/79 by an insert of 36 imperfect octapeptide repeats of unknown function (204). AKAP5 is the main AKAP that targets PKA to postsynaptic sites (88, 142, 264). Removal of its PKA anchoring site reduces the PKA content of postsynaptic densities by 70–80% (142). AKAP5 binds to the SH3 and GK domain of PSD-95 and its homolog SAP97 (48, 237) (Fig. 3). PSD-95 interacts directly with the NMDA receptor and via stargazin or its homologs with AMPA receptors (Chapter by Kennedy et al., this volume). SAP97 binds to the C terminus of AMPA receptor GluR1 subunit (123) thereby bridging AKAP5 and associated PKA with GluR1 (237, 264) (Fig. 3). Disruption of AKAP-mediated PKA anchoring leads to run-down of AMPA receptor-mediated excitatory postsynaptic currents (203). PKA-mediated phosphorylation of S845 on GluR1 requires its association with PKA via AKAP5 and SAP97 is a critical linker (237). Anchored PKA increases channel activity (12), surface expression, and postsynaptic accumulation of GluR1-containing AMPA receptors (74, 150, 175). The role of PSD-95/AKAP5–anchored PKA in NMDA receptor regulation is less clear.

AKAP5 also interacts with PKC and the Ca^{2+}/CaM-activated phosphatase calcineurin (PP2B) (45, 115). AKAP5 is critical for reversal of PKA-mediated GluR1 upregulation by PP2B (220, 237). Although PKA, PKC, and PP2B regulate NMDA receptor activity (134, 140, 195, 219) (Sections 2.2.3 and 2.3.1) it is unknown whether these effects depend partially or fully on their anchoring by AKAP5.

AKAP5 also binds to F-actin, cadherin, and phosphatidylinositol-4,5-bisphosphate (PIP_2; Fig. 3), all of which are present if not enriched at postsynaptic sites and can thereby contribute to its postsynaptic localization (59, 88, 89). It is thus difficult to precisely define whether a certain interaction is especially critical for targeting AKAP5 in general to postsynaptic sites or specifically to defined PKA substrates. Furthermore, a 3 min stimulation of Ca^{2+} influx through the NMDA receptor, which induces LTD, leads to redistribution of postsynaptic AKAP5 to the cytoplasm in a PP2B-dependent fashion (88, 102, 220). This Ca^{2+} influx also reduces the postsynaptic content of F-actin and PSD-95 but not cadherin (49, 88, 89, 220). In addition, PKC as well as Ca^{2+}/CaM inhibit AKAP5 binding to F-actin, cadherin, PIP_2, and PSD-95 (88, 89, 220). Stabilization of F-actin by jasplakinolide inhibits AKAP5 relocalization and dissociation from PSD-95 (88, 220). Furthermore, inhibition of phospholipase C to prevent PIP_2 hydrolysis blocks the Ca^{2+} influx-induced dispersal of AKAP5, F-actin, and PSD-95 (102). It thus appears that multiple interactions that foster each other are required for proper postsynaptic

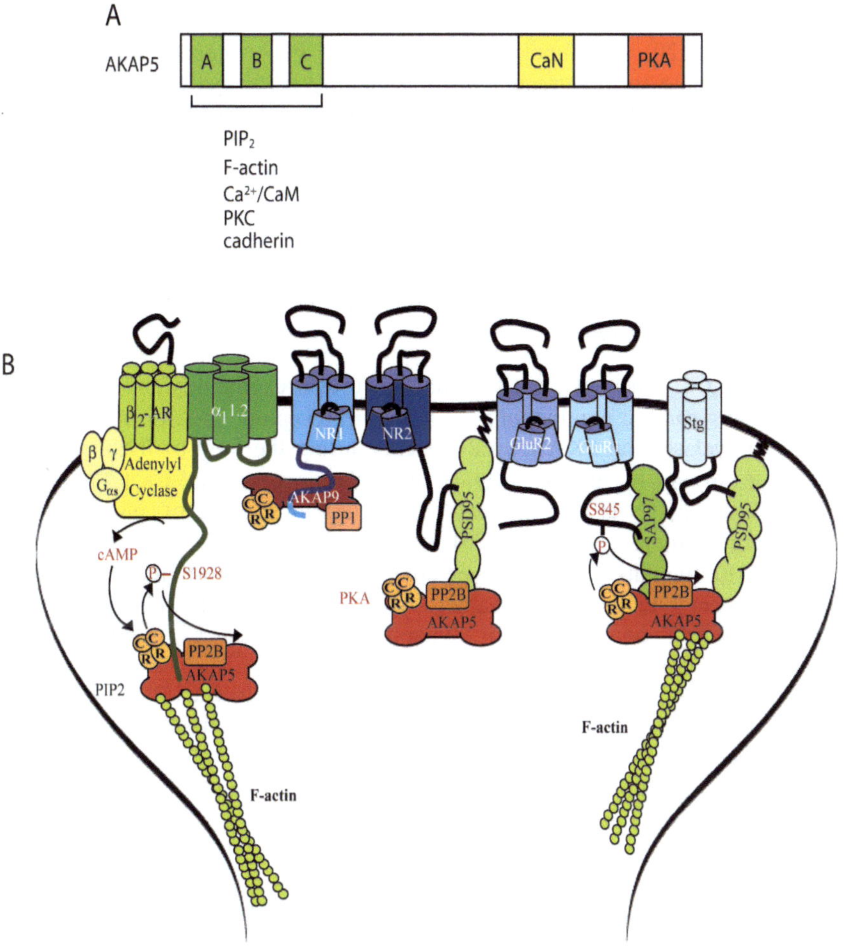

Fig. 3. Model of AKAP/PKA complexes at the postsynaptic site. (**a**) Linear model of AKAP5. The model shows the N-terminal region with three stretches rich in basic residues (*green*), which binds PIP_2, F-actin, Ca^{2+}/CaM, PKC, and cadherin. The PP2B and PKA binding sites are indicated in yellow and red, respectively. (**b**) Model of AKAP interactions at the postsynaptic site. *Right side* illustrates AKAP5-mediated anchoring of the tetrameric PKA holoenzyme (C and R) and of PP2B via SAP97 and also via PSD-95/stargazin (stg) to the AMPA receptor GluR1 subunit. SAP97 and PSD-95 bind with their C-terminal portions consisting of SH3 and GK domains to AKAP5 (*red*) but precise interaction sites have proven difficult to dissect. AKAP5 anchoring of PKA and PP2B is required for phosphorylation and dephosphorylation of S845 (*orange* P). *Center* shows the complex between the NMDA receptor, AKAP9 (present at postsynaptic sites as yotiao, the shorter splice variant of AKAP350), and AKAP5. AKAP9 binds to the C1 segment in the NR1 C-terminus. It functionally recruits not only PKA but also the antagonistic phosphatase PP1 to the NMDA receptor. AKAP5 binds to PSD-95 (or its homologs), which in turn interact with the very C-terminal ESDL-COO^- motif of NR2A and 2B. The role of AKAP5 in NMDA receptor regulation is unknown. *Left side* outlines the β_2-adrenergic receptor (AR) – adenylyl cyclase – G_s – AKAP5 – PKA – $Ca_v1.2$ signaling complex. The β_2-AR, and AKAP5 can directly bind to the pore-forming $\alpha_1 1.2$ subunit. PKA association and S1928 phosphorylation and requires AKAP5. Interaction sites for adenylyl cyclase and G_s are unknown.

AKAP5 targeting. RII but not PP2B re-distribute upon Ca^{2+} influx in parallel with AKAP5 (88, 220). The function of AKAP5 and PKA relocalization is unclear but could contribute to downregulation of signaling pathways that otherwise lead to overstimulation. The observation that dephosphorylation of GluR1 on its PKA site S845 precedes persistent AKAP5 relocation during chemically induced LTD is compatible with AKAP5 being important for linking PP2B to GluR1 (220). PKA will be displaced in parallel with AKAP5, thereby preventing GluR1 rephosphorylation in this form of LTD.

2.2.2 The β_2Adrenergic Receptor – Adenylyl Cyclase – Gs – AKAP5 – PKA – $Ca_v1.2$ Signaling Complex

The L-type Ca^{2+} channel $Ca_v1.2$ regulates neuronal excitability and gene expression and triggers cardiac contraction. β-adrenergic receptors (β-ARs) stimulate via G_s, adenylyl cyclase, and PKA $Ca_v1.2$ currents during the fight or flight response (21, 181, 196) in part by phosphorylation of S1928 in the central $Ca_v1.2$ subunit $\alpha_1 1.2$ (93, 179). $Ca_v1.2$ is concentrated at postsynaptic sites of glutamatergic and GABAergic synapses (55, 97, 173) where it contributes to Ca^{2+} influx especially upon β-AR stimulation (101). L-type channels mediate a portion of LTP induced by high frequency tetani (92) (Chapter by Lisman and Hell, this volume) and of activity-induced homeostasis of synaptic transmission by suppressing GluR1 expression (242).

AKAP5 and AKAP7 (AKAP15/18) bind directly to $Ca_v1.2$ for PKA-mediated regulation of $Ca_v1.2$ in brain (93, 179) and heart, respectively (108). This spatial confinement of PKA signaling extends in the case of $Ca_v1.2$ beyond the kinase: the $Ca_v1.2$ also contains the β_2-AR, G_s, and adenylyl cyclase (11, 55). The C-terminus of the β_2-AR directly binds to the central, pore-forming $\alpha_1 1.2$ subunit (55). It is unclear how G_s and adenylyl cyclase are linked to $Ca_v1.2$. AKAP5 binds adenylyl cyclase 5 and 6 (18) and could thereby recruit one or both isoforms to $Ca_v1.2$. During cell-attached patch clamp recording, administration of a β_2-AR- agonist increases L-type-currents by more than two fold inside, but not outside, the recording pipette electrode in neurons (55) and cardiomyocytes (42) via PKA (6, 101, 179, 268, 278). These results suggest that cAMP signaling can be restricted to sub-μm dimensions.

Mechanisms of cAMP compartmentalization are not fully understood but involve cAMP hydrolysis by phosphodiesterases (PDEs) (224, 268). Some AKAPs interact with PDEs (23) and different PDE4 isoforms are recruited to activated β_2 -ARs via β arrestin for localized breakdown of cAMP (10, 192). In the heart cGMP-inhibited cAMP PDEs (PDE3 family) and the cAMP-specific PDEs (PDE4 family) contribute to the spatial restricted regulation of $Ca_v1.2$ by the β_2-AR (111, 199). The abundant PDE4D3 isoform binds to AKAP6 (mAKAP) (64, 66), the longer splice form of AKAP9 (AKAP350) (236) and AKAP12 (gravin, AKAP250) (263). Co-localization of PDE4D with PKA by AKAPs fosters phosphorylation of PDE4D by PKA to stimulate cAMP hydrolysis (51, 66).

Like AKAP5, AKAP12 binds to the β_2-AR C-terminus for agonist-induced internalization of the β_2-AR and also recovery of the β_2-AR from this desensitization (76, 132, 216). AKAP5 and 12 also anchor of PKC and PP2B. Both AKAPs can interact with negatively charged phospholipids via three basic segments in their N-terminal regions and Ca^{2+}/CaM antagonizes these interactions (59, 88, 235, 256).

2.2.3 The Role of AKAP9 in NMDA Receptor Regulation

Yotiao is a smaller splice isoform of AKAP350/450 now called AKAP9. It binds to the C1 cassette in the C-terminus of the NR1 subunit of NMDA receptors (262). It recruits PKA to upregulate currents through the NMDA receptor (262). A recent report indicates that PKA acts mainly by selectively increasing Ca^{2+} permeability of the NMDA receptor (219). AKAP9 also links the phosphatase PP1 to the NMDA receptor to counteract the PKA effect (262). Anchoring of PKA and a phosphatase by AKAP9 is analogous to AKAP5, which links PKA and PP2B to GluR1 (220, 237) and $Ca_v1.2$ (179). However, PKA-mediated upregulation of $Ca_v1.2$ is not only counterbalanced by AKAP5-anchored PP2B but also by PP2A, which requires a direct stable association with $Ca_v1.2$ (56, 94) (Section 3.3.2).

2.3 PSD-95 as a Hub for Targeting PKC, Pyk2, Src, and Fyn

2.3.1 Regulation of the NMDA Receptor by PKC – Pyk2 – Src

G_q-coupled receptors including the metabotropic glutamate receptors mGluR1 and 5 activate phospholipase Cγ for hydrolysis of PIP_2 to inositol-1,4,5-trisphosphate (IP_3) and diacylglycerol (DAG). IP_3 induces Ca^{2+} release from intracellular stores. DAG can either alone or in conjunction with Ca^{2+} stimulate most PKC isoforms. Conventional PKCα, PKCβ, and PKCγ isoforms are activated when DAG or Ca^{2+} bind to the conserved C1 and C2 domains, respectively (164, 172). Novel isoforms (PKCδ, PKCε, PKCη, PKCθ) have no C2 domain and require only DAG but not Ca^{2+} for full activation. Atypical isoforms (human PKCι, mouse PKCλ, PKCζ) lack C1 and C2 domains and their regulation is not well understood.

PKC increases surface expression, opening probability (119), and peak currents of NMDA receptors (141). However, PKC can also decrease steady state currents by increasing inactivation (140). PKC can directly act on the NMDA receptor to reduce steady state currents and through Pyk2 and Src or related kinases (SRKs) to increase peak currents (140, 141). Inhibition of Src prevents PKC- and also Pyk2-induced increases in NMDA receptor peak currents and inhibition of Pyk2 abrogates PKC- but not Src-induced increases in NMDA receptor peak currents (106, 140, 141). These results place Pyk2 downstream of PKC and upstream of Src.

Pyk2 is a close homolog of Focal Adhesion Kinase (FAK). It consists of a Band 4.1 homology domain, a region with Y402 for autophosphorylation, which is part of the activation process, the kinase domain, a region with two proline-rich segments, and the Focal Adhesion Targeting (FAT) domain (9, 125, 209, 275). It is present in dendrites (159, 211). Pyk2 can indirectly be activated by PKC or by Ca^{2+} influx. Pyk2 links Ca^{2+} influx through NMDA receptors and Ca^{2+} channels and activation of a number of G_q coupled receptors including mGluR1 and 5 to signaling via tyrosine phosphorylation pathways (60, 63, 70, 86, 106, 125, 206). The precise mechanisms of Pyk2 activation are unclear. In contrast to FAK, Pyk2 can be activated independently of Src (186, 265) (but see (43, 222)).

Autophosphorylation of Pyk2 on Y402 allows binding of the SH2 domain of Src to promote Src activation (63, 206). Src then phosphorylates Y579 and Y580 in the activation loop to fully activity Pyk2 (8, 17, 63, 125, 128). Src activation by Pyk2 upregulates NMDA receptor peak currents likely through phosphorylation of NR2 subunits (41, 106, 116, 206, 260). This NMDA receptors regulation is important for LTP that depends on Ca^{2+} influx through the NMDA receptor (106, 143, 206, 269).

2.3.2 Binding of PKC, Pyk2, Src to PSD-95 and ND2

Postsynaptic anchoring mechanisms exist for all three kinases. PKCα has a C-terminal type I PDZ domain binding motif (E**T**A**V**- **COO⁻**; critical moieties in bold). This motif mediates PKCα binding to PICK1 and PSD-95. PICK1 is another postsynaptic structural protein that contains a PDZ domain and was first identified as a PKCα interacting protein (223). It plays a critical role in promoting AMPA receptor trafficking involving PKC, although precise mechanisms are not known (e.g., (139, 267)). PSD-95 also interacts with one ore more of its three PDZ domains with PKCα (130). PSD-95 is of special interest in the context of the PKC – Pyk2 – Src pathway because it also binds with its somewhat unusual SH3 domain to Pyk2 (211) and with its N-terminal region in a phosphotyrosine-independent manner to the SH2 domain of Src (112). Like PSD-95, Src and Pyk2 co-immunoprecipitate with the NMDA receptor (211, 276). PSD-95 might thus link PKCα, Pyk2, and Src to the NMDA receptor. A second potential linker is RACK1, initially identified as binding protein for activated PKCβ (200). RACK1 also binds Src (40) and the related Fyn (271) and thus can provide a scaffold that links these two tyrosine kinases with PKCβ. However, PSD-95 and RACK1 binding suppresses Src and Fyn activity (40, 112, 271). It is possible that RACK1 functions in anchoring Src and Fyn at postsynaptic sites and that certain stimuli relax the inhibitory interaction as occurring for PKA C subunits upon their release from AKAP-anchored R subunits by cAMP (see above). The unique domain of Src also binds to the NADH dehydrogenase subunit 2 (85). ND2 is a mitochondrial protein but remarkably is also present outside mitochondria at postsynaptic sites and required for regulation of NMDA receptors by Src (85). Binding to ND2 does not block Src activity.

3 Targeting of Phosphatases to Postsynaptic Sites

3.1 Phosphatase Classification

The human genome encodes more than 500 protein kinases but only ~140 protein phosphatases (5, 151). This ratio has been interpreted as evidence for greater promiscuity of phosphatases compared to kinases. However, PP1 and PP2A exist as multimeric enzymes composed of catalytic and variable regulatory subunits, thereby increasing the complexity of the "phosphatome" considerably.

Protein phosphatases are classified according to catalytic mechanisms as metal-based, Cys-based, or Asp-based phosphatases. Further divisions are based on sequence and structural similarities. The metal-based phosphatases are specific for phospho-Ser/Thr and belong to two groups, PPP and PPM. The PPP group includes the catalytic subunits of PP1, PP2A, PP2B (calcineurin), and PP4-7. The PPM (or PP2C) family consists of magnesium-dependent protein phosphatases, which have some structural similarity to PPP-class catalytic domains. PP1, PP2A, PP2B, and PP2C account for the majority of Ser/Thr phosphatase activity and are abundant proteins, particularly in brain. The more recently identified PP4 through PP7 are less abundant and may have relatively narrow substrate specificities.

Cys-based phosphatases feature a conserved catalytic motif (CX_5R) and are classified into four groups (5). Similar to Tyr kinases, the protein Tyr phosphatase (PTP) group can be further divided into receptor and non-receptor PTPs. The Asp-based phosphatases are characterized by a catalytic motif containing two aspartates (DXDXT). This growing group of Tyr/Ser phosphatases contains the eyes absent (Eya) transcription factors and a cofilin phosphatase.

Here we will discuss synaptic functions of the widely studied PPP group of Ser/Thr phosphatases (PP1, PP2A, PP2B). The reader is referred to excellent reviews for more information on tyrosine phosphatases in the brain (32, 188).

3.2 Protein Phosphatase PP1

Together with PP2A, PP1 contributes about 90% of phospho-Ser/Thr directed activity in most cell types. The PP1 catalytic subunit is a ~36 kD globular protein that is highly conserved from yeast to man. In mammals, there are four PP1 isoforms (α, β/δ, γ1, γ2) encoded by three genes (α-γ). PP1γ1 and PP1γ2 arise by differential splicing from the same gene. The C-termini of PP1 isoforms display the most sequence variability and may contribute to isoform-specific association with regulatory proteins (38).

Free PP1 catalytic subunits are not known to exist in significant quantities in cells; rather they form complexes with more than 50 binding partners (Fig. 4a). PP1 regulatory proteins determine the physiological functions of specific PP1 holoenzymes. They can be classified into substrate-independent activity modulators (e.g., inhibitor-1, inhibitor-2, dopamine and cAMP-dependent phosphoprotein of 32 kDa (DARPP-32)), and targeting subunits/substrate specifiers (e.g., spinophilin, neurabin). Binding of many PP1 associated proteins involves an "RVXF" motif that slots into a hydrophobic pocket on the catalytic subunit. The actual consensus sequence for PP1 binding is $[K/R]X_{0-1}[I/V]<P>[F/W]$, where X can be any residue and <P> can be any residue but proline (38).

The crystal structure of the PP1 catalytic subunit complexed with a regulatory protein, myosin phosphatase targeting subunit 1 (MYPT1), illustrates how regulatory subunits confer substrate specificity (Fig. 4b–e). The PP1/MYPT1 complex mediates smooth muscle contraction through dephosphorylation of the myosin regulatory chain, but has also a critical role in cerebellar LTD by dephosphorylating Ser880 of GluR2 (75). MYPT1 extends the substrate binding pocket of PP1 with an acidic groove, which may help attract substrates with clusters of basic residues next to the phospho-Ser/Thr (239).

Neuronal roles of the PP1 targeting/regulatory subunits spinophilin, neurabin, and DARPP-32 are well established. DARPP-32 knockout mice are largely unaffected by several drugs of abuse, underscoring the importance of this protein in psychostimulant action. However, because DARPP-32 is not believed to be enriched at synapses, the reader is referred to several excellent reviews on DARPP-32 (29, 233). In the following, we will discuss the synaptically targeted PP1 regulatory proteins spinophilin/neurabin II and neurabin.

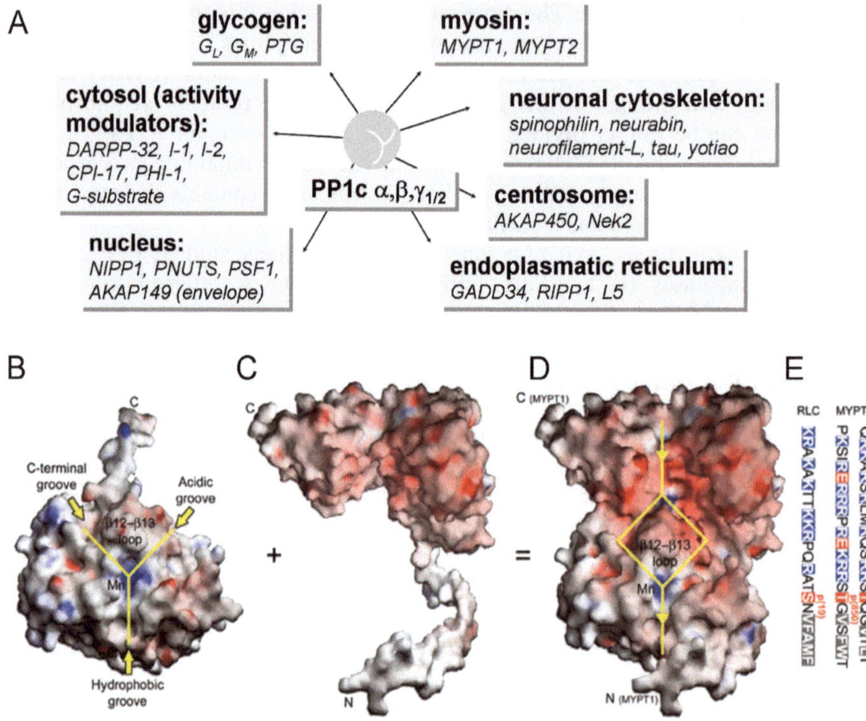

Fig. 4. Regulatory subunits determine PP1 substrate specificity. (**a**) Four closely related PP1 catalytic subunits (PP1c α, β, γ1, γ2) associate with a growing list of anchoring and regulatory proteins to dephosphorylate proteins in a variety of cellular locations (*bolded*). Most PP1 regulatory proteins contain an RVxF motif that contributes to the interaction with PP1c. (**b**) PP1c is a globular protein with a C-terminal extension. Three groves (C-terminal, acidic and hydrophobic grove) form a Y-shaped catalytic surface (*yellow*) with catalysis occurring in the center. (**c**) The N-terminal fragment of MYPT1 consists of an extended, hydrophobic N terminus that includes the RVxF motif, and an ankyrin-repeat domain with numerous acidic surface charges (*red*). (**d**) The PP1c-MYPT1 complex forms by insertion of the catalytic subunit's C-terminal tail between MYPT1's ankyrin repeats, and MYPT1's N-terminal tail wrapping halfway around PP1. This extends the substrate-binding surface (*yellow*) to include a long acidic groove on MYPT1. The acidic charges preferentially attract substrates with clustered basic residues such as myosin regulatory side chain (RLC sequence in (**e**)). (Panels b–e reproduced from Terrak et al. (2004) Nature 429:780 with permission).

3.2.1 Synaptic Targeting of PP1 by Spinophilin and Neurabin

Neurabin (neurabin I; neuronal actin binder) and spinophilin (also known as neurabin II) were initially identified in neurons as PP1 and actin-associated proteins

(3, 145, 170, 210). They are scaffold proteins that share overall 50% sequence similarity and a multi-domain architecture consisting of an N-terminal actin binding domain, a central PP1 binding motif (KI[K/H]F), a PDZ domain, and a C-terminal coiled-coil domain (Fig. 5). Spinophilin and neurabin form homo- and hetero-oligomers through their coiled-coil domains and associate with PP1 in a 1:1 stoichiometric ratio (145). In addition to PP1 and filamentous actin, spinophin/neurabin assemble complexes with several other proteins, including $p70^{S6K}$, guanyl nucleotide exchange factors (GEFs) and RGS (regulator of G protein signaling) proteins (Fig. 5; for a detailed review of spinophilin interactors see (208)). Binding of PP1 and $p70^{S6K}$ to neurabin are mutually exclusive (178). It is unknown which of the other interactions can occur simultaneously.

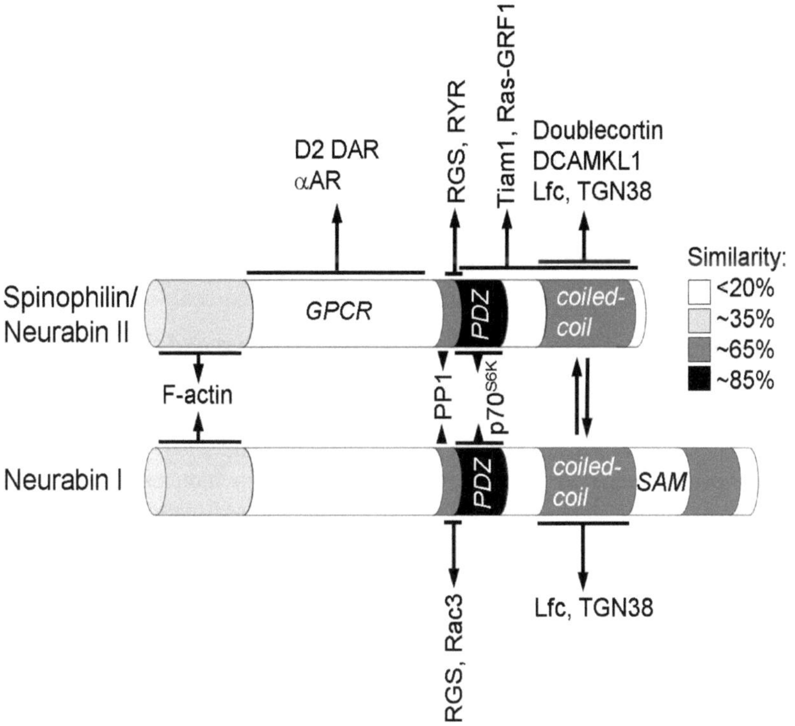

Fig. 5. Spinophilin and Neurabin: cytoskeletal PP1 scaffolds. Domain structure and interaction partners of the structurally related spinophilin and neurabin, which link PP1 and signaling enzymes to the cortical actin cytoskeleton and receptors in the membrane (see text). αAR, α adrenergic receptors; D2 DAR, D2 dopamine receptor; DCAMKL1, doublecortin and CaM kinase like 1; GPCR, G protein coupled receptor binding domain; Lfc, Lbc's (lymphoid blast crisis) first cousin; PDZ, PSD-95, disk-large, ZO-1 domain; Ras-GRF1, Ras-specific guanine nucleotide releasing factor 1; RGS, regulator of G-protein signaling; RYR, ryanodine receptor; SAM, sterile alpha motif; TGN38, trans Golgi network 38; Tiam1, T-cell lymphoma invasion and metastasis 1.

Through its unique G-protein coupled receptor (GPCR) binding domain, spinophilin associates with the third intracellular loop (3iL) of the D2 dopamine receptor and several α adrenergic receptors. Spinophilin/neurabin has therefore been proposed as an adapter complex that links the cortical cytoskeleton to receptors in the plasma membrane. Indeed, spinophilin has been shown to stabilize GPCRs at the cell surface (31), and regulates GPCR signaling by competing with arrestin for binding to the 3iL and recruiting RGS proteins to the receptor complex (257–259).

Whereas neurabin is found exclusively in neurons (170), spinophilin is expressed in many mammalian tissues (3, 210). Both are highly enriched in dendritic spines and co-fractionate with the PSD (3, 145, 170, 210). Neurabin, but not spinophilin, has also been localized to subsets of presynaptic terminals (168, 169). While both proteins are believed to target PP1 to key synaptic substrates, specific interaction partners and the differential localization of the two proteins may be responsible for the different phenotypes of spinophilin and neurabin knockout mice (see below).

A significant fraction, possibly as much as 60% of PP1 in certain brain regions exists in complexes with spinophilin and neurabin, because single knockouts decrease PP1 catalytic subunit content in homogenates of the striatum by ~30% (4). A similar decrease in PP1 content was observed in the synaptoneurosome fraction of hippocampal slice cultures infected with virus expressing a dominant-negative neurabin construct (104). Taking into account that presumably only a fraction of neurons expressed the dominant-negative construct, these experiments suggest that neurabin and spinophilin are largely responsible for targeting PP1 to dendritic spines.

3.2.2 Spinophilin and Neurabin in Cytoskeletal Dynamics and Spine Formation

The N-terminal F-actin binding domain is necessary and sufficient for the localization of spinophilin and neurabin to dendritic spines, and spine targeting of neurabin requires an intact actin cytoskeleton (13, 90, 279). Expression of the isolated F-actin binding domain of spinophilin/neurabin induces reorganization of the actin cytoskeleton and filopodial outgrowth in cultured neurons (178, 240). The neurabin N-terminus increases the density of innervated dendritic spines in cultured hippocampal slices (279). It is unclear whether these phenotypes are a non-specific consequence of the actin bundling activity of this domain (170, 210), or whether they result from a dominant-negative displacement of endogenous actin associated proteins, because overexpression of full-length spinophilin/neurabin did not have an effect.

Further evidence suggests that spinophilin and neurabin are important for the morphological differentiation of neurons. Gene knockout of spinophilin causes a transient increase in dendritic spine density in vivo and excessive filopodial extensions in vitro (77). While data on spine development in neurabin knockout mice are not available, expression of a PP1 binding-deficient neurabin mutant delays the maturation of dendritic filopodia into spines (240). Spinophilin/neurabin may also affect F-actin organization and spine morphology independent of PP1 through recruitment of the Rho GEFs kalirin-7 and Lcf (191, 205).

Spinophilin also regulates the neuronal microtubule cytoskeleton. Doublecortin, the protein product of the disease gene for the human neuronal migration disorder double cortex and X-linked lissencephaly, is a microtubule-associated protein that also interacts with the coiled-coil region of spinophilin (26, 248). By dephosphorylating doublecortin at S297, PP1-spinophilin stabilizes microtubule bundles at the transition zone between growth cone and axonal shaft. Neurons from doublecortin or spinophilin knockouts have splayed or unevenly spaced microtubules, and the double knockout exacerbated this defect and additionally prevented axons from crossing the midline to form the corpus callosum (26).

3.2.3 Spinophilin and Neurabin in Synaptic Transmission and Plasticity

PP1 is critical for the induction of LTD in the hippocampus and other brain areas. LTD-inducing stimuli promote translocation of PP1 into spines, coinciding with the dephosphorylation and internalization of AMPA receptors (71, 121, 165). Inhibition of PP1 activity or displacing PP1 from its targeting subunits (for instance with an "RVXF" motif-containing peptide) blocks LTD (165, 166).

Recent results suggested that neurabin targets PP1 to dendritic spines of CA1 hippocampal neurons to regulate surface expression of specific pools of AMPA receptors (104). Neurabin-targeted PP1 tonically stabilizes basal synaptic transmission by dephosphorylating GluR2 at S880, thereby preventing the interaction of GluR2 with PICK1 and the internalization of GluR2-containing AMPA receptors. During LTD, neurabin recruits PP1 into the vicinity of GluR1 containing AMPA-receptors, resulting in the dephosphorylation of GluR1 on S831 and S845, which in turn reduces single channel activity and surface expression of GluR1-containing AMPA receptors (Section 2). In this study, endogenous PP1-neurabin complexes were displaced with a "dominant-negative" mutant that binds to actin, but not PP1 (neurabin$_{1-490}$F460A), which may also displace PP1-spinophilin complexes from the F-actin cytoskeleton in dendritic spines. Given that spinophilin knockout mice are deficient in hippocampal LTD (77), it is possible that spinophilin and neurabin have at least partially overlapping functions at hippocampal synapses.

Knockout mice show that spinophilin and neurabin have distinct functions in dopamine-mediated plasticity (4). Deletion of neurabin blocks AMPA receptor potentiation by the D1 dopamine receptor in prefrontal cortex and striatum, whereas spinophilin deletion shows this defect only in the striatum. While neurabin knockout mice are deficient in corticostriatal LTP, spinophilin knockout mice lack LTD at this synapse. Supporting a role in AMPA receptor dephosphorylation, both knockout strains displayed enhanced induction of GluR1 S845 phosphorylation by D1-receptor activation. Interestingly, basal GluR1 S845 phosphorylation was reduced in both knockout mice, which is consistent with biochemical studies showing that spinophilin and neurabin are potent PP1 inhibitors in vitro (103, 145, 156). A factor that complicates the interpretation of these knockout studies is that both neurabin and spinophilin have been implicated in dendritic spine formation (77, 240).

3.3 PP2A and its Localization in Neurons

PP2A constitutes between 0.3 and 1% of total protein in mammalian cells. It supplies most of the soluble Ser/Thr phosphatase activity (100, 109). PP2A exists predominantly as a holoenzyme formed by three subunits. A 36 kDa catalytic or C subunit complexes with a 65 kDa scaffolding A subunit to form the AC core enzyme; the core enzyme binds a third, variable B subunit to form the ABC holoenzyme (Fig. 6). In mammals, A and C subunits are each encoded by two highly similar genes (Aα/β, Cα/β), with Aα and Cα isoforms being more abundant. Regulatory subunits are encoded by three multi-gene families, referred to as B, B', and B". The B family consists of 4 genes, Bα, Bβ, Bγ, and Bδ, that give rise to proteins with molecular masses of ~55 kDa (96, 155, 184, 226, 280). The B' family (also referred to as B56 or PR61) consists of at least 7 isoforms encoded by five genes (B'α, B'β, B'γ, B'δ, and B'ε) (52, 157, 158, 234, 238, 281), with molecular masses between 54 and 74 kDa. There are at least four members of the B" family, designated according to their size as PR48 (273), PR59 (253), and PR72/130 (98). PR72 and PR130 are splice- or promoter-use variants of the same gene. Several PP2A regulatory subunits show restricted tissue expression; in particular, Bβ1, Bβ2, and Bγ can only be detected in brain, and B'β and B'δ are enriched in brain (155, 231, 280).

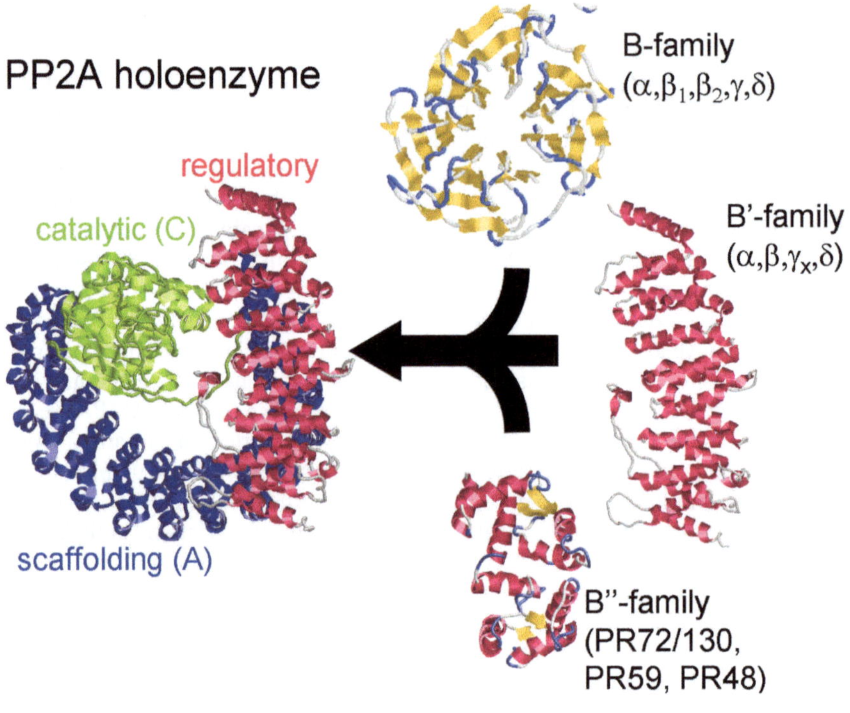

Fig. 6. The PP2A heterotrimer. Three families of regulatory subunits associate with the AC core dimer to form a spectrum of heterotrimeric PP2A holoenzymes. The heterotrimer structure was derived by X-ray crystallography (44, 270), while structures of the B- and B" family subunits are based on homology modeling.

Whereas the PP1 catalytic subunit is enriched in PSD preparations, the PP2A C subunit is found in the synaptosomal fraction, but depleted when this fraction is extracted with nonionic detergent to prepare the PSD (225). Evidence is accumulating that regulatory subunits target PP2A to specific subcellular locations and substrates in neurons. For instance, members of the B-family of regulatory subunits were localized to discrete neuronal populations and exhibited differences in dendritic, axonal, and somatic localization and subcellular fractionation profiles (231). The brain-enriched PP2A/B'β complex specifically dephosphorylates and inactivates tyrosine hydroxylase, the rate-limiting enzyme in catecholamine synthesis. PP2A/B'β was largely restricted to the somata of dopaminergic neurons in the substantia nigra and could not be detected in dopaminergic terminal fields in the striatum, which correlated with regional differences in tyrosine hydroxylase phosphorylation. B'β may curb dopamine synthesis and build-up of toxic metabolites of catecholamine at non-synaptic sites (207). Even though PP2A has been shown to regulate several pre- and postsynaptic proteins, the identity of the regulatory subunits that recruit the phosphatase holoenzyme to synapses remains to be established.

3.3.1 PP2A in Synaptic Plasticity

Compared to PP1 and PP2B, relatively little is known about the role of PP2A in synaptic function. The majority of studies implicating PP2A in synaptic transmission and plasticity employed pharmacological inhibitors such as okadaic acid or fostriecin, rather than inhibitory proteins or genetic tools. Even though okadaic acid inhibits PP2A at 10-fold lower concentration than PP1 in vitro, it is often applied at non-selective concentrations to overcome the slow uptake of the drug into cells.

Perhaps the best evidence for an involvement of PP2A derives from studying synaptic plasticity between granule and Purkinje cells in the cerebellum, which is thought to be critical for motor learning. Contrasting LTP/LTD in the CA1 region of the hippocampus, cerebellar LTD requires kinase activation and phosphatase inactivation (146), whereas LTP requires the activity of multiple phosphatases, including PP2A (24). In dissociated cerebellar cultures, specific inhibition of postsynaptic PP2A induces synaptic depression that occludes electrically-evoked LTD. Synaptic depression was use-dependent and was associated with declustering or internalization of GluR2-containing AMPA receptors(120). Available data point to a developmental switch in the phosphatases whose inhibition mediates cerebellar LTD, with PP2A playing the major role in mature cerebellar cultures (>22 DIV), and PP1 dominating in younger cultures (75, 120).

3.3.2 Complexes of PP2A with Ca^{2+} Channels and the NMDA Receptor

PP2A forms complexes with several ion channels and neurotransmitter receptors, which often contain an opposing kinase, to increase the spatial and temporal fidelity and specificity of signaling events in response to discretely localized stimuli.

Phosphorylation of the Ca^{2+} channel $Ca_V1.2$ on S1928 by PKA increases its activity. PP2A containing B' and B", but not B-family regulatory subunits interacts with a region downstream of this phosphorylation site to downregulate Cav1.2 phosphorylation and activity (94). The N-type Ca^{2+} channel ($Ca_V2.2$) mediates neurotransmitter release from nerve terminals. It binds the PP2A scaffolding A subunit, although the functional consequences of this association were not studied (114).

Intracellular Ca^{2+} release channels are also intimately associated with the phosphatase. The ryanodine receptor RyR2 contains three leucine-isoleucine zipper (LIZ) motifs. One mediates PP2A binding, possibly through the B" regulatory subunit PR130 (154). AKAP6 (mAKAP) binds to the second LIZ and spinophilin to the third LIZ motif in RyR2 (154). AKAP6-anchored PKA increases open probability of RyR2, which is opposed by PP2A and PP1. RyR1 in skeletal muscle lacks the LIZ motif that mediates PP2A binding, and therefore assembles selectively with PP1 and PKA (154). The neuronal isoform of the inositol 1,4,5-trisphosphatase receptor (IP3R1) exists as a similar supramolecular signaling complex with PP1, PP2A, and PKA (62). AKAP9 (yotiao) recruits both PKA and PP1 to the IP3R1, whereas PP2A could assemble into the complex through AKAP450, a longer splice variant of AKAP9/yotiao (249). The relative importance of the two phosphatases in intracellular Ca^{2+} channel regulation remains to be investigated.

Recruitment of PKA and PP1 to the NMDA receptor NR1 subunit via AKAP9/yotiao (262) has been described above in Section 3.3. Yeast two-hybrid screening identified the PP2A catalytic subunit as an interaction partner for the C-terminus of NR3A subunit (39, 144). Whereas NR1 and NR2 form Ca^{2+}-permeable NMDA receptors, receptors containing NR1 and NR3 function as Ca^{2+}-impermeable excitatory glycine receptors during development (185). NR3A associates with PP2A in brain lysates and thereby mediates dephosphorylation of NR1 at S897 (39, 144).

3.3.3 Regulation of CaMKII by PP2A

Autophosphorylation of CaMKII confers Ca^{2+}-independent activity and is critical for LTP (Section 2 and Chapter by Lisman and Hell, this volume). Autophosphorylation at T305 or T306 in the CaM binding region will keep CaMKII inactive by preventing its activation through Ca^{2+}/CaM binding. There are additional in vivo autophosphorylation sites (e.g., T253) with unclear functional relevance (67, 163).

PP1 and PP2A dephosphorylate CaMKII in different subcellular locations (PSD vs. cytosolic fractions, respectively) (14, 68, 225). This compartment-specific dephosphorylation is maintained when soluble or PSD-associated CaMKII are incubated with excess phosphatase in vitro, suggesting that conformational changes or local accessibility of CaMKII dictate selectivity for PP1 vs. PP2A (225, 227). Recent evidence suggests that T286 in PSD-associated CaMKII is somehow protected from dephosphorylation, and that PSD-bound PP1 dephosphorylates CaMKII at other sites (167). The identity of the CaMKII T305/306 phosphatase is less well defined, but genetic reduction of PP2A activity increases inhibitory autophosphorylation of CaMKII in *Drosophila* (138). Therefore, available evidence points to PP2A as the principal phosphatase that opposes autophosphorylation of CaMKII at both stimulatory (T286) and inhibitory (T305/306) sites.

LTP in the CA1 region of the hippocampus is associated with a decrease in PP2A activity, as shown with ^{32}P-labeled CaMKII as an in vitro substrate (79). LTP is accompanied by an increase in CaMKII-dependent phosphorylation of B'β subunit of PP2A (denoted as B'α in that report according to older nomenclature). Phosphorylation of B'β inhibited its activity towards T286-autophosphorylated CaMKII in vitro, suggesting a negative feedback mechanism in which autonomous CaMKII inactivates its modulator for persistent Ca^{2+}/CaM -independent activity during LTP (79). Recently, dimeric PP2A containing the α4 subunit was found to associate with cytosolic CaMKII, and neuron-specific α4 knockout mice displayed learning and memory deficits and increased Ca^{2+}/CaM-dependent CaMKII activity.

However, Ca^{2+}-independent activity was unchanged in the α4 knockout, leaving it unclear as to exactly how the PP2A/α4 dimer regulates the kinase (272).

The importance of phosphatases in returning CaMKII to its basal, Ca^{2+}/CaM responsive state is illustrated by the molecular etiology of Angelman syndrome, a developmental disorder characterized by mental retardation and seizures. Angelman syndrome can be caused by loss of function of the maternal copy of the UBE3A gene, which, due to brain region-specific imprinting, is the only copy expressed in the hippocampus and Purkinje cells of the cerebellum (54, 72). UBE3A codes for the E3A ubiquitin ligase. The few identified substrates have yet to be tied to the pathology. Mice with a maternally inherited UBE3A mutation display Angelman disease-like symptoms and show hyperphosphorylation of CaMKII at T286 and T305. CaMKII hyperphosphorylation correlated with reduced phosphatase activity in the hippocampus, most likely PP2A (261). Connecting dysregulated CaMKII phosphorylation to the disease, the neurological symptoms could be rescued by crossing Angelman syndrome mice with mice expressing a mutant CaMKII that cannot undergo inhibitory autophosphorylation (T305V/T306A) (251).

3.4 The Ca^{2+}/CaM-Dependent Phosphatase PP2B/Calcineurin

3.4.1 Basic Features and Functions of PP2B

PP2B is a ubiquitously expressed, Ca^{2+}/CaM activated Ser/Thr phosphatase with particularly high levels in the somatodendritic compartment of neurons (91, 152). PP2B is perhaps best known as a central mediator of T lymphocyte activation and target for the immunosuppressive drugs FK506 and cyclosporine A. The enzyme consists of a 60 kDa catalytic subunit (CaN-A), and a 19 kDa regulatory subunit (CaN-B), which has structural similarity to CaM. Three genes code for catalytic subunits, with Aα and Aβ being the predominant forms in brain. The regulatory subunit is encoded by two genes, of which only B1 is detectably expressed in neurons. Full activity requires both Ca^{2+}/CaM binding to an autoinhibitory domain on CaN-A and Ca^{2+} binding to the EF hands of CaN-B.

PP2B has a much lower threshold for activation by Ca^{2+}/CaM (0.1–1 nM) than CaMKs including CaMKII, which might leads to the preferential activation of the phosphatase during LTD-inducing low frequency stimulation (165, 166). Ca^{2+}-dependent phosphorylation events could predominate at the higher stimulation frequencies that induce LTP in part because accumulation of superoxide anions upon Ca^{2+} influx results in inactivation of PP2B through oxidation of metal ions in the active site (27, 171).

Because of steric constraints of the substrate binding region (105), PP2B has narrower substrate specificity than other Ser/Thr phosphatases. The number of PP2B-sensitive phospho-proteins is considerable because PP2B also acts indirectly by disinhibiting PP1. As discussed above, this disinhibition is accomplished through dephosphorylation of PKA sites on inhibitor-1 and DARPP-32, two cytosolic PP1 inhibitory proteins (38). Direct and indirect (PP1-mediated) PP2B actions can be distinguished by pharmacological inhibition of PP1 with tautomycin or calyculin A.

Many PP2B targets and anchoring proteins, including AKAP5 feature a PxIxIT motif (actually $Px_{1-2}IxIT$) that interacts with the CaN-A subunit (7, 30, 35, 53, 58, 179). First identified in the transcription factor NFAT, this sequence motif binds to a hydrophobic groove on CaN-A that is structurally related to the region on the PP1 catalytic subunit, where proteins containing the RVXF motif dock (126, 127).

3.4.2 Interactions of PP2B with Glutamate Receptors and Ca^{2+} Channels

PP2B is a potent modulator of excitatory neurotransmission and can downregulate ionotropic glutamate receptors by multiple mechanisms. NMDA receptor activity is negatively constrained by PP2B (129, 215, 245, 246), and this effect may be mediated by dephosphorylation of the NR2A subunit, possibly of S900 in its C-terminal tail (117, 247). PP2B further decreases surface expression and single-channel activity and of AMPA receptors, mediated through dephosphorylation (directly or indirectly via PP1) of S831 and S845 in the GluR1 subunit C-terminus (22, 121, 221, 237). Dephosphorylation of components of the endocytotic machinery provides an additional mechanism by which PP2B promotes internalization of AMPA receptors (22, 133).

As discussed above (Section 2.2.1), PP2B is recruited to the AMPA receptor GluR1 subunit through AKAP5 via SAP97 and possibly also PSD-95 (48). These interactions are crucial for PP2B to antagonize upregulation of GluR1 currents by PKA (220, 237). AKAP5 also targets PP2B to $Ca_V1.2$ via a direct interaction of AKAP5 with the C-terminus of the pore-forming α_1 subunit (179). In this complex, PP2B dominantly suppresses PKA-mediated enhancement of L-type currents. Strikingly, PP2B recruitment by AKAP5 is required for nuclear translocation of the transcription factor NFATc4/NFAT3 in response to L-type channel activation (179).

4 Concluding Remarks

Synapses not only transmit signals between neurons but also process signals and store information. Multiple signal pathways regulate the functions of central synapses in a highly complex manner. Many kinases and phosphatases have to be anchored near or at defined substrates for fast, efficient, and specific regulation of their phosphorylation. A true understanding of the numerous forms of synaptic plasticity will require the knowledge of the precise molecular interactions of kinases and phosphatases at synaptic sites. As discussed above, specific recruitment of CaMKII to postsynaptic sites that experience a period of hightened activity likely contributes to their potentiation. New sophisticated experimental approaches such as modification of kinase and phosphatase anchoring by point mutations in the mouse genes of anchoring proteins or, in a more acute setting, with membrane-penetrable peptides will help unravel the spatio-temporal organization of synaptic signaling. Numerous neurological and mental diseases are due to synaptic dysregulations (e.g., Chapters by McNamara et al., Meller and Simon, and Salter, this volume). Defining the exact organization of signaling pathways will pave the way for developing novel therapeutics for the treatment of these diseases.

Acknowledgements

We thank Yucui Chen and Ivar Stein for micrographs of CaMKII clustering in Fig. 2, J. Ulrich for excellent help with illustrations, and Mark Dell'Acqua for critical review of the manuscript. We apologize to all, whose work we were not able to cite because of space limitations. Work in the laboratory of SS is supported by NIH grants NS043254, NS056244, and NS057714 and in the laboratory of JWH by NIH grants NS035563, NS046450, and AG017502.

References

1. Adesnik H and Nicoll RA. Conservation of glutamate receptor 2-containing AMPA receptors during long-term potentiation. *J Neurosci* 27: 4598–4602, 2007.
2. Akyol Z, Bartos JA, Merrill MA, Faga LA, Jaren OR, Shea MA, and Hell JW. Apo-calmodulin binds with its C-terminal domain to the N-methyl-D-aspartate receptor NR1 C0 region. *J Biol Chem* 279: 2166–2175, 2004.
3. Allen PB, Ouimet CC, and Greengard P. Spinophilin, a novel protein phosphatase 1 binding protein localized to dendritic spines. *Proc Natl Acad Sci USA* 94: 9956–9961, 1997.
4. Allen PB, Zachariou V, Svenningsson P, Lepore AC, Centonze D, Costa C, Rossi S, Bender G, Chen G, Feng J, Snyder GL, Bernardi G, Nestler EJ, Yan Z, Calabresi P, and Greengard P. Distinct roles for spinophilin and neurabin in dopamine-mediated plasticity. *Neuroscience* 140: 897–911, 2006.
5. Alonso A, Sasin J, Bottini N, Friedberg I, Friedberg I, Osterman A, Godzik A, Hunter T, Dixon J, and Mustelin T. Protein tyrosine phosphatases in the human genome. *Cell* 117: 699–711, 2004.
6. An R, Heath BM, Higgins JP, Koch WJ, Lefkowitz RJ, and Kass RS. Beta2-adrenergic receptor overexpression in the developing mouse heart: evidence for targeted modulation of ion channels. *J Physiol* 516 (Pt 1): 19–30, 1999.
7. Aramburu J, Garcia-Cozar F, Raghavan A, Okamura H, Rao A, and Hogan PG. Selective inhibition of NFAT activation by a peptide spanning the calcineurin targeting site of NFAT. *Mol Cell* 1: 627–637, 1998.
8. Avraham H, Park SY, Schinkmann K, and Avraham S. RAFTK/Pyk2-mediated cellular signalling. *Cell Signal* 12: 123–133, 2000.
9. Avraham S, London R, Fu Y, Ota S, Hiregowdara D, Li J, Jiang S, Pasztor LM, White RA, Groopman JE, et al. Identification and characterization of a novel related adhesion focal tyrosine kinase (RAFTK) from megakaryocytes and brain. *J Biol Chem* 270: 27742–27751, 1995.
10. Baillie GS, Sood A, McPhee I, Gall I, Perry SJ, Lefkowitz RJ, and Houslay MD. beta-Arrestin-mediated PDE4 cAMP phosphodiesterase recruitment regulates beta-adrenoceptor switching from Gs to Gi. *Proc Natl Acad Sci USA* 100: 940–945, 2003.
11. Balijepalli RC, Foell JD, Hall DD, Hell JW, and Kamp TJ. From the Cover: Localization of cardiac L-type Ca2+ channels to a caveolar macromolecular signaling complex is required for beta2-adrenergic regulation. *Proc Natl Acad Sci USA* 103: 7500–7505, 2006.
12. Banke TG, Bowie D, Lee H, Huganir RL, Schousboe A, and Traynelis SF. Control of GluR1 AMPA receptor function by cAMP-dependent protein kinase. *J Neurosci* 20: 89–102, 2000.

13. Barnes AP, Smith FD, 3rd, VanDongen HM, VanDongen AM, and Milgram SL. The identification of a second actin-binding region in spinophilin/neurabin II. *Brain Res Mol Brain Res* 124: 105–113, 2004.
14. Barnes GN, Slevin JT, and Vanaman TC. Rat brain protein phosphatase 2A: an enzyme that may regulate autophosphorylated protein kinases. *J Neurochem* 64: 340–353, 1995.
15. Barria A and Malinow R. NMDA receptor subunit composition controls synaptic plasticity by regulating binding to CaMKII. *Neuron* 48: 289–301, 2005.
16. Barria A, Muller D, Derkach V, Griffith LC, and Soderling TR. Regulatory phosphorylation of AMPA-type glutamate receptors by CaM-KII during long-term potentiation. *Science* 276: 2042–2045, 1997.
17. Barsacchi R, Heider H, Girault J, and Meldolesi J. Requirement of pyk2 for the activation of the MAP kinase cascade induced by Ca(2+) (but not by PKC or G protein) in PC12 cells. *FEBS Lett* 461: 273–276, 1999.
18. Bauman AL, Soughayer J, Nguyen BT, Willoughby D, Carnegie GK, Wong W, Hoshi N, Langeberg LK, Cooper DM, Dessauer CW, and Scott JD. Dynamic regulation of cAMP synthesis through anchored PKA-adenylyl cyclase V/VI complexes. *Mol Cell* 23: 925–931, 2006.
19. Bayer KU, De Koninck P, Leonard AS, Hell JW, and Schulman H. Interaction with the NMDA receptor locks CaMKII in an active conformation. *Nature* 411: 801–805, 2001.
20. Bayer KU, LeBel E, McDonald GL, O'Leary H, Schulman H, and De Koninck P. Transition from reversible to persistent binding of CaMKII to postsynaptic sites and NR2B. *J Neurosci* 26: 1164–1174, 2006.
21. Bean BP, Nowycky MC, and Tsien RW. β-Adrenergic modulation of calcium channels in frog ventricular heart cells. *Nature* 307: 371–375, 1984.
22. Beattie EC, Carroll RC, Yu X, Morishita W, Yasuda H, von Zastrow M, and Malenka RC. Regulation of AMPA receptor endocytosis by a signaling mechanism shared with LTD. *Nat Neurosci* 3: 1291–1300, 2000.
23. Beene DL and Scott JD. A-kinase anchoring proteins take shape. *Curr Opin Cell Biol* 19: 192–198, 2007.
24. Belmeguenai A and Hansel C. A role for protein phosphatases 1, 2A, and 2B in cerebellar long-term potentiation. *J Neurosci* 25: 10768–10772, 2005.
25. Benke TA, Luthi A, Isaac JT, and Collingridge GL. Modulation of AMPA receptor unitary conductance by synaptic activity. *Nature* 393: 793–797, 1998.
26. Bielas SL, Serneo FF, Chechlacz M, Deerinck TJ, Perkins GA, Allen PB, Ellisman MH, and Gleeson JG. Spinophilin facilitates dephosphorylation of doublecortin by PP1 to mediate microtubule bundling at the axonal wrist. *Cell* 129: 579–591, 2007.
27. Bito H, Deisseroth K, and Tsien RW. CREB phosphorylation and dephosphorylation: a Ca(2+)- and stimulus duration-dependent switch for hippocampal gene expression. *Cell* 87: 1203–1214, 1996.
28. Blitzer RD, Connor JH, Brown GP, Wong T, Shenolikar S, Iyengar R, and Landau EM. Gating of CaMKII by cAMP-regulated protein phosphatase activity during LTP. *Science* 280: 1940–1942, 1998.
29. Borgkvist A and Fisone G. Psychoactive drugs and regulation of the cAMP/PKA/DARPP-32 cascade in striatal medium spiny neurons. *Neurosci Biobehav Rev* 31: 79–88, 2007.
30. Boustany LM and Cyert MS. Calcineurin-dependent regulation of Crz1p nuclear export requires Msn5p and a conserved calcineurin docking site. *Genes Dev* 16: 608–619, 2002.
31. Brady AE, Wang Q, Colbran RJ, Allen PB, Greengard P, and Limbird LE. Spinophilin stabilizes cell surface expression of alpha 2B-adrenergic receptors. *J Biol Chem* 278: 32405–32412, 2003.
32. Braithwaite SP, Paul S, Nairn AC, and Lombroso PJ. Synaptic plasticity: one STEP at a time. *Trends Neurosci* 29: 452–458, 2006.

33. Brandon EP, Idzerda RL, and McKnight GS. PKA isoforms, neural pathways, and behaviour: making the connection. *Curr Opin Neurobiol* 7: 397–403, 1997.
34. Bredt DS and Nicoll RA. AMPA receptor trafficking at excitatory synapses. *Neuron* 40: 361–379, 2003.
35. Bultynck G, Heath VL, Majeed AP, Galan J, Haguenauer-Tsapis R, and Cyert MS. Slm1 and slm2 are novel substrates of the calcineurin phosphatase required for heat stress-induced endocytosis of the yeast uracil permease. *Mol Cell Biol* 26: 4729–4745, 2006.
36. Burgin KE, Waxham MN, Rickling S, Westgate SA, Mobley WC, and Kelly PT. In situ hybridization histochemistry of Ca2+/calmodulin-dependent protein kinase in developing rat brain. *J Neurosci* 10: 1788–1798, 1990.
37. Carlisle Michel JJ and Scott JD. AKAP mediated signal transduction. *Annu Rev Pharmacol Toxicol* 42: 235–257, 2002.
38. Ceulemans H and Bollen M. Functional diversity of protein phosphatase-1, a cellular economizer and reset button. *Physiol Rev* 84: 1–39, 2004.
39. Chan SF and Sucher NJ. An NMDA receptor signaling complex with protein phosphatase 2A. *J Neurosci* 21: 7985–7992, 2001.
40. Chang BY, Conroy KB, Machleder EM, and Cartwright CA. RACK1, a receptor for activated C kinase and a homolog of the beta subunit of G proteins, inhibits activity of src tyrosine kinases and growth of NIH 3T3 cells. *Mol Cell Biol* 18: 3245–3256, 1998.
41. Chen C and Leonard JP. Protein tyrosine kinase-mediated potentiation of currents from cloned NMDA receptors. *J Neurochem* 67: 194–200, 1996.
42. Chen-Izu Y, Xiao RP, Izu LT, Cheng H, Kuschel M, Spurgeon H, and Lakatta EG. G(i)-dependent localization of beta(2)-adrenergic receptor signaling to L-type Ca(2+) channels. *Biophys J* 79: 2547–2556, 2000.
43. Cheng JJ, Chao YJ, and Wang DL. Cyclic strain activates redox-sensitive proline-rich tyrosine kinase 2 (PYK2) in endothelial cells. *J Biol Chem* 277: 48152–48157, 2002.
44. Cho US and Xu W. Crystal structure of a protein phosphatase 2A heterotrimeric holoenzyme. *Nature* 445: 53–57, 2007.
45. Coghlan VM, Perrino BA, Howard M, Langeberg LK, Hicks JB, Gallatin WM, and Scott JD. Association of protein kinase A and protein phosphatase 2B with a common anchoring protein. *Science* 267: 108–111, 1995.
46. Colbran RJ and Brown AM. Calcium/calmodulin-dependent protein kinase II and synaptic plasticity. *Curr Opin Neurobiol* 14: 318–327, 2004.
47. Colbran RJ and Soderling TR. Calcium/calmodulin-independent autophosphorylation sites of calcium/calmodulin-dependent protein kinase II. Studies on the effect of phosphorylation of threonine 305/306 and serine 314 on calmodulin binding using synthetic peptides. *J Biol Chem* 265: 11213–11219, 1990.
48. Colledge M, Dean RA, Scott GK, Langeberg LK, Huganir RL, and Scott JD. Targeting of PKA to glutamate receptors through a MAGUK-AKAP complex. *Neuron* 27: 107–119, 2000.
49. Colledge M, Snyder EM, Crozier RA, Soderling JA, Jin Y, Langeberg LK, Lu H, Bear MF, and Scott JD. Ubiquitination regulates PSD-95 degradation and AMPA receptor surface expression. *Neuron* 40: 595–607, 2003.
50. Collingridge GL, Isaac JT, and Wang YT. Receptor trafficking and synaptic plasticity. *Nat Rev Neurosci* 5: 952–962, 2004.
51. Conti M, Richter W, Mehats C, Livera G, Park JY, and Jin C. Cyclic AMP-specific PDE4 phosphodiesterases as critical components of cyclic AMP signaling. *J Biol Chem* 278: 5493–5496, 2003
52. Csortos C, Zolnierowicz S, Bako E, Durbin SD, and DePaoli-Roach AA. High complexity in the expression of the B' subunit of protein phosphatase 2A0. Evidence for the existence of at least seven novel isoforms. *J Biol Chem* 271: 2578–2588, 1996.

53. Czirjak G and Enyedi P. Targeting of calcineurin to an NFAT-like docking site is required for the calcium-dependent activation of the background K+ channel, TRESK. *J Biol Chem* 281: 14677–14682, 2006.
54. Dan B, Servais L, Boyd SG, Wagstaff J, and Cheron G. From electrophysiology to chromatin: a bottom-up approach to Angelman syndrome. *Ann NY Acad Sci* 1030: 599–611, 2004.
55. Davare MA, Avdonin V, Hall DD, Peden EM, Burette A, Weinberg RJ, Horne MC, Hoshi T, and Hell JW. A beta2 adrenergic receptor signaling complex assembled with the Ca2+ channel Cav1.2 [see comments]. [erratum appears in Science 2001 Aug 3; 293(5531): 804]. *Science* 293: 98–101, 2001.
56. Davare MA, Horne MC, and Hell JW. Protein Phosphatase 2A is associated with class C L-type calcium channels ($Ca_v1.2$) and antagonizes channel phosphorylation by cAMP-dependent protein kinase. *J Biol Chem* 275: 39710–39717, 2000.
57. De Koninck P and Schulman H. Sensitivity of CaMKII to the frequency of Ca^{2+} oscillations. *Science* 279: 227–230, 1998.
58. Dell'Acqua ML, Dodge KL, Tavalin SJ, and Scott JD. Mapping the protein phosphatase-2B anchoring site on AKAP79. Binding and inhibition of phosphatase activity are mediated by residues 315–360. *J Biol Chem* 277: 48796–48802, 2002.
59. Dell'Acqua ML, Faux MC, Thorburn J, Thorburn A, and Scott JD. Membrane-targeting sequences on AKAP79 bind phosphatidylinositol-4, 5- bisphosphate. *EMBO J* 17: 2246–2260, 1998.
60. Della Rocca GJ, van Biesen T, Daaka Y, Luttrell DK, Luttrell LM, and Lefkowitz RJ. Ras-dependent mitogen-activated protein kinase activation by G protein- coupled receptors. Convergence of Gi- and Gq-mediated pathways on calcium/calmodulin, Pyk2, and Src kinase. *J Biol Chem* 272: 19125–19132, 1997.
61. Derkach V, Barria A, and Soderling TR. Ca2+/calmodulin-kinase II enhances channel conductance of alpha-amino-3- hydroxy-5-methyl-4-isoxazolepropionate type glutamate receptors. *Proc Natl Acad Sci USA* 96: 3269–3274, 1999.
62. DeSouza N, Reiken S, Ondrias K, Yang YM, Matkovich S, and Marks AR. Protein kinase A and two phosphatases are components of the inositol 1,4,5-trisphosphate receptor macromolecular signaling complex. *J Biol Chem* 277: 39397–39400, 2002.
63. Dikic I, Tokiwa G, Lev S, Courtneidge SA, and Schlessinger J. A role for Pyk2 and Src in linking G-protein-coupled receptors with MAP kinase activation. *Nature* 383: 547–549, 1996.
64. Dodge KL, Khouangsathiene S, Kapiloff MS, Mouton R, Hill EV, Houslay MD, Langeberg LK, and Scott JD. mAKAP assembles a protein kinase A/PDE4 phosphodiesterase cAMP signaling module. *EMBO J* 20: 1921–1930, 2001.
65. Dodge-Kafka K, Langeberg L, and Scott JD. Compartmentation of cyclic nucleotide signaling in the heart: the role of A-kinase anchoring proteins. *Circ Res* 98: 993–1001, 2006.
66. Dodge-Kafka KL, Soughayer J, Pare GC, Carlisle Michel JJ, Langeberg LK, Kapiloff MS, and Scott JD. The protein kinase A anchoring protein mAKAP coordinates two integrated cAMP effector pathways. *Nature* 437: 574–578, 2005.
67. Dosemeci A, Gollop N, and Jaffe H. Identification of a major autophosphorylation site on postsynaptic density-associated Ca2+/calmodulin-dependent protein kinase. *J Biol Chem* 269: 31330–31333, 1994.
68. Dosemeci A and Reese TS. Inhibition of endogenous phosphatase in a postsynaptic density fraction allows extensive phosphorylation of the major postsynaptic density protein. *J Neurochem* 61: 550–555, 1993.
69. Dosemeci A, Tao-Cheng JH, Vinade L, Winters CA, Pozzo-Miller L, and Reese TS. Glutamate-induced transient modification of the postsynaptic density. *Proc Natl Acad Sci USA* 98: 10428–10432, 2001.

70. Earp H, Huckle W, Dawson T, Li X, Graves L, and Dy R. Angiotensin II activates at least two tyrosine kinases in rat liver epithelial cells. *J Biol Chem* 270: 28440–28447, 1995.
71. Ehlers MD. Reinsertion or degradation of AMPA receptors determined by activity-dependent endocytic sorting. *Neuron* 28: 511–525, 2000.
72. Elgersma Y. Genetic engineering cures mice of neurological deficits: prospects for treating Angelman syndrome. *Pharmacogenomics* 8: 539–541, 2007.
73. Elgersma Y, Fedorov NB, Ikonen S, Choi ES, Elgersma M, Carvalho OM, Giese KP, and Silva AJ. Inhibitory autophosphorylation of CaMKII controls PSD association, plasticity, and learning. *Neuron* 36: 493–505, 2002.
74. Esteban JA, Shi SH, Wilson C, Nuriya M, Huganir RL, and Malinow R. PKA phosphorylation of AMPA receptor subunits controls synaptic trafficking underlying plasticity. *Nat Neurosci* 6: 136–143, 2003.
75. Eto M, Bock R, Brautigan DL, and Linden DJ. Cerebellar long-term synaptic depression requires PKC-mediated activation of CPI-17, a myosin/moesin phosphatase inhibitor. *Neuron* 36: 1145–1158, 2002.
76. Fan G, Shumay E, Wang H, and Malbon CC. The scaffold protein gravin (cAMP-dependent protein kinase-anchoring protein 250) binds the beta 2-adrenergic receptor via the receptor cytoplasmic Arg-329 to Leu-413 domain and provides a mobile scaffold during desensitization. *J Biol Chem* 276: 24005–24014, 2001.
77. Feng J, Yan Z, Ferreira A, Tomizawa K, Liauw JA, Zhuo M, Allen PB, Ouimet CC, and Greengard P. Spinophilin regulates the formation and function of dendritic spines. *Proc Natl Acad Sci USA* 97: 9287–9292, 2000.
78. Fink CC, Bayer KU, Myers JW, Ferrell JE, Jr., Schulman H, and Meyer T. Selective regulation of neurite extension and synapse formation by the beta but not the alpha isoform of CaMKII. *Neuron* 39: 283–297, 2003.
79. Fukunaga K, Muller D, Ohmitsu M, Bako E, DePaoli-Roach AA, and Miyamoto E. Decreased protein phosphatase 2A activity in hippocampal long-term potentiation. *J Neurochem* 74: 807–817, 2000.
80. Gardoni F, Bellone C, Cattabeni F, and Di Luca M. Protein kinase C activation modulates alpha-calmodulin kinase II binding to NR2A subunit of N-methyl-D-aspartate receptor complex. *J Biol Chem* 276: 7609–7613, 2001.
81. Gardoni F, Caputi A, Cimino M, Pastorino L, Cattabeni F, and Di Luca M. Calcium/calmodulin-dependent protein kinase II is associated with NR2A/B subunits of NMDA receptor in postsynaptic densities. *J Neurochem* 71: 1733–1741, 1998.
82. Gardoni F, Schrama LH, Kamal A, Gispen WH, Cattabeni F, and Di Luca M. Hippocampal synaptic plasticity involves competition between Ca2+/calmodulin-dependent protein kinase II and postsynaptic density 95 for binding to the NR2A subunit of the NMDA receptor. *J Neurosci* 21: 1501–1509, 2001.
83. Gardoni F, Schrama LH, van Dalen JJ, Gispen WH, Cattabeni F, and Di Luca M. AlphaCaMKII binding to the C-terminal tail of NMDA receptor subunit NR2A and its modulation by autophosphorylation. *FEBS Letters* 456: 394–398, 1999.
84. Giese KP, Fedorov NB, Filipkowski RK, and Silva AJ. Autophosphorylation at Thr286 of the alpha calcium-calmodulin kinase II in LTP and learning. *Science* 279: 870–873, 1998.
85. Gingrich JR, Pelkey KA, Fam SR, Huang Y, Petralia RS, Wenthold RJ, and Salter MW. Unique domain anchoring of Src to synaptic NMDA receptors via the mitochondrial protein NADH dehydrogenase subunit 2. *Proc Natl Acad Sci USA* 101: 6237–6242, 2004.
86. Girault JA, Costa A, Derkinderen P, Studler JM, and Toutant M. FAK and PYK2/CAKbeta in the nervous system: a link between neuronal activity, plasticity and survival? *Trends Neurosci* 22: 257–263, 1999.

87. Gleason MR, Higashijima S, Dallman J, Liu K, Mandel G, and Fetcho JR. Translocation of CaM kinase II to synaptic sites in vivo. *Nature Neurosci* 6: 217–218, 2003.
88. Gomez LL, Alam S, Smith KE, Horne E, and Dell'Acqua ML. Regulation of A-kinase anchoring protein 79/150-cAMP-dependent protein kinase postsynaptic targeting by NMDA receptor activation of calcineurin and remodeling of dendritic actin. *J Neurosci* 22: 7027–7044, 2002.
89. Gorski JA, Gomez LL, Scott JD, and Dell'Acqua ML. Association of an A-kinase-anchoring protein signaling scaffold with cadherin adhesion molecules in neurons and epithelial cells. *Mol Biol Cell* 16: 3574–3590, 2005.
90. Grossman SD, Hsieh-Wilson LC, Allen PB, Nairn AC, and Greengard P. The actin-binding domain of spinophilin is necessary and sufficient for targeting to dendritic spines. *Neuromol Med* 2: 61–69, 2002.
91. Groth RD, Dunbar RL, and Mermelstein PG. Calcineurin regulation of neuronal plasticity. *Biochem Biophys Res Commun* 311: 1159–1171, 2003.
92. Grover LM and Teyler TJ. Two components of long-term potentiation induced by different patterns of afferent activation. *Nature* 347: 477–479, 1990.
93. Hall DD, Davare MA, Shi M, Allen ML, Weisenhaus M, McKnight GS, and Hell JW. Critical role of cAMP-dependent protein kinase anchoring to the L-type calcium channel Cav1.2 via A-kinase anchor protein 150 in neurons. *Biochemistry* 46: 1635–1646, 2007.
94. Hall DD, Feekes JA, Arachchige Don AS, Shi M, Hamid J, Chen L, Strack S, Zamponi GW, Horne MC, and Hell JW. Binding of protein phosphatase 2A to the L-type calcium channel Cav1.2 next to Ser1928, its main PKA site, is critical for Ser1928 dephosphorylation. *Biochemistry* 45: 3448–3459, 2006.
95. Hayashi Y, Shi SH, Esteban JA, Piccini A, Poncer JC, and Malinow R. Driving AMPA receptors into synapses by LTP and CaMKII: requirement for GluR1 and PDZ domain interaction. *Science* 287: 2262–2267, 2000.
96. Healy AM, Zolnierowicz S, Stapleton AE, Goebl M, DePaoli-Roach AA, and Pringle JR. CDC55, a Saccharomyces cerevisiae gene involved in cellular morphogenesis: identification, characterization, and homology to the B subunit of mammalian type 2A protein phosphatase. *Mol Cell Biol* 11: 5767–5780, 1991.
97. Hell JW, Westenbroek RE, Breeze LJ, Wang K, Chavkin C, and Catterall WA. N-methyl-D-aspartate receptor-induced proteolytic conversion of postsynaptic class C L-type calcium channels in hippocampal neurons. *Proc Natl Acad Sci USA* 93: 3362–3367, 1996.
98. Hendrix P, Mayer-Jackel RE, Cron P, Goris J, Hofsteenge J, Merlevede W, and Hemmings BA. Structure and expression of a 72-kDa regulatory subunit of protein phosphatase 2A. Evidence for different size forms produced by alternative splicing. *J Biol Chem* 268: 15267–15276, 1993.
99. Hojjati MR, van Woerden GM, Tyler WJ, Giese KP, Silva AJ, Pozzo-Miller L, and Elgersma Y. Kinase activity is not required for alphaCaMKII-dependent presynaptic plasticity at CA3-CA1 synapses. *Nat Neurosci* 10: 1125–1127, 2007.
100. Honkanen RE and Golden T. Regulators of serine/threonine protein phosphatases at the dawn of a clinical era? *Curr Med Chem* 9: 2055–2075, 2002.
101. Hoogland TM and Saggau P. Facilitation of L-type Ca2+ channels in dendritic spines by activation of beta2 adrenergic receptors. *J Neurosci* 24: 8416–8427, 2004.
102. Horne EA and Dell'Acqua ML. Phospholipase C is required for changes in postsynaptic structure and function associated with NMDA receptor-dependent long-term depression. *J Neurosci* 27: 3523–3534, 2007.
103. Hsieh-Wilson LC, Allen PB, Watanabe T, Nairn AC, and Greengard P. Characterization of the neuronal targeting protein spinophilin and its interactions with protein phosphatase-1. *Biochemistry* 38: 4365–4373, 1999.

104. Hu XD, Huang Q, Yang X, and Xia H. Differential regulation of AMPA receptor trafficking by neurabin-targeted synaptic protein phosphatase-1 in synaptic transmission and long-term depression in hippocampus. *J Neurosci* 27: 4674–4686, 2007.
105. Huai Q, Kim HY, Liu Y, Zhao Y, Mondragon A, Liu JO, and Ke H. Crystal structure of calcineurin-cyclophilin-cyclosporin shows common but distinct recognition of immunophilin-drug complexes. *Proc Natl Acad Sci USA* 99: 12037–12042, 2002.
106. Huang Y, Lu W, Ali DW, Pelkey KA, Pitcher GM, Lu YM, Aoto H, Roder JC, Sasaki T, Salter MW, and MacDonald JF. CAKbeta/Pyk2 kinase is a signaling link for induction of long-term potentiation in CA1 hippocampus. *Neuron* 29: 485–496, 2001.
107. Hudmon A and Schulman H. Structure-function of the multifunctional Ca2+/calmodulin-dependent protein kinase II. *Biochem J* 364: 593–611, 2002.
108. Hulme JT, Lin TW, Westenbroek RE, Scheuer T, and Catterall WA. Beta-adrenergic regulation requires direct anchoring of PKA to cardiac $Ca_V1.2$ channels via a leucine zipper interaction with A kinase-anchoring protein 15. *Proc Natl Acad Sci USA* 100: 13093–13098, 2003.
109. Janssens V and Goris J. Protein phosphatase 2A: a highly regulated family of serine/threonine phosphatases implicated in cell growth and signalling. *Biochem J* 353: 417–439, 2001.
110. Johnson BD, Brousal JP, Peterson BZ, Gallombardo PA, Hockerman GH, Lai Y, Scheuer T, and Catterall WA. Modulation of the cloned skeletal muscle L-type Ca^{2+} channel by anchored cAMP-dependent protein kinase. *J Neurosci* 17: 1243–1255, 1997.
111. Jurevicius J, Skeberdis VA, and Fischmeister R. Role of cyclic nucleotide phosphodiesterase isoforms in cAMP compartmentation following beta2-adrenergic stimulation of ICa,L in frog ventricular myocytes. *J Physiol* 551: 239–252, 2003.
112. Kalia LV, Pitcher GM, Pelkey KA, and Salter MW. PSD-95 is a negative regulator of the tyrosine kinase Src in the NMDA receptor complex. *EMBO J* 25: 4971–4982, 2006.
113. Kennedy MB, Bennett MK, and Erondu NE. Biochemical and immunochemical evidence that the "major postsynaptic density protein" is a subunit of a calmodulin-dependent protein kinase. *Proc Natl Acad Sci USA* 80: 7357–7361, 1983.
114. Khanna R, Zougman A, and Stanley EF. A proteomic screen for presynaptic terminal N-type calcium channel (CaV2.2) binding partners. *J Biochem Mol Biol* 40: 302–314, 2007.
115. Klauck TM, Faux MC, Labudda K, Langeberg LK, Jaken S, and Scott JD. Coordination of three signaling enzymes by AKAP79, a mammalian scaffold protein. *Science* 271: 1589–1592, 1996.
116. Kohr G and Seeburg PH. Subtype-specific regulation of recombinant NMDA receptor-channels by protein tyrosine kinases of the src family. *J Physiol* 492: 445–452, 1996.
117. Krupp JJ, Vissel B, Thomas CG, Heinemann SF, and Westbrook GL. Calcineurin acts via the C-terminus of NR2A to modulate desensitization of NMDA receptors. *Neuropharmacol* 42: 593–602, 2002.
118. Krupp JJ, Vissel B, Thomas CG, Heinemann SF, and Westbrook GL. Interactions of calmodulin and alpha-actinin with the NR1 subunit modulate Ca^{2+}-dependent inactivation of NMDA receptors. *J Neurosci* 19: 1165–1178, 1999.
119. Lan JY, Skeberdis VA, Jover T, Grooms SY, Lin Y, Araneda RC, Zheng X, Bennett MV, and Zukin RS. Protein kinase C modulates NMDA receptor trafficking and gating. *Nat Neurosci* 4: 382–390, 2001.
120. Launey T, Endo S, Sakai R, Harano J, and Ito M. Protein phosphatase 2A inhibition induces cerebellar long-term depression and declustering of synaptic AMPA receptor. *Proc Natl Acad Sci USA* 101: 676–681, 2004.
121. Lee HK, Barbarosie M, Kameyama K, Bear MF, and Huganir RL. Regulation of distinct AMPA receptor phosphorylation sites during bidirectional synaptic plasticity. *Nature* 405: 955–959, 2000.

122. Leonard AS, Bayer K-U, Merrill M, Lim I, Shea MA, Schulman H, and Hell JW. Regulation of calcium/calmodulin-dependent protein kinase II docking to N-methyl-D-aspartate receptors by calcium/calmodulin and a-actinin. *J Biol Chem* 277: 48441–48448, 2002.
123. Leonard AS, Davare MA, Horne MC, Garner CC, and Hell JW. SAP97 is associated with the a-amino-3-hydroxy-5-methylisoxazole-4-propionic acid receptor GluR1 subunit. *J Biol Chem* 273: 19518–19524, 1998.
124. Leonard AS, Lim IA, Hemsworth DE, Horne MC, and Hell JW. Calcium/calmodulin-dependent protein kinase II is associated with the N-methyl-D-aspartate receptor. *Proc Natl Acad Sci USA* 96: 3239–3244, 1999.
125. Lev S, Moreno H, Martinez R, Canoll P, Peles E, Musacchio JM, Plowman GD, Rudy B, and Schlessinger J. Protein tyrosine kinase PYK2 involved in Ca(2+)-induced regulation of ion channel and MAP kinase functions. *Nature* 376: 737–745, 1995.
126. Li H, Rao A, and Hogan PG. Structural delineation of the calcineurin-NFAT interaction and its parallels to PP1 targeting interactions. *J Mol Biol* 342: 1659–1674, 2004.
127. Li H, Zhang L, Rao A, Harrison SC, and Hogan PG. Structure of calcineurin in complex with PVIVIT peptide: portrait of a low-affinity signalling interaction. *J Mol Biol* 369: 1296–1306, 2007.
128. Li X, Dy RC, Cance WG, Graves LM, and Earp HS. Interactions between two cytoskeleton-associated tyrosine kinases: calcium-dependent tyrosine kinase and focal adhesion tyrosine kinase. *J Biol Chem* 274: 8917–8924, 1999.
129. Lieberman DN and Mody I. Regulation of NMDA channel function by endogenous Ca(2+)-dependent phosphatase. *Nature* 369: 235–239, 1994.
130. Lim IA, Hall DD, and Hell JW. Selectivity and promiscuity of the first and second PDZ domains of PSD- 95 and synapse-associated protein 102. *J Biol Chem* 277: 21697–21711, 2002.
131. Lim IA, Merrill MA, Chen Y, and Hell JW. Disruption of the NMDA receptor-PSD-95 interaction in hippocampal neurons with no obvious physiological short-term effect. *Neuropharmacol* 45: 738–754, 2003.
132. Lin F, Wang H, and Malbon CC. Gravin-mediated formation of signaling complexes in beta 2-adrenergic receptor desensitization and resensitization. *J Biol Chem* 275: 19025–19034, 2000.
133. Lin JW, Ju W, Foster K, Lee SH, Ahmadian G, Wyszynski M, Wang YT, and Sheng M. Distinct molecular mechanisms and divergent endocytotic pathways of AMPA receptor internalization. *Nat Neurosci* 3: 1282–1290, 2000.
134. Lin Y, Jover-Mengual T, Wong J, Bennett MV, and Zukin RS. PSD-95 and PKC converge in regulating NMDA receptor trafficking and gating. *Proc Natl Acad Sci USA* 103: 19902–19907, 2006.
135. Lisman J and Raghavachari S. A unified model of the presynaptic and postsynaptic changes during LTP at CA1 synapses. *Sci STKE* 2006: re11, 2006.
136. Lisman J, Schulman H, and Cline H. The molecular basis ofCaMKII function in synaptic and behavioral memory. *Nat Neurosci* 3: 175–190, 2002.
137. Lledo PM, Hjelmstad GO, Mukherji S, Soderling TR, Malenka RC, and Nicoll RA. Calcium/calmodulin-dependent kinase II and long-term potentiation enhance synaptic transmission by the same mechanism. *Proc Natl Acad Sci USA* 92: 11175–11179, 1995.
138. Lu CS, Hodge JJ, Mehren J, Sun XX, and Griffith LC. Regulation of the Ca2+/CaM-responsive pool of CaMKII by scaffold-dependent autophosphorylation. *Neuron* 40: 1185–1197, 2003.
139. Lu W and Ziff EB. PICK1 interacts with ABP/GRIP to regulate AMPA receptor trafficking. *Neuron* 47: 407–421, 2005.

140. Lu WY, Jackson MF, Bai D, Orser BA, and MacDonald JF. In CA1 pyramidal neurons of the hippocampus protein kinase C regulates calcium-dependent inactivation of NMDA receptors. *J Neurosci* 20: 4452–4461, 2000.
141. Lu WY, Xiong ZG, Lei S, Orser BA, Dudek E, Browning MD, and MacDonald JF. G-protein-coupled receptors act via protein kinase C and Src to regulate NMDA receptors. *Nat Neurosci* 2: 331–338, 1999.
142. Lu Y, Allen M, Halt AR, Weisenhaus M, Dallapiazza RF, Hall DD, Usachev YM, McKnight GS, and Hell JW. Age-dependent requirement of AKAP150-anchored PKA and GluR2-lacking AMPA Receptors in LTP. *EMBO J* 26: 4879–4890, 2007.
143. Lu YM, Roder JC, Davidow J, and Salter MW. Src activation in the induction of long-term potentiation in CA1 hippocampal neurons. *Science* 279: 1363–1367, 1998.
144. Ma OK and Sucher NJ. Molecular interaction of NMDA receptor subunit NR3A with protein phosphatase 2A. *Neuroreport* 15: 1447–1450, 2004.
145. MacMillan LB, Bass MA, Cheng N, Howard EF, Tamura M, Strack S, Wadzinski BE, and Colbran RJ. Brain actin-associated protein phosphatase 1 holoenzymes containing spinophilin, neurabin, and selected catalytic subunit isoforms. *J Biol Chem* 274: 35845–35854, 1999.
146. Malenka RC and Bear MF. LTP and LTD: an embarrassment of riches. *Neuron* 44: 5–21, 2004.
147. Malenka RC, Kauer JA, Perkel DJ, Mauk MD, Kelly PT, Nicoll RA, and Waxham MN. An essential role for postsynaptic calmodulin and protein kinase activity in long-term potentiation. *Nature* 340: 554–557, 1989.
148. Malinow R and Malenka RC. AMPA receptor trafficking and synaptic plasticity. *Annu Rev Neurosci* 25: 103–126, 2002.
149. Malinow R, Schulman H, and Tsien RW. Inhibition of postsynaptic PKC or CaMKII blocks induction but not expression of LTP. *Science* 245: 862–866, 1989.
150. Man H-Y, Sekine-Aizawa Y, and Huganir R. Regulation of alpha-amino-3-hydroxy-5-methyl-4-isoxazolepropionic acid receptor trafficking through PKA phosphorylation of the Glu receptor 1 subunit. *Proc Natl Acad Sci USA* 104: 3579–3584, 2007.
151. Manning G, Whyte DB, Martinez R, Hunter T, and Sudarsanam S. The protein kinase complement of the human genome. *Science* 298: 1912–1934, 2002.
152. Mansuy IM. Calcineurin in memory and bidirectional plasticity. *Biochem Biophys Res Commun* 311: 1195–1208, 2003.
153. Martin SJ, Grimwood PD, and Morris RG. Synaptic plasticity and memory: an evaluation of the hypothesis. *Annu Rev Neurosci* 23: 649–711, 2000.
154. Marx SO, Reiken S, Hisamatsu Y, Gaburjakova M, Gaburjakova J, Yang YM, Rosemblit N, and Marks AR. Phosphorylation-dependent regulation of ryanodine receptors: a novel role for leucine/isoleucine zippers. *J Cell Biol* 153: 699–708, 2001.
155. Mayer RE, Hendrix P, Cron P, Matthies R, Stone SR, Goris J, Merlevede W, Hofsteenge J, and Hemmings BA. Structure of the 55-kDa regulatory subunit of protein phosphatase 2A: evidence for a neuronal-specific isoform. *Biochemistry* 30: 3589–3597, 1991.
156. McAvoy T, Allen PB, Obaishi H, Nakanishi H, Takai Y, Greengard P, Nairn AC, and Hemmings HC, Jr. Regulation of neurabin I interaction with protein phosphatase 1 by phosphorylation. *Biochemistry* 38: 12943–12949, 1999.
157. McCright B, Rivers AM, Audlin S, and Virshup DM. The B56 family of protein phosphatase 2A (PP2A) regulatory subunits encodes differentiation-induced phosphoproteins that target PP2A to both nucleus and cytoplasm. *J Biol Chem* 271: 22081–22089, 1996.
158. McCright B and Virshup DM. Identification of a new family of protein phosphatase 2A regulatory subunits. *J Biol Chem* 270: 26123–26128, 1995.
159. Menegon A, Burgaya F, Baudot P, Dunlap D, Girault J, and Valtorta F. FAK and PYK2/CAKbeta, two related tyrosine kinases highly expressed in the central nervous system: similarities and differences in the expression pattern. *Eur J Neurosci* 11: 3777–3788, 1999.

160. Merrill MA, Chen Y, Strack S, and Hell JW. Activity-driven postsynaptic translocation of CaMKII. *Trends Pharmacol Sci* 26: 645–653, 2005.
161. Merrill MA, Malik Z, Akyol Z, Bartos JA, Leonard AS, Hudmon A, Shea MA, and Hell JW. Displacement of alpha-Actinin from the NMDA Receptor NR1 C0 Domain By Ca(2+)/Calmodulin Promotes CaMKII Binding. *Biochemistry* 46: 8485–8497, 2007.
162. Meyer T, Hanson PI, Stryer L, and Schulman H. Calmodulin trapping by calcium-calmodulin-dependent protein kinase. *Science* 256: 1199–1202, 1992.
163. Migues PV, Lehmann IT, Fluechter L, Cammarota M, Gurd JW, Sim AT, Dickson PW, and Rostas JA. Phosphorylation of CaMKII at Thr253 occurs in vivo and enhances binding to isolated postsynaptic densities. *J Neurochem* 98: 289–299, 2006.
164. Mochly-Rosen D and Gordon AS. Anchoring proteins for protein kinase C: a means for isozyme selectivity. *FASEB J* 12: 35–42, 1998.
165. Morishita W, Connor JH, Xia H, Quinlan EM, Shenolikar S, and Malenka RC. Regulation of synaptic strength by protein phosphatase 1. *Neuron* 32: 1133–1148, 2001.
166. Mulkey RM, Endo S, Shenolikar S, and Malenka RC. Involvement of a calcineurin/inhibitor-1 phosphatase cascade in hippocampal long-term depression. *Nature* 369: 486–488, 1994.
167. Mullasseril P, Dosemeci A, Lisman JE, and Griffith LC. A structural mechanism for maintaining the 'on-state' of the CaMKII memory switch in the post-synaptic density. *J Neurochem* 103: 357–364, 2007.
168. Muly EC, Allen P, Mazloom M, Aranbayeva Z, Greenfield AT, and Greengard P. Subcellular distribution of neurabin immunolabeling in primate prefrontal cortex: comparison with spinophilin. *Cereb Cortex* 14: 1398–1407, 2004.
169. Muly EC, Smith Y, Allen P, and Greengard P. Subcellular distribution of spinophilin immunolabeling in primate prefrontal cortex: localization to and within dendritic spines. *J Comp Neurol* 469: 185–197, 2004.
170. Nakanishi H, Obaishi H, Satoh A, Wada M, Mandai K, Satoh K, Nishioka H, Matsuura Y, Mizoguchi A, and Takai Y. Neurabin: a novel neural tissue-specific actin filament-binding protein involved in neurite formation. *J Cell Biol* 139: 951–961, 1997.
171. Namgaladze D, Hofer HW, and Ullrich V. Redox control of calcineurin by targeting the binuclear Fe(2+)-Zn(2+) center at the enzyme active site. *J Biol Chem* 277: 5962–5969, 2002.
172. Nishizuka Y. Protein kinase C and lipid signaling for sustained cellular responses. *FASEB J* 9: 484–496, 1995.
173. Obermair GJ, Szabo Z, Bourinet E, and Flucher BE. Differential targeting of the L-type Ca2+ channel alpha1C (CaV1.2) to synaptic and extrasynaptic compartments in hippocampal neurons. *Eur J Neurosci* 19: 2109–2122, 2004.
174. Oh MC and Derkach VA. Dominant role of the GluR2 subunit in regulation of AMPA receptors by CaMKII. *Nat Neurosci* 8: 853–854, 2005.
175. Oh MC, Derkach VA, Guire ES, and Soderling TR. Extrasynaptic Membrane Trafficking Regulated by GluR1 Serine 845 Phosphorylation Primes AMPA Receptors for Long-term Potentiation. *J Biol Chem* 281: 752–758, 2006.
176. Okamoto K, Nagai T, Miyawaki A, and Hayashi Y. Rapid and persistent modulation of actin dynamics regulates postsynaptic reorganization underlying bidirectional plasticity. *Nat Neurosci* 7: 1104–1112, 2004.
177. Okamoto K, Narayanan R, Lee SH, Murata K, and Hayashi Y. The role of CaMKII as an F-actin-bundling protein crucial for maintenance of dendritic spine structure. *Proc Natl Acad Sci USA* 104: 6418–6423, 2007.

178. Oliver CJ, Terry-Lorenzo RT, Elliott E, Bloomer WA, Li S, Brautigan DL, Colbran RJ, and Shenolikar S. Targeting protein phosphatase 1 (PP1) to the actin cytoskeleton: the neurabin I/PP1 complex regulates cell morphology. *Mol Cell Biol* 22: 4690–4701, 2002.
179. Oliveria SF, Dell'acqua ML, and Sather WA. AKAP79/150 Anchoring of Calcineurin Controls Neuronal L-Type Ca(2+) Channel Activity and Nuclear Signaling. *Neuron* 55: 261–275, 2007.
180. Omkumar RV, Kiely MJ, Rosenstein AJ, Min K-T, and Kennedy MB. Identification of a phosphorylation site for calcium/calmodulin-dependent protein kinase II in the NR2B subunit of the N-methyl-D-aspartate receptor. *J Biol Chem* 271: 31670–31678, 1996.
181. Osterrieder W, Brum G, Hescheler J, Trautwein W, Flockerzi V, and Hofmann F. Injection of subunits of cyclic AMP-dependent protein kinase into cardiac myocytes modulates Ca2+ current. *Nature* 298: 576–578, 1982.
182. Otey CA and Carpen O. Alpha-actinin revisited: a fresh look at an old player. *Cell Motil Cytoskel* 58: 104–111, 2004.
183. Otmakhov N, Tao-Cheng JH, Carpenter S, Asrican B, Dosemeci A, Reese TS, and Lisman J. Persistent accumulation of calcium/calmodulin-dependent protein kinase II in dendritic spines after induction of NMDA receptor-dependent chemical long-term potentiation. *J Neurosci* 24: 9324–9331, 2004.
184. Pallas DC, Weller W, Jaspers S, Miller TB, Lane WS, and Roberts TM. The third subunit of protein phosphatase 2A (PP2A), a 55-kilodalton protein which is apparently substituted for by T antigens in complexes with the 36- and 63-kilodalton PP2A subunits, bears little resemblance to T antigens. *J Virol* 66: 886–893, 1992.
185. Paoletti P and Neyton J. NMDA receptor subunits: function and pharmacology. *Curr Opin Pharmacol* 7: 39–47, 2007.
186. Park SY, Avraham HK, and Avraham S. RAFTK/Pyk2 Activation Is Mediated by Trans-acting Autophosphorylation in a Src-independent Manner. *J Biol Chem* 279: 33315–33322, 2004.
187. Patton BL, Miller SG, and Kennedy MB. Activation of type II calcium/calmodulin-dependent protein kinase by Ca2+/calmodulin is inhibited by autophosphorylation of threonine within the calmodulin-binding domain. *J Biol Chem* 265: 11204–11212, 1990.
188. Paul S and Lombroso PJ. Receptor and nonreceptor protein tyrosine phosphatases in the nervous system. *Cell Mol Life Sci* 60: 2465–2482, 2003.
189. Pawson T and Nash P. Assembly of cell regulatory systems through protein interaction domains. *Science* 300: 445–452, 2003.
190. Peng J, Kim MJ, Cheng D, Duong DM, Gygi SP, and Sheng M. Semiquantitative proteomic analysis of rat forebrain postsynaptic density fractions by mass spectrometry. *J Biol Chem* 279: 21003–21011, 2004.
191. Penzes P, Johnson RC, Sattler R, Zhang X, Huganir RL, Kambampati V, Mains RE, and Eipper BA. The neuronal Rho-GEF Kalirin-7 interacts with PDZ domain-containing proteins and regulates dendritic morphogenesis. *Neuron* 29: 229–242, 2001.
192. Perry SJ, Baillie GS, Kohout TA, McPhee I, Magiera MM, Ang KL, Miller WE, McLean AJ, Conti M, Houslay MD, and Lefkowitz RJ. Targeting of cyclic AMP degradation to beta 2-adrenergic receptors by beta-arrestins. *Science* 298: 834–836, 2002.
193. Plant K, Pelkey KA, Bortolotto ZA, Morita D, Terashima A, McBain CJ, Collingridge GL, and Isaac JT. Transient incorporation of native GluR2-lacking AMPA receptors during hippocampal long-term potentiation. *Nature Neurosci* 9: 602–604, 2006.
194. Priel A, Kolleker A, Ayalon G, Gillor M, Osten P, and Stern-Bach Y. Stargazin reduces desensitization and slows deactivation of the AMPA-type glutamate receptors. *J Neurosci* 25: 2682–2686, 2005.
195. Raman IM, Tong G, and Jahr CE. Beta-adrenergic regulation of synaptic NMDA receptors by cAMP-dependent protein kinase. *Neuron* 16: 415–421, 1996.

196. Reuter H. Calcium channel modulation by neurotransmitters, enzymes and drugs. *Nature* 301: 569–574, 1983.
197. Robison AJ, Bartlett RK, Bass MA, and Colbran RJ. Differential Modulation of Ca2+/Calmodulin-dependent Protein Kinase II Activity by Regulated Interactions with N-Methyl-D-aspartate Receptor NR2B Subunits and {alpha}-Actinin. *J Biol Chem* 280: 39316–39323, 2005.
198. Robison AJ, Bass MA, Jiao Y, Macmillan LB, Carmody LC, Bartlett RK, and Colbran RJ. Multivalent Interactions of Calcium/Calmodulin-dependent Protein Kinase II with the Postsynaptic Density Proteins NR2B, Densin-180, and {alpha}-Actinin-2. *J Biol Chem* 280: 35329–35336, 2005.
199. Rochais F, Abi-Gerges A, Horner K, Lefebvre F, Cooper DM, Conti M, Fischmeister R, and Vandecasteele G. A specific pattern of phosphodiesterases controls the cAMP signals generated by different Gs-coupled receptors in adult rat ventricular myocytes. *Circ Res* 98: 1081–1088, 2006.
200. Ron D, Chen CH, Caldwell J, Jamieson L, Orr E, and Mochly-Rosen D. Cloning of an intracellular receptor for protein kinase C: a homolog of the beta subunit of G proteins. *Proc Natl Acad Sci USA* 91: 839–843, 1994.
201. Rongo C and Kaplan JM. CaMKII regulates the density of central glutamatergic synapses in vivo. *Nature* 402: 195–199, 1999.
202. Rosenberg OS, Deindl S, Sung RJ, Nairn AC, and Kuriyan J. Structure of the autoinhibited kinase domain of CaMKII and SAXS analysis of the holoenzyme. *Cell* 123: 849–860, 2005.
203. Rosenmund C, Carr DW, Bergeson SE, Nilaver G, Scott JD, and Westbrook GL. Anchoring of protein kinase A is required for modulation of AMPA/kainate receptors on hippocampal neurons. *Nature* 368: 853–856, 1994.
204. Rubin CS. A kinase anchor proteins and the intracellular targeting of signals carried by cyclic AMP. *Biochim Biophys Acta* 1224: 467–479, 1994.
205. Ryan XP, Alldritt J, Svenningsson P, Allen PB, Wu GY, Nairn AC, and Greengard P. The Rho-specific GEF Lfc interacts with neurabin and spinophilin to regulate dendritic spine morphology. *Neuron* 47: 85–100, 2005.
206. Salter MW and Kalia LV. Src kinases: a hub for NMDA receptor regulation. *Nat Rev Neurosci* 5: 317–328, 2004.
207. Saraf A, Virshup DM, and Strack S. Differential expression of the B'beta regulatory subunit of protein phosphatase 2A modulates tyrosine hydroxylase phosphorylation and catecholamine synthesis. *J Biol Chem* 282: 573–580, 2007.
208. Sarrouilhe D, di Tommaso A, Metaye T, and Ladeveze V. Spinophilin: from partners to functions. *Biochimie* 88: 1099–1113, 2006.
209. Sasaki H, Nagura K, Ishino M, Tobioka H, Kotani K, and Sasaki T. Cloning and characterization of cell adhesion kinase beta, a novel protein-tyrosine kinase of the focal adhesion kinase subfamily. *J Biol Chem* 270: 21206–21219, 1995.
210. Satoh A, Nakanishi H, Obaishi H, Wada M, Takahashi K, Satoh K, Hirao K, Nishioka H, Hata Y, Mizoguchi A, and Takai Y. Neurabin-II/spinophilin. An actin filament-binding protein with one pdz domain localized at cadherin-based cell-cell adhesion sites. *J Biol Chem* 273: 3470–3475, 1998.
211. Seabold GK, Burette A, Lim IA, Weinberg RJ, and Hell JW. Interaction of the tyrosine kinase Pyk2 with the N-methyl-D-aspartate receptor complex via the src homology 3 domains of PSD-95 and SAP102. *J Biol Chem* 278: 15040–15048, 2003.
212. Shen K and Meyer T. Dynamic control of CaMKII Translocation in hippocampal neurons by NMDA receptor stimulation. *Science* 284: 162–166, 1999.
213. Shen K, Teruel MN, Connor JH, Shenolikar S, and Meyer T. Molecular memory by reversible translocation of calcium/calmodulin-dependent protein kinase II. *Nature Neurosci* 3: 881–886, 2000.

214. Shen K, Teruel MN, Subramanian K, and Meyer T. CaMKIIbeta functions as an F-actin targeting module that localizes CaMKIIalpha/beta heterooligomers to dendritic spines. *Neuron* 21: 593–606, 1998.
215. Shi J, Townsend M, and Constantine-Paton M. Activity-dependent induction of tonic calcineurin activity mediates a rapid developmental downregulation of NMDA receptor currents. *Neuron* 28: 103–114, 2000.
216. Shih M, Lin F, Scott JD, Wang HY, and Malbon CC. Dynamic complexes of beta2-adrenergic receptors with protein kinases and phosphatases and the role of gravin. *J Biol Chem* 274: 1588–1595, 1999.
217. Silva AJ, Paylor R, Wehner JM, and Tonegawa S. Impaired spatial learning in alpha-calcium-calmodulin kinase II mutant mice. *Science* 257: 206–211, 1992.
218. Silva AJ, Stevens C, Tonegawa S, and Wang Y. Deficient hippocampal long-term potentiation in alpha-calcium- calmodulin kinase II mutant mice. *Science* 257: 201–206, 1992.
219. Skeberdis VA, Chevaleyre V, Lau CG, Goldberg JH, Pettit DL, Suadicani SO, Lin Y, Bennett MV, Yuste R, Castillo PE, and Zukin RS. Protein kinase A regulates calcium permeability of NMDA receptors. *Nat Neurosci* 9: 501–510, 2006.
220. Smith KE, Gibson ES, and Dell'Acqua ML. cAMP-dependent protein kinase postsynaptic localization regulated by NMDA receptor activation through translocation of an A-kinase anchoring protein scaffold protein. *J Neurosci* 26: 2391–2402, 2006.
221. Snyder GL, Galdi S, Fienberg AA, Allen P, Nairn AC, and Greengard P. Regulation of AMPA receptor dephosphorylation by glutamate receptor agonists. *Neuropharmacology* 45: 703–713, 2003.
222. Sorokin A, Kozlowski P, Graves L, and Philip A. Protein-tyrosine kinase Pyk2 mediates endothelin-induced p38 MAPK activation in glomerular mesangial cells. *J Biol Chem* 276: 21521–21528, 2001.
223. Staudinger J, Lu J, and Olson EN. Specific interaction of the PDZ domain protein PICK1 with the COOH terminus of protein kinase C-alpha. *J Biol Chem* 272: 32019–32024, 1997.
224. Steinberg SF and Brunton LL. Compartmentation of G protein-coupled signaling pathways in cardiac myocytes. *Annu Rev Pharmacol Toxicol* 41: 751–773, 2001.
225. Strack S, Barban MA, Wadzinski BE, and Colbran RJ. Differential inactivation of postsynaptic density-associated and soluble Ca2+/calmodulin-dependent protein kinase II by protein phosphatases 1 and 2A. *J Neurochem* 68: 2119–2128, 1997.
226. Strack S, Chang D, Zaucha JA, Colbran RJ, and Wadzinski BE. Cloning and characterization of B delta, a novel regulatory subunit of protein phosphatase 2A. *FEBS Lett* 460: 462–466, 1999.
227. Strack S, Choi S, Lovinger DM, and Colbran RJ. Translocation of autophosphorylated calcium/calmodulin-dependent protein kinase II to the postsynaptic density. *J Biol Chem* 272: 13467–13470, 1997.
228. Strack S and Colbran RJ. Autophosphorylation-dependent targeting of calcium/calmodulin-dependent protein kinase II by the NR2B subunit of the N-methyl-D-aspartate receptor. *J Biol Chem* 273: 20689–20692, 1998.
229. Strack S, McNeill RB, and Colbran RJ. Mechanism and regulation of calcium/calmodulin-dependent protein kinase II targeting to the NR2B subunit of the N-methyl-D-aspartate receptor. *J Biol Chem* 275: 23798–23806, 2000.
230. Strack S, Robison AJ, Bass MA, and Colbran RJ. Association of calcium/calmodulin-dependent kinase II with developmentally regulated splice variants of the postsynaptic density protein densin-180. *J Biol Chem* 275: 25061–25064, 2000.
231. Strack S, Zaucha JA, Ebner FF, Colbran RJ, and Wadzinski BE. Brain protein phosphatase 2A: developmental regulation and distinct cellular and subcellular localization by B subunits. *J Compar Neurol* 392: 515–527, 1998.

232. Sun XX, Hodge JJ, Zhou Y, Nguyen M, and Griffith LC. The eag potassium channel binds and locally activates calcium/calmodulin-dependent protein kinase II. *J Biol Chem* 279: 10206–10214, 2004.
233. Svenningsson P, Nishi A, Fisone G, Girault JA, Nairn AC, and Greengard P. DARPP-32: an integrator of neurotransmission. *Annu Rev Pharmacol Toxicol* 44: 269–296, 2004.
234. Tanabe O, Nagase T, Murakami T, Nozaki H, Usui H, Nishito Y, Hayashi H, Kagamiyama H, and Takeda M. Molecular cloning of a 74-kDa regulatory subunit (B" or delta) of human protein phosphatase 2A. *FEBS Lett* 379: 107–111, 1996.
235. Tao J, Shumay E, McLaughlin S, Wang HY, and Malbon CC. Regulation of AKAP-membrane interactions by calcium. *J Biol Chem* 281: 23932–23944, 2006.
236. Tasken KA, Collas P, Kemmner WA, Witczak O, Conti M, and Tasken K. Phosphodiesterase 4D and protein kinase a type II constitute a signaling unit in the centrosomal area. *J Biol Chem* 276: 21999–22002, 2001.
237. Tavalin SJ, Colledge M, Hell JW, Langeberg LK, Huganir RL, and Scott JD. Regulation of GluR1 by the A-kinase anchoring protein 79 (AKAP79) signaling complex shares properties with long-term depression. *J Neurosci* 22: 3044–3051, 2002.
238. Tehrani MA, Mumby MC, and Kamibayashi C. Identification of a novel protein phosphatase 2A regulatory subunit highly expressed in muscle. *J Biol Chem* 271: 5164–5170, 1996.
239. Terrak M, Kerff F, Langsetmo K, Tao T, and Dominguez R. Structural basis of protein phosphatase 1 regulation. *Nature* 429: 780–784, 2004.
240. Terry-Lorenzo RT, Roadcap DW, Otsuka T, Blanpied TA, Zamorano PL, Garner CC, Shenolikar S, and Ehlers MD. Neurabin/protein phosphatase-1 complex regulates dendritic spine morphogenesis and maturation. *Mol Biol Cell* 16: 2349–2362, 2005.
241. Thalhammer A, Rudhard Y, Tigaret CM, Volynski KE, Rusakov DA, and Schoepfer R. CaMKII translocation requires local NMDA receptor-mediated Ca2+ signaling. *EMBO J* 25: 5873–5883, 2006.
242. Thiagarajan TC, Lindskog M, and Tsien RW. Adaptation to synaptic inactivity in hippocampal neurons. *Neuron* 47: 725–737, 2005.
243. Tomita S, Adesnik H, Sekiguchi M, Zhang W, Wada K, Howe JR, Nicoll RA, and Bredt DS. Stargazin modulates AMPA receptor gating and trafficking by distinct domains. *Nature* 435: 1052–1058, 2005.
244. Tomita S, Stein V, Stocker TJ, Nicoll RA, and Bredt DS. Bidirectional synaptic plasticity regulated by phosphorylation of stargazin-like TARPs. *Neuron* 45: 269–277, 2005.
245. Tong G and Jahr CE. Regulation of glycine-insensitive desensitization of the NMDA receptor in outside-out patches. *J Neurophysiol* 72: 754–761, 1994.
246. Tong G, Shepherd D, and Jahr CE. Synaptic desensitization of NMDA receptors by calcineurin. *Science* 267: 1510–1512, 1995.
247. Townsend M, Liu Y, and Constantine-Paton M. Retina-driven dephosphorylation of the NR2A subunit correlates with faster NMDA receptor kinetics at developing retinocollicular synapses. *J Neurosci* 24: 11098–11107, 2004.
248. Tsukada M, Prokscha A, Oldekamp J, and Eichele G. Identification of neurabin II as a novel doublecortin interacting protein. *Mech Dev* 120: 1033–1043, 2003.
249. Tu H, Tang TS, Wang Z, and Bezprozvanny I. Association of Type 1 Inositol 1,4,5-Trisphosphate Receptor with AKAP9 (Yotiao) and Protein Kinase A. *J Biol Chem* 279: 19375–19382, 2004.
250. Turetsky D, Garringer E, and Patneau DK. Stargazin modulates native AMPA receptor functional properties by two distinct mechanisms. *J Neurosci* 25: 7438–7448, 2005.
251. van Woerden GM, Harris KD, Hojjati MR, Gustin RM, Qiu S, de Avila Freire R, Jiang YH, Elgersma Y, and Weeber EJ. Rescue of neurological deficits in a mouse model for

Angelman syndrome by reduction of alphaCaMKII inhibitory phosphorylation. *Nat Neurosci* 10: 280–282, 2007.

252. Vigil D, Blumenthal DK, Brown S, Taylor SS, and Trewhella J. Differential Effects of Substrate on Type I and Type II PKA Holoenzyme Dissociation. *Biochemistry* 43: 5629–5636, 2004.
253. Voorhoeve PM, Hijmans EM, and Bernards R. Functional interaction between a novel protein phosphatase 2A regulatory subunit, PR59, and the retinoblastoma-related p107 protein. *Oncogene* 18: 515–524, 1999.
254. Walikonis RS, Jensen ON, Mann M, Provance DW, Jr., Mercer JA, and Kennedy MB. Identification of proteins in the postsynaptic density fraction by mass spectrometry. *J Neurosci* 20: 4069–4080, 2000.
255. Walikonis RS, Oguni A, Khorosheva EM, Jeng CJ, Asuncion FJ, and Kennedy MB. Densin-180 forms a ternary complex with the (alpha)-subunit of Ca2+/calmodulin-dependent protein kinase II and (alpha)-actinin. *J Neurosci* 21: 423–433, 2001.
256. Wang HY, Tao J, Shumay E, and Malbon CC. G-Protein-coupled receptor-associated A-kinase anchoring proteins: AKAP79 and AKAP250 (gravin). *Eur J Cell Biol* 85: 643–650, 2006.
257. Wang Q, Zhao J, Brady AE, Feng J, Allen PB, Lefkowitz RJ, Greengard P, and Limbird LE. Spinophilin blocks arrestin actions in vitro and in vivo at G protein-coupled receptors. *Science* 304: 1940–1944, 2004.
258. Wang X, Zeng W, Kim MS, Allen PB, Greengard P, and Muallem S. Spinophilin/neurabin reciprocally regulate signaling intensity by G protein-coupled receptors. *EMBO J* 26: 2768–2776, 2007.
259. Wang X, Zeng W, Soyombo AA, Tang W, Ross EM, Barnes AP, Milgram SL, Penninger JM, Allen PB, Greengard P, and Muallem S. Spinophilin regulates Ca2+ signalling by binding the N-terminal domain of RGS2 and the third intracellular loop of G-protein-coupled receptors. *Nat Cell Biol* 7: 405–411, 2005.
260. Wang YT and Salter MW. Regulation of NMDA receptors by tyrosine kinases and phosphatases. *Nature* 369: 233–235, 1994.
261. Weeber EJ, Jiang YH, Elgersma Y, Varga AW, Carrasquillo Y, Brown SE, Christian JM, Mirnikjoo B, Silva A, Beaudet AL, and Sweatt JD. Derangements of hippocampal calcium/calmodulin-dependent protein kinase II in a mouse model for Angelman mental retardation syndrome. *J Neurosci* 23: 2634–2644, 2003.
262. Westphal RS, Tavalin SJ, Lin JW, Alto NM, Fraser IDC, Langeberg LK, Sheng M, and Scott JD. Regulation of NMDA receptors by an associated phosphatase-kinase signaling complex. *Science* 285: 93–96, 1999.
263. Willoughby D, Wong W, Schaack J, Scott JD, and Cooper DM. An anchored PKA and PDE4 complex regulates subplasmalemmal cAMP dynamics. *EMBO J* 25: 2051–2061, 2006.
264. Wong W and Scott JD. AKAP signalling complexes: focal points in space and time. *Nat Rev Mol Cell Biol* 5: 959–970, 2004.
265. Wu SS, Jacamo RO, Vong SK, and Rozengurt E. Differential regulation of Pyk2 phosphorylation at Tyr-402 and Tyr-580 in intestinal epithelial cells: roles of calcium, Src, Rho kinase, and the cytoskeleton. *Cell Signal* 18: 1932–1940, 2006.
266. Wyszynski M, Lin J, Rao A, Nigh E, Beggs AH, Craig AM, and Sheng M. Competitive binding of a-actinin and calmodulin to the NMDA receptor. *Nature* 385: 439–442, 1997.
267. Xia J, Zhang X, Staudinger J, and Huganir RL. Clustering of AMPA receptors by the synaptic PDZ domain-containing protein PICK1. *Neuron* 22: 179–187, 1999.
268. Xiao RP, Cheng H, Zhou YY, Kuschel M, and Lakatta EG. Recent advances in cardiac beta(2)-adrenergic signal transduction. *Circ Res* 85: 1092–1100, 1999.

269. Xiong ZG, Pelkey KA, Lu WY, Lu YM, Roder JC, MacDonald JF, and Salter MW. Src potentiation of NMDA receptors in hippocampal and spinal neurons is not mediated by reducing zinc inhibition. *J Neurosci* 19: RC37, 1999.
270. Xu Y, Xing Y, Chen Y, Chao Y, Lin Z, Fan E, Yu JW, Strack S, Jeffrey PD, and Shi Y. Structure of the protein phosphatase 2A holoenzyme. *Cell* 127: 1239–1251, 2006.
271. Yaka R, Thornton C, Vagts AJ, Phamluong K, Bonci A, and Ron D. NMDA receptor function is regulated by the inhibitory scaffolding protein, RACK1. *Proc Natl Acad Sci USA* 99: 5710–5715, 2002.
272. Yamashita T, Inui S, Maeda K, Hua DR, Takagi K, Fukunaga K, and Sakaguchi N. Regulation of CaMKII by alpha4/PP2Ac contributes to learning and memory. *Brain Res* 1082: 1–10, 2006.
273. Yan Z, Fedorov SA, Mumby MC, and Williams RS. PR48, a novel regulatory subunit of protein phosphatase 2A, interacts with Cdc6 and modulates DNA replication in human cells. *Mol Cell Biol* 20: 1021–1029, 2000.
274. Yang S, Fletcher WH, and Johnson DA. Regulation of cAMP-dependent protein kinase: Enzyme activation without dissociation. *Biochem*: 6267–6271, 1995.
275. Yu H, Li X, Marchetto GS, Dy R, Hunter D, Calvo B, Dawson TL, Wilm M, Anderegg RJ, Graves LM, and Earp HS. Activation of a novel calcium-dependent protein-tyrosine kinase. Correlation with c-Jun N-terminal kinase but not mitogen-activated protein kinase activation. *J Biol Chem* 271: 29993–29998, 1996.
276. Yu X-M, Askalan R, Keil II GJ, and Salter MW. NMDA channel regulation by channel-associated protein tyrosine kinase Src. *Science* 275: 674–678, 1997.
277. Zhang S, Ehlers MD, Bernhardt JP, Su CT, and Huganir RL. Calmodulin mediates calcium-dependent inactivation of N-methyl-D-aspartate receptors. *Neuron* 21: 443–453, 1998.
278. Zhou YY, Cheng H, Bogdanov KY, Hohl C, Altschuld R, Lakatta EG, and Xiao RP. Localized cAMP-dependent signaling mediates beta 2-adrenergic modulation of cardiac excitation-contraction coupling. *Am J Physiol* 273: H1611–H1618, 1997.
279. Zito K, Knott G, Shepherd GM, Shenolikar S, and Svoboda K. Induction of spine growth and synapse formation by regulation of the spine actin cytoskeleton. *Neuron* 44: 321–334, 2004.
280. Zolnierowicz S, Csortos C, Bondor J, Verin A, Mumby MC, and DePaoli-Roach AA. Diversity in the regulatory B-subunits of protein phosphatase 2A: identification of a novel isoform highly expressed in brain. *Biochem* 33: 11858–11867, 1994.
281. Zolnierowicz S, Van Hoof C, Andjelkovic N, Cron P, Stevens I, Merlevede W, Goris J, and Hemmings BA. The variable subunit associated with protein phosphatase 2A0 defines a novel multimember family of regulatory subunits. *Biochem J* 317 (Pt 1): 187–194, 1996.

Long-Term Potentiation

John E. Lisman[1] and Johannes W. Hell[2]

[1]Brandeis University, Biology Department and Volen Center for Complex Systems, Waltham, MA 02454, USA, lisman@brandeis.edu
[2]University of Iowa, Department of Pharmacology, Iowa City, IA 52242, USA, johannes-hell@uiowa.edu

1 Introduction

Long-term potentiation (LTP) is a long-lasting strengthening of synaptic transmission that is triggered by brief periods of high-frequency synaptic stimulation. Other patterns of stimulation can produce a long-term weakening of synapses, a process termed long-term depression (LTD) or depotentiation. Synapses thus have a stable strength that can be bi-directionally modified in an activity-dependent manner.

There are different forms of LTP at different brain synapses. This article focuses on LTP found at glutamatergic synapses in the CA1 region of the hippocampus, the most studied form of LTP. A typical experiment starts by measuring the strength of a group of synapses. This is done by firing a single action potential in some of the axons that enter this region. These axons make synapses with pyramidal cells and generate an excitatory postsynaptic potential (EPSP). The collective strength of the synapses is defined by the magnitude of the average EPSP. LTP is then induced by stimulating the axons to fire at high frequency (typically 100 Hz for 1 s), a stimulus referred to as a tetanus. Remarkably, this brief tetanus causes a long-lasting potentiation of the strength of the synapses. The size of the EPSP typically increases by 50%–100%, but can rise by as much as 800% in some situations. In the brain-slice preparation used for most studies, potentiation persists until the slice is no longer viable (5–12 h). LTP can also be induced in living animals, where it can persist for at least a year (1).

The evidence that such changes in synaptic strength are actually involved in memory has strengthened in recent years. Studies have examined the effect of a learning experience on synapses in the hippocampus. Two groups (67, 225) have demonstrated that learning induces an LTP-like process, which reduces the subsequent ability to experimentally induce LTP at these synapses. This occlusion suggests that learning-induced and experimentally-induced potentiation affect the same

J.W. Hell, M.D. Ehlers (eds.), *Structural and Functional Organization of the Synapse*,
DOI: 10.1007/978-0-387-77232-5_17,

synaptic process. Other experiments showed that electrically inducing LTP in a large fraction of synapses after learning disrupts memory (24) as would be expected if memory is stored by the pattern of synapses that undergo LTP during learning. Another approach has examined whether the molecular events required for LTP induction occur during learning. The results of these experiments are generally supportive of a linkage. Activation of CaMKII and GluR1 phosphorylation, two events associated with LTP (see later sections) occur during learning (17). Moreover, if LTP-associated changes are prevented (NMDA receptor activation, CaMKII activation, GluR1 insertion), learning is inhibited (60, 189, 217). Finally, recent work shows that procedures that can reverse LTP after learning, can reverse memory (175, 199). Taken together, these observations make a reasonably strong case that at least some kinds of memory are stored by an LTP-like process.

2 Properties of LTP and Their Relevance

2.1 Synapse Specificity and Associativity

Two properties of LTP have made it a particularly attractive candidate for memory storage. The first is the finding that synapses can be independently modified. Initial evidence for this 'synapse specificity' came from experiments in which two sets of synapses onto the same neuron were stimulated. When a tetanus was given to one set, synapses in this but not the other set were strengthened. Two-photon uncaging of glutamate in the direct vicinity of an individual synapse can be used to induce LTP at that synapse in much the way that LTP is induced by presynaptically-released glutamate (73). Remarkably, other non-stimulated synapses only microns away do not undergo LTP. These results indicate that each synapse is independently modifiable. Because most neurons have over 10,000 excitatory synapses, the potential for information storage is vast.

The second attractive property of LTP is that it is associative, i.e. LTP cannot be triggered by activity in a single input axon, but can be triggered if that activity is associated with other active excitatory inputs (107, 140). The reason is that LTP induction requires the membrane voltage of the pyramidal cell to become depolarized by an amount much larger than that produced by a single excitatory synaptic input (~ 500 μV). This requirement has been demonstrated by artificially depolarizing the cell with positive current (95). Under these conditions, even stimulation of a single input can produce LTP. Conversely, if many inputs are stimulated, but the cell is hyperpolarized by current injection, LTP does not occur (128). Thus, a synapse will be strengthened if two conditions are met: the synapse is activated by presynaptic activity and there is substantial postsynaptic depolarization due to other excitatory inputs (and not too much inhibitory input). This dual requirement for strengthening is referred to as the 'Hebb rule' (75).

2.2 Hebbian Synaptic Modification Can Underlie Associative Memory

The theory of nerve networks provides a picture of how Hebbian modification of synapses in a network could lead to distributed storage of a memory. Consider how a visual scene might be stored. The scene would first be processed by low-level networks in which different neurons responded to different elementary visual features of an object. Memory storage is defined by the association of these features within a higher-level network. In an idealized version of such a memory network, each neuron makes a synaptic connection with every other neuron and it is at these 'recurrent' connections that the critical Hebbian modification occurs. Let us say that when the visual scene is present, a subset of cells in the memory network will fire, each representing a particular feature. This subset includes cell *X* and a group of other cells, *G*. Focusing on the connection of the *G* cells to the *X* cell, we can see that the Hebb rule will be satisfied because there is both presynaptic input and postsynaptic depolarization. These connections will therefore undergo LTP. This strengthening makes it possible for *G* cells to collectively fire the *X* cell. To see the utility of this change, consider a recall process in which the same scene is presented, but missing the feature that would normally cause the *X* cell to fire. Cell *X* will nevertheless fire because it receives strong inputs from the *G* cells. In this way, a complete memory can be recalled when the network is cued with only a part. This pattern completion process is a defining property of associative memory.

A computationally important property of the Hebb rule is that it can be executed at each synapse using information that is locally available. A synapse can 'know' that its presynaptic input is active because its receptors will be activated by the neurotransmitter released. The synapse can also 'know' about the overall depolarization of the neuron because this spreads electrotonically over the cell membrane; this voltage can be sensed by the synapse using ion channels that are sensitive to membrane voltage.

3 Induction of LTP

3.1 The Critical Role of NMDA Receptors in LTP Induction

The NMDA receptor is one of the two major ionotropic receptors for glutamate, the neurotransmitter at most excitatory synapses in the central nervous system. The other major ionotropic glutamate receptor is the AMPA receptor, which produces the bulk of the excitatory postsynaptic potential. The LTP induced by standard 100 Hz, 1 s tetanus is blocked by NMDA receptor antagonists (34, 69). This requirement for NMDA receptors has been confirmed in subsequent experiments using genetic methods to eliminate the channels (217).

The NMDA receptor has characteristics that can account for the Hebbian property of LTP induction. NMDA receptor channels require glutamate to open, but are only fully conducting if there is also substantial postsynaptic depolarization because otherwise external Mg^{2+} is obstructing the pore (138, 158). Postsynaptic depolarization electrically drives Mg^{2+} out of the pore.

3.2 Nature of the Depolarizing Event in Hebbian Plasticity

An important unanswered question about LTP induction is the nature of the dendritic depolarizing potential that allows the NMDA receptor channels to open. One candidate is action potential mediated by voltage-dependent Na^+ channels. Synaptic stimuli strong enough to induce LTP produce action potentials in the postsynaptic cell. Many action potentials are initiated in the axon hillock and actively backpropagate into the dendrites (85, 132). If subthreshold synaptic stimulation too weak to evoke LTP is repeatedly coupled with a spike produced by somatic current injection, LTP occurs (18, 133). However, the role of Na^+ channel action potentials during synaptically-induced LTP (which is what occurs under physiological conditions) has not been demonstrated. Indeed, to the contrary, if the back-propagating action potential is blocked, the LTP induced by strong synaptic inputs still occurs normally (61). Thus, although there has been an enormous amount of theoretical and experimental work on spike-timing dependent plasticity, the phenomenon must be considered hypothetical until it can be demonstrated that the back-propagating spike is necessary for synaptically-induced LTP. Indeed, in cortical pyramidal neurons, synaptically induced EPSP's that trigger single postsynaptic spikes produce LTD (205), not the LTP predicted by standard theories of spike-timing dependent plasticity (54, 222).

Other depolarizing events that could activate NMDA receptors include spikes generated by the opening of voltage-dependent Ca^{2+} channels or spikes generated by current through the NMDA receptor itself (192) or even simply a large AMPA receptor-mediated EPSP. There are several indications that a burst of several action potentials is required for LTP induction (78, 89, 180). Indeed, under some conditions even a single brief burst is sufficient to induce LTP (82). These bursts are generated in the dendrites by the activation of slow voltage-dependent conductances; the slow depolarization that underlies the burst triggers multiple fast action potentials.

3.3 Ca^{2+} Entry Through the NMDA Receptor triggers LTP

But what is it about the opening of the NMDA receptor that leads to LTP induction? NMDA receptors are approximately 100-fold more permeable to Ca^{2+} relative to Na^+ than most AMPA receptors (137). LTP can be blocked by intracellular Ca^{2+} chelators (123). These findings indicate that it is the influx of Ca^{2+} through NMDA receptors that triggers LTP.

LTP has an initial transient stage that requires GluR1 and a more slowly developing stage (~20 min) that does not (78, 113). The initial form is very sensitive to the fast Ca^{2+} buffer BAPTA but not the slow buffer EGTA, a result that suggests that the Ca^{2+} elevation responsible for the initial LTP is within a microdomain near the pore of the NMDA channel. In contrast, BAPTA and EGTA are equally effective in blocking the slower phase, suggesting that this phase is triggered by the bulk Ca^{2+} level.

The intracellular Ca^{2+} elevation that occurs during LTP has been measured with fluorescent Ca^{2+} indicators. The actual sites of excitatory synaptic input into pyramidal cells are primarily on small (~ 1 μm) protuberances from dendrites called spines (71). During tetanic stimulation spines undergo Ca^{2+} elevation to above 10 μM during tetanic stimulation used to induce LTP under Hebbian conditions (179). Similar levels of Ca^{2+} are reached during a single stimulus, provided the membrane is held at a voltage optimal for entry of Ca^{2+} through the NMDA channel (190). Biophysical analysis indicates that there are Ca^{2+} buffers in the cytoplasm of the spine, but they are immobile and at comparatively low concentration. Ca^{2+} is removed from the spine very rapidly (time constant 12 ms), primarily by a Na^{+}/ Ca^{2+} exchange mechanism (190).

The biophysical analysis of Ca^{2+} dynamics in spines provides insight into several properties of LTP/LTD. First, although the peak spine Ca^{2+} entry produced by an action potential (which opens voltage-dependent Ca^{2+} channels) is very high (190), the duration of the elevation is short because the action potential is short and because Ca^{2+} is rapidly removed. Thus, it is not possible to produced sustained high Ca^{2+} with action potentials alone, explaining why LTP cannot be induced in this way. In contrast, the slow NMDA receptor conductance produces a long-lasting Ca^{2+} elevation, even as a result of single synaptic stimulus. Moreover, because the probability of NMDA receptor opening during a single stimulus is low, presentation of multiple stimuli allows the buildup of NMDA receptor conductance and spine Ca^{2+} levels. Remarkably, the Ca^{2+} level in the spine itself can be higher than in the parent dendrite and nearby spines, even though the distances involved are very small (35). This localization of Ca^{2+} is probably important in producing the synapse specificity of LTP.

Not only is Ca^{2+} elevation necessary for LTP induction, it also appears to be sufficient: This was demonstrated by using special buffers that were injected into the cell and which release Ca^{2+} in response to bright illumination. Peak concentrations of Ca^{2+} during light were over 10 μM and had a half-time of 2.5 s (230). This elevation produced LTP.

The study of Ca^{2+} levels takes on special importance in plasticity because it may be the concentration of Ca^{2+} that determines the sign of synaptic modification. It has been proposed that moderate Ca^{2+} elevation produces LTD, whereas higher elevation induces LTP (112). The strongest support for this hypothesis comes from the study of hippocampal neurons showing that with the same (high frequency) pattern of synaptic stimulation, partial block of NMDA receptors can convert an LTP-inducing stimulus to an LTD-inducing one (36). An important form of hippocampal LTD is induced by prolonged (many minutes) low frequency (1–5 Hz) stimulation (43), and depends on Ca^{2+} entry through NMDA receptors, the same channels that produce LTP (151). It is important to understand that NMDA receptor function is not all-or-none. Rather, the relief of the Mg^{2+} block and thereby NMDA receptor-mediated Ca^{2+} elevation is graded with depolarization (over about an 8-fold range) (190). Even at resting potential is the Mg^{2+} block not complete and there is significant Ca^{2+} elevation in response to a single vesicle release. This elevation is in the submicromolar range (100). These levels, which are sufficient to induce LTD, are considerably lower than the elevation that occurs during LTP (179, 230) (but see (155)).

Blocking NMDA receptors completely blocks LTP induced by a standard 100 Hz tetanus at CA1 synapses. There are, however stimulation protocols that produce LTP that is only partially blocked by NMDA receptor antagonists, for instance a 200 Hz tetanus. The NMDA receptor-independent portion of this LTP can be blocked by inhibitors of voltage-dependent L-type Ca^{2+} channels (66). Because these inhibitors are bath-applied, it is not clear whether the Ca^{2+} channels are presynaptic or postsynaptic (27, 66). However, it is clear that postsynaptic Ca^{2+} elevation is required because LTP is blocked by postsynaptic Ca^{2+} buffer. This form of NMDA-independent LTP may well be of physiological importance, given that the bursts typical of hippocampal neurons have spike frequencies >200 Hz. The LTP that involves L-type channels is synapse specific (65), but how this specificity could be achieved remains unclear. One possibility is that spine voltage is not the same as in the dendrite and that a Ca^{2+} spike can occur in the spine head. Whatever the mechanism, it is important to keep in mind the limitations of the commonly held view that NMDA receptor activation is necessary for LTP and learning.

3.4 Ca^{2+} Sensors for LTP Induction

CaMKII is highly concentrated in spines and is a major component of the postsynaptic density (PSD), a proteinacious structure attached to the intracellular side of the postsynaptic membrane. The kinase is thus strategically located to sense Ca^{2+} entry through NMDA receptors and to control synaptic strength (e.g., by phosphorylation of AMPA receptor complexes) in a synapse-specific manner. Several lines of evidence implicate CaMKII in LTP. Introduction of active CaMKII into the postsynaptic cell potentiates synaptic transmission and prevents ('occludes') subsequent LTP induction by a tetanus, as if the two methods for enhancing transmission share a common mechanism (74, 117). Injection of CaMKII inhibitors into the postsynaptic cell (129) or genetic knockout of the protein strongly reduces LTP (201). CaMKII inhibitors also block other aspects of the potentiation process, notably the increase in spine size (25, 73) and the activation of protein synthesis (94). With certain induction protocols, postsynaptic CaMKII inhibition only blocks a portion of LTP; the residual LTP is blocked by presynaptic CaMKII inhibitor (119). It is thus possible that CaMKII is a critical sensor for both presynaptic and postsynaptic forms of LTP (see below).

On the other hand, there are indications that some forms of LTP do not depend on CaMKII and that other Ca^{2+} sensors also have important roles. In young animals, LTP is dependent on PKA and does not require CaMKII (231). There is also evidence that the ras-activating guanosine nucleotide release factor ras-GRF, which is directly activated by Ca^{2+}, and Ca^{2+}-stimulated adenylyl cyclases AC1 and AC8 are important for LTP under some conditions (108, 193, 228). Other Ca^{2+} sensitive proteins are sensors for LTD induction, including the phosphatase calcineurin/PP2B (87, 150, 238) and hippocalcin (172).

Many of the Ca^{2+} sensors important in synaptic plasticity (e.g. adenylate cyclase, calcineurin, CaMKII) do not bind Ca^{2+} directly, but depend on the Ca^{2+} binding protein, calmodulin. Interestingly, the concentration of free calmodulin appears to be controlled by presynaptic (neuromodulin/GAP-43/B50/P-57) and postsynaptic proteins (neuogranin/RC3) that bind calmodulin. Knockout of neurogranin blocks LTP induction (171). One way of explaining this finding is that the free concentration of calmodulin might be too low to activate the high concentrations of CaMKII in spines during LTP induction. Neurogranin, which is concentrated in spines, may act as a local calmodulin cache capable of supplying sufficient calmodulin to effectively activate CaMKII (237). The binding of calmodulin to neurogranin is regulated by PKC-dependent phosphorylation of neurogranin at Ser36 (80, 184). During LTP induction, this site is phosphorylated by PKCγ (186), but the functional role of this phosphorylation remains to be determined.

4 LTP Expression

4.1 Presynaptic Versus Postsynaptic

The expression of potentiated transmission during LTP could be due to changes that enhance the responsiveness of the postsynaptic cell to neurotransmitter. Alternatively, potentiation could be due to presynaptic changes that result in more transmitters being released. As LTP is induced by postsynaptic events, presynaptic expression processes would require a 'retrograde message' that carried information from the postsynaptic cell to the presynaptic cell. Determining the site of expression of LTP has been surprisingly difficult, but there is now strong evidence that postsynaptic modification occurs and specific mechanisms for producing it have been established. There is also increasing evidence for presynaptic modifications.

Evidence for postsynaptic modifications (reviewed in (156)) is as follows: 1) AMPA receptors have enhanced conductance after LTP (16, 40). 2) Miniature synaptic current responses, the amplitude of which is thought to be controlled postsynaptically, are increased after LTP (126, 131). 3) Recombinant GluR1-homomeric AMPA receptors become incorporated into the synaptic membrane during LTP in organotypic hippocampal slice cultures overexpressing GluR1 (74). 4) The two-photon Ca^{2+} uncaging method mentioned above has been used to induce LTP, as measured by the enhanced response to uncaged glutamate; because the presynaptic terminal is not involved under these conditions, the effects must be entirely postsynaptic (73).

Under some conditions, LTP appears to be expressed *exclusively* through postsynaptic processes. These experiments use a 'pairing protocol' in which the presynaptic axon is stimulated at low (1 Hz) frequency while the postsynaptic neuron is tonically depolarized under voltage-clamp. Under these conditions, LTP of AMPA receptor-mediated transmission occurs, but there is no change in the NMDA receptor-mediated component (177). The lack of effect on the NMDA receptor component would seem to rule out any increase in transmitter release (note that NMDA receptors are not saturated during a synaptic response (125)). However, this argument cannot be applied to all induction protocols because inducing LTP with a tetanus does increase the NMDA receptor component (6, 7, 153).

One line of evidence suggesting that LTP involves presynaptic changes is that synapses 'fail' less often after LTP induction. The classical interpretation of 'failures' (derived from study of the neuromuscular junction) is that a presynaptic action potential sometimes fails to produce the release of even a single synaptic vesicle. Thus, a change in 'failures' after LTP induction is consistent with a presynaptic change in vesicle release. However, there is now substantial evidence that at central synapses a reduction in 'failure' rate can arise through a postsynaptic change. Some synapses (termed silent synapses) initially have NMDA but no AMPA receptors (109) (84). Because NMDA receptors do not produce significant postsynaptic current near resting potential, the synapse would appear to be silent during synaptic transmission under normal conditions (i.e. near resting potential). After LTP, postsynaptic expression of functional AMPA receptors at these synapses would generate synaptic responses. It follows that the decrease in failure rate seen after LTP induction might be explainable as a purely postsynaptic process.

Because of the ambiguity in interpreting the results of quantal analysis, there has been the need for more direct methods to determine whether there are presynaptic changes during LTP. Direct evidence for presynaptic changes has been provided by loading the dye FM1-43 into synaptic vesicles via uptake (234). Dye release upon synaptic stimulation as detected by a change in fluorescence is enhanced after LTP indicating enhanced vesicle release after LTP. Importantly this enhancement develops slowly (30 min) after LTP induction. By comparison, postsynaptic effects occur within seconds. Furthermore, the release changes were more effectively induced by theta burst stimulation than with a single tetanus. In a related line of work, synapto-pHluorin release was used to test for presynaptic changes during LTP (13). The LTP induced by a theta-burst protocol induced a delayed increase in presynaptic release. A 200 Hz tetanus induced a similar increase but a 50 Hz tetanus did not. Importantly, presynaptic release changes were almost totally dependent on voltage-dependent Ca^{2+} channels, rather than NMDA channels. These direct measurements of presynaptic release leave little doubt that LTP has a presynaptic component. On the other hand, it is also clear that this component depends on the stimulation protocol used. This dependence may help to explain some of the discrepant data that has fueled the pre/post debate.

In addition to changes in the number of vesicles released, there are indications, albeit more preliminary, that a change in the *mode* of vesicle release may occur (30). In one mode, glutamate is released from the vesicle so slowly that AMPA receptors are not effectively activated. Such synapses are termed whispering synapses. After LTP induction, release through the fusion pore becomes fast enough to activate AMPA receptors, thereby contributing to the potentiation of transmission.

Several substances could act as retrograde messengers through which postsynaptic changes trigger presynaptic release properties. Nitric oxide (NO) and various metabolites of arachidonic acid (endocanabinoids, prostaglandins, eicosinoids) are possible candidates. Early pharmacological studies suggested that LTP could be blocked by interfering with NO (20, 161, 195), but subsequent work showed that LTP in the CA1 hippocampal region could be induced even in the absence of neuronal nitric oxide synthase, the isoform activated by synaptic activity (162, 203). However, LTP under these induction conditions may have been primarily postsynaptic. A clear demonstration of a presynaptic component of LTP dependent on NO activity has been found in the barrel cortex of mice (68). In these mice the postsynaptic component of LTP, which is dependent on GluR1-containing AMPA receptors, was eliminated genetically. The remaining LTP was completely dependent on NOS activity and this LTP had the properties of presynaptic changes in transmitter release.

4.2 Molecular Mechanisms of Postsynaptic LTP Expression Processes

Expression of LTP is defined by the set of pre- and postsynaptic processes that control the size of the synaptic response. We will concentrate here on the molecular mechanisms involved in postsynaptic expression processes because so little is known about the presynaptic processes. Progress in understanding postsynaptic expression mechanisms has built on progress in understanding AMPA receptors. There are four AMPA receptor subunits, GluR1-4, which assemble into heterotetramers. The majority of AMPA receptors contain GluR2 (224). GluR2-containing AMPA receptors have a linear current-voltage relationship and, are Ca^{2+} impermeable due to an Arg at the so-called Q/R/N site in the pore region of GluR2 (227) (other AMPA receptor subunits have Gln and NMDA receptors have Asn at this site). However, about 8% of the AMPA receptors in CA1/CA2 are GluR1 homomers, which constitute the only sizable pool of GluR2-lacking AMPA receptors in adult rats (224). These receptors are Ca^{2+} permeable. They also are inwardly rectifying because of an intracellular pore block by spermine upon depolarization (22, 88, 98).

Stargazin (also know as γ2) and its homologs (γ3, γ4, γ7, γ8) are emerging as auxiliary subunits of AMPA receptors (157). Stargazin fosters extrasynaptic surface expression of AMPA receptors independent of its C-terminus, but synaptic targeting requires the interaction of the C-terminus with PSD-95 (157, 194). Stargazin also promotes AMPA receptor opening, especially by inhibiting desensitization (157). However, other mechanisms and specifically differential protein interactions with the C-termini of AMPA receptor subunits also govern AMPA receptor trafficking and distribution (Chapter by Esteban) and are important for other forms of synaptic plasticity (e.g., LTD in cerebellar Purkinje cells) (241). In the subsequent sections we discuss how changes in the phosphorylation, subunit composition, and number of AMPA channels contributes to LTP.

4.2.1 Modulation of Single-Channel Properties

In acute hippocampal slices from P13-15 rats, pairing-induced LTP is paralleled by enhancement of single channel conductance, as detected by non-stationary noise analysis during the first 15 min (16). This potentiation could be due to phosphorylation-induced increase of AMPA receptor activity. LTP induction by tetani is correlated with phosphorylation of GluR1 on Ser831 by either CaMKII or PKC (10, 105). This phosphorylation enhances the conductance of GluR1 homomeric, but not GluR1/R2 heteromeric AMPA receptors (40, 165). However, this mechanism accounts for only a fraction of the potentation induced by constitutively active CaMKII (182) as mutating Ser831 to Ala on GluR1 has relatively little effect on the potentiation induced by CaMKII in organotypic hippocampal slices (74). Furthermore, the GluR2-lacking AMPA receptors (probably GluR1 homomers) that can be modulated by S831 phosphorylation might only be transiently present after LTP induction and only in mice of certain ages (2 weeks and more than 7 weeks) but not in 3–4 week old mice (see below) (3, 181).

Several other phosphorylation reactions may also affect AMPA receptor function. There is some evidence that phosphorylation of GluR1 by PKA on Ser845 can increases channel activity of AMPA receptors (8, 188) although some of this evidence could be explained by stimulation of GluR1 surface expression by Ser845 phosphorylation (see below). Furthermore, phosphorylation of stargazin or its homologues by CaMKII or PKC affects pairing-induced LTP (see below) (216), possibly by upregulating stargazin's positive effect on AMPA receptor gating (157).

4.2.2 Changes in AMPA Receptor Trafficking

Postsynaptic injection of *N*-ethyl-maleimide (NEM), botulinus toxin, and other reagents that interfere with vesicle fusion block LTP (118). These findings are consistent with the hypothesis that LTP involves addition of AMPA receptors to the postsynaptic plasma membrane, but it does not necessarily imply that this delivery occurs at the moment of LTP induction; vesicular delivery might be required constitutively to maintain a plasma membrane pool of AMPA receptors that are inserted into the synapse during LTP by some other mechanism.

The first evidence that AMPA receptors are rapidly delivered into the synapse after LTP induction involved overexpression of GluR1 in organotypic hippocampal slices. Under these conditions GluR1 forms predominantly homomeric receptors. Their insertion into the synapse was detectable because the GluR1 homomers rectify much more strongly than GluR2 containing receptors (74).

Various forms of synaptic plasticity appear to require temporary functional expression of GluR2-lacking AMPA receptors for initiation and establishment of increased postsynaptic AMPA receptor responses (15, 139, 210, 215). In 12–14 day old mice, GluR2-lacking AMPA receptors account for the initial increase in postsynaptic responses during the first 10–15 min of LTP, but are subsequently replaced by GluR2-containing non-rectifying AMPA receptors (120, 181). At 3 weeks of age the LTP dependency on GluR2-lacking AMPA receptors is lost (3, 64, 118). In vivo the only detectable pool of inwardly-rectifying, Ca^{2+} permeable GluR2-lacking AMPA receptors are GluR1 homomers (~8% in CA1/CA2; (224)). Collectively these findings suggest that insertion of new GluR1 homomeric receptors account for the initial phase of LTP in mice at P12-14 but not at P20-23. There is also good evidence that in mice that are older than 6 weeks GluR2-lacking receptors are required for LTP induced by a single tetanus during its early phase (120, 181). However, stronger stimulus and induction protocols (2 tetani, interface rather than submersion chamber) might circumvent this requirement (64, 120).

GluR1 knock-out mice provide another perspective on the special role of GluR1. These mice show no LTP induced by standard tetanus when 7 weeks and older (86, 236). However, a theta-burst pairing protocol produces slowly developing but robust LTP. At late times, this potentiation matches that in wt mice. These findings support the idea that GluR1 is especially critical for the early phase of LTP (78), but is not needed for the late phase. GluR1 knock-out also eliminates distance-dependent scaling of the synaptic response (more distant synapses show larger EPSCs than more proximal ones to ensure they equally weigh in on somatic potentials), but leaves a substantial component of basal transmission (5). Remarkably, GluR1 knockout completely eliminates extrasynaptic AMPA receptors (5) supporting the notion that extrasynaptic GluR1 could be the source of receptors required for early LTP and distance-dependent scaling (113). In contrast, the substantial component of transmission that remains in the GluR1 knockout is likely to be due to GluR2/3 receptors.

4.2.3 Role of Stargazin in LTP-Induced AMPA Receptor Trafficking

As described above, stargazin and its homologs (γ3, γ4, γ7, γ8) are important for synaptic AMPA receptor targeting via PSD-95 binding to their very C-termini (157). The cytosolic C-terminus of stargazing contains 9 conserved serine residues that are phosphorylation sites for CaMKII or PKC. Overexpression of stargazinΔC (which has a deletion of the last 4 residues that mediate PSD-95 binding) produces a 4–5 fold elevation of extrasynaptic receptors (as measured by the response to bath-applied AMPA) but not postsynaptic receptors (response to presynaptic stimulus), consistent with a model in which the C-terminus is not required for delivery of receptors to the extrasynaptic membrane, but is required for these receptors to enter the synapse (157, 194). Transfection of stargazin with all identified CaMKII/PKC sites mutated to aspartate to mimic phosphorylation produces a 100% increase in synaptic strength and occludes LTP produced by a pairing protocol (216). Thus, in the simplest model, the early phase of LTP is due to the phosphorylation of stargazin and the resulting enhancement of the ability of AMPA receptor complexes to bind to synaptic anchors such as PSD-95. This emphasis on 'early phase' is important because other lines of evidence suggest that stargazin is *not* involved in persistent changes in synaptic strength. Specifically, overexpression of stargazin with all identified CaMKII/PKC sites mutated to alanine (these mutations prevent phosphorylation) causes no change in basal synaptic strength. This finding indicates that there is no persistent involvement of CaMKII phosphorylation of stargazin in the control of synaptic strength. It can thus be inferred that LTP produces only transient stargazin phosphorylation, but this prediction remains to be directly demonstrated. The above results also illustrate that even a drastic increase in extrasynaptic AMPA receptors is not sufficient to drive an increase in functional AMPA receptors at postsynaptic sites, which therefore requires additional mechanisms.

The importance of stargazin's interaction with PSD-95 is supported by several lines of evidence. Quantum dot labeling and single particle tracking of GluR1 and GluR2 movements in the plasma membrane have made it possible to monitor their lateral diffusion in neuronal cultures (12). Lateral diffusion is restricted (although not abolished) after receptors enter postsynaptic sites. This effect depends on the interaction of the C-terminus of stargazin with PSD-95 (12). It will be important to determine whether phosphorylation of stargazin affects lateral mobility. Other work points to the importance of PSD-95 in the control of synaptic strength. Overexpression of PSD-95 enhances synaptic strength and spine size (46), while eliminating PSD-95 and its homologs reduces synaptic strength (47). However, it is unclear what keeps the AMPA receptor – stargazin – PSD-95 complex at the postsynaptic site.

4.2.4 Control of AMPA Receptor Trafficking by Kinases

This review introduced earlier the Ca^{2+} sensors that initiate LTP and, in the preceding paragraphs, the changes in AMPA receptors that underlie changes in postsynaptic strength. A major goal of current research is to understand the reactions that connect these two processes. This goal has not yet been achieved because multiple control systems are involved, probably affecting different pools of receptors. For instance, it appears that GluR2-containing AMPA receptors continuously cycle between intracellular stores and surface including postsynaptic sites to maintain a constant level of functionally available postsynaptic AMPA receptors in culture (11, 127, 174, 197) (but see (4)). In contrast to GluR2, GluR1-containing AMPA receptors and especially GluR1 homomers require synaptic activity, PKA, PKC, and CaMKII for surface expression and postsynaptic accumulation during pairing-induced LTP in organotypic slice cultures (11, 19, 127). Although the PKC/CaMKII site Ser831 on GluR1 is not critical, the PKC site Ser818 and the PKA site Ser845 are important for this form of LTP (19, 48). Furthermore, PKA and specifically Ser845 phosphorylation stimulates surface expression of GluR1 by inhibiting internalization and promoting recycling back to the plasma membrane after ligand-induced endocytosis (45, 130, 166, 208, 211). This mechanism fosters LTP by increasing the pool of surface AMPA receptors that are available for postsynaptic incorporation during LTP (48, 166).

The LTP dependency on PKA and the involvement of GluR2-lacking AMPA receptors are correlated; both are required for LTP in 2 but not 3–4 week old mice (120). In mice older than 6 weeks LTP induced by a single tetanus in a superfusion chamber depends on PKA and GluR2-lacking receptors (120) (although the use of two tetani in an interface chamber circumvents the dependence on GluR2-lacking receptors (64)). This correlation provides further indication that the two mechanisms might be coupled, probably because PKA fosters GluR1 surface expression and thereby provides a pool of GluR1 homomeric receptors for LTP. However, the fact that neither requirement holds true at 3–4 weeks of age indicates that LTP mechanisms can fundamentally vary with age. These findings are important because they indicate that the exact conditions including age, strain, induction protocols, and temperature (e.g., (97)) might lead to substantial differences in potentiation mechanisms. They might explain controversies surrounding many findings on LTP mechanisms.

The role of MAP kinases in stimulating delivery of AMPA receptors to the synapse has been another major line of investigation. Ca^{2+} can activate the MAP kinases p42/p44 (ERK2 and ERK1, respectively) via direct stimulation of the Ras activatory rasGRF and potentially via CaMKII-mediated inhibition of the Ras (and Rap) inhibitory protein, SynGAP (28, 102) but see (164). Although in many cell types PKA inhibits MAP kinases (e.g., (63)), in hippocampal neurons PKA can activate the MAP kinase ERK2 via Rap-1/B-Raf (2). PKC can stimulate ERK2 via Ras/Raf-1 (63). ERK1 and 2 in turn can act on a number of different levels including the transcription (83) and translation (93) that affects the later phases of LTP. Early phases of LTP can at least in part be affected by MAP kinase control of dendritic excitability; this occurs by inhibition of the dendritic K^+ channel Kv4.2 via PKA/MAP kinase and PKC/MAP kinase pathways (223, 233). During pairing-induced LTP in organotypic slices the Ras/ERK pathway is activated by CaMKII to drive GluR1-containing AMPA receptors to postsynaptic sites (239). In contrast, Rap1 and Rap2 remove postsynaptic AMPA receptors via the MAP kinases p38 and JNK during LTD and depotentiation, respectively (239, 240). MAP kinases are also regulated by BDNF and its receptor TrkB and could thus in part mediate the BDNF-dependence of late LTP described below. However, it appears that MAP kinases are mainly required for the LTP induced by the theta burst protocol, but not for the LTP induced by high-frequency tetani (e.g., (198)). The overall picture of how MAP kinases regulate postsynaptic expression of AMPA receptors is unclear.

5 LTP as a Postsynaptic and Trans-Synaptic Growth Process

Within minutes spines become larger after LTP induction (73) and smaller during LTD induction (238). F-actin is the main cytoskeletal element in spines (136) and inhibiting actin polymerization inhibits spine enlargement (73) and LTP (96, 103). Furthermore, the normal process of LTP-induced spine enlargement involves an increases of polymerized actin in spines (55, 73, 167). F-actin is thus a strong candidate for determining spine size and this appears to be related to the number of functionally available postsynaptic AMPA receptors and the size of the synapse (73, 212). One mechanism that controls F-actin involves cofilin, which, when dephosphorylated, disassembles F-actin. LTP induction activates p21-activated kinase PAK, leading to cofilin phosphorylation (29). This may account for the increase in F-actin after LTP. Conversely, inactivating cofilin by genetic removal of LIM-kinase leads to a reduction of spine size and deficits in LTP (141). It is also noteworthy that LTD processes are affected by this pathway: inhibition of the phosphatase that normally activates cofilin (by dephosphorylating it) produces deficits in LTD (148).

It is difficult to study the size changes of the synapse itself because it cannot be resolved by light microscopy. EM analysis, which is based on statistical analysis of synapse size, indicates that the average size of synapse is enlarged by LTP (70, 168). This enlargement occurs within at least an hour, but could be faster. Such slow growth is likely to contribute to the slow increase in the probability of vesicle release discussed earlier (235).

Even nearby synapses on the same dendrite differ greatly in size (over 10-fold variation). The presynaptic grid and postsynaptic density are both highly variable in size, but have exactly the same size at any given synapse; their edges are exactly in register (115). These properties strongly suggest that synapse growth is due to trans-synaptic mechanisms. These mechanisms might involve adhesion molecules that make physical connections between the presynaptic and postsynaptic sites. Consistent with this notion, LTP is inhibited by agents that interfere with adhesion molecules (31, 62, 122, 213). These molecules, including L1, NCaM, integrins and cadherin, are found in synaptic junctions and may be particularly concentrated at the puncta adherent, a specialized subregion of the synaptic region. Recent experiments show that presynaptic changes can be initiated by defined molecular changes in the postsynaptic cell and that interfering with adhesion molecules, such as integrins, cadherins, or neuroligin prevents the presynaptic changes (58, 183, 187).

6 An Integrated Model of Pre- and Postsynaptic LTP Expression Processes

As outlined in the previous sections, many processes contribute to LTP expression. A recent model (113) has attempted to provide an integrated view of how presynaptic and postsynaptic processes can function together to account for LTP. The model is described in the Fig. 1 caption, but several key findings require explanation are noteworthy. Quantal analysis of LTP shows that early LTP is associated with an increase in quantal size (104, 110, 207), which is the postsynaptic response to a single synaptic vesicle release event, but late LTP is not (202). This lack of any long-term change in quantal size is consistent with another important line of evidence having to do with AMPA receptor density at the synapse. Immuno-EM shows little variation in AMPA receptor density from synapse to synapse (159). Computer simulations show that quantal size is determined by AMPA receptor density, rather than by the number of AMPA receptors (185).These facts suggest that there is a process that can change AMPA receptor density during early LTP, but that it is transient. In the long run, there are no changes in AMPA receptor density or quantal size; synapses are stronger only because of an increase in quantal content (the average number of vesicles released). By traditional quantal analysis, this observation would imply that LTP becomes solely presynaptic. However, central synapses are too complex to be interpreted in terms of traditional theory and both presynaptic and postsynaptic changes may mediate changes in quantal content, as will be discussed below. A further finding that requires explanation is that the development of LTP (178) and LTD (160) involve a sudden discrete change in synaptic strength. The size of the current change is considerably larger than could be accounted for by a single AMPA receptor. These findings indicate that there must be changes at the synapse that involve whole groups of AMPA receptors.

The model of Fig. 1 attempts to explain these findings with the assumption that synapses are modular with respect to AMPA receptor-mediated transmission. One aspect of this is hypothesis is that the AMPA receptors activated by a vesicle are preferentially those near the site of vesicle release, i.e. in the same module where release occur (185). Another aspect is that modules can either be functional or non-functional (silent) for AMPA receptor – mediated transmission. Consistent with the idea of modules, newly developed anatomical methods show that AMPA receptors exist in multiple clusters (134). According to the model of Fig. 1, what LTP does is to convert silent modules to functional modules to yield the observed increase in quantal content. The mechanisms responsible for making a module functional may involve both a presynaptic change in the mode of vesicle release (e.g. an increase in fusion pore diameter that makes AMPA receptor activation more effective) and the postsynaptic addition of a hyperslot, a structure that has slots (binding sites) for many AMPA receptors. The addition of a hyperslot is postulated to produce the large discrete change in synaptic strength mentioned above. Not all slots may be filled and empty slots can be filled by a transient trafficking process that enhances the binding of AMPA receptors to slots, perhaps by stargazin phosphorylation. It is this process that is postulated to increase receptor density and thereby produce the transient increase in quantal size during LTP. A final aspect of the model is a slow growth of the synapse. This creates a new silent module with additional sites for vesicle release. This mechanism accounts for the slow increase in vesicle release detected by FM1-43 (234). This model appears to be the minimal model that can account for the data on quantal analysis, the evidence for both presynaptic and postsynaptic changes, and the evidence for synaptic growth.

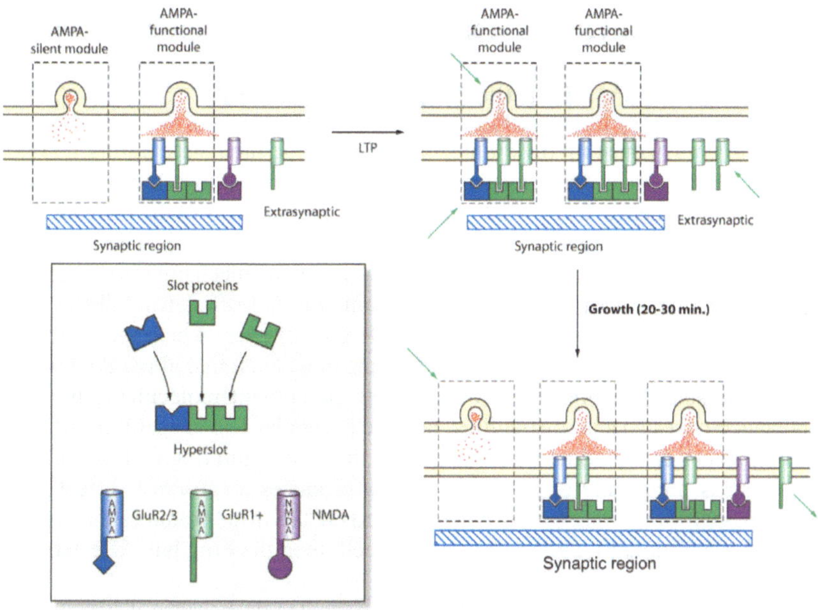

Fig. 1. Model for the transitions from basal to early LTP and from early to late LTP. In this model postsynaptic core proteins (*striped bar*) anchor transsynaptic modules represented by the dotted line rectangles. Each such module contains vesicle release site on the presynaptic side. In AMPA receptor silent modules (see *upper left* of figure), transmission fails both because of slow release of glutamate (*red dots*) and because of the absence of AMPA receptors in the module (*green* and *blue* marked cylinders). AMPA receptors are missing because their anchoring slot proteins (*green* and *blue* boxes with corresponding notches) are not bound to each other to form a 'hyperslot' (*lower left*) that would be anchored by synaptic core proteins. Under basal conditions (*upper left* synaptic region) silent modules are interspersed with functional modules. These can be identified by fast release of glutamate from vesicles (no neck and cloud of glutamate released) and the presence of postsynaptic AMPA receptors held in register by the hyperslot. AMPA receptors in the slot are in equilibrium with free AMPA receptors found in extrasynaptic membrane. Note that NMDA receptors (*purple cylinder*) are sparsely located on the periphery (*purple slot protein*) of the synapse. The small amount of glutamate that diffuses out of functional AMPA receptor modules is sufficient to activate these NMDA receptors. During early LTP (*upper right* synaptic region) transmission is enhanced by three mechanisms (*inward green* arrows): slow release of glutamate from a vesicle is converted to fast release, increasing glutamate concentration in the cleft at that module; complexes of AMPA receptors and hyperslots form at previously silent modules causing a step increase in conductance; and more AMPA receptors are inserted into the extrasynaptic membrane driving occupancy of hyperslot sites to higher levels. Late LTP (*lower right*) involves the growth of the synaptic region with the trans-synaptic formation of a new silent module (*inward green* arrow) and the relaxation of extrasynaptic AMPA receptor concentration to basal levels (*outward green* arrow). Reproduced with permission from (113).

7 LTP Maintenance: the Molecular Basis of Synaptic Memory

7.1 The Role of CaMKII in the Maintenance of LTP

Memory processes must involve molecules that change their state during LTP induction. This change must be persistent and have the power to affect the expression processes that underlie LTP. Such persistent processes constitute the maintenance processes. The most critical aspect of the maintenance process is stability: the storage must be stable for years. Thus, whatever molecular mechanisms underlie maintenance can be described as a switch with stable on and off states. Whatever molecules are involved in maintenance must undergo turnover, which generally occurs within hours or days. Thus, special mechanisms must exist that allow the state of the switch to be stable despite the turnover of its constituent molecules. The study of maintenance poses special difficulties, but one particular test is useful in distinguishing maintenance from induction processes. In this test, saturating LTP is induced and then a drug is transiently applied. If the drug produces a *reversible* effect on LTP, then an expression processes has been affected. If the drug produces an irreversible effect and if additional LTP can then be induced, then the drug has, affected a maintenance process.

CaMKII is a major candidate as a memory molecule because it has switch-like properties that make its activity persist after calcium elevation returns to baseline (145). This 'memory' occurs because Thr286 becomes autophosphorylated during calcium elevation (146, 196) and can persist for many hours after LTP induction (56). What potentially limits long-term information storage by such a process is spontaneous phosphatase activity, which could dephosphorylate Thr286. Furthermore, all proteins undergo protein turnover, so information stored by any one molecule would be slowly eliminated by this process. Theoretical work, however, points to the possibility that Thr286-phosphorylated 'on' CaMKII holoenzymes (there are 12 homologous subunits in a holoenzyme) could retain the on-state for many years (144). Dephosphorylation of Thr286 sites could be counteracted by rapid rephosphorylation of these sites by neighboring subunits. Cooperative interactions between holoenzymes could also lead to phosphorylation of holoenyzmes newly inserted in the process of protein turnover. In this way, information could be stably stored by groups of CaMKII holoenzymes (116).

Although there are indications that CaMKII is responsible for the maintenance of LTP, the evidence is not yet conclusive. The strongest experiments (191) involved the saturation of LTP and the subsequent bath-application of a membrane-permeant CaMKII inhibitor. This application reduced transmission; importantly a component of this reduction persisted after the inhibitor was removed. At this point, it was possible to induce additional LTP, indicating that the inhibitor had reversed a process critical in the maintenance of LTP.

Given the importance of CaMKII in LTP and its abundance in the PSD, it becomes critical to understand what happens to the enzyme within the PSD. CaMKII binds to a number of PSD proteins including the NMDA receptor, densin-180, and α-actinin (32, 142) (see Chapter by Strack and Hell). Binding to the NMDA receptor NR1 and NR2B subunits requires CaMKII activation by either Ca^{2+}/calmodulin

(NR2B) or Thr286 autophosphorylation (NR1, NR2B). Stimulation of postsynaptic Ca^{2+} influx through NMDA receptors augments binding of endogenous CaMKII to the NMDA receptor in acute hippocampal slices (106) and postsynaptic clustering of GFP-tagged CaMKII in primary hippocampal cultures (200). LTP produces a persistent translocation of CaMKII from the cytoplasm to the synapse (170). Such translocation is driven specifically by Ca^{2+}calmodulin influx through the NMDA receptor (214). One surprising aspect of the translocation process is that the Thr286Asp mutation, which mimics Thr286 phosphorylation and stimulates binding to NR1 and NR2B, does not automatically induce translocation, but requires synaptic activity. One possible explanation is that although capable of binding to the NMDA receptor, it cannot bind to NR1 until α-actinin is removed from NR1 by the Ca^{2+}/calmodulin elevation produced by synaptic activity (α-actinin and CaMKII compete for binding to the C0 region in the C-terminus of NR1) (143). Important evidence for the functional relevance of the association of CaMKII with the NMDA receptor comes from studies in hippocampal slice cultures: overexpression of NR2B with mutations in the CaMKII binding site that abrogate the CaMKII interaction produces an strong inhibition of LTP (9).

Once induced, binding of CaMKII to NR2B is stable even if calmodulin is removed by chelating Ca^{2+}. It can also persist if T286 autophosphorylation has not occurred (14). The complex between non-phosphorylated CaMKII and NR2B exhibits autonomous activity similar to T286-autophosphorylated CaMKII (14). Analogous constitutive CaMKII activation is observed when CaMKII is associated with the *Drosophila* eag K^+ channel (209). The CaMKII binding sites on NR2B and eag possess significant similarity with the Thr286 region. This region of CaMKII binds to the so-called T-site to foster intramolecular inhibition of CaMKII catalytic activity by placing the autohinhibitory pseudosubstrate region near the catalytic site (Chapter by Strack and Hell). When Thr286 is displaced from the T site by Ca^{2+}/calmodulin, which binds immediately downstream of Thr286, NR2B or eag associate with the T site. The pseudosubstrate region then cannot re-associate with the catalytic site, much as occurs when Thr286 CaMKII is phosphorylated. This autophosphorylation-independent autonomous activity may support continued CaMKII activity, in part by fostering Thr286 phosphorylation of neighboring subunits in the CaMKII holoenzyme (for this event the substrate subunit must bind Ca^{2+}/calmodulin, but such binding can occur at a low rate even at resting Ca^{2+} levels). Thus, NR2B binding can locally support autonomous CaMKII activity when Ca^{2+}/calmodulin is limited. Such activation of CaMKII would explain the lasting autonomous activity and apparent inaccessibility for phosphatases that CaMKII exhibits upon LTP induction (57). The maintenance of CaMKII phosphorylation in the PSD appears to be further supported by reactions that keep the phosphatase away from the Thr286 site (152, 206).

7.2 The Role of PKMζ in Maintenance of LTP

Evidence is accumulating that another kinase, the PKMζ isoform of PKC, has an important role in LTP maintenance. The concentration of this enzyme is increased during LTP and decreased during LTD (79). The increase of PKMζ after LTP induction is due to protein synthesis (76). The gene produces a protein that lacks a regulatory domain and is therefore constitutively 'on'. Inhibition of PKMζ does not affect the early phase of LTP. Rather late phase LTP will not develop or will be reversed if inhibitor is applied after LTP is already established (111). Basal synaptic transmission is not affected. Importantly, a membrane- penetrable peptide inhibitor of PKMζ, ZIP, not only reverses LTP when applied in the slice preparation, but can irreversibly destroy a memory (place avoidance), when applied in vivo after learning (175, 199). Recent work shows that CaMKII activity is necessary for the synthesis of PKMζ (94). It is thus possible that CaMKII and PKMζ are both involved in the maintenance process.

7.3 Late Phase LTP

The early phase of LTP develops within seconds (101) and does not require protein synthesis. However, late-phase LTP is inhibited by protein synthesis inhibitors. In some reports, the effects of a protein synthesis inhibitor take hours to develop (51), but others report strong effects within 30 min (204). The induction of LTP triggers an increase in protein synthesis. For instance, it is known that some forms of mRNA are translated locally in the dendrites and that local regulators of translation can themselves be controlled by NMDA receptor activation. These effects on protein synthesis can occur within minutes after LTP induction (218).

The utilization of these proteins is affected by a process called synaptic tagging (53). After inducing LTP in one pathway, subsequent application of a protein synthesis inhibitor does not prevent LTP induced in a second pathway recorded from the same neuron. LTP in this second pathway had a normal late phase, despite the presence of protein synthesis inhibitor, presumably because it was able to use the proteins whose synthesis that was triggered by the first pathway. This ability of one pathway to enhance another has a limited lifetime; if the pathways were stimulated more than a few hours apart, there was no interaction. To explain these results, it was proposed that LTP induction produces a transient synapse-specific biochemical modification that functions as a tag to incorporate proteins whose synthesis is triggered by a process that is not synapse-specific.

LTP induction causes changes in the transcription of many genes and triggers new protein synthesis. The early, transient burst of protein synthesis in the postsynaptic dendrite utilizes pre-existing mRNA transcribed earlier from what are known as intermediate-early genes (IEGs) (33). These genes encode proteins such as the transcription factors zif/268 (33), growth factors such as BDNF (26, 42, 176), and enzymes such as CaMKII (124). Another immediate early gene product, Arc/Arg3.1 interacts with dynamin and endophilin to regulate AMPA receptor endocytosis (221). Experiments in which inhibitors of transcription were found to block late LTP and the findings that some IEGs are transcription factors have suggested a role for the nucleus in supplying something needed for late LTP. In support of this hypothesis, in mutant mice lacking the transcription factor CREB, LTP is nearly normal soon after induction but is reduced at late times (21).

It was originally thought that transient elevation of cAMP could directly activate the late phase of LTP, perhaps by stimulating CREB, which is responsive to cAMP. However, it now appears that this process is not a purely biochemical process, but rather depends on presynaptic electrical activity induced by cAMP (which can lead to closure of K^+ channels). In fact, the potentiation induced by elevating cAMP requires NMDA receptor activity (147, 169).

While cAMP may not be sufficient for late LTP, it is probably necessary. This scenario may explain why dopamine modulation is necessary for late LTP. D1 receptors activate adenylate cyclase and thereby elevate cAMP levels. Experiments with dopamine antagonists (52) or knockout of D1 receptors (135) indicate that late LTP is blocked under these conditions. These findings raise the question of what causes the firing of the dopamine cells in the ventral tegmental area (VTA) that provides dopamine to the hippocampus. It now seems likely that the hippocampus itself computes the novelty of incoming information and that the resulting novelty signal stimulates the VTA. In this way, the transition to long-term information storage (late LTP) may be conditional upon the novelty of the incoming information (114).

BDNF has a major role in late LTP (23) and a single nucleotide polymorphism that impairs BDNF secretion correlates with poor episodic memory (44). The peptide is secreted from the presynaptic cell during LTP induction (234), primarily as proBDNF (149), and is converted to BDNF through the action of plasmin, a protease that itself undergoes activity-dependent secretion and activation by tpA (173) and possibly MMP-9 (154). BDNF acts on presynaptic TrkB receptors to enhance the synaptic vesicle release process during high frequency stimulation (49, 219, 220). BDNF also appears to be released in response to activity from the postsynaptic cell (72, 99). The relative contributions of pre- and post-synaptic release have not been determined.

The early phase of LTP that occurs during the first 30 min does not appear to require BDNF action, but later phases do (92, 229). Notably, interfering with TrkB function at 30 min after LTP induction produces a rapid decrease in potentiation, suggesting that BDNF action is required for a considerable time after induction to maintain LTP. Surprisingly, the synthesis of BDNF may be the *only* protein synthesis that is required for late LTP as demonstrated by the ability of BDNF to restore late LTP in the presence of protein synthesis inhibitor. This finding takes on special importance given the recent evidence on the role of test pulses in the transition from early LTP to late LTP. Specifically, if test pulses are stopped after induction and then resumed about an hour later, the system is sensitive to protein synthesis inhibitor applied at this late time (50).

These findings can be understood in terms of the model shown in Fig.2. LTP induction creates a potentiated state that is stable in the absence of stimulation, but labile in the presence of even low frequency test pulses. Normally, a product of protein synthesis, likely BDNF, helps to convert the labile state into a stable state through an activity-dependent process. If, however, a protein synthesis inhibitor is applied, a second competing reaction becomes the most likely, the phosphatase-dependent de-potentiation of the labile state (232). Thus any of the following actions will interfere with the development of the stable state: 1) Raising the frequency of test pulses because this activity attacks the labile state by a phosphatase-dependent process. 2) Factors that prevent BDNF-dependent processes including: protein synthesis inhibition (91); failure to process proBDNF to BDNF (173); lack of the BDNF receptor, TrkB (229); and interference with the phospholipase C pathway gated by TrkB (59, 90).

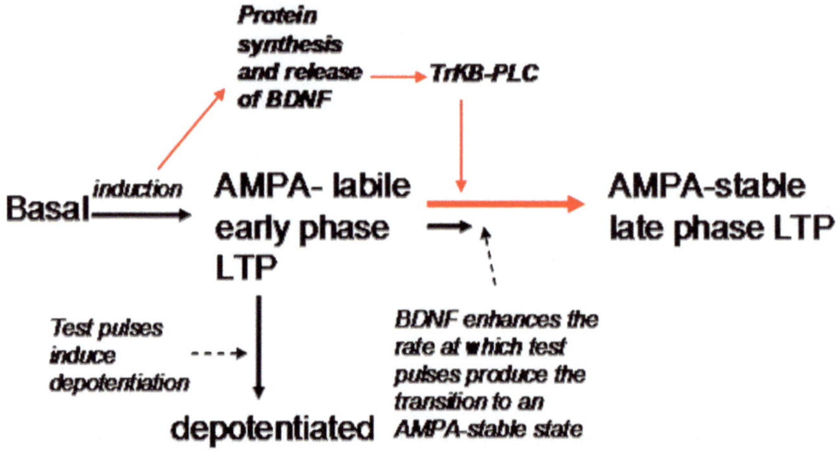

Fig. 2. Model for the BDNF-driven conversion of labile early LTP to stable late LTP. Induction of synaptic potentiation induces early LTP by increasing the amplitude of AMPA receptor-mediated responses. Activity-dependent release of BDNF has no direct effect on this step, but does induce TrkB-phospholipase-C activity, which strongly favors the activity (test pulse)-dependent transition to late LTP. In the absence of coordinated BDNF release, neuronal activity (test pulses) drives depotentiation, the first step down a distinct pathway to return to the basal state.

8 Modulatory Effects on LTP: Metaplasticity

Scores of substances affect the induction of LTP, giving the impression that synaptic modification may be an extremely complex process. However, many of these may not interfere with the biochemistry of synaptic modification itself, but rather act indirectly to affect postsynaptic depolarization. As such action will influence NMDA receptor currents, LTP will be secondarily affected. For instance, the inhibition that is also evoked by the stimulating electrode may be so strong that LTP cannot be induced. Thus, blocking inhibition can enhance LTP induction (226). GABA antagonists ease the induction of LTP by blocking the rapid GABA-A receptor iontropic current, which would otherwise reduce the depolarization that is necessary to relieve Mg^{2+} block of the NMDA receptor (41, 77). Activation of the slower metabotropic GABA-B receptors sustains the hyperpolarizing IPSP. On the other hand, there are some GABA-dependent processes that can lead to an enhancement of LTP induction. GABA can bind to presynaptic GABA-B autoreceptors on interneurons (39). Slow activation of these receptors inhibits GABA release, thereby facilitating the induction of LTP by subsequent glutamategic input. This ‘priming’ effect peaks at approximately 200 ms with the consequence that theta patterns of stimuli optimally produce LTP (37, 38).

Other work shows that activation of MAP-kinase through the insulin receptor can enhance the BK K^+ conductance in hippocampal neurons thereby reducing excitability (163). The resulting interference with synaptically induced depolarization may account for the reduction of LTP by insulin. Also of clear importance in LTP are tyrosine kinases, particular Src. Results indicate that a major role of Src and its upstream regulator, the tyrosine kinase Pyk2, is in the activity-dependent upregulation of the NMDA receptor itself (81, 121).

9 Prospects

There are many developing areas of research that will yield new insight into the mechanisms of LTP and memory. In the past, physiological studies have dealt almost exclusively with the study of populations of synapses, but techniques are now being developed to study individual synapses. Combined with imaging methods, it should be possible to watch synapses as they undergo LTP and to obtain insight into structure–function issues.

Rapid progress is also being made in understanding the molecular composition of the synapse. Once a molecule is identified, it is possible to screen for other molecules that bind to that molecule. A picture is emerging of the molecular complexes on both sides of the synapse. However, this effort is hindered by lack of direct structural information about the PSD. Refinements of EM methods using tomography offers hope in this regard.

Ultimately, the testing of hypotheses regarding the molecular basis of memory requires the ability to assay memory itself. There is increased understanding that there are many forms of memory and that they may be mediated by different mechanisms. New behavioral tests are being developed that allow particular forms of memory to be assessed. The ability of genetic methods to produce defined molecular changes at precise times and precise locations in living animals will make it possible to test whether particular forms of memory rely on particular molecules.

We are who we are largely because of our accumulated experience stored in memory. As more is learned about LTP, it will become possible to manipulate memory. It would clearly be undesirable to wipe our memory clear. But as we come to understand the biochemical basis of memory, it may become possible to affect specific memories. This may help in the treatment of unwanted memories, such as those involved in posttraumatic stress disorders. On the other hand, an understanding of memory and the process of forgetting is likely to yield drugs that will help in the treatment of memory disorders. The implications for understanding the molecular basis of memory are vast and probably largely unanticipated.

Acknowledgements

We would like to thank Roger Nicoll for his comments on the manuscript. Work by JEL has been supported by NIH grants NS027337, NS50944-04, and MH60450 and by JWH by NIH grants NS035563, NS046450, and AG017502.

References

1. Abraham WC, Logan B, Greenwood JM, and Dragunow M. Induction and experience-dependent consolidation of stable long-term potentiation lasting months in the hippocampus. *J Neurosci* 22: 9626–9634, 2002.
2. Adams JP and Sweatt JD. Molecular psychology: roles for the ERK MAP kinase cascade in memory. *Annu Rev Pharmacol Toxicol* 42: 135–163, 2002.
3. Adesnik H and Nicoll RA. Conservation of glutamate receptor 2-containing AMPA receptors during long-term potentiation. *J Neurosci* 27: 4598–4602, 2007.
4. Adesnik H, Nicoll RA, and England PM. Photoinactivation of native AMPA receptors reveals their real-time trafficking. *Neuron* 48: 977–985, 2005.
5. Andrasfalvy BK, Smith MA, Borchardt T, Sprengel R, and Magee JC. Impaired regulation of synaptic strength in hippocampal neurons from GluR1-deficient mice. *J Physiol* 552: 35–45, 2003.
6. Aniksztejn L and Ben-Ari Y. Expression of LTP by AMPA and/or NMDA receptors is determined by the extent of NMDA receptors activation during the tetanus. *J Neurophysiol* 74: 2349–2357, 1995.
7. Asztely F, Wigstrom H, and Gustafsson B. The relative contribution of NMDA receptor channels in the expression of long-term potentiation in the hippocampal CA1 region. *Eur J Neurosci* 4: 681–690, 1992.
8. Banke TG, Bowie D, Lee H, Huganir RL, Schousboe A, and Traynelis SF. Control of GluR1 AMPA receptor function by cAMP-dependent protein kinase. *J Neurosci* 20: 89–102, 2000.
9. Barria A and Malinow R. NMDA receptor subunit composition controls synaptic plasticity by regulating binding to CaMKII. *Neuron* 48: 289–301, 2005.
10. Barria A, Muller D, Derkach V, Griffith LC, and Soderling TR. Regulatory phosphorylation of AMPA-type glutamate receptors by CaM-KII during long-term potentiation. *Science* 276: 2042–2045, 1997.
11. Barry MF and Ziff EB. Receptor trafficking and the plasticity of excitatory synapses. *Curr Opin Neurobiol* 12: 279–286, 2002.
12. Bats C, Groc L, and Choquet D. The interaction between Stargazin and PSD-95 regulates AMPA receptor surface trafficking. *Neuron* 53: 719–734, 2007.
13. Bayazitov I, Richardson R, Fricke R, and Zakharenko S. Slow presynaptic and fast postsynaptic components of compound long-term potentiation. *J Neurosci* 27(43): 11510–11521, 2007.
14. Bayer KU, De Koninck P, Leonard AS, Hell JW, and Schulman H. Interaction with the NMDA receptor locks CaMKII in an active conformation. *Nature* 411: 801–805, 2001.
15. Bellone C and Luscher C. Cocaine triggered AMPA receptor redistribution is reversed in vivo by mGluR-dependent long-term depression. *Nat Neurosci* 9: 636–641, 2006.
16. Benke TA, Luthi A, Isaac JT, and Collingridge GL. Modulation of AMPA receptor unitary conductance by synaptic activity. *Nature* 393: 793–797, 1998.
17. Bevilaqua LR, Medina JH, Izquierdo I, and Cammarota M. Memory consolidation induces N-methyl-D-aspartic acid-receptor- and Ca2+/calmodulin-dependent protein kinase II-dependent modifications in alpha-amino-3-hydroxy-5-methylisoxazole-4-propionic acid receptor properties. *Neuroscience* 136: 397–403, 2005.
18. Bi GQ and Poo MM. Synaptic modifications in cultured hippocampal neurons: dependence on spike timing, synaptic strength, and postsynaptic cell type. *J Neurosci* 18: 10464–10472, 1998.
19. Boehm J, Kang MG, Johnson RC, Esteban J, Huganir RL, and Malinow R. Synaptic incorporation of AMPA receptors during LTP is controlled by a PKC phosphorylation site on GluR1. *Neuron* 51: 213–225, 2006.

20. Bohme GA, Bon C, Stutzmann JM, Doble A, and Blanchard JC. Possible involvement of nitric oxide in long-term potentiation. *Eur J Pharmacol* 199: 379–381, 1991.
21. Bourtchuladze R, Frenguelli B, Blendy J, Cioffi D, Schutz G, and Silva AJ. Deficient long-term memory in mice with a targeted mutation of the cAMP-responsive element-binding protein. *Cell* 79: 59–68, 1994.
22. Bowie D and Mayer ML. Inward rectification of both AMPA and kainate subtype glutamate receptors generated by polyamine-mediated ion channel block. *Neuron* 15: 453–462, 1995.
23. Bramham CR and Messaoudi E. BDNF function in adult synaptic plasticity: the synaptic consolidation hypothesis. *Progr Neurobiol* 76: 99–125, 2005.
24. Brun VH, Ytterbo K, Morris RG, Moser MB, and Moser EI. Retrograde amnesia for spatial memory induced by NMDA receptor-mediated long-term potentiation. *J Neurosci* 21: 356–362, 2001.
25. Buchs PA and Muller D. Induction of long-term potentiation is associated with major ultrastructural changes of activated synapses. *Proc Natl Acad Sci U S A* 93: 8040–8045, 1996.
26. Castren E, Pitkanen M, Sirvio J, Parsadanian A, Lindholm D, Thoenen H, and Riekkinen PJ. The induction of LTP increases BDNF and NGF mRNA but decreases NT-3 mRNA in the dentate gyrus. *Neuroreport* 4: 895–898, 1993.
27. Cavus I and Teyler T. Two forms of long-term potentiation in area CA1 activate different signal transduction cascades. *J Neurophysiol* 76: 3038–3047, 1996.
28. Chen H-J, Rojas-Soto M, Oguni A, and Kennedy MB. A synaptic Ras-GTPase activating protein (p135 SynGAP) inhibited by CaM kinase II. *Neuron* 20: 895–904, 1998.
29. Chen LY, Rex CS, Casale MS, Gall CM, and Lynch G. Changes in synaptic morphology accompany actin signaling during LTP. *J Neurosci* 27: 5363–5372, 2007.
30. Choi S, Klingauf J, and Tsien RW. Fusion pore modulation as a presynaptic mechanism contributing to expression of long-term potentiation. *Philos Trans R Soc Lond* 358: 695–705, 2003.
31. Chun D, Gall CM, Bi X, and Lynch G. Evidence that integrins contribute to multiple stages in the consolidation of long term potentiation in rat hippocampus. *Neuroscience* 105: 815–829, 2001.
32. Colbran RJ. Targeting of calcium/calmodulin-dependent protein kinase II. *Biochem J* 378: 1–16, 2004.
33. Cole AJ, Saffen DW, Baraban JM, and Worley PF. Rapid increase of an immediate early gene messenger RNA in hippocampal neurons by synaptic NMDA receptor activation. *Nature* 340: 474–476, 1989.
34. Collingridge GL, Kehl SJ, and McLennan H. Excitatory amino acids in synaptic transmission in the Schaffer collateral-commissural pathway of the rat hippocampus. *J Physiol* 334: 33–46, 1983.
35. Connor JA, Miller LD, Petrozzino J, and Muller W. Calcium signaling in dendritic spines of hippocampal neurons. *J Neurobiol* 25: 234–242, 1994.
36. Cummings JA, Mulkey RM, Nicoll RA, and Malenka RC. Ca2+ signaling requirements for long-term depression in the hippocampus. *Neuron* 16: 825–833, 1996.
37. Davies CH and Collingridge GL. The physiological regulation of synaptic inhibition by GABAB autoreceptors in rat hippocampus. *J Physiol* 472: 245–265, 1993.
38. Davies CH, Davies SN, and Collingridge GL. Paired-pulse depression of monosynaptic GABA-mediated inhibitory postsynaptic responses in rat hippocampus. *J Physiol* 424: 513–531, 1990.
39. Davies CH, Starkey SJ, Pozza MF, and Collingridge GL. GABA autoreceptors regulate the induction of LTP. *Nature* 349: 609–611, 1991.

40. Derkach V, Barria A, and Soderling TR. Ca2+/calmodulin-kinase II enhances channel conductance of alpha-amino-3-hydroxy-5-methyl-4-isoxazolepropionate type glutamate receptors. *Proc Natl Acad Sci U S A* 96: 3269–3274, 1999.
41. Dingledine R, Hynes MA, and King GL. Involvement of N-methyl-D-aspartate receptors in epileptiform bursting in the rat hippocampal slice. *J Physiol* 380: 175–189, 1986.
42. Dragunow M, Beilharz E, Mason B, Lawlor P, Abraham W, and Gluckman P. Brain-derived neurotrophic factor expression after long-term potentiation. *Neurosci Lett* 160: 232–236, 1993.
43. Dudek SM and Bear MF. Homosynaptic long-term depression in area CA1 of hippocampus and effects of N-methyl-D-aspartate receptor blockade. *Proc Natl Acad Sci U S A* 89: 4363–4367, 1992.
44. Egan MF, Kojima M, Callicott JH, Goldberg TE, Kolachana BS, Bertolino A, Zaitsev E, Gold B, Goldman D, Dean M, Lu B, and Weinberger DR. The BDNF val66met polymorphism affects activity-dependent secretion of BDNF and human memory and hippocampal function. *Cell* 112: 257–269, 2003.
45. Ehlers MD. Reinsertion or degradation of AMPA receptors determined by activity-dependent endocytic sorting. *Neuron* 28: 511–525, 2000.
46. El-Husseini AE, Schnell E, Chetkovich DM, Nicoll RA, and Bredt DS. PSD-95 involvement in maturation of excitatory synapses. *Science* 290: 1364–1368, 2000.
47. Elias GM, Funke L, Stein V, Grant SG, Bredt DS, and Nicoll RA. Synapse-specific and developmentally regulated targeting of AMPA receptors by a family of MAGUK scaffolding proteins. *Neuron* 52: 307–320, 2006.
48. Esteban JA, Shi SH, Wilson C, Nuriya M, Huganir RL, and Malinow R. PKA phosphorylation of AMPA receptor subunits controls synaptic trafficking underlying plasticity. *Nat Neurosci* 6: 136–143, 2003.
49. Figurov A, Pozzo-Miller LD, Olafsson P, Wang T, and Lu B. Regulation of synaptic responses to high-frequency stimulation and LTP by neurotrophins in the hippocampus. *Nature* 381: 706–709, 1996.
50. Fonseca R, Nagerl UV, and Bonhoeffer T. Neuronal activity determines the protein synthesis dependence of long-term potentiation. *Nat Neurosci* 9: 478–480, 2006.
51. Frey U, Krug M, Reymann KG, and Matthies H. Anisomycin, an inhibitor of protein synthesis, blocks late phases of LTP phenomena in the hippocampal CA1 region in vitro. *Brain Res* 452: 57–65, 1988.
52. Frey U, Matthies H, Reymann KG, and Matthies H. The effect of dopaminergic D1 receptor blockade during tetanization on the expression of long-term potentiation in the rat CA1 region in vitro. *Neurosci Lett* 129: 111–114, 1991.
53. Frey U and Morris RG. Synaptic tagging and long-term potentiation. *Nature* 385: 533–536, 1997.
54. Froemke RC and Dan Y. Spike-timing-dependent synaptic modification induced by natural spike trains. *Nature* 416: 433–438, 2002.
55. Fukazawa Y, Saitoh Y, Ozawa F, Ohta Y, Mizuno K, and Inokuchi K. Hippocampal LTP is accompanied by enhanced F-actin content within the dendritic spine that is essential for late LTP maintenance in vivo. *Neuron* 38: 447–460, 2003.
56. Fukunaga K, Muller D, and Miyamoto E. Increased phosphorylation of Ca2+/calmodulin-dependent protein kinase II and its endogenous substrates in the induction of long-term potentiation. *J Biol Chem* 270: 6119–6124, 1995.
57. Fukunaga K, Stoppini L, Miyamoto E, and Muller D. Long-term potentiation is associated with an increased activity of Ca2+/calmodulin-dependent protein kinase II. *J Biol Chem* 268: 7863–7867, 1993.

58. Futai K, Kim MJ, Hashikawa T, Scheiffele P, Sheng M, and Hayashi Y. Retrograde modulation of presynaptic release probability through signaling mediated by PSD-95-neuroligin. *Nat Neurosci* 10: 186–195, 2007.
59. Gartner A, Polnau DG, Staiger V, Sciarretta C, Minichiello L, Thoenen H, Bonhoeffer T, and Korte M. Hippocampal long-term potentiation is supported by presynaptic and post-synaptic tyrosine receptor kinase B-mediated phospholipase Cgamma signaling. *J Neurosci* 26: 3496–3504, 2006.
60. Giese KP, Fedorov NB, Filipkowski R, and Silva AJ. Autophosphorylation at Thr286 of the alpha calcium-calmodulin kinase II in LTP and learning. *Science* 279: 870–873, 1998.
61. Golding NL, Staff NP, and Spruston N. Dendritic spikes as a mechanism for cooperative long-term potentiation. *Nature* 418: 326–331, 2002.
62. Gorski JA, Gomez LL, Scott JD, and Dell'Acqua ML. Association of an A-kinase-anchoring protein signaling scaffold with cadherin adhesion molecules in neurons and epithelial cells. *Mol Biol Cell* 16: 3574–3590, 2005.
63. Graves LM, Bornfeldt KE, Raines EW, Potts BC, Macdonald SG, Ross R, and Krebs EG. Protein kinase A antagonizes platelet-derived growth factor-induced signaling by mitogen-activated protein kinase in human arterial smooth muscle cells. *Proc Natl Acad Sci U S A* 90: 10300–10304, 1993.
64. Gray EE, Fink AE, Sarinana J, Vissel B, and O'Dell TJ. LTP in the hippocampal CA1 region does not require insertion and activation of GluR2-lacking AMPA receptors. *J Neurophysiol* 98: 2488–2492, 2007.
65. Grover LM and Teyler TJ. N-methyl-D-aspartate receptor-independent long-term potentiation in area CA1 of rat hippocampus: input-specific induction and preclusion in a non-tetanized pathway. *Neuroscience* 49: 7–11, 1992.
66. Grover LM and Teyler TJ. Two components of long-term potentiation induced by different patterns of afferent activation. *Nature* 347: 477–479, 1990.
67. Gruart A, Munoz MD, and Delgado-Garcia JM. Involvement of the CA3-CA1 synapse in the acquisition of associative learning in behaving mice. *J Neurosci* 26: 1077–1087, 2006.
68. Hardingham N and Fox K. The role of nitric oxide and GluR1 in presynaptic and postsynaptic components of neocortical potentiation. *J Neurosci* 26: 7395–7404, 2006.
69. Harris EW, Ganong AH, and Cotman CW. Long-term potentiation in the hippocampus involves activation of N-methyl-D-aspartate receptors. *Brain Res* 323: 132–137, 1984.
70. Harris KM, Fiala JC, and Ostroff L. Structural changes at dendritic spine synapses during long-term potentiation. *Philos Trans R Soc Lond* 358: 745–748, 2003.
71. Harris KM and Stevens JK. Dendritic spines of CA 1 pyramidal cells in the rat hippocampus: serial electron microscopy with reference to their biophysical characteristics. *J Neurosci* 9: 2982–2997, 1989.
72. Hartmann M, Heumann R, and Lessmann V. Synaptic secretion of BDNF after high-frequency stimulation of glutamatergic synapses. *EMBO J* 20: 5887–5897, 2001.
73 Harvey CD, Svoboda K. Locally dynamic synaptic learning rules in pyramidal neuron dendrites. *Nature*. 450(7173): 1195–200, 2007.
74. Hayashi Y, Shi SH, Esteban JA, Piccini A, Poncer JC, and Malinow R. Driving AMPA receptors into synapses by LTP and CaMKII: requirement for GluR1 and PDZ domain interaction. *Science* 287: 2262–2267, 2000.
75. Hebb D. *The Organization of Behavior*. New York: Wiley, 1949.
76. Hernandez AI, Blace N, Crary JF, Serrano PA, Leitges M, Libien JM, Weinstein G, Tcherapanov A, and Sacktor TC. Protein kinase M zeta synthesis from a brain mRNA encoding an independent protein kinase C zeta catalytic domain. Implications for the molecular mechanism of memory. *J Biol Chem* 278: 40305–40316, 2003.

77. Herron CE, Williamson R, and Collingridge GL. A selective N-methyl-D-aspartate antagonist depresses epileptiform activity in rat hippocampal slices. *Neurosci Lett* 61: 255–260, 1985.
78. Hoffman DA, Sprengel R, and Sakmann B. Molecular dissection of hippocampal theta-burst pairing potentiation. *Proc Natl Acad Sci U S A* 99: 7740–7745, 2002.
79. Hrabetova S and Sacktor TC. Bidirectional regulation of protein kinase M zeta in the maintenance of long-term potentiation and long-term depression. *J Neurosci* 16: 5324–5333, 1996.
80. Huang KP, Huang FL, and Chen HC. Characterization of a 7.5-kDa protein kinase C substrate (RC3 protein, neurogranin) from rat brain. *Arch Biochem Biophys* 305: 570–580, 1993.
81. Huang Y, Lu W, Ali DW, Pelkey KA, Pitcher GM, Lu YM, Aoto H, Roder JC, Sasaki T, Salter MW, and MacDonald JF. CAKbeta/Pyk2 kinase is a signaling link for induction of long-term potentiation in CA1 hippocampus. *Neuron* 29: 485–496., 2001.
82. Huerta PT and Lisman JE. Bidirectional synaptic plasticity induced by a single burst during cholinergic theta oscillation in CA1 in vitro. *Neuron* 15: 1053–1063, 1995.
83. Impey S, Obrietan K, Wong ST, Poser S, Yano S, Wayman G, Deloulme JC, Chan G, and Storm DR. Cross talk between ERK and PKA is required for Ca2+ stimulation of CREB-dependent transcription and ERK nuclear translocation. *Neuron* 21: 869–883, 1998.
84. Isaac JT, Nicoll RA, and Malenka RC. Evidence for silent synapses: implications for the expression of LTP. *Neuron* 15: 427–434, 1995.
85. Jaffe DB, Johnston D, Lasser-Ross N, Lisman JE, Miyakawa H, and Ross WN. The spread of Na+ spikes determines the pattern of dendritic Ca2+ entry into hippocampal neurons. *Nature* 357: 244–246, 1992.
86. Jensen V, Kaiser KM, Borchardt T, Adelmann G, Rozov A, Burnashev N, Brix C, Frotscher M, Andersen P, Hvalby O, Sakmann B, Seeburg PH, and Sprengel R. A juvenile form of postsynaptic hippocampal long-term potentiation in mice deficient for the AMPA receptor subunit GluR-A. *J Physiol* 553: 843–856, 2003.
87. Jouvenceau A, Billard JM, Haditsch U, Mansuy IM, and Dutar P. Different phosphatase-dependent mechanisms mediate long-term depression and depotentiation of long-term potentiation in mouse hippocampal CA1 area. *Eur J Neurosci* 18: 1279–1285, 2003.
88. Kamboj SK, Swanson G, and Cull-Candy SG. Intracellular spermine confers rectification on rat calcium-permeable AMPA and kainate receptors. *J Physiol* 486: 297–303, 1995.
89. Kampa BM, Letzkus JJ, and Stuart GJ. Requirement of dendritic calcium spikes for induction of spike-timing-dependent synaptic plasticity. *J Physiol* 574: 283–290, 2006.
90. Kang H and Schuman EM. Intracellular Ca(2+) signaling is required for neurotrophin-induced potentiation in the adult rat hippocampus. *Neurosci Lett* 282: 141–144, 2000.
91. Kang H and Schuman EM. A requirement for local protein synthesis in neurotrophin-induced hippocampal synaptic plasticity. *Science* 273: 1402–1406, 1996.
92. Kang H, Welcher AA, Shelton D, and Schuman EM. Neurotrophins and time: different roles for TrkB signaling in hippocampal long-term potentiation. *Neuron* 19: 653–664, 1997.
93. Kelleher RJ, 3rd, Govindarajan A, Jung HY, Kang H, and Tonegawa S. Translational control by MAPK signaling in long-term synaptic plasticity and memory. *Cell* 116: 467–479, 2004.
94. Kelly MT, Crary JF, and Sacktor TC. Regulation of protein kinase M: synthesis by multiple kinases in long-term potentiation. *J Neurosci* 27: 3439–3444, 2007.
95. Kelso SR, Ganong AH, and Brown TH. Hebbian synapses in hippocampus. *Proc Natl Acad Sci U S A* 83: 5326–5330, 1986.

96. Kim CH and Lisman JE. A role of actin filament in synaptic transmission and long-term potentiation. *J Neurosci* 19: 4314–4324., 1999.
97. Klyachko VA and Stevens CF. Temperature-dependent shift of balance among the components of short-term plasticity in hippocampal synapses. *J Neurosci* 26: 6945–6957, 2006.
98. Koh DS, Burnashev N, and Jonas P. Block of native Ca(2+)-permeable AMPA receptors in rat brain by intracellular polyamines generates double rectification. *J Physiol* 486: 305–312, 1995.
99. Kojima M, Takei N, Numakawa T, Ishikawa Y, Suzuki S, Matsumoto T, Katoh-Semba R, Nawa H, and Hatanaka H. Biological characterization and optical imaging of brain-derived neurotrophic factor-green fluorescent protein suggest an activity-dependent local release of brain-derived neurotrophic factor in neurites of cultured hippocampal neurons. *J Neurosci Res* 64: 1–10, 2001.
100. Kovalchuk Y, Eilers J, Lisman J, and Konnerth A. NMDA receptor-mediated subthreshold Ca(2+) signals in spines of hippocampal neurons. *J Neurosci* 20: 1791–1799, 2000.
101. Kovalchuk Y, Hanse E, Kafitz KW, and Konnerth A. Postsynaptic induction of BDNF-mediated long-term potentiation. *Science* 295: 1729–1734, 2002.
102. Krapivinsky G, Medina I, Krapivinsky L, Gapon S, and Clapham DE. SynGAP-MUPP1-CaMKII synaptic complexes regulate p38 MAP kinase activity and NMDA receptor-dependent synaptic AMPA receptor potentiation. *Neuron* 43: 563–574, 2004.
103. Krucker T, Siggins GR, and Halpain S. Dynamic actin filaments are required for stable long-term potentiation (LTP) in area CA1 of the hippocampus. *Proc Natl Acad Sci U S A* 97: 6856–6861., 2000.
104. Kullmann DM and Nicoll RA. Long-term potentiation is associated with increases in quantal content and quantal amplitude. *Nature* 357: 240–244, 1992.
105. Lee HK, Barbarosie M, Kameyama K, Bear MF, and Huganir RL. Regulation of distinct AMPA receptor phosphorylation sites during bidirectional synaptic plasticity. *Nature* 405: 955–959, 2000.
106. Leonard AS, Lim IA, Hemsworth DE, Horne MC, and Hell JW. Calcium/calmodulin-dependent protein kinase II is associated with the N-methyl-D-aspartate receptor. *Proc Natl Acad Sci U S A* 96: 3239–3244, 1999.
107. Levy WB and Steward O. Synapses as associative memory elements in the hippocampal formation. *Brain Res* 175: 233–245, 1979.
108. Li S, Tian X, Hartley DM, and Feig LA. Distinct roles for Ras-guanine nucleotide-releasing factor 1 (Ras-GRF1) and Ras-GRF2 in the induction of long-term potentiation and long-term depression. *J Neurosci* 26: 1721–1729, 2006.
109. Liao D, Hessler NA, and Malinow R. Activation of postsynaptically silent synapses during pairing-induced LTP in CA1 region of hippocampal slice. *Nature* 375: 400–404, 1995.
110. Liao D, Jones A, and Malinow R. Direct measurement of quantal changes underlying long-term potentiation in CA1 hippocampus. *Neuron* 9: 1089–1097, 1992.
111. Ling DS, Benardo LS, Serrano PA, Blace N, Kelly MT, Crary JF, and Sacktor TC. Protein kinase Mzeta is necessary and sufficient for LTP maintenance. *Nat Neurosci* 5: 295–296, 2002.
112. Lisman J. A mechanism for the Hebb and the anti-Hebb processes underlying learning and memory. *Proc Natl Acad Sci U S A* 86: 9574–9578, 1989.
113. Lisman J and Raghavachari S. A unified model of the presynaptic and postsynaptic changes during LTP at CA1 synapses. *Sci STKE* 2006: re11, 2006.
114. Lisman JE and Grace AA. The hippocampal-VTA loop: controlling the entry of information into long-term memory. *Neuron* 46: 703–713, 2005.
115. Lisman JE and Harris KM. Quantal analysis and synaptic anatomy–integrating two views of hippocampal plasticity. *Trends Neurosci* 16: 141–147, 1993.

116. Lisman JE and Zhabotinsky AM. A model of synaptic memory: a CaMKII/PP1 switch that potentiates transmission by organizing an AMPA receptor anchoring assembly. *Neuron* 31: 191–201, 2001.
117. Lledo PM, Hjelmstad GO, Mukherji S, Soderling TR, Malenka RC, and Nicoll RA. Calcium/calmodulin-dependent kinase II and long-term potentiation enhance synaptic transmission by the same mechanism. *Proc Natl Acad Sci U S A* 92: 11175–11179, 1995.
118. Lledo PM, Zhang X, Sudhof TC, Malenka RC, and Nicoll RA. Postsynaptic membrane fusion and long-term potentiation. *Science* 279: 399–403, 1998.
119. Lu FM and Hawkins RD. Presynaptic and postsynaptic Ca(2+) and CamKII contribute to long-term potentiation at synapses between individual CA3 neurons. *Proc Natl Acad Sci U S A* 103: 4264–4269, 2006.
120. Lu Y, Allen M, Halt AR, Weisenhaus M, Dallapiazza RF, Hall DD, Usachev YM, McKnight GS, and Hell JW. Age-dependent requirement of AKAP150-anchored PKA and GluR2-lacking AMPA receptors in LTP. *EMBO J* 26: 4879–4890, 2007.
121. Lu YM, Roder JC, Davidow J, and Salter MW. Src activation in the induction of long-term potentiation in CA1 hippocampal neurons. *Science* 279: 1363–1367, 1998.
122. Luthl A, Laurent JP, Figurov A, Muller D, and Schachner M. Hippocampal long-term potentiation and neural cell adhesion molecules L1 and NCAM. *Nature* 372: 777–779, 1994.
123. Lynch G, Larson J, Kelso S, Barrionuevo G, and Schottler F. Intracellular injections of EGTA block induction of hippocampal long-term potentiation. *Nature* 305: 719–721, 1983.
124. Mackler SA, Brooks BP, and Eberwine JH. Stimulus-induced coordinate changes in mRNA abundance in single postsynaptic hippocampal CA1 neurons. *Neuron* 9: 539–548, 1992.
125. Mainen ZF, Malinow R, and Svoboda K. Synaptic calcium transients in single spines indicate that NMDA receptors are not saturated. *Nature* 399: 151–155, 1999.
126. Malinow R. Transmission between pairs of hippocampal slice neurons: quantal levels, oscillations, and LTP. *Science* 252: 722–724, 1991.
127. Malinow R and Malenka RC. AMPA receptor trafficking and synaptic plasticity. *Annu Rev Neurosci* 25: 103–126, 2002.
128. Malinow R and Miller JP. Postsynaptic hyperpolarization during conditioning reversibly blocks induction of long-term potentiation. *Nature* 320: 529–530, 1986.
129. Malinow R, Schulman H, and Tsien RW. Inhibition of postsynaptic PKC or CaMKII blocks induction but not expression of LTP. *Science* 245: 862–866, 1989.
130. Man H-Y, Sekine-Aizawa Y, and Huganir R. Regulation of alpha-amino-3-hydroxy-5-methyl-4-isoxazolepropionic acid receptor trafficking through PKA phosphorylation of the Glu receptor 1 subunit. *Proc Natl Acad Sci U S A* 104: 3579–3584, 2007.
131. Manabe T, Renner P, and Nicoll RA. Postsynaptic contribution to long-term potentiation revealed by the analysis of miniature synaptic currents. *Nature* 355: 50–55, 1992.
132. Markram H, Helm PJ, and Sakmann B. Dendritic calcium transients evoked by single back-propagating action potentials in rat neocortical pyramidal neurons. *J Physiol* 485 (Pt 1): 1–20, 1995.
133. Markram H, Lubke J, Frotscher M, and Sakmann B. Regulation of synaptic efficacy by coincidence of postsynaptic APs and EPSPs. *Science* 275: 213–215, 1997.
134. Masugi-Tokita M, Tarusawa E, Watanabe M, Molnar E, Fujimoto K, and Shigemoto R. Number and density of AMPA receptors in individual synapses in the rat cerebellum as revealed by SDS-digested freeze-fracture replica labeling. *J Neurosci* 27: 2135–2144, 2007.

135. Matthies H, Becker A, Schroeder H, Kraus J, Hollt V, and Krug M. Dopamine D1-deficient mutant mice do not express the late phase of hippocampal long-term potentiation. *Neuroreport* 8: 3533–3535, 1997.
136. Matus A. Growth of dendritic spines: a continuing story. *Curr Opin Neurobiol* 15: 67–72, 2005.
137. Mayer ML and Westbrook GL. Permeation and block of N-methyl-D-aspartic acid receptor channels by divalent cations in mouse cultured central neurones. *J Physiol* 394: 501–527, 1987.
138. Mayer ML, Westbrook GL, and Guthrie PB. Voltage-dependent block by Mg2+ of NMDA responses in spinal cord neurones. *Nature* 309: 261–263, 1984.
139. McCormack SG, Stornetta RL, and Zhu JJ. Synaptic AMPA receptor exchange maintains bidirectional plasticity. *Neuron* 50: 75–88, 2006.
140. McNaughton BL, Douglas RM, and Goddard GV. Synaptic enhancement in fascia dentata: cooperativity among coactive afferents. *Brain Res* 157: 277–293, 1978.
141. Meng Y, Zhang Y, Tregoubov V, Janus C, Cruz L, Jackson M, Lu WY, MacDonald JF, Wang JY, Falls DL, and Jia Z. Abnormal spine morphology and enhanced LTP in LIMK-1 knockout mice. *Neuron* 35: 121–133, 2002.
142. Merrill MA, Chen Y, Strack S, and Hell JW. Activity-driven postsynaptic translocation of CaMKII. *Trends Pharmacol Sci* 26: 645–653, 2005.
143. Merrill MA, Malik Z, Akyol Z, Bartos JA, Leonard AS, Hudmon A, Shea MA, and Hell JW. Displacement of alpha-Actinin from the NMDA receptor NR1 C0 domain by Ca(2+)/calmodulin promotes CaMKII binding. *Biochem* 46: 8485–8497, 2007.
144. Miller P, Zhabotinsky AM, Lisman JE, and Wang XJ. The stability of a stochastic CaMKII switch: dependence on the number of enzyme molecules and protein turnover. *PLoS Biol* 3: e107, 2005.
145. Miller SG and Kennedy MB. Regulation of brain type II Ca2+/calmodulin-dependent protein kinase by autophosphorylation: a Ca2+-triggered molecular switch. *Cell* 44: 861–870, 1986.
146. Miller SG, Patton BL, and Kennedy MB. Sequences of autophosphorylation sites in neuronal type II CaM kinase that control Ca2(+)-independent activity. *Neuron* 1: 593–604, 1988.
147. Mockett BG, Brooks WM, Tate WP, and Abraham WC. Dopamine D1/D5 receptor activation fails to initiate an activity-independent late-phase LTP in rat hippocampus. *Brain Res* 1021: 92–100, 2004.
148. Morishita W, Marie H, and Malenka RC. Distinct triggering and expression mechanisms underlie LTD of AMPA and NMDA synaptic responses. *Nat Neurosci* 8: 1043–1050, 2005.
149. Mowla SJ, Pareek S, Farhadi HF, Petrecca K, Fawcett JP, Seidah NG, Morris SJ, Sossin WS, and Murphy RA. Differential sorting of nerve growth factor and brain-derived neurotrophic factor in hippocampal neurons. *J Neurosci* 19: 2069–2080, 1999.
150. Mulkey RM, Endo S, Shenolikar S, and Malenka RC. Involvement of a calcineurin/inhibitor-1 phosphatase cascade in hippocampal long-term depression. *Nature* 369: 486–488, 1994.
151. Mulkey RM and Malenka RC. Mechanisms underlying induction of homosynaptic long-term depression in area CA1 of the hippocampus. *Neuron* 9: 967–975, 1992.
152. Mullasseril P, Dosemeci A, Lisman JE, and Griffith LC. A structural mechanism for maintaining the 'on-state' of the CaMKII memory switch in the post-synaptic density. *J Neurochem* 103: 357–364, 2007.
153. Muller D, Arai A, and Lynch G. Factors governing the potentiation of NMDA receptor-mediated responses in hippocampus. *Hippocampus* 2: 29–38, 1992.

154. Nagy V, Bozdagi O, Matynia A, Balcerzyk M, Okulski P, Dzwonek J, Costa RM, Silva AJ, Kaczmarek L, and Huntley GW. Matrix metalloproteinase-9 is required for hippocampal late-phase long-term potentiation and memory. *J Neurosci* 26: 1923–1934, 2006.
155. Nevian T and Sakmann B. Spine Ca2+ signaling in spike-timing-dependent plasticity. *J Neurosci* 26: 11001–11013, 2006.
156. Nicoll RA. Expression mechanisms underlying long-term potentiation: a postsynaptic view. *Philos Trans R Soc Lond* 358: 721–726, 2003.
157. Nicoll RA, Tomita S, and Bredt DS. Auxiliary subunits assist AMPA-type glutamate receptors. *Science* 311: 1253–1256, 2006.
158. Nowak L, Bregestovski P, Ascher P, Herbet A, and Prochiantz A. Magnesium gates glutamate-activated channels in mouse central neurones. *Nature* 307: 462–465, 1984.
159. Nusser Z, Lujan R, Laube G, Roberts JD, Molnar E, and Somogyi P. Cell type and pathway dependence of synaptic AMPA receptor number and variability in the hippocampus. *Neuron* 21: 545–559, 1998.
160. O'Connor DH, Wittenberg GM, and Wang SS. Graded bidirectional synaptic plasticity is composed of switch-like unitary events. *Proc Natl Acad Sci U S A* 102: 9679–9684, 2005.
161. O'Dell TJ, Hawkins RD, Kandel ER, and Arancio O. Tests of the roles of two diffusible substances in long-term potentiation: evidence for nitric oxide as a possible early retrograde messenger. *Proc Natl Acad Sci U S A* 88: 11285–11289, 1991.
162. O'Dell TJ, Huang PL, Dawson TM, Dinerman JL, Snyder SH, Kandel ER, and Fishman MC. Endothelial NOS and the blockade of LTP by NOS inhibitors in mice lacking neuronal NOS. *Science* 265: 542–546, 1994.
163. O'Malley D, Shanley LJ, and Harvey J. Insulin inhibits rat hippocampal neurones via activation of ATP-sensitive K+ and large conductance Ca2+-activated K+ channels. *Neuropharmacol* 44: 855–863, 2003.
164. Oh JS, Manzerra P, and Kennedy MB. Regulation of the neuron-specific Ras GTPase-activating protein, synGAP, by Ca2+/calmodulin-dependent protein kinase II. *J Biol Chem*279: 17980–17988, 2004.
165. Oh MC and Derkach VA. Dominant role of the GluR2 subunit in regulation of AMPA receptors by CaMKII. *Nat Neurosci* 8: 853–854, 2005.
166. Oh MC, Derkach VA, Guire ES, and Soderling TR. Extrasynaptic membrane trafficking regulated by GluR1 serine 845 phosphorylation primes AMPA receptors for long-term potentiation. *J Biol Chem* 281: 752–758, 2006.
167. Okamoto K, Nagai T, Miyawaki A, and Hayashi Y. Rapid and persistent modulation of actin dynamics regulates postsynaptic reorganization underlying bidirectional plasticity. *Nat Neurosci* 7: 1104–1112, 2004.
168. Ostroff LE, Fiala JC, Allwardt B, and Harris KM. Polyribosomes redistribute from dendritic shafts into spines with enlarged synapses during LTP in developing rat hippocampal slices. *Neuron* 35: 535–545, 2002.
169. Otmakhov N, Khibnik L, Otmakhova N, Carpenter S, Riahi S, Asrican B, and Lisman J. Forskolin-induced LTP in the CA1 hippocampal region is NMDA receptor dependent. *J Neurophysiol* 91: 1955–1962, 2004.
170. Otmakhov N, Tao-Cheng JH, Carpenter S, Asrican B, Dosemeci A, Reese TS, and Lisman J. Persistent accumulation of calcium/calmodulin-dependent protein kinase II in dendritic spines after induction of NMDA receptor-dependent chemical long-term potentiation. *J Neurosci* 24: 9324–9331, 2004.
171. Pak JH, Huang FL, Li J, Balschun D, Reymann KG, Chiang C, Westphal H, and Huang KP. Involvement of neurogranin in the modulation of calcium/calmodulin-dependent protein kinase II, synaptic plasticity, and spatial learning: a study with knockout mice. *Proc Natl Acad Sci U S A* 97: 11232–11237, 2000.

172. Palmer CL, Lim W, Hastie PG, Toward M, Korolchuk VI, Burbidge SA, Banting G, Collingridge GL, Isaac JT, and Henley JM. Hippocalcin functions as a calcium sensor in hippocampal LTD. *Neuron* 47: 487–494, 2005.
173. Pang PT, Teng HK, Zaitsev E, Woo NT, Sakata K, Zhen S, Teng KK, Yung WH, Hempstead BL, and Lu B. Cleavage of proBDNF by tPA/plasmin is essential for long-term hippocampal plasticity. *Science* 306: 487–491, 2004.
174. Passafaro M, Piech V, and Sheng M. Subunit-specific temporal and spatial patterns of AMPA receptor exocytosis in hippocampal neurons. *Nat Neurosci* 4: 917–926, 2001.
175. Pastalkova E, Serrano P, Pinkhasova D, Wallace E, Fenton AA, and Sacktor TC. Storage of spatial information by the maintenance mechanism of LTP. *Science* 313: 1141–1144, 2006.
176. Patterson SL, Grover LM, Schwartzkroin PA, and Bothwell M. Neurotrophin expression in rat hippocampal slices: a stimulus paradigm inducing LTP in CA1 evokes increases in BDNF and NT-3 mRNAs. *Neuron* 9: 1081–1088, 1992.
177. Perkel DJ and Nicoll RA. Evidence for all-or-none regulation of neurotransmitter release: implications for long-term potentiation. *J Physiol* 471: 481–500, 1993.
178. Petersen CC, Malenka RC, Nicoll RA, and Hopfield JJ. All-or-none potentiation at CA3-CA1 synapses. *Proc Natl Acad Sci U S A* 95: 4732–4737, 1998.
179. Petrozzino JJ, Pozzo Miller LD, and Connor JA. Micromolar Ca2+ transients in dendritic spines of hippocampal pyramidal neurons in brain slice. *Neuron* 14: 1223–1231, 1995.
180. Pike FG, Meredith RM, Olding AW, and Paulsen O. Rapid report: postsynaptic bursting is essential for 'Hebbian' induction of associative long-term potentiation at excitatory synapses in rat hippocampus. *J Physiol* 518 (Pt 2): 571–576, 1999.
181. Plant K, Pelkey KA, Bortolotto ZA, Morita D, Terashima A, McBain CJ, Collingridge GL, and Isaac JT. Transient incorporation of native GluR2-lacking AMPA receptors during hippocampal long-term potentiation. *Nat Neurosci* 9: 602–604, 2006.
182. Poncer JC, Esteban JA, and Malinow R. Multiple mechanisms for the potentiation of AMPA receptor-mediated transmission by alpha-Ca2+/calmodulin-dependent protein kinase II. *J Neurosci* 22: 4406–4411, 2002.
183. Pratt KG, Watt AJ, Griffith LC, Nelson SB, and Turrigiano GG. Activity-dependent remodeling of presynaptic inputs by postsynaptic expression of activated CaMKII. *Neuron* 39: 269–281, 2003.
184. Prichard L, Deloulme JC, and Storm DR. Interactions between neurogranin and calmodulin in vivo. *J Biol Chem* 274: 7689–7694, 1999.
185. Raghavachari S and Lisman JE. Properties of quantal transmission at CA1 synapses. *J Neurophysiol* 92: 2456–2467, 2004.
186. Ramakers GM, Pasinelli P, van Beest M, van der Slot A, Gispen WH, and De Graan PN. Activation of pre- and postsynaptic protein kinase C during tetraethylammonium-induced long-term potentiation in the CA1 field of the hippocampus. *Neurosci Lett* 286: 53–56, 2000.
187. Regalado MP, Terry-Lorenzo RT, Waites CL, Garner CC, and Malenka RC. Transsynaptic signaling by postsynaptic synapse-associated protein 97. *J Neurosci* 26: 2343–2357, 2006.
188. Roche KW, O'Brien RJ, Mammen AL, Bernhardt J, and Huganir RL. Characterization of multiple phosphorylation sites on the AMPA receptor GluR1 subunit. *Neuron* 16: 1179–1188, 1996.
189. Rumpel S, LeDoux J, Zador A, and Malinow R. Postsynaptic receptor trafficking underlying a form of associative learning. *Science* 308: 83–88, 2005.
190. Sabatini BL, Oertner TG, and Svoboda K. The life cycle of Ca(2+) ions in dendritic spines. *Neuron* 33: 439–452, 2002.

191. Sanhueza M, McIntyre CC, and Lisman JE. Reversal of synaptic memory by Ca2+/calmodulin-dependent protein kinase II inhibitor. *J Neurosci* 27: 5190–5199, 2007.
192. Schiller J, Major G, Koester HJ, and Schiller Y. NMDA spikes in basal dendrites of cortical pyramidal neurons. *Nature* 404: 285–289, 2000.
193. Schmitt JM, Guire ES, Saneyoshi T, and Soderling TR. Calmodulin-dependent kinase kinase/calmodulin kinase I activity gates extracellular-regulated kinase-dependent long-term potentiation. *J Neurosci* 25: 1281–1290, 2005.
194. Schnell E, Sizemore M, Karimzadegan S, Chen L, Bredt DS, and Nicoll RA. Direct interactions between PSD-95 and stargazin control synaptic AMPA receptor number. *Proc Natl Acad Sci U S A* 99: 13902–13907, 2002.
195. Schuman EM and Madison DV. A requirement for the intercellular messenger nitric oxide in long-term potentiation. *Science* 254: 1503–1506, 1991.
196. Schworer CM, Colbran RJ, Keefer JR, and Soderling TR. Ca2+/calmodulin-dependent protein kinase II. Identification of a regulatory autophosphorylation site adjacent to the inhibitory and calmodulin-binding domains. *J Biol Chem* 263: 13486–13489, 1988.
197. Sekine-Aizawa Y and Huganir RL. Imaging of receptor trafficking by using alpha-bungarotoxin-binding-site-tagged receptors. *Proc Natl Acad Sci U S A* 101: 17114–17119, 2004.
198. Shalin SC, Hernandez CM, Dougherty MK, Morrison DK, and Sweatt JD. Kinase suppressor of Ras1 compartmentalizes hippocampal signal transduction and subserves synaptic plasticity and memory formation. *Neuron* 50: 765–779, 2006.
199. Shema R, Sacktor TC, and Dudai Y. Rapid erasure of long-term memory associations in the cortex by an inhibitor of PKM zeta. *Science* 317: 951–953, 2007.
200. Shen K and Meyer T. Dynamic control of CaMKII translocation and localization in hippocampal neurons by NMDA receptor stimulation. *Science* 284: 162–166, 1999.
201. Silva AJ, Stevens CF, Tonegawa S, and Wang Y. Deficient hippocampal long-term potentiation in alpha-calcium-calmodulin kinase II mutant mice. *Science* 257: 201–206, 1992.
202. Sokolov MV, Rossokhin AV, Astrelin AV, Frey JU, and Voronin LL. Quantal analysis suggests strong involvement of presynaptic mechanisms during the initial 3 h maintenance of long-term potentiation in rat hippocampal CA1 area in vitro. *Brain Res* 957: 61–75, 2002.
203. Son H, Hawkins RD, Martin K, Kiebler M, Huang PL, Fishman MC, and Kandel ER. Long-term potentiation is reduced in mice that are doubly mutant in endothelial and neuronal nitric oxide synthase. *Cell* 87: 1015–1023, 1996.
204. Stanton PK and Sarvey JM. Blockade of long-term potentiation in rat hippocampal CA1 region by inhibitors of protein synthesis. *J Neurosci* 4: 3080–3088, 1984.
205. Staubli UV and Ji ZX. The induction of homo- vs. heterosynaptic LTD in area CA1 of hippocampal slices from adult rats. *Brain Res* 714: 169–176, 1996.
206. Strack S and Colbran RJ. Autophosphorylation-dependent targeting of calcium/calmodulin-dependent protein kinase II by the NR2B subunit of the N-methyl-D-aspartate receptor. *J Biol Chem* 273: 20689–20692, 1998.
207. Stricker C, Field AC, and Redman SJ. Changes in quantal parameters of EPSCs in rat CA1 neurones in vitro after the induction of long-term potentiation. *J Physiol* 490 (Pt 2): 443–454, 1996.
208. Sun X, Zhao Y, and Wolf ME. Dopamine receptor stimulation modulates AMPA receptor synaptic insertion in prefrontal cortex neurons. *J Neurosci* 25: 7342–7351, 2005.
209. Sun XX, Hodge JJ, Zhou Y, Nguyen M, and Griffith LC. The eag potassium channel binds and locally activates calcium/calmodulin-dependent protein kinase II. *J Biol Chem* 279: 10206–10214, 2004.

210. Sutton MA, Ito HT, Cressy P, Kempf C, Woo JC, and Schuman EM. Miniature neurotransmission stabilizes synaptic function via tonic suppression of local dendritic protein synthesis. *Cell* 125: 785–799, 2006.
211. Swayze RD, Lise MF, Levinson JN, Phillips A, and El-Husseini A. Modulation of dopamine mediated phosphorylation of AMPA receptors by PSD-95 and AKAP79/150. *Neuropharmacol* 47: 764–778, 2004.
212. Takumi Y, Ramirez-Leon V, Laake P, Rinvik E, and Ottersen OP. Different modes of expression of AMPA and NMDA receptors in hippocampal synapses. *Nat Neurosci* 2: 618–624., 1999.
213. Tang L, Hung CP, and Schuman EM. A role for the cadherin family of cell adhesion molecules in hippocampal long-term potentiation. *Neuron* 20: 1165–1175, 1998.
214. Thalhammer A, Rudhard Y, Tigaret CM, Volynski KE, Rusakov DA, and Schoepfer R. CaMKII translocation requires local NMDA receptor-mediated Ca2+ signaling. *EMBO J* 25: 5873–5883, 2006.
215. Thiagarajan TC, Lindskog M, and Tsien RW. Adaptation to synaptic inactivity in hippocampal neurons. *Neuron* 47: 725–737, 2005.
216. Tomita S, Stein V, Stocker TJ, Nicoll RA, and Bredt DS. Bidirectional synaptic plasticity regulated by phosphorylation of stargazin-like TARPs. *Neuron* 45: 269–277, 2005.
217. Tsien JZ, Huerta PT, and Tonegawa S. The essential role of hippocampal CA1 NMDA receptor-dependent synaptic plasticity in spatial memory. *Cell* 87: 1327–1338, 1996.
218. Tsokas P, Grace EA, Chan P, Ma T, Sealfon SC, Iyengar R, Landau EM, and Blitzer RD. Local protein synthesis mediates a rapid increase in dendritic elongation factor 1A after induction of late long-term potentiation. *J Neurosci* 25: 5833–5843, 2005.
219. Tyler WJ and Pozzo-Miller LD. BDNF enhances quantal neurotransmitter release and increases the number of docked vesicles at the active zones of hippocampal excitatory synapses. *J Neurosci* 21: 4249–4258, 2001.
220. Tyler WJ, Zhang XL, Hartman K, Winterer J, Muller W, Stanton PK, and Pozzo-Miller L. BDNF increases release probability and the size of a rapidly recycling vesicle pool within rat hippocampal excitatory synapses. *J Physiol* 574: 787–803, 2006.
221. Tzingounis AV and Nicoll RA. Arc/Arg3.1: linking gene expression to synaptic plasticity and memory. *Neuron* 52: 403–407, 2006.
222. Wang HX, Gerkin RC, Nauen DW, and Bi GQ. Coactivation and timing-dependent integration of synaptic potentiation and depression. *Nat Neurosci* 8: 187–193, 2005.
223. Watanabe S, Hoffman DA, Migliore M, and Johnston D. Dendritic K+ channels contribute to spike-timing dependent long-term potentiation in hippocampal pyramidal neurons. *Proc Natl Acad Sci U S A* 99: 8366–8371, 2002.
224. Wenthold RJ, Petralia RS, Blahos J, II, and Niedzielski AS. Evidence for multiple AMPA receptor complexes in hippocampal CA1/CA2 neurons. *J Neurosci* 16: 1982–1989., 1996.
225. Whitlock JR, Heynen AJ, Shuler MG, and Bear MF. Learning induces long-term potentiation in the hippocampus. *Science* 313: 1093–1097, 2006.
226. Wigstrom H and Gustafsson B. Facilitated induction of hippocampal long-lasting potentiation during blockade of inhibition. *Nature* 301: 603–604, 1983.
227. Wisden W and Seeburg PH. Mammalian ionotropic glutamate receptors. *Curr Opin Neurobiol* 3: 291–298, 1993.
228. Wong ST, Athos J, Figueroa XA, Pineda VV, Schaefer ML, Chavkin CC, Muglia LJ, and Storm DR. Calcium-stimulated adenylyl cyclase activity is critical for hippocampus-dependent long-term memory and late phase LTP. *Neuron* 23: 787–798, 1999.
229. Xu B, Gottschalk W, Chow A, Wilson RI, Schnell E, Zang K, Wang D, Nicoll RA, Lu B, and Reichardt LF. The role of brain-derived neurotrophic factor receptors in the ma-

ture hippocampus: modulation of long-term potentiation through a presynaptic mechanism involving TrkB. *J Neurosci* 20: 6888–6897, 2000.
230. Yang SN, Tang YG, and Zucker RS. Selective induction of LTP and LTD by postsynaptic [Ca2+]i elevation. *Journal of neurophysiology* 81: 781–787, 1999.
231. Yasuda H, Barth AL, Stellwagen D, and Malenka RC. A developmental switch in the signaling cascades for LTP induction. *Nat Neurosci* 6: 15–16, 2003.
232. Young JZ, Isiegas C, Abel T, and Nguyen PV. Metaplasticity of the late-phase of long-term potentiation: a critical role for protein kinase A in synaptic tagging. *Eur J Neurosci* 23: 1784–1794, 2006.
233. Yuan LL, Adams JP, Swank M, Sweatt JD, and Johnston D. Protein kinase modulation of dendritic K+ channels in hippocampus involves a mitogen-activated protein kinase pathway. *J Neurosci* 22: 4860–4868, 2002.
234. Zakharenko SS, Patterson SL, Dragatsis I, Zeitlin SO, Siegelbaum SA, Kandel ER, and Morozov A. Presynaptic BDNF required for a presynaptic but not postsynaptic component of LTP at hippocampal CA1-CA3 synapses. *Neuron* 39: 975–990, 2003.
235. Zakharenko SS, Zablow L, and Siegelbaum SA. Altered presynaptic vesicle release and cycling during mGluR-dependent LTD. *Neuron* 35: 1099–1110, 2002.
236. Zamanillo D, Sprengel R, Hvalby O, Jensen V, Burnashev N, Rozov A, Kaiser KM, Koster HJ, Borchardt T, Worley P, Lubke J, Frotscher M, Kelly PH, Sommer B, Andersen P, Seeburg PH, and Sakmann B. Importance of AMPA receptors for hippocampal synaptic plasticity but not for spatial learning. *Science* 284: 1805–1811, 1999.
237. Zhabotinsky AM, Camp RN, Epstein IR, and Lisman JE. Role of the neurogranin concentrated in spines in the induction of long-term potentiation. *J Neurosci* 26: 7337–7347, 2006.
238. Zhou Q, Homma KJ, and Poo MM. Shrinkage of dendritic spines associated with long-term depression of hippocampal synapses. *Neuron* 44: 749–757, 2004.
239. Zhu J, Qin Y, Zhao M, Van Aelst L, and Malinow R. Ras and rap control AMPA receptor trafficking during synaptic plasticity. *Cell* 110: 443., 2002.
240. Zhu Y, Pak D, Qin Y, McCormack SG, Kim MJ, Baumgart JP, Velamoor V, Auberson YP, Osten P, van Aelst L, Sheng M, and Zhu JJ. Rap2-JNK removes synaptic AMPA receptors during depotentiation. *Neuron* 46: 905–916, 2005.
241. Ziff EB. TARPs and the AMPA receptor trafficking paradox. *Neuron* 53: 627–633, 2007.

Homeostatic Synaptic Plasticity

Gina Turrigiano

Brandeis University, Department of Biology and Center for Behavioral Genomics, Waltham, MA 02454, USA, turrigiano@brandeis.edu

Abstract. Homeostatic synaptic plasticity mechanisms provide a means for neurons and circuits to maintain stable function in the face of perturbations such as developmental or activity-dependent changes in synapse number or strength. These forms of plasticity use negative feedback signaling to adjust synaptic properties to keep activity close to some internal set point value. Recent work suggests that there are likely multiple forms of synaptic homeostasis, mediated by distinct signaling pathways and with distinct expression mechanisms. Both excitatory and inhibitory synapses are subject to homeostatic regulation, and the form of plasticity present at a particular synapse likely depends on its function within a neuronal circuit.

1 Introduction

"Somehow the unstable stuff of which we are composed has learned the trick of maintaining stability."

- Walter Cannon, The Wisdom of the Body, 1932

Nervous systems are faced with a fundamental problem: how to allow plastic mechanisms to shape their output and function, without compromising the stability and integrity of the underlying circuits that drive behavior. Many of the plastic changes that underlie developmental and learning-related adaptations are thought to require synapse-specific changes in synaptic strengths, yet these processes exert a powerful destabilizing influence on network function (1, 56). "Hebbian" (after Donald Hebb) forms of synaptic plasticity, such as long-term potentiation (LTP) and long-term depression (LTD), strengthen synaptic inputs that are effective at depolarizing the postsynaptic neuron, and weaken inputs whose firing is not correlated with the postsynaptic neuron. Such correlation-based mechanisms are problematic: by analogy to unconstrained capitalism, the rich continually get richer, and the poor get poorer. For example, in the synaptic arena this happens because strong synapses that are effective at depolarizing the postsynaptic neuron will continue to be strengthened, and will thus become better able to depolarize the postsynaptic neuron, initiating an unconstrained positive feedback cycle.

J.W. Hell, M.D. Ehlers (eds.), *Structural and Functional Organization of the Synapse*, DOI: 10.1007/978-0-387-77232-5_18,

A related problem is the synapse-specificity of Hebbian learning rules can break down because of this positive feedback. As correlated activity drives strengthening of specific synapses, the postsynaptic neuron will be driven more strongly, and synapses that initially were only poorly correlated with postsynaptic firing will be better able to fire the postsynaptic neuron, and they too can become strengthened even without a triggering environmental stimulus. This positive feedback aspect of Hebbian learning rules has been appreciated since these rules were first formalized by theoreticians many decades ago (1, 7, 55, 56). The exact nature of the stability problem depends on how the learning rules are formalized and the many forms of Hebbian plasticity that have now been described biologically are each likely to carry with them their own unique destabilizing influences (1).

Many general solutions to this stability problem have been proposed (7, 51, 84), and this issue is of such fundamental importance that there appear to be many biological mechanisms for promoting stability, that operate over different spatial and temporal scales. Following Walter Cannon, who first formalized the idea of physiological homeostasis (9), these mechanisms have collectively been termed "homeostatic plasticity". Homeostatic synaptic plasticity mechanisms are emerging as important complements to Hebbian forms of plasticity in the activity-dependent refinement of synaptic connectivity (14, 87, 1). Here we will define a "homeostatic" form of synaptic plasticity as one that acts to stabilize the activity of a neuron or neuronal circuit in the face of perturbations that alter excitability (Fig. 1). Plasticity phenomena that conform to this definition of homeostatic plasticity include the activity-dependent regulation of intrinsic neuronal firing properties (50, 98); pre and postsynaptic forms of excitatory synaptic plasticity, such as synaptic scaling, that adjust all of a neuron's excitatory synapses up or down in the right direction to stabilize firing (14, 87); the balancing of excitation and inhibition within neuronal networks (23, 45); compensatory changes in synapse number (36, 97); and meta-plastic mechanisms that adjust the relative ease of inducing LTP and LTD (2, 7). In this book on the synapse, I will focus on homeostatic forms of synaptic plasticity.

The classic concept of homeostasis in physiological systems is of a negative feedback system that is composed of a **sensor** of a variable, an **internal set point** for that variable, a **comparator** that can determine how far away the system is from the set point, and an **effector** that acts to move the variable back toward the set point value (9). To implement synaptic homeostasis neurons must sense some aspect of activity, integrate this measure over time, compare this to a set-point value, and adjust synaptic properties to minimize the difference between the actual signal and this set-point (14, 71). For most forms of synaptic homeostasis we are still unclear on the nature of the "activity signal", whether neurons indeed possess an "integrator" of this signal, and if so what the integration time-step is. It is not known whether activity "set-points" are rigid or can themselves be modified by development or experience. Further, for several forms of synaptic homeostasis, it is not clear whether the process is cell-autonomous or requires altered function of entire networks, and/or whether synaptic homeostasis acts globally on all of a neuron's synapses or can be synapse-specific. There have been a number of excellent recent reviews on homeostatic plasticity of intrinsic and synaptic properties (14, 49, 87), so here I will concentrate on recent findings that are beginning to shed light on these outstanding questions in the field.

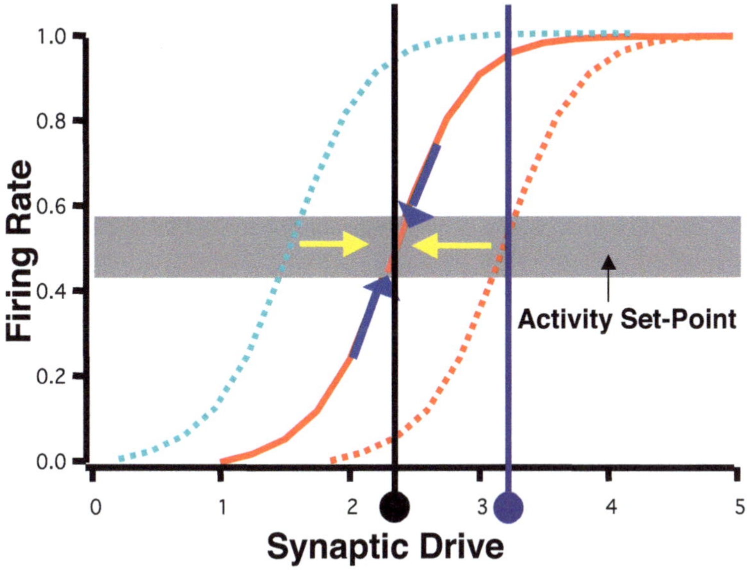

Fig. 1. Cell-autonomous forms of homeostatic plasticity. Diagrams of possible input-output (I-O) functions of a neuron (*blue dashed, red*, and *red dashed lines*), showing firing rate on the *Y* axis as a function of synaptic drive on the *X* axis. Neurons are postulated to have an activity set-point (*gray shaded area*) and to regulate synaptic and/or intrinsic properties to maintain their average activity within this set-point. **Homeostatic Intrinsic plasticity**: if a neuron has the blue dashed I-O curve, then the level of synaptic drive indicated by the black vertical line will drive the neuron's activity much higher than the set-point. Neurons might respond by regulating their intrinsic properties to shift their I-O curves to the right (*yellow arrow*). **Homeostatic Synaptic Plasticity**: If a neuron has the red I-O function, then average activity will fall within the set point for the level of synaptic drive indicated by the black vertical line. However, if average synaptic drive were to increase to the level indicated by the blue vertical line (through LTP or some other mechanisms) then the activity of the neuron would be pushed well outside the set point. Neurons might respond by reducing the strength of all excitatory synaptic inputs until average activity decreases back to the set point (*blue arrow*).

2 Homeostatic Plasticity at the Neuromuscular Junction (NMJ)

In mammals each muscle fiber is innervated by only one motoneuron. The stability problem at the neuromuscular junction is thus fairly simple: as a muscle fiber grows, the motor neuron must remain capable of bringing it over the threshold for contraction, either because there are more presynaptic release sites or more postsynaptic receptors at the endplate (or both). This matching of motoneuron and muscle fiber properties is accomplished through an active process that requires signaling between motoneuron and muscle, and can keep the gain of neuromuscular coupling remarkably constant. Some of the first reports of "homeostatic" regulation of excitability in the nervous system came from the denervation supersensitivity literature (6, 43), where it was observed that loss of synaptic inervation initiated an increase in muscle excitability. More recently work at the rodent and drosophila NMJ has shown that perturbations in presynaptic function lead to compensatory changes in postsynaptic excitability, and vice-versa (14, 15). In Drosophila the ease of genetic manipulations has made it possible to perturb both the presynaptic and postsynaptic side of the synapse and observe quite precise compensation, so that the gain of transmission remains remarkably constant. For example, genetic reductions in glutamate receptor function, or chronic hyperpolarization of the muscle, lead to compensatory increases in transmitter release which restore evoked transmission to control levels (63).

A recent study has shown that these compensatory changes in transmission can occur very quickly when induced by a pharmacological blockade of postsynaptic glutamate receptors, and does not require muscle contraction, but instead occurs in response to subtle changes in the amplitude of excitatory postsynaptic potentials (mEPSPs, the response to individual vesicles of glutamate) (20). This form of homeostatic plasticity requires a presynaptic calcium channel for its expression, implicating a retrograde or transsynaptic signal from the postsynaptic muscle to the presynaptic terminal in the induction of synaptic compensation.

3 Excitatory Homeostatic Plasticity at Vertebrate CNS Synapses

Many central neurons are embedded in complex and highly recurrent microcircuits composed of excitatory and inhibitory inputs. Maintaining stable function in such complicated networks is not trivial, and small changes in the balance between excitation and inhibition can have highly destabilizing effects on network function. It is thus not surprising that once people started looking for stabilizing forms of plasticity in such networks they encountered a rich variety of such mechanisms, operating over various temporal and spatial scales, and utilizing different expression mechanisms (86, 87). An important function of homeostatic synaptic plasticity in central neurons is to compensate for developmental or learning-induced changes in synapse strength, and this compensation could, in theory, occur through a variety of presynaptic and postsynaptic changes. For example, synaptic strength could be modified through changes in postsynaptic receptors, or through changes in presynaptic neurotransmitter release. The locus of change will have important consequences for circuit function, since (for example) presynaptic changes in release probability will strongly affect short term plasticity and thus information transfer across the synapses, while postsynaptic changes in receptor number will tend to scale up or down postsynaptic responsiveness without affecting the short-term dynamics of synaptic transmission (1).

The first reports of synaptic homeostasis at central synapses found that activity blockade increased excitatory synaptic transmission through a simple change in the accumulation of postsynaptic glutamate receptors (42, 62, 85), with no changes in presynaptic function. For example, blockade of spiking in neocortical cultures with TTX increased the amplitude but not the frequency of quantal events as determined by measuring the amplitude of AMPA-mediated miniature excitatory postsynaptic currents (mEPSCs) which probe the postsynaptic response to single vesicles of glutamate. Subsequent studies found no changes in release probability, synapse number, or short-term plasticity at excitatory synapses onto pyramidal neurons (85, 96). In contrast, several other in vitro studies reported that activity-deprivation induced changes in presynaptic release, either with or without complementary changes in postsynaptic receptors (8, 58, 81, 82). The genesis of this difference in expression locus between studies was not clear until the demonstration that cortical and hippocampal neurons respond differently to the same activity deprivation paradigm (2 days of TTX treatment) after different periods of time in vitro. Neurons transition from a purely postsynaptic response to activity deprivation to a mixed presynaptic and postsynaptic response when in culture for more than 3 weeks (97). It is not clear whether this transition is due to synaptic maturation, or to some artifact of maintenance in vitro. Whichever of these proves to be the case, the literature strongly suggests that there are several distinct homeostatic processes at central synapses that can be separately engaged by activity deprivation, and likely involve different signaling pathways.

An important aspect of these compensatory changes in synaptic transmission is that they act to stabilize network activity and are thus "homeostatic" by definition. In culture systems this has been directly demonstrated by comparing the average firing rates of pyramidal neurons before, during, and after various perturbations to activity. For example, blocking spiking with TTX for 1–2 days and then washing out the TTX generates hyperexcitable networks with strongly elevated, epileptic-like bursts of activity (13, 67, 74, 83), indicating that these cultured networks have regulated their properties to become more excitable. In TTX no amount of regulation can restore activity, so a better way to probe for "homeostasis" is to perturb firing rates up or down, and look for a restoration of firing rates over time. Two studies have performed this experiment in different ways. One blocked a fraction of inhibition and found that this initially raised firing rates, but that over time firing rates were restored close to baseline values (85). A second study hyperpolarized individual neurons by overexpressing a K channel and found that firing rates were initially reduced, but again, over time they returned close to baseline values (8). These studies demonstrate that, at least in vitro, homeostatic plasticity mechanisms are in place that function to keep activity relatively constant in the face of even major perturbations. Compensatory changes in activity have now been demonstrated in a variety of networks and systems in vivo (see section below on in vivo homeostasis).

4 Presynaptic Forms of Excitatory Synaptic Homeostasis

Activity-deprivation in older neuronal cultures has been reported to increase mEPSC frequency, presynaptic vesicle recycling, vesicle release probability during evoked synaptic transmission, and the expression of vesicular glutamate transporters – although not all effects have been observed in the same preparation (8, 58, 81, 97, 16, 19). Thus under some conditions chronic inactivity at central synapses can induce compensatory presynaptic changes in synaptic function. Most of these studies also found that inactivity induces complementary postsynaptic changes in receptor accumulation (81, 82, 97). These studies suggest that under some conditions, changes in postsynaptic glutamate receptor number and in presynaptic release may cooperate to homeostatically regulate synaptic transmission.

The signaling mechanisms that generate presynaptic homeostasis at central synapses are not understood. Changes in the level of postsynaptic depolarization can induce changes in mEPSC frequency (presumably due to presynaptic changes in synaptic function (8), but it is not clear whether local dendritic depolarization, or prevention of postsynaptic firing, is the critical factor in signaling back to the presynaptic terminal. It is also unclear whether postsynaptic hyperpolarization is signaled back to the presynaptic terminal through a classic retrograde signaling pathway of some sort, or whether it is mediated through some kind of transsynaptic signaling or even through a simple synaptic growth process, which could generate a coordinated expansion of the presynaptic and postsynaptic terminals (along with the associated release and receptor machinery). This has been suggested by studies showing that prolonged (> 3 days) activity blockade produces an expansion of synaptic terminals in culture (58), as does sensory deprivation in visual cortex (91). Since homeostatic plasticity can occur before there is any perceptible expansion of the presynaptic terminal, this growth may represent a later phase of homeostatic plasticity (78, 85). In addition, since postsynaptic changes in transmission can occur independently of presynaptic changes (96, 97), it is clear that mechanisms other than a simple expansion of the existing synaptic machinery must play a role in some forms of homeostatic plasticity.

5 Excitatory Synaptic Scaling

Pharmacological manipulations of activity induce bidirectional compensatory changes in mEPSC amplitude at glutamatergic synapses (87). These changes operate uniformly on the entire distribution of quantal amplitudes, in effect scaling synaptic strength up or down so that each synapse is changed proportionally (85). This scaling mechanism has the computational advantage of allowing neurons to regulate their excitability while preserving the relative differences between synapses (such as might be used to store information). Because of this proportional change in postsynaptic strength, this form of homeostatic plasticity has been termed "synaptic scaling" (85). The major expression mechanism is through changes in glutamate receptor accumulation in the postsynaptic membrane: bidirectional changes in activity lead to bidirectional changes in the accumulation of AMPA receptors (42, 62, 85, 96), and these changes are sufficiently large to account for the majority of the effect on mEPSC amplitude (but see prior section on regulation of VGLUT levels). Interestingly, not all postsynaptic changes are confined to changes in receptor accumulation. At cortical synapses changes in quantal amplitude are accompanied by changes in evoked transmission (as one would expect); however the increase in evoked transmission is proportionally larger than the effect on mEPSCs. This is not due to changes in quantal content. Rather, there is an increase in the boosting of synaptic potentials by dendritic sodium channels, suggesting that sodium channel accumulation in the dendrites is also increased in concert with changes in AMPA receptors, and these two changes act synergistically to increase evoked synaptic transmission (96). This suggests that activity-deprivation could affect processes such as synaptic integration and the ability of spikes to propagate backwards into dendrites, possibilities that have not yet been explored.

While studies agree that synaptic scaling is induced by changes in AMPA receptor accumulation, there is less agreement on the subunit composition of the newly accumulated receptors. In spinal and neocortical neurons there are proportional changes in GluR1 and GluR2 subunits of the AMPAR (62, 96), whereas several studies on hippocampal neurons have reported enhanced GluR1 accumulation with smaller or absent changes in GluR2 (32, 78, 81). This may represent real differences in the trafficking rules for glutamate receptors in these different cell types, or some other methodological difference between studies (such as method of activity block). A recent study found that when activity was globally blocked with TTX, and then AMPA and NMDA receptors were locally blocked at a subset of synapses, there was a rapid increase in GluR1 (but not GluR2) at the site of local block (78). This raises the possibility that there are two distinct forms of postsynaptic homeostatic plasticity present at central synapses: one that is induced by block or reduction of postsynaptic firing and leads to insertion of GluR1/2 heteromeric AMPAR; and one that is initiated locally when (in addition) postsynaptic receptor activation by mEPSCs is prevented, and leads to insertion of GluR1 homomers in a local-protein synthesis-dependent manner (32, 78). Changes in AMPA subunit composition can affect the rectification, kinetics, and calcium permeability of AMPAR receptors, so that activity-dependent changes in this composition could have interesting consequences for the function and plasticity of central synapses beyond a role in stabilizing activity (30, 80).

6 Signaling Pathways Underlying Synaptic Scaling

How does altered neuronal activity lead to homeostatic changes in AMPA receptor trafficking? What are the activity signals, how are these signals detected, and how do they lead to changes in AMPAR number at synapses? To date in vitro studies of postsynaptic forms of homeostatic plasticity (synaptic scaling) have used global pharmacological manipulations (either TTX or glutamate receptor blockers) to alter activity, so it is unknown whether the relevant activity signal for synaptic scaling is postsynaptic changes in firing, presynaptic changes in release, or local dendritic changes in receptor activation and/or calcium influx. Postsynaptic hyperpolarization is sufficient to induce presynaptic forms of homeostatic plasticity (8), but it is not clear if this is also true for synaptic scaling. Blockade of NMDA receptors alone does not induce synaptic scaling (41, 81, 85, 94) (although NMDAR blockade can accelerate scaling under some conditions (78)), so the signaling pathways are likely to be distinct from those that underlie many forms of LTP and LTD. Understanding the mechanism(s) and function(s) of synaptic scaling will depend critically on working out the precise activity requirements for its induction.

A second important issue is how changes in activity are translated into a signal that can be detected by neurons and/or synapses. Several molecular pathways have now been proposed to play a role in synaptic scaling. The first molecule suggested to be a homeostatic "activity" signal was brain-derived neurotrophic factor (BDNF). BDNF is thought to be released by cortical pyramidal neurons in an activity-dependent manner, and exogenous BDNF can prevent the effects of activity-deprivation. Further, preventing activation of endogenous BDNF receptors mimics the effects of activity blockade (74). Together these data suggest that activity-blockade increases excitatory synaptic strengths by reducing the amount of BDNF released by pyramidal neurons. BDNF also can homeostatically regulate inhibitory synaptic strengths (see below), raising the attractive possibility that BDNF coordinately regulates several classes of synapse in a homeostatic manner. However, in some studies BDNF has been reported to *enhance* mEPSC amplitude onto excitatory neurons (12, 54), suggesting that the effects of chronic BDNF may depend on the amount of BDNF released, on brain region/cell type, or possibly on developmental stage.

Several recent studies have shown that application of the cytokine TNF-alpha rapidly increases mEPSC amplitude and surface AMPAR number (5, 77). Conditioned medium from cultures treated for 2 days with TTX increased mEPSC amplitude through a TNF-alpha-dependent mechanism (77), and interestingly the TNF-alpha originated from glia rather than neurons (77). This indicates that TNF-alpha-mediated synaptic scaling is not cell-autonomous but requires some interaction between neurons and glia. Both this study and studies on BDNF raise the possibility that network-wide changes in activity increase or decrease the amount of TNF-alpha and/or BDNF that is released, which then regulates surface AMPAR on neurons in a network-level homeostatic process.

Finally, two recent papers have suggested that activity-dependent expression of the immediate early gene product Arc is intimately involved in synaptic scaling (70, 75). Arc protein levels are affected by the chronic changes in activity used to induce synaptic scaling in vitro, and overexpression of Arc decreases AMPAR-mediated currents and prevents the increase in mEPSC amplitude induced by chronic TTX. Conversely knockdown of Arc elevates AMPA-mediated transmission and occludes the effects of TTX on mEPSCs (75). The effects of Arc appear to be mediated through endocytosis of AMPAR (11), although there is disagreement on which subunits (GluR1 or GluR2/3) are targeted (70, 75). These studies suggest that environmental stimuli that strongly activate neurons and thus generate strong Arc expression will reduce surface AMPAR number, thus producing a "homeostatic" reduction in synaptic strength. Since Arc can be induced in vivo in a cell-autonomous way (24, 90), this Arc-mediated plasticity, in contrast to TNF-alpha-mediated (and possibly BDNF mediated) plasticity, is likely cell-autonomous. It is not clear whether Arc expression in vivo is a graded function of neuronal activity, or whether Arc is only induced following strong stimuli. If the later is the case, then this pathway may only be engaged by extreme stimuli such as seizures, rather than being a mechanism for the precise and graded compensation for small perturbations to activity.

It is difficult to fit all of the data described above into a unified view of the signaling pathways underlying synaptic scaling. BDNF can induce Arc expression under some conditions (3, 69, 95), suggesting one possible link between these two pathways, but whether this pathway is activated in response to physiological changes in activity is not known. Whether BDNF, TNF-alpha, and Arc prove to be part of a single core signaling pathway for synaptic scaling, or are eventually demoted to the status of modulators of synaptic function and the synaptic scaling machinery, remains to be seen. Another possibility is that each of these signaling pathways (BDNF, glial-derived TNF-alpha, and Arc) act independently to induce functionally distinct forms of homeostatic synaptic plasticity.

The study of homeostatic synaptic plasticity is still relatively young, and it is likely that the cast of molecular players thought to be involved will rapidly accumulate over the next few years. Given the wide range of phenomena currently sheltering under the "homeostatic" umbrella, it will be crucial for progress for investigators to determine which form is under study, since the mechanisms will clearly differ: compare, for example, what is known about rapid presynaptic homeostasis at the NMJ with what is known about synaptic scaling. Several of the molecular players linked to synaptic homeostasis (notably Arc and BDNF) also influence other forms of plasticity such as LTP, LTD, and inhibitory and intrinsic plasticity (73, 18, 79, 70, 88, 31). In some cases this likely reflects a convergence of multiple plasticity pathways onto the same molecular effectors, but in other cases it may be that a change in synaptic function induced by one plasticity mechanism is having indirect effects on the expression of others. A final and related point is that it will be essential to differentiate modulators of synaptic homeostasis from core mediators. A glance through the LTP literature suggests that this is no easy task (46–48), and as the number of molecules implicated in synaptic homeostasis proliferate, things are likely to get darker before the dawn.

7 Homeostatic Plasticity of NMDA Receptor-Mediated Currents

Central glutamatergic synapses typically cluster both AMPAR and NMDAR. NMDAR can contribute to information flow, and in addition flux calcium and can initiate several forms of synaptic plasticity (46). Like AMPAR, NMDAR are trafficked into and out of synapses, and the number of NMDAR clustered at synaptic sites can itself be regulated activity and experience (46, 64), as can the subunit composition of the receptors (65, 66). These changes would be predicted to alter NMDAR-dependent calcium influx induced by synaptic activation, and so in turn could modify the ability of the synapse to undergo NMDAR-dependent forms of plasticity. This plasticity of plasticity has been termed "metaplasticity" (2), and may contribute to the synaptic changes underlying experience-dependent development.

In vitro studies have found that NMDA current and/or receptor localization can be bidirectionally modified by activity in a homeostatic manner (57, 68, 72, 94). In cortical cultures, synapses that have both AMPA and NMDA currents have a tightly conserved ratio of AMPA to NMDA current (89, 92, 94), and AMPA and NMDA currents are scaled up and down proportionally by changes in activity, so that the ratio of current through the two receptor types is maintained at a remarkably constant level (94). In contrast, some manipulations that scale up AMPA currents at hippocampal synapses do not appear to effect NMDA currents (42), so it is not clear how general this proportional scaling is across synapse types. Interestingly, the ratio of AMPA to NMDA current at some central synapses (28, 59) is quite constant after the first few weeks of postnatal development, suggesting that plasticity mechanisms that perturb this ratio (65, 66) must be counterbalanced by mechanisms that act to restore it. Several forms of LTP can transiently perturb the AMPA-to-NMDA ratio by rapidly and selectively increasing AMPA-mediated transmission (46). At cortical synapses rapid potentiation of AMPA currents is followed after one to two hours by a slower potentiation of NMDA currents, and this delayed NMDA-potentiation again acts to restore the AMPA-to-NMDA ratio (93).

The mechanisms underlying homeostatic plasticity of NMDA currents are poorly understood (64). Activity-dependent alternative splicing of NMDAR controls export of receptors from the endoplasmic reticulum, and can exert one level of control over the surface expression of NMDA receptor (57). A second level of control has been demonstrated in spinal neurons (which normally have very low levels of synaptic NMDAR), where synaptic activity has been shown to inhibit the surface redistribution of NMDAR into synaptic sites (72). Whether this mechanism also operates at cortical and hippocampal synapses is not known. The slower NMDA potentiation during LTP at neocortical synapses depends upon AMPA potentiation and can be blocked by manipulations that prevent insertion of AMPAR, suggesting that accumulating more AMPA receptors at synaptic sites eventually leads to the accumulation of more NMDA receptors (93), possibly through protein-protein interactions within the postsynaptic density.

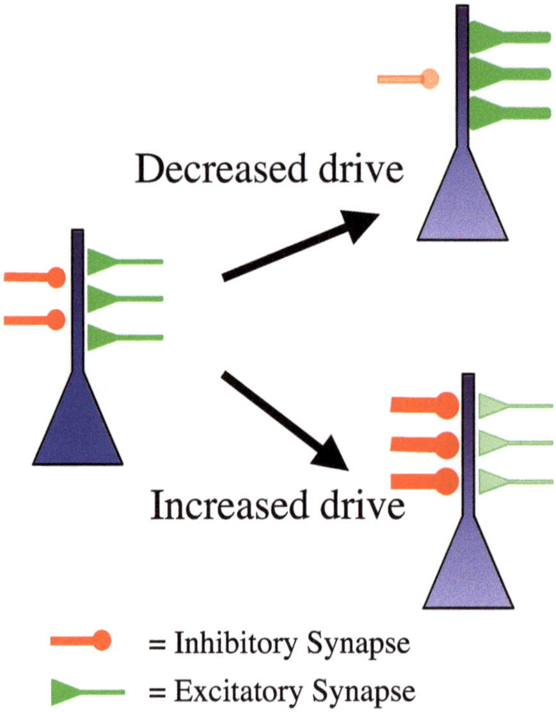

Fig. 2. Activity-dependent regulation of the excitation/inhibition balance. Homeostatic plasticity of excitatory (*green*) and inhibitory (*red*) synapses cooperate to adjust neuronal activity. When activity falls too low, excitation is increased and inhibition is decreased, and vice versa.

At many synapses information is carried both by AMPA and NMDA currents. Functionally, maintaining a relatively constant ratio of AMPA to NMDA current may ensure that plasticity mechanisms like LTP and synaptic scaling can change synaptic strength without fundamentally altering the information content of synaptic transmission. On the other hand, increasing or decreasing NMDA current may strongly affect NMDAR-dependent forms of synaptic plasticity. Given that high levels of calcium influx through NMDAR tend to induce LTP, while lower levels tend to induce LTD, the likelihood of inducing potentiation or depression by any given pattern of pre and postsynaptic co-activation will be altered by scaling NMDA currents up or down. If neurons are inactive and scale up synaptic strengths, this should promote LTP and the expense of LTD, and vice-versa. Scaling of NMDA currents may thus generate the kind of "sliding plasticity threshold" that has been postulated to operate in sensory cortex, and can contribute to the generation of stability within plastic neuronal networks (4).

8 Homeostatic Plasticity of Inhibitory Synapses

Homeostatic plasticity at excitatory synapses is probably not sufficient to maintain network stability in many central networks. For example, cortical networks have extensive positive feedback connections between excitatory pyramidal neurons both within and between cortical layers, which are kept in check by feedback and feedforward inhibition mediated by several types of inhibitory interneurons. Small changes in the balance between excitation and inhibition in such networks can disrupt cortical function (10, 38, 60) and can affect the induction of Hebbian forms of experience-dependent plasticity (27, 29, 34, 35). This suggests that excitation and inhibition must be delicately balanced to keep CNS networks functional, In more mature networks, recent work has established that several kinds of plasticity at synapses onto and from GABAergic interneurons are present in the CNS, and like plasticity at excitatory synapses can provide both stabilizing and destabilizing influences on network function (21, 39, 40, 44, 61).

Reduced visual experience has long been known to induce compensatory changes in inhibition within primary visual cortex (26). A similar phenomenon has been found in dissociated cortical and hippocampal networks, where activity-blockade reversibly decreases GABA immunoreactivity (52, 53, 73), and reduces the amount of functional inhibition received by pyramidal neurons (73). Like homeostatic plasticity at excitatory synapses inhibition is modified through changes in quantal amplitude and receptor clustering. However, a major locus of expression is also a dramatic reduction in the number of functional inhibitory synapses (33), at a time in vitro when there are no changes in excitatory synapse number (97); conversely, inhibitory synapse number does not change in older cultures (25) (at a time when excitatory synapse number does (97)). Excitatory and inhibition synapses onto cultured cortical pyramidal neurons are therefore regulated in opposite directions, and utilize a different subset of expression mechanisms. Changes in inhibition are an important component of firing rate homeostasis within these cultured networks (45, 74), and these studies strongly suggest that the balance between excitation and inhibition is not a fixed quantity, but can be continually adjusted by ongoing activity.

GABAergic interneurons are very diverse in morphology and connectivity, and play distinct roles within cortical and hippocampal microcircuits. This raises the important question of whether all inhibitory synapse types are regulated the same by homeostatic mechanisms, or whether functionally distinct classes of interneuron respond differently to long-lasting changes in activity. This question has recently been addressed within the cortical microcircuitry of layer 4, where visual deprivation during early postnatal development was found to induce compensatory changes in network excitability through reciprocal changes in total excitation and inhibition (45). These effects look very similar to those seen in dissociated cortical networks (74). In slices from deprived and control cortex it was possible to independently evaluate changes at two distinct types of inhibitory synapse onto layer 4 star pyramidal neurons, fast-spiking (FS) basket cells and regular-spiking non-pyramidal (RSNP) neurons. FS synapses onto star pyramidal neurons behaved as expected from changes in total inhibition (i.e., reduced visual drive reduced the strength of these connections). However, RSNP synapses actually underwent an increase in synaptic strength following visual deprivation. This was accompanied by a reduction in connection probability between RSNP and star pyramidal neurons, so that the net effect at this synapse is that a given star pyramid is contacted by only half as many RSNP neurons, but each unitary connection is twice as strong (45). While the functional effects of this synaptic remodeling are not clear, this study illustrates the point that the rules for homeostatic plasticity within real microcircuits are likely to be quite complicated.

Studies on both in vivo and in vitro cortical networks show that homeostatic synaptic plasticity rules function to independently adjust excitatory and inhibitory feedback loops, in a manner predicted to preserve the level of network activity in the face of changes in drive to the network (45, 74). When activity falls too low (because, for example, sensory drive is reduced), excitation between pyramidal neurons is boosted, and feedback inhibition is reduced, which should cooperate to raise the firing rates of pyramidal neurons. Conversely, when activity is too high, excitation between pyramids is reduced (85), while inhibition is increased (Hurvitz and Turrigiano, unpublished observations), so that the activity of pyramidal neurons is reduced (74, 85). Homeostatic regulation of network activity in recurrent cortical circuits thus appears to be "pyramidal neuron-centric", operating through a coordinated set of changes that selectively adjust synaptic strengths to drive pyramidal neuron activity toward some set point range.

One important signaling pathway in the homeostatic regulation of inhibitory synapses is through the activity-dependent release of BDNF (25, 73, 74, 79). Since BDNF is produced and released by pyramidal neurons, this suggests that inhibitory synapses onto pyramidal neurons should be regulated as a function of pyramidal neuron, rather than interneuron, activity. Interestingly, although hyperpolarization of individual pyramidal neurons is capable of inducing some forms of excitatory synaptic homeostasis (8), hyperpolarization of neither individual pyramidal neurons nor individual interneurons was capable of inducing homeostatic changes in inhibition. Only global changes in network activity accomplished this, suggesting that network activity is monitored through extracellular BDNF levels, which then act to homeostatically adjust inhibitory synapses.

BDNF plays an important role in the maturation of inhibitory circuits in vivo and in vitro. In cortical cultures a chronic increase in BDNF levels has been shown to potentiate excitatory synapses onto inhibitory interneurons, as well as inhibitory synapses onto pyramidal neurons. These two effects act synergistically to strengthen feedback inhibition onto pyramidal neurons, and thus cooperate to shift the balance between excitation and inhibition within these cultured networks toward inhibition. This suggests that when pyramidal neuron activity rises too high the resulting increased release of BDNF recruits additional inhibition, which then dampens pyramidal neuron activity again. Somewhat at odds with this idea that BDNF acts as a network-wide homeostatic regulator of circuit activity is the demonstration that BDNF can acutely potentiate excitatory synapses onto hippocampal and cortical neurons under some conditions (see section on excitatory synaptic scaling above). BDNF plays many roles in synaptic function, and so understanding its specific role in homeostatic plasticity will require a better understanding of the different effects of acute and chronic changes in BDNF signaling, as well as local versus global release and action of this neurotrophin. The role of other signaling pathways in inhibitory homeostatic plasticity has not as yet been investigated.

9 Homeostatic synaptic plasticity in vivo

Microcircuits in intact brains are generally quite complex. Most are composed of many different neuron types, and have extensive feedback and feedforward connections that can be excitatory, inhibitory, or modulatory. It is now clear, from a number of studies. that homeostatic plasticity is present in the central nervous system in vivo, and plays important roles in shaping connectivity during development and in response to experience (17, 22, 23, 37, 45, 76). How homeostatic plasticity rules work within such complex networks is far from understood, but several points are becoming clear. First, like many other forms of plasticity, homeostatic synaptic plasticity is strongly developmentally regulated, and can be turned on and off in different cell types at different times in development (17, 22, 44). Second, different synapse "types" respond differently to the same activity-manipulation, indicating that homeostatic plasticity rules for a given synapse type depend on the identity of both the presynaptic and the postsynaptic neuron (45). For example, excitatory synapse onto interneurons are regulated differently than excitatory synapses onto pyramidal neurons within neocortical layer 4 of rodent visual cortex; additionally, different classes of inhibitory synapse respond differently to visual deprivation (44).

Major challenges for the field now include understanding when and why different forms of homeostatic plasticity are present in real neuronal networks, and how these homeostatic mechanisms interact with Hebbian forms of plasticity to generate flexible yet stable microcircuits. When one measures a change in synaptic properties, it is very difficult to determine which plasticity mechanism gave rise to that change. Progress in assigning cellular mechanisms to experience-dependent changes in circuit properties will require that we know the full range of plasticity mechanisms present within a given microcircuit, as well as possess a much deeper mechanistic understanding of how each change is produced. Only when we can selectively interfere with particular forms of Hebbian and homeostatic plasticity during experience-dependent circuit refinement and/or learning will we understand the precise role of each plasticity mechanism in the generation of our flexible, yet stable, brains.

References

1. Abbott LF and Nelson SB. Synaptic plasticity: taming the beast. *Nat Neurosci* 3: 1178–1183, 2000.
2. Abraham WC and Bear MF. Metaplasticity: the plasticity of synaptic plasticity. *Trends In Neurosciences* 19: 126–130, 1996.
3. Alder J, Thakker-Varia S, Bangasser DA, Kuroiwa M, Plummer MR, Shors TJ, and Black IB. Brain-derived neurotrophic factor-induced gene expression reveals novel actions of VGF in hippocampal synaptic plasticity. *J Neurosci* 23: 10800–10808, 2003.
4. Bear MF. Mechanism for a sliding synaptic modification threshold. *Neuron* 15: 1–4, 1995.
5. Beattie EC, Stellwagen D, Morishita W, Bresnahan JC, Ha BK, Von Zastrow M, Beattie MS, and Malenka RC. Control of synaptic strength by glial TNFalpha. *Science* 295: 2282–2285, 2002.
6. Berg DK and Hall ZW. Increased extrajunctional acetylcholine sensitivity produced by chronic acetylcholine sensitivity produced by chronic post-synaptic neuromuscular blockade. *J Physiol* 244: 659–676, 1975.
7. Bienenstock EL, Cooper LN, and Munro PW. Theory for the development of neuron selectivity: orientation specificity and binocular interaction in visual cortex. *J Neurosci* 2: 32–48, 1982.
8. Burrone J, O'Byrne M, and Murthy VN. Multiple forms of synaptic plasticity triggered by selective suppression of activity in individual neurons. *Nature* 420: 414–418, 2002.
9. Cannon WB. *The Wisdom of the Body*. New York: W.W. Norton Co., Inc, 1932.
10. Chagnac-Amitai Y and Connors BW. Horizontal spread of synchronized activity in neocortex and its control by GABA-mediated inhibition. *J Neurophysiol* 61: 747–758, 1989.
11. Chowdhury S, Shepherd JD, Okuno H, Lyford G, Petralia RS, Plath N, Kuhl D, Huganir RL, and Worley PF. Arc/Arg3.1 interacts with the endocytic machinery to regulate AMPA receptor trafficking. *Neuron* 52: 445–459, 2006.
12. Copi A, Jungling K, and Gottmann K. Activity – and BDNF-induced plasticity of miniature synaptic currents in ES cell-derived neurons integrated in a neocortical network. *J Neurophysiol* 94: 4538–4543, 2005.
13. Corner MA and Ramakers GJ. Spontaneous firing as an epigenetic factor in brain development – physiological consequences of chronic tetrodotoxin and picrotoxin exposure on cultured rat neocortex neurons. *Brain Res Dev Brain Res* 65: 57–64, 1992.
14. Davis GW. Homeostatic Control of Neural Activity: From Phenomenology to Molecular Design. *Annu Rev Neurosci* 29: 307–323, 2006.
15. Davis GW and Bezprozvanny I. Maintaining the stability of neural function: a homeostatic hypothesis. *Annu Rev Physiol* 63: 847–869, 2001.
16. De Gois S, Schafer MK, Defamie N, Chen C, Ricci A, Weihe E, Varoqui H, and Erickson JD. Homeostatic scaling of vesicular glutamate and GABA transporter expression in rat neocortical circuits. *J Neurosci* 25: 7121–7133, 2005.
17. Desai NS, Cudmore RH, Nelson SB, and Turrigiano GG. Critical periods for experience-dependent synaptic scaling in visual cortex. *Nat Neurosci* 5: 783–789, 2002.
18. Desai NS, Rutherford LC, and Turrigiano GG. BDNF regulates the intrinsic excitability of cortical neurons. *Learn Mem* 6: 284–291, 1999.

19. Erickson JD, De Gois S, Varoqui H, Schafer MK, and Weihe E. Activity-dependent regulation of vesicular glutamate and GABA transporters: a means to scale quantal size. *Neurochem Int* 48: 643–649, 2006.
20. Frank CA, Kennedy MJ, Goold CP, Marek KW, and Davis GW. Mechanisms underlying the rapid induction and sustained expression of synaptic homeostasis. *Neuron* 52: 663–677, 2006.
21. Gaiarsa JL, Caillard O, and Ben-Ari Y. Long-term plasticity at GABAergic and glycinergic synapses: mechanisms and functional significance. *Trends Neurosci* 25: 564–570, 2002.
22. Goel A and Lee HK. Persistence of experience-induced homeostatic synaptic plasticity through adulthood in superficial layers of mouse visual cortex. *J Neurosci* 27: 6692–6700, 2007.
23. Gonzalez-Islas C and Wenner P. Spontaneous network activity in the embryonic spinal cord regulates AMPAergic and GABAergic synaptic strength. *Neuron* 49: 563–575, 2006.
24. Guzowski JF, McNaughton BL, Barnes CA, and Worley PF. Environment-specific expression of the immediate-early gene Arc in hippocampal neuronal ensembles. *Nat Neurosci* 2: 1120–1124, 1999.
25. Hartman KN, Pal SK, Burrone J, and Murthy VN. Activity-dependent regulation of inhibitory synaptic transmission in hippocampal neurons. *Nat Neurosci* 9: 642–649, 2006.
26. Hendry SH and Jones EG. Reduction in number of immunostained GABAergic neurones in deprived-eye dominance columns of monkey area 17. *Nature* 320: 750–753, 1986.
27. Hensch TK, Fagiolini M, Mataga N, Stryker MP, Baekkeskov S, and Kash SF. Local GABA circuit control of experience-dependent plasticity in developing visual cortex. *Science* 282: 1504–1508, 1998.
28. Hohnke CD, Oray S, and Sur M. Activity-dependent patterning of retinogeniculate axons proceeds with a constant contribution from AMPA and NMDA receptors. *J Neurosci* 20: 8051–8060, 2000.
29. Huang ZJ, Kirkwood A, Pizzorusso T, Porciatti V, Morales B, Bear MF, Maffei L, and Tonegawa S. BDNF regulates the maturation of inhibition and the critical period of plasticity in mouse visual cortex. *Cell* 98: 739–755, 1999.
30. Isaac JT, Ashby M, and McBain CJ. The Role of the GluR2 Subunit in AMPA Receptor Function and Synaptic Plasticity. *Neuron* 54: 859–871, 2007.
31. Jankowsky JL and Patterson PH. Cytokine and growth factor involvement in long-term potentiation. *Mol Cell Neurosci* 14: 273–286, 1999.
32. Ju W, Morishita W, Tsui J, Gaietta G, Deerinck TJ, Adams SR, Garner CC, Tsien RY, Ellisman MH, and Malenka RC. Activity-dependent regulation of dendritic synthesis and trafficking of AMPA receptors. *Nat Neurosci*, 2004.
33. Kilman V, van Rossum MC, and Turrigiano GG. Activity deprivation reduces miniature IPSC amplitude by decreasing the number of postsynaptic GABA(A) receptors clustered at neocortical synapses. *J Neurosci* 22: 1328–1337, 2002.
34. Kirkwood A and Bear MF. Elementary forms of synaptic plasticity in the visual cortex. *Biol Res* 28: 73–80, 1995.
35. Kirkwood A and Bear MF. Hebbian synapses in visual cortex. *J Neurosci* 14: 1634–1645, 1994.
36. Kirov SA, Sorra KE, and Harris KM. Slices have more synapses than perfusion-fixed hippocampus from both young and mature rats. *J Neurosci* 19: 2876–2886, 1999.
37. Kotak VC, Fujisawa S, Lee FA, Karthikeyan O, Aoki C, and Sanes DH. Hearing loss raises excitability in the auditory cortex. *J Neurosci* 25: 3908–3918, 2005.

38. Kriegstein AR. Synaptic responses of cortical pyramidal neurons to light stimulation in the isolated turtle visual system. *J Neurosci* 7: 2488–2492, 1987.
39. Lamsa K, Heeroma JH, and Kullmann DM. Hebbian LTP in feed-forward inhibitory interneurons and the temporal fidelity of input discrimination. *Nat Neurosci* 8: 916–924, 2005.
40. Lamsa KP, Heeroma JH, Somogyi P, Rusakov DA, and Kullmann DM. Anti-Hebbian long-term potentiation in the hippocampal feedback inhibitory circuit. *Science* 315: 1262–1266, 2007.
41. Leslie KR, Nelson SB, and Turrigiano GG. Postsynaptic depolarization scales quantal amplitude in cortical pyramidal neurons. *J Neurosci* 21: RC170, 2001.
42. Lissin DV, Gomperts SN, Carroll RC, Christine CW, Kalman D, Kitamura M, Hardy S, Nicoll RA, Malenka RC, and von Zastrow M. Activity differentially regulates the surface expression of synaptic AMPA and NMDA glutamate receptors. *Proc Natl Acad Sci USA* 95: 7097–7102, 1998.
43. Lomo T and Westgaard RH. Further studies on the control of ACh sensitivity by muscle activity in the rat. *J Physiol* 252: 603–626, 1975.
44. Maffei A, Nataraj K, Nelson SB, and Turrigiano GG. Potentiation of cortical inhibition by visual deprivation. *Nature* 443: 81–84, 2006.
45. Maffei A, Nelson SB, and Turrigiano GG. Selective reconfiguration of layer 4 visual cortical circuitry by visual deprivation. *Nat Neurosci* 7: 1353–1359, 2004.
46. Malenka RC and Bear MF. LTP and LTD: an embarrassment of riches. *Neuron* 44: 5–21, 2004.
47. Malenka RC and Nicoll RA. Long-term potentiation – a decade of progress? *Science* 285: 1870–1874, 1999.
48. Malinow R and Malenka RC. AMPA receptor trafficking and synaptic plasticity. *Annu Rev Neurosci* 25: 103–126, 2002.
49. Marder E and Goaillard JM. Variability, compensation and homeostasis in neuron and network function. *Nat Rev Neurosci* 7: 563–574, 2006.
50. Marder E and Prinz AA. Current compensation in neuronal homeostasis. *Neuron* 37: 2–4, 2003.
51. Marder E and Prinz AA. Modeling stability in neuron and network function: the role of activity in homeostasis. *Bioessays* 24: 1145–1154, 2002.
52. Marty S, Berninger B, Carroll P, and Thoenen H. GABAergic stimulation regulates the phenotype of hippocampal interneurons through the regulation of brain-derived neurotrophic factor. *Neuron* 16: 565–570, 1996.
53. Marty S, Berzaghi Mda P, and Berninger B. Neurotrophins and activity-dependent plasticity of cortical interneurons. *Trends Neurosci* 20: 198–202, 1997.
54. McLean Bolton M, Pittman AJ, and Lo DC. Brain-derived neurotrophic factor differentially regulates excitatory and inhibitory synaptic transmission in hippocampal cultures. *J Neurosci* 20: 3221–3232, 2000.
55. Miller KD. Synaptic economics: competition and cooperation in synaptic plasticity. *Neuron* 17: 371–374, 1996.
56. Miller KD and MacKay DJC. The role of constraints in Hebbian learning. *Neural Comput* 6: 100–124, 1994.
57. Mu Y, Ostuka T, Horton AC, Scott DB, and Ehlers MD. Activity-dependent mRNA splicing controls ER export and synaptic delivery of NMDA receptors. *Neuron* 40: 581–594, 2003.
58. Murthy VN, Schikorski T, Stevens CF, and Zhu Y. Inactivity produces increases in neurotransmitter release and synapse size. *Neuron* 32: 673–682, 2001.
59. Myme CI, Sugino K, Turrigiano GG, and Nelson SB. The NMDA-to-AMPA ratio at synapses onto layer 2/3 pyramidal neurons is conserved across prefrontal and visual cortices. *J Neurophysiol* 90: 771–779, 2003.

60. Nelson SB. Temporal interactions in the cat visual system. III. Pharmacological studies of cortical suppression suggest a presynaptic mechanism. *J Neurosci* 11: 369–380, 1991.
61. Nugent FS, Penick EC, and Kauer JA. Opioids block long-term potentiation of inhibitory synapses. *Nature* 446: 1086–1090, 2007.
62. O'Brien RJ, Kamboj S, Ehlers MD, Rosen KR, Fischbach GD, and Huganir RL. Activity-dependent modulation of synaptic AMPA receptor accumulation. *Neuron* 21: 1067–1078, 1998.
63. Paradis S, Sweeney ST, and Davis GW. Homeostatic control of presynaptic release is triggered by postsynaptic membrane depolarization. *Neuron* 30: 737–749, 2001.
64. Perez-Otano I and Ehlers MD. Homeostatic plasticity and NMDA receptor trafficking. *Trends Neurosci* 28: 229–238, 2005.
65. Philpot BD, Sekhar AK, Shouval HZ, and Bear MF. Visual experience and deprivation bidirectionally modify the composition and function of NMDA receptors in visual cortex. *Neuron* 29: 157–169, 2001.
66. Quinlan EM, Philpot BD, Huganir RL, and Bear MF. Rapid, experience-dependent expression of synaptic NMDA receptors in visual cortex in vivo. *Nat Neurosci* 2: 352–357, 1999.
67. Ramakers GJ, Corner MA, and Habets AM. Development in the absence of spontaneous bioelectric activity results in increased stereotyped burst firing in cultures of dissociated cerebral cortex. *Exp Brain Res* 79: 157–166, 1990.
68. Rao A and Craig AM. Activity regulates the synaptic localization of the NMDA receptor in hippocampal neurons. *Neuron* 19: 801–812, 1997.
69. Rao VR, Pintchovski SA, Chin J, Peebles CL, Mitra S, and Finkbeiner S. AMPA receptors regulate transcription of the plasticity-related immediate-early gene Arc. *Nat Neurosci* 9: 887–895, 2006.
70. Rial Verde EM, Lee-Osbourne J, Worley PF, Malinow R, and Cline HT. Increased expression of the immediate-early gene arc/arg3.1 reduces AMPA receptor-mediated synaptic transmission. *Neuron* 52: 461–474, 2006.
71. Rich MM and Wenner P. Sensing and expressing homeostatic synaptic plasticity. *Trends Neurosci* 30: 119–125, 2007.
72. Rosen KM, Moghekar A, and O'Brien RJ. Activity dependent localization of synaptic NMDA receptors in spinal neurons. *Mol Cell Neurosci* 34: 578–591, 2007.
73. Rutherford LC, DeWan A, Lauer HM, and Turrigiano GG. Brain-derived neurotrophic factor mediates the activity-dependent regulation of inhibition in neocortical cultures. *J Neurosci* 17: 4527–4535, 1997.
74. Rutherford LC, Nelson SB, and Turrigiano GG. BDNF has opposite effects on the quantal amplitude of pyramidal neuron and interneuron excitatory synapses. *Neuron* 21: 521–530, 1998.
75. Shepherd JD, Rumbaugh G, Wu J, Chowdhury S, Plath N, Kuhl D, Huganir RL, and Worley PF. Arc/Arg3.1 mediates homeostatic synaptic scaling of AMPA receptors. *Neuron* 52: 475–484, 2006.
76. Stacy RC, Demas J, Burgess RW, Sanes JR, and Wong RO. Disruption and recovery of patterned retinal activity in the absence of acetylcholine. *J Neurosci* 25: 9347–9357, 2005.
77. Stellwagen D and Malenka RC. Synaptic scaling mediated by glial TNF-alpha. *Nature* 440: 1054–1059, 2006.
78. Sutton MA, Ito HT, Cressy P, Kempf C, Woo JC, and Schuman EM. Miniature neurotransmission stabilizes synaptic function via tonic suppression of local dendritic protein synthesis. *Cell* 125: 785–799, 2006.

79. Swanwick CC, Murthy NR, and Kapur J. Activity-dependent scaling of GABAergic synapse strength is regulated by brain-derived neurotrophic factor. *Mol Cell Neurosci* 31: 481–492, 2006.
80. Thiagarajan TC, Lindskog M, Malgaroli A, and Tsien RW. LTP and adaptation to inactivity: overlapping mechanisms and implications for metaplasticity. *Neuropharmacology* 52: 156–175, 2007.
81. Thiagarajan TC, Lindskog M, and Tsien RW. Adaptation to synaptic inactivity in hippocampal neurons. *Neuron* 47: 725–737, 2005.
82. Thiagarajan TC, Piedras-Renteria ES, and Tsien RW. alpha- and betaCaMKII. Inverse regulation by neuronal activity and opposing effects on synaptic strength. *Neuron* 36: 1103–1114, 2002.
83. Trasande CA and Ramirez JM. Activity deprivation leads to seizures in hippocampal slice cultures: is epilepsy the consequence of homeostatic plasticity? *J Clin Neurophysiol* 24: 154–164, 2007.
84. Turrigiano GG. Homeostatic plasticity in neuronal networks: the more things change, the more they stay the same. *Trends Neurosci* 22: 221–227, 1999.
85. Turrigiano GG, Leslie KR, Desai NS, Rutherford LC, and Nelson SB. Activity-dependent scaling of quantal amplitude in neocortical neurons. *Nature* 391: 892–896, 1998.
86. Turrigiano GG and Nelson SB. Hebb and homeostasis in neuronal plasticity. *Curr Opin Neurobiol* 10: 358–364, 2000.
87. Turrigiano GG and Nelson SB. Homeostatic plasticity in the developing nervous system. *Nat Rev Neurosci* 5: 97–107, 2004.
88. Tzingounis AV and Nicoll RA. Arc/Arg3.1: linking gene expression to synaptic plasticity and memory. *Neuron* 52: 403–407, 2006.
89. Umemiya M, Senda M, and Murphy TH. Behaviour of NMDA and AMPA receptor-mediated miniature EPSCs at rat cortical neuron synapses identified by calcium imaging. *J Physiol* 521: 113–122, 1999.
90. Vazdarjanova A, McNaughton BL, Barnes CA, Worley PF, and Guzowski JF. Experience-dependent coincident expression of the effector immediate-early genes arc and Homer 1a in hippocampal and neocortical neuronal networks. *J Neurosci* 22: 10067–10071, 2002.
91. Wallace W and Bear MF. A morphological correlate of synaptic scaling in visual cortex. *J Neurosci* 24: 6928–6938, 2004.
92. Wang S, Jia Z, Roder J, and Murphy TH. AMPA receptor-mediated miniature synaptic calcium transients in GluR2 null mice. *J Neurophysiol* 88: 29–40, 2002.
93. Watt AJ, Sjostrom PJ, Hausser M, Nelson SB, and Turrigiano GG. A proportional but slower NMDA potentiation follows AMPA potentiation in LTP. *Nat Neurosci*, 2004.
94. Watt AJ, van Rossum MC, MacLeod KM, Nelson SB, and Turrigiano GG. Activity coregulates quantal AMPA and NMDA currents at neocortical synapses. *Neuron* 26: 659–670, 2000.
95. Wibrand K, Messaoudi E, Havik B, Steenslid V, Lovlie R, Steen VM, and Bramham CR. Identification of genes co-upregulated with Arc during BDNF-induced long-term potentiation in adult rat dentate gyrus in vivo. *Eur J Neurosci* 23: 1501–1511, 2006.
96. Wierenga CJ, Ibata K, and Turrigiano GG. Postsynaptic expression of homeostatic plasticity at neocortical synapses. *J Neurosci* 25: 2895–2905, 2005.
97. Wierenga CJ, Walsh MF, and Turrigiano GG. Temporal Regulation of the Expression Locus of Homeostatic Plasticity. *J Neurophysiol*, 2006.
98. Zhang W and D.J. L. The other side of the engram: experience-driven changes in neuronal intrinsic excitability. *Nat Rev Neurosci* 4: 885–900, 2003.

Ubiquitin and Protein Degradation in Synapse Function

Thomas D. Helton and Michael D. Ehlers

Howard Hughes Medical Institute, Department of Neurobiology, Duke University Medical Center, Durham, NC 27710, USA, helton@neuro.duke.edu, ehlers@neuro.duke.edu

1 Introduction

The remodeling of synapses is a fundamental mechanism for information storage and processing in the brain. During brain development, and in response to learning-related activity, synapses undergo remarkable structural changes, including growth, shrinkage, and elimination. Such structural plasticity provides a physical basis for enduring changes in neural circuits that are mediated by alterations in the molecular composition of the synapse. Indeed, stabilization or removal of neurotransmitter receptors, scaffold proteins, and signaling molecules from the synapse has been proposed to account for long-term changes in synaptic strength. Patterns of molecular changes in large sets of synaptic proteins, which ultimately encode the history of activity at the synapse, represent a level of complexity that we are only now beginning to understand.

Long-lasting changes in the molecular content of synapses arise by two general mechanisms: the incorporation of new proteins and the selective removal of existing synaptic proteins. For much of the past two decades, the prevailing model for enduring changes in synapse function and structure has been stimulus-dependent gene expression and protein synthesis. Indeed, substantial evidence indicates that transcriptional events are critical for long-term activity-dependent plasticity (135, 184). In addition, local translation of mRNAs is thought to orchestrate long-lasting forms of learning-related synaptic plasticity (269). On the other hand, considerably less attention has been given to the contribution of protein turnover to long-term structural and functional changes at synapses.

What controls the turnover and replacement of synaptic proteins? For most cellular proteins, the major pathway of degradation occurs by ubiquitin conjugation and subsequent targeting of ubiquitin-conjugated proteins to the proteasome. Here we will discuss the role of the ubiquitin–proteasome system (UPS) in the structural and

J.W. Hell, M.D. Ehlers (eds.), *Structural and Functional Organization of the Synapse*,
DOI: 10.1007/978-0-387-77232-5_19, © Springer Science+Business Media, LLC 2008

functional regulation of synapses. In addition to targeting proteins to the proteasome, post-translational modification by ubiquitination can trigger the assembly of modular protein complexes that recognize ubiquitin moieties in a manner similar to protein phosphorylation (146). In this case, ubiquitination can regulate protein function and protein trafficking (198). We will further discuss recent work revealing ubiquitin-dependent protein modification in synaptic signaling.

Far from being immutable structures, synapses contain diverse interconnected protein networks within which reside the molecular traces of experience and memory. Central to how synapses change and how synaptic molecules store information is how neurons regulate protein degradation and replacement at the synapse. Our focus in this chapter will be to summarize current knowledge of ubiquitination and protein degradation at synapses, a topic that promises to provide a new conceptual framework for understanding both the plasticity and persistence of neural circuits.

2 Components of the Ubiquitin Proteasome System

2.1 Ubiquitin, E1, E2, and E3 Enzymes

Ubiquitin activation, conjugation, ligation, protein degradation, and ubiquitin recycling are the five general steps associated with the UPS-dependent degradative pathway (Fig. 1). Ubiquitin is a highly conserved 76-amino acid protein that is covalently attached via an isopeptide linkage to lysine side chains of protein substrates targeted for proteolytic degradation (48, 49). In rare instances, non-lysine moieties of protein targets can also serve as reactive groups (38, 47). The ubiquitin moiety is attached to its targeted substrate by the concerted action of ubiquitin-activating enzymes (E1s), ubiquitin-conjugating enzymes (E2s), and ubiquitin ligases (E3s) which orchestrate ubiquitin chain elongation on a substrate through sequential isopeptide linkages at internal lysine residues (e.g., K48, K63) of ubiquitin itself. The E1 serves to prime ubiquitin at the onset of the pathway in an ATP-dependent manner (Fig. 1). Ubiquitin molecules activated by the E1 enzyme are accepted by E2 ubiquitin-conjugating enzymes, which in mammals consist of more than 30 genes whose protein products are characterized by conserved ubiquitin-conjugating domains and a conserved catalytic cysteine residue (227, 285). E2 enzymes can interact with E3 ubiquitin ligases to form specific topological-binding surfaces for substrates allowing for efficient transfer of ubiquitin (Fig. 1) (120). E2 enzymes have a varied domain architecture which accommodates interactions with distinct physiological substrates (133).

E3 ubiquitin ligases facilitate the transfer of ubiquitin from E2-activating enzymes to lysine residues of target substrates (Fig. 1). Two major classes of E3 ligases have been identified and are distinguished by containing either a conserved region *h*omologous to *E*6-AP *c*arboxyl *t*erminus (HECT) or *r*eally *i*nteresting *n*ew *g*ene (RING) domains (Fig. 2). HECT domains contain a critical conserved cysteine residue that is essential for E3 ligase function (123). The RING domain family of E3 ligases consists of several hundred genes characterized by a common 40- to 100-amino acid RING domain containing conserved cysteine and histidine residues that coordinate two zinc ions in a cross-braced manner (25, 84, 129, 131) (Fig. 2).

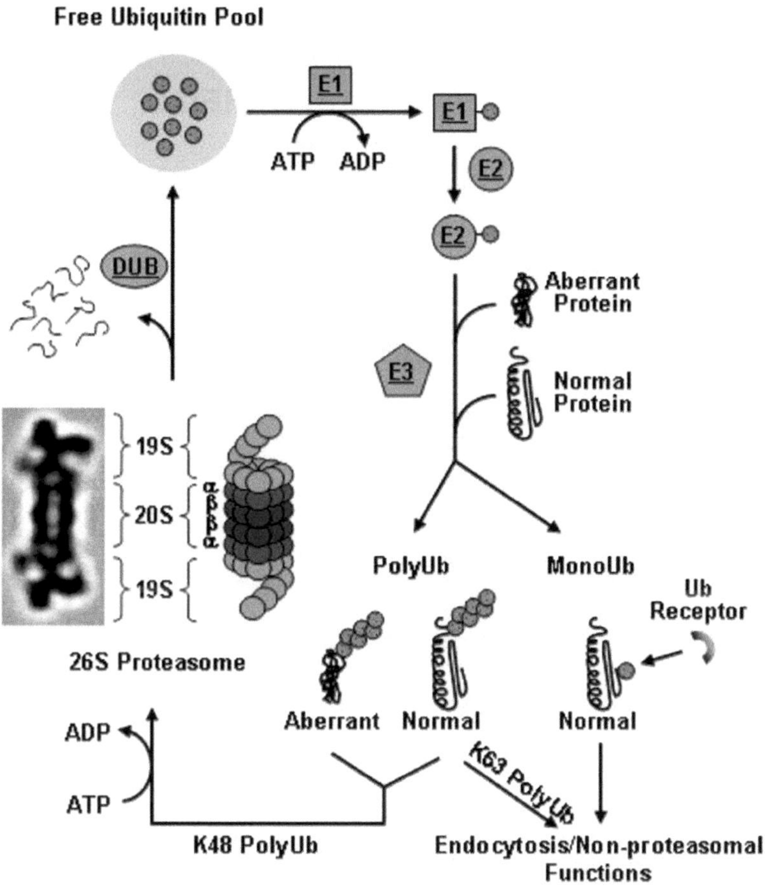

Fig. 1. Overview of the UPS.
Ubiquitin (Ub) monomers are activated in an ATP-dependent reaction by the Ub-activating enzyme (E1) and subsequently transferred to a Ub-conjugating enzyme (E2). Ub ligases (E3) facilitate the transfer of Ub to normal or abnormal target proteins. The covalent transfer of Ub monomers to lysine (K) acceptor residues of the previously attached Ub results in the formation of a poly-Ub chain. Poly-Ub chains linked through lysine 48 (K48) target proteins for degradation at the 26S proteasome. The residual poly-Ub chains are recycled into Ub monomers by deubiquitinating (DUB) enzymes. DUBs can also trim or remove ubiquitin from ubiquitin-conjugated proteins. Monoubiquitination and K63-linked polyubiquitination direct non-proteasomal functions including endocytosis, protein trafficking, DNA repair, and transcription. Proteins possessing ubiquitin-binding domains (UBDs) act as ubiquitin receptors. Recognition of ubiquitin moieties signals key regulatory events in biological pathways. Note that other lysine linkages on Ub can occur but are not indicated.

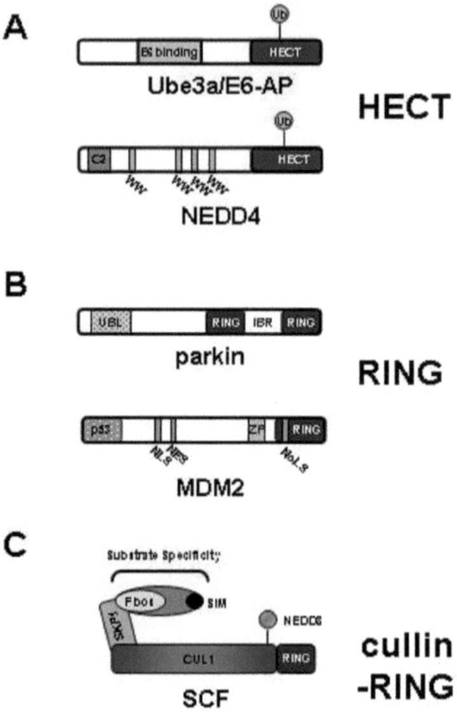

Fig. 2. E3 ligase subtype structure.
Two principal classes of E3 ligases. (**A**) Two examples of HECT (homologous to E6-AP carboxyl terminus) domain E3 ligases are shown: Ube3a/E6-AP and NEDD4. Ube3a/E6-AP contains an E6 substrate-binding domain and a C-terminal HECT domain. NEDD4 has an N-terminal C2 domain and several WW domains conferring cellular localization and substrate recognition, and a C-terminal HECT domain. In both cases, ubiquitin moities are covalently coupled to the HECT domain prior to being transferred to substrate lysine residues. (**B**) RING domain E3 ligase. Two examples of unit RING E3 ligases are shown. Parkin has an N-terminal UBL domain and two RING-binding domains separated by an IBR (in between RING) domain. Mdm2 has a single C-terminal RING domain along with an N-terminal p53-binding domain, a zinc finger (ZF) domain, a nuclear localization signal (NLS), a nuclear exclusion signal (NES), and a nucleolar localization signal (NoLS). RING domains recognize substrates and, by positioning substrates near cognate E2s, mediate the transfer of Ub from E2 enzymes to target proteins. (**C**) SCF (Skp1-Cul1-F-box) is a member of a multisubunit RING E3 ligase class where target proteins are recognized through members of the F-box family of proteins which are linked to cullin scaffolds by a Skp1 adaptor. A key feature of the cullin-RING E3 ligases is that each cullin can assemble with any number of substrate receptors to form ubiquitin ligase complexes containing a common structural core but different substrate specificities.

The identification of HECT domain ubiquitin ligases ensued from studies of tumors arising from infection by the human papilloma virus. During infection, the ubiquitin ligase E6-AP, also known as Ube3a (Fig. 2, Table 1), complexes with the E6 viral protein, causing the degradation of the p53 tumor suppressor and other pathological substrates (247). Sequence alignment revealed conservation between the

Table 1. UPS proteins identified in neuronal function

Protein	Neuronal charaterization	References
E2/E3 enzymes		
Ube3a(E6-AP)	Maternal copy disrupted in Angelman syndrome, defects in CaMKII regulation and hippocampal LTP	Kishino et al. (147); Matsuura et al. (191); Jiang et al. (130); Weeber et al. (293); van Woerden et al. (281)
APC	Axon growth, synaptogenesis, glutamate receptor clustering	Juo and Kaplan (134); Konishi et al. (156); van Roessel et al. (280)
Rnf6	Axon formation and growth	Tursun et al. (278)
Parkin	Loss of function associated with familial Parkinson's disease	Kitada et al.(150)
Pam/Highwire/RPM-1	Synapse formation in *Drosophila* NMJ and synapse development in *C. elegans*	Wan et al. (290); Liao et al. (174); Nakata et al. (204); Wu et al. (304)
Mdm2	PSD remodeling through PSD-95 degradation	Colledge et al. (51)
FSN-1	F-box compared of SCF^{FSN-1} ligase complex; synapse development in *C. elegans*	Liao et al. (174)
β-TrCP/LIN-23	F-box component of $SCF^{\beta\text{-}TrCP}$; glutamate receptor clustering in C. *elegans*. Axonal growth and targeting	Kipreos et al. (145); Mehta et al. (193); Dreier et al. (66)
Fbx2	F-box component of SCF^{Fbx2}; NRI degradation	Kato et al. (137)
Nedd4/Siah	Axon growth, growth cone steering	Hu et al. (118); Myat et al. (203)
Nedd4-2	DAT endocytosis and DA signaling. TrkA receptor and neuronal survival	Sorkina et al. (259); Arévalo et al. (319)
Smurf2	Neuronal polarity, axon/dendrite differentiation	Schwambom et al. (248)
UbcD1/DIAPI	E2/E3 involved in *Drosophila* dendritic pruning through caspase regulation	Muro et al. (201); Wilson et al. (299); Kuo et al. (159)
PDZRN3	Synaptic growth and maturation at the NMJ	Lu et al. (181)
SKR-1/SEL-10	SCF compontents. Regulate synapse elimination in *C. elegans*	Ding et al. (65)
KEL-8	CUL-3-associated protein. Regulation of GLR-1 levels at postsynaptic clusters	Schaefer and Rongo (245)
SCRAPPER	RIM1 degradation. Regulation of presynaptic vesicle release	Yao et al. (308)
Mind Bomb 1	Inhibition of neurite outgrowth in *Drosphila.* Regulation of neuronal morphogenesis	Choe et al. (320)
Deubiquitinating enzymes		
Ap-uch	Long-term facilitation in *Aplysia*	Hegde et al. (106)
UCH-L1	Loss of function in gad mice; familial Parkinson's disease	Leroy et al. (171); Saigoh et al. (321)
Fat facets/FAM	Synapse formation in Drosophila eye and NMJ; presynaptic membrane dynamics	Huang et al. (121); Cadavid et.al. (37); DiAntonio et al. (63); Chen et al. (44)
USP14	Loss of function in Ataxia mouse; ubiquitin recycling	D'Amato and Hicks (57); Wilson et al. (300); Kawaguchi et al. (322)

C-terminus of E6-AP, including the critical catalytic cysteine, and several seemingly unrelated proteins (123), defining the HECT class of E3 ligases. Most non-HECT domain E3 enzymes possess a RING domain (81, 132, 173, 178) (Fig. 2). The RING class of E3 enzymes is further subdivided into the plant homeobox domain/leukemia-associated protein and U-box families, both characterized by small modifications in their RING domains. It is interesting to note that ubiquitin fusion degradation 2, the first identified U-box protein, is also classified as an E4 enzyme (154). E4 enzymes are a novel class of proteins that coordinate with E3 ligases and facilitate ubiquitin chain elongation.

Another family of RING ubiquitin ligases is the cullin-RING ligases (82, 256). Cullins are scaffold proteins with a distinct globular C-terminal domain (cullin homology domain) and a series of N-terminal repeats of a five-helix bundle (cullin repeats) (316). In mammals, cullin-RING ligases are composed of one of seven known cullin proteins (CUL1, 2, 3, 4a, 4b, 5, and 7) and a catalytic RING protein. The core complex is linked to specific targets by various adaptor proteins, most notably F-box proteins. A well-studied group of cullin-RING ligases is the SCF family. The SCF ubiquitin ligases are multisubunit E3 ligases consisting of a CUL1/RING core and SKP1 and F-box proteins that confer substrate specificity (Fig. 2). The combination of the several identified classes of E3 ligases along with numerous possible subunit combinations, such as the cullin proteins and diverse F-box proteins, affords a capacity to recognize a wide array of target proteins for ubiquitination.

2.2 The 26S Proteasome

The 26S proteasome is a 2.5 MDa protein machine made up of two copies of at least 32 different subunits that are highly conserved among eukaryotes (93). The proteasome is an abundant complex making up approximately 1% of total cellular protein (168). The 26S proteasome has been localized to nearly every subcellular compartment including the nucleus, nucleolus, and cytoplasm (302). The 26S proteasome itself is composed of two stable, multimeric complexes, the catalytic 20S particle and the 19S regulatory complex (14, 289) (Fig. 1). The 20S component is composed of seven α and seven β subunits that assemble to form two α and two β heptameric rings. The rings are stacked to form a cylinder with two β rings bordered on each side by an α ring (Fig. 1) (14, 208). In eukaryotes, the two β heptameric rings contain multiple proteolytic sites. There are at least two trypsin-like, two chymotrypsin-like, and two caspase-like proteolytic regions that act cooperatively to allow the proteasome to cleave target proteins at distinct peptide linkages. The base of the 19S component is composed of eight subunits and a ring of six AAA family ATPases that gate the lumen of the proteasome and regulate the energy-dependent unfolding of target proteins (30, 94, 155, 206). Within the 19S complex, the S5a subunit (known as Rpn10 or Pus1 in budding yeast and fission yeast, respectively), Rad23, and one of the ATPases have been shown to interact with polyubiquitin chains and ubiquitin-like protein domains (74, 75, 161, 283, 310), thereby participating in the targeting of polyubiquitinated proteins to the proteasome.

2.3 Deubiquitinating Enzymes

Just as the addition of ubiquitin is carried out by ubiquitin ligases, there are more than 70 deubiquitinating enzymes (DUBs) in two major classes, both of which cleave the G76 isopeptide bond of ubiquitin (258). The ubiquitin C-terminal hydrolases (UCH) process ubiquitinated precursor molecules into their mature forms, as well as interact with monoubiquitinated proteins (212, 295), and recycle ubiquitin from polyubiquitinated proteins at the proteasome to maintain a free ubiquitin pool (Fig. 1, Table 1) (4, 26, 74, 102, 170). Ubiquitin-specific proteases (USP) have conserved 350-amino acid core catalytic domains that disassemble polyubiquitinated protein chains (270).

2.4 Degradative Pathway of Ubiquitinated Substrates

2.4.1. Ubiquitin Recognition

Target proteins typically contain at least four K48-linked ubiquitin molecules that are able to bind the proteasome via S5a/Rpn10 (274), the major proteasome subunit known to bind ubiquitin chains (see Section 2.2). Rpn10 contains a ubiquitin-binding site within its C-terminus in a hydrophobic region containing a LALAL motif (86). The carboxy termini of Rpn10 homologs (p54 from *Drosophila* and mammalian S5a) contain two such hydrophobic repeats within their ubiquitin-interacting motifs (UIM), and both have been shown to bind ubiquitin (103, 310). While Rpn10/p54/S5a is the primary proteasome subunit with an affinity for polyubiquitin chains, Rad23 can be co-precipitated with substrates that contain short multiubiquitin chains and contain a ubiquitin-associated domain (UBA) at the carboxy terminus of the protein that can directly interact with ubiquitin (18, 114). Rad23 and its human counterparts hHR23a and hHR23b bind to proteasomes via an amino terminal ubiquitin-like (UBL) domain which shares homology with ubiquitin (112, 246). While these proteins have been shown to aid in ubiquitin recognition, it is likely that more proteins on the proteasome are involved in this process.

2.4.2 Substrate Binding

Rpn1 and Rpn2 comprise the proteasome base along with the six AAA ATPases. Rpn1 and Rpn2 contain multiple leucine-rich repeats, regions that promote protein–protein interactions with substrates (182). In order for substrate binding to occur, the target protein must be properly positioned to be unfolded by the base and translocated into the proteasome core. The distal positioning of the proteasome lid likely ensures that ubiquitin chains conjugated to substrates do not occlude access of the target protein to the channel. The proteasome base can also bind to non-ubiquitinated, unfolded substrates and promote their folding (30, 268).

2.4.3 Unfolding and Translocation

Proteolysis of proteins by the proteasome is ATP-dependent (62, 108, 113). ATP hydrolysis cycles the regulatory 19S subunit between high- and low-affinity states, alternately binding and releasing substrates. The subsequent conformational changes associated with ATPase activity lead to gating of the proteolytic channel, substrate unfolding, and threading of the linearized substrate through the channel into the catalytic lumen of the 20S core particle.

Protein unfolding is required for proteolysis because the estimated inner diameter of the 20S subunit lumen is too narrow for proteins to enter in their native state. ATP hydrolysis alters 19S subunit affinities for substrate proteins, thus stabilizing them in an unfolded state. The location of the proteasome base points to its role in promoting substrate unfolding. The base, with its six ATPases, covers the entry ports into the 20S core particle. The archaea proteasome base homolog, PAN, unfolds proteins before translocation to the 20S subunit (205, 318). Similar to the 19S regulatory particle, the 20S subunit also binds unfolded proteins in an ATP-dependent manner, releasing them in vitro (30, 268). Thus, substrates are translocated from one domain within the proteasome (19S) to another (20S) in an ATP-dependent manner.

2.4.4 Gating

The N-termini of α-subunits in the α-rings obstruct access to the proteolytic chamber of the 20S core particle (see Fig. 1), indicating that the proteasome channel is gated (99, 100). The N-terminus of α3 is distinct from the other α-subunits in that it points directly across the α-ring surface toward the center of the lumen contacting every other α-subunit. For substrates to enter and exit the proteolytic core, the blocking N-terminal residues of these α-subunits must be altered. The formation of the well-defined closed configuration of the gate requires dramatic changes in the structure of the core particle. A possible mechanism for gating activation is the attachment of the 19S regulatory particle to the core particle. Attachment of the 19S regulatory particle to the α-ring forms competing interactions with tyrosine, aspartate, or arginine residues in the α-subunits themselves. This realignment then allows the passage of unfolded substrates into the core particle. It has been shown that the Rpt2-α3 (one of the six AAA ATPases and one of the 19S α-subunits, respectively) pairing is involved in the gating of the channel into the core particle. However, it is possible that gating is controlled by more than one Rpt-α interaction (60, 85, 99, 155, 237).

2.4.5 Proteolysis

Typically, the proteasome cleaves protein substrates into peptides from 3 to 23 amino acids in length (149, 213). This process is restrictive such that a protein is hydrolyzed within the proteasome to its final products before the next substrate enters, leading to a pattern of generated peptides from a specific substrate that is stable over time (148, 210). The resulting peptide products are likely rapidly hydrolyzed by downstream proteases and aminopeptidases.

The 20S core particle has the ability to cleave peptide bonds after any amino acid, but each of the three active site-containing β-subunits preferentially cleaves after specific amino acids. The β1-subunit has been shown to cleave after acidic or small hydrophobic amino acids, β2 cleaves after small hydrophobic or basic amino acids, while β5 hydrolyzes the peptide bond after hydrophobic residues (64). Interestingly, the 20S core particle and the proteasome holoenzyme can generate different patterns of cleavage products from the same target protein (76, 149). This implies that the 19S regulatory particle can affect the behavior of the core particle.

2.4.6 Deubiquitination

Free ubiquitin is also a product of the proteasome. After proteolysis, ubiquitin, or ubiquitin attached to residual peptide chains, is released from the proteasome and recycled back into the ubiquitin pathway (Fig. 1, Table 1) (116, 270). In order to dismantle polyubiquitin chains, a deubiquitinating enzyme must work in conjunction with the 19S regulatory particle to replenish the monoubiquitin pool. The 19S regulatory particle from various sources has been shown to possess ubiquitin hydrolase activity leading to editing or complete removal of ubiquitin from protein substrates (78, 160, 162, 167). The 19S regulatory particle p37a is thought to be responsible for this deubiquitinating activity in *Drosophila*. Doa4 has been suggested as the deubiquitinating subunit of the 19S regulatory particle in *Saccharomyces cerevisiae* (218), and the 19S lid subunit Rpn11/POH1 has been shown to have deubiquitinating activity in mammals (228). The deubiquitinating enzyme USP14 (the mammalian homolog to yeast Ubp6p) has also been shown to associate with mammalian proteasomes (26).

2.5 Protein Domains That Bind or Mimic Ubiquitin

Various adaptor proteins that regulate protein degradation and trafficking possess protein domains that bind or mimic ubiquitin (110, 146). These domains include ubiquitin-associated (UBA), ubiquitin-like (UBL), and ubiquitin E2-variant (UEV) domains, as well as ubiquitin-interacting motifs (UIM), and CUE domains. The structure of UPS adaptor proteins points to the general function of these domains. For example, the DNA repair protein Rad23 in yeast contains two UBA domains and an N-terminal UBL domain (232). This protein arrangement indicates that Rad23 has the ability to bind polyubiquitinated proteins via its UBA domain while simultaneously docking with the proteasome through its UBL domain.

A number of proteins associated with the endocytic pathway have UIM domains including Eps15, Eps15R, epsins, and Hrs (151, 229, 233, 253). Eps15 and Hrs form protein complexes that mediate sorting of ubiquitinated membrane cargo in the lumen of endocytic vesicles that are destined for lysosomal degradation. The Hrs-UIM has recently been shown to simultaneously bind two ubiquitin molecules (111). This double ubiquitin-interacting motif (DUIM)-binding ability is thought to make Hrs-ubiquitin interaction more efficient, assisting in endosomal sorting. As another example, Tsg101 (the mammalian homolog to yeast Vps23), a component of the ESCRT-I complex, contains a UEV domain that participates in sorting ubiquitinated cargo in late endocytic compartments (further discussed in the next section). On the other hand, CUE domains are structurally related to UBA domains. The ubiquitin-binding CUE domain of the yeast protein Vps9 and Vps9 domain-containing proteins in mammals have conserved functions in intracellular receptor trafficking through the endocytic pathway (40). These proteins are discussed in further detail in the following sections.

2.6 Ubiquitination in the Endocytic Pathway

In all eukaryotic cells, a highly organized network of endosomes facilitates protein trafficking between the Golgi apparatus, lysosomes, and the plasma membrane. Ubiquitination and ubiquitin adaptor proteins regulate cargo endocytosis from the plasma membrane and subsequent sorting in the endocytic pathway (37, 132, 151, 192, 229, 233, 253, 261, 298). Often, the addition of ubiquitin itself can trigger surface protein internalization of a variety of transmembrane proteins, including ion channels, G protein-coupled receptors, and receptor tyrosine kinases (RTKs) (109, 195, 252, 264).

The epidermal growth factor receptor (EGFR) is an extensively studied model of cell signaling and receptor internalization through the endocytic pathway. EGFRs are RTKs involved in controlling the proliferation, differentiation, migration, adhesion, and survival of numerous mammalian cell types including neurons. EGF binding induces the dimerization of EGFRs which form docking sites for Src-homology-2 (SH2) domain effector proteins (Fig. 3). Upon ligand binding, the internalization of EGFRs is mediated by the SH2 domain-containing E3 RING ubiquitin ligase Cbl. At the cell surface, activated, dimerized EGFRs enter clathrin-coated pits where the multiubiquitinated receptor-Cbl complex associates with Cbl-interacting protein of 85 kDa (CIN85) and endophilins (Fig. 3). At this point, EGFRs autophosphorylate and induce the monoubiquitination of EGFR-pathway substrate-15 (Eps15) and epsin, the latter of which interacts with the plasma membrane. As indicated in the previous section, Eps15 and epsin contain UIMs. These domains promote interactions between EGFRs and Eps15/epsin via ubiquitin binding. Interestingly, monoubiquitinated epsin loses its ability to interact with AP2 and clathrin indicating that ubiquitination may promote the release of internalized EGFRs from clathrin adaptor proteins (187).

On early endosomes, the hepatocyte growth factor-regulated tyrosine kinase substrate (Hrs) binds to phosphatidylinositol-3-phosphate (PtdIns3P) where it assembles in a ternary complex with signal transduction adaptor molecule (STAM) and Eps15 (Fig. 3) (11). Once internalized into early endosomes, EGFRs are either recycled back to the cell surface or sorted to multivesicular endosomes (MVE)/multivesicular bodies (MVB) where Hrs interacts with Tsg101 (Fig. 3), a component of the ESCRT-I complex. The mammalian ESCRT-I complex, consisting of Tsg101, Vps28, and one of four isotypes of Vps37(A-D), binds to ubiquitin through the N-terminal UEV domain of Tsg101 increasing the efficiency of ubiquitinated cargo sorting (19).

The next step in MVE/MVB sorting includes the association of ESCRT-II with vesicular cargo (Fig. 3). ESCRT-II consists of Vps22, Vps25, and Vps36. The N-terminus of Vps36 has been termed a GLUE (GRAM-like ubiquitin binding in EAP45) domain which binds ubiquitin in vitro (257). In yeast, ESCRT-II binds ESCRT-I with high affinity through an interaction with the Vps36 GLUE domain and the Vps28 subunit of ESCRT-I (92). In yeast and *Drosophila*, deletion of ESCRT-II components results in missorting of vacuolar hydrolases and surface accumulation of Notch, respectively (273). In mammalian cells, siRNA-mediated depletion of ESCRT-II inhibits the downregulation of specific cell surface receptors (163).

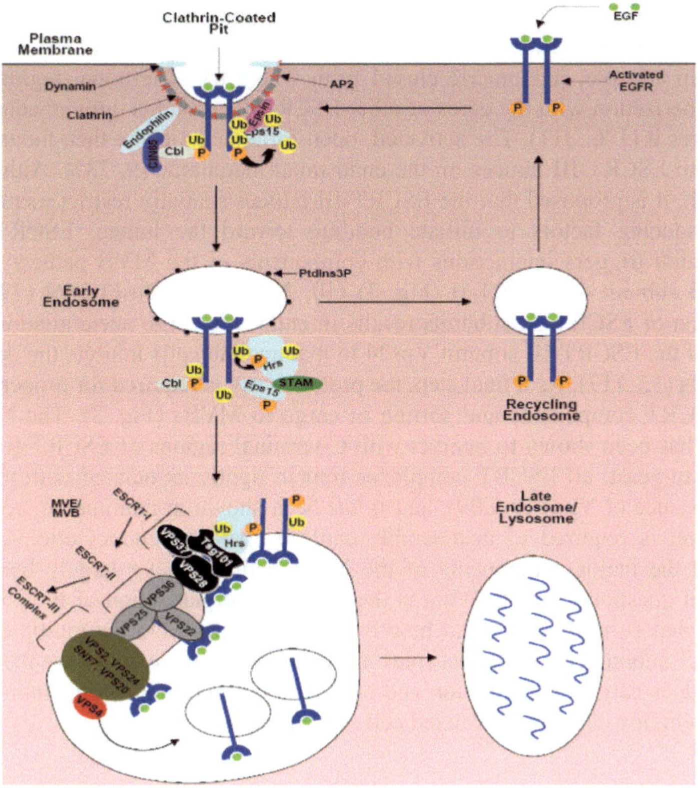

Fig. 3. Ubiquitin in receptor trafficking.

The EGF receptor is a model for ubiquitin-dependent receptor trafficking. Upon binding ligand, EGFRs dimerize and autophosphorylate their tyrosine residues. Activated EGFRs then enter clathrin-coated pits where the phosphorylated receptors recruit the E3 ubiquitin ligase Cbl, which multiubiquitinates the EGFR leading to an association with the Cbl-interacting protein of 85 kDa (CIN85) and endophilins. The clathrin adaptors epsin and Eps15 form a complex and bind to monoubiquitinated EGFRs through their ubiquitin-interacting motifs (UIMs). Within early endosomes, Hrs binds to phosphatidylinositol-3-phosphate (PtdIns3P) forming a multiprotein complex with signal transduction adaptor molecule (STAM) and Eps15 which in turn interacts with EGFRs. Hrs and STAM bind ubiquitin directly through their UIM domains, and Hrs undergoes tyrosine phosphorylation after receptor tyrosine kinase activation. In early endosomes, EGFRs can be either recycled back to the cell surface or directed to the multivesicular endosome/multivesicular body (MVE/MVB) for degradation. At the MVE/MVB, Hrs binds to Tsg101, a UIM-containing component of the ESCRT-I complex. This interaction shuttles endocytosed EGFRs to ESCRT-II and ESCRT-III complexes and then into the intralumenal vesicles of MVEs/MVBs. Following invagination into MVEs/MVBs, EGFRs are then degraded by delivery to the lysosome.

The yeast ESCRT-III complex consists of Vps2, Vps24, Snf7, and Vps20 which form two subcomplexes (Fig. 3) (9). Studies have suggested that ESCRT-III proteins exist in a stable, monomeric closed form where the C-terminal region blocks heterodimerization with the cores of other ESCRT-III subunits until an activation signal removes it (176, 311). The activated, open-form subunits are then thought to assemble into ESCRT-III lattices on the endosomal membrane (9, 287). Although yet unproven, it is proposed that the ESCRT-III lattices spatially restrict membrane curvature-inducing factors to initiate budding toward the lumen. ESCRT-III subunit activation triggers interactions with components of the MVB pathway such as the Vps25 subunit of ESCRT-II (Fig. 3) (10, 272) and PtdIns(3,5)P2 (294). In yeast, deletion of ESCRT-III subunits results in endocytic cargo accumulation, and depletion of the ESCRT-III subunit Vps24 in mammalian cells inhibits the degradation of EGFRs (12, 117). As a final step, the protein Vps4 is required for proper disassembly of ESCRT components and sorting of cargo to MVBs (Fig. 3). The N-terminus of Vps4 has been shown to interact with C-terminal regions of ESCRT-III proteins (9, 176). In yeast, all ESCRT complexes remain tightly associated with membranes in the absence of Vps4 (9, 209), and it has been shown in mammalian cells that Vps4 function is required to disassemble clathrin coats on endocytotic vesicles (239). While the major components of the MVB pathway have largely been identified, several questions remain. What is the molecular organization of the ESCRT-III lattice, what is its function, and how is it disassembled? Do the multiple mammalian ESCRT subunits have distinct roles in the endocytic pathway or are they redundant? How does cargo ubiquitination and endosomal sorting regulate membrane composition in highly compartmentalized cells such as neurons?

3 The Ubiquitin–Proteasome System in Synapse Development

3.1 Axon Growth, Steering, and Pruning

Ubiquitin-dependent protein trafficking has been implicated in early axonal growth cone migration. Two of the better characterized proteins in axon guidance are the roundabout (Robo) and DCC/Frazzled (Fra) receptors that respond to extracellular migration signals. DCC/FRA is an attractant receptor that responds to the secreted protein Netrin-1, while the repellent receptor Robo responds to secreted Slit protein ligand (32, 61). Growth cone surface expression of either DCC/FRA or Robo is involved in proper steering of the growth cone. During early growth cone development, Robo expression is controlled by an association with Commisureless (Comm), preventing Robo cell surface expression by facilitating Robo targeting to the lysosome for degradation (89, 140). At this stage, the absence of Robo ensures that the growth cone will continue migrating toward the midline. The HECT domain E3 ligase Nedd4 recognizes Comm and the subsequent ubiquitination of Comm/Robo regulates its delivery to lysosomes for degradation (Table 1) (203). Once the growth cone has crossed the midline, transcriptional regulation mechanisms diminish Comm expression leading to Robo expression on the surface of the growth cone (140). In contrast to Robo, DCC/FRA expression on the surface of the growth cone leads to steering toward the midline. DCC/FRA is polyubiquitinated by the RING domain E3 ligase Siah and subsequently degraded (Table 1) (118). Robo, Comm, and DCC/Fra have also been implicated in dendritic growth, indicating that ubiquitin-dependent regulation of guidance molecule receptors is not restricted to axons (87).

Interestingly, as Robo signaling directs the growth cone toward its synaptic target, formation of the synapse requires an elimination of Comm surface expression (303). Without Comm removal, synaptogenesis fails. Comm localizes to intracellular vesicles in muscle and endocytosis of Comm requires binding to dNedd4 via its two PY motifs and accompanying ubiquitination of Comm (125). Comm has also been shown to regulate the delivery of Robo to the growth cone (141). Muscle surface accumulation of Comm and innervation defects are observed when dNedd4 is knocked down by RNAi in the muscle or in muscles overexpressing catalytically inactive dNedd4. Expression of Comm mutants fused to a single ubiquitin which cannot be polyubiquitinated prevents the defects in both Comm endocytosis and synaptogenesis, suggesting that monoubiquitination is sufficient for Comm endocytosis in muscles (125). Expression of Comm mutants after synaptic innervation has no effect.

Another regulator of axon growth is the anaphase-promoting complex (APC) (Table 1). The APC is a multimeric ubiquitin ligase complex that is well known to mediate progression through the cell cycle (222, 223). In the brain, the APC immunopurifies as a catalytically active ubiquitin ligase from brain extracts and is present in postmitotic neurons (91). Knocking down the APC co-activator Cdh1 in dissociated cerebellar granule neurons using RNA interference (RNAi) increased axonal growth rates while leaving dendritic growth unaltered (156). Dominant negative forms of the catalytic APC11 subunit (172, 271), as well as the APC inhibitor Emi1 (234, 235), likewise increase axonal growth rates. Similarly, Cdh1-depleted neurons in vitro grow over myelinated substrates and exhibit random growth patterns when plated on cerebellar slices, while in vivo, the parallel fibers of cerebellar granule neurons transfected with Cdh1-specific RNAi defasciculate and move away from normal anatomic pathways (156).

The APC^{Cdh1} also targets inhibitor of DNA binding 2 (Id2) and the transcriptional corepressor SnoN for degradation (Fig. 4) (164, 265). Id2 is a member of the Id family of proteins linked to cell proliferation and tumor progression. Id2 inhibits basic helix–loop–helix (bHLH) transcription factors that limit axonal growth. Activation of APC^{Cdh1} targets Id2 for degradation through a destruction box motif (D box) motif conserved in Id1 and Id4. Id2 D box mutants, which are resistant to APC^{Cdh1} degradation, enhance axonal growth in cerebellar granule neurons and overcome the myelin inhibitory signals for growth (164). Conversely, bHLH activation induces genes with strong axonal inhibitory functions including the Nogo receptor, a known effector of myelin inhibition (164). The APC^{Cdh1} has also been shown to operate in the nucleus and inhibit axonal growth and patterning in the mammalian brain (265). The APC^{Cdh1} has been shown to physically interact with SnoN initiating the ubiquitin-dependent proteasomal degradation of SnoN in rat cerebellar granule neurons in vitro. RNAi knockdown of SnoN reduces axonal growth and suppresses APC^{Cdh1} RNAi-enhanced axon outgrowth (265). Furthermore, in vivo knockdown of SnoN in developing rat cerebellum significantly impairs the development of granule neuron parallel fibers by inhibiting their growth and/or stabilization (265). These findings point to the APC^{Cdh1} as an important mediator in the control of axonal morphogenesis.

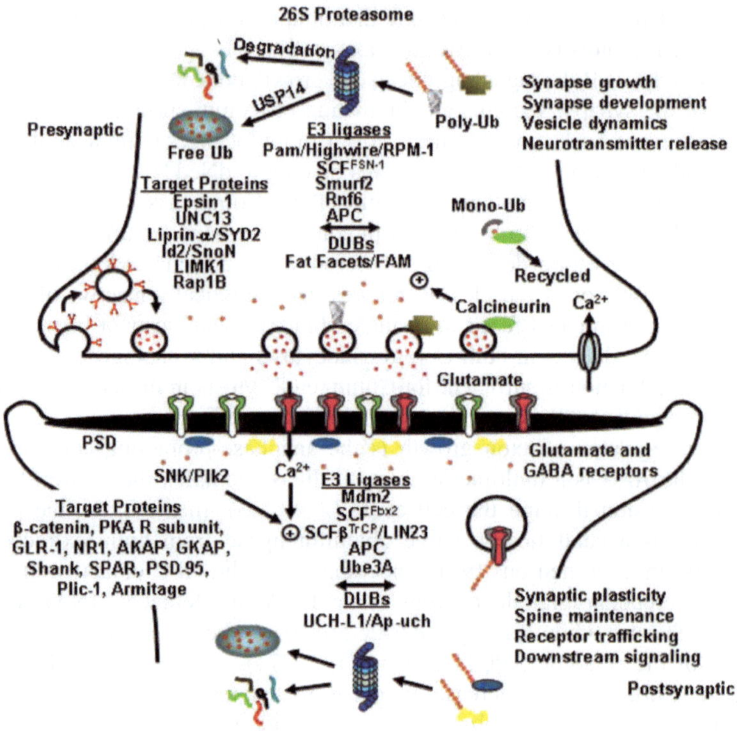

Fig. 4. The UPS regulates many aspects of synaptic physiology.
Presynaptic monoubiquitination (Mono-Ub) modulates endocytosis and interactions of endocytic adaptors. Polyubiquitination (Poly-Ub) and proteasomal degradation alter the abundance of synaptic proteins. Presynaptic effects include (1) the regulation of neurotransmitter release by proteolysis of Munc-13, (2) regulation of vesicle dynamics by Ca^{2+} and calcineurin-dependent modulation of fat facets/FAM activity on epsin 1, (3) regulation of synapse growth and development by Highwire/RPM-1, FSN-1, and the deubiquitinating enzyme (DUB) fat facets/FAM, and (4) APC^{Cdh1} regulation of the axonal transcription factor Id2. Usp14 helps maintain the cellular pool of free Ub monomers by recycling them after protein degradation. The UPS also has profound influences on PSD proteins. These include (1) spine elimination through UPS-dependent degradation of the actin-organizing molecule SPAR, (2) activity-dependent remodeling of the PSD through degradation of PSD scaffolds AKAP79/150 (AKAP), GKAP, and Shank, (3) modulation of downstream signaling by $SCF^{\beta TrCP}$-dependent degradation of β-catenin, (4) regulation of glutamate receptor abundance by the APC and SCF^{Fbx2}, (5) degradation of PSD-95 by Mdm2, and (6) regulation of CaMKII mRNA by the RISC protein armitage. See text for details.

Another ubiquitin ligase involved in axon growth and targeting is LIN-23 (Table 1). LIN-23 is part of the F-box family of proteins, which recognize substrates as part of SCF ubiquitin ligase complexes (224, 256). In *Caenorhabditis elegans*, SCF^{LIN-23} mutants exhibit excessive embryonic cell proliferation (145), while animals with a

P610S point mutation in the C-terminus of LIN-23 exhibit axonal sprouting defects in sensory neurons and interneurons as well as abnormal motor neuron axon outgrowth, but do not demonstrate alterations in cell division (193). These results show that LIN-23 activity is important for multiple stages of nervous system development.

Further evidence from Xenopus retinal ganglion neurons demonstrates that the UPS is important for growth cone steering. Isolated growth cones from Xenopus retinal ganglion neurons are highly enriched with ubiquitin, proteasomes, and E1 enzymes. In addition, pharmacological blockade of the proteasome leads to a decrease in growth cone sensitivity to extracellular guidance molecules (39, 282). These findings show that the presence of a functional UPS is essential for proper axonal growth and guidance.

Further supporting a role for the UPS in axon guidance is the finding that Rnf6, a RING domain ubiquitin ligase, targets LIM kinase 1 in mammalian neurons (Table 1, Fig. 4) (278). LIM kinase 1 is a Rho GTPase effector that modulates cellular morphology via phosphorylation of the actin depolymerizing factor cofilin (7, 306). Cofilin phosphorylation disrupts its association with actin, thereby increasing filamentous actin in the cell. Proteasomal inhibition in hippocampal neurons leads to increased levels of LIM kinase 1 while Rnf6, which polyubiquitinates LIM kinase 1 in vitro, has been shown to coimmunoprecipitate along with LIM kinase 1 (278). Rnf6 localizes to growth cones, and overexpression of an Rnf6 mutant lacking its RING domain and RNAi knockdown of Rnf6 both increase axon growth (278). In contrast, wild-type Rnf6 overexpression suppresses axon growth (278). It is widely accepted that local actin polymerization and depolymerization are important contributors to axon formation and growth (29), and thus localized proteasomal activity may be involved in modulating actin cytoskeletal changes that induce directional extension of neurites.

The UPS also influences neuronal differentiation by modulating the levels and activity of small GTPases such as Rap1B and Rho family members. Sequential Rap1B-Cdc42 activity is needed for establishment of hippocampal neuron polarity. Furthermore, Rap1B is initially present at the tips of all neurites of unpolarized early stage 2 neurons, but is then later restricted to a single late stage 2 neurite (249). The confinement of Rap1B is required to establish neuronal polarity, and recent evidence points to selective degradation of Rap1B by the UPS as a mechanism to discriminate which neurite becomes an axon. The HECT domain E3 ligase Smurf2 regulates neurite extension and polarity through Rap1B (Table 1, Fig. 4). Smurf2 ubiquitinates inactive Rap1B, initiating its degradation via the UPS. In cultured, stage 3 rat hippocampal neurons, overexpression of wild-type Smurf2 leads to a loss of neuronal polarity while overexpression of a catalytically inactive Smurf2 mutant results in neurons with supernumerary axons and a significant reduction in minor neurites (248). Thus, the E3 ligase Smurf2 modulates axonal outgrowth by ubiquitinating Rap1B, initiating its UPS degradation. The degradation of Rap1B then allows the restriction of axonal outgrowth to a single neurite. Intriguingly, in non-neuronal cells, Par6 phosphorylation recruits Smurf1 which in turn degrades RhoA to locally regulate cell polarity, migration, and motility (214, 240, 292). Similar mechanisms have been shown to control neurite outgrowth (34).

The UPS is also involved in axonal degeneration. Axonal degeneration is a common pathological characteristic of many neurodegenerative diseases. In studies of rat transected axons, termed Wallerian degeneration, pharmacological and genetic inhibition of the UPS markedly delays axon degeneration both in vitro and in vivo (312). Also, similar to transected axons, degeneration of axons from nerve growth factor (NGF)-deprived sympathetic neurons is also suppressed by proteasome inhibitors (312). In a further study of both developmental and Wallerian axonal degeneration, Wlds mutant mice, which overexpress a chimeric protein consisting of the first 70 amino acids of the UPS protein UFD2/E4 and the full-length nicotinamide mononucleotide adenylyltransferase (Nmnat), an enzyme that facilitates NAD synthesis, protects against degeneration of injury-induced retinal ganglion cells (115). In contrast, overexpression of the Wlds protein does not effect the naturally occurring degeneration of these same axonal projections during the identical developmental time period (115). These data indicate that naturally occurring developmental degenerative axonal pruning and injury-induced axon degeneration differ significantly in early steps, but may ultimately converge onto a common pathway which may involve the UPS.

3.2 The UPS in Synapse Formation and Elimination

3.2.1 Pam/Highwire/RPM-1

One of the first links discovered between the UPS and synapses was the isolation of the *Drosophila* mutant highwire (hiw). The hiw mutant results from a disruption in the function of a RING domain E3 ligase (290, 304), resulting in a variety of morphological and signaling defects at neuromuscular synapses (Table 1, Fig. 5) (304). Specifically, highwire loss-of-function mutations result in excessive synaptic bouton formation and reductions in both quantal size and content (290, 304). Interestingly, hiw loss-of-function mutants are phenocopied by the gain of function in a deubiquitinating enzyme called fat facets (Table 1, Fig. 4) (37, 63, 121). Such a phenotypic similarity shows that a proper balance between ubiquitination and deubiquitination events is critical for synaptic development.

Interestingly, the conserved Pam/Highwire/RPM-1 class of proteins, consisting of *Drosophila* highwire, its *C. elegans* ortholog RPM-1, and the mammalian orthologs Phr1 and Pam are all involved in synapse development (36, 101, 244, 290, 314). As discussed above, highwire mutants in *Drosophila* display normal neuromuscular junction motor axon targeting and synaptogenesis, but synapse growth is excessive with increases in both the number of boutons and the overall length of axonal branches (Fig. 5) (290). Similarly, RPM-1 loss-of-function *C. elegans* mutants display retracted synaptic branches, ectopically extended axons, and altered NMJ phenotypes including motor neuron overgrowth which often manifests as a single large presynaptic terminal with multiple active zones (Fig. 5) (244, 314). Moreover, mice lacking the RPM-1 ortholog Phr1 exhibit incomplete diaphragm innervation by the phrenic nerve, and at a more detailed level, nerve terminal morphologies are disrupted as NMJs show excessive postsynaptic sprouting of nerve terminals (36).

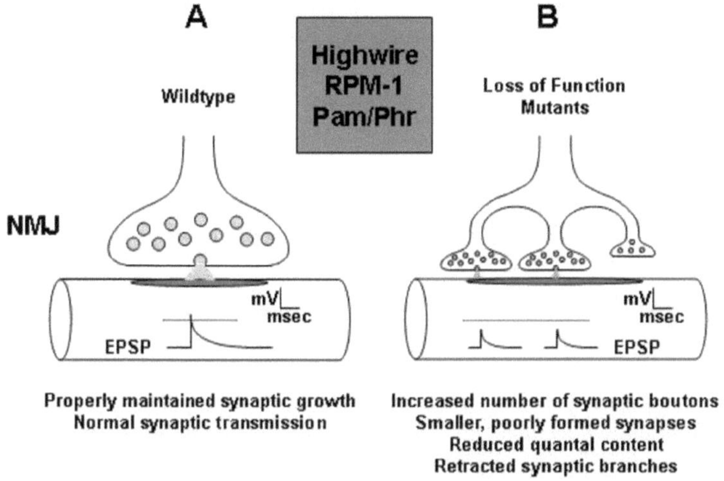

Fig. 5. The Pam/Highwire/RPM-1 (PHR) family of ubiquitin ligases regulates synapse formation and function. In flies (highwire), worms (RPM-1), and mammals (Pam/Phr), PHR family ubiquitin ligases control axonal sprouting, synapse morphology, and maintenance of synaptic transmission. Loss-of-function mutations in the Highwire/RPM-1/Pam E3 ligase result in synaptic defects including excessive synaptic bouton formation and reductions in both quantal size and content, retracted synaptic branches, ectopically extended axons, and excessive post-synaptic accumulation of nerve terminals. Phenotypes vary between worms, flies, and mammals, but have been best studied at the *Drosophila* neuromuscular synapse, and mechanosensory and motor synapses in *C. elegans*.

In addition to the conserved function of the Pam/Highwire/RPM-1 class of proteins, the signaling components of the highwire pathway are also conserved. A genetic screen in *Drosophila* for modulators of the highwire phenotype identified mutations in wallenda, a mitogen-activated protein (MAP) kinase kinase kinase (MAPK.KK) of the p38 pathway (52).

While mutations in highwire lead to neuromuscular synapse defects, this phenotype is suppressed by loss-of-function mutations in wallenda, which is consistent with the idea that wallenda is negatively regulated by highwire (52). In *C. elegans*, screens for suppressors of the RPM-1 loss-of-function phenotype identified the wallenda homolog DLK-1 (204). DLK-1 loss-of-function mutants show no significant changes in synaptic growth, but DLK-1 loss of function in an RPM-1 mutant background suppresses the RPM-1 mutant synaptic overgrowth phenotype (204). Although the complete pathway remains to be elucidated, these data indicate that Highwire/RPM-1 modulates a key signaling pathway involving the p38 MAPK pathway that promotes synaptic growth. Intriguingly, the T10H9.2 receptor tyrosine kinase, a homolog of the mammalian brain-specific anaplastic lymphoma kinase (127), is regulated by RPM-1 and the F-box protein FSN-1 in *C. elegans* (174, 204). RPM-1 and FSN-1 loss-of-function phenotypes exhibit varying degrees of under-

and overdeveloped synapses including presynaptic densities with single, presynaptic vesicle clusters or undifferentiated synaptic varicosities with single clusters of synaptic vesicles (Fig. 3, right) (244, 314). RPM-1 is part of a cullin-RING ubiquitin ligase complex where the F-box component, such as FSN-1, is interchangeable and determines substrate specificity (Table 1) (5). The multiple effects of RPM-1/FSN-1 loss-of-function mutants may indicate the existence of synapse-specific cullin-RING complexes that direct the growth and maturation of distinct synapses by targeting discrete substrates, such as DLK-1, in certain synaptic populations.

3.2.2 Anaphase-Promoting Complex (APC)

In addition to its role in axonal outgrowth, the APC also promotes differentiation of presynaptic boutons and formation of functional synapses at *Drosophila* neuromuscular synapses (Table 1). Mutations in components of the APC lead to alterations in neuromuscular synapse development. Flies expressing a truncated form of the cullin domain APC component protein APC2 survive to early pupal stages and display expanded neuromuscular synapse size (280). This augmentation in synaptic size results from a twofold increase in the number of presynaptic terminals per endplate without an increase in the size of individual boutons (280). These malformations correspond to enhanced spontaneous and evoked excitatory postsynaptic currents and an increase in postsynaptic clustering of glutamate receptors. Mutations in APC subunits in *C. elegans* increase the abundance of GLR-1 in the ventral nerve cord resulting in locomotor defects consistent with increased synaptic strength (134). Mutations that disrupt clathrin-mediated endocytosis inhibit the effects of the APC mutations, indicating that the APC regulates GLR-1 endocytosis or endocytic recycling (134).

Another putative neuronal substrate for the APC, liprin-α, was identified through database homology screens using the conserved APC destruction box motif RXXLXXXXN as a template (Fig. 4) (95, 144, 280). Liprin-α has been shown to affect NMJ bouton numbers in flies (139) while its *C. elegans* ortholog, SYD2, regulates synapse size and function (315). Furthermore, the multi-PDZ domain protein GRIP/ABP binds to liprin-α and, through an association with LAR receptor protein tyrosine phosphatases (LAR-RPTPs), clusters AMPA receptors at mammalian excitatory synapses (Table 1) (305). Loss-of-function APCmr3 (the *Drosophila* ortholog of APC2) mutant flies show a 33% increase in presynaptic levels of liprin-α, while a simultaneous expression of loss function liprin-α rescues the APCmr3 phenotype (280). Furthermore, immunoprecipitation experiments have shown that liprin-α is in a monoubiquitinated form in vivo (280). These data indicate that the APC can modulate liprin-α activity during synapse development.

Recent studies at the mammalian neuromuscular junction have identified a PDZ domain containing RING finger 3 (PDZRN3) as a synapse-associated E3 ubiquitin ligase that regulates the surface expression of the muscle-specific receptor tyrosine kinase (MuSK) (Table 1), the key organizer of postsynaptic NMJ development (181). PDZRN3 binds to MuSK and promotes its ubiquitination, leading to endocytic removal of MuSK from the muscle membrane. Gain- and loss-of-function studies in cultured myotubes and transgenic mice show that downregulation of MuSK by

PDZRN3 plays an important role in MuSK-mediated nicotinic acetylcholine receptor clustering and synapse maturation.

Similar to synapse formation, the UPS also regulates synapse elimination. The spine-associated Rap GTPase-activating protein (SPAR) (Fig. 4), an actin-binding positive regulator of spine growth that localizes to the PSD via its interaction with postsynaptic density protein-95 (PSD-95) (217), undergoes proteasomal degradation after phosphorylation by serum-inducible kinase (SNK, Table 1), also known as polo-like kinase 2 (Plk2) (216). SPAR overexpression creates enlarged spines with increased PSD-95 content in cultured hippocampal neurons. Conversely, SNK/Plk2 overexpression with simultaneous phosphorylation of SPAR eliminates spines (216, 217). SNK and SPAR are thus thought to play a role in homeostatic synaptic plasticity by dampening synaptic signaling during chronic increases in activity (see chapter by Turrigiano, this issue).

The UPS is also involved in synapse elimination in *C. elegans*. In adult worms, the hermaphrodite-specific motor neuron (HSNL) interfaces with its vulval muscle and VC motor neuron targets via a cluster of synapses in the primary synapse region (PSR). Experiments have shown that stereotypical developmental elimination of HSNL synapses is mediated by an SCF E3 ligase complex consisting of SKR-1 and the F-box protein SEL-10 (65). RNAi knockdown of SKR-1 and another obligatory component of the SCF complex CUL-1 resulted in a marked decrease in synapse elimination in the PSR over control animals (65). Furthermore, the immunoglobulin superfamily synaptic adhesion molecule SYG-1 (NEPH1 and IrreC in mouse and *Drosophila*, respectively), which is exclusively expressed in the PSR during developmental stages, binds to SKR-1 (65) and inhibits SCF complex assembly. This inhibition results in protection of local synapses in the PSR (65). These data indicate that the localization and regulation of ubiquitin-mediated protein degradation can contribute to maintaining detailed synaptic connectivity via selective synapse elimination.

3.2.3 Dendritic Pruning

In addition to the elimination of individual synapses, entire dendrite segments undergo developmental retraction and loss in a manner regulated by the UPS. *Drosophila* larvae class IV dendritic arborization (C4da) neurons undergo extensive dendritic pruning during metamorphosis and this pruning is regulated by UPS activity (158). Specifically, mutations in the *Drosophila* E1 enzyme uba1 and the proteasome subunit mov34 prevent pruning of C4da neuron larval dendrites during metamorphosis (158). Subsequent genetic screens identified the E2-conjugating enzyme UbcD1 and the RING E3 ligase *Drosophila* inhibitor of apoptosis protein 1 (DIAP1) as key components in the pruning of C4da dendrites (Table 1) (159). The antiapoptotic protein DIAP1 protects neurons from apoptosis by downregulating Dronc caspase activity by targeting Dronc for proteasomal degradation (201, 299). Loss of DIAP1 leads to increased Dronc activity resulting in caspase activation and cleavage of proximal dendrites in C4da neurons during metamorphosis. C4da neurons lacking Dronc have normal dendrites at early larval stages but fail to cleave their larval dendrites during metamorphosis (159). Mutations in UbcD1 also lead to retention of intact, non-severed larval dendrites 18 h after puparium formation (APF), whereas wild-type C4da neurons lose dendrites over this same period (159). Furthermore, gain-of-function DIAP1 mutants hinder larval neuron dendrite pruning. These experiments emphasize the importance of the UPS in large-scale remodeling of dendrites and postsynaptic contacts during development.

4 The UPS in Synaptic Function

4.1 UPS-Dependent Mechanisms of Presynaptic Regulation

Recent studies indicate that dynamic cycles of ubiquitination and deubiquitination in presynaptic terminals regulate neurotransmitter release. In isolated synaptosomes, acute depolarization causes a rapid decrease in ubiquitin-immunopositive proteins (44). This loss of ubiquitin conjugates is Ca^{2+} dependent and is blocked by inhibitors of the Ca^{2+}-dependent phosphatase calcineurin/PP2B (44). Furthermore, proteasome inhibitors only partially block the depolarization-induced loss of ubiquitin conjugates, indicating that the loss of ubiquitin immunoreactivity is due principally to deubiquitination rather than proteasomal degradation. Consistent with this notion, RNAi knockdown of FAM, a mammalian ortholog of the DUB fat facets, causes the UIM protein epsin-1 to remain monoubiquitinated after depolarization, in turn altering presynaptic vesicle release dynamics by altering adaptor interactions involving components of the endocytic pathway (Table 1, Fig. 4) (44).

Ubiquitin recycling and regulation of free pools of monomeric ubiquitin also play an important role in regulating neuronal function. The ataxia mouse, which displays severe tremors and undergoes premature death (57, 300), is deficient in the DUB enzyme Usp14 (Table 1), a mammalian homolog of the yeast deubiquitinating enzyme Ubp6 (300). Usp14/Ubp6 directly interacts with the proteasome and functions to remove and recycle ubiquitin molecules from ubiquitinated proteins bound to the proteasome (Fig. 4) (26, 170). Interestingly, wild-type mice have 35% greater monomeric ubiquitin levels in neural tissues compared to ataxia mice (4). This decrease in free ubiquitin seen in ataxia mice can be rescued in the brain by expression of a full-length Usp14 transgene (56). Electrophysiological characteristics of ataxia mice, including defects in hippocampal short-term plasticity as well as increased amplitudes but decreased frequencies of miniature endplate potentials at neuromuscular synapses, suggest that Usp14 defects alter presynaptic function (300).

Protein degradation after ubiquitination also has significant effects on presynaptic function. In cultured hippocampal neurons, proteasome inhibitors increased FM-dye uptake after a 1 min, 90 mM K^+ depolarization, by more than 75% over control, whereas the rate and magnitude of dye release were unaltered (296). This increase in FM-dye uptake is not dependent on protein synthesis but does require neuronal activity as it is blocked by TTX and augmented by elevating action potential frequency through blockade of inhibitory $GABA_A$ receptors (296). These observations indicate that proteasomal degradation at the presynaptic terminal negatively regulates neurotransmitter release during times of increased activity. The relevant ubiquitin ligases that control vesicle cycling in presynaptic terminals are not yet known, but may include members of the ZNRF family of RING domain E3 ligases that associate with synaptic membranes and regulate Ca^{2+}-dependent exocytosis in PC12 cells (6).

Potential presynaptic proteasome substrates include the vesicle priming protein Munc-13, which plays an important role in action potential-evoked neurotransmitter release (13, 33, 260). Pharmacological inhibition of the proteasome or genetic alteration of the proteasome was found to increase the accumulation of Munc-13, resulting in decreased neurotransmitter release (33, 260). Conversely, anisomycin, a protein synthesis inhibitor, had no effect on synaptic currents, demonstrating the specific involvement of protein degradation (260). Interestingly, RIM1, which is a binding partner for Munc-13, is degraded by the synaptic F-box E3 ligase SCRAPPER. Scrapper knockout mice display elevated levels of RIM1 and altered synaptic activity (Table 1) (308).

Ubiquitin-dependent protein trafficking also controls neurotransmitter reuptake into nerve terminals. The HECT domain E3 ligase Nedd4-2 has been shown to regulate dopamine transporter surface expression (see Table 1). The dopamine transporter (DAT) is localized to dopaminergic neurons where it functions to transport extracellular dopamine (DA) into DA neurons, leading to a termination of DA signaling. The dynamic nature of expression of DATs at the neuronal cell surface determines the duration and spatial limitation of dopaminergic neurotransmission. DAT activity, in turn, is regulated by endocytic trafficking. Large-scale RNAi experiments disrupting PKC-dependent endocytosis assays involving HA-tagged DAT (HA-DAT) in cultured rat neurons revealed that Nedd4-2, as well as adaptor proteins such as epsin, Eps15, and Eps15R, are involved in modulating surface expression of DAT. RNAi depletion of Nedd4-2 results in a dramatic reduction of PKC-dependent ubiquitination of DAT (259). Furthermore, endogenous Nedd4-2, epsin, and Eps15 coimmunoprecipitated with endogenous DAT isolated from rat striatum (259) indicating direct interactions between DAT and the endocytic machinery. These data indicate that PKC-induced ubiquitination of DAT is mediated by Nedd4-2, which leads to an interaction of DAT with adaptor proteins in coated pits accelerating DAT endocytosis.

4.2 The Ubiquitin–Proteasome System in Postsynaptic Remodeling and Plasticity

Long-term potentiation (LTP) and long-term depression (LTD) are considered cellular models for learning and memory (22, 23, 177, 185) (see chapter by Lisman and Hell, this volume). Both LTP and LTD are Ca^{2+}-dependent processes requiring the activation of kinase and phosphatase cascades, which lead to synaptic strengthening or weakening, respectively (119, 186, 199, 200). LTP and LTD result in long-lasting alterations in synaptic strength caused by enduring changes in synapse structure and protein composition. Traditionally, changes in macromolecular synthesis, including gene transcription and local protein synthesis, were thought to cause long-lasting synaptic remodeling (142, 143, 267). However, recent evidence points to an important role for ubiquitin-dependent protein degradation in enduring synaptic modification.

4.2.1 The UPS in Synaptic Plasticity

The importance of the UPS in synaptic plasticity was first noted in behavioral conditioning studies of Aplysia (28, 58, 194). During long-term facilitation (LTF), a paradigm similar to LTP, cAMP-dependent protein kinase (PKA) activity continues even after cAMP levels return to baseline concentrations (105). In the absence of cAMP, inactivated PKA exists as a holoenzyme consisting of two catalytic subunits whose activity is suppressed by a dimeric regulatory subunit. As cAMP concentrations increase, four cAMP molecules interact with the regulatory dimer, releasing the constitutively active catalytic subunits (see chapter by Hell and Strack, this volume). When cAMP levels return to baseline, the holoenzyme reforms, inactivating PKA catalytic activity. During LTF, PKA activation is accompanied by UPS degradation of its regulatory subunits, leading to persistent PKA activity without elevated cAMP levels (42, 90, 196, 231). Intriguingly, ubiquitin recycling also increases during LTF further enhancing degradative events. Specifically, the level of Aplysia ubiquitin C-terminal hydrolase Ap-uch, an ortholog of mammalian UCH-L1, is increased following the induction of LTF leading to amplified disassembly of polyubiquitin chains and thus a larger free pool of ubiquitin (Table 1, Fig. 4) (105, 106).

Protein degradation by the UPS is also important for the generation of short-term and long-term memory in concert with protein synthesis. For example, the well-defined odor-shock learning paradigm in *Drosophila* involves an increase in the expression of genes involved in mRNA transport as well as local translation (277). These events are regulated by components of the RNA interference (RISC) pathway. In particular, armitage, a component of the RISC pathway, inversely regulates synaptic CaMKII expression in *Drosophila* in response to synaptic activity. As synaptic activity increases, armitage expression decreases, leading to increased expression of CaMKII and subsequent postsynaptic enhancement (54). Armitage itself is targeted for degradation by the UPS (Fig. 4). In *Drosophila* brain, activity-dependent loss of armitage was prevented by a dominant-negative mutant of the proteasome β-subunit (DTS5) (8). Furthermore, incubation with the proteasome inhibitor lactacystin blocked nicotine-induced loss of GFP:Armi during periods of increased neuronal activity associated with memory (8).

Although it has been well documented that long-term changes in synaptic efficacy rely on protein synthesis, recent evidence indicates an equally important role for protein degradation in establishing late-phase LTP at hippocampal synapses. Both early and late phases of LTP induced by high-frequency stimulation are reduced at Schaeffer collateral CA1 synapses after the application of proteasome inhibitors (83, 136). Interestingly, coapplication of protein synthesis inhibitors along with proteasome inhibitors largely restores late LTP (83), indicating a delicate balance of production and destruction of plasticity-associated proteins. The precise identity of these plasticity-associated proteins remains to be discovered.

While the majority of UPS components and targets involved in LTP/LTD are not yet known, the HECT domain E3 ligase Ube3a/E6-AP has been shown to regulate synaptic plasticity. The involvement of Ube3a in synaptic function was first observed in mice lacking the maternal allele of Ube3a (130). These mice exhibit abnormal motor coordination and defects in spatial learning and hippocampal LTP. These characteristics are thought to be analogous to those of Angelman syndrome, a hereditary neurodevelopmental disorder where the loss of the maternal copy of Ube3a leads to cognitive, behavioral, and neurological deficits. The neuronal substrates of Ube3a are unknown; however, recent evidence indicates that Ube3a regulates the activity CaMKII, a critical plasticity molecule for LTP (see chapter by Lisman and Hell, this volume). Maternally deficient Ube3a mice show increased phosphorylation of threonines 286 and 305 of CaMKII leading to a corresponding reduction in kinase activity. This inhibition is due to the dominant effect of phosphorylation at the inhibitory threonine 305 site and reduced postsynaptic localization (281, 293). One attractive possibility is that Ube3a positively regulates CaMKII activity by degrading a yet to be determined negative regulator of a CaMKII phosphatase. The loss of Ube3a from neurons leads to increased CaMKII phosphorylation at threonine 305 and decreased CaMKII activity with corresponding deficits in synaptic plasticity.

4.2.2 UPS-Dependent Remodeling of the PSD

The PSD is a biochemically complex and persistent structure with a highly organized framework of protein scaffolds (251). Despite its biochemical stability, the PSD undergoes structural changes in response to alterations in synaptic activity (68, 69, 88, 104, 188, 202, 207, 211, 275). Long-term time-lapse imaging has revealed continuous turnover and remodeling of the PSD structural protein PSD-95 at excitatory synapses. In addition, fluorescence photobleaching experiments have revealed significant ongoing exchange of PSD proteins (97, 250, 276). Biochemical pulse-chase experiments have demonstrated that the aggregate turnover rate of PSD-associated proteins is on the order of hours (69). Evidence now indicates that these dynamic molecular changes are at least partially mediated by the UPS.

Biochemically isolated PSD fractions have long been known to be rich in ubiquitin-immunopositive components (43). Recent studies show that altering the amount of neuronal activity causes reciprocal changes in global protein composition of the PSD (Fig. 6) (69). Correlating with these protein compositional changes, decreasing synaptic activity diminishes ubiquitin-conjugated proteins in the PSD and decreases protein turnover, whereas increasing synaptic activity increases ubiquitin conjugation and accelerates protein turnover (Fig. 6) (69). These results indicate that the rate of postsynaptic protein turnover is tightly tuned by synaptic activity during molecular remodeling.

To date, few PSD-associated proteins have been shown to be directly ubiquitinated. However, among the known targets of ubiquitination are major PSD scaffold proteins including Shank, GKAP/SAPAP, AKAP79/150, and PSD-95 (51, 69). In addition, the postsynaptic spine-associated Rap GTPase and actin regulatory protein SPAR undergoes activity and phosphorylation-dependent degradation (216). Furthermore, Vesl-1S/Homer-1a, which is known as a binding partner of Shank and mGluR1/5, is also degraded by the UPS, and its synaptic localization is enhanced by the proteasome inhibitors MG132 and lactacystin (1, 2). Thus, UPS targeting of important structural PSD proteins for degradation may alter the relative ratios of critical protein subtypes at the postsynaptic membrane.

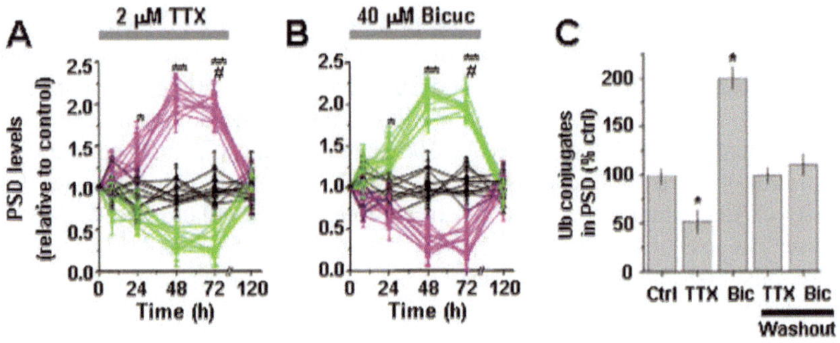

Fig. 6. Postsynaptic protein levels are regulated by activity. Quantitative analysis of selected proteins in PSD fractions isolated from cultured cortical neurons at various times after adding and then removing TTX to block activity (*left*) or bicuculline to enhance excitatory synaptic activity (*right*). Proteins indicated in *magenta* include NR1, NR2B, SAP102, Shank, GKAP, AKAP79/150, spinophilin, PP1, PKA-cat, and PKA-RIIβ. Proteins in *black* include PSD-93, SAP97, NSF, nNOS, SynGAP, α-actinin2, tubulin, PKCβ, and PKCε. Proteins indicated in *green* include NR2A, mGluR1α, PSD-95, Homer, myosin Va, CaMKIIα, CaMKIIβ, CaMKII-T286P, and PKCγ. *Gray bars* indicate duration of drug treatment. (A) Upon activity blockade by addition of TTX the abundance of certain PSD proteins are depressed (*green*), some are unchanged (*black*), and some are elevated (*magenta*). (B) Enhanced activity (Bicuc) results in an inverse pattern of protein abundance in the PSD. (C) Increases in activity (Bicuc) result in a nearly twofold increase in ubiquitin-conjugated proteins in the PSD over control, while reduced activity (TTX) results in a decrease in ubiquitin-conjugated proteins. Data represent means ± SEM of band intensities from purified PSD fractions isolated from cultured cortical neurons normalized to control values from untreated neurons (*$p < 0.05$ for all pairwise comparisons between *red* and *green* proteins; **$p < 0.01$ for all pairwise comparisons between *red*, *green*, and *black* proteins; *t*-test; $n = 3$ to 6 for each time point). Figure adapted with permission from Ehlers (2003).

Activity-dependent changes in synaptic composition correlate with alterations in the distribution of proteasomes in neurons (20). Live imaging of neurons expressing GFP-tagged 19S and 20S subunit components Rpt1 and α4, respectively, showed that these proteins are ubiquitously distributed in dendrites at resting membrane potentials. After strong, continuous KCl-induced depolarization, proteasomes are redistributed into spines (20). This proteasomal spine enrichment is sensitive to NMDA receptor inhibition by AP5, and fluorescence photobleaching experiments indicate that spines retain an increasingly immobile pool of proteasomes in response to activity (20). Correspondingly, elevated neuronal activity increases the detergent-insoluble, actin-associated fraction of neuronal proteasomes (20). These experiments indicate a model where NMDA receptor activation induces local reorganization of the actin cytoskeleton in spines leading to changes in proteasome localization.

4.3 Ubiquitin and Postsynaptic Receptor Trafficking

Forward genetic studies in *C. elegans* have been instrumental in identifying UPS adaptor proteins and ubiquitin ligases involved in postsynaptic glutamate receptor trafficking. While experiments have clearly demonstrated that glutamate receptors in *C. elegans* undergo direct ubiquitination (35), detecting ubiquitination of mammalian glutamate receptors has been more elusive. In this section, we will highlight key experiments involving glutamate receptor trafficking in both *C. elegans* and mammals and point out possible mechanistic differences between the two systems. We will also discuss the UPS-dependent effects on trafficking of $GABA_A$ and nicotinic acetylcholine receptors (nAChRs) in mammals.

4.3.1 Glutamate Receptor Regulation in C. elegans

Ubiquitination of the GLR-1 glutamate receptor at *C. elegans* synapses causes receptor elimination from the postsynaptic membrane through an endocytotic mechanism involving the clathrin adaptor protein AP180 (35). While seemingly unrelated, APC loss-of-function mutants exhibit locomotor deficits that correspond to increases in postsynaptic levels of GLR-1 and synapse size (Table 1) (134). This is similar to the findings that *Drosophila* APC mutants have enlarged synapses, increased postsynaptic glutamate receptor clustering, and deficits in synaptic transmission at the neuromuscular synapse (280). Although GLR-1 has not been shown to be a direct substrate of the APC, the APC loss-of-function phenotype is suppressed by loss of function of the unc-11/AP180 clathrin adaptor, indicating that APC activity is linked to the endocytotic pathway (134). The APC has not been localized to the synapse, although in *Drosophila*, the APC controls presynaptic size via liprin-α (280), which is known to organize the presynaptic active zone and regulate neurotransmitter release (Fig. 4).

In a previous section, we noted that LIN-23 is important in axonal outgrowth (193). Recent genetic experiments have also shown that LIN-23 is a negative regulator of synaptic GLR-1 levels in the ventral nerve cord of *C. elegans* (Table 1) (66). LIN-23 shares sequence similarity to the mammalian F-box protein β-TrCP, a key component of the Wnt signaling pathway. During Wnt signaling, β-catenin binds to the TCF/Lef family of transcription factors, thereby upregulating the expression of Wnt target genes (17, 122). Low levels of Wnt expression cause β-catenin to associate with a destruction complex where β-catenin is phosphorylated by glycogen synthase kinase-3β. Phosphorylated β-catenin is then recognized by β-TrCP and targeted for proteasomal degradation (165, 220, 301, 309). Similar to the function of β-TrCP in mammalian cells, SCF^{LIN-23} targets the *C. elegans* β-catenin ortholog BAR-1 for degradation (66). The effects of loss-of-function mutants of LIN-23 on GLR-1 clustering are rescued by an unc-11/AP180 loss of function mutant, suggesting that SCF^{LIN-23} promotes GLR-1 endocytosis. The receptor clustering phenotype in LIN-23 knockout animals can be partially rescued by loss of function of the TCF/Lef homolog POP-1. This indicates that GLR-1 is also a target gene of Wnt signaling (66) and SCF^{LIN-23} negatively regulates GLR-1 expression through both the endocytic pathway and gene transcription.

Another UPS protein shown to interact with GLR-1 is KEL-8. KEL-8 is a *b*road complex *t*ramtrack *b*ric-a-brac (BTB)-Kelch protein that directly associates with CUL-3, a scaffold component of cullin-RING family E3 ligases (224), imparting substrate specificity to the E3 ligase. KEL-8 localizes to sites adjacent to GLR-1 postsynaptic clusters along ventral cord neurites, and KEL-8 mutant worms show an increased frequency of spontaneous reversal in locomotion indicative of changes in GLR-1 levels at synapses (245). Indeed, mutations in KEL-8 lead to increases in GLR-1 levels at postsynaptic clusters, while the levels and localization of other postsynaptic proteins are unaltered (245). These results show that KEL-8 is involved in ubiquitin-mediated turnover of GLR-1 subunits, and that cullin-3 ubiquitin ligases are required for the turnover of GLR-1 receptors.

4.3.2 The UPS and Glutamate Receptor Regulation at Mammalian Synapses

One of the primary determinants of excitatory synaptic strength is the number of AMPA receptors expressed at the postsynaptic surface (see chapter by Esteban, this volume). AMPA receptors are endocytosed and sorted into various recycling or degradative pathways in an activity-dependent fashion. Under conditions of minimal synaptic activity, AMPA receptors in hippocampal cultures undergo endocytosis in as little as 14 min. However, during periods of increased neuronal activity, AMPA receptor endocytosis is accelerated by threefold (70). AMPA receptor internalization can also be triggered by direct ligand binding or by NMDA or insulin receptor activation (15, 31, 41, 70, 175, 183). While both direct and indirect activation can initiate AMPA receptor internalization, the fate of the endocytosed receptors depends on the nature of the stimuli. Activation of AMPA receptors alone directs receptors to be sorted for degradation while NMDA receptor activation alone leads to AMPA receptor recycling (70, 169).

Unlike *C. elegans* GLR-1 receptors, which undergo direct ubiquitination (35), mammalian AMPA receptors can be internalized indirectly through ubiquitination of the postsynaptic scaffolding protein PSD-95 by the RING domain ubiquitin ligase Mdm2 (Table 1, Fig. 4) (51). In hippocampal neurons, NMDA-dependent PSD-95 ubiquitination is inhibited in the absence of Ca^{2+} or in the presence of the calcineurin inhibitor ascomycin, (51) suggesting that Mdm2 and calcineurin function in the same pathway with calcineurin operating upstream of Mdm2. Furthermore, overexpression of a ubiquitin mutant that allows monoubiquitnation, but not polyubiquitination, prevents AMPA-induced receptor endocytosis (219). While it has been shown that changes in PSD-95 levels can modulate synaptic AMPA receptor content (71-73, 266), PSD-95 ubiquitination is not always detectable (21, 69). This may result from a tight regulation of PSD-95 ubiquitination or de-ubiquitination.

The degradation of NMDA-type glutamate receptors is also regulated by the UPS. The F-box protein Fbx2, an SCF complex component, mediates activity-dependent ubiquitination and degradation of NMDA receptor NR1 subunits by recognizing high mannose glycans in the N-terminal extracellular domain of NR1 (Table 1, Fig. 4) (137). Such high mannose modifications are typically present on proteins trafficking through early stages of the secretory pathway including the endoplasmic reticulum and *cis*-Golgi apparatus (107) (see chapter by Hanus and Ehlers, this volume). In neurons, Fbx2 localizes to dendritic spines, and expression of a dominant-negative form of Fbx2 leads to an increase in surface NMDA receptor expression (137), suggesting that ongoing ubiquitination of NR1 by Fbx2 in intracellular compartments limits or interrupts forward trafficking.

In a manner similar to ubiquitination, posttranslational modification of proteins by the small ubiquitin-like modifier protein (SUMO) alters protein trafficking and function. Although best known for its role in signaling within the nucleus, SUMOylation occurs at the synapse, with one key target being the kainate-type glutamate receptor GluR6 (189). SUMOylation of GluR6 is increased in response to kainate and regulates endocytosis of the receptor. Correspondingly, kainate receptor-mediated postsynaptic currents in hippocampal slices are bidirectionally regulated by SUMOylation and deSUMOylation (189). In addition to SUMOylation, GluR6 is ubiquitinated and binds actinfilin, a synaptic BTB-Kelch protein that acts as an adaptor for a Cul3 E3 ubiquitin ligase complex (242) in a manner surprisingly similar to the *C. elegans* GLR-1 receptor (245). Surface levels of GluR6 vary inversely with levels of actinfilin, suggesting that the actinfilin-Cul3 complex targets kainate receptors for membrane removal and/or degradation (242).

4.3.3 Trafficking of $GABA_A$ and Nicotinic Acetylcholine Receptors

While glutamatergic receptors are the primary determinants of excitatory synaptic strength, $GABA_A$ receptors mediate rapid inhibitory synaptic transmission (see chapter by Lorena Arancibia-Carcamo et al., this issue). Certain $GABA_A$ receptor α- and β-subunits have been shown to interact with *p*rotein *l*inking *I*AP to *c*ytoplasmic 1 (Plic-1), an adaptor protein with an N-terminal UBL domain and a C-terminal UBA domain (Fig. 4) (16). In neurons, Plic-1 colocalizes with $GABA_A$ receptors by immunofluorescent imaging while electron microscopy shows Plic-1 in proximity to $GABA_A$ receptors (16). Disruption of the Plic-1/$GABA_A$ receptor interaction reduces $GABA_A$ receptor surface expression levels by 20%, indicating that Plic-1 enhances receptor expression at the plasma membrane (16). Pulse-chase experiments in the presence of the proteasome inhibitor lactacystin showed a 30% increase in the stability of recombinant β3-containing $GABA_A$ receptors when coexpressed with Plic-1, indicating that $GABA_A$ subunit turnover is sensitive to proteasome inhibition (16). These results indicate that the Plic-1 UBA domain may bind monoubiquitinated $GABA_A$ receptors preventing polyubiquitination and subsequent degradation of the receptor.

5 The Ubiquitin–Proteasome System in Neurological Disease

A shared characteristic of several prevalent neurodegenerative disorders such as Parkinson's disease (PD), Alzheimer's disease (AD), Huntington's disease (HD), and amyotrophic lateral sclerosis (ALS) is the presence of ubiquitin-positive intracellular inclusions (179, 180, 238, 254). The significance of ubiquitin-rich aggregates associated with the onset of these neurological disorders is not well understood. Ubiquitin-conjugated protein aggregation may be a mechanism for coping with the buildup of dysfunctional proteins or a direct consequence of the disorder itself. As more information points to the UPS as an important regulator of synapse development and function, the question naturally arises as to whether aberrant UPS function could lead to neurological disease by impaired synapse regulation and plasticity (238). Furthermore, mutations in UPS-linked genes have been shown to contribute to the etiology of heritable forms of neurodegenerative disorders such as familial Parkinson's disease and neurodevelopmental diseases such as Angelman syndrome whose underlying pathogenesis may involve synaptic dysfunction.

5.1 Dysregulation of the UPS in Parkinson's Disease

Parkinson's disease (PD) is a movement disorder resulting from a gradual degeneration of dopaminergic neurons in the substantia nigra. Approximately 1% of the population older than 65 years is affected, of which only 5% of reported cases possess an inherited component (27). To date, several genes have been linked to the rare, inherited forms of PD, and the identification of these genes has generated important clues into the mechanisms underlying an otherwise mysterious disease. Proteins encoded by PARK genes include α-synuclein (PARK1) (230), parkin (PARK2, Table 1) (150), UCH-L1 (PARK5) (171), phosphatase and tensin homolog deleted on chromosome 10-induced kinase 1 or PINK-1 (PARK6) (279), DJ-1 (PARK7) (24), and leucine-rich repeat kinase 2 (PARK8) (215). In this chapter, we focus on the role of parkin (PARK2) and its role in the etiology of familial PD.

5.1.1 Parkin

Parkin is a member of the RING class of E3 ligases. Structurally, parkin has a C-terminal PDZ-binding domain (80) and a modified C-terminal in-between (IBR) RING domain flanked by two RING motifs. The N-terminus of parkin has a UBL domain that binds the Rpn10 subunit of the proteasome and is thought to aid the shuttling of ubiquitinated proteins to the proteasome for degradation (241). Numerous mutations spanning the entire parkin gene result in PD phenotypes with detectable symptoms manifesting before the age of 30 (59). Interestingly, patients with PD-linked parkin mutations lack the accumulation of Lewy bodies commonly seen in patients with sporadic forms of the disease, suggesting that Lewy body deposits are not solely responsible for PD. One possibility is that parkin activity may be involved in the conjugation of ubiquitin to Lewy body components as evidenced by ubiquitinated forms of putative parkin substrates α-synuclein (255), synphilin (46), and parkin itself (45, 313), all of which have been shown to be present in Lewy bodies (27, 171).

Parkin likely has diverse cellular roles as it localizes to both presynaptic and postsynaptic structures, including synaptic vesicles and dendritic vesicles as well as glial cell bodies (197). Parkin interacts with NMDA receptors and PSD-95 core components of the postsynaptic density at glutamatergic synapses indicating a potential role in postsynaptic function (80). Consistent with a role in synaptic transmission, parkin null mice show subtle behavioral deficits and altered excitability in the hippocampus and striatum (96, 126, 286). The diversity of disease-linked parkin mutations can lead to widely varying effects on E3 ligase activity, localization, and substrate binding (190, 262). While PD linked to parkin mutations most commonly follow a recessive inheritance trend, disease-linked mutations may lead to disruption or enhancement of parkin function (243, 262, 291), indicating the cellular effects of parkin mutations may not be simply due to loss of protein function.

Of the many possible roles parkin may play in neuronal function, one common theme is neuroprotection. Levels of the parkin substrate cyclin E, a positive regulator of neuronal apoptosis (263), is increased in patients with parkin-linked PD. Overexpression of wild-type parkin in cultured neurons causes a reduction in cyclin E levels and a greater resistance to excitotoxic insult (263). Furthermore, excitotoxicity linked to overexpression of the parkin substrates α-synuclein and the Pael-1 receptor is reduced by coexpression of wild-type parkin (124, 225, 307). Through its N-terminal UBL domain, parkin also interacts with the UIM-containing adaptor protein Eps15, and thereby regulates the internalization and trafficking of the epidermal growth factor receptor (EGFR) (53, 79). Eps15 is monoubiquitinated in the presence of EGF in a parkin-dependent manner reducing its binding to EGF receptors (79, 195). Correspondingly, RNAi-mediated parkin knockdown increases the internalization of EGF indicating that endogenous parkin prevents Eps15-dependent EGF receptor endocytosis (79). Interestingly, EGF receptor activation protects dopaminergic neurons (79), and EGFR expression is decreased in the striata of patients with PD (128). Moreover, EGFR signaling activates the downstream kinase Akt (79), which is well known to mediate neuronal survival (67, 77, 138, 157, 226). These results suggest that the neuroprotective effect of parkin may arise from its ability to increase the cellular response to EGF through an Akt-dependent pathway.

A complication in studying the function of parkin in normal neuronal physiology or in PD pathways is that few putative parkin substrates show elevated expression levels in parkin null mice. Also, these parkin knockout mice do not exhibit dopaminergic neurodegeration (96, 98, 153, 221) further clouding the link between parkin function and disease phenotypes (284). One recent important clue in this regard is that dopamine itself covalently modifies parkin inactivating its ubiquitin ligase activity (166), suggesting a link to the progressive and selective loss of parkin activity in dopaminergic neurons during sporadic Parkinson's disease.

5.1.2 Neurodevelopmental Disorders: Angelman Syndrome

Angelman syndrome (AS) is a neurodevelopmental disorder characterized by severe learning disabilities, ataxia, limited speech, inappropriate laughter, and seizures with no apparent neurodegeneration or gross brain morphology defects (297). Approximately 70% of cases of AS are caused by characteristic disruptions in the maternal chromosome 15q11-q13, leading to a microdeletion of chromosomal DNA (152). This deleted region contains genes for the $GABA_A$ receptor β3 subunit (GABRB3) and Ube3a, which encodes a HECT domain E3 ubiquitin ligase generally accepted to be the primary genetic cause of AS (Table 1, Fig. 4) (147, 191, 247). Indeed, in some patients with AS, the genetic etiology is a point mutation or small deletion within the maternal Ube3a allele (3, 55, 147). Ube3a is selectively imprinted in the brain where expression occurs only from the maternal allele. Ube3a expression shows a stereotypical pattern of imprinting in most brain regions. For example, the maternal copy of Ube3a is highly expressed in hippocampal neurons and Purkinje cells of the cerebellum while the paternal copy in these regions is not (3, 236, 288). Heterozygous mice lacking maternal copies of Ube3a have decreased levels of Ube3a mRNA in these imprinted areas (130). These animals show defects in hippocampal LTP, context-dependent learning, mild motor dysfunction, and a susceptibility to audiogenically induced seizures (130).

While the genetic component of AS has been identified, the precise physiological role of Ube3a function has not. Recent studies indicate that Ube3a regulates the activity of CamKIIα. Heterozygous mice lacking the maternal Ube3a allele (Ube3a m–/p+) show increased CaMKII phosphorylation on threonines 286 and 305 in hippocampal neurons and cerebellar Purkinje cells (281, 293). As mentioned in previous sections, inhibitory phosphorylation of CaMKII is predicted to have a significant impact on synaptic plasticity and transmission (50, 281, 293). Synaptic dysfunction has been proposed as a general mechanism for neurodevelopmental disorders (317), which in the case of AS points to the involvement of the UPS.

6 Conclusions and Perspectives

While recent work has shed light on the role of the UPS at synapses, many basic questions remain. Even though a large number of protein components comprising the UPS have been identified, few have been studied in great detail at synapses. What determines the cellular distribution of UPS components at or near synapses? How are specific PSD components modulated by the UPS while not affecting structural integrity? How does neuronal activity induce or limit the activity of UPS enzymes?

The emerging importance of protein turnover at synapses raises interesting questions about how neurons maintain their synaptic integrity. What determines synaptic protein turnover? What determines replacement of specific proteins and how are they replaced? How is protein degradation coupled to protein synthesis? How is synapse structure preserved? Are there degradation pathways for synaptic proteins relevant to synaptic plasticity similar to pathways postulated for synaptic protein synthesis? Also, a central theory of learning-based synaptic plasticity is that information is somehow stored by the composition of proteins at synapses. During ongoing activity-dependent protein turnover, how does the synapse sustain its molecular memory? While stimulating questions still remain, the multiple roles of the UPS in synapse development, signaling, and plasticity promise to offer many avenues for future study. Such efforts may be crucial in determining the link between normal synaptic physiology and dysfunction in neurological disorders involving aberrant protein degradation by the UPS.

References

1. Ageta H, Kato A, Fukazawa Y, Inokuchi K, and Sugiyama H. Effects of proteasome inhibitors on the synaptic localization of Vesl-1S/Homer-1a proteins. *Brain Res* 97: 186–189, 2001.
2. Ageta H, Kato A, Hatakeyama S, Nakayama K, Isojima Y, and Sugiyama H. Regulation of the level of Vesl-1S/Homer-1a proteins by ubiquitin-proteasome proteolytic systems. *J Biol Chem* 276: 15893–15897, 2001.

3. Albrecht U, Sutcliffe JS, Cattanach BM, Beechey CV, Armstrong D, Eichele G, and Beaudet AL. Imprinted expression of the murine Angelman syndrome gene, Ube3a, in hippocampal and Purkinje neurons. *Nat Genet* 17: 75–78, 1997.
4. Anderson C, Crimmins S, Wilson JA, Korbel GA, Ploegh HL, and Wilson SM. Loss of Usp14 results in reduced levels of ubiquitin in ataxia mice. *J Neurochem* 95: 724–731, 2005.
5. Ang XL and Harper JW. SCF-mediated protein degradation and cell cycle control. *Oncogene* 24: 2860–2870, 2005.
6. Araki T and Milbrandt J. ZNRF proteins constitute a family of presynaptic E3 ubiquitin ligases. *J Neurosci* 23: 9385–9394, 2003.
7. Arber S, Barbayannis FA, Hanser H, Schneider C, Stanyon CA, Bernard O, and Caroni P. Regulation of actin dynamics through phosphorylation of cofilin by LIM-kinase. *Nature* 393: 805–809, 1998.
8. Ashraf SI, McLoon AL, Sclarsic SM, and Kunes S. Synaptic protein synthesis associated with memory is regulated by the RISC pathway in Drosophila. *Cell* 124: 191–205, 2006.
9. Babst M, Katzmann DJ, Estepa-Sabal EJ, Meerloo T, and Emr SD. Escrt-III: an endosome-associated heterooligomeric protein complex required for mvb sorting. *Dev Cell* 3: 271–282, 2002.
10. Babst M, Katzmann DJ, Snyder WB, Wendland B, and Emr SD. Endosome-associated complex, ESCRT-II, recruits transport machinery for protein sorting at the multivesicular body. *Dev cell* 3: 283–289, 2002.
11. Bache KG, Raiborg C, Mehlum A, and Stenmark H. STAM and Hrs are subunits of a multivalent ubiquitin-binding complex on early endosomes. *J Biol Chem* 278: 12513–12521, 2003.
12. Bache KG, Stuffers S, Malerod L, Slagsvold T, Raiborg C, Lechardeur D, Walchli S, Lukacs GL, Brech A, and Stenmark H. The ESCRT-III subunit hVps24 is required for degradation but not silencing of the epidermal growth factor receptor. *Mol Biol Cell* 17: 2513–2523, 2006.
13. Basu J, Betz A, Brose N, and Rosenmund C. Munc13-1 C1 domain activation lowers the energy barrier for synaptic vesicle fusion. *J Neurosci* 27: 1200–1210, 2007.
14. Baumeister W, Walz J, Zuhl F, and Seemuller E. The proteasome: paradigm of a self-compartmentalizing protease. *Cell* 92: 367–380, 1998.
15. Beattie EC, Carroll RC, Yu X, Morishita W, Yasuda H, von Zastrow M, and Malenka RC. Regulation of AMPA receptor endocytosis by a signaling mechanism shared with LTD. *Nat Neurosci* 3: 1291–1300, 2000.
16. Bedford FK, Kittler JT, Muller E, Thomas P, Uren JM, Merlo D, Wisden W, Triller A, Smart TG, and Moss SJ. GABA(A) receptor cell surface number and subunit stability are regulated by the ubiquitin-like protein Plic-1. *Nat Neurosci* 4: 908–916, 2001.
17. Behrens J, von Kries JP, Kuhl M, Bruhn L, Wedlich D, Grosschedl R, and Birchmeier W. Functional interaction of beta-catenin with the transcription factor LEF-1. *Nature* 382: 638–642, 1996.
18. Bertolaet BL, Clarke DJ, Wolff M, Watson MH, Henze M, Divita G, and Reed SI. UBA domains of DNA damage-inducible proteins interact with ubiquitin. *Nat Struct Biol* 8: 417–422, 2001.
19. Bilodeau PS, Urbanowski JL, Winistorfer SC, and Piper RC. The Vps27p Hse1p complex binds ubiquitin and mediates endosomal protein sorting. *Nat Cell Biol* 4: 534–539, 2002.
20. Bingol B and Schuman EM. Activity-dependent dynamics and sequestration of proteasomes in dendritic spines. *Nature* 441: 1144–1148, 2006.
21. Bingol B and Schuman EM. A proteasome-sensitive connection between PSD-95 and GluR1 endocytosis. *Neuropharmacology* 47: 755–763, 2004.

22. Bliss TV and Gardner-Medwin AR. Long-lasting potentiation of synaptic transmission in the dentate area of the unanaestetized rabbit following stimulation of the perforant path. *J Physiol* 232: 357–374, 1973.
23. Bliss TV and Lomo T. Long-lasting potentiation of synaptic transmission in the dentate area of the anaesthetized rabbit following stimulation of the perforant path. *J Physiol* 232: 331–356, 1973.
24. Bonifati V, Rizzu P, van Baren MJ, Schaap O, Breedveld GJ, Krieger E, Dekker MC, Squitieri F, Ibanez P, Joosse M, van Dongen JW, Vanacore N, van Swieten JC, Brice A, Meco G, van Duijn CM, Oostra BA, and Heutink P. Mutations in the DJ-1 gene associated with autosomal recessive early-onset parkinsonism. *Science* (New York) 299: 256–259, 2003.
25. Borden KL and Freemont PS. The RING finger domain: a recent example of a sequence-structure family. *Curr Opin Struct Biol* 6: 395–401, 1996.
26. Borodovsky A, Kessler BM, Casagrande R, Overkleeft HS, Wilkinson KD, and Ploegh HL. A novel active site-directed probe specific for deubiquitylating enzymes reveals proteasome association of USP14. *The EMBO Journal* 20: 5187–5196, 2001.
27. Bossy-Wetzel E, Schwarzenbacher R, and Lipton SA. Molecular pathways to neurodegeneration. *Nat Med* 10 Suppl: S2–9, 2004.
28. Boyle MB, Klein M, Smith SJ, and Kandel ER. Serotonin increases intracellular Ca2+ transients in voltage-clamped sensory neurons of Aplysia californica. *Proc Nat Acad Sci USA* 81: 7642–7646, 1984.
29. Bradke F and Dotti CG. The role of local actin instability in axon formation. *Science* New York) 283: 1931–1934, 1999.
30. Braun BC, Glickman M, Kraft R, Dahlmann B, Kloetzel PM, Finley D, and Schmidt M. The base of the proteasome regulatory particle exhibits chaperone-like activity. *Nat Cell Biol* 1: 221–226, 1999.
31. Bredt DS and Nicoll RA. AMPA receptor trafficking at excitatory synapses. *Neuron* 40: 361–379, 2003.
32. Brose K, Bland KS, Wang KH, Arnott D, Henzel W, Goodman CS, Tessier-Lavigne M, and Kidd T. Slit proteins bind Robo receptors and have an evolutionarily conserved role in repulsive axon guidance. *Cell* 96: 795–806, 1999.
33. Brose N, Hofmann K, Hata Y, and Sudhof TC. Mammalian homologues of Caenorhabditis elegans unc-13 gene define novel family of C2-domain proteins. *J Biol Chem* 270: 25273–25280, 1995.
34. Bryan B, Cai Y, Wrighton K, Wu G, Feng XH, and Liu M. Ubiquitination of RhoA by Smurfl promotes neurite outgrowth. *FEBS Lett* 579: 1015–1019, 2005.
35. Burbea M, Dreier L, Dittman JS, Grunwald ME, and Kaplan JM. Ubiquitin and AP180 regulate the abundance of GLR-1 glutamate receptors at postsynaptic elements in C. elegans. *Neuron* 35: 107–120, 2002.
36. Burgess RW, Peterson KA, Johnson MJ, Roix JJ, Welsh IC, and O'Brien TP. Evidence for a conserved function in synapse formation reveals Phr1 as a candidate gene for respiratory failure in newborn mice. *Mol Cell Biol* 24: 1096–1105, 2004.
37. Cadavid AL, Ginzel A, and Fischer JA. The function of the Drosophila fat facets deubiquitinating enzyme in limiting photoreceptor cell number is intimately associated with endocytosis. *Development* 127: 1727–1736, 2000.
38. Cadwell K and Coscoy L. Ubiquitination on nonlysine residues by a viral E3 ubiquitin ligase. *Science* (New York) 309: 127–130, 2005.
39. Campbell DS and Holt CE. Chemotropic responses of retinal growth cones mediated by rapid local protein synthesis and degradation. *Neuron* 32: 1013–1026, 2001.
40. Carney DS, Davies BA, and Horazdovsky BF. Vps9 domain-containing proteins: activators of Rab5 GTPases from yeast to neurons. *Trends Cell Biol* 16: 27–35, 2006.

41. Carroll RC, Lissin DV, von Zastrow M, Nicoll RA, and Malenka RC. Rapid redistribution of glutamate receptors contributes to long-term depression in hippocampal cultures. *Nat Neurosci* 2: 454–460, 1999.
42. Chain DG, Casadio A, Schacher S, Hegde AN, Valbrun M, Yamamoto N, Goldberg AL, Bartsch D, Kandel ER, and Schwartz JH. Mechanisms for generating the autonomous cAMP-dependent protein kinase required for long-term facilitation in Aplysia. *Neuron* 22: 147–156, 1999.
43. Chapman AP, Courtney SC, Smith SJ, Rider CC, and Beesley PW. Ubiquitin immunoreactivity of multiple polypeptides in rat brain synaptic membranes. *Biochem Soc Trans* 20: 155S, 1992.
44. Chen H, Polo S, Di Fiore PP, and De Camilli PV. Rapid Ca2+-dependent decrease of protein ubiquitination at synapses. *Proc Nat Acad Sci USA* 100: 14908–14913, 2003.
45. Choi P, Ostrerova-Golts N, Sparkman D, Cochran E, Lee JM, and Wolozin B. Parkin is metabolized by the ubiquitin/proteosome system. *Neuroreport* 11: 2635–2638, 2000.
46. Chung KK, Zhang Y, Lim KL, Tanaka Y, Huang H, Gao J, Ross CA, Dawson VL, and Dawson TM. Parkin ubiquitinates the alpha-synuclein-interacting protein, synphilin-1: implications for Lewy-body formation in Parkinson disease. *Nat Med* 7: 1144–1150, 2001.
47. Ciechanover A and Ben-Saadon R. N-terminal ubiquitination: more protein substrates join in. *Trends Cell Biol* 14: 103–106, 2004.
48. Ciechanover A, Elias S, Heller H, Ferber S, and Hershko A. Characterization of the heat-stable polypeptide of the ATP-dependent proteolytic system from reticulocytes. *J Biol Chem* 255: 7525–7528, 1980.
49. Ciehanover A, Hod Y, and Hershko A. A heat-stable polypeptide component of an ATP-dependent proteolytic system from reticulocytes. *Biochem Biophys Res Comm* 81: 1100–1105, 1978.
50. Colbran RJ and Brown AM. Calcium/calmodulin-dependent protein kinase II and synaptic plasticity. *Curr Opin Neurobiol* 14: 318–327, 2004.
51. Colledge M, Snyder EM, Crozier RA, Soderling JA, Jin Y, Langeberg LK, Lu H, Bear MF, and Scott JD. Ubiquitination regulates PSD-95 degradation and AMPA receptor surface expression. *Neuron* 40: 595–607, 2003.
52. Collins CA, Wairkar YP, Johnson SL, and DiAntonio A. Highwire restrains synaptic growth by attenuating a MAP kinase signal. *Neuron* 51: 57–69, 2006.
53. Confalonieri S, Salcini AE, Puri C, Tacchetti C, and Di Fiore PP. Tyrosine phosphorylation of Eps15 is required for ligand-regulated, but not constitutive, endocytosis. *J Cell Biol* 150: 905–912, 2000.
54. Cook HA, Koppetsch BS, Wu J, and Theurkauf WE. The Drosophila SDE3 homolog armitage is required for oskar mRNA silencing and embryonic axis specification. *Cell* 116: 817–829, 2004.
55. Cooper EM, Hudson AW, Amos J, Wagstaff J, and Howley PM. Biochemical analysis of Angelman syndrome-associated mutations in the E3 ubiquitin ligase E6-associated protein. *J Biol Chem* 279: 41208–41217, 2004.
56. Crimmins S, Jin Y, Wheeler C, Huffman AK, Chapman C, Dobrunz LE, Levey A, Roth KA, Wilson JA, and Wilson SM. Transgenic rescue of ataxia mice with neuronal-specific expression of ubiquitin-specific protease 14. *J Neurosci* 26: 11423–11431, 2006.
57. D'Amato CJ and Hicks SP. Neuropathologic alterations in the ataxia (paralytic) mouse. *Arch Pathol* 80: 604–612, 1965.
58. Dale N, Schacher S, and Kandel ER. Long-term facilitation in Aplysia involves increase in transmitter release. *Science* (New York) 239: 282–285, 1988.
59. Dauer W and Przedborski S. Parkinson's disease: mechanisms and models. *Neuron* 39: 889–909, 2003.

60. Davy A, Bello P, Thierry-Mieg N, Vaglio P, Hitti J, Doucette-Stamm L, Thierry-Mieg D, Reboul J, Boulton S, Walhout AJ, Coux O, and Vidal M. A protein-protein interaction map of the Caenorhabditis elegans 26S proteasome. *EMBO Rep* 2: 821–828, 2001.
61. de la Torre JR, Hopker VH, Ming GL, Poo MM, Tessier-Lavigne M, Hemmati-Brivanlou A, and Holt CE. Turning of retinal growth cones in a netrin-1 gradient mediated by the netrin receptor DCC. *Neuron* 19: 1211–1224, 1997.
62. DeMartino GN, Moomaw CR, Zagnitko OP, Proske RJ, Chu-Ping M, Afendis SJ, Swaffield JC, and Slaughter CA. PA700, an ATP-dependent activator of the 20 S proteasome, is an ATPase containing multiple members of a nucleotide-binding protein family. *J Biol Chem* 269: 20878–20884, 1994.
63. DiAntonio A, Haghighi AP, Portman SL, Lee JD, Amaranto AM, and Goodman CS. Ubiquitination-dependent mechanisms regulate synaptic growth and function. *Nature* 412: 449–452, 2001.
64. Dick TP, Nussbaum AK, Deeg M, Heinemeyer W, Groll M, Schirle M, Keilholz W, Stevanovic S, Wolf DH, Huber R, Rammensee HG, and Schild H. Contribution of proteasomal beta-subunits to the cleavage of peptide substrates analyzed with yeast mutants. *J Biol Chem* 273: 25637–25646, 1998.
65. Ding M, Chao D, Wang G, and Shen K. Spatial regulation of an E3 ubiquitin ligase directs selective synapse elimination. *Science* (New York) 317: 947–951, 2007.
66. Dreier L, Burbea M, and Kaplan JM. LIN-23-mediated degradation of beta-catenin regulates the abundance of GLR-1 glutamate receptors in the ventral nerve cord of C. elegans. *Neuron* 46: 51–64, 2005.
67. Dudek H, Datta SR, Franke TF, Birnbaum MJ, Yao R, Cooper GM, Segal RA, Kaplan DR, and Greenberg ME. Regulation of neuronal survival by the serine-threonine protein kinase Akt. *Science* (New York) 275: 661–665, 1997.
68. Ebihara T, Kawabata I, Usui S, Sobue K, and Okabe S. Synchronized formation and remodeling of postsynaptic densities: long-term visualization of hippocampal neurons expressing postsynaptic density proteins tagged with green fluorescent protein. *J Neurosci* 23: 2170–2181, 2003.
69. Ehlers MD. Activity level controls postsynaptic composition and signaling via the ubiquitin-proteasome system. *Nat Neurosci* 6: 231–242, 2003a.
70. Ehlers MD. Reinsertion or degradation of AMPA receptors determined by activity-dependent endocytic sorting. *Neuron* 28: 511–525, 2000.
71. Ehrlich I and Malinow R. Postsynaptic density 95 controls AMPA receptor incorporation during long-term potentiation and experience-driven synaptic plasticity. *J Neurosci* 24: 916–927, 2004.
72. El-Husseini AE, Schnell E, Chetkovich DM, Nicoll RA, and Bredt DS. PSD-95 involvement in maturation of excitatory synapses. *Science* New York 290: 1364–1368, 2000.
73. El-Husseini Ael D, Schnell E, Dakoji S, Sweeney N, Zhou Q, Prange O, Gauthier-Campbell C, Aguilera-Moreno A, Nicoll RA, and Bredt DS. Synaptic strength regulated by palmitate cycling on PSD-95. *Cell* 108: 849–863, 2002.
74. Elsasser S, Chandler-Militello D, Muller B, Hanna J, and Finley D. Rad23 and Rpn10 serve as alternative ubiquitin receptors for the proteasome. *J Biol Chem* 279: 26817–26822, 2004.
75. Elsasser S, Gali RR, Schwickart M, Larsen CN, Leggett DS, Muller B, Feng MT, Tubing F, Dittmar GA, and Finley D. Proteasome subunit Rpn1 binds ubiquitin-like protein domains. *Nat Cell Biol* 4: 725–730, 2002.
76. Emmerich NP, Nussbaum AK, Stevanovic S, Priemer M, Toes RE, Rammensee HG, and Schild H. The human 26 S and 20 S proteasomes generate overlapping but different sets of peptide fragments from a model protein substrate. *J Biol Chem* 275: 21140–21148, 2000.

77. Eves EM, Xiong W, Bellacosa A, Kennedy SG, Tsichlis PN, Rosner MR, and Hay N. Akt, a target of phosphatidylinositol 3-kinase, inhibits apoptosis in a differentiating neuronal cell line. *Mol Cell Biol* 18: 2143–2152, 1998.
78. Eytan E, Armon T, Heller H, Beck S, and Hershko A. Ubiquitin C-terminal hydrolase activity associated with the 26 S protease complex. *J Biol Chem* 268: 4668–4674, 1993.
79. Fallon L, Belanger CM, Corera AT, Kontogiannea M, Regan-Klapisz E, Moreau F, Voortman J, Haber M, Rouleau G, Thorarinsdottir T, Brice A, van Bergen En Henegouwen PM, and Fon EA. A regulated interaction with the UIM protein Eps15 implicates parkin in EGF receptor trafficking and PI(3)K-Akt signalling. *Nat Cell Biol* 8: 834–842, 2006.
80. Fallon L, Moreau F, Croft BG, Labib N, Gu WJ, and Fon EA. Parkin and CASK/LIN-2 associate via a PDZ-mediated interaction and are co-localized in lipid rafts and postsynaptic densities in brain. *J Biol Chem* 277: 486–491, 2002.
81. Fang S, Jensen JP, Ludwig RL, Vousden KH, and Weissman AM. Mdm2 is a RING finger-dependent ubiquitin protein ligase for itself and p53. *J Biol Chem* 275: 8945–8951, 2000.
82. Feldman RM, Correll CC, Kaplan KB, and Deshaies RJ. A complex of Cdc4p, Skp1p, and Cdc53p/cullin catalyzes ubiquitination of the phosphorylated CDK inhibitor Sic1p. *Cell* 91: 221–230, 1997.
83. Fonseca R, Vabulas RM, Hartl FU, Bonhoeffer T, and Nagerl UV. A balance of protein synthesis and proteasome-dependent degradation determines the maintenance of LTP. *Neuron* 52: 239–245, 2006.
84. Freemont PS. RING for destruction? *Curr Biol* 10: R84–87, 2000.
85. Fu H, Reis N, Lee Y, Glickman MH, and Vierstra RD. Subunit interaction maps for the regulatory particle of the 26S proteasome and the COP9 signalosome. *EMBO J* 20: 7096–7107, 2001.
86. Fu H, Sadis S, Rubin DM, Glickman M, van Nocker S, Finley D, and Vierstra RD. Multiubiquitin chain binding and protein degradation are mediated by distinct domains within the 26 S proteasome subunit Mcb1. *J Biol Chem* 273: 1970–1981, 1998.
87. Furrer MP, Kim S, Wolf B, and Chiba A. Robo and Frazzled/DCC mediate dendritic guidance at the CNS midline. *Nat Neurosci* 6: 223–230, 2003.
88. Ganeshina O, Berry RW, Petralia RS, Nicholson DA, and Geinisman Y. Synapses with a segmented, completely partitioned postsynaptic density express more AMPA receptors than other axospinous synaptic junctions. *Neuroscience* 125: 615–623, 2004.
89. Georgiou M and Tear G. Commissureless is required both in commissural neurones and midline cells for axon guidance across the midline. *Development* 129: 2947–2956, 2002.
90. Ghirardi M, Braha O, Hochner B, Montarolo PG, Kandel ER, and Dale N. Roles of PKA and PKC in facilitation of evoked and spontaneous transmitter release at depressed and nondepressed synapses in Aplysia sensory neurons. *Neuron* 9: 479–489, 1992.
91. Gieffers C, Peters BH, Kramer ER, Dotti CG, and Peters JM. Expression of the CDH1-associated form of the anaphase-promoting complex in postmitotic neurons. *Proc Nat Acad Sci USA* 96: 11317–11322, 1999.
92. Gill DJ, Teo H, Sun J, Perisic O, Veprintsev DB, Emr SD, and Williams RL. Structural insight into the ESCRT-I/-II link and its role in MVB trafficking. *EMBO J* 26: 600–612, 2007.
93. Glickman MH and Ciechanover A. The ubiquitin-proteasome proteolytic pathway: destruction for the sake of construction. *Physiol Rev* 82: 373–428, 2002.
94. Glickman MH, Rubin DM, Fried VA, and Finley D. The regulatory particle of the Saccharomyces cerevisiae proteasome. *Mol Cell Biol* 18: 3149–3162, 1998.
95. Glotzer M, Murray AW, and Kirschner MW. Cyclin is degraded by the ubiquitin pathway. *Nature* 349: 132–138, 1991.

96. Goldberg MS, Fleming SM, Palacino JJ, Cepeda C, Lam HA, Bhatnagar A, Meloni EG, Wu N, Ackerson LC, Klapstein GJ, Gajendiran M, Roth BL, Chesselet MF, Maidment NT, Levine MS, and Shen J. Parkin-deficient mice exhibit nigrostriatal deficits but not loss of dopaminergic neurons. *J Biol Chem* 278: 43628–43635, 2003.
97. Gray NW, Weimer RM, Bureau I, and Svoboda K. Rapid redistribution of synaptic PSD-95 in the neocortex in vivo. *PLoS Biol* 4: e370, 2006.
98. Greene JC, Whitworth AJ, Kuo I, Andrews LA, Feany MB, and Pallanck LJ. Mitochondrial pathology and apoptotic muscle degeneration in Drosophila parkin mutants. *Proc Nat Acad Sci USA* 100: 4078–4083, 2003.
99. Groll M, Bajorek M, Kohler A, Moroder L, Rubin DM, Huber R, Glickman MH, and Finley D. A gated channel into the proteasome core particle. *Nat Struct Biol* 7: 1062–1067, 2000.
100. Groll M, Ditzel L, Lowe J, Stock D, Bochtler M, Bartunik HD, and Huber R. Structure of 20S proteasome from yeast at 2.4 A resolution. *Nature* 386: 463–471, 1997.
101. Guo Q, Xie J, Dang CV, Liu ET, and Bishop JM. Identification of a large Myc-binding protein that contains RCC1-like repeats. *Proc Nat Acad Sci USA* 95: 9172–9177, 1998.
102. Guterman A and Glickman MH. Complementary roles for Rpn11 and Ubp6 in deubiquitination and proteolysis by the proteasome. *J Biol C* 279: 1729–1738, 2004.
103. Haracska L and Udvardy A. Mapping the ubiquitin-binding domains in the p54 regulatory complex subunit of the Drosophila 26S protease. *FEBS Lett* 412: 331–336, 1997.
104. Harris KM, Fiala JC, and Ostroff L. Structural changes at dendritic spine synapses during long-term potentiation. *Philosophical Trans Royal Soc Lon* 358: 745–748, 2003.
105. Hegde AN, Goldberg AL, and Schwartz JH. Regulatory subunits of cAMP-dependent protein kinases are degraded after conjugation to ubiquitin: a molecular mechanism underlying long-term synaptic plasticity. *Proc Nat Acad Sci USA* 90: 7436–7440, 1993.
106. Hegde AN, Inokuchi K, Pei W, Casadio A, Ghirardi M, Chain DG, Martin KC, Kandel ER, and Schwartz JH. Ubiquitin C-terminal hydrolase is an immediate-early gene essential for long-term facilitation in Aplysia. *Cell* 89: 115–126, 1997.
107. Helenius A and Aebi M. Intracellular functions of N-linked glycans. *Science* (New York) 291: 2364–2369, 2001.
108. Hershko A, Leshinsky E, Ganoth D, and Heller H. ATP-dependent degradation of ubiquitin-protein conjugates. *Proc Natl Acad Sci USA* 81: 1619–1623, 1984.
109. Hicke L and Dunn R. Regulation of membrane protein transport by ubiquitin and ubiquitin-binding proteins. *Annu Rev Cell Dev Biol* 19: 141–172, 2003.
110. Hicke L, Schubert HL, and Hill CP. Ubiquitin-binding domains. *Nat Rev* 6: 610–621, 2005.
111. Hirano S, Kawasaki M, Ura H, Kato R, Raiborg C, Stenmark H, and Wakatsuki S. Double-sided ubiquitin binding of Hrs-UIM in endosomal protein sorting. *Nat Struc Mol Biol* 13: 272–277, 2006.
112. Hiyama H, Yokoi M, Masutani C, Sugasawa K, Maekawa T, Tanaka K, Hoeijmakers JH, and Hanaoka F. Interaction of hHR23 with S5a. The ubiquitin-like domain of hHR23 mediates interaction with S5a subunit of 26 S proteasome. *J Biol Chem* 274: 28019–28025, 1999.
113. Hoffman L and Rechsteiner M. Activation of the multicatalytic protease. The 11 S regulator and 20 S ATPase complexes contain distinct 30-kilodalton subunits. *J Biol Chem* 269: 16890–16895, 1994.
114. Hofmann K and Bucher P. The UBA domain: a sequence motif present in multiple enzyme classes of the ubiquitination pathway. *Trends Biochem Sci* 21: 172–173, 1996.
115. Hoopfer ED, McLaughlin T, Watts RJ, Schuldiner O, O'Leary DD, and Luo L. Wlds protection distinguishes axon degeneration following injury from naturally occurring developmental pruning. *Neuron* 50: 883–895, 2006.

116. Hough R, Pratt G, and Rechsteiner M. Ubiquitin-lysozyme conjugates. Identification and characterization of an ATP-dependent protease from rabbit reticulocyte lysates. *J Biol Chem* 261: 2400–2408, 1986.
117. Howard TL, Stauffer DR, Degnin CR, and Hollenberg SM. CHMP1 functions as a member of a newly defined family of vesicle trafficking proteins. *J Cell Sci* 114: 2395–2404, 2001.
118. Hu G, Zhang S, Vidal M, Baer JL, Xu T, and Fearon ER. Mammalian homologs of seven in absentia regulate DCC via the ubiquitin-proteasome pathway. *Genes Dev* 11: 2701–2714, 1997.
119. Hu GY, Hvalby O, Walaas SI, Albert KA, Skjeflo P, Andersen P, and Greengard P. Protein kinase C injection into hippocampal pyramidal cells elicits features of long term potentiation. *Nature* 328: 426–429, 1987.
120. Huang L, Kinnucan E, Wang G, Beaudenon S, Howley PM, Huibregtse JM, and Pavletich NP. Structure of an E6AP-UbcH7 complex: insights into ubiquitination by the E2-E3 enzyme cascade. *Science* (New York) 286: 1321–1326, 1999.
121. Huang Y, Baker RT, and Fischer-Vize JA. Control of cell fate by a deubiquitinating enzyme encoded by the fat facets gene. *Science* (New York) 270: 1828–1831, 1995.
122. Huber O, Korn R, McLaughlin J, Ohsugi M, Herrmann BG, and Kemler R. Nuclear localization of beta-catenin by interaction with transcription factor LEF-1. *Mech Dev* 59: 3–10, 1996.
123. Huibregtse JM, Scheffner M, Beaudenon S, and Howley PM. A family of proteins structurally and functionally related to the E6-AP ubiquitin-protein ligase. *Proc Nat Acad Sci USA* 92: 2563-2567, 1995.
124. Imai Y, Soda M, Inoue H, Hattori N, Mizuno Y, and Takahashi R. An unfolded putative transmembrane polypeptide, which can lead to endoplasmic reticulum stress, is a substrate of Parkin. *Cell* 105: 891–902, 2001.
125. Ing B, Shteiman-Kotler A, Castelli M, Henry P, Pak Y, Stewart B, Boulianne GL, and Rotin D. Regulation of Commissureless by the ubiquitin ligase DNedd4 is required for neuromuscular synaptogenesis in Drosophila melanogaster. *Mol Cell Biol* 27: 481–496, 2007.
126. Itier JM, Ibanez P, Mena MA, Abbas N, Cohen-Salmon C, Bohme GA, Laville M, Pratt J, Corti O, Pradier L, Ret G, Joubert C, Periquet M, Araujo F, Negroni J, Casarejos MJ, Canals S, Solano R, Serrano A, Gallego E, Sanchez M, Denefle P, Benavides J, Tremp G, Rooney TA, Brice A, and Garcia de Yebenes J. Parkin gene inactivation alters behaviour and dopamine neurotransmission in the mouse. *Hum Mol Genet* 12: 2277–2291, 2003.
127. Iwahara T, Fujimoto J, Wen D, Cupples R, Bucay N, Arakawa T, Mori S, Ratzkin B, and Yamamoto T. Molecular characterization of ALK, a receptor tyrosine kinase expressed specifically in the nervous system. *Oncogene* 14: 439–449, 1997.
128. Iwakura Y, Piao YS, Mizuno M, Takei N, Kakita A, Takahashi H, and Nawa H. Influences of dopaminergic lesion on epidermal growth factor-ErbB signals in Parkinson's disease and its model: neurotrophic implication in nigrostriatal neurons. *J Neurochem* 93: 974–983, 2005.
129. Jackson PK, Eldridge AG, Freed E, Furstenthal L, Hsu JY, Kaiser BK, and Reimann JD. The lore of the RINGs: substrate recognition and catalysis by ubiquitin ligases. *Trends Cell Biol* 10: 429–439, 2000.
130. Jiang YH, Armstrong D, Albrecht U, Atkins CM, Noebels JL, Eichele G, Sweatt JD, and Beaudet AL. Mutation of the Angelman ubiquitin ligase in mice causes increased cytoplasmic p53 and deficits of contextual learning and long-term potentiation. *Neuron* 21: 799–811, 1998.
131. Joazeiro CA and Weissman AM. RING finger proteins: mediators of ubiquitin ligase activity. *Cell* 102: 549–552, 2000.

132. Joazeiro CA, Wing SS, Huang H, Leverson JD, Hunter T, and Liu YC. The tyrosine kinase negative regulator c-Cbl as a RING-type, E2-dependent ubiquitin-protein ligase. *Science* (New York) 286: 309–312, 1999.
133. Johnston JA and Madura K. Rings, chains and ladders: ubiquitin goes to work in the neuron. *Prog Neurobiol* 73: 227–257, 2004.
134. Juo P and Kaplan JM. The anaphase-promoting complex regulates the abundance of GLR-1 glutamate receptors in the ventral nerve cord of C. elegans. *Curr Biol* 14: 2057–2062, 2004.
135. Kandel ER. The molecular biology of memory storage: a dialogue between genes and synapses. *Science* (New York) 294: 1030–1038, 2001.
136. Karpova A, Mikhaylova M, Thomas U, Knopfel T, and Behnisch T. Involvement of protein synthesis and degradation in long-term potentiation of Schaffer collateral CA1 synapses. *J Neurosci* 26: 4949–4955, 2006.
137. Kato A, Rouach N, Nicoll RA, and Bredt DS. Activity-dependent NMDA receptor degradation mediated by retrotranslocation and ubiquitination. *Proc Nat Acad Sci USA* 102: 5600–5605, 2005.
138. Kauffmann-Zeh A, Rodriguez-Viciana P, Ulrich E, Gilbert C, Coffer P, Downward J, and Evan G. Suppression of c-Myc-induced apoptosis by Ras signalling through PI(3)K and PKB. *Nature* 385: 544–548, 1997.
139. Kaufmann N, DeProto J, Ranjan R, Wan H, and Van Vactor D. Drosophila liprin-alpha and the receptor phosphatase Dlar control synapse morphogenesis. *Neuron* 34: 27–38, 2002.
140. Keleman K, Rajagopalan S, Cleppien D, Teis D, Paiha K, Huber LA, Technau GM, and Dickson BJ. Comm sorts robo to control axon guidance at the Drosophila midline. *Cell* 110: 415–427, 2002.
141. Keleman K, Ribeiro C, and Dickson BJ. Comm function in commissural axon guidance: cell-autonomous sorting of Robo in vivo. *Nat Nneurosci* 8: 156–163, 2005.
142. Kelleher RJ, 3rd, Govindarajan A, Jung HY, Kang H, and Tonegawa S. Translational control by MAPK signaling in long-term synaptic plasticity and memory. *Cell* 116: 467–479, 2004.
143. Kelleher RJ, 3rd, Govindarajan A, and Tonegawa S. Translational regulatory mechanisms in persistent forms of synaptic plasticity. *Neuron* 44: 59–73, 2004.
144. King RW, Glotzer M, and Kirschner MW. Mutagenic analysis of the destruction signal of mitotic cyclins and structural characterization of ubiquitinated intermediates. *Mol Biol Cell* 7: 1343–1357, 1996.
145. Kipreos ET, Gohel SP, and Hedgecock EM. The C. elegans F-box/WD-repeat protein LIN-23 functions to limit cell division during development. *Development* 127: 5071–5082, 2000.
146. Kirkin V and Dikic I. Role of ubiquitin- and Ubl-binding proteins in cell signaling. *Curr Opin Cell Biol* 19: 199–205, 2007.
147. Kishino T, Lalande M, and Wagstaff J. UBE3A/E6-AP mutations cause Angelman syndrome. *Nat Genet* 15: 70–73, 1997.
148. Kisselev AF, Akopian TN, and Goldberg AL. Range of sizes of peptide products generated during degradation of different proteins by archaeal proteasomes. *J Biol Chem* 273: 1982–1989, 1998.
149. Kisselev AF, Akopian TN, Woo KM, and Goldberg AL. The sizes of peptides generated from protein by mammalian 26 and 20 S proteasomes. Implications for understanding the degradative mechanism and antigen presentation. *J Biol Chem* 274: 3363–3371, 1999.

150. Kitada T, Asakawa S, Hattori N, Matsumine H, Yamamura Y, Minoshima S, Yokochi M, Mizuno Y, and Shimizu N. Mutations in the parkin gene cause autosomal recessive juvenile parkinsonism. *Nature* 392: 605–608, 1998.
151. Klapisz E, Sorokina I, Lemeer S, Pijnenburg M, Verkleij AJ, and van Bergen en Henegouwen PM. A ubiquitin-interacting motif (UIM) is essential for Eps15 and Eps15R ubiquitination. *J Biol Chem* 277: 30746–30753, 2002.
152. Knoll JH, Nicholls RD, Magenis RE, Graham JM, Jr., Lalande M, and Latt SA. Angelman and Prader-Willi syndromes share a common chromosome 15 deletion but differ in parental origin of the deletion. *Am J Med Genet* 32: 285–290, 1989.
153. Ko HS, von Coelln R, Sriram SR, Kim SW, Chung KK, Pletnikova O, Troncoso J, Johnson B, Saffary R, Goh EL, Song H, Park BJ, Kim MJ, Kim S, Dawson VL, and Dawson TM. Accumulation of the authentic parkin substrate aminoacyl-tRNA synthetase cofactor, p38/JTV-1, leads to catecholaminergic cell death. *J Neurosci* 25: 7968–7978, 2005.
154. Koegl M, Hoppe T, Schlenker S, Ulrich HD, Mayer TU, and Jentsch S. A novel ubiquitination factor, E4, is involved in multiubiquitin chain assembly. *Cell* 96: 635–644, 1999.
155. Kohler A, Cascio P, Leggett DS, Woo KM, Goldberg AL, and Finley D. The axial channel of the proteasome core particle is gated by the Rpt2 ATPase and controls both substrate entry and product release. *Mol Cell* 7: 1143–1152, 2001.
156. Konishi Y, Stegmuller J, Matsuda T, Bonni S, and Bonni A. Cdh1-APC controls axonal growth and patterning in the mammalian brain. *Science* (New York) 303: 1026–1030, 2004.
157. Kulik G and Weber MJ. Akt-dependent and -independent survival signaling pathways utilized by insulin-like growth factor I. *Mol Cell Biol* 18: 6711–6718, 1998.
158. Kuo CT, Jan LY, and Jan YN. Dendrite-specific remodeling of Drosophila sensory neurons requires matrix metalloproteases, ubiquitin-proteasome, and ecdysone signaling. *Proc Nat Acad Sci USA* 102: 15230–15235, 2005.
159. Kuo CT, Zhu S, Younger S, Jan LY, and Jan YN. Identification of E2/E3 ubiquitinating enzymes and caspase activity regulating Drosophila sensory neuron dendrite pruning. *Neuron* 51: 283–290, 2006.
160. Lam YA, DeMartino GN, Pickart CM, and Cohen RE. Specificity of the ubiquitin isopeptidase in the PA700 regulatory complex of 26 S proteasomes. *J Biol Chem* 272: 28438–28446, 1997.
161. Lam YA, Lawson TG, Velayutham M, Zweier JL, and Pickart CM. A proteasomal ATPase subunit recognizes the polyubiquitin degradation signal. *Nature* 416: 763–767, 2002.
162. Lam YA, Xu W, DeMartino GN, and Cohen RE. Editing of ubiquitin conjugates by an isopeptidase in the 26S proteasome. *Nature* 385: 737–740, 1997.
163. Langelier C, von Schwedler UK, Fisher RD, De Domenico I, White PL, Hill CP, Kaplan J, Ward D, and Sundquist WI. Human ESCRT-II complex and its role in human immunodeficiency virus type 1 release. *J Virol* 80: 9465–9480, 2006.
164. Lasorella A, Stegmuller J, Guardavaccaro D, Liu G, Carro MS, Rothschild G, de la Torre-Ubieta L, Pagano M, Bonni A, and Iavarone A. Degradation of Id2 by the anaphase-promoting complex couples cell cycle exit and axonal growth. *Nature* 442: 471–474, 2006.
165. Latres E, Chiaur DS, and Pagano M. The human F box protein beta-Trcp associates with the Cul1/Skp1 complex and regulates the stability of beta-catenin. *Oncogene* 18: 849–854, 1999.
166. LaVoie MJ, Ostaszewski BL, Weihofen A, Schlossmacher MG, and Selkoe DJ. Dopamine covalently modifies and functionally inactivates parkin. *Nat Med* 11: 1214–1221, 2005.

167. Layfield R, Franklin K, Landon M, Walker G, Wang P, Ramage R, Brown A, Love S, Urquhart K, Muir T, Baker R, and Mayer RJ. Chemically synthesized ubiquitin extension proteins detect distinct catalytic capacities of deubiquitinating enzymes. *Anal Biochem* 274: 40–49, 1999.
168. Lee DH and Goldberg AL. Proteasome inhibitors: valuable new tools for cell biologists. *Trends Cell Biol* 8: 397–403, 1998.
169. Lee SH, Simonetta A, and Sheng M. Subunit rules governing the sorting of internalized AMPA receptors in hippocampal neurons. *Neuron* 43: 221–236, 2004.
170. Leggett DS, Hanna J, Borodovsky A, Crosas B, Schmidt M, Baker RT, Walz T, Ploegh H, and Finley D. Multiple associated proteins regulate proteasome structure and function. *Mol Cell* 10: 495–507, 2002.
171. Leroy E, Boyer R, Auburger G, Leube B, Ulm G, Mezey E, Harta G, Brownstein MJ, Jonnalagada S, Chernova T, Dehejia A, Lavedan C, Gasser T, Steinbach PJ, Wilkinson KD, and Polymeropoulos MH. The ubiquitin pathway in Parkinson's disease. *Nature* 395: 451–452, 1998.
172. Leverson JD, Joazeiro CA, Page AM, Huang H, Hieter P, and Hunter T. The APC11 RING-H2 finger mediates E2-dependent ubiquitination. *Mol Biol Cell* 11: 2315–2325, 2000.
173. Levkowitz G, Waterman H, Ettenberg SA, Katz M, Tsygankov AY, Alroy I, Lavi S, Iwai K, Reiss Y, Ciechanover A, Lipkowitz S, and Yarden Y. Ubiquitin ligase activity and tyrosine phosphorylation underlie suppression of growth factor signaling by c-Cbl/Sli-1. *Mol Cell* 4: 1029–1040, 1999.
174. Liao EH, Hung W, Abrams B, and Zhen M. An SCF-like ubiquitin ligase complex that controls presynaptic differentiation. *Nature* 430: 345–350, 2004.
175. Lin JW, Ju W, Foster K, Lee SH, Ahmadian G, Wyszynski M, Wang YT, and Sheng M. Distinct molecular mechanisms and divergent endocytotic pathways of AMPA receptor internalization. *Nat Neurosci* 3: 1282–1290, 2000.
176. Lin Y, Kimpler LA, Naismith TV, Lauer JM, and Hanson PI. Interaction of the mammalian endosomal sorting complex required for transport (ESCRT) III protein hSnf7-1 with itself, membranes, and the AAA+ ATPase SKD1. *J Biol Chem* 280: 12799–12809, 2005.
177. Lisman JE, Raghavachari S, and Tsien RW. The sequence of events that underlie quantal transmission at central glutamatergic synapses. *Nat Rev Neurosci* 8: 597–609, 2007.
178. Lorick KL, Jensen JP, Fang S, Ong AM, Hatakeyama S, and Weissman AM. RING fingers mediate ubiquitin-conjugating enzyme (E2)-dependent ubiquitination. *Proc Nat Acad Sci USA* 96: 11364–11369, 1999.
179. Lowe J, Blanchard A, Morrell K, Lennox G, Reynolds L, Billett M, Landon M, and Mayer RJ. Ubiquitin is a common factor in intermediate filament inclusion bodies of diverse type in man, including those of Parkinson's disease, Pick's disease, and Alzheimer's disease, as well as Rosenthal fibres in cerebellar astrocytomas, cytoplasmic bodies in muscle, and mallory bodies in alcoholic liver disease. *J Pathol* 155: 9–15, 1988.
180. Lowe J, McDermott H, Landon M, Mayer RJ, and Wilkinson KD. Ubiquitin carboxyl-terminal hydrolase (PGP 9.5) is selectively present in ubiquitinated inclusion bodies characteristic of human neurodegenerative diseases. *J Pathol* 161: 153–160, 1990.
181. Lu Z, Je HS, Young P, Gross J, Lu B, and Feng G. Regulation of synaptic growth and maturation by a synapse-associated E3 ubiquitin ligase at the neuromuscular junction. *J Cell Biol* 177: 1077–1089, 2007.
182. Lupas A, Baumeister W, and Hofmann K. A repetitive sequence in subunits of the 26S proteasome and 20S cyclosome (anaphase-promoting complex). *Trends Biochem Sci* 22: 195–196, 1997.

183. Luscher C, Xia H, Beattie EC, Carroll RC, von Zastrow M, Malenka RC, and Nicoll RA. Role of AMPA receptor cycling in synaptic transmission and plasticity. *Neuron* 24: 649–658, 1999.
184. Lynch MA. Long-term potentiation and memory. *Physiol Rev* 84: 87–136, 2004.
185. Malenka RC and Bear MF. LTP and LTD: an embarrassment of riches. *Neuron* 44: 5–21, 2004.
186. Malenka RC, Madison DV, and Nicoll RA. Potentiation of synaptic transmission in the hippocampus by phorbol esters. *Nature* 321: 175–177, 1986.
187. Marmor MD and Yarden Y. Role of protein ubiquitylation in regulating endocytosis of receptor tyrosine kinases. *Oncogene* 23: 2057–2070, 2004.
188. Marrs GS, Green SH, and Dailey ME. Rapid formation and remodeling of postsynaptic densities in developing dendrites. *Nat Neurosci* 4: 1006–1013, 2001.
189. Martin S, Nishimune A, Mellor JR, and Henley JM. SUMOylation regulates kainate-receptor-mediated synaptic transmission. *Nature* 447: 321–325, 2007.
190. Matsuda N, Kitami T, Suzuki T, Mizuno Y, Hattori N, and Tanaka K. Diverse effects of pathogenic mutations of Parkin that catalyze multiple monoubiquitylation in vitro. *J Biol Chem* 281: 3204–3209, 2006.
191. Matsuura T, Sutcliffe JS, Fang P, Galjaard RJ, Jiang YH, Benton CS, Rommens JM, and Beaudet AL. De novo truncating mutations in E6-AP ubiquitin-protein ligase gene (UBE3A) in Angelman syndrome. *Nat Genet* 15: 74–77, 1997.
192. McCullough J, Clague MJ, and Urbe S. AMSH is an endosome-associated ubiquitin isopeptidase. *J Cell Biol* 166: 487–492, 2004.
193. Mehta N, Loria PM, and Hobert O. A genetic screen for neurite outgrowth mutants in Caenorhabditis elegans reveals a new function for the F-box ubiquitin ligase component LIN-23. *Genetics* 166: 1253–1267, 2004.
194. Montarolo PG, Goelet P, Castellucci VF, Morgan J, Kandel ER, and Schacher S. A critical period for macromolecular synthesis in long-term heterosynaptic facilitation in Aplysia. *Science* (New York) 234: 1249–1254, 1986.
195. Mori S, Claesson-Welsh L, Okuyama Y, and Saito Y. Ligand-induced polyubiquitination of receptor tyrosine kinases. *Biochem Biophys Res Comm* 213: 32–39, 1995.
196. Morton ME, Street VA, and Nathanson NM. Selective regulation of Gi alpha 1 expression and function in PC12 cells by cAMP. *J Neurosci* 12: 1839–1846, 1992.
197. Mouatt-Prigent A, Muriel MP, Gu WJ, El Hachimi KH, Lucking CB, Brice A, and Hirsch EC. Ultrastructural localization of parkin in the rat brainstem, thalamus and basal ganglia. *J Neural Transm* 111: 1209–1218, 2004.
198. Mukhopadhyay D and Riezman H. Proteasome-independent functions of ubiquitin in endocytosis and signaling. *Science* (New York) 315: 201–205, 2007.
199. Mulkey RM, Endo S, Shenolikar S, and Malenka RC. Involvement of a calcineurin/inhibitor-1 phosphatase cascade in hippocampal long-term depression. *Nature* 369: 486–488, 1994.
200. Mulkey RM, Herron CE, and Malenka RC. An essential role for protein phosphatases in hippocampal long-term depression. *Science* (New York) 261: 1051–1055, 1993.
201. Muro I, Hay BA, and Clem RJ. The Drosophila DIAP1 protein is required to prevent accumulation of a continuously generated, processed form of the apical caspase DRONC. *J Biol Chem* 277: 49644–49650, 2002.
202. Murthy VN, Schikorski T, Stevens CF, and Zhu Y. Inactivity produces increases in neurotransmitter release and synapse size. *Neuron* 32: 673–682, 2001.
203. Myat A, Henry P, McCabe V, Flintoft L, Rotin D, and Tear G. Drosophila Nedd4, a ubiquitin ligase, is recruited by Commissureless to control cell surface levels of the roundabout receptor. *Neuron* 35: 447–459, 2002.

204. Nakata K, Abrams B, Grill B, Goncharov A, Huang X, Chisholm AD, and Jin Y. Regulation of a DLK-1 and p38 MAP kinase pathway by the ubiquitin ligase RPM-1 is required for presynaptic development. *Cell* 120: 407–420, 2005.
205. Naumann M, Bech-Otschir D, Huang X, Ferrell K, and Dubiel W. COP9 signalosome-directed c-Jun activation/stabilization is independent of JNK. *J Biol Chem* 274: 35297–35300, 1999.
206. Navon A and Goldberg AL. Proteins are unfolded on the surface of the ATPase ring before transport into the proteasome. *Mol Cell* 8: 1339–1349, 2001.
207. Nicholson DA, Yoshida R, Berry RW, Gallagher M, and Geinisman Y. Reduction in size of perforated postsynaptic densities in hippocampal axospinous synapses and age-related spatial learning impairments. *J Neurosci* 24: 7648–7653, 2004.
208. Nickell S, Mihalache O, Beck F, Hegerl R, Korinek A, and Baumeister W. Structural analysis of the 26S proteasome by cryoelectron tomography. *Biochem Biophys Res Comm* 353: 115–120, 2007.
209. Nickerson DP, West M, and Odorizzi G. Did2 coordinates Vps4-mediated dissociation of ESCRT-III from endosomes. *J Cell Biol* 175: 715–720, 2006.
210. Nussbaum AK, Dick TP, Keilholz W, Schirle M, Stevanovic S, Dietz K, Heinemeyer W, Groll M, Wolf DH, Huber R, Rammensee HG, and Schild H. Cleavage motifs of the yeast 20S proteasome beta subunits deduced from digests of enolase 1. *Proc Nat Acad Sci USA* 95: 12504–12509, 1998.
211. Okabe S, Kim HD, Miwa A, Kuriu T, and Okado H. Continual remodeling of postsynaptic density and its regulation by synaptic activity. *Nat Neurosci* 2: 804–811, 1999.
212. Osaka H, Wang YL, Takada K, Takizawa S, Setsuie R, Li H, Sato Y, Nishikawa K, Sun YJ, Sakurai M, Harada T, Hara Y, Kimura I, Chiba S, Namikawa K, Kiyama H, Noda M, Aoki S, and Wada K. Ubiquitin carboxy-terminal hydrolase L1 binds to and stabilizes monoubiquitin in neuron. *Hum Mol Genet* 12: 1945–1958, 2003.
213. Osterlund MT, Hardtke CS, Wei N, and Deng XW. Targeted destabilization of HY5 during light-regulated development of Arabidopsis. *Nature* 405: 462–466, 2000.
214. Ozdamar B, Bose R, Barrios-Rodiles M, Wang HR, Zhang Y, and Wrana JL. Regulation of the polarity protein Par6 by TGFbeta receptors controls epithelial cell plasticity. *Science* (New York) 307: 1603–1609, 2005.
215. Paisan-Ruiz C, Jain S, Evans EW, Gilks WP, Simon J, van der Brug M, de Munain AL, Aparicio S, Gil AM, Khan N, Johnson J, Martinez JR, Nicholl D, Carrera IM, Pena AS, de Silva R, Lees A, Marti-Masso JF, Perez-Tur J, Wood NW, and Singleton AB. Cloning of the gene containing mutations that cause PARK8-linked Parkinson's disease. *Neuron* 44: 595–600, 2004.
216. Pak DT and Sheng M. Targeted protein degradation and synapse remodeling by an inducible protein kinase. *Science* (New York) 302: 1368–1373, 2003.
217. Pak DT, Yang S, Rudolph-Correia S, Kim E, and Sheng M. Regulation of dendritic spine morphology by SPAR, a PSD-95-associated RapGAP. *Neuron* 31: 289–303, 2001.
218. Papa FR, Amerik AY, and Hochstrasser M. Interaction of the Doa4 deubiquitinating enzyme with the yeast 26S proteasome. *Mol Biol Cell* 10: 741–756, 1999.
219. Patrick GN, Bingol B, Weld HA, and Schuman EM. Ubiquitin-mediated proteasome activity is required for agonist-induced endocytosis of GluRs. *Curr Biol* 13: 2073–2081, 2003.
220. Peifer M, Pai LM, and Casey M. Phosphorylation of the Drosophila adherens junction protein Armadillo: roles for wingless signal and zeste-white 3 kinase. *Dev Biol* 166: 543–556, 1994.
221. Pesah Y, Pham T, Burgess H, Middlebrooks B, Verstreken P, Zhou Y, Harding M, Bellen H, and Mardon G. Drosophila parkin mutants have decreased mass and cell size and increased sensitivity to oxygen radical stress. *Development* 131: 2183–2194, 2004.

222. Peters JM. The anaphase-promoting complex: proteolysis in mitosis and beyond. *Mol Cell* 9: 931–943, 2002.
223. Peters JM. The anaphase promoting complex/cyclosome: a machine designed to destroy. *Nat Rev* 7: 644–656, 2006.
224. Petroski MD and Deshaies RJ. Function and regulation of cullin-RING ubiquitin ligases. *Nat Rev* 6: 9–20, 2005.
225. Petrucelli L, O'Farrell C, Lockhart PJ, Baptista M, Kehoe K, Vink L, Choi P, Wolozin B, Farrer M, Hardy J, and Cookson MR. Parkin protects against the toxicity associated with mutant alpha-synuclein: proteasome dysfunction selectively affects catecholaminergic neurons. *Neuron* 36: 1007–1019, 2002.
226. Philpott KL, McCarthy MJ, Klippel A, and Rubin LL. Activated phosphatidylinositol 3-kinase and Akt kinase promote survival of superior cervical neurons. *J Cell Biol* 139: 809–815, 1997.
227. Pickart CM. Mechanisms underlying ubiquitination. *Annu Rev Biochem* 70: 503–533, 2001.
228. Pickart CM and Cohen RE. Proteasomes and their kin: proteases in the machine age. *Nat Rev* 5: 177–187, 2004.
229. Polo S, Sigismund S, Faretta M, Guidi M, Capua MR, Bossi G, Chen H, De Camilli P, and Di Fiore PP. A single motif responsible for ubiquitin recognition and monoubiquitination in endocytic proteins. *Nature* 416: 451–455, 2002.
230. Polymeropoulos MH, Lavedan C, Leroy E, Ide SE, Dehejia A, Dutra A, Pike B, Root H, Rubenstein J, Boyer R, Stenroos ES, Chandrasekharappa S, Athanassiadou A, Papapetropoulos T, Johnson WG, Lazzarini AM, Duvoisin RC, Di Iorio G, Golbe LI, and Nussbaum RL. Mutation in the alpha-synuclein gene identified in families with Parkinson's disease. *Science* (New York) 276: 2045–2047, 1997.
231. Qi M, Zhuo M, Skalhegg BS, Brandon EP, Kandel ER, McKnight GS, and Idzerda RL. Impaired hippocampal plasticity in mice lacking the Cbeta1 catalytic subunit of cAMP-dependent protein kinase. *Proc Nat Acad Sci USA* 93: 1571–1576, 1996.
232. Raasi S and Pickart CM. Rad23 ubiquitin-associated domains (UBA) inhibit 26 S proteasome-catalyzed proteolysis by sequestering lysine 48-linked polyubiquitin chains. *J Biol Chem* 278: 8951–8959, 2003.
233. Raiborg C, Bache KG, Gillooly DJ, Madshus IH, Stang E, and Stenmark H. Hrs sorts ubiquitinated proteins into clathrin-coated microdomains of early endosomes. *Nat Cell Biol* 4: 394–398, 2002.
234. Reimann JD, Freed E, Hsu JY, Kramer ER, Peters JM, and Jackson PK. Emi1 is a mitotic regulator that interacts with Cdc20 and inhibits the anaphase promoting complex. *Cell* 105: 645–655, 2001.
235. Reimann JD, Gardner BE, Margottin-Goguet F, and Jackson PK. Emi1 regulates the anaphase-promoting complex by a different mechanism than Mad2 proteins. *Genes Dev* 15: 3278–3285, 2001.
236. Rougeulle C, Glatt H, and Lalande M. The Angelman syndrome candidate gene, UBE3A/E6-AP, is imprinted in brain. *Nat Genet* 17: 14–15, 1997.
237. Rubin DM, Glickman MH, Larsen CN, Dhruvakumar S, and Finley D. Active site mutants in the six regulatory particle ATPases reveal multiple roles for ATP in the proteasome. *EMBO J* 17: 4909–4919, 1998.
238. Rubinsztein DC. The roles of intracellular protein-degradation pathways in neurodegeneration. *Nature* 443: 780–786, 2006.
239. Sachse M, Strous GJ, and Klumperman J. ATPase-deficient hVPS4 impairs formation of internal endosomal vesicles and stabilizes bilayered clathrin coats on endosomal vacuoles. *J Cell Sci* 117: 1699–1708, 2004.

240. Sahai E, Garcia-Medina R, Pouyssegur J, and Vial E. Smurfl regulates tumor cell plasticity and motility through degradation of RhoA leading to localized inhibition of contractility. *J Cell Biol* 176: 35–42, 2007.
241. Sakata E, Yamaguchi Y, Kurimoto E, Kikuchi J, Yokoyama S, Yamada S, Kawahara H, Yokosawa H, Hattori N, Mizuno Y, Tanaka K, and Kato K. Parkin binds the Rpn10 subunit of 26S proteasomes through its ubiquitin-like domain. *EMBO Reports* 4: 301–306, 2003.
242. Salinas GD, Blair LA, Needleman LA, Gonzales JD, Chen Y, Li M, Singer JD, and Marshall J. Actinfilin is a Cul3 substrate adaptor, linking GluR6 kainate receptor subunits to the ubiquitin-proteasome pathway. *J Biol Chem* 281: 40164–40173, 2006.
243. Sang TK, Chang HY, Lawless GM, Ratnaparkhi A, Mee L, Ackerson LC, Maidment NT, Krantz DE, and Jackson GR. A Drosophila model of mutant human parkin-induced toxicity demonstrates selective loss of dopaminergic neurons and dependence on cellular dopamine. *J Neurosci* 27: 981–992, 2007.
244. Schaefer AM, Hadwiger GD, and Nonet ML. rpm-1, a conserved neuronal gene that regulates targeting and synaptogenesis in C. elegans. *Neuron* 26: 345–356, 2000.
245. Schaefer H and Rongo C. KEL-8 Is a Substrate Receptor for CUL3-dependent Ubiquitin Ligase That Regulates Synaptic Glutamate Receptor Turnover. *Mol Biol Cell* 17: 1250–60, 2006.
246. Schauber C, Chen L, Tongaonkar P, Vega I, Lambertson D, Potts W, and Madura K. Rad23 links DNA repair to the ubiquitin/proteasome pathway. *Nature* 391: 715–718, 1998.
247. Scheffner M, Huibregtse JM, and Howley PM. Identification of a human ubiquitin-conjugating enzyme that mediates the E6-AP-dependent ubiquitination of p53. *Proc Nat Acad Sci USA* 91: 8797–8801, 1994.
248. Schwamborn JC, Muller M, Becker AH, and Puschel AW. Ubiquitination of the GTPase Rap1B by the ubiquitin ligase Smurf2 is required for the establishment of neuronal polarity. *EMBO J* 26: 1410–1422, 2007.
249. Schwamborn JC and Puschel AW. The sequential activity of the GTPases Rap1B and Cdc42 determines neuronal polarity. *Nat Neurosci* 7: 923–929, 2004.
250. Sharma K, Fong DK, and Craig AM. Postsynaptic protein mobility in dendritic spines: long-term regulation by synaptic NMDA receptor activation. *Mol Cell Neurosci* 31: 702–712, 2006.
251. Sheng M and Hoogenraad CC. The postsynaptic architecture of excitatory synapses: a more quantitative view. *Annu Rev Biochem* 76: 823-847, 2007.
252. Shenoy SK, McDonald PH, Kohout TA, and Lefkowitz RJ. Regulation of receptor fate by ubiquitination of activated beta 2-adrenergic receptor and beta-arrestin. *Science* (New York) 294: 1307–1313, 2001.
253. Shih SC, Katzmann DJ, Schnell JD, Sutanto M, Emr SD, and Hicke L. Epsins and Vps27p/Hrs contain ubiquitin-binding domains that function in receptor endocytosis. *Nat Cell Biol* 4: 389–393, 2002.
254. Shimura H, Hattori N, Kubo S, Yoshikawa M, Kitada T, Matsumine H, Asakawa S, Minoshima S, Yamamura Y, Shimizu N, and Mizuno Y. Immunohistochemical and subcellular localization of Parkin protein: absence of protein in autosomal recessive juvenile parkinsonism patients. *Ann Neurol* 45: 668–672, 1999.
255. Shimura H, Schlossmacher MG, Hattori N, Frosch MP, Trockenbacher A, Schneider R, Mizuno Y, Kosik KS, and Selkoe DJ. Ubiquitination of a new form of alpha-synuclein by parkin from human brain: implications for Parkinson's disease. *Science* (New York) 293: 263–269, 2001.
256. Skowyra D, Craig KL, Tyers M, Elledge SJ, and Harper JW. F-box proteins are receptors that recruit phosphorylated substrates to the SCF ubiquitin-ligase complex. *Cell* 91: 209–219, 1997.

257. Slagsvold T, Aasland R, Hirano S, Bache KG, Raiborg C, Trambaiolo D, Wakatsuki S, and Stenmark H. Eap45 in mammalian ESCRT-II binds ubiquitin via a phosphoinositide-interacting GLUE domain. *J Biol Chem* 280: 19600–19606, 2005.
258. Soboleva TA and Baker RT. Deubiquitinating enzymes: their functions and substrate specificity. *Curr Protein & Peptide Sci* 5: 191–200, 2004.
259. Sorkina T, Miranda M, Dionne KR, Hoover BR, Zahniser NR, and Sorkin A. RNA interference screen reveals an essential role of Nedd4-2 in dopamine transporter ubiquitination and endocytosis. *J Neurosci* 26: 8195–8205, 2006.
260. Speese SD, Trotta N, Rodesch CK, Aravamudan B, and Broadie K. The ubiquitin proteasome system acutely regulates presynaptic protein turnover and synaptic efficacy. *Curr Biol* 13: 899–910, 2003.
261. Springael JY, Galan JM, Haguenauer-Tsapis R, and Andre B. NH4+-induced down-regulation of the Saccharomyces cerevisiae Gap1p permease involves its ubiquitination with lysine-63-linked chains. *J Cell Sci* 112 (Pt 9): 1375–1383, 1999.
262. Sriram SR, Li X, Ko HS, Chung KK, Wong E, Lim KL, Dawson VL, and Dawson TM. Familial-associated mutations differentially disrupt the solubility, localization, binding and ubiquitination properties of parkin. *Hum Mol Genet* 14: 2571–2586, 2005.
263. Staropoli JF, McDermott C, Martinat C, Schulman B, Demireva E, and Abeliovich A. Parkin is a component of an SCF-like ubiquitin ligase complex and protects postmitotic neurons from kainate excitotoxicity. *Neuron* 37: 735–749, 2003.
264. Staub O, Gautschi I, Ishikawa T, Breitschopf K, Ciechanover A, Schild L, and Rotin D. Regulation of stability and function of the epithelial Na+ channel (ENaC) by ubiquitination. *EMBO J* 16: 6325–6336, 1997.
265. Stegmuller J, Konishi Y, Huynh MA, Yuan Z, Dibacco S, and Bonni A. Cell-intrinsic regulation of axonal morphogenesis by the Cdh1-APC target SnoN. *Neuron* 50: 389–400, 2006.
266. Stein V, House DR, Bredt DS, and Nicoll RA. Postsynaptic density-95 mimics and occludes hippocampal long-term potentiation and enhances long-term depression. *J Neurosci* 23: 5503–5506, 2003.
267. Steward O and Schuman EM. Protein synthesis at synaptic sites on dendrites. *Annu Rev Neurosci* 24: 299–325, 2001.
268. Strickland E, Hakala K, Thomas PJ, and DeMartino GN. Recognition of misfolding proteins by PA700, the regulatory subcomplex of the 26 S proteasome. *J Biol Chem* 275: 5565–5572, 2000.
269. Sutton MA and Schuman EM. Dendritic protein synthesis, synaptic plasticity, and memory. *Cell* 127: 49–58, 2006.
270. Swaminathan S, Amerik AY, and Hochstrasser M. The Doa4 deubiquitinating enzyme is required for ubiquitin homeostasis in yeast. *Molr Biol Cell* 10: 2583–2594, 1999.
271. Tang Z, Li B, Bharadwaj R, Zhu H, Ozkan E, Hakala K, Deisenhofer J, and Yu H. APC2 Cullin protein and APC11 RING protein comprise the minimal ubiquitin ligase module of the anaphase-promoting complex. *Mol Biol Cell* 12: 3839–3851, 2001.
272. Teo H, Veprintsev DB, and Williams RL. Structural insights into endosomal sorting complex required for transport (ESCRT-I) recognition of ubiquitinated proteins. *J Biol Chem* 279: 28689–28696, 2004.
273. Thompson BJ, Mathieu J, Sung HH, Loeser E, Rorth P, and Cohen SM. Tumor suppressor properties of the ESCRT-II complex component Vps25 in Drosophila. *Dev Cell* 9: 711–720, 2005.
274. Thrower JS, Hoffman L, Rechsteiner M, and Pickart CM. Recognition of the polyubiquitin proteolytic signal. *EMBO J* 19: 94–102, 2000.

275. Toni N, Buchs PA, Nikonenko I, Povilaitite P, Parisi L, and Muller D. Remodeling of synaptic membranes after induction of long-term potentiation. *J Neurosci* 21: 6245–6251, 2001.
276. Tsuriel S, Geva R, Zamorano P, Dresbach T, Boeckers T, Gundelfinger ED, Garner CC, and Ziv NE. Local sharing as a predominant determinant of synaptic matrix molecular dynamics. *PLoS Biol* 4: e271, 2006.
277. Tully T, Preat T, Boynton SC, and Del Vecchio M. Genetic dissection of consolidated memory in Drosophila. *Cell* 79: 35–47, 1994.
278. Tursun B, Schluter A, Peters MA, Viehweger B, Ostendorff HP, Soosairajah J, Drung A, Bossenz M, Johnsen SA, Schweizer M, Bernard O, and Bach I. The ubiquitin ligase Rnf6 regulates local LIM kinase 1 levels in axonal growth cones. *Genes Dev* 19: 2307–2319, 2005.
279. Valente EM, Abou-Sleiman PM, Caputo V, Muqit MM, Harvey K, Gispert S, Ali Z, Del Turco D, Bentivoglio AR, Healy DG, Albanese A, Nussbaum R, Gonzalez-Maldonado R, Deller T, Salvi S, Cortelli P, Gilks WP, Latchman DS, Harvey RJ, Dallapiccola B, Auburger G, and Wood NW. Hereditary early-onset Parkinson's disease caused by mutations in PINK1. *Science* (New York) 304: 1158–1160, 2004.
280. van Roessel P, Elliott DA, Robinson IM, Prokop A, and Brand AH. Independent regulation of synaptic size and activity by the anaphase-promoting complex. *Cell* 119: 707–718, 2004.
281. van Woerden GM, Harris KD, Hojjati MR, Gustin RM, Qiu S, de Avila Freire R, Jiang YH, Elgersma Y, and Weeber EJ. Rescue of neurological deficits in a mouse model for Angelman syndrome by reduction of alphaCaMKII inhibitory phosphorylation. *Nat Neurosci* 10: 280–282, 2007.
282. Verma P, Chierzi S, Codd AM, Campbell DS, Meyer RL, Holt CE, and Fawcett JW. Axonal protein synthesis and degradation are necessary for efficient growth cone regeneration. *J Neurosci* 25: 331–342, 2005.
283. Verma R, Oania R, Graumann J, and Deshaies RJ. Multiubiquitin chain receptors define a layer of substrate selectivity in the ubiquitin-proteasome system. *Cell* 118: 99–110, 2004.
284. Vila M and Przedborski S. Genetic clues to the pathogenesis of Parkinson's disease. *Nat Med* 10 Suppl: S58–62, 2004.
285. von Arnim AG. A hitchhiker's guide to the proteasome. *Sci STKE* 2001: PE2, 2001.
286. Von Coelln R, Thomas B, Savitt JM, Lim KL, Sasaki M, Hess EJ, Dawson VL, and Dawson TM. Loss of locus coeruleus neurons and reduced startle in parkin null mice. *Proc Nat Acad Sci USA* 101: 10744–10749, 2004.
287. von Schwedler UK, Stuchell M, Muller B, Ward DM, Chung HY, Morita E, Wang HE, Davis T, He GP, Cimbora DM, Scott A, Krausslich HG, Kaplan J, Morham SG, and Sundquist WI. The protein network of HIV budding. *Cell* 114: 701–713, 2003.
288. Vu TH and Hoffman AR. Imprinting of the Angelman syndrome gene, UBE3A, is restricted to brain. *Nat Genet* 17: 12–13, 1997.
289. Walz J, Erdmann A, Kania M, Typke D, Koster AJ, and Baumeister W. 26S proteasome structure revealed by three-dimensional electron microscopy. *J Struct Biol* 121: 19–29, 1998.
290. Wan HI, DiAntonio A, Fetter RD, Bergstrom K, Strauss R, and Goodman CS. Highwire regulates synaptic growth in Drosophila. *Neuron* 26: 313–329, 2000.
291. Wang C, Lu R, Ouyang X, Ho MW, Chia W, Yu F, and Lim KL. Drosophila overexpressing parkin R275W mutant exhibits dopaminergic neuron degeneration and mitochondrial abnormalities. *J Neurosci* 27: 8563–8570, 2007.

292. Wang HR, Zhang Y, Ozdamar B, Ogunjimi AA, Alexandrova E, Thomsen GH, and Wrana JL. Regulation of cell polarity and protrusion formation by targeting RhoA for degradation. *Science* (New York) 302: 1775–1779, 2003.
293. Weeber EJ, Jiang YH, Elgersma Y, Varga AW, Carrasquillo Y, Brown SE, Christian JM, Mirnikjoo B, Silva A, Beaudet AL, and Sweatt JD. Derangements of hippocampal calcium/calmodulin-dependent protein kinase II in a mouse model for Angelman mental retardation syndrome. *J Neurosci* 23: 2634–2644, 2003.
294. Whitley P, Reaves BJ, Hashimoto M, Riley AM, Potter BV, and Holman GD. Identification of mammalian Vps24p as an effector of phosphatidylinositol 3,5-bisphosphate-dependent endosome compartmentalization. *J Biol Chem* 278: 38786–38795, 2003.
295. Wilkinson CR, Ferrell K, Penney M, Wallace M, Dubiel W, and Gordon C. Analysis of a gene encoding Rpn10 of the fission yeast proteasome reveals that the polyubiquitin-binding site of this subunit is essential when Rpn12/Mts3 activity is compromised. *J Bioll Chem* 275: 15182–15192, 2000.
296. Willeumier K, Pulst SM, and Schweizer FE. Proteasome inhibition triggers activity-dependent increase in the size of the recycling vesicle pool in cultured hippocampal neurons. *J Neurosci* 26: 11333–11341, 2006.
297. Williams CA, Beaudet AL, Clayton-Smith J, Knoll JH, Kyllerman M, Laan LA, Magenis RE, Moncla A, Schinzel AA, Summers JA, and Wagstaff J. Angelman syndrome 2005: updated consensus for diagnostic criteria. *Am J Medical Genet* 140: 413–418, 2006.
298. Williams RL and Urbe S. The emerging shape of the ESCRT machinery. *Nat Rev* 8: 355–368, 2007.
299. Wilson R, Goyal L, Ditzel M, Zachariou A, Baker DA, Agapite J, Steller H, and Meier P. The DIAP1 RING finger mediates ubiquitination of Dronc and is indispensable for regulating apoptosis. *Nat Cell Biol* 4: 445–450, 2002.
300. Wilson SM, Bhattacharyya B, Rachel RA, Coppola V, Tessarollo L, Householder DB, Fletcher CF, Miller RJ, Copeland NG, and Jenkins NA. Synaptic defects in ataxia mice result from a mutation in Usp14, encoding a ubiquitin-specific protease. *Nat Genet* 32: 420–425, 2002.
301. Winston JT, Strack P, Beer-Romero P, Chu CY, Elledge SJ, and Harper JW. The SCFbeta-TRCP-ubiquitin ligase complex associates specifically with phosphorylated destruction motifs in IkappaBalpha and beta-catenin and stimulates IkappaBalpha ubiquitination in vitro. *Genes Dev* 13: 270–283, 1999.
302. Wojcik C and DeMartino GN. Intracellular localization of proteasomes. *Int J Biochem Cell Biol* 35: 579–589, 2003.
303. Wolf B, Seeger MA, and Chiba A. Commissureless endocytosis is correlated with initiation of neuromuscular synaptogenesis. *Development* 125: 3853–3863, 1998.
304. Wu C, Wairkar YP, Collins CA, and DiAntonio A. Highwire function at the Drosophila neuromuscular junction: spatial, structural, and temporal requirements. *J Neurosci* 25: 9557–9566, 2005.
305. Wyszynski M, Kim E, Dunah AW, Passafaro M, Valtschanoff JG, Serra-Pages C, Streuli M, Weinberg RJ, and Sheng M. Interaction between GRIP and liprin-alpha/SYD2 is required for AMPA receptor targeting. *Neuron* 34: 39–52, 2002.
306. Yang N, Higuchi O, Ohashi K, Nagata K, Wada A, Kangawa K, Nishida E, and Mizuno K. Cofilin phosphorylation by LIM-kinase 1 and its role in Rac-mediated actin reorganization. *Nature* 393: 809–812, 1998.
307. Yang Y, Nishimura I, Imai Y, Takahashi R, and Lu B. Parkin suppresses dopaminergic neuron-selective neurotoxicity induced by Pael-R in Drosophila. *Neuron* 37: 911–924, 2003.

308. Yao I, Takagi H, Ageta H, Kahyo T, Sato S, Hatanaka K, Fukuda Y, Chiba T, Morone N, Yuasa S, Inokuchi K, Ohtsuka T, Macgregor GR, Tanaka K, and Setou M. SCRAPPER-Dependent Ubiquitination of Active Zone Protein RIM1 Regulates Synaptic Vesicle Release. *Cell* 130: 943–957, 2007.
309. Yost C, Torres M, Miller JR, Huang E, Kimelman D, and Moon RT. The axis-inducing activity, stability, and subcellular distribution of beta-catenin is regulated in Xenopus embryos by glycogen synthase kinase 3. *Genes Dev* 10: 1443–1454, 1996.
310. Young P, Deveraux Q, Beal RE, Pickart CM, and Rechsteiner M. Characterization of two polyubiquitin binding sites in the 26 S protease subunit 5a. *J Biol Chem* 273: 5461–5467, 1998.
311. Zamborlini A, Usami Y, Radoshitzky SR, Popova E, Palu G, and Gottlinger H. Release of autoinhibition converts ESCRT-III components into potent inhibitors of HIV-1 budding. *Proc Nat Acad Sci USA* 103: 19140–19145, 2006.
312. Zhai Q, Wang J, Kim A, Liu Q, Watts R, Hoopfer E, Mitchison T, Luo L, and He Z. Involvement of the ubiquitin-proteasome system in the early stages of wallerian degeneration. *Neuron* 39: 217–225, 2003.
313. Zhang Y, Gao J, Chung KK, Huang H, Dawson VL, and Dawson TM. Parkin functions as an E2-dependent ubiquitin- protein ligase and promotes the degradation of the synaptic vesicle-associated protein, CDCrel-1. *Proc Nat Acad Sci USA* 97: 13354–13359, 2000.
314. Zhen M, Huang X, Bamber B, and Jin Y. Regulation of presynaptic terminal organization by C. elegans RPM-1, a putative guanine nucleotide exchanger with a RING-H2 finger domain. *Neuron* 26: 331–343, 2000.
315. Zhen M and Jin Y. The liprin protein SYD-2 regulates the differentiation of presynaptic termini in C. elegans. *Nature* 401: 371–375, 1999.
316. Zheng N, Schulman BA, Song L, Miller JJ, Jeffrey PD, Wang P, Chu C, Koepp DM, Elledge SJ, Pagano M, Conaway RC, Conaway JW, Harper JW, and Pavletich NP. Structure of the Cul1-Rbx1-Skp1-F boxSkp2 SCF ubiquitin ligase complex. *Nature* 416: 703–709, 2002.
317. Zoghbi HY. Postnatal neurodevelopmental disorders: meeting at the synapse? *Science* (New York) 302: 826–830, 2003.
318. Zwickl P, Ng D, Woo KM, Klenk HP, and Goldberg AL. An archaebacterial ATPase, homologous to ATPases in the eukaryotic 26 S proteasome, activates protein breakdown by 20 S proteasomes. *J Biol Chem* 274: 26008–26014, 1999.
319. Arevalo JC, Waite J, Rajagopal R, Beyna M, Chen ZY, Lee FS, and Chao MV. Cell survival through Trk neurotrophin receptors is differentially regulated by ubiquitination. *Neuron* 50: 549–559, 2006.
320. Choe EA, Liao L, Zhou JY, Cheng D, Duong DM, Jin P, Tsai LH, and Peng J. Neuronal morphogenesis is regulated by the interplay between cyclin-dependent kinase 5 and the ubiquitin ligase mind bomb 1. *J Neurosci* 27: 9503–9512, 2007.
321. Saigoh K, Wang YL, Suh JG, Yamanishi T, Sakai Y, Kiyosawa H, Harada T, Ichihara N, Wakana S, Kikuchi T, and Wada K. Intragenic deletion in the gene encoding ubiquitin carboxy-terminal hydrolase in gad mice. *Nat Genet* 23: 47–51, 1999.
322. Kawaguchi Y, Okamoto T, Taniwaki M, Aizawa M, Inoue M, Katayama S, Kawakami H, Nakamura S, Nishimura M, Akiguchi I, and et al. CAG expansions in a novel gene for Machado-Joseph disease at chromosome 14q32.1. *Nat Genet* 8: 221–228, 1994.

Signaling from Synapse to Nucleus

Carrie L. Heusner and Kelsey C. Martin

Department of Biological Chemistry and Department of Psychiatry and Biobehavioral Sciences, Brain Research Institute, Semel Institute for Neuroscience and Human Behavior, UCLA, Los Angeles, CA 90095, USA, kcmartin@mednet.ucla.edu

1 Introduction

Neurons, like all other eukaryotic cells, persistently alter their structure and/or function in response to extracellular signals through stimulus-induced changes in gene expression. While signal transduction pathways that serve to convey signals from the plasma membrane to the nucleus have been well characterized in non-neuronal cells, the transduction of signals from cell surface to nucleus poses a unique set of challenges in neurons because the distances between the sites receiving the signal and the nucleus can be extremely large (Fig. 1). The polarized morphology of nerve cells, with processes extending distances that can exceed the diameter of the cell body by orders of magnitude, has thus necessitated the development of specialized mechanisms for retrograde transport of information received at the cell surface to the soma.

Long-distance retrograde transport of signals is essential to the survival, growth and adaptability of neurons in the developing and the adult nervous system. During development, signals are transported from distal growth cones to the nucleus while neurons change their transcriptional profiles as axonal growth cones navigate and contact potential synaptic partners (41). In the adult brain, long-lasting forms of synaptic plasticity, including those that underlie learning and memory, depend on new transcription, indicating that stimuli received at synaptic sites are relayed to the nucleus to alter gene expression (2, 42). Retrograde signaling from distal sites also occurs in mature neurons following axonal injury, when injury-induced signals are transported to the nucleus to trigger transcription-dependent axonal regeneration (28, 61). During each of these processes, signals received at distal compartments elicit transcriptional changes in the nucleus. While many studies have identified genes whose expression is altered by specific stimuli, the mechanisms whereby signals are transduced from the site of stimulation to the nucleus are just beginning to be elucidated.

J.W. Hell, M.D. Ehlers (eds.), *Structural and Functional Organization of the Synapse*, DOI: 10.1007/978-0-387-77232-5_20,

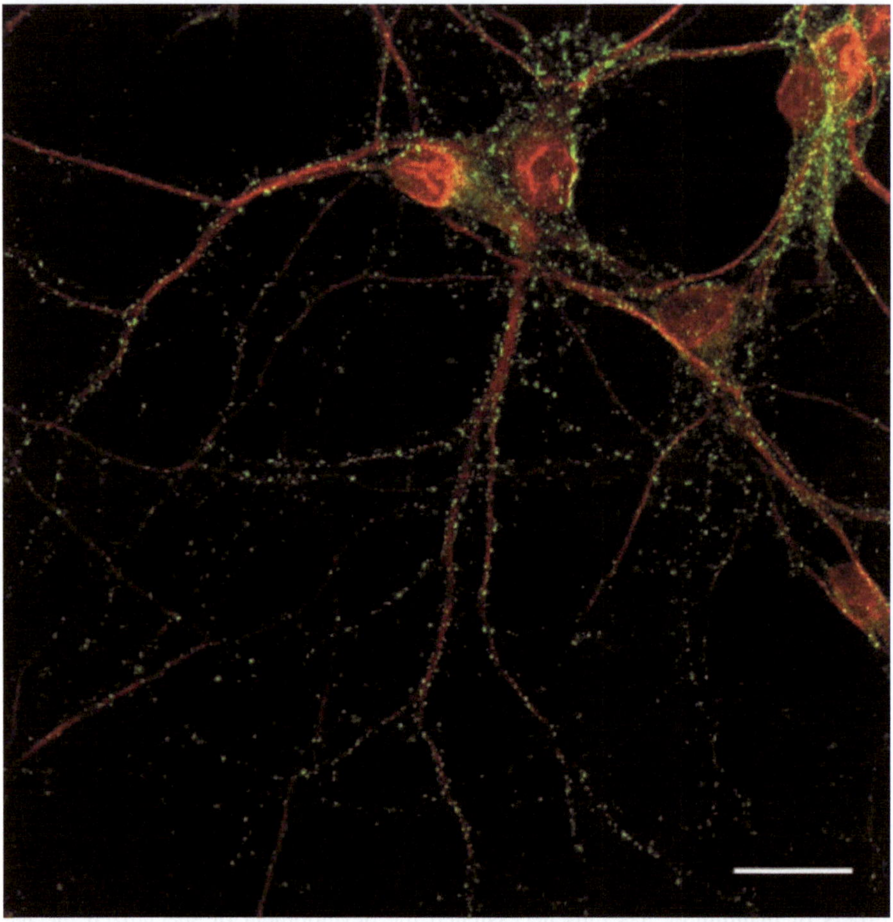

Fig. 1. Given the highly polarized morphology of neurons, signals received at distal synapses must travel significant distances to reach the cell body and the nucleus. Shown are cultured rat hippocampal neurons (14 days in vitro), stained for the synaptic marker PSD-95 (*green*) and the somatodendritic marker MAP-2 (*red*). Many of the PSD-95 immunoreactive synapses are present at significant distances from the cell soma, raising the question of how synaptically generated signals can be efficiently transmitted to the nucleus. Scale bar = 25 microns. Photomicrograph courtesy of M.M. Poon.

A central question, given the significant distances that exist between distal compartments and the soma in neurons, is how retrograde signaling can occur in a manner in which the transcriptional response is temporally coupled to the stimulus. Nerve cells are specialized for rapid signaling between compartments (Fig. 2). Thus, depolarization at growth cones or synapses can travel by passive electrotonic

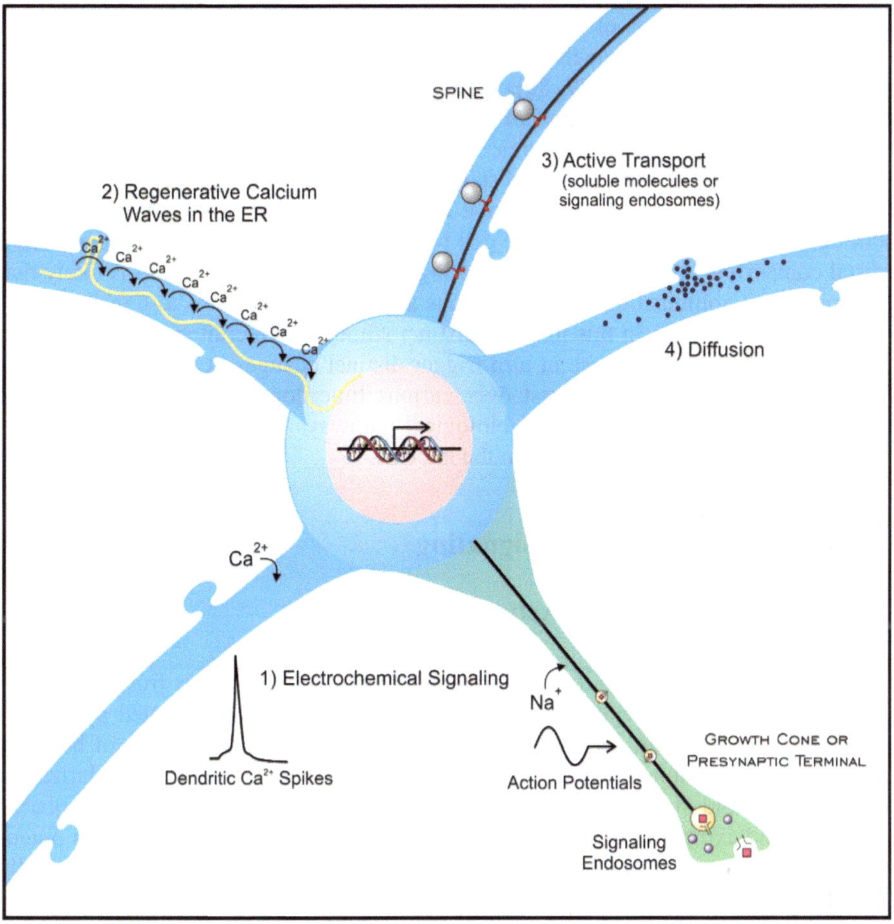

Fig. 2. Neurons possess several cell biological mechanisms for transmitting signals from distal compartments in axons (shown in *green*) and dendrites (shown in *blue*) to the nucleus (shown in *pink*). These include (1) rapid electrochemical signaling, in which action potentials (in axons) or dendritic spikes (in dendrites) rapidly propagate from distal sites to the soma where they open voltage-gated calcium channels. The resulting elevation of somatic calcium concentrations triggers calcium-dependent changes in gene expression in the nucleus. (2) Synaptic signals can also cause a release of calcium stored in the ER (represented by the *yellow line*), leading to regenerative calcium waves mediated by inositol triphosphate and ryanodine receptors. Since the ER extends into axons and dendrites and is continuous with the nuclear membrane, the regenerative calcium wave can be rapidly propagated to the nucleus, where they initiate calcium-dependent changes in gene expression. (3) Signals can be actively transported from synapses along cytoskeletal components (*black line*) by molecular motors (*red*). These signals can consist of soluble proteins or signaling endosomes. In the nucleus, these signals function to alter transcription. (4) Both second messengers and soluble proteins can passively diffuse from distal compartments toward the cell body and into the nucleus.

spread, action potentials or calcium spikes to the cell soma, where the opening of voltage-gated calcium channels in the plasma membrane rapidly transmits signals to the nucleus (8, 18, 24, 66, 79, 80). Rapid signaling can also occur by regenerative calcium waves mediated by inositol triphosphate receptors ($InsP_3Rs$) and ryanodine receptors (RYRs) in the endoplasmic reticulum (ER), which is continuous with the nuclear envelope yet extends into distal neuronal processes (12). Slower retrograde signaling pathways include the passive or active transport of signaling molecules through the axon and dendrite to the cell body (58) and the transport of signals from distal compartments in the form of signaling endosomes (22, 33, 36). It is likely that neurons use a combination of these various signaling mechanisms to transmit information to the nucleus. In this manner, distinct stimuli can recruit specific retrograde signaling pathways, and these in turn trigger distinct changes in transcriptional profiles that are initiated and persist over various time domains. In this chapter, we introduce each of the various cell biological pathways neurons can use to convey information received at distal sites to the nucleus.

2 Rapid Synapse-Nucleus Signaling

2.1 Signaling Mediated by Action Potentials

Electrochemical signaling allows neurons to rapidly deliver signals from distal sites to the nucleus. Local depolarization of the synaptic terminal, mediated by ligand-gated ion channels, can activate voltage-gated ion channels. The influx of both sodium and calcium ions further depolarizes the membrane, and if this depolarization reaches threshold at the axon hillock, it triggers action potentials, which depolarize the soma and axon (and potentially into dendrites via backpropagating action potentials) (10). Synaptic potentials generated at distal dendritic sites can also be rapidly conveyed to the cell soma via forward-propagating dendritic spikes (80). Once they reach the soma, action potentials and dendritic spikes both result in the opening of voltage-gated calcium channels, triggering a large increase in intracellular calcium. The increase in somatic calcium can in turn activate calcium-sensitive signaling cascades that function to alter gene expression in the nucleus. Calcium-sensitive signaling molecules can be activated directly within the nucleus via increases in nuclear calcium concentrations, or they can be activated in the somatic cytoplasm and then imported into the nucleus. These signaling molecules activate transcription factors and thereby couple neuronal activity to changes in gene expression (1, 8).

Whether calcium-dependent signaling cascades are directly activated in the nucleus or in the somatic cytoplasm, action potential or spike-mediated signaling from synapse to nucleus is extremely rapid. Electrochemical signals propagate along the length of the neuron within milliseconds. The propagation of calcium transients to the nucleus occurs within seconds, and the transport of signaling molecules (e.g. activated kinases) from somatic cytoplasm to the nucleus occurs within minutes. As such, transcriptional changes that result from action potentials or dendritic spikes could occur very quickly (within minutes) after a signal is received at the synapse.

An additional advantage of electrochemical signaling between synapse and nucleus is that an influx of calcium occurs in the cell body. This increases the likelihood that a calcium-dependent intracellular signal will be efficiently conveyed to the nucleus, and not be diminished as it travels the entire length of the dendrite. As a result, the amount of signal that reaches the nucleus will be significantly greater than it would be if the signal were transported directly from the synapse.

Two other features of signaling to the nucleus via action potentials should be considered. First, signaling by action potentials spreads throughout the cell, and thereby can activate signaling cascades throughout the cell. As such, it has been proposed that action potentials are not "synapse to nucleus" signals, but rather global, cell-wide signals initiated by synaptic stimulation (1). Presumably, in order to mediate signaling underlying synapse-specific changes in function, the products of gene expression induced by action potentials must combine with local synaptic tags that are generated by the initial synaptic stimulation (20, 51). Synaptic tags are localized changes that persist at synapses following stimulation and that function together with the products of gene expression to produce persistent changes in synaptic structure and function. In fact, any transcriptional process should be considered to be a "global," cell-wide change, and likely needs to combine with a synaptic tag to have synapse-specific effects. A second consideration about electrochemical signaling is that to function as a mechanism mediating activity-dependent changes in gene expression, it is likely that a threshold number of action potentials or spikes occur within a certain time window in order to activate transcriptional changes, with distinct patterns of activity triggering differential patterns of transcription.

2.2 Signaling Via the Endoplasmic Reticulum

Rapid calcium-dependent signaling from synapse to nucleus can also occur via the ER, a major internal store of calcium. In neurons, the ER is continuous with the nuclear membrane and extends into both axons and dendrites. Within the axon, soma and initial dendritic segment, electron microscopy studies reveal that the ER can come into close proximity to the plasma membrane (for review see (12)). Within the dendrite, the ER extends into spines where it forms the spine apparatus (71).

The ER contains two types of receptors that function to release calcium from the ER lumen into the cytosol: $InsP_3Rs$ (19) and RYRs (84). $InsP_3Rs$ are found throughout the ER, including in synapses and spines as well as in regions where the ER is in close proximity to the plasma membrane. $InsP_3Rs$ are activated by inositol triphosphate (IP_3), a second messenger generated by activation of G-protein coupled receptors that are coupled to phospholipase C, such as metabotropic glutamate receptors at excitatory synapses. Their activity is critically modulated by cytosolic calcium, such that calcium release is significantly potentiated by increased concentrations of cytosolic calcium. This sensitivity to calcium allows $InsP_3Rs$ to function as calcium-induced calcium release (CICR) channels. Notably, recent studies have shown that $InsP_3Rs$ can also be activated in the absence of IP_3, through the binding of a family of calcium binding proteins (77, 83). While RYRs are found primarily in the cell body, they are also present at more distal sites and can colocalize with $InsP_3Rs$. RYRs are activated when cytosolic calcium concentrations are between 1-10 μM; binding of calcium to the RYR promotes release of calcium from the ER lumen into the cytosol (84). As such, RYRs also function as CICR channels.

The sensitivity of both $InsP_3Rs$ and RYRs to calcium underlies their ability to generate regenerative calcium waves. Thus, calcium released from the internal stores in the ER is capable of activating neighboring $InsP_3Rs$ or RYRs, allowing the signal to continue along the ER. Because the ER can extend to the most distal sites in axons and dendrites, yet is connected with the nuclear envelope, activation of neighboring receptors provides a means for a continuous signal to be rapidly propagated all the way to the nucleus. Receptors located at synapses and/or near the plasma membrane are activated by calcium influx through plasma membrane receptors; calcium released from internal stores then promotes propagation of a calcium wave from the initial site of activation through the neuronal process, to the nuclear envelope (9, 12, 56, 74). Release of calcium from the ER amplifies the effects of calcium influx through plasma membrane calcium channels. Michael Berridge (12) has thus proposed that the ER acts as a "neuron-within-a-neuron." While the outer membrane responds quickly to external stimuli through the opening of ligand-gated ion channels and activation of G protein-coupled receptors, the ER responds to the internal signals generated in response to external stimuli.

The influx of extracellular calcium through ion channels and calcium released from intracellular stores in the ER both cause an increase in calcium concentrations in the cell body. The increases in somatic calcium concentrations can in turn activate calcium-dependent signaling cascades in the somatic cytoplasm and/or in the nucleus, both of which can ultimately trigger calcium-dependent changes in gene expression. While signaling from synapse to nucleus via regenerative calcium waves in the ER is somewhat slower than electrochemical signals, since it depends on both metabotropic and ionotropic receptors, it is nonetheless a rapid mechanism of conveying signals from distal sites to the soma. Further, the regenerative nature of the calcium waves in the ER indicates that the signal is not diminished as it travels from the site of initiation to the soma.

3 Physical Translocation of Signaling Molecules from Distal Compartments to the Nucleus

The physical translocation of molecules through the axon or dendrite to the cell soma provides another mechanism whereby neurons can signal from distal compartments to the nucleus. This type of signaling includes the retrograde transport of specialized organelles known as signaling endosomes as well as transport of soluble molecules via either passive or active diffusion. Transport of signals via these pathways is much slower than signaling by action potentials or by regenerative calcium waves in the ER. As a result, the coupling between stimulation at a distal site and the transcriptional response is slower and may persist over a longer time period. Notably, unlike nuclear signaling by action potentials or regenerative calcium waves in the ER, the physical translocation of molecules from distal sites to the nucleus does not necessarily depend on calcium. This thus provides a mechanism whereby stimuli that do not cause depolarization or calcium influx (e.g. some modulatory inputs or growth factors) can nonetheless elicit changes in gene expression.

Before considering specific aspects of retrograde transport via signaling endosomes or active nuclear import pathways, it is useful to briefly discuss the role of the cytoskeleton and of motor proteins in the retrograde transport of molecules to the neuronal soma. The neuronal cytoskeleton includes microtubules, actin and intermediate filaments. In addition to their structural roles in neurons, microtubules and actin filaments serve as tracks for the directed movements of molecules within neuronal processes (27, 47). Microtubules are particularly relevant to the transport of signals from distal synapses to the soma because they provide long-range tracks for transport. In axons, microtubules are uniformly oriented with the plus ends at the synapse and minus ends at the cell body. In dendrites, the polarity of microtubules with respect to the cell body is mixed. Active transport along microtubules is mediated by molecular motors, of which there are two major classes: kinesin motor proteins, which generally transport from minus to plus ends of microtubules, and dynein motor proteins, which transport from plus to minus ends of microtubules. In axons, anterograde transport from the cell body to the synapse thus occurs via plus-end-directed kinesins and retrograde transport from the terminals to the soma occurs via dynein, and possibly minus-end-directed kinesins. The directionality of motor-protein driven transport is less clear in dendrites because of the mixed polarity of microtubules, although microtubules in distal dendrites have their plus ends oriented distally indicating that dynein is required for retrograde transport of cargoes from these sites (68).

A central question in the field of molecular motors concerns the specificity and regulation of motor-cargo binding (27). This question is relevant to signaling from synapse to nucleus in neurons since the regulated binding of cargoes to specific motors will have obvious effects on the directionality of transport in axons and/or dendrites. Cargo binding to kinesins, either directly or through adaptor or scaffolding proteins relieves auto-inhibition of the N-terminal tail of the motor protein by C-terminal sequences, allowing the cargo-containing kinesin to travel along microtubules. Cargo binding to dynein is frequently mediated by a large dynein-binding protein called dynactin (34). Recent studies have indicated that post-translational modifications of tubulin can also influence transport in neurons by modulating the association of motor proteins with microtubules (14). Specifically, acetylation of tubulin has been shown to increase the association of kinesin-1 to microtubules and the transport of c-Jun N-terminal kinase-interacting proteins (JIPs, a family of adaptor proteins for kinesin cargoes) into distal neurites (64), and to increase the trafficking of BDNF-containing vesicles along microtubules in neurons (17). Thus, transport along microtubules in neuronal processes may be regulated by the initial binding of cargoes to molecular motors and by the binding of molecular motors to microtubules. The latter may be specifically regulated by post-translational modifications of tubulin, which include acetylation, tyrosination and polyglutamylation (76).

3.1 Retrograde Signaling by Signaling Endosomes

One way of transmitting information received at distal compartments to the soma is via the signaling endosome pathway (22, 33, 36). In this pathway, ligand-bound receptors are internalized and transported back to the cell body in signaling endosomes, functioning as activated receptors as they travel to the soma. This process has been best characterized in neurons in terms of the binding of nerve growth factor (NGF) to the tyrosine kinase receptor TrkA at the axon terminal. Binding of NGF to TrkA receptors causes dimerization of the receptor and autophosphorylation of specific tyrosine residues in the cytoplasmic domain. The binding of the ligand and subsequent phosphorylation leads to an "activated" ligand-receptor complex. This entire complex can be internalized through an endocytotic pathway that travels along microtubules in a dynein-dependent manner. The ligand-receptor complex serves as the message that is transported from the nerve terminal to the soma in a "signaling endosome." In the soma, the ligand-bound receptor activates signaling cascades that are then transmitted into the nucleus where they can alter transcription. A proposed advantage of the signaling endosome pathway in neurons is that it is "regenerative," in the sense that the ligand remains bound to the receptor and thus continuously recruits signaling cascades as it travels through the axon to the nucleus (37). Many of the experiments elucidating the signaling endosome pathway have been in compartmentalized chambers in which the cell bodies of sympathetic neurons are separated from their axons by barriers, a powerful system that allows manipulation and discrimination of trafficking in distal axons from events in cell bodies (15).

While there is considerable evidence for an endosome-mediated transport process for transmitting neurotrophic signaling from the nerve terminal to the nucleus, there are alternative hypotheses for how activation of distal Trk receptors transmits information to the nucleus. One such alternative idea is that activation of a Trk receptor initiates a ligand-independent cascade of phosphorylation of Trk receptors down the length of the axon (67). A second possibility is that activation of Trk receptors at the nerve terminal leads to an increase in intracellular calcium (11), which as discussed above, can be rapidly propagated to the nucleus.

3.2 Translocation of Soluble Molecules Through the Process to the Cell Soma

Another mechanism of signaling from distal compartments to the nucleus is via the transport of soluble proteins through the neuronal process to the cell soma. This transport could occur by passive diffusion or by active transport. Several theoretical studies have indicated that passive diffusion of proteins through the cytoplasmic milieu is not efficient for long-distance signaling (35), arguing that active, facilitated transport of proteins is required for efficient signaling from synapse to soma. Depending on the size of the soluble molecule, once it reaches the soma, it may either diffuse through the nuclear membrane (passive transport) or use specialized transport systems and energy-dependent processes (active transport) to enter the nucleus.

The nucleus is separated from the cytoplasm by two membranes, forming the nuclear envelope. The outer membrane is continuous with the ER, and the space between the inner and outer membranes is continuous with the lumen of the ER. The nuclear envelope is perforated with nuclear pores, through which molecules can pass between the cytoplasm and the nucleus. Nuclear pores are formed by structures known as nuclear pore complexes, octagonally-shaped channels, approximately 9 nm in diameter, and composed of more than 100 different proteins called nucleoporins

(48). The size of the channel formed by the nuclear pore complex limits the size of molecules that can readily diffuse through the nuclear pore. Small molecules, such as ions and second messengers can rapidly diffuse through the pore; larger molecules, including proteins up to approximately 40 kDa in size, can also diffuse through the pore, although at a much slower rate; and proteins greater than approximately 40–60 kDa in size must be transported through the pore by facilitated, active nuclear import.

3.3 The Active Nuclear Import Pathway

Proteins greater than 40-60 kDa, as well as some smaller proteins, are transported into the nucleus through active, energy-dependent nuclear import, facilitated by one of a family of nuclear transport receptors called importins (or karyopherins). The largest class of nuclear transporters is the importin βs (32), which bind directly to their cargoes and to the nucleoporins. More than 20 members of the importin β superfamily have been identified in humans, each of which transports a unique set of cargo proteins (38). In the classical nuclear import pathway, importin β1 binds its cargo through one of a family of nuclear import adaptors called importin αs. Six importin αs have been identified in humans, and five in mice (38).

Cargo proteins are recognized by importins through a short amino acid sequence known as a nuclear localization signal (NLS). The classical NLS, first described for the large-T antigen of SV40, is a short stretch of basic amino acids, PKKKRKV (40). Another type of classical NLS is bipartite, with two stretches of basic amino acids separated by a short linker of nonbasic amino acids (the prototype being the NLS of nucleoplasmin, KRPAATKKKAGQAKKKK) (65). Although putative NLSs can be inferred from primary sequence data, they are defined empirically either by fusing the sequence to a non-nuclear protein (which is then transported into the nucleus) or by mutational analysis. The part of importin α that binds to importin β1 is an NLS-like domain in the N-terminus. In the absence of importin β1, this NLS-like domain is autoinhibitory and masks the NLS-binding portion of importin α. Thus, with importin β1 bound, the affinity of importin α for its cargo increases from the micromolar to the nanomolar range. The high affinity cargo-importin α-importin β1 complex is transported through the nuclear pore in an energy-dependent process in a mechanism that is not entirely understood. Inside the nucleus, the complex is disassembled when the small GTP-binding protein Ran, in its GTP-bound form, binds to importin β1, dissociating it from importin α and the NLS-bearing proteins. Importin α and importin β1 then shuttle back into the cytoplasm, leaving the cargo in the nucleus.

Importin-mediated transport functions to carry any protein larger than ~40-60 kD from the cytoplasm into the nucleus. It thus is involved in the nuclear import of signaling proteins activated by calcium transients in the somatic cytoplasm. Recent studies have indicated that importin-mediated transport may additionally play a specialized role in neurons in which it also functions to transport signals from distal compartments to the nucleus (58, 61). Importins have been localized to distal axons and dendrites of a variety of types of neurons, including *Aplysia* sensory and motor neurons, rodent hippocampal neurons, and rodent sensory neurons (29, 73). In rodent motor neurons, importin-mediated transport from sites of injury to the nucleus has been shown to be required for transcription-dependent phases of regeneration following axonal crush injury (29). In rodent hippocampal and *Aplysia* sensory neurons, importins have been shown to translocate from distal sites to the nucleus during transcription-dependent forms of learning-related synaptic plasticity, and in *Aplysia* neurons, blockade of active nuclear import has been shown to block long-term plasticity (73).

In *Aplysia* neurons, Richard Ambron and colleagues (5) found that when a large protein (human serum albumin) was coupled to the SV40 NLS and microinjected into a distal growth cone, it translocated to the nucleus in a manner that required an intact microtubule network. These studies indicated both that the nuclear import machinery (importins) was present in distal growth cones and that transport of importins and their cargoes occurred along microtubules. More recently, Hanz et al. described an interaction between importin α and dynein motor proteins in the motor neuron axon (29). In a second study, they found that importin β1 interacts with vimentin in motor neurons, and that vimentin serves to couple importin β1 bound to a cargo (in this case, mitogen-activated protein kinase 42/44) to the dynein motor protein complex. While the details of this pathway have not been elucidated, these results are consistent with the idea that importins retrogradely transport their protein cargoes from distal sites to the soma along microtubule tracks in axons and dendrites, in a dynein-motor protein-dependent manner. Interestingly, Giri et al. (23) recently found that the intracellular domain of the ErbB2 tyrosine kinase receptor interacts with importin β1, and that this interaction is required for its transport into the soma and signaling to the nucleus. Since this interaction occurred when ErbB2 was intact, it suggests that importins may also be involved in the retrograde transport of signaling endosomes.

One important consideration regarding importin-mediated transport from synapse to nucleus has is that there is no amplification of the signal from the site of initiation to the nucleus. Unlike action potentials or retrograde calcium signals, the number of molecules being transmitted is limited by the amount of that material at the stimulated synapse. Adams and Dudek (1) have referred to this as a problem of stoichiometry, in which the small amount of protein transported from stimulated synapses has to compete with a large amount of protein in the nucleus, and is thus unlikely to make significant contributions to transcriptional change. Essentially, this is a signal to noise problem. This problem is less significant if the signal is an enzyme, since the enzymatic activity serves to amplify the signal. Additionally, it is not clear whether a threshold number of transcription factors is required to trigger significant changes in transcription and gene expression.

4 Examples of Signaling Between Distal Compartments and the Nucleus in Neurons

Retrograde signaling is required for the growth, survival and adaptability of neurons throughout development and adult life. Insights into the cell biological pathways utilized by neurons to mediate long-distance transport to the soma have emerged from studies in a variety of systems. Together, these studies indicate that each of the various pathways outlined above are used to transduce signals received at distal compartments to the nucleus in response to distinct stimuli.

4.1 Retrograde Transport of Signals from Distal Compartments to the Nucleus During Development of the Synapse

A number of studies have indicated that synaptogenesis and synapse differentiation are accompanied by transcriptional changes pre- and post-synaptically (reviewed in (41)). Although the signaling cascades and factors that initiate transcriptional changes during synapse formation are not well understood, a number of studies have begun to elucidate potential candidates. These include signaling by neurotrophins as well the Wingless-Int (wnt) and Transforming Growth Factor β (TGFβ) signaling pathways. In each of these pathways, signaling from synapse to nucleus appears to occur via signaling endosomes.

Studies of the role of target-derived growth factors in promoting transcriptional changes in the nuclei of innervating neurons have provided the bulk of evidence for the signaling endosome pathway outlined above (22, 33, 36). Briefly, growth factors bind to their cognate Trk receptors and are internalized in endosomes that are retrogradely transported the distance of the axon to the cell soma. In the cell body, the ligand-bound Trk receptors retain their catalytic activity, and signal to downstream signaling pathways, including the mitogen-activated protein kinase (MAPK), phosphotidylinositol 3 kinase (PI3 kinase) and phospholipase C-γ pathways. The end result is the transduction of target-derived signals (initially received at great distances from the soma) to the nucleus where they alter transcription.

Budnik and colleagues have shown in flies that wingless is secreted from presynaptic terminals during synaptogenesis and binds to the *Drosophila* frizzled2 receptor (dFrizzled2). dFrizzled2 is internalized into endosomes in the presynaptic terminal and travels back, in a dynein-dependent manner, to the soma of the postsynaptic cell, where the cytoplasmic tail is cleaved and then translocated into the nucleus where it may regulate transcription (52).Interestingly, the retrograde transport of dFrizzled2 was also shown to depend on *Drosophila* glutamate receptor interacting protein (dGRIP), a synaptically localized PDZ-domain containing protein (7).

Studies of synapse formation at the neuromuscular junction of *Drosophila* have also revealed that a TGFβ protein called Gbb is released from the muscle and binds to a TGFβ receptor (53, 54, 63). Ligand binding to this receptor in turn leads to phosphorylation of the cytoplasmic protein mammalian mothers against decapentaplegic, SMAD, which in turn associates with a co-SMAD protein to translocate into the nucleus and regulate transcription. Several findings indicate that this retrograde transport involves signaling endosomes: TGFβ ligands and receptors are transported in vesicles along axons (39); and nuclear phospho-SMAD as well as the functional effects of TGFβ signaling are both blocked by inhibitors of the dynein motor protein pathway (3, 54).

Studies in *Drosophila* have also revealed a role for importin-mediated signaling during axon guidance and synapse formation. Thus, Kumar et al. (45) expressed a truncated importin β, which acts as a general inhibitor of all importin-mediated nuclear import, in the *Drosophila* eye disc at a time when photoreceptors project their axons to the brain. Photoreceptors initially projected their axons correctly but later failed to enter the optic stalk, leading to an extensive network of misguided axons. Disruption of importin function also led to adhesion defects that caused neurons to descend below the appropriate layer. The authors concluded that axon guidance and cell adhesion in the eye rely on importin β-mediated nuclear signaling.

4.2 Transport of Signals from Synapse to Nucleus During Synaptic Plasticity

Long-lasting forms of memory and of learning-related synaptic plasticity have been shown to depend on transcription (2, 42), indicating that signals received at distal synapses are transmitted to the nucleus. Studies in a variety of systems have revealed roles for rapid action potential mediated signaling, importin-mediated active nuclear import as well as for signaling endosome-mediated transport during transcription-dependent plasticity. Many of these studies have focused on the activity-dependent induction of transcription mediated by the cAMP-responsive element binding protein (CREB). This approach is based on research showing that long-term facilitation (LTF) of *Aplysia* sensory-motor synapses, long-term potentiation (LTP) of hippocampal synapses, and long-term hippocampal-mediated memories in rodents all require cAMP-responsive element (CRE)-driven gene expression (43, 49). The transcriptional activity of CREB is activated by phosphorylation at a particular site, serine 133, and antibodies that specifically recognize CREB phosphorylated at this site (21) have been very useful in identifying pathways whereby activity leads to CREB phosphorylation in the nucleus during long-term synaptic plasticity. Such studies have shown that both calcium and cAMP signaling cascades trigger CREB phosphorylation, and have defined roles for a number of specific kinases, including calcium-calmodulin dependent kinases (CaMKs), MAPKs, and the cAMP-dependent kinase (PKA).

Studies of plasticity at mammalian hippocampal synapses have revealed roles for rapid calcium-dependent signaling to the nucleus during transcription-dependent LTP. Thus, Dudek and Fields (18) have demonstrated that somatic action potentials are sufficient to induce CREB phosphorylation and to induce specific immediate early genes in hippocampal neurons. Based on these findings, Dudek has argued that a synapse-to-nucleus signal is not required to induce transcription during the late phase of hippocampal LTP (1). Consistent with this idea, Bading and colleagues (31) have shown that depolarization and electrical stimulation of cultured hippocampal neurons can lead to CREB phosphorylation in a manner that depends entirely on calcium waves in the nucleus and is independent of any import of cytoplasmic proteins into the nucleus. These studies indicate that depolarizing stimuli trigger action potentials and/or regenerative calcium waves in the ER, and that the increased calcium in the nucleus in turn activates nuclear kinases that phosphorylate CREB and activate CRE-driven gene expression.

Earlier studies from Richard Tsien and colleagues (16) had shown that brief bursts of activity in cultured hippocampal neurons triggered CREB phosphorylation within one minute, and that this was mediated by the translocation of calmodulin from the somatic cytoplasm into the nucleus, where it activated CamKIV. The Tsien group (82) also demonstrated that strongly depolarizing stimuli recruited a second pathway that leads to CREB phosphorylation. Activation of this pathway, the ras/MAPK pathway resulted in a slower and more persistent phosphorylation of CREB. Bading and colleagues (30) also described a pathway whereby bursts of activity in cultured hippocampal neurons led to calcium increases near NMDA receptors, which in turn activated MAPK. MAPK then translocated into the nucleus, where it activated both serum response element (SRE)-driven gene expression as well as CREB phosphorylation. Notably, the phosphorylation of CREB mediated by MAPK nuclear translocation was more sustained than CREB phosphorylation mediated by nuclear calcium increases. Together, these studies indicate that depolarizing stimuli can activate cytosolic signaling molecules that are subsequently transported into the nucleus where they function to alter transcription. They further reveal that signaling to the nucleus can occur in a more rapid and transient manner or in more sustained and persistent manner. What remained unclear from these experiments, however, was whether or not these signaling molecules were transported from distal synapses to the nucleus or whether the signaling molecules were generated in the soma in response to depolarization, and subsequently transported into the nucleus.

That signals can indeed be transported from distal synapses to the nucleus during learning-related synaptic plasticity is clear from studies of long-term facilitation of *Aplysia* sensory-motor synapses. At this synapse, five spaced applications of the modulatory neurotransmitter serotonin (5HT) produce LTF, which is dependent on both mRNA and protein synthesis (42). Localized application of 5HT to distal synaptic sites is sufficient to recruit transcription-dependent LTF (50, 69), indicating that signals can in fact travel from a distal synpase to the nucleus. 5HT stimulation of sensory neurons increases cAMP concentrations in the sensory neuron, leading to activation of PKA and MAPK (62). Importantly, 5HT does not increase intracellular calcium in sensory neurons (13) and does not depolarize sensory neurons, indicating that the signal is likely to travel as a soluble signal through the cytoplasm, and not as an electrical signal or in the form of an intracellular calcium wave.

Thompson et al. (73) have provided evidence that importins function to carry signals from distal *Aplysia* synapses to the nucleus during LTF. Thus, they found that importin α was present at the synapse of *Aplysia* sensory-motor neurons, and further showed that stimuli that elicited LTF triggered importin translocation into the nucleus of the sensory neuron. Inhibition of importin-mediated transport into the nucleus by microinjection of anti-nuclear pore antibodies into the sensory neuron blocked LTF without affecting basal synaptic transmission or short term, transcription-independent plasticity. In the same study, Thompson and colleagues also showed that importins were present at synapses of rodent hippocampal neurons, and that importin α was present in post-synaptic density fractions. Activation of the NMDA receptor in dissociated hippocampal neurons triggered translocation of importins into the nucleus. In hippocampal slices, stimuli that elicited late-phase, transcription-dependent LTP also induced an NMDA receptor-dependent increase in the concentration of importin in the nuclei of CA1 pyramidal neurons. Together, these findings indicate that importins function to carry signals from the synapse to the nucleus during transcription-dependent forms of learning-related synaptic plasticity (58).

What are the cargoes of importins that travel from the synapse to the nucleus during synaptic plasticity? A number of interesting candidate molecules exist, although direct demonstration that importins do transport these molecules from synapse to nucleus during plasticity is lacking. Some of these cargoes are transcription factors that localize constitutively to the cytoplasm, and travel to the nucleus following stimulation. One such cargo is the nuclear factor-kappa b (NF-κB), which has been shown to translocate from synapse to nucleus of hippocampal neurons following glutamatergic stimulation and to play a role in learning and memory (55, 75). A recent study demonstrated that the retrograde transport of NFκB in hippocampal neurons occurs along microtubules, and that it requires dynein motor proteins and an intact NLS in NFκB (57). This study thus strongly suggests that importins do function to carry NFκB from synapse to nucleus. Nuclear factor of activated T-cells (NFATc4) is another transcription factor that has been shown to undergo stimulus-dependent translocation from dendrites to nucleus in hippocampal neurons (25, 26). It also contains an NLS and is known to require importins for its nuclear transport (81), suggesting that it may translocate to the nucleus in an importin-dependent manner. In *Aplysia* neurons, a recent study by Kaang and colleagues (46) discovered that during LTF of sensory-motor synapses, a cell-adhesion molecule associated protein (CAMAP) is phosphorylated by PKA and translocates from the synapse to the nucleus where it functions as a transcriptional co-activator. CAMAP was found to interact with importin α, suggesting that it may be transported through the neurite in an importin-dependent manner. Studies in a number of systems have revealed roles for nuclear translocation of kinases, including PKA and MAPK, during transcription-dependent forms of plasticity (42, 72), and these signaling molecules are thus also potential cargoes of importins for trafficking from distal synapses to the nucleus (although, see (78)).

Studies of the role of brain-derived growth factor (BDNF) suggest a role for signaling endosomes in transmitting signals from synapse to nucleus during transcription-dependent plasticity. Thus, BDNF binding to the TrkB receptor has been shown to be critical to long-lasting forms of long-term potentiation at hippocampal synapses (44, 59). Since BDNF-bound TrkB receptors are internalized into endosomes and subsequently travel to the nucleus in signaling endosomes (86), it is possible that BDNF recruits a signaling endosome pathway to transmit signals to the nucleus during learning-related synaptic plasticity.

4.3 Transport of Signals from Distal Axons to the Nucleus During Injury and Regeneration

In the peripheral nervous system, regeneration following axonal injury requires new transcription, indicating that signals are transmitted from the site of injury to the cell body (for review see (28, 70)). The transcriptional component includes (1) a rapid component that depends on injury-induced action potentials that signal from the site of injury to the soma; and (2) a slower component that depends both on a decrease in the normal trafficking of trophic factors to the soma and on the transport of injury-induced signals from the site of injury to the nucleus (28, 61). Studies of injury-induced growth in *Aplysia* neurons has provided clear evidence that injury-induced signals are transported to the nucleus. Thus, Richard Ambron and colleagues (6, 85) have shown that axoplasm isolated from lesioned neurons elicits neuronal growth when microinjected into uninjured neurons. They found that many of these proteins contained NLSs, suggesting that they used importin-mediated transport mechanisms to be transported from the site of injury to the nucleus (4).

Studies in lesioned dorsal root ganglion (DRG) neurons and in sciatic nerve have provided strong evidence that importins play a central role in carrying signals from sites of injury to the nucleus (29, 60). In these studies, Michael Fainzilber and colleagues showed that importin α was present in axons. They further showed that importin β1 mRNA was present in axons where it was locally translated following axonal injury. Saturation of the active import pathway with excess NLS peptides delayed regenerative outgrowth in culture and inhibited the growth-enhancing effects of an in vivo conditioning lesion. These findings are consistent with importin-mediated signaling functioning to transport injury signals to the nucleus to initiate transcription-dependent regeneration.

5 Concluding Remarks

Sites of signal initiation in neurons can occur at great distances from the nucleus. In the case of dendrites, synapses can be hundreds of microns away from the soma, and in the case of axons, synapses can be as far as a meter from the soma. These distances pose a significant challenge to the transport of signals from the site of initiation to the nucleus. Given that many processes in neurons depend on stimulus-induced changes in gene expression, the question of how signals are transported from distal sites to the nucleus is of great interest.

A number of mechanisms exist whereby signals can be delivered from their site of initiation to the nucleus. Rapid signaling mechanisms include (1) electrochemical signaling to the soma via action potentials or dendritic spikes and (2) regenerative calcium waves in the ER. Slower signaling mechanisms include (1) active transport of signaling molecules from distal compartments to the nucleus, either as soluble proteins (e.g. via importin-mediated retrograde transport) or as signaling endosomes, as well as (2) passive diffusion of signaling molecules to the nucleus. When and how neurons use each of these cell biological pathways is only beginning to be elucidated. The data indicate that distinct stimuli likely recruit distinct synapse-to-nucleus signaling pathways, and that each of these functions to produce changes over various time domains. Cell biological investigations into the mechanisms of signal transduction between synapse and nucleus, and between growth cone and nucleus, promise to provide insight into the many physiological and pathological processes in the brain that depend on stimulus-induced changes in gene expression.

Acknowledgements

The authors thank Besim Uzgil and other members of the Martin lab for comments and helpful discussions, Donna Crandall for graphic work and Michael Poon for providing the photomicrograph in Fig. 1. Work on signaling from synapse to nucleus in the Martin lab is funded by grants from the NIMH (KCM), W.M. Keck Foundation (KCM), EJLB Foundation (KCM), and NARSAD (CLH).

References

1. Adams JP and Dudek SM. Late-phase long-term potentiation: getting to the nucleus. *Nat Rev Neurosci* 6: 737–743, 2005.
2. Alberini CM. Genes to remember. *J Exp Biol* 202: 2887–2891, 1999.
3. Allan DW, St Pierre SE, Miguel-Aliaga I, and Thor S. Specification of neuropeptide cell identity by the integration of retrograde BMP signaling and a combinatorial transcription factor code. *Cell* 113: 73–86, 2003.
4. Ambron RT, Dulin MF, Zhang XP, Schmied R, and Walters ET. Axoplasm enriched in a protein mobilized by nerve injury induces memory-like alterations in *Aplysia* neurons. *J Neurosci* 15: 3440–3446, 1995.
5. Ambron RT, Schmied R, Huang CC, and Smedman M. A signal sequence mediates the retrograde transport of proteins from the axon periphery to the cell body and then into the nucleus. *J Neurosci* 12: 2813–2818, 1992.
6. Ambron RT and Walters ET. Priming events and retrograde injury signals. A new perspective on the cellular and molecular biology of nerve regeneration. *Mol Neurobiol* 13: 61–79, 1996.
7. Ataman B, Ashley J, Gorczyca D, Gorczyca M, Mathew D, Wichmann C, Sigrist SJ, and Budnik V. Nuclear trafficking of Drosophila Frizzled-2 during synapse development requires the PDZ protein dGRIP. *Proc Natl Acad Sci U S A* 103: 7841–7846, 2006.
8. Bading H. Transcription-dependent neuronal plasticity the nuclear calcium hypothesis. *Eur J Biochem* 267: 5280–5283, 2000.
9. Bardo S, Cavazzini MG, and Emptage N. The role of the endoplasmic reticulum Ca2+ store in the plasticity of central neurons. *Trends Pharmacol Sci* 27: 78–84, 2006.
10. Bean BP. The action potential in mammalian central neurons. *Nat Rev Neurosci* 8: 451–465, 2007.
11. Berninger B, Garcia DE, Inagaki N, Hahnel C, and Lindholm D. BDNF and NT-3 induce intracellular Ca2+ elevation in hippocampal neurones. *Neuroreport* 4: 1303–1306, 1993.
12. Berridge MJ. Neuronal calcium signaling. *Neuron* 21: 13–26, 1998.
13. Blumenfeld H, Spira ME, Kandel ER, and Siegelbaum SA. Facilitatory and inhibitory transmitters modulate calcium influx during action potentials in *Aplysia* sensory neurons. *Neuron* 5: 487–499, 1990.
14. Bulinski JC. Microtubule modification: acetylation speeds anterograde traffic flow. *Curr Biol* 17: R18–20, 2007.
15. Campenot RB. Local control of neurite development by nerve growth factor. *Proc Natl Acad Sci U S A* 74: 4516–4519, 1977.
16. Deisseroth K, Heist EK, and Tsien RW. Translocation of calmodulin to the nucleus supports CREB phosphorylation in hippocampal neurons. *Nature* 392: 198–202, 1998.
17. Dompierre JP, Godin JD, Charrin BC, Cordelieres FP, King SJ, Humbert S, and Saudou F. Histone deacetylase 6 inhibition compensates for the transport deficit in Huntington's disease by increasing tubulin acetylation. *J Neurosci* 27: 3571–3583, 2007.
18. Dudek SM and Fields RD. Somatic action potentials are sufficient for late-phase LTP-related cell signaling. *Proc Natl Acad Sci U S A* 99: 3962–3967, 2002.
19. Foskett JK, White C, Cheung KH, and Mak DO. Inositol trisphosphate receptor Ca2+ release channels. *Physiol rev* 87: 593–658, 2007.

20. Frey U and Morris RG. Synaptic tagging: implications for late maintenance of hippocampal long-term potentiation. *Trends Neurosci* 21: 181–188, 1998.
21. Ginty DD, Kornhauser JM, Thompson MA, Bading H, Mayo KE, Takahashi JS, and Greenberg ME. Regulation of CREB phosphorylation in the suprachiasmatic nucleus by light and a circadian clock. *Science* 260: 238–241, 1993.
22. Ginty DD and Segal RA. Retrograde neurotrophin signaling: Trk-ing along the axon. *Curr Opin Neurobiol* 12: 268–274, 2002.
23. Giri DK, Ali-Seyed M, Li LY, Lee DF, Ling P, Bartholomeusz G, Wang SC, and Hung MC. Endosomal transport of ErbB-2: mechanism for nuclear entry of the cell surface receptor. *Mol Cell Biol* 25: 11005–11018, 2005.
24. Golding NL, Staff NP, and Spruston N. Dendritic spikes as a mechanism for cooperative long-term potentiation. *Nature* 418: 326–331, 2002.
25. Graef IA, Mermelstein PG, Stankunas K, Neilson JR, Deisseroth K, Tsien RW, and Crabtree GR. L-type calcium channels and GSK-3 regulate the activity of NF-ATc4 in hippocampal neurons. *Nature* 401: 703–708, 1999.
26. Groth RD and Mermelstein PG. Brain-derived neurotrophic factor activation of NFAT (nuclear factor of activated T-cells)-dependent transcription: a role for the transcription factor NFATc4 in neurotrophin-mediated gene expression. *J Neurosci* 23: 8125–8134, 2003.
27. Guzik BW and Goldstein LS. Microtubule-dependent transport in neurons: steps towards an understanding of regulation, function and dysfunction. *Curr Opin Cell Biol* 16: 443–450, 2004.
28. Hanz S and Fainzilber M. Retrograde signaling in injured nerve–the axon reaction revisited. *J Neurochem* 99: 13–19, 2006.
29. Hanz S, Perlson E, Willis D, Zheng JQ, Massarwa R, Huerta JJ, Koltzenburg M, Kohler M, van-Minnen J, Twiss JL, and Fainzilber M. Axoplasmic importins enable retrograde injury signaling in lesioned nerve. *Neuron* 40: 1095–1104, 2003.
30. Hardingham GE, Arnold FJ, and Bading H. A calcium microdomain near NMDA receptors: on switch for ERK-dependent synapse-to-nucleus communication. *Nat Neurosci* 4: 565–566, 2001.
31. Hardingham GE, Arnold FJ, and Bading H. Nuclear calcium signaling controls CREB-mediated gene expression triggered by synaptic activity. *Nat Neurosci* 4: 261–267, 2001.
32. Harel A and Forbes DJ. Importin beta: conducting a much larger cellular symphony. *Mol Cell* 16: 319–330, 2004.
33. Heerssen HM and Segal RA. Location, location, location: a spatial view of neurotrophin signal transduction. *Trends Neurosci* 25: 160–165, 2002.
34. Hirokawa N and Takemura R. Molecular motors and mechanisms of directional transport in neurons. *Nat Rev Neurosci* 6: 201–214, 2005.
35. Howe CL. Modeling the signaling endosome hypothesis: why a drive to the nucleus is better than a (random) walk. *Theor Biol Med Model* 2: 43, 2005.
36. Howe CL and Mobley WC. Long-distance retrograde neurotrophic signaling. *Curr Opin Neurobiol* 15: 40–48, 2005.
37. Howe CL and Mobley WC. Signaling endosome hypothesis: A cellular mechanism for long distance communication. *J Neurobiol* 58: 207–216, 2004.
38. Jans DA, Xiao CY, and Lam MH. Nuclear targeting signal recognition: a key control point in nuclear transport? *Bioessays* 22: 532–544, 2000.
39. Jiang Y, McLennan IS, Koishi K, and Hendry IA. Transforming growth factor-beta 2 is anterogradely and retrogradely transported in motoneurons and up-regulated after nerve injury. *Neuroscience* 97: 735–742, 2000.
40. Kalderon D, Roberts BL, Richardson WD, and Smith AE. A short amino acid sequence able to specify nuclear location. *Cell* 39: 499–509, 1984.

41. Kalinovsky A and Scheiffele P. Transcriptional control of synaptic differentiation by retrograde signals. *Curr Opin Neurobiol* 14: 272–279, 2004.
42. Kandel ER. The molecular biology of memory storage: a dialog between genes and synapses. *Biosci Rep* 21: 565–611, 2001.
43. Kornhauser JM, Cowan CW, Shaywitz AJ, Dolmetsch RE, Griffith EC, Hu LS, Haddad C, Xia Z, and Greenberg ME. CREB transcriptional activity in neurons is regulated by multiple, calcium-specific phosphorylation events. *Neuron* 34: 221–233, 2002.
44. Korte M, Kang H, Bonhoeffer T, and Schuman E. A role for BDNF in the late-phase of hippocampal long-term potentiation. *Neuropharmacology* 37: 553–559, 1998.
45. Kumar JP, Wilkie GS, Tekotte H, Moses K, and Davis I. Perturbing nuclear transport in Drosophila eye imaginal discs causes specific cell adhesion and axon guidance defects. *Dev Biol* 240: 315–325, 2001.
46. Lee SH, Lim CS, Park H, Lee JA, Han JH, Kim H, Cheang YH, Lee SH, Lee YS, Ko HG, Jang DH, Kim H, Miniaci MC, Bartsch D, Kim E, Bailey CH, Kandel ER, and Kaang BK. Nuclear translocation of CAM-associated protein activates transcription for long-term facilitation in *Aplysia*. *Cell* 129: 801–812, 2007.
47. Levy JR and Holzbaur EL. Cytoplasmic dynein/dynactin function and dysfunction in motor neurons. *Int J Dev Neurosci* 24: 103–111, 2006.
48. Lim RY and Fahrenkrog B. The nuclear pore complex up close. *Currt Opin Cell Biol* 18: 342–347, 2006.
49. Lonze BE and Ginty DD. Function and regulation of CREB family transcription factors in the nervous system. *Neuron* 35: 605–623, 2002.
50. Martin KC, Casadio A, Zhu H, Yaping E, Rose JC, Chen M, Bailey CH, and Kandel ER. Synapse-specific, long-term facilitation of *Aplysia* sensory to motor synapses: a function for local protein synthesis in memory storage. *Cell* 91: 927–938, 1997.
51. Martin KC and Kosik KS. Synaptic tagging – who's it? *Nat Rev Neurosci* 3: 813–820, 2002.
52. Mathew D, Ataman B, Chen J, Zhang Y, Cumberledge S, and Budnik V. Wingless signaling at synapses is through cleavage and nuclear import of receptor DFrizzled2. *Science* 310: 1344–1347, 2005.
53. McCabe BD, Hom S, Aberle H, Fetter RD, Marques G, Haerry TE, Wan H, O'Connor MB, Goodman CS, and Haghighi AP. Highwire regulates presynaptic BMP signaling essential for synaptic growth. *Neuron* 41: 891–905, 2004.
54. McCabe BD, Marques G, Haghighi AP, Fetter RD, Crotty ML, Haerry TE, Goodman CS, and O'Connor MB. The BMP homolog Gbb provides a retrograde signal that regulates synaptic growth at the Drosophila neuromuscular junction. *Neuron* 39: 241–254, 2003.
55. Meffert MK, Chang JM, Wiltgen BJ, Fanselow MS, and Baltimore D. NF-kappa B functions in synaptic signaling and behavior. *Nat Neurosci* 6: 1072–1078, 2003.
56. Meldolesi J. Rapidly exchanging Ca2+ stores in neurons: molecular, structural and functional properties. *Prog Nneurobiol* 65: 309–338, 2001.
57. Mikenberg I, Widera D, Kaus A, Kaltschmidt B, and Kaltschmidt C. Transcription Factor NF-kappaB Is Transported to the Nucleus via Cytoplasmic Dynein/Dynactin Motor Complex in Hippocampal Neurons. *PLoS ONE* 2: e589, 2007.
58. Otis KO, Thompson KR, and Martin KC. Importin-mediated nuclear transport in neurons. *Curr Opin Neurobiol* 16: 329–335, 2006.
59. Patterson SL, Pittenger C, Morozov A, Martin KC, Scanlin H, Drake C, and Kandel ER. Some forms of cAMP-mediated long-lasting potentiation are associated with release of BDNF and nuclear translocation of phospho-MAP kinase. *Neuron* 32: 123–140, 2001.

60. Perlson E, Hanz S, Ben-Yaakov K, Segal-Ruder Y, Seger R, and Fainzilber M. Vimentin-dependent spatial translocation of an activated MAP kinase in injured nerve. *Neuron* 45: 715–726, 2005.
61. Perlson E, Hanz S, Medzihradszky KF, Burlingame AL, and Fainzilber M. From snails to sciatic nerve: Retrograde injury signaling from axon to soma in lesioned neurons. *J Neurobiol* 58: 287–294, 2004.
62. Pittenger C and Kandel ER. In search of general mechanisms for long-lasting plasticity: *Aplysia* and the hippocampus. *Philos Trans R Soc Lond B Biol Sci* 358: 757–763, 2003.
63. Rawson JM, Lee M, Kennedy EL, and Selleck SB. Drosophila neuromuscular synapse assembly and function require the TGF-beta type I receptor saxophone and the transcription factor Mad. *J Neurobiol* 55: 134–150, 2003.
64. Reed NA, Cai D, Blasius TL, Jih GT, Meyhofer E, Gaertig J, and Verhey KJ. Microtubule acetylation promotes kinesin-1 binding and transport. *Curr Biol* 16: 2166–2172, 2006.
65. Robbins J, Dilworth SM, Laskey RA, and Dingwall C. Two interdependent basic domains in nucleoplasmin nuclear targeting sequence: identification of a class of bipartite nuclear targeting sequence. *Cell* 64: 615–623, 1991.
66. Sabatini BL, Maravall M, and Svoboda K. Ca(2+) signaling in dendritic spines. *Curr Opin Neurobiol* 11: 349–356, 2001.
67. Senger DL and Campenot RB. Rapid retrograde tyrosine phosphorylation of trkA and other proteins in rat sympathetic neurons in compartmented cultures. *J Cell Biol* 138: 411–421, 1997.
68. Setou M, Hayasaka T, and Yao I. Axonal transport versus dendritic transport. *J Neurobiol* 58: 201–206, 2004.
69. Sherff CM and Carew TJ. Coincident induction of long-term facilitation in *Aplysia*: cooperativity between cell bodies and remote synapses. *Science* 285: 1911–1914, 1999.
70. Snider WD, Zhou FQ, Zhong J, and Markus A. Signaling the pathway to regeneration. *Neuron* 35: 13–16, 2002.
71. Spacek J and Harris KM. Three-dimensional organization of smooth endoplasmic reticulum in hippocampal CA1 dendrites and dendritic spines of the immature and mature rat. *J Neurosci* 17: 190–203, 1997.
72. Sweatt JD. Mitogen-activated protein kinases in synaptic plasticity and memory. *Curr Opin Neurobiol* 14: 311–317, 2004.
73. Thompson KR, Otis KO, Chen DY, Zhao Y, O'Dell TJ, and Martin KC. Synapse to nucleus signaling during long-term synaptic plasticity; a role for the classical active nuclear import pathway. *Neuron* 44: 997–1009, 2004.
74. Verkhratsky A. The endoplasmic reticulum and neuronal calcium signaling. *Cell Calcium* 32: 393–404, 2002.
75. Wellmann H, Kaltschmidt B, and Kaltschmidt C. Retrograde transport of transcription factor NF-kappa B in living neurons. *J Biol Chem* 276: 11821–11829, 2001.
76. Westermann S and Weber K. Post-translational modifications regulate microtubule function. *Nat Rev Mol Cell Biol* 4: 938–947, 2003.
77. White C, Yang J, Monteiro MJ, and Foskett JK. CIB1, a ubiquitously expressed Ca2+-binding protein ligand of the InsP3 receptor Ca2+ release channel. *J Biol Chem* 281: 20825–20833, 2006.
78. Whitehurst AW, Wilsbacher JL, You Y, Luby-Phelps K, Moore MS, and Cobb MH. ERK2 enters the nucleus by a carrier-independent mechanism. *Proc Natl Acad Sci U S A* 99: 7496–7501, 2002.
79. Williams SR and Stuart GJ. Dependence of EPSP efficacy on synapse location in neocortical pyramidal neurons. *Science* 295: 1907–1910, 2002.

80. Williams SR and Stuart GJ. Role of dendritic synapse location in the control of action potential output. *Trends Neurosci* 26: 147–154, 2003.
81. Willingham AT, Orth AP, Batalov S, Peters EC, Wen BG, Aza-Blanc P, Hogenesch JB, and Schultz PG. A strategy for probing the function of noncoding RNAs finds a repressor of NFAT. *Science* 309: 1570–1573, 2005.
82. Wu GY, Deisseroth K, and Tsien RW. Activity-dependent CREB phosphorylation: convergence of a fast, sensitive calmodulin kinase pathway and a slow, less sensitive mitogen-activated protein kinase pathway. *Proc Natl Acad Sci U S A* 98: 2808–2813, 2001.
83. Yang J, McBride S, Mak DO, Vardi N, Palczewski K, Haeseleer F, and Foskett JK. Identification of a family of calcium sensors as protein ligands of inositol trisphosphate receptor Ca(2+) release channels. *Proc Natl Acad Sci U S A* 99: 7711–7716, 2002.
84. Zalk R, Lehnart SE, and Marks AR. Modulation of the Ryanodine Receptor and Intracellular Calcium. *Annu Rev Biochem* 76: 367–385, 2007.
85. Zhang XP and Ambron RT. Positive injury signals induce growth and prolong survival in *Aplysia* neurons. *J Neurobiol* 45: 84–94, 2000.
86. Zhou P, Porcionatto M, Pilapil M, Chen Y, Choi Y, Tolias KF, Bikoff JB, Hong EJ, Greenberg ME, and Segal RA. Polarized signaling endosomes coordinate BDNF-Induced Chemotaxis of Cerebellar Precursors. *Neuron* 55: 53–68, 2007.

Molecular Organization of the Postsynaptic Membrane at Inhibitory Synapses

I. Lorena Arancibia-Carcamo[1], Antoine Triller[2] and Josef T. Kittler[3]

[1] Department of Pharmacology, University College London, Gower Street, London, WC1E 6BT, UK, l.carcamo@ucl.ac.uk
[2] Biologie Cellulaire de la Synapse N&P, Inserm UR497, Ecole Normale Supérieure, 46, rue d'Ulm 75005 Paris
[3] Department of Physiology, University College London, Gower Street, London, WC1E 6BT, UK, j.kittler@ucl.ac.uk

Abstract

The majority of fast inhibitory neurotransmission in the mammalian central nervous system (CNS) is mediated by γ-aminobutyric acid type A receptors ($GABA_A$ receptors), whereas in the brain stem and spinal cord fast inhibition is primarily mediated by glycine receptors where they can be associated with $GABA_A$ receptors at mixed synapses. $GABA_A$ receptors and glycine receptors are pentameric hetero-oligomeric ligand gated ion channels, the activation of which leads to the opening of an integral chloride permeable channel (Fig. 1). When chloride is low intracellularly, as in most adult neurons, this results in a chloride influx into the cell, resulting in cell hyperpolarization, moving the membrane potential away from the spike threshold for action potential generation. $GABA_A$ receptors and glycine receptors play a key role in regulating neuronal excitability, therefore modulating their function has important implications for regulating cell and network activity in the nervous system. In addition $GABA_A$ receptors are drug targets for many clinically relevant compounds including benzodiazepines, barbiturates, ethanol, neurosteroids and some anesthetics. Alterations in $GABA_A$ receptor and glycine receptor function have been implicated in an increasing number of neurological and neuropsychiatric diseases including: anxiety, depression, schizophrenia, epilepsy, stroke, substance abuse, neuropathic pain and hyperekplexia/startle disease.

Correct synaptic inhibition is dependent on the formation and maintenance of inhibitory postsynaptic domains where postsynaptic GABA and/or glycine receptors can be recruited and stabilized below release sites for their cognate neurotransmitter. The formation and maintenance of inhibitory synapses depends on the dynamic recruitment of cell adhesion molecules and scaffold proteins to sites opposing inhibitory presynaptic terminals. In addition, the correct function of inhibitory synapses is

J.W. Hell, M.D. Ehlers (eds.), *Structural and Functional Organization of the Synapse*, DOI: 10.1007/978-0-387-77232-5_21, © Springer Science+Business Media, LLC 2008

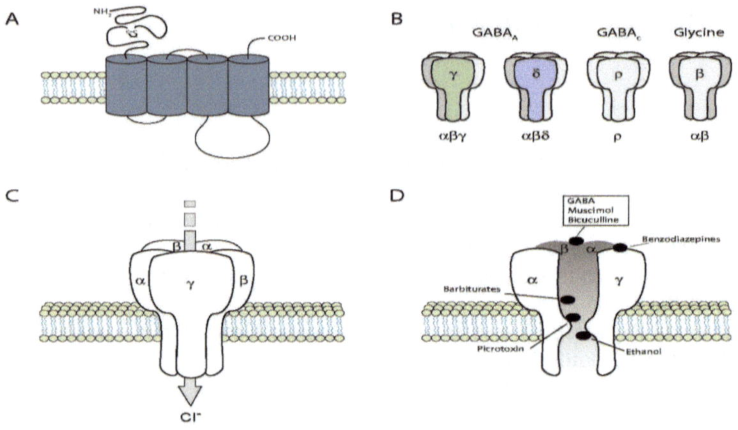

Fig. 1. Structure of inhibitory ligand gated channels.
(**a**) Membrane topology of ligand gated ion channel subunits. Each subunit consists of a large extracellular N-terminus, four transmembrane domains (TM1-4) and a large intracellular loop between TM3 and TM4. (**b**) Proposed pentameric structures of $GABA_A$, $GABA_C$ and glycine receptor subtypes. Most $GABA_A$ receptors are believed to be composed of 2α:2β:1γ subunits, whereas extrasynaptic receptors contain 2α:2β:1δ subunit combination. $GABA_C$ receptors are constructed from ρ1–3 subunits either as heteromeric or homomeric pentameric assemblies. Glycine receptors are pentamers constructed from α and β subunits in a ratio of 2:3. All these subunit combinations allow for the formation of a chloride permeable channel (**c**). (**d**) Site of action of pharmacological agents on $GABA_A$ receptors. The neurotransmitter GABA binds $GABA_A$ receptors on a site formed by the N-terminus of α and β subunits. $GABA_A$ receptors can be potentiated by binding of benzodiazepines to a site created at the interface of the α and γ subunit. In addition, $GABA_A$ receptors can also be modulated by barbiturates, ethanol and picrotoxin. The site of action for these drugs is believed to be inside the chloride permeable channel.

dependent on the membrane trafficking of inhibitory receptors into and out of inhibitory postsynaptic domains, both by diffusion within the membrane and by membrane transport between surface and intracellular compartments. This process is regulated by interaction of receptors with proteins in the cytosol. The strength of inhibition can therefore be regulated by modulation of both the function and number of synaptic GABA and glycine receptors, controlled by receptor associated proteins that facilitate receptor activity, transport, and synaptic confinement.

1 Pharmacology of $GABA_A$ and Glycine Receptors

The three major GABA receptor types in the brain were initially identified based on their pharmacological and electrophysiological properties. GABA type A receptors ($GABA_A$ receptors) are ligand-gated ion channels, activated by muscimol and inhibited by bicuculline (141). A second GABA receptor (termed $GABA_B$ receptor) is insensitive to these agents, but is activated by baclofen, inhibited by 2-hydroxysaclofen and produces a slower inhibitory response (20, 91). $GABA_B$ receptors are G protein-coupled receptors that mediate their inhibitory effects by coupling to ion channels (for a detailed review of $GABA_B$ receptors see (45)). In addition to $GABA_A$ and $GABA_B$ receptors, a third GABA receptor type ($GABA_C$ receptor), insensitive to both bicuculline and baclofen, but sensitive to the GABA analogue cis-4-aminocrotonic acid and picrotoxin, has also been identified (56, 162, 180). $GABA_C$ receptors are ligand-gated ion channels, homologous in structure to the $GABA_A$ receptor but are primarily expressed in the retina (56, 162). Glycine receptors are primarily expressed in the brain stem and spinal cord where they play a crucial role in regulating inhibitory tone (49, 228). Glycine receptors are similar in structure to $GABA_A$ and $GABA_C$ receptors, are activated by glycine and also potentially taurine, but are insensitive to picrotoxin and instead are inhibited by the plant alkaloid strychnine (139).

A range of compounds, several of which are clinically relevant therapeutic agents, can allosterically modulate $GABA_A$ receptor and glycine receptor function. $GABA_A$ receptor function is modulated by benzodiazepines, barbiturates, steroids, anesthetics and ethanol (141, 157, 181) (Fig.1). Benzodiazepines (such as diazepam) potentiate GABA responses by increasing $GABA_A$ channel opening frequency (141, 209). Diazepam has also been reported to increase channel conductance of some native receptors (58), which may be due to receptor interactions with $GABA_A$ receptor associated proteins (62). Barbiturates, such as pentobarbital, enhance GABA responses by increasing the open probability of GABA-activated channels and have also been reported to enhance single channel conductance of some native receptors, and directly activate $GABA_A$ receptors at high concentrations (11, 18, 59, 141, 144, 209). General anesthetics such as isoflurane and halothane increase GABA-induced currents by potentiating the action of GABA or by directly activating the channel (for reviews see (66, 81)). Ethanol has also been shown to have a potentiating effect on $GABA_A$ receptors (for reviews see (51, 141, 171, 189)). Recently, elegant work from several groups have reported that $GABA_A$ receptors containing the δ subunit are enhanced by low alcohol concentrations that are thought to mediate alcohol effects experienced during social drinking, although several controversies with respect to these findings remain (for detailed discussions see (158, 171, 189)). $GABA_A$ receptors are also subject to regulation by several endogenous modulators including steroids, protons and Zn^{2+} (9, 93, 141). At low nanomolar concentrations which occur during stress, alcohol intoxication and pregnancy/oestrous, neurosteroids, such as allopregnanolone and tetrahydrodeoxycorticosteroneprogesterone, can potentiate GABA responses. At higher concentrations (submicromolar to micromolar), which may occur during parturition, neurosteroids can directly activate the receptor (for detailed reviews see (9, 93, 124, 157)). $GABA_A$ receptors are also inhibited by a number of pharmacological agents. Picrotoxin, a plant convulsant, is a noncompetitive $GABA_A$ receptor inhibitor that is thought to act by binding to a site within the channel (241) leading to stabilization of receptors in an agonist bound desensitized state (164). In addition, $GABA_A$ receptors lacking the γ subunit, and therefore likely extrasynaptic, are inhibited by Zn^{2+} (201). Although the pharmacology of glycine receptors is not as rich as that of GABA receptors, glycine receptor function is enhanced by several of the same agents that potentiate GABA receptors, such as alcohols, anesthetics, and neurosteroids (30, 81, 87, 128, 153). Glycine receptors are also sensitive to Zn^{2+} which potentiates receptor activation at submicromolar Zn^{2+} concentrations, but causes inhibition at concentrations greater than 10 μM (13, 201). In a physiological context, low nanomolar basal Zn^{2+} concentrations are sufficient to prolong the decay phase of glycinergic inhibitory postsynaptic currents (128). In contrast, an equivalent of benzodiazepines is lacking for glycine receptors.

2 Molecular Identification of $GABA_A$ Receptors and Glycine Receptors

$GABA_A$ receptors and glycine receptors are members of the Cys-loop ligand gated ion channel family of receptors, which also include nicotinic acetylcholine receptors (nACh receptors), $GABA_C$ receptors and serotonin ($5HT_3$) receptors (44, 192, 222). The glycine receptor was the first neurotransmitter receptor to be isolated from the mammalian brain using aminostrychnine-agarose affinity chromatography, taking advantage of the high affinity interaction (K_D 1–10 nM) of the receptor with strychnine (175). The receptor appeared to be a complex of three proteins: a 48 kDa α subunit, 58 kDa β subunit, and an additional 93 kDa protein (80). Several biochemical studies of the purified form of glycine receptors have established that the integral membrane α and β subunit glycoproteins represent the constitutive subunits of the receptor (122). Cross-linking techniques showed that α and β subunits assemble to form the channel-containing transmembrane core of the glycine receptor. The size of the complex (250 kDa) suggested a pentameric assembly of the subunits giving a quaternary structure that is now well established for all members of the Cys-loop family of receptors (Fig. 1). In contrast to the receptor, the co-purifying 93-kDa protein, gephyrin, is a nonglycosylated polypeptide that can interact reversibly with the α/β pentamer. Gephyrin is a cytoplasmic extrinsic membrane protein that plays a key role in regulating glycine receptor localization as discussed below.

3 Molecular Diversity and Distribution Pattern of Glycine Receptor Subunit Isoforms

Initial biochemical and immunochemical studies during spinal cord development established the notion of glycine receptor subtypes (128). A comparison of glycine receptor expression in neonatal and adult membranes revealed that a distinct form of glycine receptor predominated around birth (neonatal receptor, GlyRN). It is characterized pharmacologically by its relatively low strychnine-binding affinity. It is composed of pentamers of 49 kDa polypetides denominated α2 and it is found mainly during the fetal and neonatal stages. This supported the notion of a developmental switch from GlyRN to the adult form, GlyRA, corresponding to the formation of α1/β heteromers. This switch occurs in the spinal cord within 3 weeks after birth, and is not as complete in other regions of the nervous system. One current view is that postsynaptic glycine receptors correspond to a mixture of α2 pentamers and α2/β heteromers in immature neurons, while α1/β heteromers predominate in mature synapses. Currently, the greater diversity of glycine receptor subunits is established to include not only α1 and α2 but also α3 and α2* subunits (139) and in addition an α4 subunit has also been identified in mouse, which is absent in rat and human (145). These highly homologous isoforms can form glycine-gated chloride channels of comparable strychnine sensitivity, except α2*, which is 99% identical to human α2 shows a 500-fold lower sensitivity to strychnine (123). Further, alternative exon usage increases glycine receptor diversity. For the α1 subunit, eight additional amino acids can be inserted in the M3–M4 loop. This sequence contains a serine residue, which is a potential phosphorylation site allowing a functional modulation of the α1 subunit. Alternative splicing has also been identified for the α2 and α3 subunits (139).

In situ hybridization (122, 182) revealed that the distribution of individual glycine receptor subunits was not identical for all variants. The mRNAs encoding the α1 and α3 subunits are mainly transcribed in spinal cord and brain stem at later postnatal stages. The α3 subunit mRNA is present in the infralimbic system, the hippocampal complex, and the cerebellar granular layer. Expression levels of the α2 subunit, by contrast, are high in the embryonic and perinatal stages, but barely detectable in the adult brain, with some expression in higher cortical regions. The α4 subunit, which is expressed at low levels in the adult brain and spinal cord, can form functional glycine receptors, which are restricted to the spinal cord and the sympathetic nervous system. Compared to the α subunit transcripts, the β subunit mRNA is more widely expressed throughout the embryonic and adult nervous system (103) and is found in brain loci devoid of strychnine-binding sites or glycine receptor immunoreactivity. The physiological significance of this widespread β subunit mRNA expression is not understood, and there is no evidence that glycine receptor β subunits assemble with subunits of other Cys-loop receptor family members.

4 Molecular Heterogeneity of $GABA_A$ Receptors

A total of 16 genes encoding $GABA_A$ receptor subunits have been identified in the mammalian nervous system. The $GABA_A$ receptor was initially purified using GABA/benzodiazepine affinity chromatography. SDS-PAGE analysis revealed two receptor subunits, α and β, with molecular weights of 53 kDa and 58 kDa respectively (199, 200). Oligonucleotides, designed based on peptide microsequencing results, were used for screening bovine cDNA libraries and allowed for the cloning of both α and β subunits. Analysis of the amino acid sequences of α and β subunits revealed that these receptor subunits showed significant homology with each other and with other members of the Cys-loop receptors such as the nAChRs. Based on amino acid sequence homology six α subunits as well as three β subunits have been identified (96, 98, 133, 134, 137, 142, 238). Co-expression of α and β subunits in heterologous systems, however, formed receptors that lacked benzodiazepine sensitivity, a pharmacological trait observed in neuronal $GABA_A$ receptors. A novel $GABA_A$ receptor subunit that when co-expressed with α and β subunits inferred benzodiazepine sensitivity on the assembled receptors was then identified (179). This subunit, termed γ2, shows 42% and 35% identity to the α1 and β1 subunits respectively. Three γ subunits have been identified in total (179, 197, 231, 239). In addition to α, β and γ subunits other receptor subunits have been identified: δ (197), ε (50), π (90) and θ (17). In addition some of these genes, including the α6 (121), β2 (146), β3 (101), ε (230) and γ2 subunits (229), undergo alternative splicing resulting in a long (L) and short (S) version of the gene product. However, a complete understanding of the physiological relevance of these splice variants remains to be addressed.

In situ hybridization and immunohistochemical studies have provided a detailed picture of the regional and cellular expression patterns of $GABA_A$ receptor subunit mRNA and corresponding receptor subunit protein, in the nervous system (67, 82, 126, 127, 159, 174, 176, 198, 232, 240). The α1, β2/3, and γ2 subunits are the most abundant and broadly expressed $GABA_A$ receptor subunits in the brain and spinal cord (126, 174, 176, 232). In agreement with this, gene deletion studies have revealed that α1 and β2 subunits account for up to half of all $GABA_A$ receptors in the CNS (212). The α1 subunit is found to be present, at high levels, in most regions of the adult brain (88, 176) but not in embryonic or neonatal brain, where it is expressed at low levels (127). The α2 subunit is found in most brain regions, except the thalamus and globus pallidum, although at lower levels than the α1 subunit (176). The α3 and α4 are the least abundant α subunits in the adult brain. The α3 subunit is highly expressed in early developmental stages, but at low levels in the adult brain (176, 240) where it is concentrated in the cerebral cortex, mainly on monoaminergic neurons (72), the basal forebrain, on cholinergic neurons (73), and the thalamic reticular nucleus (69). In contrast the α4 subunit is restricted to the thalamus, the striatum and the molecular layer of the dentate gyrus (176, 232). The α5 subunit is evenly expressed throughout development and enriched in the CA1 region of the hippocampus, although expression levels show a continuous decrease with aging (240). In contrast the α6 subunit is found almost exclusively, postnatally in the granular layer of the cerebellum (126, 127, 176).

Despite a wide distribution of all three β subunits, β1 is expressed at much lower levels compared to β2 and β3 subunits (176). Although there is considerable overlap in the expression of β2 and β3 subunits, there are some regions where higher expression of one β subunit in certain brain regions comes at the expense of another. For example, β2 subunits are highly expressed in the adult thalamus in comparison to β1 and β3 subunits (176, 194, 232), although the β3 subunit is expressed at higher levels than the β2 in the fetal and neo-natal thalamus (127). The β3 subunit is expressed at high levels in the striatum where very low levels of the β2 subunit are found. In addition, β1 and β3 subunits are found at higher concentrations than the β2 subunit in the hippocampus (232). Furthermore, in the hippocampus, the β2 subunit is found concentrated in non-pyramidal neurons, whereas β1 and β3 subunits are found in primary pyramidal neurons (155, 161, 176, 232).

Of the γ subunit isoforms, the γ2 subunit is, by far, the most abundant. However, all three γ subunits are found widely distributed, with the γ1 subunit expressed at higher concentrations in the pallidum and substantia nigra whereas the γ3 subunit is slightly more concentrated in the cerebral cortex (126, 127, 176, 232). Similar to other $GABA_A$ receptor subunits, the expression of γ1 and γ3 is altered during development, with these two subunits being expressed at higher levels pre-natally (127). Furthermore, the γ2 subunit also shows a small but significant reduction in expression levels with aging (240). Importantly, the γ1 and γ3 subunits are unable to fully substitute for the essential γ2 subunit in γ2 knockout mice, which show a lethal phenotype (4, 61). In contrast to the γ subunits, the δ subunit of $GABA_A$ receptors is most highly expressed in the thalamus and in the granular layer of the cerebellum where it is thought to assemble into receptors which contain either α4 or α6 subunits, respectively (176). The ε, π and θ subunits are the least abundant $GABA_A$ receptor subunits expressed in the CNS. The ε subunit has been observed in the hypothalamus, amygdala and brainstem (160). Interestingly, the θ subunit has also been observed in all these regions as well as the substantia nigra and hippocampus (160). The π subunit appears to be expressed mainly in the uterus, although low expression levels have been observed in the hippocampus (90).

5 $GABA_A$ and Glycine Receptor Structure

$GABA_A$ receptors, glycine receptors and the other members of the Cys-loop receptor family are polytopic type I membrane proteins that share a common subunit structure (44). Each subunit encompasses a large extended N-terminal domain, bearing potential glycosylation sites, four highly conserved transmembrane domains (deduced by hydropathy analysis), and a large intracellular loop of significantly lower homology between TM3 and TM4 which protrudes into the cytoplasm (TM3–TM4 loop; Fig. 1). The four membrane-spanning domains form α helices (TM2) and/or β strands (TM1, TM3, TM4). Based on extensive biochemical and electron microscopy analysis of the nicotinic acetylcholine receptor, members of this family are assumed to be pentameric in structure where subunits are arranged around a central aqueous pore (222). The tertiary and quaternary structure of the soluble pentameric acetylcholine-binding protein (AChBP), which shares approximately 20% sequence homology with nAChR, GABA and glycine receptors, has been useful in further elucidating structure–function parameters for receptors of the family (21, 202). In addition to improving our understanding of the binding of agonists, antagonists and modulators, (1, 35, 92, 93, 201), the AChBP crystal structure provides a powerful model of the N-terminal domain of the Cys-loop receptor family. AChBP is mainly a sandwich of antiparallel β sheets positioning conserved residues in order to stabilize the protomers, whereas variable residues are at their interface (21, 202). The topology of the transmembrane domains (Fig. 1) delineates the ion permeation pathway away from the hydrophobic core of the phospholipid bilayer. For all Cys-loop receptors, the subunit's α-helical TM2 domain lines the central water-filled pore, while TM1, TM3, and TM4 form the interface with the lipids and isolate TM2 from a hydrophobic environment (44). Recent studies on the GlyR α1 subunit have challenged the four-helix model and provide evidence that whereas TM2 and TM4 are entirely helical, TM1 and TM3 also contain β strands.

6 Glycine Receptor Assembly

The α1 and β subunits assemble as a pentameric complex of 2α:3β stoichiometry (84, 102). Short amino acid sequences, named assembly boxes, all located in the N-terminal domain of the β subunit and corresponding to three diverging motifs in α and β subunits, have been identified (83, 102, 154). They have a role in determining receptor assembly. Replacement of these motifs in the β subunit by the corresponding α1 motifs results in the loss of the subunit ratio in α/β oligomers, suggesting that different amino acid positions are determinants in the early step of subunit–subunit interaction. These residues impose a mutually exclusive mode of assembly, either in complexes of invariant α/β stoichiometry or in homo-oligomers. A 2α–3β stoichiometry has recently been established following affinity purification of expressed engineered tandem subunits (84). These experiments indicated that the β subunit contributes to the agonist binding properties of hetero-oligomeric α1/β glycine receptors. In fact, ionic interactions at the α/β interface are required to stabilize glycine in its binding pocket. This feature, with an agonist binding shared between two adjacent subunits is characteristic of the Cys-loop receptor family and also holds true for $GABA_A$ receptors (102, 162).

7 $GABA_A$ Receptor Assembly and Composition of Plasma Membrane Receptors

The large diversity of $GABA_A$ receptor subunits generates the potential for a bewildering heterogeneity of receptor structure. However, a number of functional, biochemical and immunocytochemical studies have revealed that only a limited number of receptor subunit combinations are likely to exist on the neuronal cell surface (162, 198). Restriction on $GABA_A$ receptor molecular heterogeneity in the brain is due to a number of factors. $GABA_A$ receptor subunit composition is in part restricted by the regional and temporal selectivity in subunit expression (69, 176, 198). In addition, a number of assembly rules further restrict the heterogeneity of native $GABA_A$ receptors (44). Expression studies indicated that individual $GABA_A$ receptor subunits do not result in the formation of GABA gated channels. Expression of β1 and β3 subunits alone results in the formation of chloride permeable channels that are sensitive to picrotoxin and barbiturates, however, these channels are insensitive to application of GABA. In contrast, co-expression of α and β subunits results in the production of functional GABA gated chloride channels sensitive to picrotoxin, bicuculline, barbiturates and Zn^{2+} (55, 133, 192). This is in agreement with the binding site for GABA lying at the interface of α and β subunits (14, 15, 141). However, the co-expression of α and β subunits does not form channels that are sensitive to benzodiazepine modulation, which represents the pharmacology of most native receptors. It is only when γ subunits are additionally co-expressed with α and β subunits that GABA gated, benzodiazepine sensitive channels are formed (239) (179), which are in addition insensitive to Zn^{2+} modulation (201).

As shown with fluorescence, only certain $GABA_A$ receptor subunit combinations can access the cell surface whereas other subunit combinations are retained within the endoplasmic reticulum (ER) (15, 43, 44, 162). This suggested the existence of molecular mechanisms to restrict the surface expression of receptors with given compositions. With exception of the β1, β3 and γ2S subunits, individual $GABA_A$ receptor subunits are mostly retained in the ER (43). Moreover, pulse chase experiments demonstrated that α1 and β2 subunits are rapidly degraded when expressed alone (77), suggesting that these proteins are targeted for degradation from the ER. Importantly, $GABA_A$ receptor subunits have been shown to associate with BiP and calnexin (43, 77), two chaperone molecules which assist in protein quality control in the ER (114). Initial studies suggested that assembly was dependent on the N-terminus of receptor subunits. Using a chimeric approach, four amino acids in the N-terminal domain of the β3 subunit have been shown to mediate functional cell surface expression of this subunit compared to β2 (215, 216). Introduction of these four amino acids into the N-terminus of the β2 subunit is sufficient to enable β2 homomerization and ER exit. These four amino acids are also important for the oligomerization of β subunits with γ subunits but not with α subunits. Interestingly, mutation of these four amino acids within the β3 subunit abolished its ability to form homomers but not its ability to oligomerize with γ2L suggesting that at least one alternative signal for the assembly of β and γ subunits must exist (215, 216). A conserved domain in the N-terminus of α subunits has also been identified to play a role in the oligomerization of α with β but not γ subunits (15, 16, 215, 216). Furthermore, conserved glutamine and arginine residues have been shown to independently play a key role in determining the assembly of α with β subunits (15, 16, 215, 216), whereas a conserved arginine in α, β and γ subunits has been shown to be essential for subunit oligomerization (85). Finally, two regions within the N-terminus of the γ2 subunit have been shown to mediate assembly of this subunit with α and β subunits (111, 112). The above results together with biochemical experiments suggest that the most prevalent $GABA_A$ receptor subunit composition in the brain consists of α, β and γ subunits with a majority of receptors containing 2α, 2β and 1γ subunit isoform (16, 68, 138, 141, 162).

8 Role of $GABA_A$ Receptor Subunit Composition in Determining Subcellular Localization

Immunofluorescence and EM immunogold studies have revealed that α1, α2, α3, α6, β2/3 and γ2 subunits are enriched at postsynaptic domains of inhibitory synapses in many brain regions including cortex, hippocampus, globus pallidus and cerebellum (61, 64, 67–69, 166, 167, 198). These synaptically targeted receptor subunits can also be found extrasynaptically (64). Immunocytochemical and immunogold electron microscopy studies have revealed that $GABA_A$ receptor subunit combinations can be targeted to different subcellular domains (67, 68, 166–168). For example, although the α1–3 and α5 subunits are all expressed in hippocampal and cerebellar cells, the α2 subunit is found concentrated in the axon initial segment (AIS) of the majority of cells where it colocalizes with the inhibitory synaptic marker gephyrin (25, 168). In contrast, the α1 subunit is found throughout the cell and shows both diffuse and clustered staining suggesting both a synaptic and extrasynaptic localization (25). Interestingly, the α1 subunit cannot be found in the AIS on its own and is always co-localized there with the α2 subunit suggesting that subunit composition of $GABA_A$ receptors in the AIS consists of one copy of α1 and one of α2 (25). Recent functional experiments in hippocampal pyramidal neurons have revealed that fast phasic responses are mediated by synaptic α2 subunit containing $GABA_A$ receptors on the cell soma but synaptic α1 containing $GABA_A$ receptors on dendrites (177). Receptors containing α3 are differentially targeted depending on the cell type where they are expressed. In pyramidal cells, the α3 subunit is found in clusters at postsynaptic sites, whereas in a subset of hippocampal cells characterized by a round cell body and numerous short dendrites, α3 containing receptors show a diffuse expression pattern across the membrane and are not found at synaptic sites (25). α5 subunits have been found highly expressed in extrasynaptic locations on hippocampal dendrites where it has been proposed they may contribute to tonic inhibition (25), but have also been localized to synapses both by immunofluorescence and EM postembedding immunogold (40, 195).

In addition to the α subunit, receptor subcellular localization is also determined by the γ and δ subunits. In the cerebellum, immunocytochemical studies on electron microscopic sections demonstrated that synaptic and extrasynaptic $GABA_A$ receptors differed in their subunit composition. Synapses in cerebellar granule neurons are positive for the $GABA_A$ receptor γ2 subunit, whereas the δ subunit was found exclusively in extrasynaptic membranes (165). In the hippocampus δ subunits are localized perisynaptically and not at synapses (227). Interestingly, in the cerebellum phasic and tonic $GABA_A$ receptor mediated inhibition has been observed and this has been attributed to synaptic and extrasynaptic $GABA_A$ receptors respectively (22, 23). Furthermore, in addition to immunocytochemical studies that show γ2 and δ subunits being respectively targeted to synaptic and extrasynaptic sites, electrophysiological studies have confirmed that the γ2 subunit mediates phasic inhibition whereas the δ subunit plays a role in mediating tonic inhibition (23, 24, 206). In agreement with this γ2 subunit $GABA_A$ receptor knockout (KO) experiments have demonstrated the critical role of the γ2 subunit in mediating the synaptic targeting of $GABA_A$ receptors which is critical for correct animal behavior (2, 47, 61). Overall, it is proposed based on work from immunofluorescent, EM, functional and gene deletion studies, that α1, α2 or α3 subunits, co assembled with the γ2 subunit and β subunit variants, are the major receptor subtypes localized to inhibitory synapses and contributing to phasic inhibition (64). In contrast, α5 subunit containing receptors assembled with the γ2 subunit and β subunit variants, and α4 or α6 subunits co assembled with the δ subunit and β subunit variants are primarily localized extrasynaptically and mediate tonic inhibition (29, 64). Recently it has been demonstrated that in some cell types such as hippocampal interneurons, the α1 subunit may also assemble with the δ subunit and β subunit variants to form extrasynaptic receptors mediating tonic inhibition suggesting that novel $GABA_A$ receptor subunit partnerships may yet be identified (76).

9 Components of the Inhibitory Postsynaptic Domain and Their Role in the Formation and Maintenance of Inhibitory Synapses

Specific mechanisms must exist to specify the formation and maturation of synapses and the correct apposition of presynaptic terminals with postsynaptic domains containing the correct cognate neurotransmitter receptors. In addition mechanisms must exist to direct and retain inhibitory receptors from intracellular or extrasynaptic compartments within the inhibitory postsynaptic apparatus. Several components of the inhibitory postsynaptic domains are proposed to play a key role in these processes, either as components of the inhibitory synaptic scaffold or by regulating transport of receptors within the cell (Figs. 2 and 4).

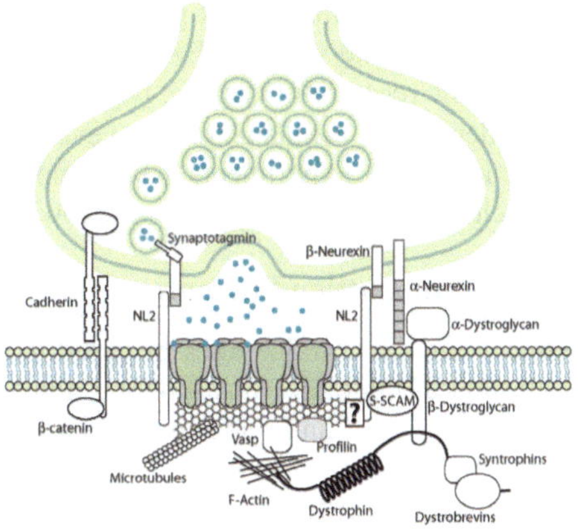

Fig. 2. Schematic representation of a central inhibitory terminal.
The specification and validation of inhibitory synapses and the correct apposition of inhibitory presynaptic terminals with inhibitory postsynaptic domains is thought to be determined in part by the co-ordinated action of adhesion molecules including cadherins, the neurexin-neuroligin 2 complex and the dystroglycan complex. Inhibitory neurotransmitter receptors ($GABA_A$ and glycine) are confined to these inhibitory specializations by scaffold molecules, most notably gephyrin, which can form a sub-synaptic clustered lattice important for synaptic receptor retention. Either directly or via intermediate binding partners, gephyrin can interact with cytoskeletal elements including tubulin and actin. The dystroglycan complex can interact with cytoplasmic binding partners including dystrobrevin and S-SCAM, which may also play a role in regulating receptor clustering and trafficking and furthermore, may also facilitate the co-clustering of both neurexin-neuroligin 2 and neurexin-dystrolgycan complexes.

9.1 A Critical Role for Gephyrin in the Molecular Organization of Inhibitory Synaptic Domains

One of the first proteins to be found enriched at inhibitory synapses is the ubiquitously expressed 93 kDa protein gephyrin (221). Initially isolated as a protein co-purifying with glycine receptors from rat spinal cord (175), gephyrin was found to be present in front of GABAergic terminal boutons (220) and was then demonstrated to play a central role in the organization of GABAergic synapses. Immunofluorescence and EM studies have conclusively shown that gephyrin is enriched at inhibitory GABAergic and glycinergic synapses (reviewed in (162)). However, several key differences remain between gephyrin regulation of glycinergic and GABAergic synapses. Firstly, whereas a direct interaction between gephyrin and glycine receptors has been demonstrated between a 14 residue stretch in the glycine receptor β-subunit loop and gephyrin (151), the exact molecular mechanism whereby gephyrin is recruited to $GABA_A$ receptors and whether or not gephyrin can interact either directly or indirectly via a bridging molecule with $GABA_A$ receptors remains unclear. Furthermore, in contrast to glycine receptor clustering, not all $GABA_A$ receptor subtypes are clustered by gephyrin dependent mechanisms. However, gephyrin and γ2 subunit $GABA_A$ receptor mouse knockout experiments have emphasized the relationship between GABA receptor clustering and gephyrin (see below) (2, 61).

9.2 Gephyrin Structure and Lattice Formation

Several recent biochemical and structural studies have led to a model for the ability of gephyrin to forms clusters at inhibitory postsynaptic domains (7, 125). Gephyrin has a modular structure consisting of an N-terminal G-domain linked by a 170 residue central region to a C-terminal E-domain (99, 204). This structure originates from the fusion of two genes of bacterial origin MogA and Moe, important for the biosynthesis of molybdenum co-factor and homologous to gephyrin G- and E-domains respectively (3, 204, 236). Structural and biochemical data have revealed that the gephyrin N-terminal G-domain and the C-terminal E-domain can form trimers and dimers respectively (99, 203), suggesting a mechanism for the formation of a hexagonal submembrane lattice onto which inhibitory receptors can be sequestered (Fig. 3). In agreement with this, experiments using Blue Native-PAGE of affinity purified gephyrin expressed in *Xenopus* oocytes found gephyrin to run as hexamers, possibly dimers of trimers, in addition to some higher order complexes (188). In contrast oligomerization mutants no longer formed hexamers in this expression system. The gephyrin hexamers may be a natural intermediate of gephyrin lattice formation. Initial structural and mapping studies have located the GlyR β-loop binding to the C-terminal E-domain of gephyrin (89, 193). More recently atomic resolution structural data define the GlyR β-loop as binding to each E-domain monomer in a pocket adjacent to the dimer interface. Complementary mutagenesis experiments reveal that β-loop binding is mediated by a hydrophobic interaction between phenylalanine 330 of gephyrin and two residues of the glycine receptor β-loop (phenylalanine 398, isoleucine 400) that are essential for this interaction. In contrast, the mechanisms of gephyrin recruitment to $GABA_A$ receptors or $GABA_A$ receptor recruitment to gephyrin, remain unknown.

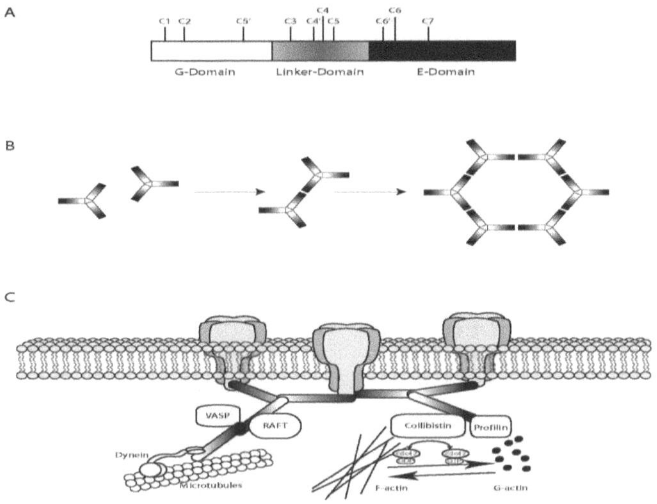

Fig. 3. Gephyrin structure and binding partners.
(**a**) Domain and splicing map of gephyrin. The 93kDa protein is divided into three domains: An N-terminal G-domain (homologous to the bacterial gene product MogA), a linker region (L-domain) and a C-terminal E-domain (homologous to the bacterial gene product MoeA). Gephyrin can exist in several splice variants termed C1-C7 and C4'-C6', the sites where cassettes are subject to alternative splicing are shown. (**b**) Gephyrin may assemble as a hexagonal lattice by the formation of dimers of trimers. G-domains can come together to form a trimeric structure, which in turn can form dimers via their E-domains. (**c**) Molecular partners of the glycine receptor/gephyrin complex. Gephyrin interacts with the intracellular loop of glycine β-subunits via its E-domain. This domain is also responsible for the interaction of gephyrin with other binding partners such as collybistin, profilin and VASP. RAFT-1 has been shown to interact with the E- and L-domains of gephyrin. Gephyrin has also been shown to bind microtubules via its G-domain as well as the light chain of the microtubule motor dynein.

Interestingly, gephyrin has several splice variants (147–149, 172, 173, 186) which may generate functional diversity and direct specific roles in regulating $GABA_AR$ function (Fig. 3). Gephyrin splicing appears to have no effect on high affinity interaction with the GlyR β-subunit but whether splicing may underlie gephyrin's interaction with $GABA_A$ receptor clustering remains less clear (147–149, 172, 173). On the other hand, gephyrin splicing within its central domain affects its ability to interact with various gephyrin associated proteins (102). Of greatest interest, splicing also regulates the ability of gephyrin to form oligomers. Two groups have recently demonstrated that insertion of a gephyrin splice cassette C5’ into the gephyrin N-terminal G-domain interferes with N-terminal trimerization (7, 188).

Several studies (e.g. (7)) have now demonstrated that interfering with gephyrin dimerization and trimerization inhibits gephyrin’s ability to form clusters in neurons. Gephyrin expression constructs containing substitutions at oligomerization interfaces have confirmed the essential role of G- and E-domains for clustering gephyrin at synapses. Furthermore, expression of recombinant gephyrin G- or E- domains, or dimerization or trimerization defective gephyrin mutants, disrupts gephyrin clusters in cultured neurons. Furthermore this causes disruption of both $GABA_A$ receptor and glycine receptor clusters at inhibitory postsynaptic sites supporting a critical role of gephyrin lattice formation in the maintenance of inhibitory postsynaptic domains.

9.3 Gephyrin Associated Proteins

Several gephyrin interacting partners have been identified that tie gephyrin functionally to cytoskeletal transport and anchoring processes (Fig. 3). Key gephyrin associated proteins identified so far include: the cdc42 guanylate exchange factor (GEF) collybistin (100), tubulin (178), the motor protein component dynein light chain (Dlc) (71), Mena/VASP (75), profilin (143) and RAFT (187). Gephyrin binding to tubulin provides a physical linkage between gephyrin and microtubules. Gephyrin also interacts directly with the actin monomer binding proteins profilin I and profilin II, and the actin microfilament adaptors Mena and VASP, providing a link to the actin cytoskeleton (5, 75, 143). Gephyrin can form complexes with profilin and Mena dependent on the gephyrin E-domain and competes with G-actin and phospholipids for the same binding region on profilin (75). In cell lines or cultured hippocampal neurons, profilin and Mena/VASP partially colocalize with inhibitory synaptic markers and gephyrin clusters and may contribute to a link between gephyrin dependent receptor clustering mechanisms and the microfilament system to regulate the dynamics of receptor localization at inhibitory synapses (163).

Of particular interest was the identification by yeast two-hybrid screening of a direct interaction of gephyrin with guanine nucleotide exchange factor (GEF) collybistin (hPEM-2 in humans), which can accelerate the GDP-GTP exchange on small GTPases (100). Collybistin is a member of the Dbl family of GEFs which are composed of tandem Dbl-homology (DH) and pleckstrin-homology (PH) domains and are specific for the Rho GTPases (Cdc42, Rac and Rho and their isoforms) which can regulate, among other processes, the reorganization of the actin cytoskeleton. Collybistin exists in a number of splice variants with three alternatively spliced C-terminal isoforms combined with presence or absence of an N-terminal SH3 domain. Collybistin 2 (the shortest version) recruits gephyrin to submembranous clusters which recruit glycine receptors (89, 100). Collybistin is a selective GEF for the GTPase Cdc42, which can regulate the reorganization of actin filaments. Since gephyrin can interact with components of the cytoskeleton, collybistin may regulate gephyrin function and receptor clustering by controlling local actin dynamics around the gephyrin lattice. Recent structural and biochemical results indicate that gephyrin binding has an inhibitory function on collybistin GEF activity (235). Collybistin may therefore play a role in terminating Cdc42 signaling during the initial stages of inhibitory synapse formation. In addition to control of cytoskeletal dynamics, the ability of gephyrin to associate with mTOR/RAFT1 (rapamycin and FKB12 target protein) suggests that gephyrin may also be able to participate in translational control at the synapse (187). This is in line with the demonstration that glycine receptor alpha subunit mRNAs can be found localized in the vicinity of the postsynaptic membrane (183) where a protein synthesis machinery is also present (74).

Gephyrin also interacts with dynein light chain 1 (also known as dynein light chain 8) and its homologue dynein light chain 2 (71). Dynein light chain interacts with a 63 amino acid binding domain within the central linker region of gephyrin and this interaction allows the recruitment of gephyrin and glycine receptors to dynein motor complexes for retrograde transport in neurons (71, 140). Whether gephyrin can similarly recruit $GABA_A$ receptor complexes to dynein motors remains unknown.

9.4 Functional Role of Gephyrin in Regulating the Formation and Maintenance of Inhibitory Glycinergic and GABAergic Synapses

Evidence supports a role for gephyrin in regulating the formation of inhibitory synapses and the recruitment of receptors to these postsynaptic specializations. Loss of function studies using antisense or gephyrin KO mice (104) (65, 104, 132) demonstrated that this protein is essential for the synaptic recruitment and/or clustering of all synaptic glycine receptors, both in the brain and spinal cord. Furthermore, gephyrin knockout mice die soon after birth, exhibiting a rigid hyperextended posture similar to animals treated with strychnine. In contrast the consequences of gephyrin depletion on the function of GABAergic synapses have remained more controversial. Antisense treatment against gephyrin in cultured hippocampal neurons causes a dramatic reduction in the density of clusters for $GABA_A$ receptor $\alpha2$ and $\gamma2$ subunits (61). Similarly RNAi mediated gephyrin knockdown resulted in a significant reduction but did not completely abolish the clustering of $\alpha2$ and $\gamma2$ containing $GABA_A$ receptors (94). A complete absence of $\alpha2$ and $\gamma2$ containing clusters in hippocampal cultures from gephyrin knockout mice was reported in one study (117). In contrast other studies have shown that, whereas surface and synaptic clusters of $\alpha2$ and $\gamma2$ subunit containing $GABA_A$ receptors are significantly reduced, many clusters could still be detected, and furthermore $\alpha1$ subunit clusters remain unaffected in gephyrin knockout neurons (116, 131, 132). Therefore in most neuronal types, gephyrin is not essential for $GABA_A$ receptor clustering but does appear to contribute to the aggregation of $\alpha2$, $\alpha3$ and $\gamma2$ but not $\alpha1$ or $\alpha5$ subunits. In agreement with this, GABAergic miniature inhibitory postsynaptic currents (mIPSCs), although reduced in amplitude, are present in hippocampal neurons lacking gephyrin, suggesting that gephyrin independent mechanisms of $GABA_A$ receptor clustering exist and may compensate during inhibitory synapse development in the absence of gephyrin (132).

Intriguingly, the clustering of gephyrin at GABAergic synapses is dependent on $GABA_A$ receptors. Gephyrin does not form clusters in the absence of $GABA_A$ receptors in cultured neurons and gephyrin clustering is disrupted *in vivo* in neurons from $GABA_A$ receptor $\alpha1$, $\alpha3$ and $\gamma2$ subunit knockout mice (2, 61, 125, 207). For example, in $\alpha3$ knockout neurons of the reticular nucleus of the thalamus, the absence of

$GABA_A$ receptors results in gephyrin forming large intracellular aggregates rather than synaptic clusters (207). Similarly in α1 knockout mice gephyrin clustering at postsynaptic sites is disrupted, leading to intracellular aggregates (125). In agreement with these results, artificial aggregation of $GABA_A$ receptors induces co-clustering of gephyrin (132). These results suggest that clustered synaptic $GABA_A$ receptors may recruit gephyrin to synapses. Interestingly, the glycinergic or GABAergic mixed phenotype of the presynaptic element determines the postsynaptic accumulation of specific cognate receptors but not of gephyrin and the postsynaptic accumulation of gephyrin alone cannot account for the formation of glycine receptor rich microdomains (130). A developmental analysis revealed that, at mixed glycine-GABAergic synapses, $GABA_A$ receptors formed clusters first, followed by gephyrin and then by glycine receptors (57). Gephyrin may then form a submembranous lattice that could serve to further stabilize newly recruited $GABA_A$ receptors at these sites. In agreement with this notion, RNAi mediated gephyrin knockdown combined with fluorescence imaging of GFP-labeled $GABA_A$ receptors found that gephyrin regulates $GABA_A$ receptor cell surface dynamics by reducing the mobility of $GABA_A$ receptor clusters (94). This would provide a feedback loop to promote and validate the formation of inhibitory postsynaptic specializations.

9.5 GRIP1

The glutamate receptor interacting protein 1 (GRIP1) is one of the few proteins to be found enriched at both excitatory and inhibitory synapses (52, 53). GRIP1 and its homologue GRIP2 are 7 PDZ domain containing proteins originally identified to interact with glutamate receptors. In addition, a second GRIP family member, ABP (for AMPA receptor binding protein) is a shorter splice variant derived from the same gene as GRIP2 but lacking the seventh PDZ domain. GRIP1 is encoded by a separate gene which can produce several GRIP1 isoforms by alternative splicing (GRIP1a, GRIP1b and GRIP1c4-7) (31, 32). GRIP1 associates with several proteins including the C-termini of AMPA receptor GluR2/3 and 4c subunits, ephrins and Eph receptors, the proteoglycan NG2, the putative extracellular matrix (ECM) protein Fraser syndrome protein (Fras1), GRASPs (for GRIP associated proteins) and liprin α family members (32, 52, 53, 205, 213, 233, 237). This suggests that GRIP may be important for regulating the function of other proteins in addition to glutamate receptors and has a more general role in membrane transport processes. In addition GRIP1 associates with several signaling molecules and proteins implicated in cytoskeletal transport processes including kinesin motor proteins and the microtubule associated protein GABARAP. Surprisingly, several groups have reported that GRIP1 associates with the postsynaptic domains of inhibitory synapses. Immunofluorescence experiments in cultured neurons and brain slices reveal that GRIP-1 and the 4-PDZ domain splice form GRIP-1c (PDZ 4-7) co-localize with several inhibitory postsynaptic markers such as gephyrin and $GABA_A$ receptors (27, 32, 53, 135) (234). Furthermore, using immunoelectron microscopy, GRIP1 was found to localize at inhibitory synapses in intact brain where it localizes to both the presynaptic and postsynaptic side of inhibitory synapses (31). The role of GRIP-1 at inhibitory synapses remains unknown, although its ability to interact with cytoskeletal and trafficking proteins suggest that it may be involved in the regulation of $GABA_A$ receptor function or trafficking (105).

10 Cell Adhesion Molecules at the Inhibitory Synapse

The exact mechanisms that underlie the precise alignment of presynaptic terminals from innervating neurons with postsynaptic domains containing the cognate neurotransmitter receptor remain unclear. It has become clear however that in addition to postsynaptic scaffold proteins a critical role also exists for cell adhesion molecules that can span the synapse linking pre- and postsynaptic membranes. Several cell adhesion molecules such as cadherins (10), the dystroglycan complex and neurexin/neuroligin complexes have been localized to inhibitory synapses and proposed to play an important role in their formation and maintenance (Fig. 2).

10.1 Neurexins and Neuroligins: Key Molecules for the Induction of Inhibitory Synapses

The neuroligins, and their presynaptic counterparts neurexins, have emerged as key trans-synaptic organizing molecules at both excitatory and inhibitory synapses. Neurexins were originally identified as a binding partner for the black widow spider venom α-latrotoxin (223). Using biochemical techniques, a number of endogenous neurexin binding partners were then identified such as neuroligins, neurexophilins and dystroglycan (46). The most well characterized neurexin complex is that with neuroligins. Neurexins and neuroligins form cell adhesion complexes that span the synaptic cleft, with neurexins localized on the presynaptic side and the neuroligins located in the postsynaptic domain (46) (see Chapter by El-Husseini, this volume). A key role of the neurexin/neuroligin complex for generating synaptic connections became clear when it was shown that neuroligins expressed in a human embryonic kidney (HEK) cell line could induce glutamatergic axons from co-cultured neurons to form functional release sites onto these non-neuronal cells (191). In a complementary set of co-culture experiments, it was also found that when neurexins were expressed alone on non-neuronal cells they could trigger postsynaptic differentiation and clustering of postsynaptic receptors and scaffolds in contacting dendrites (79).

Neurexins are cell surface proteins that contain a large extracellular N-terminal domain, transmembrane region and short intracellular C-terminus. There are three vertebrate genes for Neurexins that contain two independent promoters, which allow the formation of either α- or (the shorter) β-neurexins. Neurexins can be alternatively spliced at five different sites allowing for the expression of hundreds of alternatively spliced isoforms (46). The N-terminal domain of α-neurexins contains six laminin-neurexin-sex hormone-binding globulin (LNS) domains interspersed by three EGF-like regions whereas β-neurexins contain only a single LNS domain. Neurexins have relatively short intracellular domains which bind to synaptotagmin and the PDZ domains of CASK, syntenin and Mint linking them to synaptic vesicles and the vesicle fusion machinery (46). Extracellularly, α- and β-neurexins bind to neuroligins, which are concentrated in the postsynaptic domains of inhibitory and excitatory synapses.

There are believed to be at least four main neuroligin isoforms in mammals (NLGN1-NLGN4). Neuroligins comprise a large extracellular domain similar in structure to that of acetylcholinesterase (but lacking cholinesterase activity), which interacts with neurexins. In addition, neuroligins can homo-multimerise through their AChE-like domain. In contrast to their large extracellular domains, the intracellular domains of neuroligins are relatively short and terminate with a PDZ domain ligand. The PDZ domain ligand is believed to link them to several PDZ domain containing postsynaptic proteins such as PSD-95 and S-SCAM family members in addition to Shank, PICK1, GOPC and SPAR (46, 152). The identification of protein interactions between neuroligins and components of the excitatory postsynaptic scaffold suggested a mechanism allowing them to co-ordinate the formation of excitatory synapses. However, the importance of neuroligins for the formation of inhibitory synapses was suggested when it was reported that neuroligin-2 is enriched at inhibitory (but not excitatory) synapses (225).

Substantial evidence has now accumulated to support the ability of neurexin-neuroligin complexes to trigger presynaptic/postsynaptic differentiation and/or stabilization of inhibitory synapses. Graf et al. (2004) showed that neurexin presented alone on the surface of HEK cells or on beads is sufficient to induce the localized clustering of glutamatergic and GABAergic postsynaptic receptors and scaffolds (79). They also went on to demonstrate that, whereas neuroligins-1, -3 and -4 localize to excitatory postsynaptic domains, neuroligin-2 localizes primarily to inhibitory synapses supporting the idea that neuroligin-2 is a key factor for the differentiation of inhibitory synapses. In agreement with this, aggregating neuroligins-1, -3 and -4 resulted in clustering of only excitatory postsynaptic components, whereas aggregating neuroligin-2 resulted in redistribution of both excitatory and inhibitory postsynaptic proteins. Although it is now well established that neuroligin-2 can selectively link gephyrin and $GABA_A$ receptors, the mechanisms of this linkage remain unclear. Importantly, the presynaptic contacts induced by neuroligins appear to exhibit many of the release characteristics of normal synapses. This has most recently been elegantly demonstrated by experiments showing that HEK cells co-expressing neuroligins with either AMPA receptors or $GABA_A$ receptors and co-cultured with neurons produce miniature excitatory or inhibitory postsynaptic like currents (mIPSCs or mEPSCs) (54, 70).

Additional functional evidence for a critical role of neuroligins comes from RNA interference studies which showed that knockdown of neuroligins-1, -3 in hippocampal cultures reduced the density of inhibitory and excitatory synapses identified by immunofluorescence with inhibitory and excitatory presynaptic markers (38). Furthermore knockdown also resulted in a large reduction in the amplitude and frequency of mIPSCs. The synaptogenic activity of neurexins/neuroligins appears to be in part dependent on alternative splicing in splice site B of the neuroligin AChE-domain and at site 4 in β-neurexins. It has been proposed that neuroligin lacking a splice insert in splice B together with β-neurexins containing an S4 insert selectively promote differentiation of GABAergic synapses, whereas β-neurexins lacking S4 but containing splice insert B selectively promote differentiation of glutamatergic synapses (19, 39, 46).

It had generally been assumed that postsynaptic neuroligins interact trans-synaptically with presynaptic neurexins. Using immunoelectron microscopy, it was revealed that in addition to being localized pre-synaptically, neurexins were also found to be abundantly expressed in the postsynaptic density suggesting that cis-neurexin-neuroligin complexes may also form postsynaptically (214). Furthermore cis-expression of recombinant neurexin was found to block the synaptogenic activity of neuroligin-1. Postsynaptic cis-interactions between neuroligins and neurexins could provide a novel mechanism for silencing the activity of trans-synaptic complexes. Whether postsynaptic expression of neurexins can also mediate an inhibitory effect on synaptogenic activity of neuroligin-2 at inhibitory synapses remains unclear. However, these recent findings suggest a mechanism for silencing neuroligin function to regulate synapse formation or disassembly (214).

The increase in synapse formation observed *in vitro* by neuroligins and neurexins has been proposed to be due to the synaptogenic activity promoting the differentiation of new synapses. In contrast, *in vivo* although neuroligin function appears essential for synaptic function and animal survival it does not appear to be required for synapse formation *per se* (46, 224). Recent work reconciling these apparently contradictory observations suggests that rather than inducing the differentiation of new synapses, neurexin/neuroligin complexes act to specify and validate already formed synapses by an activity-dependent mechanism (42). In this model, activity-dependent postsynaptic signaling mechanisms converge on either neuroligin-1 or neuroligin-2 to validate the formation of excitatory versus inhibitory synapses, respectively (42).

10.2 Dystrophin Glycoprotein Complex (DGC)

There are accumulating localization and functional studies suggesting that the dystrophin-glycoprotein complex (DGC), which in non-neuronal cells links the extracellular matrix to the intracellular cytoskeleton, is another adhesion complex important for the proper maturation and function of a subset of inhibitory synapses. The DGC is a large complex containing at least 10 proteins (e.g. dystrophin, dystroglycan, utrophin, syntrophins, sarcoglycans and dystrobrevins) (12). Dystroglycan is a major component of the DGC and is composed of an extracellular α-subunit and transmembrane β-subunit, both derived from proteolytic cleavage of a single pre-cursor protein, and which remain bound to each other on the cell surface. Through its extracellular domain dystroglycan binds matrix proteins agrin, laminin and perlcan and the presynaptic proteins neurexins. Through its intracellular domain dystroglycan binds dystrophin (a large > 400 kDa cytoskeletal protein of the α-actinin/β-spectrin family) and utrophin (12). Dystrophin interacts directly with dystrobrevins, syntrophin and the actin cytoskeleton. Mutations in dystrophin result in Duchene and Becker muscular dystrophies in humans. Similarly, mutations in many of the genes encoding components of the DGC (e.g. α-, β-, γ- and δ-sarcoglycan, α-dystrobrevin, laminin α2, and four enzymes that glycosylate dystroglycan) lead to muscular dystrophies in human or animal models (12).

A number of studies support a role for the DGC at inhibitory synapses (26, 119, 120, 131). Several groups have localized components of the DGC (including α- and β-dystroglycan, long and short forms of dystrophin, syntrophin and α- and β-dystrobrevin) to a subset of inhibitory synapses on neuronal somata and dendrites (131). Although the role of the DGC at inhibitory synapses remains unclear, the demonstration that dystrophin mutant *mdx* mice or DGC knockout mice have a large reduction in $GABA_A$ receptor clusters (but not gephyrin) has lead to the proposal that the DGC may be important for regulating the stability of $GABA_A$ receptors at DGC containing synapses (120). More recently, cerebellar Purkinje cells from double α- and β-dystrobrevin knockout mice have been found to have depleted synaptic dystrophin levels, reduced $GABA_A$ receptor clusters and altered sensorimotor behaviors similar to those seen in dystrophin-deficient (*mdx*) mice suggesting that other DGC components in addition to dystrophin can affect inhibitory synaptic structure (78). In addition, observations of the critical requirement for the DGC for the proper maturation and function of a subset of inhibitory cerebellar synapses suggests that the motor deficits observed in muscular dystrophy patients may reflect not only peripheral dysfunctions but in addition alterations of synaptic inhibition in the CNS (78).

The major extracellular matrix components that bind the DGC in non-neuronal cells (laminin, perlecan, collagen and agrin) are not as abundant in brain and the extracellular space between neurons in adult brain lacks a basal lamina. This suggests that other neuro-specific ligands of the DGC may exist. In agreement with this it has been demonstrated that in brain, the DGC forms a stoichiometric complex with the presynaptic neurexins (210). Both α- and β-neurexin can interact with dystroglycan via the second and sixth LNS domain of α-neurexin and the single LNS domain in β-neurexin, suggesting that they may be the main neuronal DGC receptors. Intriguingly, in a hippocampal cell culture model of inhibitory synapse formation, gephyrin and $GABA_A$Rs can be mistargeted to glutamatergic terminals when there is insufficient inhibitory presynaptic terminal innervation (41, 184, 208). In contrast, components of the DGC are not mistargeted to glutamatergic synapses due to reduced inhibitory presynaptic input or in the absence of $GABA_A$ receptor γ2 subunits suggesting independent targeting mechanisms which may rely on interaction with specific presynaptic neurexins present at inhibitory synapses (26, 210). Recently it has been demonstrated that the synaptic scaffolding molecule S-SCAM can be found localized to inhibitory synapses (211). S-SCAM contains 5 or 6 PDZ domains, a guanylate kinase domain and two WW domains. S-SCAM can interact with β-dystroglycan via its WW domains and neuroligin-2 via its WW domains and second PDZ domain. Furthermore, S-SCAM, β-dystroglycan and neuroligin-2 form a tripartite complex *in vitro*, co-immunoprecipitate from rat brain and partially co-localize in rat hippocampal neurons. These results suggest that S-SCAM functions as a linker between the DGC/neurexin complex and the neuroligin/neurexin complex.

11 Membrane Trafficking and Dynamics of $GABA_A$ Receptors and Glycine Receptors

Although postsynaptic specializations can be viewed as relatively fixed entities, the individual components of the inhibitory postsynaptic domain themselves are dynamic. The number of surface and synaptic GABA and glycine receptors is determined by rates of receptor cycling between surface and intracellular compartments and the exchange of synaptic and extrasynaptic surface receptors. This is determined by association with receptor associated proteins that regulates exocytosis and endocytosis of plasma membrane receptors and also by the relationship of receptors with their associated scaffold proteins.

11.1 $GABA_A$ Receptor Associated Proteins in Receptor Transport, Localization, and Internalization

Given the selective cellular localization of $GABA_A$ receptors and glycine receptors, it is clear that the trafficking, targeting and clustering of these receptors is a tightly controlled and regulated process. A number of $GABA_A$ receptor associated proteins that interact directly with $GABA_A$ receptor intracellular domains have been recently implicated in receptor membrane trafficking and localization (Fig. 4). Several of these proteins are not highly enriched at inhibitory synapses, but rather are thought to be important for receptor transport and maturation throughout the secretory pathway (see Chapter by Hanus and Ehlers, this volume). $GABA_A$ receptor α- and β-subunits are components of essentially all native $GABA_A$ receptors and play key roles in controlling receptor trafficking. Several proteins interact with the intracellular domains of these receptors. The ubiquitin related protein Plic-1 interacts with $GABA_A$ receptor α- and β-subunits (8). Plic-1 is a 67 kDa protein with a ubiquitin like N-terminal (UBL) domain and a carboxy-terminal ubiquitin associated domain (UBA) (113). Plic proteins regulate ubiquitin-dependent protein degradation in the proteasome by their ability to bind ubiquitin ligases and components of the proteasome (113) (see Chapter by Helton and Ehlers, this volume). Yeast two hybrid screens and GST affinity purification assays have shown that Plic-1 interacts with all α (1–6) and β (1–3) subunits of $GABA_A$ receptors indicating that Plic-1 function may be relevant for the majority of $GABA_A$ receptor subtypes expressed in the brain (8). Immunofluorescence studies found Plic-1 to be mainly expressed in intracellular compartments and not at synapses (8). Significantly, a number of $GABA_A$ receptors colocalize with Plic-1 beneath the plasma membrane at subsynaptic membranes (8). Interestingly binding to the intracellular loops of α and β subunits is mediated by the UBA domain of Plic-1 and functional studies in non-neuronal HEK293 cells and hippocampal brain slices revealed that blockade of the interaction between Plic-1 and $GABA_A$ receptors results in reduced receptor cell surface expression (8). Consistent with these results, overexpression of Plic-1 in recombinant systems results in an increase in surface expressed $GABA_A$ receptors without affecting internalization rates (8). Plic-1 may thus act to modulate $GABA_A$ receptor cell surface numbers by inhibiting receptor degradation in the proteasome, providing initial evidence that $GABA_A$ receptor function and synaptic receptor number may be regulated by ubiquitin dependent degradation processes. The role of ubiquitination in regulating glycine receptor surface stability and membrane trafficking has also been demonstrated, however, the molecular mechanisms involved in this process remain unknown (28).

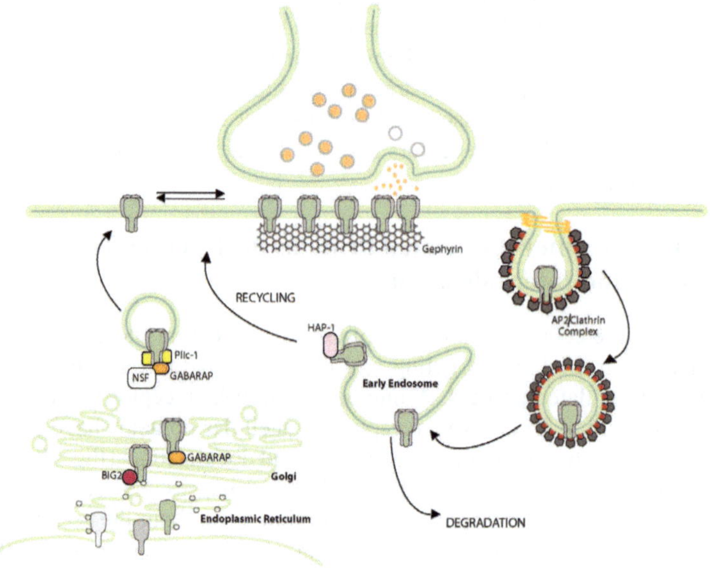

Fig. 4. Trafficking of $GABA_A$ receptors.
Assembled $GABA_A$ receptors in the endoplasmic reticulum are delivered to the Golgi, possibly with aid of the guanidine exchange factor BIG2. GABARAP can associate with the intracellular loop of the $\gamma 2$ subunit in the Golgi as well as with NSF, which may facilitate transportation of the receptor complexes to the cell surface. In addition the ubiquitin like protein Plic-1 is localized in intracellular compartments where it binds α and β subunits. This interaction is thought to regulate membrane targeting by preventing degradation in the proteasome. $GABA_A$ receptors inserted at extrasynaptic sites can diffuse along the plasma membrane into postsynaptic sites where they are stabilized by an undefined interaction with the scaffolding protein gephyrin. $GABA_A$ receptors are internalized via clathrin mediated pathway, where $GABA_A$ receptors move out of synaptic sites and are recruited to clathrin pits by the association of β and γ subunits with the clathrin adaptor protein AP2. The large GTPase dynamin releases the clathrin-coated vesicle which delivers the internalized receptor to the endosomal system. In early endosomes, $GABA_A$ receptors can be targeted for degradation or recycle back to the plasma membrane, a process that is mediated by an interaction with the Huntingtin associated protein HAP1.

The guanine nucleotide exchange factor BIG2 (brefeldin A-inhibited GDP/GTP exchange factor 2) has also been identified in a yeast two-hybrid screen as a $GABA_A$ receptor β subunit interacting protein (33). BIG2 is mainly found concentrated in the trans Golgi network although a small proportion is found in vesicle-like structures along dendrites and near post-synaptic sites (33). BIG2 has previously been implicated in vesicular transport by its ability to catalyse the GDP-GTP exchange on ARF GTPases. Interestingly, coexpression of BIG2 with the $GABA_A$ receptor β3 subunit results in an increase in β3 exit from the ER suggesting that BIG2 is involved in the post-Golgi vesicular trafficking of $GABA_A$ receptors (33). Other $GABA_A$ receptor β-subunit interacting proteins that have been identified include GRIF-1 (6) and gC1qR (190) whereas radixin has been shown to interact with receptor α-subunits (136). Neither GRIF-1 nor gC1qR have been found enriched at synapses and their functional significance remains to be established. The ERM family member radixin was identified by yeast two hybrid to interact with extrasynaptic $GABA_A$ receptor α5 subunits. Radixin's ability to bind to actin suggests that it may act to link $GABA_A$ receptors to the actin cytoskeleton and to the formation of extrasynaptic α5 subunit containing $GABA_A$ receptors (136).

As outlined above, the $GABA_A$ receptor γ2 subunit is essential for the synaptic localization, clustering and function of $GABA_A$ receptors. Several yeast two hybrid screens using the γ2 subunit have identified γ2 subunit associated proteins that are important for the selective synaptic targeting mediated by this subunit. The first to be identified was the 17 kDa $GABA_A$ receptor associated protein (GABARAP), which participates in the trafficking of $GABA_A$ receptors (226). GABARAP binds γ subunits both *in vitro* and *in vivo*. Immunocytochemical and immunoelectron microscopy studies revealed that it is not found at inhibitory synapses but instead colocalizes with $GABA_A$ receptors in intracellular compartments, namely at the Golgi compartment and subsynaptic cisternae (109, 226). GABARAP is a member of a family of homologous small microtubule binding proteins that also includes GABARAP like1 (GBRL1, GEC1), GABARAP like 2 (GBRL2, GATE-16) and the light chain-3A and B (LC3A, LC3B) subunits of MAP-1A and 1B (226). GABARAP is able to bind microtubules as well as N-ethylmaleimide sensitive factor (NSF) supporting a role for GABARAP in $GABA_A$ receptor intracellular transport (109, 169, 226). Recent immunochemical and electrophysiological studies have provided evidence to support a functional role for GABARAP in the trafficking of $GABA_A$ receptors to the plasma membrane and in inhibitory synaptic plasticity. Coexpression of GABARAP and $GABA_A$ receptors in COS-7 cells and neurons results in an increase in the levels of $GABA_A$ receptors expressed at the cell surface as well as an increase of GABA mediated currents in oocytes (36, 37, 129). Several GABARAP associated proteins have been identified that may also be important for $GABA_A$ receptor function. Although GABARAP itself is not localized to inhibitory synaptic domains, yeast two hybrid screens and *in vitro* binding studies have demonstrated that two other proteins localized to inhibitory postsynaptic domains, GRIP-1 and gephyrin interact with GABARAP, suggesting that GABARAP may be important for the transport of other components of the inhibitory postsynaptic domain in addition to $GABA_A$ receptors (105, 118). GABARAP has also been shown to associate with the phospholipase-C related inactive protein type 1 (PRIP-1). PRIP-1 appears to compete with the γ2 subunit for GABARAP binding (95). Interestingly PRIP-1 knockout mice showed an impairment in $GABA_A$ receptor modulation by γ2 selective pharmacological agents suggesting that PRIP-1 plays a role in the regulation of $GABA_A$ receptor trafficking by GABARAP ensuring that only mature αβγ receptors are delivered to the plasma membrane (95, 156). Gene deletion studies have shown that *in vivo* GABARAP is dispensable for postsynaptic trafficking of $GABA_A$ receptors. However whether this is due to compensation by the other closely related GABARAP family members remains to be established (170).

In addition to GABARAP, γ2 subunits have also been identified by yeast two hybrid to interact with the Golgi-specific palmitoyl transferase GODZ and to co-localize with this protein in Golgi compartments (63, 97). The γ2 subunit is the only $GABA_A$ receptor subunit family member that contains cysteine residues within its intracellular domain which have been shown to serve as substrates for palmitoylation, a post-translational modification important for correct surface delivery and synaptic targeting of $GABA_A$ receptors (63, 97, 185). GODZ is the major palmitoylating enzyme for $GABA_A$ receptors and RNAi mediated knockdown of this protein disrupts the formation of normal whole cell and synaptic inhibitory currents. Palmitoylation of $GABA_A$ receptor γ2 subunits by GODZ (and to a lesser extent by a GODZ paralogue, SERZ beta) therefore is important for receptor transport through the Golgi apparatus and the selective formation and normal function of inhibitory synapses (63, 97). Other $GABA_A$ receptor subunits, including γ1, γ3, θ, ε and π also contain cysteine residues within their intracellular domains, although whether they are substrates for palmitoylation remains to be addressed.

The internalization of $GABA_A$ receptors from the cell surface by clathrin and dynamin dependent mechanisms has also been demonstrated to be an important mechanism for regulating surface and synaptic receptor number (108). $GABA_A$ receptors have been found localized to clathrin coated pits and receptor β- and γ- subunits can interact directly with the clathrin adaptor protein AP2 which facilitates the recruitment of these receptors into the endocytic pathway (107). The interaction of $GABA_A$ receptor β-subunits with AP2 has been mapped to a phospho-dependent basic amino acid motif that also contains the major site for β-subunit phosphorylation (106). Blocking $GABA_A$ receptor internalization, by either targeting dynamin function or by blocking $GABA_A$ receptor phospho-dependent interaction with AP2 inhibits dopaminergic modulation of inhibitory synapse strength in the striatum suggesting that signaling dependent alterations in AP2-dependent $GABA_A$ receptor endocytosis may be a critical mechanism underlying functional modulation of inhibition (34). Internalized $GABA_A$ receptors can be either targeted for degradation in the lysosome or rapidly recycled in a process that is dependent on a direct interaction (also identified by yeast two hybrid) of $GABA_A$ receptors with the huntingtin associated protein 1 (HAP1) (110). Interestingly, decreasing HAP1 levels in the hypothalamus using siRNA leads to reduced levels and activity of $GABA_A$ receptors and altered feeding suggesting that HAP1 dependent $GABA_A$ receptor trafficking may underlie important aspects of normal animal behavior (110, 196).

11.2 Glycine Receptor Associated Proteins

With the exception of gephyrin (see above), virtually no proteins have yet been demonstrated to interact directly with glycine receptor subunits. It remains to be tested whether any of the identified $GABA_A$ receptor associated proteins may also be important for glycine receptor function and membrane trafficking (analogous to the overlapping role of gephyrin for regulating both these Cys-loop receptors). Interestingly, gephyrin itself has been implicated in receptor transport and trafficking in addition to its well characterized role in receptor anchoring at the synapse (60, 86, 115). Gephyrin associates with the vesicular compartments used for intracellular trafficking of glycine receptors and has also been demonstrated to act as a mediator of microtubule-dependent insertion of newly synthesized receptor molecules into the plasma membrane (86). It has also been suggested that non-synaptic glycine receptor/gephyrin complexes can be recruited to microtubules by dynein motors for retrograde transport along neuronal processes (115), and diffusing extrasynaptic receptors could also bind gephyrin (60). The demonstration of intracellular and extrasynaptic glycine receptor/gephyrin complexes suggests a dual function for gephyrin in both trafficking and assembly of postsynaptic scaffolds, as has similarly been suggested for proteins of excitatory synapses (218). Whether gephyrin also plays a role in the intracellular transport of $GABA_A$ receptors remains to be determined.

11.3 Diffusional Properties of Cell Surface $GABA_A$ Receptors and Glycine Receptors

Lateral diffusion within the plane of the plasma membrane has been demonstrated for both $GABA_A$ receptors and glycine receptors. This concept has been extended to almost all excitatory and inhibitory receptors so far studied (see Chapter by Choquet and Triller, this volume). The best estimates of inhibitory receptor diffusion rates in the plasma membrane have been derived from single particle tracking (SPT) of glycine receptors labeled with antibodies to extracellular domains (48, 150). These studies have shown that receptor movements display interspersed periods of high and low diffusion rates in the extrasynaptic and synaptic membrane, respectively. Whereas periods of fast diffusion correspond to Brownian movement, those of slow mobility result from transient association of glycine receptor with clusters of gephyrin. This association, which reduces diffusion coefficients from up to 0.5 $\mu m^2\ s^{-1}$ to less than 0.01 $\mu m^2\ s^{-1}$, imposes a restricted area of exploration (confinement) similar to that generated for other proteins by insertion into lipid rafts or by contact with protein fences or obstacles. The fact that glycine receptors are not irreversibly retained in gephyrin domains suggests that any postsynaptic cluster of gephyrin can behave as donor and/or acceptor for glycine receptors (219). The diffusion rates of synaptic and extrasynaptic $GABA_A$ receptors have not yet been directly measured. However, studies using electrophysiologically tagged $GABA_A$ receptors indicate that significant lateral mobility of $GABA_A$ receptors in the plasma membrane also occurs (217). Furthermore, combined imaging of GFP-tagged $GABA_A$ receptors and RNAi-mediated gephyrin knockdown demonstrate that synaptic $GABA_A$ receptors have lower levels of lateral mobility than their extrasynaptic counterparts and that gephyrin plays a specific role in limiting receptor diffusion (94). The dynamic properties of gephyrin interactions with glycine receptors and $GABA_A$ receptors is therefore also a means whereby the number of synaptic receptors can be regulated. Therefore, the interactions between receptors and scaffold molecules is likely to be one of the targets for the regulation of receptors at inhibitory synapses during synaptogenesis and in synaptic plasticity, or in homeostatic mechanisms setting the level of inhibition.

12 Concluding Remarks

Since their initial identification some two decades ago, significant progress has been made in our understanding of the mechanisms that underlie the biology of inhibitory ligand-gated neurotransmitter receptors. Now that all subunits are cloned, we have a clear picture of the extent of receptor subunit heterogeneity and their specific regional and temporal expression profiles in the brain and spinal cord. In addition, a detailed picture of the assembly mechanisms that specify the formation of native receptor subunit combinations and the molecular mechanisms underlying recognition of receptors by agonists, antagonists and allosteric modulators has emerged. Molecular, biochemical, structural and genetic approaches have significantly contributed to developing an in depth understanding of the mechanisms that underlie gephyrin-dependent recruitment and clustering of GABA and glycine receptors to inhibitory synapses. More recently the critical role of neuroligin family cell adhesion molecules for receptor clustering and the formation and validation of inhibitory synaptic sites has also become apparent. Finally, the importance of intracellular membrane trafficking and surface membrane diffusion of receptors has also emerged as a key mechanism for regulating the strength and plasticity of inhibitory synapses. Much work lies ahead in establishing a detailed model of how the varying processes of receptor membrane dynamics and the structural establishment of inhibitory postsynaptic domains are coordinated, both during development and in the adult, to underlie the formation and synaptic remodeling of inhibitory synapses. Another challenge is to understand potential cross talk in the processes regulating excitatory and inhibitory receptors in a same cell. An in depth understanding of how these processes may be altered could shed new light on neurological diseases where homeostatic regulation of synaptic inhibition and excitation may be dysregulated.

References

1. Absalom NL, Lewis TM, Kaplan W, Pierce KD, and Schofield PR. Role of charged residues in coupling ligand binding and channel activation in the extracellular domain of the glycine receptor. *J Biol Chem* 278: 50151–50157, 2003.
2. Alldred MJ, Mulder-Rosi J, Lingenfelter SE, Chen G, and Luscher B. Distinct gamma2 subunit domains mediate clustering and synaptic function of postsynaptic GABAA receptors and gephyrin. *J Neurosci* 25: 594–603, 2005.
3. Bader G, Gomez-Ortiz M, Haussmann C, Bacher A, Huber R, and Fischer M. Structure of the molybdenum-cofactor biosynthesis protein MoaB of Escherichia coli. *Acta Crystallogr D Biol Crystallogr* 60: 1068–1075, 2004.

4. Baer K, Essrich C, Balsiger S, Wick MJ, Harris RA, Fritschy JM, and Luscher B. Rescue of gamma2 subunit-deficient mice by transgenic overexpression of the GABAA receptor gamma2S or gamma2L subunit isoforms. *Eur J Neurosci* 12: 2639–2643, 2000.
5. Bausen M, Fuhrmann JC, Betz H, and O'Sullivan G A. The state of the actin cytoskeleton determines its association with gephyrin: role of ena/VASP family members. *Mol Cell Neurosci* 31: 376–386, 2006.
6. Beck M, Brickley K, Wilkinson HL, Sharma S, Smith M, Chazot PL, Pollard S, and Stephenson FA. Identification, molecular cloning, and characterization of a novel GABAA receptor-associated protein, GRIF-1. *J Biol Chem* 277: 30079–30090, 2002.
7. Bedet C, Bruusgaard JC, Vergo S, Groth-Pedersen L, Eimer S, Triller A, and Vannier C. Regulation of gephyrin assembly and glycine receptor synaptic stability. *J Biol Chem* 281: 30046–30056, 2006.
8. Bedford FK, Kittler JT, Muller E, Thomas P, Uren JM, Merlo D, Wisden W, Triller A, Smart TG, and Moss SJ. GABA(A) receptor cell surface number and subunit stability are regulated by the ubiquitin-like protein Plic-1. *Nat Neurosci* 4: 908–916, 2001.
9. Belelli D and Lambert JJ. Neurosteroids: endogenous regulators of the GABA(A) receptor. *Nat Rev Neurosci* 6: 565–575, 2005.
10. Benson DL and Tanaka H. N-cadherin redistribution during synaptogenesis in hippocampal neurons. *J Neurosci* 18: 6892–6904, 1998.
11. Birnir B, Eghbali M, Everitt AB, and Gage PW. Bicuculline, pentobarbital and diazepam modulate spontaneous GABA(A) channels in rat hippocampal neurons. *Br J Pharmacol* 131: 695–704, 2000.
12. Blake DJ, Weir A, Newey SE, and Davies KE. Function and genetics of dystrophin and dystrophin-related proteins in muscle. *Physiol Rev* 82: 291–329, 2002.
13. Bloomenthal AB, Goldwater E, Pritchett DB, and Harrison NL. Biphasic modulation of the strychnine-sensitive glycine receptor by Zn2+. *Mol Pharmacol* 46: 1156–1159, 1994.
14. Boileau AJ, Evers AR, Davis AF, and Czajkowski C. Mapping the agonist binding site of the GABAA receptor: evidence for a beta-strand. *J Neurosci* 19: 4847–4854, 1999.
15. Bollan K, King D, Robertson LA, Brown K, Taylor PM, Moss SJ, and Connolly CN. GABA(A) receptor composition is determined by distinct assembly signals within alpha and beta subunits. *J Biol Chem* 278: 4747–4755, 2003.
16. Bollan K, Robertson LA, Tang H, and Connolly CN. Multiple assembly signals in gamma-aminobutyric acid (type A) receptor subunits combine to drive receptor construction and composition. *Biochem Soc Trans* 31: 875–879, 2003.
17. Bonnert TP, McKernan RM, Farrar S, le Bourdelles B, Heavens RP, Smith DW, Hewson L, Rigby MR, Sirinathsinghji DJ, Brown N, Wafford KA, and Whiting PJ. theta, a novel gamma-aminobutyric acid type A receptor subunit. *Proc Natl Acad Sci USA* 96: 9891–9896, 1999.
18. Bormann J and Kettenmann H. Patch-clamp study of gamma-aminobutyric acid receptor Cl- channels in cultured astrocytes. *Proc Natl Acad Sci USA* 85: 9336–9340, 1988.
19. Boucard AA, Chubykin AA, Comoletti D, Taylor P, and Sudhof TC. A splice code for trans-synaptic cell adhesion mediated by binding of neuroligin 1 to alpha- and beta-neurexins. *Neuron* 48: 229–236, 2005.
20. Bowery NG, Doble A, Hill DR, Hudson AL, Turnbull MJ, and Warrington R. Structure/activity studies at a baclofen-sensitive, bicuculline-insensitive GABA receptor. *Adv Biochem Psychopharmacol* 29: 333–341, 1981.
21. Brejc K, van Dijk WJ, Klaassen RV, Schuurmans M, van Der Oost J, Smit AB, and Sixma TK. Crystal structure of an ACh-binding protein reveals the ligand-binding domain of nicotinic receptors. *Nature* 411: 269–276, 2001.

22. Brickley SG, Cull-Candy SG, and Farrant M. Development of a tonic form of synaptic inhibition in rat cerebellar granule cells resulting from persistent activation of GABAA receptors. *J Physiol* 497 (Pt 3): 753–759, 1996.
23. Brickley SG, Cull-Candy SG, and Farrant M. Single-channel properties of synaptic and extrasynaptic GABAA receptors suggest differential targeting of receptor subtypes. *J Neurosci* 19: 2960–2973, 1999.
24. Brickley SG, Revilla V, Cull-Candy SG, Wisden W, and Farrant M. Adaptive regulation of neuronal excitability by a voltage-independent potassium conductance. *Nature* 409: 88–92, 2001.
25. Brunig I, Scotti E, Sidler C, and Fritschy JM. Intact sorting, targeting, and clustering of gamma-aminobutyric acid A receptor subtypes in hippocampal neurons in vitro. *J Comp Neurol* 443: 43–55, 2002.
26. Brunig I, Suter A, Knuesel I, Luscher B, and Fritschy JM. GABAergic terminals are required for postsynaptic clustering of dystrophin but not of GABA(A) receptors and gephyrin. *J Neurosci* 22: 4805–4813, 2002.
27. Burette A, Wyszynski M, Valtschanoff JG, Sheng M, and Weinberg RJ. Characterization of glutamate receptor interacting protein-immunopositive neurons in cerebellum and cerebral cortex of the albino rat. *J Comp Neurol* 411: 601–612, 1999.
28. Buttner C, Sadtler S, Leyendecker A, Laube B, Griffon N, Betz H, and Schmalzing G. Ubiquitination precedes internalization and proteolytic cleavage of plasma membrane-bound glycine receptors. *J Biol Chem* 276: 42978–42985, 2001.
29. Caraiscos VB, Elliott EM, You-Ten KE, Cheng VY, Belelli D, Newell JG, Jackson MF, Lambert JJ, Rosahl TW, Wafford KA, MacDonald JF, and Orser BA. Tonic inhibition in mouse hippocampal CA1 pyramidal neurons is mediated by alpha5 subunit-containing gamma-aminobutyric acid type A receptors. *Proc Natl Acad Sci USA* 101: 3662–3667, 2004.
30. Celentano JJ, Gibbs TT, and Farb DH. Ethanol potentiates GABA- and glycine-induced chloride currents in chick spinal cord neurons. *Brain Res* 455: 377–380, 1988.
31. Charych EI, Li R, Serwanski DR, Li X, Miralles CP, Pinal N, and De Blas AL. Identification and characterization of two novel splice forms of GRIP1 in the rat brain. *J Neurochem* 97: 884–898, 2006.
32. Charych EI, Yu W, Li R, Serwanski DR, Miralles CP, Li X, Yang BY, Pinal N, Walikonis R, and De Blas AL. A four PDZ domain-containing splice variant form of GRIP1 is localized in GABAergic and glutamatergic synapses in the brain. *J Biol Chem* 279: 38978–38990, 2004.
33. Charych EI, Yu W, Miralles CP, Serwanski DR, Li X, Rubio M, and De Blas AL. The brefeldin A-inhibited GDP/GTP exchange factor 2, a protein involved in vesicular trafficking, interacts with the beta subunits of the GABA receptors. *J Neurochem* 90: 173–189, 2004.
34. Chen G, Kittler JT, Moss SJ, and Yan Z. Dopamine D3 receptors regulate GABAA receptor function through a phospho-dependent endocytosis mechanism in nucleus accumbens. *J Neurosci* 26: 2513–2521, 2006.
35. Chen Z, Dillon GH, and Huang R. Molecular determinants of proton modulation of glycine receptors. *J Biol Chem* 279: 876–883, 2004.
36. Chen ZW, Chang CS, Leil TA, Olcese R, and Olsen RW. GABAA receptor-associated protein regulates GABAA receptor cell-surface number in Xenopus laevis oocytes. *Mol Pharmacol* 68: 152–159, 2005.
37. Chen ZW, Chang CS, Leil TA, and Olsen RW. C-terminal modification is required for GABARAP-mediated GABA(A) receptor trafficking. *J Neurosci* 27: 6655–6663, 2007.
38. Chih B, Engelman H, and Scheiffele P. Control of excitatory and inhibitory synapse formation by neuroligins. *Science* 307: 1324–1328, 2005.

39. Chih B, Gollan L, and Scheiffele P. Alternative splicing controls selective trans-synaptic interactions of the neuroligin-neurexin complex. *Neuron* 51: 171–178, 2006.
40. Christie SB and de Blas AL. alpha5 Subunit-containing GABA(A) receptors form clusters at GABAergic synapses in hippocampal cultures. *Neuroreport* 13: 2355–2358, 2002.
41. Christie SB, Miralles CP, and De Blas AL. GABAergic innervation organizes synaptic and extrasynaptic GABAA receptor clustering in cultured hippocampal neurons. *J Neurosci* 22: 684–697, 2002.
42. Chubykin AA, Atasoy D, Etherton MR, Brose N, Kavalali ET, Gibson JR, and Sudhof TC. Activity-dependent validation of excitatory versus inhibitory synapses by neuroligin-1 versus neuroligin-2. *Neuron* 54: 919–931, 2007.
43. Connolly CN, Krishek BJ, McDonald BJ, Smart TG, and Moss SJ. Assembly and cell surface expression of heteromeric and homomeric gamma-aminobutyric acid type A receptors. *J Biol Chem* 271: 89–96, 1996.
44. Connolly CN and Wafford KA. The Cys-loop superfamily of ligand-gated ion channels: the impact of receptor structure on function. *Biochem Soc Trans* 32: 529–534, 2004.
45. Couve A, Moss SJ, and Pangalos MN. GABAB receptors: a new paradigm in G protein signaling. *Mol Cell Neurosci* 16: 296–312, 2000.
46. Craig AM and Kang Y. Neurexin-neuroligin signaling in synapse development. *Curr Opin Neurobiol* 17: 43–52, 2007.
47. Crestani F, Lorez M, Baer K, Essrich C, Benke D, Laurent JP, Belzung C, Fritschy JM, Luscher B, and Mohler H. Decreased GABAA-receptor clustering results in enhanced anxiety and a bias for threat cues. *Nat Neurosci* 2: 833–839, 1999.
48. Dahan M, Levi S, Luccardini C, Rostaing P, Riveau B, and Triller A. Diffusion dynamics of glycine receptors revealed by single-quantum dot tracking. *Science* 302: 442–445, 2003.
49. Davidoff RA, Shank RP, Graham LT, Jr., Aprison MH, and Werman R. Association of glycine with spinal interneurones. *Nature* 214: 680–681, 1967.
50. Davies PA, Hanna MC, Hales TG, and Kirkness EF. Insensitivity to anaesthetic agents conferred by a class of GABA(A) receptor subunit. *Nature* 385: 820–823, 1997.
51. Deitrich RA, Dunwiddie TV, Harris RA, and Erwin VG. Mechanism of action of ethanol: initial central nervous system actions. *Pharmacol Rev* 41: 489–537, 1989.
52. Dong H, O'Brien RJ, Fung ET, Lanahan AA, Worley PF, and Huganir RL. GRIP: a synaptic PDZ domain-containing protein that interacts with AMPA receptors. *Nature* 386: 279–284, 1997.
53. Dong H, Zhang P, Liao D, and Huganir RL. Characterization, expression, and distribution of GRIP protein. *Ann N Y Acad Sci* 868: 535–540, 1999.
54. Dong N, Qi J, and Chen G. Molecular reconstitution of functional GABAergic synapses with expression of neuroligin-2 and GABAA receptors. *Mol Cell Neurosci* 35: 14–23, 2007.
55. Draguhn A, Verdorn TA, Ewert M, Seeburg PH, and Sakmann B. Functional and molecular distinction between recombinant rat GABAA receptor subtypes by Zn2+. *Neuron* 5: 781–788, 1990.
56. Drew CA, Johnston GA, and Weatherby RP. Bicuculline-insensitive GABA receptors: studies on the binding of (-)-baclofen to rat cerebellar membranes. *Neurosci Lett* 52: 317–321, 1984.
57. Dumoulin A, Levi S, Riveau B, Gasnier B, and Triller A. Formation of mixed glycine and GABAergic synapses in cultured spinal cord neurons. *Eur J Neurosci* 12: 3883–3892, 2000.
58. Eghbali M, Curmi JP, Birnir B, and Gage PW. Hippocampal GABA(A) channel conductance increased by diazepam. *Nature* 388: 71–75, 1997.

59. Eghbali M, Gage PW, and Birnir B. Pentobarbital modulates gamma-aminobutyric acid-activated single-channel conductance in rat cultured hippocampal neurons. *Mol Pharmacol* 58: 463–469, 2000.
60. Ehrensperger MV, Hanus C, Vannier C, Triller A, and Dahan M. Multiple association states between glycine receptors and gephyrin identified by SPT analysis. *Biophys J* 92: 3706–3718, 2007.
61. Essrich C, Lorez M, Benson JA, Fritschy JM, and Luscher B. Postsynaptic clustering of major GABAA receptor subtypes requires the gamma 2 subunit and gephyrin. *Nat Neurosci* 1: 563–571, 1998.
62. Everitt AB, Luu T, Cromer B, Tierney ML, Birnir B, Olsen RW, and Gage PW. Conductance of recombinant GABA (A) channels is increased in cells co-expressing GABA(A) receptor-associated protein. *J Biol Chem* 279: 21701–21706, 2004.
63. Fang C, Deng L, Keller CA, Fukata M, Fukata Y, Chen G, and Luscher B. GODZ-mediated palmitoylation of GABA(A) receptors is required for normal assembly and function of GABAergic inhibitory synapses. *J Neurosci* 26: 12758–12768, 2006.
64. Farrant M and Nusser Z. Variations on an inhibitory theme: phasic and tonic activation of GABA(A) receptors. *Nat Rev Neurosci* 6: 215–229, 2005.
65. Feng G, Tintrup H, Kirsch J, Nichol MC, Kuhse J, Betz H, and Sanes JR. Dual requirement for gephyrin in glycine receptor clustering and molybdoenzyme activity. *Science* 282: 1321–1324, 1998.
66. Franks NP. Molecular targets underlying general anaesthesia. *Br J Pharmacol* 147 Suppl 1: S72–S81, 2006.
67. Fritschy JM, Benke D, Mertens S, Oertel WH, Bachi T, and Mohler H. Five subtypes of type A gamma-aminobutyric acid receptors identified in neurons by double and triple immunofluorescence staining with subunit-specific antibodies. *Proc Natl Acad Sci USA* 89: 6726–6730, 1992.
68. Fritschy JM and Brunig I. Formation and plasticity of GABAergic synapses: physiological mechanisms and pathophysiological implications. *Pharmacol Ther* 98: 299–323, 2003.
69. Fritschy JM and Mohler H. GABAA-receptor heterogeneity in the adult rat brain: differential regional and cellular distribution of seven major subunits. *J Comp Neurol* 359: 154–194, 1995.
70. Fu Z, Washbourne P, Ortinski P, and Vicini S. Functional excitatory synapses in HEK293 cells expressing neuroligin and glutamate receptors. *J Neurophysiol* 90: 3950–3957, 2003.
71. Fuhrmann JC, Kins S, Rostaing P, El Far O, Kirsch J, Sheng M, Triller A, Betz H, and Kneussel M. Gephyrin interacts with Dynein light chains 1 and 2, components of motor protein complexes. *J Neurosci* 22: 5393–5402, 2002.
72. Gao B, Fritschy JM, Benke D, and Mohler H. Neuron-specific expression of GABAA-receptor subtypes: differential association of the alpha 1- and alpha 3-subunits with serotonergic and GABAergic neurons. *Neuroscience* 54: 881–892, 1993.
73. Gao B, Hornung JP, and Fritschy JM. Identification of distinct GABAA-receptor subtypes in cholinergic and parvalbumin-positive neurons of the rat and marmoset medial septum-diagonal band complex. *Neuroscience* 65: 101–117, 1995.
74. Gardiol A, Racca C, and Triller A. Dendritic and postsynaptic protein synthetic machinery. *J Neurosci* 19: 168–179, 1999.
75. Giesemann T, Schwarz G, Nawrotzki R, Berhorster K, Rothkegel M, Schluter K, Schrader N, Schindelin H, Mendel RR, Kirsch J, and Jockusch BM. Complex formation between the postsynaptic scaffolding protein gephyrin, profilin, and Mena: a possible link to the microfilament system. *J Neurosci* 23: 8330–8339, 2003.

76. Glykys J, Peng Z, Chandra D, Homanics GE, Houser CR, and Mody I. A new naturally occurring GABA(A) receptor subunit partnership with high sensitivity to ethanol. *Nat Neurosci* 10: 40–48, 2007.
77. Gorrie GH, Vallis Y, Stephenson A, Whitfield J, Browning B, Smart TG, and Moss SJ. Assembly of GABAA receptors composed of alpha1 and beta2 subunits in both cultured neurons and fibroblasts. *J Neurosci* 17: 6587–6596, 1997.
78. Grady RM, Wozniak DF, Ohlemiller KK, and Sanes JR. Cerebellar synaptic defects and abnormal motor behavior in mice lacking alpha- and beta-dystrobrevin. *J Neurosci* 26: 2841–2851, 2006.
79. Graf ER, Zhang X, Jin SX, Linhoff MW, and Craig AM. Neurexins induce differentiation of GABA and glutamate postsynaptic specializations via neuroligins. *Cell* 119: 1013–1026, 2004.
80. Graham D, Pfeiffer F, Simler R, and Betz H. Purification and characterization of the glycine receptor of pig spinal cord. *Biochemistry* 24: 990–994, 1985.
81. Grasshoff C, Drexler B, Rudolph U, and Antkowiak B. Anaesthetic drugs: linking molecular actions to clinical effects. *Curr Pharm Des* 12: 3665–3679, 2006.
82. Greferath U, Grunert U, Fritschy JM, Stephenson A, Mohler H, and Wassle H. GABAA receptor subunits have differential distributions in the rat retina: in situ hybridization and immunohistochemistry. *J Comp Neurol* 353: 553–571, 1995.
83. Griffon N, Buttner C, Nicke A, Kuhse J, Schmalzing G, and Betz H. Molecular determinants of glycine receptor subunit assembly. *Embo J* 18: 4711–4721, 1999.
84. Grudzinska J, Schemm R, Haeger S, Nicke A, Schmalzing G, Betz H, and Laube B. The beta subunit determines the ligand binding properties of synaptic glycine receptors. *Neuron* 45: 727–739, 2005.
85. Hales TG, Tang H, Bollan KA, Johnson SJ, King DP, McDonald NA, Cheng A, and Connolly CN. The epilepsy mutation, gamma2(R43Q) disrupts a highly conserved intersubunit contact site, perturbing the biogenesis of GABAA receptors. *Mol Cell Neurosci* 29: 120–127, 2005.
86. Hanus C, Vannier C, and Triller A. Intracellular association of glycine receptor with gephyrin increases its plasma membrane accumulation rate. *J Neurosci* 24: 1119–1128, 2004.
87. Harrison NL, Kugler JL, Jones MV, Greenblatt EP, and Pritchett DB. Positive modulation of human gamma-aminobutyric acid type A and glycine receptors by the inhalation anesthetic isoflurane. *Mol Pharmacol* 44: 628–632, 1993.
88. Hartig W, Brauer K, Fritschy JM, Bruckner G, and Bigl V. Regional and cellular expression sites of the alpha 1 subunit of GABAA receptors in the rat basal forebrain: a cytochemical study with glutamic acid decarboxylase, choline acetyltransferase, calcium-binding proteins and nitric oxide synthase as second markers. *Brain Res* 692: 215–226, 1995.
89. Harvey K, Duguid IC, Alldred MJ, Beatty SE, Ward H, Keep NH, Lingenfelter SE, Pearce BR, Lundgren J, Owen MJ, Smart TG, Luscher B, Rees MI, and Harvey RJ. The GDP-GTP exchange factor collybistin: an essential determinant of neuronal gephyrin clustering. *J Neurosci* 24: 5816–5826, 2004.
90. Hedblom E and Kirkness EF. A novel class of GABAA receptor subunit in tissues of the reproductive system. *J Biol Chem* 272: 15346–15350, 1997.
91. Hill DR and Bowery NG. 3H-baclofen and 3H-GABA bind to bicuculline-insensitive GABA B sites in rat brain. *Nature* 290: 149–152, 1981.
92. Hosie AM, Wilkins ME, da Silva HM, and Smart TG. Endogenous neurosteroids regulate GABAA receptors through two discrete transmembrane sites. *Nature* 444: 486–489, 2006.

93. Hosie AM, Wilkins ME, and Smart TG. Neurosteroid binding sites on GABA(A) receptors. *Pharmacol Ther* 116: 7–19, 2007.
94. Jacob TC, Bogdanov YD, Magnus C, Saliba RS, Kittler JT, Haydon PG, and Moss SJ. Gephyrin regulates the cell surface dynamics of synaptic GABAA receptors. *J Neurosci* 25: 10469–10478, 2005.
95. Kanematsu T, Jang IS, Yamaguchi T, Nagahama H, Yoshimura K, Hidaka K, Matsuda M, Takeuchi H, Misumi Y, Nakayama K, Yamamoto T, Akaike N, Hirata M, and Nakayama K. Role of the PLC-related, catalytically inactive protein p130 in GABA(A) receptor function. *Embo J* 21: 1004–1011, 2002.
96. Kato K. Novel GABAA receptor alpha subunit is expressed only in cerebellar granule cells. *J Mol Biol* 214: 619–624, 1990.
97. Keller CA, Yuan X, Panzanelli P, Martin ML, Alldred M, Sassoe-Pognetto M, and Luscher B. The gamma2 subunit of GABA(A) receptors is a substrate for palmitoylation by GODZ. *J Neurosci* 24: 5881–5891, 2004.
98. Khrestchatisky M, MacLennan AJ, Chiang MY, Xu WT, Jackson MB, Brecha N, Sternini C, Olsen RW, and Tobin AJ. A novel alpha subunit in rat brain GABAA receptors. *Neuron* 3: 745–753, 1989.
99. Kim EY, Schrader N, Smolinsky B, Bedet C, Vannier C, Schwarz G, and Schindelin H. Deciphering the structural framework of glycine receptor anchoring by gephyrin. *Embo J* 25: 1385–1395, 2006.
100. Kins S, Betz H, and Kirsch J. Collybistin, a newly identified brain-specific GEF, induces submembrane clustering of gephyrin. *Nat Neurosci* 3: 22–29, 2000.
101. Kirkness EF and Fraser CM. A strong promoter element is located between alternative exons of a gene encoding the human gamma-aminobutyric acid-type A receptor beta 3 subunit (GABRB3). *J Biol Chem* 268: 4420–4428, 1993.
102. Kirsch J. Glycinergic transmission. *Cell Tissue Res* 326: 535–540, 2006.
103. Kirsch J and Betz H. Widespread expression of gephyrin, a putative glycine receptor-tubulin linker protein, in rat brain. *Brain Res* 621: 301–310, 1993.
104. Kirsch J, Wolters I, Triller A, and Betz H. Gephyrin antisense oligonucleotides prevent glycine receptor clustering in spinal neurons. *Nature* 366: 745–748, 1993.
105. Kittler JT, Arancibia-Carcamo IL, and Moss SJ. Association of GRIP1 with a GABA(A) receptor associated protein suggests a role for GRIP1 at inhibitory synapses. *Biochem Pharmacol* 68: 1649–1654, 2004.
106. Kittler JT, Chen G, Honing S, Bogdanov Y, McAinsh K, Arancibia-Carcamo IL, Jovanovic JN, Pangalos MN, Haucke V, Yan Z, and Moss SJ. Phospho-dependent binding of the clathrin AP2 adaptor complex to GABAA receptors regulates the efficacy of inhibitory synaptic transmission. *Proc Natl Acad Sci USA* 102: 14871–14876, 2005.
107. Kittler JT, Delmas P, Jovanovic JN, Brown DA, Smart TG, and Moss SJ. Constitutive endocytosis of GABAA receptors by an association with the adaptin AP2 complex modulates inhibitory synaptic currents in hippocampal neurons. *J Neurosci* 20: 7972–7977, 2000.
108. Kittler JT and Moss SJ. Modulation of GABAA receptor activity by phosphorylation and receptor trafficking: implications for the efficacy of synaptic inhibition. *Curr Opin Neurobiol* 13: 341–347, 2003.
109. Kittler JT, Rostaing P, Schiavo G, Fritschy JM, Olsen R, Triller A, and Moss SJ. The subcellular distribution of GABARAP and its ability to interact with NSF suggest a role for this protein in the intracellular transport of GABA(A) receptors. *Mol Cell Neurosci* 18: 13–25, 2001.
110. Kittler JT, Thomas P, Tretter V, Bogdanov YD, Haucke V, Smart TG, and Moss SJ. Huntingtin-associated protein 1 regulates inhibitory synaptic transmission by modulating

gamma-aminobutyric acid type A receptor membrane trafficking. *Proc Natl Acad Sci USA* 101: 12736–12741, 2004.
111. Klausberger T, Fuchs K, Mayer B, Ehya N, and Sieghart W. GABA(A) receptor assembly. Identification and structure of gamma(2) sequences forming the intersubunit contacts with alpha(1) and beta(3) subunits. *J Biol Chem* 275: 8921–8928, 2000.
112. Klausberger T, Sarto I, Ehya N, Fuchs K, Furtmuller R, Mayer B, Huck S, and Sieghart W. Alternate use of distinct intersubunit contacts controls GABAA receptor assembly and stoichiometry. *J Neurosci* 21: 9124–9133, 2001.
113. Kleijnen MF, Shih AH, Zhou P, Kumar S, Soccio RE, Kedersha NL, Gill G, and Howley PM. The hPLIC proteins may provide a link between the ubiquitination machinery and the proteasome. *Mol Cell* 6: 409–419, 2000.
114. Kleizen B and Braakman I. Protein folding and quality control in the endoplasmic reticulum. *Curr Opin Cell Biol* 16: 343–349, 2004.
115. Kneussel M. Dynamic stabilization: Structural plasticity at inhibitory postsynaptic sites. *Traffic* 7: 1604–1606, 2006.
116. Kneussel M, Brandstatter JH, Gasnier B, Feng G, Sanes JR, and Betz H. Gephyrin-independent clustering of postsynaptic GABA(A) receptor subtypes. *Mol Cell Neurosci* 17: 973–982, 2001.
117. Kneussel M, Brandstatter JH, Laube B, Stahl S, Muller U, and Betz H. Loss of postsynaptic GABA(A) receptor clustering in gephyrin-deficient mice. *J Neurosci* 19: 9289–9297, 1999.
118. Kneussel M, Haverkamp S, Fuhrmann JC, Wang H, Wassle H, Olsen RW, and Betz H. The gamma-aminobutyric acid type A receptor (GABAAR)-associated protein GABARAP interacts with gephyrin but is not involved in receptor anchoring at the synapse. *Proc Natl Acad Sci USA* 97: 8594–8599, 2000.
119. Knuesel I, Bornhauser BC, Zuellig RA, Heller F, Schaub MC, and Fritschy JM. Differential expression of utrophin and dystrophin in CNS neurons: an in situ hybridization and immunohistochemical study. *J Comp Neurol* 422: 594–611, 2000.
120. Knuesel I, Mastrocola M, Zuellig RA, Bornhauser B, Schaub MC, and Fritschy JM. Short communication: altered synaptic clustering of GABAA receptors in mice lacking dystrophin (mdx mice). *Eur J Neurosci* 11: 4457–4462, 1999.
121. Korpi ER, Kuner T, Kristo P, Kohler M, Herb A, Luddens H, and Seeburg PH. Small N-terminal deletion by splicing in cerebellar alpha 6 subunit abolishes GABAA receptor function. *J Neurochem* 63: 1167–1170, 1994.
122. Kuhse J, Betz H, and Kirsch J. The inhibitory glycine receptor: architecture, synaptic localization and molecular pathology of a postsynaptic ion-channel complex. *Curr Opin Neurobiol* 5: 318–323, 1995.
123. Kuhse J, Laube B, Magalei D, and Betz H. Assembly of the inhibitory glycine receptor: identification of amino acid sequence motifs governing subunit stoichiometry. *Neuron* 11: 1049–1056, 1993.
124. Lambert JJ, Belelli D, Peden DR, Vardy AW, and Peters JA. Neurosteroid modulation of GABAA receptors. *Prog Neurobiol* 71: 67–80, 2003.
125. Lardi-Studler B and Fritschy JM. Matching of pre- and postsynaptic specializations during synaptogenesis. *Neuroscientist* 13: 115–126, 2007.
126. Laurie DJ, Seeburg PH, and Wisden W. The distribution of 13 GABAA receptor subunit mRNAs in the rat brain. II. Olfactory bulb and cerebellum. *J Neurosci* 12: 1063–1076, 1992.
127. Laurie DJ, Wisden W, and Seeburg PH. The distribution of thirteen GABAA receptor subunit mRNAs in the rat brain. III. Embryonic and postnatal development. *J Neurosci* 12: 4151–4172, 1992.
128. Legendre P. The glycinergic inhibitory synapse. *Cell Mol Life Sci* 58: 760–793, 2001.

129. Leil TA, Chen ZW, Chang CS, and Olsen RW. GABAA receptor-associated protein traffics GABAA receptors to the plasma membrane in neurons. *J Neurosci* 24: 11429–11438, 2004.
130. Levi S, Chesnoy-Marchais D, Sieghart W, and Triller A. Synaptic control of glycine and GABA(A) receptors and gephyrin expression in cultured motoneurons. *J Neurosci* 19: 7434–7449, 1999.
131. Levi S, Grady RM, Henry MD, Campbell KP, Sanes JR, and Craig AM. Dystroglycan is selectively associated with inhibitory GABAergic synapses but is dispensable for their differentiation. *J Neurosci* 22: 4274–4285, 2002.
132. Levi S, Logan SM, Tovar KR, and Craig AM. Gephyrin is critical for glycine receptor clustering but not for the formation of functional GABAergic synapses in hippocampal neurons. *J Neurosci* 24: 207–217, 2004.
133. Levitan ES, Blair LA, Dionne VE, and Barnard EA. Biophysical and pharmacological properties of cloned GABAA receptor subunits expressed in Xenopus oocytes. *Neuron* 1: 773–781, 1988.
134. Levitan ES, Schofield PR, Burt DR, Rhee LM, Wisden W, Kohler M, Fujita N, Rodriguez HF, Stephenson A, Darlison MG, and et al. Structural and functional basis for GABAA receptor heterogeneity. *Nature* 335: 76–79, 1988.
135. Li RW, Serwanski DR, Miralles CP, Li X, Charych E, Riquelme R, Huganir RL, and de Blas AL. GRIP1 in GABAergic synapses. *J Comp Neurol* 488: 11–27, 2005.
136. Loebrich S, Bahring R, Katsuno T, Tsukita S, and Kneussel M. Activated radixin is essential for GABAA receptor alpha5 subunit anchoring at the actin cytoskeleton. *Embo J* 25: 987–999, 2006.
137. Luddens H, Pritchett DB, Kohler M, Killisch I, Keinanen K, Monyer H, Sprengel R, and Seeburg PH. Cerebellar GABAA receptor selective for a behavioural alcohol antagonist. *Nature* 346: 648–651, 1990.
138. Luscher B and Fritschy JM. Subcellular localization and regulation of GABAA receptors and associated proteins. *Int Rev Neurobiol* 48: 31–64, 2001.
139. Lynch JW. Molecular structure and function of the glycine receptor chloride channel. *Physiol Rev* 84: 1051–1095, 2004.
140. Maas C, Tagnaouti N, Loebrich S, Behrend B, Lappe-Siefke C, and Kneussel M. Neuronal cotransport of glycine receptor and the scaffold protein gephyrin. *J Cell Biol* 172: 441–451, 2006.
141. Macdonald RL and Olsen RW. GABAA receptor channels. *Annu Rev Neurosci* 17: 569–602, 1994.
142. Malherbe P, Sigel E, Baur R, Persohn E, Richards JG, and Mohler H. Functional expression and sites of gene transcription of a novel alpha subunit of the GABAA receptor in rat brain. *FEBS Lett* 260: 261–265, 1990.
143. Mammoto A, Sasaki T, Asakura T, Hotta I, Imamura H, Takahashi K, Matsuura Y, Shirao T, and Takai Y. Interactions of drebrin and gephyrin with profilin. *Biochem Biophys Res Commun* 243: 86–89, 1998.
144. Mathers DA and Barker JL. (-)Pentobarbital opens ion channels of long duration in cultured mouse spinal neurons. *Science* 209: 507–509, 1980.
145. Matzenbach B, Maulet Y, Sefton L, Courtier B, Avner P, Guenet JL, and Betz H. Structural analysis of mouse glycine receptor alpha subunit genes. Identification and chromosomal localization of a novel variant. *J Biol Chem* 269: 2607–2612, 1994.
146. McKinley DD, Lennon DJ, and Carter DB. Cloning, sequence analysis and expression of two forms of mRNA coding for the human beta 2 subunit of the GABAA receptor. *Brain Res Mol Brain Res* 28: 175–179, 1995.
147. Meier J, De Chaldee M, Triller A, and Vannier C. Functional heterogeneity of gephyrins. *Mol Cell Neurosci* 16: 566–577, 2000.

148. Meier J and Grantyn R. A gephyrin-related mechanism restraining glycine receptor anchoring at GABAergic synapses. *J Neurosci* 24: 1398–1405, 2004.
149. Meier J and Grantyn R. Preferential accumulation of GABAA receptor gamma 2L, not gamma 2S, cytoplasmic loops at rat spinal cord inhibitory synapses. *J Physiol* 559: 355–365, 2004.
150. Meier J, Vannier C, Serge A, Triller A, and Choquet D. Fast and reversible trapping of surface glycine receptors by gephyrin. *Nat Neurosci* 4: 253–260, 2001.
151. Meyer G, Kirsch J, Betz H, and Langosch D. Identification of a gephyrin binding motif on the glycine receptor beta subunit. *Neuron* 15: 563–572, 1995.
152. Meyer G, Varoqueaux F, Neeb A, Oschlies M, and Brose N. The complexity of PDZ domain-mediated interactions at glutamatergic synapses: a case study on neuroligin. *Neuropharmacology* 47: 724–733, 2004.
153. Mihic SJ, Ye Q, Wick MJ, Koltchine VV, Krasowski MD, Finn SE, Mascia MP, Valenzuela CF, Hanson KK, Greenblatt EP, Harris RA, and Harrison NL. Sites of alcohol and volatile anaesthetic action on GABA(A) and glycine receptors. *Nature* 389: 385–389, 1997.
154. Minier F and Sigel E. Techniques: Use of concatenated subunits for the study of ligand-gated ion channels. *Trends Pharmacol Sci* 25: 499–503, 2004.
155. Miralles CP, Li M, Mehta AK, Khan ZU, and De Blas AL. Immunocytochemical localization of the beta(3) subunit of the gamma-aminobutyric acid(A) receptor in the rat brain. *J Comp Neurol* 413: 535–548, 1999.
156. Mizokami A, Kanematsu T, Ishibashi H, Yamaguchi T, Tanida I, Takenaka K, Nakayama KI, Fukami K, Takenawa T, Kominami E, Moss SJ, Yamamoto T, Nabekura J, and Hirata M. Phospholipase C-related inactive protein is involved in trafficking of gamma2 subunit-containing GABA(A) receptors to the cell surface. *J Neurosci* 27: 1692–1701, 2007.
157. Mody I. Extrasynaptic GABA(A) receptors in the crosshairs of hormones and ethanol. *Neurochem Int*, 52(1–2):60–4, 1: 2008.
158. Mody I, Glykys J, and Wei W. A new meaning for "Gin & Tonic": tonic inhibition as the target for ethanol action in the brain. *Alcohol* 41: 145–153, 2007.
159. Mohler H, Knoflach F, Paysan J, Motejlek K, Benke D, Luscher B, and Fritschy JM. Heterogeneity of GABAA-receptors: cell-specific expression, pharmacology, and regulation. *Neurochem Res* 20: 631–636, 1995.
160. Moragues N, Ciofi P, Tramu G, and Garret M. Localisation of GABA(A) receptor epsilon-subunit in cholinergic and aminergic neurones and evidence for co-distribution with the theta-subunit in rat brain. *Neuroscience* 111: 657–669, 2002.
161. Moreno JI, Piva MA, Miralles CP, and De Blas AL. Immunocytochemical localization of the beta 2 subunit of the gamma-aminobutyric acidA receptor in the rat brain. *J Comp Neurol* 350: 260–271, 1994.
162. Moss SJ and Smart TG. Constructing inhibitory synapses. *Nat Rev Neurosci* 2: 240–250, 2001.
163. Neuhoff H, Sassoe-Pognetto M, Panzanelli P, Maas C, Witke W, and Kneussel M. The actin-binding protein profilin I is localized at synaptic sites in an activity-regulated manner. *Eur J Neurosci* 21: 15–25, 2005.
164. Newland CF and Cull-Candy SG. On the mechanism of action of picrotoxin on GABA receptor channels in dissociated sympathetic neurones of the rat. *J Physiol* 447: 191–213, 1992.
165. Nusser Z, Hajos N, Somogyi P, and Mody I. Increased number of synaptic GABA(A) receptors underlies potentiation at hippocampal inhibitory synapses. *Nature* 395: 172–177, 1998.

166. Nusser Z, Roberts JD, Baude A, Richards JG, Sieghart W, and Somogyi P. Immunocytochemical localization of the alpha 1 and beta 2/3 subunits of the GABAA receptor in relation to specific GABAergic synapses in the dentate gyrus. *Eur J Neurosci* 7: 630–646, 1995.
167. Nusser Z, Roberts JD, Baude A, Richards JG, and Somogyi P. Relative densities of synaptic and extrasynaptic GABAA receptors on cerebellar granule cells as determined by a quantitative immunogold method. *J Neurosci* 15: 2948–2960, 1995.
168. Nusser Z, Sieghart W, Benke D, Fritschy JM, and Somogyi P. Differential synaptic localization of two major gamma-aminobutyric acid type A receptor alpha subunits on hippocampal pyramidal cells. *Proc Natl Acad Sci U S A* 93: 11939–11944, 1996.
169. Nymann-Andersen J, Wang H, Chen L, Kittler JT, Moss SJ, and Olsen RW. Subunit specificity and interaction domain between GABA(A) receptor-associated protein (GABARAP) and GABA(A) receptors. *J Neurochem* 80: 815–823, 2002.
170. O'Sullivan GA, Kneussel M, Elazar Z, and Betz H. GABARAP is not essential for GABA receptor targeting to the synapse. *Eur J Neurosci* 22: 2644–2648, 2005.
171. Olsen RW, Hanchar HJ, Meera P, and Wallner M. GABAA receptor subtypes: the "one glass of wine" receptors. *Alcohol* 41: 201–209, 2007.
172. Paarmann I, Saiyed T, Schmitt B, and Betz H. Gephyrin: does splicing affect its function? *Biochem Soc Trans* 34: 45–47, 2006.
173. Paarmann I, Schmitt B, Meyer B, Karas M, and Betz H. Mass spectrometric analysis of glycine receptor-associated gephyrin splice variants. *J Biol Chem* 281: 34918–34925, 2006.
174. Persohn E, Malherbe P, and Richards JG. In situ hybridization histochemistry reveals a diversity of GABAA receptor subunit mRNAs in neurons of the rat spinal cord and dorsal root ganglia. *Neuroscience* 42: 497–507, 1991.
175. Pfeiffer F, Graham D, and Betz H. Purification by affinity chromatography of the glycine receptor of rat spinal cord. *J Biol Chem* 257: 9389–9393, 1982.
176. Pirker S, Schwarzer C, Wieselthaler A, Sieghart W, and Sperk G. GABA(A) receptors: immunocytochemical distribution of 13 subunits in the adult rat brain. *Neuroscience* 101: 815–850, 2000.
177. Prenosil GA, Schneider Gasser EM, Rudolph U, Keist R, Fritschy JM, and Vogt KE. Specific subtypes of GABAA receptors mediate phasic and tonic forms of inhibition in hippocampal pyramidal neurons. *J Neurophysiol* 96: 846–857, 2006.
178. Prior P, Schmitt B, Grenningloh G, Pribilla I, Multhaup G, Beyreuther K, Maulet Y, Werner P, Langosch D, Kirsch J, and et al. Primary structure and alternative splice variants of gephyrin, a putative glycine receptor-tubulin linker protein. *Neuron* 8: 1161–1170, 1992.
179. Pritchett DB, Sontheimer H, Shivers BD, Ymer S, Kettenmann H, Schofield PR, and Seeburg PH. Importance of a novel GABAA receptor subunit for benzodiazepine pharmacology. *Nature* 338: 582–585, 1989.
180. Qian H and Dowling JE. Novel GABA responses from rod-driven retinal horizontal cells. *Nature* 361: 162–164, 1993.
181. Rabow LE, Russek SJ, and Farb DH. From ion currents to genomic analysis: recent advances in GABAA receptor research. *Synapse* 21: 189–274, 1995.
182. Racca C, Gardiol A, and Triller A. Cell-specific dendritic localization of glycine receptor alpha subunit messenger RNAs. *Neuroscience* 84: 997–1012, 1998.
183. Racca C, Gardiol A, and Triller A. Dendritic and postsynaptic localizations of glycine receptor alpha subunit mRNAs. *J Neurosci* 17: 1691–1700, 1997.
184. Rao A, Cha EM, and Craig AM. Mismatched appositions of presynaptic and postsynaptic components in isolated hippocampal neurons. *J Neurosci* 20: 8344–8353, 2000.

185. Rathenberg J, Kittler JT, and Moss SJ. Palmitoylation regulates the clustering and cell surface stability of GABAA receptors. *Mol Cell Neurosci* 26: 251–257, 2004.
186. Rees MI, Harvey K, Ward H, White JH, Evans L, Duguid IC, Hsu CC, Coleman SL, Miller J, Baer K, Waldvogel HJ, Gibbon F, Smart TG, Owen MJ, Harvey RJ, and Snell RG. Isoform heterogeneity of the human gephyrin gene (GPHN), binding domains to the glycine receptor, and mutation analysis in hyperekplexia. *J Biol Chem* 278: 24688–24696, 2003.
187. Sabatini DM, Barrow RK, Blackshaw S, Burnett PE, Lai MM, Field ME, Bahr BA, Kirsch J, Betz H, and Snyder SH. Interaction of RAFT1 with gephyrin required for rapamycin-sensitive signaling. *Science* 284: 1161–1164, 1999.
188. Saiyed T, Paarmann I, Schmitt B, Haeger S, Sola M, Schmalzing G, Weissenhorn W, and Betz H. Molecular basis of gephyrin clustering at inhibitory synapses: role of G- and E-domain interactions. *J Biol Chem* 282: 5625–5632, 2007.
189. Santhakumar V, Wallner M, and Otis TS. Ethanol acts directly on extrasynaptic subtypes of GABAA receptors to increase tonic inhibition. *Alcohol* 41: 211–221, 2007.
190. Schaerer MT, Kannenberg K, Hunziker P, Baumann SW, and Sigel E. Interaction between GABA(A) receptor beta subunits and the multifunctional protein gC1q-R. *J Biol Chem* 276: 26597–26604, 2001.
191. Scheiffele P, Fan J, Choih J, Fetter R, and Serafini T. Neuroligin expressed in nonneuronal cells triggers presynaptic development in contacting axons. *Cell* 101: 657–669, 2000.
192. Schofield PR, Darlison MG, Fujita N, Burt DR, Stephenson FA, Rodriguez H, Rhee LM, Ramachandran J, Reale V, Glencorse TA, and et al. Sequence and functional expression of the GABA A receptor shows a ligand-gated receptor super-family. *Nature* 328: 221–227, 1987.
193. Schrader N, Kim EY, Winking J, Paulukat J, Schindelin H, and Schwarz G. Biochemical characterization of the high affinity binding between the glycine receptor and gephyrin. *J Biol Chem* 279: 18733–18741, 2004.
194. Schwarzer C, Berresheim U, Pirker S, Wieselthaler A, Fuchs K, Sieghart W, and Sperk G. Distribution of the major gamma-aminobutyric acid(A) receptor subunits in the basal ganglia and associated limbic brain areas of the adult rat. *J Comp Neurol* 433: 526–549, 2001.
195. Serwanski DR, Miralles CP, Christie SB, Mehta AK, Li X, and De Blas AL. Synaptic and nonsynaptic localization of GABAA receptors containing the alpha5 subunit in the rat brain. *J Comp Neurol* 499: 458–470, 2006.
196. Sheng G, Chang GQ, Lin JY, Yu ZX, Fang ZH, Rong J, Lipton SA, Li SH, Tong G, Leibowitz SF, and Li XJ. Hypothalamic huntingtin-associated protein 1 as a mediator of feeding behavior. *Nat Med* 12: 526–533, 2006.
197. Shivers BD, Killisch I, Sprengel R, Sontheimer H, Kohler M, Schofield PR, and Seeburg PH. Two novel GABAA receptor subunits exist in distinct neuronal subpopulations. *Neuron* 3: 327–337, 1989.
198. Sieghart W and Sperk G. Subunit composition, distribution and function of GABA(A) receptor subtypes. *Curr Top Med Chem* 2: 795–816, 2002.
199. Sigel E and Barnard EA. A gamma-aminobutyric acid/benzodiazepine receptor complex from bovine cerebral cortex. Improved purification with preservation of regulatory sites and their interactions. *J Biol Chem* 259: 7219–7223, 1984.
200. Sigel E, Stephenson FA, Mamalaki C, and Barnard EA. A gamma-aminobutyric acid/benzodiazepine receptor complex of bovine cerebral cortex. *J Biol Chem* 258: 6965–6971, 1983.
201. Smart TG, Hosie AM, and Miller PS. Zn2+ ions: modulators of excitatory and inhibitory synaptic activity. *Neuroscientist* 10: 432–442, 2004.

202. Smit AB, Syed NI, Schaap D, van Minnen J, Klumperman J, Kits KS, Lodder H, van der Schors RC, van Elk R, Sorgedrager B, Brejc K, Sixma TK, and Geraerts WP. A glia-derived acetylcholine-binding protein that modulates synaptic transmission. *Nature* 411: 261–268, 2001.
203. Sola M, Bavro VN, Timmins J, Franz T, Ricard-Blum S, Schoehn G, Ruigrok RW, Paarmann I, Saiyed T, O'Sullivan GA, Schmitt B, Betz H, and Weissenhorn W. Structural basis of dynamic glycine receptor clustering by gephyrin. *Embo J* 23: 2510–2519, 2004.
204. Sola M, Kneussel M, Heck IS, Betz H, and Weissenhorn W. X-ray crystal structure of the trimeric N-terminal domain of gephyrin. *J Biol Chem* 276: 25294–25301, 2001.
205. Stegmuller J, Werner H, Nave KA, and Trotter J. The proteoglycan NG2 is complexed with alpha-amino-3-hydroxy-5-methyl-4-isoxazolepropionic acid (AMPA) receptors by the PDZ glutamate receptor interaction protein (GRIP) in glial progenitor cells. Implications for glial-neuronal signaling. *J Biol Chem* 278: 3590–3598, 2003.
206. Stell BM, Brickley SG, Tang CY, Farrant M, and Mody I. Neuroactive steroids reduce neuronal excitability by selectively enhancing tonic inhibition mediated by delta subunit-containing GABAA receptors. *Proc Natl Acad Sci U S A* 100: 14439–14444, 2003.
207. Studer R, von Boehmer L, Haenggi T, Schweizer C, Benke D, Rudolph U, and Fritschy JM. Alteration of GABAergic synapses and gephyrin clusters in the thalamic reticular nucleus of GABAA receptor alpha3 subunit-null mice. *Eur J Neurosci* 24: 1307–1315, 2006.
208. Studler B, Fritschy JM, and Brunig I. GABAergic and glutamatergic terminals differentially influence the organization of GABAergic synapses in rat cerebellar granule cells in vitro. *Neuroscience* 114: 123–133, 2002.
209. Study RE and Barker JL. Diazepam and (–)-pentobarbital: fluctuation analysis reveals different mechanisms for potentiation of gamma-aminobutyric acid responses in cultured central neurons. *Proc Natl Acad Sci U S A* 78: 7180–7184, 1981.
210. Sugita S, Saito F, Tang J, Satz J, Campbell K, and Sudhof TC. A stoichiometric complex of neurexins and dystroglycan in brain. *J Cell Biol* 154: 435–445, 2001.
211. Sumita K, Sato Y, Iida J, Kawata A, Hamano M, Hirabayashi S, Ohno K, Peles E, and Hata Y. Synaptic scaffolding molecule (S-SCAM) membrane-associated guanylate kinase with inverted organization (MAGI)-2 is associated with cell adhesion molecules at inhibitory synapses in rat hippocampal neurons. *J Neurochem* 100: 154–166, 2007.
212. Sur C, Wafford KA, Reynolds DS, Hadingham KL, Bromidge F, Macaulay A, Collinson N, O'Meara G, Howell O, Newman R, Myers J, Atack JR, Dawson GR, McKernan RM, Whiting PJ, and Rosahl TW. Loss of the major GABA(A) receptor subtype in the brain is not lethal in mice. *J Neurosci* 21: 3409–3418, 2001.
213. Takamiya K, Kostourou V, Adams S, Jadeja S, Chalepakis G, Scambler PJ, Huganir RL, and Adams RH. A direct functional link between the multi-PDZ domain protein GRIP1 and the Fraser syndrome protein Fras1. *Nat Genet* 36: 172–177, 2004.
214. Taniguchi H, Gollan L, Scholl FG, Mahadomrongkul V, Dobler E, Limthong N, Peck M, Aoki C, and Scheiffele P. Silencing of neuroligin function by postsynaptic neurexins. *J Neurosci* 27: 2815–2824, 2007.
215. Taylor PM, Connolly CN, Kittler JT, Gorrie GH, Hosie A, Smart TG, and Moss SJ. Identification of residues within GABA(A) receptor alpha subunits that mediate specific assembly with receptor beta subunits. *J Neurosci* 20: 1297–1306, 2000.
216. Taylor PM, Thomas P, Gorrie GH, Connolly CN, Smart TG, and Moss SJ. Identification of amino acid residues within GABA(A) receptor beta subunits that mediate both homomeric and heteromeric receptor expression. *J Neurosci* 19: 6360–6371, 1999.
217. Thomas P, Mortensen M, Hosie AM, and Smart TG. Dynamic mobility of functional GABAA receptors at inhibitory synapses. *Nat Neurosci* 8: 889, 2005.

218. Triller A and Choquet D. Surface trafficking of receptors between synaptic and extra-synaptic membranes: and yet they do move! *Trends Neurosci* 28: 133–139, 2005.
219. Triller A and Choquet D. Synaptic structure and diffusion dynamics of synaptic receptors. *Biol Cell* 95: 465–476, 2003.
220. Triller A, Cluzeaud F, and Korn H. gamma-Aminobutyric acid-containing terminals can be apposed to glycine receptors at central synapses. *J Cell Biol* 104: 947–956, 1987.
221. Triller A, Cluzeaud F, Pfeiffer F, Betz H, and Korn H. Distribution of glycine receptors at central synapses: an immunoelectron microscopy study. *J Cell Biol* 101: 683–688, 1985.
222. Unwin N. Nicotinic acetylcholine receptor at 9 A resolution. *J Mol Biol* 229: 1101–1124, 1993.
223. Ushkaryov YA, Petrenko AG, Geppert M, and Sudhof TC. Neurexins: synaptic cell surface proteins related to the alpha-latrotoxin receptor and laminin. *Science* 257: 50–56, 1992.
224. Varoqueaux F, Aramuni G, Rawson RL, Mohrmann R, Missler M, Gottmann K, Zhang W, Sudhof TC, and Brose N. Neuroligins determine synapse maturation and function. *Neuron* 51: 741–754, 2006.
225. Varoqueaux F, Jamain S, and Brose N. Neuroligin 2 is exclusively localized to inhibitory synapses. *Eur J Cell Biol* 83: 449–456, 2004.
226. Wang H, Bedford FK, Brandon NJ, Moss SJ, and Olsen RW. GABA(A)-receptor-associated protein links GABA(A) receptors and the cytoskeleton. *Nature* 397: 69–72, 1999.
227. Wei W, Zhang N, Peng Z, Houser CR, and Mody I. Perisynaptic localization of delta subunit-containing GABA(A) receptors and their activation by GABA spillover in the mouse dentate gyrus. *J Neurosci* 23: 10650–10661, 2003.
228. Werman R, Davidoff RA, and Aprison MH. Inhibition of motoneurones by iontophoresis of glycine. *Nature* 214: 681–683, 1967.
229. Whiting P, McKernan RM, and Iversen LL. Another mechanism for creating diversity in gamma-aminobutyrate type A receptors: RNA splicing directs expression of two forms of gamma 2 phosphorylation site. *Proc Natl Acad Sci USA* 87: 9966–9970, 1990.
230. Whiting PJ, McAllister G, Vassilatis D, Bonnert TP, Heavens RP, Smith DW, Hewson L, O'Donnell R, Rigby MR, Sirinathsinghji DJ, Marshall G, Thompson SA, Wafford KA, and Vasilatis D. Neuronally restricted RNA splicing regulates the expression of a novel GABAA receptor subunit conferring atypical functional properties [corrected; erratum to be published]. *J Neurosci* 17: 5027–5037, 1997.
231. Wilson-Shaw D, Robinson M, Gambarana C, Siegel RE, and Sikela JM. A novel gamma subunit of the GABAA receptor identified using the polymerase chain reaction. *FEBS Lett* 284: 211–215, 1991.
232. Wisden W, Laurie DJ, Monyer H, and Seeburg PH. The distribution of 13 GABAA receptor subunit mRNAs in the rat brain. I. Telencephalon, diencephalon, mesencephalon. *J Neurosci* 12: 1040–1062, 1992.
233. Wyszynski M, Kim E, Dunah AW, Passafaro M, Valtschanoff JG, Serra-Pages C, Streuli M, Weinberg RJ, and Sheng M. Interaction between GRIP and liprin-alpha/SYD2 is required for AMPA receptor targeting. *Neuron* 34: 39–52, 2002.
234. Wyszynski M, Valtschanoff JG, Naisbitt S, Dunah AW, Kim E, Standaert DG, Weinberg R, and Sheng M. Association of AMPA receptors with a subset of glutamate receptor-interacting protein in vivo. *J Neurosci* 19: 6528–6537, 1999.
235. Xiang S, Kim EY, Connelly JJ, Nassar N, Kirsch J, Winking J, Schwarz G, and Schindelin H. The crystal structure of Cdc42 in complex with collybistin II, a gephyrin-interacting guanine nucleotide exchange factor. *J Mol Biol* 359: 35–46, 2006.

236. Xiang S, Nichols J, Rajagopalan KV, and Schindelin H. The crystal structure of Escherichia coli MoeA and its relationship to the multifunctional protein gephyrin. *Structure* 9: 299–310, 2001.
237. Ye B, Liao D, Zhang X, Zhang P, Dong H, and Huganir RL. GRASP-1: a neuronal RasGEF associated with the AMPA receptor/GRIP complex. *Neuron* 26: 603–617, 2000.
238. Ymer S, Draguhn A, Kohler M, Schofield PR, and Seeburg PH. Sequence and expression of a novel GABAA receptor alpha subunit. *FEBS Lett* 258: 119–122, 1989.
239. Ymer S, Draguhn A, Wisden W, Werner P, Keinanen K, Schofield PR, Sprengel R, Pritchett DB, and Seeburg PH. Structural and functional characterization of the gamma 1 subunit of GABAA/benzodiazepine receptors. *Embo J* 9: 3261–3267, 1990.
240. Yu ZY, Wang W, Fritschy JM, Witte OW, and Redecker C. Changes in neocortical and hippocampal GABAA receptor subunit distribution during brain maturation and aging. *Brain Res* 1099: 73–81, 2006.
241. Zhang HG, ffrench-Constant RH, and Jackson MB. A unique amino acid of the Drosophila GABA receptor with influence on drug sensitivity by two mechanisms. *J Physiol* 479 (Pt 1): 65–75, 1994.

Acid-Sensing Ion Channels (ASICs) and pH in Synapse Physiology

John A. Wemmie[1], Xiang-ming Zha[2], and Michael J. Welsh[3]
University of Iowa, Roy J. and Lucille A. Carver College of Medicine

[1] Department of Psychiatry, Neuroscience Program, and Department of Veterans Affairs Medical Center, Iowa City, IA,USA, john-wemmie@uiowa.edu
[2] Department of Internal Medicine and Howard Hughes Medical Institute, Iowa City, IA 52242, USA, xiangming-zha@uiowa.edu
[3] Departments of Internal Medicine and Molecular Physiology, and Howard Hughes Medical Institute, Iowa City, IA 52242, USA, michael-welsh@uiowa.edu

Abstract. Although brain pH is tightly controlled, it can be more dynamic than commonly appreciated. Physiological fluctuations in extracellular pH provide ample opportunity for protons to influence synaptic signaling. A number of synaptic proteins are modified by extracellular pH. Of these, the acid sensing ion channels (ASICs) are gated by extracellular protons and thus they may be particularly well suited to respond to synaptic pH. Here we review extracellular pH changes that accompany neural activity, the synaptic localization of ASICs, and their known effects on synapse function. Results from manipulating ASICs in mice suggest important roles for synaptic ASICs in behavior and neurological disease.

1 Introduction

In general, the brain's extracellular pH is controlled within a narrow range (1). However, neural activity causes pH alterations that can vary in location and with time. In addition, disease can generate sustained deviations of physiologic pH. Here, we consider some of the changes in brain pH, and some possible molecular targets, focusing particularly on the acid sensing ion channels.

2 Extracellular pH at the Synapse

2.1 Proton Release During Neurotransmission Lowers Synaptic pH

Vacuolar-H^+ATPases pump protons into synaptic vesicles lowering vesicle pH to ~5.2–5.7 (2–5). The resulting proton gradient energizes neurotransmitter uptake; it may also serve an important signaling role during neurotransmission (Fig. 1a). In addition to free protons, at low pH other molecules in the vesicle are protonated, creating a source of releasable protons. Thus, when synaptic vesicles fuse with the presynaptic membrane, free protons and protonated acids are released into the synaptic cleft (5). At the release site, pH probably falls very quickly, but the

J.W. Hell, M.D. Ehlers (eds.), *Structural and Functional Organization of the Synapse*,
DOI: 10.1007/978-0-387-77232-5_22, © Springer Science+Business Media, LLC 2008

reduction is likely to be short lived (milliseconds or less) due to rapid H^+ buffering and diffusion (Fig. 1b). The speed and spatial characteristics of this phenomenon make it technically challenging to monitor. However, by electrophysiological methods and by pH sensitive fluorophores, extracellular acidification at the presynaptic side has been detected (5–8). Based on these measurements, pH in the retinal ribbon synapse is estimated to drop 0.2–0.6 pH units during neurotransmission (6, 8). Faster and more sensitive detection strategies may improve our ability to measure and understand this phenomenon.

As a consequence of synaptic proton release, pH-sensitive voltage-gated Ca^{2+}-channels (VGCC) are inhibited in the presynaptic membrane of retinal ribbon synapses (6–8). Other proteins near the synaptic cleft could also be affected. For example, extracellular acidosis is known to inhibit NMDA receptors (9–12) and AMPA receptors (13). $GABA_A$-receptors are also pH sensitive; some subunits are activated (14) and some inhibited by extracellular protons (15). As will be discussed later (Sections 3 and 4), acid-sensing ion channels (ASICs) in the post-synaptic membrane are well positioned to respond to rapid acid transients.

2.2 Alkalosis Follows Neurotransmission

Slower changes in extracellular pH follow bursts of synaptic activity. A transient alkalosis begins ~20 ms after a 100 Hz stimulus train and can last hundreds of milliseconds (Fig. 1b) (16, 17). This increase in extracellular pH has been detected in the hippocampus and elsewhere in the brain (16). pH-sensitive microelectrodes in extracellular fluid detected a rise of ~0.05 pH units (18, 19). However, judging by the effects on synaptic transmission (18, 19), the magnitude of the alkalosis at the synapse is probably significantly larger. As described above, additional high-resolution measurements would help the field. The mechanisms underlying the net loss of extracellular protons are not clear. Nor is it clear whether the alkalosis is related to the more rapid acidification that occurs during neurotransmission. Suspected mechanisms for the alkalosis include glutamate receptor activated H^+ transport into cells (17), $GABA_A$-receptor mediated extrusion of HCO_3^- (17, 20, 21), and Ca^{2+}- H^+ exchange by cell surface ATPase (22–24). Carbonic anhydrase also plays an important role, since carbonic anhydrase inhibitors magnified the alkaline transient (25), and adding exogenous carbonic anhydrase to brain slices attenuated the alkalinization (19).

The alkaline transients can have important physiological consequences. As a general rule, interstitial alkalosis tends to increase excitability in most central neurons (26). Because the NMDA receptor is partially blocked by protons at physiological pH (~7.35), alkalosis would be expected to relieve proton inhibition. Consistent with this prediction, the alkaline transient boosted NMDA receptor activation and increased Ca^{2+} entry (18). As a consequence, depolarization was prolonged following EPSC trains and excitability was increased (27). The effects of the alkaline transient on other pH sensitive channels, and on behavior and disease are not yet clear.

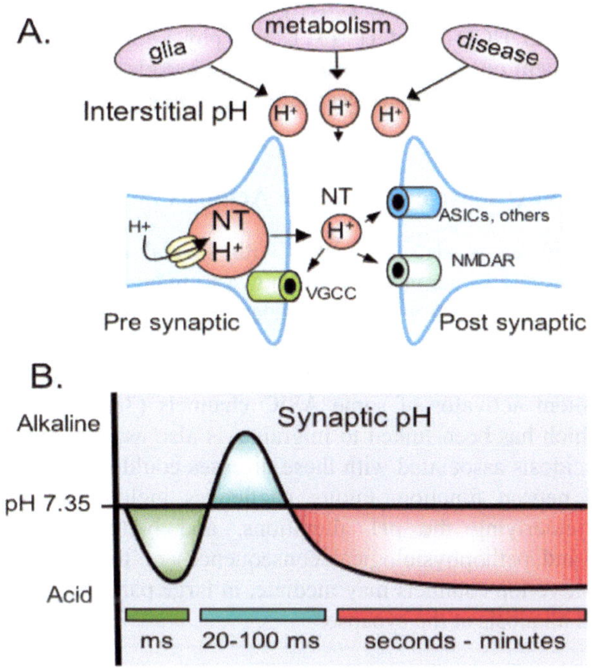

Fig. 1. Model of pH fluctuations at the synapse. (**a**) Synaptic vesicles release protons and neurotransmitter (NT) into the synaptic cleft during neurotransmission, which may lower pH in the synaptic cleft and modulate pH sensitive channels and other proteins in the pre- and post-synaptic membrane. Illustrated are voltage-gated Ca^{2+}-channels (VGCC), NMDA receptors, and ASICs, although other pH sensitive proteins may also be present. Glia, metabolism, and disease processes lower interstitial pH and may also affect synaptic pH and physiology. (**b**) At lease three transient pH fluctuations may occur at the synapse in response to neural activity. Rapid acidification of the cleft (*green*) may occur with vesicle release and probably lasts milliseconds (ms) or less. A slower alkalosis (*blue*) has been detected in the interstitial space within tens of milliseconds. An even slower acidosis (*red*) can follow intense neural activity and last for seconds. Improved techniques are needed to better characterize the degree and duration of these pH fluctuations at the synapse.

2.3 Intense Neural Activity Causes Acidosis

On an even slower timescale, intense neural activity produces a wave of acidification (Fig. 1b, red waveform)(17). The acidosis occurs within seconds to minutes of stimulation and can last for minutes or longer. The degree of acidification and its duration appear to depend on the magnitude of neural and metabolic activity. With extreme activity, such as during a seizure, interstitial pH can fall well below 7.0; raising the chance of both synaptic and extra-synaptic consequences. Metabolic mechanisms may contribute to the acidification. Mitochondria take up Ca^{2+} and extrude H^{+}, which in turn can be transported out of the cell (22–24). Lactate is also produced and extruded from cells (17). Glia move acid across the cell membrane (28, 29), and glial extrusion of lactate may provide an important energy source for active neurons (30). Increased metabolic activity also produces more CO_2, which can be rapidly hydrated to HCO_3^- and H^+ (31). The physiological consequences of this slower acid transient on synaptic transmission are not clear, although the effects are likely to be mainly inhibitory because prolonged acidosis reduces the excitability of most central neurons (26).

2.4 Neurological Disease can Produce Acidosis

Extracellular acidosis is also associated with several neurological diseases. For example, seizures and ischemia are well known to reduce extracellular pH (32–35). Inflammatory diseases, such as multiple sclerosis, and neurodegenerative diseases, such as Huntington's disease, impair energy metabolism (36, 37). A resulting accumulation of lactate may contribute to acidosis. Interestingly, lactic acid is a particularly potent activator of some ASIC channels (38, 39). Cortical spreading depression, which has been linked to migraine, is also associated with acidosis (32, 40–42). The acidosis associated with these diseases could readily influence synaptic signaling and neuron function. Future challenges include better elucidating the mechanisms underlying the pH alterations, and better understanding of the physiological and pathophysiological consequences of low pH. At the molecular level, pH-sensitive ion channels may mediate, in large part, the physiological effects of acidosis and alkalosis at the synapse.

3 Acid Sensing Ion Channels (ASICs)

3.1 ASICs are Well Suited to Respond to Synaptic pH

A number of synaptic channels are sensitive to pH modulation; including glutamate receptors, GABA-receptors, and voltage-gated Ca^{2+} channels (9–15, 43). These ion channels likely contribute to the physiological effects of fluctuating extracellular pH (9, 44). Unlike these other channels, in which pH modulates the response to a ligand or voltage, extracellular protons directly gate acid sensing ion channels (ASICs). ASICs are members of the DEG/ENaC family located at synapses where they may play an important role in synaptic signaling. They are preferentially permeable to Na^+, but to a lesser extent can conduct other cations (Ca^{2+}, K^+, Li^+, H^+). Thus, ASICs may be well suited to mediate effects of acute pH changes at the synapse.

3.2 ASICs are Comprised of Multiple Subunits

Five ASIC subunits have been identified in mammalian neurons: ASIC1a, -1b, -2a, -2b, and -3, (a and b refer to alternatively spliced isoforms). The ASIC subunits combine to form both homomultimeric and heteromultimeric channel complexes (45–47). ASIC2b by itself does not form an acid-activated channel, but modifies currents from other acid-activated subunits (45). The related proteins, ASIC4 and BLINaC, share sequence homology with ASICs, but these mammalian proteins have not been shown to produce or modify acid-activated currents (48, 49), although an ASIC4 homolog in zebrafish is acid-activated (50). The recently determined crystal structure of ASIC1a indicates a unique trimeric assembly (Fig. 2b), which suggests that three subunits are required to form a channel (55). Interestingly, previous studies using electrophysiological, biochemical, and fluorescence resonance energy transfer approaches had suggested that 4–9 subunits may combine to form a channel (51–54). Although the explanation for this difference is unknown, understanding the reason for the discrepancy may give the field insight into the structure and function of these channels.

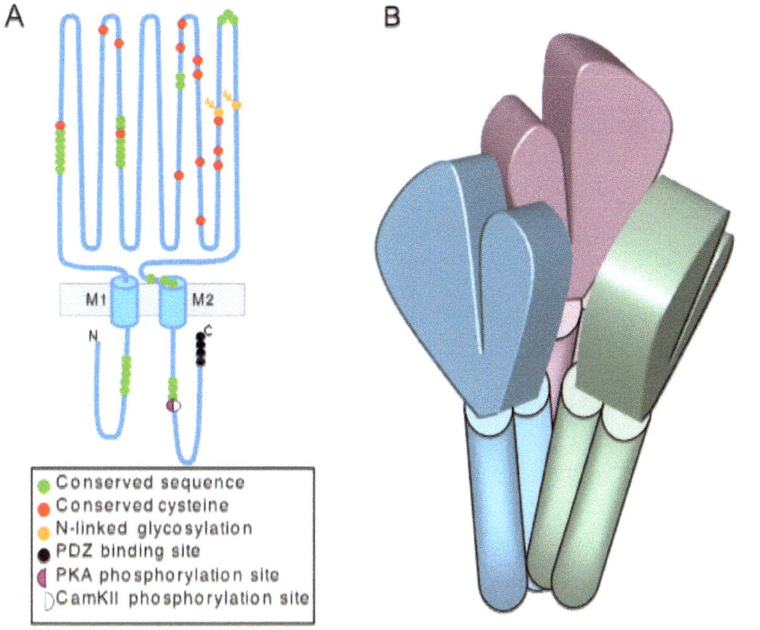

Fig. 2. Topology and trimeric assembly of ASICs. (**a**) ASIC subunits are comprised of two membrane-spanning domains, a large cysteine-rich extracellular domain, and several small conserved amino acid motifs (for recent reviews see (62–64)). Adapted with permission, from (62). (**b**) Trimeric assembly of ASIC/DEG/ENaC family proteins based on the recently determined crystal structure of chicken ASIC1a minus portions of the amino- and carboxy-termini (55). Individual ASIC1a subunits separated by color.

3.3 Subunit Composition Dictates Channel Properties

Homomultimeric ASIC channels vary in activation and desensitization kinetics, and in pH sensitivity (Fig. 3a). For example, one study reported the half-maximal activation ($pH_{0.5}$) of mouse ASIC subunits as 6.8 (ASIC1a), 6.2 (ASIC1b), 4.9 (ASIC2a), and 6.6 (ASIC3) (46). Slightly different values have also been reported (47), which may to some degree reflect differences between species, for example rat ASIC1a may be less sensitive ($pH_{0.5}$ = 6.2–5.8) (47, 56) than mouse ASIC1a (46). Properties can also vary considerably when subunits are co-expressed to form heteromultimers. For example,

ASIC1a homomultimers are permeable to Ca^{2+}, whereas when ASIC1a heteromultimerizes with ASIC2a the channel loses Ca^{2+} permeability (57). Also ASIC1a, -2a, and -3 heteromultimers desensitize faster than any of the individual subunits alone (Fig. 3) (46, 47). This result indicates that the properties of heteromultimeric channels are not simply the sum of properties from the individual subunits. Instead, subunits combine to confer new properties on channels.

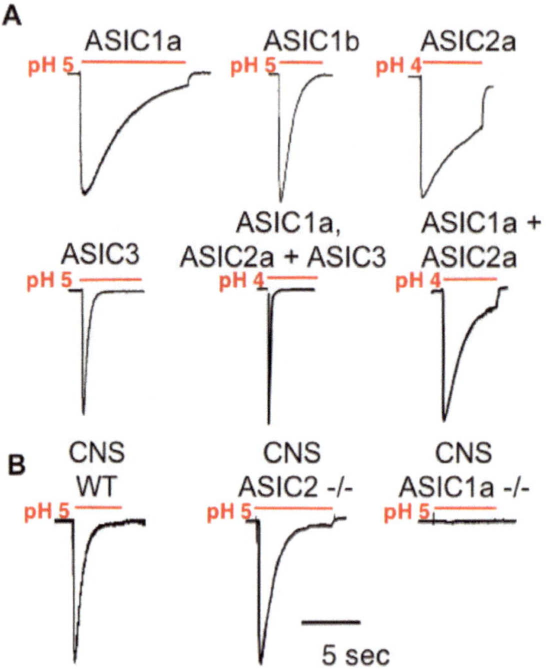

Fig. 3. Representative acid-evoked currents. (**a**) Currents evoked when indicated subunits were transfected into heterologous (COS-7) cells. Note the effects of subunit combination on rate of desensitization. (**b**) Acid-evoked current in hippocampal neurons cultured from mice of indicated genotypes. Disrupting ASIC1a eliminated pH 5-evoked current. Adapted from (62) with permission.

3.4 ASIC1a is Critical for Acid-Evoked Currents in the Brain

In the brain, ASIC1a, -2a, and -2b are the most prominently expressed subunits (58–61). Of these, ASIC1a plays a critical role. Disrupting the ASIC1a gene eliminates current evoked by lowering extracellular pH to 5.0 in cultured hippocampal, cortical, and amygdala neurons (Fig. 3b). Conversely, disrupting ASIC2 slightly increased pH 5-evoked current (Fig. 3b) (61) (65–67). In addition, an antagonist specific for ASIC1a homomultimers (PcTx1, from tarantula venom) inhibited a significant proportion of the acid-activated current in cultured hippocampal and cortical neurons (60, 67, 68). Thus, much of the acid-activated current in the brain is mediated by ASIC1a homomultimers and ASIC1a-containing heteromultimers. Consequently, relative to the other ASICs, we currently know the most about ASIC1a in the brain.

3.5 Fluctuating Brain pH Could Activate or Desensitize ASIC1a

ASIC1a homomultimeric channels are activated as pH falls below 7.2, with a $pH_{0.5}$ of about 6.8–6.2 (46), (56). This pH sensitivity indicates that ASIC1a is well within the range of acidosis recorded during both physiological and pathophysiological states. Because ASIC currents are largely transient and densensitize quickly, some have considered them best suited to respond to rapid, transient pH fluctuations such as those thought to occur at the synapse. However, ASICs can also manifest a persistent current (45, 46, 69), particularly when activated in the presence of modulatory neuropeptides such as NPFF or FMRFamide (70). Moreover, it is possible that ASIC activation could produce sustained effects in second messenger pathways, by raising intracellular Ca^{2+}, for example (57, 67, 71).

Another way that pH regulates ASICs is by desensitization; sustained exposure to extracellular protons inactivates the channels. Desensitization begins at pH < 8.0 in cultured sensory neurons, with half-maximal desensitization occurring at ~pH 7.2 (Fig. 4) (72). Thus, similar to the NMDA receptor, when extracellular pH is held at

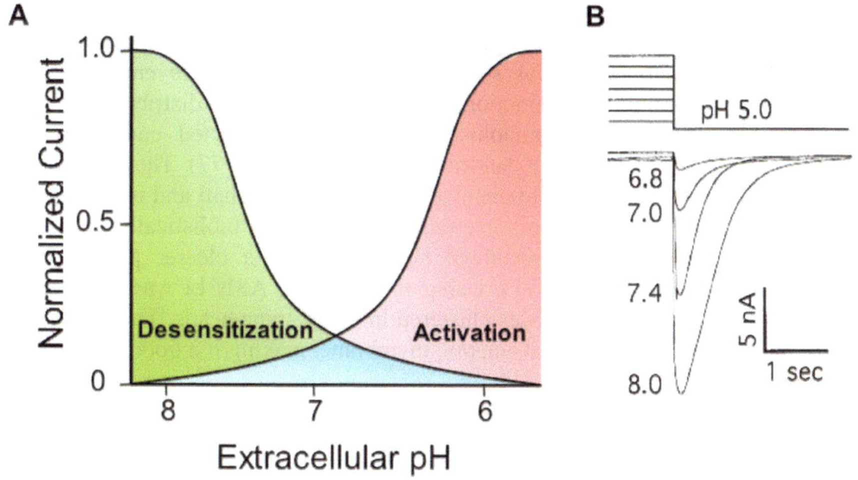

Fig. 4. Extracellular pH regulates ASIC activation and steady-state inactivation (desensitization). (**a**) Illustration of desensitization and activation sensitivity to extracellular pH in sensory neurons. Desensitization curve represents the effects of variations in sustained pH on the subsequent pH 5-evoked current (protocol illustrated in b). Activation curve illustrates current evoked by stepping from pH 8 to lower pH levels. Depending on the rate of recovery from inactivation, the overlapping area beneath the two curves (*blue*) could provide a window of steady state current; evidence from sensory neurons supports this possibility (68). (**b**) Desensitization protocol. Due to channel desensitization, stepping from pH 7 to pH 5 evokes smaller current than stepping from pH 7.4 to pH 5. Similarly, stepping from pH 8 to pH 5 evokes larger ASIC current. Adapted from (72) with permission.

7.35, a significant proportion of ASIC channels are tonically desensitized. As sustained pH becomes more alkaline, desensitization is reduced and transient acidosis evokes greater ASIC current. Similarly, as sustained pH falls, desensitization increases and transient acidosis evokes smaller ASIC current. These dynamic effects of pH raise the possibility that both acid and alkaline pH could regulate ASIC activity in vivo. Moreover, the effects of pH on channel activation and desensitization could generate seemingly paradoxical effects. For example, a generalized alkaline pH might increase the effect of a transient pH drop during neurotransmission. The degree of channel activity may also depend on the speed of onset, magnitude, and duration of the pH change, as well as the presence of modulators in the extracellular milieu (such as FMRFamide (70), Zn^{2+} (73, 74), lactate (38), arachidonic acid (75) and others (62)). However, because of this complexity, it is difficult to precisely predict the amount of ASIC channel activity in vivo.

4 ASIC1a Modulates Synapse Physiology

4.1 ASIC1a is Localized to Central Synapses

Data from multiple experimental approaches place ASIC1a in dendritic spines. By subcellular fractionation, several studies found that ASIC1a is enriched in the synaptosome-containing brain fraction, suggesting a synaptic distribution (65, 76, 77). In cultured neurons, immunohistochemistry also detected endogenous and overexpressed ASIC1a protein in dendritic spines (65, 71, 76, 77). These studies also revealed ASIC1a to be present at puncta along the dendritic shaft and in the cell body (Fig. 5). The improved signal-to-noise ratio obtained from biolistically transfecting ASIC1a into organotypic hippocampal slices revealed a clearer picture of the protein's distribution (Fig. 5) (71). Using this method, ASIC1a was visualized in almost all dendritic spines, and was clustered in many spine heads. This localization pattern places ASIC1a at the postsynaptic membrane, making it a good candidate for sensing protons there.

4.2 ASIC1a is Not Abundant in Axons

Unlike the somatodendritic distribution of ASIC1a, there is controversy about its presence in axons. One study reported ASIC1a immunostaining in axon-like branches of cultured CNS neurons (77). In contrast, others failed to detect endogenous ASIC1a in axons (65). In addition, those studies failed to detect overexpressed ASIC1a in tau-positive branches of cultured neurons. Moreover, no detectable ASIC1a immunofluorescence was seen in axons from biolistically transfected slice neurons co-labeled with an axonal marker, VAMP2/synaptobrevin (71). Together, these data do not rule out the presence of ASIC1a in axons, but suggest it is probably not abundant there. The subcellular distribution of other ASIC subunits in brain neurons remains unclear, although ASIC2 has been suggested to localize to synapses in the cerebellum (78) and the retina (79).

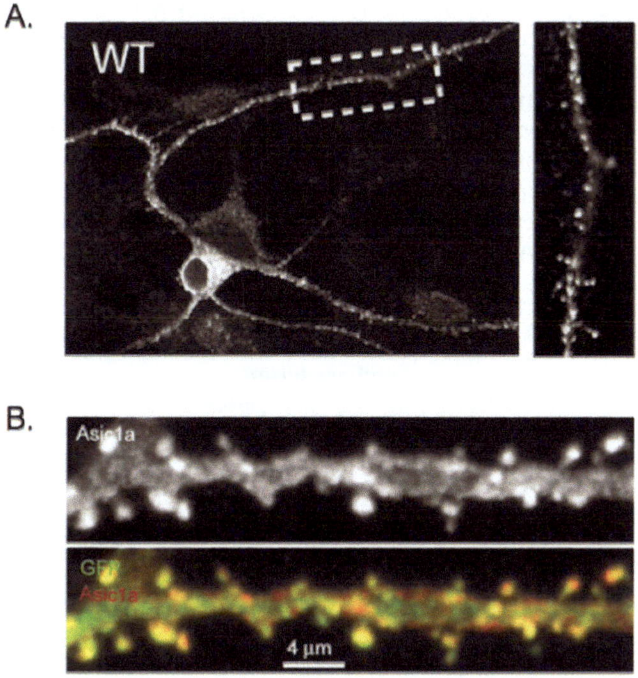

Fig. 5. ASIC1a is localized to dendritic spines. (**a**) Immunofluorescence (IF) of endogenous ASIC1a in dissociated cortical neurons. Shown is a cortical neuron from a wild-type culture. ASIC1a IF is visible in the cell body and dendrites. Little or no fluorescence was observed in cultures from ASIC1a knockout mice (71). Right is the high-magnification view of the *boxed area*. (**b**) IF of ASIC1a in transfected hippocampal slice neurons. Shown is an enlarged view of an apical dendrite of a CA1 pyramidal neuron. *Top panel* shows ASIC1a IF. *Bottom panel* is the merged image of ASIC1a (*red*) and eGFP (*green*). In both endogenous staining and in hippocampal slices, ASIC1a IF presents in a clustered pattern in spines and dendrites. Reproduced with permission from (71).

4.3 ASIC1a Associates with Post-Synaptic Scaffolding Proteins

Consistent with the speculation that ASIC1a functions in the postsynaptic membrane, ASIC1a associates with at least two synaptic scaffolding proteins, PICK1 and AKAP150 (Fig. 6) (80–82). Both PICK1 and AKAP150 interact with ASIC1a biochemically and colocalize with ASIC1a in neurons. In addition to these interactions, ASIC1a co-localizes with PSD-95 in dendritic spines (65, 76). This observation is intriguing, although there is currently no biochemical evidence for a direct interaction between ASIC1a and PSD-95. However, other ASIC family members interact with PSD-95 (83), raising the possibility of an indirect association between ASIC1a and PSD-95. Functionally, ASIC1a associates with several other synaptic signaling molecules including calmodulin-dependent protein kinase II (CaMKII), protein kinase A (PKA) and NMDA receptors (71, 82, 84, 85). In mice, disrupting ASIC1a reduced CaMKII phosphorylation, and overexpressing ASIC1a increased CaMKII phosphorylation (71). Similarly, CaMKII-dependent phosphorylation affected ASIC1a function; in response to NMDA receptor activation, CaMKII phosphorylated the intracellular C-terminus of human ASIC1a at serines 478 and 479, which potentiated proton-elicited current and acid-induced neuron damage (84). Hippocampal slice recordings also suggest a functional interaction between ASIC1a and the NMDA receptor (Section 4.5). PKA also phosphorylates mouse ASIC1a at serine 478 and disrupts its interaction and co-localization with PICK1 in heterologous cells (85). This event is likely to be functionally important, because PKA binding with AKAP150 reduced ASIC1a current in heterologous cells and in cultured neurons (82). Together, these studies implicate ASIC1a in synaptic signaling networks.

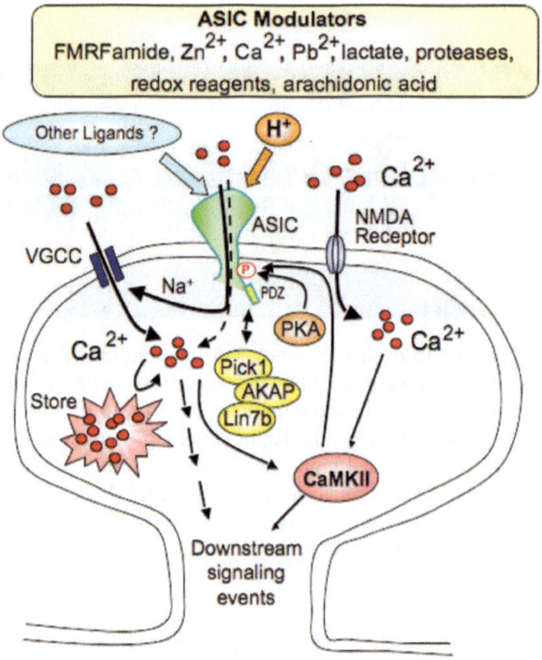

Fig. 6. Model of ASICs and interacting molecules at the synapse. ASICs are activated by protons, possibly from pre-synaptic release or other sources. Additional molecules can modify acid-evoked ASIC currents, suggesting the possibility of other ligands not yet identified (62). Upon activation, ASIC1a increases intracellular calcium $[Ca^{2+}]_i$, largely through voltage-gated calcium channels (VGCCs) and the release of intracellular Ca^{2+} stores (71). The resulting rise in $[Ca^{2+}]_i$ initiates downstream signaling events, including CaMKII activation. ASIC1a channel activity may be regulated by interactions with CaMKII, PKA, AKAP150, calcineurin, PICK1 and possibly other PDZ-domain proteins (81–85).

4.4 ASIC1a Confers Acid Sensitivity to Dendritic Spines

The presence of ASIC1a at dendritic spines raised the question about its physiological effects there. To explore this question, one study tested whether

ASIC1a influenced the intracellular Ca^{2+} concentration ($[Ca^{2+}]_i$) at dendritic spines by transfecting a ratiometric Ca^{2+} reporter, cameleon YC3.60, into hippocampal slices (71). Acute acid application induced $[Ca^{2+}]_i$ transients in wild-type spines (Fig. 7). Knocking down ASIC1a with siRNA attenuated the number of acid-sensitive spines. Conversely ASIC1a overexpression increased the number of acid-sensitive spines. The Ca^{2+} signal was also observed in the dendritic shaft and the cell body. These data combined with the ASIC1a localization indicate that ASIC1a confers acid sensitivity to dendritic spines and that ASIC1a can function as a postsynaptic proton receptor.

ASIC1a homomultimeric channels conduct both Na^+ and Ca^{2+}, thus channel activation could increase $[Ca^{2+}]_i$ directly, or indirectly through voltage-gated Ca^{2+} channels (VGCC) and by releasing from intracellular Ca^{2+} stores. In heterologous cells, Ca^{2+} influx through ASIC1a is responsible for acid-induced increase in $[Ca^{2+}]_i$ (57). In contrast, in hippocampal slices, direct Ca^{2+} influx through ASIC1a made only a small contribution; most of the $[Ca^{2+}]_i$ increase came from VGCCs and from intracellular stores (71). These observations illustrate the multiple mechanisms by which ASIC1a may influence $[Ca^{2+}]_i$ at synapses.

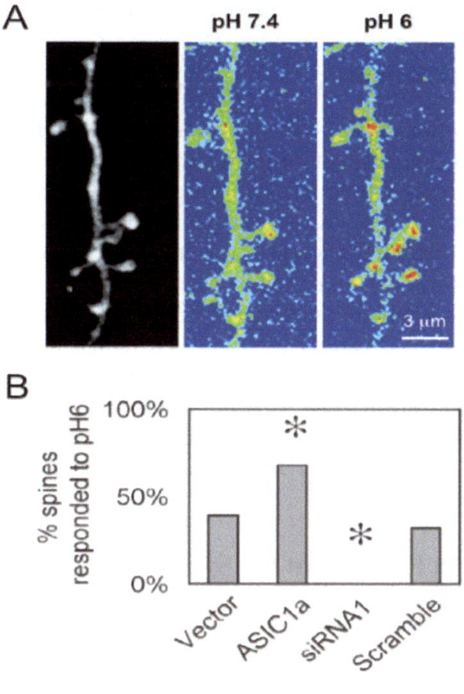

Fig. 7. ASIC1a confers pH sensitivity to dendritic spines. (**a**) Image on *left* shows a segment of an apical dendrite from a neuron cotransfected with ASIC1a and cameleon. Two images on the *right* show the YFP/CFP fluorescence ratio (*blue* indicates a low ratio and *red* indicates a high ratio) obtained at baseline (pH 7.4) and during acid stimulation (pH 6). Note that pH 6 increased Ca^{2+} levels in most spines. (**b**) Acid-induced calcium increase in spines was ASIC1a-dependent. ASIC1a overexpression increased while ASIC1a siRNA reduced the % of spines responding to acid stimulation. Adapted from (71) with permission.

4.5 Disrupting ASIC1a Impairs Synaptic Transmission and Plasticity

The pH changes at the synapse, the subcellular distribution of ASIC1a, and its effects on dendritic [Ca^{2+}], hinted that ASIC1a may play an important role in synaptic physiology and plasticity. Results from ASIC1a knockout mice support this possibility. An analysis of synaptic function in hippocampal slices found that long term potentiation (LTP) evoked by Schaffer collateral fiber stimulation, was impaired by the loss of ASIC1a (65). High frequency stimulation (HFS-100 Hz) produced potentiation in wild-type mice lasting beyond 40 minutes. However, within this time frame, potentiation of excitatory post-synaptic potentials (EPSP) decayed to baseline in ASIC1a knockout slices. Prior to LTP induction, paired-pulse facilitation and single evoked EPSPs were normal, suggesting ASIC1a disruption did not impair presynaptic vesicle release. However, loss of ASIC1a attenuated the summation of EPSPs during the HFS. Interestingly, the NMDA receptor antagonist D-APV inhibited EPSP summation in wild-type slices but not in ASIC1a-null slices suggesting that the loss of ASIC1a impaired NMDA-receptor function. These data support the possible connection between ASIC1a and the NMDA receptor, however the mechanism for the suggested crosstalk remains uncertain.

4.6 ASICs at Retinal Synapses

Glutamate release at retinal ribbon synapses is accompanied by acidosis, raising the possibility that ASICs could affect neurotransmission in the retina. Recent studies investigated this possibility and found ASIC1a protein located in the retina near the synaptic cleft in bipolar cells. ASIC1a-antisense RNA and PcTx1 significantly decreased a- and b-electroretinogram measurements, supporting the idea that ASIC1a contributes to synaptic signaling in the retina (86). While ASIC1a inhibition decreased electroretinogram waves, knocking out ASIC2 increased the a- and b-electroretinogram waves and caused light-induced retinal degeneration (79). Thus, disrupting ASIC2 produced an effect essentially opposite to that of disrupting ASIC1a. Further studies are needed to better understand the roles of ASIC1a and ASIC2 at the retinal synapse. However, the retina may be an excellent location for studying synaptic ASICs because of large size of retinal synapses and the proton release recorded there (6–8).

4.7 Model of ASIC1a Function at the Synapse

The available data support a model of ASIC1a at the synapse (Figs. 1 and 6). In the post-synaptic membrane, ASICs seem well positioned to respond to protons released from presynaptic neurotransmitter (NT)-containing vesicles and also from other sources. In this model, various extracellular modulators and intracellular ASIC-interacting proteins might influence the response, which would be expected to depolarize the membrane potential and raise intracellular Ca^{2+} concentration, perhaps influencing other receptors and signaling proteins. An important prediction from this model is that ASIC currents will be activated during neurotransmission. However, thus far, studies in brain slices (65) and cultured neurons (77) have not detected them. There are many possible explanations for these negative results, but they hint that the story could be more complex. Perhaps the appropriate experimental conditions have not been identified. Interestingly however, a recent study discovered spontaneous ASIC activation in HEK293 cells (87). Presumably, protons released by vesicles evoked the ASIC current, because blocking vesicle acidification with bafilomycin prevented the spontaneous currents. This finding represents a potentially important step in identifying how endogenous ASICs may be activated.

4.8 ASIC1a Affects Synaptic Spine Density

Regardless of how ASIC1a acts at the synapse, evidence for some potentially important consequences has been obtained. For example, it was recently found that knocking down ASIC1a expression with RNAi in hippocampal slice neurons reduced the number of dendritic spines (71). An ASIC1a dominant-negative construct had a similar effect. In contrast, overexpressing ASIC1a in CA1 neurons had the opposite effect and increased the number of dendritic spines. Possibly, the ability of ASIC1a to affect spine number is related to its effects on synaptic $[Ca^{2+}]$ (57, 67, 71, 84) and LTP (65), although the precise mechanisms have not yet been established. Nevertheless, these observations suggest ASICs may influence spine remodeling in disease. Unlike the effects of acutely changing ASIC1a expression, disrupting ASIC1a throughout development in the knockout mice did not change spine number, suggesting developmental mechanisms might compensate for some effects of ASIC1a disruption (71). Together, these functional effects on synaptic physiology suggest that genetically or pharmacologically altering ASIC1a might have important physiological and behavioral consequences.

5 ASIC1a Regulates Behavior and May Contribute to Neurological disease

5.1 Disrupting ASIC1a Impaired Cerebellum-Dependent Learning

The location of ASIC1a in mouse brain provided important clues about the potential contribution of the channel to behavior. By immunohistochemistry and protein blotting, it was found that ASIC1a protein was enriched in gray matter and in brain regions with high synaptic density (65, 66, 88). One region where ASIC1a was abundant was the molecular layer of the cerebellar cortex (Fig. 8a) (65, 66). In addition, Purkinje cells which populate this layer of the cerebellum have large acid-evoked currents (75). Consistent with a role in cerebellum-dependent learning, the loss of ASIC1a significantly impaired classical delay eyeblink conditioning (Fig. 8b) (65). Mice normally learn that a tone predicts a periorbital shock and acquire the ability to defensively blink before the shock occurs (i.e. a conditioned response). After 10 days of training, wild-type mice produced conditioned responses in more than 80% of trials, whereas the ASIC1a-null mice produced a conditioned response in only 40% of trials. Disrupting ASIC1a did not alter shock sensitivity (65), hearing (88), or the blink reflex (65), suggesting a learning-related deficit. Although the mechanism has not been determined, the loss of ASIC1a may have impaired plasticity in the cerebellar cortex.

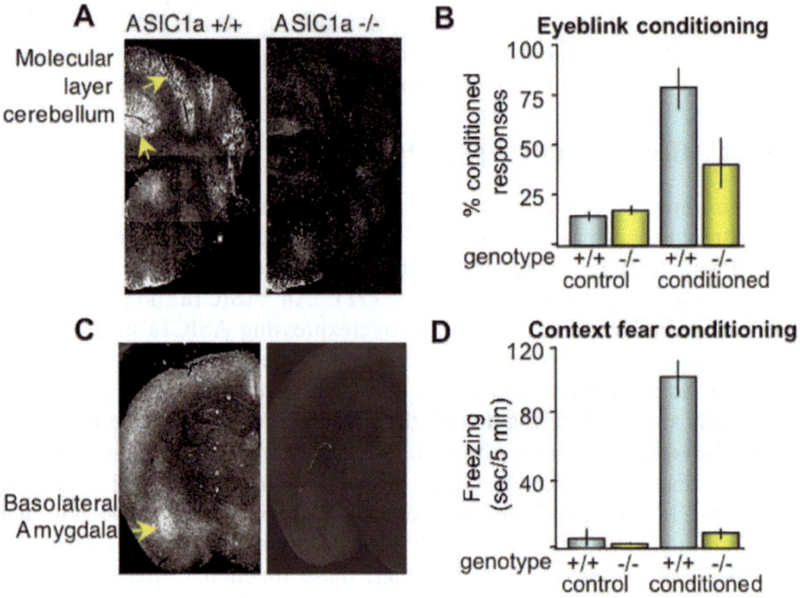

Fig. 8. Localization of ASIC1a in brain structures underlying Pavlovian conditioning. (**a**) Immunofluorescent labeling of ASIC1a protein in ASIC1a+/+ relative to ASIC1a–/– mice revealed abundant ASIC1a expression in the molecular layer of the cerebellar cortex (*arrowheads*). (**b**) Consistent with a role in cerebellum-dependent learning, loss of ASIC1a significantly impaired delay eyeblink conditioning. When the tone and shock were not paired (control), neither group developed conditioned responses. (**c**) In the forebrain, ASIC1a immunolabeling was particularly abundant in the basolateral amygdala of wild-type mice (*arrowheads*). (**d**) Suggesting a role in amygdala-dependent learning, ASIC1a–/– disruption severely impaired single-shock context fear conditioning. Twenty-four hours after receiving a single 0.5 mA footshock, wild-type mice froze when returned to the training chamber. In contrast, ASIC1a-null mice froze very little. Mice not receiving a footshock during training (control) froze little or none during testing. Panels (a), (b), and (c) reproduced with permission from (65, 66, 88).

5.2 ASIC1a Contributes to Pavlovian Fear Conditioning

Outside of the cerebellum, ASIC1a is particularly abundant in structures that underlie fear behaviors including the amygdala complex, bed nucleus of the stria terminalis, lateral hypothalamus, habenula, cingulate cortex, and periacqueductal gray (56, 59, 66, 77, 88, 89). In the amygdala, multiple nuclei express ASIC1a including the lateral, central, and medial nuclei, although ASIC1a immuno-labeling was especially pronounced in the basolateral amygdala (88, 89).

The distribution of ASIC1a in fear circuit structures and the effects of ASIC1a on synaptic function, led to the question of whether ASIC1a influences Pavlovian fear conditioning. Wild-type and ASIC1a-null mice were tested in their ability to learn the association between a fear-producing footshock and either the environmental context or an audible tone (66). Disrupting ASIC1a significantly impaired both context fear conditioning (Fig. 8d) and tone conditioning (66). Conversely, overexpressing ASIC1a in the mouse brain increased fear conditioning (76). The effects were not due to sensory dysfunction because both hearing (88) and footshock sensitivity (66) were unaffected by ASIC1a gene disruption.

5.3 Interrupting ASIC1a Reduces Unconditioned Fear

Effects of ASIC1a on unconditioned fear responses have also been tested (67). The loss of ASIC1a reduced unconditioned fear in the open field test, during acoustic startle, and in response to predator odor. Curiously, the loss of ASIC1a did not reduce undonditioned fear in the elevated plus maze, suggesting that this fear behavior is not ASIC1a-dependent (65). The effects of ASIC1a disruption on unconditioned fear are probably not due to developmental abnormalities, because acutely inhibiting ASIC1a in the brain post-developmentally with PcTx1 venom reduced TMT-evoked fear in wild-type mice, but not in ASIC1a-null mice (67).

The observation that ASIC1a contributes to fear learning, suggests that ASIC1a contributes to plasticity in the fear circuit. In addition, the finding that ASIC1a affects unconditioned fear suggests that ASIC1a can regulate fear circuit activity independent of long-lasting plasticity. These effects of ASIC1a resemble those of AMPA and NMDA receptors, the inhibition of which reduces both fear learning and the expression of learned fear. However, further studies are needed to better undestand how ASIC1a affects neurotransmission in the amygdala and elsewhere in the fear circuit. Nevertheless, the ability of ASIC1a inhibition to reduce multiple fear behaviors, raises the possibility that ASIC1a antagonists could reduce anxiety in humans and might benefit other psychiatric diseases. Anti-anxiety medications with novel mechanisms of action could offer advantages over current medications, which can cause a number of unwanted side effects.

5.4 Ischemic Stroke

Central acidosis exacerbates neurodegenerative diseases including ischemic stroke. Because Ca^{2+} overload causes toxicity in the ischemic brain, and because ASIC1a activation raises intracellular Ca^{2+}, it was hypothesized that ASIC1a activation might increase ischemic cell death. Enhancement of ASIC1a function by activation of NMDA receptors (84), lack of oxygen or glucose (67), or accumulation of lactate or arachidonic acid (75) might further exacerbate ASIC1a-mediated neurotoxicity. A possible role for ASIC1a in ischemic toxicity was initially supported by experiments in cells heterologously expressing ASIC1a and in cultured hippocampal neurons (57). Neurons lacking ASIC1a and cells treated with amiloride or PcTx1 resisted acidosis-induced injury (57, 67). Moreover, PcTx1 diminished the effects of NMDA-induced cell death (67). Establishing a strong case for a role in ischemic disease, disrupting ASIC1a in mice reduced infarct volume by 60% following experimental stroke due to middle cerebral artery (MCA) occlusion (67). Interestingly, pH remains low for hours after a stroke, thus opening a window of therapeutic opportunity. Even administering PcTx1 five hours after MCA occlusion reduced infarct volume by >50% (90). These encouraging results endorse ASIC1a inhibitors as a potential new treatment for stroke.

5.5 Central Pain Regulation

In peripheral neurons, ASICs contribute to acid-evoked nociception (reviewed in (62)). Recently, ASICs in the central nervous system (CNS) were also implicated in pain control (91). Inhibiting ASIC1a in the CNS by injecting PcTx1 or ASIC1a antisense RNA into the cerebrospinal fluid reduced inflammatory and neuropathic pain. The analgesic effects were linked to elevated met-enkephalin levels. The possible relationship between ASIC1a function at the synapse and met-enkephalin expression is not yet clear, although it is interesting to speculate that blocking ASIC1a could attenuate previously observed anti-opiod effects of FMRFamide related peptides (92). Thexse observations suggest a promising opportunity for targeting ASICs in pain control.

6 Concluding Remarks

The identification of ASIC1a as synaptic proton receptors and an increasing appreciation of pH dynamics in the CNS suggest new possibilities for proton mediated cellular signaling. Although much remains to be learned about the ASICs, the available data indicate an intriguing role in synapse function. Perhaps like a neurotransmitter, protons cross the synapse to activate or modulate ASICs and other proteins. In addition, at the cell surface and at synapses, ASICs may be readily accessible to pharmacological modification, suggesting fertile ground for novel drug discovery. Studies in mice point to possible therapeutic benefits of ASIC1a antagonists in psychiatric disease, neurodegeneration, and pain.

Acknowledgements

We thank Theresa Mayhew and Ashley Small for assistance preparing the manuscript, and Kelsey Coulter for help preparing the figures. We also thank Drs. Chris Benson, Candice Askwith, and Mikael Schnizler for examples of electrophysiological traces. JW is supported by a Department of Veteran's Affairs Advanced Career Development Award, a NARSAD young investigator award, and a beginning grant in aid award from the American Heart Association. XZ is an Associate and MJW is an Investigator of the Howard Hughes Medical Institute.

References

1. Kaila, K. and B.R. Ransom, Concept of pH and its importance in neurobiology, in *pH and Brain Function*, K. Kaila and B.R. Ransom, Editors. 1998, Wiley-Liss, Inc. p. 1–10.
2. Morel, N., Neurotransmitter release: the dark side of the vacuolar-H+ATPase. *Biol. Cell.*, 2003. **95**(7): p. 453–457.
3. Fuldner, H.H. and H. Stadler, 31P-NMR analysis of synaptic vesicles. Status of ATP and internal pH. *Eur. J. Biochem.*, 1982. **121**(3): p. 519–524.
4. Michaelson, D.M. and I. Angel, Determination of delta pH in cholinergic synaptic vesicles: its effect on storage and release of acetylcholine. *Life Sci.*, 1980. **27**(1): p. 39–44.
5. Miesenbock, G., D.A. De Angelis, and J.E. Rothman, Visualizing secretion and synaptic transmission with pH-sensitive green fluorescent proteins. *Nature*, 1998. **394**: p. 192–195.
6. DeVries, S.H., Exocytosed protons feedback to suppress the Ca(2+) current in mammalian cone photoreceptors. *Neuron*, 2001. **32**(6): p. 1107–1117.
7. Vessey, J.P., et al., Proton-mediated feedback inhibition of presynaptic calcium channels at the cone photoreceptor synapse. *J. Neurosci.*, 2005. **25**(16): p. 4108–4117.
8. Palmer, M.J., et al., Synaptic cleft acidification and modulation of short-term depression by exocytosed protons in retinal bipolar cells. *J. Neurosci.*, 2003. **23**(36): p. 11332–11341.
9. Traynelis, S.F., pH modulation of ligand-gated ion channels, in *pH and Brain Function*, K. Kaila and B.R. Ransom, Editors. 1998, Wiley-Liss, Inc.
10. Traynelis, S.F. and S.G. Cull-Candy, Proton inhibition of N-methyl-D-aspartate receptors in cerebellar neurons. *Nature*, 1990. **345**(6273): p. 347–350.
11. Vyklicky, L.J., V. Vlachova, and J. Krusek, The effect of external pH changes on responses to excitatory amino acids in mouse hippocampal neurones. *J. Physiol. (Lond)*, 1990. **430**: p. 497–517.
12. Tang, C.M., M. Dichter, and M. Morad, Modulation of the N-methyl-D-aspartate channel by extracellular H+. *Proc. Natl. Acad. Sci. U. S. A.*, 1990. **87**(16): p. 6445–6449.
13. Ihle, E.C. and D.K. Patneau, Modulation of ∂-amino-3-hydroxy-5-methyl-4-isoxazolepropionic acid receptor desensitization by extracellular protons. *Mol. Pharmacol.*, 2000. **58**(6): p. 1204–1212.
14. Pasternack, M., S. Smirnov, and K. Kaila, Proton modulation of functionally distinct GABAA receptors in acutely isolated pyramidal neurons of rat hippocampus. *Neuropharmacology*, 1996. **35**(9–10): p. 1279–1288.
15. Huang, R.Q., Z. Chen, and G.H. Dillon, Molecular basis for modulation of recombinant alpha1beta2gamma2 GABAA receptors by protons. *J. Neurophysiol.*, 2004. **92**(2): p. 883–894.
16. Chesler, M., Regulation and modulation of pH in the brain. *Physiol. Rev.*, 2003. **83**(4): p. 1183–1221.
17. Kaila, K. and M. Chesler, Activity-evoked changes in extracellular pH, in *pH and Brain Function*, K. Kaila and B.R. Ransom, Editors. 1998, Wiley-Liss, Inc. p. 309.
18. Fedirko, N., et al., Regulation of postsynaptic Ca^{2+} influx in hippocampal CA1 pyramidal neurons via extracellular carbonic anhydrase. *J. Neurosci.*, 2007. **27**(5): p. 1167–1175.
19. Makani, S. and M. Chesler, Endogenous alkaline transients boost postsynaptic NMDA receptor responses in hippocampal CA1 pyramidal neurons. *J. Neurosci.*, 2007. **27**(28): p. 7438–7446.
20. Kaila, K. and J. Voipio, Postsynaptic fall in intracellular pH induced by GABA-activated bicarbonate conductance. *Nature*, 1987. **330**(6144): p. 163–165.

21. Kaila, K., J. Saarikoski, and J. Voipio, Mechanism of action of GABA on intracellular pH and on surface pH in crayfish muscle fibres. *J. Physiol.*, 1990. **427**: p. 241–260.
22. Schwiening, C.J., H.J. Kennedy, and R.C. Thomas, Calcium-hydrogen exchange by the plasma membrane Ca-ATPase of voltage-clamped snail neurons. *Proc. R. Soc. Lond. Biol. Sci.*, 1993. **253**(1338): p. 285–289.
23. Schwiening, C.J. and R.C. Thomas, pH consequences of calcium regulation, in *pH and Brain Function*, K. Kaila and B. Ransom, Editors. 1998, Wiley-Liss, Inc. p. 277.
24. Thomas, R.C., Proton channels in snail neurones. Does calcium entry mimic the effects of proton influx? *Ann. N. Y. Acad. Sci.*, 1989. **574**: p. 287–293.
25. Chen, J.C. and M. Chesler, pH transients evoked by excitatory synaptic transmission are increased by inhibition of extracellular carbonic anhydrase. *Proc. Natl. Acad. Sci. U. S. A.*, 1992. **89**(16): p. 7786–7790.
26. Somjen, G.G. and G.C. Tombaugh, pH modulation of neuronal excitability and central nervous system functions, in *pH and Brain Function*, K. Kaila and B.R. Ransom, Editors. 1998, Wiley-Leiss Inc: New York. p. 373–393.
27. Taira, T., et al., Relative contributions of excitatory and inhibitory neuronal activity to alkaline transients evoked by stimulation of Schaffer collaterals in the rat hippocampal slice. *J. Neurophysiol.*, 1995. **74**(2): p. 643–649.
28. Chesler, M. and R.P. Kraig, Intracellular pH of astrocytes increases rapidly with cortical stimulation. *Am. J. Physiol.*, 1987. **253**(4 Pt 2): p. R666–R6670.
29. Chesler, M. and R.P. Kraig, Intracellular pH transients of mammalian astrocytes. *J. Neurosci.*, 1989. **9**(6): p. 2011–2019.
30. Pellerin, L., et al., Activity-dependent regulation of energy metabolism by astrocytes: An update. *Glia*, 2007. **55**(12): p. 1251–1262.
31. Voipio, J. and K. Kaila, Interstitial PCO2 and pH in rat hippocampal slices measured by means of a novel fast CO2/H(+)-sensitive microelectrode based on a PVC-gelled membrane. *Pflugers Arch.*, 1993. **423**(3–4): p. 193–201.
32. Somjen, G.G., Acidification of interstitial fluid in hippocampal formation caused by seizures and by spreading depression. *Brain Res.*, 1984. **311**(1): p. 186–188.
33. Siesjo, B.K., et al., Extra- and intracellular pH in the brain during seizures and in the recovery period following the arrest of seizure activity. *J. Cereb. Blood Flow Metab.*, 1985. **5**(1): p. 47–57.
34. Siesjo, B.K., Acidosis and ischemic brain damage. *Neurochem. Pathol.*, 1988. **9**: p. 31–88.
35. Wang, R.I.H. and S. R.R., PH of cerebral cortex during induced convulsions. *J Neurophysiol.*, 1955. **18**(2): p. 130–137.
36. Waxman, S.G., Axonal conduction and injury in multiple sclerosis: the role of sodium channels. *Nat. Rev. Neurosci.*, 2006. **7**(12): p. 932–941.
37. Browne, S.E. and M.F. Beal, The energetics of Huntington's disease. *Neurochem. Res.*, 2004. **29**(3): p. 531–546.
38. Immke, D.C. and E.W. McCleskey, Lactate enhances the acid-sensing Na+ channel on ischemia-sensing neurons. *Nat. Neurosci.*, 2001. **4**(9): p. 869–870.
39. Immke, D.C. and E.W. McCleskey, Protons open acid-sensing ion channels by catalyzing relief of Ca^{2+} blockade. *Neuron*, 2003. **37**(1): p. 75–84.
40. Parsons, A.A. and P.J. Strijbos, The neuronal versus vascular hypothesis of migraine and cortical spreading depression. *Curr. Opin. Pharmacol.*, 2003. **3**(1): p. 73–77.
41. Sanchez-Del-Rio, M., U. Reuter, and M.A. Moskowitz, New insights into migraine pathophysiology. *Curr. Opin. Neurol.*, 2006. **19**(3): p. 294–298.
42. Csiba, L., W. Paschen, and G. Mies, Regional changes in tissue pH and glucose content during cortical spreading depression in rat brain. *Brain Res.*, 1985. **336**(1): p. 167–170.

43. Tombaugh, G.C. and G.G. Somjen, Effects of extracellular pH on voltage-gated Na+, K+ and Ca2+ currents in isolated rat CA1 neurons. *J. Physiol.*, 1996. **493(Pt 3)**: p. 719–732.
44. Tombaugh, G.C. and G.G. Somjen, pH modulation of voltage-gated ion channels, in *pH and Brain Function*, K. Kaila and B.R. Ransom, Editors. 1998, Wiley-Liss, Inc. p. 395.
45. Linguegila, E., et al., A modulatory subunit of acid sensing ion channels in brain and dorsal root ganglion cells. *J. Biol. Chem.*, 1997. **272**(47): p. 29778–29783.
46. Benson, C.J., et al., Heteromultimerics of DEG/ENaC subunits form H^+-gated channels in mouse sensory neurons. *Proc. Natl. Acad. Sci. U. S. A.*, 2002. **99**(4): p. 2338–2343.
47. Hesselager, M., D.B. Timmermann, and P.K. Ahring, pH-dependency and desensitization kinetics of heterologously expressed combinations of ASIC subunits. *J. Biol. Chem.*, 2004. **279**(12): p. 11006–11015.
48. Akopian, A.N., et al., A new member of the acid-sensing ion channel family. *Neuroreport*, 2000. **11**(10): p. 2217–2222.
49. Sakai, H., et al., Cloning and functional expression of a novel degenerin-like Na^+ channel gene in mammals. *J. Physiol. (Lond)*, 1999. **519**(2): p. 323–333.
50. Paukert, M., et al., A family of acid-sensing ion channels from the zebrafish: widespread expression in the central nervous system suggests a conserved role in neuronal communication. *J. Biol. Chem.*, 2004. **279**(18): p. 18783–18791.
51. Gao, Y., et al., Fluorescence resonance energy transfer analysis of subunit assembly of the ASIC channel. *Biochem. Biophys. Res. Commun.*, 2007. **359**(1): p. 143–150.
52. Snyder, P.M., et al., Electrophysiological and biochemical evicence that DEG/ENaC cation channels are composed of nine subunits. *J. Biol. Chem.*, 1998. **273**(2): p. 681–684.
53. Firsov, D., et al., The heterotetrameric architecture of the epithelial sodium channel (ENaC). *EMBO J.*, 1998. **17**(2): p. 344–352.
54. Eskandari, S., et al., Number of subunits comprising the epithelial sodium channel. *J. Biol. Chem.*, 1999. **274**(38): p. 27281–27286.
55. Jasti, J., et al., Structure of acid-sensing ion channel 1 at 1.9 A resolution and low pH. *Nature*, 2007. **449**(7160): p. 316–323.
56. Waldmann, R., et al., A proton-gated cation channel involved in acid-sensing. *Nature*, 1997. **386**: p. 173–177.
57. Yermolaieva, O., et al., Extracellular acidosis increases neuronal cell calcium by activating acid-sensing ion channel 1a. *Proc. Natl. Acad. Sci. U. S. A.*, 2004. **101**(17): p. 6752–6757.
58. Waldmann, R., Proton-gated cation channels–neuronal acid sensors in the central and peripheral nervous system. *Adv. Exp. Med. Biol.*, 2001. **502**: p. 293–304.
59. García-Añoveros, J., et al., BNaC1 and BNaC2 constitute a new family of human neuronal sodium channels related to degenerins and epithelial sodium channels. *Proc. Natl. Acad. Sci. U. S. A.*, 1997. **94**: p. 1459–1464.
60. Baron, A., R. Waldmann, and M. Lazdunski, ASIC-like, proton-activated currents in rat hippocampal neurons. *J. Physiol.*, 2002. **539**(Pt 2): p. 485–494.
61. Askwith, C.C., et al., ASIC2 modulates ASIC1 H^+-activated currents in hippocampal neurons. *J. Biol. Chem.*, 2003. **279**(18): p. 18296–18305.
62. Wemmie, J.A., M.P. Price, and M.J. Welsh, Acid-sensing ion channels: advances, questions and therapeutic opportunities. *Trends Neurosci.*, 2006. **29**(10): p. 578–586.
63. Krishtal, O., The ASICs: signaling molecules? Modulators? *Trends Neurosci.*, 2003. **26**(9): p. 477–483.
64. Lingueglia, E., E. Deval, and M. Lazdunski, FMRFamide-gated sodium channel and ASIC channels: a new class of ionotropic receptors for FMRFamide and related peptides. *Peptides*, 2006. **27**(5): p. 1138–1152.

65. Wemmie, J.A., et al., The acid-activated ion channel ASIC contributes to synaptic plasticity, learning, and memory. *Neuron*, 2002. **34**: p. 463–477.
66. Wemmie, J.A., et al., Acid-sensing ion channel 1 is localized in brain regions with high synaptic density and contributes to fear conditioning. *J. Neurosci.*, 2003. **23**(13): p. 5496–5502.
67. Xiong, Z.G., et al., Neuroprotection in ischemia: blocking calcium-permeable acid-sensing ion channels. *Cell*, 2004. **118**(6): p. 687–698.
68. Coryell, M.W., Ziemann, A.E., Westmoreland, P.J., Haenfler, J.M., Kurjakovic, Z., Zha, X.M., Price, M., Schinzler, M.K., Wemmie, J.A. Targeting ASIC1a reduces innate fear and alters neuronal acitivity in the fear circuit. Biol. Psychiatry., 2007. **62** (10): P. 1140–1148.
69. Yagi, J., et al., Sustained currents through ASIC3 ion channels at the modest pH changes that occur during myocardial ischemia. *Circ. Res.*, 2006. **99**(5): p. 501–509.
70. Askwith, C.C., et al., Neuropeptide FF and FMRFamide potentiate acid-evoked currents from sensory neurons and proton-gated DEG/ENaC channels. *Neuron*, 2000. **26**: p. 133–141.
71. Zha, X.-M., J.A. Wemmie, and M.J. Welsh, ASIC1a is a postsynaptic proton receptor that influences the density of dendritic spines. *Proc. Natl. Acad. Sci. U. S. A.*, 2006. **103**(44): p. 16556–16561.
72. Benson, C.J., S.P. Eckert, and E.W. McCleskey, Acid-evoked currents in cardiac sensory neurons: a possible mediator of myocardial ischemic sensation. *Circ. Res.*, 1999. **84**(8): p. 921–928.
73. Chu, X.P., et al., Subunit-dependent high-affinity zinc inhibition of acid-sensing ion channels. *J. Neurosci.*, 2004. **24**(40): p. 8678–8689.
74. Baron, A., et al., Zn^{2+} and H^{+}, coactivators of acid sensing ion channels (ASIC). *J. Biol. Chem.*, 2001. **276**: p. 35361–35367.
75. Allen, N.J. and D. Attwell, Modulation of ASIC channels in rat cerebellar Purkinje neurons by ischaemia-related signals. *J. Physiol.*, 2002. **543**(2): p. 521–529.
76. Wemmie, J., Coryell M, Askwith C, Lamani E, Leonard S, Sigmund C, Welsh M, Overexpression of acid-sensing ion channel 1a in transgenic mice increases fear-related behavior. *Proc. Natl. Acad. Sci. U. S. A.*, 2004. **101**(10): p. 3621–3626.
77. Alvarez de la Rosa, D., et al., Distribution, subcellular localization and ontogeny of ASIC1 in the mammalian central nervous system. *J. Physiol.*, 2003. **546**(Pt 1): p. 77–87.
78. Jovov, B., et al., Immunolocalization of the acid-sensing ion channel 2a in the rat cerebellum. *Histochem. Cell Biol.*, 2003. **119**(6): p. 437.
79. Ettaiche, M., et al., Acid-sensing ion channel 2 is important for retinal function and protects against light-induced retinal degeneration. *J. Neurosci.*, 2004. **24**(5): p. 1005–1012.
80. Hruska-Hageman, A.M., et al., Interaction of the synaptic protein PICK1 (protein interacting with C kinase 1) with the non-voltage gated sodium channels BNC1 (brain Na^{+} channel 1) and ASIC (acid-sensing ion channel). *Biochem. J.*, 2002. **361**(Pt 3): p. 443–450.
81. Duggan, A., J. Garcia-Anoveros, and D.P. Corey, The PDZ domain protein PICK1 and the sodium channel BNaC1 interact and localize at mechanosensory terminals of dorsal root ganglion neurons and dendrites of central neurons. *J. Biol. Chem.*, 2002. **277**(7): p. 5203–5208.
82. Chai, S., et al., A Kinase-anchoring Protein 150 and Calcineurin Are Involved in Regulation of Acid-sensing Ion Channels ASIC1a and ASIC2a. *J. Biol. Chem.*, 2007. **282**(31): p. 22668–22677.
83. Hruska-Hageman, A.M., et al., PSD-95 and Lin-7b interact with acid-sensing ion channel-3 and have opposite effects on H+- gated current. *J. Biol. Chem.*, 2004. **279**(45): p. 46962–46968.

84. Gao, J., et al., Coupling between NMDA receptor and acid-sensing ion channel contributes to ischemic neuronal death. *Neuron*, 2005. **48**(4): p. 635–646.
85. Leonard, A.S., et al., cAMP-dependent protein kinase phosphorylation of the acid-sensing ion channel-1 regulates its binding to the protein interacting with C-kinase-1. *Proc. Natl. Acad. Sci. U. S. A.*, 2003. **100**(4): p. 2029–2034.
86. Ettaiche, M., et al., Silencing acid-sensing ion channel 1a alters cone-mediated retinal function. *J. Neurosci.*, 2006. **26**(21): p. 5800–5809.
87. Lalo, U., et al., Spontaneous autocrine release of protons activates ASIC-mediated currents in HEK293 cells. *J. Cell. Physiol.*, 2007. **212**(2): p. 473–480.
88. Coryell, M.W., et al., Targeting ASIC1a Reduces Innate Fear and Alters Neuronal Activity in the Fear Circuit. *Biol. Psychiatry*, 2007: p. Epub ahead of Print.
89. Olson, T.H., et al., An acid sensing ion channel (ASIC) localizes to small primary afferent neurons in rats. *Neuron*, 1998. **9**: p. 1109–1113.
90. Pignataro, G., R.P. Simon, and Z.G. Xiong, Prolonged activation of ASIC1a and the time window for neuroprotection in cerebral ischaemia. *Brain*, 2007. **130**(Pt 1): p. 151–158.
91. Mazzuca, M., et al., A tarantula peptide against pain via ASIC1a channels and opioid mechanisms. *Nat. Neurosci.*, 2007. **10**(8): p. 943–945.
92. Roumy, M. and J.M. Zajac, Neuropeptide FF, pain and analgesia. *Eur. J. Pharmacol.*, 1998. **345**: p. 1–11.

Glia as Active Participants in the Development and Function of Synapses

Cagla Eroglu[1], Ben A. Barres[2] and Beth Stevens[3]

[1]Department of Cell Biology, Duke University Medical Center, Durham, NC 27710, USA, ceroglu@stanfordmedalumni.org
[2]Department of Neurobiology, Stanford University School of Medicine, Stanford, CA 94305, USA, barres@stanford.edu
[3]Department of Neurology, Harvard Medical School, Children's Hospital, Boston, Massachusetts 02115, USA, beths@stanfordmedalumni.org

1 Introduction

Long before the discovery of green fluorescent protein (GFP), Ramon y Cajal recognized glia's intimate association with synapses in the brain. Though their function was at that time a mystery, he predicted that glia must be doing more than simply filling the spaces between neurons. Not surprisingly, Cajal was correct. Although incapable of firing action potentials, glia are highly excitable cells, communicating with neurons and other glia primarily through chemical signals. While it is well established that glia have many important support functions in the nervous system, emerging evidence reveals a far more dynamic role for glial cells at the synapse.

We now appreciate that glia express many of the same voltage-gated ion channels and neurotransmitter receptors that neurons do and are thus well equipped to receive and transmit neuroactive signals. Through these mechanisms, glia can signal to pre and postsynaptic neurons and affect synaptic excitability and plasticity. Glia also actively communicate with one another in the form of calcium waves. The realization that dynamic bi-directional signaling between neurons and glia modulates synaptic transmission has led to re-definition of the synapse. The "tripartite synapse" is a term that was recently coined to incorporate glia as the third participant at the synapse, actively communicating with both the pre- and the post-synaptic neuron (2) (Fig. 1c).

Glia also play an important role in the development and maintenance of synapses. Their appearance at synapses in the postnatal brain and developing neuromuscular junction coincides with periods of developmental plasticity. A growing body of research indicates a pivotal role for glial-derived signals in the formation of structural and functional synapses. In addition, glia appear to secrete multiple signals that control the remodeling and stabilization of synapses, as well as the modulation of synaptic plasticity.

J.W. Hell, M.D. Ehlers (eds.), *Structural and Functional Organization of the Synapse*, DOI: 10.1007/978-0-387-77232-5_23,

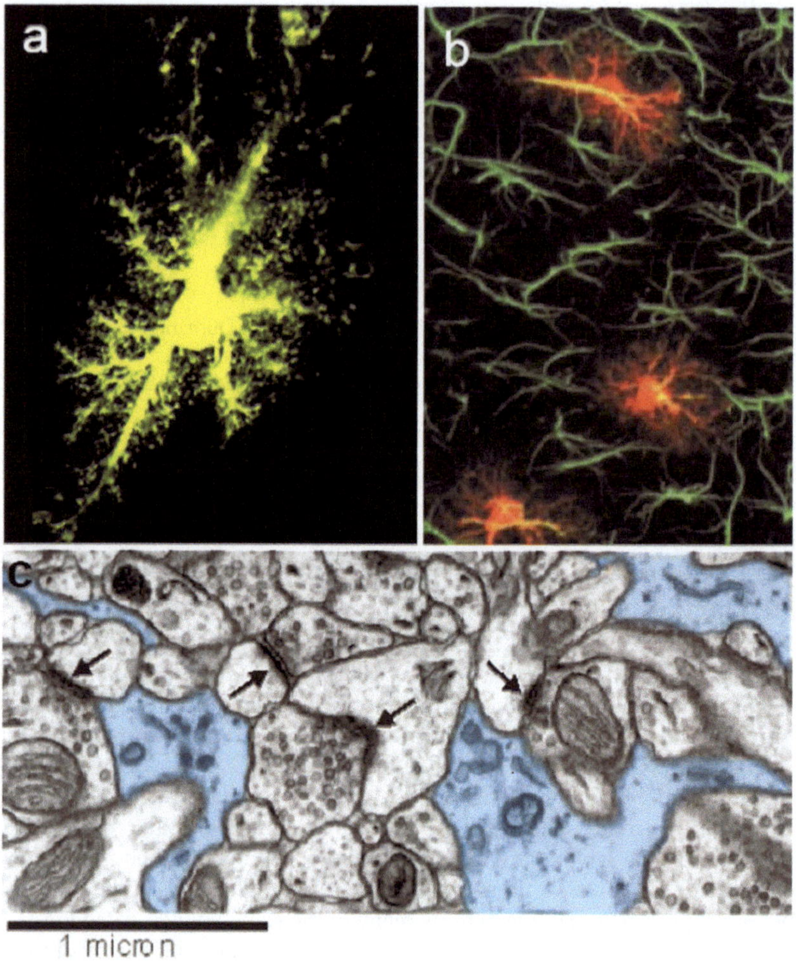

Fig. 1. Protoplasmic astrocytes are intimately associated with synapses.
(**a**) Protoplasmic astrocyte of CA1 region of rat hippocampus filled with Lucifier *Yellow* dye revealed thousands of dense spongiform processes that ensheath synapses (14). (**b**) Immunostaining with traditional astrocytic markers clearly underestimates the complexity and anatomical organization of astrocytic processes, as shown by GFAP immunostaining (*green*) of dye-filled astrocytes (*red*) in CA1 stratum radiatum (14). (**c**) Electron Micrograph of the "Tripartite Synapse" in the rodent brain show several synapses (*arrows*) ensheathed by astrocytic processes (*blue*). From http://synapse-web.org/anatomy/astrocyte/Astrocyte.stm (Dr. Kristin Harris).

It is becoming increasingly clear that glia can no longer be regarded as passive bystanders at the synapse. Neuroscientists are now appreciating that glial cells not only support essential neuronal functions, but also actively communicate with both neurons and one another to influence synaptic structure and function. This chapter reviews recent evidence for the role of glia in synapse development and synaptic transmission and plasticity.

2 Types of Glial Cells Associated with Synapses

In the human brain, glial cells outnumber neurons by a factor of ten, and numerous glial subtypes are easily identified based on their unique morphological, physiological, and biochemical features. Major types of glia in the nervous system include: myelinating glia (oligodendrocytes and Schwann cells), astrocytes, and microglia. In the CNS, glia that communicate with synapses include astrocytes, oligodendrocyte progenitor cells and possibly microglia. In the PNS, perisynaptic Schwann cells (PSCs) ensheath synapses at the neuromuscular junction.

Classical neurohistologists divided astrocytes into two main classes, which are distinguished by morphology and location, and perhaps function as well. Protoplasmic astrocytes are found in grey matter and their processes are closely associated with synapses as well as blood vessels. Fibrillary (or fibrous) astrocytes in white matter contact nodes of Ranvier. In addition, there are specialized astrocytes including Müller glia in the retina and Bergmann glia in the cerebellum.

Although most roles of protoplasmic astrocytes remain a great mystery, there is little doubt that they buffer ions and neurotransmitters in the extracellular space. Astrocytic processes intimately contact and ensheath pre- and postsynaptic terminals (115). Until recently, our understanding of astrocyte morphology was based on classical immunostaining for a widely used astrocytic marker GFAP, (glial fibrillary acidic protein), which grossly underestimates the complexity of astrocytes and their interactions with neurons and other cells (Fig. 1b). Recent *in vivo* imaging and three dimensional reconstructions of dye filled astrocytes revealed that adjacent astrocytes extend thousands of dense spongiform processes and are organized into large, non-overlapping anatomical domains. In fact, it has been estimated that a single astrocyte can associate with an average of four neurons (40), and contact over 100,000 synapses (14) (Fig. 1). Astrocytes express receptors for most neurotransmitters, and can release a variety of neuroactive, synaptogenic and trophic factors. Together these findings place astrocytes in an important position to actively exchange signals with neurons and other glial cells to coordinate synaptic networks.

Recent research has revealed a surprising new role for oligodendrocyte precursor cells (OPCs) at the synapse. These cells represent a third class of glia in the adult CNS and have morphological and physiological features distinct from oligodendrocytes and astrocytes (19, 49). These small, stellate shaped glial cells express the chondroitin sulfate proteoglycan, NG2. Based on their anatomical and physiological properties, it is likely that many of the cells originally classified as “complex cells” or “immature astrocytes” are actually NG2+ OPCs (NG2 cells). While the function of this new class of glial cells is still unknown, recent electrophysiological and anatomical studies revealed that NG2 cells express an array of voltage gated channels, as well as receptors for neurotransmitters. However unlike ensheathing perisynaptic glia, which respond to “spillover” of chemical transmitters from presynaptic terminals, NG2 cells receive direct synaptic inputs from neurons (81). These findings challenge current dogma that functional synapses are a unique feature of neurons.

Microglia comprise about 10% of CNS glia, and exist in multiple morphological states in the healthy and damaged brain. Like other glial cell types, much of their function remains mysterious. They act as internal “sensors” of the extracellular environment, rapidly responding to and communicating changes in the levels of extracellular ions and signaling molecules to neurons and glia. Recent *in vivo* time-lapse imaging has revealed dynamic interactions between microglia and neurons in the brain following lesion or injury (23, 73). Much like immune cells, microglia can elicit inflammatory responses and phagocytose cellular debris in response to injury, but new research indicates that microglia may also play a role in synapse development in the healthy brain.

In the peripheral nervous system (PNS), perisynaptic Schwann cells are specialized non-myelinating Schwann cells that ensheath the synaptic junction between motor neuron and muscle fibers PSCs are important for the formation, function and maintenance of neuromuscular junctions and play a pivotal role in synaptic repair after injury.

3 Role of Glia in Synapse Formation and Development

The traditional assumption that the ability of neurons to form synapses is intrinsic led early studies on synaptic development to focus on neuronal signals and surface molecules. However, in the last decade, mounting evidence from a variety of studies has demonstrated a role for glia in the formation of both the central and peripheral nervous system synapses. In this section we will discuss these studies and their influence on our current perception of glia as an active partner in the development of the nervous system.

3.1 Role of Astrocytes in Synapse Formation in the Central Nervous System

Astrocytes are the most abundant cell type in the brain. Through their intricate processes, astrocytes ensheath synapses and are known to listen and respond to synaptic signals. Do they also play a role in the establishment of the synaptic contacts in the developing brain? Interestingly, there is a strong correlation between the timing of the birth and maturation of astrocytes and the formation of synapses in the CNS. For example, in the visual system, retinal ganglion cells (RGCs) reach their thalamic targets in the superior colliculus beginning around embryonic day 16 and most RGC axons innervate the colliculus by the day of birth, postnatal day 0 (54). Despite the arrival of axons in their target area, there is a one week delay before the majority of synapses are formed in the superior colliculus. Coincident with this delay, the birth and proliferation of astrocytes in the superior colliculus also occurs during the first postnatal week. This spatiotemporal correlation between synaptogenesis and astrocyte development strongly suggests that signals from astrocytes might regulate the initiation of synapse formation (114). The presence of this additional regulatory step in the timing of synaptogenesis might be necessary to ensure that all axons reach their targets before the competition for synaptic territory begins.

Do astrocytes regulate the formation of synapses? Evidence supporting the hypothesis that astrocytes secrete extracellular signals that induce synapse formation has emerged from *in virto* studies with purified RGCs. RGCs can be purified from rodent retina and can be cultured in defined serum-free media with the support of several neurotrophic factors in the absence other cell types. Under these conditions, RGCs have equally high rates of survival and neurite outgrowth, regardless of whether or not they are cultured in the presence of astrocytes (62). Using these cells Pfrieger and Barres found that RGCs cultured in the absence of astrocytes exhibited very little spontaneous synaptic activity when excitatory postsynaptic currents were measured by patch-clamp. In contrast, the RGCs exhibited high levels of synaptic activity when they were cultured in the presence of a feeding layer of astrocytes or astrocyte-conditioned medium (ACM) (86) (Fig. 2). They have concluded that the

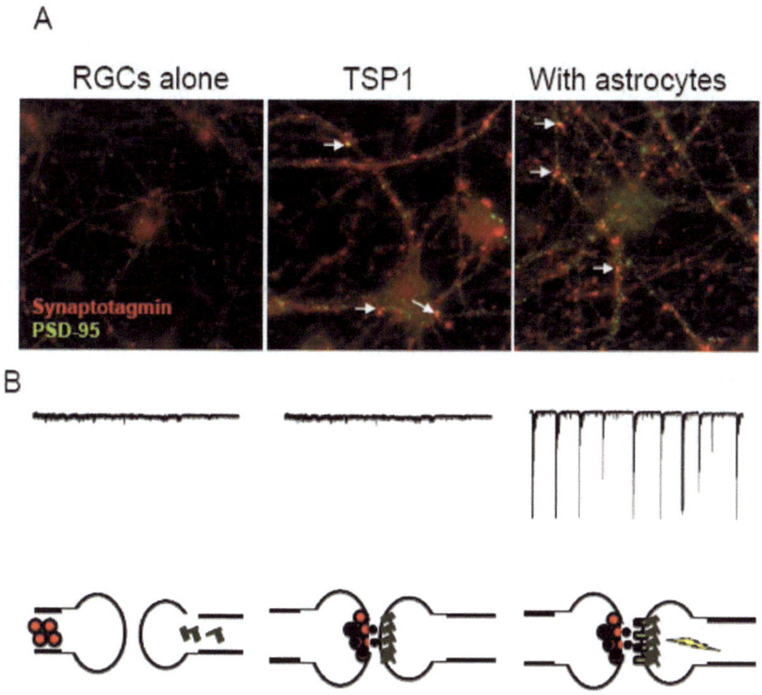

Fig. 2. Astrocytes promote synapse formation via the release of secreted factors. (**a**) RGCs require astrocyte-secreted factors to form synapses. Immunostaining of RGCs for colocalization of presynaptic Synaptotagmin (*red*) and postsynaptic PSD-95 (*green*) shows few co-localized synaptic puncta in the absence of astrocytes indicating that RGCs alone did not form many synapses. Many synapses (colocalized puncta) were present in RGCs cultured with a feeding layer of astrocytes or with purified TSP. (**b**) Astrocytes induce synapse formation in two steps. First TSP and/or other synaptogenic factors induce formation of structural synapses that have pre and post synaptic specializations and presynaptically active but postsynaptically silent as no significant action potentials can be recorded. A second factor(s) activates these postsynaptically silent synapses by facilitating insertion of AMPARs to the postsynaptic site.

ability of astrocytes to enhance synaptic activity depends on astrocyte-secreted proteins. This activity was mimicked by Schwann cells, and oligodendrocytes, but not by fibroblasts or microglia (86, 112). Similar to RGCs, purified spinal motor neurons exhibited little synaptic activity unless astrocytes or Schwann cells were present (113). Taken together, these findings reveal that astrocytes and other perisynaptic glia, such as Schwann cells, release soluble signals that profoundly enhance synaptic activity by nearly 100-fold.

To determine whether this increase in synaptic activity could be attributed to an increase in either synapse number or synaptic efficacy, Ullian and colleagues examined the effects of astrocytes on the number of synapses that formed between RGCs *in vitro*. They found that astrocytes induce a 7-fold increase in synapse number between RGCs, as measured by immunostaining with pre- and postsynaptic markers, as well as by electron microscopy (114). These synapses were ultrastructurally normal and they were both pre- and postsynaptically functional. Patch-clamp recordings from neurons that have been grown at clonal density and allowed to synapse upon themselves (a.k.a. autaptic neurons), showed a close correlation between the number of synapses observed by immunostaining and the number of synapses estimated by measuring the quantal content of the evoked response. Taken together, these findings show that astrocytes profoundly increase the number of synapses that form between CNS neurons *in vitro*. This was the first evidence that the number of synapses a neuron can form is not solely determined by the intrinsic properties of that neuron but instead can be powerfully controlled by extracellular signals. The 7-fold increase in synapse number, however, is not as large as the nearly 100-fold increase in synaptic activity induced by astrocytes suggesting that astrocytes can also enhance synaptic function.

What are the astrocyte-secreted factors that induce synapse formation? Interestingly, Christopherson and colleagues identified Thrombospondins (TSP) as the signals coming from astrocytes that can induce an increase in synapse number (20). They showed that purified TSP1 can increase synapse number in RGC cultures to a similar degree to ACM (Fig. 2). TSPs are a family of very large (450kDa) oligomeric extracellular matrix proteins and they are normally expressed in brain by glia and are present in ACM. TSP2, a closely related TSP family member, was also found to induce synapse formation at levels similar to TSP1, and was the major TSP type in the ACM. Immunodepletion of TSPs from ACM decreased its synaptogenic effect down to control levels showing that TSPs are the key synaptogenic component of ACM (20).

The TSP-induced synapses are ultrastructurally normal with pre- and postsynaptic specializations similar to those in RGC cultures grown with an astrocyte feeding layer. TSP treatment did not affect the total levels of synaptic proteins but instead induced synapse formation by promoting the localization of synaptic proteins to synapses and the alignment of pre and postsynaptic sites (20). Interestingly, while the TSP-induced synapses are structurally normal and presynaptically active, they are postsynaptically silent, due to the lack of surface AMPA receptors at postsynaptic sites (20). In contrast, astrocyte feeding layers and ACM can each induce both pre- and postsynaptic activity (Fig. 2). These data suggest that the regulation of synaptogenesis by astrocytes occurs in at least two steps (112). Astrocytes provide TSPs and/or other synaptogenic molecules to drive the assembly of synapses that have pre- and postsynaptic specializations but lack postsynaptic AMPA receptors thus are postsynaptically silent (Fig. 2B, middle panel). Astrocytes also secrete a second signal that converts these silent synapses into functional units (Fig. 2B, right panel). This silent to active synapse conversion through the activation of AMPA receptor-mediated responses resembles the AMPA receptor-dependent enhancement of synaptic strength observed following the induction of long-term potentiation (32) suggesting that astrocytes also secrete signals that play important roles in synaptic plasticity.

Do TSPs promote synapse formation *in vivo*? Interestingly, TSP1 and TSP2 are expressed in the brain during the early postnatal stages when the majority of synapses are forming, and both proteins are significantly down-regulated in the adult when rate of synaptogenesis is significantly reduced (20). In agreement with the *in vitro* data, TSP1/TSP2 double mutant mice exhibited a 25–30% reduction in the number of synapses formed during first and third postnatal weeks (20). In conclusion, TSPs are the key synaptogenic signal in ACM, TSP1 and TSP2 expression coincides with synaptogenesis *in vivo*, and loss of TSP1/2 function significantly decreases synapse formation.

How do TSPs promote synapse formation? Most likely TSPs provide a permissive signal for synapse formation within the developing brain. However, if TSPs are permissive and are required for synapse maintenance, they should also be present in the adult CNS where synapses are stably maintained. Infact there are five TSP isoforms in mammals and recent developmental gene expression analysis of astrocytes in rodent brain revealed an interesting spatial and temporal expression pattern for TSPs (25). All TSP isoforms except TSP5 are expressed by astrocytes. In agreement with previous observations, TSP1 and 2 are expressed by young astrocytes in developing brain during the first two postnatal weeks, when the bulk of synaptogenesis takes place. As the animal matures their expression goes down. In contrast, TSP4 expression is only detected in mature astrocytes (after postnatal day 17). These results suggest that TSP4 is the adult form of TSP in the CNS and might play a role in synaptogenesis and synaptic plasticity in the adult brain.

Recent data indicate that all TSP isoforms are synaptogenic, but that they differ in their effects on neurite outgrowth. TSP1 promotes RGC neurite extension, while TSP4 inhibits outgrowth and branching *in vitro* (25). This differential effect on neurite outgrowth might explain the reason for this developmental isoform switch. Early in development neurons are required to both extend their neurites and form new synapses. Correspondingly TSP1 and TSP2, which promote both of these processes, are expressed. As the animal matures, the major neural circuits are established and the formation of new synapses does not require extensive outgrowth; thus synapse formation and maintenance function is taken over by TSP4 in the adult brain. Surprising recent work comparing gene expression patterns in brains of humans and other nonhuman primates reveals that in humans TSP4 and TSP2 expression levels are dramatically increased (15). Does this mean that humans, because of TSPs, have a greater capacity to form synapses, and that this capability underlies our increased brain function?

Is TSP the only synaptogenic signal secreted by astrocytes or just the tip of the iceberg? Further study of astrocyte-secreted proteins and their role in synapse formation is necessary to answer this question. Astrocytes might secrete other signals that are capable of inducing formation of structural synapses similar to TSPs or they may also secrete signals that inhibit synapse formation. Secretion of both positive and negative regulators of synapse formation would give astrocytes the ability to regulate the timing and location of synapse formation with greater precision. Additionally, while identification of TSP significantly improved our understanding of the role of astrocytes in synapse formation, discovery and characterization of the neuronal TSP receptor will help clarify the cellular mechanism by which TSP establishes synaptic contact and might help us to better understand molecular basis of synapse formation in the CNS.

In addition to astrocyte-secreted proteins, Mauch et al. identified cholesterol in Apolipoprotein E (ApoE) particles coming from astrocytes as a potential signal that induces the formation of functional synapses between RGCs in culture (58). They observed that cholesterol enhances synaptic activity by increasing the quantal content of evoked responses in autoptic RGCs. This increase can be explained by an increase in the number of release sites (36). In addition, cholesterol is a limiting factor in synaptic vesicle formation and is required for vesicle turn-over events due to its regulation of vesicle curvature (109). Cholesterol is also necessary for lipid raft formation, and its membrane concentration can be critical for correct function of signaling processes as well as the activity of postsynaptic glutamate receptors that regulate synaptic function (43). One caveat is that most of these experiments were performed on autaptic neurons cultured at clonal density and thus these neurons were possibly in highly cholesterol-starved condition. Thus, future studies are necessary to determine the importance of astrocyte-derived cholesterol for synaptogenesis *in vivo*. Interestingly, in higher density cultures, cholesterol enhanced presynaptic function while it had no effect on postsynaptic function, and its effects in increasing synapse number were negligible (20).

Astrocytes also regulate the formation of inhibitory synapses between hippocampal neurons *in vitro* (24). When hippocampal neurons were either co-cultured with astrocytes or treated with astrocyte-conditioned medium (ACM), Elmariah et al. observed a significant increase in the number of inhibitory pre- and postsynaptic terminals, as well as an increase in the frequency of miniature IPSCs . They further explored whether the neurotrophin, Brain Derived Neurotrophic Factor (BDNF) and its receptor Tyrosine receptor kinase B (TrkB), which are potent modulators of inhibitory synaptic structure and function, mediate the effects of astrocytes on inhibitory synapses. Astrocytes deficient in either BDNF or TrkB maintained the ability to increase the number of inhibitory pre- and postsynaptic specializations in wild type neurons. However when neurons from these knockout mice were treated with wild type ACM, no increase was observed in the number of postsynaptic GABA (A) receptor clusters even though the increase in the number of inhibitory presynaptic terminals persisted. These results suggest that astrocytes may modulate inhibitory postsynaptic development by stimulating Trk signaling between neurons, but that the neurons themselves are the source of TrkB ligand BDNF. It is noteworthy that astrocytes modulate inhibitory synapse formation via distinct pre- and postsynaptic mechanisms, quite similar to the two step model of astrocytes regulation of excitatory synaptogenesis.

In addition to their role in development, astrocytes might also modulate synapse formation in the adult CNS. Two consequent studies by Song et al. on neuronal stem cells elegantly show that astrocytes indeed play an active role in synaptogenesis in adult brain. Stem cells from the adult hippocampus require a signal to differentiate into neurons. One source of this signal is hippocampal astrocytes (102). Once the stem cells have differentiated into their appropriate neuronal cell types, however, the neurons need an additional signal from astrocytes in order to integrate into synaptic circuits (102, 103). The second astrocyte signal appears to be necessary to promote an increase in synapse number in fully differentiated neurons. This work suggests that astrocytes continue to play a role in synaptogenesis in the adult brain and are involved in integration of newly born neurons into already existing circuits in the hippocampus. Whether TSPs or other known astrocyte derived signals are involved in this process remains to be discovered.

Besides astrocyte-secreted factors that induce synapse formation, astrocytes might also regulate synapse formation through contact mediated mechanisms. A study by Hama et al. provided evidence that astrocytes affect neuronal synaptogenesis by the process of adhesion (41). They showed that local contact with astrocytes via integrin receptors activates protein kinase C (PKC) in individual dissociated neurons. They observed that PKC activation, while initially focal, eventually spread throughout the entire neuron and facilitated excitatory synaptogenesis. Thus, they conclude that propagation of PKC signaling represents an underlying mechanism for global neuronal maturation following local astrocyte adhesion (41). One interesting difference between this clonal density hippocampal neuron culture system and RGCs is that treatment with ACM alone is able to induce RGC synapse formation, while hippocampal neurons require astrocytic contact. One possible reason for this discrepancy could be the age difference between the hippocampal neurons (isolated from embryonic day 17 brains) and RGCs (isolated from postnatal day 5 retinas). There might be an age dependent switch in neurons in which their ability to respond to soluble synaptogenic signals from astrocytes is induced by another environmental cue, such as contact with astrocytes themselves. Indeed, Ullian et al., found that astrocyte contact induced embryonic (E17) RGCs to become receptive to synaptogenic signals secreted from astrocytes (E. Ullian and B. Barres, in preparation).

3.2 Role of Perisynaptic Schwann Cells in Synapse Formation in the PNS

Parallel to astrocytes' role in CNS synapse formation, Schwann cells (SCs), the perisynaptic glia at the neuromuscular junction (NMJ), also actively participate in PNS synaptogenesis and synaptic function.

The classical doctrine of NMJ formation states that motor neuron derived factors such as Agrin induce the clustering of postsynaptic acetylcholine receptors (AChRs) (60). Is there also a SC component to NMJ formation? Indeed early studies on the formation of NMJ pointed to the presence of a factor that was able to induce upregulation of postsynaptic AChR subunits. This factor was then cloned and identified as Neuregulin, also known as AChR inducing activity (ARIA). Interestingly, Neuregulins are growth/differentiation factors essential for SC lineage differentiation and survival. In addition to their effects on SCs, Neuregulins have been associated with NMJ formation. A decade ago Trachtenberg and Thompson (111) found that the addition of the Neuregulin to developing NMJs resulted in SC process extension and migration away from the synapse. Coinciding with SC migration, the nerve terminals retracted from the synapse and regrew along with the SCs away from postsynaptic sites. The motor neuron (MN) axons were unable to make synapses onto muscle until the Neuregulin was removed. They suggested that SCs directly instruct the synaptic remodeling that occurs *in vivo* and they concluded that SCs play an important role in the formation of the NMJ synapse. However, it is been hard to pinpoint the cellular origin of Neuregulin function as both SCs and MNs express the receptors for Neuregulin. It was suggested that either Neuregulin independently induces both the SCs and MN axons to leave the synapse and reenter a plastic growth state, or Neuregulin affects only SCs by inducing them to become synaptogenic.

If SCs regulate synaptogenesis at the NMJ then lack of SCs should affect NMJ formation and function. A conditional knockdown of Neuregulin1 in motor and sensory neurons led to the total loss of Schwann cells in the peripheral nerves. However there were no significant changes in the AChR gene expression (120). On the other hand mice deficient in ErbB2, a Neuregulin receptor, also failed to develop the Schwann cell lineage, a defect that was accompanied by impaired development of the junctional folds of NMJ synapses (52). Similarly, an elegant study by Reddy and colleagues showed that when perisynaptic Schwann cells were ablated by complement lysis in live frog muscles, formation of synapses was dramatically reduced and existing synapses underwent retraction (90).

SCs strongly induce the formation of functional synapses between spinal MNs (113) or between *Xenopus* MNs and muscle cells in culture (82). SC-conditioned medium (SC-CM) mimics these effects, indicating the presence of soluble SC-secreted factors that promote synapse formation in PNS. How do SCs increase synapse number in MN-muscle co-cultures? Peng and colleagues proposed that SCs promote synaptogenesis by instructing neurons to synthesize a factor necessary for NMJ formation and showed that SC-CM induced an increase in Agrin expression by cultured MNs. This way SCs may mediate a switch in MNs from a growth state to a synaptogenic state (82). This is quite reminiscent of the permissive role of TSPs in the CNS. It is possible to postulate that glia act as switches both in PNS and CNS, converting axons from growth state to synaptogenic state. Feng and Ko identified transforming growth factor beta (TGFβ) as a component of SC-CM that is able to induce formation of synapses in *Xenopus* nerve-muscle cultures (26). They showed that TGF-β1 protein was present in SC-CM and that treatment of nerve-muscle cultures with purified TGF-β1, in the presence of trophic stimulation, induced neuronal Agrin expression and synaptogenesis along nerve-muscle contacts. Immunodepletion of TGF-β1 with a specific antibody abolished the synaptogenic effect of SC-CM. Latency-associated protein (LAP), which binds to TGF-β1 and prevents it from binding to TGF-β1 receptors, also blocked the effect of SC-CM (26). Taken together, these results indicate that TGF-β1 in SC-CM may be the glial signal that instructs neurons to switch from a "growth state" to a "synaptogenic state".

Recently Cao and Ko characterized SC-CM's effect on synaptic function (16). They found that SC-CM increased the frequency of spontaneous synaptic currents

150-fold at developing neuromuscular synapses. This was a presynaptic effect independent of neuronal excitability and required the influx of Ca2+, most likely through L and N-type voltage gated calcium channels potentiating spontaneous transmitter release. In contrast, SC-CM suppressed the evoked transmitter release. The molecular weight of functional factor(s) in SC-CM was within 500 and 5000 Da range and the SC-CM effect was not attributable to currently known factors that modulate synaptic efficacy, such as neurotrophins, glutamate, or ATP (16). These results suggest that SCs release small molecules that enhance spontaneous synaptic activity acutely and potently at developing neuromuscular synapses; however it is not yet clear whether this small factor also contributes to the increase in synapse number observed by Ullian et al. and Peng et al. Taken together, these studies show that SCs, like astrocytes, secrete synapse-promoting factors that can increase the number of MN synapses *in vitro* and profoundly affect NMJ formation and function.

3.3 Role of Glia in Synapse Elimination and Refinement of Neural Circuits

During development, the formation of mature neural circuits requires the selective elimination or pruning of inappropriate synaptic connections. The specific molecular mechanisms that drive synapse elimination remain a mystery but it is has been hypothesized that synapses are removed by either the retraction of presynaptic axons or the degeneration of axons and their subsequent phagocytosis by glial cells. Recent evidence suggests that glia may not only be engulfing axonal debris, but may also be actively involved in the process of developmental pruning.

In the *Drosophila* nervous system, larval mushroom body (MB) γ neurons initially extend axon branches into the dorsal and medial MB axon lobes that later undergo fragmentation and pruning. Awasaki and colleagues recently used a combination of high resolution imaging and molecular approaches to label MB axons to show that glial cells are recruited to MB axon lobes before axonal fragmentation and that they engulf MB axonal varicosities prior to their degeneration. Importantly, glial invasion and axon pruning was significantly impaired when the ability of glia to phagocytose was perturbed by inhibitors of glial endocytosis (5). In addition, Watts et al. showed that MB γ neuron terminals were incorporated within the lysosomal compartments of neighboring glial cells, further suggesting a role for phagocytosis in axon pruning (119). It was later shown that glial cells recognize and engulf degenerating MB axons through the function of Draper, the glial engulfment receptor (6, 33). Together these findings strongly suggest that glia actively contribute to developmental pruning in Drosophila.

A similar process of axon removal by SCs, termed "axosome shedding", occurs during synapse elimination at the mammalian NMJ junction. Postsynaptic muscle cells are initially innervated by multiple motor axons, but by the second postnatal week, a process of activity-dependent competition promotes the permanent elimination of the majority of these immature inputs, allowing only one input to be maintained and strengthened (95). As the losing axon withdraws, its distal ends form a retraction bulb (34, 92). Until recently, it was not known how these retraction bulbs formed or how retracted axons were eliminated. A combination of time-lapse imaging and serial electron microscopy (EM) demonstrated that as inappropriate axons disappeared, they shed small membrane bound particles termed "axosomes" that contain intact synaptic vesicles and organelles. Importantly, axosomes were formed by the engulfment of the tips of retreating axons by neighboring SCs (10). It was suggested that this phagocytic process is a novel means of transferring synaptic information between the cytoplasmic compartments of neurons and glial cells. While it is not known what molecules drive this phagocytic process, this study raises the intriguing possibility that glia actively contribute to synapse elimination in mammals.

Moreover, a recent study has provided evidence that the classical complement cascade, part of the innate immune system, helps to mediate synapse elimination in the developing CNS. In the developing mammalian visual system, astrocytes induce neuronal expression of C1q, the initiating protein of this cascade. C1q's best known role in the innate immune system is to opsonize or "tag" unwanted cells or debris for elimination. Recent work shows that C1q is expressed by postnatal retinal ganglion neurons in response to immature astrocytes and is then localized to synapses throughout the postnatal brain and retina (105). Mice deficient in complement protein C1q or the downstream complement cascade protein C3 exhibit large sustained defects in CNS synapse elimination, as shown by the failure of anatomical refinement of retinogeniculate connections and the retention of excess retinal innervation by lateral geniculate neurons. Microglia, the resident phagocytes of the brain, express receptors for C1q and C3, and have been shown to phagoctose synaptic terminals of motor neurons following injury, in a process known as synaptic stripping. Together these findings add to the growing evidence that immune system molecules are crucial for patterning of neural circuits (11, 45), and support a model in which unwanted synapses are tagged by complement proteins for elimination by phagocytic cells. As C1q levels become highly elevated in the adult brain in the setting of neurodegenerative disease, these findings also raise the interesting possibility that the classical complement cascade mediates synapse loss in diseases such as Alzheimer's disease, ALS, and glaucoma.

3.4 Role of Glia in Synaptic Maintenance and Structural Plasticity

3.4.1 Glia in Synaptic Maintenance

Glia have traditionally been viewed as the support cells that provide trophic support to neurons and maintain the homeostatic environment necessary for proper synaptic function. Are glia also necessary for the maintenance of the synaptic adhesions? To find out, Ullian and colleagues (114) examined whether the glial-induced synapses would be maintained in culture if the glial cells were removed. First synapses were allowed to form by co-culturing purified RGCs with a feeder layer of astrocytes cultured on a removable insert. After 1 week, as expected many synapses had formed. They then removed the astrocyte insert and cultured RGCs alone for an additional week. When they examined the synapses formed by these RGC neurons by immunostaining and patch clamp recording, they found that the majority of synapses were lost when the astrocytes were removed (114). These findings provide evidence that astrocytes are required for synapse stability, ensuring synaptic maintenance. This also raises the possibility that astrocytes may increase synapse number by enhancing synapse stability. However, dissection of synapse maintenance from other possible mechanisms such as enhancement of the rate of formation of synapses by astrocytes is difficult. The relative contribution of astrocytes to synapse formation and stabilization need to be investigated further.

Are perisynaptic Schwann cells also required for synaptic maintenance and function? Reddy et al. showed that in addition to their role in the establishment of the NMJ, SCs play a very important role in the maintenance of synaptic adhesion and synaptic function in mature NMJ (90). Using the frog NMJ as a model system, they investigated the effects of removing synaptic glia from a mature synaptic contact. Perisynaptic SCs in an intact adult frog were selectively labeled with a monoclonal antibody that specifically recognizes these cells. Once the antibody labeled cells were exposed to complement and lysed via the complement cascade, the presynaptic MN terminals and the postsynaptic muscle cell were intact but "naked", lacking their synaptic partner the perisynaptic SC. Interestingly, one week post ablation the presynaptic function at the naked synapses decreased by half and there was a ten-fold increase in the retraction of presynaptic terminals from the muscle. Taken together these studies suggest that SCs in PNS and astrocytes in CNS play important roles in stabilizing mature synaptic contacts.

3.4.2 Glia Regulates Structural Plasticity in the Developing and Mature Nervous System

So far we have reviewed the function of glia in the formation and maintenance of synapses as well as the elimination of unwanted ones. Capability of astrocytes and Schwann cells to control the establishment and breakdown of synapses gives them the ability to sculpt our nervous system, begging the question of whether glial signals also regulate structural plasticity in the brain.

Dendritic spines are small protrusions visible on dendrites of neurons that serve as postsynaptic sites for excitatory inputs. They are highly dynamic structures and a large body of research has linked changes in spine number and morphology to changes in neuronal activity and experience (53). During development spines are highly dynamic whereas in the adult brain they become more stable; hence a correlation exists between age dependent spine dynamics and the plasticity of the brain. The decrease in spine motility in adult could also be attributed to the close association of astrocytes with synapses. Astrocytes might provide both physical constraints that inhibit spine movement as well as molecular interactions that stabilize spines. Indeed an exciting study by Murai et al. showed that astrocytes help to control dendritic spine morphology *in vivo* through bi-directional Ephrin/EphA signaling (68). They observed that in the hippocampus, astrocytes and their processes express Ephrin A3, while neurons express the Ephrin receptor EphA4. Interestingly, soluble Ephrin A3 was sufficient to cause spine retraction in hippocampal slice cultures and transfecting neurons with a kinase inactive EphA4 resulted in longer spine extension. Together, these experiments imply that this receptor-ligand interaction regulates spine shape by inhibiting spine extension. An *ephA4* knockout mouse showed extension of spines in hippocampal slice cultures similar to dominant negative EphA4 transfected neurons, indicating that this receptor is required to maintain spine morphology (68). This work suggests that, in addition to soluble factors released by astrocytes, there might be bi-directional, contact-mediated, signaling between astrocytes and neurons that is important for maintenance of synaptic structure and number. Recently, timelapse imaging in mature hippocampal slices has shown that astrocyte processes and their neigbouring spines move cooperatively, further signifying a direct link between these two structures (39). In addition, a similar study using two-photon time-lapse imaging of astrocytic and dendritic processes revealed that dendritic protrusions with astrocytic contacts had longer lifetime and were morphological more mature (75). They showed that both motility of astrocytic processes and the Ephrin/EphA4 signaling between spines and astrocytes were necessary for astrocyte-dependent stabilization of newly generated dendritic protrusions. They proposed a model in which astrocytic processes are key local regulators of both the stabilization of individual dendritic protrusions and their subsequent maturation into spines.

A classical study supporting an active role for glia in developmental synaptic plasticity investigated the role of astrocytes in ocular dominance plasticity in the primary visual cortex (66). During visual system development in animals with binocular vision such as cats or humans, inputs corresponding to each eye converge and establish permanent connections in segregated columns in the visual cortex. This division of cortex into ocular dominance columns is an activity-dependent event, requiring active inputs from both eyes. If the input from one eye is prevented by either lid suture (monocular deprivation) or tetrodotoxin injection, the columns from the open eye expand extensively into the territory of the closed eye. This plasticity of synaptic inputs, termed ocular dominance plasticity, occurs only during the first 4–7 weeks after birth in cat visual cortex. In addition, if the deprived eye is reopened anytime after 7 weeks after birth, the deprived eye can no longer regain its lost territory; in contrast, if the eye is reopened before the end of this critical period, it can. These events are known to depend on neuronal activity, NMDA receptors, and inhibitory input (42).

Several lines of evidence suggest that astrocytes may also play a role in ocular dominance plasticity. First, there is a temporal correlation between the critical window for ocular dominance plasticity and the immature state of astrocytes in the cortex. In addition, dark rearing that is known to prolong ocular dominance plasticity also prolongs astrocyte immaturity, as assessed by delay of expression of GFAP, a marker of older astrocytes (65). Amazingly, when Muller and Best injected immature astrocytes from newborn kitten into the visual cortex of adult cats after the termination of the critical period, ocular dominance plasticity period was reopened (66). The implication of these results is that astrocytes play two roles in ocular dominance plasticity: immature astrocytes are instructive for synaptic plasticity, while mature astrocytes may help limit synaptic plasticity.

Could maturation of astrocytes be a general signal for the end of critical period plasticity in the CNS? A study by Pizzorusso et al. suggests that chondroitan sulfate proteoglycan (CSPG), an extracellular matrix (ECM) molecule that is secreted by mature astrocytes, might be involved in the regulation of plasticity in the mature visual cortex (88). They found that the appearance of CSPGs in the visual cortex is temporally correlated with the end of the critical period and that, like astrocyte maturation, their appearance can be delayed by dark rearing. Could immature astrocytes breakdown the limiting CSPG matrix thus reopen critical period plasticity? Indeed when Pizzorusso et al. dissolved the CSPG network in the brain enzymatically, the critical period was reopened, mimicking the effect of immature astrocyte transplantation observed by Muller and Best. This work implies that astrocytes can use secreted molecules to signal the end of the critical period and alter synaptic plasticity. Are there other molecules involved in astrocyte-regulation of synaptic plasticity? And what are the changes that astrocytes go through as they mature? Answering these questions will surely broaden our understanding of CNS development.

There is also evidence that myelination may help signal the end of critical period plasticity in the brain (30). Myelin proteins, such as Nogo, myelin associated glycoprotein (MAG) and oligodendrocyte-myelin glycoprotein (OMgp), are strong inhibitors of axon outgrowth and regeneration after injury. In part, they have been found to inhibit axon growth through their interaction with the Nogo-66 receptor (NgR), which is expressed by neurons. Interestingly, the period of ocular dominance plasticity is dramatically prolonged in mice deficient in NgR (59). These findings suggest that physiological NgR signaling is involved in ending the critical period, however they fall short of demonstrating a role for myelin-associated NgR ligands as there is emerging evidence that these ligands are also localized to developing synapses.

In addition to their role in developmental plasticity, astrocytes have also been linked to regulation of synaptic plasticity of the hypothalamic supraoptic nucleus in the adult. Activation of this system by release of oxytocin due to physiological conditions such as parturition or lactation leads to a fast coordinated withdrawal of astrocytic processes that normally lie between neuronal cell bodies and innervating dendrites and the somatodendritic synapses (42). After astrocyte withdrawal there is a significant increase in the number of GABAergic, glutamatergic, and noradrenergic synapses as postsynaptic target sites, which were previously shielded by astrocytes, become available (107). In addition since the astrocytes have withdrawn from synaptic sites, glutamate concentration at the synapse increases, leading to a dramatic increase in neuronal excitability (87). It is possible that astrocyte-secreted ECM molecules that regulate cell adhesion such as PSA-NCAM and Tenascin-C are involved in these structural changes at the synapse (108). These studies also suggest that astrocytes, most likely through their blood vessel-associated end-feet, can detect and respond to somatic stimuli that may be critical for the initiation of important structural and functional plasticity in the adult CNS.

In conclusion, astrocytes and oligodendrocytes can regulate developmental structural plasticity of the CNS through modulation of dendritic spine dynamics, ECM and myelination. Whether the inability of the CNS to regenerate after injury is actually a side effect of the glial regulation of plasticity remains an intriguing and important question.

4. Role of Glia in Synaptic Transmission in the Mature Nervous System

4.1 Activity-Dependent Neuron-Glia Communication at the Synapse

There is little question that glia perform several regulatory roles essential for normal neurotransmission and that there are many mechanisms by which glia indirectly regulate synaptic function. Astrocytic processes surround pre- and postsynaptic terminals where they rapidly clear and recycle ions and neurotransmitters from the synaptic cleft. Astrocytes express high levels of glutamate transporters, GLAST (glutamate/aspartate transporter) and GLT1 (glutamate transporter1), which actively clear the synapse of excess transmitter. During neurotransmission, glutamate is taken up by astrocytes, converted into glutamine, and then released into the extracellular space. Glutamine uptake by neurons is subsequently used to synthesize glutamate and GABA, the excitatory and inhibitory neurotransmitters. Importantly, astrocytes also protect neurons from excessive glutamate and other forms of excitotoxicity by both regulating pH and buffering potassium and other ions released during neurotransmission.

It is becoming increasingly clear, however, that glia also play a more direct role in modulating synaptic transmission. The notion that glial cells are active, excitable cells emerged with the advent of new techniques with which to study glia. The first clue that the close structural relationship between glia and synapses played an important role in modulating synaptic transmission came from the realization that perisynaptic glia, like their neuronal partners, express membrane receptors for most neurotransmitters (43). Although incapable of firing action potentials, glia can respond to neuronal activity via the activation of metabotropic neurotransmitter receptors. When activated, these neurotransmitter receptors promote the production of second messengers and the release of calcium from intracellular stores. Glial calcium signaling is thus a measurable indicator of cell excitability, providing a powerful tool to study the activity of glial cells in action.

Do glia signal back to synapses? Activity-dependent activation of glial receptors can induce the synthesis and release of several neuroactive substances, including ATP (and its byproduct, adenosine), glutamate, and D-serine. Activity-dependent release of "gliotransmitters" affects synaptic activity in neurons and modulate synaptic efficacy and plasticity. Time-lapse calcium imaging has demonstrated that glia surrounding synapses respond to neuronal activity with increases in intracellular calcium and that activity-dependent communication between glia and neurons is bidirectional. Chemical or mechanical stimulation of glia can in turn elicit calcium responses in nearby neurons. Glia also actively communicate with each other through intercellular calcium waves. Stimulation of a single astrocyte can trigger a local elevation in astrocytic calcium that can subsequently spread to neighboring astrocytes in the form of a calcium wave. This section highlights examples of activity-dependent intercellular communication between neurons and glial cells in the CNS and PNS (Fig. 3).

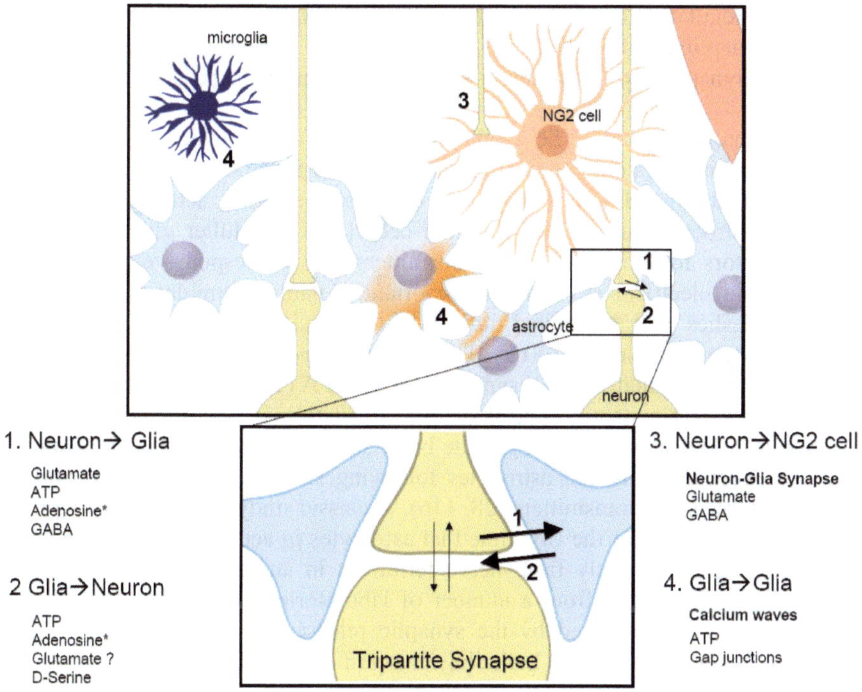

Fig. 3. Multiple forms of activity-dependent neuron-glia synaptic communication in the brain. Schematic illustrating several forms of neuron-glia communication at the synapse. The "tripartite synapse" in the CNS (*Box*) is composed of presynaptic neuron, postsynaptic neurons, and the processes of protoplasmic astrocyte, which ensheath synaptic terminals. Neuron-glia communication is bi-directional and include neuron to glia signaling (1) and glia to neuron signaling (2). Neurons form direct synaptic contacts onto NG2 cells in the brain (3). Astrocytic calcium waves (shaded) propagate between neighboring astrocytes is a well studied example of glia-glia communication. Several of the key signals and mechanisms mediating each form (1–4) of synaptic communication are highlighted. *note: adenosine is generated from the rapid degradation of ATP by ectonucleotidases.

4.2 Neuron to Glia Communication

Are glia actively listening to neurons at synapses? Activity-dependent neuron-glia communication has been extensively studied at NMJ. Elegant electrophysiological and imaging studies performed at the intact frog and mouse NMJ have demonstrated that perisynaptic SCs (PSCs) respond to high frequency stimulation of motor nerves with increases in intracellular calcium (91, 94). Nerve-evoked calcium responses in PSCs are mediated the by the activation of purinergic and muscarinic receptors following the release of ATP and acetylcholine from the presynaptic terminal (94). In addition, classic neuromodulators such as neuropeptides and nitric oxide (NO) can fine-tune neurotransmitter-evoked calcium signaling in PSCs just as they do in neurons. One of the functional consequences of the synaptic activation of transmitter receptors in PSCs may be the activation of downstream intracellular signaling pathways and subsequent regulation of glial gene expression. As we'll discuss later, activation of PSCs by synaptic activity also leads to the regulated release of gliotransmitters that can act presynaptically to modulate synaptic transmission and short term synaptic plasticity.

The retina has proven to be a valuable model system in which to study activity-dependent neuron glia communication in intact CNS tissue. Müller cells, specialized radial glia that expand almost the entire depth of the retina, actively communicate with both astrocytes and retinal ganglion cells (RGCs). Müller glia express membrane receptors for ATP and other transmitters and are responsive to mechanical, chemical and electrical stimulation. Importantly, natural stimulation of RGCs by repetitive flashes of light elicited rapid calcium transients in Müller cells that were blocked by tetrodotoxin (TTX)(69). This activity-dependent response in Müller glia was shown to be mediated by presynaptic release of ATP by neurons (69).

It appears that synaptic control of glial excitation is widespread in the brain. Electrophysiological studies using acute brain slices have revealed active communication between neurons and astrocytes following stimulation of neurons and their subsequent release of transmitters (28, 116). A classic study by Porter and McCarthy (1996) demonstrated for the first time that astrocytes in acute brain slices respond to glutamate released directly from nerve terminals in an activity-dependent manner (89). Subsequent studies from a number of laboratories have shown that glial calcium increases are mediated by the synaptic release of transmitters that act on a variety of astrocytic receptors including muscarinic receptors (1) and GABA receptors (47). Importantly, activity-dependent neuron-glial synaptic signaling also occurs *in vivo* under physiological conditions. A recent study by Wang et al. provided compelling evidence that astrocytes respond to sensory input *in vivo*. Two photon imaging in anesthetized mice showed that whisker stimulation evoked increases in astrocyte calcium levels within layer 2 of barrel cortex of adult mice (117). Sensory-evoked increases in astrocytic calcium were frequency-dependent and were mediated in part by ectopic glutamate release, which is in agreement with previous *ex-vivo* studies.

Bergmann glia, the specialized astrocytes of the cerebellum, are exquisitely sensitive to synaptic activity. Stimulation of cerebellar granule cells in acute slices induced rapid calcium transients in associated Bergmann glia (48). Similarly, electrical stimulation of parallel fibers elicited calcium signaling in Bergmann glia that was mediated by nitric oxide release (37, 57). Interestingly, Bergmann glia have hundreds of subcellular compartments termed 'microdomains' that wrap around synapses and respond with increases in intracellular calcium following electrical stimulation (37). Bergmann glia, like neurons, are sensitive to the pattern and frequency of neuronal activity. Single pulse stimulation of parallel fibers evoked local calcium fluxes in microdomains, whereas multiple trains of pulses evoked much larger and sustained calcium increases in Bergman glia cell bodies and processes. Frequency-dependent neuron-glia signaling has also been demonstrated in hippocampal slices (80) and at the NMJ (46, 91, 93). That glia are sensitive to the pattern of neuronal firing suggests that glia are capable of sensing and integrating information from neurons according to functional requirements of neural circuits.

4.2.1 Synapses Between Neurons and Glia

Glial cells have recently challenged the dogma that fast synaptic transmission occurs exclusively between neurons. Direct activity-dependent neuron-glia communication was recently demonstrated between NG2- positive oligodendrocyte progenitor cells (NG2 cells) and neurons. These small stellate-shaped cells are the newest subtype of CNS glia and they have attracted significant attention due to their widespread distribution in the brain and their persistence in the developing and mature nervous system.

In contrast to perisynaptic astrocytes and Schwann cells, which ensheath synapses and respond to the ectopic release of chemical transmitters, electrophysiological and anatomical evidence indicates that NG2 cells form direct synaptic junctions with both glutamatergic and GABAeric neurons. Whole cell patch-clamp recordings of NG2 cells in acute hippocampal slices revealed that NG2 cells responded to afferent stimulation with the extremely rapid kinetics traditionally associated with neuron-neuron synaptic communication. In addition to activating calcium signaling in NG2 cells, quantal release of neurotransmitter from neurons triggered rapid activation of AMPA receptors and evoked postsynaptic currents (mEPSCs) in NG2 cells (8). These findings suggested that fusion of synaptic vesicles in neurons induced the binding of glutamate to AMPA receptors on NG2 cells. Indeed, single NG2 cells processes were often located in direct contact with synaptic junctions (8). Although it is possible that ionotropic receptors on NG2 cells were activated by ectopic release of transmitter, electron microscopy analysis revealed that nerve terminals containing synaptic vesicles formed synaptic junctions with the processes of NG2 cells. NG2 cells also express ionotropic GABA (A) receptors and direct GABAergic signaling between neurons and NG2 cells has also been observed (50).

Direct rapid synaptic communication between NG2 cells and neurons is not limited to the hippocampus. NG2 cells are found in both grey and white matter throughout the developing and adult CNS (19, 49). In the molecular layer of the adult cerebellum, climbing fibers from the inferior olive form direct synaptic connections with NG2 cells, in addition to Purkinje cells. In fact, it has been estimated that NG2 cells can receive up to 70 inputs from a single climbing fiber. In addition, unlike Purkinje neurons, NG2 cells are often innervated by multiple climbing fibers, which suggests a major role for NG2 cells in cerebellar circuitry. Indeed, activation of climbing fiber inputs resulted in excitatory AMPA synaptic currents in NG2 cells (51). Given their widespread distribution, it is likely that direct signaling between NG2 cells and neurons is a general phenomenon. What is the functional significance of this rapid neuron-glia signaling? This remains an important and open question. It may enable neurons to rapidly regulate gene expression in NG2 cells according to functional requirements. Neurons could also rapidly signal NG2 cells to release gliotransmitters and signal back to pre- and postsynaptic neurons, as has been described for astrocytes and perisynaptic Schwann cells.

4.3 Glia to Glia Signaling

Active communication between glial cells contributes an additional layer of complexity to synaptic signaling in the CNS. Neurotransmitters released by presynaptic neurons not only evoke local calcium oscillations in nearby glial cells, but can also trigger calcium waves through glial networks, which might allow for the transmission of synaptic signals over extensive cortical areas.

In culture, stimulation of a single astrocyte can trigger a local elevation of astrocytic calcium that can subsequently spread to neighboring astrocytes in the form of a calcium wave. Calcium waves can propagate at a rate of 100 μm/sec (18) and over a distance of hundreds of micrometers (96, 28). Intracellular and intercellular calcium waves involve the activation of G-coupled receptors, activation of phospholipase C, and the production of the second messenger IP3, which triggers calcium release from intracellular ER stores. Although these waves can occur spontaneously, the frequency and extent of calcium signaling can also be potentiated by mechanical, electrical and chemical stimulation.

What mechanisms underlie the propagation of calcium waves? Waves can be transmitted to neighboring astrocytes via either the local diffusion of second messengers through gap junction channels or the release of "gliotransmitters" such as ATP along astrocytic networks. Recent evidence suggests that these two mechanisms are likely to work together to coordinate signaling between glial cells. Gap junction mediated propagation of calcium waves was the first such mechanism described in astrocytes (31), and there is now convincing evidence from a number of groups that disruption of gap junction channels impairs the propagation of calcium waves *in vitro* and *in vivo* (83, 96). In addition, extracellular ATP released from stimulated astrocytes has been shown to propagate long-range calcium signaling via the activation purinergic receptors (P2Y) on neighboring astrocytes (38). There is also evidence for a "point source ATP release" mechanism in which calcium waves are propagated via the release and diffusion of ATP from the stimulated cell and the activation of purinergic receptors on neighboring astrocytes (4). The mechanism of ATP release from astrocytes is still not clear. Possible mechanisms include calcium-dependent exocytosis and ATP release through connexin hemichannels (116).

An important question is whether propagating calcium waves through glia normally occur *in vivo*. Currently there is little evidence that intercellular astrocytic calicium waves occur under conditions considered to be physiological. Spontaneous and evoked calcium waves occur between Müller glia and astrocytes in the intact retina and other systems (28, 71), but there are clear differences in the extent and magnitude of waves in different tissue preparations, which has led to confusion in the field. Curiously, electrical stimulation generally fail to elicit large propagating astrocytic waves in most brain slice preparations (37, 48, 89) or *in vivo* (44, 74). Instead, calcium increases are typically observed in a few neighboring astrocytes in response to local stimulation. This, together with the realization that astrocytes are organized into distinct non-overlapping domains, suggests the existence of local astrocytic circuits (14, 106).

Are calcium waves physiological or pathological? It is clear that intercellular waves can be induced in culture and *in situ* under more pathological conditions. Propagating waves have been observed in corpus collosum glia (99) and rat cortical slices (85) in association with spreading depression associated with ischemia and injury. Schipke and colleagues demonstrated that ATP and activity- induced astrocytic calcium waves can propagate over hundreds of micrometers *in situ* and can trigger responses in microglial cells (99). Recent advances in high resolution imaging and molecular approaches to study neuron-glia communication in anesthetized and awake behaving animals will hopefully shed light on these important issues (73, 74).

An important function of rapid intercellular communication among groups of astrocytes could be to modulate and coordinate the activity of neural circuits. Retinal ganglion cells actively respond to waves of calcium that ripple through neighboring Müller glia (72) *in situ*. Increases in neuronal spontaneous and evoked activity occur in response to calcium waves in neuron-astrocyte co-cultures (3) and in intact tissue slices (28). Astrocytic calcium signaling also have important consequences for the cerebral vasculature. In addition to ensheathing thousands of synapses, astrocytic end feet are also in direct contact with blood vessels. Recent time-lapse imaging experiments provide convincing evidence that calcium increases in astrocytic endfeet can regulate blood flow in associated blood vessels (67, 122). In the retina, stimulation of glia in the intact retina can evoke dilation or constriction of arterioles (61). Although the mechanisms underlying astrocyte-vessel signaling are still unclear, these recent findings point toward a possible role for astrocytes in controlling vasculature function and possibly dysfunction in pathophysiological conditions and disease.

4.4 Glia-Neuronal Signaling at the Synapse

Glia rapidly respond to synaptic activity, but can glia signal back to the synaptic terminal to influence synaptic function? This question has been investigated extensively over the past decade in several model systems and more recently *in vivo*. It is now well established that activated glia can release neuroactive molecules including ATP (21, 70, 118), glutamate (77), inflammatory cytokines (9), vasoactive molecules (123) and D-Serine (64) that could signal back to pre- and postsynaptic terminals to modulate synaptic activity and function.

Perhaps the best example of glial modulation of synaptic transmission *in vivo* occurs at the mammalian NMJ. Perisynaptic SCs at the mammalian and amphibian NMJs respond to the release of Acetylcholine and ATP from the presynaptic terminal with robust increases in intracellular calcium (91, 46). It is well established that repetitive stimulation of motor neurons induces synaptic depression by decreasing presynaptic transmitter release. Recent research implicates the perisynaptic Schwann cell (PSC) as a major player in this form of synaptic depression. Activation of PSCs induced release of a chemical signal from PSCs that stimulated nitric oxide production at the NMJ. This glial induced NO release mediates synaptic depression by decreasing the probability of presynaptic acetylcholine release (110).

What is the functional consequence of this activity-dependent neuron-glia communication? Perisynaptic Schwann cells can both positively and negatively influence synaptic transmission, depending on the types of glial signaling pathways activated during synaptic transmission. Sustained IP3- dependent calcium influxes in perisynaptic Schwann cells can regulate synaptic strength (17); conversely, synaptic activation of G-protein coupled signaling cascades in Schwann cells can affect synaptic depression as described above. Together these findings suggest that activity-dependent activation of specific signaling cascades in terminal Schwann cells can integrate synaptic activity and directly influence neuronal activity and synaptic efficacy.

The first evidence that astrocytes can directly modulate the activity of adjacent neurons came from studies using neuron-glial co-cultures. Araque and colleagues found that elevations in intracellular calcium in astrocytes (by electrical or mechanical stimulation) resulted in a corresponding slow inward current in hippocampal neurons. They went on to show that calcium-dependent release of glutamate from astrocytes enhanced the frequency of postsynaptic currents in neurons by acting on extrasynaptic NMDA receptors (3, 77). Glutamate was later shown to mediate astrocyte-neuron signaling in acute brain slices. Fiacco and McCarthy (27) recorded from CA1 pyramidal cells in acute brain slices while evoking Ca^{2+} increases in the adjacent stratum radiatum astrocytes. They used whole-cell patch clamping to deliver caged IP3 and a Ca^{2+} indicator dye to astrocytes. Uncaging IP3 in astrocytes produced a propagating calcium wave that spread into fine distal astrocytic processes that perfectly coincided with a significant increase in the frequency of AMPA spontaneous excitatory post synaptic currents (sEPSCs) in CA1 neurons. Recent evidence also suggests that astrocytes may be a necessary mediator in activity-dependent modulation of inhibitory synapses in the hippocampus. Stimulation of interneurons elicited a GABA(B)-receptor-mediated elevation of calcium in surrounding astrocytes, which in turn potentiated inhibitory transmission (47).

ATP is a key signal mediating the propagation of astrocytic calcium waves, raising the question of whether astrocytic ATP release might regulate synaptic transmission. In the retina, electrical activity of retinal ganglion neurons (RGCs) is depressed in response to glial calcium waves that propagate through the retinal ganglion cell layer (72). It turned out that Müller glia inhibited RGC firing via the release of ATP, which was converted to adenosine by ectotidonucleases and activated adenosine receptors on RGCs (70). A very similar process has been recently demonstrated in acute hippocampal slices. Glial-derived adenosine acts on presynaptic A1 receptors in neurons leading to inhibition of calcium channels, a decreased probability of vesicular release and reduced frequency of sEPSCs (79). Glial mediated synaptic depression has also been demonstrated in cerebellar slices in the case of depression of Purkinje neurons by Bergmann glia (12). Interestingly, a recent study demonstrated that ATP release from glia can also positively regulate synaptic transmission. ATP activation of postsynaptic ionotropic P2X receptors increases postsynaptic intracellular calcium levels, which is thought to regulate insertion of AMPA receptors, thereby increasing excitatory synaptic transmission (35).

It has been proposed that astrocytes may normally release vesicle-stored glutamate *in vivo* at synapse like contacts onto neurons, but recent studies suggest that this is unlikely. Regulated glutamate release by astrocytes has been measured in culture and several mechanisms of release have been proposed including: calcium-dependent exocytosis, volume-sensitive ion channels, gap junction hemichannels, and the large pore of the ionotropic P2X7 ATP receptor (78). There is little evidence, however, that glutamate can normally accumulate in astrocytes *in vivo*. To directly address the question of whether stimulation of calcium waves in astrocytes induces them to secrete signals that are sufficient to activate nearby neurons, Fiacco et al., (2007) constructed transgenic mice that express a Gq-coupled receptor specifically in astrocytes (29). This enabled them to use an agonist that would activate calcium waves specifically in astrocytes. Despite widespread calcium elevations in astrocytes, nearby neuronal activation could not be detected. These new findings call into question the popular notion that calcium-dependent release of glutamate or other gliotransmitters by astrocytes directly affects neuronal synaptic activity *in vivo.*

5. Role of Glia in Synaptic Plasticity

As we discussed in the previous section, glia have the ability signal to neurons through neuromodulators and neurotransmitters that affect pre- and/or postsynaptic function. This signaling between glia and synapses establishes glia as a major player in the regulation of the functional plasticity at the synapse. Indeed, recent literature shows that several factors secreted by glia play an active role in activity-dependent modulation of synaptic efficacy, a process that is known to contribute to neural circuit development and experience-dependent plasticity.

The most well known forms of synaptic plasticity are long-term potentiation (LTP) and long-term depression (LTD), which involve rapid adjustments in the strength of individual synapses in response to specific patterns of correlated synaptic activity. The postsynaptic ionotropic glutamate receptors, AMPA receptors (AMPAR) and NMDA receptors (NMDAR), are thought to be central players in these plasticity processes through the regulation of their surface delivery, retention and calcium channel properties (22, 100). NMDARs contain a glutamate binding site as well as a site for glycine, a coagonist that is required for the opening of the calcium channel. In the past decade an unlikely amino acid D-serine emerged as a potent agonist for this site. Interestingly, D-serine is exclusively produced by the glial enzyme serine racemase, which converts L-serine to D-serine (63). D-serine is secreted by astrocytes through a vesicular release process that is dependent on AMPAR activation in glial cells. This indicates that the glutamate, which is released onto perisynaptic astrocyte processes by neurons during synaptic transmission, could be a signal that induces D-serine secretion by astrocytes (97, 98).

Interestingly, neuronal activity-dependent release of D-serine might be necessary for hippocampal LTP (121). When Yang and colleagues cultured hippocampal neurons in contact with astrocytes, the neurons were capable of undergoing LTP; however neurons grown in ACM without direct astrocytic contact were not. The addition of exogenous D-serine restored the ability of neurons grown in ACM to undergo LTP. These data suggest that astrocytes are actively involved in synaptic plasticity by regulating the availability of D-serine, released at perisynaptic astrocyte processes. Recently Panatier and colleagues looked at the function of D-serine mediated NMDAR signaling in the hypothalamic supraoptic nucleus (76), a system in which structural changes in astrocyte association with synapses are known to be important for the oxytocin-mediated responses at synapses. They provided compelling evidence that in this hypothalamic structure the endogenous co-agonist of NMDA receptors is D-serine rather than glycine. Therefore, the degree of astrocytic coverage of neurons governs the level of glycine site occupancy on the NMDA receptor, thereby affecting their availability for activation and thus the activity dependence of LTP/LTD (76). In conclusion, D-serine released by astrocytes is a major regulator of NMDA function thus a potent regulator of synaptic plasticity.

In addition to regulation of NMDAR function, surface delivery of AMPARs is thought to be an important mechanism in the regulation of LTP and LTD. Beattie et al. showed that the pro-inflammatory cytokine TNF-alpha released by glia enhanced synaptic efficacy by increasing surface expression of AMPA receptors (7). Blocking TNF-alpha signaling had opposite effects. The authors suggested that the continuous presence of TNF-alpha is crucial for the preservation of synaptic strength at excitatory synapses. Interestingly, even though TNF-alpha increased surface AMPAR expression, LTP and LTD were not affected in its absence (104).

Homeostatic synaptic scaling, a second form of synaptic plasticity, involves uniform adjustments in the strength of all synapses on a cell in response to prolonged changes in the cell's synaptic activity such as blockade of synaptic function. This regulation might be critical for preventing neural networks from becoming unstable and dysfunctional (13). The mechanisms responsible for synaptic scaling have generally remained a mystery, with the exception of evidence that regulation of receptor content at synapses might be involved (84). Stellwagen and Malenka showed that TNF-alpha mediates synaptic scaling in response to prolonged activity blockade (104). Using mixtures of wild-type and TNF-alpha-deficient neurons and glia, they were able to illustrate that glia are the source of the TNF-alpha involved in this synaptic scaling. Since TNF alpha is secreted by microglia in response to injury, TNF-alpha mediated synaptic plasticity might modulate neural responses to injury and neurodegeneration such as Alzheimer disease. Further studies will be necessary to address whether such TNF-alpha signaling is present *in vivo* under normal developmental conditions or whether it is an injury mediated adaptation mechanism. Characterization of the underlying molecular mechanisms through which TNF-alpha can induce AMPAR surface expression will also improve our understanding of this phenomenon.

Another synaptic plasticity mechanism that requires coordinated regulation of synaptic function is heterosynaptic depression. This phenomenon was first noticed in the retina and was later shown to be mediated by the ability of an active synaptic pathway to laterally regulate the relative strength of neighboring synapses (55). Such a mechanism is also present in hippocampus and it entails the specific depression of an unstimulated pathway in the presence of another independent pathway in which LTP is induced (56). This inter-synaptic communication is regulated by adenosine through A1 receptors (56). Recently, vesicular release by astrocytes was reported to be crucial for heterosynaptic depression (79). To do so they created a transgenic mouse line in which a dominant negative SNARE domain that is known to inhibit vesicular exocytosis was specifically expressed in astrocytes using a GFAP promoter. Heterosynaptic depression was prevented in these mice despite the fact that LTP could still be potentiated. They also showed that this effect was mediated by secreted ATP, which is converted to adenosine and accumulates at the synapse to mediate its inhibitory function through A1 receptors. These results provide evidence that ATP released by astrocytes is required for heterosynaptic depression. An important caveat is that the conclusion that the vesicular ATP derives from astrocytes depends on the absolute specificity of dominant negative SNARE expression in astrocytes. This specificity might be difficult to achieve with the GFAP promoter, as GFAP is expressed in some neural stem cells, which could thus give rise to neurons that also express the dominant negative SNARE construct.

To investigate the trigger of glial ATP release, Serrano and colleagues tested whether Ca^{2+} activation of glial cells was necessary for heterosynaptic depression (101). Indeed, a rise of intracellular Ca^{2+} in glial cells, induced by activation of glial GABA (B) receptors, was required for heterosynaptic depression. This effect could be inhibited by blocking ATP degradation, indicating that ATP-adenosine signaling was downstream of Ca^{2+} signaling in glia (101). Taken together these results indicate that GABA released by inhibitory neurons activates Ca^{2+} signaling in astrocytes, which leads to secretion of ATP. This ATP is then converted to adenosine in the extracellular space, where it activates A1 receptors on the post-synaptic excitatory neuron, inducing synaptic depression. The presence of this regulatory network modulating heterosynaptic depression is an elegant example of how neuron-glia/glia-neuron interactions mediate complex plasticity processes in the brain.
In conclusion, there is overwhelming evidence that glia use their ability to respond to neurotransmitters and secrete neuromodulators to actively regulate a number of processes involving synaptic plasticity.

6. Concluding Remarks

It is becoming increasingly clear that glia play a crucial and dynamic role at the synapse and influence functions that have long been thought to be strictly under neuronal control. The inclusion of glial cells as active participants at the synapse is gradually shifting long-standing neurocentric perspectives on how the brain works. Glial signals help promote both the formation and elimination of synapses as the brain wires up during development. Once the appropriate synaptic connections are formed, glial cells then influence their maintenance and stabilization in the mature CNS and PNS. We now appreciate that glia are well equipped to actively communicate with pre and postsynaptic neurons at the tripartite synapse. This dynamic two-way dialogue can directly influence synaptic transmission and several well studied mechanisms of synaptic plasticity, including LTP and synaptic scaling. Over the past decade, an exciting new field has emerged, and neuroscientists are gradually beginning to appreciate a more "gliocentric" view of the synapse (Fig. 4).

The recent work reviewed here lays a strong foundation for understanding the many roles that glia play in synapse development and function, and paves the way for many future studies. Several important and open questions remain including: Do astrocytes (and other glia) control the location and timing of critical period plasticity? What are NG2+ progenitor cells doing in the adult brain and what is the function of their direct synaptic contacts with neurons? Why are astrocytes organized into non-overlapping anatomical domains? Could glia actively influence neural circuits to control information processing? Do glia release gliotransmitters and directly control synaptic function *in vivo*? As glia are crucial for the normal structure and function of synapses, it has been difficult to dissect out their normal role in synapse function and behavior *in vivo.* Recent advances in the power of live imaging has allowed us to observe glia in action, but there is still a need for better molecular tools to study and manipulate specific types of glial cells at the right place and time *in vivo.* Regardless, the exciting new findings highlighted here lay a strong foundation in new area of neuroscience, in which there is much to explore and re-explore.

Fig. 4. The "Gliocentric" View of the Synapse.
In the last decade overwhelming evidence indicate that glial cells, which were once ignored and viewed as "support cells", are in fact central players in the formation, function and plasticity of the synapse.

Acknowledgements

We thank Alissa Winzeler for helpful discussions and for criticalreview of this manuscript. CE is funded by Human Frontiers Scientific Program Long term Fellowship, and BS is funded by a Larry H Hillblom Fellowship.

References

1. Araque A, Martin ED, Perea G, Arellano JI, and Buno W. Synaptically released acetylcholine evokes Ca2+ elevations in astrocytes in hippocampal slices. *J Neurosci* 22: 2443–2450, 2002.

2. Araque A, Parpura V, Sanzgiri RP, and Haydon PG. Tripartite synapses: glia, the unacknowledged partner. *Trends Neurosci* 22: 208–215, 1999.
3. Araque A, Sanzgiri RP, Parpura V, and Haydon PG. Calcium elevation in astrocytes causes an NMDA receptor-dependent increase in the frequency of miniature synaptic currents in cultured hippocampal neurons. *J Neurosci* 18: 6822–6829, 1998.
4. Arcuino G, Lin JH, Takano T, Liu C, Jiang L, Gao Q, Kang J, and Nedergaard M. Intercellular calcium signaling mediated by point-source burst release of ATP. *Proc Natl Acad Sci U S A* 99: 9840–9845, 2002.
5. Awasaki T and Ito K. Engulfing action of glial cells is required for programmed axon pruning during Drosophila metamorphosis. *Curr Biol* 14: 668–677, 2004.
6. Awasaki T, Tatsumi R, Takahashi K, Arai K, Nakanishi Y, Ueda R, and Ito K. Essential role of the apoptotic cell engulfment genes draper and ced-6 in programmed axon pruning during Drosophila metamorphosis. *Neuron* 50: 855–867, 2006.
7. Beattie EC, Stellwagen D, Morishita W, Bresnahan JC, Ha BK, Von Zastrow M, Beattie MS, and Malenka RC. Control of synaptic strength by glial TNFalpha. *Science* 295: 2282–2285, 2002.
8. Bergles DE, Roberts JD, Somogyi P, and Jahr CE. Glutamatergic synapses on oligodendrocyte precursor cells in the hippocampus. *Nature* 405: 187–191, 2000.
9. Bezzi P, Domercq M, Brambilla L, Galli R, Schols D, De Clercq E, Vescovi A, Bagetta G, Kollias G, Meldolesi J, and Volterra A. CXCR4-activated astrocyte glutamate release via TNFalpha: amplification by microglia triggers neurotoxicity. *Nat Neurosci* 4: 702–710, 2001.
10. Bishop DL, Misgeld T, Walsh MK, Gan WB, and Lichtman JW. Axon branch removal at developing synapses by axosome shedding. *Neuron* 44: 651–661, 2004.
11. Boulanger LM and Shatz CJ. Immune signalling in neural development, synaptic plasticity and disease. *Nat Rev Neurosci* 5: 521–531, 2004.
12. Brockhaus J and Deitmer JW. Long-lasting modulation of synaptic input to Purkinje neurons by Bergmann glia stimulation in rat brain slices. *J Physiol* 545: 581–593, 2002.
13. Buckby LE, Jensen TP, Smith PJ, and Empson RM. Network stability through homeostatic scaling of excitatory and inhibitory synapses following inactivity in CA3 of rat organotypic hippocampal slice cultures. *Mol Cell Neurosci* 31: 805–816, 2006.
14. Bushong EA, Martone ME, Jones YZ, and Ellisman MH. Protoplasmic astrocytes in CA1 stratum radiatum occupy separate anatomical domains. *J Neurosci* 22: 183–192, 2002.
15. Caceres M, Suwyn C, Maddox M, Thomas JW, and Preuss TM. Increased Cortical Expression of Two Synaptogenic Thrombospondins in Human Brain Evolution. *Cereb Cortex*, 2006.
16. Cao G and Ko CP. Schwann cell-derived factors modulate synaptic activities at developing neuromuscular synapses. *J Neurosci* 27: 6712–6722, 2007.
17. Castonguay A and Robitaille R. Differential regulation of transmitter release by presynaptic and glial Ca2+ internal stores at the neuromuscular synapse. *J Neurosci* 21: 1911–1922, 2001.
18. Charles AC, Merrill JE, Dirksen ER, and Sanderson MJ. Intercellular signaling in glial cells: calcium waves and oscillations in response to mechanical stimulation and glutamate. *Neuron* 6: 983–992, 1991.
19. Chittajallu R, Aguirre A, and Gallo V. NG2-positive cells in the mouse white and grey matter display distinct physiological properties. *J Physiol* 561: 109–122, 2004.
20. Christopherson KS, Ullian EM, Stokes CC, Mullowney CE, Hell JW, Agah A, Lawler J, Mosher DF, Bornstein P, and Barres BA. Thrombospondins are astrocyte-secreted proteins that promote CNS synaptogenesis. *Cell* 120: 421–433, 2005.

21. Coco S, Calegari F, Pravettoni E, Pozzi D, Taverna E, Rosa P, Matteoli M, and Verderio C. Storage and release of ATP from astrocytes in culture. *J Biol Chem* 278: 1354–1362, 2003.
22. Cummings JA, Mulkey RM, Nicoll RA, and Malenka RC. Ca2+ signaling requirements for long-term depression in the hippocampus. *Neuron* 16: 825–833, 1996.
23. Davalos D, Grutzendler J, Yang G, Kim JV, Zuo Y, Jung S, Littman DR, Dustin ML, and Gan WB. ATP mediates rapid microglial response to local brain injury in vivo. *Nat Neurosci* 8: 752–758, 2005.
24. Elmariah SB, Oh EJ, Hughes EG, and Balice-Gordon RJ. Astrocytes regulate inhibitory synapse formation via Trk-mediated modulation of postsynaptic GABAA receptors. *J Neurosci* 25: 3638–3650, 2005.
25. Eroglu C, Winzeler, A., Annis, D., Lawler,J., Mosher,D.F., Bornstein, P., Barres, B.A. How does thrombospondin induce CNS synaptogenesis? Program No 6012 2005 Abstract Viewer/Itinerary Planner Washington, DC: Society for Neuroscience, 2005.
26. Feng Z, Koirala S, and Ko CP. Synapse-glia interactions at the vertebrate neuromuscular junction. *Neuroscientist* 11: 503–513, 2005.
27 Fiacco TA and McCarthy KD. Intracellular astrocyte calcium waves in situ increase the frequency of spontaneous AMPA recaptor currents in CA1 pyramidal neurons. *J Neurosci* 24: 722–732, 2004.
28. Fiacco TA and McCarthy KD. Astrocyte calcium elevations: properties, propagation, and effects on brain signaling. *Glia* 54: 676–690, 2006.
29. Fiacco TA, Agulhon C, Taves SR, Petravicz J, Casper KB, Dong X, Chen J, and McCarthy KD. Selective stimulation of astrocyte calcium in situ does not affect neuronal excitatory synaptic activity. *Neuron* 54: 611–626, 2007.
30. Fields RD. Myelination: an overlooked mechanism of synaptic plasticity? *Neuroscientist* 11: 528–531, 2005.
31. Finkbeiner S. Calcium waves in astrocytes-filling in the gaps. *Neuron* 8: 1101–1108, 1992.
32. Freeman MR. Glial control of synaptogenesis. *Cell* 120: 292–293, 2005.
33. Freeman MR. Sculpting the nervous system: glial control of neuronal development. *Curr Opin Neurobiol* 16: 119–125, 2006.
34. Gan WB and Lichtman JW. Synaptic segregation at the developing neuromuscular junction. *Science* 282: 1508–1511, 1998.
35. Gordon GR, Baimoukhametova DV, Hewitt SA, Rajapaksha WR, Fisher TE, and Bains JS. Norepinephrine triggers release of glial ATP to increase postsynaptic efficacy. *Nat Neurosci* 8: 1078–1086, 2005.
36. Goritz C, Mauch DH, and Pfrieger FW. Multiple mechanisms mediate cholesterol-induced synaptogenesis in a CNS neuron. *Mol Cell Neurosci* 29: 190–201, 2005.
37. Grosche J, Matyash V, Moller T, Verkhratsky A, Reichenbach A, and Kettenmann H. Microdomains for neuron-glia interaction: parallel fiber signaling to Bergmann glial cells. *Nat Neurosci* 2: 139–143, 1999.
38. Guthrie PB, Knappenberger J, Segal M, Bennett MV, Charles AC, and Kater SB. ATP released from astrocytes mediates glial calcium waves. *J Neurosci* 19: 520–528, 1999.
39. Haber M, Zhou L, and Murai KK. Cooperative astrocyte and dendritic spine dynamics at hippocampal excitatory synapses. *J Neurosci* 26: 8881–8891, 2006.
40. Halassa MM, Fellin T, Takano H, Dong JH, and Haydon PG. Synaptic islands defined by the territory of a single astrocyte. *J Neurosci* 27: 6473–6477, 2007.
41. Hama H, Hara C, Yamaguchi K, and Miyawaki A. PKC signaling mediates global enhancement of excitatory synaptogenesis in neurons triggered by local contact with astrocytes. *Neuron* 41: 405–415, 2004.

42. Hatton GI. Function-related plasticity in hypothalamus. *Annu Rev Neurosci* 20: 375–397, 1997.
43. Haydon PG. GLIA: listening and talking to the synapse. *Nat Rev Neurosci* 2: 185–193, 2001.
44. Hensch TK. Controlling the critical period. *Neurosci Res* 47: 17–22, 2003.
45. Hering H, Lin CC, and Sheng M. Lipid rafts in the maintenance of synapses, dendritic spines, and surface AMPA receptor stability. *J Neurosci* 23: 3262–3271, 2003.
46. Hirase H, Qian L, Bartho P, and Buzsaki G. Calcium dynamics of cortical astrocytic networks in vivo. *PLoS Biol* 2: E96, 2004.
47. Huh GS, Boulanger LM, Du H, Riquelme PA, Brotz TM, and Shatz CJ. Functional requirement for class I MHC in CNS development and plasticity. *Science* 290: 2155–2159, 2000.
48. Jahromi BS, Robitaille R, and Charlton MP. Transmitter release increases intracellular calcium in perisynaptic Schwann cells in situ. *Neuron* 8: 1069–1077, 1992.
49. Kang J, Jiang L, Goldman SA, and Nedergaard M. Astrocyte-mediated potentiation of inhibitory synaptic transmission. *Nat Neurosci* 1: 683–692, 1998.
50. Kulik A, Haentzsch A, Luckermann M, Reichelt W, and Ballanyi K. Neuron-glia signaling via alpha(1) adrenoceptor-mediated Ca(2+) release in Bergmann glial cells in situ. *J Neurosci* 19: 8401–8408, 1999.
51. Lin SC and Bergles DE. Physiological characteristics of NG2-expressing glial cells. *J Neurocytol* 31: 537–549, 2002.
52. Lin SC and Bergles DE. Synaptic signaling between GABAergic interneurons and oligodendrocyte precursor cells in the hippocampus. *Nat Neurosci* 7: 24–32, 2004.
53. Lin SC, Huck JH, Roberts JD, Macklin WB, Somogyi P, and Bergles DE. Climbing fiber innervation of NG2-expressing glia in the mammalian cerebellum. *Neuron* 46: 773–785, 2005.
54. Lin W, Sanchez HB, Deerinck T, Morris JK, Ellisman M, and Lee KF. Aberrant development of motor axons and neuromuscular synapses in erbB2-deficient mice. *Proc Natl Acad Sci U S A* 97: 1299–1304, 2000.
55. Lippman J and Dunaevsky A. Dendritic spine morphogenesis and plasticity. *J Neurobiol* 64: 47–57, 2005.
56. Lund RD. Synaptic patterns in the superficial layers of the superior colliculus of the monkey, Macaca mulatta. *Exp Brain Res* 15: 194–211, 1972.
57. Lynch GS, Dunwiddie T, and Gribkoff V. Heterosynaptic depression: a postsynaptic correlate of long-term potentiation. *Nature* 266: 737–739, 1977.
58. Manzoni OJ, Manabe T, and Nicoll RA. Release of adenosine by activation of NMDA receptors in the hippocampus. *Science* 265: 2098–2101, 1994.
59. Matyash V, Filippov V, Mohrhagen K, and Kettenmann H. Nitric oxide signals parallel fiber activity to Bergmann glial cells in the mouse cerebellar slice. *Mol Cell Neurosci* 18: 664–670, 2001.
60. Mauch DH, Nagler K, Schumacher S, Goritz C, Muller EC, Otto A, and Pfrieger FW. CNS synaptogenesis promoted by glia-derived cholesterol. *Science* 294: 1354–1357, 2001.
61. McGee AW, Yang Y, Fischer QS, Daw NW, and Strittmatter SM. Experience-driven plasticity of visual cortex limited by myelin and Nogo receptor. *Science* 309: 2222–2226, 2005.
62. McMahan UJ. The agrin hypothesis. *Cold Spring Harb Symp Quant Biol* 55: 407–418, 1990.
63. Metea MR and Newman EA. Glial cells dilate and constrict blood vessels: a mechanism of neurovascular coupling. *J Neurosci* 26: 2862–2870, 2006.

64. Meyer-Franke A, Kaplan MR, Pfrieger FW, and Barres BA. Characterization of the signaling interactions that promote the survival and growth of developing retinal ganglion cells in culture. *Neuron* 15: 805–819, 1995.
65. Miller RF. D-Serine as a glial modulator of nerve cells. *Glia* 47: 275–283, 2004.
66. Mothet JP, Pollegioni L, Ouanounou G, Martineau M, Fossier P, and Baux G. Glutamate receptor activation triggers a calcium-dependent and SNARE protein-dependent release of the gliotransmitter D-serine. *Proc Natl Acad Sci U S A* 102: 5606–5611, 2005.
67. Muller CM. Dark-rearing retards the maturation of astrocytes in restricted layers of cat visual cortex. *Glia* 3: 487–494, 1990.
68. Muller CM and Best J. Ocular dominance plasticity in adult cat visual cortex after transplantation of cultured astrocytes. *Nature* 342: 427–430, 1989.
69. Mulligan SJ and MacVicar BA. Calcium transients in astrocyte endfeet cause cerebrovascular constrictions. *Nature* 431: 195–199, 2004.
70. Murai KK, Nguyen LN, Irie F, Yamaguchi Y, and Pasquale EB. Control of hippocampal dendritic spine morphology through ephrin-A3/EphA4 signaling. *Nat Neurosci* 6: 153–160, 2003.
71. Newman EA. Calcium increases in retinal glial cells evoked by light-induced neuronal activity. *J Neurosci* 25: 5502–5510, 2005.
72. Newman EA. Glial cell inhibition of neurons by release of ATP. *J Neurosci* 23: 1659–1666, 2003.
73. Newman EA. Glial modulation of synaptic transmission in the retina. *Glia* 47: 268–274, 2004.
74. Newman EA and Zahs KR. Modulation of neuronal activity by glial cells in the retina. *J Neurosci* 18: 4022–4028, 1998.
75. Nimmerjahn A, Kirchhoff F, and Helmchen F. Resting microglial cells are highly dynamic surveillants of brain parenchyma in vivo. *Science* 308: 1314–1318, 2005.
76. Nimmerjahn A, Kirchhoff F, Kerr JN, and Helmchen F. Sulforhodamine 101 as a specific marker of astroglia in the neocortex in vivo. *Nat Methods* 1: 31–37, 2004.
77. Nishida H and Okabe S. Direct astrocytic contacts regulate local maturation of dendritic spines. *J Neurosci* 27: 331–340, 2007.
78. Panatier A, Theodosis DT, Mothet JP, Touquet B, Pollegioni L, Poulain DA, and Oliet SH. Glia-derived D-serine controls NMDA receptor activity and synaptic memory. *Cell* 125: 775–784, 2006.
79. Parpura V, Basarsky TA, Liu F, Jeftinija K, Jeftinija S, and Haydon PG. Glutamate-mediated astrocyte-neuron signalling. *Nature* 369: 744–747, 1994.
80. Parpura V, Scemes E, and Spray DC. Mechanisms of glutamate release from astrocytes: gap junction "hemichannels", purinergic receptors and exocytotic release. *Neurochem Int* 45: 259–264, 2004.
81. Pascual O, Casper KB, Kubera C, Zhang J, Revilla-Sanchez R, Sul JY, Takano H, Moss SJ, McCarthy K, and Haydon PG. Astrocytic purinergic signaling coordinates synaptic networks. *Science* 310: 113–116, 2005.
82. Pasti L, Volterra A, Pozzan T, and Carmignoto G. Intracellular calcium oscillations in astrocytes: a highly plastic, bidirectional form of communication between neurons and astrocytes in situ. *J Neurosci* 17: 7817–7830, 1997.
83. Paukert M and Bergles DE. Synaptic communication between neurons and NG2+ cells. *Curr Opin Neurobiol* 16: 515–521, 2006.
84. Peng HB, Yang JF, Dai Z, Lee CW, Hung HW, Feng ZH, and Ko CP. Differential effects of neurotrophins and schwann cell-derived signals on neuronal survival/growth and synaptogenesis. *J Neurosci* 23: 5050–5060, 2003.
85. Perea G and Araque A. Glial calcium signaling and neuron-glia communication. *Cell Calcium* 38: 375–382, 2005.

86. Perez-Otano I and Ehlers MD. Homeostatic plasticity and NMDA receptor trafficking. *Trends Neurosci* 28: 229–238, 2005.
87. Peters O, Schipke CG, Hashimoto Y, and Kettenmann H. Different mechanisms promote astrocyte Ca2+ waves and spreading depression in the mouse neocortex. *J Neurosci* 23: 9888–9896, 2003.
88. Pfrieger FW and Barres BA. Synaptic efficacy enhanced by glial cells in vitro. *Science* 277: 1684–1687, 1997.
89. Piet R, Poulain DA, and Oliet SH. Modulation of synaptic transmission by astrocytes in the rat supraoptic nucleus. *J Physiol Paris* 96: 231–236, 2002.
90. Pizzorusso T, Medini P, Berardi N, Chierzi S, Fawcett JW, and Maffei L. Reactivation of ocular dominance plasticity in the adult visual cortex. *Science* 298: 1248–1251, 2002.
91. Porter JT and McCarthy KD. Hippocampal astrocytes in situ respond to glutamate released from synaptic terminals. *J Neurosci* 16: 5073–5081, 1996.
92. Reddy LV, Koirala S, Sugiura Y, Herrera AA, and Ko CP. Glial cells maintain synaptic structure and function and promote development of the neuromuscular junction in vivo. *Neuron* 40: 563–580, 2003.
93. Reist NE and Smith SJ. Neurally evoked calcium transients in terminal Schwann cells at the neuromuscular junction. *Proc Natl Acad Sci U S A* 89: 7625–7629, 1992.
94. Riley DA. Ultrastructural evidence for axon retraction during the spontaneous elimination of polyneuronal innervation of the rat soleus muscle. *J Neurocytol* 10: 425–440, 1981.
95. Rochon D, Rousse I, and Robitaille R. Synapse-glia interactions at the mammalian neuromuscular junction. *J Neurosci* 21: 3819–3829, 2001.
96. Rousse I and Robitaille R. Calcium signaling in Schwann cells at synaptic and extrasynaptic sites: active glial modulation of neuronal activity. *Glia* 54: 691–699, 2006.
97. Sanes JR and Lichtman JW. Development of the vertebrate neuromuscular junction. *Annu Rev Neurosci* 22: 389–442, 1999.
98. Scemes E and Giaume C. Astrocyte calcium waves: what they are and what they do. *Glia* 54: 716–725, 2006.
99. Schell MJ, Brady RO, Jr., Molliver ME, and Snyder SH. D-serine as a neuromodulator: regional and developmental localizations in rat brain glia resemble NMDA receptors. *J Neurosci* 17: 1604–1615, 1997.
100. Schell MJ, Molliver ME, and Snyder SH. D-serine, an endogenous synaptic modulator: localization to astrocytes and glutamate-stimulated release. *Proc Natl Acad Sci U S A* 92: 3948–3952, 1995.
101. Schipke CG, Boucsein C, Ohlemeyer C, Kirchhoff F, and Kettenmann H. Astrocyte Ca2+ waves trigger responses in microglial cells in brain slices. *Faseb J* 16: 255–257, 2002.
102. Selig DK, Hjelmstad GO, Herron C, Nicoll RA, and Malenka RC. Independent mechanisms for long-term depression of AMPA and NMDA responses. *Neuron* 15: 417–426, 1995.
103. Serrano A, Haddjeri N, Lacaille JC, and Robitaille R. GABAergic network activation of glial cells underlies hippocampal heterosynaptic depression. *J Neurosci* 26: 5370–5382, 2006.
104. Song H, Stevens CF, and Gage FH. Astroglia induce neurogenesis from adult neural stem cells. *Nature* 417: 39–44, 2002.
105. Song HJ, Stevens CF, and Gage FH. Neural stem cells from adult hippocampus develop essential properties of functional CNS neurons. *Nat Neurosci* 5: 438–445, 2002.
106. Stellwagen D and Malenka RC. Synaptic scaling mediated by glial TNF-alpha. *Nature* 440: 1054–1059, 2006.

107. Stevens B, Allen NJ, Vazquez LE, Howell GR, Christopherson KS, Nouri N, Micheva KD, Mehalow AK, Huberman AD, Stafford B, et al. The classical complement cascade mediates CNS synapse elimination. *Cell* 131: 1164–1178, 2007.
108. Sul JY, Orosz G, Givens RS, and Haydon PG. Astrocytic Connectivity in the Hippocampus. *Neuron Glia Biol* 1: 3–11, 2004.
109. Theodosis DT and Poulain DA. Maternity leads to morphological synaptic plasticity in the oxytocin system. *Prog Brain Res* 133: 49–58, 2001.
110. Theodosis DT, Schachner M, and Neumann ID. Oxytocin neuron activation in NCAM-deficient mice: anatomical and functional consequences. *Eur J Neurosci* 20: 3270–3280, 2004.
111. Thiele C, Hannah MJ, Fahrenholz F, and Huttner WB. Cholesterol binds to synaptophysin and is required for biogenesis of synaptic vesicles. *Nat Cell Biol* 2: 42–49, 2000.
112. Thomas S and Robitaille R. Differential frequency-dependent regulation of transmitter release by endogenous nitric oxide at the amphibian neuromuscular synapse. *J Neurosci* 21: 1087–1095, 2001.
113. Trachtenberg JT and Thompson WJ. Nerve terminal withdrawal from rat neuromuscular junctions induced by neuregulin and Schwann cells. *J Neurosci* 17: 6243–6255, 1997.
114. Ullian EM, Christopherson KS, and Barres BA. Role for glia in synaptogenesis. *Glia* 47: 209–216, 2004.
115. Ullian EM, Harris BT, Wu A, Chan JR, and Barres BA. Schwann cells and astrocytes induce synapse formation by spinal motor neurons in culture. *Mol Cell Neurosci* 25: 241–251, 2004.
116. Ullian EM, Sapperstein SK, Christopherson KS, and Barres BA. Control of synapse number by glia. *Science* 291: 657–661, 2001.
117. Ventura R and Harris KM. Three-dimensional relationships between hippocampal synapses and astrocytes. *J Neurosci* 19: 6897–6906, 1999.
118. Volterra A and Meldolesi J. Astrocytes, from brain glue to communication elements: the revolution continues. *Nat Rev Neurosci* 6: 626–640, 2005.
119. Wang X, Lou N, Xu Q, Tian GF, Peng WG, Han X, Kang J, Takano T, and Nedergaard M. Astrocytic Ca2+ signaling evoked by sensory stimulation in vivo. *Nat Neurosci* 9: 816–823, 2006.
120. Wang Z, Haydon PG, and Yeung ES. Direct observation of calcium-independent intercellular ATP signaling in astrocytes. *Anal Chem* 72: 2001–2007, 2000.
121. Watts RJ, Schuldiner O, Perrino J, Larsen C, and Luo L. Glia engulf degenerating axons during developmental axon pruning. *Curr Biol* 14: 678–684, 2004.
122. Yang X, Arber S, William C, Li L, Tanabe Y, Jessell TM, Birchmeier C, and Burden SJ. Patterning of muscle acetylcholine receptor gene expression in the absence of motor innervation. *Neuron* 30: 399–410, 2001.
123. Yang Y, Ge W, Chen Y, Zhang Z, Shen W, Wu C, Poo M, and Duan S. Contribution of astrocytes to hippocampal long-term potentiation through release of D-serine. *Proc Natl Acad Sci U S A* 100: 15194–15199, 2003.
124. Zonta M, Angulo MC, Gobbo S, Rosengarten B, Hossmann KA, Pozzan T, and Carmignoto G. Neuron-to-astrocyte signaling is central to the dynamic control of brain microcirculation. *Nat Neurosci* 6: 43–50, 2003.
125. Zonta M, Sebelin A, Gobbo S, Fellin T, Pozzan T, and Carmignoto G. Glutamate-mediated cytosolic calcium oscillations regulate a pulsatile prostaglandin release from cultured rat astrocytes. *J Physiol* 553: 407–414, 2003.

Plasticity of Dentate Granule Cell Mossy Fiber Synapses: A Putative Mechanism of Limbic Epileptogenesis

James O. McNamara[1]*, Yang Z. Huang[2], and Enhui Pan[3]

[1] Duke University, Department of Neurobiology, Durham, NC 27710, USA, jmc@neuro.duke.edu
[2] Duke University, Department of Neurobiology, Durham, NC 27710, USA, huang@neuro.duke.edu
[3] Duke University, Department of Neurobiology, Durham, NC 27710, USA, enhui@neuro.duke.edu

1 Introduction

The epilepsies constitute the third most common serious neurological disorder. Among the more than 40 different types of epilepsy, limbic epilepsy is the single most common and devastating form. Antiseizure drugs provide symptomatic relief, in that they suppress seizures, yet limbic seizures persist in 30–40% of patients and unwanted effects of drugs are common. No disease modifying therapies are available. Insight into the underlying mechanisms in cellular and molecular terms may provide novel molecular targets for developing preventive or curative therapies. The availability of animal models of limbic epilepsy has provided an opportunity to elucidate the mechanisms. Plasticity of excitatory synapses of principal neurons has been advanced as one cellular mechanism underlying the hyperexcitability of the epileptic brain (14).

Among the diverse populations of neurons within the limbic system, the dentate granule cells of the hippocampal formation have emerged as a rate limiting step in limbic seizure propagation. Moreover, this "gatekeeper" function of the granule cells is compromised in some models of limbic epilepsy. This chapter reviews evidence implicating the granule cells as gatekeepers in a normal brain, as well as, evidence of compromise of the gatekeeper function in an epilepsy model. We further discuss the functional connectivity of the granule cells with their principal cell targets in CA3, the seizure prone CA3 pyramidal cells. We propose that plasticity of dentate granule cell mossy fiber synapses is a putative mechanism of limbic epileptogenesis.

J.W. Hell, M.D. Ehlers (eds.), *Structural and Functional Organization of the Synapse*,
DOI: 10.1007/978-0-387-77232-5_24,

2 The Clinical Problem

The epilepsies, disorders of recurrent seizures, affect approximately 1% of the population worldwide. Available therapy is symptomatic, in that drugs inhibit seizures, yet no effective preventive measures or cure has been identified, except for surgery in a small number of cases. The term epileptogenesis refers to the process by which a normal brain becomes epileptic. Understanding the mechanisms of epileptogenesis in cellular and molecular terms may help identify molecular targets for development of drugs to prevent or reduce the severity of epilepsy in individuals at high risk.

Approximately 60% of all epilepsies are partial in nature, meaning that they are thought to arise from some localized area of hyperexcitable cerebral cortex. The most devastating form of partial epilepsy is termed limbic epilepsy, a form of epilepsy that arises most frequently from the temporal lobe. Limbic epilepsy is the single most common form of epilepsy, accounting for approximately 40% of all cases of adult epilepsy. Limbic seizures impair consciousness, thus limiting driving and leaving the individual susceptible to bodily harm. Roughly 30% of patients suffer recurrent limbic seizures despite optimal contemporary therapy.

Interestingly, refractoriness to medication often develops over time; that is, longitudinal studies of children reveal that seizures may be responsive to medication initially, but become unresponsive years later (7). Moreover, this is a feature particularly of limbic seizures and was not observed in seizures arising from frontal, parietal, or occipital lobes (7).

3 Seizures Beget Seizures: Pathological Neuronal Activity Contributes to Progression of Epileptic Condition

While the factors underlying development of refractoriness to medication are incompletely understood, an astute British neurologist, William Gowers, suggested more than a century ago that "seizures beget seizures" (17). That is, Gowers' careful clinical observations led him to propose that the occurrence of seizures per se may increase the likelihood of subsequent seizures, regardless of the inciting factor that led to the initial seizure. Approximately 80 years later, Graham Goddard established an experimental model that supported Gowers' prediction (15). In this model, termed kindling, brief, low intensity electrical stimulations are applied focally to a limbic structure (e.g. amygdala or hippocampus); the stimulus triggers a brief focal seizure that can be detected by EEG, but is not associated with overt change in behavior. Repeated elicitation of these focal seizures results in progressive intensification of evoked seizures consisting of limbic and eventually tonic-clonic seizures after perhaps a dozen stimulations. Interestingly, this robust, strikingly progressive intensification of evoked seizures can be triggered by stimulation of limbic sites, but not of frontal, parietal, or occipital cortex, a pattern coinciding with the subtype of focal epilepsies that become refractory to medication (7). If 80 or more stimulations are administered to the animal, spontaneous recurrent seizures, not simply stimulation induced seizures, arise and become progressively more intense and often fatal (39). Importantly, the repeated focal seizures are the critical causal factor; triggering these seizures with either electrical or chemical stimuli induces the progressive intensification of seizure activity. Focal electrical stimulations that fail to trigger seizures do not induce kindling. The requirement for focal seizures implies that this is a model of a pathological activity-determined plasticity.

Evidence implicating pathological neuronal activity as a critical cellular event underlying epileptogenesis was subsequently demonstrated in other models of limbic epilepsy such as those induced by "status epilepticus", a state of continuous seizure activity. Status epilepticus can be produced experimentally by chemical convulsants or by sustained electrical stimulations administered through intracerebral electrodes, the common feature being continuous limbic and motor seizures manifest by tonic and clonic contractions of limb and facial muscles that persist for hours (40). In these experimental models, an episode of status epilepticus lasting a few hours is followed by a seizure-free "latent period" of one to several weeks, after which spontaneous seizures emerge (24, 28). This mimics the human condition in which an otherwise normal individual undergoes an episode of status epilepticus and develops recurrent seizures following recovery. In the animal models of status epilepticus, pathologic neuronal activity in the form of electrographic seizure is a prerequisite for epileptogenesis because anticonvulsant treatments administered 1–2 hours after onset of status epilepticus can prevent late onset epilepsy (41). In contrast to the kindling model described above, the pathological neuronal activity in the status epilepticus models occurs continuously over several hours rather than being distributed into brief episodes lasting tens of seconds that are evoked at daily intervals.

4 Potentiation of Excitatory Synapses: A Putative Cellular Mechanism of Epileptogenesis

A rich diversity of cellular mechanisms almost certainly contribute to the hyperexcitability of medically refractory limbic epilepsy in different settings (32). For example, loss of inhibition due to the death or reduced firing of inhibitory interneurons likely contributes in some instances (48). Recent evidence implicates plasticity of intrinsic properties of principal neurons of hippocampus as a causal factor (8). Death of susceptible neurons followed by formation of aberrant excitatory networks has been demonstrated in some models (36). Each of these putative mechanisms has been identified in status epilepticus models at a stage in which extensive death of neurons has occurred.

Enhanced function of excitatory synapses on principal neurons has emerged as yet another putative mechanism contributing to increased excitability. Interestingly, enhanced function of these synapses has been identified in early stages of epileptogenesis in which no death of neurons has been detected (46). For example, development of hyperexcitability in the early stages of the kindling model is characterized by propagation of seizures into progressively wider populations of neurons within the limbic system. This same progression of behavioral seizure severity reflecting recruitment of wider populations of neurons is also evident during the first several initial seizures in status epilepticus models, well before death of neurons is evident. These populations of neurons are almost certainly connected mainly by excitatory synapses, most of which use glutamate as a transmitter. The fact that high frequency firing of neurons promotes more widespread invasion of synaptically connected populations of neurons by seizures led to the idea that potentiation of the efficacy of excitatory synapses similar to that of LTP contributed to the cellular mechanisms of epileptogenesis (14). In support of this idea, prior application of electrical stimulations that induce LTP, but not focal seizures, is sufficient to increase the rate of epileptogenesis in the kindling model (47). Moreover, a stimulation-induced focal seizure is sufficient to induce LTP (46). Enhanced function of excitatory synapses has been identified in intracellular recordings of CA1 pyramidal cells in an in vitro model of epilepsy (44). Together these findings support the idea that one mechanism of epileptogenesis involves enhanced function of excitatory synapses.

5 The Dentate Granule

5.1 Dentate Granule Cells Limit Invasion of Hippocampus by Seizures

Among the many populations of principal neurons at which the function of excitatory synapses might be studied, the dentate granule cells of hippocampus have emerged as particularly important with respect to mechanisms of limbic epilepsy. That is, the granule cells of the normal brain have been touted as "gatekeepers" that resist invasion of hippocampus by seizure activity. This gatekeeper function appears to be compromised in an epileptic brain. A brief review of the connectivity of the entorhinal-hippocampal circuitry will facilitate understanding the experimental evidence that supports these ideas.

Extensive anatomical and physiological studies of the connectivity of this circuitry lead to predictions of the pattern of propagation of seizure activity from entorhinal cortex into hippocampus (Fig. 1). The entorhinal cortex is a major site of convergence of excitatory projections from many cortical and subcortical brain regions and provides the major afferent input to hippocampus. Thus, the entorhinal cortex serves as a crucial focal point for the relay of seizure activity into hippocampus. Stellate cells of layer II entorhinal cortex project excitatory afferents via the perforant path to the dentate granule cells, accounting for >80% of excitatory synapses on the outer two thirds of the apical dendrites of the granule cells (2). These entorhinal afferents also innervate interneurons within the dentate gyrus (DG), providing disynaptic feedforward inhibition to the granule cells. At the next step of the trisynaptic circuitry, the mossy fiber axons of the dentate granule cells provide monosynaptic excitatory input to the CA3 pyramidal cells, as well as, interneurons in stratum lucidum, the net effect of the latter providing feedforward inhibition of the CA3 pyramids. In addition to these inputs, CA3 pyramidal cells also receive a portion of the perforant path projection which bypasses dentate gyrus and contacts distal apical dendrites directly (45, 52). At the third step of the trisynaptic circuitry, CA3 pyramidal cells provide monosynaptic excitatory input to CA1 pyramidal cells through the Schaffer collateral projection. In addition to this indirect route to CA1, stellate cells of layer III of entorhinal cortex can directly excite CA1 pyramidal cells through the temporoammonic projection independently of the perforant path projection. Axons traveling via the temporoammonic projection contact the outermost extent of the CA1 apical dendrites in stratum lacunosum-moleculare (51). Excitatory afferents of CA1 pyramids project to principal neurons in subiculum, which in turn project back to entorhinal cortex, thus providing pathways for reverberatory activity within the entorhinal-hippocampal circuitry.

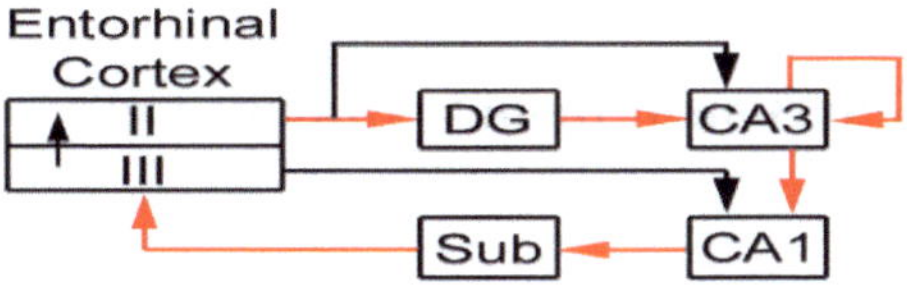

Fig. 1. Circuitry diagram illustrating connections among neuronal populations within the trisynaptic circuit.
Red arrows indicate the hypothesized pathway through which seizures induced in entorhinal cortex propagate through this circuitry.

The idea that the dentate granule cells limit invasion of hippocampal circuitry by seizures in entorhinal cortex emerged from a study using 2-deoxyglucose and EEG (11). In this study, focal seizures were evoked by microinjections of penicillin or picrotoxin locally into the entorhinal cortex and the resulting epileptiform activity was monitored by behavioral observations and EEG in both entorhinal cortex and dorsal hippocampus. Following EEG recordings, animals were infused with [^{14}C] deoxyglucose and deoxyglucose uptake, assessed by autoradiography, was used as a surrogate measure of neuronal activity. When low doses of the chemoconvulsant were injected, the investigators observed no change in behavior or very subtle limbic seizures; analyses of EEG revealed epileptiform activity in entorhinal cortex but not in hippocampus. Deoxyglucose autoradiography of these animals revealed striking increases of uptake in the dentate gyrus but not in CA3 or CA1 of hippocampus. In contrast, when higher doses of chemoconvulsants were injected, moderate convulsive behavior resulted and epileptiform activity was now evident in EEG recordings from both entorhinal cortex and hippocampus. Similarly, deoxyglucose uptake revealed prominent increases in CA3 and CA1 of hippocampus in addition to dentate gyrus. The authors proposed that the dentate granule cells serve as a gatekeeper in that they resist seizure propagation from entorhinal cortex into hippocampus, a resistance overwhelmed by intense seizure activity in entorhinal cortex induced by higher concentrations of convulsants. This gatekeeper function is likely subserved in part by resistance of the granule cells to fire repetitively in response to excitatory synaptic input, the resistance due in part to the high resting membrane potential and powerful recurrent GABA mediated inhibition (13). Analyses of granule cell firing in awake, freely moving rats performing a memory task revealed that the granule cells fire single spikes in some instances and brief bursts in others (23). A single action potential in a dentate granule cell rarely triggers an action potential in CA3 pyramidal cell (20), likely due in part to the low release probability of a granule cell's mossy fiber axon synapse on a CA3 pyramidal cell (26, 50). Limiting excitatory synaptic input to the CA3 pyramidal cells seems critical because the CA3 pyramidal cells are particularly prone to seizure activity, a consequence of their intrinsic burst firing properties together with the extensive recurrent excitatory synapses among CA3 pyramids (49).

5.2 Gatekeeper Function of Dentate Granule Cells is Compromised in Epileptic Brain

These studies of focal seizures in entorhinal cortex in an otherwise normal brain led to the idea that the gatekeeper function of the granule cells may be compromised in an epileptic brain. To test this idea, Behr and colleagues (6) attempted to examine the integrity of the "dentate gate" in entorhinal-hippocampal slices isolated from animals following induction of kindling. Local field potentials were measured in entorhinal cortex, dentate gyrus and CA3 in slices isolated from control or kindled animals. In slices from control animals, focal application of a small amount of a convulsant to entorhinal cortex induced epileptiform activity that propagated to dentate gyrus but not CA3; by contrast, in slices from kindled animals, the same amount of this convulsant induced epileptiform activity that propagated to dentate and CA3. The authors concluded that the dentate granule cells normally limit invasion of hippocampus by epileptiform activity in entorhinal cortex and that this gating function is compromised after kindling (6).

5.3 Compromise of the Dentate Gate: Potential Cellular Mechanisms

This raises the question as to the cellular mechanism by which the gatekeeper function is compromised in an epileptic brain. A diversity of candidates have been advanced and it seems plausible that diverse mechanisms may contribute in different settings. For example, limbic seizures induce acceleration of neurogenesis of the granule cells and these newborn granule cells are functionally integrated into the local circuitry of the dentate gyrus (10, 35). Whether these promote or inhibit the excitability of the entorhinal-hippocampal circuitry remains to be established. Formation of aberrant recurrent excitatory circuits among dentate granule cells likewise occurs in these models, circuitry formed by sprouting of both axons and dendrites of the granule cells (36). Again the extent to which this underlies the increased excitability in these circuits remains to be established. Yet another possibility involves reduction of inhibitory synaptic input to the granule cells, mediated by either death of inhibitory interneurons or reduced activity of these neurons (12, 42, 34). With respect to the kindling model in particular, this seems unlikely because the function of the inhibitory GABAergic input to the granule cells seems to be increased, not decreased (38, 34).

5.4 Plasticity of Mossy Fiber Synapses: A Putative Contributor to Compromise of Gatekeeper Function

Another potential cellular mechanism contributing to compromise of the dentate gate might involve the synaptic output of the granule cells, that is, enhancing the functional connectivity of the granule cells with their targets in CA3. Our interest in the functional connectivity of the granule cells and CA3 pyramidal cells grew out of studies examining molecular mechanisms of epileptogenesis in the kindling model. Work from multiple laboratories including our own has investigated the idea that neurotrophins, such as brain derived neurotrophic factor (BDNF), promote limbic epileptogenesis. Interestingly, the cellular consequences of BDNF in simple models are strikingly similar to structural and functional plasticities thought to underlie epileptogenesis, including regulation of synaptic plasticity, sprouting of axons and dendrites, and others. Moreover, focal seizures, the event necessary for epileptogenesis in multiple models, induces striking increases of BDNF expression. This led to the discovery that conditional deletion of the BDNF receptor, TrkB, from subsets of CNS neurons eliminated behavioral evidence of epileptogenesis in the kindling model (18). That is, repeated induction of focal seizures induced by initially subconvulsive electrical stimulations resulted in seizures of progressively increased intensity and duration in the wild type (WT) but not *TrkB* null mutant mice (18). This is the only genetic or pharmacological perturbation that eliminates epileptogenesis in this model. Importantly, similar intensity of seizures were induced by electroshock in wild type and *TrkB* mutant mice (18), demonstrating that the plastic response of epileptogenesis was selectively eliminated in the *TrkB* null mutant mice.

The fact that eliminating the TrkB gene prevented epileptogenesis in the kindling model raised the possibility that enhanced signaling through TrkB may promote epileptogenesis in WT mice. If so, knowing where and when this occurred might provide a valuable clue as to the nature of the cellular consequences of TrkB activation that led to hyperexcitability in this model. BDNF binds to the ectodomain of TrkB and induces its activation by a mechanism that involves dimerization, increased intrinsic kinase activity, autophosphorylation of the receptor, and initiation of downstream signaling pathways (21). Because tyrosine phosphorylation of TrkB (pTrkB) correlates with its activation, the phosphorylated state of TrkB provides a surrogate measure of TrkB activity. The availability of antibodies that recognize phosphorylated, but not non-phosphorylated Trks, provides a valuable tool for biochemical and histochemical studies of TrkB activation (43). This led to the demonstration of enhanced pTrkB content in multiple models of limbic epileptogenesis (9, 19). Interestingly, immunohistochemical studies localized the enhanced pTrkB content to the mossy fiber pathway of hippocampus in each of these models (9, 19). Preliminary (unpublished) evidence reveals that at least some of the pTrkB localizes within the mossy fiber terminals of the dentate granule cells.

The localization of increased TrkB activation to the mossy fiber pathway in multiple models provided an attractive clue to potential mechanisms of limbic epileptogenesis. That is, among principal neurons within the hippocampus, CA3 pyramidal cells are renowned for their propensity to exhibit epileptiform activity. The basis of this propensity lies in their intrinsic bursting ability together with their recurrent excitatory synaptic connections, properties which together enable the neurons in this local circuit to fire synchronously in a bursting pattern – the hallmark of epileptiform activity (49). Remarkably, activation of a single neuron within isolated segments of the CA3 region in the presence of the $GABA_A$ receptor antagonist, picrotoxin, is capable of entraining the entire population of CA3 pyramids into synchronized bursting (33). Given this propensity, controlling synaptic input from the dentate granule cells is critical to the normal function of hippocampal circuitry.

5.5 Mossy Fiber Projection to CA3: Anatomical and Physiological Studies

Given the importance of controlling the functional connectivity between dentate granule cells and CA3 pyramidal cells, it is not surprising that an elaborate local circuitry has evolved to subserve this function. A brief overview of the structure and function of the granule cell afferents to CA3 will provide a useful context for understanding how these connections function in a normal brain and how they might be modified in an epileptic brain to compromise dentate gatekeeper function.

The mossy fiber axons of the dentate granule cells are unique among CNS principal neurons in that they exhibit three distinct presynaptic terminal specializations, mossy fiber giant boutons, *en passant* terminals and filipodial extensions (3, 1). Remarkably, these distinct specializations innervate distinct targets. Intracellular labeling of granule cells in vivo followed by immunochemistry and electron microscopy enabled Acsady et al (1) to demonstrate that the giant boutons of the mossy fibers almost exclusively innervate CA3 pyramids whereas the filopodia and *en passant* terminals primarily innervate interneurons in stratum lucidum, a subdomain of CA3 between the CA3 pyramidal cells and stratum radiatum. A giant bouton of a mossy fiber has as many as 35 release sites apposed to a specialized spine (termed thorny excrescence) of the proximal dendrite of a single CA3 pyramidal cell, thereby optimally positioning this massive specialization to trigger action potentials in its target. By contrast, a filipodial extension of the giant bouton typically forms a single active zone on an interneuron in stratum lucidum. However, interneurons in stratum lucidum receive at least tenfold more synapses from mossy fibers than do CA3 pyramidal cells, a single mossy fiber typically innervating >40 interneurons (1). Because each inhibitory interneuron can innervate hundreds of CA3 pyramidal cells (1), these interneurons are ideally positioned to mediate feedforward inhibition whereby granule cells excite inhibitory neurons, which in turn inhibit CA3 pyramids (26). This feedforward inhibition exerts powerful regulatory control of the CA3 pyramidal cells, limiting both their excessive firing and the time at which action potentials can be generated to a narrow window between arrival of a monosynaptic EPSP and a disynaptic IPSP.

Each of these mossy fiber terminal specializations releases glutamate and the simplistic prediction would be that the properties of synapses at CA3 pyramids and interneurons would be similar. In retrospect, the elaborate anatomical arrangement suggests a far more complicated physiology of these synapses, a suggestion that has proven to the case. That is, the synapses formed by these distinct terminals of the mossy fiber axon with distinct targets exhibit strikingly different properties, with the postsynaptic target of the presynaptic terminal specifying the mechanisms of plasticity. With respect to short term plasticity, the mossy fiber synapse with CA3 pyramids is notable among CNS synapses for its dramatic paired pulse facilitation (~ 3 fold), as well as, its striking frequency facilitation in which the synaptic gain can exceed 20 fold (37). By contrast, mossy fiber synapses formed with interneurons exhibit variable short term plasticity ranging from depression to modest facilitation (37). Equally dramatic differences between these synapses have been identified with respect to long term plasticities. Like many cortical synapses, high frequency stimulation of the mossy fibers induces LTP of the mossy fiber-CA3 pyramid synapse (Fig. 2). By contrast, this same high frequency stimulation of the mossy fibers induces LTD of the mossy fiber-interneuron synapse (Fig. 2). Thus it seems that invasion of the same high frequency train of action potentials into a giant bouton triggers LTP of the giant bouton's synapse with a CA3 pyramidal cell but LTD of the bouton's filopodial extension with an interneuron. Notably, whereas induction of LTD at most CNS synapses requires sustained application of low frequency stimulation of the afferents, induction of LTD of the mossy fiber-interneuron synapses is produced by the same HFS stimulus of the mossy fibers that produces LTP of the mossy fiber-CA3 pyramid synapse (29).

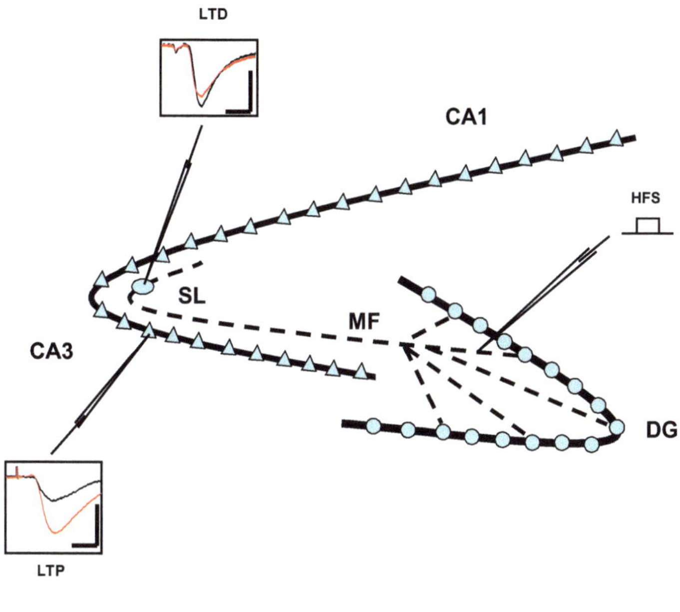

Fig. 2. Schematic diagram illustrating plasticity of hippocampal mossy fiber synapses. Stimulating electrode is placed at the junction of the dentate granule cell (*circles*) layer and hilus near the midpoint of the suprapyramidal blade of the dentate gyrus (DG). The mossy fiber (MF)-evoked excitatory post-synaptic currents recorded from CA3 pyramid (*triangle*) and stratum lucidum interneuron (*oval*), respectively. High frequency stimulation (HFS) of MF induces long-term potentiation (LTP) of MF-CA3 pyramid EPSC and long-term depression (LTD) of MF-interneuron EPSC simultaneously (*black*: representative trace before HFS; *red*: representative trace after HFS). Scale bar, 5 ms and 50 pA.

The mechanisms underlying the plasticity of the mossy fiber synapses with CA3 pyramids contrast sharply with mechanisms of plasticity of the Schaffer collateral synapse with CA1 pyramids and most other cortical synapses. Whereas induction of LTP of the Schaffer collateral synapse with a CA1 pyramid requires NMDA receptors, induction of LTP of the mossy fiber-CA3 pyramid synapse is not dependent upon NMDA receptors. Moreover, whereas expression of NMDA receptor-dependent LTP is mediated by insertion of AMPA receptors into the postsynaptic membrane of the CA1 pyramids (31), the expression of NMDA receptor-independent LTP at the mossy fiber-CA3 pyramid synapse is mediated by increased release of glutamate from the mossy fiber terminals (37).

At least two distinct mechanisms underlie LTD of the mossy fiber synapses with interneurons in stratum lucidum. Common to each mechanism of LTD is the requirement for elevation of calcium postsynaptically (27). At synapses containing calcium-permeable AMPA receptors, expression of LTD resides presynaptically, mediated by a reduction of glutamate release. At synapses containing calcium-impermeable AMPA receptors, expression is mediated by removal of AMPA receptors from the postsynaptic membrane as occurs in LTD of Schaffer collateral synapses with CA1 pyramidal cells (27).

The striking divergence of HFS-induced plasticity at distinct terminals of the mossy fiber axons implies remarkable compartmentalization of the molecular signaling events underlying these plasticities. Whereas LTP of the mossy fiber-CA3 pyramid synapse requires cAMP dependent signaling cascades, neither form of LTD of the mossy fiber synapses with interneurons involves cAMP signaling.

5.6 Plasticity of Mossy Fiber Synapses in Animal Models of Epileptogenesis

The fact that high frequency stimulation of the mossy fibers in vitro produces these plasticities of mossy fiber synapses suggests that recruitment of the dentate granule cells into high frequency firing during limbic seizures in vivo may induce similar plasiticities of mossy fiber synapses. That is, field potential recordings reveal that the dentate granule cells fire synchronously at high frequency during a seizure in vivo (25), a firing pattern predicted to result in LTP of the mossy fiber-CA3 pyramid synapse as well as LTD of the mossy fiber-interneuron synapses.

Analyses of mossy fiber-CA3 pyramid synapses in hippocampal slices isolated several weeks after chemoconvulsant (kainic acid) induced status epilepticus provide evidence that the mossy fiber-CA3 pyramid synapse had indeed undergone LTP in vivo (16). The amplitude of mossy fiber evoked fEPSPs in CA3 was greater in slices from animals following status epilepticus compared to controls, evidence that the synapse had undergone LTP in the epileptic animals. High frequency stimulation of the mossy fibers induced LTP in slices from control animals, but not from animals following status epilepticus, implying that LTP had been occluded and further supporting the idea that the synapse had already been potentiated in the status epilepticus treated animals. Likewise, paired pulse facilitation and frequency facilitation were impaired in slices following status epilepticus compared to control, further implying that the synapse had undergone potentiation in vivo during status epilepticus. Finally, biochemical studies revealed increased release of glutamate in mossy fiber synaptosomes prepared from the epileptic compared to control animals. Together with in vitro studies of the effects of high frequency stimulation of mossy fibers, it seems likely that the high frequency firing of the granule cells and their mossy fiber axons during status epilepticus induced LTP of the mossy fiber-CA3 pyramid synapse.

In contrast to the mossy fiber-CA3 pyramid synapse, studies of the mossy fiber-interneuron synapses in slices isolated from an animal model of limbic epilepsy have not been reported. Interestingly, interneurons in stratum lucidum are among the most, if not the most, sensitive CNS neurons to death induced by either status epilepticus or global ischemia (30, 5). Thus these neurons may have been absent from the hippocampal slices isolated from kainic acid treated animals in which LTP of the mossy fiber-CA3 pyramid synapse was identified (16). This subpopulation of stratum lucidum interneurons possesses a very high density of dendritic spines and expresses both the calcium binding protein, calretinin, as well as somatostatin (5). Together with other types of interneurons within stratum lucidum, this subpopulation is thought to provide feedforward inhibition to the CA3 pyramidal cells.

The induction of LTP of the mossy fiber-CA3 pyramid synapse, if combined with death of stratum lucidum interneurons and the presumed loss of feedforward inhibition, would be expected to facilitate functional connectivity of granule cells with CA3 pyramidal cells and compromise the dentate gate. Such effects on this local circuit could contribute to a lowered threshold for triggering seizure activity in CA3 pyramids with subsequent propagation through limbic circuitry and may contribute to emergence of spontaneous recurrent seizures in the kainic acid model.

5.7 TrkB and Plasticity of Mossy Fiber Synapses

The question arises as to whether signaling through TrkB might contribute to LTP of the mossy fiber-CA3 pyramid synapse during limbic epileptogenesis. As noted above, evidence of enhanced activation of TrkB has been identified in the mossy fiber pathway in multiple models of limbic epileptogenesis (9, 19). Among potential cellular consequences of TrkB activation that might promote limbic epileptogenesis, LTP of excitatory synapses on principal neurons is an attractive possibility. We therefore asked whether TrkB activation may be required for LTP of the mossy fiber-CA3 pyramid synapse. To address this question, we analyzed the mossy fiber-CA3 pyramid synapse in hippocampal slices acutely isolated from WT and *TrkB* conditional null mutant mice (22). Analyses of the mf-CA3 pyramid synapse in slices from *TrkB* conditional null mutants revealed paired pulse facilitation (PPF) of the fEPSP similar to that found in slices of WT mice. By contrast, the increase of the fEPSP induced by high frequency stimulation of the mossy fibers was reduced by approximately 60% in slices from *TrkB* conditional null mutant mice compared to WT littermate controls, thereby demonstrating a requirement for TrkB in LTP of the mossy fiber-CA3 pyramid synapse. Preventing LTP of the mossy fiber-CA3 pyramid synapse may be one mechanism by which the conditional deletion of *TrkB* prevents epileptogenesis in the kindling model (18).

6 Perspective and Questions Arising

It seems likely that the "gatekeeper" function assigned to the dentate granule cells of hippocampal formation does serve to limit seizure propagation through the normal entorhinal-hippocampal circuitry. It further seems plausible that plasticity of the granule cell synapses contributes to the impaired gatekeeper function evident in epilepsy models. We suspect that the enhanced functional connectivity involves LTP of the mossy fiber synapse with CA3 pyramids as well as a compromise of the mossy fiber-evoked feedforward inhibition of the CA3 pyramids. With respect to the latter, the compromise of the feedforward inhibition is likely mediated by death of the spiny interneurons in stratum lucidum in some models of epilepsy (e.g. status epilepticus models). In other models of limbic epilepsy where these interneurons survive, we suspect that LTD of the mossy fiber synapse with these interneurons may contribute.

That said, the perspective advanced above raises a series of important questions. Starting with plasticity of these synapses, what is the signaling cascade triggered by TrkB activation that leads to LTP of the mossy fiber-CA3 pyramidal cell synapse? What is the cellular locale at which the TrkB must be activated, a mossy fiber bouton or CA3 dendritic spine? Is BDNF the key ligand activating TrkB? How does TrkB signaling interact with identified molecules required for LTP of the mossy fiber-CA3 pyramid synapse including GluR6, EphB2 receptor, Rab3a, PKA, RIM1α, and other signaling proteins (37)? Also unclear is whether TrkB signaling is required for LTD of the mossy fiber synapses with interneurons in stratum lucidum. With respect to the gatekeeper function and its proposed defects, the in vitro studies provide a strong rationale for testing the integrity of the "dentate gate" in vivo in animal models of limbic epilepsy. Addressing this question is entirely feasible using multiple electrodes recording local field potentials and single unit activity in entorhinal cortex and various hippocampal regions in chronic models. This will permit testing not only the integrity or lack thereof of the dentate gate but also the potential role of temporoammonic projections directly to CA1 in propagation of seizure from entorhinal cortex (4). Finally, developing causal inference as to the role of the granule cells and their synapses in epileptogenesis in vivo will be facilitated by engineering mice in which gene expression can be selectively and conditionally regulated in the dentate granule cells of the mature animal. This will permit a determination of how regulating activity of granule cells as well as plasticity of their synapses impacts epileptogenesis in vivo.

Acknowledgments

This work was supported by NS056217 from the National Institute of Neurological Disease and Stroke (J.O.M.) and an Epilepsy Foundation postdoctoral fellowship (Y.Z.H).

References

1. Acsady, L., Kamondi, A., Sik, A., Freund, T., and Buzsaki, G. (1998). GABAergic cells are the major postsynaptic targets of mossy fibers in the rat hippocampus. J Neurosci *18*, 3386–3403.
2. Amaral, D.G. (1993). Emerging principles of intrinsic hippocampal organization. Curr Opin Neurobiol *3*, 225–229.
3. Amaral, D.G., and Dent, J.A. (1981). Development of the mossy fibers of the dentate gyrus: I. A light and electron microscopic study of the mossy fibers and their expansions. J Comp Neurol *195*, 51–86.
4. Ang, C.W., Carlson, G.C., and Coulter, D.A. (2006). Massive and specific dysregulation of direct cortical input to the hippocampus in temporal lobe epilepsy. J Neurosci *26*, 11850–11856.
5. Arabadzisz, D., and Freund, T.F. (1999). Changes in excitatory and inhibitory circuits of the rat hippocampus 12–14 months after complete forebrain ischemia. Neuroscience *92*, 27–45.
6. Behr, J., Lyson, K.J., and Mody, I. (1998). Enhanced propagation of epileptiform activity through the kindled dentate gyrus. J Neurophysiol *79*, 1726–1732.
7. Berg, A.T., Vickrey, B.G., Testa, F.M., Levy, S.R., Shinnar, S., DiMario, F., and Smith, S. (2006). How long does it take for epilepsy to become intractable? A prospective investigation. Ann Neurol *60*, 73–79.
8. Bernard, C., Anderson, A., Becker, A., Poolos, N.P., Beck, H., and Johnston, D. (2004). Acquired dendritic channelopathy in temporal lobe epilepsy. Science *305*, 532–535.
9. Binder, D.K., Routbort, M.J., and McNamara, J.O. (1999). Immunohistochemical evidence of seizure-induced activation of trk receptors in the mossy fiber pathway of adult rat hippocampus. J Neurosci *19*, 4616–4626.
10. Bonde, S., Ekdahl, C.T., and Lindvall, O. (2006). Long-term neuronal replacement in adult rat hippocampus after status epilepticus despite chronic inflammation. Eur J Neurosci *23*, 965–974.
11. Collins, R.C., Tearse, R.G., and Lothman, E.W. (1983). Functional anatomy of limbic seizures: focal discharges from medial entorhinal cortex in rat. Brain Res *280*, 25–40.
12. Doherty, J., and Dingledine, R. (2001). Reduced excitatory drive onto interneurons in the dentate gyrus after status epilepticus. J Neurosci *21*, 2048–2057.
13. Fricke, R.A., and Prince, D.A. (1984). Electrophysiology of dentate gyrus granule cells. J Neurophysiol *51*, 195–209.
14. Goddard, G.V., and Douglas, R.M. (1975). Does the engram of kindling model the engram of normal long term memory? Can J Neurol Sci *2*, 385–394.
15. Goddard, G.V., McIntyre, D.C., and Leech, C.K. (1969). A permanent change in brain function resulting from daily electrical stimulation. Exp Neurol *25*, 295–330.
16. Goussakov, I.V., Fink, K., Elger, C.E., and Beck, H. (2000). Metaplasticity of mossy fiber synaptic transmission involves altered release probability. J Neurosci *20*, 3434–3441.
17. Gowers WR (1881). Epilepsy and Other Chronic Convulsive Diseases. London: Churchill.
18. He, X.P., Kotloski, R., Nef, S., Luikart, B.W., Parada, L.F., and McNamara, J.O. (2004). Conditional deletion of TrkB but not BDNF prevents epileptogenesis in the kindling model. Neuron *43*, 31–42.
19. He, X.P., Minichiello, L., Klein, R., and McNamara, J.O. (2002). Immunohistochemical evidence of seizure-induced activation of trkB receptors in the mossy fiber pathway of adult mouse hippocampus. J Neurosci *22*, 7502–7508.

20. Henze, D.A., Wittner, L., and Buzsaki, G. (2002). Single granule cells reliably discharge targets in the hippocampal CA3 network in vivo. Nat Neurosci *5*, 790–795.
21. Huang, E.J., and Reichardt, L.F. (2003). Trk receptors: roles in neuronal signal transduction. Annu Rev Biochem *72*, 609–642.
22. Huang, Y.Z., Pan, E., Xiong, Z.Q., and McNamara, J.O. (2008). Zinc-mediated transactivation of TrkB potentiates the hippocampal mossy fiber-CA3 pyramid synapse. *57*, 546–558.
23. Jung, M.W., and McNaughton, B.L. (1993) Spatial selectivity of unit activity in the hippocampal granular layer. Hippocampus *3*, 165–182
24. Klitgaard, H., Matagne, A., Vanneste-Goemaere, J., and Margineanu, D.G. (2002). Pilocarpine-induced epileptogenesis in the rat: impact of initial duration of status epilepticus on electrophysiological and neuropathological alterations. Epilepsy Res *51*, 93–107.
25. Labiner, D.M., Butler, L.S., Cao, Z., Hosford, D.A., Shin, C., and McNamara, J.O. (1993). Induction of c-fos mRNA by kindled seizures: complex relationship with neuronal burst firing. J Neurosci *13*, 744–751.
26. Lawrence, J.J., and McBain, C.J. (2003). Interneuron diversity series: containing the detonation–feedforward inhibition in the CA3 hippocampus. Trends Neurosci *26*, 631–640.
27. Lei, S., and McBain, C.J. (2004). Two Loci of expression for long-term depression at hippocampal mossy fiber-interneuron synapses. J Neurosci *24*, 2112–2121.
28. Lemos, T., and Cavalheiro, E.A. (1995). Suppression of pilocarpine-induced status epilepticus and the late development of epilepsy in rats. Exp Brain Res *102*, 423–428.
29. Maccaferri G., Toth K., and McBain C.J. (1998). Target-specific expression of presynaptic mossy fiber plasticity. Science *279*(5355), 1368–70.
30. Magloczky, Z., and Freund, T.F. (1993). Selective neuronal death in the contralateral hippocampus following unilateral kainate injections into the CA3 subfield. Neuroscience *56*, 317–335.
31. Malenka, R.C., and Nicoll, R.A. (1999). Long-term potentiation–a decade of progress? Science *285*, 1870–1874.
32. McNamara, J.O., Huang, Y.Z., and Leonard, A.S. (2006). Molecular signaling mechanisms underlying epileptogenesis. Sci STKE *2006*, re12.
33. Miles, R., and Wong, R.K. (1983). Single neurones can initiate synchronized population discharge in the hippocampus. Nature *306*, 371–373.
34. Mody, I. (1999). Synaptic plasticity in kindling. Adv Neurol *79*, 631–643.
35. Mohapel, P., Ekdahl, C.T., and Lindvall, O. (2004). Status epilepticus severity influences the long-term outcome of neurogenesis in the adult dentate gyrus. Neurobiol Dis *15*, 196–205.
36. Nadler, J.V. (2003). The recurrent mossy fiber pathway of the epileptic brain. Neurochem Res *28*, 1649–1658.
37. Nicoll, R.A., and Schmitz, D. (2005). Synaptic plasticity at hippocampal mossy fibre synapses. Nat Rev Neurosci *6*, 863–876.
38. Nusser, Z., Hajos, N., Somogyi, P., and Mody, I. (1998). Increased number of synaptic GABA(A) receptors underlies potentiation at hippocampal inhibitory synapses. Nature *395*, 172–177.
39. Pinel, J.P., and Rovner, L.I. (1978). Experimental epileptogenesis: kindling-induced epilepsy in rats. Exp Neurol *58*, 190–202.
40. Pitkanen, A., Schwartzkroin, P.A., and Moshe, S.L. (2005). Models of Seizures and Epilepsy. San Diego: Academic Press.
41. Prasad, A., Williamson, J.M., and Bertram, E.H. (2002). Phenobarbital and MK-801, but not phenytoin, improve the long-term outcome of status epilepticus. Ann Neurol *51*, 175–181.

42. Prince, D.A., Deisz, R.A., Thompson, S.M., and Chagnac-Amitai, Y. (1992). Functional alterations in GABAergic inhibition during activity. Epilepsy Res Suppl *8*, 31–38.
43. Segal, R.A., Bhattacharyya, A., Rua, L.A., Alberta, J.A., Stephens, R.M., Kaplan, D.R., and Stiles, C.D. (1996). Differential utilization of Trk autophosphorylation sites. J Biol Chem *271*, 20175–20181.
44. Stelzer, A., Slater, N.T., and ten Bruggencate, G. (1987). Activation of NMDA receptors blocks GABAergic inhibition in an in vitro model of epilepsy. Nature *326*, 698–701.
45. Steward, O. (1976). Topographic organization of the projections from the entorhinal area to the hippocampal formation of the rat. J Comp Neurol *167*, 285–314.
46. Sutula, T., and Steward, O. (1986). Quantitative analysis of synaptic potentiation during kindling of the perforant path. J Neurophysiol *56*, 732–746.
47. Sutula, T., and Steward, O. (1987). Facilitation of kindling by prior induction of long-term potentiation in the perforant path. Brain Res *420*, 109–117.
48. Tower, D.B. (1976). GABA in Nervous System Function. New York: Raven.
49. Traub, R.D., Miles, R., and Wong, R.K. (1987). Models of synchronized hippocampal bursts in the presence of inhibition. I. Single population events. J Neurophysiol *58*, 739–751.
50. von Kitzing, E., Jonas, P., and Sakmann, B. (1994). Quantal analysis of excitatory post-synaptic currents at the hippocampal mossy fiber-CA3 pyramidal cell synapse. Adv Second Messenger Phosphoprotein Res *29*, 235–260.
51. Witter, M.P., Van Hoesen, G.W., and Amaral, D.G. (1989). Topographical organization of the entorhinal projection to the dentate gyrus of the monkey. J Neurosci *9*, 216–228.
52. Yeckel, M.F., and Berger, T.W. (1990). Feedforward excitation of the hippocampus by afferents from the entorhinal cortex: redefinition of the role of the trisynaptic pathway. Proc Natl Acad Sci USA *87*, 5832–583.

Stroke – A Synaptic Perspective

Robert Meller and Roger P. Simon

RS Dow Neurobiology Laboratory, Legacy Clinical Research and Technology Center, 1225 NE 2nd Ave, Portland, OR, 97232, USA. Rsimon@downeurobiology.org

1 Introduction

Evolving data implicate the synapse in the pathogenesis of ischemic brain injury and stroke. Ischemic conditions in the brain induce profound changes in synaptic function and synaptic morphology, which may account for early functional loss and deficits observed following stroke. A better understanding of the effects of ischemia on the synapse may help guide novel therapeutic approaches to reduce brain damage and aid in the recovery of function following a stroke.

2 Introduction to Stroke

2.1 The Challenge of Stroke

Stroke accounts for one in every 16 deaths in the United States (2004), making it the third most common cause of death after heart disease and cancer. Stroke is the leading cause of long-term disability, because 76% of people survive their stroke (14). The economic burden of stroke is huge, $63 billion dollars in 2007. Part of the challenge facing stroke research is the multifaceted etiology of the disorder in humans, with many cerebrovascular, cardiac and metabolite risk factors contributing to the condition.

2.2 Classification of Stroke

Stroke is the clinical syndrome resulting from impaired blood flow in the brain due to local disturbances in cerebral perfusion, produced by large vessel (thrombotic) or small vessel (lacunar) in situ vascular occlusion or occlusion by embolism from a

J.W. Hell, M.D. Ehlers (eds.), *Structural and Functional Organization of the Synapse*,
DOI: 10.1007/978-0-387-77232-5_25,

remote source (9, 103). Transient symptoms (TIA: transient ischemic attack) or permanent symptoms (stroke) result from these occlusions. The clinical manifestation of the stroke is defined by the region of brain infarcted: unilateral weakness with or without impairment of language function being the commonest impairment. A focal stroke syndrome may also result from rupture of a blood vessel with in the parenchyma of the brain usually as the result of chronic hypertension. The basal ganglia is the most common location and unilateral paralysis is the most common presentation (46).

Global brain ischemia occurs when the blood supply to the entire brain is reduced, for example following cardiac arrest, or due to raised intracranial pressure resulting from the rupture of an intracranial aneurysm (46). Global ischemia produces damage in metabolically sensitive and selectively vulnerable brain regions, such as the hippocampus, basal ganglia and cortical lamina. Damage to the hippocampus has been associated with memory deficits in patients recovering from global ischemia (46).

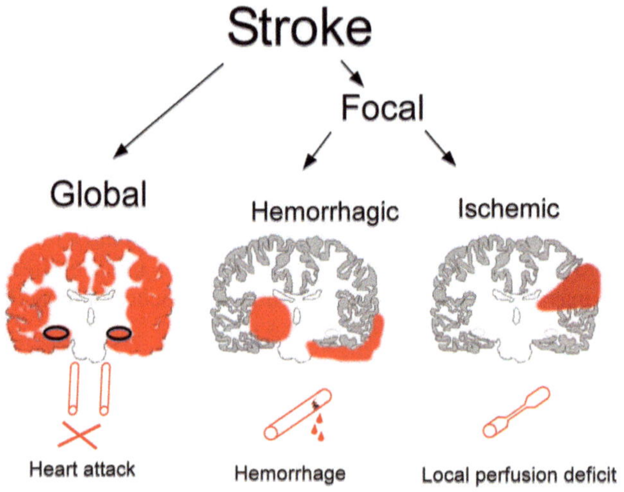

Fig. 1. Different types of stroke result in different patterns of damage in the brain. Stoke can be caused by a global reduction in global brain perfusion, such as a heart attach or during heart bypass surgery. (*Left*) Global ischemia results in damage to selective metabolically brain regions such as the hippocampus (*black circled region*) and the cortical lamina. Focal ischemia can be induced by either a hemorrhage or by local blockade of perfusion, such as results from an embolism. Focal stoke results in local damage to the brain resulting in functional deficits based on that brain region's function. (*Middle*) Following a hemorrhage in the brain parenchyma small ischemic regions occur, where as a sub-arachnoid bleed will result in the cortical lamina being subject to ischemic conditions. (*Right*) Ischemia induced by embolic stroke in a blood vessel result in wedge shaped pattern of damage to the cortical and sub cortical regions which receive their greatest blood flow from the occluded vessel.

2.3 Therapeutic Management of Stroke

Each type of ischemic event requires different clinical management: symptomatic ischemia of thrombotic, embolic, or lacunar etiology may improve clinically by administration of thrombolytic therapy (tissue plasminogen activator: tPA) begun within three hours of the onset of symptoms (1). This is a so-called "clot-busting drug", which will aid the dissolution of a clot. This treatment is counter indicated in hemorrhagic stroke. No neuroprotective drugs to preserve or protect brain parenchyma are currently available and approved by the FDA. Unfortunately with the recent failure of the SAINT II trial (*S*troke *A*cute *I*schemic *N*XY-059 *T*reatment), interest in spin trap reagents (free radical scavengers) has been reduced (101, 102). Given the increasing prevalence of stroke in the community, which may be related to the increasing population age, it seems imperative that additional therapeutic options be identified.

2.4 Experimental Models to Study Stroke and Ischemia – In vivo Models of Stroke

There are multiple experimental models used to study stroke, because of this caution is needed in interpreting sometimes conflicting results from different models. Different studies may employ different techniques of inducing ischemia, different species or strain of experimental animals, different brain regions for study and differing techniques to determine infarction area/volume and functional effects. Experimental ischemia can be permanent, or transient (temporary) whereby the cause of reduced perfusion is removed, i.e. withdrawal of an occluding filament.

The nature of damage following focal and global ischemia, as well as the development of damage (progressive vs. completed) is determined by the choice of model used to study ischemia. Global ischemia models are performed to reduce blood flow to the entire brain. Common models are the 2-vessel occlusion (whereby the bilateral carotid arteries are clamped and systemic blood pressure reduced) or 4-vessel occlusion (most commonly used in rat) whereby the vertebral arteries are permanently occluded by ligature or cauterizing and then the carotid arteries are transiently occluded by aneurysm clips (95, 113). Some studies have employed a cardiac arrest model whereby the heart is stopped for a varing duration (by KCl injection), and then resuscitated (chest compressions, intravenous epinephrine, and ventilation with 100% O_2).

Focal ischemic models attempt to recapitulate the situation of disrupted local blood flow to a specific area of the brain. Probably the most common model is the filament occlusion of the middle cerebral artery. While commonly used, due to the ease of controlling the reperfusion of the tissue, filament insertion may exacerbate injury following ischemia due to damage to the vascular endothelium (25). Other focal models include electro-coagulation and chemical-coagulation of blood vessels (112) on the surface of the brain. While similar to human stroke the duration of ischemia is more variable in these models, and distal thrombosis may induce secondary infarctions. As an alternative, endothelin-1 has been used to induce constriction in the cerebral blood vessels, which can be reversed by an antagonist, thereby reducing filament damage to the vasculature endothelium (54). (For a review of ischemia models see (14, 113)).

2.5 Experimental Models to Study Stroke and Ischemia – In vitro Models of Ischemia

Most studies of in vitro ischemia utilize primary cultures of cortical or hippocampal cells (18), although immortalized neuronal and neuronal like cells have been used (55, 56). These cultures however vary between study groups, regarding age of cultures, growth media, species (mouse vs. rat) and date of tissue preparation (embryonic vs. post-natal). Most studies focus on the effects of ischemia on neuronal cell populations, however in vitro systems can be used to study astrocyte and oligodendrocyte responses (4, 40). Of note while in general astrocytes tend to show higher resilience to ischemia than neurons, protoplasmic astrocytes may have a similar vulnerability to ischemia as neurons (40). Organotypic slices have also been used for the study of ischemia.

There are a number of methods to mimic ischemic conditions. Hypoxia/anoxia can be induced by placing cells in a solution where oxygen is removed or lowered by bubbling the solution with an anaerobic gas mixture or by placing cells into a hypobaric/anoxic chamber (80). Hypoglycemia can be modeled by the removal of glucose, or its replacement with a non-metabolized variant (2-deoxy glucose) (115). Mitochondrial inhibitors, for example potassium cyanide or 3-nitopropionic acid (115, 117), have been used to model the metabolic inhibition that occurs following ischemia,. Ischemia also induces glutamate release and excitotoxicity, hence many studies have focused on the effects of high concentrations of glutamate or the selective agonist N- methyl-D Aspartate (NMDA).

2.6 Physiology and Biochemistry of Brain Ischemia

Ischemic conditions in the brain rapidly deplete neurons, and to a lesser extent astrocytes, of the metabolites oxygen and glucose that are necessary for energy production. Energy loss results in a rundown of the electrochemical gradients of ions across membranes, notably Ca^{2+} and K^{+}, leading to depolarization of cells and neurotransmitter release (59). The rapid rise in intracellular calcium, via influx from voltage and ligand-gated calcium ion channels can activate calcium sensitive proteases such as calpain and cathepsin resulting in protein degradation.

The brain, when subject to focal ischemia, can be divided into two regions. The portion of the brain subject to the most extreme ischemic conditions exhibits rapid cell death and is termed the ischemic core. The region surrounding the ischemic core is termed the penumbra. The penumbra is also vulnerable to ischemia-induced cell damage but cell death evolves over a slower time course. The slower progression of cell death enables the penumbra to be salvaged by attenuating neurochemical cell death processes. In this regard, ischemia activates a complex series of biochemical and molecular processes in the core and penumbra, which have been covered extensively in a number of reviews (28, 34, 45, 69, 76); the following is a brief overview.

In the ischemic core, the combined loss of energy and ion gradients results in cell swelling, compromising the cell membrane and leading to the release of cell constituents and cell death. The mitochondria of cells in the ischemic core show swelling and rupture of contents. Associated with this rapid passive cell death is an accompanying inflammatory response. Released cell constituents (K^+ and glutamate) may then cause a cascade effect resulting in further damage to adjacent cells. Raised extracellular K^+ can also induce the release of glutamate from astrocytes via reversal of uptake transporters (24). Extracellular glutamate may evoke calcium mediated glutamate release from astrocytes (104). Hence once these conditions of high extracellular glutamate and potassium are obtained, they can overwhelm compensatory mechanisms leading to excitotoxic "necrotic" cell death.

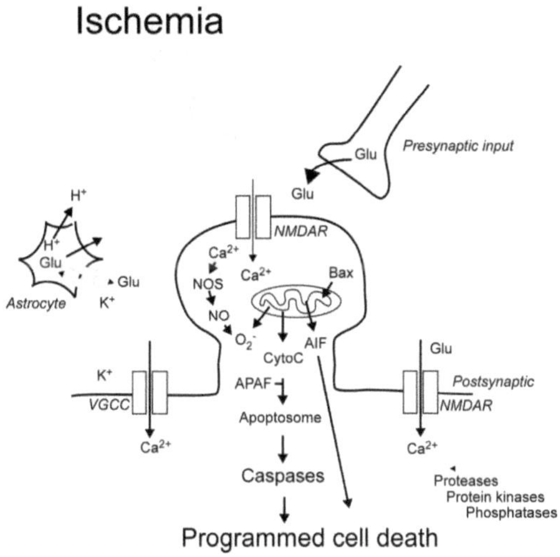

Fig. 2. Ischemia induced excitotoxicity. Glutamate excitotoxicty is implicated in the cell damage following ischemic conditions. Excessive extracellular glutamate results in prolonged NMDA receptor activation leading to calcium overload in the cells. Calcium can activate Nitric oxide synthase leading to NO production and the generation of oxygen radicals and peroxynitrite. Calcium can also lead to the activation of proteases, protein kinases and phosphatases. Once a critical threashold is reached, bcl-2 proteins are activated and translocate to the mitochondria, resulting in a loss of mitochondrial membrane potential, and release of apoptogenic factors cytochrome C and AIF. Cytochrome C can then interact with APAF-1 to initiate the caspase cascade which ultimately results in cell death.

In contrast to the ischemic core, cells in the penumbra show transient damage and can die by programmed cell death mechanisms. Following reperfusion to these cells, oxidative stress and the release of free radicals can drive lipid peroxidation and membrane damage. Stress-activated protein kinases initiate the activation of intracellular signaling cascades associated with apoptosis/programmed cell death. The activation of the pro-cell death protein Bax compromises the mitochondria membrane resulting in the release of cytochrome c and other apoptosis-inducing factors, including AIF (apoptosis inducing factor) and SMAC/Diablo (second mitochondria-derived activator of caspase/direct IAP-binding protein with low PI) which neutralizes IAPs (inhibitor of apoptosis protein) (13, 41, 63, 68). Released mitochondrial cytochrome C interacts with APAF-1 (Apoptotic Protease Activating Factor-1) and ATP (adenosine 5' triphosphate) to form the apoptosome, which catalyzes the cleavage and subsequent activation of the caspase family of proteolytic enzymes.

The initiator caspase, caspase 9 is directly activated by the apoptosome and can then cleave and activate caspase 3, which can further cleave and activate executioner caspases (caspases 6, 7 and 8). Caspase 3 also cleaves ICAD (inhibitor of CAD), resulting in the activation of CAD (caspase activated deoxynuclease), which cleaves DNA into regular sized packages (50–300-kb). These biochemical cascades and subsequent DNA damage render the cells committed to die. The complex nature of this cell death cascade reveals many potential therapeutic targets, although to date none have been used successfully in clinical trials (76).

3 Evidence Implicating the Synapse in Mediating Ischemia-Induced Brain Damage

3.1 Synaptically Released Glutamate Mediates Excitotoxicity Following Ischemia

Early work on the retina showed that excessive levels of the excitatory amino acid aspartate resulted in brain damage and neurodegeneration termed excitotoxicity (87). Excitotoxicity due to excessive brain glutamate levels was suggested as a potential mechanism of injury in stroke and other acute/chronic neurological disorders (77). Two pieces of evidence strongly support the role of the synapse in mediating the effect of ischemia. First, it was shown that cutting the pre-synaptic inputs to the CA1 region of the hippocampus blocks global ischemia induced neurodegeneration (11) and that ischemia induced cell death in vitro is attenuated by blocking synaptic release of glutamate with tetanus toxin (85) Second, glutamate receptor antagonists attenuate ischemia-induced brain damage in animal models of global ischemia and stroke (78). Indeed, this latter observation initiated the large interest in NMDA and glutamate receptor acting drugs as potential therapeutics for stroke.

3.2 NMDA Receptor Antagonists as a Therapy for Stroke: Potential and Pitfalls

Since these initial experiments, glutamate receptor blocking agents have been shown to be effective in reducing ischemia induced brain damage in a variety of species (rat, mouse, hamster, gerbil, cat, primate) and ischemia models (focal, global temporary permanent and in vitro models). However, it should be noted that some of the neuroprotective effects of MK801 may be due to its ability to reduce body temperature (61). Yet it is of note that even though it is over thirty years since the initial experiments proving that blockade of excitatory amino acid receptors blocks ischemia induced brain damage, not a single compound has to date been successful enough in clinical trials to be approved for general clinical use (although the NMDA antagonist memantine was approved for treatment of Alzheimer's disease by the FDA in 2003) (61).

The reasons behind this apparent failure of glutamate antagonists in clinical stroke trials may be due to multiple factors, including the time window of efficacy, pH modulation of the ligand gated ion channel (low pH in ischemia may reduce channel function) and acute toxicity of the compounds. Furthermore some NMDA blocking compounds show unacceptable side effects such a psychosis (ketamine) or vacuolation of the brain (MK801: dizocilpine) (51). One significant issue is the that time window for therapeutic NMDAR blockade, at least in rodents is slender, up to one hour following the ischemic event (97). Such a short therapeutic time window makes it unlikely that a patient presenting with stroke symptoms will be given an NMDA blocking compound within the effective time range. Indeed this may also be a reason for the failure of the SAINT II trial of free radical blocking agents in stroke (101). However, one potential solution is to identify therapeutic combinations that extend the therapeutic time window of NMDA receptor antagonists. It has recently been shown that combination of an ASIC1 receptor blocker PcTX (see previous chapter) with the NMDA antagonist memantine enhanced the therapeutic time window of the NMDA blocker following ischemia (92). This suggests that combinations of neuroprotective agents may provide more effective therapeutic options for treating stroke than a single compound alone. Indeed this approach has proved effective in cancer therapy, whereby multiple cytotoxic compounds are combined.

The failure in clinical trials of NMDA receptor antagonists and the SAINT trial also highlight some of the limitations of studies assessing neuroprotective potential of an agents using in vivo models of stroke. For example, in in vivo models the test agent is frequently administered at the time of ischemia induction, rather than at time points more equivalent to clinical response times (3–6 hours post ischemia). As such, numerous recommendations have been made from the review of failed clinical trials of neuroprotective agents with recommendations for future studies (27, 35, 101, 102).

3.3 Physiological and Pathological NMDA Receptor Signaling: Cell Fate in the Balance

Another compelling reason for the apparent failure of NMDA antagonists concerns the apparent Yin/Yang effects of NMDA receptor signaling. On the one hand, activation of NMDA receptors has been shown to mediate pro-survival physiological effects, such as CREB transcription factor activation and BDNF expression (50, 116). However, excessive activation of NMDA receptors promotes neurodegeneration and cell death. In a series of experiments, Biegnon showed that the prolonged blockade of NMDA receptors is detrimental if administered hours following traumatic brain injury. Indeed, administration of a low dose of NMDA agonist (NMDA) results in less brain damage than the administration of the NMDA antagonist MK801 in the TBI-injured mice (6).

The results of the Beignon study leads to two intriguing suggestions. First, that a weak partial agonist of NMDA receptors may actually be a more therapeutically effective therapy. A weak partial agonist would antagonize excessive endogenous activation of a given receptor, yet maintain a low level of tonic activation of the receptor if endogenous activation falters. This approach is currently in clinical trials; the NMDA antagonist memantine may be a weak partial agonist.

The second suggestion regards the nature of NMDA receptors and their contribution to physiological and pathological signaling. Hardingham showed that following blockade of synaptic receptors on neurons, further NMDA stimulation resulted in toxicity, which was attributed to the activation of extra-synaptic receptors (49, 50). Other groups have shown that NMDA toxicity is not blocked by loss of spines (100), e.g. following the removal of dendritic spines with latrunculin (an actin de-stabilizing agent). These data suggest that toxic NMDA signaling exists when either synaptic NMDA receptors are blocked or following removal of dendritic spines (and presumably the synaptic receptors). This implies that toxic NMDA signaling is mediated through extra-synaptic NMDA receptors.

However, ischemia induced excitotoxicity may exert its toxic signaling effects through synaptic NMDA receptors, and not extrasynaptic receptors. For example, removal of dendritic spine heads, following exposure of neurons to the actin depolymerizing agent latrunculin, blocks ischemia-induced cell death in vitro (100). This suggests that ischemia induced cell death may be reliant on synaptic signaling mechanisms and synaptic NMDA receptors that are activated by endogenous synaptic glutamate that is released during energy failure. This concept is supported by the observation that cutting pre-synaptic inputs to the CA1 neurons of the hippocampus blocks ischemia-induced neurodegeneration (11). Furthermore blocking synaptically released glutamate with tetrodotoxin blocks oxygen glucose modeled ischemia induced cell death in vitro (85). However what is not clear from the experiments of Sattler et al. (100) is what happens to synaptic NMDA receptor subunits following latrunculin treatment, and whether latrunculin induces changes to all NMDA receptors, or whether its effects are subtype specific (NR2A vs. NR2B).

In a recent study it was suggested that blockade of NR2B subunits, irrespective of synaptic location mediates neurotoxic NMDA-receptor mediated signaling, but NR2A subunits were pro-survival. This suggests that selective enhancement of NR2A, but reduction of NR2B mediated signaling may offer a new therapeutic mechanism for stroke. Unfortunatly, the polyamine site NR2B glutamate receptor subunit selective antagonist Eliprodil was withdrawn from clinical trials in 1997, reducing the enthusiasm of NR2B subunit selective NMDA receptor antagonists as a therapeutic option for stroke (see (120)). Hence the challenge for one future direction of neurprotective stroke research is to devise a mechanism whereby physiological signaling can be maintained, but toxic NMDA signaling can be selectively attenuated. Interestingly, endogenous protective mechanisms may indeed mimic this situation, and are considered later on in the chapter.

Taken together the role of the synapse in mediating ischemia induced damage to neurons is implicated by a number of studies, but ischemia can also result in profound changes to the synapse. These mechanisms may be part of a neuroprotective response to the ischemia, or a result of excitotoxic signaling.

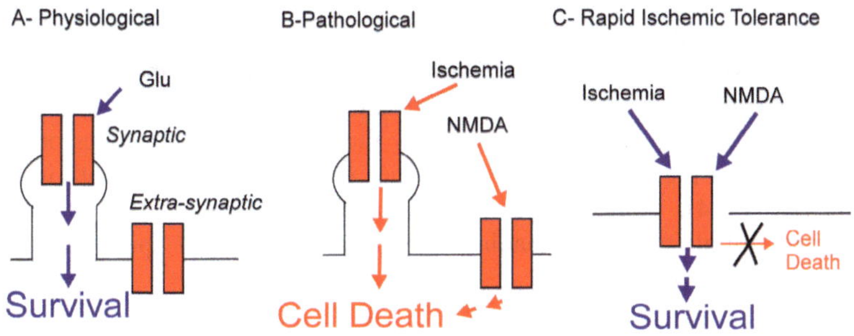

Fig. 3. Different signaling modes of NMDA receptors. Synaptic NMDA receptors appear essential for physiological roles of NMDA in neurotransmission. Synaptic NMDA receptors may mediate ischemia induced cell death, in contrast to NMDA, which activates extrasynaptic NMDA receptors. Following ischemic preconditioning, NMDA receptors show reduced toxic signaling, but maintain physiological function.

4 Effect of Ischemia on Synaptic Morphology

The physical loss of neurons following an injurious ischemic episode will result defective brain function. However, synaptic changes are evident in areas subject to non-harmful ischemic where the loss of function can be temporary or reversible. The form and function of the synapse are intertwined, for simplicity we consider the effect of ischemia on synaptic morphological and then the effect of ischemia on synaptic function.

4.1 Methods to Study Synaptic Structure Following Ischemia

Structural studies have been performed using both in vitro and in vivo stroke models to determine the morphological effects of ischemia. In vitro studies have allowed the "real time" assessment of dendritic structures through the application of fluorescence and confocal microscopy. Typically immunocytochemistry, fluorescent proteins, tagged actin or lipophilic fluorescent membrane dyes are used in these studies. In vivo studies have commonly used fixed sections to assess morphology either by confocal imaging, or electron microscopy. However, recently "real time" imaging of live neurons expressing yellow fluorescent protein (YFP) in layer V of the mouse cortex was reported (124).

4.2 Ischemia Induces Rapid Dendritic Spine Loss and Morphological Changes to Neurons

Long periods of ischemia induces neurons and other cells in the brain to die, whereas briefer ischemic periods may not be toxic. The initial response of neurons to both lethal and sub-lethal ischemia is similar regardless of the technique and model system used: ischemia induces a rapid loss of dendritic spines and an increase in varicosity formation along the dendrite. Spine loss can occur as quickly at 10 minutes following the onset of ischemia (124). The spine loss is recoverable if the ischemia is transient and non-harmful, but is persistent when toxic levels of ischemia occur (88, 124). It should be noted however that neurons in the penumbral region that recover spines may yet be lost at a later time point. The recovery of spines takes approximately 2–4 hours (88). Interestingly the recovery of the spines occurs at the same location on the dendrites that the spines were originally lost both in vitro (52) and in vivo (124). Spine loss is similar when toxic and non-toxic ischemia or excitotoxic stimuli are used, leading to the suggestion that spine loss may be a protective phenotype (see later).

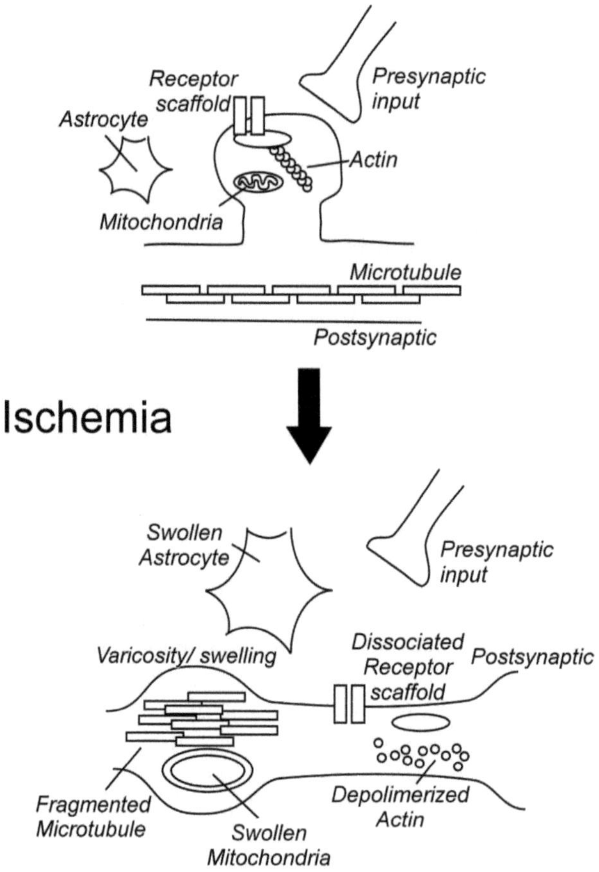

Fig. 4. Overview of effect of ischemia on synapse structure. Following ischemia various modifications to synaptic structrure have been noted, including, temporary swelling of pre-synaptic boutons and astrocytes. The post-synaptic regions show more profound changes, including loss of dendrites, disassembly of actin cytoskeleton and receptor scaffolds. Microtubules are cleaved and accumulate in varicosities along with swollen mitochondria.

The mechanisms regulating spine loss following ischemia have not been thoroughly studied, rather many studies have focused on the mechanisms of NMDA-induced morphological changes. NMDA activates protein phosphatase 2B, calcineurin, which causes actin depolymerization and may be responsible in part for spine loss. However, actin depolymerization alone is insufficient to remove spines from dendrites. Latrunculin treatment of hippocampal cells results in a loss of the spine head, but the shaft is still evident (100). In contrast, incubation of hippocampal cells with NMDA results in loss of the entire spine (44, 52, 53). This suggests additional mechanisms are responsible for the spine loss and rather than just actin remodeling. Interestingly, latrunculin treatment does prevent spine recovery following NMDA treatment (52).

NMDA may regulate spines via the activation of proteolytic enzymes. NMDA-mediated activation of the proteolytic enzyme cathepsin D degrades MARCKS: a protein, which is responsible for anchoring the actin cytoskeleton to the plasma membrane (44). Calpain however does not regulate spine loss, but calpain blockers slow the recovery of spines following NMDA treatment (32). The ubiquitin-proteasome system may also play a role in the loss of MARCKs following brief ischemia, and contribute to synaptic remodeling (Meller: unpublished results). This suggests that brief ischemia may remodel the synapse via the selective degradation of actin-associated elements.

4.3 Varicosity Formation in Neuronal Dendrites Following Ischemia

A second feature following ischemia is the appearance of beaded varicosities along the dendrites. These varicosities have been reported in response to excitotoxic stimuli and were originally considered cell death indicators resulting from excitotoxicity (86). The varicosities occur in the same time period as spine loss is observed (88). Interestingly, more varicosities are observed on proximal dendrites following ischemia, than those in close proximity to the cell soma (Meller; unpublished observations). Microtubule associated proteins (MAP2A/MAP2B etc.) have been shown to be cleaved following ischemia, and they accumulate into the varicosites (12). The formation of varicosities following ischemia is not inhibited by a calpain inhibtor, MDL 28, 170, however the recovery of varicosities following NMDA treatment was slowed in the presence of MDL28,170(32). More recent data show that varicosity formation and excitotoxic cell death occur through independent mechanisms; death is Ca^{+2} mediated while varicosities are mediated by Na^{+2} flux and AMPA internalization (64). The functional consequence of varicosity formation is not clear, but may also serve as a protective function, as they may help compartmentalize calcium to reduce calcium overload (64). Indeed, varicosity formation by NMDA induces a transient, reversible neuroprotective state by attenuating excitatory neurotransmission (64). Thus the varicosity formation as well as spine loss following ischemia may be part of a protective response.

5 Ultrastructural Changes in the Brain Following Ischemia

A number of ultrastructural studies have helped to shed light on the effect of global ischemia on selectively vulnerable neurons, especially damage to the CA1 pyramidal neurons of the hippocampus. Most of these studies involve observations made on brain sections prepared for electron microscopy.

5.1 Presynaptic Membranes are Tolerant of Ischemia

Pre-synaptic membranes are not as affected by global ischemia as the post synaptic surface. The pre-synaptic membrane and glia show temporary swelling, which is reversible within one hour (31, 57). This is consistent with the lack of permanent cell damage to astrocytes and pre-synaptic neurons in global ischemia models (67, 82). One common observation is that astrocyte end feet are swollen, and appear to reduce the extracellular space (31), which may have implications for synaptic transmission. Following ischemia, vesicle aggregation is observed in pre-synaptic terminals, and some synaptic boutons show a reduction in vesicle number (31, 118). This may correlate to ischemia/reperfusion induced vesicular release of glutamate, and other transmitters. The morphological changes to the pre-synaptic terminal and astrocytes appear temporary and return to relatively normal levels within one to three hours following the ischemic insult.

Further support for the robustness of the pre-synaptic neurons in global ischemia comes from multiple experiments, and not just ultra-structural studies. Global ischemia induces cell death in the CA1 field of the hippocampus but not the innervating cells. Following global ischemia, potassium evoked glutamate release can still be measured using microdialysis a few days after the ischemic insult (82). Histological analysis of these brains show that the post-synaptic cells were lost, but the pre-synaptic axons appear to withstand the ischemic shock. In a study by Hasbani, the pre-synaptic labeled cells appear relatively stable, following exposure to NMDA (52). In comparison the post synaptic dendritic spine shrinks, but reappears following wash out of the NMDA. This apparent robustness of pre-synaptic boutons to ischemia may also serve to guide dendritic spines back to their original contacts. However, one week following global ischemia pre-synaptic bouton numbers are reduced, which may be due to a lack of target sites in fields where neurons have degenerated (58).

5.2 Postsynaptic Membranes are Vulnerable to Ischemia

Ischemia induces more morphological defects in the post-synaptic region, compared to the pre-synaptic terminal. Following ischemia, modifications of the post-synaptic density (PSD) have been reported, including projections of electron dense material (the consequence of this is as yet unclear). The PSD becomes irregular following the ischemic insult. The PSD is thicker and discontinuous 20–60 minutes following reperfusion (31). Apparent breaks in the PSD correspond to the breaks in the post-synaptic membrane at 1 hour. A disruption or compromise of the post-synaptic membrane is consistent with the observation that post-synaptic sites take up exogenous HRP (horse radish peroxidase) following ischemia (although the HRP is associated with cell membranes and so may be due to endocytotic mechanisms rather than cell lysis) (26). The potential of post fixation to induce the membrane breaks should also be considered. One unusual observation (albeit with unknown significance) is the formation of pockets in the post-synaptic membrane, which are filled by the pre-synaptic membrane, but no PSD is present in the "buckets " (31).

One hour following ischemia, many studies report the swelling of post-synaptic membranes and changes in the morphology of internal organelles: mitochondria appear swollen with a loss of the cristae, the endoplasmic reticulum appears swollen, vacuoles form in the cytoplasm and poly-ribosomes are disintegrated into single units (31, 57, 118). Some cells also start to show signs of chromatin clumping following ischemia. Microtubules in axons and dendrites also appear to show kinking and breaks following ischemia, which may correspond to the formation of varicosities (see above).

The thickening of the post synaptic density in ischemic-vulnerable CA1 hippocampal neurons can last 4–24 hours following global ischemia, where as in the dentate gyrus (which is less vulnerable to ischemia) changes in the PSD thickness reverse 24 hours following the ischemic insult (74). This thickening of the PSD may correspond to an enhanced NMDA receptor mediated facilitation (signaling) following ischemia in CA1 neurons or be associated with ischemic damage to the post-synaptic region. Hence the persistent modification of the PSD may be a marker or a cause of the delayed cell loss in the CA1 following global ischemia.

The thickening of the PSD following ischemia may also involve the denaturation of proteins. Ubiquitinated protein aggregates have been reported to occur following ischemia, especially in synaptic preparations (70). If proteins are present but denatured (62), this may lead to a decrease in signaling function of synaptic proteins. Since multiple proteins in the post-synaptic density are targets for ubiquitination (30), this may affect post synaptic receptor signaling following ischemia.

Fewer studies have investigated long term changes to the structure of the brain following harmful ischemia. There is evidence to support reinnervation of some areas lost following an ischemic insult. One week following ischemia there is a persistent decrease in synaptic number in the hippocampus. In particular there is a reduced number of asymmetrical synapses, which correlate with excitatory synapses (58). However there is an increase in glutamate in the remaining synaptic boutons of the hippocampus, which may be a compensatory mechanism for the initial loss of cells following ischemia (58). In the short-term, presynaptic markers levels (GAP43) do not change following ischemia, but at 2 weeks changes were apparent (increased protein expression) (47, 71). This may coincide with reinnervation of damaged areas or a recovery in function. Recently, realtime in vivo two-photon imaging has demonstrated up to 5-fold higher rates of spine formation in neurons near the infarct borders (10). It is notable that the brain is capable of recovering function following the loss of discrete brain regions due to ischemia.

6 Functional Effects of Ischemia on Synaptic Transmission

Many studies have been performed using global ischemia to determine the effects of harmful ischemia on synaptic transmission (123). Such studies are performed using the global model of ischemia, where the maturation of injury in the vulnerable CA1 region of the can be compared to dentate gyrus or CA3 cells, which are resistant to the effects of ischemia. While it is assumed that regions of the brain exposed to hypoxic conditions will show greatest functional deficits following ischemia, it is of note that in a receptor study Zhang and Murphy showed that functional deficits extended on average up to 400 μm from areas with damaged dendrites indicating synaptic function may even be compromised in the structurally intact penumbra (125).

6.1 Rapid Onset of Synaptic Depression During Ischemia

When cells are subject to hypoxia there is an initial depression or failure of synaptic transmission (108). This may be due to the rapid energy deletion in the cells following the onset of ischemic conditions. The synaptic depression is reversible if the duration of ischemia is short. The CA1 cells of the hippocampus show a rapid decrease in population spike response to stimulation following ischemia, compared to dentate gyrus cells.

6.2 Anoxia-Induced Spreading Depression

If ischemia is prolonged the cells suffer a spreading depression type event, which has some similarities to normoxic spreading depression (108). Interestingly while some drugs are able to delay the time of onset of spreading depression in brain slices, no drugs have been shown to produce blockade (108). The consequence of this period of depolarization is not fully understood but may play a role in the post ischemic enhancement of synaptic transmission. The ability of the brain to recover from this depolarization is inversely proportional to the time exposed to spreading depression (108). The timing of the spreading depression may contribute to two features of post ischemic responses. First the spreading depression may result in excessive release of the excitatory neurotransmitter glutamate, which may activate NMDA receptors leading to enhanced intracellular calcium levels. Second, the spreading depression may also account for some of the post-ischemic abnormalities in synaptic transmission.

Following recovery from ischemia the neuronal responses may appear normal or enhanced, but drop off between 8 and 24 hours following ischemia, a time interval that corresponds to that of CA1 neuronal loss. It must be noted that this mechanism of cells dieing off after a period of re-perfusion in global ischemic models is distinct from the cell death following irreversible immediate hypoxic damage usually observed in the core of a focal ischemic infarct (commonly referred to apoptosis vs. necrosis, but see (106)). Following global ischemia, there is a temporary decrease in inhibitory synaptic transmission, but this is not as prolonged as the enhancement of excitatory transmission (15).

6.3 Enhanced Postsynaptic Responses Following Ischemia

Prolonged post-synaptic glutamate receptor activation may result in an apparent increase in EPSP amplitude up to 12 hours following an ischemic episode (21, 22, 37, 38, 83). Enhanced NMDA responses have been termed post ischemic long term potentiation, in that it shows some similarities to long term potentiation induced by a tetanic stimulation under normoxic conditions. The potentiation is due to the increased synaptic efficiency in post ischemic hippocampal CA1 neurons following ischemia. Post ischemic LTP can be blocked by NMDA receptor antagonists, which may account for the ability of some glutamate antagonist studies to show a reduction in cell damage when administered following global ischemia (84). Calcium chelating agents can also block post ischemic LTP, but unlike tetanic LTP AMPA antagonists are not effective (21).

A number of mechanisms may account for the enhanced response of NMDA receptors following ischemia. Ischemia decreases ATP, changes pH, increases glycine release, activates protein kinases (PKC) and results in reducing conditions, all of which have been shown to enhance NMDA receptor currents (21). Some studies suggest that the increased expression of the NMDA receptor 2C subunit (NR2C) following ischemia may account for enhanced NMDA responses (66, 107). NR2C subunits have a reduced magnesium blockade of the channel resulting in enhanced calcium influx and prolonged opening of the channel. Hypoxic LTP requires a strong depolarizing event to be induced, and one potential candidate could be anoxic spreading depression, which is observed following ischemia.

The enhanced NMDA response in CA1 regions may also lead to additional complications in the brain following stroke. Following ischemia there is an increased occurrence of epileptic seizures. NMDA receptors regulate burst firing in the CA1 post ischemic, but in some recent studies the CA3 was shown to have a lowered seizure threshold following global ischemia (19). Enhanced seizure susceptibility occurs 2–3 weeks following the primary ischemic insult.

6.4 Changes in NMDA Receptors Following Ischemia

Following global ischemia there is a reduced expression of NMDA receptor subunits (123) which probably correlates with the loss of cells from the CA1 following injury. These decreases are blocked by administration of NMDA receptor antagonists at the time of ischemia (again supporting the view that ischemia induced cell death involves excitotoxicity). Following ischemia there is also a calpain sensitive cleavage of the NR2A and NR2B receptor subunit C termini (39, 121), which may also enhance receptor signaling. Interestingly the post ischemic induction of LTP, cell death and reduction of NR2B subunits is blocked by the NMDA receptor 2B subunit selective antagonist ifenprodil (91). Other studies have reported that NR2B mediated signaling is toxic (50).

The interaction of the NMDA receptor with the postsynaptic scaffold may also result in enhanced NMDA function. Following ischemia there is an increased phosphorylation of NMDA receptor subunits by the tyrosine kinases Src Fyn, and Pyk (17). The phosphorylation of NR2A and NR2B receptor subunits has been linked to hyper-excitability of the cell and cell death. Disruption of PSD-95 interactions with NMDA receptors, or reduction in its expression by antisense oligonucleotides reduced NMDA receptor subunit phosphorylation and ischemia induced cell death (2, 60). Hence the downstream intracellular signaling components of the NMDA receptor scaffold may offer new targets for anti-stroke therapies

7 Role of the Synapse in Ischemic Tolerance

Tolerance is the biological phenomenon whereby a brief sub toxic exposure to a stimuli (preconditioning) renders the cell, organ or organism protected against a normal toxic exposure to the same stimuli. In ischemic tolerance, brief exposure to subtoxic ischemia (albeit sufficient to elicit a response in the tissue) renders the brain (or heart) tolerant to subsequent normally injurious levels of ischemia (29). The protection induced by preconditioning can render the system protected from other harmful stimuli as well, for example endotoxin (lipo-poly saccharide¨LPS) can induce tolerance to ischemia, and seizures or spreading depression can also induce ischemic tolerance (29). Tolerance to ischemia can occur over two time frames, rapid tolerance occurs within minutes and lasts approximately one hour, delayed tolerance require 24–72 hours to be maximal and is lost after 7 days (29).

7.1 Multiple Intracellular Mechanisms Regulate Ischemic Tolerance

Studies of neuroprotection in ischemic tolerance usually revolve around the study of an induced protective gene following preconditioning. Central to this is the observation that long term ischemic tolerance is blocked by protein synthesis inhibitors, which suggests that it requires de-novo protein synthesis (5, 79, 80). Long-term ischemic tolerance is also blocked by RNA synthesis inhibitor actinomycin D (Meller, unpublished observation). Ischemic tolerance following preconditioning with ischemia is inhibited by NMDA receptor antagonists (7, 111), but this may be specific to the nature of ischemic preconditioning (i.e. requiring a synaptic event).

Ischemic tolerance induced by ischemic preconditioning is blocked by pharmacological and genomic inhibition of Protein Kinase A (PKA), p42/44 mitogen-activated protein kinase (MAPK), calcium/calmodulin activated protein kinase (CAMK) and AKT (42, 80). Each of these protein kinases have been shown to be activated following synaptic activation of NMDA receptors, although the exact mechanism linking these protein kinases remains to be determined. Activation of these protein kinases has been shown to drive multiple transcription factors and CREB activation has been shown by a number of groups to regulate ischemic tolerance. Blocking CREB phosphorylation, interaction with CBP or CRE promoters has been shown to block ischemic tolerance (48, 80). Interestingly, the prosurvival protein bcl-2 contains a CRE promoter (36, 94, 119) and blocking CREB function blocks bcl-2 expression and ischemic tolerance (48, 72, 80).

However, apart from bcl-2, numerous genes and proteins have been touted as mediating protection in tolerance paradigms, including heat shock proteins, other anti-apoptotic proteins e.g. Bcl-w, changes in expression of glutamate transport proteins EAAT2/3 and erythropoietin (23, 75, 81, 98, 99). When one considers a mechanism for ischemic tolerance, one must take into account the nature of the preconditioning agent as well the nature of the harmful event, as this may have lead to misleading generalizations regarding common mechanisms of ischemic tolerance. For example, induction of heat shock proteins have been shown as a potential mechanism of ischemic tolerance (16, 122), but ischemic tolerance is induced at a lower insult threashold than required to induce HSP expression (3). This does not rule out heat shock proteins as a relevant mechanism of inducing tolerance to ischemia, but suggests other mechanisms may be more relevant for spreading depression induced protection.

Many agents may induce long term (protein synthesis dependent) tolerance, however the mechanisms by which this tolerance is acquired may be quite different. Comparison of the genomic profile of ischemic preconditioning is quite different from other preconditioning agents, and no one common mechanism appears to result in protection. One potential explanation for this is not that preconditioning agents cause the same induction of protective genes, but rather they cause a reprogramming in the genomic response to normally harmful ischemia, thereby changing the cells fate. This has indeed been shown in ischemic tolerance using microarray profiling. Stenzel Poore et al. showed that the genomic overlap of infarcted cortex and tolerant cortex subject to harmful ischemia show no common regulated genes (110). Thus the genomic response to ischemia was “reprogrammed” by preconditioning the brain by brief ischemia and that this reprogramming produced protection. What remains to be determined is whether the different preconditioning agents also induce the same reprogramming effect or whether each stimuli not only induces its own pattern of protective gene expression but also its own re-programming effect, based of the genes expressed by the preconditioning agent. Stenzel-Poore has suggested that the tolerizing agent (ischemia, LPS) dictates the mechanism of protection to subsequent ischemia (109). Ischemic preconditioning induces a hibernation-like phenotype, the down regulation of high energy utilizers such as transporters and channels (110). In contract LPS down regulated inflammatory mediators and upregulates anti inflammatory mediators (109).

7.2 Is Synaptic Function Changed in Delayed Ischemic Tolerance?

Since ischemia tolerance induced by ischemic preconditioning can be blocked by NMDA receptor antagonists (43, 111), this would suggest that tolerance pathways are activated following ischemia-induced activation of NMDA receptors. However, no clear pattern of expression of synaptic proteins is induced by preconditioning or the tolerant state in microarray studies (110).

However, preconditioning ischemia may induce changes to synaptic transmission via indirect mechanisms. Microinjection of Bcl-XL into pre-synaptic terminals enhances post-synaptic responses and the rate of recovery from synaptic depression (65). In contrast an N-terminal cleaved Bcl-XL, which is pro-apoptotic, reduced post-synaptic responses. The effect of bcl-2 family proteins on the cell may depend on the developmental stage of the neuron. Loss of Bak results in reduced

GABA-ergic inhibition and enhanced seizure-like activity in young mice exposed to the excitatory amino acid analogue, kainite (33). However, in mature mice, loss of Bak renders cells protected against ischemia (33). Therefore in the case if ischemic tolerance, where Bcl-2 expression is increased (8, 80, 105), a change in the balance of pro-survival vs. pro cell death proteins may result in changes in mitochondrial function and therefore synaptic function.

Ischemic preconditioning may induce long-term changes to the synapse. A recent study reported that 3–30 days following preconditioning ischemia there is a higher number of dendritic spines compared to control (20). Increased spine number is usually associated with an increase in vulnerability to ischemia and other excitotoxic stimuli. This suggests that long-term changes occur in the brain following preconditioning, out lasting the neuroprotective period.

7.3 Rapid Ischemic Tolerance

Neurons also become tolerant to harmful ischemia one hour following preconditioning ischemia: a process termed rapid ischemic tolerance (79, 89, 90, 96). This form of tolerance is mediated by biochemical mechanisms in the brain. Blocking protein synthesis does not block rapid ischemic tolerance (79), suggesting it is not dependent on new protein synthesis. A complement of protein kinases appear to mediate rapid ischemic tolerance including p42/44 mitogen-activated protein kinase (MAPK), protein kinase C (PKC) and AKT (79, 96), but not protein kinase A (PKA) or calcium calmodulin dependent protein kinase (CAMK) (Meller unpublished results).

7.4 Proteasomal Degradation of Structural Proteins in Rapid Ischemic Tolerance Leads to Dendritic Spine Remodeling

Rapid ischemic tolerance is blocked by inhibition of the ubiquitin-proteasome system (79), which is implicated in the regulation of synaptic signaling (30). Our recent studies have identified a number of proteins implicated in post-synaptic density regulation and signaling as being ubiquitinated, and potentially degraded one hour following preconditioning ischemia. Of particular focus, we identified the actin binding proteins MARCKS and fascin as being ubiquitinated and degraded. The loss of these proteins appears to result in re-organization of the actin cytoskeleton, although rapid acting calcium mediated processes following NMDA receptor activation, such as calpain, may also account for some of the rapid re-organization of actin.

A direct consequence of actin reorganization is the loss of dendritic spines on the neurons, which are therefore unable to make synaptic contacts. The loss of dendritic spines following ischemia has been reported in neurons in vitro following ischemia (53), but the role of the ubiquitin proteasome system has not yet been investigated. We showed that the proteasome inhibitor reduces the number of spines lost at one hour. Whether the proteasome directly mediates spine loss or speeds up the spine recovery is currently under investigation.

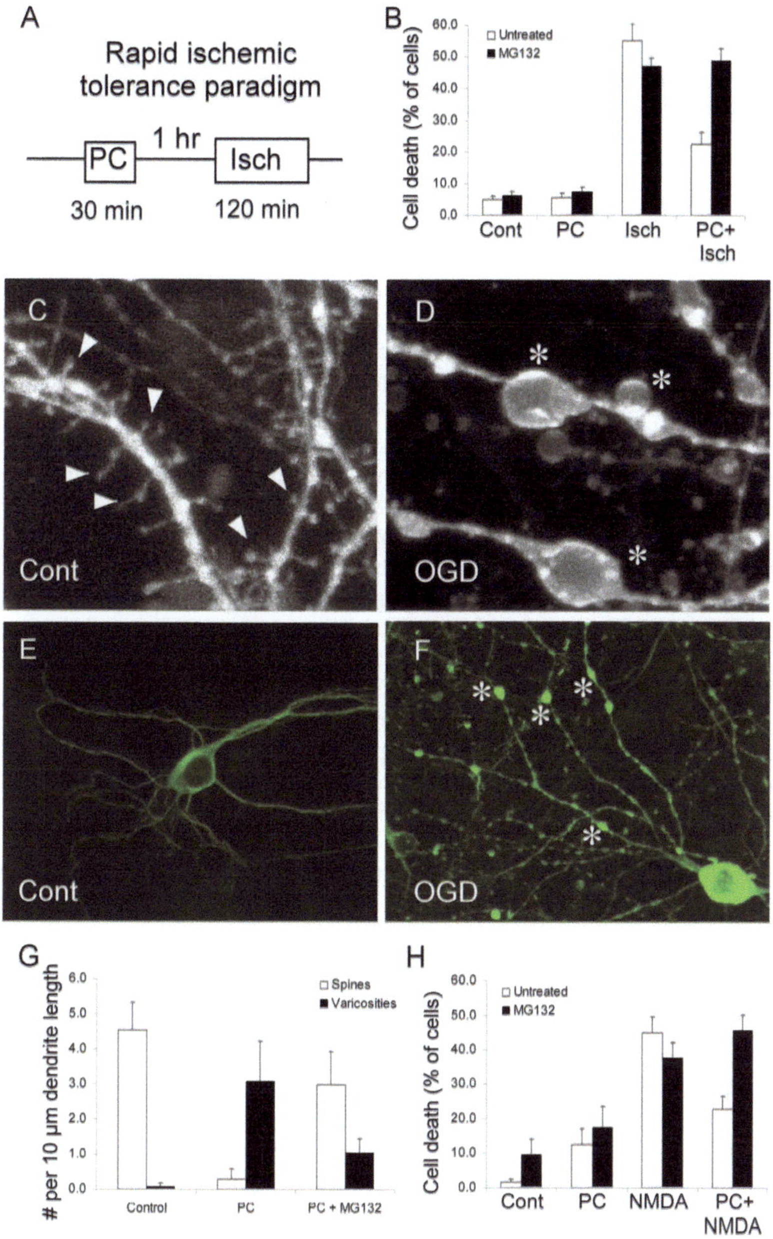

Fig. 5. Rapid ischemic tolerance induced protection is mediated by changes to the synapse. (**a**) Ischemic tolerance is induced by pairing a preconditioning brief ischemic stimulus (30 min OGD: PC) with a prolonged harmful ischemic challenge (120 min OGD: Isch). (**b**) The neuroprotection in ischemia tolerance is reversed by the proteasome inhibitor MG132. Compared to untreated cells, cells treated with the proteasome inhibitor MG132 do not aquire tolerance to harmful ischemia. (**c–f**). Compared to control (c), following preconditioning

The loss of dendritic spines may be a feature of endogenous protective mechanisms resulting in tolerance to ischemia. A reduction in dendritic spine number has been associated with the ischemia tolerant effect of hibernation in artic ground squirrels and European hamsters (73, 93, 114). Hence the remodeling of dendritic spines following transient sub-toxic ischemia appears to be a protective mechanism.

7.5 Reduced Excitotoxicity of NMDA in Rapid Ischemic Tolerance

The remodeling of the post-synaptic membrane may not just be mediated by changes to the actin cytoskeleton. The loss of dendritic spines one hour following preconditioning ischemia, would no doubt have an effect on synaptic transmission. We have also discovered that at the time of spine loss in preconditioned cells, NMDA-mediated cell death is reduced. The tolerance to NMDA excitotoxicity is blocked by a proteasome inhibitor, which also blocks the reduction in spine number at one hour following preconditioning ischemia (See Fig. 5). This suggest that preconditioning ischemia results in a selective inhibition of post-synaptic toxic glutamate signaling in cells, via the ubiquitin-proteasome pathway, yet NMDA function (CREB phosphorylation) and receptor subunit expression levels were not effected by the preconditioning ischemia (not shown).

The loss of NMDA toxicity following preconditioning ischemia, suggest a selective loss in pathological NMDA signaling that may account for the tolerance to ischemia. These studies agree with that of Aarts, who showed that disruption of the interaction of PSD-95 with NMDA2A/2B receptors blocks ischemia induced cell death, but not NMDA electrophysiological responses (2). Further this result agrees with Sattler whereby remodeling of dendritic spines results in protection to ischemia (100). However in that study, dendritic remodeling using latrunculin (an actin depolymerizing agent) only protected against ischemia and not NMDA toxicity. In contrast ischemic preconditioning protects against toxicity induced by both ischemia and NMDA. Our data suggest that preconditioning induced protection, involves more than just actin re-organization, rather the ubiquitination of post-synaptic density associated proteins results in a protective phenotype. Hence preconditioning cells with ischemia results in protection to toxic NMDA signaling from extra-synaptic sites as well as synaptically derived ischemia – induced neurotoxicity.

Summary and Concluding Remarks

Ischemia induces profound changes to the synapse, both in terms of mediating the death of cells following toxic ischemia, and the remodeling of synapses in response

Fig. 5. (Continued) ischemia a reduction in dendritic spine number (arrow heads) and increases in varicosity (*) formation is observed (d). Compared to control cells (e), varicosity formation (*) is highlighted by an accumulation of MAP2 following preconditioning ischemia (f). (**g**) The loss of spines and the formation of varicosities is blocked by a proteasome inhibitor MG132. (**h**) NMDA tolerance induced by preconditioning (PC) is also blocked by the proteasome inhibitor MG132. Taken together we suggest that following preconditioning ischemia the protein degradation mediated by the proteasome results in morphological changes in the spines, resulting in endogenous protection to excitotoxicity.

to sub-toxic ischemia. By further investigating the mechanisms that separate physiological from pathological NMDA signaling, potential therapeutic options may be identified with significant therapeutic protection from stroke potential and a more favorable therapeutic index than currently available investigational tools. Such an intervention may be of use post-stroke as well as in situations whereby ischemic brain conditions can be predicted, for example heart bypass surgery.

Acknowledgements

Thanks to Simon Thompson Ph.D for images of MAP2 staining in cells. We apologize to those whose work we could not cite due to space limitations. Work in the Robert Stone Dow Neurobiology Laboratories is supported by the National Institute for Health (NINDS), the American Heart Association, the Medical Research Foundation of Oregon and The Good Samaritan Foundation.

References

1. Tissue plasminogen activator for acute ischemic stroke. The National Institute of Neurological Disorders and Stroke rt-PA Stroke Study Group. *N Engl J Med* 333: 1581–1587, 1995.
2. Aarts M, Liu Y, Liu L, Besshoh S, Arundine M, Gurd JW, Wang YT, Salter MW, and Tymianski M. Treatment of ischemic brain damage by perturbing NMDA receptor-PSD-95 protein interactions. *Science* 298: 846–850, 2002.
3. Abe H and Nowak TS, Jr. Gene expression and induced ischemic tolerance following brief insults. *Acta Neurobiol Exp* 56: 3–8, 1996.
4. Back SA, Han BH, Luo NL, Chricton CA, Xanthoudakis S, Tam J, Arvin KL, and Holtzman DM. Selective vulnerability of late oligodendrocyte progenitors to hypoxia-ischemia. *J Neurosci* 22: 455–463, 2002.
5. Barone FC, White RF, Spera PA, Ellison J, Currie RW, Wang X, and Feuerstein GZ. Ischemic preconditioning and brain tolerance: temporal histological and functional outcomes, protein synthesis requirement, and interleukin-1 receptor antagonist and early gene expression. *Stroke* 29: 1937–1950; discussion 1950–1931, 1998.
6. Biegon A, Fry PA, Paden CM, Alexandrovich A, Tsenter J, and Shohami E. Dynamic changes in N-methyl-D-aspartate receptors after closed head injury in mice: Implications for treatment of neurological and cognitive deficits. *Proc Natl Acad Sci USA* 101: 5117–5122, 2004.
7. Bond A, Lodge D, Hicks CA, Ward MA, and O'Neill MJ. NMDA receptor antagonism, but not AMPA receptor antagonism attenuates induced ischaemic tolerance in the gerbil hippocampus. *Eur J Pharmacol* 380: 91–99, 1999.
8. Bossenmeyer C, Chihab R, Muller S, Schroeder H, and Daval JL. Hypoxia/reoxygenation induces apoptosis through biphasic induction of protein synthesis in cultured rat brain neurons. *Brain Res* 787: 107–116, 1998.
9. Brott T and Bogousslavsky J. Treatment of acute ischemic stroke. *N Engl J Med* 343: 710–722, 2000.
10. Brown CE, Li P, Boyd JD, Delaney KR, and Murphy TH. Extensive turnover of dendritic spines and vascular remodeling in cortical tissues recovering from stroke. *J Neurosci* 27: 4101–4109, 2007.

11. Buchan AM and Pulsinelli WA. Septo-hippocampal deafferentation protects CA1 neurons against ischemic injury. *Brain Res* 512: 7–14, 1990.
12. Buddle M, Eberhardt E, Ciminello LH, Levin T, Wing R, DiPasquale K, and Raley-Susman KM. Microtubule-associated protein 2 (MAP2) associates with the NMDA receptor and is spatially redistributed within rat hippocampal neurons after oxygen-glucose deprivation. *Brain Res* 978: 38–50, 2003.
13. Cao G, Clark RS, Pei W, Yin W, Zhang F, Sun FY, Graham SH, and Chen J. Translocation of apoptosis-inducing factor in vulnerable neurons after transient cerebral ischemia and in neuronal cultures after oxygen-glucose deprivation. *J Cereb Blood Flow Metab* 23: 1137–1150, 2003.
14. Carmichael ST. Rodent models of Focal Stroke: Size, Mechanism and Purpose. *NeuroRx* 2: 396–409, 2005.
15. Centonze D, Saulle E, Pisani A, Bernardi G, and Calabresi P. Adenosine-mediated inhibition of striatal GABAergic synaptic transmission during in vitro ischaemia. *Brain* 124: 1855–1865, 2001.
16. Chen J, Graham SH, Zhu RL, and Simon RP. Stress proteins and tolerance to focal cerebral ischemia. *J Cereb Blood Flow Metab* 16: 566–577, 1996.
17. Cheung HH, Takagi N, Teves L, Logan R, Wallace MC, and Gurd JW. Altered association of protein tyrosine kinases with postsynaptic densities after transient cerebral ischemia in the rat brain. *J Cereb Blood Flow Metab* 20: 505–512, 2000.
18. Choi DW. Calcium-mediated neurotoxicity: relationship to specific channel types and role in ischemic damage. *Trends Neurosci* 11: 465–469, 1988.
19. Congar P, Gaiarsa JL, Popovici T, Ben-Ari Y, and Crepel V. Permanent reduction of seizure threshold in post-ischemic CA3 pyramidal neurons. *J Neurophysiol* 83: 2040–2046, 2000.
20. Corbett D, Giles T, Evans S, McLean J, and Biernaskie J. Dynamic changes in CA1 dendritic spines associated with ischemic tolerance. *Exp Neurol* 202: 133–138, 2006.
21. Crepel V, Congar P, Aniksztejn L, Gozlan H, Hammond C, and Ben-Ari Y. Synaptic plasticity in ischemia: role of NMDA receptors. *Prog Brain Res* 116: 273–285, 1998.
22. Crepel V, Epsztein J, and Ben-Ari Y. Ischemia induces short- and long-term remodeling of synaptic activity in the hippocampus. *J Cell Mol Med* 7: 401–407, 2003.
23. Currie RW, Ellison JA, White RF, Feuerstein GZ, Wang X, and Barone FC. Benign focal ischemic preconditioning induces neuronal Hsp70 and prolonged astrogliosis with expression of Hsp27. *Brain Res* 863: 169–181, 2000.
24. Dawson LA, Djali S, Gonzales C, Vinegra MA, and Zaleska MM. Characterization of transient focal ischemia-induced increases in extracellular glutamate and aspartate in spontaneously hypertensive rats. *Brain Res Bull* 53: 767-776, 2000.
25. del Zoppo GJ. Microvascular changes during cerebral ischemia and reperfusion. *Cerebrovasc Brain Metab Rev* 6: 47–96, 1994.
26. Diemer NH and Ekstrom von Lubitz DK. Cerebral ischaemia in the rat: increased permeability of post-synaptic membranes to horseradish peroxidase in the early post-ischaemic period. *Neuropathol Appl Neurobiol* 9: 403–414, 1983.
27. Dirnagl U. Bench to bedside: the quest for quality in experimental stroke research. *J Cereb Blood Flow Metab* 26: 1465–1478, 2006.
28. Dirnagl U, Iadecola C, and Moskowitz MA. Pathobiology of ischaemic stroke: an integrated view. *Trends Neurosci* 22: 391–397, 1999.
29. Dirnagl U, Simon RP, and Hallenbeck JM. Ischemic tolerance and endogenous neuroprotection. *Trends Neurosci* 26: 248–254, 2003.
30. Ehlers MD. Activity level controls postsynaptic composition and signaling via the ubiquitin-proteasome system. *Nat Neurosci* 6: 231–242, 2003.

31. Ekstrom von Lubitz DK and Diemer NH. Complete cerebral ischaemia in the rat: an ultrastructural and stereological analysis of the distal stratum radiatum in the hippocampal CA-1 region. *Neuropathol Appl Neurobiol* 8: 197–215, 1982.
32. Faddis BT, Hasbani MJ, and Goldberg MP. Calpain activation contributes to dendritic remodeling after brief excitotoxic injury in vitro. *J Neurosci* 17: 951–959, 1997.
33. Fannjiang Y, Kim CH, Huganir RL, Zou S, Lindsten T, Thompson CB, Mito T, Traystman RJ, Larsen T, Griffin DE, Mandir AS, Dawson TM, Dike S, Sappington AL, Kerr DA, Jonas EA, Kaczmarek LK, and Hardwick JM. BAK Alters Neuronal Excitability and Can Switch from Anti- to Pro-Death Function during Postnatal Development. *Dev Cell* 4: 575–585, 2003.
34. Ferrer I and Planas AM. Signaling of cell death and cell survival following focal cerebral ischemia: life and death struggle in the penumbra. *J Neuropathol Exp Neurol* 62: 329–339, 2003.
35. Fisher M, Hanley DF, Howard G, Jauch EC, and Warach S. Recommendations from the STAIR V meeting on acute stroke trials, technology and outcomes. *Stroke* 38: 245–248, 2007.
36. Freeland K, Boxer LM, and Latchman DS. The cyclic AMP response element in the Bcl-2 promoter confers inducibility by hypoxia in neuronal cells. *Brain Res Mol Brain Res* 92: 98–106, 2001.
37. Gao TM, Howard EM, and Xu ZC. Transient neurophysiological changes in CA3 neurons and dentate granule cells after severe forebrain ischemia in vivo. *J Neurophysiol* 80: 2860–2869, 1998.
38. Gao TM, Pulsinelli WA, and Xu ZC. Prolonged enhancement and depression of synaptic transmission in CA1 pyramidal neurons induced by transient forebrain ischemia in vivo. *Neuroscience* 87: 371–383, 1998.
39. Gascon S, Sobrado M, Roda JM, Rodriguez-Pena A, and Diaz-Guerra M. Excitotoxicity and focal cerebral ischemia induce truncation of the NR2A and NR2B subunits of the NMDA receptor and cleavage of the scaffolding protein PSD-95. *Mol Psychiatry*, 13: 99–14, 2007.
40. Giffard RG and Swanson RA. Ischemia-induced programmed cell death in astrocytes. *Glia* 50: 299–306, 2005.
41. Gomez-Lazaro M, Galindo MF, Melero-Fernandez de Mera RM, Fernandez-Gomez FJ, Concannon CG, Segura MF, Comella JX, Prehn JH, and Jordan J. Reactive oxygen species and p38 mitogen-activated protein kinase activate Bax to induce mitochondrial cytochrome c release and apoptosis in response to malonate. *Mol Pharmacol* 71: 736–743, 2007.
42. Gonzalez-Zulueta M, Feldman AB, Klesse LJ, Kalb RG, Dillman JF, Parada LF, Dawson TM, and Dawson VL. Requirement for nitric oxide activation of p21(ras)/extracellular regulated kinase in neuronal ischemic preconditioning. *Proc Natl Acad Sci USA* 97: 436–441, 2000.
43. Grabb MC and Choi DW. Ischemic tolerance in murine cortical cell culture: critical role for NMDA receptors. *J Neurosci* 19: 1657–1662, 1999.
44. Graber S, Maiti S, and Halpain S. Cathepsin B-like proteolysis and MARCKS degradation in sub-lethal NMDA-induced collapse of dendritic spines. *Neuropharmacology* 47: 706–713, 2004.
45. Graham SH and Chen J. Programmed cell death in cerebral ischemia. *J Cereb Blood Flow Metab* 21: 99–109, 2001.
46. Greenberg DA, Aminoff MJ, and R.P.Simon. Stroke. In: *Clinical Neurology*, edited by Greenberg DA, Aminoff MJ, and R.P.Simon: Lange medical Books/McGraw-Hill, pp. 282–316, 2002.

47. Gregersen R, Christensen T, Lehrmann E, Diemer NH, and Finsen B. Focal cerebral ischemia induces increased myelin basic protein and growth-associated protein-43 gene transcription in peri-infarct areas in the rat brain. *Exp Brain Res* 138: 384–392, 2001.
48. Hara T, Hamada J, Yano S, Morioka M, Kai Y, and Ushio Y. CREB is required for acquisition of ischemic tolerance in gerbil hippocampal CA1 region. *J Neurochem* 86: 805–814, 2003.
49. Hardingham GE and Bading H. Coupling of extrasynaptic NMDA receptors to a CREB shut-off pathway is developmentally regulated. *Biochim Biophys Acta* 1600: 148–153, 2002.
50. Hardingham GE, Fukunaga Y, and Bading H. Extrasynaptic NMDARs oppose synaptic NMDARs by triggering CREB shut-off and cell death pathways. *Nat Neurosci* 5: 405–414, 2002.
51. Hargreaves RJ, Hill RG, and Iversen LL. Neuroprotective NMDA antagonists: the controversy over their potential for adverse effects on cortical neuronal morphology. *Acta Neurochir Suppl (Wien)* 60: 15–19, 1994.
52. Hasbani MJ, Schlief ML, Fisher DA, and Goldberg MP. Dendritic spines lost during glutamate receptor activation reemerge at original sites of synaptic contact. *J Neurosci* 21: 2393–2403, 2001.
53. Hasbani MJ, Viquez NM, and Goldberg MP. NMDA receptors mediate hypoxic spine loss in cultured neurons. *Neuroreport* 12: 2731–2735, 2001.
54. Henshall DC, Butcher SP, and Sharkey J. A rat model of endothelin-3-induced middle cerebral artery occlusion with controlled reperfusion. *Brain Res* 843: 105–111, 1999.
55. Hillion JA, Li Y, Maric D, Takanohashi A, Klimanis D, Barker JL, and Hallenbeck JM. Involvement of Akt in preconditioning-induced tolerance to ischemia in PC12 cells. *J Cereb Blood Flow Metab* 26: 1323–1331, 2006.
56. Hillion JA, Takahashi K, Maric D, Ruetzler C, Barker JL, and Hallenbeck JM. Development of an ischemic tolerance model in a PC12 cell line. *J Cereb Blood Flow Metab* 25: 154–162, 2005.
57. Hills CP. The Ultrastructure of Anoxic-Ischaemic Lesions in the Cerebral Cortex of the Adult Rat Brain. *Guys Hosp Rep* 113: 333–348, 1964.
58. Horner CH, Davies HA, and Stewart MG. Hippocampal synaptic density and glutamate immunoreactivity following transient cerebral ischaemia in the chick. *Eur J Neurosci* 10: 3913–3917, 1998.
59. Hossmann KA. Pathophysiology and therapy of experimental stroke. *Cell Mol Neurobiol* 26: 1057–1083, 2006.
60. Hou XY, Zhang GY, and Zong YY. Suppression of postsynaptic density protein 95 expression attenuates increased tyrosine phosphorylation of NR2A subunits of N-methyl-D-aspartate receptors and interactions of Src and Fyn with NR2A after transient brain ischemia in rat hippocampus. *Neurosci Lett* 343: 125–128, 2003.
61. Hoyte L, Barber PA, Buchan AM, and Hill MD. The rise and fall of NMDA antagonists for ischemic stroke. *Curr Mol Med* 4: 131–136, 2004.
62. Hu BR, Martone ME, Jones YZ, and Liu CL. Protein aggregation after transient cerebral ischemia. *J Neurosci* 20: 3191–3199, 2000.
63. Hu XL, Olsson T, Johansson IM, Brannstrom T, and Wester P. Dynamic changes of the anti- and pro-apoptotic proteins Bcl-w, Bcl-2, and Bax with Smac/Diablo mitochondrial release after photothrombotic ring stroke in rats. *Eur J Neurosci* 20: 1177–1188, 2004.
64. Ikegaya Y, Kim JA, Baba M, Iwatsubo T, Nishiyama N, and Matsuki N. Rapid and reversible changes in dendrite morphology and synaptic efficacy following NMDA receptor activation: implication for a cellular defense against excitotoxicity. *J Cell Sci* 114: 4083–4093, 2001.

65. Jonas EA, Hoit D, Hickman JA, Brandt TA, Polster BM, Fannjiang Y, McCarthy E, Montanez MK, Hardwick JM, and Kaczmarek LK. Modulation of synaptic transmission by the BCL-2 family protein BCL-xL. *J Neurosci* 23: 8423–8431, 2003.
66. Kadotani H, Namura S, Katsuura G, Terashima T, and Kikuchi H. Attenuation of focal cerebral infarct in mice lacking NMDA receptor subunit NR2C. *Neuroreport* 9: 471–475, 1998.
67. Kirino T, Tamura A, and Sano K. Chronic maintenance of presynaptic terminals in gliotic hippocampus following ischemia. *Brain Res* 510: 17–25, 1990.
68. Lin CH, Lu YZ, Cheng FC, Chu LF, and Hsueh CM. Bax-regulated mitochondria-mediated apoptosis is responsible for the in vitro ischemia induced neuronal cell death of Sprague Dawley rat. *Neurosci Lett* 387: 22–27, 2005.
69. Liou AKF, Clark RS, Henshall DC, Yin XM, and Chen J. To die or not to die for neurons in ischemia, traumatic brain injury and epilepsy: a review on the stress-activated signaling pathways and apoptotic pathways. *Prog Neurobiol* 69: 103–142, 2003.
70. Liu CL, Martone ME, and Hu BR. Protein ubiquitination in postsynaptic densities after transient cerebral ischemia. *J Cereb Blood Flow Metab* 24: 1219–1225, 2004.
71. Luque JM, Puig N, Martinez JM, Gonzalez-Garcia C, and Cena V. Glutamate N-methyl-D-aspartate receptor blockade prevents induction of GAP-43 after focal ischemia in rats. *Neurosci Lett* 305: 87–90, 2001.
72. Mabuchi T, Kitagawa K, Kuwabara K, Takasawa K, Ohtsuki T, Xia Z, Storm D, Yanagihara T, Hori M, and Matsumoto M. Phosphorylation of cAMP response element-binding protein in hippocampal neurons as a protective response after exposure to glutamate in vitro and ischemia in vivo. *J Neurosci* 21: 9204–9213, 2001.
73. Magarinos AM, McEwen BS, Saboureau M, and Pevet P. Rapid and reversible changes in intrahippocampal connectivity during the course of hibernation in European hamsters. *Proc Natl Acad Sci USA* 103: 18775–18780, 2006.
74. Martone ME, Hu BR, and Ellisman MH. Alterations of hippocampal postsynaptic densities following transient ischemia. *Hippocampus* 10: 610–616, 2000.
75. McLaughlin B, Hartnett KA, Erhardt JA, Legos JJ, White RF, Barone FC, and Aizenman E. Caspase 3 activation is essential for neuroprotection in preconditioning. *Proc Natl Acad Sci USA* 100: 715–720, 2003.
76. Mehta SL, Manhas N, and Raghubir R. Molecular targets in cerebral ischemia for developing novel therapeutics. *Brain Res Rev* 54: 34–66, 2007.
77. Meldrum B. Excitotoxicity and epileptic brain damage. *Epilepsy Res* 10: 55–61, 1991.
78. Meldrum BS, Evans MC, Swan JH, and Simon RP. Protection against hypoxic/ischaemic brain damage with excitatory amino acid antagonists. *Med Biol* 65: 153–157, 1987.
79. Meller R, Cameron JA, Torrey DJ, Clayton CE, Ordonez AN, Henshall DC, Minami M, Schindler CK, Saugstad JA, and Simon RP. Rapid degradation of Bim by the ubiquitin-proteasome pathway mediates short-term ischemic tolerance in cultured neurons. *J Biol Chem* 281: 7429–7436, 2006.
80. Meller R, Minami M, Cameron JA, Impey S, Chen D, Lan JQ, Henshall DC, and Simon RP. CREB-mediated Bcl-2 protein expression after ischemic preconditioning. *J Cereb Blood Flow Metab* 25: 234–246, 2005.
81. Minami M, Jin KL, Li W, Nagayama T, Henshall DC, and Simon RP. Bcl-w expression is increased in brain regions affected by focal cerebral ischemia in the rat. *Neurosci Lett* 279: 193–195, 2000.
82. Mitani A, Matsuda S, Yamamoto H, Sakanaka M, and Kataoka K. The role of remaining presynaptic terminals in the hippocampal CA1 after selective neuronal death: ischemia-induced glutamate efflux. *Acta Neuropathol (Berl)* 91: 41–46, 1996.
83. Mitani A, Namba S, Ikemune K, Yanase H, Arai T, and Kataoka K. Postischemic enhancements of N-methyl-D-aspartic acid (NMDA) and non-NMDA receptor-mediated

responses in hippocampal CA1 pyramidal neurons. *J Cereb Blood Flow Metab* 18: 1088–1098, 1998.
84. Miyazaki S, Katayama Y, Furuichi M, Kinoshita K, Kawamata T, and Tsubokawa T. Impairment of hippocampal long-term potentiation following transient cerebral ischaemia in rat: effects of bifemelane, a potent inhibitor of ischaemia-induced acetylcholine release. *Neurol Res* 15: 249–252, 1993.
85. Monyer H, Giffard RG, Hartley DM, Dugan LL, Goldberg MP, and Choi DW. Oxygen or glucose deprivation-induced neuronal injury in cortical cell cultures is reduced by tetanus toxin. *Neuron* 8: 967–973, 1992.
86. Olney JW. Glutamate-induced neuronal necrosis in the infant mouse hypothalamus. An electron microscopic study. *J Neuropathol Exp Neurol* 30: 75-90, 1971.
87. Olney JW and Ho OL. Brain damage in infant mice following oral intake of glutamate, aspartate or cysteine. *Nature* 227: 609–611, 1970.
88. Park JS, Bateman MC, and Goldberg MP. Rapid alterations in dendrite morphology during sublethal hypoxia or glutamate receptor activation. *Neurobiol Dis* 3: 215–227, 1996.
89. Perez-Pinzon MA and Born JG. Rapid preconditioning neuroprotection following anoxia in hippocampal slices: role of the K+ ATP channel and protein kinase C. *Neuroscience* 89: 453–459, 1999.
90. Perez-Pinzon MA, Vitro TM, Dietrich WD, and Sick TJ. The effect of rapid preconditioning on the microglial, astrocytic and neuronal consequences of global cerebral ischemia. *Acta Neuropathol (Berl)* 97: 495–501, 1999.
91. Picconi B, Tortiglione A, Barone I, Centonze D, Gardoni F, Gubellini P, Bonsi P, Pisani A, Bernardi G, Di Luca M, and Calabresi P. NR2B subunit exerts a critical role in postischemic synaptic plasticity. *Stroke* 37: 1895–1901, 2006.
92. Pignataro G, Simon RP, and Xiong ZG. Prolonged activation of ASIC1a and the time window for neuroprotection in cerebral ischaemia. *Brain* 130: 151–158, 2007.
93. Popov VI and Bocharova LS. Hibernation-induced structural changes in synaptic contacts between mossy fibres and hippocampal pyramidal neurons. *Neuroscience* 48: 53–62, 1992.
94. Pugazhenthi S, Nesterova A, Sable C, Heidenreich KA, Boxer LM, Heasley LE, and Reusch JE. Akt/protein kinase B up-regulates Bcl-2 expression through cAMP- response element-binding protein. *J Biol Chem* 275: 10761–10766, 2000.
95. Pulsinelli WA and Buchan AM. The four-vessel occlusion rat model: method for complete occlusion of vertebral arteries and control of collateral circulation. *Stroke* 19: 913–914, 1988.
96. Reshef A, Sperling O, and Zoref-Shani E. Activation and inhibition of protein kinase C protect rat neuronal cultures against ischemia-reperfusion insult. *Neurosci Lett* 238: 37–40, 1997.
97. Rod MR and Auer RN. Pre- and post-ischemic administration of dizocilpine (MK-801) reduces cerebral necrosis in the rat. *Can J Neurol Sci* 16: 340–344, 1989.
98. Romera C, Hurtado O, Botella SH, Lizasoain I, Cardenas A, Fernandez-Tome P, Leza JC, Lorenzo P, and Moro MA. In vitro ischemic tolerance involves upregulation of glutamate transport partly mediated by the TACE/ADAM17-tumor necrosis factor-alpha pathway. *J Neurosci* 24: 1350–1357, 2004.
99. Ruscher K, Freyer D, Karsch M, Isaev N, Megow D, Sawitzki B, Priller J, Dirnagl U, and Meisel A. Erythropoietin is a paracrine mediator of ischemic tolerance in the brain: evidence from an in vitro model. *J Neurosci* 22: 10291–10301, 2002.
100. Sattler R, Xiong Z, Lu WY, MacDonald JF, and Tymianski M. Distinct roles of synaptic and extrasynaptic NMDA receptors in excitotoxicity. *J Neurosci* 20: 22–33, 2000.

101. Savitz SI. A critical appraisal of the NXY-059 neuroprotection studies for acute stroke: a need for more rigorous testing of neuroprotective agents in animal models of stroke. *Exp Neurol* 205: 20–25, 2007.
102. Savitz SI and Fisher M. Future of neuroprotection for acute stroke: in the aftermath of the SAINT trials. *Ann Neurol* 61: 396–402, 2007.
103. Selim M. Perioperative stroke. *N Engl J Med* 356: 706–713, 2007.
104. Shelton MK and McCarthy KD. Mature hippocampal astrocytes exhibit functional metabotropic and ionotropic glutamate receptors in situ. *Glia* 26: 1–11, 1999.
105. Shimizu S, Nagayama T, Jin KL, Zhu L, Loeffert JE, Watkins SC, Graham SH, and Simon RP. bcl-2 Antisense treatment prevents induction of tolerance to focal ischemia in the rat brain. *J Cereb Blood Flow Metab* 21: 233–243, 2001.
106. Sloviter RS. Apoptosis: a guide for the perplexed. *Trends Pharmacol Sci* 23: 19–24, 2002.
107. Small DL, Poulter MO, Buchan AM, and Morley P. Alteration in NMDA receptor subunit mRNA expression in vulnerable and resistant regions of in vitro ischemic rat hippocampal slices. *Neurosci Lett* 232: 87–90, 1997.
108. Somjen GG, Aitken PG, Czeh G, Jing J, and Young JN. Cellular physiology of hypoxia of the mammalian central nervous system. *Res Publ Assoc Res Nerv Ment Dis* 71: 51–65, 1993.
109. Stenzel-Poore MP, Stevens SL, King JS, and Simon RP. Preconditioning reprograms the response to ischemic injury and primes the emergence of unique endogenous neuroprotective phenotypes: a speculative synthesis. *Stroke* 38: 680–685, 2007.
110. Stenzel-Poore MP, Stevens SL, Xiong Z, Lessov NS, Harrington CA, Mori M, Meller R, Rosenzweig HL, Tobar E, Shaw TE, Chu X, and Simon RP. Effect of ischaemic preconditioning on genomic response to cerebral ischaemia: similarity to neuroprotective strategies in hibernation and hypoxia-tolerant states. *Lancet* 362: 1028–1037, 2003.
111. Sydserff SG, Borrelli AR, Palmer GC, and Cross AJ. Ischaemic tolerance is blocked by NMDA antagonism. *Soc Neurosci Abstr* 24: P87.86, 1998.
112. Takamatsu H, Kondo K, Ikeda Y, and Umemura K. Neuroprotective effects depend on the model of focal ischemia following middle cerebral artery occlusion. *Eur J Pharmacol* 362: 137–142, 1998.
113. Traystman RJ. Animal models of focal and global cerebral ischemia. *Ilar J* 44: 85–95, 2003.
114. von der Ohe CG, Darian-Smith C, Garner CC, and Heller HC. Ubiquitous and temperature-dependent neural plasticity in hibernators. *J Neurosci* 26: 10590–10598, 2006.
115. Vornov JJ, Tasker RC, and Coyle JT. Delayed protection by MK-801 and tetrodotoxin in a rat organotypic hippocampal culture model of ischemia. *Stroke* 25: 457–464; discussion 464–455, 1994.
116. Walton M, Woodgate AM, Muravlev A, Xu R, During MJ, and Dragunow M. CREB phosphorylation promotes nerve cell survival. *J Neurochem* 73: 1836–1842, 1999.
117. Weih M, Bergk A, Isaev NK, Ruscher K, Megow D, Riepe M, Meisel A, Victorov IV, and Dirnagl U. Induction of ischemic tolerance in rat cortical neurons by 3-nitropropionic acid: chemical preconditioning. *Neurosci Lett* 272: 207–210, 1999.
118. Williams V and Grossman RG. Ultrastructure of cortical synapses after failure of presynaptic activity in ischemia. *Anat Rec* 166: 131–141, 1970.
119. Wilson BE, Mochon E, and Boxer LM. Induction of bcl-2 expression by phosphorylated CREB proteins during B- cell activation and rescue from apoptosis. *Mol Cell Biol* 16: 5546–5556, 1996.
120. Wood PL and Hawkinson JE. N-methyl-D-aspartate antagonists for stroke and head trauma. *Expert Opin Investig Drugs* 6: 389–397, 1997.

121. Wu HY, Yuen EY, Lu YF, Matsushita M, Matsui H, Yan Z, and Tomizawa K. Regulation of N-methyl-D-aspartate receptors by calpain in cortical neurons. *J Biol Chem* 280: 21588–21593, 2005.
122. Yagita Y, Kitagawa K, Ohtsuki T, Tanaka S, Hori M, and Matsumoto M. Induction of the HSP110/105 family in the rat hippocampus in cerebral ischemia and ischemic tolerance. *J Cereb Blood Flow Metab* 21: 811–819, 2001.
123. Zhang L, Hsu JC, Takagi N, Gurd JW, Wallace MC, and Eubanks JH. Transient global ischemia alters NMDA receptor expression in rat hippocampus: correlation with decreased immunoreactive protein levels of the NR2A/2B subunits, and an altered NMDA receptor functionality. *J Neurochem* 69: 1983–1994, 1997.
124. Zhang S, Boyd J, Delaney K, and Murphy TH. Rapid reversible changes in dendritic spine structure in vivo gated by the degree of ischemia. *J Neurosci* 25: 5333–5338, 2005.
125. Zhang S and Murphy TH. Imaging the impact of cortical microcirculation on synaptic structure and sensory-evoked hemodynamic responses in vivo. *PLoS Biol* 5: e119, 2007.

Neuroplasticity and Pathological Pain

Michael W. Salter

Program in Neurosciences & Mental Health, The Hospital for Sick Children, and The University of Toronto Centre for the Study of Pain, Toronto, ON, Canada M5G 1X8, mike.salter@utoronto.ca

1 Introduction and Pathological Pain

There is great survival value in appropriately responding to environmental stimuli that damage, or threaten to damage tissue – noxious stimuli. In higher organisms, the detection of noxious stimuli is done through a specialized system of noxious stimuli-detecting, i.e. nociceptive, neurons, and in higher organisms, nociception leads to acute nociceptive pain. Nociceptive pain is a crucial defensive mechanism that warns an individual of recent or imminent damage to the body, and is produced as the physiological outcome of normal functioning of peripheral and central nervous systems. By contrast, chronic pain, which are inflammatory or neuropathic, are the result of aberrant functioning of peripheral or central nervous systems that have been pathologically modified (95, 97). These chronic pains typically are not directly related to nociception, and may persist long after any tissue damage, which may have initiated nociceptive pain, has subsided. There are no known defensive, or other helpful, functions of chronic pain. Rather, chronic pain may be insidious and destructive to the life of the individual who is suffering. Over the past few years there has been increasing recognition that pathological alterations underlie and amplify chronic pain occur in the peripheral nervous system or numerous sites within the central nervous system leading to the concept that chronic pain is a group of mechanistically separable nervous system disorders produced by one or more abnormal cellular signaling mechanisms.

In this chapter I focus on synaptic and cellular signaling pathways within the dorsal horn of the spinal cord underlying inflammatory and neuropathic pain. In the dorsal horn, nociceptive inputs from primary afferents are primarily mediated at fast glutamatergic synapses onto second order neurons in the dorsal horn of the spinal cord through activation of AMPA/kainate and NMDA receptor subtypes of ionotropic glutamate receptors. At these glutamatergic synapses several forms of

J.W. Hell, M.D. Ehlers (eds.), *Structural and Functional Organization of the Synapse*,
DOI: 10.1007/978-0-387-77232-5_26, © Springer Science+Business Media, LLC 2008

short-lasting and long-lasting enhancement of synaptic transmission are known. Enhancement of excitatory synaptic transmission in nociceptive pathways is a key neural substrate underlying chronic pain, and thus the cellular and molecular mechanisms producing this enhancement represent potential targets for developing novel forms of therapeutics. Central to the mechanisms for pain hypersensitivity is the NMDA receptor, the activity of which is facilitated by convergent intracellular biochemical cascades in dorsal horn neurons. Recently, it has been increasingly recognized that cellular changes are not restricted to neurons in the dorsal horn, however, and there is growing evidence for involvement of glia, and of glia-neuronal interactions, in initiating and sustaining enhancement of nociceptive transmission. In particular, a role has emerged for microglia in pain hypersensitivity following nerve injury. Expanding our understanding of synaptic and cellular signaling mechanisms in the dorsal horn leads to a basis for creating new types of strategies for management of chronic pain.

Within the dorsal horn there is a complex nociceptive processing network through which inputs from the periphery are transduced and modulated (97). The network includes local, as well as descending, inhibitory control mechanisms, and the output of this network is transmitted to other areas of the CNS involved in sensory, emotional, autonomic and motor processing. The ultimate effect of the intracellular signaling pathways focused on here is alteration of the activity of neurons in the dorsal horn nociceptive processing network such that the output of this network, for a given input, is increased. This increased output may arise from modifications of the intrinsic currents in dorsal horn neurons, enhancement of excitatory inputs or depression of inhibitory control mechanisms, or a combination of these. These diverse mechanisms have been subject to several recent reviews (34, 35, 42, 97). Here I restrict the discussion to mechanisms of persistent enhancement of excitatory synaptic transmission which leads to increased nociceptive output by facilitating the input of primary afferent neurons onto nociceptive neurons in the dorsal horn, or between dorsal horn neurons themselves.

2 Synaptic Transmission in the Dorsal Horn

2.1 Primary Afferent Transmission onto Dorsal Horn Neurons is Excitatory

Synaptic input from all primary sensory afferents onto second order neurons in the dorsal horn is excitatory. Like the vast majority of synapses in the CNS, this excitation is mediated by glutamate released from the pre-synaptic terminals which produces an excitatory post-synaptic potential (EPSP) in the dorsal horn neurons. With low frequency action potential discharge of the primary afferents, the EPSP is mediated mainly by activation of the AMPA (31) and kainate (43) subtypes of glutamate receptor. Unlike most regions of the CNS, a large portion of synaptic transmission at primary afferent synapses is mediated by calcium-permeable AMPA receptors (86) providing the opportunity for these receptors to participate in synaptic plasticity. Under normal circumstances the NMDA subtype of glutamate receptor, which is also localized at excitatory primary afferent synapses and is a key to pain neuroplasticity (see below), contributes little to the post-synaptic responses to low-frequency presynaptic action potentials. The lack of NMDAR contribution arises as a consequence of tonic inhibition of current flow by pore blockade by extracellular Mg^{2+} at the resting membrane potential current flow and because the activity *per se* of NMDARs is tonically downregulated by biasing of kinase/phosphatase regulatory systems towards dephosphorylation. It is important to note that even with primary afferents that may be considered 'slow', from the perspective of axonal conduction velocity and which are predominantly nociceptors, the synaptic transmission to second order neurons is the means of this fast glutamatergic transmission (54, 95).

Each individual primary afferent makes few synapses with any given dorsal horn neuron. Thus, in order to generate action potentials in the second order neurons there must be summation of EPSPs from multiple primary afferent nociceptors, or nociceptors together with non-nociceptors in the case of dorsal horn neurons with multi-convergent inputs often referred to as wide-dynamic range neurons. Acute noxious peripheral stimulation, that is normally experienced as nociceptive pain, produces action potential activity of primary afferent nociceptors that signal the onset, duration and intensity of the stimulation. Within the nociceptive neurons in the dorsal horn the EPSPs from these primary afferents, and the resultant action potentials, reflect this afferent input such that discharge rate encodes the intensity of the stimulation. However, there are two main classes of dorsal horn nociceptive neurons: nociceptive-specific neurons, in which action potential discharge is driven only by noxious peripheral stimulation, and the wide-dynamic range neurons which encode innocuous as well noxious stimuli in their action potential discharge frequency.

2.2 Inhibitory Synaptic Transmission in the Dorsal Horn

Although action potential discharge by dorsal horn neurons is driven by glutamatergic EPSPs, the activity of these neurons is powerfully suppressed by inhibitory inputs that act pre- or post-synaptically (16, 47, 95). The predominant inhibition is that produced post-synaptically by GABA and/or glycine, released from dorsal horn inhibitory interneurons which themselves are excited by primary afferent inputs, and as well as by inputs from descending neurons. These neurons evoke fast inhibitory post-synaptic potentials (IPSPs) mediated by $GABA_A$ and glycine receptors, respectively. As is well-known, these receptors are ligand-gated Cl^- channels, the opening of which normally suppresses generation of action potentials by hyperpolarizing the membrane and by "shunting" the post-synaptic cell (63). Post-synaptic inhibition may also be produced by activation of metabotropic receptors, i.e. G-protein coupled receptors, such as $GABA_B$, adenosine and opioid receptors, which typically produce post-synaptic hyperpolarization by means of activating K^+ channels. Discharge of dorsal horn neurons is also inhibited pre-synaptically by suppression of the release of glutamate from primary afferent terminals. This pre-synaptic inhibition involves many of the same chemical mediators that cause post-synaptic inhibition, with receptors localized on the pre-synaptic terminals of primary afferents (96).

2.3 Integration of Excitation and Inhibition Controls Nociceptive Output from the Dorsal Horn

The efficacy of excitatory synaptic transmission and of the coupling between EPSPs and generation of action potential generation are fundamentally important for controlling the activity of neurons in the nociceptive network within the dorsal horn. There are various cellular and molecular mechanisms for increasing, or decreasing, the strength of glutamatergic synapses and the efficiency by which this synaptic input produces action potentials. It is the increase in action potential discharge in dorsal horn nociceptive neurons that is key to pain hypersensitivity. As described below, these mechanisms range from those that are very rapid in onset, and typically also rapidly reversible, to those that are more slowly developing and persistent. The integration of these various mechanisms, based on the short- and long-term past electrical and biochemical history of the dorsal horn neurons, is a major means for controlling the output of the nociceptive dorsal horn network. Understanding the molecular details of these mechanisms also provides the potential for developing new strategies for reversing or inhibiting persistent enhancement of nociceptive transmission.

3 Synaptic Plasticity in the Dorsal Horn Nociceptive Network

Synaptic transfer of information from the periphery to dorsal horn neurons is governed by the nature and amount of the transmitter released by different primary afferents, the density and identity of postsynaptic receptors (ionotropic and metabotropic), the kinetics of receptor activation and ion-channel opening and closing, and the uptake or breakdown of the transmitter. Each of these factors is subject to multiple modulatory influences.

Nociceptive neurons in the dorsal horn exhibit mechanistically distinct forms of activity-dependent enhancement of responsiveness to nociceptive inputs that are often referred to together as "central sensitization" (93). One form is windup, an activity-dependent progressive increase in the response of neurons over the course of a train of inputs (53), another is a heterosynaptic plasticity that outlasts the initiating stimulus for tens of minutes (classic central sensitization) (84, 85) and a third is a homosynaptic long-term potentiation elicited by brief high frequency inputs (66, 69). In addition, there are pre- and postsynaptic transcription-dependent changes in synaptic function which take longer to manifest (hours) but which last for prolonged periods (days) (97). Finally, there are reductions in the level of tonic and phasic inhibition which also facilitate sensory flow (12, 55).

Importantly, the processes producing central sensitization are initiated by inputs of peripheral nociceptors whereas central sensitization is not evoked by inputs from primary afferents that are non-nociceptive. Nociceptive inputs initiate intracellular signaling cascades leading to a concerted modification of neuronal network behaviour consisting of enhancement of excitatory post-synaptic responses. One consequence of this signaling is an increase in the gain of nociceptive output neurons, resulting in increased responses to the same input from primary afferents. A second consequence of sensitization of dorsal horn neurons is unmasking of subthreshold inputs and as such neurons are recruited to become responsive to stimuli that were previously subliminal (96). Thus, when the dorsal horn nociceptive processing system is sensitized, a given noxious peripheral stimulus will evoke increased responses of individual nociceptive dorsal neurons and, as well, there is an increase in the overall number of nociceptive output neurons activated.

3.1 Rapid-Onset and Reversible Enhancement of Excitatory Transmission – Windup

At low rates of discharge primary afferent nociceptors primarily release glutamate. However, noxious peripheral stimulation that is more intense or sustained induces primary afferent nociceptors to discharge at higher frequencies resulting in release of peptide neuromodulators such as substance P and CGRP, together with glutamate,from central nociceptor terminals. These peptides act on their cognate G-protein coupled receptors to produce depolarizing synaptic potentials lasting up to tens of seconds (14). These slow EPSPs can greatly facilitate temporal summation of fast EPSPs with the cumulative depolarization of the dorsal horn neurons that is accentuated by feedforward recruitment of NMDAR current through relief of the Mg^{2+} blockade of the channels (74). The current flow through NMDARs leads to a rise in intracellular Ca^{2+}levels. Moreover, the sustained depolarization may also recruit voltage-gated Ca^{2+} channels, causing a further boost in the level of intracellular Ca^{2+} and triggering plateau potentials mediated by Ca^{2+}-activated non-selective cation channels (56). The end result of these processes is a progressive increase in the action potential discharge in dorsal horn nociceptive neurons during a train of nociceptive peripheral stimuli (95) that is commonly referred to as "windup" (52). The summation leading to windup is reversed within seconds when the initiating input from primary afferent nociceptors ends and there appear to be no long-term consequences to windup. Nevertheless, in conjunction with persisting forms of plasticity, windup can greatly enhance the nociceptive output of the dorsal horn.

3.2 Heterosynaptic Potentiation in Dorsal Horn Nociceptive Neurons

Central sensitization was originally described as a rapid-onset, activity-dependent increase in the excitability of nociceptive dorsal horn neurons as a result of, and outlasting a short barrage nociceptor afferent input for tens of minutes (10, 98). After this form of central sensitization was induced by a brief (as short as 10–20 seconds) intense nociceptor-conditioning stimulus, normally subliminal inputs become sufficient to activate dorsal horn neurons. A striking feature of the increase in synaptic efficacy in dorsal horn neurons is that while it includes those nociceptor central terminal synapses activated by the conditioning or initiating stimulus (a homosynaptic potentiation) it also occurs for synapses not activated by the conditioning/initiating stimulus, such as those made by low-threshold mechanosensitive primary afferents on the dorsal horn neurons (heterosynaptic potentiation) (73). Activity-dependent heterosynaptic central potentiation may manifest within seconds of an appropriate nociceptive conditioning stimulus and if the stimulus is maintained, even at low levels, the central sensitization persists up to several hours or even longer.

As a result of heterosynaptic potentiation, low-threshold sensory fibers activated by innocuous stimuli such as light touch can, after the induction of the heterosynaptic central sensitization activate normally high threshold nociceptive neurons, a prominent feature of pain hypersensitivity known as "allodynia". Thus, although evoked by a non-noxious stimulus in the periphery, allodynia essentially arises as a result of changes in sensory processing within the CNS. This is prominent in the neurons in the most superficial dorsal horn, lamina I, where the functional phenotype of neurons switches from purely nociceptive to nociceptive/non-nociceptive (37).

Heterosynaptic potentiation in nociceptive dorsal horn neurons is realized as a decrease in discharge threshold due to the recruitment of previously subthreshold non-nociceptive afferent inputs; as an increase in the responsiveness of dorsal horn neurons, i.e. an increase in the number of action potentials elicited by a suprathreshold input; and as an expansion of the excitatory cutaneous or deep receptive fields. The unmasking of subthreshold inputs causes the neurons to become sensitive to stimuli in surrounding regions of the periphery producing a spread of sensitivity beyond the site of the tissue damage which may be a neural substrate for secondary hyperalgesia – the phenomena whereby stimuli outside an area of injury are experienced as painful.

Heterosynaptic potentiation is produced as a consequence of the engagement of multiple intracellular signaling cascades in dorsal horn neurons, cascades that were effectively dormant before activation by fast neurotransmitters such as glutamate or modulators such as substance P, BDNF, or Eph-B. This leads to an orchestrated modification of neuronal behaviour consisting of enhanced excitatory post-synaptic responses and depressed inhibition (34, 95). Two major classes of mechanisms contribute to the resultant increased synaptic efficacy: i) post-translational processing of ion channel/receptors or associated regulatory proteins altering intrinsic functional properties, and ii) cell-surface expression and trafficking of channels in primary sensory and dorsal horn neurons due to the trafficking of receptors to the membrane.

Calcium influx through NMDARs is the principal, but not sole means, by which heterosynaptic potentiation is induced in dorsal horn neurons (95). Activation of voltage-gated calcium channels can enhance excitatory transmission through mechanisms not involving NMDA receptors (39). Neurotrophins, such as BDNF, acting through their cognate trk receptors facilitate synaptic transmission (38, 50) through mechanisms that are in part dependent on post-synaptic NMDA receptors, as further described in the section on microglia-neuron signaling. The intracellular pathways that lead to central sensitization involve interactions of multiple serine/threonine and tyrosine kinase signaling cascades. Activation of several of these protein kinases leads to the phosphorylation of ionotropic glutamate receptors (NMDA and AMPA) increasing synaptic efficacy by altering channel open time, increasing bursting, suppression of the Mg^{2+} channel blockade and promoting the trafficking of receptors to the synaptic membrane (95).

Deletion of genes for isoforms of adenyl cyclase, PKA, and PKC all impair the development of pain hypersensitivity in transgenic mice (49, 91). Noxious stimuli and inflammation induce phosphorylation of NR1 and NR2A/B NMDA receptor subunits in dorsal horn neurons (106). A major mediator of NMDA receptor tyrosine phosphorylation is the non-receptor tyrosine kinase Src, which is activated by Gq coupled GPCRs (29) as well as by TrkB and EphB receptors, and subsequently enhances NMDA receptor currents (see below). That the NMDA receptor plays a key role in heterosynaptic potentiation has been revealed both pharmacologically and more recently by a conditional knockout of the NR1 NMDA receptor subunit which eliminates both NMDA synaptic currents and a behavioral manifestation of acute central sensitization (76). There is emerging evidence that additional serine/threonine and tyrosine kinases are activated and control phosphorylation of NMDA and AMPA receptors in dorsal horn neurons in response to different nociceptor inputs. One major modulator may be the MAP kinase ERK (32, 33). ERK is activated (phosphorylated) in dorsal horn neurons following nociceptive afferent input and inhibition of its activation suppresses behavioral manifestations of central sensitization. Excitability and EPSC-action potential coupling in dorsal horn neurons can also be controlled through ERK regulation of Kv4.2 channels, the major contributor to A-type potassium currents (28), and it is likely that there are other substrates downstream of ERK including NMDA and AMPA receptors (34). Moreover, the balance between metabotropic glutamate (mGluR) and GABA(B) receptor activation can switch the intrinsic firing properties of deep dorsal horn neurons from a tonic to a plateau or even an endogenous bursting pattern, through modulation of the inwardly rectifying potassium channel (Kir3) and this contributes to persistent heterosynaptic potentiation (15).

3.3 Long-Term Potentiation in Dorsal Horn Nociceptive Neurons

A conceptually simple means to sensitize central pain transmission neurons is to homosynaptically increase the efficacy of the excitatory primary afferent inputs onto these neurons. Brief duration, high-frequency primary afferent stimulation does induce a potentiation of AMPA-receptor mediated responses at synapses onto second order neurons (34, 69). The potentiation is prevented by pharmacological blockade of NMDA receptors, and persists for as long as experimentally observable, up to many hours. The lasting enhancement of monosynaptic excitatory synaptic responses at primary afferent-second order synapses in pain pathways shares common signaling cascades with the NMDA receptor-dependent form of LTP of excitatory synaptic transmission observed in many regions of the CNS, which resemble many aspects of the signaling responsible for heterosynaptic central sensitization described above.

The mechanisms of NMDA receptor-dependent LTP have been examined in most detail for Schaffer collateral synapses onto CA1 neurons in the hippocampus, as described in detail in other chapters. At Schaffer collateral-CA1 synapses LTP requires calcium influx through NMDA receptors during the tetanic stimulation, which is accomplished by temporal summation of EPSPs, thereby diminishing Mg^{2+} suppression of NMDA channel activity and by enhancement of NMDA channel function by the tyrosine kinase Src (46). The activation of Src results from stimulation of another tyrosine kinase, CAKß/Pyk2 (29). The effect of Src activation counters tonic depression of NMDA channel activity by the phosphotyrosine phosphatase, STEP (61). Tetanic stimulation of Schaffer collaterals has been demonstrated to increase the level of tyrosine phosphorylation of the NMDA receptor subunit NR2B in CA1 neurons, predominantly on Y1472 of NR2B, indicating effects through direct phosphorylation of the NMDA receptor itself. A coincident rise in post-synaptic sodium concentration may additionally contribute to boosting NMDA receptor activity (101). The function of NMDARs can be enhanced by increasing the activity of Src which can be directly achieved through the kinase CAKβ (29) which binds to the SH2 domain of Src, relieving autoinhibition. This autoinhibition can also be relieved by the phosphatase PTPα (41) and can be enhanced by the kinase Csk (99), which increase and decrease NMDAR function, respectively. Thus, PTPα and Csk may also participate in LTP induction.

It is hypothesized that the multiplicative boost in NMDA receptor function and resultant dramatic enhancement influx of calcium sets off a cascade leading to activation of calcium/calmodulin dependent kinase II [CAMKII] and phosphorylation of the AMPA receptor subunit protein GluR1 which causes AMPA channels to open in a high conductance state (48, 72, 75). Phosphorylation of AMPA receptors may also cause increased cell surface expression of AMPA receptors and allows conversion of "silent synapses", those lacking AMPA receptors, into active ones.

The general form of this core signaling cascade – NMDA receptor activation leading to post-synaptic enhancement of AMPA receptor function or cell-surface expression – is likely applicable in spinal nociceptive transmission neurons (Fig. 1). NMDA receptors are expressed in the dorsal horn and are necessary for inducing central sensitization. The key tyrosine kinases Src and CAKß/Pyk2, as well as the phosphotyrosine phosphatase STEP, are also all expressed in spinal cord, and are physically associated with spinal NMDA receptors (61, 68, 101). In dorsal horn neurons, NMDAR function is enhanced by CAKß/Pyk2 – Src signaling and is depressed by STEP and tyrosine phosphorylation of the NR2B subunit of the NMDAR has been shown in the dorsal horn (23). Upregulation of NMDAR function by signaling pathways mediated by Src appears to be crucial for the initiation and maintenance of the enhanced responsiveness of nociceptive neurons in the dorsal horn of the spinal cord that occurs in experimental pain models(101). In vivo, tyrosine phosphorylation of NR2B in the spinal cord increases with the development and maintenance of hyperalgesia in a rat model of inflammation(70), in which hyperalgesia depends on NMDARs. Inhibition of Src in vivo delays the onset of hyperalgesia, and inhibition of Src, PKC or group I mGluRs prevents the increase in NR2B tyrosine phosphorylation (23), indicating that a GPCR signaling cascade upstream of SFK-mediated NMDAR upregulation might be required for pain-related maladaptive changes in synaptic transmission. Activation of EphB in the spinal cord with ephrinB2 results in prolonged hyperalgesia that depends on NMDARs and is associated with an increase in tyrosine phosphorylation of EphB-associated Src (6). Thus, the convergence of multiple signaling pathways on Src allows plasticity in the dorsal horn, which is likely to be mediated through upregulation of NMDARs by these kinases. Moreover, there is evidence for silent synapses in dorsal horn neurons

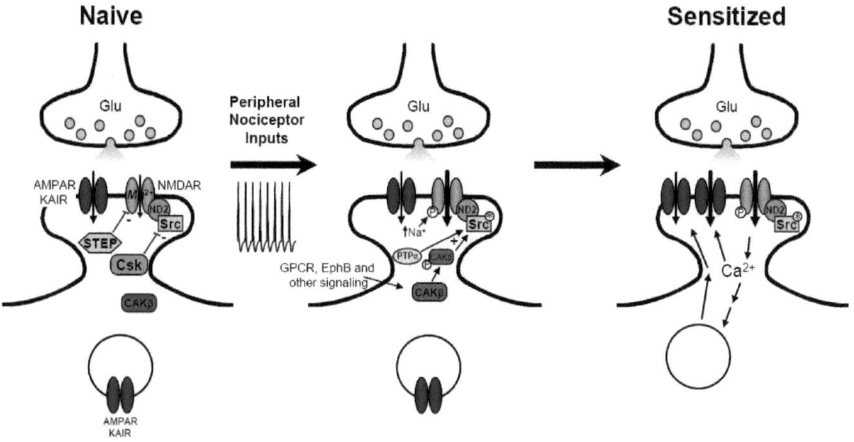

Fig. 1. A model for the role of synaptic potentiation in nociceptive dorsal horn neurons in pain hypersensitivity. *Left*; under basal conditions NMDAR activity is suppressed by partial blockade of the channel by Mg^{2+} and by the activity of the PTP STEP and the kinase Csk. KAIR, kainate receptor. Middle; nociceptive input increases current through NMDARs by relief of Mg^{2+} inhibition, by activation of Src (Src∗) via the actions of PTPα and activated CAKß (CAKß-P) which overcomes the suppression by STEP, and by sensitizing the NMDARs to raised intracellular [Na^{+}]. Right; upregulation of NMDAR function allows a large boost in entry of Ca^{2+} which binds to calmodulin (CaM) causing activation of CaMKII, not illustrated. The enhancement of glutamatergic transmission is ultimately expressed through increased number of AMPA/KAI receptors in the postsynaptic membrane and/or enhanced AMPA/KAI receptor activity. Adapted from Kalia et al, 2004.

and for conversion of these to active synapses, a process requiring PDZ domain interactions of AMPA receptors (44). As in many regions of the CNS, silent synapses in the dorsal horn are most prominent at early developmental stages (5). Thus, in the adult dorsal horn LTP may be expressed by an NMDA-receptor mediated enhanced single-channel conductance of AMPA channels and enhanced cell-surface expression of AMPA receptors (34).

The applicability of the entire signaling cascade described in CA1 is likely limited to a subpopulation of neurons in the dorsal horn. Administering exogenous CAMKII has been shown to enhance AMPA responses of dorsal horn neurons and pharmacological inhibition of CAMKII inhibits neuronal and behavioural responses to intradermal capsaicin. However, expression of endogenous CAMKII is highly restricted within the dorsal horn. A protein kinase which might substitute for CAMKII is protein kinase C (PKC) which potentiates synaptic transmission in dorsal horn neurons. In CA1, PKC has been implicated in initiating LTP but evidence indicates that it is likely upstream of Src and its effects mediated via the protein tyrosine kinase CAKß/Pyk2 (29). In the dorsal horn PKC could play a dual role, phosphorylating AMPA receptors and stimulating CAKß/Pyk2-Src signaling, or alternatively, phosphorylation of AMPA receptors may be produced by an as yet unidentified serine/threonine kinase.

Another kinase signaling cascade that appears required for induction of LTP in CA1 is the mitogen-activated protein kinase (MAPK) pathway. MAPK is activated upon phosphorylation by MEK and inhibitors of MEK block induction of LTP. Importantly, the early phase of LTP is prevented by MEK inhibition at times too early to be accounted for by changes in gene expression. In the superficial dorsal horn, MAPK phosphorylation increases following nociceptive stimulation and inhibiting MEK suppresses the second phase of the formalin test, indicative of suppression of central responsiveness. Thus, it has been hypothesized that the MAPK pathway is necessary for amplification in spinal nociceptive pathways but it is not known if it has a role in spinal LTP (34).

An additional mechanism for lasting enhancement of excitatory transmission is through activation of AMPA receptors lacking the edited form of the GluR2 subunit. While AMPA receptors are generally impermeable to Ca^{2+}, by virtue of containing the edited form of the GluR2 subunit (17, 71), AMPA receptors lacking edited GluR2 do have a high permeability to Ca^{2+}, which provides a source for post-synaptic Ca^{2+} that may mediate synaptic plasticity. Neurons expressing Ca^{2+}-permeable AMPA receptors are preferentially localized in the superficial dorsal horn (2). Lasting enhancement of synaptic transmission mediated by GluR2-less AMPA receptors by a post-synaptic mechanism has been demonstrated at dorsal horn synapses (22) and may participate in enhancing nociceptive transmission (25). Such LTP may contribute to certain types of amplification of responsiveness of neurons in pain pathways.

In addition to the post-synaptic mechanisms described above, pre-synaptic increase in the release of glutamate might also result in sustained increase in the gain of pain pathways. This could be produced by direct facilitation of transmitter release or by suppression of tonic presynaptic inhibition. Release of transmitter could be enhanced by stimulating receptors on primary afferent terminals, including P2X3 receptors and NMDA autoreceptors. However, the effects of such receptor stimulation may be relatively short-lived.

The lasting potentiation of synaptic transmission at Schaffer collateral synapses onto CA1 neurons in the hippocampus is a form of homosynaptic potentiation, i.e. the potentiation is restricted to synapses that are activated at the time of stimulation. Homosynaptic potentiation of AMPA-receptor responses at synapses on dorsal horn neurons can occur experimentally in response to brief duration, high-frequency nociceptor stimulation (66). The potentiation is restricted to the activated synapse, and is persistent. Nociceptors do not usually fire at high frequencies, and therefore homosynaptic potentiation was thought to be limited to very intense stimuli producing a spatially and modality constrained, if long lasting, facilitation. However, it has been recently demonstrated that homosynaptic LTP may occur at primary afferent firing rates that match those produced by peripheral noxious stimulation (69). Nevertheless, homosynaptic LTP cannot account for the spatial spread of pain hypersensitivity nor for the facilitation of non-nociceptive inputs in nociceptive dorsal horn neurons, these common features of clinical chronic pain can only be accounted for by heterosynaptic potentiation.

4 Microglia-Neuron Signaling Mediates Enhanced Transmission after Peripheral Nerve Injury

Although outnumbering neurons in the CNS glial cells have traditionally been viewed as cells that serve primarily housekeeping roles. However, this perspective has changed dramatically over the past decade and glia are known to play key roles in regulating synaptic transmission and participating in synaptic plasticity (4, 20, 59, 103). In particular for pain resulting from peripheral nerve injury, there is a growing body of evidence indicating that hyperalgesia, allodynia and ongoing pain involve active participation of glia in the spinal dorsal horn (88, 92). Of the three types of CNS glia – astrocytes, oligodendrocytes and microglia – it is the microglia that have emerged as a central player in enhanced pain behaviours after peripheral nerve-injury (88, 89).

Under basal conditions microglia are considered to be in a resting state. The cells can respond to various stimuli that threaten physiological homeostasis, which cause the microglia to undergo a variety of stimulus-dependent changes in morphology, gene expression, function and number (62). The responding microglia typically change their morphology from a resting, ramified shape into an active, amoeboid shape (40, 57, 77, 79). In addition, responding microglia upregulate expression of a variety of cell-surface molecules including the complement receptor 3 [CR3; also known as CD11b/CD18 or Mac-1; comprised of integrin subunits αM (CD11b) and β2 (CD18)(18), which is recognized by the antibody OX-42 (67). Activated microglia also express immunomolecules such as major histocompatibility complex (MHC) class I and II (40, 77, 79) which play a role in antigen presentation to T-lymphocytes. Activated microglia produce and release various chemical mediators, including proinflammatory cytokines, that can produce immunological actions and can also act on neurons to alter their function (24, 57, 77).

4.1 Microglia as Mediators of Pain Hypersensitivity Following Peripheral Nerve Injury

Microglia in the spinal cord have been examined in numerous models of peripheral nerve injury (7). A common feature of these models, that is often considered to be clinically relevant, is that after the nerve injury withdrawal behaviours are evoked by stimuli that are innocuous, like gentle mechanical stimulation. The potential clinical relevance of the behaviour change is that allodynia is a common, and often very severe, component of neuropathic pain in humans (30, 94, 105). Following peripheral nerve injury there is a series of changes in microglia within spinal dorsal horn (13, 19, 45, 83), although there may be some variability between the models in terms of the extent and time course of the changes (8, 9). Within hours after PNI, the microglia start to become activated as evidenced by the small soma becoming hypertrophic and the long and thin processes withdrawing(19). Subsequently, microglia proliferate, with this peaking about 3 days after nerve injury (21). The microglia show an increased level of a number of activation 'marker' proteins including CR3 (13, 19, 45, 89), toll-like receptor 4 (TLR4) (83), cluster determinant14 (CD14) (83), CD4 (82) and MHC class II protein(81, 82). Many reports have shown a correlation between signs of activation of microglia and peripheral nerve injury-induced pain hypersensitivity. But that microglia have a causal role in these nerve injury-evoked pain behaviours was demonstrated in studies implicating $P2X_4$ receptors (89) and p38 MAP kinase (36).

$P2X_4$ receptors form a subtype of the P2X family of ligand-gated ion channels activated by ATP (58). These receptors were implicated in mechanical allodynia following spinal nerve ligation because the behavioural sensitization is acutely reversed by means of intrathecally administering blockers of $P2X_4$ receptors (89). Moreover, inhibiting $P2X_4$ receptor expression prevents the development of allodynia. Surprisingly at the time, $P2X_4$ receptors in the dorsal horn are expressed by microglia rather than neurons or astrocytes; this expression is low in the naïve spinal cord but progressively increases in the days following nerve injury paralleling the development of mechanical allodynia. In addition, in otherwise naïve animals, mechanical allodynia develops following the intrathecal administration of microglia in which $P2X_4$ receptors are stimulated in vitro. Unstimulated microglia do not cause allodynia nor does administering vehicle or ATP-only controls. Thus, stimulating $P2X_4$ receptors in microglia responding to peripheral nerve injury appears to be both necessary and sufficient for producing mechanical allodynia.

Inhibiting p38 MAP kinase pharmacologically, by means of an inhibitor administered intrathecally, reverses mechanical allodynia following spinal nerve ligation (36). Infusing the p38 inhibitor beginning prior to the nerve injury prevented allodynia from developing. The nerve lesion leads to persistent activation of p38 MAP kinase, as assessed by phospho-p38 MAP kinase labeling, which is restricted to microglia.

Together the findings from these two studies suggest that $P2X_4$ receptors and p38 MAP kinase in dorsal horn microglia are central to allodynia and neuropathic pain following nerve injury. Because $P2X_4$ receptors are permeable to Ca^{2+}, it is possible that stimulation of these receptors leads to Ca^{2+} influx that indirectly activates p38 MAP kinase, but this remains to be investigated. Activated microglia are functionally and phenotypically heterogeneous with various microglial proteins altered in different time courses and amounts, depending upon the stimulus that produces the activation (40, 57, 77, 78). The heterogeneity observed in the responses of spinal microglia to stimulation in the periphery may depend upon the type of nociceptive stimulus. For example, expression of CR3 is elevated in the spinal cord following various types of peripheral nerve injury (see above and (64, 102)). Increases in CR3 levels are also observed following stimulation that produces inflammation peripherally (65, 80), although this is not a uniform finding as there are reports that peripheral inflammatory stimulation may not lead to an increase in CR3 (27, 102) [48,67]. TLR4 and CD14 are also increased following peripheral nerve injury [24] or peripheral inflammation by complete Freund's adjuvant [66]. In contrast, expression of $P2X_4R$(89), CCR2 (1), CB2 (102) and MHC Class II (82) is increased in spinal microglia following peripheral nerve injury but not following peripheral inflammation. Thus, while there are some similarities in the spinal microglial proteins upregulated following peripheral nerve injury compared with inflammation, there are also differences in the responses of the microglia that depend upon the type of stimulus in the periphery.

4.2 Microglia Signaling to Neurons: Role of BDNF

Pain hypersensitivity following nerve injury depends upon increased responsiveness of neurons in the dorsal horn of the spinal cord and thus to effect a change in the nociceptive processing network the microglia must signal to neurons within the dorsal horn. While this enhanced responsiveness could potentially arise from alteration of intrinsic passive or active membrane conductances, increased excitatory drive or suppressed inhibition, it is known that after peripheral nerve injury there is a prominent suppression of inhibition in dorsal horn neurons, that is critical for producing mechanical allodynia (12, 55). In particular, allodynia following peripheral nerve injury involves reduction in the expression of the K^+-Cl^- co-transporter KCC2, normally responsible for the extrusion of Cl^- from the cell, in neurons of spinal lamina I (12). As a consequence, after peripheral nerve injury there is a rise in intracellular Cl^- in lamina I dorsal horn neurons, resulting in a depolarizing shift in the anion reversal potential (E_{anion}) and hyperexcitability by means of dramatically reducing $GABA_A$-ergic and glycinergic inhibition. In about one-third of lamina I neurons, GABA produces frank excitation rather than inhibition. Local blockade or knock-down of spinal KCC2 in naïve rats markedly reduces nociceptive threshold, confirming that disruption of anion homeostasis in lamina I neurons is sufficient to produce neuropathic pain behaviours. Although sensitization after peripheral nerve injury is displayed by many cells in both the superficial and deep laminae of the dorsal horn, the impact in terms of pain hypersensitivity appears to be most important for lamina I spinothalamic or spinoparabrachial projection neurons, particularly those expressing the NK1 receptor (3, 26).

The possibility that stimulating $P2X_4$ receptors on microglia engages this mechanism was shown by Coull et al. (2005), who found that intrathecal injection of $P2X_4$ receptor-stimulated cultured microglia causes a rise in intracellular Cl^- in lamina I neurons that reduces inhibition and converts $GABA_A$-ergic responses to excitation. Moreover, pharmacological blockade of $P2X_4$ receptors acutely reverses the rise in intracellular Cl^- in lamina I neurons after peripheral nerve injury. These findings unify, seemingly disparate observations that $P2X_4$ receptors on microglia and raised intracellular Cl^- concentrations in lamina I neurons are each necessary for maintaining pain behaviours after peripheral nerve injury.

Thus, the issue of how microglia signal to neurons reduces to how microglia signal to lamina I neurons to effect the shift in E_{anion} that contributes to enhancing neuronal responses. This signaling could occur through the release of one or more diffusible chemical messengers from microglia upon stimulation of $P2X_4$ receptors. Indeed, activated microglia are known to secrete a number of diffusible chemical messengers including interleukin-1β, interleukin-6, and tumor necrosis factor-α. Release of these factors in the spinal cord has been hypothesized to contribute to nerve injury-induced pain hypersensitivity (51, 88, 90).

Microglia also express and release neurotrophic factors such as BDNF, which in hippocampal neurons mediates anion gradient shifts leading to increased Cl^- (60). In addition, it has been reported that BDNF heterozygous null mutant mice show attenuated pain hypersensitivity following nerve injury as compared to wild-type (100). Thus, these lines of evidence suggest that BDNF is a primary candidate signaling molecule between microglia and neurons. Consistent with this, intrathecal application of BDNF mimicked the mechanical allodynia and alteration in E_{anion} caused by peripheral nerve injury or by administering $P2X_4$ receptor-stimulated microglia, demonstrating that BDNF is sufficient to cause tactile allodynia and to raise intracellular Cl^- in lamina I neurons. The mechanical allodynia produced by peripheral nerve injury is reversed by a function-blocking antibody against TrkB, the cognate receptor for BDNF, and by a BDNF sequestering fusion protein (TrkB-Fc), implying that BDNF-TrkB signaling is necessary for neuropathic pain hypersensitivity.

The function-blocking TrkB antibody and the TrkB-Fc prevented mechanical allodynia evoked by administering $P2X_4$ receptor-stimulated microglia. Moreover, treatment of the cultured microglia with double-stranded interference RNA against BDNF prevented the effects of intrathecally administered microglia on lamina I neurons and on the microglia-elicited pain behaviours suggesting that the requisite BNDF may derive from the microglia rather than from some other source. $P2X_4$ receptor stimulation by ATP caused release of BDNF from microglia in culture, a response prevented by TNP-ATP or by knocking down expression of BDNF with siRNA. Collectively, the results provide the basis for a conceptual model whereby $P2X_4$ receptor stimulation of microglia causes release of BDNF, which then acts on dorsal horn neurons to raise intracellular Cl^- and resultant enhancement of neuronal responsiveness (Fig. 2). It remains to be shown directly that native microglia in the spinal dorsal horn respond to peripheral nerve injury by releasing BDNF. Another potential source of BDNF in the dorsal horn is small diameter sensory afferents. However, selectively eliminating BDNF expression in these sensory neurons, which blocks inflammatory pain hypersensitivity, has no effect on peripheral nerve injury-induced mechanical allodynia (104). Thus, while BDNF-TrkB signaling is required for maintaining neuropathic pain hypersensitivity (11), BDNF expression by primary sensory neurons is not required, consistent with microglia as the source of BDNF.

4.3 Microglia-Induced Disinhibition and NMDAR Synaptic Transmission

GABA- and glycinergic inhibition within the dorsal horn strongly suppresses polysynaptic synaptic inputs from primary afferents to lamina I neurons, such that these neurons mainly discharge only in response to inputs from primary afferent nociceptors. Pharmacological blockade of $GABA_A$ and glycine receptors unmasks polysynaptic excitatory synaptic transmission (87), from inputs which largely arise from non-nociceptive primary afferents, switching the functional phenotype of lamina I neurons from nociceptive only to nociceptive/non-nociceptive (37). Acute inhibition of KCC2 or applying ATP-stimulated microglia to the spinal cord in vivo similarly unmasks excitatory non-nociceptive synaptic inputs to lamina I neurons (37). In vitro, the unmasking of polysynaptic inputs by blocking GABAA and glycine receptors depends upon NMDARs (87). Thus, by causing disinhibition in the dorsal horn, microglia-neuron signaling following peripheral nerve injury may facilitate NMDAR-dependent polysynaptic transmission. As such, it is anticipated that the intracellular biochemical cascades that regulate NMDARs in dorsal horn neurons may be critical for pain hypersensitivity induced by peripheral nerve injury.

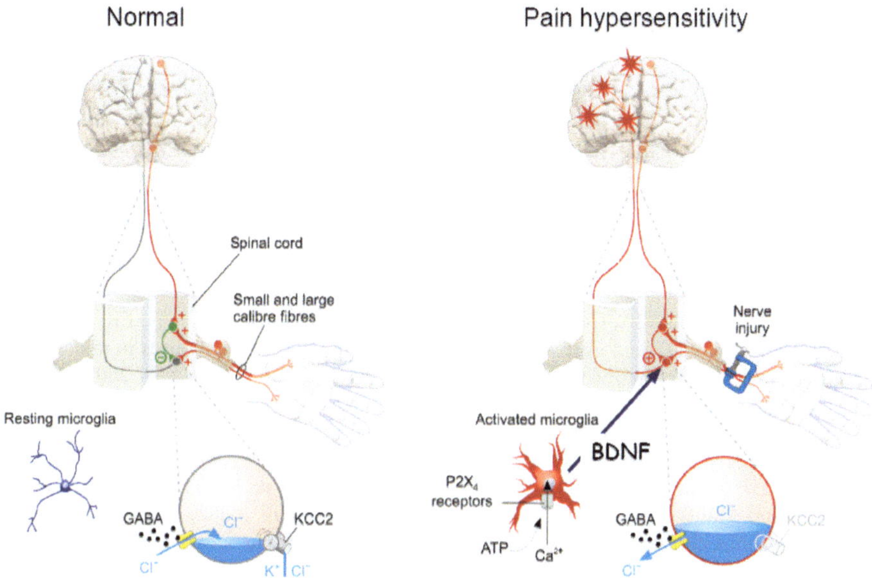

Fig. 2. A model for pain hypersensitivity induced by peripheral nerve injury. *Left*; Under normal conditions, intracellular Cl^- concentration is sufficiently low that opening $GABA_A$, or glycine, receptors Cl^- flows into the dorsal horn nociceptive neurons causing hyperpolarizing inhibition that suppresses transmission. *Right*; After peripheral nerve injury microglia in the dorsal horn become activated and upregulate expression of $P2X_4$ receptors, stimulation of which causes the microglia to release BDNF. BDNF, acting on TrkB receptors, causes downregulation of KCC2 causing intracellular Cl^- concentration to increase thereby decreasing inhibition of the neurons.

5 Conclusions and Future Directions

Overall, it is evident that enhancement of excitatory synaptic responses of projection neurons within the nociceptive network within the dorsal horn, directly or through suppression of inhibition, is one of the key processes which increases the gain of nociceptive transmission and leads to pain hypersensitivity. The elucidation of the molecular mechanisms underlying this enhanced transmission has led to potential targets for new types of analgesics. It has become apparent that the key molecular mechanisms are not restricted to neurons but also involve glial cells in the spinal cord, with most of the existing evidence implicating microglia. Thus, understanding how microglia-neuron signaling impacts synaptic transmission and plasticity may lead to additional new strategies for the diagnosis and management of neuropathic pain, strategies not previously anticipated by a view of pain plasticity focused solely on neurons.

Acknowledgments

The work of the author is supported by grants from the Canadian Institutes of Health Research and from Neuroscience Canada. MWS holds a Canada Research Chair (Tier I), in Neuroplasticity and Pain, and is an International Scholar of the Howard Hughes Medical Institute. Thanks to Yves De Koninck for assistance with Fig. 2.

References

1. Abbadie C, Lindia JA, Cumiskey AM, Peterson LB, Mudgett JS, Bayne EK, DeMartino JA, MacIntyre DE and Forrest MJ. Impaired neuropathic pain responses in mice lacking the chemokine receptor CCR2. *Proc Natl Acad Sci USA* 100: 7947–7952, 2003.
2. Albuquerque C, Lee CJ, Jackson AC and MacDermott AB. Subpopulations of GABAergic and non-GABAergic rat dorsal horn neurons express Ca2+-permeable AMPA receptors. *Eur J Neurosci* 11: 2758–2766, 1999.
3. Allen BJ, Rogers SD, Ghilardi JR, Menning PM, Kuskowski MA, Basbaum AI, Simone DA and Mantyh PW. Noxious cutaneous thermal stimuli induce a graded release of endogenous substance P in the spinal cord: imaging peptide action in vivo. *J Neurosci* 17: 5921–5927, 1997.
4. Auld DS and Robitaille R. Glial cells and neurotransmission: an inclusive view of synaptic function. *Neuron* 40: 389–400, 2003.
5. Baba H, Doubell TP, Moore KA and Woolf CJ. Silent NMDA receptor-mediated synapses are developmentally regulated in the dorsal horn of the rat spinal cord. *J Neurophysiol* 83: 955–962, 2000.
6. Battaglia AA, Sehayek K, Grist J, McMahon SB and Gavazzi I. EphB receptors and ephrin-B ligands regulate spinal sensory connectivity and modulate pain processing. *Nat Neurosci* 6: 339–340, 2003.
7. Beggs S and Salter MW. Neuropathic pain: symptoms, models and mechanisms. *Drug Dev Res* 67: 289–301, 2006.
8. Colburn RW, DeLeo JA, Rickman AJ, Yeager MP, Kwon P and Hickey WF. Dissociation of microglial activation and neuropathic pain behaviors following peripheral nerve injury in the rat. *J Neuroimmunol* 79: 163–175, 1997.
9. Colburn RW, Rickman AJ and DeLeo JA. The effect of site and type of nerve injury on spinal glial activation and neuropathic pain behavior. *Exp Neurol* 157: 289–304, 1999.
10. Cook AJ, Woolf CJ, Wall PD and McMahon SB. Dynamic receptive field plasticity in rat spinal cord dorsal horn following C-primary afferent input. *Nature* 325: 151–153, 1987.
11. Coull JA, Beggs S, Boudreau D, Boivin D, Tsuda M, Inoue K, Gravel C, Salter MW and De Koninck Y. BDNF from microglia causes the shift in neuronal anion gradient underlying neuropathic pain. *Nature* 438: 1017–1021, 2005.
12. Coull JA, Boudreau D, Bachand K, Prescott SA, Nault F, Sik A, De Koninck P and De Koninck Y. Trans-synaptic shift in anion gradient in spinal lamina I neurons as a mechanism of neuropathic pain. *Nature* 424: 938–942, 2003.
13. Coyle DE. Partial peripheral nerve injury leads to activation of astroglia and microglia which parallels the development of allodynic behavior. *Glia* 23: 75–83, 1998.
14. De Koninck Y and Henry JL. Substance P-mediated slow excitatory postsynaptic potential elicited in dorsal horn neurons in vivo by noxious stimulation. *Proc Natl Acad Sci USA* 88: 11344–11348, 1991.

15. Derjean D, Bertrand S, Le MG, Landry M, Morisset V and Nagy F. Dynamic balance of metabotropic inputs causes dorsal horn neurons to switch functional states. *Nat Neurosci* 6: 274–281, 2003.
16. Dickenson AH, Chapman V and Green GM. The pharmacology of excitatory and inhibitory amino acid-mediated events in the transmission and modulation of pain in the spinal cord. *Gen Pharmacol* 28: 633–638, 1997.
17. Dingledine R, Borges K, Bowie D and Traynelis SF. The glutamate receptor ion channels. *Pharmacol Rev* 51: 7–61, 1999.
18. Ehlers MR. CR3: a general purpose adhesion-recognition receptor essential for innate immunity. *Microbes Infect* 2: 289–294, 2000.
19. Eriksson NP, Persson JK, Svensson M, Arvidsson J, Molander C and Aldskogius H. A quantitative analysis of the microglial cell reaction in central primary sensory projection territories following peripheral nerve injury in the adult rat. *Exp Brain Res* 96: 19–27, 1993.
20. Fields RD and Stevens-Graham B. New insights into neuron-glia communication. *Science* 298: 556–562, 2002.
21. Gehrmann J and Banati RB. Microglial turnover in the injured CNS: Activated microglia undergo delayed DNA fragmentation following peripheral nerve injury. *J Neuropathol Exp Neurol* 54: 680–688, 1995.
22. Gu JG, Albuquerque C, Lee CJ and MacDermott AB. Synaptic strengthening through activation of Ca2+-permeable AMPA receptors. *Nature* 381: 793–796, 1996.
23. Guo W, Zou S, Guan Y, Ikeda T, Tal M, Dubner R and Ren K. Tyrosine Phosphorylation of the NR2B Subunit of the NMDA Receptor in the Spinal Cord during the Development and Maintenance of Inflammatory Hyperalgesia. *J Neurosci* 22: 6208–6217, 2002.
24. Hanisch UK. Microglia as a source and target of cytokines. *Glia* 40: 140–155, 2002.
25. Hartmann B, Ahmadi S, Heppenstall PA, Lewin GR, Schott C, Borchardt T, Seeburg PH, Zeilhofer HU, Sprengel R and Kuner R. The AMPA receptor subunits GluR-A and GluR-B reciprocally modulate spinal synaptic plasticity and inflammatory pain. *Neuron* 44: 637–650, 2004.
26. Honor P, Menning PM, Rogers SD, Nichols ML, Basbaum AI, Besson JM and Mantyh PW. Spinal substance P receptor expression and internalization in acute, short-term, and long-term inflammatory pain states. *J Neurosci* 19: 7670–7678, 1999.
27. Honore P, Rogers SD, Schwei MJ, Salak-Johnson JL, Luger NM, Sabino MC, Clohisy DR and Mantyh PW. Murine models of inflammatory, neuropathic and cancer pain each generates a unique set of neurochemical changes in the spinal cord and sensory neurons. *Neuroscience* 98: 585–598, 2000.
28. Hu HJ, Carrasquillo Y, Karim F, Jung WE, Nerbonne JM, Schwarz TL and Gereau RW. The kv4.2 potassium channel subunit is required for pain plasticity. *Neuron* 50: 89–100, 2006.
29. Huang Y, Lu W, Ali DW, Pelkey KA, Pitcher GM, Lu YM, Aoto H, Roder JC, Sasaki T, Salter MW and MacDonald JF. CAKß/Pyk2 kinase is a signaling link for induction of long-term potentiation in CA1 hippocampus. *Neuron* 29: 485–496, 2001.
30. Jensen TS and Baron R. Translation of symptoms and signs into mechanisms in neuropathic pain. *Pain* 102: 1–8, 2003.
31. Jessell TM and Jahr CE. Fast and Slow Excitatory Transmitters at Primary Afferent Synapses in the Dorsal Horn of the Spinal-Cord. In: Advances in Pain Research and Therapy, New York: Raven Press, p. 31–39, 1985.
32. Ji RR, Baba H, Brenner GJ and Woolf CJ. Nociceptive-specific activation of ERK in spinal neurons contributes to pain hypersensitivity. *Nat Neurosci* 2: 1114–1119, 1999.

33. Ji RR, Befort K, Brenner GJ and Woolf CJ. ERK MAP kinase activation in superficial spinal cord neurons induces prodynorphin and NK-1 upregulation and contributes to persistent inflammatory pain hypersensitivity. *J Neurosci* 22: 478–485, 2002.
34. Ji RR, Kohno T, Moore KA and Woolf CJ. Central sensitization and LTP: do pain and memory share similar mechanisms? *Trends Neurosci* 26: 696–705, 2003.
35. Ji RR and Strichartz G. Cell signaling and the genesis of neuropathic pain. *Sci STKE* 2004: reE14, 2004.
36. Jin SX, Zhuang ZY, Woolf CJ and Ji RR. p38 mitogen-activated protein kinase is activated after a spinal nerve ligation in spinal cord microglia and dorsal root ganglion neurons and contributes to the generation of neuropathic pain. *J Neurosci* 23: 4017–4022, 2003.
37. Keller, A. F., Beggs, S., Salter, M. W., and De Koninck, Y. Disrupting anion homeostasis in the spinal dorsal horn replicates the enhanced excitability of Lamina I projection neurons observed following peripheral nerve injury. *Soc Neurosci Abs* 31. 2005.
38. Kerr BJ, Bradbury EJ, Bennett DL, Trivedi PM, Dassan P, French J, Shelton DB, McMahon SB and Thompson SW. Brain-derived neurotrophic factor modulates nociceptive sensory inputs and NMDA-evoked responses in the rat spinal cord. *J Neurosci* 19: 5138–5148, 1999.
39. Koltzenburg M, Lundberg LE and Torebjork HE. Dynamic and static components of mechanical hyperalgesia in human hairy skin. *Pain* 51: 207–219, 1992.
40. Kreutzberg GW. Microglia: a sensor for pathological events in the CNS. *Trends Neurosci* 19: 312–318, 1996.
41. Lei G, Xue S, Chery N, Liu Q, Xu J, Kwan CL, Fu YP, Lu YM, Liu M, Harder KW and Yu XM. Gain control of N-methyl-D-aspartate receptor activity by receptor-like protein tyrosine phosphatase alpha. *EMBO J* 21: 2977–2989, 2002.
42. Lewin GR, Lu Y and Park TJ. A plethora of painful molecules. *Curr Opin Neurobiol* 14: 443–449, 2004.
43. Li P, Wilding TJ, Kim SJ, Calejesan AA, Huettner JE and Zhuo M. Kainate-receptor-mediated sensory synaptic transmission in mammalian spinal cord. *Nature* 397: 161–164, 1999.
44. Li P and Zhuo M. Silent glutamatergic synapses and nociception in mammalian spinal cord. *Nature* 393: 695–698, 1998.
45. Liu L, Tornqvist E, Mattsson P, Eriksson NP, Persson JK, Morgan BP, Aldskogius H and Svensson M. Complement and clusterin in the spinal cord dorsal horn and gracile nucleus following sciatic nerve injury in the adult rat. *Neuroscience* 68: 167–179, 1995.
46. Lu YM, Roder JC, Davidow J and Salter MW. Src activation in the induction of long-term potentiation in CA1 hippocampal neurons. *Science* 279: 1363–1368, 1998.
47. Malcangio M and Bowery NG. GABA and its receptors in the spinal cord. *Trends Pharmacol Sci* 17: 457–462, 1996.
48. Malinow R and Malenka RC. AMPA receptor trafficking and synaptic plasticity. *Annu Rev Neurosci* 25: 103–126, 2002.
49. Malmberg AB, Chen C, Tonegawa S and Basbaum AI. Preserved acute pain and reduced neuropathic pain in mice lacking PKCgamma. *Science* 278: 279–283, 1997.
50. Mannion RJ, Costigan M, Decosterd I, Amaya F, Ma Q, Holstege JC, Ji R, Acheson A, Lindsay RM, Wilkinson GA and Woolf CJ. Neurotrophins: peripherally and centrally acting modulators of tactile stimulus-induced inflammatory pain hypersensitivity. *Proc Natl Acad Sci USA* 96: 9385–9390, 1999.
51. Marchand F, Perretti M and McMahon SB. Role of the immune system in chronic pain. *Nat Rev Neurosci* 6: 521–532, 2005.
52. Mendell LM. Modifiability of spinal synapses. *Physiol Rev* 64: 260–324, 1984.

53. Mendell LM and Wall PD. Responses of single dorsal cord cells to peripheral cutaneous unmyelinated fibres. *Nature* 206: 97–99, 1965.
54. Moore KA, Baba H and Woolf CJ. Synaptic transmission and plasticity in the superficial dorsal horn. *Prog Brain Res* 129: 63–80, 2000.
55. Moore KA, Kohno T, Karchewski LA, Scholz J, Baba H and Woolf CJ. Partial peripheral nerve injury promotes a selective loss of GABAergic inhibition in the superficial dorsal horn of the spinal cord. *J Neurosci* 22: 6724–6731, 2002.
56. Morisset V and Nagy F. Ionic basis for plateau potentials in deep dorsal horn neurons of the rat spinal cord. *J Neurosci* 19: 7309–7316, 1999.
57. Nakajima K and Kohsaka S. Microglia: activation and their significance in the central nervous system. *J Biochem (Tokyo)* 130: 169–175, 2001.
58. North RA. Molecular physiology of P2X receptors. *Physiol Rev* 82: 1013–1067, 2002.
59. Pascual O and Haydon PG. Synaptic inhibition mediated by glia. *Neuron* 40: 873–875, 2003.
60. Payne JA, Rivera C, Voipio J and Kaila K. Cation-chloride co-transporters in neuronal communication, development and trauma. *Trends Neurosci* 26: 199–206, 2003.
61. Pelkey KA, Askalan R, Paul S, Kalia LV, Nguyen TH, Pitcher GM, Salter MW and Lombroso PJ. Tyrosine phosphatase STEP is a tonic brake on induction of long-term potentiation. *Neuron* 34: 127–138, 2002.
62. Perry VH. Modulation of microglia phenotype. *Neuropathol Appl Neurobiol* 20: 177, 1994.
63. Prescott SA and De Koninck Y. Gain control of firing rate by shunting inhibition: roles of synaptic noise and dendritic saturation. *Proc Natl Acad Sci USA* 100: 2076–2081, 2003.
64. Raghavendra V, Tanga F and DeLeo JA. Inhibition of microglial activation attenuates the development but not existing hypersensitivity in a rat model of neuropathy. *J Pharmacol Exp Ther* 306: 624–630, 2003.
65. Raghavendra V, Tanga FY and DeLeo JA. Complete Freunds adjuvant-induced peripheral inflammation evokes glial activation and proinflammatory cytokine expression in the CNS. *Eur J Neurosci* 20: 467–473, 2004.
66. Randic M, Jiang MC and Cerne R. Long-term potentiation and long-term depression of primary afferent neurotransmission in the rat spinal cord. *J Neurosci* 13: 5228–5241, 1993.
67. Robinson AP, White TM and Mason DW. Macrophage heterogeneity in the rat as delineated by two monoclonal antibodies MRC OX-41 and MRC OX-42, the latter recognizing complement receptor type 3. *Immunology* 57: 239–247, 1986.
68. Salter MW and Kalia LV. Src kinases: a hub for NMDA receptor regulation. *Nat Rev Neurosci* 5: 317–328, 2004.
69. Sandkuhler J. Understanding LTP in pain pathways. *Mol Pain* 3: 9, 2007.
70. Sato E, Takano Y, Kuno Y, Takano M and Sato I. Involvement of spinal tyrosine kinase in inflammatory and N-methyl-D-aspartate-induced hyperalgesia in rats. *Eur J Pharmacol* 468: 191–198, 2003.
71. Seeburg PH, Higuchi M and Sprengel R. RNA editing of brain glutamate receptor channels: mechanism and physiology. *Brain Res Brain Res Rev* 26: 217–229, 1998.
72. Sheng M and Kim MJ. Postsynaptic signaling and plasticity mechanisms. *Science* 298: 776–780, 2002.
73. Simone DA, Baumann TK, Collins JG and LaMotte RH. Sensitization of cat dorsal horn neurons to innocuous mechanical stimulation after intradermal injection of capsaicin. *Brain Res* 486: 185–189, 1989.

74. Sivilotti LG, Thompson SW and Woolf CJ. Rate of rise of the cumulative depolarization evoked by repetitive stimulation of small-caliber afferents is a predictor of action potential windup in rat spinal neurons in vitro. *J Neurophysiol* 69: 1621–1631, 1993.
75. Song I and Huganir RL. Regulation of AMPA receptors during synaptic plasticity. *Trends Neurosci* 25: 578–588, 2002.
76. South SM, Kohno T, Kaspar BK, Hegarty D, Vissel B, Drake CT, Ohata M, Jenab S, Sailer AW, Malkmus S, Masuyama T, Horner P, Bogulavsky J, Gage FH, Yaksh TL, Woolf CJ, Heinemann SF and Inturrisi CE. A conditional deletion of the NR1 subunit of the NMDA receptor in adult spinal cord dorsal horn reduces NMDA currents and injury-induced pain. *J Neurosci* 23: 5031–5040, 2003.
77. Stoll G and Jander S. The role of microglia and macrophages in the pathophysiology of the CNS. *Prog Neurobiol* 58: 233–247, 1999.
78. Streit WJ and Graeber MB. Heterogeneity of microglial and perivascular cell populations: insights gained from the facial nucleus paradigm. *Glia* 7: 68–74, 1993.
79. Streit WJ, Walter SA and Pennell NA. Reactive microgliosis. *Prog Neurobiol* 57: 563–581, 1999.
80. Sweitzer SM, Colburn RW, Rutkowski M and DeLeo JA. Acute peripheral inflammation induces moderate glial activation and spinal IL-1beta expression that correlates with pain behavior in the rat. *Brain Res* 829: 209–221, 1999.
81. Sweitzer SM and DeLeo JA. The active metabolite of leflunomide, an immunosuppressive agent, reduces mechanical sensitivity in a rat mononeuropathy model. *J Pain* 3: 360–368, 2002.
82. Sweitzer SM, White KA, Dutta C and DeLeo JA. The differential role of spinal MHC class II and cellular adhesion molecules in peripheral inflammatory versus neuropathic pain in rodents. *J Neuroimmunol* 125: 82–93, 2002.
83. Tanga FY, Raghavendra V and DeLeo JA. Quantitative real-time RT-PCR assessment of spinal microglial and astrocytic activation markers in a rat model of neuropathic pain. *Neurochem Int* 45: 397–407, 2004.
84. Thompson SW, Woolf CJ and Sivilotti LG. Small-caliber afferent inputs produce a heterosynaptic facilitation of the synaptic responses evoked by primary afferent A-fibers in the neonatal rat spinal cord in vitro. *J Neurophysiol* 69: 2116–2128, 1993.
85. Thompson SWN, King AE and Woolf CJ. Activity-dependent changes in rat ventral horn neurones in vitro; summation of prolonged afferent evoked postsynaptic depolarizations produce a d-APV sensitive windup. *Eur J Neurosci* 2: 638–649, 1990.
86. Tong CK and MacDermott AB. Both Ca2+-permeable and -impermeable AMPA receptors contribute to primary synaptic drive onto rat dorsal horn neurons. *J Physiol* 575: 133–144, 2006.
87. Torsney C and MacDermott AB. Disinhibition opens the gate to pathological pain signaling in superficial neurokinin 1 receptor-expressing neurons in rat spinal cord. *J Neurosci* 26: 1833–1843, 2006.
88. Tsuda M, Inoue K and Salter MW. Neuropathic pain and spinal microglia: a big problem from molecules in 'small' glia. *Trends Neurosci* 28: 101–107, 2005.
89. Tsuda M, Shigemoto-Mogami Y, Koizumi S, Mizokoshi A, Kohsaka S, Salter MW and Inoue K. P2X4 receptors induced in spinal microglia gate tactile allodynia after nerve injury. *Nature* 424: 778–783, 2003.
90. Watkins LR and Maier SF. Glia: a novel drug discovery target for clinical pain. *Nat Rev Drug Discov* 2: 973–985, 2003.
91. Wei F, Qiu CS, Kim SJ, Muglia L, Maas JW, Pineda VV, Xu HM, Chen ZF, Storm DR, Muglia LJ and Zhuo M. Genetic elimination of behavioral sensitization in mice lacking calmodulin-stimulated adenylyl cyclases. *Neuron* 36: 713–726, 2002.

92. Wieseler-Frank J, Maier SF and Watkins LR. Glial activation and pathological pain. *Neurochem Int* 45: 389–395, 2004.
93. Woolf CJ. Evidence for a central component of post-injury pain hypersensitivity. *Nature* 306: 686–688, 1983.
94. Woolf CJ. Dissecting out mechanisms responsible for peripheral neuropathic pain: implications for diagnosis and therapy. *Life Sci* 74: 2605–2610, 2004.
95. Woolf CJ and Salter MW. Neuronal plasticity: increasing the gain in pain. *Science* 288: 1765–1769, 2000.
96. Woolf CJ and Salter MW. Cellular and molecular mechanisms of central sensitization. In: The Neurobiology of Pain. Edited by Hunt sp and koltzenburg M. Oxford University Press, 2005, p. 95–114.
97. Woolf CJ and Salter MW. Plasticity and pain: role of the dorsal horn. In: Melzack and Wall's Textbook of Pain. 5th Edition, edited by McMahon SB and Koltzenburg M. Elsevier, 2006, p. 91–106.
98. Woolf CJ and Wall PD. Relative effectiveness of C primary afferent fibers of different origins in evoking a prolonged facilitation of the flexor reflex in the rat. *J Neurosci* 6: 1433–1442, 1986.
99. Xu J, Xue S, Lei G, Kwan CL and Yu XM. C-terminal Src kinase (CSK) depresses NMDA receptor tyrosine phosphorylation and channel activity through targeting the C-tail tyrosine of Src. *Soc Neurosci Abs* 30: 49.3, 2004.
100. Yajima Y, Narita M, Usui A, Kaneko C, Miyatake M, Narita M, Yamaguchi T, Tamaki H, Wachi H, Seyama Y and Suzuki T. Direct evidence for the involvement of brain-derived neurotrophic factor in the development of a neuropathic pain-like state in mice. *J Neurochem* 93: 584–594, 2005.
101. Yu XM and Salter MW. Src, a molecular switch governing gain control of synaptic transmission mediated by N-methyl-D-aspartate receptors. *Proc Natl Acad Sci USA* 96: 7697–7704, 1999.
102. Zhang J, Hoffert C, Vu HK, Groblewski T, Ahmad S and O'Donnell D. Induction of CB2 receptor expression in the rat spinal cord of neuropathic but not inflammatory chronic pain models. *Eur J Neurosci* 17: 2750–2754, 2003.
103. Zhang Q and Haydon PG. Roles for gliotransmission in the nervous system. *J Neural Transm* 112: 121–125, 2005.
104. Zhao J, Seereeram A, Nassar MA, Levato A, Pezet S, Hathaway G, Morenilla-Palao C, Stirling C, Fitzgerald M, McMahon SB, Rios M and Wood JN. Nociceptor-derived brain-derived neurotrophic factor regulates acute and inflammatory but not neuropathic pain. *Mol Cell Neurosci* 31: 539–548, 2006.
105. Zimmermann M. Pathobiology of neuropathic pain. *Eur J Pharmacol* 429: 23–37, 2001.
106. Zou X, Lin Q and Willis WD. Enhanced phosphorylation of NMDA receptor 1 subunits in spinal cord dorsal horn and spinothalamic tract neurons after intradermal injection of capsaicin in rats. *J Neurosci* 20: 6989–6997, 2000.

Index

MIX
Papier aus verantwortungsvollen Quellen
Paper from responsible sources
FSC® C105338

If you have any concerns about our products,
you can contact us on
ProductSafety@springernature.com

In case Publisher is established outside the EU,
the EU authorized representative is:
Springer Nature Customer Service Center GmbH
Europaplatz 3, 69115 Heidelberg, Germany

Printed by Libri Plureos GmbH
in Hamburg, Germany